4.1   *Sum or Difference Rule*   If $f(x) = u(x) \pm v(x)$, then

$$f'(x) = u'(x) \pm v'(x).$$

4.2   *Product Rule*   If $f(x) = u(x) \cdot v(x)$, then

$$f'(x) = u(x) \cdot v'(x) + v(x) \cdot u'(x).$$

4.2   *Quotient Rule*   If $f(x) = \dfrac{u(x)}{v(x)}$, and $v(x) \neq 0$, then

$$f'(x) = \frac{v(x) \cdot u'(x) - u(x) \cdot v'(x)}{[v(x)]^2}.$$

4.3   *Chain Rule*   If y is a function of u, say $y = f(u)$, and if u is a function of x, say $u = g(x)$, then $y = f(u) = f[g(x)]$, and

$$\frac{dy}{dx} = \frac{dy}{du} \cdot \frac{du}{dx}.$$

4.3   *Chain Rule (Alternate Form)*   If $y = f[g(x)]$, then $dy/dx = f'[g(x)] \cdot g'(x)$.

4.3   *Generalized Power Rule*   Let $g(x)$ be a function of x, and let $y = [g(x)]^n$ for any real number n. Then

$$dy/dx = n[g(x)]^{n-1} \cdot g'(x).$$

4.4   *Exponential Function*

$$D_x[a^{g(x)}] = (\ln a)a^{g(x)}g'(x)$$

$$D_x[e^{g(x)}] = e^{g(x)}g'(x)$$

4.5   *Logarithmic Function*

$$D_x\Big[\log_a|g(x)|\Big] = \frac{1}{\ln a} \cdot \frac{g'(x)}{g(x)}$$

$$D_x\Big[\ln |g(x)|\Big] = \frac{g'(x)}{g(x)}$$

4.6   *Trigonometric Functions*

$$D_x(\sin x) = \cos x \qquad D_x(\cos x) = -\sin x$$

$$D_x(\tan x) = \sec^2 x \qquad D_x(\cot x) = -\csc^2 x$$

5.2 First Derivative Test   Let c be a critical number for a function f. Suppose that f is differentiable on (a, b) except possibly at c, and that c is the only critical number for f in (a, b).

1. $f(c)$ is a relative maximum of f if the derivative $f'(x)$ is positive in the interval (a, c) and negative in the interval (c, b).

2. $f(c)$ is a relative minimum of f if the derivative $f'(x)$ is negative in the interval (a, c) and positive in the interval (c, b).

5.3 Second Derivative Test   Let f'' exist on some open interval containing c, and let $f'(c) = 0$.

1. If $f''(c) > 0$, then $f(c)$ is a relative minimum.

2. If $f''(c) < 0$, then $f(c)$ is a relative maximum.

3. If $f''(c) = 0$, then the test gives no information about extrema.

CALCULUS
WITH
APPLICATIONS
FOR THE
LIFE SCIENCES

CALCULUS WITH APPLICATIONS FOR THE LIFE SCIENCES

- **Raymond N. Greenwell**
  Hofstra University

- **Nathan P. Ritchey**
  Youngstown State University

- **Margaret L. Lial**
  American River College

Addison
Wesley

Boston San Francisco New York
London Toronto Sydney Tokyo Singapore Madrid
Mexico City Munich Paris Cape Town Hong Kong Montreal

Sponsoring Editor	William Hoffman
Executive Project Manager	Christine O'Brien
Production Supervisor	Julie LaChance
Production Coordination and Text Design	Elm Street Publishing Services, Inc.
Marketing Coordinator	Weslie Lewis
Senior Prepress Supervisor	Caroline Fell
Senior Manufacturing Buyer	Evelyn Beaton
Media Producer	Ruth Berry
Editorial Project Assistant	Joanne Ha
Senior Software Engineer	Marty Wright
Software Editor	Malcolm Litowitz
Compositor	The Beacon Group, Inc.
Illustrations	Precision Graphics
Cover Design	Barbara T. Atkinson
Cover Photos	© Art Wolfe/Getty Images, PhotoDisc

For permission to use copyrighted material, grateful acknowledgment is made to the copyright holders on page A-49, which is hereby made part of this copyright page.

Library of Congress Cataloging-in-Publication Data
Greenwell, Raymond N.
    Calculus for the life sciences / Raymond N. Greenwell, Nathan P. Ritchey, Margaret L. Lial—1st ed.
        p. cm.
    Includes bibliographical references and index.
    ISBN 0-201-74582-8 (alk. paper)
    1. Calculus.   2. Life sciences—Mathematics.   I. Ritchey, Nathan P.   II. Lial, Margaret L.   III. Title.

    QA303.2.G74 2002
    570'.1'515—dc15   2002031168

7   8   9   10—QWD—08

CONTENTS

Preface   xi

CHAPTER R   ALGEBRA REFERENCE

R.1   Polynomials   xvi
R.2   Factoring   xix
R.3   Rational Expressions   xxii
R.4   Equations   xxv
R.5   Inequalities   xxxi
R.6   Exponents   xxxvi
R.7   Radicals   xli

CHAPTER 1   FUNCTIONS

1.1   Lines and Linear Functions   2
1.2   The Least Squares Line   19
1.3   Properties of Functions   31
1.4   Quadratic Functions; Translation and Reflection   44
1.5   Polynomial and Rational Functions   54
Review Exercises   68
Extended Application: Using Extrapolation to Predict Life Expectancy   75

CHAPTER 2   EXPONENTIAL, LOGARITHMIC, AND TRIGONOMETRIC FUNCTIONS

2.1   Exponential Functions   77
2.2   Logarithmic Functions   89
2.3   Applications: Growth and Decay   103
2.4   Trigonometric Functions   109
Summary of Important Functions   126
Review Exercises   128
Extended Application: Characteristics of the Monkeyface Prickleback   133

CHAPTER 3   THE DERIVATIVE

3.1   Limits   135
3.2   Continuity   154
3.3   Rates of Change   162
3.4   Definition of the Derivative   173
3.5   Graphical Differentiation   190
          Review Exercises   197

CHAPTER 4   CALCULATING THE DERIVATIVE

4.1   Techniques for Finding Derivatives   203
4.2   Derivatives of Products and Quotients   217
4.3   The Chain Rule   223
4.4   Derivatives of Exponential Functions   234
4.5   Derivatives of Logarithmic Functions   242
4.6   Derivatives of Trigonometric Functions   250
          Rules for Derivatives Summary   259
          Review Exercises   260

CHAPTER 5   GRAPHS AND THE DERIVATIVE

5.1   Increasing and Decreasing Functions   266
5.2   Relative Extrema   278
5.3   Higher Derivatives, Concavity, and the Second Derivative Test   290
5.4   Curve Sketching   306
          Review Exercises   316

CHAPTER 6   APPLICATIONS OF THE DERIVATIVE

6.1   Absolute Extrema   322
6.2   Applications of Extrema   331
6.3   Implicit Differentiation   342
6.4   Related Rates   349
6.5   Differentials: Linear Approximation   356
          Review Exercises   363
          Extended Application: A Total Cost Model for a Training Program   367

■ **CHAPTER 7   INTEGRATION**

7.1   Antiderivatives   369
7.2   Substitution   380
7.3   Area and the Definite Integral   389
7.4   The Fundamental Theorem of Calculus   403
7.5   Integrals of Trigonometric Functions   414
7.6   The Area Between Two Curves   419
Integration Summary   426
Review Exercises   428
Extended Application: Estimating Depletion Dates for Minerals   433

■ **CHAPTER 8   FURTHER TECHNIQUES AND APPLICATIONS OF INTEGRATION**

8.1   Numerical Integration   437
8.2   Integration by Parts   446
8.3   Volume and Average Value   456
8.4   Improper Integrals   463
Review Exercises   469
Extended Application: Flow Systems   472

■ **CHAPTER 9   MULTIVARIABLE CALCULUS**

9.1   Functions of Several Variables   476
9.2   Partial Derivatives   489
9.3   Maxima and Minima   500
9.4   Total Differentials and Approximations   509
9.5   Double Integrals   514
Review Exercises   527
Extended Application: Optimization for a Predator   531

■ **CHAPTER 10   MATRICES**

10.1   Solution of Linear Systems   534
10.2   Addition and Subtraction of Matrices   552
10.3   Multiplication of Matrices   561

10.4   Matrix Inverses   573
10.5   Eigenvalues and Eigenvectors   584
Review Exercises   595
Extended Application: Contagion   601

CHAPTER 11   DIFFERENTIAL EQUATIONS

11.1   Solutions of Elementary and Separable Differential Equations   603
11.2   Linear First-Order Differential Equations   616
11.3   Euler's Method   623
11.4   Linear Systems of Differential Equations   630
11.5   Nonlinear Systems of Differential Equations   639
11.6   Applications of Differential Equations   646
Review Exercises   652
Extended Application: Pollution of the Great Lakes   656

CHAPTER 12   PROBABILITY

12.1   Sets   659
12.2   Introduction to Probability   674
12.3   Conditional Probability; Independent Events; Bayes' Theorem   691
12.4   Discrete Random Variables; Applications to Decision Making   709
Probability Summary   721
Review Exercises   722
Extended Application: Medical Diagnosis   727

CHAPTER 13   PROBABILITY AND CALCULUS

13.1   Continuous Probability Models   729
13.2   Expected Value and Variance of Continuous Random Variables   738
13.3   Special Probability Density Functions   746
Review Exercises   760
Extended Application: Exponential Waiting Times   765

TABLES

Table 1   Formulas from Geometry   A-1
Table 2   Area Under a Normal Curve   A-2
Table 3   Integrals   A-4
Table 4   Integrals Involving Trigonometric Functions   A-5

Answers to Selected Exercises   A-7

Photo Acknowledgments   A-49

Index of Applications   I-1

Index   I-7

SPECIAL TOPICS TO ACCOMPANY CALCULUS WITH APPLICATIONS FOR THE LIFE SCIENCES*

Chapter 14   Discrete Dynamical Systems
          14.1 Sequences
          14.2 Equilibrium Points
          14.3 Determining Stability

Markov Chains

---

*These supplements are available on the Web site for this text: www.aw.com/greenwell.

PREFACE

Calculus with Applications for the Life Sciences emphasizes those aspects of calculus most relevant to the life sciences. The application examples and exercises are drawn predominately from biology, medicine, ecology, and other life sciences. A prerequisite of two years of high school algebra is assumed.

■ FEATURES

Pedagogical Features

- careful explanation of the mathematics
- fully developed examples with explanatory comments in color on the side
- an algebra reference (Chapter R), designed to be used either in class or by individual students
- thought-provoking questions that open most sections, which are answered in an application within the section or in the section exercises
- "just in time" margin reviews giving short explanations or comments reminding students of skills or techniques learned earlier that are needed at this point
- common student difficulties and errors highlighted under the heading "Caution"
- important treatments and asides highlighted with the heading "Notes"
- summaries of rules or formulas for chapters where students may have trouble deciding which of several techniques to use
- an index of applications showing the abundant variety of real-data applications used in the text and allowing direct reference to particular topics

Exercises

- exercises arranged according to the material in the section, with the more challenging exercises placed near the end
- applied exercises (labeled "Applications") that are grouped by subject, with titles indicating the specific topic
- writing exercises, labeled with ✎, to provide students with an opportunity to explain important mathematical ideas
- connections exercises, labeled with ⟲, that integrate topics presented in different sections or chapters

■ SPECIAL FEATURES

Use of Referenced Real-Data Applications   This text includes a large number of application exercises and examples using real data, with references to articles appearing in newspapers, books, and journals. For example, see pages 100, 123, 153, and 214–215. These examples will help students learn how extensively mathematics is applied to the life sciences, as well as motivate students who wonder how the mathematics they learn relates to their other courses and their future careers. We believe that the quantity and quality of real-data applications set this book apart from others at this level.

Multiple Methods of Solution   We emphasize multiple representations of a topic, whenever possible, by examining each topic symbolically, numerically, graphically, and verbally, and offer multiple methods of solution to various examples. Some of these alternative methods involve technology, while others represent a different way of performing a computation. For example, see pages 20–22, 394–395, and 447–448. We believe that students will better understand a topic when they learn more than one approach. Furthermore, the multiple methods will give instructors greater flexibility to emphasize the methods they prefer.

Use of Graphing Calculators and Spreadsheets   We emphasize graphing calculators and spreadsheets. Although students reading this book do not need a graphing calculator or a spreadsheet program, using either or both will enhance their learning. Exercises requiring technology are labeled with ◤; many of these can be done with either a spreadsheet or a graphing calculator. For example, see pages 27–31 and 86–88. We have used the TI-83 for our graphing calculator examples and Microsoft® Excel for our spreadsheet examples, but students can succeed in the course with any appropriate graphing calculator or spreadsheet program. Inclusion of both graphing calculators and spreadsheets in the text gives instructors the ability to incorporate technology into their course as they wish.

Extended Applications   We include in-depth applied exercises (labeled "Extended Applications") at the end of most sections to stimulate student interest and for use as group projects. Additional extended applications are available at the Web site for *Calculus with Applications:* www.aw.com/LGR.

Full-Color Graphics and Photography   To make this book easy, lively, and fun to read, we have used full-color graphics. Occasionally, color in the text enhances the exposition, such as on page 4. More often, we have used color to clarify the different parts of a figure, such as on pages 308 and 323. We also have photographs illustrating many of the applications.

■ SUPPLEMENTS

For the Instructor   The *Instructor's Solutions Manual* (ISBN 0-201-77017-2) contains solutions to even-numbered exercises and a set of teaching tips.

MyMathLab is a complete online course available with this text and is perfect for helping students succeed in their course. This site offers a wide variety of resources, from dynamic multimedia—video clips, animations, audio explanations, and more—to an online gradebook that allows students to track their performance on homework assignments and tests. With MyMathLab students can increase comprehension and get help with concepts they don't understand.

TestGen-EQ with QuizMaster-EQ (dual platform for Windows and Macintosh, ISBN 0-201-79418-7) is a computerized test generator with algorithmically defined problems organized specifically for this textbook. Its user-friendly graphical interface enables instructors to select, view, edit, and add test items, then print tests in a variety of fonts and forms. A built-in question editor gives the user the power to create graphs, import graphics, insert mathematical symbols and templates, and insert variable numbers or text. An "Export to HTML" feature lets instructors create practice tests that can be posted to a Web site. Tests created with TestGen-EQ can also be displayed on Web pages when the free TestGen-EQ plugin has been installed with Internet Explorer or Netscape Navigator. Tests created with TestGen-EQ can be used with QuizMaster-EQ, which enables students to take exams on a computer network. QuizMaster-EQ automatically grades the exams, stores results on disk, and allows the instructor to view or print a variety of reports for individual students, classes, or courses. Contact your Addison-Wesley sales consultant.

For the Student   The *Student's Solutions Manual* (ISBN 0-201-77016-4) provides solutions to odd-numbered exercises and sample chapter tests with answers.

www.aw.com/greenwell is a free site available to everyone with World Wide Web access. Among the items provided on the Web site are the following:

- *The Graphing Calculator Manual for Calculus for the Life Sciences* provides detailed information on using some of the more popular graphing calculators to work through examples in the text. Listings of some programs for graphing calculators are also included.

- *The Excel Spreadsheet Manual for Calculus for the Life Sciences* provides detailed information on using the Excel spreadsheet program to work through examples in the text.

- The supplementary chapter *Discrete Dynamical Systems* is available for classes that wish to cover this topic.

- The supplementary section *Markov Chains* is also available for classes interested in the topic.

Just-in-Time Algebra for Students of Calculus in Management and the Life Sciences, Second Edition (ISBN 0-201-74611-5) lets students review key algebra topics. The user-friendly table of contents has key prerequisite topics arranged in the order in which students will need to review them as they study applied calculus.

The Addison-Wesley Math Tutor Center is staffed by qualified mathematics instructors who provide students with tutoring on examples and exercises answered at the back of the textbook. Tutoring is available via toll-free telephone, fax, e-mail, or whiteboard technology—which allows tutors and students to actually see the problems worked while they "talk" in real time over the Internet.

This service is available five days a week, seven hours a day. For more information, go to www.aw.com/tutorcenter.

InterAct MathXL® (www.mathxl.com) is a Web site that provides diagnostic testing and tutorial help, all online, using InterAct Math tutorial software and TestGen-EQ testing software. Students can take chapter tests correlated to this textbook, receive individualized study plans based on those test results, work practice problems and receive tutorial instruction for areas in which they need improvement, and take further tests to gauge their progress. Instructors can customize tests and track all student test results, study plans, and practice work. The site is free when an access code is bundled with a new text.

Acknowledgments   We wish to thank the following professors for their contributions in reviewing portions of this text.

David Carlson, *San Diego State University*
Edward Crotty, *University of Pennsylvania*
John Daughtry, *East Carolina State University*
Richard Driver, *Washburn University*
Daniela Ferrero, *Southwest Texas State University*
David Goldberg, *Colorado State University*
Kat Gustafson, *University of Alaska, Fairbanks*
Philip Gustafson, *Mesa State College*
Jocelyn Krebs, *University of Alaska, Anchorage*
Melvin Lax, *California State University, Long Beach*
Thomas Riedel, *University of Louisville*
Maria Schonbek, *University of California, Santa Cruz*
Christopher Sogge, *Johns Hopkins University*
Larry Taylor, *North Dakota State University*

We are grateful to LaurelTech Integrated Publishing Services for doing an excellent job coordinating the *Student's Solutions Manual* and the *Instructor's Solutions Manual*, an enormous and time-consuming task. We also thank Sheri Minkner and Judy Martinez for typesetting these manuals. Paul Van Erden, retired from American River College, has created an accurate and complete index for us, and Becky Troutman has compiled the extensive index of applications. For their invaluable help in maintaining high standards of accuracy in the answer section, we thank Timothy Comar, Jennifer Crawford, Cathy Ferrer, and Lauri Semarne. We also want to thank Karla Harby and Mary Ann Ritchey for their editorial assistance. We especially appreciate the staff at Addison-Wesley, whose contributions have been very important in bringing this project to a successful conclusion: Greg Tobin, Christine O'Brien, Bill Hoffman, Joanne Ha, Julie LaChance, and Barbara Atkinson. Our sincere thanks go to Susan Gallier of Elm Street Publishing Services for a great job as project editor. Finally, we are especially grateful to Laurie Rosatone, former editor at Addison-Wesley, for inspiring us to write this book and giving us wonderful guidance during her tenure as editor.

Raymond N. Greenwell
Nathan P. Ritchey
Margaret L. Lial

CHAPTER

R

Algebra Reference

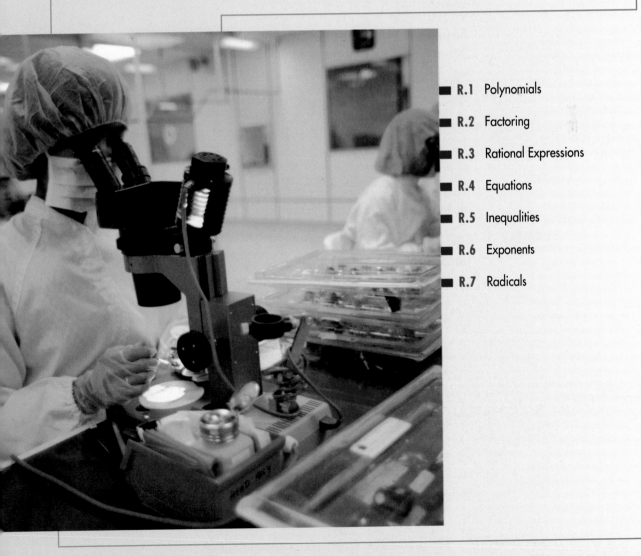

■ **R.1** Polynomials

■ **R.2** Factoring

■ **R.3** Rational Expressions

■ **R.4** Equations

■ **R.5** Inequalities

■ **R.6** Exponents

■ **R.7** Radicals

In this chapter, we will review the most important topics in algebra. Knowing algebra is a fundamental prerequisite to success in higher mathematics. This algebra reference is designed for self-study; study it all at once or refer to it when needed throughout the course. Since this is a review, answers to all exercises are given in the answer section at the back of the book.

■ R.1 POLYNOMIALS

An expression such as $9p^4$ is a **term;** the number 9 is the **coefficient,** p is the **variable,** and 4 is the **exponent.** The expression p^4 means $p \cdot p \cdot p \cdot p$, while p^2 means $p \cdot p$, and so on. Terms having the same variable and the same exponent, such as $9x^4$ and $-3x^4$, are **like terms.** Terms that do not have both the same variable and the same exponent, such as m^2 and m^4, are **unlike terms.**

A **polynomial** is a term or a finite sum of terms in which all variables have whole number exponents, and no variables appear in denominators. Examples of polynomials include

$$5x^4 + 2x^3 + 6x, \qquad 8m^3 + 9m^2n - 6mn^2 + 3n^3, \qquad 10p, \qquad \text{and} \qquad -9.$$

Adding and Subtracting Polynomials   The following properties of real numbers are useful for performing operations on polynomials.

> **PROPERTIES OF REAL NUMBERS**
>
> For all real numbers a, b, and c,
>
> **1.** $a + b = b + a$;   **Commutative properties**
> $ab = ba$;
>
> **2.** $(a + b) + c = a + (b + c)$;   **Associative properties**
> $(ab)c = a(bc)$;
>
> **3.** $a(b + c) = ab + ac$.   **Distributive property**

EXAMPLE 1   Properties of Real Numbers

(a) $2 + x = x + 2$ Commutative property of addition

(b) $x \cdot 3 = 3x$ Commutative property of multiplication

(c) $(7x)x = 7(x \cdot x) = 7x^2$ Associative property of multiplication

(d) $3(x + 4) = 3x + 12$ Distributive property   ■

The distributive property is used to add or subtract polynomials. Only like terms may be added or subtracted. For example,

$$12y^4 + 6y^4 = (12 + 6)y^4 = 18y^4,$$

and

$$-2m^2 + 8m^2 = (-2 + 8)m^2 = 6m^2,$$

but the polynomial $8y^4 + 2y^5$ cannot be further simplified. To subtract polynomials, use the facts that $-(a + b) = -a - b$ and $-(a - b) = -a + b$. In the next example, we show how to add and subtract polynomials.

EXAMPLE 2 Adding and Subtracting Polynomials
Add or subtract as indicated.

(a) $(8x^3 - 4x^2 + 6x) + (3x^3 + 5x^2 - 9x + 8)$

Solution Combine like terms.
$$(8x^3 - 4x^2 + 6x) + (3x^3 + 5x^2 - 9x + 8)$$
$$= (8x^3 + 3x^3) + (-4x^2 + 5x^2) + (6x - 9x) + 8$$
$$= 11x^3 + x^2 - 3x + 8$$

(b) $(-4x^4 + 6x^3 - 9x^2 - 12) + (-3x^3 + 8x^2 - 11x + 7)$

Solution Combining like terms as before yields
$$-4x^4 + 3x^3 - x^2 - 11x - 5.$$

(c) $(2x^2 - 11x + 8) - (7x^2 - 6x + 2)$

Solution Distributing the minus sign yields
$$(2x^2 - 11x + 8) + (-7x^2 + 6x - 2)$$
$$= -5x^2 - 5x + 6.$$

Multiplying Polynomials The distributive property is also used to multiply polynomials, along with the fact that $a^m \cdot a^n = a^{m+n}$. For example,

$$x \cdot x = x^1 \cdot x^1 = x^2 \qquad \text{and} \qquad x^2 \cdot x^5 = x^7.$$

EXAMPLE 3 Multiplying Polynomials
Multiply.

(a) $8x(6x - 4)$

Solution
$$8x(6x - 4) = 8x(6x) - 8x(4)$$
$$= 48x^2 - 32x$$

(b) $(3p - 2)(p^2 + 5p - 1)$

Solution
$$(3p - 2)(p^2 + 5p - 1)$$
$$= 3p(p^2 + 5p - 1) - 2(p^2 + 5p - 1)$$
$$= 3p(p^2) + 3p(5p) + 3p(-1) - 2(p^2) - 2(5p) - 2(-1)$$
$$= 3p^3 + 15p^2 - 3p - 2p^2 - 10p + 2$$
$$= 3p^3 + 13p^2 - 13p + 2$$

(c) $(x + 2)(x + 3)(x - 4)$

Solution
$$(x + 2)(x + 3)(x - 4)$$
$$= [(x + 2)(x + 3)](x - 4)$$
$$= (x^2 + 2x + 3x + 6)(x - 4)$$
$$= (x^2 + 5x + 6)(x - 4)$$
$$= x^3 + 5x^2 + 6x - 4x^2 - 20x - 24$$
$$= x^3 + x^2 - 14x - 24$$

A **binomial** is a polynomial with exactly two terms, such as $2x + 1$ or $m + n$. When two binomials are multiplied, the FOIL method (First, Outer, Inner, Last) is used as a memory aid.

EXAMPLE 4   Multiplying Polynomials

Find $(2m - 5)(m + 4)$ using the FOIL method.

Solution

$$\begin{array}{cccc} \text{F} & \text{O} & \text{I} & \text{L} \end{array}$$

$$(2m - 5)(m + 4) = (2m)(m) + (2m)(4) + (-5)(m) + (-5)(4)$$
$$= 2m^2 + 8m - 5m - 20$$
$$= 2m^2 + 3m - 20$$

EXAMPLE 5   Multiplying Polynomials

Find $(2k - 5)^2$.

Solution Use FOIL.
$$(2k - 5)^2 = (2k - 5)(2k - 5)$$
$$= 4k^2 - 10k - 10k + 25$$
$$= 4k^2 - 20k + 25$$

Notice that the product of the square of a binomial is the square of the first term, $(2k)^2$, plus twice the product of the two terms, $(2)(2k)(-5)$, plus the square of the last term, $(-5)^2$.

CAUTION Avoid the common error of writing $(x + y)^2 = x^2 + y^2$. As Example 5 shows, the square of a binomial has three terms, so

$$(x + y)^2 = x^2 + 2xy + y^2.$$

Furthermore, higher powers of a binomial also result in more than two terms. For example, verify by multiplication that

$$(x + y)^3 = x^3 + 3x^2y + 3xy^2 + y^3.$$

Remember, for any value of $n \neq 1$,

$$(x + y)^n \neq x^n + y^n.$$

R.1 EXERCISES

Perform the indicated operations.

1. $(2x^2 - 6x + 11) + (-3x^2 + 7x - 2)$

2. $(-4y^2 - 3y + 8) - (2y^2 - 6y - 2)$

3. $-3(4q^2 - 3q + 2) + 2(-q^2 + q - 4)$

4. $2(3r^2 + 4r + 2) - 3(-r^2 + 4r - 5)$

5. $(0.613x^2 - 4.215x + 0.892) - 0.47(2x^2 - 3x + 5)$

6. $0.83(5r^2 - 2r + 7) - (7.12r^2 + 6.423r - 2)$

7. $-9m(2m^2 + 3m - 1)$

8. $(6k - 1)(2k - 3)$

9. $(5r - 3s)(5r + 4s)$

10. $(9k + q)(2k - q)$

11. $\left(\dfrac{2}{5}y + \dfrac{1}{8}z\right)\left(\dfrac{3}{5}y + \dfrac{1}{2}z\right)$

12. $\left(\dfrac{3}{4}r - \dfrac{2}{3}s\right)\left(\dfrac{5}{4}r + \dfrac{1}{3}s\right)$

13. $(12x - 1)(12x + 1)$

14. $(6m + 5)(6m - 5)$

15. $(3p - 1)(9p^2 + 3p + 1)$

16. $(2p - 1)(3p^2 - 4p + 5)$

17. $(2m + 1)(4m^2 - 2m + 1)$

18. $(k + 2)(12k^3 - 3k^2 + k + 1)$

19. $(m - n + k)(m + 2n - 3k)$

20. $(r - 3s + t)(2r - s + t)$

21. $(x + 1)(x + 2)(x + 3)$

22. $(x - 1)(x + 2)(x - 3)$

23. $(3a + b)^2$

24. $(x - 2y)^3$

■ R.2 FACTORING

Multiplication of polynomials relies on the distributive property. The reverse process, where a polynomial is written as a product of other polynomials, is called **factoring.** For example, one way to factor the number 18 is to write it as the product $9 \cdot 2$; both 9 and 2 are **factors** of 18. Usually, only integers are used as factors of integers. The number 18 can also be written with three integer factors as $2 \cdot 3 \cdot 3$.

The Greatest Common Factor   To factor the algebraic expression $15m + 45$, first note that both $15m$ and 45 are divisible by 15; $15m = 15 \cdot m$ and $45 = 15 \cdot 3$. By the distributive property,

$$15m + 45 = 15 \cdot m + 15 \cdot 3 = 15(m + 3).$$

Both 15 and $m + 3$ are factors of $15m + 45$. Since 15 divides into both terms of $15m + 45$ (and is the largest number that will do so), 15 is the **greatest common factor** for the polynomial $15m + 45$. The process of writing $15m + 45$ as $15(m + 3)$ is often called **factoring out** the greatest common factor.

EXAMPLE 1   Factoring
Factor out the greatest common factor.

(a) $12p - 18q$

Solution Both $12p$ and $18q$ are divisible by 6. Therefore,

$$12p - 18q = 6 \cdot 2p - 6 \cdot 3q = 6(2p - 3q).$$

(b) $8x^3 - 9x^2 + 15x$

Solution Each of these terms is divisible by x.

$$8x^3 - 9x^2 + 15x = (8x^2) \cdot x - (9x) \cdot x + 15 \cdot x$$
$$= x(8x^2 - 9x + 15) \qquad \text{or} \qquad (8x^2 - 9x + 15)x \quad ■$$

One can always check factorization by finding the product of the factors and comparing it to the original expression.

CAUTION When factoring out the greatest common factor in an expression like $2x^2 + x$, be careful to remember the 1 in the second term.

$$2x^2 + x = 2x^2 + 1x = x(2x + 1), \text{ not } x(2x).$$

Factoring Trinomials   A polynomial that has no greatest common factor (other than 1) may still be factorable. For example, the polynomial $x^2 + 5x + 6$ can be factored as $(x + 2)(x + 3)$. To see that this is correct, find the product $(x + 2)(x + 3)$; you should get $x^2 + 5x + 6$. A polynomial such as this with three terms is called a **trinomial.** To factor the trinomial $x^2 + 5x + 6$, where the coefficient of x^2 is 1, we use FOIL backwards.

EXAMPLE 2   Factoring a Trinomial

Factor $y^2 + 8y + 15$.

Solution   Since the coefficient of y^2 is 1, factor by finding two numbers whose *product* is 15 and whose *sum* is 8. Since the constant and the middle term are positive, the numbers must both be positive. Begin by listing all pairs of positive integers having a product of 15. As you do this, also form the sum of each pair of numbers.

Products	*Sums*
$15 \cdot 1 = 15$	$15 + 1 = 16$
$5 \cdot 3 = 15$	$5 + 3 = 8$

The numbers 5 and 3 have a product of 15 and a sum of 8. Thus, $y^2 + 8y + 15$ factors as

$$y^2 + 8y + 15 = (y + 5)(y + 3).$$

The answer also can be written as $(y + 3)(y + 5)$.   ▪

If the coefficient of the squared term is *not* 1, work as shown below.

EXAMPLE 3   Factoring a Trinomial

Factor $2x^2 + 9xy - 5y^2$.

Solution   The factors of $2x^2$ are $2x$ and x; the possible factors of $-5y^2$ are $-5y$ and y, or $5y$ and $-y$. Try various combinations of these factors until one works (if, indeed, any work). For example, try the product $(2x + 5y)(x - y)$.

$$(2x + 5y)(x - y) = 2x^2 - 2xy + 5xy - 5y^2$$
$$= 2x^2 + 3xy - 5y^2$$

This product is not correct, so try another combination.

$$(2x - y)(x + 5y) = 2x^2 + 10xy - xy - 5y^2$$
$$= 2x^2 + 9xy - 5y^2$$

Since this combination gives the correct polynomial,

$$2x^2 + 9xy - 5y^2 = (2x - y)(x + 5y).$$ ▪

Special Factorizations   Four special factorizations occur so often that they are listed here for future reference.

SPECIAL FACTORIZATIONS

$x^2 - y^2 = (x + y)(x - y)$	**Difference of two squares**
$x^2 + 2xy + y^2 = (x + y)^2$	**Perfect square**
$x^3 - y^3 = (x - y)(x^2 + xy + y^2)$	**Difference of two cubes**
$x^3 + y^3 = (x + y)(x^2 - xy + y^2)$	**Sum of two cubes**

A polynomial that cannot be factored is called a **prime polynomial.**

EXAMPLE 4 Factoring

Factor each of the following.

(a) $64p^2 - 49q^2 = (8p)^2 - (7q)^2 = (8p + 7q)(8p - 7q)$

(b) $x^2 + 36$ is a prime polynomial.

(c) $x^2 + 12x + 36 = (x + 6)^2$

(d) $9y^2 - 24yz + 16z^2 = (3y - 4z)^2$

(e) $y^3 - 8 = y^3 - 2^3 = (y - 2)(y^2 + 2y + 4)$

(f) $m^3 + 125 = m^3 + 5^3 = (m + 5)(m^2 - 5m + 25)$

(g) $8k^3 - 27z^3 = (2k)^3 - (3z)^3 = (2k - 3z)(4k^2 + 6kz + 9z^2)$

CAUTION In factoring, always look for a common factor first. Since $36x^2 - 4y^2$ has a common factor of 4,

$$36x^2 - 4y^2 = 4(9x^2 - y^2) = 4(3x + y)(3x - y).$$

It would be incomplete to factor it as

$$36x^2 - 4y^2 = (6x + 2y)(6x - 2y)$$

since each factor can be factored still further. To *factor* means to factor completely, so that each polynomial factor is prime.

R.2 EXERCISES

Factor each of the following. If a polynomial cannot be factored, write prime. Factor out the greatest common factor as necessary.

1. $8a^3 - 16a^2 + 24a$

2. $3y^3 + 24y^2 + 9y$

3. $25p^4 - 20p^3q + 100p^2q^2$

4. $60m^4 - 120m^3n + 50m^2n^2$

5. $m^2 + 9m + 14$

6. $x^2 + 4x - 5$

7. $z^2 + 9z + 20$

8. $b^2 - 8b + 7$

9. $a^2 - 6ab + 5b^2$

10. $s^2 + 2st - 35t^2$

11. $y^2 - 4yz - 21z^2$

12. $6a^2 - 48a - 120$

13. $3m^3 + 12m^2 + 9m$

14. $2x^2 - 5x - 3$

15. $3a^2 + 10a + 7$

16. $2a^2 - 17a + 30$

17. $15y^2 + y - 2$

18. $21m^2 + 13mn + 2n^2$

19. $24a^4 + 10a^3b - 4a^2b^2$

20. $32z^5 - 20z^4a - 12z^3a^2$

21. $x^2 - 64$

22. $9m^2 - 25$

23. $121a^2 - 100$

24. $9x^2 + 64$

25. $z^2 + 14zy + 49y^2$

26. $m^2 - 6mn + 9n^2$

27. $9p^2 - 24p + 16$

28. $a^3 - 216$

29. $8r^3 - 27s^3$

30. $64m^3 + 125$

31. $x^4 - y^4$

32. $16a^4 - 81b^4$

R.3 RATIONAL EXPRESSIONS

Many algebraic fractions are **rational expressions,** which are quotients of polynomials with nonzero denominators. Examples include

$$\frac{8}{x-1}, \qquad \frac{3x^2 + 4x}{5x - 6}, \qquad \text{and} \qquad \frac{2y + 1}{y^2}.$$

Properties for working with rational expressions are summarized next.

PROPERTIES OF RATIONAL EXPRESSIONS

For all mathematical expressions P, Q, R, and S, with Q and $S \neq 0$:

$$\frac{P}{Q} = \frac{PS}{QS} \qquad\qquad \textbf{Fundamental property}$$

$$\frac{P}{Q} + \frac{R}{Q} = \frac{P + R}{Q} \qquad\qquad \textbf{Addition}$$

$$\frac{P}{Q} - \frac{R}{Q} = \frac{P - R}{Q} \qquad\qquad \textbf{Subtraction}$$

$$\frac{P}{Q} \cdot \frac{R}{S} = \frac{PR}{QS} \qquad\qquad \textbf{Multiplication}$$

$$\frac{P}{Q} \div \frac{R}{S} = \frac{P}{Q} \cdot \frac{S}{R} \quad (R \neq 0) \qquad \textbf{Division}$$

When using the fundamental property to write a rational expression in lowest terms, we may need to use the fact that $\dfrac{a^m}{a^n} = a^{m-n}$. For example,

$$\frac{x^2}{3x} = \frac{1x^2}{3x} = \frac{1}{3} \cdot \frac{x^2}{x} = \frac{1}{3}x.$$

EXAMPLE 1 Reducing Rational Expressions

Write each rational expression in lowest terms, that is, reduce the expression as much as possible.

(a) $\dfrac{8x + 16}{4} = \dfrac{8(x + 2)}{4} = \dfrac{4 \cdot 2(x + 2)}{4} = 2(x + 2)$

Factor both the numerator and denominator in order to identify any common factors, which have a quotient of 1. The answer could also be written as $2x + 4$.

(b) $\dfrac{k^2 + 7k + 12}{k^2 + 2k - 3} = \dfrac{(k + 4)(k + 3)}{(k - 1)(k + 3)} = \dfrac{k + 4}{k - 1}$

The answer cannot be further reduced.

CAUTION One of the most common errors in algebra involves incorrect use of the fundamental property of rational expressions. Only common *factors* may be divided or "canceled." It is essential to factor rational expressions before writing them in lowest terms. In Example 1(b), for instance, it is not correct to "cancel" k^2 (or cancel k, or divide 12 by -3) because the additions and subtraction must be performed first. Here they cannot be performed, so it is not possible to divide. After factoring, however, the fundamental property can be used to write the expression in lowest terms.

EXAMPLE 2 Combining Rational Expressions
Perform each operation.

(a) $\dfrac{3y + 9}{6} \cdot \dfrac{18}{5y + 15}$

Solution Factor where possible, then multiply numerators and denominators and reduce to lowest terms.

$$\frac{3y + 9}{6} \cdot \frac{18}{5y + 15} = \frac{3(y + 3)}{6} \cdot \frac{18}{5(y + 3)}$$

$$= \frac{3 \cdot 18(y + 3)}{6 \cdot 5(y + 3)}$$

$$= \frac{3 \cdot 6 \cdot 3(y + 3)}{6 \cdot 5(y + 3)} = \frac{3 \cdot 3}{5} = \frac{9}{5}$$

(b) $\dfrac{m^2 + 5m + 6}{m + 3} \cdot \dfrac{m}{m^2 + 3m + 2}$

Solution Factor where possible.

$$\frac{(m + 2)(m + 3)}{m + 3} \cdot \frac{m}{(m + 2)(m + 1)}$$

$$= \frac{m(m + 2)(m + 3)}{(m + 3)(m + 2)(m + 1)} = \frac{m}{m + 1}$$

(c) $\dfrac{9p - 36}{12} \div \dfrac{5(p - 4)}{18}$

Solution Use the division property of rational expressions.

$$\frac{9p - 36}{12} \cdot \frac{18}{5(p - 4)} \quad \text{Invert and multiply.}$$

$$= \frac{9(p - 4)}{6 \cdot 2} \cdot \frac{6 \cdot 3}{5(p - 4)} = \frac{27}{10}$$

(d) $\dfrac{4}{5k} - \dfrac{11}{5k}$

Solution As shown in the list of properties, to subtract two rational expressions that have the same denominators, we subtract the numerators while keeping the same denominator.

$$\frac{4}{5k} - \frac{11}{5k} = \frac{4 - 11}{5k} = -\frac{7}{5k}$$

(e) $\dfrac{7}{p} + \dfrac{9}{2p} + \dfrac{1}{3p}$

Solution These three fractions cannot be added until their denominators are the same. A **common denominator** into which p, $2p$, and $3p$ all divide is $6p$. Note that $12p$ is also a common denominator, but $6p$ is the **least common denominator.** Use the fundamental property to rewrite each rational expression with a denominator of $6p$.

$$\frac{7}{p} + \frac{9}{2p} + \frac{1}{3p} = \frac{6 \cdot 7}{6 \cdot p} + \frac{3 \cdot 9}{3 \cdot 2p} + \frac{2 \cdot 1}{2 \cdot 3p}$$

$$= \frac{42}{6p} + \frac{27}{6p} + \frac{2}{6p}$$

$$= \frac{42 + 27 + 2}{6p}$$

$$= \frac{71}{6p}$$

(f) $\dfrac{x + 1}{x^2 + 5x + 6} - \dfrac{5x - 1}{x^2 - x - 12}$

Solution To find the least common denominator, first factor each denominator. Then change each fraction so they all have the same denominator, being careful to multiply only by quotients that equal 1.

$$\frac{x + 1}{x^2 + 5x + 6} - \frac{5x - 1}{x^2 - x - 12}$$

$$= \frac{x + 1}{(x + 2)(x + 3)} - \frac{5x - 1}{(x + 3)(x - 4)}$$

$$= \frac{x + 1}{(x + 2)(x + 3)} \cdot \frac{(x - 4)}{(x - 4)} - \frac{5x - 1}{(x + 3)(x - 4)} \cdot \frac{(x + 2)}{(x + 2)}$$

$$= \frac{(x^2 - 3x - 4) - (5x^2 + 9x - 2)}{(x + 2)(x + 3)(x - 4)}$$

$$= \frac{-4x^2 - 12x - 2}{(x + 2)(x + 3)(x - 4)}$$

$$= \frac{-2(2x^2 + 6x + 1)}{(x + 2)(x + 3)(x - 4)}$$

Because the numerator cannot be factored further, we leave our answer in this form. We could also multiply out the denominator, but factored form is usually more useful.

R.3 EXERCISES

Write each rational expression in lowest terms.

1. $\dfrac{7z^2}{14z}$

2. $\dfrac{25p^3}{10p^2}$

3. $\dfrac{8k + 16}{9k + 18}$

4. $\dfrac{3(t + 5)}{(t + 5)(t - 3)}$

5. $\dfrac{8x^2 + 16x}{4x^2}$

6. $\dfrac{36y^2 + 72y}{9y}$

7. $\dfrac{m^2 - 4m + 4}{m^2 + m - 6}$

8. $\dfrac{r^2 - r - 6}{r^2 + r - 12}$

9. $\dfrac{x^2 + 3x - 4}{x^2 - 1}$

10. $\dfrac{z^2 - 5z + 6}{z^2 - 4}$

11. $\dfrac{8m^2 + 6m - 9}{16m^2 - 9}$

12. $\dfrac{6y^2 + 11y + 4}{3y^2 + 7y + 4} \cdot$

Perform the indicated operations.

13. $\dfrac{9k^2}{25} \cdot \dfrac{5}{3k}$

14. $\dfrac{15p^3}{9p^2} \div \dfrac{6p}{10p^2}$

15. $\dfrac{a + b}{2p} \cdot \dfrac{12}{5(a + b)}$

16. $\dfrac{a - 3}{16} \div \dfrac{a - 3}{32}$

17. $\dfrac{2k + 8}{6} \div \dfrac{3k + 12}{2}$

18. $\dfrac{9y - 18}{6y + 12} \cdot \dfrac{3y + 6}{15y - 30}$

19. $\dfrac{4a + 12}{2a - 10} \div \dfrac{a^2 - 9}{a^2 - a - 20}$

20. $\dfrac{6r - 18}{9r^2 + 6r - 24} \cdot \dfrac{12r - 16}{4r - 12}$

21. $\dfrac{k^2 - k - 6}{k^2 + k - 12} \cdot \dfrac{k^2 + 3k - 4}{k^2 + 2k - 3}$

22. $\dfrac{m^2 + 3m + 2}{m^2 + 5m + 4} \div \dfrac{m^2 + 5m + 6}{m^2 + 10m + 24}$

23. $\dfrac{2m^2 - 5m - 12}{m^2 - 10m + 24} \div \dfrac{4m^2 - 9}{m^2 - 9m + 18}$

24. $\dfrac{6n^2 - 5n - 6}{6n^2 + 5n - 6} \cdot \dfrac{12n^2 - 17n + 6}{12n^2 - n - 6}$

25. $\dfrac{a + 1}{2} - \dfrac{a - 1}{2}$

26. $\dfrac{3}{p} + \dfrac{1}{2}$

27. $\dfrac{2}{y} - \dfrac{1}{4}$

28. $\dfrac{1}{6m} + \dfrac{2}{5m} + \dfrac{4}{m}$

29. $\dfrac{1}{m - 1} + \dfrac{2}{m}$

30. $\dfrac{6}{r} - \dfrac{5}{r - 2}$

31. $\dfrac{8}{3(a - 1)} + \dfrac{2}{a - 1}$

32. $\dfrac{2}{5(k - 2)} + \dfrac{3}{4(k - 2)}$

33. $\dfrac{2}{x^2 - 2x - 3} + \dfrac{5}{x^2 - x - 6}$

34. $\dfrac{2y}{y^2 + 7y + 12} - \dfrac{y}{y^2 + 5y + 6}$

35. $\dfrac{3k}{2k^2 + 3k - 2} - \dfrac{2k}{2k^2 - 7k + 3}$

36. $\dfrac{4m}{3m^2 + 7m - 6} - \dfrac{m}{3m^2 - 14m + 8}$

37. $\dfrac{2}{a + 2} + \dfrac{1}{a} + \dfrac{a - 1}{a^2 + 2a}$

38. $\dfrac{5x + 2}{x^2 - 1} + \dfrac{3}{x^2 + x} - \dfrac{1}{x^2 - x}$

R.4 EQUATIONS

Linear Equations Equations that can be written in the form $ax + b = 0$, where a and b are real numbers, with $a \neq 0$, are **linear equations.** Examples of linear equations include $5y + 9 = 16$, $8x = 4$, and $-3p + 5 = -8$. Equations that are *not* linear include absolute value equations such as $|x| = 4$. The following properties are used to solve linear equations.

PROPERTIES OF EQUALITY

For all real numbers a, b, and c:

1. If $a = b$, then $a + c = b + c$.   **Addition property of equality**
(The same number may be added to both sides of an equation.)

2. If $a = b$, then $ac = bc$.   **Multiplication property of**
(Both sides of an equation may be   **equality**
multiplied by the same number.)

EXAMPLE 1   Solving Linear Equations

(a) If $x - 2 = 3$, then $x = 2 + 3 = 5$ Addition property of equality

(b) If $x/2 = 3$, then $x = 2 \cdot 3 = 6$ Multiplication property of equality   ▰

The following example shows how these properties are used to solve linear equations. Of course, the solutions should always be checked by substitution in the original equation.

EXAMPLE 2   Solving Linear Equations

Solve $2x - 5 + 8 = 3x + 2(2 - 3x)$.

Solution

$$2x - 5 + 8 = 3x + 4 - 6x \quad \text{Distributive property}$$

$$2x + 3 = -3x + 4 \quad \text{Combine like terms.}$$

$$5x + 3 = 4 \quad \text{Add } 3x \text{ to both sides.}$$

$$5x = 1 \quad \text{Add } -3 \text{ to both sides.}$$

$$x = \frac{1}{5} \quad \text{Multiply both sides by } \tfrac{1}{5}.$$

Check by substituting in the original equation. The left side becomes $2(1/5) - 5 + 8$ and the right side becomes $3(1/5) + 2(2 - 3(1/5))$. Verify that both of these expressions simplify to $17/5$.   ▰

Quadratic Equations   An equation with 2 as the highest exponent of the variable is a *quadratic equation*. A **quadratic equation** has the form $ax^2 + bx + c = 0$, where a, b, and c are real numbers and $a \neq 0$. A quadratic equation written in the form $ax^2 + bx + c = 0$ is said to be in **standard form.**

The simplest way to solve a quadratic equation, but one that is not always applicable, is by factoring. This method depends on the **zero-factor property.**

ZERO-FACTOR PROPERTY

If a and b are real numbers, with $ab = 0$, then

$$a = 0, b = 0, \text{ or both.}$$

EXAMPLE 3   Solving Quadratic Equations

Solve $6r^2 + 7r = 3$.

Solution First write the equation in standard form.

$$6r^2 + 7r - 3 = 0$$

Now factor $6r^2 + 7r - 3$ to get

$$(3r - 1)(2r + 3) = 0.$$

By the zero-factor property, the product $(3r - 1)(2r + 3)$ can equal 0 if and only if

$$3r - 1 = 0 \quad \text{or} \quad 2r + 3 = 0.$$

Solve each of these equations separately to find that the solutions are $1/3$ and $-3/2$. Check these solutions by substituting them in the original equation.   ▰

▌**CAUTION** Remember, the zero-factor property requires that the product of two (or more) factors be equal to *zero,* not some other quantity. It would be incorrect to use the zero-factor property with an equation in the form $(x + 3)(x - 1) = 4$, for example.

If a quadratic equation cannot be solved easily by factoring, use the *quadratic formula.* (The derivation of the quadratic formula is given in most algebra books.)

QUADRATIC FORMULA

The solutions of the quadratic equation $ax^2 + bx + c = 0$, where $a \neq 0$, are given by

$$x = \frac{-b \pm \sqrt{b^2 - 4ac}}{2a}.$$

EXAMPLE 4 The Quadratic Formula
Solve $x^2 - 4x - 5 = 0$ by the quadratic formula.

Solution The equation is already in standard form (it has 0 alone on one side of the equals sign), so the values of a, b, and c from the quadratic formula are easily identified. The coefficient of the squared term gives the value of a; here, $a = 1$. Also, $b = -4$ and $c = -5$. (Be careful to use the correct signs.) Substitute these values into the quadratic formula.

$$x = \frac{-(-4) \pm \sqrt{(-4)^2 - 4(1)(-5)}}{2(1)} \qquad \text{Let } a = 1, b = -4, c = -5.$$

$$x = \frac{4 \pm \sqrt{16 + 20}}{2} \qquad (-4)^2 = (-4)(-4) = 16$$

$$x = \frac{4 \pm 6}{2} \qquad \sqrt{16 + 20} = \sqrt{36} = 6$$

The $\pm$ sign represents the two solutions of the equation. To find both of the solutions, first use $+$ and then use $-$.

$$x = \frac{4 + 6}{2} = \frac{10}{2} = 5 \qquad \text{or} \qquad x = \frac{4 - 6}{2} = \frac{-2}{2} = -1$$

The two solutions are 5 and -1.

▌**CAUTION** Notice in the quadratic formula that the square root is added to or subtracted from the value of $-b$ *before* dividing by $2a$.

EXAMPLE 5 Quadratic Formula
Solve $x^2 + 1 = 4x$.

Solution First, add $-4x$ on both sides of the equals sign in order to get the equation in standard form.

$$x^2 - 4x + 1 = 0$$

Now identify the letters a, b, and c. Here $a = 1$, $b = -4$, and $c = 1$. Substitute these numbers into the quadratic formula.

$$x = \frac{-(-4) \pm \sqrt{(-4)^2 - 4(1)(1)}}{2(1)}$$

$$= \frac{4 \pm \sqrt{16 - 4}}{2}$$

$$= \frac{4 \pm \sqrt{12}}{2}$$

Simplify the solutions by writing $\sqrt{12}$ as $\sqrt{4 \cdot 3} = \sqrt{4} \cdot \sqrt{3} = 2\sqrt{3}$. Substituting $2\sqrt{3}$ for $\sqrt{12}$ gives

$$x = \frac{4 \pm 2\sqrt{3}}{2}$$

$$= \frac{2(2 \pm \sqrt{3})}{2} \qquad \text{Factor } 4 \pm 2\sqrt{3}.$$

$$= 2 \pm \sqrt{3}. \qquad \text{Reduce to lowest terms.}$$

The two solutions are $2 + \sqrt{3}$ and $2 - \sqrt{3}$.

The exact values of the solutions are $2 + \sqrt{3}$ and $2 - \sqrt{3}$. The $\sqrt{}$ key on a calculator gives decimal approximations of these solutions (to the nearest thousandth):

$$2 + \sqrt{3} \approx 2 + 1.732 = 3.732*$$

$$2 - \sqrt{3} \approx 2 - 1.732 = 0.268$$

NOTE Sometimes the quadratic formula will give a result with a negative number under the radical sign, such as $3 \pm \sqrt{-5}$. A solution of this type is not a real number. Since this text deals only with real numbers, such solutions cannot be used.

Equations with Fractions When an equation includes fractions, first eliminate all denominators by multiplying both sides of the equation by a common denominator, a number that can be divided (with no remainder) by each denominator in the equation. When an equation involves fractions with variable denominators, it is *necessary* to check all solutions in the original equation to be sure that no solution will lead to a zero denominator.

EXAMPLE 6 Solving Rational Equations
Solve each equation.

(a) $\dfrac{r}{10} - \dfrac{2}{15} = \dfrac{3r}{20} - \dfrac{1}{5}$

Solution The denominators are 10, 15, 20, and 5. Each of these numbers can be divided into 60, so 60 is a common denominator. Multiply both sides of the equation by 60 and use the distributive property. (If a common denomi-

---

*The symbol $\approx$ means "is approximately equal to."

nator cannot be found easily, all the denominators in the problem can be multiplied together to produce one.)

$$\frac{r}{10} - \frac{2}{15} = \frac{3r}{20} - \frac{1}{5}$$

$$60\left(\frac{r}{10} - \frac{2}{15}\right) = 60\left(\frac{3r}{20} - \frac{1}{5}\right) \qquad \text{Multiply by the common denominator.}$$

$$60\left(\frac{r}{10}\right) - 60\left(\frac{2}{15}\right) = 60\left(\frac{3r}{20}\right) - 60\left(\frac{1}{5}\right) \qquad \text{Distributive property}$$

$$6r - 8 = 9r - 12$$

Add $-9r$ and 8 to both sides.

$$6r - 8 + (-9r) + 8 = 9r - 12 + (-9r) + 8$$

$$-3r = -4$$

$$r = \frac{4}{3} \qquad \text{Multiply each side by } -\tfrac{1}{3}.$$

Check by substituting into the original equation.

(b) $\dfrac{3}{x^2} - 12 = 0$

Solution Begin by multiplying both sides of the equation by x^2 to get $3 - 12x^2 = 0$. This equation could be solved by using the quadratic formula with $a = -12$, $b = 0$, and $c = 3$. Another method, which works well for the type of quadratic equation in which $b = 0$, is shown below.

$$3 - 12x^2 = 0$$

$$3 = 12x^2 \qquad \text{Add } 12x^2.$$

$$\frac{1}{4} = x^2 \qquad \text{Multiply by } \tfrac{1}{12}.$$

$$\pm\frac{1}{2} = x \qquad \text{Take square roots.}$$

Verify that there are two solutions, $-1/2$ and $1/2$.

(c) $\dfrac{2}{k} - \dfrac{3k}{k+2} = \dfrac{k}{k^2 + 2k}$

Solution Factor $k^2 + 2k$ as $k(k + 2)$. The least common denominator for all the fractions is $k(k + 2)$. Multiplying both sides by $k(k + 2)$ gives the following.

$$2(k + 2) - 3k(k) = k$$

$$2k + 4 - 3k^2 = k \qquad \text{Distributive property}$$

$$-3k^2 + k + 4 = 0 \qquad \text{Add } -k; \text{ rearrange terms.}$$

$$3k^2 - k - 4 = 0 \qquad \text{Multiply by } -1.$$

$$(3k - 4)(k + 1) = 0 \qquad \text{Factor.}$$

$$3k - 4 = 0 \qquad \text{or} \qquad k + 1 = 0$$

$$k = \frac{4}{3} \qquad\qquad\qquad k = -1$$

Verify that the solutions are $4/3$ and -1.

> **CAUTION** It is possible to get, as a solution of a rational equation, a number that makes one or more of the denominators in the original equation equal to zero. That number is not a solution, so it is *necessary* to check all potential solutions of rational equations. These introduced solutions are called **extraneous solutions.**

EXAMPLE 7 Solving Rational Equations

Solve $\dfrac{2}{x-3} + \dfrac{1}{x} = \dfrac{6}{x(x-3)}$.

Solution The common denominator is $x(x-3)$. Multiply both sides by $x(x-3)$ and solve the resulting equation.

$$\frac{2}{x-3} + \frac{1}{x} = \frac{6}{x(x-3)}$$
$$2x + x - 3 = 6$$
$$3x = 9$$
$$x = 3$$

Checking this potential solution by substitution in the original equation shows that 3 makes two denominators 0. Thus 3 cannot be a solution, so there is no solution for this equation.

R.4 EXERCISES

Solve each equation.

1. $0.2m - 0.5 = 0.1m + 0.7$

2. $\dfrac{5}{6}k - 2k + \dfrac{1}{3} = \dfrac{2}{3}$

3. $2x + 8 = x - 4$

4. $5x + 2 = 8 - 3x$

5. $3r + 2 - 5(r + 1) = 6r + 4$

6. $5(a + 3) + 4a - 5 = -(2a - 4)$

7. $2[m - (4 + 2m) + 3] = 2m + 2$

8. $4[2p - (3 - p) + 5] = -7p - 2$

Solve each of the following equations by factoring or by using the quadratic formula. If the solutions involve square roots, give both the exact solutions and the approximate solutions to three decimal places.

9. $x^2 + 5x + 6 = 0$

10. $x^2 = 3 + 2x$

11. $m^2 + 16 = 8m$

12. $2k^2 - k = 10$

13. $6x^2 - 5x = 4$

14. $m(m - 7) = -10$

15. $9x^2 - 16 = 0$

16. $z(2z + 7) = 4$

17. $12y^2 - 48y = 0$

18. $3x^2 - 5x + 1 = 0$

19. $2m^2 = m + 4$

20. $p^2 + p - 1 = 0$

21. $k^2 - 10k = -20$

22. $2x^2 + 12x + 5 = 0$

23. $2r^2 - 7r + 5 = 0$

24. $2x^2 - 7x + 30 = 0$

25. $3k^2 + k = 6$

26. $5m^2 + 5m = 0$

Solve each of the following equations.

27. $\dfrac{3x - 2}{7} = \dfrac{x + 2}{5}$

28. $\dfrac{x}{3} - 7 = 6 - \dfrac{3x}{4}$

29. $\dfrac{4}{x - 3} - \dfrac{8}{2x + 5} + \dfrac{3}{x - 3} = 0$

30. $\dfrac{5}{2p + 3} - \dfrac{3}{p - 2} = \dfrac{4}{2p + 3}$

31. $\dfrac{2}{m} + \dfrac{m}{m + 3} = \dfrac{3m}{m^2 + 3m}$

32. $\dfrac{2y}{y - 1} = \dfrac{5}{y} + \dfrac{10 - 8y}{y^2 - y}$

33. $\dfrac{1}{x - 2} - \dfrac{3x}{x - 1} = \dfrac{2x + 1}{x^2 - 3x + 2}$ 　　**34.** $\dfrac{5}{a} + \dfrac{-7}{a + 1} = \dfrac{a^2 - 2a + 4}{a^2 + a}$ 　　**35.** $\dfrac{2b^2 + 5b - 8}{b^2 + 2b} + \dfrac{5}{b + 2} = -\dfrac{3}{b}$

36. $\dfrac{2}{x^2 - 2x - 3} + \dfrac{5}{x^2 - x - 6} = \dfrac{1}{x^2 + 3x + 2}$ 　　**37.** $\dfrac{2}{y^2 + 7y + 12} - \dfrac{1}{y^2 + 5y + 6} = \dfrac{5}{y^2 + 6y + 8}$

■ R.5 INEQUALITIES

To write that one number is greater than or less than another number, we use the following symbols.

> **INEQUALITY SYMBOLS**
>
> $<$ means *is less than* 　　　　$\leq$ means *is less than or equal to*
> $>$ means *is greater than* 　　　$\geq$ means *is greater than or equal to*

Linear Inequalities　An equation states that two expressions are equal; an **inequality** states that they are unequal. A **linear inequality** is an inequality that can be simplified to the form $ax < b$. (Properties introduced in this section are given only for $<$, but they are equally valid for $>$, $\leq$, or $\geq$.) Linear inequalities are solved with the following properties.

> **PROPERTIES OF INEQUALITY**
>
> For all real numbers a, b, and c:
>
> **1.** If $a < b$, then $a + c < b + c$.
> **2.** If $a < b$ and if $c > 0$, then $ac < bc$.
> **3.** If $a < b$ and if $c < 0$, then $ac > bc$.

Pay careful attention to property 3; it says that if both sides of an inequality are multiplied by a negative number, the direction of the inequality symbol must be reversed.

EXAMPLE 1　Solving Linear Inequalities
Solve $4 - 3y \leq 7 + 2y$.

Solution　Use the properties of inequality.

$$4 - 3y + (-4) \leq 7 + 2y + (-4) \qquad \text{Add } -4 \text{ to both sides.}$$
$$-3y \leq 3 + 2y$$

Remember that *adding* the same number to both sides never changes the direction of the inequality symbol.

$$-3y + (-2y) \leq 3 + 2y + (-2y) \qquad \text{Add } -2y \text{ to both sides.}$$
$$-5y \leq 3$$

Multiply both sides by $-1/5$. Since $-1/5$ is negative, change the direction of the inequality symbol.

$$-\frac{1}{5}(-5y) \ge -\frac{1}{5}(3)$$

$$y \ge -\frac{3}{5}$$

> **CAUTION** It is a common error to forget to reverse the direction of the inequality sign when multiplying or dividing by a negative number. For example, to solve $-4x \le 12$, we must multiply by $-1/4$ on both sides *and* reverse the inequality symbol to get $x \ge -3$.

The solution $y \ge -3/5$ in Example 1 represents an interval on the number line. **Interval notation** often is used for writing intervals. With interval notation, $y \ge -3/5$ is written as $[-3/5, \infty)$. This is an example of a **half-open interval,** since one endpoint, $-3/5$, is included. The **open interval** $(2, 5)$ corresponds to $2 < x < 5$, with neither endpoint included. The **closed interval** $[2, 5]$ includes both endpoints and corresponds to $2 \le x \le 5$.

The **graph** of an interval shows all points on a number line that correspond to the numbers in the interval. To graph the interval $[-3/5, \infty)$, for example, use a solid circle at $-3/5$, since $-3/5$ is part of the solution. To show that the solution includes all real numbers greater than or equal to $-3/5$, draw a heavy arrow pointing to the right (the positive direction). See Figure 1.

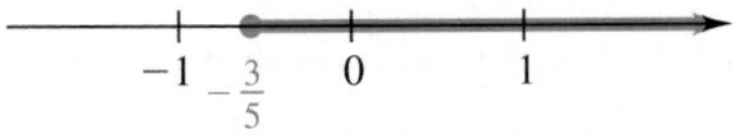

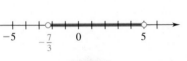

FIGURE 1

EXAMPLE 2   Graphing Linear Inequalities

Solve $-2 < 5 + 3m < 20$. Graph the solution.

Solution The inequality $-2 < 5 + 3m < 20$ says that $5 + 3m$ is *between* -2 and 20. Solve this inequality with an extension of the properties given on the previous page. Work as follows, first adding -5 to each part.

$$-2 + (-5) < 5 + 3m + (-5) < 20 + (-5)$$

$$-7 < 3m < 15$$

Now multiply each part by $1/3$.

$$-\frac{7}{3} < m < 5$$

A graph of the solution is given in Figure 2; here open circles are used to show that $-7/3$ and 5 are *not* part of the graph.*

FIGURE 2

Quadratic Inequalities   A **quadratic inequality** has the form $ax^2 + bx + c > 0$ (or $<$, or $\le$, or $\ge$). The highest exponent is 2. The next few examples show how to solve quadratic inequalities.

---

*Some textbooks use brackets in place of solid circles for the graph of a closed interval, and parentheses in place of open circles for the graph of an open interval.

EXAMPLE 3   Solving Quadratic Inequalities

Solve the quadratic inequality $x^2 - x < 12$.

Solution Write the inequality with 0 on one side, as $x^2 - x - 12 < 0$. This inequality is solved with values of x that make $x^2 - x - 12$ negative (<0). The quantity $x^2 - x - 12$ changes from positive to negative or from negative to positive at the points where it equals 0. For this reason, first solve the *equation* $x^2 - x - 12 = 0$.

$$x^2 - x - 12 = 0$$
$$(x - 4)(x + 3) = 0$$
$$x = 4 \quad \text{or} \quad x = -3$$

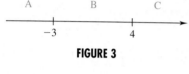

FIGURE 3

Locating -3 and 4 on a number line, as shown in Figure 3, determines three intervals A, B, and C. Decide which intervals include numbers that make $x^2 - x - 12$ negative by substituting any number from each interval in the polynomial. For example,

$$\text{choose } -4 \text{ from interval A: } (-4)^2 - (-4) - 12 = 8 > 0;$$
$$\text{choose } 0 \text{ from interval B: } 0^2 - 0 - 12 = -12 < 0;$$
$$\text{choose } 5 \text{ from interval C: } 5^2 - 5 - 12 = 8 > 0.$$

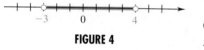

FIGURE 4

Only numbers in interval B satisfy the given inequality, so the solution is $(-3, 4)$. A graph of this solution is shown in Figure 4.

EXAMPLE 4   Solving Polynomial Inequalities

Solve the inequality $x(x - 1)(x + 3) \geq 0$.

Solution This is not a quadratic inequality. If the three factors are multiplied, the highest-degree term is x^3. However, it can be solved in the same way as a quadratic inequality because it is in factored form. First solve the corresponding equation.

$$x(x - 1)(x + 3) = 0$$
$$x = 0 \quad \text{or} \quad x - 1 = 0 \quad \text{or} \quad x + 3 = 0$$
$$x = 1 \qquad\qquad x = -3$$

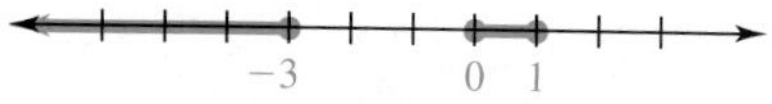

These three solutions determine four intervals on the number line: $(-\infty, -3)$, $(-3, 0)$, $(0, 1)$, and $(1, \infty)$. Substitute a number from each interval into the original inequality to determine that the solution includes the numbers less than or equal to -3 and the numbers that are equal to or between 0 and 1. See Figure 5. In interval notation, the solution is

FIGURE 5

$$(-\infty, -3] \cup [0, 1].*$$

Inequalities with Fractions

Inequalities with fractions are solved in a similar manner as quadratic inequalities.

EXAMPLE 5   Solving Rational Inequalities

Solve $\dfrac{2x - 3}{x} \geq 1$.

---

*The symbol $\cup$ indicates the *union* of two sets, which includes all elements in either set.

Solution First solve the corresponding equation.

$$\frac{2x - 3}{x} = 1$$

$$2x - 3 = x$$

$$x = 3$$

The solution, $x = 3$, determines the intervals on the number line where the fraction may change from greater than 1 to less than 1. This change also may occur on either side of a number that makes the denominator equal 0. Here, the x-value that makes the denominator 0 is $x = 0$. Test each of the three intervals determined by the numbers 0 and 3.

$$\text{For } (-\infty, 0), \text{ choose } -1: \frac{2(-1) - 3}{-1} = 5 \geq 1.$$

$$\text{For } (0, 3), \quad \text{choose} \quad 1: \frac{2(1) - 3}{1} = -1 \not\geq 1.$$

$$\text{For } (3, \infty), \quad \text{choose} \quad 4: \frac{2(4) - 3}{4} = \frac{5}{4} \geq 1.$$

The symbol $\not\geq$ means "is *not* greater than or equal to." Testing the endpoints 0 and 3 shows that the solution is $(-\infty, 0) \cup [3, \infty)$.

> **CAUTION** A common error is to try to solve the inequality in Example 5 by multiplying both sides by x. The reason this is wrong is that we don't know in the beginning whether x is positive or negative. If x is negative, the $\geq$ would change to $\leq$ according to the third property of inequality listed at the beginning of this section.

EXAMPLE 6 Solving Rational Inequalities

Solve $\dfrac{(x - 1)(x + 1)}{x} \leq 0.$

Solution Solve the corresponding equation.

$$\frac{(x - 1)(x + 1)}{x} = 0$$

$$(x - 1)(x + 1) = 0 \qquad \text{Multiply both sides by } x.$$

$$x = 1 \qquad \text{or} \qquad x = -1 \qquad \text{Use the zero-factor property.}$$

Setting the denominator equal to 0 gives $x = 0$, so the intervals of interest are $(-\infty, -1)$, $(-1, 0)$, and $(0, \infty)$. Testing a number from each region in the original inequality and checking the endpoints, we find the solution is

$$(-\infty, -1] \cup (0, 1].$$

> **CAUTION** Remember to solve the equation formed by setting the *denominator* equal to zero. Any number that makes the denominator zero always creates two intervals on the number line. For instance, in Example 6, 0 makes the denominator of the rational inequality equal to 0, so we know that there may be a sign change from one side of 0 to the other (as was indeed the case).

EXAMPLE 7 Solving Rational Inequalities

Solve $\dfrac{x^2 - 3x}{x^2 - 9} < 4$.

Solution Solve the corresponding equation.

$$\frac{x^2 - 3x}{x^2 - 9} = 4$$

$$x^2 - 3x = 4x^2 - 36 \qquad \text{Multiply by } x^2 - 9.$$

$$0 = 3x^2 + 3x - 36 \qquad \text{Get 0 on one side.}$$

$$0 = x^2 + x - 12 \qquad \text{Multiply by } \tfrac{1}{3}.$$

$$0 = (x + 4)(x - 3) \qquad \text{Factor.}$$

$$x = -4 \qquad \text{or} \qquad x = 3$$

Now set the denominator equal to 0 and solve that equation.

$$x^2 - 9 = 0$$

$$(x - 3)(x + 3) = 0$$

$$x = 3 \qquad \text{or} \qquad x = -3$$

The intervals determined by the three (different) solutions are $(-\infty, -4)$, $(-4, -3)$, $(-3, 3)$, and $(3, \infty)$. Testing a number from each interval in the given inequality shows that the solution is

$$(-\infty, -4) \cup (-3, 3) \cup (3, \infty).$$

For this example, none of the endpoints are part of the solution because $x = 3$ and $x = -3$ force the denominator to be zero and $x = -4$ produces an equality.

R.5 EXERCISES

Write each expression in interval notation. Graph each interval.

1. $x < 0$

2. $x \geq -3$

3. $1 \leq x < 2$

4. $-5 < x \leq -4$

5. $-9 > x$

6. $6 \leq x$

Using the variable x, write each interval in Exercises 7–14 as an inequality.

7. $(-4, 3)$

8. $[2, 7)$

9. $(-\infty, -1]$

10. $(3, \infty)$

11. (number line with points -2, 0, 6)

12. (number line with points 0, 8)

13. (number line with points -4, 0, 4)

14. (number line with points 0, 3)

Solve each inequality and graph the solution.

15. $-3p - 2 \geq 1$

16. $6k - 4 < 3k - 1$

17. $m - (4 + 2m) + 3 < 2m + 2$

18. $-2(3y - 8) \geq 5(4y - 2)$

19. $3p - 1 < 6p + 2(p - 1)$

20. $x + 5(x + 1) > 4(2 - x) + x$

21. $-7 < y - 2 < 4$

22. $8 \leq 3r + 1 \leq 13$

23. $-4 \le \dfrac{2k - 1}{3} \le 2$

24. $-1 \le \dfrac{5y + 2}{3} \le 4$

25. $\dfrac{3}{5}(2p + 3) \ge \dfrac{1}{10}(5p + 1)$

26. $\dfrac{8}{3}(z - 4) \le \dfrac{2}{9}(3z + 2)$

Solve each of the following quadratic inequalities. Graph each solution.

27. $(m + 2)(m - 4) < 0$

28. $(t + 6)(t - 1) \ge 0$

29. $y^2 - 3y + 2 < 0$

30. $2k^2 + 7k - 4 > 0$

31. $q^2 - 7q + 6 \le 0$

32. $2k^2 - 7k - 15 \le 0$

33. $6m^2 + m > 1$

34. $10r^2 + r \le 2$

35. $2y^2 + 5y \le 3$

36. $3a^2 + a > 10$

37. $x^2 \le 25$

38. $p^2 - 16p > 0$

Solve the following inequalities.

39. $\dfrac{m - 3}{m + 5} \le 0$

40. $\dfrac{r + 1}{r - 1} > 0$

41. $\dfrac{k - 1}{k + 2} > 1$

42. $\dfrac{a - 5}{a + 2} < -1$

43. $\dfrac{2y + 3}{y - 5} \le 1$

44. $\dfrac{a + 2}{3 + 2a} \le 5$

45. $\dfrac{7}{k + 2} \ge \dfrac{1}{k + 2}$

46. $\dfrac{5}{p + 1} > \dfrac{12}{p + 1}$

47. $\dfrac{3x}{x^2 - 1} < 2$

48. $\dfrac{8}{p^2 + 2p} > 1$

49. $\dfrac{z^2 + z}{z^2 - 1} \ge 3$

50. $\dfrac{a^2 + 2a}{a^2 - 4} \le 2$

R.6   EXPONENTS

Integer Exponents   Recall that $a^2 = a \cdot a$, while $a^3 = a \cdot a \cdot a$, and so on. In this section a more general meaning is given to the symbol a^n.

DEFINITION OF EXPONENT

If n is a natural number, then
$$a^n = a \cdot a \cdot a \cdot \ \cdots \ \cdot a,$$
where a appears as a factor n times.

In the expression a^n, n is the **exponent** and a is the **base.** This definition can be extended by defining a^n for zero and negative integer values of n.

ZERO AND NEGATIVE EXPONENTS

If a is any nonzero real number, and if n is a positive integer, then
$$a^0 = 1 \quad \text{and} \quad a^{-n} = \dfrac{1}{a^n}.$$

(The symbol 0^0 is meaningless.)

EXAMPLE 1 Exponents

(a) $6^0 = 1$

(b) $(-9)^0 = 1$

(c) $3^{-2} = \dfrac{1}{3^2} = \dfrac{1}{9}$

(d) $9^{-1} = \dfrac{1}{9^1} = \dfrac{1}{9}$

(e) $\left(\dfrac{3}{4}\right)^{-1} = \dfrac{1}{(3/4)^1} = \dfrac{1}{3/4} = \dfrac{4}{3}$

The following properties follow from the definitions of exponents given on the previous page.

PROPERTIES OF EXPONENTS

For any integers m and n, and any real numbers a and b for which the following exist:

1. $a^m \cdot a^n = a^{m+n}$ **4.** $(ab)^m = a^m \cdot b^m$

2. $\dfrac{a^m}{a^n} = a^{m-n}$ **5.** $\left(\dfrac{a}{b}\right)^m = \dfrac{a^m}{b^m}.$

3. $(a^m)^n = a^{mn}$

EXAMPLE 2 Simplifying Exponential Expressions

Use the properties of exponents to simplify each of the following. Leave answers with positive exponents. Assume that all variables represent positive real numbers.

(a) $7^4 \cdot 7^6 = 7^{4+6} = 7^{10}$ (or 282,475,249)   Property 1

(b) $\dfrac{9^{14}}{9^6} = 9^{14-6} = 9^8$ (or 43,046,721)   Property 2

(c) $\dfrac{r^9}{r^{17}} = r^{9-17} = r^{-8} = \dfrac{1}{r^8}$ Property 2

(d) $(2m^3)^4 = 2^4 \cdot (m^3)^4 = 16m^{12}$ Properties 3 and 4

(e) $(3x)^4 = 3^4 \cdot x^4 = 81x^4$ Property 4

(f) $\left(\dfrac{x^2}{y^3}\right)^6 = \dfrac{(x^2)^6}{(y^3)^6} = \dfrac{x^{2\cdot6}}{y^{3\cdot6}} = \dfrac{x^{12}}{y^{18}}$ Properties 4 and 5

(g) $\dfrac{a^{-3}b^5}{a^4 b^{-7}} = \dfrac{b^{5-(-7)}}{a^{4-(-3)}} = \dfrac{b^{5+7}}{a^{4+3}} = \dfrac{b^{12}}{a^7}$ Property 2

(h) $p^{-1} + q^{-1} = \dfrac{1}{p} + \dfrac{1}{q} = \dfrac{1}{p} \cdot \dfrac{q}{q} + \dfrac{1}{q} \cdot \dfrac{p}{p} = \dfrac{q}{pq} + \dfrac{p}{pq} = \dfrac{p+q}{pq}$

(i) $\dfrac{x^{-2} - y^{-2}}{x^{-1} - y^{-1}} = \dfrac{\dfrac{1}{x^2} - \dfrac{1}{y^2}}{\dfrac{1}{x} - \dfrac{1}{y}}$ Definition of a^{-n}

$= \dfrac{\dfrac{y^2 - x^2}{x^2 y^2}}{\dfrac{y - x}{xy}}$ Get common denominators and combine terms.

$= \dfrac{y^2 - x^2}{x^2 y^2} \cdot \dfrac{xy}{y - x}$ Invert and multiply.

$= \dfrac{(y - x)(y + x)}{x^2 y^2} \cdot \dfrac{xy}{y - x}$ Factor.

$= \dfrac{x + y}{xy}$ Simplify.

> **CAUTION** If Example 2(e) were written $3x^4$, the properties of exponents would not apply. When no parentheses are used, the exponent refers only to the factor closest to it. Also notice in Examples 2(c), 2(g), 2(h), and 2(i) that a negative exponent does *not* indicate a negative number.

Roots   For *even* values of n, the expression $a^{1/n}$ is defined to be the **positive nth root** of a or the **principal nth root** of a. For example, $a^{1/2}$ denotes the positive second root, or **square root,** of a, while $a^{1/4}$ is the positive fourth root of a. When n is *odd,* there is only one nth root, which has the same sign as a. For example, $a^{1/3}$, the **cube root** of a, has the same sign as a. By definition, if $b = a^{1/n}$, then $b^n = a$. On a calculator, a number is raised to a power using a key labeled x^y, y^x, or $\wedge$. For example, to take the fourth root of 6 on a TI-83 calculator, enter $6 \wedge (1/4)$, to get the result 1.56508458.

EXAMPLE 3   Calculations with Exponents

(a) $121^{1/2} = 11$, since 11 is positive and $11^2 = 121$.

(b) $625^{1/4} = 5$, since $5^4 = 625$.

(c) $256^{1/4} = 4$

(d) $64^{1/6} = 2$

(e) $27^{1/3} = 3$

(f) $(-32)^{1/5} = -2$

(g) $128^{1/7} = 2$

(h) $(-49)^{1/2}$ is not a real number.

Rational Exponents   In the following definition, the domain of an exponent is extended to include all rational numbers.

DEFINITION OF $a^{m/n}$

For all real numbers a for which the indicated roots exist, and for any rational number m/n,

$$a^{m/n} = (a^{1/n})^m.$$

EXAMPLE 4 Calculations with Exponents

(a) $27^{2/3} = (27^{1/3})^2 = 3^2 = 9$

(b) $32^{2/5} = (32^{1/5})^2 = 2^2 = 4$

(c) $64^{4/3} = (64^{1/3})^4 = 4^4 = 256$

(d) $25^{3/2} = (25^{1/2})^3 = 5^3 = 125$

NOTE $27^{2/3}$ could also be evaluated as $(27^2)^{1/3}$, but this is more difficult to perform without a calculator because it involves squaring 27, and then taking the cube root of this large number. On the other hand, when we evaluate it as $(27^{1/3})^2$, we know that the cube root of 27 is 3 without using a calculator, and squaring 3 is easy.

All the properties for integer exponents given in this section also apply to any rational exponent on a nonnegative real-number base.

EXAMPLE 5 Simplifying Exponential Expressions

(a) $\dfrac{y^{1/3}y^{5/3}}{y^3} = \dfrac{y^{1/3+5/3}}{y^3} = \dfrac{y^2}{y^3} = y^{2-3} = y^{-1} = \dfrac{1}{y}$

(b) $m^{2/3}(m^{7/3} + 2m^{1/3}) = m^{2/3+7/3} + 2m^{2/3+1/3} = m^3 + 2m$

(c) $\left(\dfrac{m^7n^{-2}}{m^{-5}n^2}\right)^{1/4} = \left(\dfrac{m^{7-(-5)}}{n^{2-(-2)}}\right)^{1/4} = \left(\dfrac{m^{12}}{n^4}\right)^{1/4} = \dfrac{(m^{12})^{1/4}}{(n^4)^{1/4}} = \dfrac{m^{12/4}}{n^{4/4}} = \dfrac{m^3}{n}$

In calculus, it is often necessary to factor expressions involving fractional exponents.

EXAMPLE 6 Simplifying Exponential Expressions

Factor out the smallest power of the variable, assuming all variables represent positive real numbers.

(a) $4m^{1/2} + 3m^{3/2} = m^{1/2}(4 + 3m)$

Solution To check this result, multiply $m^{1/2}$ by $4 + 3m$.

(b) $9x^{-2} - 6x^{-3}$

Solution The smallest exponent here is -3. Since 3 is a common numerical factor, factor out $3x^{-3}$.

$$9x^{-2} - 6x^{-3} = 3x^{-3}(3x^{-2-(-3)} - 2x^{-3-(-3)}) = 3x^{-3}(3x - 2)$$

Check by multiplying. The factored form can be written without negative exponents as

$$\frac{3(3x - 2)}{x^3}.$$

(c) $(x^2 + 5)(3x - 1)^{-1/2}(2) + (3x - 1)^{1/2}(2x)$.

Solution There is a common factor of 2. Also, $(3x - 1)^{-1/2}$ and $(3x - 1)^{1/2}$ have a common factor. Always factor out the quantity to the *smallest* exponent. Here $-1/2 < 1/2$, so the common factor is $2(3x - 1)^{-1/2}$ and the factored form is

$$2(3x - 1)^{-1/2}[(x^2 + 5) + (3x - 1)x] = 2(3x - 1)^{-1/2}(4x^2 - x + 5).$$

R.6 EXERCISES

Evaluate each expression. Write all answers without exponents.

1. 8^{-2} **2.** 3^{-4} **3.** 5^0 **4.** $(-12)^0$

5. $-(-3)^{-2}$ **6.** $-(-3^{-2})$ **7.** $\left(\dfrac{2}{7}\right)^{-2}$ **8.** $\left(\dfrac{4}{3}\right)^{-3}$

Simplify each expression. Assume that all variables represent positive real numbers. Write answers with only positive exponents.

9. $\dfrac{3^{-4}}{3^2}$ **10.** $\dfrac{8^9 \cdot 8^{-7}}{8^{-3}}$ **11.** $\dfrac{10^8 \cdot 10^{-10}}{10^4 \cdot 10^2}$ **12.** $\left(\dfrac{5^{-6} \cdot 5^3}{5^{-2}}\right)^{-1}$

13. $\dfrac{x^4 \cdot x^3}{x^5}$ **14.** $\dfrac{y^9 \cdot y^7}{y^{13}}$ **15.** $\dfrac{(4k^{-1})^2}{2k^{-5}}$ **16.** $\dfrac{(3z^2)^{-1}}{z^5}$

17. $\dfrac{2^{-1}x^3y^{-3}}{xy^{-2}}$ **18.** $\dfrac{5^{-2}m^2y^{-2}}{5^2m^{-1}y^{-2}}$ **19.** $\left(\dfrac{a^{-1}}{b^2}\right)^{-3}$ **20.** $\left(\dfrac{2c^2}{d^3}\right)^{-2}$

21. $\left(\dfrac{x^6y^{-3}}{x^{-2}y^5}\right)^{1/2}$ **22.** $\left(\dfrac{a^{-7}b^{-1}}{b^{-4}a^2}\right)^{1/3}$

Simplify each expression, writing the answer as a single term without negative exponents.

23. $a^{-1} + b^{-1}$ **24.** $b^{-2} - a$ **25.** $\dfrac{2n^{-1} - 2m^{-1}}{m + n^2}$

26. $\left(\dfrac{m}{3}\right)^{-1} + \left(\dfrac{n}{2}\right)^{-2}$ **27.** $(x^{-1} - y^{-1})^{-1}$ **28.** $(x^{-2} + y^{-2})^{-2}$

Write each number without exponents.

29. $81^{1/2}$ **30.** $27^{1/3}$ **31.** $32^{2/5}$ **32.** $-125^{2/3}$

33. $\left(\dfrac{4}{9}\right)^{1/2}$ **34.** $\left(\dfrac{64}{27}\right)^{1/3}$ **35.** $16^{-5/4}$ **36.** $625^{-1/4}$

37. $\left(\dfrac{27}{64}\right)^{-1/3}$ **38.** $\left(\dfrac{121}{100}\right)^{-3/2}$

Simplify each expression. Write all answers with only positive exponents. Assume that all variables represent positive real numbers.

39. $2^{1/2} \cdot 2^{3/2}$ **40.** $27^{2/3} \cdot 27^{-1/3}$ **41.** $\dfrac{4^{2/3} \cdot 4^{5/3}}{4^{1/3}}$

42. $\dfrac{3^{-5/2} \cdot 3^{3/2}}{3^{7/2} \cdot 3^{-9/2}}$ **43.** $\dfrac{7^{-1/3} \cdot 7r^{-3}}{7^{2/3} \cdot (r^{-2})^2}$ **44.** $\dfrac{12^{3/4} \cdot 12^{5/4} \cdot y^{-2}}{12^{-1} \cdot (y^{-3})^{-2}}$

45. $\dfrac{6k^{-4} \cdot (3k^{-1})^{-2}}{2^3 \cdot k^{1/2}}$

46. $\dfrac{8p^{-3} \cdot (4p^2)^{-2}}{p^{-5}}$

47. $\dfrac{a^{4/3} \cdot b^{1/2}}{a^{2/3} \cdot b^{-3/2}}$

48. $\dfrac{x^{1/3} \cdot y^{2/3} \cdot z^{1/4}}{x^{5/3} \cdot y^{-1/3} \cdot z^{3/4}}$

49. $\dfrac{k^{-3/5} \cdot h^{-1/3} \cdot t^{2/5}}{k^{-1/5} \cdot h^{-2/3} \cdot t^{1/5}}$

50. $\dfrac{m^{7/3} \cdot n^{-2/5} \cdot p^{3/8}}{m^{-2/3} \cdot n^{3/5} \cdot p^{-5/8}}$

Factor each expression.

51. $12x^2(x^2 + 2)^2 - 4x(4x^3 + 1)(x^2 + 2)$

52. $6x(x^3 + 7)^2 - 6x^2(3x^2 + 5)(x^3 + 7)$

53. $(x^2 + 2)(x^2 - 1)^{-1/2}(x) + (x^2 - 1)^{1/2}(2x)$

54. $9(6x + 2)^{1/2} + 3(9x - 1)(6x + 2)^{-1/2}$

55. $x(2x + 5)^2(x^2 - 4)^{-1/2} + 2(x^2 - 4)^{1/2}(2x + 5)$

56. $(4x^2 + 1)^2(2x - 1)^{-1/2} + 16x(4x^2 + 1)(2x - 1)^{1/2}$

■ R.7   RADICALS

We have defined $a^{1/n}$ as the positive or principal nth root of a for appropriate values of a and n. An alternative notation for $a^{1/n}$ uses radicals.

RADICALS

If n is an even natural number and $a > 0$, or n is an odd natural number, then

$$a^{1/n} = \sqrt[n]{a}.$$

The symbol $\sqrt{}$ is a **radical sign,** the number a is the **radicand,** and n is the **index** of the radical. The familiar symbol $\sqrt{a}$ is used instead of $\sqrt[2]{a}$.

EXAMPLE 1   Radical Calculations

(a) $\sqrt[4]{16} = 16^{1/4} = 2$ **(b)** $\sqrt[5]{-32} = -2$

(c) $\sqrt[3]{1{,}000} = 10$ **(d)** $\sqrt[6]{\dfrac{64}{729}} = \dfrac{2}{3}$

With $a^{1/n}$ written as $\sqrt[n]{a}$, $a^{m/n}$ also can be written using radicals.

$$a^{m/n} = \left(\sqrt[n]{a}\right)^m \qquad \text{or} \qquad a^{m/n} = \sqrt[n]{a^m}$$

The following properties of radicals depend on the definitions and properties of exponents.

PROPERTIES OF RADICALS

For all real numbers a and b and natural numbers m and n such that $\sqrt[n]{a}$ and $\sqrt[n]{b}$ are real numbers:

1. $\left(\sqrt[n]{a}\right)^n = a$ **4.** $\dfrac{\sqrt[n]{a}}{\sqrt[n]{b}} = \sqrt[n]{\dfrac{a}{b}}$ $(b \neq 0)$

2. $\sqrt[n]{a^n} = \begin{cases} |a| & \text{if } n \text{ is even} \\ a & \text{if } n \text{ is odd} \end{cases}$ **5.** $\sqrt[m]{\sqrt[n]{a}} = \sqrt[mn]{a}$

3. $\sqrt[n]{a} \cdot \sqrt[n]{b} = \sqrt[n]{ab}$

Property 3 can be used to simplify certain radicals. For example, since $48 = 16 \cdot 3$,

$$\sqrt{48} = \sqrt{16 \cdot 3} = \sqrt{16} \cdot \sqrt{3} = 4\sqrt{3}.$$

To some extent, simplification is in the eye of the beholder, and $\sqrt{48}$ might be considered as simple as $4\sqrt{3}$. In this textbook, we will consider an expression to be simpler when we have removed as many factors as possible from under the radical.

EXAMPLE 2   Radical Calculations

(a) $\sqrt{1{,}000} = \sqrt{100 \cdot 10} = \sqrt{100} \cdot \sqrt{10} = 10\sqrt{10}$

(b) $\sqrt{128} = \sqrt{64 \cdot 2} = 8\sqrt{2}$

(c) $\sqrt{108} = \sqrt{36 \cdot 3} = 6\sqrt{3}$

(d) $\sqrt[3]{54} = \sqrt[3]{27 \cdot 2} = \sqrt[3]{27} \cdot \sqrt[3]{2} = 3\sqrt[3]{2}$

(e) $\sqrt{288m^5} = \sqrt{144 \cdot m^4 \cdot 2m} = 12m^2\sqrt{2m}$

(f) $2\sqrt{18} - 5\sqrt{32} = 2\sqrt{9 \cdot 2} - 5\sqrt{16 \cdot 2}$
$$= 2\sqrt{9} \cdot \sqrt{2} - 5\sqrt{16} \cdot \sqrt{2}$$
$$= 2(3)\sqrt{2} - 5(4)\sqrt{2} = -14\sqrt{2}$$

Rationalizing Denominators   The next example shows how to *rationalize* (remove all radicals from) the denominator in an expression containing radicals.

EXAMPLE 3   Rationalizing the Denominator
Simplify each of the following expressions by rationalizing the denominator.

(a) $\dfrac{4}{\sqrt{3}}$

Solution To rationalize the denominator, multiply by $\sqrt{3}/\sqrt{3}$ (or 1) so that the denominator of the product is a rational number.

$$\frac{4}{\sqrt{3}} \cdot \frac{\sqrt{3}}{\sqrt{3}} = \frac{4\sqrt{3}}{3}$$

(b) $\dfrac{2}{\sqrt[3]{x}}$

Solution Here, we need a perfect cube under the radical sign to rationalize the denominator. Multiplying by $\sqrt[3]{x^2}/\sqrt[3]{x^2}$ gives

$$\frac{2}{\sqrt[3]{x}} \cdot \frac{\sqrt[3]{x^2}}{\sqrt[3]{x^2}} = \frac{2\sqrt[3]{x^2}}{\sqrt[3]{x^3}} = \frac{2\sqrt[3]{x^2}}{x}.$$

(c) $\dfrac{1}{1 - \sqrt{2}}$

Solution The best approach here is to multiply both numerator and denominator by the number $1 + \sqrt{2}$. The expressions $1 + \sqrt{2}$ and $1 - \sqrt{2}$ are conjugates.* Thus,

$$\frac{1}{1 - \sqrt{2}} = \frac{1(1 + \sqrt{2})}{(1 - \sqrt{2})(1 + \sqrt{2})} = \frac{1 + \sqrt{2}}{1 - 2} = -1 - \sqrt{2}. \quad \blacksquare$$

Sometimes it is advantageous to rationalize the *numerator* of a rational expression. The following example arises in calculus when evaluating a *limit*.

EXAMPLE 4   Rationalizing the Numerator
Rationalize the numerator.

(a) $\dfrac{\sqrt{x} - 3}{x - 9}$.

Solution Multiply numerator and denominator by the conjugate of the numerator, $\sqrt{x} + 3$.

$$\frac{\sqrt{x} - 3}{x - 9} \cdot \frac{\sqrt{x} + 3}{\sqrt{x} + 3} = \frac{\sqrt{x} \cdot \sqrt{x} - 3\sqrt{x} + 3\sqrt{x} - 9}{(x - 9)(\sqrt{x} + 3)}$$

$$= \frac{x - 9}{(x - 9)(\sqrt{x} + 3)}$$

$$= \frac{1}{\sqrt{x} + 3}$$

(b) $\dfrac{\sqrt{3} + \sqrt{x + 3}}{\sqrt{3} - \sqrt{x + 3}}$

Solution Multiply the numerator and denominator by the conjugate of the numerator, $\sqrt{3} - \sqrt{x + 3}$.

$$\frac{\sqrt{3} + \sqrt{x + 3}}{\sqrt{3} - \sqrt{x + 3}} \cdot \frac{\sqrt{3} - \sqrt{x + 3}}{\sqrt{3} - \sqrt{x + 3}} = \frac{3 - (x + 3)}{3 - 2\sqrt{3}\sqrt{x + 3} + (x + 3)}$$

$$= \frac{-x}{6 + x - 2\sqrt{3(x + 3)}} \quad \blacksquare$$

When simplifying a square root, keep in mind that $\sqrt{x}$ is positive by definition. Also, $\sqrt{x^2}$ is not x, but $|x|$, the **absolute value of x,** defined as

$$|x| = \begin{cases} x \text{ if } x \geq 0 \\ -x \text{ if } x < 0. \end{cases}$$

For example, $\sqrt{(-5)^2} = |-5| = 5$.

EXAMPLE 5   Simplifying by Factoring
Simplify $\sqrt{m^2 - 4m + 4}$.

Solution Factor the polynomial as $m^2 - 4m + 4 = (m - 2)^2$. Then by property 2 of radicals, and the definition of absolute value,

$$\sqrt{(m - 2)^2} = |m - 2| = \begin{cases} m - 2 & \text{if } m - 2 \geq 0 \\ -(m - 2) = 2 - m & \text{if } m - 2 < 0. \end{cases} \quad \blacksquare$$

---

*If a and b are real numbers, the *conjugate* of $a + b$ is $a - b$.

CAUTION Avoid the common error of writing $\sqrt{a^2 + b^2}$ as $\sqrt{a^2} + \sqrt{b^2}$. We must add a^2 and b^2 *before* taking the square root. For example, $\sqrt{16 + 9} = \sqrt{25} = 5$, *not* $\sqrt{16} + \sqrt{9} = 4 + 3 = 7$. This idea applies as well to higher roots. For example, in general,

$$\sqrt[3]{a^3 + b^3} \neq \sqrt[3]{a^3} + \sqrt[3]{b^3},$$

$$\sqrt[4]{a^4 + b^4} \neq \sqrt[4]{a^4} + \sqrt[4]{b^4}.$$

Also, $$\sqrt{a + b} \neq \sqrt{a} + \sqrt{b}.$$

R.7 EXERCISES

Simplify each expression by removing as many factors as possible from under the radical. Assume that all variables represent positive real numbers.

1. $\sqrt[3]{125}$

2. $\sqrt[4]{1{,}296}$

3. $\sqrt[5]{-3{,}125}$

4. $\sqrt{50}$

5. $\sqrt{2{,}000}$

6. $\sqrt{32y^5}$

7. $7\sqrt{2} - 8\sqrt{18} + 4\sqrt{72}$

8. $4\sqrt{3} - 5\sqrt{12} + 3\sqrt{75}$

9. $2\sqrt{5} - 3\sqrt{20} + 2\sqrt{45}$

10. $3\sqrt{28} - 4\sqrt{63} + \sqrt{112}$

11. $\sqrt[3]{2} - \sqrt[3]{16} + 2\sqrt[3]{54}$

12. $2\sqrt[3]{3} + 4\sqrt[3]{24} - \sqrt[3]{81}$

13. $\sqrt[3]{32} - 5\sqrt[3]{4} + 2\sqrt[3]{108}$

14. $\sqrt{2x^3y^2z^4}$

15. $\sqrt{98r^3s^4t^{10}}$

16. $\sqrt[3]{16x^8y^4z^5}$

17. $\sqrt[4]{x^8y^7z^{11}}$

18. $\sqrt{a^3b^5} - 2\sqrt{a^7b^3} + \sqrt{a^3b^9}$

19. $\sqrt{p^7q^3} - \sqrt{p^5q^9} + \sqrt{p^9q}$

Rationalize each denominator. Assume that all radicands represent positive real numbers.

20. $\dfrac{5}{\sqrt{7}}$

21. $\dfrac{-2}{\sqrt{3}}$

22. $\dfrac{-3}{\sqrt{12}}$

23. $\dfrac{4}{\sqrt{8}}$

24. $\dfrac{3}{1 - \sqrt{5}}$

25. $\dfrac{5}{2 - \sqrt{6}}$

26. $\dfrac{-2}{\sqrt{3} - \sqrt{2}}$

27. $\dfrac{1}{\sqrt{10} + \sqrt{3}}$

28. $\dfrac{1}{\sqrt{r} - \sqrt{3}}$

29. $\dfrac{5}{\sqrt{m} - \sqrt{5}}$

30. $\dfrac{y - 5}{\sqrt{y} - \sqrt{5}}$

31. $\dfrac{z - 11}{\sqrt{z} - \sqrt{11}}$

32. $\dfrac{\sqrt{x} + \sqrt{x + 1}}{\sqrt{x} - \sqrt{x + 1}}$

33. $\dfrac{\sqrt{p} + \sqrt{p^2 - 1}}{\sqrt{p} - \sqrt{p^2 - 1}}$

Rationalize each numerator. Assume that all radicands represent positive real numbers.

34. $\dfrac{1 + \sqrt{2}}{2}$

35. $\dfrac{1 - \sqrt{3}}{3}$

36. $\dfrac{\sqrt{x} + \sqrt{x + 1}}{\sqrt{x} - \sqrt{x + 1}}$

37. $\dfrac{\sqrt{p} + \sqrt{p^2 - 1}}{\sqrt{p} - \sqrt{p^2 - 1}}$

Simplify each root, if possible.

38. $\sqrt{16 - 8x + x^2}$

39. $\sqrt{4y^2 + 4y + 1}$

40. $\sqrt{4 - 25z^2}$

41. $\sqrt{9k^2 + h^2}$

1

Functions

Over short time intervals, many changes in biology are well modeled by linear functions. In an exercise in the second section of this chapter, we will examine a linear model that predicts the skeletal maturity of children given their age. Such functions are useful to pediatricians in determining the health of children.

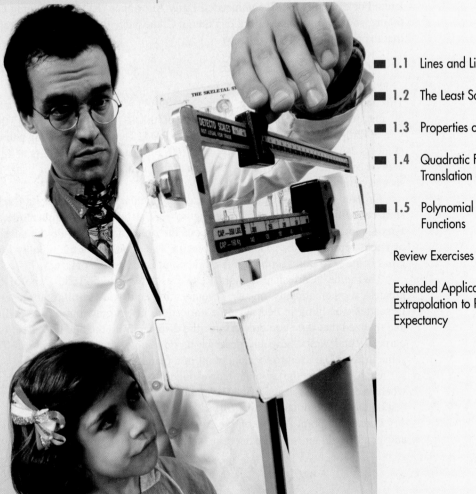

■ 1.1   Lines and Linear Functions

■ 1.2   The Least Squares Line

■ 1.3   Properties of Functions

■ 1.4   Quadratic Functions;
        Translation and Reflection

■ 1.5   Polynomial and Rational
        Functions

Review Exercises

Extended Application: Using
Extrapolation to Predict Life
Expectancy

Before using mathematics to solve a real-world problem, we usually must set up a **mathematical model,** a mathematical description of the situation. Constructing such a model requires a solid understanding of the situation to be modeled, as well as familiarity with relevant mathematical ideas and techniques.

Many mathematical models involve the concept of a *function,* which we will define and explore in this chapter. The simplest type of function is a *linear function,* so we will examine linear functions first. Despite their simplicity, or perhaps because of it, linear functions are widely used in the life sciences, as we will see in the examples and exercises. Linear functions are also a special case of polynomial functions, which we will study later in this chapter.

■ 1.1 LINES AND LINEAR FUNCTIONS

? THINK ABOUT IT   How does the amount of energy expended by an elk vary with the animal's running speed?

In Example 11 of this section we will answer this question using a linear function.

There are many situations in the life sciences in which two quantities are related. For example, the Recommended Daily Allowance (RDA) of Vitamin C is 60 mg, so if an adult takes x mg of Vitamin C, the percent of the daily allowance that the person receives is given by

$$y = 100\left(\frac{x}{60}\right), \qquad \text{or} \qquad y = \frac{5x}{3}.$$

This means that if $x = 60$ mg, then $y = 5(60)/3 = 100$; a person who takes 60 mg of Vitamin C has received 100% of the RDA. If $x = 120$ mg, then $y = 5(120)/3 = 200$; 120 mg is 200% of the RDA. These corresponding pairs of numbers can be written as **ordered pairs,** $(60, 100)$ and $(120, 200)$, pairs of numbers whose order is important. The first number denotes the value of x and the second number the value of y.

Ordered pairs are **graphed** with the perpendicular number lines of a **Cartesian coordinate system,** shown in Figure 1. The horizontal number line, or **x-axis,** represents the first components of the ordered pairs, while the vertical or **y-axis** represents the second components. The point where the number lines cross is the zero point on both lines; this point is called the **origin.**

The name "Cartesian" honors René Descartes (1596–1650), one of the greatest mathematicians of the seventeenth century. According to legend, Descartes was lying in bed when he noticed an insect crawling on the ceiling and realized that if he could determine the distance from the bug to each of two perpendicular walls, he could describe its position at any given moment. The same idea can be used to locate a point in a plane.

Each point on the *xy*-plane corresponds to an ordered pair of numbers, where the *x*-value is written first. From now on, we will refer to the point corresponding to the ordered pair (a, b) as "the point (a, b)."

Locate the point $(-2, 4)$ on the coordinate system by starting at the origin and counting 2 units to the left on the horizontal axis and 4 units upward, parallel to the vertical axis. This point is shown in Figure 1, along with several other sample points. The number -2 is the **x-coordinate** and the number 4 is the **y-coordinate** of the point $(-2, 4)$.

The x-axis and y-axis divide the plane into four parts, or **quadrants.** For example, quadrant I includes all those points whose x- and y-coordinates are both positive. The quadrants are numbered as shown in Figure 1. The points on the axes themselves belong to no quadrant. The set of points corresponding to the ordered pairs of an equation is the **graph** of the equation.

The x- and y-values of the points where the graph of an equation crosses the axes are called the **x-intercept** and **y-intercept,** respectively.* See Figure 2.

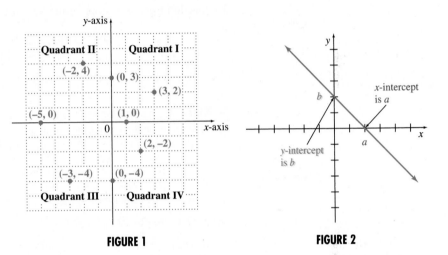

FIGURE 1   **FIGURE 2**

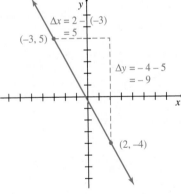

FIGURE 3

Slope of a Line

An important characteristic of a straight line is its *slope,* a number that represents the "steepness" of the line. To see how slope is defined, look at the line in Figure 3. The line goes through the points $(x_1, y_1) = (-3, 5)$ and $(x_2, y_2) = (2, -4)$. The difference in the two x-values,

$$x_2 - x_1 = 2 - (-3) = 5$$

in this example, is called the **change in x.** The symbol Δx (read "delta x") is used to represent the change in x. In the same way, Δy represents the **change in y.** In our example,

$$\Delta y = y_2 - y_1$$
$$= -4 - 5$$
$$= -9.$$

These symbols, Δx and Δy, are used in the following definition of slope.

---

*Some people prefer to define the intercepts as ordered pairs, rather than as numbers.

SLOPE OF A LINE

The **slope** of a line is defined as the vertical change (the "rise") over the horizontal change (the "run") as one travels along the line. In symbols, taking two different points (x_1, y_1) and (x_2, y_2) on the line, the slope is

$$m = \frac{\textbf{Change in } y}{\textbf{Change in } x} = \frac{\Delta y}{\Delta x} = \frac{y_2 - y_1}{x_2 - x_1},$$

where $x_1 \neq x_2$.

By this definition, the slope of the line in Figure 3 is

$$m = \frac{\Delta y}{\Delta x} = \frac{-4 - 5}{2 - (-3)} = -\frac{9}{5}.$$

The slope of a line tells how fast y changes for each unit of change in x.

NOTE Using similar triangles, it can be shown that the slope of a line is independent of the choice of points on the line. That is, the same slope will be obtained for *any* choice of two different points on the line.

EXAMPLE 1 Slope

Find the slope of the line through each of the following pairs of points.

(a) $(-7, 6)$ and $(4, 5)$

Solution Let $(x_1, y_1) = (-7, 6)$ and $(x_2, y_2) = (4, 5)$. Use the definition of slope.

$$m = \frac{\Delta y}{\Delta x} = \frac{5 - 6}{4 - (-7)} = -\frac{1}{11}$$

(b) $(5, -3)$ and $(-2, -3)$

Solution Let $(x_1, y_1) = (5, -3)$ and $(x_2, y_2) = (-2, -3)$. Then

$$m = \frac{-3 - (-3)}{-2 - 5} = \frac{0}{-7} = 0.$$

Lines with zero slope are horizontal (parallel to the x-axis).

(c) $(2, -4)$ and $(2, 3)$

Solution Let $(x_1, y_1) = (2, -4)$ and $(x_2, y_2) = (2, 3)$. Then

$$m = \frac{3 - (-4)}{2 - 2} = \frac{7}{0},$$

which is undefined. This happens when the line is vertical (parallel to the y-axis).

The slope of a horizontal line is 0.
The slope of a vertical line is undefined.

CAUTION The phrase "no slope" should be avoided; specify instead whether the slope is zero or undefined.

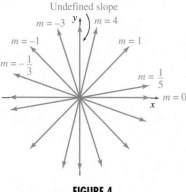

Undefined slope

FIGURE 4

In finding the slope of the line in Example 1(a) we could have let $(x_1, y_1) = (4, 5)$ and $(x_2, y_2) = (-7, 6)$. In that case,

$$m = \frac{6 - 5}{-7 - 4} = \frac{1}{-11} = -\frac{1}{11},$$

the same answer as before. The order in which coordinates are subtracted does not matter, as long as it is done consistently.

Figure 4 shows examples of lines with different slopes. Lines with positive slopes go up from left to right, while lines with negative slopes go down from left to right.

It might help you to compare slope with the percent grade of a hill. If a sign says a hill has a 10% grade uphill, this means the slope is 0.10, or 1/10, so the hill rises 1 foot for every 10 feet horizontally. A 15% grade downhill means the slope is -0.15.

FOR REVIEW ■■———

For review on solving a linear equation, see Section R.4.

Equations of a Line   An equation in two first-degree variables, such as $y = 5x - 7$, has a line as its graph, so it is called a **linear equation.** In the rest of this section, we consider various forms of the equation of a line.

EXAMPLE 2   Equation of a Line
Find the equation of the line through $(0, -3)$ with slope $3/4$.

Solution   We can use the definition of slope, letting $(x_1, y_1) = (0, -3)$ and (x, y) represent another point on the line.

$$m = \frac{y_2 - y_1}{x_2 - x_1}$$

$$\frac{3}{4} = \frac{y - (-3)}{x - 0} = \frac{y + 3}{x} \qquad \text{Substitute.}$$

$$3x = 4(y + 3) \qquad \text{Cross multiply.}$$

$$3x = 4y + 12$$

It is often useful to solve such an equation for y.

$$4y = 3x - 12 \qquad \text{Subtract 12.}$$

$$y = \frac{3}{4}x - 3 \qquad \text{Divide by 4.}$$

Notice that the slope (3/4) is the coefficient of x, and the y-intercept (-3) is the constant term.   ■

A generalization of the method of Example 2 can be used to find the equation of any line, given its y-intercept and slope. Assume that a line has y-intercept b, so that it goes through the point $(0, b)$. Let the slope of the line be represented by m. If (x, y) is any point on the line *other* than $(0, b)$, then the definition of slope can be used with the points $(0, b)$ and (x, y) to get

$$m = \frac{y - b}{x - 0}$$

$$m = \frac{y - b}{x}$$

$$mx = y - b$$

$$y = mx + b.$$

This result is called the *slope-intercept form* of the equation of a line, because b is the y-intercept of the graph of the line.

SLOPE-INTERCEPT FORM

If a line has slope m and y-intercept b, then the equation of the line in **slope-intercept form** is

$$y = mx + b.$$

When $b = 0$, we say that y is **proportional** to x.

The slope-intercept form of the equation of a line involves the slope and the y-intercept. Sometimes, however, the slope of a line is known, together with one point (perhaps *not* the y-intercept) that the line goes through. The *point-slope form* of the equation of a line is used to find the equation in this case. Let (x_1, y_1) be any fixed point on the line and let (x, y) represent any other point on the line. If m is the slope of the line, then by the definition of slope,

$$\frac{y - y_1}{x - x_1} = m,$$

or

$$y - y_1 = m(x - x_1).$$

POINT-SLOPE FORM

If a line has slope m and passes through the point (x_1, y_1), then an equation of the line is given by

$$y - y_1 = m(x - x_1),$$

the **point-slope form** of the equation of a line.

The point-slope form also can be useful to find an equation of a line if we know two different points that the line goes through. The procedure for doing this is shown in the next example.

EXAMPLE 3   Using Point-Slope Form to Find Equation
Find an equation of the line through $(5, 4)$ and $(-10, -2)$.

Solution Begin by using the definition of slope to find the slope of the line that passes through the given points.

$$\text{Slope} = m = \frac{-2 - 4}{-10 - 5} = \frac{-6}{-15} = \frac{2}{5}$$

Either $(5, 4)$ or $(-10, -2)$ can be used in the point-slope form with $m = 2/5$. If $(x_1, y_1) = (5, 4)$, then

$$y - y_1 = m(x - x_1)$$

$$y - 4 = \frac{2}{5}(x - 5) \qquad \text{Let } y_1 = 4, m = \tfrac{2}{5}, x_1 = 5.$$

$$5y - 20 = 2(x - 5) \qquad \text{Multiply both sides by 5.}$$

$$5y - 20 = 2x - 10 \qquad \text{Distributive property}$$

$$5y = 2x + 10 \qquad \text{Add 20 to both sides.}$$

$$y = \frac{2}{5}x + 2 \qquad \text{Divide by 5 to put in slope-intercept form.}$$

Check that the same result is found if $(x_1, y_1) = (-10, -2)$.

EXAMPLE 4   Horizontal Line
Find an equation of the line through $(8, -4)$ and $(-2, -4)$.

Solution Find the slope.

$$m = \frac{-4 - (-4)}{-2 - 8} = \frac{0}{-10} = 0$$

Choose, say, $(8, -4)$ as (x_1, y_1).

$$y - y_1 = m(x - x_1)$$

$$y - (-4) = 0(x - 8) \qquad \text{Let } y_1 = -4, m = 0, x_1 = 8.$$

$$y + 4 = 0 \qquad 0(x - 8) = 0$$

$$y = -4$$

Plotting the given ordered pairs and drawing a line through the points, show that the equation $y = -4$ represents a horizontal line. See Figure 5(a). Every horizontal line has a slope of zero and an equation of the form $y = k$, where k is the y-value of all ordered pairs on the line.

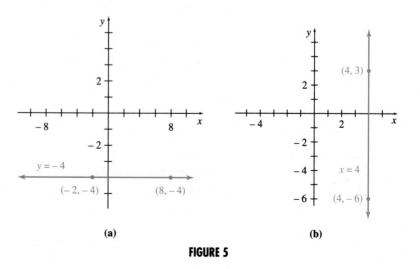

(a)   (b)

FIGURE 5

EXAMPLE 5   Vertical Line

Find an equation of the line through $(4, 3)$ and $(4, -6)$.

Solution The slope of the line is

$$m = \frac{-6 - 3}{4 - 4} = \frac{-9}{0},$$

which is undefined. Since both ordered pairs have x-coordinate 4, the equation is $x = 4$. Because the slope is undefined, the equation of this line cannot be written in the slope-intercept form.

    Again, plotting the given ordered pairs and drawing a line through them show that the graph of $x = 4$ is a vertical line. See Figure 5(b).

    The different forms of linear equations discussed in this section are summarized below. The slope-intercept and point-slope forms are equivalent ways to express the equation of a nonvertical line. The slope-intercept form is simpler for a final answer, but you may find the point-slope form easier to use when you know the slope of a line and a point through which the line passes.

EQUATIONS OF LINES

Equation	Description
$y = mx + b$	**Slope-intercept form:** slope m, y-intercept b
$y - y_1 = m(x - x_1)$	**Point-slope form:** slope m, line passes through (x_1, y_1)
$x = k$	**Vertical line:** x-intercept k, no y-intercept (except when $k = 0$), undefined slope
$y = k$	**Horizontal line:** y-intercept k, no x-intercept (except when $k = 0$), slope 0

Parallel and Perpendicular Lines   One application of slope involves deciding whether two lines are parallel. Since two parallel lines are equally "steep," they should have the same slope. Also, two lines with the same "steepness" are parallel.

PARALLEL LINES

Two lines are parallel if and only if they have the same slope, or if they are both vertical.

EXAMPLE 6   Parallel Line

Find the equation of the line that passes through the point $(3, 5)$ and is parallel to the line $2x + 5y = 4$.

Solution The slope of $2x + 5y = 4$ can be found by writing the equation in slope-intercept form.

$$2x + 5y = 4$$

$$y = -\frac{2}{5}x + \frac{4}{5}$$

This result shows that the slope is $-2/5$. Since the lines are parallel, $-2/5$ is also the slope of the line whose equation we want. This line passes through $(3, 5)$. Substituting $m = -2/5$, $x_1 = 3$, and $y_1 = 5$ into the point-slope form gives

$$y - y_1 = m(x - x_1)$$

$$y - 5 = -\frac{2}{5}x + \frac{6}{5}$$

$$y = -\frac{2}{5}x + \frac{6}{5} + 5$$

$$y = -\frac{2}{5}x + \frac{31}{5}$$

As already mentioned, two nonvertical lines are parallel if and only if they have the same slope. Two lines having slopes with a product of -1 are perpendicular. A proof of this fact, which depends on similar triangles from geometry, is given as Exercise 43 in this section.

PERPENDICULAR LINES

Two lines are perpendicular if and only if the product of their slopes is -1, or if one is vertical and the other horizontal.

EXAMPLE 7   Perpendicular Line

Find the slope of any line L perpendicular to the line having the equation $5x - y = 4$.

Solution To find the slope, write $5x - y = 4$ in slope-intercept form:

$$y = 5x - 4.$$

The slope is 5. Since the lines are perpendicular, if line L has slope m, then

$$5m = -1$$

$$m = -\frac{1}{5}.$$

Graph of a Line   A linear equation is simple to graph, as shown in the following example.

EXAMPLE 8   Graph of a Line

Graph $y = -\frac{3}{2}x + 6$.

Solution There are several ways to graph a line. To begin, note that the y-intercept is 6, so the point $(0, 6)$ is on the line. Next, by substituting $x = 2$ and $x = 4$ into the equation, we find the points $(2, 3)$ and $(4, 0)$. (We could use any values for x, but we chose even numbers so the value of y would be an integer.) These three points are plotted in Figure 6(a). A line is drawn through them in Figure 6(b). (We only need to find two points on the line. The third point gives a confirmation of our work.)

Alternatively, after finding the y-intercept, we could find the x-intercept by setting $y = 0$.

$$0 = -\frac{3}{2}x + 6$$

$$\frac{3}{2}x = 6 \qquad \text{Add } \tfrac{3}{2}x \text{ to both sides.}$$

$$x = \frac{6}{3/2} = 6\left(\frac{2}{3}\right) = 4 \qquad \text{Divide both sides by } \tfrac{3}{2}.$$

This gives the point $(4, 0)$ which we found earlier. Once the x- and y-intercepts are found, we can draw a line through them.

As a third method, observe that the slope of $-3/2$ means that every time x increases by 2, y decreases by 3. So start at $(0, 6)$ and go 2 across and 3 down to the point $(2, 3)$. Again, go 2 across and 3 down to the point $(4, 0)$. Draw the line through these points.

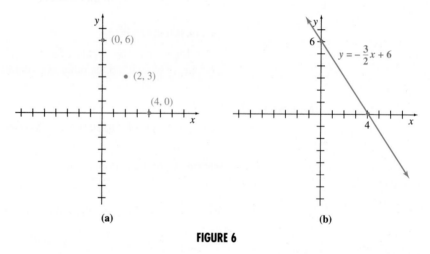

(a)   (b)

FIGURE 6

Not every line has two distinct intercepts; the graph in the next example does not cross the x-axis, and so it has no x-intercept.

EXAMPLE 9   Graph of a Horizontal Line
Graph $y = -3$.

Solution The equation $y = -3$, or equivalently, $y = 0x - 3$, always gives the same y-value, -3, for any value of x. Therefore, no value of x will make $y = 0$, so the graph has no x-intercept. As we saw in Example 4, the graph of such an equation is a horizontal line parallel to the x-axis. In this case the y-intercept is -3, as shown in Figure 7.

In general, the graph of $y = k$, where k is a real number, is the horizontal line having y-intercept k.

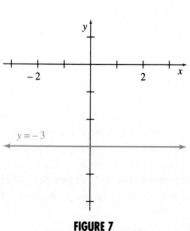

FIGURE 7

Linear Functions   When y can be expressed in terms of x in the form $y = mx + b$, such as we found in Examples 2, 3, 4, 6, 8, and 9, we say that y is a **linear function** of x. This means that for any value of x (the **independent**

variable), we can use the equation to find the corresponding value of y (the **dependent variable**). (In general, the independent variable in a linear function can be any real number, but in some applications, the independent variable may be restricted.) Examples of linear functions include $y = 2x + 3$, $y = -5$, and $2x - 3y = 7$, which can be written as $y = (2/3)x - (7/3)$. Equations in the form $x = k$, where k is a constant, are not linear functions. All other linear equations define linear functions.

Letters such as f, g, or h are often used to name functions. For example, f might be used to name the function

$$y = 5 - 3x.$$

To show that this function is named f, it is common to replace y with $f(x)$ (read "f of x") to get

$$f(x) = 5 - 3x.$$

By choosing 2 as a value of x, $f(x)$ becomes $5 - 3 \cdot 2 = 5 - 6 = -1$, written

$$f(2) = -1.$$

The corresponding ordered pair is $(2, -1)$. In a similar manner,

$$f(-4) = 5 - 3(-4) = 17, \qquad f(0) = 5, \qquad f(-6) = 23,$$

and so on.

EXAMPLE 10 Function Notation

Let $g(x) = -4x + 5$. Find $g(3)$, $g(0)$, and $g(-2)$.

Solution To find $g(3)$, substitute 3 for x.

$$g(3) = -4(3) + 5 = -12 + 5 = -7$$

Similarly,

$$g(0) = -4(0) + 5 = 0 + 5 = 5,$$
$$\text{and} \qquad g(-2) = -4(-2) + 5 = 8 + 5 = 13.$$

We summarize the discussion below.

LINEAR FUNCTION

A relationship f defined by

$$y = f(x) = mx + b,$$

for real numbers m and b, is a **linear function.**

Many real-world situations can be approximately described by a linear function. One way to find the equation of such a function is to use two typical data points from the graph and the point-slope form of the equation of a line.

EXAMPLE 11 Elk Energy

? Researchers have found that the energy expenditure (in kilocalories per kg of mass per minute) of elk calves rises linearly as a function of their speed.* A calf

---

*Robbins, Charles T., *Wildlife Feeding and Nutrition*, 2nd ed., Academic Press, 1993, p. 132.

walking 63 meters per minute expends, on average, 0.109 kilocalories per kg per minute, while a calf galloping 243 meters per minute expends, on average, 0.307 kilocalories per kg per minute.

(a) Find an equation for y as a linear function of x.

Solution The two ordered pairs representing the given information are $(63, 0.109)$ and $(243, 0.307)$. The slope of the line through these points is

$$m = \frac{0.307 - 0.109}{243 - 63} = \frac{0.198}{180} = 0.0011.$$

This means that, on average, the amount of energy expended goes up by 0.0011 kilocalories per kg per minute for each increase in speed of 1 meter per minute.

Using $m = 0.0011$ and $(x_1, y_1) = (63, 0.109)$ in the point-slope form gives the required equation.

$$y - 0.109 = 0.0011(x - 63)$$
$$y = 0.0011x + 0.0397$$

(b) Find and interpret the y-intercept.

Solution From the answer to part (a), notice that when $x = 0$, $y = 0.0397$, which is the y-intercept. This means that when the calf is at rest, it is still expending energy at a rate of 0.0397 kilocalories per kg per minute.

(c) The maximum amount of energy that a calf can expend is about 0.36 kilocalories per kg per minute. How fast is such a calf galloping?

Solution Let $y = 0.36$ in the answer to part (a).

$$0.0011x + 0.0397 = 0.36$$
$$0.0011x = 0.3203 \qquad \text{Subtract } 0.0397.$$
$$x = \frac{0.3203}{0.0011} \approx 291,^* \qquad \text{Divide by } 0.0011.$$

which indicates that the maximum speed of an elk calf is around 291 meters per minute, or about 11 miles per hour.

Another conclusion we can reach is that the equation in part (a) is only valid for $0 \leq x \leq 291$. The speed cannot be negative, and we have just found that it is unlikely to go past 291 meters per minute. Most equations are valid for some specific set of numbers. It is highly speculative to extrapolate beyond those values. On the other hand, sometimes it is necessary to make a prediction about what will happen past the data points, so a tentative conclusion based on previous data points may be better than no conclusion at all. There are also circumstances, particularly in the physical sciences, in which theoretical reasons imply that the trend will continue.

---

*The symbol $\approx$ means "is approximately equal to."

EXAMPLE 12   Antibiotic Resistance

In recent years, pathogenic bacteria have become resistant to one or many antibiotics, raising concerns about the future of infection control. The linear equation $y = 1.5457x - 3,067.7$, where x represents the year, can be used to estimate the percent of gonorrhea cases with antibiotic resistance diagnosed from 1985 to 1990.*

(a) Determine this percent in 1988.

> **Solution** Substitute 1988 for x in the equation.
>
> $$\begin{aligned} y &= 1.5457x - 3,067.7 \\ &= 1.5457(\mathbf{1988}) - 3,067.7 \\ &\approx 5.2 \end{aligned}$$
>
> This means that 5.2% of gonorrhea cases in 1988 were antibiotic resistant.

(b) Find and interpret the slope of the line.

> **Solution** The equation is given in slope-intercept form, so the slope is the coefficient of x, 1.5457. Since
>
> $$m = 1.5457 = \frac{\text{change in } y}{\text{change in } x} = \frac{1.5457}{1},$$
>
> the slope indicates the rate of change in the percent of antibiotic-resistant cases (that is, the change per year) from 1985 to 1990. Because the slope is positive, the percent of antibiotic-resistant cases increased by 1.5457% per year.

Cost Analysis

The cost of manufacturing an item commonly consists of two parts. The first is a **fixed cost** for designing the product, setting up a factory, training workers, and so on. Within broad limits, the fixed cost is constant for a particular product and does not change as more items are made. The second part is a *cost per item,* or **marginal cost,** for labor, materials, packing, shipping, and so on. The total value of this second cost *does* depend on the number of items made. Analysis of these costs is important in business. Medicine is a business, so the same ideas apply, as in the next example.

EXAMPLE 13   Cost Analysis

Suppose a hospital has to choose one of two X-ray machines to purchase. One costs $200,000 to purchase, and each image produced by the machine costs $10. The other sells for $250,000, but each image costs only $2.

(a) What is the fixed cost and marginal cost for the less expensive machine?

> **Solution** The fixed cost is $200,000 and the marginal cost is $10.

---

*Teutsch, S., and R. Churchhill, *Principles and Practice of Public Health Surveillance,* New York: Oxford University Press, 1994.

(b) How many X-ray images must be made before the cost of the two machines becomes equal?

Solution The cost function for the first machine can be written as

$$C_1(x) = 200{,}000 + 10x,$$

where x is the number of X-ray images taken, while the cost function for the second machine can be written as

$$C_2(x) = 250{,}000 + 2x.$$

The break-even point is found by setting $C_1(x) = C_2(x)$.

$$200{,}000 + 10x = 250{,}000 + 2x$$
$$8x = 50{,}000$$
$$x = \frac{50{,}000}{8} = 6{,}250$$

If the hospital purchasing agents expect the hospital to make fewer than 6,250 X-ray images over the life of the machine, they should buy the cheaper machine, assuming all else is equal. Otherwise, the more expensive machine is less expensive to use in the long run. ▮

Temperature One of the most common linear functions found in everyday situations deals with temperature. Recall that water freezes at 32° Fahrenheit and 0° Celsius, while it boils at 212° Fahrenheit and 100° Celsius.* The ordered pairs $(0, 32)$ and $(100, 212)$ are graphed in Figure 8 on axes showing Fahrenheit (F) as a function of Celsius (C). The line joining them is the graph of the function.

EXAMPLE 14 Temperature

Derive an equation relating F and C.

Solution To derive the required linear equation, first find the slope using the given ordered pairs, $(0, 32)$ and $(100, 212)$.

$$m = \frac{212 - 32}{100 - 0} = \frac{9}{5}$$

The F-intercept of the graph is 32, so by the slope-intercept form, the equation of the line is

$$F = \frac{9}{5}C + 32.$$

With simple algebra this equation can be rewritten to give C in terms of F:

$$C = \frac{5}{9}(F - 32).$$ ▮

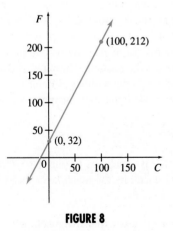

FIGURE 8

*Gabriel Fahrenheit (1686–1736), a German physicist, invented his scale with 0° representing the temperature of an equal mixture of ice and ammonium chloride (a type of salt), and 96° as the temperature of the human body. (It is often said, erroneously, that Fahrenheit set 100° as the temperature of the human body. Fahrenheit's own words are quoted in *A History of the Thermometer and Its Use in Meteorology* by W. E. Knowles, Middleton: The Johns Hopkins Press, 1966, p. 75.) The Swedish astronomer Anders Celsius (1701–1744) set 0° and 100° as the freezing and boiling points of water.

1.1 EXERCISES

Find the slope of each line that has a slope.

1. Through $(4, 5)$ and $(-1, 2)$

2. Through $(5, -4)$ and $(1, 3)$

3. Through $(8, 4)$ and $(8, -7)$

4. Through $(1, 5)$ and $(-2, 5)$

5. $y = 2x$

6. $y = 3x - 2$

7. $5x - 9y = 11$

8. $4x + 7y = 1$

9. $x = -6$

10. The x-axis

11. $y = 8$

12. $y = -4$

13. A line parallel to $2y - 4x = 7$

14. A line perpendicular to $6x = y - 3$

Find an equation in slope-intercept form (where possible) for each line in Exercises 15–34.

15. Through $(1, 3)$, $m = -2$

16. Through $(2, 4)$, $m = -1$

17. Through $(6, 1)$, $m = 0$

18. Through $(-8, 1)$, with undefined slope

19. Through $(4, 2)$ and $(1, 3)$

20. Through $(8, -1)$ and $(4, 3)$

21. Through $(1/2, 5/3)$ and $(3, 1/6)$

22. Through $(-2, 3/4)$ and $(2/3, 5/2)$

23. Through $(-8, 4)$ and $(-8, 6)$

24. Through $(-1, 3)$ and $(0, 3)$

25. x-intercept 3, y-intercept -2

26. x-intercept -2, y-intercept 4

27. Vertical, through $(-6, 5)$

28. Horizontal, through $(8, 7)$

29. Through $(-1, 4)$, parallel to $x + 3y = 5$

30. Through $(2, -5)$, parallel to $y - 4 = 2x$

31. Through $(3, -4)$, perpendicular to $x + y = 4$

32. Through $(-2, 6)$, perpendicular to $2x - 3y = 5$

33. The line with y-intercept 2 and perpendicular to $3x + 2y = 6$

34. The line with x-intercept $-2/3$ and perpendicular to $2x - y = 4$

35. Do the points $(4, 3)$, $(2, 0)$, and $(-18, -12)$ lie on the same line? (*Hint:* Find the slopes between the points.)

36. Find k so that the line through $(4, -1)$ and $(k, 2)$ is

    a. parallel to $2x + 3y = 6$

    b. perpendicular to $5x - 2y = -1$

37. Use slopes to show that the quadrilateral with vertices at $(1, 3)$, $(-5/2, 2)$, $(-7/2, 4)$, and $(2, 1)$ is a parallelogram.

38. Use slopes to show that the square with vertices at $(-2, 5)$, $(4, 5)$, $(4, -1)$, and $(-2, -1)$ has diagonals that are perpendicular.

For the lines in Exercises 39 and 40, which of the following is closest to the slope of the line? **(a)** 1   **(b)** 2   **(c)** 3   **(d)** -1 **(e)** -2 **(f)** -3

39.

40.

Estimate the slope of the lines in Exercises 41 and 42.

41.
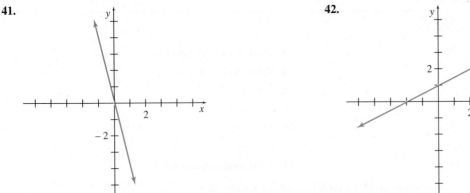

42.

43. To show that two perpendicular lines, neither of which is vertical, have slopes with a product of -1, go through the following steps. Let line L_1 have equation $y = m_1x + b_1$, and let L_2 have equation $y = m_2x + b_2$. Assume that L_1 and L_2 are perpendicular, and use right triangle *MPN* shown in the figure. Prove each of the following statements.

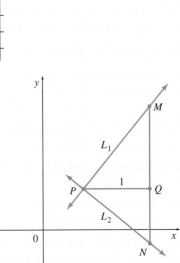

a. *MQ* has length m_1.

b. *QN* has length $-m_2$.

c. Triangles *MPQ* and *PNQ* are similar.

d. $m_1/1 = 1/(-m_2)$ and $m_1m_2 = -1$

Graph the equations for Exercises 44–59.

44. $y = x - 1$	**45.** $y = 2x + 3$	**46.** $y = -4x + 9$	**47.** $y = -6x + 12$
48. $2x - 3y = 12$	**49.** $3x - y = -9$	**50.** $3y + 4x = 12$	**51.** $4y + 5x = 10$
52. $y = -2$	**53.** $x = 4$	**54.** $x + 5 = 0$	**55.** $y - 4 = 0$
56. $y = 2x$	**57.** $y = -5x$	**58.** $x + 4y = 0$	**59.** $x - 3y = 0$

In Exercises 60–63, decide whether the statement is true or false.

60. To find the *x*-intercept of the graph of a linear function, we solve $y = f(x) = 0$, and to find the *y*-intercept, we evaluate $f(0)$.

61. The graph of $f(x) = -3$ is a vertical line.

62. The slope of the graph of a linear function cannot be undefined.

63. The graph of $f(x) = ax$ is a straight line that passes through the origin.

Applications

LIFE SCIENCES

64. *HIV Infection* The time interval between a person's initial infection with HIV and that person's eventual development of AIDS symptoms is an important issue. The method of in- fection with HIV affects the time interval before AIDS de- velops. One study of HIV patients who were infected by in- travenous drug use found that 17% of the patients had AIDS after 4 years, and 33% had developed the disease after 7 years. The relationship between the time interval and the percentage of patients with AIDS can be modeled accurately with a linear equation.*

a. Write a linear function $y = mx + b$ that models this data, using the ordered pairs $(4, 0.17)$ and $(7, 0.33)$.

b. Use your equation from part a to predict the number of years before half of these patients will have AIDS.

65. *Exercise Heart Rate* To achieve the maximum benefit for the heart when exercising, your heart rate (in beats per minute) should be in the target heart rate zone. The lower limit of this zone is found by taking 70% of the difference between 220 and your age. The upper limit is found by using 85%.[†]

a. Find formulas for the upper and lower limits (u and l) as linear functions involving the age x.

b. What is the target heart rate zone for a 20-year-old?

c. What is the target heart rate zone for a 40-year-old?

d. Two women in an aerobics class stop to take their pulse, and are surprised to find that they have the same pulse. One woman is 36 years older than the other and is working at the upper limit of her target heart rate zone. The younger woman is working at the lower limit of her target heart rate zone. What are the ages of the two women, and what is their pulse?

e. Run for 10 minutes, take your pulse, and see if it is in your target heart rate zone. (After all, this is listed as an exercise!)

66. *Ponies Trotting* A 1991 study found that the peak vertical force on a trotting Shetland pony increased linearly with the pony's speed, and that when the force reached a critical level, the pony switched from a trot to a gallop.[‡] For one pony, the critical force was 1.16 times its body weight. It experienced a force of 0.75 times its body weight at a speed of 2 meters per second, and a force of 0.93 times its body weight at 3 meters per second. At what speed did the pony switch from a trot to a gallop?

67. *Life Span* Some scientists believe there is a limit to how long humans can live.[§] One supporting argument is that during the last century, life expectancy from age 65 has increased more slowly than life expectancy from birth, so eventually these two will be equal, at which point, according to these scientists, life expectancy should increase no further. In 1900, life expectancy at birth was 46 years, and life expectancy at age 65 was 76. In 1975, these figures had risen to 75 and 80, respectively. In both cases, the increase in life expectancy has been linear. Using these assumptions and the data given, find the maximum life expectancy for humans.

68. *Deer Ticks* Deer ticks cause concern because they can carry the parasite that causes Lyme disease. One study found a relationship between the density of acorns produced in the fall and the density of deer tick larvae the following spring.[‖] The relationship can be approximated by the linear function

$$y = 34x + 230,$$

where x is the number of acorns per square meter in the fall, and y is the number of deer tick larvae per 400 square meters the following spring. According to this formula, approximately how many acorns per square meter would result in 1,000 deer tick larvae per 400 square meters?

69. *Tobacco Deaths* The U.S. Centers for Disease Control and Prevention projects that tobacco could soon be the leading cause of death in the world.[#] In 1990, 35 million years of healthy life were lost globally due to tobacco. This quantity was going up linearly at a rate of about 28 million years each decade. In contrast, 100 million years of healthy life were lost due to diarrhea, with the rate going down linearly 22 million years each decade.

a. Write the years of healthy life (y) in millions lost globally to tobacco as a linear function of the years since 1990.

b. Write the years of healthy life (y) in millions lost to diarrhea as a linear function of the years since 1990.

---

*Alcabes, P., A. Munoz, D. Vlahov, and G. Friedland, "Incubation Period of Human Immunodeficiency Virus," *Epidemiologic Review,* Vol. 15, No. 2, The Johns Hopkins University School of Hygiene and Public Health, 1993.

[†]Hockey, Robert V., *Physical Fitness: The Pathway to Healthful Living,* Times Mirror/Mosby College Publishing, 1989, pp. 85–87.

[‡]*Science,* July 19, 1991, pp. 306–8.

[§]*Science,* Nov. 15, 1991, pp. 936–38.

[‖]*Science,* Vol. 281, July 17, 1998, pp. 350–351.

[#]*Science,* Vol. 290, Nov. 17, 2000, p. 1291.

c. Using your answers to parts a and b, find in what year the amount of healthy life lost to tobacco is expected to first exceed that lost to diarrhea.

70. *Carbon Emissions* The rise in carbon emissions has caused concern because of its suspected contribution to global warming. While the total amount of carbon emitted has risen, the carbon emitted per unit of oil-equivalent energy has declined linearly from about 1.2 tons in 1855 to about 0.76 tons in 1995.*

  a. Write the tons of carbon emitted (y) as a linear function of the year (x).

  b. Assuming the linear function found in part a continues to be accurate, how much carbon will be emitted per unit of energy in the year 2020?

  c. Assuming the linear function found in part a continues to be accurate, in what year will the amount of carbon emitted per unit of energy reach 0.2 tons?

  ✎ **d.** Is it reasonable to expect the linear function found in part a to be accurate forever? Explain your answer.

71. *Ant Colonies* The number of male alates (winged, sexual ants) in an ant colony has been found to be approximately a linear function of the age of the colony, with a 5-year-old colony having 8.2 alates on average, and a 17-year-old colony having an average of 33.4 alates.[†]

  a. Write the number of male alates (y) as a linear function of the age of the colony in years (x).

  b. A one-year-old colony typically has no alates. What does the formula found in part a predict?

  c. Assuming the linear function found in part a continues to be accurate, how old would we expect a colony to be before it has about 40 male alates?

72. *Global Warming* Due to global warming, the area covered by the ice cap on Mt. Kilimanjaro has decreased linearly over the last year, from 12 square kilometers in 1912 to 2.2 square kilometers in 2000.[‡]

  a. Write the area of the ice (y) as a linear function of the year (x).

  b. In what year did the ice cover 6 square kilometers?

  c. Assuming the linear function found in part a continues, in what year will the ice cap disappear?

73. *Cost Analysis* Suppose a simple thermometer costs $10, but the cost to prepare the thermometer for the next patient costs $2. A more sophisticated thermometer costs $120 but costs only $0.75 to prepare for the next patient. How many patients would be required for the cost of the two thermometers to be equal?

74. *Cost Analysis* Acme Products sells a machine for doing a certain type of blood test for $40,000, which costs $20 for each use. Amalgamated Medical Supplies sells a similar machine for only $32,000, but it costs $30 for each use. How many times must the machines be used for the total cost to be equal?

75. *Global Warming* In 1990, the Intergovernmental Panel on Climate Change predicted that the average temperature on Earth would rise 0.3°C per decade in the absence of international controls on greenhouse emissions.[§] Let t measure the time in years since 1970, when the average global temperature was 15°C.

  a. Find a linear equation giving the average global temperature in degrees Celsius in terms of t, the number of years since 1970.

  b. Scientists have estimated that the sea level will rise by 65 cm if the average global temperature rises to 19°C. According to your answer to part a, when would this occur?

76. *Body Temperature* You may have heard that the average temperature of the human body is 98.6°. Recent experiments show that the actual figure is closer to 98.2°.[∥] The figure of 98.6 comes from experiments done by Carl Wunderlich in 1868. But Wunderlich measured the temperatures in degrees Celsius and rounded the average to the nearest degree, giving 37°C as the average temperature.[#]

  a. What is the Fahrenheit equivalent of 37°C?

  b. Given that Wunderlich rounded to the nearest degree Celsius, his experiments tell us that the actual average human body temperature is somewhere between 36.5°C and 37.5°C. Find what this range corresponds to in degrees Fahrenheit.

OTHER APPLICATIONS

77. *Immigration* In 1974, 86,821 people from other countries immigrated to the state of California. In 1996, the number of immigrants was 199,483.**

  a. If the change in foreign immigration to California is considered to be linear, write an equation expressing the

*The New York Times, Oct. 31, 1999, p. 38.

†Wagner, Diane, and Deborah M. Gordon, "Colony Age, Neighborhood Density and Reproductive Potential in Harvester Ants," *Oecologia,* Vol. 119, 1999, pp. 175–182.

‡*Science,* July 6, 2001, p. 47.

§*Science News,* June 23, 1990, p. 391.

∥*Science News,* Sept. 26, 1992, p. 195.

#*Science News,* Nov. 7, 1992, p. 399.

**Legal Immigration to California in Federal Fiscal Year 1996,* State of California Demographic Research Unit, June 1999.

number of immigrants, y, in terms of the number of years after 1974, x.

b. Use your result in part a to predict the foreign immigration to California in the year 2010.

78. *Cohabitation* The number of unmarried couples in the United States who are living together has been rising at a roughly linear rate in recent years. The number of cohabiting adults was 1.1 million in 1977 and 4.9 million in 1997.*

a. Write an equation expressing the number of cohabiting adults (in millions), y, in terms of the number of years after 1977, x.

b. Use your result in part a to predict the number of cohabiting adults in the year 2010.

79. *Older College Students* The percentage of college students who are age 35 and older has been increasing at roughly a linear rate. In 1972 the percentage was 9%, and in 1998 it was 17%.†

a. Find an equation giving the percentage of college students age 35 and older in terms of time t, where t represents the number of years since 1970.

b. If this linear trend continues, what percentage of college students will be 35 and over in 2010?

c. If this linear trend continues, in what year will the percentage of college students 35 and over reach 31%?

80. *Temperature* Use the formula for conversion between Fahrenheit and Celsius derived in Example 14 to convert each of the following temperatures.

a. 58°F to Celsius

b. −20°F to Celsius

c. 50°C to Fahrenheit

81. *Temperature* Find the temperature at which the Celsius and Fahrenheit temperatures are numerically equal.

■ 1.2 THE LEAST SQUARES LINE

? **THINK ABOUT IT**   How has the accidental death rate in the United States changed over time?

In this section, we show how to answer such questions using the method of least squares. We use past data to find trends and to make tentative predictions about the future. The only assumption we make is that the data are related linearly— that is, if we plot pairs of data, the resulting points will lie close to some line. This method cannot give exact answers. The best we can expect is that, if we are careful, we will get a reasonable approximation.

Table 1 lists the number of accidental deaths per 100,000 population in the United States through the past century.‡ If you were a manager at an insurance company, these data could be very important. You might need to make some predictions about how much you will pay out next year in accidental death benefits, and even a very tentative prediction based on past trends is better than no prediction at all.

The first step is to draw a **scatterplot,** as we have done in Figure 9. Notice that the points lie approximately along a line, which means that a linear function may give a good approximation of the data. If we select two points and find the line that passes through them, as we did in Section 1.1, we will get a different line for each pair of points, and in some cases the lines will be very different. We want to draw one line that is simultaneously close to all the points on the graph, but many such lines are possible, depending upon how we define the phrase "simultaneously close to all the points." How do we decide on the best possible line? Before going on, you might want to try drawing the line you think is best on Figure 9.

The line used most often in applications is that in which the sum of the squares of the vertical distances from the data points to the line is as small

■ Table 1

Year	Death Rate
1910	84.4
1920	71.2
1930	80.5
1940	73.4
1950	60.3
1960	52.1
1970	56.2
1980	46.5
1990	36.9

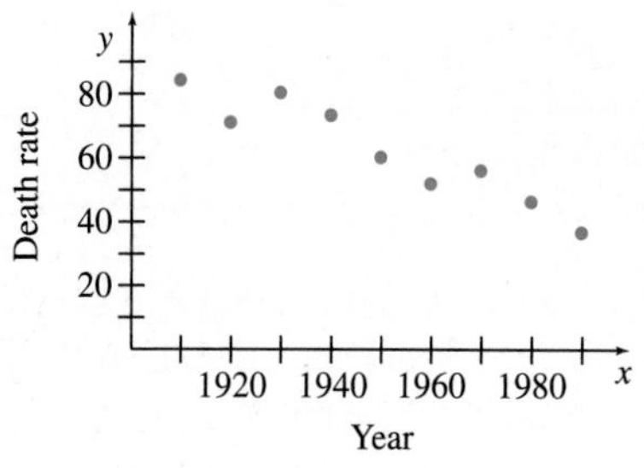

FIGURE 9

*The New York Times, Feb. 15, 2000, p. F8.
†The New York Times, Aug. 17, 1994, p. B9, and Jan. 9, 2000, p. 4A.
‡U.S. Dept. of Health and Human Services, National Center for Health Statistics.

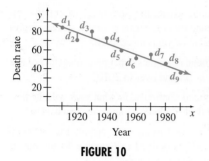

FIGURE 10

as possible. Such a line is called the **least squares line.** The least squares line for the data in Figure 9 is drawn in Figure 10. How does the line compare with the one you drew on Figure 9? It may not be exactly the same, but should appear similar.

In Figure 10, the vertical distances from the points to the line are indicated by d_1, d_2, and so on, up through d_9 (read "d-sub-one, d-sub-two, d-sub-three," and so on). For n points, corresponding to the n pairs of data, the least squares line is found by minimizing the sum $(d_1)^2 + (d_2)^2 + (d_3)^2 + \cdots + (d_n)^2$.

For the points (x_1, y_1), (x_2, y_2), ..., (x_n, y_n), if the equation of the desired line is $Y = mx + b$, then

$$d_1 = |Y_1 - y_1| = |mx_1 + b - y_1|,$$
$$d_2 = |Y_2 - y_2| = |mx_2 + b - y_2|,$$

and so on. We use Y in the equation of the line instead of y to distinguish the predicted values (Y) from the y-values of the given data points. The sum to be minimized becomes

$$(mx_1 + b - y_1)^2 + (mx_2 + b - y_2)^2 + \cdots + (mx_n + b - y_n)^2,$$

where (x_1, y_1), (x_2, y_2), ..., (x_n, y_n) are known and m and b are to be found.

The method of minimizing this sum requires advanced techniques and is not given here. The result gives equations that can be solved for the slope m and the y-intercept b of the least squares line.* In these equations, the symbol Σ, the Greek letter sigma, indicates "the sum of"; this notation is known as **summation notation.** For example, we write the sum $x_1 + x_2 + \cdots + x_n$, where n is the number of data points, as

$$x_1 + x_2 + \cdots + x_n = \Sigma x.$$

Similarly, Σxy means $x_1 y_1 + x_2 y_2 + \cdots + x_n y_n$, and so on.

❙ CAUTION Note that Σx^2 means $x_1^2 + x_2^2 + \cdots + x_n^2$, which is *not* the same as squaring Σx.

LEAST SQUARES LINE

The least squares line $Y = mx + b$ that gives the best fit to the data points (x_1, y_1), (x_2, y_2), ..., (x_n, y_n) has slope m and y-intercept b that satisfy the equations

$$nb + (\Sigma x)m = \Sigma y$$
$$(\Sigma x)b + (\Sigma x^2)m = \Sigma xy.$$

Method 1: Calculating by Hand | To find the least squares line for the given data, we first find the required sums. To reduce the size of the numbers, let x represent the years since 1900, so that, for example, $x = 10$ for the year 1910. Let y represent the death rate.

---

*Equations for m and b are derived in Exercise 4.

Table 2

x	y	xy	x^2	y^2
10	84.4	844	100	7,123.36
20	71.2	1,424	400	5,069.44
30	80.5	2,415	900	6,480.25
40	73.4	2,936	1,600	5,387.56
50	60.3	3,015	2,500	3,636.09
60	52.1	3,126	3,600	2,714.41
70	56.2	3,934	4,900	3,158.44
80	46.5	3,720	6,400	2,162.25
90	36.9	3,321	8,100	1,361.61
$\Sigma x = 450$	$\Sigma y = 561.5$	$\Sigma xy = 24{,}735$	$\Sigma x^2 = 28{,}500$	$\Sigma y^2 = 37{,}093.41$

(The column headed y^2 will be used later.) Now we can calculate m and b by solving a system of equations. Our method is to solve the first equation for b in terms of m, and substitute this into the second equation. We then solve the second equation for m. Once we have m, we put this back into the equation for b. Here, $n = 9$ (the number of data points).

$$nb + (\Sigma x)m = \Sigma y$$

$$9b + 450m = 561.5 \qquad \text{Substitute from Table 2.}$$

$$9b = 561.5 - 450m$$

$$b = (561.5 - 450m)/9 \qquad \text{Divide by 9.}$$

$$(\Sigma x)b + (\Sigma x^2)m = \Sigma xy$$

$$450(561.5 - 450m)/9 + 28{,}500m = 24{,}735 \qquad \text{Substitute.}$$

$$28{,}075 - 22{,}500m + 28{,}500m = 24{,}735 \qquad \text{Multiply.}$$

$$6{,}000m = -3{,}340 \qquad \text{Combine terms.}$$

$$m \approx -0.5566667 \approx -0.557$$

The significance of m is that the death rate per 100,000 population is tending to drop (because of the negative sign) at a rate of 0.557 per year.

Now substitute the value of m into the equation for b.

$$b = \frac{561.5 - 450(-0.5566667)}{9} \approx 90.2222 \approx 90.2$$

Substitute m and b into the least squares line equation, $Y = mx + b$; the least squares line that best fits the nine data points has equation $Y = -0.557x + 90.2$. This gives a mathematical description of the relationship between the year and the number of accidental deaths per 100,000 population. The equation can be used to predict y from a given value of x, as we will show in Example 1. As we mentioned before, however, caution must be exercised when using the least squares equation to predict data points that are far from the range of points on which the equation was modeled.

CAUTION In computing m and b, we rounded the final answer to three digits because the original data were known only to three digits. It is important, however, *not* to round any of the intermediate results (such as Σx^2) because round-off error may have a detrimental effect on the accuracy of the answer. Similarly, it is important not to use a rounded-off value of m when computing b.

Method 2: Graphing Calculator

The calculations for finding the least squares line are often tedious, even with the aid of a calculator. Fortunately, many calculators can calculate the least squares line with just a few keystrokes. For purposes of illustration, we will show how the least squares line in the previous example is found with a TI-83 graphing calculator.

We begin by entering the data into the calculator. We will be using the first two lists, called L_1 and L_2. Choosing the STAT menu, then choosing the fourth entry ClrList, we enter L_1, L_2, to indicate the lists to be cleared. Now we press STAT again and choose the first entry EDIT, which brings up the blank lists. As before, we will only use the last two digits of the year, putting the numbers in L_1. We put the death rate in L_2, giving the two screens shown in Figure 11.

FIGURE 11

Press STAT again and choose CALC instead of EDIT. Then choose item 4 LinReg $(ax + b)$ to get the values of a (the slope) and b (the y-intercept) for the least squares line, as shown in Figure 12. With a and b rounded to three decimal places, the least squares line is $Y = -0.557x + 90.222$. A graph of the data points and the line is shown in Figure 13.

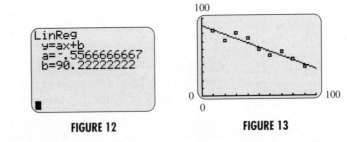

FIGURE 12   **FIGURE 13**

For more details on finding the least squares line with a graphing calculator, see *The Graphing Calculator Manual* that is available on this text's Web site.

Method 3: Spreadsheet

Many computer spreadsheet programs can also find the least squares line. Figure 14 shows the scatterplot and least squares line for the accidental death rate data using an Excel spreadsheet. The scatterplot was found using the XY(Scatter) command under Chart Wizard, and the line was found using the Add Trendline command under the Chart menu. For details, see *The Spreadsheet Manual* that is available on this text's Web site.

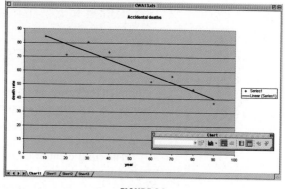

FIGURE 14

EXAMPLE 1   Least Squares Line

What do we predict the accidental death rate in 1997 to be?

Solution Use the least squares line equation given above with $x = 97$.

$$Y = -0.557x + 90.2$$
$$= -0.557(97) + 90.2$$
$$= 36.171$$

The death rate in 1997 is predicted to be about 36.2 per 100,000 population. In this case, we have the actual value for 1997. It happens to be 35.0, which is close to the predicted value.

EXAMPLE 2   Least Squares Line

In what year is the death rate predicted to drop below 26 per 100,000 population?

Solution Let $Y = 26$ in the equation above and solve for x.

$$26 = -0.557x + 90.2$$
$$-64.2 = -0.557x$$
$$x = 115.26$$

This means that after 115 years, the rate will not have quite reached 26 per 100,000, so we must wait 116 years for this to happen. This corresponds to the year 2016 (116 years after 1900), when our equation predicts the death rate to be $-0.557(116) + 90.2 = 25.6$ per 100,000 population.

Correlation   Once an equation is found for the least squares line, we need to have some way of judging just how good the equation is for predictive purposes. If the points from the data fit the line quite closely, then we have more reason to expect future data pairs to do so. But if the points are widely scattered about even the best-fitting line, then predictions are not likely to be accurate.

In order to have a quantitative basis for confidence in our predictions, we need a measure of the "goodness of fit" of the original data to the prediction line. One such measure is called the **coefficient of correlation,** denoted r.

COEFFICIENT OF CORRELATION

$$r = \frac{n(\Sigma xy) - (\Sigma x)(\Sigma y)}{\sqrt{n(\Sigma x^2) - (\Sigma x)^2} \cdot \sqrt{n(\Sigma y^2) - (\Sigma y)^2}}$$

Although the expression for r looks daunting, remember that each of the summations, Σx, Σy, Σxy, and so on, are just the totals from a table like the one we prepared for the data on accidental deaths. Also, with a calculator, the arithmetic is no problem!

The coefficient of correlation r is always equal to or between 1 and -1. Values of exactly 1 or -1 indicate that the data points lie *exactly* on the least squares line. If $r = 1$, the least squares line has a positive slope; $r = -1$ gives a negative slope. If $r = 0$, there is no linear correlation between the data points (but some *nonlinear* function might provide an excellent fit for the data). A correlation of zero may also indicate that the data fit a horizontal line. To investigate what is happening, it is always helpful to sketch a scatterplot of the data. Some scatterplots that correspond to these values of r are shown in Figure 15.

| r close to 1 | r close to -1 | r close to 0 | r close to 0 |

FIGURE 15

A value of r close to 1 or -1 indicates the presence of a linear relationship. The exact value of r necessary to conclude that there is a linear relationship depends upon n, the number of data points, as well as how confident we want to be of our conclusion. For details, consult a text on statistics.*

EXAMPLE 3   Coefficient of Correlation
Find r for the data on accidental death rates.

Solution

Method 1: Calculating by Hand   From Table 2 on page 21, $\Sigma x = 450$, $\Sigma y = 561.5$, $\Sigma xy = 24{,}735$, $\Sigma x^2 = 28{,}500$, and $\Sigma y^2 = 37{,}093.41$. Also, $n = 9$. Substituting these values into the formula for r gives

$$r = \frac{9(24{,}735) - (450)(561.5)}{\sqrt{9(28{,}500) - (450)^2} \cdot \sqrt{9(37{,}093.41) - (561.5)^2}}$$

$$= \frac{-30{,}060}{\sqrt{54{,}000} \cdot \sqrt{18{,}558.44}}$$

$$= -0.9495577064 \approx -0.950.$$

This is a high correlation, which agrees with our observation that the data fit a line quite well.

---

*For example, see *Introductory Statistics*, 6th edition, by Neil A. Weiss, Boston, Mass.: Addison Wesley, 2001.

Method 2: Graphing Calculator

```
LinReg
  y=ax+b
  a=-.5566666667
  b=90.22222222
  r²=.9016598378
  r=-.9495577064
```

FIGURE 16

Most calculators that give the least squares line will also give the coefficient of correlation. To do this on the TI-83, we press the second function CATALOG and go down the list to the entry DiagnosticOn. Press ENTER at that point, then press STAT, CALC, and choose item 4 to get the display in Figure 16. The result is the same as we got by hand. The command DiagnosticOn need only be entered once, and the coefficient of correlation will always appear in the future.

Method 3: Spreadsheet

Many computer spreadsheet programs have a built-in command to find the coefficient of correlation. For example, in Excel, use the command "= CORREL(A1:A9, B1:B9)" to find the correlation of the 9 data points stored in columns A and B. For more details, see *The Spreadsheet Manual* that is available on this text's Web site.

EXAMPLE 4   Neediest Cases

The table shows donations to the *New York Times* Neediest Cases Fund (in millions) from 1985 through 1995.*

Year	'85	'86	'87	'88	'89	'90	'91	'92	'93	'94	'95
Donation	2.7	3.0	3.8	4.0	4.2	4.3	4.9	4.3	4.4	4.4	1.9

Find the correlation coefficient, as well as the line that best fits the data.

Solution A scatterplot of the data, along with the graph of the least squares line, is shown in Figure 17. Notice that the data appear to be linear for 1985 through 1994, but the last point ruins the linearity. Figure 18 shows the result of the LinReg command on the TI-83. The coefficient of correlation is only about 0.158, indicating that the data do not fit a line. In this case, the least squares line is of little value. When we remove the 1995 value, however, we get $r = 0.828$ and $Y = 0.184x - 12.5$. (Verify this on your calculator.) Therefore, the trend from 1985 through 1994 is roughly linear, with a correlation coefficient of 0.828 and a least squares line $Y = 0.184x - 12.5$. The figure for 1995 is an anomaly that does not fit the trend, and the managers of the Neediest Case Fund undoubtedly looked for the cause.

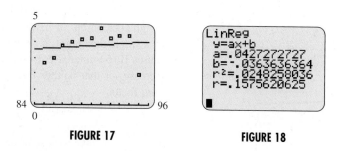

FIGURE 17   **FIGURE 18**

The final example illustrates why a plot of the data points is so important.

---

*New York Observer, Vol. 9, No. 49/50, 12/25/95–1/1/96 issue, p.1.

EXAMPLE 5 Silicone Implants

Silicones have long been used for fabricating medical devices on the presumption that they are biocompatible materials. This presumption is not entirely correct. Silicone prostheses, when implanted within the soft tissues of the breast, may evoke an inflammatory reaction. In response to silicone exposure, production of protein factors that mediate the inflammatory response (known as inflammatory mediators) was observed in experimental studies. The levels of four inflammatory mediators from ten patients with silicone breast implants after in vitro culture for 24 hours were as shown in Table 3.*

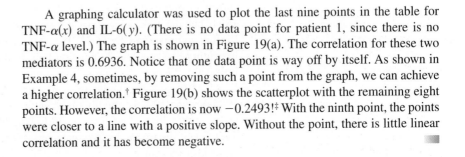

Table 3

Patient	IL-2	TNF-α	IL-6	PGE$_2$
1	48	ND	231	68
2	219	78	308,287	2,710
3	109	65	33,291	1,804
4	2,179	149	124,550	8,053
5	219	451	17,075	7,371
6	54	64	22,955	3,418
7	6	79	95,102	9,768
8	10	115	5,649	441
9	42	618	840,585	9,585
10	196	69	58,924	4,536

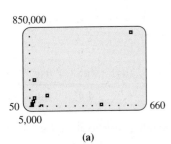

(a)

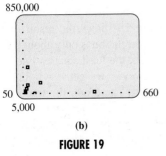

(b)

FIGURE 19

A graphing calculator was used to plot the last nine points in the table for TNF-α(x) and IL-6(y). (There is no data point for patient 1, since there is no TNF-α level.) The graph is shown in Figure 19(a). The correlation for these two mediators is 0.6936. Notice that one data point is way off by itself. As shown in Example 4, sometimes, by removing such a point from the graph, we can achieve a higher correlation.[†] Figure 19(b) shows the scatterplot with the remaining eight points. However, the correlation is now -0.2493![‡] With the ninth point, the points were closer to a line with a positive slope. Without the point, there is little linear correlation and it has become negative.

CAUTION A common error is to assume a cause-and-effect relationship from a high correlation. In Example 5, even if the high correlation were not caused by a single data point, it would be incorrect to conclude that increasing TNF-α causes IL-6 to increase. It is just as likely that increasing IL-6 causes TNF-α to increase, or that neither one affects the other, but both are affected by some third factor.

---

*Mena, E. A., et al., "Inflammatory Intermediates Produced by Tissues Encasing Silicone Breast Prostheses," *Journal of Investigative Surgery,* Vol. 8, 1995, p. 33. Copyright © 1995. Reproduced by permission of Taylor & Francis, Inc., http://www.routledge-ny.com.

[†]Before discarding a point, we should investigate the reason it is an outlier.

[‡]The observation that removing one point changes the correlation from positive to negative was made by Patrick Fleury, *Chance News* (an electronic newsletter), Vol. 4, No. 16, Dec. 1995.

1.2 EXERCISES

1. Suppose a positive linear correlation is found between two quantities. Does this mean that one of the quantities increasing causes the other to increase? If not, what does it mean?

2. Given a set of points, the least squares line formed by letting x be the independent variable will not necessarily be the same as the least squares line formed by letting y be the independent variable. Give an example to show why this is true.

*The following problem is reprinted from the November 1989 Actuarial Examination on Applied Statistical Methods.**

3. You are given

X	6.8	7.0	7.1	7.2	7.4
Y	0.8	1.2	0.9	0.9	1.5

Determine r^2, the coefficient of determination for the regression of Y on X. (*Note:* The coefficient of determination is defined as the square of the coefficient of correlation.)

a. 0.3   **b.** 0.4   **c.** 0.5   **d.** 0.6   **e.** 0.7

4. Follow the steps outlined in this section to solve the least squares line equations

$$nb + (\Sigma x)m = \Sigma y$$
$$(\Sigma x)b + (\Sigma x^2)m = \Sigma xy$$

for m and b to get

$$m = \frac{n\Sigma(xy) - (\Sigma x)(\Sigma y)}{n(\Sigma x^2) - (\Sigma x)^2}$$

$$b = \frac{\Sigma y - m(\Sigma x)}{n}.$$

Applications

LIFE SCIENCES

5. *Bird Eggs* The average length and width of various bird eggs are given in the following table.[†]

Bird Name	Width (cm)	Length (cm)
Canada goose	5.8	8.6
Robin	1.5	1.9
Turtledove	2.3	3.1
Hummingbird	1.0	1.0
Raven	3.3	5.0

a. Plot the points, putting the length on the y-axis and the width on the x-axis. Do the data appear to be linear?

b. Find the least squares line, and plot it on the same graph as the data.

c. Suppose there are birds with eggs even smaller than those of hummingbirds. Would the equation found in part b continue to make sense for all positive widths, no matter how small? Explain.

d. Find the coefficient of correlation.

6. *Skeletal Maturity* Researchers measuring skeletal maturity used a maturity score of 1 to 45 points, with 45 representing complete maturity.[‡] The average maturity score for the hip joint and pelvis of children of various ages (in years) is given in the table on the next page.

a. Find the coefficient of correlation. How closely do the data seem to fit a straight line?

b. Find the equation of the least squares line.

c. Plot the data and least squares line on the same graph. Discuss any ways in which the least squares line does not fit the data.

---

*"November 1989 Course 120 Examination Applied Statistical Methods" of the *Education and Examination Committee of The Society of Actuaries.* Reprinted by permission of The Society of Actuaries.
[†]www.nctm.org/wlme/wlme6/five.htm
[‡]Hensinger, Robert N., *Standards in Pediatric Orthopedics,* Raven Press, 1986, p. 309.

d. Assuming the linear trend continues, what would be the maturity score for an 18-year-old? Compare this with the actual score of 43.6.

Age (x)	Maturity (y)
1	8.4
2	10.7
3	12.9
4	15.2
5	17.5
6	19.1
7	20.4
8	21.9
9	23.4
10	25.0
11	27.2
12	29.9
13	32.6
14	36.7

7. *Fetal Stature* The following table gives the estimated stature (length, in cm) of human fetuses at various times (in weeks) since conception.*

Age (x)	Stature (y)
18	19.81
20	23.80
22	27.40
24	30.60
26	33.72
28	36.52
30	39.13
32	41.58
34	43.84
36	46.03
38	48.08
40	50.02

a. Find the coefficient of correlation. How closely do the data seem to fit a straight line?

b. Find the equation of the least squares line.

c. Plot the data and least squares line on the same graph. Does the line accurately fit the data?

d. According to the least squares line, how much is the fetal stature increasing per week, on average?

e. If this formula continues to be accurate, what would we expect for the stature of a 45-week-old fetus?

8. *Medical School Admissions* The number of applications to medical schools in the United States increased rapidly from 1989 to 1994 as indicated by the data in the table.[†] Years are represented by their last two digits and applications are given in thousands.

Year (x)	88	89	90	91	92	93	94
Applications (y)	27	27	29	33	37	43	45

a. Plot the data. Do the data points lie in a linear pattern?

b. Determine the least squares line for this data and graph it on the same coordinate axes. Does the line fit the data reasonably well?

c. Find the coefficient of correlation. Does it agree with your estimate of the fit in part b?

d. Explain why the coefficient of correlation is close to 1, even though the data points do not appear to be linear.

9. *Crickets Chirping* Biologists have observed a linear relationship between the temperature and the frequency with which a cricket chirps. The following data were measured for the striped ground cricket.[‡]

Temperature °F (x)	Chirps per second (y)
88.6	20.0
71.6	16.0
93.3	19.8
84.3	18.4
80.6	17.1
75.2	15.5
69.7	14.7
82.0	17.1
69.4	15.4
83.3	16.2
79.6	15.0
82.6	17.2
80.6	16.0
83.5	17.0
76.3	14.4

a. Find the equation for the least squares line for the data.

---

*Shipman, Pat, et al., *The Human Skeleton,* Harvard University Press, 1985, p. 253.
[†]*The New York Times,* Friday, Nov. 17, 1995, p. C2.
[‡]Pierce, George W., *The Songs of Insects,* Cambridge, Mass., Harvard University Press, Copyright © 1948 by the President and Fellows of Harvard College. Reprinted by permission of the publishers.

b. Use the results of part a to determine how many chirps per second you would expect to hear from the striped ground cricket if the temperature were 73°F.

c. Use the results of part a to determine what the temperature is when the striped ground crickets are chirping at a rate of 18 times per second.

d. Find the coefficient of correlation.

10. *Athletic Records* The following table shows the men's and women's world records (in seconds) in the 800-meter run.*

Year	Men's Record	Women's Record
1905	113.4	—
1915	111.9	—
1925	111.9	144.0
1935	109.7	135.6
1945	106.6	132.0
1955	105.7	125.0
1965	104.3	118.0
1975	103.7	117.48
1985	101.73	113.28
1995	101.11	113.28

a. Find the equation for the least squares line for the men's record (y) in terms of the year (x). Use 5 for 1905, 15 for 1915, and so on.

b. Find the equation for the least squares line for the women's record.

c. Suppose the men's and women's records continue to improve as predicted by the equations found in parts a and b. In what year will the women's record catch up with the men's record? Do you believe that will happen? Why or why not?

d. Calculate the coefficient of correlation for both the men's and the women's record. What do these numbers tell you?

11. *Running* If you think a marathon is a long race, consider the Hardrock 100, a 101.1 mile running race held in southwestern Colorado. The chart lists the times that the 1994 winner, Scott Hirst, arrived at various mileage points along the way.[†]

a. What was Hirst's average speed?

b. Graph the data, plotting time on the x-axis and distance on the y-axis. You will need to convert the time from hours and minutes into hours. Do the data appear to lie approximately on a straight line?

c. Find the equation for the least squares line, fitting distance as a linear function of time.

d. Calculate the coefficient of correlation. Does it indicate a good fit of the least squares line to the data?

e. Based on your answer to part d, what is a good value for Hirst's average speed? Compare this with your answer to part a. Which answer do you think is better? Explain your reasoning.

Miles	Time (hours:minutes)
0	0
12.2	2:57
19.0	5:03
28.0	7:22
33.4	9:26
36.6	10:10
43.8	11:16
59.4	16:01
69.6	21:40
78.4	24:00
85.0	26:48
91.5	28:59
101.1	32:00

OTHER APPLICATIONS

12. *Air Fares* In January 2000, American Airlines ran an ad in *The New York Times* advertising one-way air fares from New York to various cities.[‡] Fourteen of the cities are listed on the next page, with the distances from New York to the cities added.

a. Plot the data. Do the data points lie in a linear pattern?

b. Find the correlation coefficient. Combining this with your answer to part a, does the cost of a ticket tend to go up with the distance flown?

c. Find the equation of the least squares line, and use it to find the approximate marginal cost per mile to fly.

---

*Whipp, Brian J., and Susan A. Ward, "Will Women Soon Outrun Men?" *Nature,* Vol. 355 (Jan. 2, 1992), p. 25. The data are from Peter Matthews, *Track and Field Athletics: The Records,* Guinness, 1986, pp. 11 and 44, and from Robert W. Schutz and Yuanlong Liu in *Statistics in Sport,* edited by Jay Bennett, Arnold, 1998, p. 189.

†Frantz, Marny, and Sylvia Lazarnick, "Data Analysis and the Hardrock 100," *The Mathematics Teacher,* Vol. 90, No. 4, April 1997, pp. 274–276. Reprinted with permission from *The Mathematics Teacher,* copyright 1970 by the National Council of Teachers of Mathematics.

‡*The New York Times,* Jan. 7, 2000, p. A9.

✎ **d.** For similar data in an October 1993 ad, the equation of the least squares line was $y = 91.9 + 0.0313x$. Use this information and your answer to part b to compare the cost of flying American Airlines for these two time periods.

City	Distance (x) in miles	Price (y) in dollars
Boston	206	109
Chicago	802	124
Denver	1,771	154
Kansas City	1,198	144
Little Rock	1,238	144
Los Angeles	2,786	179
Minneapolis	1,207	144
Nashville	892	144
Phoenix	2,411	179
Portland	2,885	179
Reno	2,705	179
St. Louis	948	144
San Diego	2,762	179
Seattle	2,815	179

13. *Educational Expenditures* A 1991 report issued by the U.S. Department of Education listed the expenditure per pupil and the average mathematics proficiency in grade 8 for 37 states and the District of Columbia. Letting x equal the expenditure per pupil (ranging from \$2,838 in Idaho to \$7,850 in Washington, D.C.) and y equal the average mathematics proficiency score (ranging from 231 in Washington, D.C., to 281 in North Dakota), the data can be summarized as follows:*

$$\Sigma x = 175{,}878 \qquad \Sigma x^2 = 872{,}066{,}218$$
$$\Sigma y = 9{,}989 \qquad \Sigma y^2 = 2{,}629{,}701$$
$$\Sigma xy = 46{,}209{,}266 \qquad n = 38$$

Compute the coefficient of correlation for the given data. Does there appear to be a trend in the amount of money spent per pupil and the proficiency of eighth graders in mathematics?

14. *Poverty Levels* The following table lists how poverty level income cutoffs (in dollars) for a family of four have changed over time.[†] Let x be the year, with $x = 0$ corresponding to 1970, and y be the income in thousands of dollars. (*Note:* $\Sigma x = 75$, $\Sigma x^2 = 1{,}375$, $\Sigma y = 57.799$, $\Sigma y^2 = 658.405183$, $\Sigma xy = 932.88$.)

Year	Income
1970	3,968
1975	5,500
1980	8,414
1985	10,989
1990	13,359
1995	15,569

a. Sketch a graph of the data. Do the data appear to lie along a straight line?

b. Calculate the coefficient of correlation. Does your result agree with your answer to part a?

c. Find the equation of the least squares line.

d. Use your answer from part c to predict the poverty level in the year 2015.

15. *Library Budget Decline* The budget for books in New York Public Library branches has declined recently as shown in the table. Budgets are given in millions of dollars.[‡]

Year (x)	91	92	93	94	95	96
Budget (y)	7.1	6.3	7.8	7.6	7.5	6.6

a. Plot the data with a graphing calculator in the window $[90, 97]$ by $[6, 8]$. Do the data show a linear pattern?

b. Plot the data with a graphing calculator in the window $[90, 97]$ by $[4, 10]$. Do the data show a more linear pattern? Do you think a line would be a good fit?

c. Plot the data with a graphing calculator in the window $[90, 97]$ by $[0, 12]$. Is a line a good fit for this set of points?

✎ **d.** What can you conclude from a comparison of the graphs in parts a–c?

e. What is the coefficient of correlation?

✎ **f.** Which scatterplot in parts a–c describes the data most accurately? How does the coefficient of correlation support your answer?

16. *SAT Scores* At Hofstra University, all students take the math SAT before entrance, and most students take a mathematics placement test before registration. Recently, one professor collected the data on the following page for 19 students in his Finite Mathematics class.

a. Find an equation for the least squares line. Let x be the math SAT and y be the placement test score.

---

*Newsday, June 7, 1991.
†U.S. Bureau of the Census, *Current Population Reports.*
‡*New York Observer,* Feb. 12, 1996. Vol. 10, No. 6, p. 1.

Math SAT	Placement Test	Math SAT	Placement Test	Math SAT	Placement Test
540	20	580	8	440	10
510	16	680	15	520	11
490	10	560	8	620	11
560	8	560	13	680	8
470	12	500	14	550	8
600	11	470	10	620	7
540	10				

b. Use your answer from part a to predict the mathematics placement test score for a student with a math SAT score of 420.

c. Use your answer from part a to predict the mathematics placement test score for a student with a math SAT score of 620.

d. Calculate the coefficient of correlation.

e. Based on your answer to part d, what can you conclude about the relationship between a student's math SAT and mathematics placement test score?

17. *Length of a Pendulum* Grandfather clocks use pendulums to keep accurate time. The relationship between the length of a pendulum L and the time T for one complete oscillation can be determined from the data in the table.*

L (ft)	T (sec)
1.0	1.11
1.5	1.36
2.0	1.57
2.5	1.76
3.0	1.92
3.5	2.08
4.0	2.22

a. Plot the data from the table with L as the horizontal axis and T as the vertical axis.

b. Find the least squares line equation and graph it simultaneously, if possible, with the data points. Does it seem to fit the data?

c. Find the coefficient of correlation and interpret it. Does it confirm your answer to part b?[†]

■ 1.3 PROPERTIES OF FUNCTIONS

? **THINK ABOUT IT**   How does a woman's basal body temperature vary with time?

In this section we will explore this question using the concept of a *function*.

Figure 20 on the next page illustrates a function that, unlike those studied in Section 1.1, is *nonlinear*. Its graph is not a straight line. Linear functions are simple to study, and they can be used to approximate many functions over short intervals. But most functions exhibit behavior that, in the long run, does not follow a straight line. In the rest of this chapter and the next we will study some of the most common nonlinear functions.

FUNCTION

A **function** is a rule that assigns to each element from one set exactly one element from another set.

---

*Data provided by Gary Rockswold, Mankato State University, Minnesota.
[†]The actual relationship is $L = 0.81T^2$, which is not a linear equation. This illustrates that even if the relationship is not linear, a line can give a good approximation.

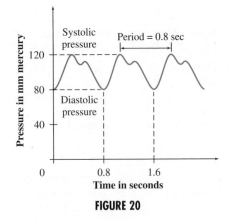

FIGURE 20

In most cases in this book, the "rule" mentioned in the box is expressed as an equation, such as $C(x) = 200,000 + 10x$. When an equation is given for a function, we say that the equation *defines* the function. Whenever x and y are used in this book to define a function, x represents the independent variable and y the dependent variable. Of course, letters other than x and y could be used and are often more meaningful. For example, if the independent variable represents the number of minutes and the dependent variable represents the number of seconds, we might write $s = 60m$.

The independent variable in a function can take on any value within a specified set of values called the *domain*.

DOMAIN AND RANGE

The set of all possible values of the independent variable in a function is called the **domain** of the function, and the resulting set of possible values of the dependent variable is called the **range.**

 A woman trying to get pregnant, or using natural methods to avoid getting pregnant, might be very interested in her basal body temperature (her temperature at rest). To predict when she is ovulating, a woman can take her temperature each day on waking.* Ovulation begins when her temperature begins to rise, usually about halfway through her cycle. A typical graph for a woman's temperature as a function of her cycle day is shown in Figure 21. Let us label

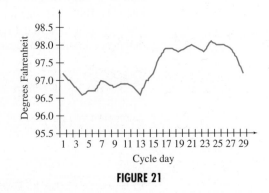

FIGURE 21

*For more information, see http://www.reproductivesurgery.com/rs11_1.htm.

this function $y = f(x)$, where y is the temperature of the woman's body and x is the cycle day. Notice that the function increases and decreases during the month, so it is not linear. Such a function, whose graph is not a straight line, is called a *nonlinear function.*

The concepts you learned in the section on linear functions apply to this and other nonlinear functions as well. The independent variable here is x, the time in days; the dependent variable is y, the temperature at any time. The domain is $\{x \mid 1 \le x \le 29\}$, or $[1, 29]$; $x = 1$ corresponds to the first day of this woman's cycle, and $x = 29$ corresponds to the last day. By looking for the lowest and highest values of the function, we estimate the range to be approximately $\{y \mid 96.6 \le y \le 98.1\}$, or $[96.6, 98.1]$. As with linear functions, the domain is mapped along the horizontal axis and the range along the vertical axis.

We do not have a formula for $f(x)$. (If a woman had such a formula, she would no longer have to take her temperature.) Instead, we can use the graph to estimate values of the function. To estimate $f(21)$, for example, we draw a vertical line from day 21, as shown in Figure 22. The y-coordinate appears to be 98.0, so we estimate $f(21) = 98.0$. Similarly, if we wanted to solve the equation $f(x) = 97.0$, we would look for points on the graph that have a y-coordinate of 97.0. As Figure 23 shows, this occurs at three points, around days 2, 7, and 14. Thus $f(x) = 97.0$ when $x = 2, 7$, and 14.

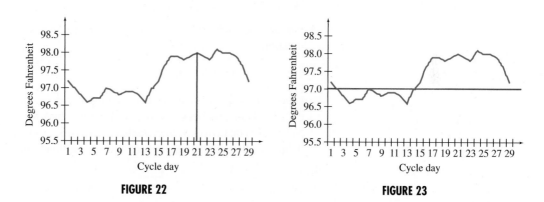

FIGURE 22   **FIGURE 23**

Table 4	
Day (x)	**Temperature (y)**
1	97.2
2	97.0
3	96.8
4	96.6
5	96.7
6	96.7
7	97.0

This function can also be given as a table. Table 4 shows the value of the function for several values of x.

Notice from the table that $f(5) = f(6) = 96.7$. This illustrates an important property of functions: Several different values of the independent variable can have the same value for the dependent variable. On the other hand, we cannot have several different y-values corresponding to the same value of x; if we did, this would not be a function.

What is $f(2.5)$? We do not know. Between days 2 and 3, the woman's temperature presumably went up and back down, as normally happens on a daily basis. Unless the woman monitors her temperature continually throughout the day, we do not have access to this information.

Functions arise in numerous applications, and an understanding of them is critical for understanding calculus. The following example shows some of the ways functions can be represented, and will help you in determining whether a relationship between two variables is a function or not.

EXAMPLE 1 Functions

Which of the following are functions?

(a)

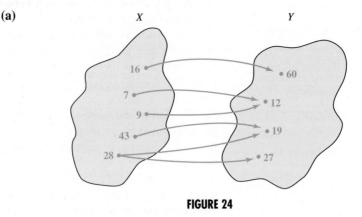

FIGURE 24

Solution Figure 24 shows that an x-value of 28 corresponds to *two* y-values, 19 and 27. In a function, each x must correspond to exactly one y, so this correspondence is not a function.

(b) The optical reader at the checkout counter in many stores that converts codes to prices

Solution For each code, the reader produces exactly one price, so this is a function.

(c) The x^2 key on a calculator

Solution This correspondence between input and output is a function because the calculator produces just one x^2 (one y-value) for each x-value entered. Notice also that two x-values, such as 3 and -3, produce the same y-value of 9, but this does not violate the definition of a function.

(d)

x	1	1	2	2	3	3
y	3	-3	5	-5	8	-8

Solution Since at least one x-value corresponds to more than one y-value, this table does not define a function.

(e) The set of ordered pairs with first elements mothers and second elements their children

Solution Here the mother is the independent variable and the child is the dependent variable. For a given mother, there may be several children, so this correspondence is not a function.

(f) The set of ordered pairs with first elements children and second elements their birth mothers

Solution In this case the child is the independent variable and the mother is the dependent variable. Since each child has only one birth mother, this is a function.

EXAMPLE 2 Functions

Decide whether each of the following equations represents a function. (Assume that x represents the independent variable here, an assumption we shall make throughout this book.) Give the domain and range of any functions.

(a) $y = 11 - 4x^2$

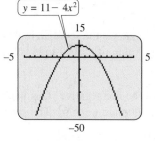

FIGURE 25

Solution For a given value of x, calculating $11 - 4x^2$ produces exactly one value of y. (For example, if $x = -7$, then $y = 11 - 4(-7)^2 = -185$, so $f(-7) = -185$.) Since one value of the independent variable leads to exactly one value of the dependent variable, $y = 11 - 4x^2$ meets the definition of a function.

Because x can take on any real-number value, the domain of this function is the set of all real numbers. Finding the range is more difficult. One way to find it would be to ask what possible values of y could come out of this function. Notice that the value of y is 11 minus a quantity that is always 0 or positive, since $4x^2$ can never be negative. There is no limit to how large $4x^2$ can be, so the range is $(-\infty, 11]$.

Another way to find the range would be to examine the graph. Figure 25 shows a graphing calculator view of this function, and we can see that the function takes on y-values of 11 or less. The calculator cannot tell us, however, whether the function continues to go down past the viewing window, or turns back up. To find out, we need to study this type of function more carefully, as we will do in the next section.

NOTE The actual graph of $y = 11 - 4x^2$ is a smooth curve. The jaggedness seen in Figure 25 is due to the limited resolution of the graphing calculator screen.

(b) $y^2 = x$

Solution Suppose $x = 36$. Then $y^2 = x$ becomes $y^2 = 36$, from which $y = 6$ or $y = -6$. Since one value of the independent variable can lead to two values of the dependent variable, $y^2 = x$ does not represent a function. ▮

The following agreement on domains is customary.

AGREEMENT ON DOMAINS

Unless otherwise stated, assume that the domain of all functions defined by an equation is the largest set of real numbers that are meaningful replacements for the independent variable.

For example, suppose

$$y = \frac{-4x}{2x - 3}.$$

Any real number can be used for x except $x = 3/2$, which makes the denominator equal 0. By the agreement on domains, the domain of this function is the set of all real numbers except $3/2$, which we denote $\{x \mid x \neq 3/2\}$, $\{x \neq 3/2\}$, or $(-\infty, 3/2) \cup (3/2, \infty)$.*

> **CAUTION** When finding the domain of a function, there are two operations to avoid: (1) dividing by zero; and (2) taking the square root (or any even root) of a negative number. Later chapters will present other functions, such as logarithms, which require further restrictions on the domain. For now, just remember these two restrictions on the domain.

EXAMPLE 3 Domain and Range

Find the domain and range for each of the functions defined as follows.

(a) $y = x^2$

Solution Any number may be squared, so the domain is the set of all real numbers, written $(-\infty, \infty)$. Since $x^2 \geq 0$ for every value of x, the range is $[0, \infty)$.

(b) $y = \sqrt{6 - x}$

Solution For y to be a real number, $6 - x$ must be nonnegative. This happens only when $6 - x \geq 0$, or $6 \geq x$, making the domain $(-\infty, 6]$. The range is $[0, \infty)$ because $\sqrt{6 - x}$ is always nonnegative.

(c) $y = \sqrt{2x^2 + 5x - 12}$

Solution The domain includes only those values of x satisfying $2x^2 + 5x - 12 \geq 0$. Using the methods for solving a quadratic inequality produces the domain

$$(-\infty, -4] \cup [3/2, \infty).$$

As in part (b), the range is $[0, \infty)$.

(d) $y = \dfrac{1}{x + 3}$

Solution Since the denominator cannot be zero, $x \neq -3$ and the domain is

$$(-\infty, -3) \cup (-3, \infty).$$

Because the numerator can never be zero, $y \neq 0$. There are no other restrictions on y, so the range is $(-\infty, 0) \cup (0, \infty)$.

FOR REVIEW ■

The Inequalities section of the Algebra Reference demonstrates the method for solving a quadratic inequality. To solve $2x^2 + 5x - 12 \geq 0$, factor the quadratic to get $(2x - 3)(x + 4) \geq 0$. Setting each factor equal to 0 gives $x = 3/2$ or $x = -4$, leading to the intervals $(-\infty, -4]$, $[-4, 3/2]$, and $[3/2, \infty)$. Testing a number from each interval shows that the solution is $(-\infty, -4] \cup [3/2, \infty)$.

To understand how a function works, think of a function f as a machine—for example, a calculator or computer—that takes an input x from the domain and uses it to produce an output $f(x)$ (which represents the y-value), as shown in Figure 26. In the basal body temperature example, when we put 21 into the machine, we get an output of 98.0, since $f(21) = 98.0$.

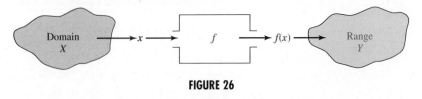

FIGURE 26

---

*The *union* of sets A and B, written $A \cup B$, is defined as the set of all elements in A or B or both.

EXAMPLE 4 Evaluating Functions

Let $g(x) = -x^2 + 4x - 5$. Find each of the following.

(a) $g(3)$

Solution Replace x with 3.

$$g(3) = -3^2 + 4 \cdot 3 - 5 = -9 + 12 - 5 = -2$$

(b) $g(a)$

Solution Replace x with a to get $g(a) = -a^2 + 4a - 5$.

This replacement of one variable with another is important in later chapters.

(c) $g(x + h) = -(x + h)^2 + 4(x + h) - 5$

$$= -(x^2 + 2xh + h^2) + 4(x + h) - 5$$

$$= -x^2 - 2xh - h^2 + 4x + 4h - 5$$

(d) $g\left(\dfrac{2}{r}\right) = -\left(\dfrac{2}{r}\right)^2 + 4\left(\dfrac{2}{r}\right) - 5 = -\dfrac{4}{r^2} + \dfrac{8}{r} - 5$

(e) Find all values of x such that $g(x) = -12$.

Solution Set $g(x)$ equal to -12, and then add 12 to both sides to make one side equal to 0.

$$-x^2 + 4x - 5 = -12$$

$$-x^2 + 4x + 7 = 0$$

This equation does not factor, but can be solved with the quadratic formula, which says that if $ax^2 + bx + c = 0$, where $a \neq 0$, then

$$x = \frac{-b \pm \sqrt{b^2 - 4ac}}{2a}.$$

In this case, with $a = -1$, $b = 4$, and $c = 7$, we have

$$x = \frac{-4 \pm \sqrt{16 - 4(-1)7}}{2(-1)}$$

$$= \frac{-4 \pm \sqrt{44}}{-2}$$

$$= 2 \pm \sqrt{11}$$

$$\approx -1.317 \quad \text{or} \quad 5.317.$$

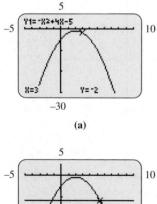

(a)

(b)

FIGURE 27

We can verify the results of parts (a) and (e) of the previous example using a graphing calculator. In Figure 27(a), after graphing $f(x) = -x^2 + 4x - 5$, we have used the "value" feature on the TI-83 to support our answer from part (a). In Figure 27(b) we have used the "intersect" feature to find the intersection of $y = g(x)$ and $y = -12$. The result is $x = 5.3166248$, which is one of our two answers to part (e). The graph clearly shows that there is another answer on the opposite side of the y-axis.

CAUTION Notice from Example 4(c) that $g(x + h)$ is *not* the same as $g(x) + h$, which equals $-x^2 + 4x - 5 + h$. There is a significant difference between applying a function to the quantity $x + h$ and applying a function to x and adding h afterwards.

If you tend to get confused when replacing x with $x + h$, as in Example 4(c), you might try replacing the x in the original function with a box, like this:

$$g\left(\boxed{}\right) = -\left(\boxed{}\right)^2 + 4\left(\boxed{}\right) - 5$$

Then, to compute $g(x + h)$, just enter $x + h$ into the box:

$$g\left(\boxed{x + h}\right) = -\left(\boxed{x + h}\right)^2 + 4\left(\boxed{x + h}\right) - 5$$

and proceed as in Example 4(c).

Notice in the basal body temperature example that to find the value of the function for a given value of x, we drew a vertical line from the value of x and found where it intersected the graph. If a graph is to represent a function, each value of x from the domain must lead to exactly one value of y. In the graph in Figure 28, the domain value x_1 leads to *two* y-values, y_1 and y_2. Since the given x-value corresponds to two different y-values, this is not the graph of a function. This example suggests the **vertical line test** for the graph of a function.

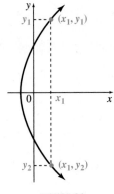

FIGURE 28

VERTICAL LINE TEST

If a vertical line intersects a graph in more than one point, the graph is not the graph of a function.

EXAMPLE 5 Vertical Line Test

Use the vertical line test to decide which of the graphs in Figure 29 are graphs of functions.

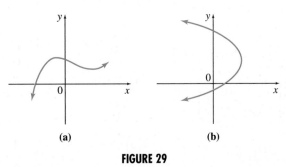

(a)   (b)

FIGURE 29

Solution

(a) Every vertical line intersects this graph in at most one point, so this is the graph of a function.

Solution

(b) It is possible for a vertical line to intersect the graph in part (b) twice. This is not the graph of a function.

Since we can substitute any expression, such as $x + h$, into a function, we can substitute one function for another. Such an operation is called *function composition*.

> **FUNCTION COMPOSITION**
>
> The **composition** of two functions f and g is the new function h, where
>
> $$h(x) = f(g(x)).$$
>
> The notation for function composition is $h = f \circ g$, or $h(x) = (f \circ g)(x).$

EXAMPLE 6 Function Composition

Given the functions $f(x) = x^2 + x$ and $g(x) = 5x - 4$, find each of the following.

(a) $h(x) = (f \circ g)(x)$

Solution

$$\begin{aligned}
h(x) &= (f \circ g)(x) \\
&= f(g(x)) \\
&= (g(x))^2 + g(x) \\
&= (5x - 4)^2 + (5x - 4) \\
&= 25x^2 - 35x + 12
\end{aligned}$$

(b) $k(x) = (g \circ f)(x)$

Solution

$$\begin{aligned}
k(x) &= (g \circ f)(x) \\
&= g(f(x)) \\
&= 5f(x) - 4 \\
&= 5(x^2 + x) - 4 \\
&= 5x^2 + 5x - 4
\end{aligned}$$

In the next example, we write an equation to model the area of a lot.

EXAMPLE 7 Area

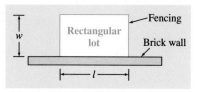

FIGURE 30

A hospital wishes to build a fence against a brick wall to form a rectangular lot, as shown in Figure 30. Only three sides of the fence need to be built, because the wall forms the fourth side. The contractor will use 200 meters of fencing. Let the length of the wall be l and the width w, as shown in Figure 30.

(a) Find the area of the lot as a function of the length l.

Solution The area formula for a rectangle is area = length × width, or

$$A = lw.$$

We want the area as a function of the length only, so we must eliminate the width. We use the fact that the total amount of fencing is the sum of the three sections, one length and two widths, so $200 = l + 2w$. Solve this for w:

$$\begin{aligned}
200 &= l + 2w \\
200 - l &= 2w \qquad \text{Subtract } l \text{ from both sides.} \\
100 - l/2 &= w. \qquad \text{Divide both sides by 2.}
\end{aligned}$$

Substituting this into the formula for area gives

$$A = l(100 - l/2).$$

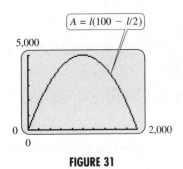

FIGURE 31

(b) Find the domain of the function in part (a).

Solution The length cannot be negative, so $l \geq 0$. Similarly, the width cannot be negative, so $100 - l/2 \geq 0$, from which we find $l \leq 200$. Therefore, the domain is $[0, 200]$.

(c) Sketch a graph of the function in part (a).

Solution The result from a graphing calculator is shown in Figure 31. Notice that at the endpoints of the domain, when $l = 0$ and $l = 200$, the area is 0. This makes sense: If the length or width is 0, the area will be 0 as well. In between, as the length increases from 0 to 100 meters, the area gets larger, and seems to reach a peak of 5,000 square meters when $l = 100$ meters. After that, the area gets smaller as the length continues to increase because the width is becoming smaller.

In the next section, we will study this type of function in more detail and determine exactly where the maximum occurs.

1.3 EXERCISES

Which of the following rules define y as a function of x?

1.

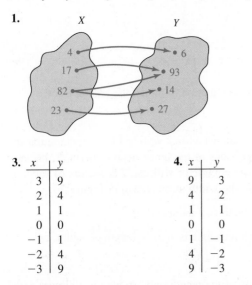

2.

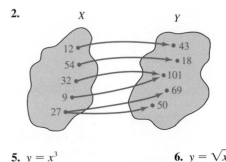

3.

x	y
3	9
2	4
1	1
0	0
−1	1
−2	4
−3	9

4.

x	y
9	3
4	2
1	1
0	0
1	−1
4	−2
9	−3

5. $y = x^3$

6. $y = \sqrt{x}$

7. $x = |y|$

8. $x = y^4 - 1$

List the ordered pairs obtained from each equation, given $\{-2, -1, 0, 1, 2, 3\}$ as the domain. Graph each set of ordered pairs. Give the range.

9. $y = 2x + 3$

10. $y = -4x + 9$

11. $2y - x = 5$

12. $6x - y = -3$

13. $y = x(x + 1)$

14. $y = (x - 2)(x - 3)$

15. $y = x^2$

16. $y = -2x^2$

17. $y = \dfrac{1}{x + 3}$

18. $y = \dfrac{-2}{x + 4}$

19. $y = \dfrac{3x - 3}{x + 5}$

20. $y = \dfrac{2x + 1}{x + 3}$

Give the domain of each function defined as follows.

21. $f(x) = 2x$

22. $f(x) = x + 2$

23. $f(x) = x^4$

24. $f(x) = (x - 2)^2$

25. $f(x) = \sqrt{16 - x^2}$

26. $f(x) = |x - 1|$

27. $f(x) = (x - 3)^{1/2}$

28. $f(x) = (3x + 5)^{1/2}$

29. $f(x) = \dfrac{2}{x^2 - 4}$

30. $f(x) = \dfrac{-8}{x^2 - 36}$

31. $f(x) = -\sqrt{\dfrac{2}{x^2 + 9}}$

32. $f(x) = -\sqrt{\dfrac{5}{x^2 + 36}}$

33. $f(x) = \sqrt{x^2 - 4x - 5}$

34. $f(x) = \sqrt{15x^2 + x - 2}$

35. $f(x) = \dfrac{1}{\sqrt{x^2 - 6x + 8}}$

36. $f(x) = \sqrt{\dfrac{x + 1}{x - 1}}$

Give the domain and the range of each function. Where arrows are drawn, assume the function continues in the indicated direction.

37.

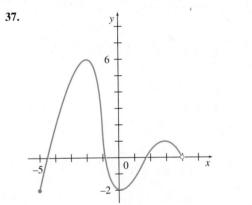

38.

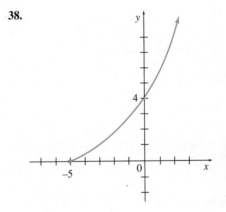

39.

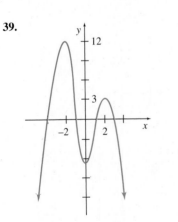

40.

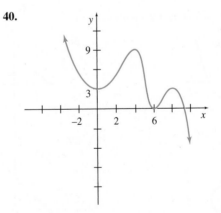

For each function, find **a.** $f(4)$, **b.** $f(-1/2)$, **c.** $f(a)$, **d.** $f(2/m)$, *and* **e.** *any values of x such that* $f(x) = 1$.

41. $f(x) = -x^2 + 5x + 1$

42. $f(x) = (x + 3)(x - 4)$

43. $f(x) = \dfrac{2x + 1}{x - 2}$

44. $f(x) = \dfrac{3x - 5}{2x + 3}$

In Exercises 45–48, give the domain and range. Then, use each graph to find
a. $f(-2)$, **b.** $f(0)$, **c.** $f(1/2)$, *and* **d.** *any values of x such that* $f(x) = 1$.

45.

46.

47.

48.

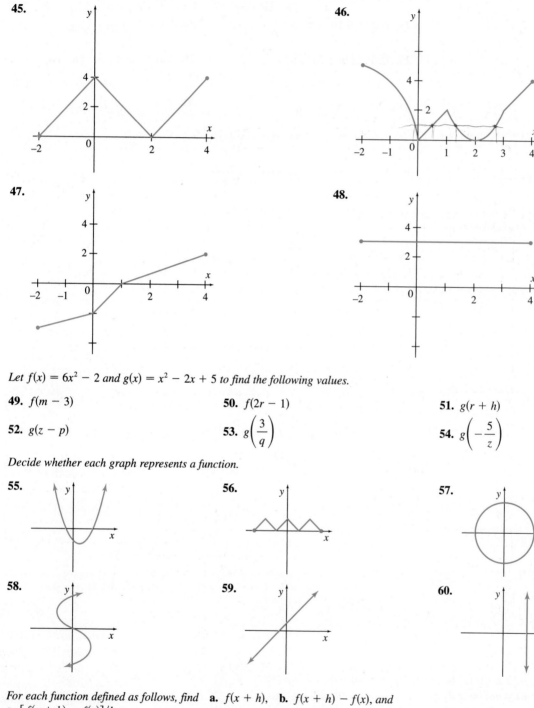

Let $f(x) = 6x^2 - 2$ *and* $g(x) = x^2 - 2x + 5$ *to find the following values.*

49. $f(m - 3)$

50. $f(2r - 1)$

51. $g(r + h)$

52. $g(z - p)$

53. $g\left(\dfrac{3}{q}\right)$

54. $g\left(-\dfrac{5}{z}\right)$

Decide whether each graph represents a function.

55.

56.

57.

58.

59.

60.

For each function defined as follows, find **a.** $f(x + h)$, **b.** $f(x + h) - f(x)$, *and*
c. $[f(x + h) - f(x)]/h$.

61. $f(x) = x^2 - 4$

62. $f(x) = 8 - 3x^2$

63. $f(x) = 2x^2 - 4x - 5$

64. $f(x) = -4x^2 + 3x + 2$

65. $f(x) = \dfrac{1}{x}$

66. $f(x) = -\dfrac{1}{x^2}$

For each pair of functions, find $(f \circ g)(x)$ and $(g \circ f)(x)$.

67. $f(x) = 2x^2 + 5x + 1$ and $g(x) = 3x - 1$

68. $f(x) = \dfrac{1}{3x + 2}$ and $g(x) = x^2 + 4x$

Applications

LIFE SCIENCES

69. *Whales Diving* The figure shows the depth of a diving sperm whale as a function of time, as recorded by researchers at the Woods Hole Oceanographic Institution in Massachusetts.*

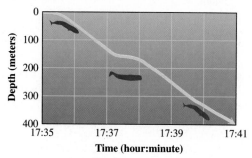

Find the depth of the whale at the following times.

a. 17 hours and 37 minutes

b. 17 hours and 39 minutes

70. *Metabolic Rate* The basal metabolic rate (in kcal/day) for large anteaters is given by

$$y = f(x) = 19.7x^{0.753},$$

where x is the anteater's weight in kg.[†][‡]

a. Find the basal metabolic rate for anteaters with the following weights.

  i. 5 kg   **ii.** 25 kg

b. Suppose the anteaters weight is given in lbs rather than kg. Given that 1 lb = 0.454 kg, find a function $x = g(z)$ giving the anteater's weight in kg if z is the animal's weight in lbs.

c. Only one of the two functions $f \circ g$ and $g \circ f$ makes physical sense. Calculate that function and describe what it does.

71. *Swimming Energy* The energy expenditure (in kcal/km) for animals swimming at the surface of the water is given by

$$y = f(x) = 0.01x^{0.88},$$

where x is the animal's weight in grams.[§]

a. Find the energy for the following animals swimming at the surface of the water.

  i. a muskrat weighing 800 g

  ii. a sea otter weighing 20,000 g

b. Suppose the animal's weight is given in kg rather than g. Given that 1 kg = 1,000 g, find a function $x = g(z)$ giving the animal's weight in g if z is the animal's weight in kgs.

c. Only one of the two functions $f \circ g$ and $g \circ f$ makes physical sense. Calculate that function and describe what it does.

72. *Energy Consumption* Over the last century, the world has shifted from using high-carbon sources of energy such as wood to lower carbon fuels such as oil and natural gas, as shown in the figure.[‖] The rise in carbon emissions during this time has caused concern because of its suspected contribution to global warming.

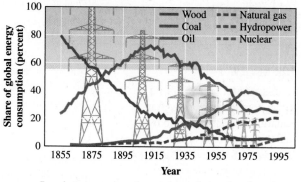

a. In what year were the percent of wood and coal use equal? What was the percent of each used in that year?

b. In what year were the percent of oil and coal use equal? What was the percent of each used in that year?

OTHER APPLICATIONS

73. *Internet Users* The table on the next page shows the estimated number of Internet users worldwide from 1995–1999.[#]

---

*Science, Vol. 291, Jan. 26, 2001, p. 577.

[†]Robbins, Charles T., *Wildlife Feeding and Nutrition,* 2nd ed., Academic Press, 1993, p. 125.

[‡]Technically, kilograms are a measure of mass, not weight. Weight is a measure of the force of gravity, which varies with the distance from the center of the earth. For objects on the surface of the earth, weight and mass are often used interchangeably, and we will do so in this text.

[§]Robbins, Charles T., *Wildlife Feeding and Nutrition,* 2nd ed., Academic Press, 1993, p. 142.

[‖]*The New York Times,* Oct. 31, 1999, p. 38.

[#]NUA Internet Surveys

Although the table shows a function with a y-value for just five x-values representing the years 1995 through 1999, the function can be defined for every x-value in the interval $1995 \leq x \leq 1999$. Let $y = f(x)$ represent the number of Internet users and x represent the years.

Worldwide Internet Users	
Year	**Millions of Users**
1995	26
1996	55
1997	98
1998	150
1999	205

a. What is the independent variable?

b. What is the dependent variable?

c. Find $f(1997)$.

d. Give the domain and range of the function.

74. *Perimeter* A rectangular field is to have an area of 500 sq m.

   a. Write the perimeter, P, of the field as a function of the width, w.

   b. Find the domain of the function in part a.

   c. Use a graphing calculator to sketch the graph of the function in part a.

   d. Describe what the graph found in part c tells you about how the perimeter of the field varies with the width.

75. *Area* A rectangular field is to have a perimeter of 6,000 ft.

   a. Write the area, A, of the field as a function of the width, w.

   b. Find the domain of the function in part a.

   c. Use a graphing calculator to sketch the graph of the function in part a.

   d. Describe what the graph found in part c tells you about how the area of the field varies with the width.

1.4   QUADRATIC FUNCTIONS; TRANSLATION AND REFLECTION

? THINK ABOUT IT   At what point in the development of a fetus is the resistance in the splenic artery at a maximum?

FOR REVIEW

In this section you will need to know how to solve a quadratic equation by factoring and by the quadratic formula, which are covered in Sections R.2 and R.4. Factoring is usually easiest; when a polynomial is set equal to zero and factored, then a solution is found by setting any one factor equal to zero. But factoring is not always possible. The quadratic formula will provide the solution to *any* quadratic equation.

In one of the exercises in this section, we will use a knowledge of *quadratic functions* to answer the question above.

A linear function is defined by

$$f(x) = ax + b,$$

for real numbers a and b. In a *quadratic function* the independent variable is squared. A quadratic function is an especially good model for many situations with a maximum or a minimum function value. Next to linear functions, they are the simplest type of function, and well worth studying thoroughly.

QUADRATIC FUNCTION

A **quadratic function** is defined by

$$f(x) = ax^2 + bx + c,$$

where a, b, and c are real numbers, with $a \neq 0$.

The simplest quadratic function has $f(x) = x^2$, with $a = 1$, $b = 0$, and $c = 0$. This function describes situations where the dependent variable y is proportional to the *square* of the independent variable x. The function can be graphed on a graphing calculator as shown in Figure 32. This graph is called a **parabola.** Every quadratic function has a parabola as its graph. The lowest (or highest) point on a

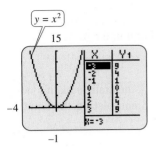

FIGURE 32

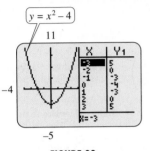

FIGURE 33

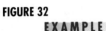

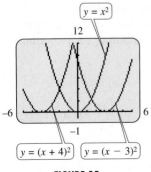

FIGURE 34

FIGURE 35

parabola is the **vertex** of the parabola. The vertex of the parabola in Figure 32 is $(0, 0)$.

If the graph in Figure 32 were folded in half along the y-axis, the two halves of the parabola would match exactly. This means that the graph of a quadratic function is *symmetric* with respect to a vertical line through the vertex; this line is the **axis** of the parabola.

There are many real-world instances of parabolas. For example, cross sections of spotlight reflectors or radar dishes form parabolas. Also, a projectile thrown in the air follows a parabolic path.

EXAMPLE 1 Graphing Quadratic Functions
Graph $y = x^2 - 4$.

Solution Each value of y will be 4 less than the corresponding value of y in $y = x^2$. The graph of $y = x^2 - 4$ has the same shape as that of $y = x^2$ but is 4 units lower. See Figure 33. The vertex of the parabola (on this parabola, the *lowest* point) is at $(0, -4)$. The x-intercepts can be found by letting $y = 0$ to get

$$0 = x^2 - 4,$$

from which $x = 2$ and $x = -2$ are the x-intercepts. The axis of the parabola is the vertical line $x = 0$.

Example 1 suggests that the effect of c in $ax^2 + bx + c$ is to lower the graph if c is negative and to raise the graph if c is positive. This is true for any function; the movement up or down is referred to as a **vertical translation** of the function.

EXAMPLE 2 Graphing Quadratic Functions
Graph $y = ax^2$ with $a = -0.5$, $a = -1$, $a = -2$, and $a = -4$.

Solution Figure 34 shows all four functions plotted on the same axes. We see that since a is negative, the graph opens downward. When the magnitude of a is less than 1 (that is, when $a = -0.5$), the graph is wider than the original graph, because the values of y are smaller in magnitude. Conversely, when the magnitude of a is greater than 1, the graph is steeper.

Example 2 shows that the sign of a in $ax^2 + bx + c$ determines whether the parabola opens upward or downward. Multiplying $f(x)$ by a negative number flips the graph of f upside down. This is called a **vertical reflection** of the graph. The magnitude of a determines how steeply the graph increases or decreases.

EXAMPLE 3 Graphing Quadratic Functions
Graph $y = (x - h)^2$ for $h = 3$, 0, and -4.

Solution Figure 35 shows a graphing calculator view of all three functions on the same axes. Notice that since the number is subtracted *before* the squaring occurs, the graph does not move up or down, but instead moves left or right. Evaluating $f(x) = (x - 3)^2$ at $x = 3$ gives the same result as evaluating $f(x) = x^2$ at $x = 0$. Therefore, when we subtract the positive number 3 from x, the graph shifts 3 units to the right, so the vertex is at $(3, 0)$. Similarly, when we subtract the negative number -4 from x—in other words, when the function becomes $f(x) = (x + 4)^2$—the graph shifts to the left 4 units.

The left or right shift of the graph illustrated in Figure 35 is called a **horizontal translation** of the function.

If a quadratic equation is given in the form $ax^2 + bx + c$, we can identify the translations and any vertical reflection by rewriting it in the form

$$y = a(x - h)^2 + k.$$

In this form, we can identify the vertex as (h, k). A quadratic equation not given in this form can be converted by a process called **completing the square.** The next example illustrates the process.

EXAMPLE 4   Graphing Quadratic Functions

Graph $f(x) = -3x^2 - 2x + 1$.

Solution To begin, factor -3 from the x-terms so the coefficient of x^2 is 1.

$$f(x) = -3\left(x^2 + \frac{2}{3}x\right) + 1 \qquad \text{Factor out } -3.$$

$$= -3\left(x^2 + \frac{2}{3}x + \frac{1}{9}\right) + 1 + 3\left(\frac{1}{9}\right) \qquad \begin{array}{l}\text{Add and subtract } -3 \text{ times}\\ \left(\frac{1}{2} \text{ the coefficient of } x\right)^2.\end{array}$$

$$= -3\left(x + \frac{1}{3}\right)^2 + \frac{4}{3} \qquad \text{Factor and combine terms.}$$

Note that in the second step we added 0. Now we can identify $a = -3$, $h = -1/3$, and $k = 4/3$, so the graph is the parabola $y = x^2$ translated $1/3$ unit to the left, flipped upside down, and translated $4/3$ units upward. This puts the vertex at $(-1/3, 4/3)$. The 3 in front of the squared term causes the parabola to be stretched vertically by a factor of 3. These results are shown in Figure 36.

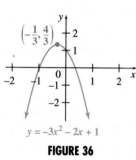

$y = -3x^2 - 2x + 1$

FIGURE 36

Instead of completing the square to find the vertex of the graph of a quadratic function given in the form $y = ax^2 + bx + c$, we can develop a formula for the vertex. By the quadratic formula, if $ax^2 + bx + c = 0$, where $a \neq 0$, then

$$x = \frac{-b \pm \sqrt{b^2 - 4ac}}{2a}.$$

Notice that this is the same as

$$x = \frac{-b}{2a} \pm \frac{\sqrt{b^2 - 4ac}}{2a} = \frac{-b}{2a} \pm Q,$$

where $Q = \sqrt{b^2 - 4ac}/(2a)$. Since a parabola is symmetric with respect to its axis, the vertex is halfway between its two roots. Halfway between $x = -b/(2a) + Q$ and $x = -b/(2a) - Q$ is $x = -b/(2a)$. Once we have the x-coordinate of the vertex, we can easily find the y-coordinate by substituting the x-coordinate into the original equation.

GRAPH OF THE QUADRATIC FUNCTION

The graph of the quadratic function $f(x) = ax^2 + bx + c$ has its vertex at

$$\left(\frac{-b}{2a}, f\left(\frac{-b}{2a}\right)\right).$$

The graph opens upward if $a > 0$ and downward if $a < 0$.

A graphing calculator does not necessarily tell us the exact value of the vertex or the *x*- or *y*-intercepts, but it gives a sketch quickly, and it allows us to verify what we have found through algebra. In many examples, the exact value of the solutions given by the quadratic formula will be irrational numbers, and a calculator is required to approximate the solutions. Another situation that may arise is the absence of any *x*-intercepts, as in the next example.

EXAMPLE 5 Graphing Quadratic Functions

Graph $y = x^2 + 4x + 6$.

Solution This does not appear to factor, so we'll try the quadratic formula.

$$x = \frac{-b \pm \sqrt{b^2 - 4ac}}{2a}$$

$$= \frac{-4 \pm \sqrt{4^2 - 4(1)(6)}}{2(1)} = \frac{-4 \pm \sqrt{-8}}{2}$$

As soon as we see the negative under the square root sign, we know there are no *x*-intercepts. Nevertheless, the vertex is still at

$$x = \frac{-b}{2a} = \frac{-4}{2} = -2.$$

Substituting this into the equation gives

$$y = (-2)^2 + 4(-2) + 6 = 2.$$

The *y*-intercept is at $(0, 6)$, which is 2 units to the right of the parabola's axis $x = -2$. Using the symmetry of the figure, we can also plot the mirror image of this point on the opposite side of the parabola's axis: at $x = -4$, 2 units to the left of the axis, *y* is also equal to 6. Plotting the vertex, the *y*-intercept, and the point $(-4, 6)$ gives the graph in Figure 37.

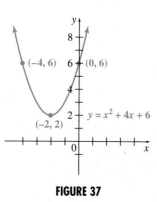

FIGURE 37

Section 1.2 showed how the equation of a line that closely approximates a set of data points is found using linear regression. Some graphing calculators with statistics capability perform other kinds of regression. For example, *quadratic regression* gives the coefficients of a quadratic equation that models a given set of points.

EXAMPLE 6 Lead Emissions

Table 5 gives lead emissions from all sources in the United States. We let $x = 0$ represent the year 1990, so $x = 2$ represents 1992, and so on, and *y* is lead emissions in thousands. A scatterplot of these data is given in Figure 38(a).

The scatterplot suggests that a quadratic function with a negative value of *a* (so the graph opens downward) would be a reasonable model for the data. Using the quadratic regression feature of a graphing calculator to model the data, we find that the quadratic function with

$$f(x) = -34.64x^2 + 298.8x + 3,347$$

approximates the data fairly well, as shown in Figure 38(b) on the next page. Since $a < 0$, the graph of $f(x)$ opens downward (as expected), and so has a maximum. Using the graphing calculator feature that locates the maximum value of the function, we find the maximum value of *y* occurs at approximately $(4.313, 3,991)$. See Figure 38(b). Thus, according to the function, maximum lead emissions of about 3,991,000 occurred in 1994. This is slightly less than the actual value of 4,043,000 given in the table.

Table 5	
Year	**Thousands of Tons**
1992	3,808
1993	3,911
1994	4,043
1995	3,924
1996	3,910

National Air Pollution Emission Trends, U.S. Environmental Protection Agency, 1900–1996.

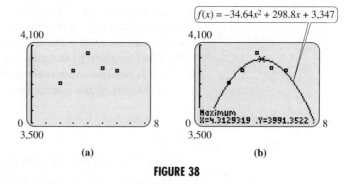

$f(x) = -34.64x^2 + 298.8x + 3,347$

(a)

(b)

FIGURE 38

TRANSLATIONS AND REFLECTIONS OF FUNCTIONS

Let f be any function, and let a, h, and k be positive constants (Figure 39).

The graph of $y = f(x) + k$ is the graph of $y = f(x)$ translated upward by an amount k (Figure 40).

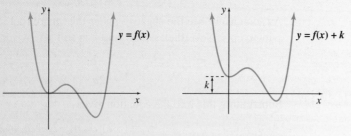

FIGURE 39   **FIGURE 40**

The graph of $y = f(x) - k$ is the graph of $y = f(x)$ translated downward by an amount k (Figure 41).

The graph of $y = f(x - h)$ is the graph of $y = f(x)$ translated to the right by an amount h (Figure 42).

The graph of $y = f(x + h)$ is the graph of $y = f(x)$ translated to the left by an amount h (Figure 43).

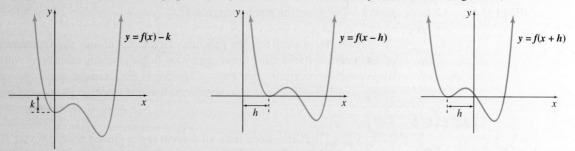

FIGURE 41   **FIGURE 42**   **FIGURE 43**

The graph of $y = -f(x)$ is the graph of $y = f(x)$ reflected vertically, that is, turned upside down (Figure 44).

The graph of $y = f(-x)$ is the graph of $y = f(x)$ reflected horizontally, that is, its mirror image (Figure 45).

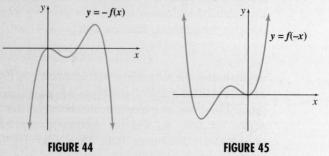

FIGURE 44   **FIGURE 45**

Multiplying x or $f(x)$ by a constant a, to get $y = f(ax)$ or $y = a \cdot f(x)$ does not change the general appearance of the graph, except to compress or stretch it. When a is negative, it also causes a reflection, as shown in the last two figures in the summary for $a = -1$.

EXAMPLE 7 Translations and Reflections of Graphs
Graph $f(x) = -\sqrt{4 - x} + 3$.

Solution Begin with the simplest possible function, then add each variation in turn. Start with the graph of $f(x) = \sqrt{x}$. As Figure 46 reveals, this is just one-half of the graph of $f(x) = x^2$ lying on its side.

Now add another component of the original function, the negative in front of the x, giving $f(x) = \sqrt{-x}$. This is a horizontal reflection of the $f(x) = \sqrt{x}$ graph, as shown in Figure 47. Next, include the 4 under the square root sign. To get $4 - x$ into the form $f(x - h)$ or $f(x + h)$, we need to factor out the negative: $\sqrt{4 - x} = \sqrt{-(x - 4)}$. Now the 4 is subtracted, so this function is a translation to the right of the function $f(x) = \sqrt{-x}$ by 4 units, as Figure 48 indicates.

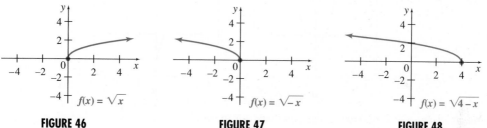

FIGURE 46   **FIGURE 47**   **FIGURE 48**

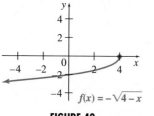

FIGURE 49

The effect of the negative in front of the radical is a vertical reflection, as in Figure 49, which shows the graph of $f(x) = -\sqrt{4 - x}$. Finally, adding the constant 3 raises the entire graph by 3 units, giving the graph of $f(x) = -\sqrt{4 - x} + 3$ in Figure 50(a).

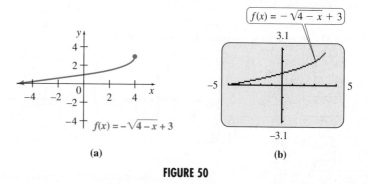

(a)   (b)

FIGURE 50

If you viewed a graphing calculator image such as Figure 50(b), you might think the function continues to go up and to the right. By realizing that $(4, 3)$ is the vertex of the sideways parabola, we see that this is the rightmost point on the graph. Another approach is to find the domain of f by setting $4 - x \geq 0$, from which we conclude that $x \leq 4$. This demonstrates the importance of knowing the algebraic techniques in order to interpret a graphing calculator image correctly.

1.4 EXERCISES

1. How does the value of a affect the graph of $y = ax^2$? Discuss the case for $a \geq 1$ and for $0 \leq a \leq 1$.

2. How does the value of a affect the graph of $y = ax^2$ if $a \leq 0$?

In Exercises 3–7, match the correct graph A, B, C, D, or E to the function without using your calculator. Then, if you have a graphing calculator, use it to check your answers. Each graph in this group shows x and y in $[-10, 10]$.

3. $y = x^2 - 3$

4. $y = (x - 3)^2 + 2$

5. $y = (x + 3)^2 + 2$

6. $y = -(3 - x)^2 + 2$

7. $y = -(x + 3)^2 + 2$

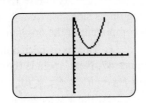

A

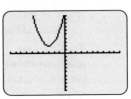

B

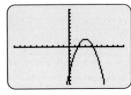

C

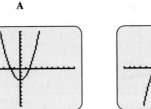

D

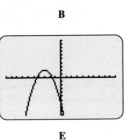

E

Graph each parabola and give its vertex, axis, x-intercepts, and y-intercept.

8. $y = x^2 + 6x + 5$

9. $y = x^2 - 10x + 21$

10. $y = -2x^2 - 12x - 16$

11. $y = -3x^2 + 12x - 11$

12. $f(x) = 2x^2 + 12x - 16$

13. $f(x) = -x^2 + 6x - 6$

14. $f(x) = 2x^2 - 4x + 5$

15. $f(x) = -3x^2 + 24x - 36$

16. $f(x) = -\frac{1}{3}x^2 + 2x + 4$

17. $f(x) = \frac{5}{2}x^2 + 10x + 8$

18. $f(x) = \frac{2}{3}x^2 - \frac{8}{3}x + \frac{5}{3}$

19. $f(x) = -\frac{1}{2}x^2 - x - \frac{7}{2}$

In Exercises 20–24, follow the directions for Exercises 3–7.

20. $y = \sqrt{x + 2} - 4$

21. $y = \sqrt{x - 2} - 4$

22. $y = \sqrt{-x + 2} - 4$

23. $y = \sqrt{-x - 2} - 4$

24. $y = -\sqrt{x + 2} - 4$

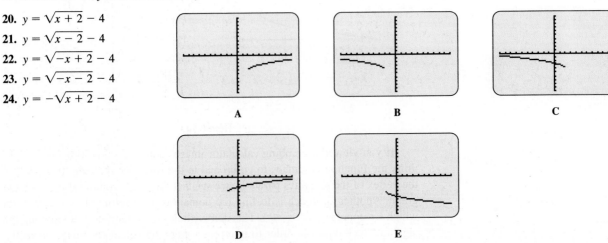

A   **B**   **C**

D   **E**

Given the following graph, sketch by hand the graph of the function described, indicating how the three points labeled on the original graph have been translated.

25. $y = -f(x)$

26. $y = f(x - 2) + 2$

27. $y = f(-x)$

28. $y = f(2 - x) + 2$

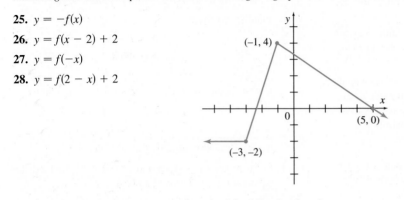

Use the ideas in this section to graph each of the following functions.

29. $f(x) = \sqrt{x - 1} + 3$

30. $f(x) = \sqrt{x + 1} - 4$

31. $f(x) = -\sqrt{-4 - x} - 2$

32. $f(x) = -\sqrt{2 - x} + 2$

Drawing a figure similar to those in the Translations and Reflections box, Figures 39–45, show the graph of $f(ax)$ where a satisfies the given condition.

33. $0 < a < 1$

34. $1 < a$

35. $-1 < a < 0$

36. $a < -1$

Drawing a figure similar to those in the Translations and Reflections box, Figures 39–45, show the graph of $af(x)$ where a satisfies the given condition.

37. $0 < a < 1$

38. $1 < a$

39. $-1 < a < 0$

40. $a < -1$

41. If r is an x-intercept of the graph of $y = f(x)$, what is an x-intercept of the graph of each of the following?

  a. $y = -f(x)$ **b.** $y = f(-x)$ **c.** $y = -f(-x)$

42. If b is the y-intercept of the graph of $y = f(x)$, what is the y-intercept of the graph of each of the following?

  a. $y = -f(x)$ **b.** $y = f(-x)$ **c.** $y = -f(-x)$

Applications

LIFE SCIENCES

43. *Splenic Artery Resistance* Blood flow to the fetal spleen is of research interest because several diseases are associated with increased resistance in the splenic artery (the artery that goes to the spleen). Researchers have found that the index of splenic artery resistance in the fetus can be described by the function

$$y = 0.057x - 0.001x^2,$$

where x is the number of weeks of gestation.*

  a. At how many weeks is the splenic artery resistance a maximum?

  b. What is the maximum splenic artery resistance?

  c. At how many weeks is the splenic artery resistance equal to 0, according to this formula? Is your answer reasonable for this function? Explain.

44. *Tooth Length* The length (in mm) of the mesiodistal crown of the first molar for human fetuses can be approximated by

$$L(t) = -0.01t^2 + 0.788t - 7.048,$$

---

*Abuhamad, A. Z., et al., "Doppler Flow Velocimetry of the Splenic Artery in the Human Fetus: Is It a Marker of Chronic Hypoxia?" *American Journal of Obstetrics and Gynecology,* Vol. 172, No. 3, March 1995, pp. 820–825.

where t is the number of weeks since conception.[*]

a. What does this formula predict for the length at 14 weeks? 24 weeks?

b. What does this formula predict for the maximum length, and when does that occur? Explain why the formula does not make sense past that time.

45. *Angioplasty* In an effort to improve angioplasty—in which a wire is threaded through a coronary artery to guide a balloon which then opens a blocked artery—researchers have modeled the guide wire using a quadratic equation[†]

$$y = px^2 + qx + r.$$

Part of the process involves taking pairs of points (x_0, y_0) and (x_1, y_1) on an X-ray image and finding the values of p, q, and r as described below.

a. Substitute the two points into the quadratic equation, and call the resulting equations E_0 and E_1. Subtract $E_1 - E_0$ to eliminate r, and solve the resulting equation for p to get

$$p = \frac{y_1 - y_0 - q(x_1 - x_0)}{x_1^2 - x_0^2}.$$

b. Substitute the result from part a into equation E_0 and solve for r to obtain

$$r = y_0 - x_0^2\left(\frac{y_1 - y_0}{x_1^2 - x_0^2}\right) + qx_0^2\left(\frac{x_1 - x_0}{x_1^2 - x_0^2}\right) - qx_0.$$

c. Substitute the values for p and r from parts a and c into the original quadratic equation to get

$$y = \left(\frac{y_1 - y_0}{x_1^2 - x_0^2}\right)x^2$$
$$+ q\left[x - x_0 - \left(\frac{x_1 - x_0}{x_1^2 - x_0^2}\right)x^2 + \left(\frac{x_1 - x_0}{x_1^2 - x_0^2}\right)x_0^2\right]$$
$$+ y_0 - \left(\frac{y_1 - y_0}{x_1^2 - x_0^2}\right)x_0^2.$$

d. Substitute

$$K = \frac{y_1 - y_0}{x_1^2 - x_0^2},$$
$$L = Kx_0^2$$
$$\text{and} \quad M = \frac{x_1 - x_0}{x_1^2 - x_0^2} = \frac{1}{x_1 + x_0}$$

into the result from part c to obtain

$$y = Kx^2 + q(x - x_0 - Mx^2 + Mx_0^2) + y_0 - L.$$

e. Finally, substitute

$$A = L - y_0,$$
$$P = Mx_0^2,$$
$$\text{and} \quad R = x_0 - P,$$

into the result from part d, and solve for q to obtain

$$q = \frac{y - Kx^2 + A}{x - R - Mx^2},$$

the final equation used by the researchers.

f. Suppose $(x_0, y_0) = (1, 2)$ and $(x_1, y_1) = (3, 5)$. Find the equation for q in terms of x and y.

46. *Length of Life* According to recent data from the Teachers Insurance and Annuity Association (TIAA), the survival function for life after 65 is approximately given by

$$S(x) = 1 - 0.058x - 0.076x^2,$$

where x is measured in decades. This function gives the probability that an individual who reaches the age of 65 will live at least x decades ($10x$ years) longer.[‡]

a. Find the median length of life for people who reach 65, that is, the age for which the survival rate is 0.50.

b. Find the age beyond which virtually nobody lives. (There are, of course, exceptions.)

47. *Medicine* Between 1992 and 1998, the percent of college freshmen who planned to get a professional degree in a medical field can be modeled by

$$f(x) = -0.2369x^2 + 1.425x + 6.905,$$

where $x = 0$ represents 1992.[§] Based on this model, in what year did the percent of freshmen planning to get a medical degree reach its maximum? What is the domain of $f(x)$?

48. *AIDS* The table[‖] on the next page lists the total (cumulative) number of AIDS cases diagnosed in the United States up to 1996. For example, a total of 22,620 AIDS cases were diagnosed between 1981 and 1985.

a. Plot the data in the table using 0 for 1980, and so on.

b. Would a linear or quadratic function model these data best? Explain.

---

[*]Harris, Edward F., Joseph D. Hicks, and Betsy D. Barcroft, "Tissue Contributions to Sex and Race: Differences in Tooth Crown Size of Deciduous Molars," *American Journal of Physical Anthropology,* Vol. 115, 2001, pp. 223–237.

[†]Palti-Wasserman, Daphna, et al., "Identifying and Tracking a Guide Wire in the Coronary Arteries During Angioplasty from X-Ray Images," *IEEE Transactions on Biomedical Engineering,* Vol. 44, No. 2, Feb. 1997, pp. 152–163.

[‡]Exercise 46 is from Ralph DeMarr, University of New Mexico.

[§]*The American Freshmen: National Norms for Fall 1992–1998,* Higher Education Research Institute, UCLA.

[‖]"Facts and Figures," Joint United Nations Programme on HIV/AIDS (UNAIDS), Feb. 1999.

Year	AIDS Cases	Year	AIDS Cases
1982	1,563	1990	193,245
1983	4,647	1991	248,023
1984	10,845	1992	315,329
1985	22,620	1993	361,509
1986	41,662	1994	441,406
1987	70,222	1995	515,586
1988	105,489	1996	584,394
1989	147,170		

c. If your graphing calculator has a regression feature, find the quadratic function that best fits the data. Graph this function on the same calculator window as the data. (On a TI-83 calculator, press the STAT key, and then select the CALC menu. QuadReg is item 5. The command `QuadReg L₁, L₂, Y₁` finds the quadratic regression equation for the data in L_1 and L_2 and stores the function in Y_1.)

d. Find a quadratic function defined by $f(x) = a(x - h)^2 + k$ that models the data using (2, 1,563) as the vertex and then choosing a second point such as (13, 361,509) to determine a.

e. Graph the function from part d on the same calculator window as the data and function from part c. Do the graphs of the two functions differ much?

OTHER APPLICATIONS

49. *Maximizing the Height of an Object* If an object is thrown upward with an initial velocity of 32 ft/sec, then its height after t sec is given by

$$h = 32t - 16t^2.$$

a. Find the maximum height attained by the object.

b. Find the number of seconds it takes the object to hit the ground.

50. *Stopping Distance* According to data from the National Traffic Safety Institute,[*] the stopping distance y in feet of a car traveling x mph can be described by the equation $y = 0.056057x^2 + 1.06657x$.

a. Find the stopping distance for a car traveling 25 mph.

b. How fast can you drive if you need to be certain of stopping within 150 ft?

51. *Highway Research* Since 1990 Congress has authorized more than $700 million for research and development of "smart" highways, with the goal of enabling cars to drive themselves.[†] The spending per year is approximated by the function

$$f(x) = -19.321x^2 + 3,608.7x - 168,310,$$

where $x = 90$ corresponds to 1990 and y is in millions of dollars.

a. In what year was the maximum amount spent?

b. What is the maximum amount spent?

c. Sketch a graph of the function.

52. *Accident Rate* According to data from the National Highway Traffic Safety Administration, the accident rate as a function of the age of the driver in years x can be approximated by the function

$$f(x) = 60.0 - 2.28x + 0.0232x^2$$

for $16 \le x \le 85$. Find the age at which the accident rate is a minimum and the minimum rate.[‡]

53. *Maximizing Area* Glenview Community College wants to construct a rectangular parking lot on land bordered on one side by a highway. It has 320 ft of fencing to use along the other three sides. What should be the dimensions of the lot if the enclosed area is to be a maximum? (*Hint:* Let x represent the width of the lot, and let $320 - 2x$ represent the length.)

54. *Maximizing Area* What would be the maximum area that could be enclosed by the college's 320 ft of fencing if it decided to close the entrance by enclosing all four sides of the lot? (See Exercise 53.)

In Exercises 55 and 56, draw a sketch of the arch or culvert on coordinate axes, with the horizontal and vertical axes through the vertex of the parabola. Use the given information to label points on the parabola. Then give the equation of the parabola and answer the question.

55. *Parabolic Arch* An arch is shaped like a parabola. It is 30 m wide at the base and 15 m high. How wide is the arch 10 m from the ground?

56. *Parabolic Culvert* A culvert is shaped like a parabola, 18 ft across the top and 12 ft deep. How wide is the culvert 8 ft from the top?

National Traffic Safety Institute Student Workbook, 1993, p. 7.
[†]IVHS America
[‡]Exercise 52 is from Ralph DeMarr, University of New Mexico.

1.5 POLYNOMIAL AND RATIONAL FUNCTIONS

When during the 1940s and 1950s was the incidence of measles in the United States the highest? How much ash does a mouse ingest?

In an example and an exercise in this section, we will explore these questions using *polynomial* and *rational functions*.

Polynomial Functions Earlier, we discussed linear and quadratic functions and their graphs. Both of these functions are special types of *polynomial functions*.

POLYNOMIAL FUNCTION

A **polynomial function** of degree n, where n is a nonnegative integer, is defined by

$$f(x) = a_n x^n + a_{n-1} x^{n-1} + \cdots + a_1 x + a_0,$$

where $a_n, a_{n-1}, \ldots, a_1$, and a_0 are real numbers, called **coefficients,** with $a_n \neq 0$. The number a_n is called the **leading coefficient.**

For $n = 1$, a polynomial function takes the form

$$f(x) = a_1 x + a_0,$$

a linear function. A linear function, therefore, is a polynomial function of degree 1. (Note, however, that a linear function of the form $f(x) = a_0$ for a real number a_0 is a polynomial function of degree 0, the constant function.) A polynomial function of degree 2 is a quadratic function.

Accurate graphs of polynomial functions of degree 3 or higher require methods of calculus to be discussed later. Meanwhile, a graphing calculator is useful for obtaining such graphs, but care must be taken in choosing a viewing window that captures the significant behavior of the function.

The simplest polynomial functions of higher degree are those of the form $f(x) = x^n$. Such a function is known as a **power function.** Figure 51 on the next page shows the graphs of $f(x) = x^3$ and $f(x) = x^5$. These functions are simple enough that they can be drawn by hand by plotting a few points and connecting them with a smooth curve. An important property of all polynomials is that their graphs are smooth curves.

The graphs of $f(x) = x^4$ and $f(x) = x^6$, shown in Figure 52, can be sketched in a similar manner. These graphs have symmetry about the y-axis, as does the graph of $f(x) = ax^2$ for a nonzero real number a. As with the graph of $f(x) = ax^2$, the value of a in $f(x) = ax^n$ affects the direction of the graph. When $a > 0$, the graph has the same general appearance as the graph of $f(x) = x^n$. However, if $a < 0$, the graph is reflected vertically.

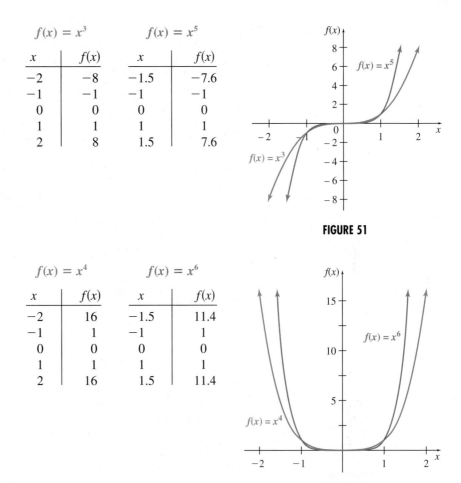

$f(x) = x^3$

x	$f(x)$
-2	-8
-1	-1
0	0
1	1
2	8

$f(x) = x^5$

x	$f(x)$
-1.5	-7.6
-1	-1
0	0
1	1
1.5	7.6

FIGURE 51

$f(x) = x^4$

x	$f(x)$
-2	16
-1	1
0	0
1	1
2	16

$f(x) = x^6$

x	$f(x)$
-1.5	11.4
-1	1
0	0
1	1
1.5	11.4

FIGURE 52

EXAMPLE 1   Translations and Reflections
Graph $f(x) = -(x - 2)^3 + 3$.

Solution Using the principles of translation and reflection from the previous section, we recognize that this is similar to the graph of $y = x^3$, but reflected vertically (because of the negative in front of $(x - 2)^3$), and with its center moved 2 units to the right and 3 units up. The result is shown in Figure 53.

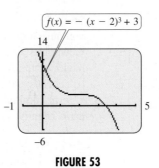

$f(x) = -(x - 2)^3 + 3$

FIGURE 53

EXAMPLE 2 Graphing Polynomials
Graph $f(x) = 8x^3 - 12x^2 + 2x + 1$.

Solution Figure 54 shows the function graphed on the x- and y-intervals $[-0.5, 0.6]$ and $[-2, 2]$. In this view, it appears similar to a parabola opening downward. Zooming out to $[-1, 2]$ by $[-8, 8]$, we see in Figure 55 that the graph goes upward as x gets large. There are also two **turning points** near $x = 0$ and $x = 1$. (In a later chapter, we will introduce another term for such turning points: *relative extrema.*) By zooming in with the graphing calculator, we can find these turning points to be at approximately $(0.09175, 1.08866)$ and $(0.90825, -1.08866)$.

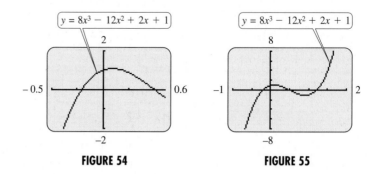

FIGURE 54   **FIGURE 55**

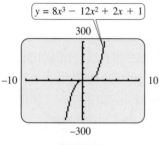

FIGURE 56

Zooming out still further, we see the function on $[-10, 10]$ by $[-300, 300]$ in Figure 56. From this viewpoint, we don't see the turning points at all, and the graph seems similar in shape to that of $y = x^3$. This is an important point: when x is large in magnitude, either positive or negative, $8x^3 - 12x^2 + 2x + 1$ behaves a lot like $8x^3$, because the other terms are small in comparison with the cubic term. So this viewpoint tells us something useful about the function, but it is less useful than the previous graph for determining the turning points.

After the previous example, you may wonder how to be sure you have the viewing window that exhibits all the important properties of a function. We will find an answer to this question in later chapters using the techniques of calculus. Meanwhile, let us consider one more example to get a better idea of what polynomials look like.

EXAMPLE 3 Graphing Polynomials
Graph $f(x) = -3x^4 + 14x^3 - 54x + 3$.

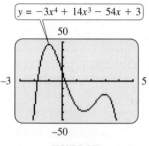

FIGURE 57

Solution Figure 57 shows a graphing calculator view on $[-3, 5]$ by $[-50, 50]$. If you have a graphing calculator, we recommend that you experiment with various viewpoints and verify for yourself that this viewpoint captures the important behavior of the function. Notice that it has three turning points. Notice also that as $|x|$ gets large, the graph turns downward. This is because as $|x|$ becomes large, the x^4 term dominates the other terms, which are small in comparison, and the x^4 term has a negative coefficient.

As suggested by the graphs above, the domain of a polynomial function is the set of all real numbers. The range of a polynomial function of odd degree is also the set of all real numbers. Some typical graphs of polynomial functions of odd degree are shown in Figure 58. These graphs suggest that for every polynomial function f of odd degree, there is at least one real value of x for which $f(x) = 0$. Such a value of x is called a **real zero** of f; these values are also the x-intercepts of the graph.

Polynomial functions of even degree have a range that takes either the form $(-\infty, k]$ or the form $[k, \infty)$ for some real number k. Figure 59 shows two typical graphs of polynomial functions of even degree.

A fifth-degree polynomial can have four turning points, as in the last graph of Figure 58, or no turning points, as in Figure 51. By examining the figures in this section, you may notice that the graph of a polynomial of degree n has at most $n - 1$ turning points. In a later chapter we will use calculus to see why this is true. Meanwhile, you can learn much about a polynomial by examining its graph. For example, if you are presented with the second graph in Figure 59 and told it is a polynomial, you know immediately that it is of even degree, because the range is of the form $(-\infty, k]$. It has five turning points, so it must be of degree 6 or higher. It could be of degree 8 or 10 or 12, etc., but you can't be sure from the graph alone. Because it goes down at the right, the leading coefficient must be negative.

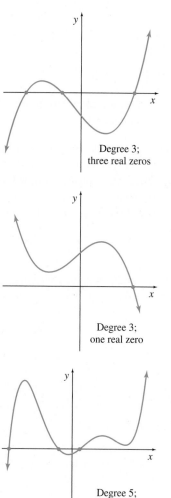

Degree 3;
three real zeros

Degree 3;
one real zero

Degree 5;
three real zeros

FIGURE 58

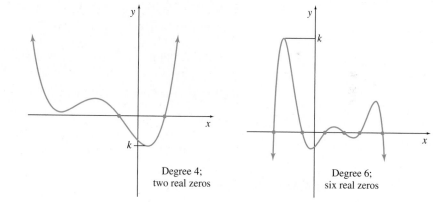

Degree 4;
two real zeros

Degree 6;
six real zeros

FIGURE 59

These ideas about polynomial functions are summarized here.

PROPERTIES OF POLYNOMIAL FUNCTIONS

1. A polynomial function of degree n can have at most $n - 1$ turning points. Conversely, if the graph of a polynomial function has n turning points, it must have degree at least $n + 1$.

2. In the graph of a polynomial function of even degree, both ends go up or both ends go down. For a polynomial function of odd degree, one end goes up and one end goes down.

3. If the graph goes up as x becomes large, the leading coefficient must be positive. If the graph goes down as x becomes large, the leading coefficient is negative.

EXAMPLE 4 Measles

The incidence of measles (per 100,000) for the period 1940–1960 can be approximated by

$$M(x) = -0.287x^3 + 8.8x^2 - 59.843x + 220.7,$$

where x is the number of years since 1940.* Use a graphing calculator to find the year in this period when the incidence of measles was highest.

Solution Figure 60 shows a graph on the window $[0, 20]$ by $[0, 500]$. By zooming in on the higher of the two turning points, or by using a special feature on the calculator (such as the maximum feature under the CALC menu on a TI-83), we find that the maximum occurs at approximately $x = 16$, where $y = 340$. Thus the incidence of measles reached a peak of about 340 per 100,000 in the year 1956.

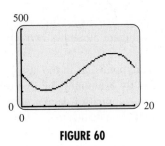

FIGURE 60

Rational Functions Many situations require mathematical models that are quotients. A common model for such situations is a *rational function*.

RATIONAL FUNCTION

A **rational function** is defined by

$$f(x) = \frac{p(x)}{q(x)},$$

where $p(x)$ and $q(x)$ are polynomial functions and $q(x) \neq 0$.

Since any values of x such that $q(x) = 0$ are excluded from the domain, a rational function often has a graph with one or more breaks.

EXAMPLE 5 Graphing Rational Functions

Graph $y = \dfrac{1}{x}$.

Solution This function is undefined for $x = 0$, since 0 is not allowed in the denominator of a fraction. For this reason, the graph of this function will not intersect the vertical line $x = 0$, which is the y-axis. Since x can take on any value except 0, the values of x can approach 0 as closely as desired from either side of 0.

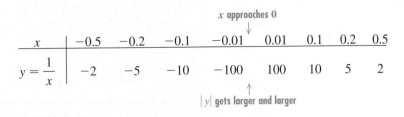

*http://www.learnnc.org/dpi/instserv.nsf/ID/IHS/$file/Alg2Indicators.pdf

The table suggests that as x gets closer and closer to 0, $|y|$ gets larger and larger. This is true in general: as the denominator gets smaller, the fraction gets larger. Thus, the graph of the function approaches the vertical line $x = 0$ (the y-axis) without ever touching it.

As $|x|$ gets larger and larger, $y = 1/x$ gets closer and closer to 0, as shown in the table below. This is also true in general: as the denominator gets larger, the fraction gets smaller.

x	-100	-10	-4	-1	1	4	10	100
$y = \dfrac{1}{x}$	-0.01	-0.1	-0.25	-1	1	0.25	0.1	0.01

The graph of the function approaches the horizontal line $y = 0$ (the x-axis). The information from both tables supports the graph in Figure 61.

FIGURE 61

In Example 5, the vertical line $x = 0$ and the horizontal line $y = 0$ are *asymptotes,* defined as follows.

ASYMPTOTES

If a number k makes the denominator 0 in a rational function but does not make the numerator 0, then the line $x = k$ is a **vertical asymptote.**

If the values of y approach a number k as $|x|$ gets larger and larger, the line $y = k$ is a **horizontal asymptote.**

EXAMPLE 6   Graphing Rational Functions

Graph $y = \dfrac{3x + 2}{2x + 4}$.

Solution   The value $x = -2$ makes the denominator 0, so the line $x = -2$ is a vertical asymptote. To find a horizontal asymptote, find y as x gets larger and larger, as in Table 6 from a graphing calculator.

Table 6 suggests that as x gets larger and larger, $(3x + 2)/(2x + 4)$ gets closer and closer to 1.5, or 3/2. In fact, when $x = 100{,}000$, $(3x + 2)/(2x + 4)$ is equal to 1.5 *to the accuracy of the calculator.* Verify that the same behavior occurs with large negative values of x. Therefore, the line $y = 3/2$ is a horizontal asymptote, with the function approaching the asymptote as x becomes large in magnitude, either positive or negative.

The intercepts should also be noted. When $x = 0$, $y = 2/4 = 1/2$ (the y-intercept). To make a fraction 0, the numerator must be 0; so to make $y = 0$, it is necessary that $3x + 2 = 0$. Solve this for x to get $x = -2/3$ (the x-intercept). We can also use these values to determine where the function is positive and where it is negative. Using the techniques described in Chapter R, verify that the function is negative on $(-2, -2/3)$ and positive on $(-\infty, -2) \cup (-2/3, \infty)$. With this information, the two asymptotes to guide us, and the fact that there are only two intercepts, we suspect the graph is as shown in Figure 62 on the next page. A graphing calculator can support this.

Table 6

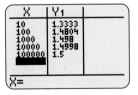

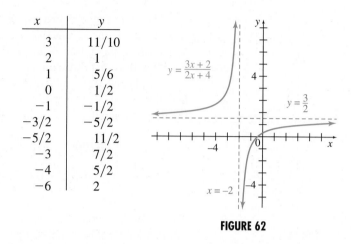

x	y
3	11/10
2	1
1	5/6
0	1/2
−1	−1/2
−3/2	−5/2
−5/2	11/2
−3	7/2
−4	5/2
−6	2

FIGURE 62

In Example 6, $y = 3/2$ was the horizontal asymptote for the rational function $y = (3x + 2)/(2x + 4)$. An equation for the horizontal asymptote can also be found by asking what happens when x gets large. If x is very large, then $3x + 2 \approx 3x$, because the 2 is very small by comparison. In other words, just keep the larger term ($3x$ in this case) and discard the smaller term. Similarly, the denominator is approximately equal to $2x$, so $y \approx 3x/2x = 3/2$. This means that the line $y = 3/2$ is a horizontal asymptote. A more precise way of approaching this idea will be seen later, when limits at infinity are discussed.

Rational functions occur often in practical applications. In many situations involving environmental pollution, much of the pollutant can be removed from the air or water at a fairly reasonable cost, but the last small part of the pollutant can be very expensive to remove. Cost as a function of the percentage of pollutant removed from the environment can be calculated for various percentages of removal, with a curve fitted through the resulting data points. This curve then leads to a mathematical model of the situation. Rational functions are often a good choice for these **cost–benefit models** because they rise rapidly as they approach a vertical asymptote.

EXAMPLE 7 Cost–Benefit Analysis

Suppose a cost–benefit model is given by

$$y = \frac{18x}{106 - x},$$

where y is the cost (in thousands of dollars) of removing x percent of a certain pollutant. The domain of x is the set of all numbers from 0 to 100 inclusive; any amount of pollutant from 0% to 100% can be removed. Find the cost to remove the following amounts of the pollutant: 100%, 95%, 90%, and 80%. Graph the function.

Solution Removal of 100% of the pollutant would cost

$$y = \frac{18(100)}{106 - 100} = 300,$$

or $300,000. Check that 95% of the pollutant can be removed for $155,000, 90% for $101,000, and 80% for $55,000. Using these points, as well as others obtained from the function, gives the graph shown in Figure 63.

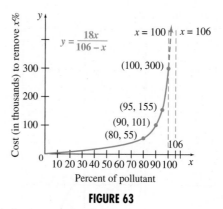

FIGURE 63

If a cost function has the form $C(x) = mx + b$, where x is the number of items produced, m is the marginal cost per item and b is the fixed cost, then the **average cost** per item is given by

$$\overline{C}(x) = \frac{C(x)}{x} = \frac{mx + b}{x}.$$

Notice that this is a rational function with a vertical asymptote at $x = 0$ and a horizontal asymptote at $y = m$. The vertical asymptote reflects the fact that, as the number of items produced approaches 0, the average cost per item becomes infinitely large, because the fixed costs are spread over fewer and fewer items. The horizontal asymptote shows that, as the number of items becomes large, the fixed costs are spread over more and more items, so most of the average cost per item is the marginal cost to produce each item. This is another example of how asymptotes give important information in real applications.

1.5 EXERCISES

1. Explain how translations and reflections can be used to graph $y = -(x - 1)^4 + 2$.

2. Describe an asymptote, and explain when a rational function will have (a) a vertical asymptote, and (b) a horizontal asymptote.

In Exercises 3–6, use the principles of the previous section with the graphs of this section to sketch a graph of the given function.

3. $f(x) = (x + 2)^3 - 5$

4. $f(x) = (x - 4)^3 + 2$

5. $f(x) = -(x - 3)^4 + 1$

6. $f(x) = -(x + 1)^4 + 3$

In Exercises 7–15, match the correct graph A–I on the next page to the function without using your calculator. Then, after you have answered all of them, if you have a graphing calculator, use your calculator to check your answers. Each graph is plotted on $[-6, 6]$ by $[-50, 50]$.

7. $y = x^3 - 7x - 9$

8. $y = -x^3 + 4x^2 + 3x - 8$

9. $y = -x^3 - 4x^2 + x + 6$

10. $y = 2x^3 + 4x + 5$

11. $y = x^4 - 5x^2 + 7$

12. $y = x^4 + 4x^3 - 20$

13. $y = -x^4 + 2x^3 + 10x + 15$

14. $y = 0.7x^5 - 2.5x^4 - x^3 + 8x^2 + x + 2$

15. $y = -x^5 + 4x^4 + x^3 - 16x^2 + 12x + 5$

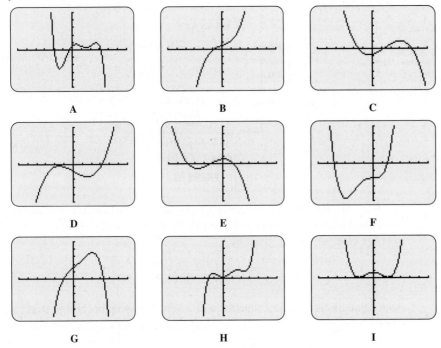

In Exercises 16–20, match the correct graph A–E to the function without using your calculator. Then, after you have answered all of them, if you have a graphing calculator, use your calculator to check your answers. Each graph in this group is plotted on $[-6, 6]$ by $[-6, 6]$. Hint: Consider the asymptotes. (These graphs are done using Dot Mode. If you try graphing them in Connected Mode, you will see some lines that are not part of the graph, but the result of the calculator connecting disconnected parts of the graph.)

16. $y = \dfrac{2x^2 + 3}{x^2 - 1}$

17. $y = \dfrac{2x^2 + 3}{x^2 + 1}$

18. $y = \dfrac{-2x^2 - 3}{x^2 - 1}$

19. $y = \dfrac{-2x^2 - 3}{x^2 + 1}$

20. $y = \dfrac{2x^2 + 3}{x^3 - 1}$

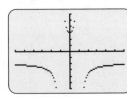
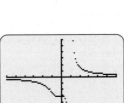
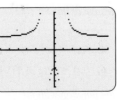
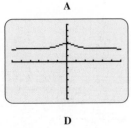
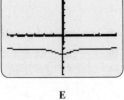

Each of the following is the graph of a polynomial function. Give the possible values for the degree of the polynomial, and give the sign (+ or −) for the x^n term.

21.

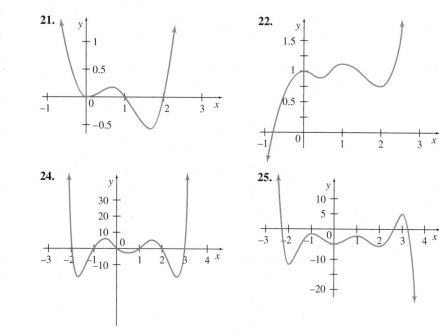

22.

23.

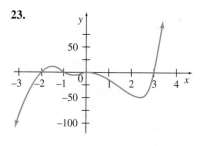

24.

25.

26.

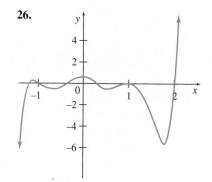

Find the horizontal and vertical asymptotes for each of the following rational functions. Draw the graph of each function, including any x- and y-intercepts.

27. $y = \dfrac{-4}{x - 3}$

28. $y = \dfrac{-1}{x + 3}$

29. $y = \dfrac{2}{3 + 2x}$

30. $y = \dfrac{4}{5 + 3x}$

31. $y = \dfrac{3x}{x - 1}$

32. $y = \dfrac{4x}{3 - 2x}$

33. $y = \dfrac{x + 1}{x - 4}$

34. $y = \dfrac{x - 3}{x + 5}$

35. $y = \dfrac{1 - 2x}{5x + 20}$

36. $y = \dfrac{6 - 3x}{4x + 12}$

37. $y = \dfrac{-x - 4}{3x + 6}$

38. $y = \dfrac{-x + 8}{2x + 5}$

39. Write an equation that defines a rational function with a vertical asymptote at $x = 1$ and a horizontal asymptote at $y = 2$.

40. Write an equation that defines a rational function with a vertical asymptote at $x = -2$ and a horizontal asymptote at $y = 0$.

41. Consider the polynomial functions defined by $f(x) = (x - 1)(x - 2)(x + 3)$, $g(x) = x^3 + 2x^2 - x - 2$, and $h(x) = 3x^3 + 6x^2 - 3x - 6$.

   a. What is the value of $f(1)$?

   b. For what values, other than 1, is $f(x) = 0$?

   c. Verify that $g(-1) = g(1) = g(-2) = 0$.

   d. Based on your answer from part c, what do you think is the factored form of $g(x)$? Verify your answer by multiplying it out and comparing with $g(x)$.

   e. Using your answer from part d, what is the factored form of $h(x)$?

   f. Based on what you have learned in this exercise, fill in the blank: If f is a polynomial and $f(a) = 0$ for some number a, then one factor of the polynomial is _____.

42. Consider the function defined by

$$f(x) = \frac{x^7 - 4x^5 - 3x^4 + 4x^3 + 12x^2 - 12}{x^7}.*$$

  a. Graph the function on $[-6, 6]$ by $[-6, 6]$. From your graph, estimate how many x-intercepts the function has and what their values are.

  b. Now graph the function on $[-1.5, -1.4]$ by $[-10^{-4}, 10^{-4}]$, and also on $[1.4, 1.5]$ by $[-10^{-5}, 10^{-5}]$. From your graphs, estimate how many x-intercepts the function has and what their values are.

  c. From your results in parts a and b, what advice would you give a friend on using a graphing calculator to find x-intercepts?

43. Consider the function defined by

$$f(x) = \frac{1}{x^5 - 2x^3 - 3x^2 + 6}.*$$

  a. Graph the function on $[-3.4, 3.4]$ by $[-3, 3]$. From your graph, estimate how many vertical asymptotes the function has and where they are located.

  b. Now graph the function on $[-1.5, -1.4]$ by $[-10, 10]$, and also on $[1.4, 1.5]$ by $[-1,000, 1,000]$. From your graphs, estimate how many vertical asymptotes the function has and where they are located.

  c. From your results in parts a and b, what advice would you give a friend on using a graphing calculator to find vertical asymptotes?

Applications

LIFE SCIENCES

44. *Brain Weight* The weight (in grams) of the human brain during the last trimester of gestation and the first two years after birth can be approximated by the function

$$w = \frac{c^3}{100} - \frac{1,500}{c},$$

where c is the circumference of the head in cm.[†]

  a. Find the approximate weight of brains with a circumference of 30, 40, or 50 cm.

  b. Clearly the formula is invalid for any values of c yielding negative values of w. For what values of c is this true?

  c. Use a graphing calculator to sketch this graph on the interval $20 \le x \le 50$.

  d. Suppose an infant brain weighs 700 g. Use features on a graphing calculator to find what the circumference of the head is expected to be.

45. *Soil Ingestion* The amount of acid-insoluble ash in scat from white-footed mice can be approximated by

$$y = \frac{b(1 - x) + cx}{1 - a(1 - x)},$$

where x is the fraction of soil in the diet, a is the digestibility of food, b is the concentration of acid-insoluble ash in the food, and c is the concentration of acid-insoluble ash in the soil.[‡]

  a. Explain in your own words what the numerator of this fraction represents.

  b. Explain in your own words what the denominator of this fraction represents.

---

*Donley, Edward, and Elizabeth Ann George, "Hidden Behavior in Graphs," *The Mathematics Teacher,* Vol. 86, No. 6, Sept. 1993.
†Dobbing, John, and Jean Sands, "Head Circumference, Biparietal Diameter and Brain Growth in Fetal and Postnatal Life," *Early Human Development,* Vol. 2, No.1, April 1978, pp. 81–87.
‡Beyer, W. Nelson, et al., "Estimates of Soil Ingestion by Wildlife," *Journal of Wildlife Management,* Vol. 58, No. 2, 1994, pp. 375–382.

c. The researchers estimated the values of the parameters as $a = 0.76$, $b = 0.025$, and $c = 0.92$. Use these values to simplify the function above.

d. Use a graphing calculator to sketch the graph of the function in part c on the interval $0 \leq x \leq 0.15$.

e. Suppose the mouse scat contains 20% ash, that is, $y = 0.20$. Use algebra to find the concentration of soil in the diet, and then verify your answer with a graphing calculator.

46. *Contact Lenses* The strength of a contact lens is given in units known as diopters, as well as in mm of arc. The following is taken from a chart used by optometrists to convert diopters to mm of arc.*

diopters	mm of arc
36.000	9.37
36.125	9.34
36.250	9.31
36.375	9.27
36.500	9.24
36.625	9.21
36.750	9.18
36.875	9.15
37.000	9.12

a. Notice that as the diopters increase, the mm of arc decrease. Find a formula of the form $a = f(d) = k/d$ giving a, the mm of arc, as a function of d, the strength in diopters. (Round k to the nearest integer. For a more accurate answer, average all the values of k given by each pair of data.)

b. An optometrist wants to order 40.50 diopter lenses for a patient. The manufacturer needs to know the strength in mm of arc. What is the strength in mm of arc?

47. *Cardiac Output* A technique for measuring cardiac output depends on the concentration of a dye after a known amount is injected into a vein near the heart. In a normal heart, the concentration of the dye at time x (in seconds) is given by the function

$$g(x) = -0.006x^4 + 0.140x^3 - 0.053x^2 + 1.79x.$$

a. Graph $g(x)$ on $[0, 6]$ by $[0, 20]$.

b. In your graph from part a, notice that the function initially increases. Considering the form of $g(x)$, do you think it can keep increasing forever? Explain.

c. Write a short paragraph about the extent to which the concentration of dye might be described by the function $g(x)$.

48. *Population Variation* During the early part of the twentieth century, the deer population of the Kaibab Plateau in Arizona experienced a rapid increase, because hunters had reduced the number of natural predators. The increase in population depleted the food resources and eventually caused the population to decline. For the period from 1905 to 1930, the deer population was approximated by

$$D(x) = -0.125x^5 + 3.125x^4 + 4,000,$$

where x is time in years from 1905.

a. Graph $D(x)$ on $0 \leq x \leq 30$.

b. From the graph, over what period of time (from 1905 to 1930) was the population increasing? relatively stable? decreasing?

49. *Alcohol Concentration* The polynomial function

$$A(x) = -0.015x^3 + 1.058x$$

gives the approximate alcohol concentration (in tenths of a percent) in an average person's bloodstream x hours after drinking about eight ounces of 100-proof whiskey. The function is approximately valid for x in the interval $[0, 8]$.

a. Graph $A(x)$ on $0 \leq x \leq 9$.

b. Using the graph from part a, estimate the time of maximum alcohol concentration.

c. In many states, a person is legally drunk if the blood alcohol concentration exceeds 0.08%. Use the graph from part a to estimate the period in which this average person is legally drunk.

---

*Provided courtesy of Bausch & Lomb. The original chart gave all data to 2 decimal places.

50. *Cancer* From 1930 to 1990, the rate of breast cancer was nearly constant at 30 cases per 100,000 females, whereas the rate of lung cancer in females over the same period increased. The number of lung cancer cases per 100,000 females in the year t (where $t = 0$ corresponds to 1930) can be modeled using the function defined by

$$f(t) = 2.8 \times 10^{-4}t^3 - 0.011t^2 + 0.23t + 0.93.*$$

a. Graph the rates of breast and lung cancer on $0 \leq t \leq 60$.

b. Zoom in on the intersection of the two functions graphed in part a to determine the year when rates for lung cancer first exceeded those for breast cancer.

51. *Population Biology* The function

$$f(x) = \frac{\lambda x}{1 + (ax)^b}$$

is used in population models to give the size of the next generation ($f(x)$) in terms of the current generation (x).[†]

a. What is a reasonable domain for this function, considering what x represents?

b. Graph this function for $\lambda = a = b = 1$.

c. Graph this function for $\lambda = a = 1$ and $b = 2$.

d. What is the effect of making b larger?

52. *Growth Model* The function

$$f(x) = \frac{Kx}{A + x}$$

is used in biology to give the growth rate of a population in the presence of a quantity x of food. This is called Michaelis-Menten kinetics.[‡]

a. What is a reasonable domain for this function, considering what x represents?

b. Graph this function for $K = 5$ and $A = 2$.

c. Show that $y = K$ is a horizontal asymptote.

d. What do you think K represents?

e. Show that A represents the quantity of food for which the growth rate is half of its maximum.

OTHER APPLICATIONS

53. *Batting Power* The rate at which energy is transferred from a batter to a baseball bat was found for one batter to be approximated by the function

$$y = 0.0078109 + 2.08079t - 393.385t^2 + 127{,}454t^3$$
$$- 3.64084 \times 10^6t^4 + 4.07979 \times 10^7t^5 - 1.99528$$
$$\times 10^8t^6 + 3.54337 \times 10^8t^7,$$

where t is the time into the swing in seconds and y is the power in horsepower.[§]

a. Use a graphing calculator to sketch this function.

b. From your graph in part a, estimate over what values of t this function is valid.

c. From your graph in part a, estimate at what time into the swing the power is a maximum. What is the maximum power?

54. *Cost–Benefit Model* Suppose a cost–benefit model is given by

$$y = \frac{6.7x}{100 - x},$$

where y is the cost in thousands of dollars of removing x percent of a given pollutant.

a. Find the cost of removing each percent of pollutants: 50%; 70%; 80%; 90%; 95%; 98%; 99%.

b. Is it possible, according to this function, to remove *all* the pollutant?

c. Graph the function.

55. *Cost–Benefit Model* Suppose a cost–benefit model is given by

$$y = \frac{6.5x}{102 - x},$$

where y is the cost in thousands of dollars of removing x percent of a certain pollutant.

a. Find the cost of removing each percent of pollutants: 0%; 50%; 80%; 90%; 95%; 99%; 100%.

b. Graph the function.

56. *Length of a Pendulum* A simple pendulum swings back and forth in regular time intervals. Grandfather clocks use pendulums to keep accurate time. The relationship between

---

*Valanis, B., *Epidemiology in Nursing and Health Care,* Appleton & Lange, 1992.
[†]Smith, J. Maynard, *Models in Ecology,* Oxford: Cambridge University Press, 1974.
[‡]Edelstein-Keshet, Leah, *Mathematical Models in Biology,* Random House, 1988.
[§]http://www.rose-hulman.edu/Class/CalculusProbs/Problems/BATTERUP/BATTERUP_3_0_0.html

the length of a pendulum L and the period (time) T for one complete oscillation can be expressed by the function $L = kT^n$, where k is a constant and n is a positive integer to be determined. The following data were taken for different lengths of pendulums.*

T (sec)	L (ft)
1.11	1.0
1.36	1.5
1.57	2.0
1.76	2.5
1.92	3.0
2.08	3.5
2.22	4.0

a. Find the value of k for $n = 1, 2,$ and 3, using the data for the 4 ft pendulum.

b. Use a graphing calculator to plot the data in the table and to graph the function $L = kT^n$ for the three values of k (and their corresponding values of n) found in part a. Which function best fits the data?

c. Use the best-fitting function from part a to predict the period of a pendulum having a length of 5 ft.

d. If the length of the pendulum doubles, what happens to the period?

e. If you have a graphing calculator or computer program with a quadratic regression feature, use it to find a quadratic function that approximately fits the data. How does this answer compare with the answer to part b?

57. *Coal Consumption* The table gives United States coal consumption for selected years, using the last two digits of each year.[†]

Year	Millions of Short Tons
50	494.1
60	398.1
70	523.2
80	702.7
85	818.0
90	895.5
95	962.0
96	1,005.6
97	1,027.1

a. Draw a scatterplot.

b. Use the quadratic regression feature of a graphing calculator to get a quadratic function that approximates the data.

c. Graph the function from part b on the same window as the scatterplot.

d. Use cubic regression to get a cubic function that approximates the data.

e. Graph the cubic function from part d on the same window as the scatterplot.

f. Which of the two functions in parts b and d appears to be a better fit for the data? Explain your reasoning.

58. *Abandoned Cars* The number of abandoned cars in New York City has dropped in recent years because of reduced auto theft and increasing scrap metal prices.[‡] The following table, based on estimates from a graph in *The New York Times,* gives the number of abandoned cars as a function of the year.

Year	Abandoned Cars (1,000s)
1986	82
1987	120
1988	140
1989	148
1990	135
1991	95
1992	75
1993	51
1994	38
1995	27
1996	20

a. Plot the points from the table, using 0 for 1986, and so on.

b. If your graphing calculator has a cubic and quartic regression feature, find the third- and fourth-degree polynomials that best fit the data according to the least squares method. Plot these polynomials on the same calculator window as the data.

c. Discuss whether a third- or fourth-degree polynomial better fits the data.

---

*Data provided by Gary Rockswold, Mankato State University, Mankato, Minnesota. See Exercise 17, Section 1.2.
[†]*Annual Energy Review,* U.S. Department of Energy, 1997.
[‡]*The New York Times,* Feb. 15, 1997, p. 25.

◼ CHAPTER SUMMARY

In this chapter, we have studied various types of functions: linear, quadratic, polynomial, and rational. Linear functions are the simplest; we can find the equation of such a function when given a point and the slope or when given two points. The method of least squares allows us to derive linear functions that approximately describe sets of data. Quadratic functions are next in order of complication; they, too, can be analyzed completely without calculus. Linear and quadratic functions are special cases of polynomial functions. We have briefly introduced polynomial functions of higher degree and rational functions; a complete analysis requires techniques discussed in later chapters. The functions in this chapter have a broad range of applications, as demonstrated in the applications exercises. Other types of functions will be studied in the next chapter.

◼ KEY TERMS

To understand the concepts presented in this chapter, you should know the meaning and use of the following terms. For easy reference, the section in the chapter where a term was first used is provided.

mathematical model	**slope-intercept form**	**1.3 function**	**completing the square**
1.1 ordered pair	**proportional**	**domain**	**1.5 polynomial function**
Cartesian coordinate	**point-slope form**	**range**	**coefficient**
system	**linear function**	**vertical line test**	**leading coefficient**
axes	**independent variable**	**function composition**	**power function**
origin	**dependent variable**	**1.4 quadratic function**	**turning point**
coordinates	**fixed cost**	**parabola**	**real zero**
quadrants	**marginal cost**	**vertex**	**rational function**
graph	**1.2 scatterplot**	**axis**	**vertical asymptote**
intercepts	**least squares line**	**vertical translation**	**horizontal asymptote**
slope	**summation notation**	**horizontal translation**	**cost–benefit model**
linear equation	**coefficient of correlation**	**vertical reflection**	**average cost**

◼ CHAPTER 1 REVIEW EXERCISES

1. What is marginal cost? Fixed cost?

2. What six quantities are needed to compute a coefficient of correlation?

Find the slope for each line in Exercises 3–12 that has a slope.

3. Through $(-2, 5)$ and $(4, 7)$

4. Through $(4, -1)$ and $(3, -3)$

5. Through the origin and $(11, -2)$

6. Through the origin and $(0, 7)$

7. $2x + 3y = 15$

8. $4x - y = 7$

9. $y + 4 = 9$

10. $3y - 1 = 14$

11. $y = -3x$

12. $x = 5y$

Find an equation in the form $y = mx + b$ (where possible) for each line.

13. Through $(5, -1)$; slope $= 2/3$

14. Through $(8, 0)$; slope $= -1/4$

15. Through $(5, -2)$ and $(1, 3)$

16. Through $(2, -3)$ and $(-3, 4)$

17. Through $(-1, 4)$; undefined slope

18. Through $(-2, 5)$; slope $= 0$

19. Through $(2, -1)$, parallel to $3x - y = 1$

20. Through $(0, 5)$, perpendicular to $8x + 5y = 3$

21. Through $(2, -10)$, perpendicular to a line with undefined slope

22. Through $(3, -5)$, parallel to $y = 4$

23. Through $(-7, 4)$, perpendicular to $y = 8$

Graph each linear equation defined as follows.

24. $y = 4x + 3$

25. $y = 6 - 2x$

26. $3x - 5y = 15$

27. $2x + 7y = 14$

28. $x + 2 = 0$

29. $y = 1$

30. $y = 2x$

31. $x + 3y = 0$

32. What is a function? A linear function? A quadratic function? A rational function?

33. How do you find a vertical asymptote? A horizontal asymptote?

34. What can you tell about the graph of a polynomial function of degree n before you plot any points?

List the ordered pairs obtained from each of the following if the domain of x for each exercise is $\{-3, -2, -1, 0, 1, 2, 3\}$. Graph each set of ordered pairs. Give the range.

35. $y = (2x + 1)(x - 1)$

36. $y = \dfrac{2}{x^2 + 1}$

In Exercises 37 and 38, find   (a) $f(6)$,   (b) $f(-2)$,   (c) $f(-4)$, and   (d) $f(r + 1)$.

37. $f(x) = -x^2 + 2x - 4$

38. $f(x) = 8 - x - x^2$

39. Let $f(x) = 5x^2 - 3$ and $g(x) = -x^2 + 4x + 1$. Find each of the following.

    a. $f(-2)$ **b.** $g(3)$ **c.** $f(-k)$ **d.** $g(3m)$

    e. $f(x + h)$ **f.** $g(x + h)$ **g.** $\dfrac{f(x + h) - f(x)}{h}$ **h.** $\dfrac{g(x + h) - g(x)}{h}$

Find the domain of each function defined as follows.

40. $y = \dfrac{\sqrt{x - 2}}{2x + 3}$

41. $y = \dfrac{3x - 4}{x}$

Graph each of the following.

42. $y = 3x^2 + 6x - 2$

43. $y = -\dfrac{1}{4}x^2 + x + 2$

44. $y = x^2 - 4x + 2$

45. $y = -3x^2 - 12x - 1$

46. $f(x) = x^3 + 5$

47. $f(x) = 1 - x^4$

48. $y = -(x - 1)^3 + 4$

49. $y = -(x + 2)^4 - 2$

50. $f(x) = \dfrac{8}{x}$

51. $f(x) = \dfrac{2}{3x - 1}$

52. $f(x) = \dfrac{4x - 2}{3x + 1}$

53. $f(x) = \dfrac{6x}{x + 2}$

For each pair of functions, find $(f \circ g)(x)$ and $(g \circ f)(x)$.

54. $f(x) = 4x^2 + 3x$ and $g(x) = \dfrac{x}{x + 1}$

55. $f(x) = 3x + 4$ and $g(x) = x^2 - 6x - 7$

Applications

LIFE SCIENCES

56. *Eating Behavior* In a study of eating behavior, the cumulative intake (in grams) of one individual was found to vary with time according to the formula

$$I = 27 + 72t - 1.5t^2,$$

where t is the time in minutes.[*]

a. At what time is the intake for this individual a maximum?

b. What is the maximum intake?

c. Over what domain is this function valid? Explain why the function makes no sense outside of this domain.

57. *Ecosystem Diversity* A study at the Cedar Creek Natural History Area of Minnesota found that an average local diversity of x plant species per 0.5 square meters required an average regional diversity given by

$$y = \frac{x - 1.1}{0.124}$$

in species per 0.5 hectare.[†]

a. What is the slope of this line? Describe in words what your answer means.

b. What average regional diversity is required for an average local diversity of 6 species per 0.5 square meters?

c. What average local diversity will lead to an average regional diversity of 70 species per hectare?

d. The coefficient of correlation for this study is given as $r = 0.82$. Describe in your own words what this means.

58. *Endangered Species* The Committee on the Status of Endangered Wildlife in Canada (COSEWIC) began proposing endangered species for protection in 1978 with a list of 17 species.[‡] As COSEWIC continued to investigate and designate species suspected of being at risk, the size of the list grew at a fairly constant rate to 380 in 2001.

a. Find an equation giving the number of proposed listings as a linear function of the years since 1978.

b. If the rate of listings were to continue, how many species would be listed in 2015?

c. If the rate of listings were to continue, in what year would the number of listings surpass 800?

59. *Blood Volume* A formula proposed by Hurley[§] for the red cell volume (RCV) in milliliters for males is

$$RCV = 1,486S^2 - 4,106S + 4,514,$$

where S is the surface area in square meters. A formula given by Pearson, et al.,[‖] is

$$RCV = 1,486S - 825.$$

a. Verify that these two formulas never give the same answer

i. using the quadratic formula;

ii. using a graphing calculator.

b. The formula for plasma volume for females given by Hurley is

$$PV = 1,278S^{1.289},$$

while the formula given by Pearson, et al., is

$$PV = 1,395S,$$

where PV is measured in milliliters and S in square meters. Find the values of S for which these two formulas give the same answer. What is the predicted plasma volume at each of these values of S?

60. *Respiratory Rate* Researchers have found that the 95th percentile (the value at which 95% of the data is at or below) for respiratory rates (in breaths per minute) during the first three years of infancy are given by

$$y = 10^{1.82411 - 0.0125995x + 0.00013401x^2}$$

for awake infants and

$$y = 10^{1.72858 - 0.0139928x + 0.00017646x^2}$$

for sleeping infants, where x is the age in months.[#]

a. What is the domain for each function?

b. For each respiratory rate, is the rate decreasing or increasing over the first three years of life? (*Hint:* Is the

---

[*]Kissileff, H. R., and J. L. Guss, "Microstructure of Eating Behavior in Humans," *Appetite,* Vol. 36, No. 1, Feb. 2001, pp. 70–78.

[†]*Science,* Vol. 286, Nov. 5, 1999, p. 1099.

[‡]*Science,* Vol. 293, Aug. 24, 2001, p. 1417.

[§]Hurley, Peter J., "Red cell and plasma volumes in normal adults," *Journal of Nuclear Medicine,* Vol. 16, 1975, pp. 46–52.

[‖]Pearson, T. C., et al., "Interpretation of measured red cell mass and plasma volume in adults," *British Journal of Haematology,* Vol. 89, 1995, pp. 748–756.

[#]Rusconi, Franca, et al., "Reference Values for Respiratory Rate in the First 3 Years of Life," *Pediatrics,* Vol. 94, No. 3, Sept. 1994, pp. 350–355.

graph of the quadratic in the exponent opening upwards or downwards? Where is the vertex?)

    c. Verify your answer to part b using a graphing calculator.

    d. For a one-year-old infant in the 95th percentile, how much higher is the waking respiratory rate than the sleeping respiratory rate?

61. *World Health* In general, people tend to live longer in countries that have a greater supply of food. Listed below is the 1992 daily calorie supply and 1994 life expectancy at birth for 10 randomly selected countries.*

Country	Calories (x)	Life Expectancy (y)
Afghanistan	1,523	44
Belize	2,670	74
Cuba	2,833	76
France	3,465	79
India	2,395	61
Mexico	3,181	72
North Korea	2,834	71
Peru	1,883	67
Sweden	2,960	78
United States	3,671	76

    a. Find the coefficient of correlation. Do the data seem to fit a straight line?

    b. Draw a scatterplot of the data. Combining this with your results from part a, do the data seem to fit a straight line?

    c. Find the equation for the least squares line.

    d. Use your answer from part c to predict the life expectancy in the United Kingdom, which has a daily calorie supply of 3,149. Compare your answer with the actual value of 77 years.

    e. Briefly speculate why countries with a higher daily calorie supply might tend to have a longer life expectancy.

    f. (For the ambitious!) Find the coefficient of correlation and least squares line using the data for a larger sample of countries, as found in an almanac or other reference. Is the result in general agreement with the previous results?

62. *Blood Sugar and Cholesterol Levels* The following data show the connection between blood sugar levels and cholesterol levels for 8 different patients.

Patient	Blood Sugar Level (x)	Cholesterol Level (y)
1	130	170
2	138	160
3	142	173
4	159	181
5	165	201
6	200	192
7	210	240
8	250	290

For the data given in the preceding table, $\Sigma x = 1{,}394$, $\Sigma y = 1{,}607$, $\Sigma xy = 291{,}990$, $\Sigma x^2 = 255{,}214$, and $\Sigma y^2 = 336{,}155$.

    a. Find the equation of the least squares line, $Y = mx + b$.

    b. Predict the cholesterol level for a person whose blood sugar level is 190.

    c. Find r.

63. *Fever* A certain viral infection causes a fever that typically lasts 6 days. A model of the fever (in °F) on day x, $1 \le x \le 6$, is

$$F(x) = -\frac{2}{3}x^2 + \frac{14}{3}x + 96.$$

According to the model, on what day should the maximum fever occur? What is the maximum fever?

64. *AIDS* HIV is the virus responsible for AIDS. The following graph shows an estimation of the HIV infection rate in the United States.[†]

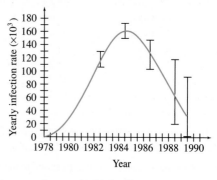

    a. When was the rate the highest?

    b. What was the maximum rate?

    c. What is the range of the function?

*The New York Times 2000 Almanac, pp. 490–492.
†Brookmeyer, Dr. Ronald, "Reconstruction and Future Trends of the AIDS Epidemic in the United States," *Science,* Vol. 253, July 5, 1991, p. 37. Copyright © 1991 by the American Association for the Advancement of Science. Reprinted by permission of the AAAS and Dr. Brookmeyer.

65. *Sunscreen* An article in a medical journal says that a sunscreen with a sun protection factor (SPF) of 2 provides 50% protection against ultraviolet B (UVB) radiation, an SPF of 4 provides 75% protection, and an SPF of 8 provides 87.5% protection (which the article rounds to 87%).*

  a. 87.5% protection means that 87.5% of the UVB radiation is screened out. Write as a fraction the amount of radiation that is let in, and then describe how this fraction, in general, relates to the SPF rating.

  b. Plot UVB percent protection (y) against x, where $x = 1/\text{SPF}$.

  c. Based on your graph from part b, give an equation relating UVB protection to SPF rating.

  d. An SPF of 8 has double the chemical concentration of an SPF 4. Find the increase in the percent protection.

  e. An SPF of 30 has double the chemical concentration of an SPF 15. Find the increase in the percent protection.

  f. Based on your answers from parts d and e, what happens to the increase in the percent protection as the SPF continues to double?

66. *AIDS Deaths* The following table lists the number of AIDS deaths per 100,000 people for all races in the resident population in the United States.[†]

Year	Deaths (per 100,000)
1987	12.7
1989	20.5
1990	23.2
1991	26.5
1992	29.9
1993	32.9
1994	36.7
1995	36.9
1996	25.9
1997	13.2

  a. If the population of the United States was 267 million in 1997, estimate the number of AIDS deaths that year.

  b. Plot the data, letting $x = 0$ correspond to the year 1987.

  c. Would a quadratic or cubic function model this data best? Explain.

  d. Using the regression feature on your calculator, find a quadratic and a cubic function that models this data.

  e. Plot the two functions with the data on the same coordinate axes.

  f. What happens if we try to use either of the above functions to predict the number of AIDS deaths in the future? What trend do you expect AIDS deaths to follow in the coming years?

67. *Population Growth* In 1960 in an article in *Science* magazine, H. Van Forester, P. M. Mora, and W. Amiot predicted that world population would be infinite in the year 2026. Their projection was based on the rational function defined by

$$p(t) = \frac{1.79 \cdot 10^{11}}{(2,026.87 - t)^{0.99}},$$

where $p(t)$ gives population in year t.[‡] This function has provided a relatively good fit to the population so far.

  a. Estimate world population in 1999 using this function, and compare it with the estimate given in *Statistical Abstract of the United States 1999,* of 5.996 billion.

  b. What does the function predict for world population in 2020? 2025?

  c. Discuss why this function is not realistic, despite its good fit to past data.

68. *Carbon Emissions* Carbon emissions are a serious problem due to their suspected contribution to global warming. A recent study comparing per capita emissions in terms of per capita gross domestic product (GDP) in 1997 U.S. dollars calculated the least squares line through selected countries.[§] At a per capita GDP of $5,000, the expected per capita emission of carbon is 1 metric ton, while at $25,000, it is 3.3 metric tons.

  a. Find the equation of the least squares line giving expected per capita emission of carbon in metric tons as a function of per capita GDP.

  b. The U.S. has a per capita GDP of $30,500 and emits 5.5 metric tons of carbon per capita. Is this more or less than the expected value?

  c. China has a per capita GDP of $3,500 and emits 0.8 metric tons of carbon per capita. Is this more or less than the expected value?

---

*Family Practice, May 17, 1993, p. 55.
[†]National Center for Health Statistics, Health, United States, 1999, Health With Aging Chartbook, Hyattsville, Maryland, 1999, Table 433, p. 183.
[‡]Chance News 7.01, Jan. 1998.
[§]The New York Times, June 24, 2001, p. WK2.

69. *Pollution* The cost to remove x percent of a pollutant is

$$y = \frac{7x}{100 - x},$$

in thousands of dollars. Find the cost of removing each of the following percents of the pollutant.

a. 80% **b.** 50% **c.** 90%

d. Graph the function.

e. Can all of the pollutant be removed?

70. *Red Meat Consumption* The per capita consumption of red meat in the United States decreased from 131.7 pounds in 1970 to 114.1 pounds in 1992.* Assume a linear function describes the decrease. Write a linear equation defining the function. Let x represent the number of years since 1900 and y represent the number of pounds of red meat consumed.

71. *Carbon Monoxide Levels* The 8-hour maximum carbon monoxide levels (in parts per million) for the United States from 1982–1992 can be modeled by the function

$$f(x) = -0.012053x^2 - 0.046607x + 9.125,$$

where $x = 0$ corresponds to 1982.†

a. Discuss the general trend in these carbon monoxide levels.

b. Find a function $g(x)$ that models the same carbon monoxide levels except that x is the actual year between 1982 and 1992. For example, $g(1985) = f(3)$. (*Hint:* Use a horizontal translation.)

OTHER APPLICATIONS

72. *Marital Status* More people are staying single longer in the United States. In 1970, the number of never-married adults, age 18 and over, was 21.4 million. By 1993, it was 42.3 million.‡ Assume the data increase linearly, and write

an equation that defines a linear function for this data. Let x represent the number of years since 1900.

73. *Governors' Salaries* In general, the larger a state's population, the more its governor earns. Listed below are the estimated 1998 populations (in millions) and the salary of the governor (in thousands of dollars) for 8 randomly selected states.§

a. Find the coefficient of correlation. Do the data seem to fit a straight line?

b. Draw a scatterplot of the data. Compare this with your answer from part a.

c. Find the equation for the least squares line.

d. Based on your answer to part c, how much does a governor's salary increase, on average, for each additional million in population?

e. Use your answer from part c to predict the governor's salary in your state. Based on your answers from parts a and b, would this prediction be very accurate? Compare with the actual salary, as listed in an almanac or other reference.

f. (For the ambitious!) Find the coefficient of correlation and least squares line using the data for all 50 states, as found in an almanac or other reference. Is the result in general agreement with the previous results?

State	AZ	DE	MD	MA	NY	PA	TN	WY
Population (x)	4.67	0.74	5.14	6.15	18.18	12.00	5.43	0.48
Governor's Salary (y)	75	107	120	75	130	105	85	95

74. *Average Speed* Suppose a plane flies from one city to another that is a distance d miles away. The plane flies at a constant speed v relative to the wind, but there is a constant wind speed of w. Therefore, the speed in one direction is $v + w$, and the speed in the other direction is $v - w$.‖

a. Using the formula distance = rate × time, show that the time to make the round trip is

$$\frac{d}{v + w} + \frac{d}{v - w}.$$

---

*U.S. Department of Agriculture, Economic Research Service, *Food Consumption, Price, and Expenditures,* annual.

†U.S. Environmental Protection Agency, 1992.

‡U.S. Bureau of the Census, *1970 Census of Population,* Vol. 1, Part 1, and *Current Population Reports,* pp. 20, 450.

§*The New York Times 2000 Almanac,* pp. 183, 220.

‖This exercise is based on a letter by Tom Blazey to *The Mathematics Teacher,* Vol. 86, No. 2, Feb. 1993, p. 178.

b. Using the result from part a, show that the average speed for the round trip is

$$v_{aver} = \frac{2d}{\dfrac{d}{v+w} + \dfrac{d}{v-w}}.$$

c. Simplify your result from part b to get

$$v_{aver} = v - \frac{w^2}{v}.$$

d. Consider v in the result in part c to be a constant. What wind speed results in the greatest average speed? Explain why your result makes sense.

75. *Average Speed* Suppose the plane in the previous exercise makes the trip one way at a speed of v, and the return trip at a speed xv.*

a. Explain in words what $x = 0.9$ and $x = 1.1$ represent.

b. Show that the average velocity can be written as

$$v_{aver} = \left(\frac{2x}{x+1}\right)v.$$

(*Hint:* Use the steps in parts a–c of the previous exercise.)

c. With v held constant, the equation in part b defines v_{aver} as a function of x. Discuss the behavior of this function. In particular, consider the horizontal asymptote and what it says about the average velocity.

76. *Planets* The following table contains the average distance D from the sun for the first eight planets and their period P of revolution around the sun in years.[†]

Planet	Distance (D)	Period (P)
Mercury	0.39	0.24
Venus	0.72	0.62
Earth	1	1
Mars	1.52	1.89
Jupiter	5.20	11.9
Saturn	9.54	29.5
Uranus	19.2	84.0
Neptune	30.1	164.8

The distances are given in astronomical units (A.U.); 1 A.U. is the average distance from Earth to the sun. For example, since Jupiter's distance is 5.2 A.U., its distance from the sun is 5.2 times farther than Earth's.

a. Find functions of the form $P = kD^n$ for $n = 1$, 1.5, and 2 that fit the data at Neptune.

b. Use a graphing calculator to plot the data in the table and to graph the three functions found in part a. Which function best fits the data?

c. Use the best-fitting function from part b to predict the period of the planet Pluto, which has a distance from the sun of 39.5 A.U. Compare your answer to the true value of 248.5 years.

d. If you have a graphing calculator or computer program with a power regression feature, use it to find a power function (a function of the form $P = kD^n$) that approximately fits the data. How does this answer compare with the answer to part b?

---

*This exercise is based on the article "Problems Whose Solutions Lie on a Hyperbola," by Steven Schwartzman, *The AMATYC Review,* Vol. 14, No. 2, Spring 1993, pp. 27–36.
[†]Ronan, C., *The Natural History of the Universe,* Macmillan, 1991.

EXTENDED APPLICATION: Using Extrapolation to Predict Life Expectancy

One reason for developing a mathematical model is to make predictions. If your model is a least squares line, you can predict the y value corresponding to some new x by substituting this x into an equation of the form $Y = mx + b$. (We use a capital Y to remind us that we're getting a predicted value rather than an actual data value.) Data analysts distinguish two very different kinds of prediction, *interpolation* and *extrapolation*. An interpolation uses a new x inside the x range of your original data. For example, if you have inflation data at five-year intervals from 1950 to 2000, estimating the rate of inflation in 1957 is an interpolation problem. But if you use the same data to estimate what the inflation rate was in 1920, or what it will be in 2020, you are extrapolating.

In general, interpolation is much safer than extrapolation, because data that are approximately linear over a short interval may be nonlinear over a larger interval. One way to detect nonlinearity is to look at *residuals,* which are the differences between the actual data values and the values predicted by the line of best fit. Here is a simple example:

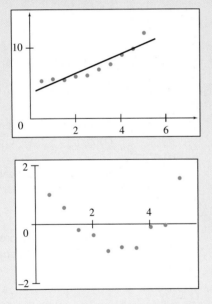

The regression equation for the linear fit on the top is $Y = 3.431 + 1.334x$. Since the r value for this regression line is 0.93, our linear model fits the data very well. But we might notice that the predictions are a bit low at the ends and high in the middle. We can get a better look at this pattern by plotting the residuals. To find them, we put each value of the independent variable into the regression equation, calculate the predicted value Y, and subtract it from the actual y value. The residual plot is below the linear fit graph, with the vertical axis rescaled to

exaggerate the pattern. The residuals indicate that our data has a nonlinear, U-shaped component that isn't captured by the linear fit. Extrapolating from this data set is probably not a good idea; our linear prediction for the value of y when x is 10 may be much too low.

Exercises

The following table gives the life expectancy at birth of females born in the United States in various years from 1950 to 1995.

Year of birth	Life expectancy in years
1950	71.3
1960	73.1
1970	74.7
1980	77.4
1985	78.2
1990	78.8
1995	78.9

1. Find an equation for the least squares line for this data, using year of birth as the independent variable.

2. Use your regression equation to guess a value for the life expectancy of females born in 1900.

3. Compare your answer with the actual life expectancy for females born in 1900, which was 48.3 years. Are you surprised?

4. Find the life expectancy predicted by your regression equation for each year in the table, and subtract it from the actual value in the second column. This gives you a table of residuals. Plot your residuals as points on a graph.

5. Now look at the residuals as a fresh data set and see if you can sketch the graph of a smooth function that fits the residuals well. How easy do you think it will be to predict the life expectancy at birth of females born in 2010?

6. What will happen if you try linear regression on the *residuals?* If you're not sure, use your calculator or software to find the regression equation for the residuals. Why does this result make sense?

7. Since most of the females born in 1985 are still alive, how did the Public Health Service come up with a life expectancy of 78.2 years for these women?

*Health, United States, 1998, Centers for Disease Control.

CHAPTER 2

Exponential, Logarithmic, and Trigonometric Functions

Ecologists want to know how long it takes for a population of animals to double. Microbiologists want to know the same thing about a population of bacteria. Oncologists want to know how long it will be until a tumor doubles in size. As we will see in Section 2 of this chapter, all of these questions can be answered with the same mathematical methods involving exponential functions and logarithms.

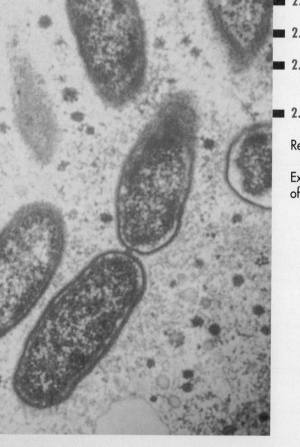

■ 2.1 Exponential Functions

■ 2.2 Logarithmic Functions

■ 2.3 Applications: Growth and Decay

■ 2.4 Trigonometric Functions

Review Exercises

Extended Application: Characteristics of the Monkeyface Prickleback

In the previous chapter, we studied *algebraic functions:* those that can be formed by addition, subtraction, multiplication, division, and raising to a numerical power. But some of the most important functions in the life sciences do not fall into this category; they are *transcendental,* because they transcend, or go beyond, the algebraic functions. In this chapter, we will study three types of transcendental functions. The first, exponential functions, arise whenever a population grows at a rate proportional to the size of the population. The second, logarithmic functions, are the inverse of exponential functions and are just as important. The third, trigonometric functions, are useful for describing periodic phenomena, such as a heartbeat or seasonal migration.

2.1 EXPONENTIAL FUNCTIONS

? **THINK ABOUT IT**   What is the oxygen consumption of yearling salmon?

Later in this section, in Example 6, we will see that the answer to this question depends on *exponential functions.*

In the previous chapter we discussed functions involving expressions such as x^2, $(2x + 1)^3$, or x^{-1}, where the variable or variable expression is the base of an exponential expression, and the exponent is a constant. In an exponential function, the variable is in the exponent and the base is a constant.

EXPONENTIAL FUNCTION

An **exponential function** with base a is defined as

$$f(x) = a^x, \text{ where } a > 0 \text{ and } a \neq 1.$$

(If $a = 1$, the function is the constant function $f(x) = 1$.)

Exponential functions may be the single most important type of functions used in practical applications. They are used to describe growth and decay, which are important ideas in management, social science, and biology.

Figure 1 shows a graph of the exponential function defined by $f(x) = 2^x$ and a table of integer values from $x = -3$ to $x = 3$ of $f(x)$. This graph is typical of the graphs of exponential functions of the form $y = a^x$, where $a > 1$. Notice that as x gets larger and larger, the function also gets larger. As x gets more and more negative, the function becomes smaller and smaller, approaching but never reaching 0. Therefore, the x-axis is a horizontal asymptote, but the function only approaches the left side of the asymptote. In contrast, rational functions approach both the left and right sides of the asymptote. The graph suggests that the domain is the set of all real numbers and the range is the set of all positive numbers.

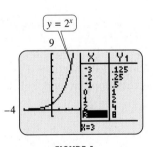

FIGURE 1

FOR REVIEW

Recall from the section on Quadratic Functions; Translation and Reflection, that the graph of $f(-x)$ is the reflection of the graph of $f(x)$ about the y-axis.

EXAMPLE 1   Graphing Exponential Functions
Graph $f(x) = 2^{-x}$.

Solution   The graph, shown in Figure 2 on the next page, is the horizontal reflection of the graph of $f(x) = 2^x$ given in Figure 1. Since $2^{-x} = 1/2^x = (1/2)^x$, this graph is typical of the graphs of exponential functions of the form $y = a^x$ where

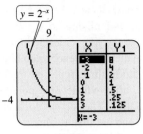

FIGURE 2

$0 < a < 1$. The domain includes all real numbers and the range includes all positive numbers. Notice that this function, with $f(x) = 2^{-x} = (1/2)^x$, is decreasing over its domain.

In the definition of an exponential function, notice that the base a is restricted to positive values, with negative or zero bases not allowed. For example, the function $y = (-4)^x$ could not include such numbers as $x = 1/2$ or $x = 1/4$ in the domain. The resulting graph would be at best a series of separate points having little practical use.

EXAMPLE 2 Graphing Exponential Functions
Graph $f(x) = -2^x + 3$.

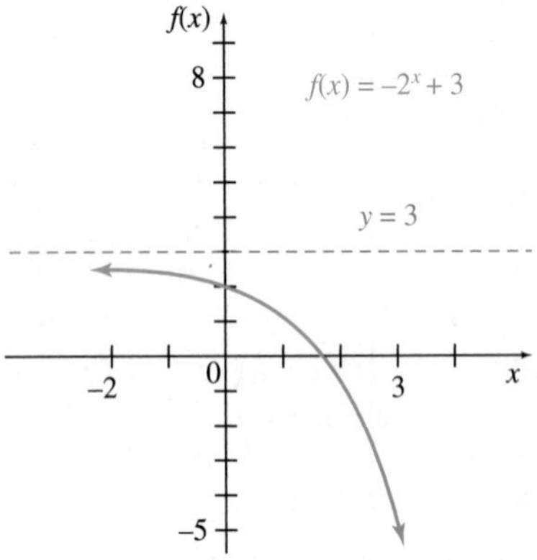

Solution The graph of $y = -2^x$ is the vertical reflection of the graph of $y = 2^x$, so this is a decreasing function. (Notice that -2^x is not the same as $(-2)^x$. In -2^x, we raise 2 to the x power and then take the negative.) The 3 indicates that the graph should be translated vertically 3 units, as compared to the graph of $y = -2^x$. Since $y = -2^x$ would have y-intercept $(0, -1)$, this function has y-intercept $(0, 2)$, which is up 3 units. For negative values of x, the graph approaches the line $y = 3$, which is a horizontal asymptote. The graph is shown in Figure 3.

FIGURE 3

Exponential Equations In Figures 1 and 2, which are typical graphs of exponential functions, a given value of x leads to exactly one value of a^x. Because of this, an equation with a variable in the exponent, called an **exponential equation,** often can be solved using the following property.

> If $a > 0$, $a \neq 1$, and $a^x = a^y$, then $x = y$. Also, if $x = y$, then $a^x = a^y$.

(Both bases must be the same.) The value $a = 1$ is excluded since $1^2 = 1^3$, for example, even though $2 \neq 3$. To solve $2^{3x} = 2^7$ using this property, work as follows.

$$2^{3x} = 2^7$$
$$3x = 7$$
$$x = \frac{7}{3}$$

EXAMPLE 3 Solving Exponential Equations

(a) Solve $9^x = 27$.

Solution First rewrite both sides of the equation so the bases are the same. Since $9 = 3^2$ and $27 = 3^3$,

$$(3^2)^x = 3^3$$
$$3^{2x} = 3^3$$
$$2x = 3$$
$$x = \frac{3}{2}.$$

FOR REVIEW

Recall from the Algebra Reference that $(a^m)^n = a^{mn}$.

(b) Solve $32^{2x-1} = 128^{x+3}$.

Solution Since the bases must be the same, write 32 as 2^5 and 128 as 2^7, giving

$$32^{2x-1} = 128^{x+3}$$
$$(2^5)^{2x-1} = (2^7)^{x+3}$$
$$2^{10x-5} = 2^{7x+21}.$$

Now use the property above to get

$$10x - 5 = 7x + 21$$
$$3x = 26$$
$$x = \frac{26}{3}.$$

Verify this solution in the original equation.

Compound Interest To understand the most fundamental type of growth function in biology, it will help to start with an example from the mathematics of finance. As we shall see in the next section, money and populations sometimes grow in a similar way.

The cost of borrowing money or the return on an investment is called **interest.** The amount borrowed or invested is the **principal,** P. The **rate of interest** r is given as a percent per year, and t is the **time,** measured in years.

SIMPLE INTEREST

The product of the principal P, rate r, and time t gives **simple interest,** I:

$$I = Prt.$$

With **compound interest,** interest is charged (or paid) on interest as well as on principal. To find a formula for compound interest, first suppose that P dollars, the principal, is deposited at a rate of interest r per year. The interest earned during the first year is found using the formula for simple interest.

$$\text{First-year interest} = P \cdot r \cdot 1 = Pr.$$

At the end of one year, the amount on deposit will be the sum of the original principal and the interest earned, or

$$P + Pr = P(1 + r). \tag{1}$$

If the deposit earns compound interest, the interest earned during the second year is found from the total amount on deposit at the end of the first year. Thus, the interest earned during the second year (again found by the formula for simple interest), is

$$[P(1 + r)](r)(1) = P(1 + r)r, \tag{2}$$

so the total amount on deposit at the end of the second year is the sum of amounts from (1) and (2) above, or

$$P(1 + r) + P(1 + r)r = P(1 + r)(1 + r) = P(1 + r)^2.$$

In the same way, the total amount on deposit at the end of three years is

$$P(1 + r)^3.$$

After t years, the total amount on deposit, called the **compound amount,** is $P(1 + r)^t$.

When interest is compounded more than once a year, the compound interest formula is adjusted. For example, if interest is to be paid quarterly (four times a year), $1/4$ of the interest rate is used each time interest is calculated, so the rate becomes $r/4$, and the number of compounding periods in t years becomes $4t$. Generalizing from this idea gives the following formula.

COMPOUND AMOUNT

If P dollars is invested at a yearly rate of interest r per year, compounded m times per year for t years, the compound amount is

$$A = P\left(1 + \frac{r}{m}\right)^{tm} \text{ dollars.}$$

EXAMPLE 4   Compound Interest

Françoise Dykler invests a bonus of $9,000 at 6% annual interest compounded semiannually for 4 years. How much interest will she earn?

Solution   Use the formula for compound interest with $P = 9,000$, $r = 0.06$, $m = 2$, and $t = 4$.

$$A = P\left(1 + \frac{r}{m}\right)^{tm}$$

$$= 9,000\left(1 + \frac{0.06}{2}\right)^{4(2)}$$

$$= 9,000(1.03)^8$$

$$\approx 11,400.93 \qquad \text{Use a calculator.}$$

The investment plus the interest is $11,400.93. The interest amounts to $11,400.93 − $9,000 = $2,400.93.

> **NOTE**   When using a calculator to compute the compound interest, store each partial result in the calculator and avoid rounding off until the final answer.

The Number e   Perhaps the single most useful base for an exponential function is the number e, an irrational number that occurs often in practical applications. The letter e was chosen to represent this number in honor of the Swiss mathematician Leonhard Euler (pronounced "oiler") (1707–1783). To see how the number e occurs in an application, begin with the formula for compound interest,

$$P\left(1 + \frac{r}{m}\right)^{tm}.$$

Suppose that a lucky investment produces annual interest of 100%, so that $r = 1.00 = 1$. Suppose also that you can deposit only $1 at this rate, and for only one year. Then $P = 1$ and $t = 1$. Substituting these values into the formula for compound interest gives

$$P\left(1 + \frac{r}{m}\right)^{t(m)} = 1\left(1 + \frac{1}{m}\right)^{1(m)} = \left(1 + \frac{1}{m}\right)^{m}.$$

As interest is compounded more and more often, m gets larger and the value of this expression will increase. For example, if $m = 1$ (interest is compounded annually),

$$\left(1 + \frac{1}{m}\right)^{m} = \left(1 + \frac{1}{1}\right)^{1} = 2^1 = 2,$$

so that your \$1 becomes \$2 in one year. Using a graphing calculator, we produced Table 1 (where m is represented by X and $(1 + 1/m)^m$ by Y_1) to see what happens as m becomes larger and larger. A spreadsheet can also be used to produce this table.

The table suggests that as m increases, the value of $(1 + 1/m)^m$ gets closer and closer to a fixed number, called e. As we shall see in the next chapter, this is an example of a limit.

Table 1

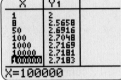

X=100000

DEFINITION OF e

As m becomes larger and larger, $\left(1 + \frac{1}{m}\right)^{m}$ becomes closer and closer to the number e, whose approximate value is 2.718281828.

The value of e is approximated here to nine decimal places. Euler approximated e to 23 decimal places using this definition. Many calculators give values of e^x, usually with a key labeled e^x. Some require two keys, either INV LN or 2nd LN. (We will define $\ln x$ in the next section.) In Figure 4, the functions $y = 2^x$, $y = e^x$, and $y = 3^x$ are graphed for comparison. Notice that e^x is between 2^x and 3^x, because e is between 2 and 3. For $x > 0$, the graphs show that $3^x > e^x > 2^x$. All three functions have y-intercept $(0, 1)$. It is difficult to see from the graph, but when $x < 0$, $3^x < e^x < 2^x$.

The number e is often used as the base in an exponential equation because it provides a good model for many natural, as well as economic, phenomena. In the exercises for this section, we will look at several examples of such applications.

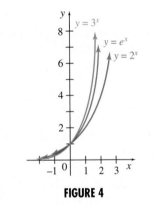

FIGURE 4

Continuous Compounding

In economics, the formula for **continuous compounding** is a good example of an exponential growth function. Recall the formula for compound amount

$$A = P\left(1 + \frac{r}{m}\right)^{tm},$$

where m is the number of times annually that interest is compounded. As m becomes larger and larger, the compound amount also becomes larger, but not without bound. Recall that as m becomes larger and larger, $(1 + 1/m)^m$ becomes closer and closer to e. Similarly,

$$\left(1 + \frac{1}{(m/r)}\right)^{m/r}$$

becomes closer and closer to e. Let us rearrange the formula for compound amount to take advantage of this fact.

$$A = P\left(1 + \frac{r}{m}\right)^{tm}$$

$$= P\left(1 + \frac{1}{(m/r)}\right)^{tm}$$

$$= P\left[\left(1 + \frac{1}{(m/r)}\right)^{m/r}\right]^{rt}$$

This last expression becomes closer and closer to Pe^{rt} as m becomes larger and larger, which describes what happens when interest is compounded continuously. Essentially, the number of times annually that interest is compounded becomes infinitely large. We thus have the following formula for the compound amount when interest is compounded continuously.

CONTINUOUS COMPOUNDING

If a deposit of P dollars is invested at a rate of interest r compounded continuously for t years, the compound amount is

$$A = Pe^{rt} \text{ dollars.}$$

EXAMPLE 5 Continuous Compound Interest

Assuming continuous compounding, if the inflation rate averaged 6% per year for 5 years, how much would a $1 item cost at the end of the 5 years?

Solution In the formula for continuous compounding, let $P = 1$, $t = 5$, and $r = 0.06$ to get

$$A = 1e^{5(0.06)} = e^{0.3} \approx 1.34986.$$

An item that costs $1 at the beginning of the 5-year period would cost $1.35 at the end of the period, an increase of 35%, or about 1/3. ▪

In situations that involve growth or decay of a population, the size of the population at a given time t often is determined by an exponential function of t. The next example illustrates a typical application of this kind.

EXAMPLE 6 Oxygen Consumption

?

Biologists studying salmon have found that the oxygen consumption of yearling salmon (in appropriate units) increases exponentially with the speed of swimming according to the function defined by

$$f(x) = 100e^{0.6x},$$

where x is the speed in feet per second. Find each of the following.

(a) The oxygen consumption when the fish are still

Solution When the fish are still, their speed is 0. Substitute 0 for x:

$$f(0) = 100e^{(0.6)(0)} = 100e^0$$
$$= 100 \cdot 1 = 100. \qquad e^0 = 1$$

When the fish are still, their oxygen consumption is 100 units.

FOR REVIEW ■

Refer to the discussion on linear regression in Section 1.2. A similar process is used to fit data points to other types of functions. Many of the functions in this chapter's applications were determined in this way, including that given in Example 6.

(b) The oxygen consumption at a speed of 2 feet per second

> **Solution** Find $f(2)$ as follows.
>
> $$f(2) = 100e^{(0.6)(2)} = 100e^{1.2} \approx 332$$
>
> At a speed of 2 feet per second, oxygen consumption is about 332 units.

NOTE In Example 6(b), we rounded the answer to the nearest integer. Because the function is only an approximation of the real situation, further accuracy is not realistic.

EXAMPLE 7 CO_2 Levels

The International Panel on Climate Change (IPCC) in 1990 published data showing that if current trends of burning fossil fuel and deforestation continue, then future amounts of atmospheric carbon dioxide in parts per million (ppm) will increase as shown in Table 2.*

Table 2

Year	Carbon Dioxide
1990	353
2000	375
2075	590
2175	1,090
2275	2,000

(a) Plot the data. Do the carbon dioxide levels appear to grow linearly or exponentially?

> **Solution** Figure 5 shows a graphing calculator plot of the data, which suggests that carbon dioxide levels are growing exponentially.

(b) Find an exponential function in the form $C(x) = C_0 a^{x-1990}$ that models this data, where x is the year and $C(x)$ is the amount of carbon dioxide in the atmosphere. Use the data for 1990 and 2275.

> **Solution** Since $C(1990) = C_0 a^0 = C_0$, we have $C_0 = 353$. Using $x = 2275$, we have

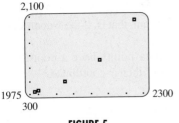

FIGURE 5

$$C(2275) = 353a^{285} = 2{,}000$$

$$a^{285} = \frac{2{,}000}{353} \qquad \text{Divide by 353.}$$

$$a = \left(\frac{2{,}000}{353}\right)^{1/285} \qquad \text{Take the 285th root.}$$

$$\approx 1.0061.$$

Thus $C(x) = 353(1.0061)^{x-1990}$. Figure 6 shows that this function fits the data well.

(c) Determine the expected annual percentage increase in carbon dioxide levels during this time period.

> **Solution** Since a is 1.0061, the amount of carbon dioxide each year is 1.0061 times its value the previous year, for a rate of increase of $0.0061 = 0.61\%$ per year.

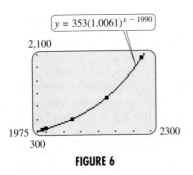

FIGURE 6

(d) Graph C and estimate the years when future levels of carbon dioxide will double and triple over the preindustrial level of 280 ppm.

---

*International Panel on Climate Change (IPCC), 1990.

Solution Figure 7(a) shows the graphs of $y = C(x)$ and $y = 2 \cdot 280 = 560$. Figure 7(b) shows the graphs of $y = C(x)$ and $y = 3 \cdot 280 = 840$ on the same coordinate axes. Their graphs intersect at approximately 2066 and 2133. Carbon dioxide levels will double by 2066 and triple by 2133. ∎

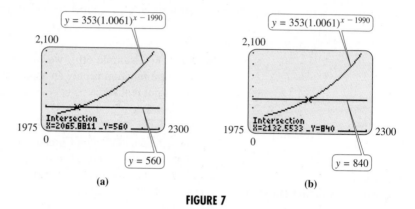

(a)   **(b)**

FIGURE 7

NOTE Another way to check whether an exponential function fits the data is to see if points whose x-coordinates are equally spaced have y-coordinates with a constant ratio. This must be true for an exponential function because if $f(x) = a \cdot b^x$, then $f(x_1) = a \cdot b^{x_1}$ and $f(x_2) = a \cdot b^{x_2}$, so

$$\frac{f(x_2)}{f(x_1)} = \frac{a \cdot b^{x_2}}{a \cdot b^{x_1}} = b^{x_2 - x_1}.$$

This last expression is constant if $x_2 - x_1$ is constant, that is, if the x-coordinates are equally spaced.

In the previous example, the last three data points have x-coordinates 100 years apart, so we can compare the ratios of their y-coordinates:

$$\frac{1090}{590} = 1.847$$

$$\frac{2000}{1090} = 1.835.$$

These ratios are fairly close to each other, so an exponential function fits the data quite well.

Another way to find an exponential function that fits a set of data is to use a graphing calculator or computer program with an exponential regression feature. This fits an exponential function through a set of points using the least squares method, introduced in Section 1.2 for fitting a line through a set of points. On a TI-83, for example, enter the year into the list L_1 and the carbon dioxide into L_2. For simplicity, subtract 1990 from each year, so that 1990 corresponds to $x = 0$. Selecting ExpReg from the STAT CALC menu yields $y = 352.65(1.0061)^x$, which is essentially the same function as we found in Example 7(b). In other cases the results may be a little different.

2.1 EXERCISES

A ream of 20-pound paper contains 500 sheets and is about 2 inches high. Suppose you take one sheet, fold it in half, then fold it in half again, continuing in this way as long as possible.*

1. Complete the table.

number of folds	1	2	3	4	5	...	10	...	50
layers of paper									

2. After folding 50 times (if this were possible), what would be the height (in miles) of the folded paper?

For Exercises 3–11, match the correct graph A–F to the function without using your calculator. Notice that there are more functions than graphs; some of the functions are equivalent. After you have answered all of them, use a graphing calculator to check your answers. Each graph in this group is plotted on the window $[-2,2]$ by $[-4,4]$.

3. $y = 3^x$

4. $y = 3^{-x}$

5. $y = \left(\dfrac{1}{3}\right)^{1-x}$

6. $y = 3^{x+1}$

7. $y = 3(3)^x$

8. $y = \left(\dfrac{1}{3}\right)^x$

9. $y = 2 - 3^{-x}$

10. $y = -2 + 3^{-x}$

11. $y = 3^{x-1}$

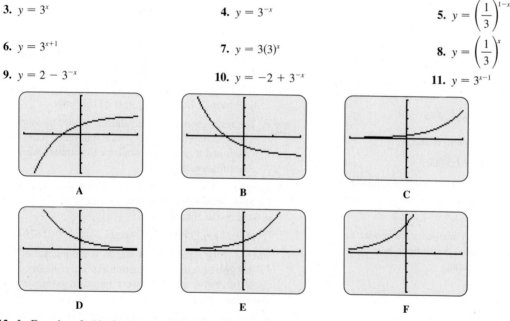

A	B	C

D	E	F

12. In Exercises 3–11, there were more formulas for functions than there were graphs. Explain how this is possible.

Solve each equation.

13. $2^x = \dfrac{1}{8}$

14. $4^x = 64$

15. $e^x = \dfrac{1}{e^2}$

16. $4^x = 8^{x+1}$

17. $25^x = 125^{x-2}$

18. $16^{-x+1} = 8^x$

19. $16^{x+2} = 64^{2x-1}$

20. $(e^4)^{-2x} = e^{-x+1}$

21. $e^{-x} = (e^2)^{x+3}$

22. $2^{|x|} = 16$

23. $5^{-|x|} = \dfrac{1}{25}$

24. $2^{x^2-4x} = \left(\dfrac{1}{16}\right)^{x-4}$

---

25. $5^{x^2+x} = 1$ **26.** $8^{x^2} = 2^{5x+2}$

27. In our definition of exponential function, we ruled out negative values of a. The author of a textbook on mathematical economics, however, obtained a "graph" of $y = (-2)^x$ by plotting the following points and drawing a smooth curve through them.

x	-4	-3	-2	-1	0	1	2	3
y	$1/16$	$-1/8$	$1/4$	$-1/2$	1	-2	4	-8

The graph oscillates very neatly from positive to negative values of y. Comment on this approach. (This exercise shows the dangers of point plotting when drawing graphs.)

28. Explain why the exponential equation $4^x = 6$ cannot be solved using the method described in Example 3.

29. Explain why $3^x > e^x > 2^x$ when $x > 0$, but $3^x < e^x < 2^x$ when $x < 0$.

30. A friend claims that as x becomes large, the expression $1 + 1/x$ gets closer and closer to 1, and 1 raised to any power is still 1. Therefore, $f(x) = (1 + 1/x)^x$ gets closer and closer to 1 as x gets larger. Use a graphing calculator to graph f on $0.1 \le x \le 50$. How might you use this graph to explain to the friend why $f(x)$ does not approach 1 as x becomes large? What does it approach?

Applications

LIFE SCIENCES

31. *Plasma Volume* A formula for plasma volume for males proposed by Hurley[*] is

$$PV = 995e^{0.6085S}.$$

A formula proposed by Pearson, et al.,[†] is

$$PV = 1{,}578S,$$

where PV is measured in milliliters and S in square meters. Use a graphing calculator to verify that these two formulas never give the same answer.

32. *Cortisone Concentration* A mathematical model for the concentration of administered cortisone in humans over a 24-hour period uses the function

$$C = \frac{D \times a}{V(a - b)}(e^{-bt} - e^{-at}),$$

where C is the concentration, D is the dose given at time $t = 0$, V is the volume of distribution (volume divided by bioavailability), a is the absorption rate, b is the elimination rate, and t is the time in hours.[‡]

a. What is the value of C at $t = 0$? Explain why this makes sense.

b. What happens to the concentration as a large amount of time passes? Explain why this makes sense.

c. The researchers used the values $D = 500$ micrograms, $a = 8.5$, $b = 0.09$, and $V = 3{,}700$ liters. Use these values and a graphing calculator to estimate when the concentration is greatest.

33. *Bacteria in Sausage* The number of *Enterococcus faecium* bacteria contaminating bologna sausage at 32°C can be described by the equation

$$N = N_0 \exp\{9.8901 \exp[-\exp(2.5420 - 0.2167t)]\},$$

where t is the time in hours and N_0 is the number at time $t = 0$.[§] (We have used the notation exp x to represent e^x because otherwise all the powers raised to powers would make the expression difficult to read.) Suppose $N_0 = 100$.

a. Find the number of bacteria after 10, 20, 30, 40, and 50 hours.

b. Use a graphing calculator to graph the function on the interval $[0, 50]$.

*Hurley, Peter J., "Red Cell and Plasma Volumes in Normal Adults," *Journal of Nuclear Medicine,* Vol. 16, 1975, pp. 46–52.
†Pearson, T. C., et al., "Interpretation of Measured Red Cell Mass and Plasma Volume in Adults," *British Journal of Haematology,* Vol. 89, 1995, pp. 748–756.
‡Chakraborty, Abhijit, et al., "Mathematical Modeling of Circadian Cortisol Concentrations Using Indirect Response Models: Comparison of Several Methods," *Journal of Pharmacokinetics and Biopharmaceutics,* Vol. 27, No. 1, 1999, pp. 23–43.
§Zanoni, B., C. Garzardi, S. Anselmi, and G. Rondinini, "Modeling the Growth of *Enterococcus Faecium* in Bologna Sausage," *Applied and Environmental Microbiology,* Vol. 59, No. 10, Oct. 1993, pp. 3411–3417.

c. Using your answers to parts a and b, what seems to happen to the number of bacteria after a long period of time?

34. *Heart Attack Risk* The following data give the 10-year risk of a heart attack for men based on their score on a test of risk factors.* (For example, a man increases his score by 2 points if his HDL cholesterol is less than 40 mg/dL.)

Score (x)	Percent Risk (y)
0	1
5	2
10	6
15	20
16	25
17	30

a. Graph the function

$$f(x) = 0.8454e^{0.2081x}$$

on the same axes as the data. How well does this function describe the data?

b. Using the function in part a, find the risk of a man with a score of 3.

c. A man aged 20 to 34 increases his score by 8 points if he is a smoker. If his score was 3 without considering his smoking, by what percent does his risk go up if he smokes?

d. Redo part c under the assumption that the man's score, without considering his smoking, was 6.

e. The formula for women is

$$g(x) = 0.1210e^{0.2249x}.$$

Answer part b for a woman.

f. A woman aged 20 to 34 increases her score by 9 points if she is a smoker. If her score was 3 without considering her smoking, by what percent does her risk go up if she smokes?

g. Redo part f under the assumption that the woman's score, without considering her smoking, was 6.

35. *Population Growth* Since 1950, the growth in world population (in millions) closely fits the exponential function defined by

$$A(t) = 2,600e^{0.018t},$$

where t is the number of years since 1950.

a. World population was about 3,700 million in 1970. How closely does the function approximate this value?

b. Use the function to approximate world population in 1990. (The actual 1990 population was about 5,320 million.)

c. Estimate world population in the year 2010.

36. *Growth of Bacteria* Salmonella bacteria, found on almost all chicken and eggs, grow rapidly in a nice warm place. If just a few hundred bacteria are left on the cutting board when a chicken is cut up, and they get into the potato salad, the population begins compounding. Suppose the number present in the potato salad after x hours is given by

$$f(x) = 500 \cdot 2^{3x}.$$

a. If the potato salad is left out on the table, how many bacteria are present 1 hour later?

b. How many were present initially?

c. How often do the bacteria double?

d. How quickly will the number of bacteria increase to 32,000?

37. *Chlorofluorocarbons* Chlorofluorocarbons (CFCs) are gases used in refrigeration units, foaming agents, and aerosols. They have great potential for destroying the ozone layer of the atmosphere. The following graph displays approximate concentrations (in parts per billion, or ppb) of CFC-11.[†]

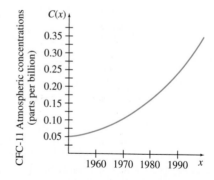

a. Letting $C(x)$ be the concentration of CFC-11 in the atmosphere from 1950 to 2000 in ppb, where x is the year, find $C(1975)/C(1950)$ and $C(2000)/C(1975)$. Based on these two ratios, does an exponential function seem to fit the data fairly well?

b. Use the values of the function at 1950 and 2000 to find an exponential function with $C(x) = C_0 a^{x-1950}$ that models the concentration of CFC-11 in the atmosphere from 1950 to 2000.

---

*The New York Times, May 16, 2001, p. A18.
[†]Nilsson, A., *Greenhouse Earth*, New York, John Wiley & Sons, 1992.

c. Graph $C(x)$, and describe how closely the graph resembles the original function. In particular, calculate $C(1975)$ and compare it with the actual concentration in 1975.

d. Approximate the average annual percentage increase of CFC-11 during this time period.

e. Use a graphing calculator to find the year in which the concentration would reach 0.50 ppb if the increase in CFC-11 were to continue.

OTHER APPLICATIONS

38. *Radioactive Decay* Suppose the quantity (in grams) of a radioactive substance present at time t is

$$Q(t) = 1,000(5^{-0.3t}),$$

where t is measured in months.

a. How much will be present in 6 months?

b. How long will it take to reduce the substance to 8 g?

39. *Atmospheric Pressure* The atmospheric pressure (in millibars) at a given altitude (in meters) is listed in the table.*

Altitude	Pressure
0	1,013
1,000	899
2,000	795
3,000	701
4,000	617
5,000	541
6,000	472
7,000	411
8,000	357
9,000	308
10,000	265

a. Find functions of the form $P = ae^{kx}$, $P = mx + b$, and $P = 1/(ax + b)$ that fit the data at $x = 0$ and $x = 10,000$, where P is the pressure and x is the altitude.

b. Plot the data in the table and graph the three functions found in part a. Which function best fits the data?

c. Use the best-fitting function from part b to predict pressure at 1,500 m and 11,000 m. Compare your answers to the true values of 846 millibars and 227 millibars, respectively.

d. If you have a graphing calculator or computer program with an exponential regression feature, use it to find

an exponential function that approximately fits the data. How does this answer compare with the answer to part b?

40. *Computer Chips* The power of personal computers has increased dramatically as a result of the ability to place an increasing number of transistors on a single processor chip. The table below lists the number of transistors on some popular computer chips made by Intel.[†]

Year	Chip	Transistors
1971	4004	2,300
1986	386DX	275,000
1989	486DX	1,200,000
1993	Pentium	3,300,000
1995	P6	5,500,000
1997	Pentium III	9,500,000

a. Let x be the year, where $x = 0$ corresponds to 1971, and y be the number of transistors. Find functions of the form $y = mx + b$, $y = ax^2 + b$, and $y = ab^x$ that fit the data at 1971 and 1997.

b. Use a graphing calculator to plot the data in the table and to graph the three functions found in part a. Which function best fits the data?

c. Use the best-fitting function from part b to predict the number of transistors on a chip in the year 2002.

d. If you have a graphing calculator or computer program with an exponential regression feature, use it to find an exponential function that approximately fits the data. How does this answer compare with the answer to part b?

41. *Internet Users* The number of Internet users was estimated to be 26 million in 1995 and 205 million in 1999.[‡] The growth can be approximated by an exponential function.

a. Find an exponential function with $f(x) = f_0 a^x$ that models the number of Internet users, where $x = 0$ (in years) corresponds to 1995, and $f(x)$ gives the number of users in millions.

b. Graph the function and use it to estimate the number of users in 2002.

c. Approximate the average yearly percent increase in users during this time period.

d. Use your graph to estimate the year when there were 122 million users.

---

*Miller, A., and J. Thompson, *Elements of Meteorology*, Charles Merrill, 1975.
[†]Data provided by Intel.
[‡]NUA Internet Surveys

42. *Interest* Find the interest earned on $10,000 invested for 5 yr at 6% interest compounded as follows.

    a. Annually   **b.** Semiannually (twice a year)

    c. Quarterly   **d.** Monthly

43. *Interest* Suppose $26,000 is borrowed for 3 yr at 12% interest. Find the interest paid over this period if the interest is compounded as follows.

    a. Annually   **b.** Semiannually

    c. Quarterly   **d.** Monthly

44. *Interest* David Glenn needs to choose between two investments: one pays 8% compounded semiannually, and the other pays 7½% compounded monthly. If he plans to invest $18,000 for 1½ yr, which investment should he choose? How much extra interest will he earn by making the better choice?

45. *Interest* Find the interest rate required for an investment of $5,000 to grow to $8,000 in 4 yr if interest is compounded as follows:

    a. annually   **b.** quarterly.

46. *Interest* Laurie Rosatone plans to invest $500 into a money market account. Find the interest rate that is needed for the money to grow to $1,200 in 14 yr if the interest is computed using each of the following methods.

    a. compounded quarterly

    b. compounded continuously

47. *Interest* Karen Guardino puts $10,500 into an account to save money to buy a car in 12 yr. She expects the car of her dreams to cost $30,000 by then. Find the interest rate that is necessary if the interest is computed using each of the following methods.

    a. compounded quarterly

    b. compounded continuously

2.2 LOGARITHMIC FUNCTIONS

? THINK ABOUT IT

If a population of squirrels grows 5% per year, how long will it take for the population to double?

The amount of time it will take for a population to double under given conditions is called the **doubling time.** To analyze how 400 squirrels become 800 in t years, assuming 5% annual growth, we can use the exact same equation that we used for money growing at 5% interest compounded annually. In mathematical symbols,

$$800 = 400(1.05)^t.$$

Dividing by 400 gives

$$2 = 1\left(1 + \frac{0.05}{1}\right)^{1(t)}$$

or

$$2 = (1.05)^t.$$

This equation would be easier to solve if the variable were not in the exponent. **Logarithms** are defined for just this purpose. In Example 7, we will use logarithms to answer the question posed above.

DEFINITION OF LOGARITHM

For $a > 0$, $a \neq 1$, and $x > 0$,

$$y = \log_a x \qquad \text{means} \qquad a^y = x.$$

(Read $y = \log_a x$ as "y is the logarithm of x to the base a.") For example, the exponential statement $2^4 = 16$ can be translated into the logarithmic statement

$4 = \log_2 16$. Also, in the problem discussed above, $(1.05)^t = 2$ can be rewritten with this definition as $t = \log_{1.05} 2$. A logarithmic is an exponent: $\log_a x$ is the exponent used with the base a to get x.

EXAMPLE 1 Equivalent Expressions

This example shows the same statements written in both exponential and logarithmic forms.

Exponential Form	*Logarithmic Form*
(a) $3^2 = 9$	$\log_3 9 = 2$
(b) $(1/5)^{-2} = 25$	$\log_{1/5} 25 = -2$
(c) $10^5 = 100{,}000$	$\log_{10} 100{,}000 = 5$
(d) $4^{-3} = 1/64$	$\log_4 1/64 = -3$
(e) $2^{-4} = 1/16$	$\log_2 1/16 = -4$
(f) $e^0 = 1$	$\log_e 1 = 0$

Logarithmic Functions For a given positive value of x, the definition of logarithm leads to exactly one value of y, so $y = \log_a x$ defines the **logarithmic function** of base a (the base a must be positive, with $a \neq 1$).

> **LOGARITHMIC FUNCTION**
>
> If $a > 0$ and $a \neq 1$, then the logarithmic function of base a is defined by
>
> $$f(x) = \log_a x$$
>
> for $x > 0$.

The graphs of the exponential function with $f(x) = 2^x$ and the logarithmic function with $g(x) = \log_2 x$ are shown in Figure 8. The graphs show that $f(3) = 2^3 = 8$, while $g(8) = \log_2 8 = 3$. Thus, $f(3) = 8$ and $g(8) = 3$. Also, $f(2) = 4$ and $g(4) = 2$. In fact, for any number m, if $f(m) = p$, then $g(p) = m$. Functions related in this way are called *inverses* of each other. The graphs also show that the domain of the exponential function (the set of real numbers) is the range of the logarithmic function. Also, the range of the exponential function (the set of positive real numbers) is the domain of the logarithmic function. Every logarithmic function is the inverse of some exponential function. This means that we can graph logarithmic functions by rewriting them as exponential functions using the definition of logarithm. The graphs in Figure 8 show a characteristic of a pair of inverse functions: their graphs are mirror images about the line $y = x$. Notice that because the exponential function has the x-axis as a horizontal asymptote, the logarithmic function has the y-axis as a vertical asymptote. A more complete discussion of inverse functions is given in most standard intermediate algebra and college algebra books.

The graph of $\log_2 x$ is typical of logarithms with bases $a > 1$. When $0 < a < 1$, the graph is the vertical reflection of the logarithm graph in Figure 8. Because logarithms with bases less than 1 are rarely used, we will not explore them here.

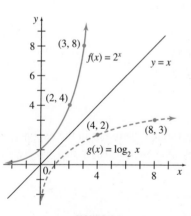

FIGURE 8

> **CAUTION** The domain of $\log_a x$ includes all $x > 0$. In other words, you cannot take the logarithm of zero or a negative number. This also means that in a function such as $g(x) = \log_a(x - 2)$, the domain is given by $x - 2 > 0$, or $x > 2$.

Properties of Logarithms The usefulness of logarithmic functions depends in large part on the following **properties of logarithms.**

PROPERTIES OF LOGARITHMS

Let x and y be any positive real numbers and r be any real number. Let a be a positive real number, $a \neq 1$. Then

a. $\log_a xy = \log_a x + \log_a y$

b. $\log_a \dfrac{x}{y} = \log_a x - \log_a y$

c. $\log_a x^r = r \log_a x$

d. $\log_a a = 1$

e. $\log_a 1 = 0$

f. $\log_a a^r = r.$

To prove property (a), let $m = \log_a x$ and $n = \log_a y$. Then, by the definition of logarithm,

$$a^m = x \qquad \text{and} \qquad a^n = y.$$

Hence,

$$a^m a^n = xy.$$

By a property of exponents, $a^m a^n = a^{m+n}$, so

$$a^{m+n} = xy.$$

Now use the definition of logarithm to write

$$\log_a xy = m + n.$$

Since $m = \log_a x$ and $n = \log_a y$,

$$\log_a xy = \log_a x + \log_a y.$$

Proofs of properties (b) and (c) are left for the exercises. Properties (d) and (e) depend on the definition of a logarithm. Property (f) follows from properties (c) and (d).

EXAMPLE 2 Properties of Logarithms

If all the following variable expressions represent positive numbers, then for $a > 0$, $a \neq 1$, the statements in (a)–(c) are true.

(a) $\log_a x + \log_a(x - 1) = \log_a x(x - 1)$

(b) $\log_a \dfrac{x^2 - 4x}{x + 6} = \log_a(x^2 - 4x) - \log_a(x + 6)$

(c) $\log_a(9x^5) = \log_a 9 + \log_a(x^5) = \log_a 9 + 5 \cdot \log_a x$

Evaluating Logarithms The invention of logarithms is credited to John Napier (1550–1617), who first called logarithms "artificial numbers." Later he joined the Greek words *logos* (ratio) and *arithmos* (number) to form the word used today. The development of logarithms was motivated by a need for faster computation. Tables of logarithms and slide rule devices were developed by Napier, Henry Briggs (1561–1631), Edmund Gunter (1581–1626), and others.

For many years logarithms were used primarily to assist in involved calculations. Current technology has made this use of logarithms obsolete, but logarithmic functions play an important role in many applications of mathematics. Since our number system has base 10, logarithms to base 10 were most convenient for numerical calculations and so base 10 logarithms were called **common logarithms.** Common logarithms are still useful in other applications. For simplicity,

$$\log_{10} x \text{ is abbreviated } \log x.$$

Most practical applications of logarithms use the number e as base. (Recall that to seven decimal places, $e = 2.7182818$.) Logarithms to base e are called **natural logarithms,** and

$$\log_e x \text{ is abbreviated } \ln x$$

(read "el-en x"). A graph of $f(x) = \ln x$ is given in Figure 9.

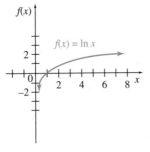

FIGURE 9

> **NOTE** Keep in mind that $\ln x$ is a logarithmic function. Therefore, all of the properties of logarithms given previously are valid when $\log_a$ is replaced with $\ln$ and a is replaced with e.

Although common logarithms may seem more "natural" than logarithms to base e, there are several good reasons for using natural logarithms instead. The most important reason is discussed later in the section on Derivatives of Logarithmic Functions.

A calculator can be used to find both common and natural logarithms. For example, using a calculator and four decimal places, we get the following values.

$$\log 2.34 = 0.3692, \quad \log 594 = 2.7738, \quad \text{and} \quad \log 0.0028 = -2.5528.$$
$$\ln 2.34 = 0.8502, \quad \ln 594 = 6.3869, \quad \text{and} \quad \ln 0.0028 = -5.8781.$$

Notice that logarithms of numbers less than 1 are negative when the base is greater than 1. A look at the graph of $y = \log_2 x$ or $y = \ln x$ will show why.

Sometimes it is convenient to use logarithms to bases other than 10 or e. For example, some computer science applications use base 2. In such cases, the following theorem is useful for converting from one base to another.

CHANGE-OF-BASE THEOREM FOR LOGARITHMS
If x is any positive number and if a and b are positive real numbers, $a \neq 1$, $b \neq 1$, then

$$\log_a x = \frac{\log_b x}{\log_b a}.$$

To prove this result, use the definition of logarithm to write $y = \log_a x$ as $x = a^y$ or $x = a^{\log_a x}$ (for positive x and positive a, $a \neq 1$). Now take base b logarithms of both sides of this last equation.

$$\log_b x = \log_b a^{\log_a x}$$

$$\log_b x = (\log_a x)(\log_b a), \qquad \log_a x^r = r \log_a x$$

$$\log_a x = \frac{\log_b x}{\log_b a} \qquad\qquad \text{Solve for } \log_a x.$$

If the base b is equal to e, then by the change-of-base theorem,

$$\log_a x = \frac{\log_e x}{\log_e a}.$$

Using $\ln x$ for $\log_e x$ gives the special case of the theorem using natural logarithms.

For any positive numbers a and x, $a \neq 1$,

$$\log_a x = \frac{\ln x}{\ln a}.$$

The change-of-base theorem is also useful when graphing $y = \log_b x$ on a graphing calculator for a base b other than e or 10. For example, to graph $y = \log_2 x$, let $y = \ln x / \ln 2$.

EXAMPLE 3 Evaluating Logarithms

Use natural logarithms to find each value. Round to the nearest hundredth.

(a) $\log_5 27$

Solution Let $x = 27$ and $a = 5$. Using the second form of the theorem gives

$$\log_5 27 = \frac{\ln 27}{\ln 5}.$$

Now use a calculator.

$$\log_5 27 \approx \frac{3.2958}{1.6094} \approx 2.05$$

To check, use a calculator, along with the definition of logarithm, to verify that $5^{2.05} \approx 27$.

(b) $\log_2 0.1$

Solution

$$\log_2 0.1 = \frac{\ln 0.1}{\ln 2} \approx -3.32$$

CAUTION As mentioned earlier, when using a calculator, do not round off intermediate results. Keep all numbers in the calculator until you have the final answer. In Example 3(a), we showed the rounded intermediate values of $\ln 27$ and $\ln 5$, but we used the unrounded quantities when doing the division.

Logarithmic Equations Equations involving logarithms are often solved by using the fact that exponential functions and logarithmic functions are inverses, so a logarithmic equation can be rewritten (with the definition of logarithm) as an exponential equation. In other cases, the properties of logarithms may be useful in simplifying a **logarithmic equation.**

EXAMPLE 4   Solving Logarithmic Equations
Solve each equation.

(a) $\log_x \dfrac{8}{27} = 3$

Solution Using the definition of logarithm, write the expression in exponential form. To solve for x, take the cube root on both sides.

$$x^3 = \frac{8}{27}$$

$$x = \frac{2}{3}$$

(b) $\log_4 x = \dfrac{5}{2}$

Solution In exponential form, the given statement becomes

$$4^{5/2} = x$$
$$(4^{1/2})^5 = x$$
$$2^5 = x$$
$$32 = x.$$

(c) $\log_2 x - \log_2 (x - 1) = 1$

Solution By a property of logarithms,

$$\log_2 x - \log_2 (x - 1) = \log_2 \frac{x}{x - 1},$$

so the original equation becomes

$$\log_2 \frac{x}{x - 1} = 1.$$

Now write this equation in exponential form, and solve.

$$\frac{x}{x - 1} = 2^1 = 2$$

Solve this equation.

$$\frac{x}{x - 1}(x - 1) = 2(x - 1)$$
$$x = 2(x - 1)$$
$$x = 2x - 2$$

$$-x = -2$$
$$x = 2$$

(d) $\log_3 \dfrac{1}{9} = x$

Solution Using the definition of logarithm, write the expression in exponential form. To solve for x, first rewrite in terms of 3 to some power. Thus, $3^x = 1/9 = (1/3)^2 = 3^{-2}$ and $x = -2$. It is important to check solutions when solving equations involving logarithms because $\log_a u$, where u is an expression in x, has domain given by $u > 0$. ■

Exponential Equations   In the previous section exponential equations like $(1/3)^x = 81$ were solved by writing each side of the equation as a power of 3. That method cannot be used to solve an equation such as $3^x = 5$, however, since 5 cannot easily be written as a power of 3. Such equations can be solved approximately with a graphing calculator, but an algebraic method is also useful, particularly when the equation involves variables such as a and b rather than just numbers such as 3 and 5. A general method for solving these equations depends on the following property of logarithms, which is supported by the graphs of logarithmic functions (Figures 8 and 9).

> For $x > 0$, $y > 0$, $b > 0$, and $b \neq 1$,
>
> $$\text{if } x = y, \text{ then } \log_b x = \log_b y,$$
>
> and $\quad\quad\quad$ if $\log_b x = \log_b y$, then $x = y$.

EXAMPLE 5   Solving Exponential Equations
Solve each equation.

(a) $3^{2x} = 4^{x+1}$

Solution Taking natural logarithms (logarithms to any base could be used) on both sides gives

$$\ln 3^{2x} = \ln 4^{x+1}$$
$$2x \ln 3 = (x + 1) \ln 4 \quad\quad \ln u^r = r \ln u$$
$$(2 \ln 3)x = (\ln 4)x + \ln 4$$
$$(2 \ln 3)x - (\ln 4)x = \ln 4$$
$$(2 \ln 3 - \ln 4)x = \ln 4$$
$$x = \frac{\ln 4}{2 \ln 3 - \ln 4}$$

Use a calculator to evaluate the logarithms, then divide, to get

$$x \approx \frac{1.3863}{2(1.0986) - 1.3863} \approx 1.710.$$

(b) $5e^{0.01x} = 9$

Solution

$$e^{0.01x} = \frac{9}{5} = 1.8 \qquad \text{Divide both sides by 5.}$$

$$\ln e^{0.01x} = \ln 1.8 \qquad \text{Take natural logarithms on both sides.}$$

$$0.01x = \ln 1.8 \qquad \ln e^u = u$$

$$x = \frac{\ln 1.8}{0.01} \approx 58.779$$

Just as $\log_a x$ can be written as a base e logarithm, any exponential function $y = a^x$ can be written as an exponential function with base e. For example, there exists a real number k such that

$$2 = e^k.$$

Raising both sides to the power x gives

$$2^x = e^{kx},$$

so that powers of 2 can be found by evaluating appropriate powers of e. To find the necessary number k, solve the equation $2 = e^k$ for k by first taking logarithms on both sides.

$$2 = e^k$$
$$\ln 2 = \ln e^k$$
$$\ln 2 = k \ln e$$
$$\ln 2 = k \qquad \ln e = 1$$

Thus, $k = \ln 2$. In the section on Derivatives of Exponential Functions, we will see why this change of base is useful. A general statement can be drawn from this example.

CHANGE-OF-BASE THEOREM FOR EXPONENTIALS

For every positive real number a,

$$a^x = e^{(\ln a)x}.$$

EXAMPLE 6   Evaluating Exponentials

Evaluate each exponential by first writing it as a power of e.

(a) $2^{3.57}$

Solution

$$2^{3.57} = e^{(\ln 2)(3.57)} \approx 11.88$$

(b) $5^{-1.38} = e^{(\ln 5)(-1.38)} \approx 0.1085$

Of course, expressions such as $2^{3.57}$ can be evaluated directly on a calculator without using e. As we shall see later, the change-of-base theorem for exponentials is more useful when converting expressions such as 2^x to $e^{(\ln 2)x} \approx e^{0.6931x}$. For many purposes, e is the simplest base to work with, and so it is convenient to be able to convert any exponential function into one with base e.

The final examples in this section illustrate the use of logarithms in practical problems.

EXAMPLE 7 Doubling Time

?

Complete the solution of the problem posed at the beginning of this section.

Solution Recall that if a population will double after t years at an annual growth rate of 5%, t is given by the equation

$$2 = (1.05)^t.$$

We solve this equation by first taking natural logarithms on both sides.

$$\ln 2 = \ln(1.05)^t$$

$$\ln 2 = t \ln 1.05 \qquad\qquad \ln x^r = r \ln x$$

$$t = \frac{\ln 2}{\ln 1.05} \approx 14.2$$

It will take a little over 14 years for the population to double.

The problem solved in Example 7 can be generalized for the growth equation

$$y = y_0(1 + r)^t.$$

Solving for t as in Example 7 (with $y = 2$ and $y_0 = 1$) gives the doubling time in years as $t = (\ln 2)/(\ln(1 + r))$.

DOUBLING TIME

For a population growing at an annual rate of $100r\%$ per year, the years for the population to double is given by

$$t = \frac{\ln 2}{\ln(1 + r)}.$$

We will study growth equations more in the next section.

EXAMPLE 8 Index of Diversity

One measure of the diversity of the species in an ecological community is given by the *index of diversity H,* where

$$H = -[P_1 \ln P_1 + P_2 \ln P_2 + \cdots + P_n \ln P_n],$$

and $P_1, P_2, \ldots, P_n$ are the proportions of a sample belonging to each of n species found in the sample.* For example, in a community with two species, where there are 90 of one species and 10 of the other, $P_1 = 90/100 = 0.9$ and $P_2 = 10/100 = 0.1$, with

$$H = -[0.9 \ln 0.9 + 0.1 \ln 0.1].$$

Using a calculator, we find

$$\ln 0.9 \approx -0.1054, \qquad \text{and} \qquad \ln 0.1 \approx -2.3026.$$

---

*Ludwig, John, and James Reynolds, *Statistical Ecology: A Primer on Methods and Computing,* New York, Wiley, 1988, p. 92.

Therefore,

$$H \approx -[(0.9)(-0.1054) + (0.1)(-2.3026)]$$
$$\approx 0.325.$$

Verify that $H \approx 0.673$ if there are 60 of one species and 40 of the other. As the proportions of n species get closer to $1/n$ each, the index of diversity increases to a maximum of $\ln n$.

2.2 EXERCISES

Write each exponential equation in logarithmic form.

1. $2^3 = 8$

2. $5^2 = 25$

3. $3^4 = 81$

4. $6^3 = 216$

5. $3^{-2} = \dfrac{1}{9}$

6. $\left(\dfrac{4}{3}\right)^{-2} = \dfrac{9}{16}$

Write each logarithmic equation in exponential form.

7. $\log_2 128 = 7$

8. $\log_3 81 = 4$

9. $\log_{25} \dfrac{1}{25} = -1$

10. $\log_2 \dfrac{1}{8} = -3$

11. $\log 10{,}000 = 4$

12. $\log 0.00001 = -5$

Evaluate each logarithm without using a calculator.

13. $\log_5 25$

14. $\log_9 81$

15. $\log_4 64$

16. $\log_6 216$

17. $\log_2 \dfrac{1}{4}$

18. $\log_3 \dfrac{1}{27}$

19. $\log_2 \sqrt[3]{\dfrac{1}{4}}$

20. $\log_8 \sqrt[4]{\dfrac{1}{2}}$

21. $\ln e$

22. $\ln e^2$

23. $\ln e^{5/3}$

24. $\ln 1$

25. Is the "logarithm to the base 3 of 4" written as $\log_4 3$ or $\log_3 4$?

26. Write a few sentences describing the relationship between e^x and $\ln x$.

Use the properties of logarithms to write each expression as a sum, difference, or product.

27. $\log_9(7m)$

28. $\log_5(8p)$

29. $\log_3 \dfrac{3p}{5k}$

30. $\log_7 \dfrac{11p}{13y}$

31. $\log_3 \dfrac{5\sqrt{2}}{\sqrt[3]{7}}$

32. $\log_2 \dfrac{9\sqrt[3]{5}}{\sqrt{3}}$

Suppose $\log_b 2 = a$ and $\log_b 3 = c$. Use the properties of logarithms to find each of the following.

33. $\log_b 8$

34. $\log_b 24$

35. $\log_b(72b)$

36. $\log_b(4b^2)$

Use natural logarithms to evaluate each logarithm to the nearest hundredth.

37. $\log_5 20$

38. $\log_{12} 170$

39. $\log_{1.2} 0.55$

40. $\log_{2.8} 0.12$

Solve each equation in Exercises 41–59. Round decimal answers to the nearest hundredth.

41. $\log_x 25 = -2$

42. $\log_9 27 = m$

43. $\log_8 4 = z$

44. $\log_y 8 = \dfrac{3}{4}$

45. $\log_r 7 = \dfrac{1}{2}$

46. $\log_3(5x + 1) = 2$

47. $\log_5(9x - 4) = 1$

48. $\log_4 x - \log_4(x + 3) = -1$

49. $\log_9 m - \log_9(m - 4) = -2$

50. $\log(x + 5) + \log(x + 2) = 1$

51. $\log_3(x - 2) + \log_3(x + 6) = 2$

52. $3^x = 5$

53. $4^x = 12$

54. $e^{k-1} = 4$

55. $e^{2y} = 12$

56. $2e^{5a+12} = 8$

57. $10e^{3z-7} = 5$

58. $5(0.10)^x = 4(0.12)^x$

59. $1.5(1.05)^x = 2(1.01)^x$

Find the domain for each of the following functions.

60. $f(x) = \log(3 - x)$

61. $f(x) = \ln(x^2 - 4)$

62. Lucky Larry was faced with solving

$$\log(2x + 1) - \log(3x - 1) = 0.$$

Larry just dropped the logs and proceeded:

$$(2x + 1) - (3x - 1) = 0$$
$$-x + 2 = 0$$
$$x = 2.$$

Although Lucky Larry is wrong in dropping the logs, his procedure will always give the correct answer to an equation of the form

$$\log A - \log B = 0,$$

where A and B are any two expressions in x. Prove that this last equation leads to the equation $A - B = 0$, which is what you get when you drop the logs.*

63. Prove: $\log_a\left(\dfrac{x}{y}\right) = \log_a x - \log_a y.$

64. Prove: $\log_a x^r = r \log_a x.$

◼ Applications

LIFE SCIENCES

65. *Population Growth* Suppose a population grows at each of the following annual rates. Find the time it would take for the population to double.

   a. 3%   **b.** 6%   **c.** 8%

66. *Population Growth* Suppose each spring a population of house sparrows produces a new brood so that the population is 6% larger.

   a. How many years are required for the population to at least double? (Note that the population only increases in size each spring.)

   b. In how many years will the population at least triple?

67. *Population Growth* Two species of squirrels inhabit a national park. There are currently 4,500 gray squirrels, whose population increases at a rate of 4% a year. There are only 3,000 black squirrels, but they increase at a rate of 6% a year. In how many years will the black squirrels outnumber the gray squirrels? Use the algebra of logarithms to solve this problem, and support your answer by using a graphing calculator to see where the two population functions intersect.

68. *Tumor Growth* An article on growth rates of cancer gives a formula for the doubling time of a tumor, when the volume of the tumor at two different times is known.[†] In this exercise we will derive the formula.

   a. Suppose at time t_1 the tumor has volume V_1 and at time t_2 it has volume V_2. Put this information into the equation

$$V = V_0(1 + r)^t$$

---

*Based on Lucky Larry #16 by Joan Page, *The AMATYC Review,* Vol. 16, No. 1, Fall 1994, p. 67.
†Jansen, Jeroen C., et al., "Estimation of Growth Rate in Patients with Head and Neck Paragangliomas Influences the Treatment Proposal," *Cancer,* Vol. 88, No. 12, June 15, 2000, pp. 2811–2816.

to get two equations. Then divide the second equation by the first to get

$$\frac{V_2}{V_1} = (1 + r)^{t_2 - t_1}.$$

b. By taking the natural logarithm of the answer to part a and simplifying, show that

$$\ln(1 + r) = \frac{\ln V_2 - \ln V_1}{t_2 - t_1}.$$

c. Put the result from part b into the equation for doubling time given in the text after Example 7 to derive the formula for doubling time given in the paper:

$$t = \frac{(t_2 - t_1) \ln 2}{\ln V_2 - \ln V_1}.$$

d. In the paper, the average follow-up time for a carotid body tumor was 4.5 years, after which the tumor had grown 55% on average. Letting $t_1 = 0$ and $V_2 = 1.55V_1$, find the doubling time for the tumor.

69. *Body Surface Area* A formula used for the body surface area (in square cm) of infants is

$$A = 4.688 w^{0.8168 - 0.0154 \log w},$$

where w is the weight in gm.*† Find the surface area of an infant with each of the following weights.

a. 4,000 gm

b. 8,000 gm

c. Use a graphing calculator to find the weight of an infant with a surface area of 4,000 square cm.

70. *Insect Species* An article in *Science* stated that the number of insect species of a given mass is proportional to $m^{-0.6}$, where m is the mass in grams.‡ A graph accompanying the article shows the common logarithm of the mass on the horizontal axis and the common logarithm of the number of species on the vertical axis. Explain why the graph is a straight line. What is the slope of the line?

Index of Diversity For Exercises 71–73, refer to Example 8.

71. Suppose a sample of a small community shows two species with 50 individuals each.

a. Find the index of diversity H.

b. What is the maximum value of the index of diversity for two species?

c. Does your answer for part a equal ln 2? Explain why.

72. A virgin forest in northwestern Pennsylvania has 4 species of large trees with the following proportions of each: hemlock, 0.521; beech, 0.324; birch, 0.081; and maple, 0.074. Find the index of diversity H.

73. Find the value of the index of diversity for populations with n species and $1/n$ of each if

a. $n = 3$;

b. $n = 4$.

c. Verify that your answers for parts a and b equal ln 3 and ln 4, respectively.

74. *Allometric Growth* The allometric formula is used to describe a wide variety of growth patterns. It says that $y = nx^m$, where x and y are variables, and n and m are constants. For example, the famous biologist J. S. Huxley used this formula to relate the weight of the large claw of the fiddler crab to the weight of the body without the claw.§ Show that if x and y are given by the allometric formula, then $X = \log_b x$, $Y = \log_b y$, and $N = \log_b n$ are related by the linear equation

$$Y = mX + N.$$

75. *Drug Concentration* When a pharmaceutical drug is injected into the bloodstream, its concentration at time t can be approximated by $C(t) = C_0 e^{-kt}$, where C_0 is the concentration at $t = 0$. Suppose the drug is ineffective below a concentration C_1 and harmful above a concentration C_2. Then it can be shown that the drug should be given at intervals of time T, where

$$T = \frac{1}{k} \ln \frac{C_2}{C_1}.\|$$

A certain drug is harmful at a concentration five times the concentration below which it is ineffective. At noon an injection of the drug results in a concentration of 2 mg per

---

*Sharkey, I., et al., "Body Surface Area Estimation in Children Using Weight Alone: Application in Paediatric Oncology," *British Journal of Cancer,* Vol. 85, No. 1, 2001, pp. 23–28.
†Technically, grams are a measure of mass, not weight. Weight is a measure of the force of gravity, which varies with the distance from the center of the earth. For objects on the surface of the earth, weight and mass are often used interchangeably, and we will do so in this text.
‡*Science,* Vol. 284, June 18, 1999, p. 1937.
§Huxley, J. S., *Problems of Relative Growth,* Dover, 1968.
‖Horelick, Brindell, and Sinan Koont, "Applications of Calculus to Medicine: Prescribing Safe and Effective Dosage," *UMAP Module 202,* 1977.

liter of blood. Three hours later the concentration is down to 1 mg per liter. How often should the drug be given?

The graph for Exercise 76* is plotted on a logarithmic scale *where differences between successive measurements are not always the same. Data that do not plot in a linear pattern on the usual Cartesian axes often form a linear pattern when plotted on a logarithmic scale. Notice that on the vertical scale, the distance from 1 to 2 is not the same as the distance from 2 to 3, and so on. This is characteristic of a graph drawn with logarithmic scales.*

76. *Oxygen Consumption* The accompanying graph gives the rate of oxygen consumption for resting guinea pigs of various sizes. This rate is proportional to body mass raised to the power 0.67.

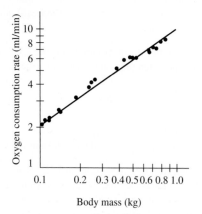

Body mass (kg)

a. Estimate the oxygen consumption for a guinea pig with body mass of 0.3 kg. Do the same for one with body mass of 0.7 kg.

b. Verify that if the relationship between x and y is of the form $y = ax^b$, then there will be a linear relationship between $\ln x$ and $\ln y$. (*Hint:* Apply $\ln$ to both sides of $y = ax^b$.)

c. If a function of the form $y = ax^b$ contains the points (x_1, y_1) and (x_2, y_2), then values for a and b can be found by dividing $y_1 = ax_1^b$ by $y_2 = ax_2^b$, solving the resulting equation for b, and putting the result back into either equation to solve for a. Use this procedure and the results from part a to find an equation of the form $y = ax^b$ that gives the oxygen consumption rate as a function of body mass.

d. Use the result of part c to predict the oxygen consumption of a guinea pig weighing 0.5 kg.

77. *Population Growth* In July 1994 the population of New York state was 18.2 million and increasing at a rate of 0.1% per year. The population of Florida was 14.0 million and increasing at a rate of 1.7% per year.[†] If this trend continues, estimate in what year the population of Florida will exceed the population of New York. Use the algebra of logarithms to solve this problem, and verify your answer by using a graphing calculator or spreadsheet to see where the two population functions intersect.

78. *Toronto's Jewish Population* The table gives the population of Toronto's Jewish community at various times.[‡]

Year	Population
1901	3,103
1911	18,294
1921	34,770
1931	46,751
1941	52,798
1951	66,773
1961	85,000
1971	97,000
1981	128,650
1991	162,605

a. Plot the population on the y-axis against the year on the x-axis. Let x represent the years since 1900. Do the data appear to lie along a straight line?

b. Plot the natural logarithm of the population against the year. Does the graph appear to be more linear than the graph in part a?

c. Find an equation for the least squares line for the data plotted in part b.

d. If your graphing calculator has an exponential regression feature, find the exponential function that best fits the given data according to the least squares method.

e. Take the natural logarithm of the equation found in part d, and verify that the result is the same as the equation found in part c. (In the section on Solutions of Elementary and Separable Differential Equations, we will see another type of function that is a better fit to this data.)

*McMahon, Thomas A., and John Tyler Bonner, *On Size and Life,* Copyright © 1983 by Thomas A. McMahon and John Tyler Bonner. Reprinted by permission of W. H. Freeman and Company.
†U.S. Census Bureau
‡*The Globe and Mail,* Feb. 17, 1995. The data were quoted by Ron Lancaster and Charlie Marion, *The Mathematics Teacher,* Vol. 90, No. 2, Feb. 1997.

OTHER APPLICATIONS

79. *Evolution of Languages* The number of years $N(r)$ since two independently evolving languages split off from a common ancestral language is approximated by

$$N(r) = -5,000 \ln r,$$

where r is the proportion of the words from the ancestral language that are common to both languages now. Find each of the following.

a. $N(0.9)$ **b.** $N(0.5)$ **c.** $N(0.3)$

d. How many years have elapsed since the split if 70% of the words of the ancestral language are common to both languages today?

e. If two languages split off from a common ancestral language about 1,000 years ago, find r.

For Exercises 80–82, recall that log x represents the common (base 10) logarithm of x.

80. *Intensity of Sound* The loudness of sounds is measured in a unit called a *decibel*. To do this, a very faint sound, called the *threshold sound,* is assigned an intensity I_0. If a particular sound has intensity I, then the decibel rating of this louder sound is

$$10 \log \frac{I}{I_0}.$$

Find the decibel ratings of the following sounds having intensities as given. Round answers to the nearest whole number.

a. Whisper, $115I_0$

b. Busy street, $9,500,000I_0$

c. Heavy truck, 20 m away, $1,200,000,000I_0$

d. Rock music concert, $895,000,000,000I_0$

e. Jetliner at takeoff, $109,000,000,000,000I_0$

f. In a noise ordinance instituted in Stamford, Connecticut, the threshold sound I_0 was defined as 0.0002 micro-bars.* Use this definition to express the sound levels in parts c and d in microbars.

81. *Earthquake Intensity* The magnitude of an earthquake, measured on the Richter scale, is given by

$$R(I) = \log \frac{I}{I_0},$$

where I is the amplitude registered on a seismograph located 100 km from the epicenter of the earthquake, and I_0 is the amplitude of a certain small size earthquake. Find the Richter scale ratings of earthquakes with the following amplitudes.

a. $1,000,000I_0$ **b.** $100,000,000I_0$

c. On June 16, 1999, the city of Puebla in central Mexico was shaken by an earthquake that measured 6.7 on the Richter scale. Express this reading in terms of I_0.[†]

d. On September 19, 1985, Mexico's largest recent earthquake, measuring 8.1 on the Richter scale, killed about 9,500 people. Express the magnitude of an 8.1 reading in terms of I_0.[†]

e. Compare your answers to parts c and d. How much greater was the force of the 1985 earthquake than the 1999 earthquake?

f. The relationship between the energy E of an earthquake and the magnitude on the Richter scale is given by

$$R(E) = \frac{2}{3} \log \frac{E}{E_0},$$

where E_0 is the energy of a certain small earthquake. Compare the energies of the 1999 and 1985 earthquakes.

g. According to a newspaper article, "Scientists say such an earthquake of magnitude 7.5 could release 15 times as much energy as the magnitude 6.7 trembler that struck the Northridge section of Los Angeles"[‡] in 1994. Using the formula from part f, verify this quote by computing the magnitude of an earthquake with 15 times the energy of a magnitude 6.7 earthquake.

82. *Acidity of a Solution* A common measure for the acidity of a solution is its pH. It is defined by pH $= -\log[H^+]$, where H^+ measures the concentration of hydrogen ions in the solution. The pH of pure water is 7. Solutions that are more acidic than pure water have a lower pH, while solutions that are less acidic (referred to as basic solutions) have a higher pH.

a. Acid rain sometimes has a pH as low as 4. How much greater is the concentration of hydrogen ions in such rain than in pure water?

b. A typical mixture of laundry soap and water for washing clothes has a pH of about 11, while black coffee has a pH of about 5. How much greater is the concentration of hydrogen ions in black coffee than in the laundry mixture?

---

*The New York Times, June 6, 1999, p. 41.
[†]Times Picayune
[‡]The New York Times, Jan. 13, 1995.

2.3 APPLICATIONS: GROWTH AND DECAY

This is one of many situations that occur in biology, economics, and the social sciences, in which a quantity changes at a rate proportional to the amount of the quantity present. In such cases the amount present at time t is a function of t called the **exponential growth and decay function.** (The derivation of this equation is presented in a later section on Differential Equations.)

> **EXPONENTIAL GROWTH AND DECAY FUNCTION**
>
> Let y_0 be the amount or number of some quantity present at time $t = 0$. Then, under certain conditions, the amount present at any time t is given by
>
> $$y = y_0 e^{kt},$$
>
> where k is a constant.

If $k > 0$, then k is called the **growth constant;** if $k < 0$, then k is called the **decay constant.** A common example is the growth of microorganisms in a culture. The more microorganisms present, the faster the population increases.

EXAMPLE 1   Yeast Production

Yeast in a sugar solution is growing at a rate such that 1 gram becomes 1.5 grams after 20 hours. Find the growth function, assuming exponential growth.

Solution   The values of y_0 and k in the exponential growth function $y = y_0 e^{kt}$ must be found. Since y_0 is the amount present at time $t = 0$, $y_0 = 1$. To find k, substitute $y = 1.5$, $t = 20$, and $y_0 = 1$ into the equation.

$$y = y_0 e^{kt}$$
$$1.5 = 1 e^{k(20)}$$

Now take the natural logarithm of both sides and use the power rule for logarithms and the fact that $\ln e = 1$.

$$1.5 = e^{20k}$$
$$\ln 1.5 = \ln e^{20k}$$
$$\ln 1.5 = 20k \qquad \ln e^x = x$$
$$\frac{\ln 1.5}{20} = k$$
$$k \approx 0.02 \text{ (to the nearest hundredth)}$$

The exponential growth function is $y = e^{0.02t}$, where y is the number of grams of yeast present after t hours.

The decline of a population or decay of a substance may also be described by the exponential growth function. In this case the decay constant k is negative, since an increase in time leads to a decrease in the quantity present. Radioactive substances provide a good example of exponential decay. By definition, the **half-life** of a radioactive substance is the time it takes for exactly half of the initial quantity to decay.

EXAMPLE 2 Carbon Dating

Carbon 14 is a radioactive form of carbon that is found in all living plants and animals. After a plant or animal dies, the carbon 14 disintegrates. Scientists determine the age of the remains by comparing its carbon 14 with the amount found in living plants and animals. The amount of carbon 14 present after t years is given by the exponential equation

$$A(t) = A_0 e^{kt},$$

with $k = -[(\ln 2)/5{,}600]$.

(a) Find the half-life of carbon 14.

Solution Let $A(t) = (1/2)A_0$ and $k = -[(\ln 2)/5{,}600]$.

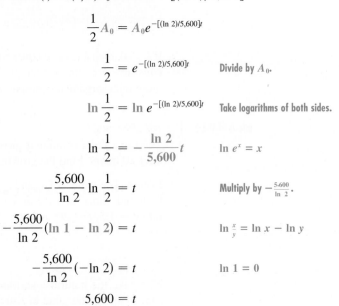

$$\frac{1}{2}A_0 = A_0 e^{-[(\ln 2)/5{,}600]t}$$

$$\frac{1}{2} = e^{-[(\ln 2)/5{,}600]t} \qquad \text{Divide by } A_0.$$

$$\ln \frac{1}{2} = \ln e^{-[(\ln 2)/5{,}600]t} \qquad \text{Take logarithms of both sides.}$$

$$\ln \frac{1}{2} = -\frac{\ln 2}{5{,}600}t \qquad \ln e^x = x$$

$$-\frac{5{,}600}{\ln 2} \ln \frac{1}{2} = t \qquad \text{Multiply by } -\frac{5{,}600}{\ln 2}.$$

$$-\frac{5{,}600}{\ln 2}(\ln 1 - \ln 2) = t \qquad \ln \frac{x}{y} = \ln x - \ln y$$

$$-\frac{5{,}600}{\ln 2}(-\ln 2) = t \qquad \ln 1 = 0$$

$$5{,}600 = t$$

The half-life is 5,600 years.

(b) Charcoal from an ancient fire pit on Java had $1/4$ the amount of carbon 14 found in a living sample of wood of the same size. Estimate the age of the charcoal.

Solution Let $A(t) = (1/4)A_0$ and $k = -[(\ln 2)/5{,}600]$.

$$\frac{1}{4}A_0 = A_0 e^{-[(\ln 2)/5{,}600]t}$$

$$\frac{1}{4} = e^{-[(\ln 2)/5{,}600]t}$$

$$\ln \frac{1}{4} = \ln e^{-[(\ln 2)/5{,}600]t}$$

$$\ln \frac{1}{4} = -\frac{\ln 2}{5{,}600} t$$

$$-\frac{5{,}600}{\ln 2} \ln \frac{1}{4} = t$$

$$t = 11{,}200$$

The charcoal is about 11,200 years old.

By following the steps in Example 2, we get the general equation giving the half-life T in terms of the decay constant k as

$$T = -\frac{\ln 2}{k}.$$

For example, the decay constant for potassium-40, where t is in billions of years, is approximately -0.5545, so its half-life is

$$T = -\frac{\ln 2}{(-0.5545)}$$

$$\approx 1.25 \text{ billion years.}$$

We can rewrite the growth and decay function as

$$y = y_0 e^{kt} = y_0 (e^k)^t = y_0 a^t,$$

where $a = e^k$. This is sometimes a helpful way to look at an exponential growth or decay function.

EXAMPLE 3 Radioactive Decay

Rewrite the function for radioactive decay of carbon 14 in the form $A(t) = A_0 a^{f(t)}$.

Solution From the previous example, we have

$$A(t) = A_0 e^{kt} = A_0 e^{-[(\ln 2)/5{,}600]t}$$

$$= A_0 (e^{\ln 2})^{-t/5{,}600}$$

$$= A_0 2^{-t/5{,}600} = A_0 \left(\frac{1}{2}\right)^{t/5{,}600}.$$

This last expression shows clearly that every time t increases by 5,600 years, the amount of carbon 14 decreases by a factor of $1/2$.

Limited Growth Functions The exponential growth functions discussed so far all continued to grow without bound. More realistically, many populations grow exponentially for a while, but then the growth is slowed by some external constraint that eventually limits the growth. For example, an animal population may grow to the point where its habitat can no longer support the population and the growth rate begins to dwindle until a stable population size is reached. Models that reflect this pattern are called **limited growth functions.** The next example discusses a function of this type.

EXAMPLE 4 Sculpin Growth

A fish in the Bering sea known as the threaded sculpin has been found to grow according to a limited growth function known as a **von Bertalanffy growth curve.*** The length of a male sculpin (in mm) is approximated by

$$L(t) = 165.5(1 - e^{-0.375299(t+0.0704311)}),$$

where t is the age of the fish in years.

(a) What happens to the length of the fish as t gets larger and larger?

Solution As t gets larger, $e^{-0.375299(t+0.0704311)}$ becomes closer to 0, so $L(x)$ approaches 165.5. The limit on the length of a male sculpin is 165.5 mm. Note that this limit represents a horizontal asymptote on the graph of L shown in Figure 10.

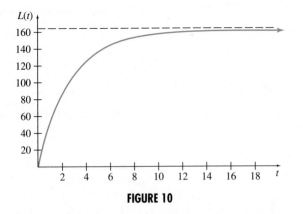

FIGURE 10

(b) How many years will it take for a male sculpin to be 20 times its length when it was hatched?

Solution The length when the fish was hatched is $L(0)$. Using a calculator, we find that $L(0) = 4.3173$, and $20L(0) = 86.346$. Let $L(t) = 86.346$ and solve for t.

$$L(t) = 165.5(1 - e^{-0.375299(t+0.0704311)})$$

$$86.346 = 165.5(1 - e^{-0.375299(t+0.0704311)})$$

$$0.52173 = 1 - e^{-0.375299(t+0.0704311)} \qquad \text{Divide by 165.5.}$$

$$-0.47827 = -e^{-0.375299(t+0.0704311)} \qquad \text{Subtract 1.}$$

$$0.47827 = e^{-0.375299(t+0.0704311)}$$

$$\ln 0.47827 = \ln e^{-0.375299(t+0.0704311)} \qquad \text{Take natural logarithms.}$$

$$-0.73758 = -0.375299(t + 0.0704311) \qquad \ln e^u = u$$

$$1.9653 = t + 0.0704311 \qquad \text{Divide by } -0.375299.$$

$$t \approx 1.89$$

In about 1.89 years the fish will be 20 times as long as when it was hatched.

---

*Hoff, Gerald R., "Biology and Ecology of Threaded Sculpin, *Gymnocanthus Pistilliger,* in the Eastern Bering Sea," *Fishery Bulletin,* Vol. 98, No. 4, Oct. 2000, pp. 711–722.

2.3   EXERCISES

1. In the exponential growth or decay function $y = y_0 e^{kt}$, what does y_0 represent? What does k represent?

2. In the exponential growth or decay function, explain the circumstances that cause k to be positive or negative.

3. What is meant by the half-life of a quantity?

4. Show that if a radioactive substance has a half-life of T, then the corresponding constant k in the exponential decay function is given by $k = -(\ln 2)/T$.

5. Show that if a radioactive substance has a half-life of T, then the corresponding exponential decay function can be written as $y = y_0(1/2)^{(t/T)}$.

Applications

LIFE SCIENCES

6. *Giardia* When a person swallows giardia cysts, stomach acids and pancreatic enzymes cause the cysts to release trophozoites, which divide every 12 hours.*

   a. Suppose the number of trophozoites at time $t = 0$ is y_0. Write a function giving the number after t hours.

   b. The article quoted above said that a single trophozoite can multiply to a million in just 10 days and a billion in 15 days. Verify this fact.

7. *Melanoma* An article on skin cancer said that the number of diagnosed melanoma cases is growing by 4% a year.[†] It also said that the chance of a U.S. resident getting cancer increased from 1 in 250 in 1980 to 1 in 84 in 1997. Is this a 4% increase annually? If not, what percent increase annually is this?

8. *Population Growth* The population of the world in the year 1650 was 470 million, and in the year 1999 was 5,996 million.[‡]

   a. Assuming that the population of the world grows exponentially, find the equation for the population $P(t)$ in millions in the year t.

   b. Use your answer from part a to find the population of the world in the year 1.

   c. Is your answer to part b reasonable? What does this tell you about how the population of the world grows?

9. *Growth of Bacteria* A culture contains 25,000 bacteria, with the population increasing exponentially. The culture contains 40,000 bacteria after 10 hr.

   a. Write an exponential equation to express the growth function y in terms of time t in hours.

   b. How long will it be until there are 60,000 bacteria?

10. *Decrease in Bacteria* When an antibiotic is introduced into a culture of 50,000 bacteria, the number of bacteria decreases exponentially. After 9 hr, there are only 20,000 bacteria.

    a. Write an exponential equation to express the growth function y in terms of time t in hours.

    b. In how many hours will half the number of bacteria remain?

11. *Growth of Bacteria* The growth of bacteria in food products makes it necessary to time-date some products (such as milk) so that they will be sold and consumed before the bacteria count is too high. Suppose for a certain product that the number of bacteria present is given by

$$f(t) = 500e^{0.1t},$$

under certain storage conditions, where t is time in days after packing of the product and the value of $f(t)$ is in millions.

    a. If the product cannot be safely eaten after the bacteria count reaches 3,000 million, how long will this take?

    b. If $t = 0$ corresponds to January 1, what date should be placed on the product?

12. *Cancer Research* An article on cancer treatment contains the following statement: A 37% five-year survival rate for women with ovarian cancer yields an estimated annual mortality rate of 0.1989.[§] The authors of this article

---

*Brody, Jane, "Sly Parasite Menaces Pets and their Owners," *The New York Times,* Dec. 21, 1999, p. F7.

[†]Seppa, N., "Skin Cancer Makes Unexpected Reappearance," *Science News,* June 21, 1997, p. 383.

[‡]*Statistical Abstract of the United States,* 1999

[§]Speroff, Theodore, et al., "A Risk-Benefit Analysis of Elective Bilateral Oophorectomy: Effect of Changes in Compliance with Estrogen Therapy on Outcome," *American Journal of Obstetrics and Gynecology,* Vol. 164, Jan. 1991, pp. 165–74.

assume that the number of survivors is described by the exponential decay function given at the beginning of this section, where y is the number of survivors and k is the mortality rate. Verify that the given survival rate leads to the given mortality rate.

13. *Chromosomal Abnormality* The graph below shows how the risk of chromosomal abnormality in a child rises with the age of the mother.*

   a. Read from the graph the risk of chromosomal abnormality (per 1,000) at ages 20, 35, 42, and 49.

   b. Assuming the graph to be of the form $y = Ce^{kt}$, find k using $t = 20$ and $t = 35$.

   c. Still assuming the graph to be of the form $y = Ce^{kt}$, find k using $t = 42$ and $t = 49$.

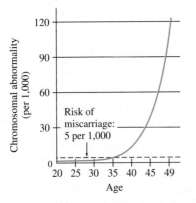

   d. Based on your results from parts a–c, is it reasonable to assume the graph is of the form $y = Ce^{kt}$? Explain.

   e. In situations such as parts a–c, where an exponential function does not fit because different data points give different values for the growth constant k, it is often appropriate to describe the data using an equation of the form $y = Ce^{kt^n}$. Parts b and c show that $n = 1$ results in a smaller constant using the interval $[20, 35]$ than using the interval $[42, 49]$. Repeat parts b and c using $n = 2, 3$, etc., until the interval $[20, 35]$ yields a larger value of k than the interval $[42, 49]$, and then estimate what n should be.

14. *Botany* A group of Tasmanian botanists have claimed that a King's holly shrub, the only one of its species in the world, is also the oldest living plant.[†] Using carbon 14 dating of charcoal found along with fossilized leaf fragments, they arrived at an age of 43,000 years for the plant, whose exact location in southwest Tasmania is being kept a secret. What percent of the original carbon 14 in the charcoal was present?

15. *Radioactive Decay* Potassium 40, with a half-life of 1.25 billion years, has been used by geochronologists trying to sort out the mass extinction of 250 million years ago.[‡] What percent of the potassium 40 remains from a creature that died 250 million years ago?

16. *Decay of Radioactivity* A large cloud of radioactive debris from a nuclear explosion has floated over the Pacific Northwest, contaminating much of the hay supply. Consequently, farmers in the area are concerned that the cows who eat this hay will give contaminated milk. (The tolerance level for radioactive iodine in milk is 0.) The percent of the initial amount of radioactive iodine still present in the hay after t days is approximated by $P(t)$, which is given by the mathematical model

$$P(t) = 100e^{-0.1t}.$$

   a. Find the percent remaining after 4 days.

   b. Find the percent remaining after 10 days.

   c. Some scientists feel that the hay is safe after the percent of radioactive iodine has declined to 10% of the original amount. Solve the equation $10 = 100e^{-0.1t}$ to find the number of days before the hay may be used.

   d. Other scientists believe that the hay is not safe until the level of radioactive iodine has declined to only 1% of the original level. Find the number of days that this would take.

OTHER APPLICATIONS

17. *Carbon Dating* Refer to Example 2. A sample from a refuse deposit near the Strait of Magellan had 60% of the carbon 14 found in a contemporary living sample. How old was the sample?

Half-Life Find the half-life of the following radioactive substances. See Example 2.

18. Plutonium 241; $A(t) = A_0 e^{-0.053t}$

19. Radium 226; $A(t) = A_0 e^{-0.00043t}$

20. *Half-Life* The half-life of plutonium 241 is approximately 13 years.

   a. How much of a sample weighing 2.0 g will remain after 100 years?

   b. How much time is necessary for a sample weighing 2.0 g to decay to 0.1 g?

21. *Half-Life* The half-life of radium 226 is approximately 1,620 years.

*The New York Times, Feb. 5, 1994, p. 24.
†Science, Vol. 277, July 25, 1997, p. 483.
‡Science, Vol. 286, Oct. 29, 1999, p. 883.

a. How much of a sample weighing 2.0 g will remain after 100 years?

b. How much time is necessary for a sample weighing 2.0 g to decay to 0.1 g?

22. *Radioactive Decay* 500 g of iodine 131 is decaying exponentially. After 3 days 386 g of iodine 131 is left.

a. Write an exponential equation to express the decay function y in terms of t in days.

b. Use your answer from part a to find the half-life of iodine 131.

23. *Radioactive Decay* 25.0 g of polonium 210 is decaying exponentially. After 50 days 19.5 g of polonium 210 is left.

a. Write an exponential equation to express the decay function y in terms of t in days.

b. Use your answer from part a to find the half-life of polonium 210.

24. *Nuclear Energy* Nuclear energy derived from radioactive isotopes can be used to supply power to space vehicles. The output of the radioactive power supply for a certain satellite is given by the function $y = 40e^{-0.004t}$, where y is in watts and t is the time in days.

a. How much power will be available at the end of 180 days?

b. How long will it take for the amount of power to be half of its original strength?

c. Will the power ever be completely gone? Explain.

25. *Chemical Dissolution* The amount of chemical that will dissolve in a solution increases exponentially as the temperature is increased. At 0°C 10 g of the chemical dissolves, and at 10°C 11 g dissolves.

a. Write an equation to express the amount of chemical dissolved, y, in terms of temperature, t, in degrees Celsius.

b. At what temperature will 15 g dissolve?

Newton's Law of Cooling *Newton's law of cooling says that the rate at which a body cools is proportional to the difference in temperature between the body and an environment into which it is introduced. This leads to an equation where the temperature $f(t)$ of the body at time t after being introduced into an environment having constant temperature T_0 is*

$$f(t) = T_0 + Ce^{-kt},$$

where C and k are constants. Use this result in Exercises 26–28.

26. Find the temperature of an object when $t = 9$ if $T_0 = 18$, $C = 5$, and $k = 0.6$.

27. If $C = 100$, $k = 0.1$, and t is time in minutes, how long will it take a hot cup of coffee to cool to a temperature of 25° Celsius in a room at 20° Celsius?

28. If $C = -14.6$ and $k = 0.6$ and t is time in hours, how long will it take a frozen pizza to thaw to 10° Celsius in a room at 18° Celsius?

29. *Rent Increase* A newspaper article contained the following quotes: "Next year's . . . legal rent increase has been pegged at 4.9%. Coming on the heels of this year's 6 percent increase and 5.4 percent in 1991, it means rents have gone up 16.3 percent in three years."* Is this figure correct? If not, what is the correct figure?

30. *Electric Rates* In 1987, a New York State Assemblywoman wrote a letter to her constituents about the electric rates of the Long Island Lighting Company (LILCO). She wrote: "If the [Public Service] Commission grants this increase, it will mean that between 1970 and 1987, LILCO's electric rates will have increased by more than 478 percent. That's an average rise of over 27 percent a year!" Is this correct? If not, what annual percent increase over 17 years leads to a total increase of 478 percent?

■ 2.4 TRIGONOMETRIC FUNCTIONS

? THINK ABOUT IT   How can the cortisone concentration in a human be modeled?

In Example 8 of this section, we will answer this question using trigonometry.

The angle is one of the basic concepts of trigonometry. The definition of an angle depends on that of a ray: a **ray** is the portion of a line that starts at a given point and continues indefinitely in one direction. Figure 11 shows a line through the two points A and B. The portion of the line AB that starts at A and continues through and past B is called ray AB. Point A is the **endpoint** of the ray.

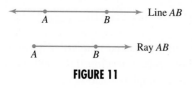

FIGURE 11

Line AB

Ray AB

———————

*Toronto Star, Aug. 2, 1992.

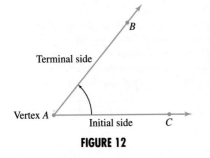

FIGURE 12

An **angle** is formed by rotating a ray about its endpoint. The initial position of the ray is called the **initial side** of the angle, and the endpoint of the ray is called the **vertex** of the angle. The location of the ray at the end of its rotation is called the **terminal side** of the angle. See Figure 12.

An angle can be named by its vertex. For example, the angle in Figure 12 can be called angle *A*. An angle also can be named by using three letters, with the vertex letter in the middle. For example, the angle in Figure 12 could be named angle *BAC* or angle *CAB*.

An angle is in **standard position** if its vertex is at the origin of a coordinate system and if its initial side is along the positive *x*-axis. The angles in Figures 13 and 14 are in standard position. An angle in standard position is said to be in the quadrant of its terminal side. For example, the angle in Figure 13(a) is in quadrant I, while the angle in Figure 13(b) is in quadrant II.

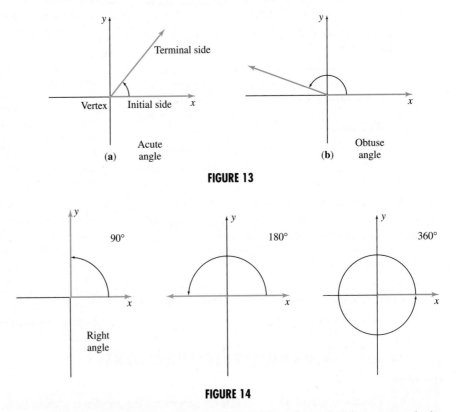

FIGURE 13

FIGURE 14

Notice that the angles in Figures 13 and 14 are formed with a counterclockwise rotation from the positive *x*-axis. This is true for any positive angle. A negative angle is measured clockwise from the positive *x*-axis, as we shall see in Example 5.

Degree Measure   The sizes of angles are often indicated in *degrees*. **Degree measure** has remained unchanged since the Babylonians developed it 4,000 years ago. In degree measure, 360 degrees represents a complete rotation of a ray. *One degree,* written 1°, is 1/360 of a rotation. Also, 90° is 90/360 or 1/4 of a rotation, and 180° is 180/360 or 1/2 of a rotation. See Figure 14.

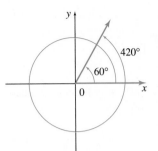

FIGURE 15

An angle having a degree measure between $0°$ and $90°$ is called an **acute angle.** An angle of $90°$ is a **right angle.** An angle having measure more than $90°$ but less than $180°$ is an **obtuse angle,** while an angle of $180°$ is a **straight angle.** See Figures 13 and 14.

A complete rotation of a ray results in an angle of measure $360°$. But there is no reason why the rotation need stop at $360°$. By continuing the rotation, angles of measure larger than $360°$ can be produced. The angles in Figure 15 have measures $60°$ and $420°$. These two angles have the same initial side and the same terminal side, but different amounts of rotation.

Radian Measure While degree measure is best for some applications, this system of angle measurement is not the best for calculus. To keep the formulas for derivatives as simple as possible, it is better to measure angles with **radian measure.** To see how this alternative system for measuring angles is obtained, look at angle θ (the Greek letter *theta*) in Figure 16(a). The angle θ is in standard position; Figure 16(a) also shows a circle of radius r, centered at the origin.

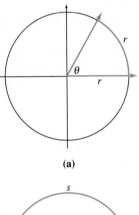

(a)

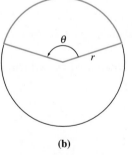

(b)

FIGURE 16

The vertex of θ is at the center of the circle in Figure 16(a). Angle θ cuts a piece of the circle called an **arc.** *One radian is the measure of an angle that has its vertex at the center of a circle and that cuts an arc on the circle equal in length to the radius of the circle.* Thus, the measure of this angle θ is defined to be 1 radian. The term *radian* comes from the phrase *radial angle.* Two nineteenth-century scientists, mathematician Thomas Muir and physicist James Thomson, are credited with the development of the radian as a unit of angular measure.

Generalizing, the radian measure of a central angle θ (see Figure 16(b)) cutting off an arc of length s in a circle of radius r is defined as follows:

$$\text{Radian measure of } \theta = \frac{\text{Length of arc}}{\text{Radius}} = \frac{s}{r}.$$

In this definition, the units of measure of the length of the arc and the radius cancel, leaving a quotient without units. For this reason, a real number can be thought of as the radian measure of some angle.

Since the circumference of a circle is 2π times the radius of the circle, the radius could be marked off 2π times around the circle. Therefore, an angle of $360°$—that is, a complete circle—cuts off an arc equal in length to 2π times the radius of the circle, or

$$360° = 2\pi \text{ radians.}$$

This result gives a basis for comparing degree and radian measure.

Since an angle of $180°$ is half the size of an angle of $360°$, an angle of $180°$ would have half the radian measure of an angle of $360°$, or

$$180° = \frac{1}{2}(2\pi) \text{ radians} = \pi \text{ radians.}$$

$$180° = \pi \text{ radians}$$

Since π radians $= 180°$, divide both sides by π to find the degree measure of 1 radian.

1 RADIAN

$$1 \text{ radian} = \left(\frac{180°}{\pi}\right)$$

This quotient is approximately $57.29578°$. Since $180° = \pi$ radians, we can find the radian measure of $1°$ by dividing by $180°$ on both sides.

1 DEGREE

$$1° = \frac{\pi}{180} \text{ radians}$$

One degree is approximately equal to 0.0174533 radians.

Graphing calculators and many scientific calculators have the capability of changing from degree to radian measure or from radian to degree measure. If your calculator has this capability, you can practice using it with the angle measures in Example 1. *The most important thing to remember when using a calculator to work with angle measures is to be sure the calculator mode is set for degrees or radians, as appropriate.*

EXAMPLE 1   Equivalent Angles
Convert degree measures to radians and radian measures to degrees.

(a) $45°$

Solution Since $1° = \pi/180$ radians,

$$45° = 45\left(\frac{\pi}{180}\right) \text{ radians} = \frac{45\pi}{180} \text{ radians} = \frac{\pi}{4} \text{ radians}.$$

The word *radian* is often omitted, so the answer could be written as just $45° = \pi/4$.

(b) $\dfrac{9\pi}{4}$

Solution Since 1 radian $= 180°/\pi$,

$$\frac{9\pi}{4} \text{ radians} = \frac{9\pi}{4}\left(\frac{180°}{\pi}\right) = 405°.$$

The following chart shows the equivalent radian and degree measure for several angles that we will encounter frequently.

Degrees	0°	30°	45°	60°	90°	180°	270°	360°
Radians	0	$\pi/6$	$\pi/4$	$\pi/3$	$\pi/2$	π	$3\pi/2$	2π

The Trigonometric Functions   To define the six basic trigonometric functions, we start with an angle θ in standard position, as shown in Figure 17. Next, we choose an arbitrary point P having coordinates (x, y), located on the terminal side of angle θ. (The point P must not be the vertex of θ.)

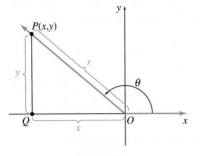

FIGURE 17

Drawing a line segment perpendicular to the x-axis from P to point Q sets up a right triangle having vertices at O (the origin), P, and Q. The distance from P to O is r. Since the distance from P to O can never be negative, $r > 0$. The six **trigonometric functions** of angle θ are defined as follows.

TRIGONOMETRIC FUNCTIONS

Let (x, y) be a point other than the origin on the terminal side of an angle θ in standard position. Let r be the distance from the origin to (x, y). Then

$$\text{sine } \theta = \sin \theta = \frac{y}{r} \qquad\qquad \text{cosecant } \theta = \csc \theta = \frac{r}{y}$$

$$\text{cosine } \theta = \cos \theta = \frac{x}{r} \qquad\qquad \text{secant } \theta = \sec \theta = \frac{r}{x}$$

$$\text{tangent } \theta = \tan \theta = \frac{y}{x} \qquad\qquad \text{cotangent } \theta = \cot \theta = \frac{x}{y}.$$

From these definitions, it is easy to prove the following elementary trigonometric identities.

ELEMENTARY TRIGONOMETRIC IDENTITIES

$$\csc \theta = \frac{1}{\sin \theta} \qquad \sec \theta = \frac{1}{\cos \theta} \qquad \cot \theta = \frac{1}{\tan \theta} \qquad \tan \theta = \frac{\sin \theta}{\cos \theta}$$

$$\cot \theta = \frac{\cos \theta}{\sin \theta} \qquad \sin^2\theta + \cos^2\theta = 1$$

EXAMPLE 2   Values of Trigonometric Functions

The terminal side of an angle α (the Greek letter *alpha*) goes through the point $(8, 15)$. Find the values of the six trigonometric functions of angle α.

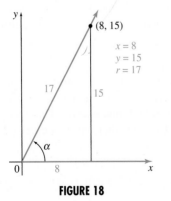

FIGURE 18

Solution Figure 18 shows angle α and the triangle formed by dropping a perpendicular from the point $(8, 15)$. To reach the point $(8, 15)$, begin at the origin and go 8 units to the right and 15 units up, so that $x = 8$ and $y = 15$. To find the radius r, use the Pythagorean theorem*: In a triangle with a right angle, if the longest side of the triangle is r and the shorter sides are x and y, then

$$r^2 = x^2 + y^2,$$

or

$$r = \sqrt{x^2 + y^2}.$$

(Recall that $\sqrt{a}$ represents the *positive* square root of a.)

Substituting the known values $x = 8$ and $y = 15$ in the equation gives

$$r = \sqrt{8^2 + 15^2} = \sqrt{64 + 225} = \sqrt{289} = 17.$$

---

*Although one of the most famous theorems in mathematics is named after the Greek mathematician Pythagoras, there is much evidence that the relationship between the sides of a right triangle was known long before his time. The Babylonian mathematical tablet identified as *Plimpton 322* has been determined to be essentially a list of *Pythagorean triples*—sets of three numbers a, b, and c that satisfy the equation $a^2 + b^2 = c^2$.

We have $x = 8$, $y = 15$, and $r = 17$. The values of the six trigonometric functions of angle α are found by using the definitions given on the previous page.

$$\sin \alpha = \frac{y}{r} = \frac{15}{17} \qquad \tan \alpha = \frac{y}{x} = \frac{15}{8} \qquad \sec \alpha = \frac{r}{x} = \frac{17}{8}$$

$$\cos \alpha = \frac{x}{r} = \frac{8}{17} \qquad \cot \alpha = \frac{x}{y} = \frac{8}{15} \qquad \csc \alpha = \frac{r}{y} = \frac{17}{15}$$

EXAMPLE 3 Values of Trigonometric Functions

Find the values of the six trigonometric functions for an angle of $\pi/2$.

Solution Select any point on the terminal side of an angle of measure $\pi/2$ radians (or 90°). See Figure 19. Selecting the point $(0, 1)$ gives $x = 0$ and $y = 1$. Check that $r = 1$ also. Then

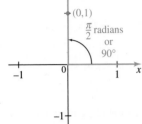

$$\sin \frac{\pi}{2} = \frac{1}{1} = 1 \qquad \tan \frac{\pi}{2} = \frac{1}{0} \text{ (undefined)} \qquad \sec \frac{\pi}{2} = \frac{1}{0} \text{ (undefined)}$$

$$\cos \frac{\pi}{2} = \frac{0}{1} = 0 \qquad \cot \frac{\pi}{2} = \frac{0}{1} = 0 \qquad \csc \frac{\pi}{2} = \frac{1}{1} = 1.$$

FIGURE 19

Methods similar to the procedure in Example 3 can be used to find the values of the six trigonometric functions for the angles with measure 0, π, and $3\pi/2$. These results are summarized in Table 3, which shows that the results for 2π are the same as those for 0.

Table 3

θ (in radians)	θ (in degrees)	$\sin \theta$	$\cos \theta$	$\tan \theta$	$\cot \theta$	$\sec \theta$	$\csc \theta$
0	0°	0	1	0	Undefined	1	Undefined
$\pi/2$	90°	1	0	Undefined	0	Undefined	1
π	180°	0	-1	0	Undefined	-1	Undefined
$3\pi/2$	270°	-1	0	Undefined	0	Undefined	-1
2π	360°	0	1	0	Undefined	1	Undefined

Special Angles The values of the trigonometric functions for most angles must be found by using a calculator with trigonometric keys. For a few angles called **special angles,** however, the function values can be found exactly. These values are found with the aid of two kinds of right triangles that will be described in this section.

30°–60°–90° TRIANGLE

In a right triangle having angles of 30°, 60°, and 90°, the hypotenuse is always twice as long as the shortest side, and the middle side has a length that is $\sqrt{3}$ times as long as that of the shortest side. Also, the shortest side is opposite the 30° angle.

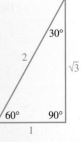

EXAMPLE 4 Values of Trigonometric Functions

Find the values of the trigonometric functions for an angle of $\pi/6$ radians.

Solution Since $\pi/6$ radians $= 30°$, find the necessary values by placing a $30°$ angle in standard position, as in Figure 20. Choose a point P on the terminal side of the angle so that $r = 2$. From the description of $30°$–$60°$–$90°$ triangles, P will have coordinates $(\sqrt{3}, 1)$, with $x = \sqrt{3}$, $y = 1$, and $r = 2$. Using the definitions of the trigonometric functions gives the following results.

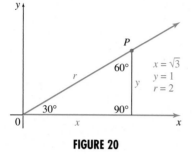

FIGURE 20

$$\sin\frac{\pi}{6} = \frac{1}{2} \qquad \tan\frac{\pi}{6} = \frac{1}{\sqrt{3}} = \frac{\sqrt{3}}{3} \qquad \sec\frac{\pi}{6} = \frac{2}{\sqrt{3}} = \frac{2\sqrt{3}}{3}$$

$$\cos\frac{\pi}{6} = \frac{\sqrt{3}}{2} \qquad \cot\frac{\pi}{6} = \sqrt{3} \qquad \csc\frac{\pi}{6} = 2$$

We can find the trigonometric function values for $45°$ angles by using the properties of a right triangle having two sides of equal length.

$45°$–$45°$–$90°$ TRIANGLE

In a $45°$–$45°$ right triangle, the hypotenuse has a length that is $\sqrt{2}$ times as long as the length of either of the shorter (equal) sides.

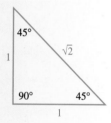

EXAMPLE 5 Values of Trigonometric Functions

Find the trigonometric function values for an angle of $-\pi/4$.

Solution Place an angle of $-\pi/4$ radians, or $-45°$, in standard position, as in Figure 21. Choose point P on the terminal side so that $r = \sqrt{2}$. By the description of $45°$–$45°$–$90°$ triangles, P has coordinates $(1, -1)$, with $x = 1$, $y = -1$, and $r = \sqrt{2}$.

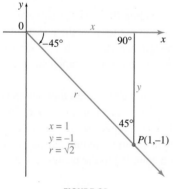

FIGURE 21

$$\sin\left(-\frac{\pi}{4}\right) = -\frac{1}{\sqrt{2}} = -\frac{\sqrt{2}}{2} \qquad \tan\left(-\frac{\pi}{4}\right) = -1 \qquad \sec\left(-\frac{\pi}{4}\right) = \sqrt{2}$$

$$\cos\left(-\frac{\pi}{4}\right) = \frac{1}{\sqrt{2}} = \frac{\sqrt{2}}{2} \qquad \cot\left(-\frac{\pi}{4}\right) = -1 \qquad \csc\left(-\frac{\pi}{4}\right) = -\sqrt{2}$$

For angles other than the special angles of $30°$, $45°$, $60°$, and their multiples, a calculator should be used. Many calculators have keys labeled sin, cos, and tan. To get the other trigonometric functions, use the fact that $\sec x = 1/\cos x$, $\csc x = 1/\sin x$, and $\cot x = 1/\tan x$. (The x^{-1} key is also useful here.)

CAUTION Whenever you use a calculator to compute trigonometric functions, check whether the calculator is set on radians or degrees. If you want one and your calculator is set on the other, you will get erroneous answers. Most calculators have a way of switching back and forth; check the calculator manual for details. On the TI-83, press the MODE button and then select Radian or Degree.

EXAMPLE 6 Values of Trigonometric Functions

Use a calculator to verify the following results.

(a) $\sin 10° \approx 0.1736$

(b) $\cos 48° \approx 0.6691$

(c) $\tan 82° \approx 7.1154$

(d) $\sin 0.2618 \approx 0.2588$

(e) $\cot 1.2043 = 1/\tan 1.2043 \approx 1/2.6053 \approx 0.3838$

(f) $\sec 0.7679 = 1/\cos 0.7679 \approx 1/0.71937 \approx 1.3901$

Graphs of the Trigonometric Functions   Because of the way the trigono-metric functions are defined (using a circle), the same function values will be obtained for any two angles that differ by 2π radians (or 360°). For example,

$$\sin(x + 2\pi) = \sin x \qquad \text{and} \qquad \cos(x + 2\pi) = \cos x$$

for any value of x. Because of this property, the trigonometric functions are *periodic functions*.

PERIODIC FUNCTION

A function $y = f(x)$ is **periodic** if there exists a positive real number a such that

$$f(x) = f(x + a)$$

for all values of x in the domain of the function. The smallest positive value of a is called the **period** of the function.*

When graphing, it is customary to use x (rather than θ) for the domain elements, as we did with earlier functions, and to write $y = \sin x$ instead of $y = \sin \theta$. Because sine is periodic, with period 2π, we can graph $y = \sin x$ by finding y for values of x between 0 and 2π. This portion of the graph can then be repeated as many times as necessary.

Think of a point moving counterclockwise around a circle, tracing out an arc for angle x. The value of $\sin x$ gradually increases from 0 to 1 as x increases from 0 to $\pi/2$. The value of $\sin x$ then decreases back to 0 as x goes from $\pi/2$ to π. For $\pi < x < 2\pi$, $\sin x$ is negative. A few typical values from these intervals are given in the following table, where decimals have been rounded to the nearest tenth.

x	0	$\pi/4$	$\pi/2$	$3\pi/4$	π	$5\pi/4$	$3\pi/2$	$7\pi/4$	2π
$\sin x$	0	0.7	1	0.7	0	−0.7	−1	−0.7	0

Plotting the points from the table of values and connecting them with a smooth curve gives the solid portion of the graph in Figure 22 on the next page. Since $y = \sin x$ is periodic, the graph continues in both directions indefinitely, as suggested by the dashed lines.

---

*Some authors define a period of a function as any value of a that satisfies $f(x) = f(x + a)$.

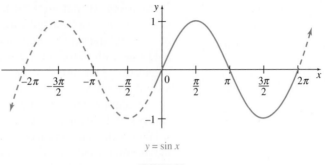

y = sin x

FIGURE 22

The graph of $y = \cos x$ in Figure 23 can be found in much the same way. Again, the period is 2π. (These graphs could also be drawn using a graphing calculator or a computer.)

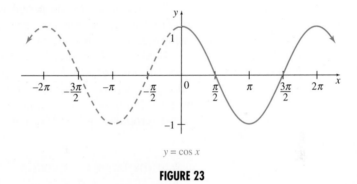

y = cos x

FIGURE 23

Finally, Figure 24 shows the graph of $y = \tan x$. Since $\tan x$ is undefined (because of zero denominators) for $x = \pi/2, 3\pi/2, -\pi/2$, and so on, the graph has vertical asymptotes at these values. As the graph suggests, the tangent function is periodic, with a period of π.

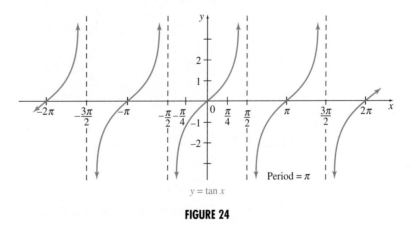

y = tan x

FIGURE 24

The graphs of the secant, cosecant, and cotangent functions are not used as often as these three, so they are not given here.

Translating Graphs of Sine and Cosine Functions In an earlier section we saw that the graph of the function $y = f(x - c) + d$ was simply the graph of $y = f(x)$ translated c units horizontally and d units vertically. The same facts hold true with trigonometric functions. The constants a, b, c, and d affect the graphs of the functions $y = a \sin(bx - c) + d$ and $y = a \cos(bx - c) + d$ in a similar manner.

In addition, the constants a, b, and c have particular properties. Since the sine and cosine functions range between -1 and 1, the value of a, whose absolute value is called the **amplitude,** can be interpreted as half the difference between the maximum and minimum values of the function.

The period of the function is determined by the constant b, which we will assume to be greater than 0. Recall that the period of both $\sin x$ and $\cos x$ is 2π. The value of $b > 0$ will increase or decrease the period, depending on its value. A similar phenomenon occurs when $b < 0$, but it is not covered in this textbook. Thus, the graph of $y = \sin(bx)$ will look like that of $y = \sin x$, but with a period of $2\pi/b$. The results are similar for $y = \cos(bx)$.

The reciprocal of the period is the **frequency.** That is, the frequency is $b/(2\pi)$. The period tells how long one cycle is, while the frequency tells how many cycles occur per unit of x.

Because $y = a \sin(bx - c) = a \sin[b(x - c/b)]$, the quantity c/b determines the number of units that the graph of $\sin bx$ (or $\cos bx$) is shifted horizontally. The quantity c/b is known as the **phase shift.** The constant d determines the vertical shift of $\sin x$ or $\cos x$. In the life sciences, c/d and d are often referred to as the **acrophase** and the **MESOR** (for Midline-Estimating Statistic of Rhythm).

TRANSFORMATION OF TRIGONOMETRIC FUNCTIONS

For the function $y = a \sin(bx - c) + d$ or $y = a \cos(bx - c) + d$,

$$a = \text{the amplitude,}$$

$$\frac{2\pi}{b} = \text{the period,}$$

$$\frac{b}{2\pi} = \text{the frequency,}$$

$$\frac{c}{b} = \text{the phase shift, or acrophase,}$$

$$d = \text{the vertical shift, or MESOR.}$$

EXAMPLE 7 Graphing Trigonometric Functions
Graph each function.

(a) $y = \sin 3x$

Solution The graph of this function has amplitude $a = 1$ and no vertical or horizontal shifts. The period of this function is $T = 2\pi/3$. Hence, the graph of $y = \sin 3x$ is the same as $y = \sin x$ except that the period is different. See Figure 25.

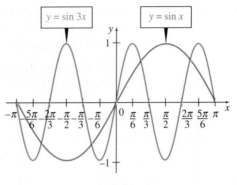

FIGURE 25

(b) $f(x) = 4 \cos\left(\dfrac{1}{2}x + \dfrac{\pi}{4}\right) - 1$

The amplitude is $a = 4$. The graph of $f(x)$ is shifted down 1 unit vertically. Since $\cos[(1/2)x + \pi/4] = \cos[(1/2)x - (-\pi/4)]$, the phase shift is $c/b = (-\pi/4)/(1/2) = -\pi/2$. This shifts the graph $\pi/2$ units to the left, relative to the graph $g(x) = \cos[(1/2)x]$. The period of $f(x)$ is $2\pi/(1/2) = 4\pi$. Making these translations on $y = \cos x$ leads to Figure 26.

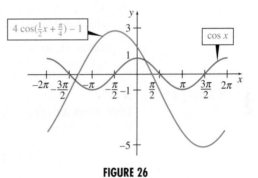

FIGURE 26

EXAMPLE 8   Cortisone Concentration

Researchers have found that the average baseline cortisone concentration in volunteers over a 24-hour period could be modeled by a sum of trigonometric functions.* They used the functions

$$R_1(t) = 67.75 + 47.72 \cos\left(\frac{\pi t}{12}\right) + 11.79 \sin\left(\frac{\pi t}{12}\right)$$

$$R_2(t) = R_1(t) + 14.29 \cos\left(\frac{\pi t}{6}\right) - 21.09 \sin\left(\frac{\pi t}{6}\right)$$

$$R_3(t) = R_2(t) - 2.86 \cos\left(\frac{\pi t}{4}\right) - 14.31 \sin\left(\frac{\pi t}{4}\right),$$

*Chakraborty, Abhijit, et al., "Mathematical Modeling of Circadian Cortisol Concentrations Using Indirect Response Models: Comparison of Several Methods," *Journal of Pharmocokinetics and Biopharmaceutics,* Vol. 27, No. 1, 1999, pp. 23–43.

where each function approximates the cortisol concentration in nanograms per milliliter, and t is the time in hours. Notice that all three functions are periodic over a 24-hour period.

(a) Plot each of these functions on the window $0 \leq t \leq 24$, $0 \leq y \leq 160$. What happens as the function contains more terms?

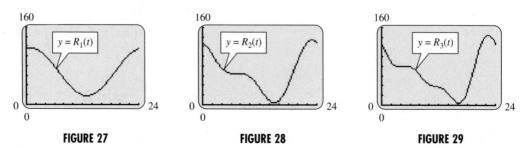

FIGURE 27   FIGURE 28   FIGURE 29

Solution The function R_1, shown in Figure 27, is a simple sinusoidal function. Figures 28 and 29 show the functions R_2 and R_3. Notice that as more terms are added, the function has greater complexity. By adding more trigonometric terms, periodic functions of arbitrary complexity can be approximated. The method is known as *Fourier analysis* and is covered in more advanced mathematics courses.

(b) The actual concentration at 12 hours is about 42 nanograms per milliliter. What value does each of the functions give? Which is most accurate?

Solution Using a calculator, we find that $R_1(12) = 20.0$, $R_2(12) = 34.3$, and $R_3(12) = 37.2$. The function R_3 gives the most accurate value.

EXAMPLE 9 Sunrise

Table 4 lists the approximate number of minutes after midnight, Eastern Standard Time, that the sun rises in Boston for specific days of the year.*

Table 4

Day of the Year	Sunrise (minutes after midnight)
21	428
52	393
81	345
112	293
142	257
173	247
203	266
234	298
265	331
295	365
326	403
356	431

*Thomas, Robert, *The Old Farmer's Almanac*, 2000.

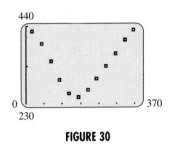

FIGURE 30

(a) Plot the data. Is it reasonable to assume that the times of sunrise are periodic?

Solution Figure 30 shows a graphing calculator plot of the data. Because of the cyclical nature of the days of the year, it is reasonable to assume that the data are periodic.

(b) Find a trigonometric function of the form $s(x) = a \sin(bx + c) + d$ that models this data when x is the day of the year and $s(x)$ is the number of minutes past midnight, Eastern Standard Time, that the sun rises. Use the data from Table 3.

Solution The function $s(x)$, derived by a TI-83 using the sine regression function under the STAT-CALC menu, is given by

$$s(x) = 92.1414 \sin(0.016297x + 1.80979) + 342.934.$$

Figure 31 shows that this function fits the data well.

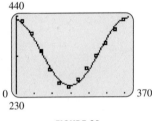

FIGURE 31

(c) Estimate the time of sunrise for days 30, 90, and 240. Round answers to the nearest minute.

Solution

$$s(30) = 92.1414 \sin(0.016297(30) + 1.80979) + 342.934$$
$$= 412 \text{ minutes} = 412/60 \text{ hours} = 6.867 \text{ hours}$$
$$= 6 \text{ hours} + 0.867(60) \text{ minutes} = 6:52 \text{ A.M.}$$

Similarly,

$$s(90) = 333 \text{ minutes} = 5:33 \text{ A.M.}$$
$$s(240) = 288 \text{ minutes} + 60 \text{ minutes (daylight savings)}$$
$$= 5:48 \text{ A.M.}$$

(d) Estimate the days of the year that the sun rises at 5:45 A.M.

Solution Figure 32 shows the graphs of $s(x)$ and $y = 345$ (corresponding to a sunrise of 5:45 A.M.). These graphs first intersect on day 80. However, because of daylight savings time, to find the second value we find where the graphs of $s(x)$ and $y = 345 - 60 = 285$ intersect. These graphs intersect on day 233. Thus, the sun rises at approximately 5:45 A.M. on the 80th and 233rd days of the year.

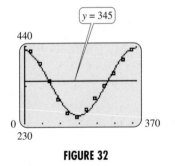

FIGURE 32

2.4 EXERCISES

Convert the following degree measures to radians. Leave answers as multiples of π.

1. 60°   **2.** 90°   **3.** 150°   **4.** 135°

5. 210°   **6.** 300°   **7.** 390°   **8.** 480°

Convert the following radian measures to degrees.

9. $\dfrac{7\pi}{4}$ **10.** $\dfrac{2\pi}{3}$ **11.** $\dfrac{11\pi}{6}$ **12.** $-\dfrac{\pi}{4}$

13. $\dfrac{8\pi}{5}$ **14.** $\dfrac{7\pi}{10}$ **15.** $\dfrac{4\pi}{15}$ **16.** 5π

Find the values of the six trigonometric functions for the angles in standard position having the points in Exercises 17–20 on their terminal sides.

17. $(-3, 4)$ **18.** $(-12, -5)$ **19.** $(6, 8)$ **20.** $(-7, 24)$

In quadrant I, x, y, and r are all positive, so that all six trigonometric functions have positive values. In quadrant II, x is negative and y is positive (r is always positive). Thus, in quadrant II, sine is positive, cosine is negative, and so on. For Exercises 21–24, complete the following table of values for the signs of the trigonometric functions.

	Quadrant of θ	$\sin\theta$	$\cos\theta$	$\tan\theta$	$\cot\theta$	$\sec\theta$	$\csc\theta$
21.	I	+					
22.	II						
23.	III						
24.	IV						

*For Exercises 25–32, complete the following chart. Use the 30°–60°–90° and 45°–45°–90° triangles. Do **not** use a calculator.*

	θ	$\sin\theta$	$\cos\theta$	$\tan\theta$	$\cot\theta$	$\sec\theta$	$\csc\theta$
25.	30°	$1/2$	$\sqrt{3}/2$			$2\sqrt{3}/3$	
26.	45°			1	1		
27.	60°		$1/2$	$\sqrt{3}$		2	
28.	120°	$\sqrt{3}/2$		$-\sqrt{3}$			$2\sqrt{3}/3$
29.	135°	$\sqrt{2}/2$	$-\sqrt{2}/2$			$-\sqrt{2}$	$\sqrt{2}$
30.	150°		$-\sqrt{3}/2$	$-\sqrt{3}/3$			2
31.	210°	$-1/2$		$\sqrt{3}/3$	$\sqrt{3}$		-2
32.	240°	$-\sqrt{3}/2$	$-1/2$			-2	$-2\sqrt{3}/3$

Find the following function values without using a calculator.

33. $\sin\dfrac{\pi}{3}$ **34.** $\cos\dfrac{\pi}{6}$ **35.** $\tan\dfrac{\pi}{4}$ **36.** $\cot\dfrac{\pi}{3}$

37. $\sec\dfrac{\pi}{6}$ **38.** $\sin\dfrac{\pi}{2}$ **39.** $\cos 3\pi$ **40.** $\sec\pi$

41. $\sin\dfrac{4\pi}{3}$ **42.** $\tan\dfrac{3\pi}{4}$ **43.** $\csc\dfrac{5\pi}{4}$ **44.** $\cos 5\pi$

45. $\tan-\dfrac{\pi}{3}$ **46.** $\cot-\dfrac{2\pi}{3}$ **47.** $\sin-\dfrac{7\pi}{6}$ **48.** $\cos-\dfrac{\pi}{6}$

Use a calculator to find the following function values.

49. $\sin 39°$ **50.** $\cos 58°$ **51.** $\tan 82°$ **52.** $\tan 54°$

53. $\sin 0.4014$ **54.** $\tan 1.0123$ **55.** $\cos 1.4137$ **56.** $\sin 1.5359$

Find the amplitude (a) and period (T) of each of the following functions.

57. $f(x) = \cos(3x)$ **58.** $g(t) = 5\sin\left(\dfrac{\pi}{6}t - 2\right)$ **59.** $s(x) = 3\sin(880\pi t - 7)$

Graph each function defined as follows over a two-period interval.

60. $y = 2 \sin x$ **61.** $y = 2 \cos x$ **62.** $y = -\sin x$ **63.** $y = -\dfrac{1}{2} \cos x$

64. $y = 3 \cos\left(2x - \dfrac{\pi}{2}\right) + 1$ **65.** $y = 4 \sin\left(\dfrac{1}{2}x + \pi\right) + 2$ **66.** $y = \dfrac{1}{2} \tan x$ **67.** $y = -3 \tan x$

Applications

LIFE SCIENCES

68. *Transylvania Hypothesis* The "Transylvania hypothesis" claims that the full moon has an effect on health-related behavior. A study investigating this effect found a significant relationship between the phase of the moon and the number of general practice consultations nationwide, given by

$$y = 100 + 1.8 \cos\left(\frac{(x - 6)\pi}{14.77}\right),$$

where y is the number of consultations as a percentage of the daily mean and x is the days since the last full moon.[*]

  a. What is the period of this function? What is the significance of this period?

  b. There was a full moon on August 4, 2001. On what day in August 2001 does this formula predict the maximum number of consultations? What percent increase would be predicted for that day?

  c. What does the formula predict for August 17, 2001?

69. *Monkey Eyes* In a study of how monkeys' eyes pursue a moving object, an image was moved sinusoidally through a monkey's field of vision with an amplitude of 2 degrees and a period of 0.350 seconds.[†]

  a. Find an equation giving the position of the image in degrees as a function of time in seconds.

  b. After how many seconds does the image reach its maximum amplitude?

  c. What is the position of the object after 2 seconds?

70. *Alzheimer's Disease* A study on the circadian rhythms of patients with Alzheimer's disease found that the body temperature of patients could be described by a function of the form

$$T = T_0 + a \cos\left(\frac{2\pi(t - k)}{24}\right),$$

where t is the time in hours since midnight.[‡] For the patients without Alzheimer's, the average values of T_0 (the MESOR), a (the amplitude), and k (the acrophase) were 36.91 degrees Celsius, 0.32 degrees Celsius, and 14.92 hours, while for the patients with the disease, the values were 37.29 degrees Celsius, 0.46 degrees Celsius, and 16.37 hours.

  a. Graph the functions giving the temperature for each of the two groups using a graphing calculator. Do these two functions ever cross?

  b. At what time is the temperature highest for the patients without Alzheimer's?

  c. At what time is the temperature highest for the patients with Alzheimer's?

71. *Femoral Angles* The true angle of torsion θ of the femur is given by

$$\tan \theta = (\tan \theta_2)(\cos a - \cot B_2 \sin a),$$

where θ_2 is the measured angle of torsion, a is the angle of abduction, and B_2 is the measured angle of inclination.[§] Find the true angle of torsion for each of the following values of the other angles. (*Hint:* After finding the value of the right-hand side, use the $\tan^{-1}$ button on your calculator to

[*]Neal, R. D., and M. Colledge, "The Effect of the Full Moon on General Practice Consultation Rates," *Family Practice,* Vol. 17, No. 6, Dec. 2000, pp. 472–474.
[†]Churchland, M. M., and S. G. J. Lisberger, "Experimental and Computational Analysis of Monkey Smooth Pursuit Eye Movements," *Journal of Neurophysiology,* Vol. 86, No. 2, Aug. 2001, pp. 741–759.
[‡]Volicer, L., et al., "Sundowning and Circadian Rhythms in Alzheimer's Disease," *American Journal of Psychiatry,* Vol. 158, No. 5, May 2001, pp. 704–11.
[§]Hensinger, Robert N., *Standards in Pediatric Orthopedics,* Raven Press, 1986, p. 51.

find the value of θ. Remember to set your calculator on degrees for this exercise.)

a. $\theta_2 = 23°$, $a = 10°$, $B_2 = 170°$

b. $\theta_2 = 20°$, $a = 10°$, $B_2 = 160°$

72. Air Pollution The amount of pollution in the air fluctuates with the seasons. It is lower after heavy spring rains and higher after periods of little rain. In addition to this seasonal fluctuation, the long-term trend is upward. An idealized graph of this situation is shown in the figure. Trigonometric functions can be used to describe the fluctuating part of the pollution levels. Powers of the number e can be used to show the long-term growth. In fact, the pollution level in a certain area might be given by

$$P(t) = 7(1 - \cos 2\pi t)(t + 10) + 100e^{0.2t},$$

where t is time in years, with $t = 0$ representing January 1 of the base year. Thus, July 1 of the same year would be represented by $t = 0.5$, while October 1 of the following year would be represented by $t = 1.75$. Find the pollution levels on the following dates.

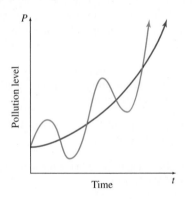

a. January 1, base year

b. July 1, base year

c. January 1, following year

d. July 1, following year

73. Air Pollution Using a computer or a graphing calculator, sketch the function for air pollution given in Exercise 72 over the interval $[0, 6]$.

OTHER APPLICATIONS

Light Rays *When a light ray travels from one medium, such as air, to another medium, such as water or glass, the speed of the light changes, and the direction that the ray is traveling changes. (This is why a fish under water is in a different* *position from the place at which it appears to be.) These changes are given by Snell's law,*

$$\frac{c_1}{c_2} = \frac{\sin \theta_1}{\sin \theta_2},$$

where c_1 is the speed in the first medium, c_2 is the speed in the second medium, and θ_1 and θ_2 are the angles shown in the figure.

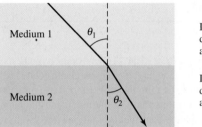

In Exercises 74 and 75, assume that $c_1 = 3 \times 10^8$ meters per second, and find the speed of light in the second medium.

74. $\theta_1 = 39°$, $\theta_2 = 28°$

75. $\theta_1 = 46°$, $\theta_2 = 31°$

Sound *Pure sounds produce single sine waves on an oscilloscope. Find the period of each sine wave in the photographs in Exercises 76 and 77. On the vertical scale each square represents 0.5, and on the horizontal scale each square represents 30°.*

76.

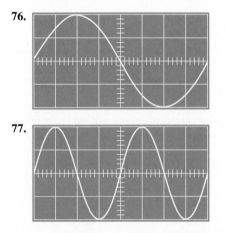

77.

78. Temperature The maximum afternoon temperature (in degrees Fahrenheit) in a given city is approximated by

$$T(x) = 60 - 30 \cos(x/2),$$

where x represents the month, with $x = 0$ representing January, $x = 1$ representing February, and so on. Use a calculator to find the maximum afternoon temperature for each of the following months.

a. January   **b.** March   **c.** October

d. June   **e.** August

79. *Temperature* A mathematical model for the temperature in Fairbanks is

$$T(x) = 37 \sin\left[\frac{2\pi}{365}(x - 101)\right] + 25,$$

where $T(x)$ is the temperature (in degrees Celsius) on day x, with $x = 0$ corresponding to January 1 and $x = 365$ corresponding to December 31.*

Use a calculator to estimate the temperature on each of the following days for a–d.

a. March 1 (Day 60)   **b.** April 1 (Day 91)

c. Day 101   **d.** Day 150

e. Find maximum and minimum values of T.

f. Find the period, T.

80. *Sunset* The number of minutes after noon, Eastern Standard Time, that the sun sets in Boston for specific days of the year is approximated in the following table.[†]

Day of the Year	Sunset (minutes after noon)
21	283
52	323
81	358
112	393
142	425
173	445
203	434
234	396
265	343
295	292
326	257
356	255

a. Plot the data. Is it reasonable to assume that the times of sunset are periodic?

b. Use a calculator with trigonometric regression to find a trigonometric function of the form $s(x) = a \sin(bx + c) + d$ that models this data when x is the day of the year and $s(x)$ is the number of minutes past noon, Eastern Standard Time, that the sun sets.

c. Estimate the time of sunset for days 60, 120, 240. Round answers to the nearest minute. (*Hint:* Don't forget about daylight savings time.)

d. Use part b to estimate the days of the year that the sun sets at 6:00 P.M. In reality, the days are close to 82 and 290.

81. *Measurement* A surveyor standing 48 m from the base of a building measures the angle to the top of the building and finds it to be 37.4°. (See the figure.) Use trigonometry to find the height of the building.

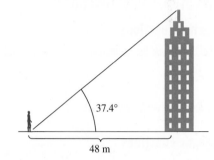

82. *Measurement* Elizabeth Linton stands on a cliff at the edge of a canyon. On the opposite side of the canyon is another cliff equal in height to the one she is on. (See the figure below.) By dropping a rock and timing its fall, she determines that it is 80 ft to the bottom of the canyon. She also determines that the angle to the base of the opposite cliff is 24°. How far is it to the opposite side of the canyon?

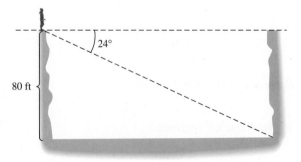

---

*Lando, Barbara, and Clifton Lando, "Is the Graph of Temperature Variation a Sine Curve?" *The Mathematics Teacher,* Vol. 70, Sept. 1977, pp. 534–37.
[†]Thomas, Robert, *The Old Farmer's Almanac,* 2000.

In this chapter we have studied three more families of functions: exponential, logarithmic, and trigonometric. When added to the repertoire of functions in the last chapter, they give us a substantial number of mathematical tools for analyzing real-life applications. We now know when the use of one function is more appropriate than another, and we can predict how the dependent variable will react to changes in the independent variable. In particular, we have applied this knowledge to the study of growth and decay. In the next chapters, we will build on this knowledge.

SUMMARY OF IMPORTANT FUNCTIONS

Exponential Functions

Exponential Growth and Decay Function   If y_0 is the number present at time $t = 0$,

$$y = y_0 e^{kt}.$$

Change-of-Base Theorem for Exponentials $a^x = e^{(\ln a)x}$

Logarithmic Functions

Properties of Logarithms   Let x and y be any positive real numbers and r be any real number. Let a be a positive real number, $a \neq 1$. Then

(a) $\log_a xy = \log_a x + \log_a y$

(b) $\log_a \dfrac{x}{y} = \log_a x - \log_a y$

(c) $\log_a x^r = r \log_a x$

(d) $\log_a a = 1$

(e) $\log_a 1 = 0$

(f) $\log_a a^r = r$.

Change-of-Base Theorem for Logarithms $\log_a x = \dfrac{\log_b x}{\log_b a} = \dfrac{\ln x}{\ln a}$

Trigonometric Functions

Let (x, y) be a point other than the origin on the terminal side of an angle θ in standard position. Let r be the distance from the origin to (x, y). Then

$$\sin \theta = \frac{y}{r} \qquad\qquad \cos \theta = \frac{x}{r} \qquad\qquad \tan \theta = \frac{y}{x} = \frac{\sin \theta}{\cos \theta}$$

$$\csc \theta = \frac{1}{\sin \theta} = \frac{r}{y} \qquad \sec \theta = \frac{1}{\cos \theta} = \frac{r}{x} \qquad \cot \theta = \frac{x}{y} = \frac{\cos \theta}{\sin \theta}$$

$$\sin^2 \theta + \cos^2 \theta = 1.$$

Graphs of Basic Functions

Quadratic	Absolute Value	Square Root	Rational

$y = x^2$

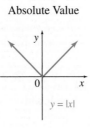

$y = |x|$

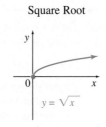

$y = \sqrt{x}$

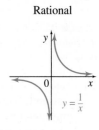

$y = \dfrac{1}{x}$

Exponential	Logarithmic

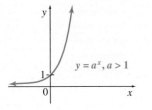

$y = a^x, a > 1$

$y = \log_a x, a > 1$

Trigonometric

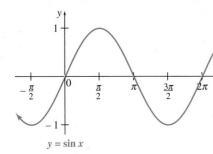

$y = \sin x$

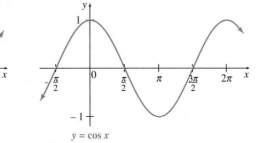

$y = \cos x$

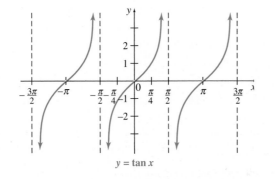

$y = \tan x$

KEY TERMS

To understand the concepts presented in this chapter, you should know the meaning and use of the following terms. For easy reference, the section in the chapter where a term was first used is provided.

2.1 exponential function
exponential equation
simple interest
compound interest
rate of interest
time
principal
compound amount
e
continuous compounding
2.2 doubling time
logarithm
logarithmic function
properties of logarithms
common logarithms

natural logarithms
change-of-base theorem
for logarithms
logarithmic equation
change-of-base theorem
for exponentials
Index of Diversity
2.3 exponential growth and
decay function
growth constant
decay constant
half-life
limited growth function
von Bertalanffy growth
curve

2.4 ray
endpoint
angle
vertex
initial side
terminal side
standard position
degree measure
acute angle
right angle
obtuse angle
straight angle
arc
radian measure
trigonometric functions

sine
cosine
tangent
cotangent
secant
cosecant
special angles
periodic functions
period
amplitude
frequency
phase shift
acrophase
MESOR

CHAPTER 2 REVIEW EXERCISES

1. Describe in words what a logarithm is.

Find the domain of each function defined as follows.

2. $y = \ln(x^2 - 9)$

3. $y = \dfrac{1}{e^x - 1}$

4. $y = \dfrac{1}{\sin x - 1}$

5. $y = \tan x$

Graph each of the following.

6. $y = 5^x$

7. $y = 5^{-x} + 1$

8. $y = \left(\dfrac{1}{5}\right)^{2x-3}$

9. $y = \left(\dfrac{1}{2}\right)^{x-1}$

10. $y = \log_2(x - 1)$

11. $y = 1 + \log_3 x$

12. $y = -\log_3 x$

13. $y = 2 - \ln x^2$

Solve each equation.

14. $2^{3x} = \dfrac{1}{8}$

15. $\left(\dfrac{9}{16}\right)^x = \dfrac{3}{4}$

16. $9^{2y-1} = 27^y$

17. $\dfrac{1}{2} = \left(\dfrac{b}{4}\right)^{1/4}$

Write the following equations using logarithms.

18. $2^6 = 64$

19. $3^{1/2} = \sqrt{3}$

20. $e^{0.09} = 1.09417$

21. $10^{1.07918} = 12$

Write the following equations using exponents.

22. $\log_2 32 = 5$

23. $\log_{10} 100 = 2$

24. $\ln 82.9 = 4.41763$

25. $\log 15.46 = 1.18921$

Evaluate each expression without using a calculator. Then support your work using a calculator and the change-of-base theorem for logarithms.

26. $\log_3 81$

27. $\log_{32} 16$

28. $\log_{25} 5$

29. $\log_{100} 1{,}000$

Simplify each expression using the properties of logarithms.

30. $\log_5 3k + \log_5 7k^3$ **31.** $\log_3 2y^3 - \log_3 8y^2$ **32.** $2 \log_2 x - 3 \log_2 m$ **33.** $5 \log_4 r - 3 \log_4 r^2$

Solve each equation. If necessary, round each answer to the nearest thousandth.

34. $8^p = 19$ **35.** $3^{z-2} = 11$ **36.** $2^{1-m} = 7$ **37.** $15^{-k} = 9$

38. $e^{-5-2x} = 5$ **39.** $e^{3x-1} = 12$ **40.** $\left(1 + \dfrac{m}{3}\right)^5 = 10$ **41.** $\left(1 + \dfrac{2p}{5}\right)^2 = 3$

42. $\log_k 64 = 6$ **43.** $\log_3(2x + 5) = 2$ **44.** $\log(4p + 1) + \log p = \log 3$

45. $\log_2(5m - 2) - \log_2(m + 3) = 2$

46. Give the following properties of the exponential function $f(x) = a^x;\, a > 0, a \neq 1$.

  a. Domain   **b.** Range   **c.** y-intercept

  d. Discontinuities   **e.** Asymptote(s)   **f.** Increasing if a is _____

  g. Decreasing if a is _____

47. Give the following properties of the logarithmic function $f(x) = \log_a x;\, a > 0, a \neq 1$.

  a. Domain   **b.** Range   **c.** x-intercept

  d. Discontinuities   **e.** Asymptote(s)   **f.** Increasing if a is _____

  g. Decreasing if a is _____

48. Compare your answers for Exercises 46 and 47. What similarities do you notice? What differences?

49. What is the relationship between the degree measure and the radian measure of an angle?

50. Under what circumstances should radian measure be used instead of degree measure? Degree measure instead of radian measure?

51. Describe in words how each of the six trigonometric functions is defined.

52. At what angles (given as rational multiples of π) can you determine the exact values for the trigonometric functions?

Convert the following degree measures to radians. Leave answers as multiples of π.

53. $90°$ **54.** $120°$ **55.** $210°$ **56.** $270°$ **57.** $360°$ **58.** $420°$

Convert the following radian measures to degrees.

59. 7π **60.** $\dfrac{3\pi}{4}$ **61.** $\dfrac{9\pi}{20}$ **62.** $\dfrac{7\pi}{15}$ **63.** $\dfrac{13\pi}{20}$ **64.** $\dfrac{11\pi}{15}$

Find each function value without using a calculator.

65. $\sin 60°$ **66.** $\tan 120°$ **67.** $\cos(-30°)$ **68.** $\sec 45°$ **69.** $\csc 120°$

70. $\cot 300°$ **71.** $\sin \dfrac{\pi}{6}$ **72.** $\cos \dfrac{2\pi}{3}$ **73.** $\sec \dfrac{5\pi}{4}$ **74.** $\csc \dfrac{7\pi}{3}$

Find each function value.

75. $\sin 47°$ **76.** $\cos 59°$ **77.** $\tan 81°$ **78.** $\sin(-32°)$

79. $\sin 1.4661$ **80.** $\cos 0.3142$ **81.** $\cos 0.5934$ **82.** $\tan 1.2915$

Graph one period of each function.

83. $y = 3 \cos x$ **84.** $y = \dfrac{1}{2} \tan x$ **85.** $y = -\tan x$ **86.** $y = -2 \sin x$

Applications

LIFE SCIENCES

87. *World Grain Production* World grain harvests have been increasing linearly over the last half century.* The harvest was about 1.1 billion metric tons in 1966 and about 2.1 billion metric tons in 1997.

    a. Write the world grain harvest (y) in billions of metric tons as a function of the years since 1900.

    b. World population in billions can be described by the function

$$p(t) = \frac{9.803}{1 + 27.28e^{-0.03391x}} + 0.99,$$

    where x is the number of years since 1900. Per capita grain production can be described as world grain production divided by the population. Use a graphing calculator to plot world per capita grain production over the last half of the twentieth century. Describe any general trends.

88. *Mutation* Researchers studying the relationship between the generation time of a species and the mutation rate for genes that cause deleterious effects gathered the following data.[†]

Species	Generation Time (x) in Years	Genomic Mutation Rate (y) per Generation
D. melanogaster/D. pseudoobscura	0.1	0.070
D. melanogaster/D. simulans	0.1	0.058
D. picticornis/D. silvestris	0.2	0.071
Mouse/rat	0.5	0.50
Chicken/old world quail	2	0.49
Dog/cat	4	1.6
Sheep/cow	6	0.90
Macaque/New World monkey	11	1.9
Human/chimpanzee	25	3.0

    a. The researchers plotted the common logarithm of y against the common logarithm of x. Do so with a graphing calculator. Do the data follow a linear trend?

    b. Find the least squares line of the common logarithm of y against the common logarithm of x. Plot this line on the same graph as in part a.

    c. Use your answer from part b to find an equation describing y in terms of x.

    d. Find the coefficient of correlation.

89. *Cancer Growth* Cancer cells often have a diameter of about 20 microns, or $2(10)^{-5}$ m.

    a. Although cancer cells are not exactly spheres, they are close enough to spheres to make this simplification. What is the volume of a cancer cell?

    b. Suppose a cancer cell doubles every day. Starting with one cancer cell, find a formula for the total volume of the cancer cells after t days.

    c. Assume that a tumor of cancer cells is roughly a sphere. Also assume that about 74% of the tumor is made up of cancerous cells, while the rest consists of blood vessels and space between the cells. How many days must pass before the single cancer cell grows to a tumor with a diameter of 1 cm (10^{-2} m)?

90. *Polar Bear Mass* One formula for estimating the mass (in kg) of a polar bear is given by

$$m(g) = e^{0.02 + 0.062g - 0.000165g^2},$$

    where g is the axillary girth in cm.[‡] It seems reasonable that as girth increases, so does the mass. What is the largest girth for which this formula gives a reasonable answer? What is the predicted mass of a polar bear with this girth?

91. *Population Growth* A population of 15,000 small deer in a specific region has grown exponentially to 17,000 in 4 years.

    a. Write an exponential equation to express the population growth y in terms of time t in years.

    b. At this rate, how long will it take for the population to reach 45,000?

92. *Intensity of Light* The intensity of light (in appropriate units) passing through water decreases exponentially with the depth it penetrates beneath the surface according to the function

$$I(x) = 10e^{-0.3x},$$

where x is the depth in meters. A certain water plant requires light of an intensity of 1 unit. What is the greatest depth of water in which it will grow?

93. *Drug Concentration* The concentration of a certain drug in the bloodstream at time t (in minutes) is given by

$$c(t) = e^{-t} - e^{-2t}.$$

*Science, Vol. 283, Jan. 15, 1999, p. 310.

[†]Keightley, Peter D., and Adam Eyre-Walker, "Deleterious Mutations, and the Evolution of Sex," Science, Vol. 290, Oct. 13, 2000, p. 331–333.

[‡]Cattet, Marc R. L., et al., "Predicting Body Mass in Polar Bears: Is Morphometry Useful?" Journal of Wildlife Management, Vol. 61, No. 4, 1997, pp. 1083–1090.

Use a graphing calculator to find the maximum concentration and the time when it occurs.

94. *Glucose Concentration* When glucose is infused into a person's bloodstream at a constant rate of c grams per minute, the glucose is converted and removed from the bloodstream at a rate proportional to the amount present. The amount of glucose in the bloodstream at time t (in minutes) is given by

$$g(t) = \frac{c}{a} + \left(g_0 - \frac{c}{a}\right)e^{-at},$$

where a is a positive constant. Assume $g_0 = 0.08$, $c = 0.1$, and $a = 1.3$.

a. At what time is the amount of glucose a maximum? What is the maximum amount of glucose in the bloodstream?

b. When is the amount of glucose in the bloodstream 0.1 gram?

c. What happens to the amount of glucose in the bloodstream after a very long time?

95. *Pollution* The cost to remove x percent of a pollutant is

$$y = \frac{7x}{100 - x},$$

in thousands of dollars. Find the cost of removing each of the following percents of the pollutant.

a. 80%   **b.** 50%   **c.** 90%

d. Graph the function.

e. Can all of the pollutant be removed?

96. *Circadian Testosterone* The following model has been used to describe the circadian rhythms in testosterone concentration:

$$C(t) = a + (C_1 - a)e^{3[\cos(b(t-t_1))-1]} + (C_0 - a)e^{3[\cos(b(t-t_0))-1]}$$

where C is the concentration in nanograms per deciliter, t is the time in hours, and a, b, C_0, C_1, t_0, and t_1 are constants.*

a. The value of b is $2\pi/24$. What is the period of this function?

b. The values of a, C_0, C_1, t_0, and t_1 for healthy young men are estimated to be 546, 511, 634, 20.27, and 6.05, respectively. Graph this function using a graphing calculator.

c. What is the value of $C(t_0)$? What is noteworthy about this answer?

d. What is the value of $C(t_1)$? What is noteworthy about this answer?

e. Is $C(t_0)$ exactly equal to C_0 or $C(t_1)$ exactly equal to C_1? Explain.

97. *Blood Pressure* A person's blood pressure at time t (in seconds) is given by

$$P(t) = 90 + 15 \sin 144\pi t.$$

Find the maximum and minimum values of P on the interval $[0, 1/72]$. Graph one period of $y = P(t)$.

OTHER APPLICATIONS

Interest Find the amount of interest earned by each of the following deposits.

98. $6,902 at 12% compounded semiannually for 8 yr

99. $2,781.36 at 8% compounded quarterly for 6 yr

100. How long will it take for $1,000 deposited at 6% compounded semiannually to double? to triple?

101. How long will it take for $2,100 deposited at 4% compounded quarterly to double? to triple?

Interest Find the compound amount if $12,104 is invested at 8% compounded continuously for each of the following periods.

102. 2 yr   **103.** 4 yr

Interest Find the compound amounts for the following deposits if interest is compounded continuously.

104. $1,500 at 10% for 9 yr

105. $12,000 at 5% for 8 yr

106. *Dating Rocks* Geologists sometimes measure the age of rocks by using "atomic clocks." By measuring the amounts of potassium 40 and argon 40 in a rock, the age t of the specimen (in years) is found with the formula

$$t = (1.26 \times 10^9)\frac{\ln[1 + 8.33(A/K)]}{\ln 2},$$

where A and K, respectively, are the numbers of atoms of argon 40 and potassium 40 in the specimen.

a. How old is a rock in which $A = 0$ and $K > 0$?

b. The ratio A/K for a sample of granite from New Hampshire is 0.212. How old is the sample?

c. Let $A/K = r$. What happens to t as r gets larger? smaller?

107. *Pace of Life* In an attempt to measure how the pace of city life is related to the size of the city, two researchers measured the mean speed of pedestrians in 15 cities by measuring the mean time it took them to walk 50 ft.[†]

---

*Gupat, Suneel K., et al., "Modeling of Circadian Testosterone in Healthy Men and Hypogonadal Men," *Journal of Clinical Pharmacology,* Vol. 40, No. 7, July 2000, pp. 731–738.
[†]*Nature,* Feb. 19, 1976, pp. 557–59.

a. Plot the original pairs of numbers. The pattern should be nonlinear.

b. Compute the coefficient of correlation for the data.

c. Plot y against $\log x$, using a calculator to compute $\log x$. Are the data more linear now than in part a?

d. Compute the coefficient of correlation for y against $\log x$. Is r closer to 1 than in part b?

e. Compute the least squares line for y against $\log x$.

City	Population (x)	Speed (ft/sec) (y)
Brno, Czechoslovakia	341,948	4.81
Prague, Czechoslovakia	1,092,759	5.88
Corte, France	5,491	3.31
Bastia, France	49,375	4.90
Munich, Germany	1,340,000	5.62
Psychro, Crete	365	2.67
Itea, Greece	2,500	2.27
Iráklion, Greece	78,200	3.85
Athens, Greece	867,023	5.21
Safed, Israel	14,000	3.70
Dimona, Israel	23,700	3.27
Netanya, Israel	70,700	4.31
Jerusalem, Israel	304,500	4.42
New Haven, Conn., U.S.A	138,000	4.39
Brooklyn, N.Y., U.S.A	2,602,000	5.05

108. *Temperature* The table lists the average monthly temperatures in Vancouver, Canada.*

Month	Jan	Feb	Mar	Apr	May	June
Temperature	36	39	43	48	55	59

Month	July	Aug	Sept	Oct	Nov	Dec
Temperature	64	63	57	50	43	39

These average temperatures cycle yearly and change only slightly over many years. Because of the repetitive nature of temperatures from year to year, they can be modeled with a sine function. Some graphing calculators have a sine regression feature. If the table is entered into a calculator, the points can be plotted automatically, as shown in the early chapters of this book with other types of functions.

a. Use a graphing calculator to plot the ordered pairs (month, temperature) in the interval $[0, 12]$ by $[30, 70]$.

b. Use a graphing calculator with a sine regression feature to find an equation of the sine function that models this data.

c. Graph the equation from part b.

---

*Miller, A., and J. Thompson, *Elements of Meteorology,* Columbus, Ohio, Charles E. Merrill Publishing Company, 1975.

EXTENDED APPLICATION: Characteristics of the Monkeyface Prickleback*

The monkeyface prickleback (*Cebidichthys violaceus*), known to anglers as the monkeyface "eel," is found in rocky intertidal and subtidal habitats ranging from San Quintin Bay, Baja, California, to Brookings, Oregon. Pricklebacks are prime targets of the few sports anglers who "poke pole" in the rocky intertidal zone at low tide. Little is known about the life history of this species. The results of a study of the length, weight, and age of this species are discussed in this case.

Data on standard length (*SL*) and total length (*TL*) were collected. Early in the study only *TL* was measured, so a conversion to *SL* was necessary. The equation relating the two lengths, calculated from 177 observations for which both lengths had been measured, is

$$SL = TL(0.931) + 1.416.$$

Ages (determined by standard aging techniques) were used to estimate parameters of the von Bertalanffy growth model

$$L_t = L_x(1 - e^{-kt}) \tag{1}$$

where L_t = length at age t,
  L_x = asymptotic age of the species,
  k = growth completion rate, and
  t_0 = theoretical age at zero length.

The constants a and b in the model

$$W = aL^b \tag{2}$$

where W = weight in g,
  L = standard length in cm,

were determined using 139 fish ranging from 27 cm and 145 g to 60 cm and 195 g.

Growth curves giving length as a function of age are shown in Figure 33. For the data marked opercle, the lengths were computed from the ages using equation (1).

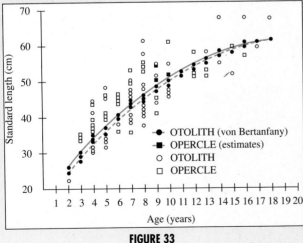

FIGURE 33

Estimated length from equation (1) at a given age was larger for males than females after age eight. See the table. Weight/length relationships found with Equation (2) are shown in Figure 34, along with data from other studies.

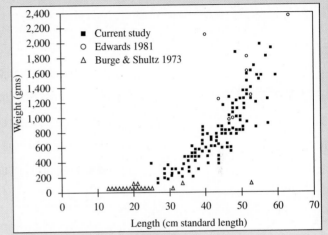

FIGURE 34

Structure/ Sex	Age (yr)	Length (cm)	L_x	k	t_0	n
Otolith						
Est.	2–18	23–67	72	0.10	−1.89	91
S.D.			8	0.03	1.08	
Opercle						
Est.	2–18	23–67	71	0.10	−2.63	91
S.D.			8	0.04	1.31	
Opercle– Females						
Est.	0–18	15–62	62	0.14	−1.95	115
S.D.			2	0.02	0.28	
Opercle– Males						
Est.	0–18	13–67	70	0.12	−1.91	74
S.D.			5	0.02	0.29	

Exercises

1. Use Equation (1) to estimate the lengths at ages 4, 11, and 17. Let $L_x = 71.5$ and $k = 0.1$. Compare your answers with the results in Figure 33. What do you find?

2. Use Equation (2) with $a = 0.01289$ and $b = 2.9$ to estimate the weights for lengths of 25 cm, 40 cm, and 60 cm. Compare with the results in Figure 34. Are your answers reasonable compared to the curve?

*From Marshall, William H., and Tina Wyllie Echeverria, "Characteristics of the Monkeyface Prickleback," *California Fish & Game,* Vol. 78, No. 2, Spring 1992. Copyright 1992, American Association for the Advancement of Science. Reprinted with permission.

CHAPTER

3

The Derivative

Controlling environmental pollution means controlling rates like kilograms of sulfur emitted per day by a power plant or micrograms per liter of pollutant flowing into a river. These rates are examples of derivatives, the subject of this chapter. Rates also govern the economics of pollution, as we will see in an exercise in Section 3 exploring the rate of increase in fines paid by polluters.

3.1 Limits

3.2 Continuity

3.3 Rates of Change

3.4 Definition of the Derivative

3.5 Graphical Differentiation

Review Exercises

In earlier chapters we began to see how mathematics can be used to study a wide range of situations. We saw how to answer questions like the following:

- In how many years will a particular population reach 10,000?
- How wide is an arch 10 meters from the ground?
- What is the oxygen consumption of yearling salmon?

In this chapter, we will build upon this knowledge by introducing calculus. With calculus, we will be able to solve problems that involve changing situations.

- What is the maximum size of a female arctic fox?
- How fast is the car moving after 20 seconds?
- At what rate is the population growing?
- At what time after a meal does the maximum thermal effect of food occur?

Each of these problems can be solved by calculating a *derivative*. Since the definition of derivative involves the idea of a *limit*, we begin with limits.

■ 3.1 LIMITS

? THINK ABOUT IT   **What happens to the oxygen concentration in a pond over the long run?**

FOR REVIEW ■

Evaluating function notation, such as $f(1) = 3$, was discussed in Section 1.1. Verify that for

$$f(x) = \frac{x^2 - 4}{x - 2},$$

$f(0) = 2$, $f(3) = 5$, and $f(-1) = 1$.

In this section we will find an answer to this question using the concept of limit. We can find the value of the function defined by

$$f(x) = \frac{x^2 - 4}{x - 2}$$

when $x = 1$ by substitution:

$$f(1) = \frac{1^2 - 4}{1 - 2} = 3.$$

It is also true that when x is a number *very close* to 1 (on either side of 1), then $f(x)$ is a number *very close* to 3, as the following table shows.

	x approaches 1 from the left.			↓	x approaches 1 from the right.					
x	0.8	0.9	0.99	0.9999	**1**	1.0000001	1.0001	1.001	1.01	1.05	1.1
$f(x)$	2.8	2.9	2.99	2.9999	**3**	3.0000001	3.0001	3.001	3.01	3.05	3.1
	f(x) approaches 3.			↑	f(x) approaches 3.						

At $x = 2$ the situation is different: $f(2)$ is not defined because the denominator is 0 when $x = 2$. But we can still ask, what happens to $f(x)$ when x is a number *very close* to (but not equal to) 2? The table on the next page provides an answer.

			x approaches 2 from the left.		↓	*x* approaches 2 from the right.				
x	1.8	1.9	1.99	1.9999	**2**	2.0000001	2.00001	2.001	2.05	2.1
f(x)	3.8	3.9	3.99	3.9999	**4**	4.0000001	4.00001	4.001	4.05	4.1
			f(x) approaches 4.		↑	*f(x)* approaches 4.				

The table suggests that, as x gets closer and closer to 2 from either side, $f(x)$ gets closer and closer to 4. In fact, by experimenting with a calculator you can convince yourself that the values of $f(x)$ can be made as close as you want to 4 by taking values of x close enough to 2. In such a case, we say "the limit of $f(x)$ as x approaches 2 is 4," which is written as

$$\lim_{x \to 2} f(x) = 4.$$

In the first example, we found that

$$\lim_{x \to 1} f(x) = 3,$$

because the values of $f(x)$ got closer and closer to 3 as x got closer and closer to 1, *from either side* of 1.

The phrase "x approaches 1 from the left" is written $x \to 1^-$. Similarly, "x approaches 1 from the right" is written $x \to 1^+$. These expressions are used to write **one-sided limits.** The **limit from the left** (as x approaches 1 from the negative direction) is written

$$\lim_{x \to 1^-} f(x) = 3,$$

and the **limit from the right** (as x approaches 1 from the positive direction) is written

$$\lim_{x \to 1^+} f(x) = 3.$$

A **two-sided limit,** such as

$$\lim_{x \to 1} f(x) = 3,$$

exists only if both one-sided limits exist and are the same; that is, if $f(x)$ approaches the same number as x approaches a given number from *either* side.

The examples suggest the following informal definition.

LIMIT OF A FUNCTION

Let f be a function and let a and L be real numbers. If

1. as x takes values closer and closer (but not equal) to a on both sides of a, the corresponding values of $f(x)$ get closer and closer (and perhaps equal) to L; and

2. the value of $f(x)$ can be made as close to L as desired by taking values of x close enough to a;

then L is the **limit** of $f(x)$ as x approaches a, written

$$\lim_{x \to a} f(x) = L.$$

This definition is informal because the expressions "closer and closer to" and "as close as desired" have not been defined. A more formal definition would be needed to prove the rules for limits given later in this section.*

> **NOTE** The definition of a limit describes what happens to $f(x)$ when x is near a. It is not affected by whether $f(a)$ is defined. Also, the definition implies that the function values cannot approach two different numbers, so that if a limit exists, it is unique.

Figure 1 shows the graph of the function in the previous example drawn with a graphing calculator. Notice that the function has a small gap at the point $(2, 4)$, which agrees with our previous observation that the function is undefined at $x = 2$, where the limit is 4. (Due to the limitations of the graphing calculator, this gap may vanish when the viewing window is changed very slightly.) Furthermore, notice that for other values of x, the graph of the function appears to be a straight line. This is because when $x \neq 2$, the function can be simplified:

$$f(x) = \frac{x^2 - 4}{x - 2}$$

$$f(x) = \frac{(x - 2)(x + 2)}{x - 2} \quad \text{Using } a^2 - b^2 = (a - b)(a + b)$$

$$= x + 2.$$

The graph of $y = x + 2$ is a straight line, and it seems reasonable that as x gets closer and closer to 2, $x + 2$ gets closer and closer to 4.

The previous discussion shows a second way of finding a limit: Use algebra to simplify the expression so the limit is easy to find.

A third way of finding a limit is to use the TRACE feature on a graphing calculator. For the example above, the result after pressing the TRACE key is shown in Figure 2. The cursor is already located at $x = 2$; if it were not, we could use the right or left arrow key to move the cursor there. The calculator does not give a y-value because the function is undefined at $x = 2$. Moving the cursor back a step gives $x = 1.95$, $y = 3.95$. Moving the cursor forward two steps gives $x = 2.05$, $y = 4.05$. It seems that as x approaches 2, y approaches 4, or at least something close to 4. Zooming in on the point $(2, 4)$ (such as using $[1.9, 2.1]$ by $[3.9, 4.1]$) allows the limit to be estimated more accurately and helps ensure that the graph has no unexpected behavior very close to $x = 2$.

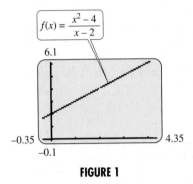

FIGURE 1

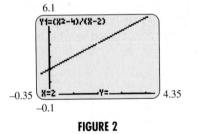

FIGURE 2

EXAMPLE 1 Limits

Find $\lim_{x \to 2} g(x)$, where $g(x) = \frac{x^2 + 4}{x - 2}$.

Solution Using the TABLE feature on a TI-83, we produce the table of numbers shown in Figure 3, where Y_1 represents the function $g(x)$. Figure 4 shows the

*The limit is the key concept from which all the ideas of calculus flow. Calculus was independently discovered by the English mathematician Isaac Newton (1642–1727) and the German mathematician Gottfried Wilhelm Leibniz (1646–1716). For the next century, supporters of each accused the other of plagiarism, resulting in a lack of communication between mathematicians in England and on the European continent. Neither Newton nor Leibniz developed a mathematically rigorous definition of the limit (and we have no intention of doing so here). More than 100 years passed before the French mathematician Augustin-Louis Cauchy (1789–1857) accomplished this feat.

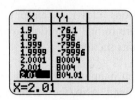

FIGURE 3

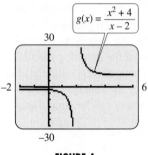

FIGURE 4

graph of the function. Both the table and the corresponding graph can be easily generated using a spreadsheet. Consult *The Spreadsheet Manual* that is available on this text's Web site for details.

Both the table and the graph suggest that as $x \to 2$ from the left, $g(x)$ gets more and more negative, becoming larger and larger in magnitude. This is indicated by writing

$$\lim_{x \to 2^-} g(x) = -\infty.$$

Because $-\infty$ is not a real number, the limit in this case does not exist. The symbol $-\infty$ simply indicates that as $x \to 2^-$, $g(x)$ becomes more and more negative without bound. Whenever we say that a limit has the value of ∞ or $-\infty$, this implies that the limit does not exist.

In the same way, the behavior of the function as $x \to 2$ from the right is indicated by writing

$$\lim_{x \to 2^+} g(x) = \infty.$$

Since there is no real number that $g(x)$ approaches as $x \to 2$ (from either side), nor does $g(x)$ approach either ∞ or $-\infty$, we simply say

$$\lim_{x \to 2} \frac{x^2 + 4}{x - 2} \text{ does not exist.}$$

We have shown three methods for determining limits: (1) using a table of numbers, (2) using algebraic simplification, and (3) tracing the graph on a graphing calculator. Which method you choose depends on the complexity of the function and the accuracy required by the application. Algebraic simplification gives the exact answer, but it can be difficult or even impossible to use in some situations. Calculating a table of numbers or tracing the graph may be easier when the function is complicated, but be careful, because the results could be inaccurate, inconclusive, or misleading. A graphing calculator does not tell us what happens between or beyond the points that are plotted.

EXAMPLE 2 Limits

Find $\lim\limits_{x \to 0} \dfrac{|x|}{x}$.

Solution

Method 1: Algebraic Approach

The function $f(x) = |x|/x$ is not defined when $x = 0$. When $x > 0$, the definition of absolute value shows that $f(x) = |x|/x = x/x = 1$. When $x < 0$, then $|x| = -x$ and $f(x) = -x/x = -1$.

Method 2: Graphing Calculator Approach

A graphing calculator graph of f is shown in Figure 5 on the next page.

As x approaches 0 from the right, x is always positive and the corresponding value of $f(x)$ is 1, so

$$\lim_{x \to 0^+} f(x) = 1.$$

But as x approaches 0 from the left, x is always negative and the corresponding value of $f(x)$ is -1, so

$$\lim_{x \to 0^-} f(x) = -1.$$

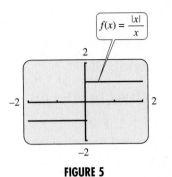

FIGURE 5

Thus, as x approaches 0 from either side, the corresponding values of $f(x)$ do not get closer and closer to a *single* real number. Therefore, the limit of $|x|/x$ as x approaches 0 does not exist because the limits from the left and from the right are not equal.

The discussion up to this point can be summarized as follows.

EXISTENCE OF LIMITS

The limit of f as x approaches a may not exist.

1. If $f(x)$ becomes infinitely large in magnitude (positive or negative) as x approaches the number a from either side, we write $\lim_{x \to a} f(x) = \infty$ or $\lim_{x \to a} f(x) = -\infty$. In either case, the limit does not exist.

2. If $f(x)$ becomes infinitely large in magnitude (positive) as x approaches a from one side and infinitely large in magnitude (negative) as x approaches a from the other side, then $\lim_{x \to a} f(x)$ does not exist.

3. If $\lim_{x \to a^-} f(x) = L$ and $\lim_{x \to a^+} f(x) = M$, and $L \neq M$, then $\lim_{x \to a} f(x)$ does not exist.

Figure 6 illustrates these three facts.

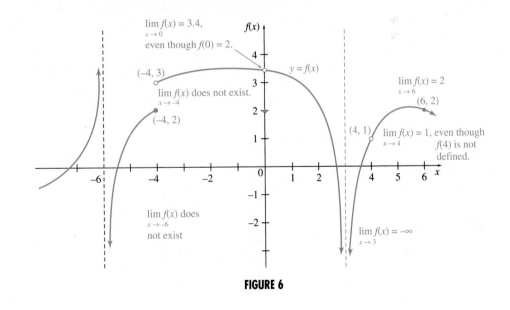

FIGURE 6

Rules for Limits   As shown by the preceding examples, tables and graphs can be used to find limits. However, it is usually more efficient to use the rules for limits given on the next page. (Proofs of these rules require a formal definition of limit, which we have not given.)

RULES FOR LIMITS

Let a, A, and B be real numbers, and let f and g be functions such that

$$\lim_{x \to a} f(x) = A \qquad \text{and} \qquad \lim_{x \to a} g(x) = B.$$

1. If k is a constant, then $\lim_{x \to a} k = k$ and $\lim_{x \to a}[k \cdot f(x)] = k \cdot \lim_{x \to a} f(x) = k \cdot A$.

2. $\lim_{x \to a}[f(x) \pm g(x)] = \lim_{x \to a} f(x) \pm \lim_{x \to a} g(x) = A \pm B$
   (The limit of a sum or difference is the sum or difference of the limits.)

3. $\lim_{x \to a}[f(x) \cdot g(x)] = [\lim_{x \to a} f(x)] \cdot [\lim_{x \to a} g(x)] = A \cdot B$
   (The limit of a product is the product of the limits.)

4. $\lim_{x \to a} \dfrac{f(x)}{g(x)} = \dfrac{\lim_{x \to a} f(x)}{\lim_{x \to a} g(x)} = \dfrac{A}{B}$ if $B \neq 0$
   (The limit of a quotient is the quotient of the limits, provided the limit of the denominator is not zero.)

5. If $p(x)$ is a polynomial, then $\lim_{x \to a} p(x) = p(a)$.

6. For any real number k, $\lim_{x \to a}[f(x)]^k = [\lim_{x \to a} f(x)]^k = A^k$, provided this limit exists.*

7. $\lim_{x \to a} f(x) = \lim_{x \to a} g(x)$ if $f(x) = g(x)$ for all $x \neq a$.

8. For any real number $b > 0$, $\lim_{x \to a} b^{f(x)} = b^{[\lim_{x \to a} f(x)]} = b^A$.

9. For any real number b such that $0 < b < 1$ or $1 < b$,
   $\lim_{x \to a}[\log_b f(x)] = \log_b[\lim_{x \to a} f(x)] = \log_b A$ if $A > 0$.

This list may seem imposing, but these limit rules, once understood, agree with common sense. For example, Rule 3 says that if $f(x)$ becomes close to A as x approaches a, and if $g(x)$ becomes close to B, then $f(x) \cdot g(x)$ should become close to $A \cdot B$, which seems plausible.

EXAMPLE 3 Rules for Limits

Suppose $\lim_{x \to 2} f(x) = 3$ and $\lim_{x \to 2} g(x) = 4$. Use the limit rules to find the following limits.

(a) $\lim_{x \to 2}[f(x) + 5g(x)]$

Solution

$$\lim_{x \to 2}[f(x) + 5g(x)] = \lim_{x \to 2} f(x) + \lim_{x \to 2} 5g(x) \qquad \text{Rule 2}$$

$$= \lim_{x \to 2} f(x) + 5 \lim_{x \to 2} g(x) \qquad \text{Rule 1}$$

$$= 3 + 5(4)$$

$$= 23$$

*This limit does not exist, for example, when $A < 0$ and $k = 1/2$, or when $A = 0$ and $k \leq 0$.

(b) $\lim\limits_{x \to 2} \dfrac{[f(x)]^2}{\ln\,g(x)}$

Solution

$$\lim_{x \to 2} \frac{[f(x)]^2}{\ln\,g(x)} = \frac{\lim\limits_{x \to 2}[f(x)]^2}{\lim\limits_{x \to 2}\ln\,g(x)} \qquad \text{Rule 4}$$

$$= \frac{[\lim\limits_{x \to 2} f(x)]^2}{\ln[\lim\limits_{x \to 2} g(x)]} \qquad \text{Rule 6 and Rule 9}$$

$$= \frac{3^2}{\ln\,4}$$

$$\approx \frac{9}{1.38629} \approx 6.4921$$

EXAMPLE 4 Limits

Find $\lim\limits_{x \to 3} \dfrac{x^2 - x - 1}{\sqrt{x + 1}}$.

Solution

$$\lim_{x \to 3} \frac{x^2 - x - 1}{\sqrt{x + 1}} = \frac{\lim\limits_{x \to 3}(x^2 - x - 1)}{\lim\limits_{x \to 3}\sqrt{x + 1}} \qquad \text{Rule 4}$$

$$= \frac{3^2 - 3 - 1}{\sqrt{\lim\limits_{x \to 3}(x + 1)}} \qquad \text{Rule 5 and Rule 6 } (\sqrt{a} = a^{1/2})$$

$$= \frac{5}{\sqrt{4}} \qquad \text{Rule 5}$$

$$= \frac{5}{2}$$

As Examples 3 and 4 suggest, the rules for limits actually mean that many limits can be found simply by evaluation. This process is valid for polynomials, rational functions, exponential functions, logarithmic functions, and roots and powers, so long as this does not involve an illegal operation, such as division by 0 or taking the logarithm of a negative number. Division by 0 presents particular problems that can often be solved by algebraic simplification, as the following example shows.

EXAMPLE 5 Limits

Find $\lim\limits_{x \to 2} \dfrac{x^2 + x - 6}{x - 2}$.

Solution Rule 4 cannot be used here, since

$$\lim_{x \to 2}(x - 2) = 0.$$

The numerator also approaches 0 as x approaches 2, and 0/0 is meaningless. For $x \ne 2$, we can, however, simplify the function by rewriting the fraction as

$$\frac{x^2 + x - 6}{x - 2} = \frac{(x + 3)(x - 2)}{x - 2} = x + 3.$$

Now Rule 7 can be used.

$$\lim_{x \to 2} \frac{x^2 + x - 6}{x - 2} = \lim_{x \to 2}(x + 3) = 2 + 3 = 5$$

EXAMPLE 6 Limits

Find $\lim\limits_{x \to 4} \dfrac{\sqrt{x} - 2}{x - 4}$.

Solution As $x \to 4$, the numerator approaches 0 and the denominator also approaches 0, giving the meaningless expression $0/0$. Algebra can be used to rationalize the numerator by multiplying both the numerator and the denominator by $\sqrt{x} + 2$. This gives

$$\frac{\sqrt{x} - 2}{x - 4} \cdot \frac{\sqrt{x} + 2}{\sqrt{x} + 2} = \frac{\sqrt{x} \cdot \sqrt{x} - 2\sqrt{x} + 2\sqrt{x} - 4}{(x - 4)(\sqrt{x} + 2)}$$

$$= \frac{x - 4}{(x - 4)(\sqrt{x} + 2)} = \frac{1}{\sqrt{x} + 2}$$

if $x \neq 4$. Now use rules for limits.

$$\lim_{x \to 4} \frac{\sqrt{x} - 2}{x - 4} = \lim_{x \to 4} \frac{1}{\sqrt{x} + 2} = \frac{1}{\sqrt{4} + 2} = \frac{1}{2 + 2} = \frac{1}{4}$$

You can support this result by using a graphing calculator to make a table of values as we did in Example 1, or by graphing the function and using the TRACE feature to investigate that near $x = 4$, y is close to 0.25, although at $x = 4$, the function is undefined.

Examples 5 and 6 suggest the following principle: **To calculate the limit of $f(x)/g(x)$ as x approaches a, where $f(a) = g(a) = 0$, you should attempt to factor $x - a$ from both the numerator and the denominator.**

CAUTION Simply because the expression in a limit is approaching $0/0$, as in Examples 5 and 6, does *not* mean that the limit is 0 or that the limit does not exist. For such a limit, try to simplify the expression using algebra.

EXAMPLE 7 Limits

Find $\lim\limits_{x \to 1} \dfrac{x + 1}{x^2 - 1}$.

Solution

Method 1: Algebraic Approach Again, Rule 4 cannot be used since $\lim\limits_{x \to 1} x^2 - 1 = 0$. If $x \neq 1$, the function can be rewritten as

$$\frac{x + 1}{x^2 - 1} = \frac{x + 1}{(x + 1)(x - 1)} = \frac{1}{x - 1}.$$

Then

$$\lim_{x \to 1} \frac{x + 1}{x^2 - 1} = \lim_{x \to 1} \frac{1}{x - 1}$$

by Rule 7. None of the rules can be used to find

$$\lim_{x \to 1} \frac{1}{x - 1},$$

but as x approaches 1, the denominator approaches 0 while the numerator stays at 1, making the result larger and larger in magnitude. If $x > 1$, both the numerator and denominator are positive, so $\lim\limits_{x \to 1^+} 1/(x - 1) = \infty$. If $x < 1$, the denominator is negative, so $\lim\limits_{x \to 1^-} 1/(x - 1) = -\infty$. Therefore,

$$\lim_{x \to 1} \frac{1}{x - 1} \text{ does not exist.}$$

Another way to understand the behavior of this function near $x = 1$ is to recall from the section on Polynomial and Rational Functions that a rational function often has a vertical asymptote at a value of x where the denominator is 0, although it may not if the numerator there is also 0. In this example, we see after simplifying that the function has a vertical asymptote at $x = 1$ because that would make the denominator of $1/(x - 1)$ equal to 0, while the numerator is 1.

Method 2: Graphing Calculator Approach   Figure 7 shows a graphing calculator view of $y = 1/(x - 1)$ on $[0, 2]$ by $[-10, 10]$. The behavior of the function indicates a vertical asymptote at $x = 1$, with the limit approaching $-\infty$ from one side and ∞ from the other, so

$$\lim_{x \to 1} \frac{1}{x - 1} \text{ does not exist.}$$

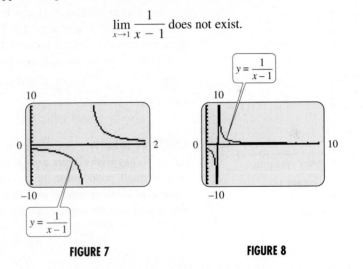

FIGURE 7   **FIGURE 8**

CAUTION A graphing calculator can give a deceptive view of a function. Figure 8 shows the result if we graph the previous function on $[0, 10]$ by $[-10, 10]$. Near $x = 1$, the graph appears to be a steep line connecting the two pieces. The graph in Figure 7 is more representative of the function near $x = 1$. When using a graphing calculator, you may need to experiment with the viewing window, guided by what you have learned about functions and limits, to get a good picture of a function. On many calculators, extraneous lines connecting parts of the graph can be avoided by using Dot mode rather than Connected mode.

EXAMPLE 8   Postage

Figure 9 on the next page shows how the postage required to mail a letter in the United States has changed with time.*

The New York Times, March 13, 1994, p. 2.

The Cost of Mailing a Letter
Postage
(in cents)

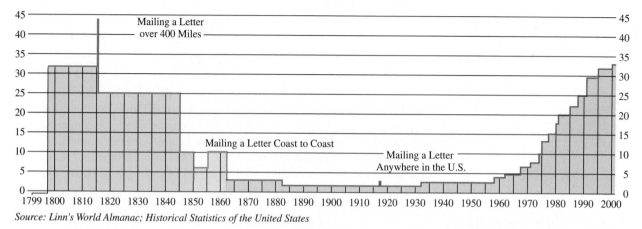

Source: Linn's World Almanac; Historical Statistics of the United States

FIGURE 9

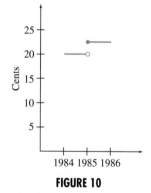

FIGURE 10

Let $C(t)$ be the cost of a letter in the year t. Consider the behavior of the function near $t = 1985$, when postage jumped from 20¢ to 22¢. The graph as drawn does not pass the vertical line test, so we have redrawn this section of the graph in Figure 10 using the notation in this text. Notice that $\lim_{t \to 1985^-} C(t) = 20$ and $\lim_{t \to 1985^+} C(t) = 22$, so $\lim_{t \to 1985} C(t)$ does not exist. Notice also that $C(1985)$ does exist and is equal to 22.

Limits at Infinity

Sometimes it is useful to examine the behavior of the values of $f(x)$ as x gets larger and larger (or smaller and smaller). For example, suppose a small pond normally contains 12 units of dissolved oxygen in a fixed volume of water. Suppose also that at time $t = 0$ a quantity of organic waste is introduced into the pond, with the oxygen concentration t weeks later given by

$$f(t) = \frac{12t^2 - 15t + 12}{t^2 + 1}.$$

As time goes on, what will be the ultimate concentration of oxygen? Will it return to 12 units?

After 2 weeks, the pond contains

$$f(2) = \frac{12 \cdot 2^2 - 15 \cdot 2 + 12}{2^2 + 1} = \frac{30}{5} = 6$$

units of oxygen, and after 4 weeks, it contains

$$f(4) = \frac{12 \cdot 4^2 - 15 \cdot 4 + 12}{4^2 + 1} \approx 8.5$$

units. Choosing several values of t and finding the corresponding values of $f(t)$, or using a graphing calculator or computer, leads to the graph in Figure 11.

The graph suggests that, as time goes on, the oxygen level gets closer and closer to the original 12 units. If so, the line $y = 12$ is a horizontal asymptote. We

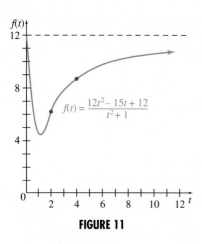

FIGURE 11

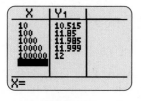

FIGURE 12

FOR REVIEW ■

In the section on Polynomial and Rational Functions, we saw a way to find horizontal asymptotes by considering the behavior of the function as x (or t) gets large. For large t, $12t^2 - 15t + 12 \approx 12t^2$, because the t term and the constant term are small compared with the t^2 term when t is large. Similarly, $t^2 + 1 \approx t^2$. Thus, for large t, $f(t) = \dfrac{12t^2 - 15t + 12}{t^2 + 1} \approx$

$\dfrac{12t^2}{t^2} = 12$. Thus the function f has a horizontal asymptote at $y = 12$.

can use the TABLE feature on a graphing calculator to investigate the behavior for large values of t. Figure 12 shows the result using a TI-83.

The table suggests that

$$\lim_{t \to \infty} f(t) = 12,$$

where $t \to \infty$ means that t increases without bound. (Similarly, $t \to -\infty$ means that t *decreases* without bound; that is, t becomes more and more negative.) Thus, the oxygen concentration will approach 12, but it will never be *exactly* 12.

The preceding example illustrates a **limit at infinity.** The phrase "t approaches infinity" (symbolically, $t \to \infty$) is simply convenient shorthand to express the fact that t becomes larger and larger without bound. Similarly, the phrase "t approaches negative infinity" (symbolically, $t \to -\infty$) means that t becomes more and more negative without bound (such as -10, $-1,000$, $-10,000$, etc.).

As we saw in the previous example, limits at infinity or negative infinity, if they exist, correspond to horizontal asymptotes of the graph of the function. In the previous chapter, we saw one way to find horizontal asymptotes. We will now show a more precise way, based upon some simple limits at infinity. The graphs of $f(x) = 1/x$ (in red) and $g(x) = 1/x^2$ (in blue) shown in Figure 13, as well as the table there, indicate that $\lim_{x \to \infty} 1/x = 0$, $\lim_{x \to -\infty} 1/x = 0$, $\lim_{x \to \infty} 1/x^2 = 0$, and $\lim_{x \to -\infty} 1/x^2 = 0$, suggesting the following rule.

x	$\dfrac{1}{x}$	$\dfrac{1}{x^2}$
-100	-0.01	0.0001
-10	-0.1	0.01
-1	-1	1
1	1	1
10	0.1	0.01
100	0.01	0.0001

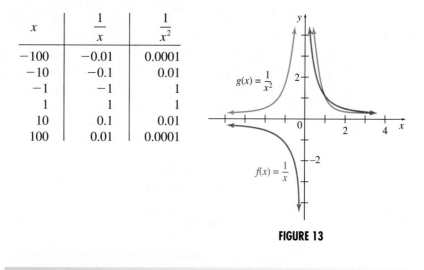

FIGURE 13

LIMITS AT INFINITY

For any positive real number n,

$$\lim_{x \to \infty} \frac{1}{x^n} = 0 \qquad \text{and} \qquad \lim_{x \to -\infty} \frac{1}{x^n} = 0.^*$$

The rules for limits given earlier remain unchanged when a is replaced with ∞ or $-\infty$.

---

*If x is negative, x^n does not exist for certain values of n, so the second limit is undefined.

To evaluate the limit at infinity of a rational function, divide the numerator and denominator by the largest power of the variable, t^2 here, and then use these results. In the previous example, we find that

$$\lim_{t \to \infty} \frac{12t^2 - 15t + 12}{t^2 + 1} = \lim_{t \to \infty} \frac{\dfrac{12t^2}{t^2} - \dfrac{15t}{t^2} + \dfrac{12}{t^2}}{\dfrac{t^2}{t^2} + \dfrac{1}{t^2}}$$

$$= \lim_{t \to \infty} \frac{12 - 15 \cdot \dfrac{1}{t} + 12 \cdot \dfrac{1}{t^2}}{1 + \dfrac{1}{t^2}}.$$

Now apply the limit rules and the fact that $\lim\limits_{t \to \infty} 1/t^n = 0$.

$$\frac{\lim\limits_{t \to \infty}\left(12 - 15 \cdot \dfrac{1}{t} + 12 \cdot \dfrac{1}{t^2}\right)}{\lim\limits_{t \to \infty}\left(1 + \dfrac{1}{t^2}\right)}$$

$$= \frac{\lim\limits_{t \to \infty} 12 - \lim\limits_{t \to \infty} 15 \cdot \dfrac{1}{t} + \lim\limits_{t \to \infty} 12 \cdot \dfrac{1}{t^2}}{\lim\limits_{t \to \infty} 1 + \lim\limits_{t \to \infty} \dfrac{1}{t^2}} \qquad \text{Rules 4 and 2}$$

$$= \frac{12 - 15\left(\lim\limits_{t \to \infty} \dfrac{1}{t}\right) + 12\left(\lim\limits_{t \to \infty} \dfrac{1}{t^2}\right)}{1 + \lim\limits_{t \to \infty} \dfrac{1}{t^2}} \qquad \text{Rule 1}$$

$$= \frac{12 - 15 \cdot 0 + 12 \cdot 0}{1 + 0} = 12. \qquad \text{Limits at infinity}$$

EXAMPLE 9 Limits at Infinity
Find each limit.

(a) $\lim\limits_{x \to \infty} \dfrac{8x + 6}{3x - 1}$

Solution We can use the rule $\lim\limits_{x \to \infty} 1/x^n = 0$ to find this limit by first dividing numerator and denominator by x, as follows.

$$\lim_{x \to \infty} \frac{8x + 6}{3x - 1} = \lim_{x \to \infty} \frac{\dfrac{8x}{x} + \dfrac{6}{x}}{\dfrac{3x}{x} - \dfrac{1}{x}} = \lim_{x \to \infty} \frac{8 + 6 \cdot \dfrac{1}{x}}{3 - \dfrac{1}{x}} = \frac{8 + 0}{3 - 0} = \frac{8}{3}$$

(b) $\lim\limits_{x \to \infty} \dfrac{3x + 2}{4x^3 - 1} = \lim\limits_{x \to \infty} \dfrac{3 \cdot \dfrac{1}{x^2} + 2 \cdot \dfrac{1}{x^3}}{4 - \dfrac{1}{x^3}} = \dfrac{0 + 0}{4 - 0} = \dfrac{0}{4} = 0$

Here, the highest power of x is x^3, which is used to divide each term in the numerator and denominator.

(c) $\lim\limits_{x\to\infty} \dfrac{3x^2 + 2}{4x - 3} = \lim\limits_{x\to\infty} \dfrac{3 + \dfrac{2}{x^2}}{\dfrac{4}{x} - \dfrac{3}{x^2}} = \dfrac{3 + 0}{0 - 0} = \dfrac{3}{0}$

Division by 0 is undefined, so this limit does not exist. We can actually say more: by examining what happens as larger and larger values of x are put into the function, verify that

$$\lim\limits_{x\to\infty} \dfrac{3x^2 + 2}{4x - 3} = \infty.$$

The method used in Example 9 is a useful way to rewrite expressions with fractions so that the rules for limits at infinity can be used.

FINDING LIMITS AT INFINITY

If $f(x) = p(x)/q(x)$, for polynomials $p(x)$ and $q(x)$, $q(x) \neq 0$, $\lim\limits_{x\to-\infty} f(x)$ and $\lim\limits_{x\to\infty} f(x)$ can be found as follows.

1. Divide $p(x)$ and $q(x)$ by the highest power of x in either polynomial.

2. Use the rules for limits, including the rules for limits at infinity,

$$\lim\limits_{x\to\infty} \frac{1}{x^n} = 0 \quad \text{and} \quad \lim\limits_{x\to-\infty} \frac{1}{x^n} = 0,$$

to find the limit of the result from Step 1.

EXAMPLE 10 Arctic Foxes

The age-weight relationship of female Arctic foxes caught in Svalbard, Norway, can be estimated by the function

$$M(t) = 3{,}102e^{-e^{-0.022(t-56)}},$$

where t is the age of the fox in days and $M(t)$ is the weight of the fox in grams.*†

(a) Use $M(t)$ to estimate the largest size that a female fox can attain.

Solution Since $\lim\limits_{t\to\infty} e^{-0.022(t-56)} = 0$, $\lim\limits_{t\to\infty} M(t) = 3{,}102e^0 = 3{,}102$. Thus, the maximum weight of female Arctic foxes predicted by this function is 3,102 g.

(b) Estimate the age of a female fox when it has reached 80% of its maximum weight.

Solution To answer this question we set $(0.8)(3{,}102) = 2{,}481.6$ equal to $M(t)$ and then solve for t. Thus,

---

*Presrud, Pal, and Kuell Nilssen, "Growth, Size, and Sexual Dimorphism in Arctic Foxes," *Journal of Mammalogy,* Vol. 76, No. 2, May 1995, pp. 522–530.

†Technically, grams are a measure of mass, not weight. Weight is a measure of the force of gravity, which varies with the distance from the center of the earth. For objects on the surface of the earth, weight and mass are often used interchangeably, and we will do so in this text.

$$2{,}481.6 = 3{,}102e^{-e^{-0.022(t-56)}}$$

$$\frac{2{,}481.6}{3{,}102} = e^{-e^{-0.022(t-56)}} \qquad \text{Divide both sides by } 3{,}102.$$

$$\ln\left(\frac{2{,}481.6}{3{,}102}\right) = \ln(e^{-e^{-0.022(t-56)}}) = -e^{-0.022(t-56)} \qquad \text{Take natural log of both sides.}$$

$$-\ln\left(\frac{2{,}481.6}{3{,}102}\right) = e^{-0.022(t-56)} \qquad \text{Multiply both sides by } -1.$$

$$\ln\left(-\ln\left(\frac{2{,}481.6}{3{,}102}\right)\right) = -0.022(t-56) \qquad \text{Take natural log of both sides.}$$

$$t - 56 = \frac{-1}{0.022}\ln\left(-\ln\left(\frac{2{,}481.6}{3{,}102}\right)\right) \approx 68.2 \qquad \text{Divide both sides by } -0.022.$$

$$t \approx 124 \text{ days} \qquad \text{Solve for } t.$$

3.1 EXERCISES

In Exercises 1–4, choose the best answer for each limit.

1. If $\lim\limits_{x \to 2^-} f(x) = 5$ and $\lim\limits_{x \to 2^+} f(x) = 6$, then $\lim\limits_{x \to 2} f(x)$

   a. is 5.

   b. is 6.

   c. does not exist.

   d. is infinite.

2. If $\lim\limits_{x \to 2^-} f(x) = \lim\limits_{x \to 2^+} f(x) = -1$, but $f(2) = 1$, then $\lim\limits_{x \to 2} f(x)$

   a. is -1.

   b. does not exist.

   c. is infinite.

   d. is 1.

3. If $\lim\limits_{x \to 4^-} f(x) = \lim\limits_{x \to 4^+} f(x) = 6$, but $f(4)$ does not exist, then $\lim\limits_{x \to 4} f(x)$

   a. does not exist.

   b. is 6.

   c. is $-\infty$.

   d. is ∞.

4. If $\lim\limits_{x \to 1^-} f(x) = -\infty$ and $\lim\limits_{x \to 1^+} f(x) = -\infty$, then $\lim\limits_{x \to 1} f(x)$

   a. is ∞.

   b. is $-\infty$.

   c. does not exist.

   d. is 1.

Decide whether each limit exists. If a limit exists, find its value.

5. $\lim\limits_{x\to 3} f(x)$

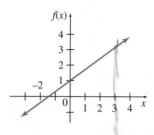

6. $\lim\limits_{x\to 2} F(x)$

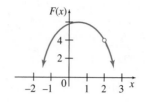

7. $\lim\limits_{x\to 0} f(x)$

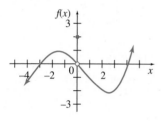

8. $\lim\limits_{x\to 3} g(x)$

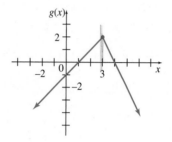

In Exercises 9 and 10, use the graph to find **(i)** $\lim\limits_{x\to a^-} f(x)$, **(ii)** $\lim\limits_{x\to a^+} f(x)$, **(iii)** $\lim\limits_{x\to a} f(x)$, *and* **(iv)** $f(a)$ *if it exists.*

9. a. $a = -2$ **b.** $a = -1$

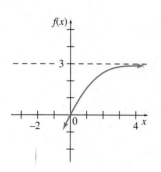

10. a. $a = 1$ **b.** $a = 2$

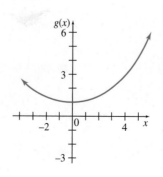

Decide whether the following limits exist. If a limit exists, find its value.

11. $\lim\limits_{x\to\infty} f(x)$

12. $\lim\limits_{x\to -\infty} g(x)$

13. Explain why $\lim\limits_{x \to 2} F(x)$ in Exercise 6 exists, but $\lim\limits_{x \to -2} f(x)$ in Exercise 9 does not.

14. In Exercise 10, why does $\lim\limits_{x \to 1} f(x) = 1$, even though $f(1) = 2$?

15. Use the table of values below to estimate $\lim\limits_{x \to 1} f(x)$.

x	0.9	0.99	0.999	0.9999	1.0001	1.001	1.01	1.1
$f(x)$	1.9	1.99	1.999	1.9999	2.0001	2.001	2.01	2.1

Complete the tables and use the results to find the indicated limits.

16. If $f(x) = 2x^2 - 4x + 3$, find $\lim\limits_{x \to 1} f(x)$.

x	0.9	0.99	0.999	1.001	1.01	1.1
$f(x)$			1.000002	1.000002		

17. If $k(x) = \dfrac{x^3 - 2x - 4}{x - 2}$, find $\lim\limits_{x \to 2} k(x)$.

x	1.9	1.99	1.999	2.001	2.01	2.1
$k(x)$						

18. If $f(x) = \dfrac{2x^3 + 3x^2 - 4x - 5}{x + 1}$, find $\lim\limits_{x \to -1} f(x)$.

x	-1.1	-1.01	-1.001	-0.999	-0.99	-0.9
$f(x)$						

19. If $h(x) = \dfrac{\sqrt{x} - 2}{x - 1}$, find $\lim\limits_{x \to 1} h(x)$.

x	0.9	0.99	0.999	1.001	1.01	1.1
$h(x)$						

20. If $f(x) = \dfrac{\sqrt{x} - 3}{x - 3}$, find $\lim\limits_{x \to 3} f(x)$.

x	2.9	2.99	2.999	3.001	3.01	3.1
$f(x)$						

Let $\lim\limits_{x \to 4} f(x) = 16$ and $\lim\limits_{x \to 4} g(x) = 8$. Use the limit rules to find the following limits.

21. $\lim\limits_{x \to 4}[f(x) - g(x)]$

22. $\lim\limits_{x \to 4}[g(x) \cdot f(x)]$

23. $\lim\limits_{x \to 4} \dfrac{f(x)}{g(x)}$

24. $\lim\limits_{x \to 4} \log_2 f(x)$

25. $\lim\limits_{x \to 4} \sqrt{f(x)}$

26. $\lim\limits_{x \to 4} \sqrt[3]{g(x)}$

27. $\lim\limits_{x \to 4} 2^{g(x)}$

28. $\lim\limits_{x \to 4}[1 + f(x)]^2$

29. $\lim\limits_{x \to 4} \dfrac{f(x) + g(x)}{2g(x)}$

30. $\lim\limits_{x \to 4} \dfrac{5g(x) + 2}{1 - f(x)}$

31. $\lim\limits_{x \to 4}\left[\sin\left(\dfrac{\pi}{16} \cdot g(x)\right)\right]$

32. $\lim\limits_{x \to 4}\left[\cos\left(\pi \cdot \dfrac{f(x)}{g(x)}\right)\right]$

Use the properties of limits to help decide whether the following limits exist. If a limit exists, find its value.

33. $\lim\limits_{x \to 3} \dfrac{x^2 - 9}{x - 3}$

34. $\lim\limits_{x \to -2} \dfrac{x^2 - 4}{x + 2}$

35. $\lim\limits_{x \to 1} \dfrac{5x^2 - 7x + 2}{x^2 - 1}$

36. $\lim\limits_{x \to -3} \dfrac{x^2 - 9}{x^2 + x - 6}$

37. $\lim\limits_{x \to -2} \dfrac{x^2 - x - 6}{x + 2}$

38. $\lim\limits_{x \to 5} \dfrac{x^2 - 3x - 10}{x - 5}$

39. $\lim\limits_{x \to 0} \dfrac{[1/(x + 3)] - 1/3}{x}$

40. $\lim\limits_{x \to 0} \dfrac{[-1/(x + 2)] + 1/2}{x}$

41. $\lim\limits_{x \to 25} \dfrac{\sqrt{x} - 5}{x - 25}$

42. $\lim\limits_{x \to 36} \dfrac{\sqrt{x} - 6}{x - 36}$

43. $\lim\limits_{h \to 0} \dfrac{(x + h)^2 - x^2}{h}$

44. $\lim\limits_{h \to 0} \dfrac{(x + h)^3 - x^3}{h}$

45. $\lim\limits_{x \to \infty} \dfrac{3x}{5x - 1}$

46. $\lim\limits_{x \to -\infty} \dfrac{8x + 2}{2x - 5}$

47. $\lim\limits_{x \to \infty} \dfrac{x^2 + 2x}{2x^2 - 2x + 1}$

48. $\lim\limits_{x \to \infty} \dfrac{x^2 + 2x - 5}{3x^2 + 2}$

49. $\lim\limits_{x \to \infty} \dfrac{3x^3 + 2x - 1}{2x^4 - 3x^3 - 2}$

50. $\lim\limits_{x \to \infty} \dfrac{2x^2 - 1}{3x^4 + 2}$

51. $\lim\limits_{x \to \infty} \dfrac{2x^3 - x - 3}{6x^2 - x - 1}$

52. $\lim\limits_{x \to \infty} \dfrac{x^4 - x^3 - 3x}{7x^2 + 9}$

53. $\lim\limits_{x \to 0} \dfrac{1 - \cos^2 x}{\sin^2 x}$

54. $\lim\limits_{x \to \pi/2} \dfrac{\tan^2 x + 1}{\sec^2 x}$

55. Let $F(x) = \dfrac{3x}{(x + 2)^3}$.

   a. Find $\lim\limits_{x \to -2} F(x)$.

   b. Find the vertical asymptote of the graph of $F(x)$.

   c. Compare your answers for parts a and b. What can you conclude?

56. Let $G(x) = \dfrac{-6}{(x - 4)^2}$.

   a. Find $\lim\limits_{x \to 4} G(x)$.

   b. Find the vertical asymptote of the graph of $G(x)$.

   c. Compare your answers for parts a and b. Are they related? How?

57. How can you tell that the graph in Figure 7 is more representative of the function $f(x) = 1/(x - 1)$ than the graph in Figure 8?

58. A friend who is confused about limits wonders why you investigate the value of a function closer and closer to a point, instead of just finding the value of a function at the point. How would you respond?

59. Use a graph of $f(x) = e^x$ to answer the following questions.

   a. Find $\lim\limits_{x \to -\infty} e^x$.

   b. Where does the function e^x have a horizontal asymptote?

60. Use a graphing calculator to answer the following questions.

   a. From a graph of $y = xe^{-x}$, what do you think is the value of $\lim\limits_{x \to \infty} xe^{-x}$? Support this by evaluating the function for several large values of x.

   b. Repeat part a, this time using the graph of $y = x^2 e^{-x}$.

   c. Based on your results from parts a and b, what do you think is the value of $\lim\limits_{x \to \infty} x^n e^{-x}$, where n is a positive integer? Support this by experimenting with other positive integers n.

61. Use a graph of $f(x) = \ln x$ to answer the following questions.

   a. Find $\lim\limits_{x \to 0^+} \ln x$.

   b. Where does the function $\ln x$ have a vertical asymptote?

62. Use a graphing calculator to answer the following questions.

   a. From a graph of $y = x \ln x$, what do you think is the value of $\lim\limits_{x \to 0^+} x \ln x$? Support this by evaluating the function for several small values of x.

   b. Repeat part a, this time using the graph of $y = x(\ln x)^2$.

c. Based on your results from parts a and b, what do you think is the value of $\lim_{x \to 0^+} x(\ln x)^n$, where n is a positive integer? Support this by experimenting with other positive integers n.

63. Explain in your own words why the rules for limits at infinity should be true.

64. Explain in your own words what Rule 4 for limits means.

Find each of the following limits (**a**) by investigating values of the function near the x-value where the limit is taken, and (**b**) using a graphing calculator to view the function near that value of x.

65. $\lim_{x \to 1} \dfrac{x^4 + 4x^3 - 9x^2 + 7x - 3}{x - 1}$

66. $\lim_{x \to 2} \dfrac{x^4 + x - 18}{x^2 - 4}$

67. $\lim_{x \to -1} \dfrac{x^{1/3} + 1}{x + 1}$

68. $\lim_{x \to 4} \dfrac{x^{3/2} - 8}{x + x^{1/2} - 6}$

Use a graphing calculator to graph the function. (**a**) Determine the limit from the graph. (**b**) Explain how your answer could be determined from the expression for f(x).

69. $\lim_{x \to \infty} \dfrac{\sqrt{9x^2 + 5}}{2x}$

70. $\lim_{x \to -\infty} \dfrac{\sqrt{9x^2 + 5}}{2x}$

71. $\lim_{x \to -\infty} \dfrac{\sqrt{36x^2 + 2x + 7}}{3x}$

72. $\lim_{x \to \infty} \dfrac{\sqrt{36x^2 + 2x + 7}}{3x}$

73. $\lim_{x \to \infty} \dfrac{(1 + 5x^{1/3} + 2x^{5/3})^3}{x^5}$

74. $\lim_{x \to -\infty} \dfrac{(1 + 5x^{1/3} + 2x^{5/3})^3}{x^5}$

75. Explain why the following rules can be used to find $\lim_{x \to \infty} p(x)/q(x)$:

a. If the degree of $p(x)$ is less than the degree of $q(x)$, the limit is 0.

b. If the degree of $p(x)$ is equal to the degree of $q(x)$, the limit is A/B, where A and B are the leading coefficients of $p(x)$ and $q(x)$, respectively.

c. If the degree of $p(x)$ is greater than the degree of $q(x)$, the limit is ∞ or $-\infty$.

Applications

LIFE SCIENCES

76. *Consumer Demand* When the price of an essential commodity (such as healthcare) decreases rapidly, consumption increases slowly at first. If the price continues to drop, however, a "tipping" point may be reached, at which consumption takes a sudden substantial increase. Suppose that the accompanying graph shows the consumption of a certain antibiotic, $C(t)$, in millions of units. We assume that the price is decreasing rapidly. Here t is time in months after the price began decreasing. Use the graph to find the following.

a. $\lim_{t \to 12} C(t)$ b. $\lim_{t \to 16} C(t)$

c. $C(16)$ d. the tipping point (in months)

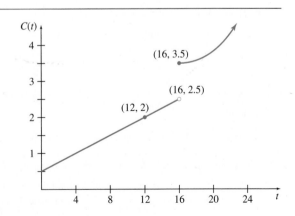

77. *Dialysis Patients* Because of new patients, transplantations, and deaths, the number of patients receiving dialysis during some part of each month in a small clinic varies over a one-year period as indicated by the following graph. Let $N(t)$ represent the number of patients in month t. Find the following.

a. $\lim\limits_{t \to 4} N(t)$ **b.** $\lim\limits_{t \to 6^+} N(t)$

c. $\lim\limits_{t \to 6^-} N(t)$ **d.** $\lim\limits_{t \to 6} N(t)$

e. $N(6)$

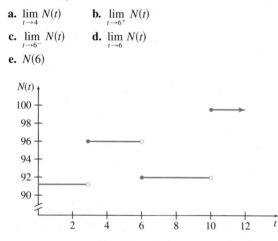

78. *Average Cost* The cost (in dollars) to a certain hospital for a particular medical test is $C(n) = 15{,}000 + 60n$, where n is the number of tests given. The average cost per test, denoted by $\overline{C}(n)$, is found by dividing $C(n)$ by n. Find and interpret $\lim\limits_{n \to \infty} \overline{C}(n)$.

79. *Drug Concentration* The concentration of a drug in a patient's bloodstream h hours after it was injected is given by

$$A(h) = \frac{0.17h}{h^2 + 2}.$$

Find and interpret $\lim\limits_{h \to \infty} A(h)$.

80. *Alligator Teeth* Researchers have developed a mathematical model that can be used to estimate the number of teeth $N(t)$ at time t (days of incubation) for *Alligator mississippiensis*,* where

$$N(t) = 71.8e^{-8.96e^{(-0.0685t)}}.$$

a. Find $N(65)$, the number of teeth of an alligator that hatched after 65 days.

b. Find $\lim\limits_{t \to \infty} N(t)$ and use this value as an estimate of the number of teeth of a newborn alligator. Does this estimate differ significantly from the estimate of part a?

81. *Sediment* To develop strategies to manage water quality in polluted lakes, biologists must determine the depths of sediments and the rate of sedimentation. It has been determined that the depth of sediment $D(t)$ (in centimeters) with respect to time (in years before 1990) for Lake Coeur d'Alene, Idaho, can be estimated by the equation

$$D(t) = 155(1 - e^{-0.0133t}).^\dagger$$

a. Find $D(20)$ and interpret.

b. Find $\lim\limits_{t \to \infty} D(t)$ and interpret.

82. *Cell Surface Receptors* In an article on the local clustering of cell surface receptors, the researcher analyzed limits of the function

$$C = F(S) = \frac{1 - S}{S(1 + kS)^{f-1}},$$

where C is the concentration of free ligand in the medium, S is the concentration of free receptors, f is the number of functional groups of cells, and k is a positive constant.[‡] Find each of the following limits.

a. $\lim\limits_{S \to 1} F(S)$ **b.** $\lim\limits_{S \to 0} F(S)$

83. *Nervous System* In a model of the nervous system, the intensity of excitation of a nerve pathway is given by

$$I = E[1 - e^{-a(S-h)/E}],$$

where E is the maximum possible excitation, S is the intensity of a stimulus, h is a threshold stimulus, and a is a constant.[§] Find $\lim\limits_{S \to \infty} I$.

OTHER APPLICATION

84. *Legislative Voting* Members of a legislature often must vote repeatedly on the same bill. As time goes on, members may change their votes. Suppose that p_0 is the probability that an individual legislator favors an issue before the first roll call vote, and suppose that p is the probability of a change in position from one vote to the next. Then the probability that the legislator will vote "yes" on the nth roll

---

*Kulesa, P., G. Cruywagen, et al., "On a Model Mechanism for the Spatial Patterning of Teeth Primordia in the Alligator," *Journal of Theoretical Biology,* Vol. 180, 1996, pp. 287–296.

†Nord, Gail, and John Nord, "Sediment in Lake Coeur d'Alene, Idaho," *The Mathematics Teacher,* Vol. 91, No. 4, April 1998, pp. 292–295.

‡Perelson, Alan S., "Receptor Clustering on a Cell Surface," *Mathematical Biosciences,* Vol. 53, No. 1/2, Feb. 1981, pp. 1–39.

§Rashevsky, Nicolas, *Mathematical Biology of Social Behavior,* rev. ed., Chicago, The University of Chicago Press, 1959, p. 7.

call is given by

$$p_n = \frac{1}{2} + \left(p_0 - \frac{1}{2}\right)(1 - 2p)^n.*$$

For example, the chance of a "yes" on the third roll call vote is

$$p_3 = \frac{1}{2} + \left(p_0 - \frac{1}{2}\right)(1 - 2p)^3.$$

Suppose that there is a chance of $p_0 = 0.7$ that Congressman Stephens will favor the budget appropriation bill before the first roll call, but only a probability of $p = 0.2$ that he will change his mind on the subsequent vote. Find and interpret the following.

a. p_2 **b.** p_4

c. p_8 **d.** $\lim_{n \to \infty} p_n$

■ 3.2 CONTINUITY

? **THINK ABOUT IT**   How does the recommended dosage of medication change with respect to weight?

? Figure 14 shows how the recommended dosage of Children's TYLENOL® Soft Chews† changes with respect to a child's weight. The weight is given in pounds and the dosage is given as the number of tablets recommended. Notice that whenever the dosage changes, the height of the graph jumps to a higher level. To study the behavior of this function at these points, we will use the idea of limits, presented in the previous section. Let us denote this function $g(w)$, where w is the weight, in pounds, of the child.

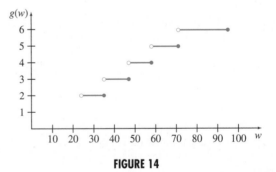

FIGURE 14

Notice from the graph that $\lim_{w \to 59^-} g(w) = 4$ and that $\lim_{w \to 59^+} g(w) = 5$, so $\lim_{w \to 59} g(w)$ does not exist. Notice also that $g(59) = 4$. A point such as this, where a function has a sudden sharp break, is a point where the function is *discontinuous*.

Intuitively speaking, a function is *continuous* at a point if you can draw the graph of the function near that point without lifting your pencil from the paper. As you can see, this would not be possible in Figure 14. Conversely, a function is

---

*Bishir, John W., and Donald W. Drewes, *Mathematics in the Behavioral and Social Sciences,* New York, Harcourt Brace Jovanovich, 1970, p. 538.
†Children's TYLENOL® Soft Chews, www.tylenol.com. Under 24 lbs: Please consult a physician.

discontinuous at any *x*-value where the pencil *must* be lifted from the paper in order to draw the graph on both sides of the point. A more precise definition is as follows.

DEFINITION OF CONTINUITY AT $x = c$

A function *f* is **continuous** at $x = c$ if the following three conditions are satisfied:

    1. $f(c)$ is defined,   **2.** $\lim\limits_{x \to c} f(x)$ exists, and   **3.** $\lim\limits_{x \to c} f(x) = f(c)$.

If *f* is not continuous at *c*, it is **discontinuous** there.

The following example shows the various ways a function can be discontinuous.

EXAMPLE 1   Continuity

Tell why the functions are discontinuous at the indicated *x*-values.

(a) $f(x)$ in Figure 15 at $x = 3$

    Solution The open circle on the graph of Figure 15 at the point where $x = 3$ means that $f(3)$ is not defined. Because of this, part 1 of the definition fails.

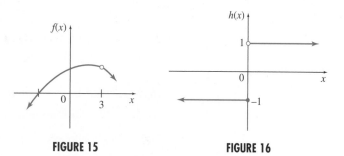

    FIGURE 15   **FIGURE 16**

(b) $h(x)$ in Figure 16 at $x = 0$

    Solution According to the graph of Figure 16, $h(0) = -1$. Also, as *x* approaches 0 from the left, $h(x)$ is -1. As *x* approaches 0 from the right, however, $h(x)$ is 1. In other words,

$$\lim_{x \to 0^-} h(x) = -1,$$

while

$$\lim_{x \to 0^+} h(x) = 1.$$

    Since no single number is approached by the values of $h(x)$ as *x* approaches 0, $\lim\limits_{x \to 0} h(x)$ does not exist, and part 2 of the definition fails.

(c) $g(x)$ in Figure 17 (on the next page) at $x = 4$

    Solution In Figure 17, the heavy dot above 4 shows that $g(4)$ is defined. In fact, $g(4) = 1$. The graph also shows, however, that

$$\lim_{x \to 4} g(x) = -2,$$

so $\lim\limits_{x \to 4} g(x) \neq g(4)$, and part 3 of the definition fails.

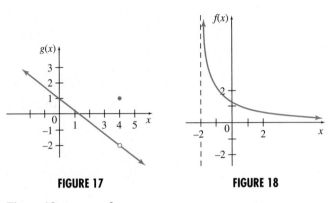

FIGURE 17   **FIGURE 18**

(d) $f(x)$ in Figure 18 at $x = -2$

 Solution The function f graphed in Figure 18 is not defined at $x = -2$, and $\lim\limits_{x \to -2} f(x)$ does not exist there. Either of these reasons is sufficient to show that f is not continuous at -2. (Function f is continuous at any value of x greater than -2, however.)

   A function is said to be **continuous on an open interval** if it is continuous at every x-value in the interval. Continuity on a closed interval is slightly more complicated because we must decide what to do with the endpoints. We will say that a function f is **continuous from the right** at $x = c$ if $\lim\limits_{x \to c^+} f(x) = f(c)$. A function f is **continuous from the left** at $x = c$ if $\lim\limits_{x \to c^-} f(x) = f(c)$. With these ideas, we can now define continuity on a closed interval.

> **CONTINUITY ON A CLOSED INTERVAL**
>
> A function is **continuous on a closed interval** $[a, b]$ if
>
> **1.** it is continuous on the open interval (a, b),
> **2.** it is continuous from the right at $x = a$, and
> **3.** it is continuous from the left at $x = b$.

   For example, the function $f(x) = \sqrt{1 - x^2}$, shown in Figure 19, is continuous on the closed interval $[-1, 1]$. By defining continuity on a closed interval in this way, we need not worry about the fact that $\sqrt{1 - x^2}$ does not exist to the left of $x = -1$ or to the right of $x = 1$.

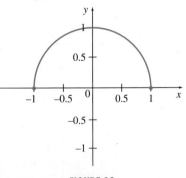

FIGURE 19

The next table lists some key functions and tells where each is continuous.

Continuous Functions

Type of Function	Where It Is Continuous	Graphic Example
Polynomial Function $y = a_nx^n + a_{n-1}x^{n-1} + \cdots + a_1x + a_0$, where a_n, $a_{n-1}, \ldots, a_1, a_0$ are real numbers, not all 0	For all x	
Rational Function $y = \dfrac{p(x)}{q(x)}$, where $p(x)$ and $q(x)$ are polynomials, with $q(x) \neq 0$	For all x where $q(x) \neq 0$	
Root Function $y = \sqrt{ax + b}$, where a and b are real numbers, with $a \neq 0$ and $ax + b \geq 0$	For all x where $ax + b \geq 0$	
Exponential Function $y = a^x$ where $a > 0$	For all x	
Logarithmic Function $y = \log_a x$ where $a > 0$	For all $x > 0$	

Continuous functions are nice to work with because finding $\lim\limits_{x \to c} f(x)$ is simple if f is continuous: just evaluate $f(c)$.

When a function is given by a graph, any discontinuities are clearly visible. When a function is given by a formula, it is usually continuous at all x-values except those where the function is undefined, or possibly where there is a change in the defining formula for the function, as in the following example.

EXAMPLE 2 Continuity

Find all values of x where the following function is discontinuous.

$$f(x) = \begin{cases} x + 1 & \text{if } x < 1 \\ x^2 - 3x + 4 & \text{if } 1 \leq x \leq 3. \\ 5 - x & \text{if } x > 3 \end{cases}$$

Solution A function defined by two or more cases is called a *piecewise function*. The only x-values where f might be discontinuous here are 1 and 3. We investigate at $x = 1$ first. From the left,

$$\lim_{x \to 1^-} f(x) = \lim_{x \to 1^-} (x + 1) = 1 + 1 = 2.$$

From the right,

$$\lim_{x \to 1^+} f(x) = \lim_{x \to 1^+} (x^2 - 3x + 4) = 1^2 - 3 + 4 = 2.$$

Furthermore, $f(1) = 1^2 - 3 + 4 = 2$, so $\lim_{x \to 1} f(x) = f(1) = 2$. Thus f is continuous at $x = 1$.

Now let us investigate $x = 3$. From the left,

$$\lim_{x \to 3^-} f(x) = \lim_{x \to 3^-} (x^2 - 3x + 4) = 3^2 - 3(3) + 4 = 4.$$

From the right,

$$\lim_{x \to 3^+} f(x) = \lim_{x \to 3^+} (5 - x) = 5 - 3 = 2.$$

Because $\lim_{x \to 3^-} f(x) \neq \lim_{x \to 3^+} f(x)$, $\lim_{x \to 3} f(x)$ does not exist, so f is discontinuous at $x = 3$, regardless of the value of $f(3)$. ▪

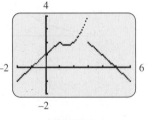

FIGURE 20

The graph of the function in Example 2 can be drawn by considering each of the three parts separately. For example, for the first part, the line $y = x + 1$ is drawn using the techniques of Chapter 1, but including only the section of the line to the left of $x = 1$. The other two parts are drawn similarly.

Alternatively, some graphing calculators have the ability to draw piecewise functions. On the TI-83, letting $Y_1 = (X + 1)(X < 1) + (X^2 - 3X + 4)(1 \leq X)$ $(X \leq 3) + (5 - X)(X > 3)$ produces the graph shown in Figure 20. It is important here that the graphing mode be set on "Dot" rather than "Connected." Otherwise, the calculator will show a line segment at $x = 3$ connecting the parabola to the line, although such a segment does not really exist.

One application of continuity is the **Intermediate Value Theorem,** which says that if a function is continuous on a closed interval $[a, b]$, the function takes on every value between $f(a)$ and $f(b)$. For example, if $f(1) = -3$ and $f(2) = 5$, then f must take on every value between -3 and 5 as x varies over the interval $[1, 2]$. In particular (in this case), there must be a value of x in the interval $(1, 2)$ such that $f(x) = 0$. If f were discontinuous, however, this conclusion would not necessarily be true. Before searching for a solution to $f(x) = 0$ in $[1, 2]$, we would like to know that a solution exists.

3.2 EXERCISES

In Exercises 1–6, find all values x = a where the function is discontinuous.
For each point of discontinuity, give **(a)** $\lim\limits_{x \to a^-} f(x)$, **(b)** $\lim\limits_{x \to a^+} f(x)$, **(c)** $\lim\limits_{x \to a} f(x)$, *and*
(d) $f(a)$ *if it exists.*

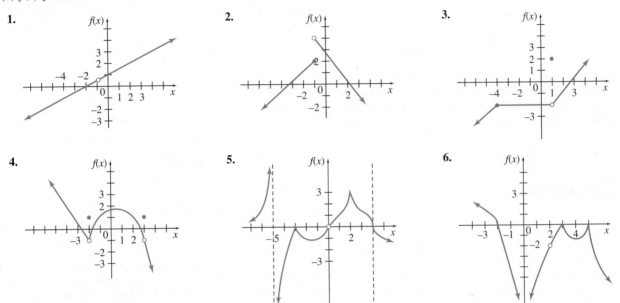

1.   **2.**   **3.**

4.   **5.**   **6.**

Find all values x = a where the function is discontinuous. For each value of x, give the
limit of the function as x approaches a.

7. $f(x) = \dfrac{5 + x}{x(x - 2)}$

8. $f(x) = \dfrac{-2x}{(2x + 1)(3x + 6)}$

9. $f(x) = \dfrac{x^2 - 4}{x - 2}$

10. $f(x) = \dfrac{x^2 - 25}{x + 5}$

11. $p(x) = x^2 - 4x + 11$

12. $q(x) = -3x^3 + 2x^2 - 4x + 1$

13. $p(x) = \dfrac{|x + 2|}{x + 2}$

14. $r(x) = \dfrac{|5 - x|}{x - 5}$

15. $f(x) = \sin\left(\dfrac{x}{x + 2}\right)$

16. $g(x) = \tan(\pi x)$

17. $k(x) = e^{\sqrt{x - 1}}$

18. $j(x) = e^{1/x}$

19. $r(x) = \ln\left(\dfrac{x}{x - 1}\right)$

20. $r(x) = \ln\left(\dfrac{x + 2}{x - 1}\right)$

In Exercises 21–26, **(a)** *graph the given function,* **(b)** *find all values of x where the*
function is discontinuous, and **(c)** *find the limit from the left and from the right at any*
values of x found in part b.

21. $f(x) = \begin{cases} 1 & \text{if } x < 2 \\ x + 3 & \text{if } 2 \le x \le 4 \\ 7 & \text{if } x > 4 \end{cases}$

22. $f(x) = \begin{cases} x - 1 & \text{if } x < 1 \\ 0 & \text{if } 1 \le x \le 4 \\ x - 2 & \text{if } x > 4 \end{cases}$

23. $g(x) = \begin{cases} 11 & \text{if } x < -1 \\ x^2 + 2 & \text{if } -1 \leq x \leq 3 \\ 11 & \text{if } x > 3 \end{cases}$

24. $g(x) = \begin{cases} 0 & \text{if } x < 0 \\ x^2 - 5x & \text{if } 0 \leq x \leq 5 \\ 5 & \text{if } x > 5 \end{cases}$

25. $h(x) = \begin{cases} 4x + 4 & \text{if } x \leq 0 \\ x^2 - 4x + 4 & \text{if } x > 0 \end{cases}$

26. $h(x) = \begin{cases} x^2 + x - 12 & \text{if } x \leq 1 \\ 3 - x & \text{if } x > 1 \end{cases}$

27. Explain in your own words what the Intermediate Value Theorem says and why it seems plausible.

28. Explain why $\lim\limits_{x \to 2}(3x^2 + 8x)$ can be evaluated by substituting $x = 3$.

In Exercises 29–30, (a) use a graphing calculator to tell where the rational function $P(x)/Q(x)$ is discontinuous, and (b) verify your answer from part (a) by using the graphing calculator to plot $Q(x)$ and determine where $Q(x) = 0$. You will need to choose the viewing window carefully.

29. $f(x) = \dfrac{x^2 + x + 2}{x^3 - 0.9x^2 + 4.14x - 5.4}$

30. $f(x) = \dfrac{x^2 + 3x - 2}{x^3 - 0.9x^2 + 4.14x + 5.4}$

Applications

LIFE SCIENCES

31. *Tumor Growth* A very aggressive tumor is growing according to the function $N(t) = 2^t$, where $N(t)$ represents the number of cells at time t (in months). The tumor continues to grow for 40 months, at which time the tumor is diagnosed.

a. How many tumor cells are there at diagnosis?

b. Suppose the patient receives chemotherapy immediately after the tumor has grown for 40 months, and 99.9% of the cells are instantaneously killed. Graph the function both before and after the chemotherapy. (*Hint:* To draw a manageable graph, use $\log_2 N(t)$ for the vertical axis.)

c. Is the function described in the graph in part b continuous? If not, then find the value(s) of t where the function is discontinuous.

32. *Pregnancy* During pregnancy, a woman's weight naturally increases during the course of the event. When she delivers, her weight immediately decreases by the approximate weight of the child. Suppose that a 120 lb woman gains 27 lbs during pregnancy, delivers a 7 lb baby, and then, through diet and exercise, loses the remaining weight during the next 20 weeks.

a. Graph the weight gain and loss during the pregnancy and the 20 weeks following the birth of the baby. Assume that the pregnancy lasts 40 weeks, that delivery occurs immediately after this time interval, and that the weight-gain/loss before and after birth is linear.

b. Is this a continuous function? If not, then find the value(s) of t where the function is discontinuous.

33. *Pharmacology* When a patient receives an injection, the amount of the injected substance in the body immediately goes up. Comment on whether the function which describes the amount of the drug with respect to time is a continuous function.

34. *Pharmacology* When a person takes a pill, the amount of the drug contained in the pill is immediately inside the body. Comment on whether the function which describes the amount of the drug inside the body with respect to time is a continuous function.

35. *Poultry Farming* Researchers at Iowa State University and the University of Arkansas have developed a piecewise function that can be used to estimate the body weight (in grams) of a male broiler during the first 56 days of life according to

$$W(t) = \begin{cases} 48 + 3.64t + 0.6363t^2 + 0.00963t^3, & 1 \leq t \leq 28, \\ -1{,}004 + 65.8t, & 28 < t \leq 56, \end{cases}$$

where t is the age of the chicken (in days).*

a. Determine the weight of a male broiler that is 25 days old.

b. Is $W(t)$ a continuous function?

c. Use a graphing calculator to graph $W(t)$ on $[1, 56]$ by $[0, 3{,}000]$. Comment on the accuracy of the graph.

d. Comment on why researchers would use two different

---

*Xin, H., I. Berry, T. Barton, and G. Tabler, "Feed and Water Consumption, Growth, and Mortality of Male Broilers," *Poultry Science*, Vol. 73, No. 5, May 1994, pp. 610–616.

types of functions to estimate the weight of a chicken at various ages.

36. *Electrocardiogram* Electrocardiograms (EKGs) are used by physicians to study the relative health of a patient's heart. The figures below indicate part of a complete EKG for a particular healthy patient.* The *x*-axis is time and the *y*-axis is voltage.

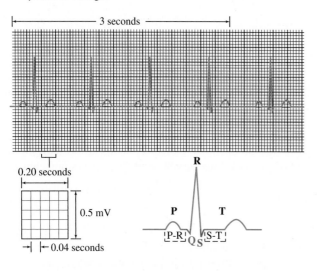

a. A patient's heart-rate can be determined by the formula $R = 300/N$, where N is the number of large squares between QRS complexes, or cycles of the heart. Find the heart rate for this particular patient.

b. Comment on whether the graph above actually represents a continuous function.

37. *Production* The graph shows the profit from the daily production of *x* thousand kilograms of an industrial disinfectant. Use the graph to find the following limits.

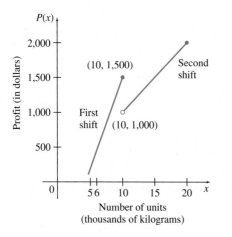

a. $\lim_{x \to 6} P(x)$

b. $\lim_{x \to 10^-} P(x)$

c. $\lim_{x \to 10^+} P(x)$

d. $\lim_{x \to 10} P(x)$

e. Where is the function discontinuous? What might account for such a discontinuity?

f. Use the graph to estimate the number of units of the chemical that must be produced before the second shift is as profitable as the first.

38. *Cost Analysis* The cost of ambulance transport depends on the distance, *x*, in miles that the patient is moved. Let $C(x)$ represent the cost to transport a patient *x* miles. One firm charges as follows.

Cost per Mile	Distance in Miles
$4.00	$0 < x \leq 150$
$3.00	$150 < x \leq 400$
$2.50	$400 < x$

Find the cost to transport a patient the following distances.

a. 130 mi

b. 150 mi

c. 210 mi

d. 400 mi

e. 500 mi

f. Where is *C* discontinuous?

OTHER APPLICATIONS

39. *Cost Analysis* A company charges $1.20 per pound for a certain fertilizer on all orders not over 100 lb, and $1 per pound for orders over 100 lb. Let $F(x)$ represent the cost for buying *x* lb of the fertilizer. Find the cost of buying the following.

a. 80 lb

b. 150 lb

c. 100 lb

d. Where is *F* discontinuous?

40. *Postage* In 2000, to send an international airmail letter to most countries from the United States, it cost $0.60 for the first half ounce, and $0.40 for each additional half ounce or

*Posey, Andrea, "Basic EKG Dysrhythmia Identification," http://www.rnceus.com/ekg/ekgcontent. html, 2001

fraction of a half ounce. Let $C(x)$ represent the postage for a first-class letter weighing x oz. Find each of the following.

a. $\lim\limits_{x \to 1.5^-} C(x)$

b. $\lim\limits_{x \to 1.5^+} C(x)$

c. $\lim\limits_{x \to 1.5} C(x)$

d. $C(1.5)$

e. Find all values of x on the interval $(0, 4)$ where the function C is discontinuous.

f. Sketch a graph of $y = C(x)$ on the interval $(0, 4]$.

3.3 RATES OF CHANGE

? THINK ABOUT IT   What is the relationship between the age of a moose and its weight?

This question will be answered in Example 1 of this section as we develop a method for finding the rate of change of one variable with respect to a unit change in another variable.

Average Rate of Change   One of the main applications of calculus is determining how one variable changes in relation to another. A marketing manager wants to know how profit changes with respect to the amount spent on advertising, while a physician wants to know how a patient's reaction to a drug changes with respect to the dose.

For another example, suppose we take a trip from San Francisco driving south. Every half-hour we note how far we have traveled, with the following results for the first three hours.

Time in Hours	0	0.5	1	1.5	2	2.5	3
Distance in Miles	0	20	48	80	104	126	150

If s is the function whose rule is

$$s(t) = \text{distance from San Francisco at time } t,$$

then the table shows, for example, that $s(0) = 0$, $s(1) = 48$, $s(2.5) = 126$, and so on. The distance traveled during, say, the second hour can be calculated by $s(2) - s(1) = 104 - 48 = 56$ miles.

Distance equals time multiplied by rate (or speed); so the distance formula is $d = rt$. Solving for rate gives $r = d/t$, or

$$\text{Average speed} = \frac{\text{Distance}}{\text{Time}}.$$

For example, the average speed over the time interval from $t = 0$ to $t = 3$ is

$$\text{Average speed} = \frac{s(3) - s(0)}{3 - 0} = \frac{150 - 0}{3} = 50,$$

or 50 miles per hour. We can use this formula to find the average speed for any interval of time during the trip, as shown on the next page.

FOR REVIEW ■

Recall the formula for the slope of a line through two points (x_1, y_1) and (x_2, y_2):

$$\frac{y_2 - y_1}{x_2 - x_1}.$$

Find the slopes of the lines through the following points.

(0.5, 20)	and	(1, 48)
(0.5, 20)	and	(1.5, 80)
(1, 48)	and	(2, 104)

Compare your answers to the average speeds shown in the chart.

Time Interval	Average Speed $= \dfrac{\text{Distance}}{\text{Time}}$
$t = 0.5$ to $t = 1$	$\dfrac{s(1) - s(0.5)}{1 - 0.5} = \dfrac{28}{0.5} = 56$
$t = 0.5$ to $t = 1.5$	$\dfrac{s(1.5) - s(0.5)}{1.5 - 0.5} = \dfrac{60}{1} = 60$
$t = 1$ to $t = 2$	$\dfrac{s(2) - s(1)}{2 - 1} = \dfrac{56}{1} = 56$
$t = 1$ to $t = 3$	$\dfrac{s(3) - s(1)}{3 - 1} = \dfrac{102}{2} = 51$
$t = a$ to $t = b$	$\dfrac{s(b) - s(a)}{b - a}$

The analysis of the average speed or **average rate of change** of distance s with respect to t can be applied to any function defined by $f(x)$ to get a formula for the average rate of change of f with respect to x.

AVERAGE RATE OF CHANGE

The average rate of change of $f(x)$ with respect to x for a function f as x changes from a to b is

$$\frac{f(b) - f(a)}{b - a}.$$

NOTE The formula for the average rate of change is the same as the formula for the slope of the line through $(a, f(a))$ and $(b, f(b))$. This connection between slope and rate of change will be examined more closely in the next section.

We will sometimes refer to the quantity $(f(b) - f(a))/(b - a)$ as the **difference quotient**.

EXAMPLE 1 Alaskan Moose

?

Researchers from the Moose Research Center in Soldotna, Alaska, have developed a mathematical relationship between the age of a captive female moose and its weight. The function is

$$M(t) = 369(0.93)^t t^{0.36},$$

where M is the weight of the moose (in kg) and t is the age (in years) of the moose.* Find the average rate of change in the weight of a female moose between the ages of two and three years.

Solution On the interval from $t = 2$ to $t = 3$, the average rate of change is

$$\frac{M(3) - M(2)}{3 - 2} = \frac{369(0.93)^3(3)^{0.36} - 369(0.93)^2(2)^{0.36}}{3 - 2} \approx 31.194.$$

*Schwartz, C., and Kris Hundertmark, "Reproductive Characteristics of Alaskan Moose," *Journal of Wildlife Management,* Vol. 57, No. 3, July 1993, pp. 454–468.

Therefore, the average amount of weight gained by a captive female moose be-tween the ages of two and three is 31.194 kg per year.

EXAMPLE 2 Drug Control

The graph in Figure 21 shows federal spending $P(t)$, measured in millions of dol-lars, as a function of time t, measured in years, on drug control programs.* The average rate of change of *price* with respect to *time* on some interval is defined as the change in price divided by the change in time on the interval. Prices must be estimated from the graph; we estimate that the amount spent (in millions of dollars) in 1995 was \$13,250, the amount spent in 1997 was \$15,000, the amount spent in 1999 was \$17,900, the amount spent in 2000 was \$17,800, the amount spent in 2001 was \$18,100, and the amount spent in 2002 was \$18,800. (Verify these figures from the graph.)

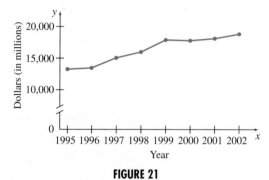

FIGURE 21

(a) From 1995 to 1997, the average rate of change in the amount of federal drug control spending with respect to time is given by

$$\frac{P(1997) - P(1995)}{1997 - 1995} = \frac{15,000 - 13,250}{2} = \frac{1,750}{2} = \$875.$$

On average, the amount the federal government spent to control drugs in-creased by 875 million dollars per year in this time interval.

(b) From 1999 to 2002, the average rate of change in the amount of federal drug control spending with respect to time is given by

$$\frac{P(2002) - P(1999)}{2002 - 1999} = \frac{18,800 - 17,900}{3} = \frac{900}{3} = \$300.$$

On average, the amount the federal government spent to control drugs in-creased by 300 million dollars per year in this time interval.

(c) From the amount spent in 1999 to the amount spent in 2000, the average rate of change in the amount of federal drug control spending with respect to time is

$$\frac{P(2000) - P(1999)}{2000 - 1999} = \frac{17,800 - 17,900}{1} = -\frac{100}{1} = -\$100.$$

*National Drug Control Strategy, FY 2003 Budget Summary, The White House, www.whitehousedrugpolicy.gov/publications/.

On average, the amount the federal government spent to control drugs decreased by 100 million dollars in this time interval.

Instantaneous Rate of Change

Finding the average rate of change of a function over a large interval can sometimes lead to answers that are not very helpful. The results often are more useful if the average rate of change is found over a fairly narrow interval. Finding the exact rate of change at a particular *x*-value requires a continuous function. For example, suppose a car starts from a stop sign and moves along a straight road. Assume that the distance in feet traveled in *t* seconds is given by $s(t) = t^2 + 3t$. After 5 seconds, the car has traveled $s(5) = 5^2 + 3(5) = 40$ feet. The speed of the car at exactly 10 seconds after starting can be estimated by finding the average speed over shorter and shorter time intervals.

Interval	Average Speed	
$t = 10$ to $t = 10.1$	$\dfrac{s(10.1) - s(10)}{10.1 - 10} = \dfrac{132.31 - 130}{0.1}$	$= 23.1$
$t = 10$ to $t = 10.01$	$\dfrac{s(10.01) - s(10)}{10.01 - 10} = \dfrac{130.2301 - 130}{0.01}$	$= 23.01$
$t = 10$ to $t = 10.001$	$\dfrac{s(10.001) - s(10)}{10.001 - 10} = \dfrac{130.023001 - 130}{0.001}$	$= 23.001$

The results in the chart suggest that the exact speed at $t = 10$ seconds is 23 feet per second.

This example can be easily generalized to any function *f*. Let *a* be a specific *x*-value, such as 10 in the example. Let *h* be a (small) number. The average rate of change of *f* as *x* changes from *a* to $a + h$ is

$$\frac{f(a + h) - f(a)}{(a + h) - a} = \frac{f(a + h) - f(a)}{h}.$$

The exact rate of change of *f* at $x = a$, called the **instantaneous rate of change of *f* at *x* = *a*,** is the limit of this quotient.

INSTANTANEOUS RATE OF CHANGE

The instantaneous rate of change for a function *f* when $x = a$ is

$$\lim_{h \to 0} \frac{f(a + h) - f(a)}{h},$$

provided this limit exists.

CAUTION Remember that $f(x + h) \neq f(x) + f(h)$. To find $f(x + h)$, replace *x* with $x + h$ in the expression for $f(x)$. For example, if $f(x) = x^2$,

$$f(x + h) = (x + h)^2 = x^2 + 2xh + h^2,$$

but

$$f(x) + f(h) = x^2 + h^2.$$

For the next example, we need to make a subtle distinction between speed, the term we have been using to describe how fast someone is moving, and velocity, which also incorporates the direction of the motion. In any motion along a straight line, one direction is arbitrarily labeled as positive, so the other direction is negative. Velocity in the negative direction is considered negative. Speed is always positive; it is the absolute value of velocity.

EXAMPLE 3 Velocity

The distance in feet of an object from a starting point is given by $s(t) = 2t^2 - 5t + 40$, where t is time in seconds.

(a) Find the average velocity of the object from 2 seconds to 4 seconds.

Solution The average velocity is

$$\frac{s(4) - s(2)}{4 - 2} = \frac{52 - 38}{2} = \frac{14}{2} = 7$$

feet per second.

(b) Find the instantaneous velocity at 4 seconds.

Solution For $t = 4$, the instantaneous velocity is

$$\lim_{h \to 0} \frac{s(4 + h) - s(4)}{h}$$

feet per second.

$$s(4 + h) = 2(4 + h)^2 - 5(4 + h) + 40$$
$$= 2(16 + 8h + h^2) - 20 - 5h + 40$$
$$= 32 + 16h + 2h^2 - 20 - 5h + 40$$
$$= 2h^2 + 11h + 52$$

Also,

$$s(4) = 2(4)^2 - 5(4) + 40 = 52.$$

Therefore, the instantaneous velocity at $t = 4$ is

$$\lim_{h \to 0} \frac{(2h^2 + 11h + 52) - 52}{h} = \lim_{h \to 0} \frac{2h^2 + 11h}{h}$$
$$= \lim_{h \to 0} (2h + 11) = 11,$$

or 11 feet per second.

EXAMPLE 4 Mass of Bighorn Yearlings

Researchers have determined that the body mass of yearling bighorn sheep on Ram Mountain in Alberta, Canada, can be estimated by

$$M(t) = 27.5 + 0.3t - 0.001t^2$$

where $M(t)$ is the mass of the sheep (in kg) and t is the number of days since May 25.*

---

*Jorgenson, J., M. Festa-Bianchet, M. Lucherini, and W. Wishart, "Effects of Body Size, Population Density, and Maternal Characteristics on Age at First Reproduction of Bighorn Ewes," *Canadian Journal of Zoology*, Vol. 71, No. 12, Dec. 1993, pp. 2509–2517.

(a) Find the average rate of change of the weight of a bighorn yearling between 70 and 75 days past May 25.

Solution Use the formula for average rate of change. The weight of the yearling 70 days after May 25 is

$$M(70) = 27.5 + 0.3(70) - 0.001(70)^2 = 43.6 \text{ kg}.$$

Similarly, the weight of the yearling 75 days after May 25 is

$$M(75) = 27.5 + 0.3(75) - 0.001(75)^2 = 44.375 \text{ kg}.$$

The average rate of change of weight in this time interval is

$$\frac{M(75) - M(70)}{75 - 70} = \frac{44.375 - 43.6}{75 - 70} = 0.155.$$

Thus, on the average, the weight increases at the rate of 0.155 kg/day as the age of the sheep increases from 70 to 75 days past May 25.

(b) Find the instantaneous rate of change of weight for a yearling sheep whose age is 70 days past May 25.

Solution The instantaneous rate of change for $t = 70$ is given by

$$\lim_{h \to 0} \frac{M(70 + h) - M(70)}{h}$$

$$= \lim_{h \to 0} \frac{[27.5 + 0.3(70 + h) - 0.001(70 + h)^2] - [27.5 + 0.3(70) - 0.001(70)^2]}{h}$$

$$= \lim_{h \to 0} \frac{27.5 + 21 + 0.3h - 4.9 - 0.14h - 0.001h^2 - 43.6}{h}$$

$$= \lim_{h \to 0} \frac{0.16h - 0.001h^2}{h} \qquad \text{Combine terms.}$$

$$= \lim_{h \to 0} \frac{h(0.16 - 0.001h)}{h} \qquad \text{Divide by } h.$$

$$= \lim_{h \to 0}(0.16 - 0.001h) = 0.16. \qquad \text{Calculate the limit.}$$

When 70 days have passed since May 25, the weight of the yearling bighorn sheep is increasing at the rate of 0.16 kg/day. We can use this notion to predict that the sheep will weigh $43.6 + 0.16 = 43.76$ kg on day 71. Another name for this instantaneous rate is the *marginal value*. The concept of marginal analysis is quite important in the field of economics where people are interested in calculating marginal cost, marginal demand, marginal profit, and marginal revenue. These values are obtained in the same manner as described above.

EXAMPLE 5 Alaskan Moose

Estimate the instantaneous rate of change at the beginning of year two for a captive female moose.

Solution We saw in Example 1 that the estimated weight of a captive female moose (in kg) is given by $M(t) = 369(0.93)^t t^{0.36}$, where t is the age (in years) of the moose.* Unlike the previous example, in which the function was a polynomial, the function in this example is exponential, making it harder to compute the limit directly using the formula for instantaneous rate of change. Instead, we will approximate the instantaneous rate of change at $t = 2$ by using smaller and smaller values of h. This can be done using the TABLE feature on a TI-83 graphing calculator by entering Y_1 as the function above, and $Y_2 = (Y_1(2 + X) - Y_1(2))/X$. (The graphing calculator requires us to use X in place of h in the formula for instantaneous rate of change.) The result is shown in Figure 22. This table can also be generated using a spreadsheet.

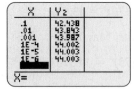

FIGURE 22

The limit seems to be approaching 44.003. Thus, the instantaneous rate of change in the weight of a captive two-year-old female moose is about 44 kg per year.

When we have data, rather than a formula, we can find the average velocity or rate of change, but we can only approximate the instantaneous velocity or rate of change.

EXAMPLE 6 Velocity

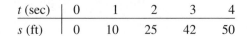

One day Musk, the friendly pit bull, escaped from the yard and ran across the street to see a neighbor, who was 50 ft away. An estimate of the distance Musk ran as a function of time is given by the following table.

t (sec)	0	1	2	3	4
s (ft)	0	10	25	42	50

(a) Find Musk's average velocity during her 4-second trip.

Solution The total distance she traveled is 50 ft, and the total time is 4 seconds, so her average velocity is $50/4 = 12.5$ feet per second.

(b) Estimate Musk's velocity at 2 seconds.

Solution We could estimate her velocity by taking the short time interval from 2 to 3 seconds, for which the velocity is

$$\frac{42 - 25}{1} = 17 \text{ feet per second.}$$

Alternatively, we could estimate her velocity by taking the short time interval from 1 to 2 seconds, for which the velocity is

$$\frac{25 - 10}{1} = 15 \text{ feet per second.}$$

A better estimate is found by averaging these two values, to get

$$\frac{17 + 15}{2} = 16 \text{ feet per second.}$$

Another way to get this same answer is to take the time interval from 1 to 3 seconds, for which the velocity is

$$\frac{42 - 10}{2} = 16 \text{ feet per second.}$$

*Schwartz, C., and Kris Hundertmark, "Reproductive Characteristics of Alaskan Moose," *Journal of Wildlife Management,* Vol. 57, No. 3, July 1993, pp. 454–468.

This answer is reasonable if we assume Musk's velocity changes at a fairly steady rate, and does not increase or decrease drastically from one second to the next. It is impossible to calculate Musk's exact velocity without knowing her position at times arbitrarily close to 2 seconds, or without a formula for her position as a function of time, or without a radar gun or speedometer on her. (In any case, she was very happy when she reached the neighbor.)

3.3 EXERCISES

Find the average rate of change for each function over the given interval.

1. $y = x^2 + 2x$ between $x = 0$ and $x = 3$

2. $y = -4x^2 - 6$ between $x = 2$ and $x = 5$

3. $y = 2x^3 - 4x^2 + 6x$ between $x = -1$ and $x = 1$

4. $y = -3x^3 + 2x^2 - 4x + 1$ between $x = 0$ and $x = 1$

5. $y = \sqrt{x}$ between $x = 1$ and $x = 4$

6. $y = \sqrt{3x - 2}$ between $x = 1$ and $x = 2$

7. $y = \dfrac{1}{x - 1}$ between $x = -2$ and $x = 0$

8. $y = \dfrac{-5}{2x - 3}$ between $x = 2$ and $x = 4$

Suppose the position of an object moving in a straight line is given by $s(t) = t^2 + 5t + 2$. Find the instantaneous velocity at each of the following times.

9. $t = 6$ 10. $t = 1$

Suppose the position of an object moving in a straight line is given by $s(t) = t^3 + 2t + 9$. Find the instantaneous velocity at each of the following times.

11. $t = 1$ 12. $t = 4$

Find the instantaneous rate of change for each function at the given value.

13. $f(x) = x^2 + 2x$ at $x = 0$ 14. $s(t) = -4t^2 - 6$ at $t = 2$ 15. $g(t) = 1 - t^2$ at $t = -1$ 16. $F(x) = x^2 + 2$ at $x = 0$

 *Use the formula for instantaneous rate of change, approximating the limit by using smaller and smaller values of h, to find the instantaneous rate of change for each function at the given value.*

17. $f(x) = x^x$ at $x = 2$ 18. $f(x) = x^x$ at $x = 3$ 19. $f(x) = x^{\ln x}$ at $x = 2$ 20. $f(x) = x^{\ln x}$ at $x = 3$

21. Explain the difference between the average rate of change of y as x changes from a to b, and the instantaneous rate of change of y at $x = a$.

22. If the instantaneous rate of change of $f(x)$ with respect to x is positive when $x = 1$, is f increasing or decreasing there?

Applications

LIFE SCIENCES

23. *Molars* As we saw in the section on quadratic functions, the mesiodistal crown length (as shown on the next page)

of deciduous mandibular first molars in fetuses is related to the postconception age of the tooth as

$$L(t) = -0.01t^2 + 0.788t - 7.048,$$

where $L(t)$ is the crown length, in millimeters, of the molar t weeks after conception.*

a. Find the average rate of growth in mesiodistal crown length during weeks 22 through 28.

b. Find the instantaneous rate of growth in mesiodistal crown length when the tooth is exactly 22 weeks of age.

c. Graph the function on $[0, 50]$ by $[0, 9]$. Does a function that increases and then begins to decrease make sense for this particular application? What do you suppose is happening during the first 11 weeks? Does this function accurately model crown length during those weeks?

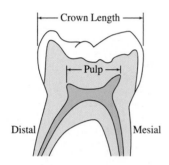

Crown Length

Pulp

Distal   Mesial

24. *Thermic Effect of Food* The metabolic rate of a person who has just eaten a meal tends to go up and then, after some time has passed, returns to a resting metabolic rate. This phenomenon is known as the thermic effect of food. Researchers have indicated that the thermic effect of food (in kJ/hr) for one person is

$$F(t) = -10.28 + 175.9te^{-t/1.3},$$

where t is the number of hours that have elapsed since eating a meal.[†]

a. Graph the function on $[0, 6]$ by $[-20, 100]$.

b. Find the average rate of change of the thermic effect of food during the first hour after eating.

c. Use a graphing calculator to find the instantaneous rate of change of the thermic effect of food exactly one hour after eating.

d. Use a graphing calculator to estimate when the function stops increasing and begins to decrease.

25. *Flu Epidemic* Epidemiologists in College Station, Texas, estimate that t days after the flu begins to spread in town, the percent of the population infected by the flu is approximated by

$$p(t) = t^2 + t$$

for $0 \le t \le 5$.

a. Find the average rate of change of p with respect to t over the interval from 1 to 4 days.

b. Find the instantaneous rate of change of p with respect to t at $t = 3$.

26. *World Population Growth* The future size of the world population depends on how soon it reaches replacement-level fertility, the point at which each woman bears on average about 2.1 children. The graph shows projections for reaching that point in different years.[‡]

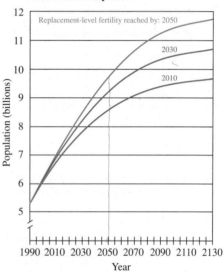

Ultimate World Population Size Under Different Assumptions

a. Estimate the average rate of change in population for each projection from 1990 to 2050. Which projection shows the smallest rate of change in world population?

b. Estimate the average rate of change in population from 2090 to 2130 for each projection. Interpret your answer.

27. *Bacteria Population* The graph on the next page shows the population in millions of bacteria t minutes after an antibiotic is introduced into a culture. Find and interpret the average rate of change of population with respect to time for the following time intervals.

a. 1 to 2   **b.** 2 to 3   **c.** 3 to 4   **d.** 4 to 5

*Harris, E. F., J. D. Hicks, and B. D. Barcroft, "Tissue Contributions to Sex and Race: Differences in Tooth Crown Size of Deciduous Molars," *American Journal of Physical Anthropology,* Vol. 115, 2001, pp. 223–237.
†Reed, G., and J. Hill, "Measuring the Thermic Effect of Food," *American Journal of Clinical Nutrition,* Vol. 63, 1996, pp. 164–169.
‡Carl Haub, Population Reference Bureau, 2000.

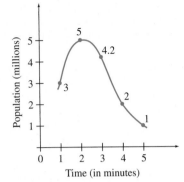

e. How long after the antibiotic was introduced did the population begin to decrease?

f. At what time did the rate of decrease of the population begin to slow down?

28. *Mass of Bighorn Yearlings* As indicated in Example 4, the body mass of yearling bighorn sheep on Ram Mountain in Alberta, Canada, can be estimated by

$$M(t) = 27.5 + 0.3t - 0.001t^2$$

where $M(t)$ is measured in kg and t is days since May 25.*

a. Find the average rate of change of the weight of a bighorn yearling between 105 and 115 days past May 25.

b. Find the instantaneous rate of change of weight for a bighorn yearling sheep whose age is 105 days past May 25.

c. Graph the function $M(t)$ on $[5, 125]$ by $[25, 65]$.

d. Does the behavior of the function past 125 days accurately model the mass of the sheep? Why or why not?

29. *Drug Trade* The U.S. government estimates of the worldwide potential net drug production (in metric tons) for the years 1994–1998 are given in the graph below.[†]

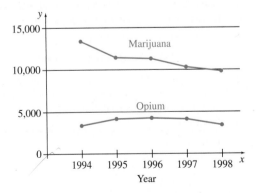

a. Estimate the average rate of change in the net production of marijuana and opium from 1994 to 1996 and from 1996 to 1998.

b. Which drug had the greatest change in net production during this time?

c. List some possible reasons for the observed decreases in marijuana production and the observed behavior of opium production during this time.

30. *High School Drug Use* The U.S. government estimates of the percentage of high school seniors who have tried illegal drugs at least once in their lives are shown in the graph below.[†]

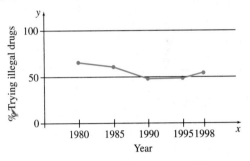

a. Estimate the average rate of change in the percentage of high school seniors who have tried illegal drugs from 1980 to 1990 and from 1990 to 1998.

b. Estimate the average rate of change in the percentage of high school seniors who have tried illegal drugs from 1980 to 1998 and compare your answer to those found in part a.

c. List some possible reasons for the observed decrease and then increase in the percentage of high school seniors who have tried illegal drugs.

31. *Medicare Trust Fund* The graph shows the money remaining in the Medicare Trust Fund at the end of the calendar year, adjusted for inflation in 2000 dollars.[‡]

Medicare Trust Fund

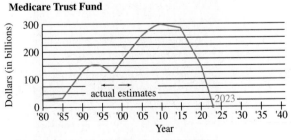

Using the Consumer Price Index for Urban Wage Earners and Clerical Workers

---

*Jorgenson, J., M. Festa-Bianchet, M. Lucherini, and W. Wishart, "Effects of Body Size, Population Density, and Maternal Characteristics on Age at First Reproduction of Bighorn Ewes," *Canadian Journal of Zoology,* Vol. 71, No. 12, Dec. 1993, pp. 2509–2517.
[†]"ONDCP Drug Policy Information Clearinghouse, Drug Data Summary," Executive Office of President, Office of National Drug Control Policy, NCJ-175050, June 1999, p. 5.
[‡]Social Security Administration; Department of Health and Human Services.

Find the approximate average rate of change in the trust fund for each time period.

a. From 1994 (the peak) to 1998 (the low point)

b. From 1998 to the estimated value for 2010

c. From 1990 to 1998

32. *Human Strength* According to research done by T. Curetun and Barry Ak at the University of Illinois Sports Fitness School, the distance (in inches) that typical boys of various ages can throw an 8 lb shotput can be estimated by the following graph.* Use the graph to estimate and interpret the average rate of increase in shotput distance for boys from 9 to 11 years old.

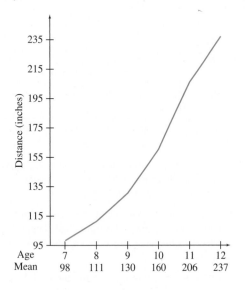

Age	7	8	9	10	11	12
Mean	98	111	130	160	206	237

OTHER APPLICATIONS

33. *New Technology versus Old* The following graph shows the ratio of users of new technology to users of old technology for two new technologies—Internet messages and PCs.[†]

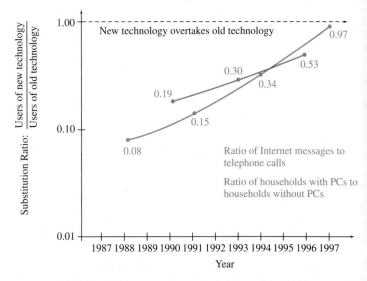

a. Which curve has a greater rate of change from 1991 to 1997? Interpret your answer.

b. The graphs indicate that one curve is approximately linear, while the other could be modeled by an exponential function. Which ratio has an increasing, rather than constant, rate of change?

34. *Temperature* The graph on the next page shows the temperature T in degrees Celsius as a function of the altitude h (in feet) when an inversion layer is over Southern California. (An inversion layer is formed when air at a higher altitude, say 3,000 ft, is warmer than air at sea level, even though air normally is cooler with increasing altitude.) Estimate and interpret the average rate of change in temperature for the following changes in altitude.

a. 1,000 to 3,000 ft   **b.** 1,000 to 5,000 ft

c. 3,000 to 9,000 ft   **d.** 1,000 to 9,000 ft

e. At what altitude at or below 7,000 ft is the temperature highest? lowest? How would your answer change if 7,000 ft is changed to 10,000 ft?

f. At what altitude is the temperature the same as it is at 1,000 ft?

*Hensinger, Robert, *Standards in Pediatric Orthopedics: Tables, Charts, and Graphs Illustrating Growth,* New York, Raven Press, 1986, p. 378.
†Data from May 1999 survey by Mercer Management Consulting in Washington, D.C.

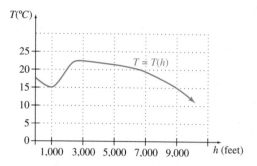

35. *Velocity* A car is moving along a straight test track. The position in feet of the car, $s(t)$, at various times t is measured, with the following results.

t (seconds)	0	2	4	6	8	10
$s(t)$ (feet)	0	10	14	20	30	36

Find and interpret the average velocities for the following changes in t.

a. 0 to 2 sec   **b.** 2 to 4 sec

c. 4 to 6 sec   **d.** 6 to 8 sec

e. Estimate the instantaneous velocity at 4 seconds
   i. by using the formula for estimating instantaneous rate (with $h = 2$), and
   ii. by averaging the answers for the average velocity in the two seconds before and the two seconds afterwards (that is, the answers to parts b and c).

f. Estimate the instantaneous velocity at 6 seconds using the two methods in part e.

g. Notice in parts e and f that your two answers are the same. Discuss whether this will always be the case, and why.

3.4 DEFINITION OF THE DERIVATIVE

? THINK ABOUT IT   How does the risk of chromosomal abnormality in a child change with the mother's age?

We will answer this question in Example 3, using the concept of the derivative. In the previous section, the formula

$$\lim_{h \to 0} \frac{f(a + h) - f(a)}{h}$$

was used to calculate the instantaneous rate of change of a function f at the point where $x = a$. Now we will give a geometric interpretation of this limit.

The Tangent Line   In geometry, a tangent line to a circle is defined as a line that touches the circle at only one point, as at the point P in Figure 23 on the next page (which shows the top half of a circle). If you think of this half-circle as part of a curving road on which you are driving at night, then the tangent line indicates the direction of the light beam from your headlights as you pass through the point P. Intuitively, the tangent line to an arbitrary curve at a point P on the curve should touch the curve at P, but not at any points nearby, and should indicate the direction of the curve. In Figure 24, for example, the lines through P_1 and P_3 are tangent lines, while the lines through P_2 and P_5 are not. The tangent lines just touch the curve, while the other lines pass through it. To decide about the line at P_4, we need to define the idea of a tangent line to the graph of a function more carefully.

To see how we might define the slope of a line tangent to the graph of a function f at a given point, let R be a fixed point with coordinates $(a, f(a))$ on the graph of a function $y = f(x)$, as in Figure 25. Choose a different point S on the graph

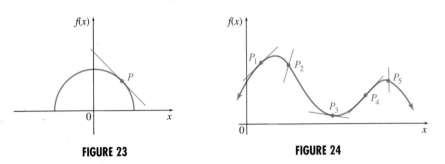

FIGURE 23   **FIGURE 24**

and draw the line through R and S; this line is called a **secant line.** If S has coordinates $(a + h, f(a + h))$, then by the definition of slope, the slope of the secant line RS is given by

$$\text{Slope of secant} = \frac{\Delta y}{\Delta x} = \frac{f(a + h) - f(a)}{a + h - a} = \frac{f(a + h) - f(a)}{h}.$$

This slope corresponds to the average rate of change of y with respect to x over the interval from a to $a + h$. As h approaches 0, point S will slide along the curve, getting closer and closer to the fixed point R. See Figure 26, which shows successive positions S_1, S_2, S_3, and S_4 of the point S. If the slopes of the corresponding secant lines approach a limit as h approaches 0, then this limit is defined to be the slope of the tangent line at point R.

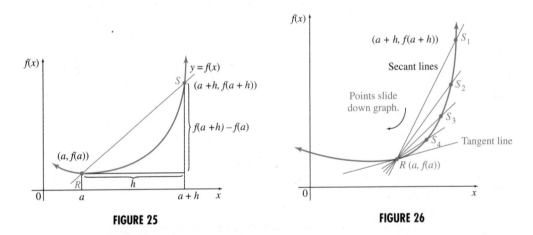

FIGURE 25   **FIGURE 26**

SLOPE OF THE TANGENT LINE

The **tangent line** of the graph of $y = f(x)$ at the point $(a, f(a))$ is the line through this point having slope

$$\lim_{h \to 0} \frac{f(a + h) - f(a)}{h},$$

provided this limit exists. If this limit does not exist, then there is no tangent at the point.

The slope of this line at a point is also called the **slope of the curve** at the point and corresponds to the instantaneous rate of change of y with respect to x at the point.

In certain applications of mathematics it is necessary to determine the equation of a line tangent to the graph of a function at a given point, as in the next example.

EXAMPLE 1 Tangent Line

Find the slope of the tangent line to the graph of $f(x) = x^2 + 2$ at $x = -1$. Find the equation of the tangent line.

Solution Use the definition given above, with $f(x) = x^2 + 2$ and $a = -1$. The slope of the tangent line is given by

$$\text{Slope of tangent} = \lim_{h \to 0} \frac{f(a + h) - f(a)}{h}$$

$$= \lim_{h \to 0} \frac{[(-1 + h)^2 + 2] - [(-1)^2 + 2]}{h}$$

$$= \lim_{h \to 0} \frac{[1 - 2h + h^2 + 2] - [1 + 2]}{h}$$

$$= \lim_{h \to 0} \frac{-2h + h^2}{h}$$

$$= \lim_{h \to 0}(-2 + h) = -2.$$

The slope of the tangent line at $(-1, f(-1)) = (-1, 3)$ is -2.

The equation of the tangent line can be found with the point-slope form of the equation of a line from Chapter 1.

$$y - y_1 = m(x - x_1)$$

$$y - 3 = -2[x - (-1)]$$

$$y - 3 = -2(x + 1)$$

$$y - 3 = -2x - 2$$

$$y = -2x + 1$$

Figure 27 shows a graph of $f(x) = x^2 + 2$, along with a graph of the tangent line at $x = -1$.

FOR REVIEW ▣

In Section 1.1, we saw that the equation of a line can be found with the point-slope form $y - y_1 = m(x - x_1)$, if the slope m and the coordinates (x_1, y_1) of a point on the line are known. Use the point-slope form to find the equation of the line with slope 3 that goes through the point $(-1, 4)$.

Let $m = 3$, $x_1 = -1$, $y_1 = 4$. Then

$$y - y_1 = m(x - x_1)$$

$$y - 4 = 3(x - (-1))$$

$$y - 4 = 3x + 3$$

$$y = 3x + 7.$$

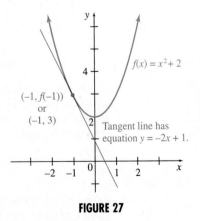

FIGURE 27

NOTE Some graphing calculators have a TANLN feature which draws the tangent line and gives its equation. If your calculator has this feature, use it to duplicate Figure 27.

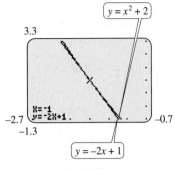

FIGURE 28

Figure 28 shows the result of using a graphing calculator to zoom in on the point $(-1, 3)$ in Figure 27. Notice that in this closeup view, the graph and its tangent line appear virtually identical. This gives us another interpretation of the tangent line. Suppose, as we zoom in on a function, the graph appears to become a straight line. Then this line is the tangent line to the graph at that point. In other words, the tangent line captures the behavior of the function very close to the point under consideration. (This assumes, of course, that the function when viewed close up is approximately a straight line. As we will see later in this section, this may not occur.)

> If it exists, the tangent line at $x = a$ is a good approximation of the graph of a function near $x = a$.

Consequently, another way to approximate the slope of the curve is to zoom in on the function until it appears to be a straight line (the tangent line). Then find the slope using any two points on that line.

EXAMPLE 2 Slope

Use a graphing calculator to find the slope of the graph of $f(x) = x^x$ at $x = 1$.

Solution The slope would be challenging to evaluate algebraically using the limit definition. Instead, using a graphing calculator on the window $[0, 2]$ by $[0, 2]$, we see the graph in Figure 29. Zooming in gives the view in Figure 30. Using the TRACE key, we find two points on the line to be $(1, 1)$ and $(1.0021277, 1.0021322)$. Therefore, the slope is approximately

$$\frac{1.0021322 - 1}{1.0021277 - 1} \approx 1.$$

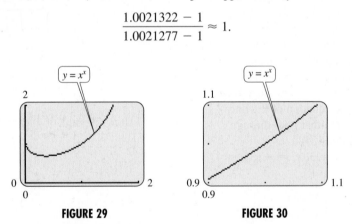

FIGURE 29   **FIGURE 30**

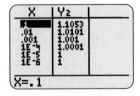

FIGURE 31

Rather than using a graph, we could use the graphing calculator to create a table, as we did in the previous section to estimate the instantaneous rate of change. Letting $Y_1 = X \wedge X$ and $Y_2 = (Y_1(1 + X) - Y_1(1))/X$ results in the table shown in Figure 31. Based on this table, we estimate that the slope of the graph of $f(x) = x^x$ at $x = 1$ is 1.

EXAMPLE 3 Genetics

Figure 32 on the next page shows how the risk of chromosomal abnormality in a child increases with the age of the mother.* Find the rate that the risk is rising when the mother is 40 years old.

*The New York Times, Feb. 5, 1994, p. 24.

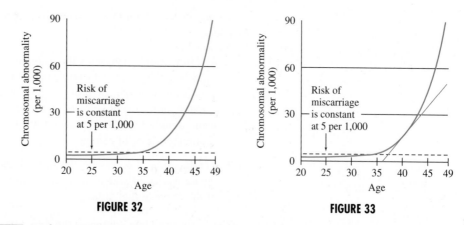

FIGURE 32   **FIGURE 33**

Solution In Figure 33, we have added the tangent line to the graph at the point where the age of the mother is 40. At that point, the risk is approximately 15 per 1,000. Extending the line, we estimate that when the age is 45, the y-coordinate of the line is roughly 35. Thus, the slope of the line is

$$\frac{35 - 15}{45 - 40} = \frac{20}{5} = 4.$$

Therefore, at the age of 40, the risk of chromosomal abnormality in the child is increasing at the rate of about 4 per 1,000 for each additional year of the mother's age.

The Derivative If $y = f(x)$ is a function and a is a number in its domain, then we shall use the symbol $f'(a)$ to denote the special limit

$$\lim_{h \to 0} \frac{f(a + h) - f(a)}{h},$$

provided that it exists. This means that for each number a we can assign the number $f'(a)$ found by calculating this limit. This assignment defines an important new function.

DERIVATIVE

The **derivative** of the function f at x is defined as

$$f'(x) = \lim_{h \to 0} \frac{f(x + h) - f(x)}{h},$$

provided this limit exists.

The notation $f'(x)$ is read "f-prime of x." The function $f'(x)$ is called the derivative of f with respect to x. If x is a value in the domain of f and if $f'(x)$ exists, then f is **differentiable** at x. The process that produces f' is called **differentiation.**

NOTE The derivative is a *function of x,* since $f'(x)$ varies as x varies. This differs from both the slope of the tangent line and the instantaneous rate of change, either of which is represented by the number $f'(a)$ that corresponds to a number a.

The derivative function has several interpretations, two of which we have discussed.

1. The function f' represents the *instantaneous rate of change* of y with respect to x. This instantaneous rate of change could be interpreted as marginal cost, revenue, or profit (if the original function represented cost, revenue, or profit) or velocity (if the original function described displacement along a line). From now on we will use *rate of change* to mean *instantaneous* rate of change.

2. The function f' represents the *slope* of the graph of $f(x)$ at any point x. If the derivative is evaluated at the point $x = a$, then it represents the slope of the curve at that point.

The following table compares the different interpretations of the difference quotient and the derivative.

The Difference Quotient and the Derivative	
Difference Quotient	**Derivative**
$$\dfrac{f(b) - f(a)}{b - a}$$	$$\lim_{h \to 0} \dfrac{f(x + h) - f(x)}{h}$$
■ Slope of the secant line	■ Slope of the tangent line
■ Average rate of change	■ Instantaneous rate of change
■ Average velocity	■ Instantaneous velocity

The next few examples show how to use the definition to find the derivative of a function by means of a four-step procedure.

EXAMPLE 4   Derivative

Let $f(x) = x^2$.

(a) Find the derivative.

Solution By definition, for all values of x where the following limit exists, the derivative is given by

$$f'(x) = \lim_{h \to 0} \frac{f(x + h) - f(x)}{h}.$$

Use the following sequence of steps to evaluate this limit.

Step 1 Find $f(x + h)$.

Replace x with $x + h$ in the equation for $f(x)$. Simplify the result.

$$f(x) = x^2$$
$$f(x + h) = (x + h)^2$$
$$= x^2 + 2xh + h^2$$

(Note that $f(x + h) \neq f(x) + h$, since $f(x) + h = x^2 + h$.)

Step 2 Find $f(x + h) - f(x)$.
   Since $f(x) = x^2$,

$$f(x + h) - f(x) = (x^2 + 2xh + h^2) - x^2 = 2xh + h^2.$$

Step 3 Find and simplify the quotient $\dfrac{f(x + h) - f(x)}{h}$.

$$\frac{f(x + h) - f(x)}{h} = \frac{2xh + h^2}{h} = \frac{h(2x + h)}{h} = 2x + h$$

Step 4 Finally, find the limit as h approaches 0. In this step, h is the variable and x is fixed.

$$f'(x) = \lim_{h \to 0} \frac{f(x + h) - f(x)}{h}$$

$$= \lim_{h \to 0}(2x + h)$$

$$= 2x + 0 = 2x$$

(b) Calculate and interpret $f'(3)$. Use the function defined by $f'(x) = 2x$.

Solution

Method 1: Algebraic Method

$$f'(3) = 2 \cdot 3 = 6$$

The number 6 is the slope of the tangent line to the graph of $f(x) = x^2$ at the point where $x = 3$, that is, at $(3, f(3)) = (3, 9)$. See Figure 34(a).

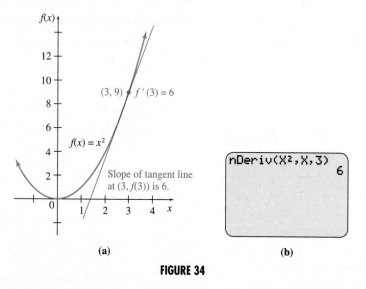

(a)   (b)

FIGURE 34

Method 2: Graphing Calculator Some graphing calculators can calculate the value of the derivative at a given x-value. For example, the TI-83 uses the `nDeriv` command as shown in Figure 34(b), with the expression for $f(x)$, the variable, and the value of a entered in the parentheses, to find $f'(3)$ for $f(x) = x^2$.

CAUTION

1. In Example 4(a) notice that $f(x + h)$ is *not* equal to $f(x) + h$. In fact,

$$f(x + h) = (x + h)^2 = x^2 + 2xh + h^2,$$

but

$$f(x) + h = x^2 + h.$$

2. In Example 4(b), do not confuse $f(3)$ and $f'(3)$. The value $f(3)$ is the y-value that corresponds to $x = 3$. It is found by substituting 3 for x in $f(x)$; $f(3) = 3^2 = 9$. On the other hand, $f'(3)$ is the slope of the tangent line to the curve at $x = 3$; as Example 4(b) shows, $f'(3) = 2 \cdot 3 = 6$.

FINDING $f'(x)$ FROM THE DEFINITION OF DERIVATIVE

The four steps used to find the derivative $f'(x)$ for a function $y = f(x)$ are summarized here.

1. Find $f(x + h)$.
2. Find and simplify $f(x + h) - f(x)$.
3. Divide by h to get $\dfrac{f(x + h) - f(x)}{h}$.
4. Let $h \to 0$; $f'(x) = \lim\limits_{h \to 0} \dfrac{f(x + h) - f(x)}{h}$ if this limit exists.

EXAMPLE 5 Derivative

Let $f(x) = 2x^3 + 4x$. Find $f'(x)$, $f'(2)$, and $f'(-3)$.

Solution Go through the four steps to find $f'(x)$.

Step 1 Find $f(x + h)$ by replacing x with $x + h$.

$$
\begin{aligned}
f(x + h) &= 2(x + h)^3 + 4(x + h) \\
&= 2(x^3 + 3x^2h + 3xh^2 + h^3) + 4(x + h) \\
&= 2x^3 + 6x^2h + 6xh^2 + 2h^3 + 4x + 4h
\end{aligned}
$$

Step 2 $f(x + h) - f(x) = 2x^3 + 6x^2h + 6xh^2 + 2h^3 + 4x + 4h - 2x^3 - 4x$
$$= 6x^2h + 6xh^2 + 2h^3 + 4h$$

Step 3 $\dfrac{f(x + h) - f(x)}{h} = \dfrac{6x^2h + 6xh^2 + 2h^3 + 4h}{h}$

$$= \dfrac{h(6x^2 + 6xh + 2h^2 + 4)}{h}$$

$$= 6x^2 + 6xh + 2h^2 + 4$$

Step 4 Now use the rules for limits to get

$$
\begin{aligned}
f'(x) &= \lim_{h \to 0} \frac{f(x + h) - f(x)}{h} \\
&= \lim_{h \to 0}(6x^2 + 6xh + 2h^2 + 4) \\
&= 6x^2 + 6x(0) + 2(0)^2 + 4 \\
f'(x) &= 6x^2 + 4.
\end{aligned}
$$

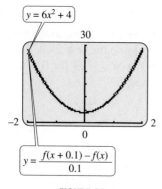

$y = 6x^2 + 4$

$$y = \frac{f(x + 0.1) - f(x)}{0.1}$$

FIGURE 35

Use this result to find $f'(2)$ and $f'(-3)$.

$$f'(2) = 6 \cdot 2^2 + 4$$
$$= 28$$
$$f'(-3) = 6 \cdot (-3)^2 + 4 = 58$$

One way to support this result is to plot $(f(x + h) - f(x))/h$ on a graphing calculator with a small value of h. Figure 35 shows a graphing calculator screen of $y = (f(x + 0.1) - f(x))/0.1$, where f is the function $f(x) = 2x^3 + 4x$, and $y = 6x^2 + 4$, which was just found to be the derivative of f. The two functions, plotted on the window $[-2, 2]$ by $[0, 30]$, appear virtually identical. If $h = 0.01$ had been used, the two functions would be indistinguishable.

EXAMPLE 6   Derivative

Let $f(x) = \dfrac{4}{x}$. Find $f'(x)$.

Solution

Step 1 $f(x + h) = \dfrac{4}{x + h}$

Step 2 $f(x + h) - f(x) = \dfrac{4}{x + h} - \dfrac{4}{x}$

$$= \frac{4x - 4(x + h)}{x(x + h)} \qquad \text{Find a common denominator.}$$

$$= \frac{4x - 4x - 4h}{x(x + h)} \qquad \text{Simplify the numerator.}$$

$$= \frac{-4h}{x(x + h)}$$

Step 3 $\dfrac{f(x + h) - f(x)}{h} = \dfrac{\dfrac{-4h}{x(x + h)}}{h}$

$$= \frac{-4h}{x(x + h)} \cdot \frac{1}{h} \qquad \text{Invert and multiply.}$$

$$= \frac{-4}{x(x + h)}$$

Step 4 $f'(x) = \lim\limits_{h \to 0} \dfrac{f(x + h) - f(x)}{h}$

$$= \lim\limits_{h \to 0} \frac{-4}{x(x + h)}$$

$$= \frac{-4}{x(x + 0)}$$

$$f'(x) = \frac{-4}{x(x)} = \frac{-4}{x^2}$$

Notice that in Example 6 neither $f(x)$ nor $f'(x)$ is defined when $x = 0$. Look at a graph of $f(x) = 4/x$ to see why this is true.

EXAMPLE 7 Weight Gain

Lindsay Branson finds that, after introducing her dog Buddy to a new brand of food, Buddy's weight begins to increase. After x weeks on the new food, Buddy's weight (in pounds) is approximately given by $w(x) = \sqrt{x} + 17$ for $0 \le x \le 6$. Find the rate of change of Buddy's weight after x weeks.

Solution

Step 1 $w(x + h) = \sqrt{x + h} + 17$

Step 2 $w(x + h) - w(x) = \sqrt{x + h} + 17 - \left(\sqrt{x} + 17\right)$
$$= \sqrt{x + h} - \sqrt{x}$$

Step 3 $\dfrac{w(x + h) - w(x)}{h} = \dfrac{\sqrt{x + h} - \sqrt{x}}{h}$

At this point, in order to be able to divide by h, multiply both numerator and denominator by $\sqrt{x + h} + \sqrt{x}$; that is, rationalize the *numerator.*

$$\frac{w(x + h) - w(x)}{h} = \frac{\sqrt{x + h} - \sqrt{x}}{h} \cdot \frac{\sqrt{x + h} + \sqrt{x}}{\sqrt{x + h} + \sqrt{x}}$$

$$= \frac{\left(\sqrt{x + h}\right)^2 - \left(\sqrt{x}\right)^2}{h\left(\sqrt{x + h} + \sqrt{x}\right)}$$

$$= \frac{x + h - x}{h\left(\sqrt{x + h} + \sqrt{x}\right)}$$

$$= \frac{1}{\sqrt{x + h} + \sqrt{x}}$$

Step 4 $w'(x) = \lim\limits_{h \to 0} \dfrac{1}{\sqrt{x + h} + \sqrt{x}} = \dfrac{1}{\sqrt{x} + \sqrt{x}} = \dfrac{1}{2\sqrt{x}}$

This tells us, for example, that after 4 weeks, when Buddy's weight is $w(4) = \sqrt{4} + 17 = 19$ lb, her weight is increasing at a rate of $w'(4) = 1/\left(2\sqrt{4}\right) = 1/4$ pound per week.

EXAMPLE 8 Eating Behavior

As we saw in the review exercises for Chapter 1, studies on the eating behavior of humans revealed that for one subject, the cumulative intake of food during a meal can be described by

$$I(t) = 27 + 72t - 1.5t^2,$$

where t is the number of minutes since the meal began, and $I(t)$ represents the amount, in grams, that the person has eaten at time t.* Find the rate of change of the intake of food for this particular person 1 minute into a meal and 10 minutes into a meal.

Solution The rate of change of intake is given by the derivative of the intake function,

---

*Kissileff, H. R., and J. L. Guss, "Microstructure of Eating Behaviors in Humans," *Appetite,* Vol. 36, No. 1, Feb. 2001, pp. 70–78.

$$I'(t) = \lim_{h \to 0} \frac{I(t + h) - I(t)}{h}.$$

Going through the steps for finding $I'(t)$ gives

$$I'(t) = 72 - 3t.$$

When $t = 1$,

$$I'(1) = 72 - 3(1) = 69.$$

This rate of change of intake per minute gives the *marginal value* of the function at $t = 1$, which means that the person will eat approximately 69 grams of food during the second minute of the meal.

When 10 minutes have passed, the marginal value of the function is

$$I'(10) = 72 - 3(10) = 42,$$

or 42 g/min. This indicates that, as the meal progresses, the rate of intake of food decreases. In fact, the rate of intake is zero 24 minutes after the meal began. ∎

We can use the notation for the derivative to write the equation of the tangent line. Using the point-slope form, $y - y_1 = m(x - x_1)$, and letting $y_1 = f(x_1)$ and $m = f'(x_1)$, we have the following formula.

EQUATION OF THE TANGENT LINE

The tangent line to the graph of $y = f(x)$ at the point $(x_1, f(x_1))$ is given by the equation

$$y - f(x_1) = f'(x_1)(x - x_1),$$

provided $f'(x)$ exists.

EXAMPLE 9 Tangent Line

Find the equation of the tangent line to the graph of $f(x) = 4/x$ at $x = 2$.

Solution From the answer to Example 6, we have $f'(x) = -4/x^2$, so $f'(x_1) = f'(2) = -4/2^2 = -1$. Also $f(x_1) = f(2) = 4/2 = 2$. Then the equation of the tangent line is

$$y - 2 = (-1)(x - 2),$$

or

$$y = -x + 4$$

after simplifying. ∎

Existence of the Derivative

The definition of the derivative included the phrase "provided this limit exists." If the limit used to define the derivative does not exist, then of course the derivative does not exist. For example, a derivative cannot exist at a point where the function itself is not defined. If there is no function value for a particular value of x, there can be no tangent line for that value. This was the case in Example 6—there was no tangent line (and no derivative) when $x = 0$.

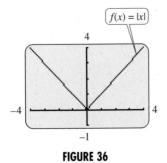

FIGURE 36

Derivatives also do not exist at "corners" or "sharp points" on a graph. For example, the function graphed in Figure 36 is the *absolute value function,* defined previously as

$$f(x) = \begin{cases} x & \text{if } x \geq 0 \\ -x & \text{if } x < 0, \end{cases}$$

and written $f(x) = |x|$. By the definition of derivative, the derivative at any value of x is given by

$$f'(x) = \lim_{h \to 0} \frac{f(x + h) - f(x)}{h},$$

provided this limit exists. To find the derivative at 0 for $f(x) = |x|$, replace x with 0 and $f(x)$ with $|0|$ to get

$$f'(0) = \lim_{h \to 0} \frac{|0 + h| - |0|}{h} = \lim_{h \to 0} \frac{|h|}{h}.$$

In Example 2 in the first section of this chapter, we showed that

$$\lim_{h \to 0} \frac{|h|}{h} \text{ does not exist;}$$

therefore, the derivative does not exist at 0. However, the derivative does exist for all values of x other than 0.

In Figure 37, we have zoomed in on the origin in Figure 36. Notice that the graph looks essentially the same. The corner is still sharp, and the graph does not resemble a straight line any more than it originally did. As we observed earlier, the derivative only exists at a point when the function more and more resembles a straight line as we zoom in on the point.

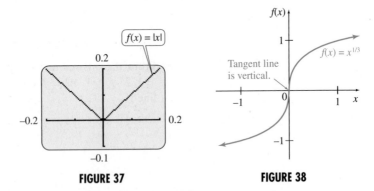

FIGURE 37   **FIGURE 38**

A graph of the function $f(x) = x^{1/3}$ is shown in Figure 38. As the graph suggests, the tangent line is vertical when $x = 0$. Since a vertical line has an undefined slope, the derivative of $f(x) = x^{1/3}$ cannot exist when $x = 0$. Use the fact that $\lim_{h \to 0} h^{1/3}/h = \lim_{h \to 0} 1/h^{2/3}$ does not exist and the definition of the derivative to verify that $f'(0)$ does not exist for $f(x) = x^{1/3}$.

Figure 39 on the next page summarizes various ways that a derivative can fail to exist. Notice in Figure 39 that at a point where the function is discontinuous, such as x_3, x_4, and x_6, the derivative does not exist. A function must be continuous at a point for the derivative to exist there. But just because a function is continuous at a point does not mean the derivative necessarily exists; for example, observe the behavior of the function in Figure 39 at x_1, x_2, and x_5, or recall that $f(x) = |x|$ is continuous at $x = 0$ but $f'(0)$ does not exist.

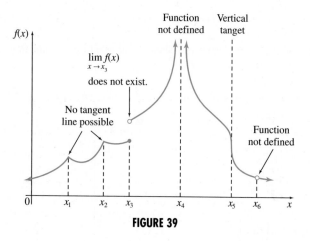

FIGURE 39

EXAMPLE 10   Astronomy

A nova is a star whose brightness suddenly increases and then gradually fades. The cause of the sudden increase in brightness is thought to be an explosion of some kind. The intensity of light emitted by a nova as a function of time is shown in Figure 40.* Notice that although the graph is a continuous curve, it is not differentiable at the point of the explosion.

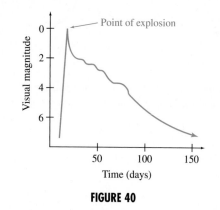

FIGURE 40

3.4 EXERCISES

1. By considering, but not calculating, the slope of the tangent line, give the derivative of each of the following.

   a. $f(x) = 5$ **b.** $f(x) = x$ **c.** $f(x) = -x$ **d.** the line $x = 3$

   e. the line $y = mx + b$

*Kaufmann, III, William J., "The Light Curve of a Nova," *Astronomy: The Structure of the Universe*, New York, Macmillan 1977. Reprinted by permission of William J. Kaufmann, III.

2. a. Suppose $g(x) = \sqrt[3]{x}$. Use the graph of $g(x)$ to find $g'(0)$.

b. Explain why the derivative of a function does not exist at a point where the tangent line is vertical.

3. If $f(x) = \dfrac{x^2 - 1}{x + 2}$, where is f not differentiable?

4. If the rate of change of $f(x)$ is zero when $x = a$, what can be said about the tangent line to the graph of $f(x)$ at $x = a$?

Estimate the slope of the tangent line to each curve at the given point (x, y).

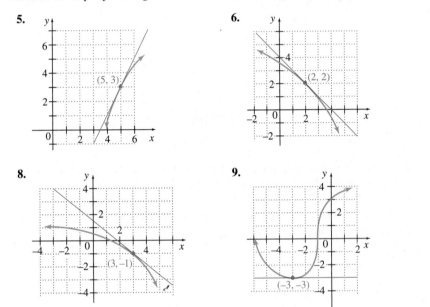

5. $(5, 3)$

6. $(2, 2)$

7. $(-2, 2)$

8. $(3, -1)$

9. $(-3, -3)$

10.

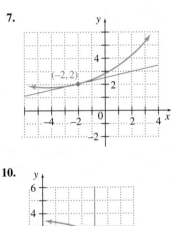

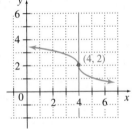
$(4, 2)$

Using the definition of the derivative, find $f'(x)$. Then find $f'(-2)$, $f'(0)$, and $f'(3)$.
(Hint for Exercises 15 and 16: *In Step 3, multiply numerator and denominator by* $\sqrt{x + h} + \sqrt{x}$.)

11. $f(x) = -4x^2 + 11x$

12. $f(x) = 6x^2 - 4x$

13. $f(x) = -2/x$

14. $f(x) = 6/x$

15. $f(x) = \sqrt{x}$

16. $f(x) = -3\sqrt{x}$

Find the equation of the tangent line to each curve when x has the given value.

17. $f(x) = x^2 + 2x; \quad x = 3$

18. $f(x) = 6 - x^2; \quad x = -1$

19. $f(x) = 5/x; \quad x = 2$

20. $f(x) = -3/(x + 1); \quad x = 1$

21. $f(x) = 4\sqrt{x}; \quad x = 9$

22. $f(x) = \sqrt{x}; \quad x = 25$

Use a graphing calculator to find $f'(2)$, $f'(16)$, and $f'(-3)$ for each of the following.

23. $f(x) = -4x^2 + 11x$

24. $f(x) = 6x^2 - 4x$

25. $f(x) = 8x + 6$

26. $f(x) = -9x - 5$

27. $f(x) = -\dfrac{2}{x}$

28. $f(x) = \dfrac{6}{x}$

29. $f(x) = \sqrt{x}$

30. $f(x) = -3\sqrt{x}$

Find the x-values where the following do not have derivatives.

31.

32.

33.

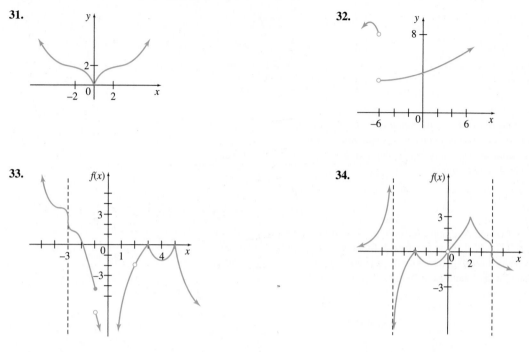

34.

35. For the function shown in the sketch, give the intervals or points on the *x*-axis where the rate of change of $f(x)$ with respect to x is

    a. positive   **b.** negative   **c.** zero

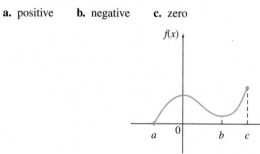

*In Exercises 36 and 37, tell which graph, **a** or **b**, represents velocity and which represents distance from a starting point. (Hint: Consider where the derivative is zero, positive, or negative.)*

36. a.

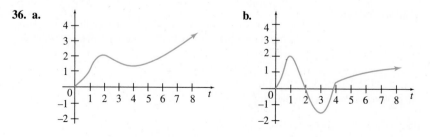

                b.

37. a.

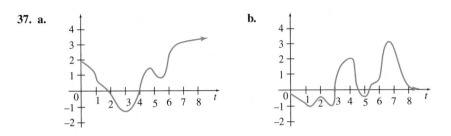

In Exercises 38–41, find the derivative of the function at the given point.

a. Approximate the definition of the derivative with small values of h.

b. Use a graphing calculator to zoom in on the function until it appears to be a straight line, and then find the slope of that line.

38. $f(x) = x^x, a = 2$

39. $f(x) = x^x, a = 3$

40. $f(x) = x^{1/x}, a = 2$

41. $f(x) = x^{1/x}, a = 3$

42. For each function in Column A, graph $(f(x + h) - f(x))/h$ for a small value of h on the window $[-2, 2]$ by $[-2, 8]$. Then graph each function in Column B on the same window. Compare the first set of graphs with the second set to choose from Column B the derivative of each of the functions in Column A.

Column A	Column B
$\ln |x|$	e^x
e^x	$3x^2$
x^3	$\dfrac{1}{x}$

Applications

LIFE SCIENCES

43. *Water Snakes* Scientists in Arkansas have collected data which reveals that there is a linear relationship between the length (snout to vent) of female water snakes and the length of their catfish prey such that

$$l(x) = -3.6 + 0.17x,$$

where x is the snout-vent length (cm) of the snake and $l(x)$ is the predicted length (cm) of the catfish.[*]

a. Use the function above to estimate the size of catfish that a snake which has a snout-vent length of 70 cm will choose for prey.

b. Calculate the derivative of this function and interpret the results.

c. Estimate the snout-vent length of the smallest snake for which the function above makes sense.

44. *Beaked Sea Snake* Scientists in Malaysia have collected data which reveals that there is a linear relationship between the length (snout to vent) of the beaked sea snake and the length of its catfish prey such that

$$l(x) = 2.41 + 0.12x,$$

where x is the snout-vent length (cm) of the snake and $l(x)$ is the predicted length (cm) of the catfish.[†]

a. Use the function above to estimate the size of catfish that a snake which has a snout-vent length of 70 cm will choose for prey.

b. Calculate the derivative of this function and interpret the results.

45. *Eating Behavior* As indicated in Example 8, the eating behavior of a typical human during a meal can be described by

[*]Plummer, M., and J. Goy, "Ontogenetic Dietary Shift of Water Snakes in a Fish Hatchery," *Copeia*, No. 2, 1984, pp. 550–552.

[†]Voris, H., and M. Moffett, "Size and Proportion Relationship between the Beaked Sea Snake and its Prey," *Biotropica*, Vol. 13, No. 1, 1981, pp. 15–19.

$$I(t) = 27 + 72t - 1.5t^2,$$

where t is the number of minutes since the meal began, and $I(t)$ represents the amount (in grams) that the person has eaten at time t.*

a. Find the rate of change of the intake of food for this particular person 5 minutes into a meal and interpret.

b. Verify that the rate in which food is consumed is zero 24 minutes after the meal starts.

✎ **c.** Comment on the assumptions and usefulness of this function after 24 minutes. Given this fact, determine a logical range for this function.

46. *Quality Control of Cheese* It is often difficult to evaluate the quality of products that undergo a ripening or maturation process. Researchers have successfully used ultrasonic velocity to determine the maturation time of Mahon cheese. The age can be determined by

$$M(v) = 0.0312443v^2 - 101.39v + 82,264, \quad v \geq 1,620,$$

where $M(v)$ is the estimated age of the cheese (in days) for a velocity v (m/s).†

a. If Mahon cheese ripens in 150 days determine the velocity of the ultrasound that one would expect to measure. (*Hint:* Set $M(v) = 150$ and solve for v.)

b. Determine the derivative of this function when $v = 1,700$ m/s and interpret.

47. *Shellfish Population* In one research study, the population of a certain shellfish in an area at time t was closely approximated by the graph below. Estimate and interpret the derivative at each of the marked points.

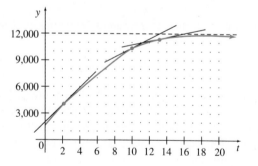

OTHER APPLICATIONS

48. *Temperature* The graph shows the temperature in degrees Celsius as a function of the altitude h in feet when an inversion layer is over Southern California. (See Exercise 34 in the previous section.) Estimate and interpret the derivatives of $T(h)$ at the marked points.

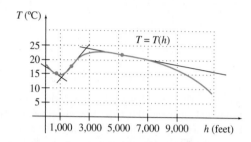

49. *Oven Temperature* The graph shows the temperature in an oven during a self-cleaning cycle.‡ (*Note:* The circles on the graph do not represent points of discontinuity, but merely the times when the thermal door lock turns on and off.) Let $T(x)$ be the temperature after x hours.

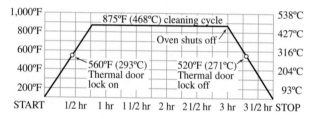

a. Find all x-values where the derivative does not exist.

b. Find and interpret $T'(0.5)$.

c. Find and interpret $T'(2)$.

d. Find and interpret $T'(3.5)$.

50. *Baseball* The graph on the next page shows how the distance a baseball travels varies with the weight of the bat.§ Estimate and interpret the derivative for a 24 oz and a 51 oz bat under the assumption that the bat plus the batter have a constant energy.

---

*Kissileff, H. R., and J. L. Guss, "Microstructure of Eating Behavior in Humans," *Appetite,* Vol. 36, No. 1, Feb. 2001, pp. 70–78.
†Benedito, J., J. Carcel, M. Gisbert, and A. Mulet, "Quality Control of Cheese Maturation and Defects Using Ultrasonics," *Journal of Food Science,* Vol. 66, No. 1, 2001, pp. 100–104.
‡*Whirlpool Use and Care Guide, Self-Cleaning Electric Range,* Whirlpool Corporation. Reprinted with permission.
§Exercises 50 and 51 are from Adair, Robert K., *The Physics of Baseball,* HarperCollins, © 1990, p. 82. Reprinted with permission.

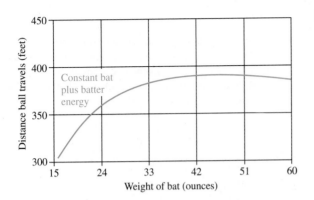

Distance ball travels (feet) vs. Weight of bat (ounces)

Constant bat plus batter energy

51. *Baseball* The graph below shows how the velocity of the hands and the baseball bat vary with the time of the swing. Estimate and interpret the derivative for the hands and the bat at the time when the velocity of the two are equal. (*Hint:* The rate of change of velocity is called acceleration.)

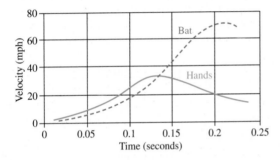

Velocity (mph) vs. Time (seconds)

Bat

Hands

52. *Social Security Assets* The table below gives actual and projected year-end assets in Social Security trust funds, in billions of current dollars, where Year represents the number of years since 1990.*

The polynomial function defined by

$$f(x) = -0.0142x^4 + 0.6698x^3 - 6.113x^2 + 84.05x + 203.9$$

models these data quite well.

Year	Billions of Dollars
0	214
6	550
8	800
10	1,000
20	2,500
30	3,800
40	250

a. To verify the fit of the model, find $f(10)$, $f(20)$, and $f(30)$.

b. Use a graphing calculator with a command such as nDeriv to find the slope of the tangent line to the graph of f at the following x-values: 10, 20, 30, 35.

c. Use your results in part b to describe the graph of f and interpret the corresponding changes in Social Security assets.

3.5 GRAPHICAL DIFFERENTIATION

?

THINK ABOUT IT   Given a graph of the weight of a girl as a function of age, how can we find the graph of the derivative of that function?

?

To understand how the value of a function and its derivative are affected by changes in the independent variable, it is helpful to draw graphs of both functions. In Figure 41(a) on the next page, from a book on human growth,[†] the graph shows the typical weight (in kg) of an English girl for the first 18 years of life. The graph in Figure 41(b), which represents the derivative of the curve, shows the rate of change in weight of the girl each year.

These graphs should remind us that, in the previous section, we estimated the derivative at various points of a graph by estimating the slope of the tangent line at those points. We will now extend this process to show how to sketch the graph of the derivative given the graph of the original function. This is important because, as indicated above, in many applications a graph of the function is all we have, and it is easier to find the derivative graphically than to find a formula that fits the graph and then take the derivative of that formula.

---

*Social Security Administration
[†]Sinclair, David, *Human Growth After Birth*, New York, Oxford University Press, 1985.

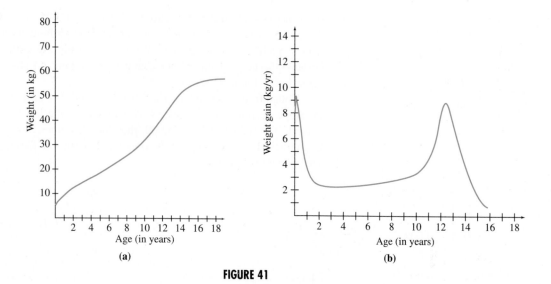

FIGURE 41

In the example above, let t refer to the age (in years) of the girl. Observe that the slope of the weight function is always positive. However, notice that as age increases, the weight gain is not constant. For example, right after birth, the slope of the function is steeper than it is from about one year to ten years. But, from about age 11 to age 13, the function increases rapidly, illustrating a second growth spurt. Notice that these growth spurts can also be seen by looking at the graph in Figure 41(b).

In this example, we have seen how the general shape of the graph of the derivative is related to the graph of the original function. To get a more accurate graph of the derivative, we need to estimate the slope of the tangent line at various points, as we did in the previous section.

EXAMPLE 1   Temperature

Figure 42 is from Exercise 48 in the previous section. It gives the temperature in degrees Celsius as a function of the altitude h in feet when an inversion layer is over Southern California. The exercise asks for an estimate of the derivative of $T(h)$ at the marked points. If you have not done that exercise yet, do the estimation now by finding two points on each tangent line and computing the slope between the two points.

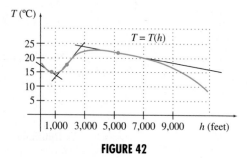

FIGURE 42

Your answers should be roughly $T'(500) = -0.005$, $T'(1,500) = 0.008$, and $T'(5,000) = -0.00125$. Your answers may be slightly different, since estimation from a picture can be inexact. We will also add the estimate $T'(3,500) = 0$, because the tangent line is horizontal there. Figure 43 shows a graph of these values of T' found so far.

Notice from the graph of $T(h)$ that the slope is largest at $h = 1,500$, and that the slope becomes more negative as we move to the left from 500. Also, as we move to the right from $h = 5,000$, the slope becomes more and more negative. Using these facts, we connect the points in the graph $T'(h)$ smoothly, with the result shown in Figure 44.

t	T'
500	−0.005
1,500	0.008
3,500	0
5,000	−0.00125

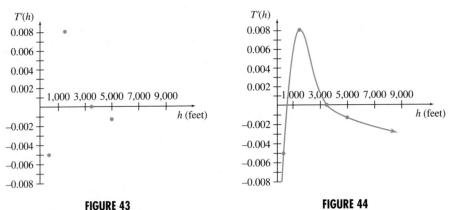

FIGURE 43　　　　　　　**FIGURE 44**

CAUTION Remember that when you graph the derivative, you are graphing the *slope* of the original function. Do not confuse the slope of the original function with the y-value of the original function.

Sometimes the original function is not smooth or even continuous, so the graph of the derivative may also be discontinuous.

EXAMPLE 2　**Graphing Derivatives**
Sketch the graph of the derivative of the function shown in Figure 45.

Solution Notice that when $x < -2$, the slope is 1, and when $-2 < x < 0$, the slope is −1. At $x = -2$, the derivative does not exist due to the sharp corner in the graph. The derivative also does not exist at $x = 0$ because the function is discontinuous there. Using this information, the graph of $f'(x)$ on $x < 0$ is shown in Figure 46.

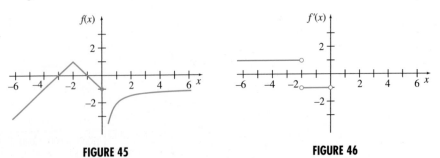

FIGURE 45　　　　　　　**FIGURE 46**

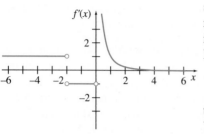

FIGURE 47

For $x > 0$, the derivative is positive. If you draw a tangent line at $x = 1$, you should find that the slope of this line is roughly 1. As x approaches 0 from the right, the derivative becomes larger and larger. As x approaches infinity, the derivative approaches 0. The resulting sketch of the graph of $y = f'(x)$ is shown in Figure 47. ∎

Finding the derivative graphically may seem difficult at first, but with practice you should be able to quickly sketch the derivative of any function graphed. Your answers to the exercises may not look exactly like those in the back of the book, because estimating the slope accurately can be difficult, but your answers should have the same general shape.

Figures 48(a), (b), and (c) show the graphs of $y = x^2$, $y = x^4$, and $y = x^{4/3}$ on a graphing calculator. When finding the derivative graphically, all three seem to have the same behavior: negative derivative for $x < 0$, 0 derivative at $x = 0$, and positive derivative for $x > 0$. Beyond these general features, however, the derivatives look quite different, as you can see from Figures 49(a), (b), and (c), which show the graphs of the derivatives. When finding derivatives graphically, detailed information can only be found by very carefully measuring the slope of the tangent line at a large number of points.

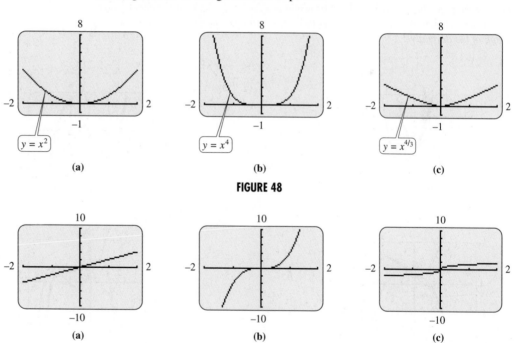

FIGURE 48

FIGURE 49

> **NOTE** On many calculators, the graph of the derivative can be plotted if a formula for the original function is known. For example, the graphs in Figure 49 were drawn on a TI-83 by defining $Y_2 = \texttt{nDeriv}\,(Y_1,\ x,\ x)$ after entering the original function into Y_1. You can use this feature to practice finding the derivative graphically. Enter a function into Y_1, sketch the graph on the graphing calculator, and use it to draw by hand the graph of the derivative. Then use $\texttt{nDeriv}$ to draw the graph of the derivative, and compare it with your sketch.

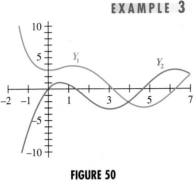

FIGURE 50

EXAMPLE 3 Graphical Differentiation

Figure 50 shows the graph of a function f and its derivative function f'. Use slopes to decide which graph is that of f and which is the graph of f'.

Solution Look at the places where each graph crosses the x-axis; that is, the x-intercepts, since x-intercepts occur on the graph of f' whenever the graph of f has a horizontal tangent line or slope of zero. Also, a decreasing graph corresponds to negative slope or a negative derivative, while an increasing graph corresponds to positive slope or a positive derivative. Y_1 has zero slope near $x = 0$, $x = 1$, and $x = 5$; Y_2 has x-intercepts near these values of x. Y_1 decreases on $(-2, 0)$ and $(1, 5)$; Y_2 is negative on those intervals. Y_1 increases on $(0, 1)$ and $(5, 7)$; Y_2 is positive there. Thus, Y_1 is the graph of f and Y_2 is the graph of f'.

3.5 EXERCISES

1. Explain how to graph the derivative of a function given the graph of the function.

2. Explain how to graph a function given the graph of the derivative function.

Each of the following graphing calculator windows shows the graphs of a function f and its derivative function f'. Decide which is the graph of the function and which is the graph of the derivative.

3.

4.

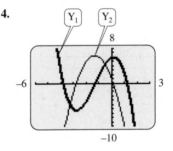

5.

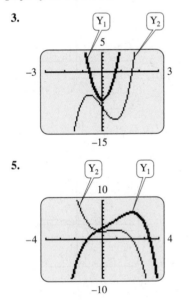

6.

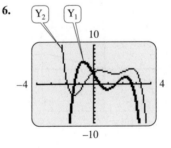

Sketch the graph of the derivative for each function shown.

7.

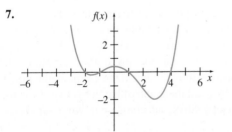

8.

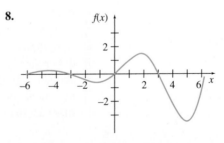

9.

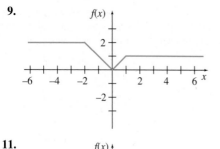

10.

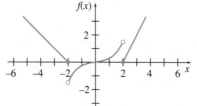

11.

12.

13.

14.

Applications

LIFE SCIENCES

15. *Shellfish Population* In one research study, the population of a certain shellfish in an area at time *t* was closely approximated by the graph below. Sketch a graph of the growth rate of the population.

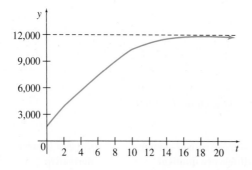

16. *Flight Speed* The graph in the next column shows the relationship between the speed of the Arctic tern in flight and the required power expended by flight muscles.* Sketch

the graph of the rate of change of the power as a function of the speed.

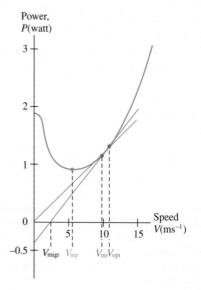

---

*Alerstam, Thomas, "Bird Flight and Optimal Migration," *Ecology and Evolution,* July 1991. Copyright © 1991 by Elsevier Trends Journals. Reprinted by permission of Elsevier Trends Journals and Thomas Alerstam.

17. *Human Growth* The growth remaining in sitting height at consecutive skeletal age levels is indicated below for boys.* Sketch a graph showing the rate of change of growth remaining for the indicated years. Use the graph and your sketch to estimate the remaining growth and the rate of change of remaining growth for a 14-year-old boy.

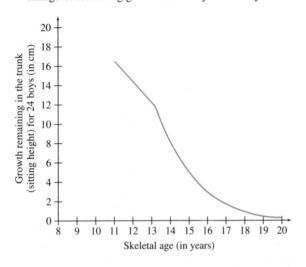

18. *Weight Gain* The graph below shows the typical weight (in kg) of an English boy for his first 18 years of life.† Sketch the graph of the rate of change of weight with respect to time.

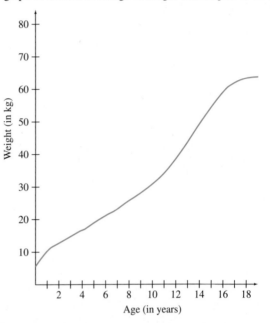

CHAPTER SUMMARY

In this chapter we introduced the ideas of limit and continuity of functions and then used these ideas to discover calculus. We investigated the relationship between instantaneous rate of change, velocity, and the definition of a derivative. We also learned how to estimate the value of the derivative using graphical differentiation. In the next chapter, we will take a closer look at the definition of a derivative to develop a set of rules in which the derivative of a wide range of functions can be quickly and easily calculated without having to directly apply the definition of a derivative each time.

KEY TERMS

To understand the concepts presented in this chapter, you should know the meaning and use of the following words. For easy reference, the section in the chapter where a word (or expression) was first used is given with each item.

3.1 limit
limit from the left/right
one-/two-sided limit
limit at infinity
3.2 continuous
discontinuous

continuous from the
right/left
continuous on an
open/closed interval
Intermediate Value
Theorem

3.3 average rate of change
difference quotient
instantaneous rate
of change
3.4 secant line
tangent line

slope of a curve
derivative
differentiable
differentiation

---

*Hensinger, Robert, *Standards in Pediatric Orthopedics: Tables, Charts, and Graphs Illustrating Growth,* New York, Raven Press, 1986, p. 192.
†Sinclair, David, *Human Growth After Birth,* New York, Oxford University Press, 1985.

CHAPTER 3 REVIEW EXERCISES

1. Is a derivative always a limit? Is a limit always a derivative? Explain.

2. Is every continuous function differentiable? Is every differentiable function continuous? Explain.

3. Describe how to tell when a function is discontinuous at the real number $x = a$.

4. Give two applications of the derivative.

$$f'(x) = \lim_{h \to 0} \frac{f(x + h) - f(x)}{h}.$$

Decide whether the limits in Exercises 5–22 exist. If a limit exists, find its value.

5. **a.** $\lim\limits_{x \to -3^-} f(x)$ **b.** $\lim\limits_{x \to -3^+} f(x)$ **c.** $\lim\limits_{x \to -3} f(x)$ **d.** $f(-3)$ 6. **a.** $\lim\limits_{x \to -1^-} g(x)$ **b.** $\lim\limits_{x \to -1^+} g(x)$ **c.** $\lim\limits_{x \to -1} g(x)$ **d.** $g(-1)$

7. **a.** $\lim\limits_{x \to 4^-} f(x)$ **b.** $\lim\limits_{x \to 4^+} f(x)$ **c.** $\lim\limits_{x \to 4} f(x)$ **d.** $f(4)$ 8. **a.** $\lim\limits_{x \to 2^-} h(x)$ **b.** $\lim\limits_{x \to 2^+} h(x)$ **c.** $\lim\limits_{x \to 2} h(x)$ **d.** $h(2)$

9. $\lim\limits_{x \to -\infty} g(x)$ 10. $\lim\limits_{x \to \infty} f(x)$

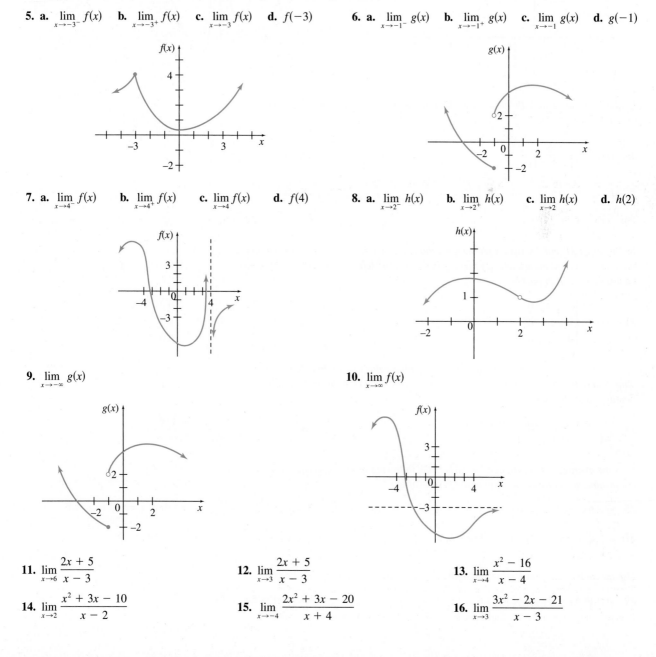

11. $\lim\limits_{x \to 6} \dfrac{2x + 5}{x - 3}$ 12. $\lim\limits_{x \to 3} \dfrac{2x + 5}{x - 3}$ 13. $\lim\limits_{x \to 4} \dfrac{x^2 - 16}{x - 4}$

14. $\lim\limits_{x \to 2} \dfrac{x^2 + 3x - 10}{x - 2}$ 15. $\lim\limits_{x \to -4} \dfrac{2x^2 + 3x - 20}{x + 4}$ 16. $\lim\limits_{x \to 3} \dfrac{3x^2 - 2x - 21}{x - 3}$

17. $\lim\limits_{x \to 9} \dfrac{\sqrt{x} - 3}{x - 9}$

18. $\lim\limits_{x \to 16} \dfrac{\sqrt{x} - 4}{x - 16}$

19. $\lim\limits_{x \to \infty} \dfrac{x^2 + 5}{5x^2 - 1}$

20. $\lim\limits_{x \to \infty} \dfrac{x^2 + 6x + 8}{x^3 + 2x + 1}$

21. $\lim\limits_{x \to -\infty} \left(\dfrac{3}{4} + \dfrac{2}{x} - \dfrac{5}{x^2} \right)$

22. $\lim\limits_{x \to -\infty} \left(\dfrac{9}{x^4} + \dfrac{1}{x^2} - 3 \right)$

Identify the x-values where f is discontinuous.

23.

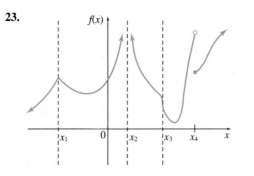

24.

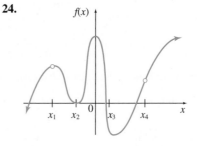

Find all x-values where the function is discontinuous. For each such value, give $f(a)$ and $\lim\limits_{x \to a} f(x)$.

25. $f(x) = \dfrac{-5}{3x(2x - 1)}$

26. $f(x) = \dfrac{2 - 3x}{(1 + x)(2 - x)}$

27. $f(x) = \dfrac{x - 6}{x + 5}$

28. $f(x) = \dfrac{x^2 - 9}{x + 3}$

29. $f(x) = x^2 + 3x - 4$

30. $f(x) = 2x^2 - 5x - 3$

In Exercises 31 and 32, (a) graph the given function, (b) find all values of x where the function is discontinuous, and (c) find the limit from the left and from the right at any values of x found in part b.

31. $f(x) = \begin{cases} 1 - x & \text{if } x < 1 \\ 2 & \text{if } 1 \le x \le 2 \\ 4 - x & \text{if } x > 2 \end{cases}$

32. $f(x) = \begin{cases} 2 & \text{if } x < 0 \\ -x^2 + x + 2 & \text{if } 0 \le x \le 2 \\ 1 & \text{if } x > 2 \end{cases}$

Find each of the following limits (a) by investigating values of the function near the point where the limit is taken, and (b) by using a graphing calculator to view the function near the point.

33. $\lim\limits_{x \to 1} \dfrac{x^4 + 2x^3 + 2x^2 - 10x + 5}{x^2 - 1}$

34. $\lim\limits_{x \to -2} \dfrac{x^4 + 3x^3 + 7x^2 + 11x + 2}{x^3 + 2x^2 - 3x - 6}$

Find the average rate of change for each of the following on the given interval. Then find the instantaneous rate of change at the first x-value.

35. $y = 6x^2 + 2$; from $x = 1$ to $x = 4$

36. $y = -2x^3 - x^2 + 5$; from $x = -2$ to $x = 6$

37. $y = \dfrac{-6}{3x - 5}$; from $x = 4$ to $x = 9$

38. $y = \dfrac{x + 4}{x - 1}$; from $x = 2$ to $x = 5$

Use the definition of the derivative to find the derivative of each of the following.

39. $f(x) = 4x + 3$

40. $f(x) = 5x^2 + 6x$

In Exercises 41 and 42, find the derivative of the function at the given point
(a) *by approximating the definition of the derivative with small values of h, and*
(b) *by using a graphing calculator to zoom in on the function until it appears to be a straight line, and then finding the slope of that line.*

41. $f(x) = (\ln x)^x$, $x_0 = 2$

42. $f(x) = x^{\ln x}$, $x_0 = 3$

Sketch the graph of the derivative for each function shown.

43.

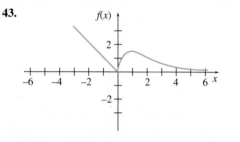

44.

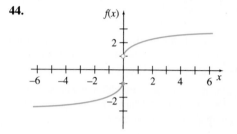

Applications

LIFE SCIENCES

45. *Cholesterol* The graph shows how the risk of coronary heart attack rises as blood cholesterol increases*. Estimate and interpret the derivative when blood cholesterol is as follows.

a. 100 mg/dL   **b.** 200 mg/dL

c. Find the average rate of change of the risk of coronary heart attack as blood cholesterol goes from 100 to 300 mg/dL.

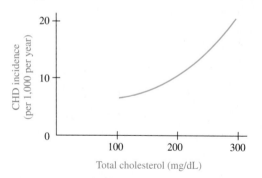

46. *Spread of a Virus* The spread of a virus is modeled by

$$V(t) = -t^2 + 6t - 4,$$

where $V(t)$ is the number of people (in hundreds) with the virus and t is the number of weeks since the first case was observed.

a. Graph $V(t)$.

b. What is a reasonable domain of t for this problem?

c. When does the number of cases reach a maximum? What is the maximum number of cases?

d. Find the rate of change function.

e. What is the rate of change in the number of cases at the maximum?

f. Give the sign (+ or −) of the rate of change up to the maximum and after the maximum.

47. *Wolf Harvest* The number of wolves harvested in Dornod Aimag, Mongolia, each year from 1964 to 1990 can be estimated by the function

$$N(t) = 1,883.41 - 16.91t,$$

where t is the number of years since 1964 and $N(t)$ is the number of wolves harvested in year t.[†]

a. Sketch the graph of this function.

b. Use the definition of the derivative to find the derivative of this function and interpret the result.

48. *TYLENOL®* The table on the next page shows the recommended dosage of Children's TYLENOL® Liquid with respect to a child's age.[‡]

---

[*]LaRosa, John C., et al., "The Cholesterol Facts: A Joint Statement by the American Heart Association and the National Heart, Lung, and Blood Institute," *Circulation,* Vol. 81, No. 5, May 1990, p. 1722. Reprinted with permission.
[†]Reading, R., H. Mix, B. Lhagvasuren, and N. Tseveenmyadag, "The Commercial Harvest of Wildlife in Dornod Aimag, Mongolia," *Journal of Wildlife Management,* Vol. 62, No. 1, 1998, pp. 59–71.
[‡]Children's TYLENOL® Liquid, www.tylenol.com.

Weight (lb)	Dose (tsps)
24–35	1
36–47	$1\frac{1}{2}$
48–59	2
60–71	$2\frac{1}{2}$
72–95	3

a. Draw a graph of this function. Assume that any fractional weight that is between the weights listed (35.5 pounds, for example) gets the lower dosage.

b. Is this function continuous for children who weigh between 24 and 95 pounds? If not, what are the weights at which the function is discontinuous?

49. *Whales Diving* The following figure, already shown in the section on Properties of Functions, shows the depth of a sperm whale as a function of time, recorded by researchers at the Woods Hole Oceanographic Institution in Massachusetts.*

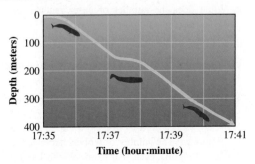

Time (hour:minute)

a. Find the rate that the whale was descending at the following times.
  i. 17 hours and 37 minutes
  ii. 17 hours and 39 minutes

b. Sketch a graph of the rate the whale was descending as a function of time.

50. *Extinction* The probability of a population going extinct by time t can be estimated by

$$p(t) = \left(\frac{a[e^{(b-a)t} - 1]}{be^{(b-a)t} - a}\right)^N,$$

where a is the death rate, b is the birth rate ($b \neq a$), and N is the number of individuals in the population at time $t = 0$.†

a. Find $\lim\limits_{t \to \infty} p(t)$
  i. if $b > a$;
  ii. if $b < a$.

b. If $a = b$, the probability of a population going extinct by time t can be estimated by

$$p(t) = \left(\frac{at}{at + 1}\right)^N.$$

Find $\lim\limits_{t \to \infty} p(t)$.

51. *Human Growth* The growth remaining in sitting height at consecutive skeletal age levels is indicated below for girls.‡ Sketch a graph showing the rate of change of growth remaining for the indicated years. Use the graph and your sketch to estimate the remaining growth and the rate of change of remaining growth for a 10-year-old girl.

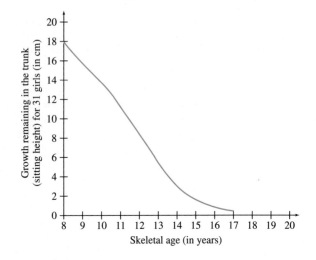

Skeletal age (in years)

OTHER APPLICATIONS

52. *Average Cost* The graph on the next page shows the total cost $C(x)$ to produce x tons of lumber. (Recall that average cost is given by total cost divided by the number produced, or $\overline{C}(x) = C(x)/x$.)

*Science, Vol. 291, Jan. 26, 2001, p. 577.
†Bailey, Norman T. J., *The Mathematical Approach to Biology and Medicine*, Wiley, 1967, p. 161.
‡Hensinger, Robert, *Standards in Pediatric Orthopedics: Tables, Charts, and Graphs Illustrating Growth*, New York, Raven Press, 1986, p. 193.

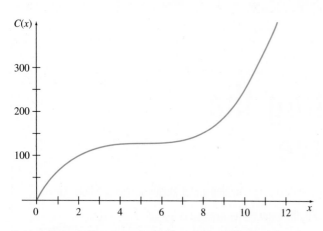

a. Draw a line through $(0,0)$ and $(5, C(5))$. Explain why the slope of this line represents the average cost per ton when 5 tons of lumber are produced.

b. Find the value of x for which the average cost is smallest.

c. What can you say about the marginal cost at the point where the average cost is smallest?

53. *Baseball* When a batter hits a baseball, the bat may not hit the center of the ball, but might hit over or under the center by various amounts. The graph shows the trajectories of balls struck by a bat swung under the ball by various amounts.* Estimate and interpret the derivative for a bat swung under the ball by 1.5 in. when the ball has traveled the following distances.

a. 100 ft   b. 200 ft

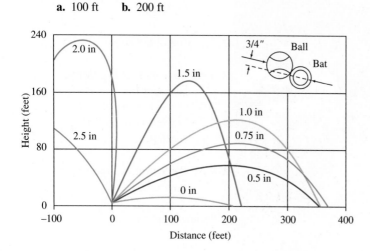

54. *Temperature* Suppose a gram of ice is at a temperature of $-100°C$. The graph shows the temperature of the ice as increasing numbers of calories of heat are applied. It takes 80 calories to melt one gram of ice at $0°C$ into water, and 540 calories to boil one gram of water at $100°C$ into steam.

a. Where is this graph discontinuous?

b. Where is this graph not differentiable?

c. Sketch the graph of the derivative.

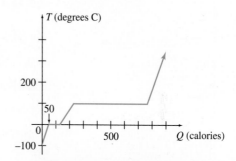

*Adair, Robert K., *The Physics of Baseball*, HarperCollins, © 1990, p. 83. Reprinted with permission.

4

Calculating the Derivative

By differentiating the function defining a mathematical model we can see how the model's output changes with the input. In an exercise in Section 2 we explore a rational-function model for the length of the rest period needed to recover from vigorous exercise such as riding a bike. The derivative indicates how the rest required changes with the work expended in kilocalories per minute.

4.1 Techniques for Finding Derivatives

4.2 Derivatives of Products and Quotients

4.3 The Chain Rule

4.4 Derivatives of Exponential Functions

4.5 Derivatives of Logarithmic Functions

4.6 Derivatives of Trigonometric Functions

Review Exercises

In the previous chapter, we found the derivative to be a useful tool for describing the rate of change, velocity, and the slope of a curve. Taking the derivative by using the definition, however, can be difficult. To take full advantage of the power of the derivative, we need faster ways of calculating the derivative. That is the goal of this chapter.

■ 4.1 TECHNIQUES FOR FINDING DERIVATIVES

? **T**HINK ABOUT IT   How fast is the number of Americans who are expected to be over 100 years old growing? What is the aortic pressure-diameter relation after a cardiac surgical procedure?

These questions can be answered by finding the derivative of an appropriate function. We shall return to them at the end of this section in Examples 6 and 7.

Using the definition to calculate the derivative of a function is a very involved process even for simple functions. In this section we develop rules that make the calculation of derivatives much easier. Keep in mind that even though the process of finding a derivative will be greatly simplified with these rules, *the interpretation of the derivative will not change.* But first, a few words about notation are in order.

Several alternative notations for the derivative are used. In the previous chapter the symbol $f'(x)$ was used to represent the derivative of $y = f(x)$. Sometimes it is important to show that the derivative is taken with respect to a particular variable; for example, if y is a function of x, the notation

$$\frac{dy}{dx} \qquad \text{or} \qquad D_x y$$

(both read "the derivative of y with respect to x") can be used for the derivative of y with respect to x. The dy/dx notation for the derivative is sometimes referred to as *Leibniz notation,* named after one of the co-inventors of the calculus, Gottfried Wilhelm von Leibniz (1646–1716). (The other was Sir Isaac Newton, 1642–1727.)

With the above notation, the derivative of $f(x) = 2x^3 + 4x$, for example, which was found in Example 5 of the section on the Definition of the Derivative to be $f'(x) = 6x^2 + 4$, would be written

$$\frac{dy}{dx} = \frac{d}{dx}[2x^3 + 4x] = 6x^2 + 4, \qquad \text{or} \qquad D_x[2x^3 + 4x] = 6x^2 + 4.$$

Either $(d/dx)[f(x)]$ or $D_x[f(x)]$ represents the derivative of the function f with respect to x.

NOTATIONS FOR THE DERIVATIVE

The derivative of $y = f(x)$ may be written in any of the following ways:

$$f'(x), \qquad \frac{dy}{dx}, \qquad \frac{d}{dx}[f(x)], \qquad \text{or} \qquad D_x[f(x)].$$

A variable other than x is often used as the independent variable. For example, if $y = f(t)$ gives population growth as a function of time, then the derivative of y with respect to t could be written

$$f'(t), \qquad \frac{dy}{dt}, \qquad \frac{d}{dt}[f(t)], \qquad \text{or} \qquad D_t[f(t)].$$

Other variables also may be used to name the function, as in $g(x)$ or $h(t)$.

Now we will use the definition

$$f'(x) = \lim_{h \to 0} \frac{f(x + h) - f(x)}{h}$$

to develop some rules for finding derivatives more easily than by the four-step process given in the previous chapter.

The first rule tells how to find the derivative of a constant function defined by $f(x) = k$, where k is a constant real number. Since $f(x + h)$ is also k, by definition $f'(x)$ is

$$f'(x) = \lim_{h \to 0} \frac{f(x + h) - f(x)}{h}$$

$$= \lim_{h \to 0} \frac{k - k}{h} = \lim_{h \to 0} \frac{0}{h} = \lim_{h \to 0} 0 = 0,$$

establishing the following rule.

CONSTANT RULE

If $f(x) = k$, where k is any real number, then

$$f'(x) = 0.$$

(The derivative of a constant is 0.)

FIGURE 1

Figure 1 illustrates this constant rule geometrically; it shows a graph of the horizontal line $y = k$. At any point P on this line, the tangent line at P is the line itself. Since a horizontal line has a slope of 0, the slope of the tangent line is 0. This agrees with the result above: the derivative of a constant is 0.

EXAMPLE 1 Derivative of a Constant

(a) If $f(x) = 9$, then $f'(x) = 0$.

(b) If $h(t) = \pi$, then $D_t[h(t)] = 0$.

(c) If $y = 2^3$, then $dy/dx = 0$.

Functions of the form $y = x^n$, where n is a real number, are very common in applications. To obtain a rule for finding the derivative of such a function, we can use the definition to work out the derivatives for various special values of n. This was done in the section on the Definition of the Derivative in Example 4 to show that for $f(x) = x^2$, $f'(x) = 2x$.

For $f(x) = x^3$, the derivative is found as follows.

$$f'(x) = \lim_{h \to 0} \frac{f(x + h) - f(x)}{h}$$

$$= \lim_{h \to 0} \frac{(x + h)^3 - x^3}{h}$$

$$= \lim_{h \to 0} \frac{(x^3 + 3x^2h + 3xh^2 + h^3) - x^3}{h}$$

The binomial theorem (discussed in most intermediate and college algebra texts) was used to expand $(x + h)^3$ in the last step. Now, the limit can be determined.

$$f'(x) = \lim_{h \to 0} \frac{3x^2h + 3xh^2 + h^3}{h}$$

$$= \lim_{h \to 0} (3x^2 + 3xh + h^2)$$

$$= 3x^2$$

The results in the following table were found in a similar way, using the definition of the derivative. (These results are modifications of some of the examples and exercises from the previous chapter.)

Function	n	Derivative
$f(x) = x^2$	2	$f'(x) = 2x = 2x^1$
$f(x) = x^3$	3	$f'(x) = 3x^2$
$f(x) = x^4$	4	$f'(x) = 4x^3$
$f(x) = x^{-1}$	-1	$f'(x) = -1 \cdot x^{-2} = \dfrac{-1}{x^2}$
$f(x) = x^{1/2}$	$1/2$	$f'(x) = \dfrac{1}{2}x^{-1/2} = \dfrac{1}{2x^{1/2}}$

These results suggest the following rule.

POWER RULE

If $f(x) = x^n$ for any real number n, then

$$f'(x) = nx^{n-1}.$$

(The derivative of $f(x) = x^n$ is found by multiplying by the exponent n and decreasing the exponent on x by 1.)

While the power rule is true for every real-number value of n, a proof is given here only for positive integer values of n. This proof follows the steps used above in finding the derivative of $f(x) = x^3$.

For any real numbers p and q, by the binomial theorem,

$$(p + q)^n = p^n + np^{n-1}q + \frac{n(n - 1)}{2}p^{n-2}q^2 + \cdots + npq^{n-1} + q^n.$$

Replacing p with x and q with h gives

$$(x + h)^n = x^n + nx^{n-1}h + \frac{n(n-1)}{2}x^{n-2}h^2 + \cdots + nxh^{n-1} + h^n,$$

from which

$$(x + h)^n - x^n = nx^{n-1}h + \frac{n(n-1)}{2}x^{n-2}h^2 + \cdots + nxh^{n-1} + h^n.$$

Dividing each term by h yields

$$\frac{(x + h)^n - x^n}{h} = nx^{n-1} + \frac{n(n-1)}{2}x^{n-2}h + \cdots + nxh^{n-2} + h^{n-1}.$$

Use the definition of derivative, and the fact that each term except the first contains h as a factor and thus approaches 0 as h approaches 0, to get

$$f'(x) = \lim_{h \to 0} \frac{(x + h)^n - x^n}{h}$$

$$= nx^{n-1} + \frac{n(n-1)}{2}x^{n-2}0 + \cdots + nx0^{n-2} + 0^{n-1}$$

$$= nx^{n-1}.$$

This shows that the derivative of $f(x) = x^n$ is $f'(x) = nx^{n-1}$, proving the power rule for positive integer values of n.

EXAMPLE 2 Power Rule

(a) If $y = x^6$, find $D_x y$.

Solution
$$D_x y = 6x^{6-1} = 6x^5$$

(b) If $y = t = t^1$, find $\dfrac{dy}{dt}$.

Solution
$$\frac{dy}{dt} = 1t^{1-1} = t^0 = 1$$

(c) If $y = 1/x^3$, find dy/dx.

Solution Use a negative exponent to rewrite this equation as $y = x^{-3}$; then
$$\frac{dy}{dx} = -3x^{-3-1} = -3x^{-4} \qquad \text{or} \qquad \frac{-3}{x^4}.$$

(d) Find $D_x(x^{4/3})$.

Solution
$$D_x(x^{4/3}) = \frac{4}{3}x^{4/3-1} = \frac{4}{3}x^{1/3}$$

(e) If $y = \sqrt{z}$, find dy/dz.

Solution Rewrite this as $y = z^{1/2}$; then
$$\frac{dy}{dz} = \frac{1}{2}z^{1/2-1} = \frac{1}{2}z^{-1/2} \qquad \text{or} \qquad \frac{1}{2z^{1/2}} \qquad \text{or} \qquad \frac{1}{2\sqrt{z}}.$$

FOR REVIEW ■

At this point you may wish to turn back to the Algebra Reference for a review of negative exponents and rational exponents. The relationship between powers, roots, and rational exponents is explained there.

The next rule shows how to find the derivative of the product of a constant and a function.

CONSTANT TIMES A FUNCTION

Let k be a real number. If $f'(x)$ exists, then

$$D_x[kf(x)] = kf'(x).$$

(The derivative of a constant times a function is the constant times the derivative of the function.)

This rule is proved with the definition of the derivative and rules for limits.

$$D_x[kf(x)] = \lim_{h \to 0} \frac{kf(x + h) - kf(x)}{h}$$

$$= \lim_{h \to 0} k\frac{[f(x + h) - f(x)]}{h} \qquad \text{Factor out } k.$$

$$= k \lim_{h \to 0} \frac{f(x + h) - f(x)}{h} \qquad \text{Limit rule 1}$$

$$= kf'(x) \qquad \text{Definition of derivative}$$

EXAMPLE 3 Derivative of a Constant Times a Function

(a) If $y = 8x^4$, find $\dfrac{dy}{dx}$.

Solution

$$\frac{dy}{dx} = 8(4x^3) = 32x^3$$

(b) If $y = -\dfrac{3}{4}x^{12}$, find dy/dx.

Solution

$$\frac{dy}{dx} = -\frac{3}{4}(12x^{11}) = -9x^{11}$$

(c) Find $D_t(-8t)$.

Solution

$$D_t(-8t) = -8(1) = -8$$

(d) Find $D_p(10p^{3/2})$.

Solution

$$D_p(10p^{3/2}) = 10\left(\frac{3}{2}p^{1/2}\right) = 15p^{1/2}$$

(e) If $y = \dfrac{6}{x}$, find $\dfrac{dy}{dx}$.

Solution Rewrite this as $y = 6x^{-1}$; then

$$\frac{dy}{dx} = 6(-1x^{-2}) = -6x^{-2} \qquad \text{or} \qquad \frac{-6}{x^2}.$$

EXAMPLE 4 Beagles

Researchers have determined that the daily energy requirements of female beagles who are at least one year old change with respect to age according to the function

$$E(t) = 753t^{-0.1321},$$

where $E(t)$ is the daily energy requirements (in kJ/W$^{0.67}$) for a dog that is t years old.*

(a) Graph this function. As a female beagle gets older, how do the daily energy requirements change? What happens to the function for very young animals?

Solution As indicated in Figure 2, the daily energy requirements are decreasing with respect to time. This indicates that older dogs do not have the same energy requirements as younger ones. The function approaches infinity for newborn beagles. This indicates that the function should not be used to predict energy requirements of very young beagles.

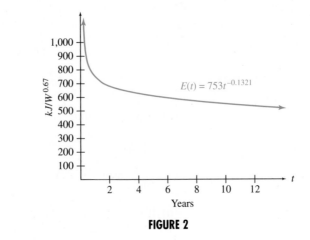

FIGURE 2

(b) Find $E'(t)$.

Solution Using the rules of differentiation we find that

$$E'(t) = -99.4713t^{-0.8679}.$$

(c) Determine the rate of change of the daily energy requirements of a two-year-old female beagle.

Solution

$$E'(2) = -99.4713(2)^{-0.8679} \approx -54.5$$

Thus, the daily energy requirements of a two-year-old female beagle are decreasing at the rate of 54.5 kJ/W$^{0.67}$/yr.

The final rule in this section is for the derivative of a function that is a sum or difference of terms.

---

*Finke, M., "Energy Requirements of Adult Female Beagles," *Journal of Nutrition,* Vol. 124, 1994, pp. 2604s–2608s.

SUM OR DIFFERENCE RULE

If $f(x) = u(x) \pm v(x)$, and if $u'(x)$ and $v'(x)$ exist, then

$$f'(x) = u'(x) \pm v'(x).$$

(The derivative of a sum or difference of functions is the sum or difference of the derivatives.)

The proof of the sum part of this rule is as follows: If $f(x) = u(x) + v(x)$, then

$$f'(x) = \lim_{h \to 0} \frac{[u(x + h) + v(x + h)] - [u(x) + v(x)]}{h}$$

$$= \lim_{h \to 0} \frac{[u(x + h) - u(x)] + [v(x + h) - v(x)]}{h}$$

$$= \lim_{h \to 0} \left[\frac{u(x + h) - u(x)}{h} + \frac{v(x + h) - v(x)}{h} \right]$$

$$= \lim_{h \to 0} \frac{u(x + h) - u(x)}{h} + \lim_{h \to 0} \frac{v(x + h) - v(x)}{h}$$

$$= u'(x) + v'(x).$$

A similar proof can be given for the difference of two functions.

EXAMPLE 5   Derivative of a Sum
Find the derivative of each function.

(a) $y = 6x^3 + 15x^2$

Solution Let $u(x) = 6x^3$ and $v(x) = 15x^2$; then $y = u(x) + v(x)$. Since $u'(x) = 18x^2$ and $v'(x) = 30x$,

$$\frac{dy}{dx} = 18x^2 + 30x.$$

(b) $p(t) = 12t^4 - 6\sqrt{t} + \dfrac{5}{t}$

Solution Rewrite $p(t)$ as $p(t) = 12t^4 - 6t^{1/2} + 5t^{-1}$; then
$$p'(t) = 48t^3 - 3t^{-1/2} - 5t^{-2}.$$

Also, $p'(t)$ may be written as $p'(t) = 48t^3 - 3/\sqrt{t} - 5/t^2$.

(c) $f(x) = \dfrac{x^3 + 3\sqrt{x}}{x}$

Solution Rewrite $f(x)$ as $f(x) = \dfrac{x^3}{x} + \dfrac{3x^{1/2}}{x} = x^2 + 3x^{-1/2}$. Then,

$$D_x f(x) = 2x - \frac{3}{2}x^{-3/2},$$

or

$$D_x f(x) = 2x - \frac{3}{2\sqrt{x^3}}.$$

(d) $f(x) = (4x^2 - 3x)^2$

Solution Rewrite $f(x)$ as $f(x) = 16x^4 - 24x^3 + 9x^2$. Then,
$$f'(x) = 64x^3 - 72x^2 + 18x.$$

> **NOTE** Some computer programs and calculators have built-in methods for taking derivatives symbolically, which is what we have been doing in this section, as opposed to approximating the derivative numerically by using a small number for h in the definition of the derivative. In the computer program Maple, we would do part (a) of Example 5 by entering
>
> ```
> > diff(6*x^3+15*x^2,x);
> ```
>
> and Maple would respond with
>
> ```
> 18*x^2+30*x.
> ```
>
> Similarly, on the TI-92, we would enter `d(6x^3+15x^2,x)` and the calculator would give "`18 · x^2+30 · x`."
>
> Other graphing calculators, such as the TI-83, do not have built-in methods for taking derivatives symbolically. As we saw in the last chapter, however, they do have the ability to calculate the derivative of a function at a particular point and to simultaneously graph a function and its derivative.
>
> Recall that, on the TI-83, we could enter `nDeriv(6x^3+15x^2,x,1)`. The number 48 will appear on the screen of the calculator, indicating the value of the derivative when $x = 1$. Figure 3 and Figure 4 indicate how to input the functions into the calculator and the corresponding graphs of both the function and its derivative. Consult *The Graphing Calculator Manual* that is available on this text's Web site for assistance.

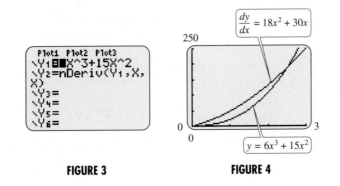

FIGURE 3   **FIGURE 4**

The rules developed in this section make it possible to find the derivative of a function more directly, so that applications of the derivative can be dealt with more effectively.

EXAMPLE 6 Centenarians

 The number of Americans (in thousands) who are expected to be over 100 years old can be approximated by the function

$$f(x) = 0.4018x^2 + 2.039x + 50.071,^*$$

_____
*U.S. Census Bureau

where x is the year, with $x = 0$ corresponding to 1994. This formula is based on estimates from 1994–2004.

(a) Find a formula giving the rate of change of the number of Americans over 100 years old.

Solution Using the techniques for finding the derivative, we have

$$f'(x) = 0.8036x + 2.039.$$

This tells us that the number of Americans over 100 years old is expected to grow at a linear rate.

(b) Find the rate of change in the number of Americans who are expected to be over 100 years old in the year 2003.

Solution The year 2003 corresponds to $x = 9$.

$$f'(9) = 0.8036(9) + 2.039 \approx 9.27$$

The number of Americans over 100 years old is expected to grow at a rate of 9.27 thousand, or about 9,300, per year in the year 2003. ▪

EXAMPLE 7 Cardiology

?

The aortic pressure-diameter relation in a single patient who underwent cardiac catheterization can be represented by

$$D(p) = 0.000002p^3 - 0.0008p^2 + 0.1141p + 16.683, \quad 55 \le p \le 130,$$

where $D(p)$ is the aortic diameter (in mm) and p is the aortic pressure (in mmHg).*

(a) Find a formula that gives the rate of change of aortic diameter with respect to aortic pressure.

Solution Using the techniques for finding the derivative, we have

$$D'(p) = 0.000006p^2 - 0.0016p + 0.1141.$$

This tells us that the rate of change in aortic diameter with respect to pressure is expected to follow a quadratic function.

(b) Find the rate of change in the aortic diameter when the pressure is 100 mmHg.

Solution A pressure of 100 mmHg corresponds to $p = 100$.

$$D'(100) = 0.000006(100)^2 - 0.0016(100) + 0.1141 = 0.0141.$$

Thus the aortic diameter is expected to increase at a rate of 0.0141 mm per mmHg when the pressure is 100 mmHg. ▪

---

*Stefanadis, C., J. Dernellis, et al., "Assessment of Aortic Line of Elasticity Using Polynomial Regression Analysis," *Circulation*, Vol. 101, No. 15, April 18, 2000, pp. 1819–1825.

4.1 EXERCISES

Find the derivative of each function defined as follows.

1. $y = 10x^3 - 9x^2 + 6x + 5$

2. $y = 3x^3 - x^2 - \dfrac{x}{12}$

3. $y = x^4 - 5x^3 + \dfrac{x^2}{9} + 5$

4. $y = 3x^4 + 11x^3 + 2x^2 - 4x$

5. $f(x) = 6x^{1.5} - 4x^{0.5}$

6. $f(x) = -2x^{2.5} + 8x^{0.5}$

7. $y = 8\sqrt{x} + 6x^{3/4}$

8. $y = -100\sqrt{x} - 11x^{2/3}$

9. $g(x) = 6x^{-5} - x^{-1}$

10. $y = 10x^{-2} + 3x^{-4} - 6x$

11. $y = x^{-5} - x^{-2} + 5x^{-1}$

12. $f(t) = \dfrac{6}{t} - \dfrac{8}{t^2}$

13. $f(t) = \dfrac{4}{t} + \dfrac{2}{t^3} + \sqrt{2}$

14. $y = \dfrac{9}{x^4} - \dfrac{8}{x^3} + \dfrac{2}{x}$

15. $y = \dfrac{3}{x^6} + \dfrac{1}{x^5} - \dfrac{7}{x^2}$

16. $p(x) = -10x^{-1/2} + 8x^{-3/2}$

17. $h(x) = x^{-1/2} - 14x^{-3/2}$

18. $y = \dfrac{6}{\sqrt[4]{x}}$

19. $y = \dfrac{-2}{\sqrt[3]{x}}$

20. $f(x) = \dfrac{x^2 + 2}{x}$

21. $g(x) = \dfrac{x^2 - 2x}{\sqrt{x}}$

22. $g(x) = (8x^2 - 4x)^2$

23. $h(x) = (x^2 - 1)^3$

24. Which of the following describes the derivative function f' of a quadratic function f?

    a. quadratic   **b.** linear   **c.** constant   **d.** cubic (third degree)

25. Explain the relationship between the slope and the derivative of $f(x)$ at $x = a$.

26. Which of the following does *not* equal $\dfrac{d}{dx}(4x^3 - 6x^{-2})$?

    a. $\dfrac{12x^2 + 12}{x^3}$ **b.** $\dfrac{12x^5 + 12}{x^3}$ **c.** $12x^2 + \dfrac{12}{x^3}$ **d.** $12x^3 + 12x^{-3}$

Find each of the following.

27. $D_x\left[9x^{-1/2} + \dfrac{2}{x^{3/2}}\right]$

28. $D_x\left[\dfrac{8}{\sqrt[4]{x}} - \dfrac{3}{\sqrt{x^3}}\right]$

29. $f'(-2)$ if $f(x) = \dfrac{x^2}{6} - 4x$

30. $f'(3)$ if $f(x) = \dfrac{x^3}{9} - 8x^2$

In Exercises 31–34, find the slope of the tangent line to the graph of the given function at the given value of x. Find the equation of the tangent line in Exercises 31 and 32.

31. $y = x^4 - 5x^3 + 2$; $x = 2$

32. $y = -2x^5 - 7x^3 + 8x^2$; $x = 1$

33. $y = -2x^{1/2} + x^{3/2}$; $x = 4$

34. $y = -x^{-3} + x^{-2}$; $x = 1$

35. Find all points on the graph of $f(x) = 9x^2 - 8x + 4$ where the slope of the tangent line is 0.

In Exercises 36–39, for each function find all values of x where the tangent line is horizontal.

36. $f(x) = 2x^3 + 9x^2 - 60x + 4$

37. $f(x) = x^3 + 15x^2 + 63x - 10$

38. $f(x) = x^3 - 4x^2 - 7x + 8$

39. $f(x) = x^3 - 5x^2 + 6x + 3$

40. At what points on the graph of $f(x) = 6x^2 + 4x - 9$ is the slope of the tangent line -2?

41. At what points on the graph of $f(x) = 2x^3 - 9x^2 - 12x + 5$ is the slope of the tangent line 12?

42. At what points on the graph of $f(x) = x^3 + 6x^2 + 21x + 2$ is the slope of the tangent line 9?

43. If $g'(5) = 10$ and $h'(5) = -4$, find $f'(5)$ for $f(x) = 3g(x) - 2h(x) + 3$.

44. If $g'(2) = 3$ and $h'(2) = 6$, find $f'(2)$ for $f(x) = \dfrac{1}{2}g(x) + \dfrac{1}{4}h(x)$.

45. Use the information given in the figure to find the following values.

    a. $f(1)$ **b.** $f'(1)$ **c.** the domain of f **d.** the range of f

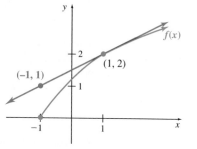

46. In Exercises 37–40 of the section on Quadratic Functions; Translation and Reflection, the effect of a when graphing $y = af(x)$ was discussed. Now describe how this relates to the fact that $D_x[af(x)] = af'(x)$.

47. Show that, for any constant k,

$$\frac{d}{dx}\left[\frac{f(x)}{k}\right] = \frac{f'(x)}{k}.$$

48. Use the differentiation feature on your graphing calculator to solve the problems (to two decimal places) below, where $f(x)$ is defined as follows:

$$f(x) = 1.25x^3 + 0.01x^2 - 2.9x + 1.$$

    a. Find $f'(4)$.

    b. Find all values of x where $f'(x) = 0$.

Applications

LIFE SCIENCES

49. *Cancer* Insulation workers who were exposed to asbestos and employed before 1960 experienced an increased likelihood of lung cancer. If a group of insulation workers has a cumulative total of 100,000 years of work experience with their first date of employment t years ago, then the number of lung cancer cases occurring within the group can be modeled using the function

$$N(t) = 0.00437t^{3.2}.*$$

Find the rate of growth of the number of workers with lung cancer in a group as described by the following first dates of employment.

    a. 5 years ago   **b.** 10 years ago

50. *Blood Sugar Level* Insulin affects the glucose, or blood sugar, level of some diabetics according to the function

$$G(x) = -0.2x^2 + 450,$$

where $G(x)$ is the blood sugar level one hour after x units of insulin are injected. (This mathematical model is only approximate, and it is valid only for values of x less than about 40.) Find the blood sugar level after the following numbers of units of insulin are injected.

    a. 0   **b.** 25

Find the rate of change of blood sugar level after injection of the following numbers of units of insulin.

    c. 10   **d.** 25

*Walker, A., *Observation and Inference: An Introduction to the Methods of Epidemiology,* Epidemiology Resources, Inc., 1991.

51. *Insect Mating Patterns* In an experiment testing methods of sexually attracting male insects to sterile females, equal numbers of males and females of a certain species are permitted to intermingle. (Such an experiment tests ways to control insect populations.) Assume that

$$M(t) = 4t^{3/2} + 2t^{1/2}$$

approximates the number of matings observed among the insects in an hour, where t is the temperature in degrees Celsius. (This formula is valid only for certain temperature ranges.) Find the number of matings at the following temperatures.

a. 16°C **b.** 25°C

c. Find the rate of change in the number of matings when the temperature is 16°C.

52. *Track and Field* In 1906 Kennelly developed a simple formula for predicting an upper limit on the fastest time that humans could ever run distances from 100 yards to 10 miles. His formula is given by

$$t = 0.0588s^{1.125},$$

where s is the distance in meters and t is the time to run that distance in seconds.*

a. Find Kennelly's estimate for the fastest mile. (*Hint:* 1 mile ≈ 1,609 meters.)

b. Find dt/ds when $s = 100$ and interpret your answer.

✎ **c.** Compare this and other estimates to the current world records. Have these estimates been surpassed?

53. *Bighorn Sheep* The cumulative horn volume for certain types of bighorn rams, found in the Rocky Mountains, can be described by the quadratic function

$$V(t) = -2,159 + 1,313t - 60.82t^2,$$

where $V(t)$ is the horn volume (in cm^3) and t is the year of growth, $2 \le t \le 9$.[†]

a. Find the horn volume for a three-year-old ram.

b. Find the rate at which the horn volume of a three-year-old ram is changing.

54. *Brain Weight* As we saw in Chapter 1, the brain weight of a human fetus during the last trimester can be accurately estimated from the circumference of the head by

$$w(c) = \frac{c^3}{100} - \frac{1,500}{c},$$

where $w(c)$ is the weight of the brain (in g[‡]) and c is the circumference (in cm) of the head.[§]

a. Estimate the brain weight of a fetus that has a head circumference of 30 cm.

b. Find the rate of change of the brain weight for a fetus that has a head circumference of 30 cm and interpret your results.

55. *Human Cough* To increase the velocity of the air flowing through the trachea when a human coughs, the body contracts the windpipe, producing a more effective cough. Tuchinsky formulated that the velocity of air that is flowing through the trachea during a cough is

$$V = C(R_0 - R)R^2,$$

where C is a constant based on individual body characteristics, R_0 is the radius of the windpipe before the cough, and R is the radius of the windpipe during the cough.[‖] It can be shown that the maximum velocity of the cough occurs when $\dfrac{dV}{dR} = 0$. Find the value of R that maximizes the velocity.[#]

56. *Body Mass Index* The body mass index (BMI) is a number that can be calculated for any individual as follows: Multiply weight (lbs) by 703 and divide by the person's height (in.) squared. That is,

$$BMI = \frac{703w}{h^2},$$

where w is in pounds and h is in inches. The National Heart, Lung and Blood Institute uses the BMI to determine whether a person is "overweight" ($25 \le BMI < 30$) or "obese" ($BMI \ge 30$).

a. Calculate the BMI for a person who weighs 220 lbs and is 6′2″ tall.

b. How much weight would the person in part a have to lose until he reaches a BMI of 24.9 and is no longer "overweight"?

---

*Kennelly, A., "An Approximate Law of Fatigue in Speeds of Racing Animals," *Proceedings of the American Academy of Arts and Sciences,* Vol. 42, 1906, pp. 275–331.
†Fitzsimmons, N., S. Buskirk, and M. Smith, "Population History, Genetic Variability, and Horn Growth in Bighorn Sheep," *Conservation Biology,* Vol. 9, No. 2, April 1995, pp. 314–323.
‡Technically, grams are a measure of mass, not weight. Weight is a measure of the force of gravity, which varies with the distance from the center of the earth. For objects on the surface of the earth, weight and mass are often used interchangeably, and we will do so in this text.
§Dobbing, John, and Jean Sands, "Head Circumference, Biparietal Diameter and Brain Growth in Fetal and Postnatal Life," *Early Human Development,* Vol. 2, No. 1, April 1978, pp. 81–87.
‖Tuchinsky, Philip, "The Human Cough," *UMAP Module 211,* Lexington, MA, COMAP, Inc., 1979, pp. 1–9.
#Interestingly, Tuchinsky also states that X-rays indicate that the body naturally contracts the windpipe to this radius during a cough.

c. For a 125-lb female, what is the rate of change of BMI with respect to height? (*Hint:* Take the derivative of the function: $f(h) = 703(125)/h^2$.)

d. Calculate and interpret the meaning of $f'(65)$.

e. Use the table function on your graphing calculator to construct a table for BMI for various weights and heights.

57. *Heart* The left ventricular length (viewed from the front of the heart) of a fetus that is at least 18 weeks old can be estimated by

$$l(x) = -2.318 + 0.2356x - 0.002674x^2,$$

where $l(x)$ is the ventricular length (in cm) and x is the age (in weeks) of the fetus.[*]

a. Determine a meaningful domain for this function.

b. Find $l'(x)$.

c. Find $l'(25)$.

58. *Velocity of Marine Organism* The typical velocity (in cm/sec) of a marine organism of length l (in cm) is given by $v = 2.69l^{1.86}$.[†] Find the rate of change of the velocity with respect to the length of the organism.

59. *Insect Species* The number of species of a particular type of insect depends on the size of the insect according to the formula $f(m) = 40m^{-0.6}$, where m is the mass in grams.[‡] Find and interpret $f'(0.01)$.

60. *Bird Eggs* Using results from Exercise 5 in Sec. 1.2 and Exercise 38 in Sec. 8.3, we can approximate the volume (in cubic cm) of a bird egg of width w (in cm) according to the formula

$$V = \frac{\pi(1.585w^3 - 0.487w^2)}{6}.$$

Find the rate of change of the volume with respect to the width for each of the following.

a. a robin, whose egg has an average width of 1.5 cm

b. a Canada goose, whose egg has an average width of 5.8 cm

61. *Popcorn* Researchers have determined that the amount of moisture present in a kernel of popcorn affects the volume of the popped corn and can be described for certain sizes of kernels by the function

$$v(x) = -35.98 + 12.09x - 0.4450x^2,$$

where x is moisture content (percent, wet basis) and $v(x)$ is the expansion volume (in cm^3/g).[§]

a. Graph this function on $[8, 20]$ by $[40, 50]$.

b. Determine the moisture content where the maximum volume occurs.

c. What is the value of the derivative of the above function at this value?

62. *Dog's Human Age* From the data printed in the following table from the *Minneapolis Star Tribune* on September 20, 1998, a dog's age when compared to a human's age can be modeled using either a linear formula or a quadratic formula as follows:

$$y_1 = 4.13x + 14.63$$
$$y_2 = -0.033x^2 + 4.647x + 13.347,$$

where y_1 and y_2 represent a dog's human age for each formula and x represents a dog's actual age.[‖]

Dog Age	Human Age
1	16
2	24
3	28
5	36
7	44
9	52
11	60
13	68
15	76

a. Find y_1 and y_2 when $x = 5$.

b. Find dy_1/dx and dy_2/dx when $x = 5$ and interpret your answers.

c. If the first three points are eliminated from the table, find the equation of a line that perfectly fits the reduced set of data. Interpret your findings.

d. Of the three formulas, which do you prefer?

---

[*]Tan, J., N. Silverman, J. Hoffman, M. Villegas, and K. Schmidt, "Cardiac Dimensions Determined by Cross-Sectional Echocardiography in the Normal Human Fetus From 18 Weeks to Term," *American Journal of Cardiology,* Vol. 70, No. 18, Dec. 1, 1992, pp. 1459–1467.

[†]Okubo, Akira, "Fantastic Voyage into the Deep: Marine Biofluid Mechanics," in *Mathematical Topics in Population Biology, Morphogenesis and Neurosciences,* edited by E. Teramoto and M. Yamaguti, Springer-Verlag, 1987, pp. 32–47.

[‡]*Science,* Vol. 284, June 18, 1999, p. 1937.

[§]Song, A., and S. Eckhoff, "Optimum Popping Moisture Content for Popcorn Kernels of Different Sizes," *Cereal Chemistry,* Vol. 71, No. 5, 1994, pp. 458–460.

[‖]Vennebush, Patrick, "Media Clips: A Dog's Human Age," *Mathematics Teacher,* Vol. 92, 1999, pp. 710–712.

OTHER APPLICATIONS

Velocity We saw in the previous chapter that if a function $s(t)$ gives the position of an object at time t, the derivative gives the velocity, that is, $v(t) = s'(t)$. For each of the position functions in Exercises 63–66, find **(a)** $v(t)$ and **(b)** the velocity when $t = 0$, $t = 5$, and $t = 10$.

63. $s(t) = 11t^2 + 4t + 2$

64. $s(t) = 25t^2 - 9t + 8$

65. $s(t) = 4t^3 + 8t^2$

66. $s(t) = -2t^3 + 4t^2 - 1$

67. *Velocity* If a rock is dropped from a 144-foot building, its position (in feet above the ground) is given by $s(t) = -16t^2 + 144$, where t is the time in seconds since it was dropped.

    a. What is its velocity 1 second after being dropped? 2 seconds after being dropped?

    b. When will it hit the ground?

    c. What is its velocity upon impact?

68. *Velocity* A ball is thrown vertically upward from the ground at a velocity of 64 feet per second. Its distance from the ground at t seconds is given by $s(t) = -16t^2 + 64t$.

    a. How fast is the ball moving 2 seconds after being thrown? 3 seconds after being thrown?

    b. How long after the ball is thrown does it reach its maximum height?

    c. How high will it go?

69. *Dead Sea* Researchers who have been studying the alarming rate at which the level of the Dead Sea has been dropping have shown that the density $d(x)$ in g/cm³ of the Dead Sea brine during evaporation can be estimated by the function

$$d(x) = 1.66 - 0.90x + 0.47x^2,$$

where x is the fraction of the remaining brine, $0 \le x \le 1$.*

    a. Estimate the density of the brine when 50% of the brine remains.

    b. Find and interpret the instantaneous rate of change of the density when 50% of the brine remains.

70. *Postal Rates* U.S. postal rates have steadily increased since 1932. Using data depicted in the table in the next column for the years 1932–1999, the cost to mail a single letter can be modeled using a quadratic formula as follows:

$$C(x) = 0.0102x^2 - 0.1901x + 2.0579,$$

where x is the number of years since 1932.†

Year	Cost
1932	3
1958	4
1963	5
1968	6
1971	8
1974	10
1975	13
1978	15
1981	18
1981	20
1985	22
1988	25
1991	29
1995	32
1999	33
2001	34

    a. Find the predicted cost of mailing a letter in 1982 and 2002 and compare these estimates with the actual rates.

    b. Find the rate of change of the postage cost for the years ending 1982 and 2002 and interpret your results.

    c. Critically evaluate possible limitations of this model and make suggestions for improvements.

71. *Money* The total amount of money in circulation for the years 1915–1999 can be closely approximated by

$$M(x) = 2.99x^3 - 382.12x^2 + 15,316.35x - 172,148.37,$$

where x represents the number of years since 1900 and $M(x)$ is in millions of dollars.‡ Find the derivative of $M(x)$ and use it to find the rate of change of money in circulation in the following years.

    a. 1930   **b.** 1935   **c.** 1940

    d. 1955   **e.** 1980   **f.** 1999

    g. What do your answers to parts a–f tell you about the amount of money in circulation in those years?

---

*Yechieli, Yoseph, Ittai Gavrieli, Brian Berkowitz, and Daniel Ronen, "Will the Dead Sea Die?" *Geology,* Vol. 26, No. 8, Aug. 1998, pp. 755–758. These researchers have predicted that the Dead Sea will not die but reach an equilibrium level.

†U.S. Postal Service

‡*The 2000 World Almanac and Book of Facts,* World Almanac Books, 2000, p. 120.

■ 4.2 DERIVATIVES OF PRODUCTS AND QUOTIENTS

? THINK ABOUT IT
How can we describe the behavior of the immune system when it is being attacked by a parasite?

We show how the derivative is used to study a problem like this in Example 5, later in this section. In the previous section we developed several rules for finding derivatives. We develop two additional rules in this section, again using the definition of the derivative.

The derivative of a sum of two functions is found from the sum of the derivatives. What about products? Is the derivative of a product equal to the product of the derivatives? For example, if

$$u(x) = 2x + 3 \quad \text{and} \quad v(x) = 3x^2,$$

then $\quad u'(x) = 2 \quad \text{and} \quad v'(x) = 6x.$

Let $f(x)$ be the product of u and v; that is, $f(x) = (2x + 3)(3x^2) = 6x^3 + 9x^2$. By the rules of the preceding section, $f'(x) = 18x^2 + 18x$. On the other hand, $u'(x) = 2$ and $v'(x) = 6x$, with the product $u'(x) \cdot v'(x) = 2(6x) = 12x \neq f'(x)$. In this example, the derivative of a product is *not* equal to the product of the derivatives, nor is this usually the case.

The rule for finding derivatives of products is now developed.

PRODUCT RULE

If $f(x) = u(x) \cdot v(x)$, and if $u'(x)$ and $v'(x)$ both exist, then

$$f'(x) = u(x) \cdot v'(x) + v(x) \cdot u'(x).$$

(The derivative of a product of two functions is the first function times the derivative of the second, plus the second function times the derivative of the first.)

FOR REVIEW ■

This proof uses several of the rules for limits given in the first section of the previous chapter. You may want to review them at this time.

To sketch the method used to prove the product rule, let

$$f(x) = u(x) \cdot v(x).$$

Then $f(x + h) = u(x + h) \cdot v(x + h)$, and, by definition, $f'(x)$ is given by

$$f'(x) = \lim_{h \to 0} \frac{f(x + h) - f(x)}{h}$$

$$= \lim_{h \to 0} \frac{u(x + h) \cdot v(x + h) - u(x) \cdot v(x)}{h}.$$

Now subtract and add $u(x + h) \cdot v(x)$ in the numerator, giving

$$f'(x) = \lim_{h \to 0} \frac{u(x + h) \cdot v(x + h) - u(x + h) \cdot v(x) + u(x + h) \cdot v(x) - u(x) \cdot v(x)}{h}$$

$$= \lim_{h \to 0} \frac{u(x + h)[v(x + h) - v(x)] + v(x)[u(x + h) - u(x)]}{h}$$

$$= \lim_{h \to 0} u(x + h) \left[\frac{v(x + h) - v(x)}{h} \right] + \lim_{h \to 0} v(x) \left[\frac{u(x + h) - u(x)}{h} \right]$$

$$= \lim_{h \to 0} u(x + h) \cdot \lim_{h \to 0} \frac{v(x + h) - v(x)}{h} + \lim_{h \to 0} v(x) \cdot \lim_{h \to 0} \frac{u(x + h) - u(x)}{h}. \quad \text{(1)}$$

If u' and v' both exist, then

$$\lim_{h \to 0} \frac{u(x + h) - u(x)}{h} = u'(x) \qquad \text{and} \qquad \lim_{h \to 0} \frac{v(x + h) - v(x)}{h} = v'(x).$$

The fact that u' exists can be used to prove

$$\lim_{h \to 0} u(x + h) = u(x),$$

and since no h is involved in $v(x)$,

$$\lim_{h \to 0} v(x) = v(x).$$

Substituting these results into equation (1) gives

$$f'(x) = u(x) \cdot v'(x) + v(x) \cdot u'(x),$$

the desired result.

EXAMPLE 1 Product Rule

Let $f(x) = (2x + 3)(3x^2)$. Use the product rule to find $f'(x)$.

Solution Here f is given as the product of $u(x) = 2x + 3$ and $v(x) = 3x^2$. By the product rule and the fact that $u'(x) = 2$ and $v'(x) = 6x$,

$$f'(x) = u(x) \cdot v'(x) + v(x) \cdot u'(x)$$
$$= (2x + 3)(6x) + (3x^2)(2)$$
$$= 12x^2 + 18x + 6x^2 = 18x^2 + 18x.$$

This result is the same as that found at the beginning of the section.

EXAMPLE 2 Product Rule

Find the derivative of $y = \left(\sqrt{x} + 3 \right)(x^2 - 5x)$.

Solution Let $u(x) = \sqrt{x} + 3 = x^{1/2} + 3$, and $v(x) = x^2 - 5x$. Then

$$\frac{dy}{dx} = u(x) \cdot v'(x) + v(x) \cdot u'(x)$$

$$= (x^{1/2} + 3)(2x - 5) + (x^2 - 5x)\left(\frac{1}{2} x^{-1/2} \right).$$

Simplify by multiplying and combining terms.

$$\frac{dy}{dx} = (2x)(x^{1/2}) + 6x - 5x^{1/2} - 15 + (x^2)\left(\frac{1}{2} x^{-1/2} \right) - (5x)\left(\frac{1}{2} x^{-1/2} \right)$$

$$= 2x^{3/2} + 6x - 5x^{1/2} - 15 + \frac{1}{2} x^{3/2} - \frac{5}{2} x^{1/2}$$

$$= \frac{5}{2} x^{3/2} + 6x - \frac{15}{2} x^{1/2} - 15$$

We could have found the derivatives above by multiplying out the original functions. The product rule then would not have been needed. In the next section, however, we shall see products of functions where the product rule is essential.

What about *quotients* of functions? To find the derivative of the quotient of two functions, use the next result.

QUOTIENT RULE

If $f(x) = u(x)/v(x)$, if all indicated derivatives exist, and if $v(x) \neq 0$, then

$$f'(x) = \frac{v(x) \cdot u'(x) - u(x) \cdot v'(x)}{[v(x)]^2}.$$

(The derivative of a quotient is the denominator times the derivative of the numerator, minus the numerator times the derivative of the denominator, all divided by the square of the denominator.)

The proof of the quotient rule is similar to that of the product rule and is left for the exercises.

> **CAUTION** Just as the derivative of a product is *not* the product of the derivatives, the derivative of a quotient is *not* the quotient of the derivatives. If you are asked to take the derivative of a product or a quotient, it is essential that you recognize that the function contains a product or quotient and then use the appropriate rule.

EXAMPLE 3 Quotient Rule

Find $f'(x)$ if $f(x) = \dfrac{2x - 1}{4x + 3}$.

FOR REVIEW ■

You may want to consult the Rational Expressions section of the Algebra Reference chapter (Section 3) to help you work with the fractions in this section.

Solution Let $u(x) = 2x - 1$, with $u'(x) = 2$. Also, let $v(x) = 4x + 3$, with $v'(x) = 4$. Then, by the quotient rule,

$$f'(x) = \frac{v(x) \cdot u'(x) - u(x) \cdot v'(x)}{[v(x)]^2}$$

$$= \frac{(4x + 3)(2) - (2x - 1)(4)}{(4x + 3)^2}$$

$$= \frac{8x + 6 - 8x + 4}{(4x + 3)^2}$$

$$= \frac{10}{(4x + 3)^2}.$$

> **CAUTION** In the second step of Example 3, we had the expression
>
> $$\frac{(4x + 3)(2) - (2x - 1)(4)}{(4x + 3)^2}.$$
>
> Students often incorrectly "cancel" the $4x + 3$ in the numerator with one factor of the denominator. Because the numerator is a *difference* of two products, however, you must multiply and combine terms *before* looking for common factors in the numerator and denominator.

EXAMPLE 4 Product and Quotient Rules

Find $D_x\left(\dfrac{(3 - 4x)(5x + 1)}{7x - 9}\right)$.

Solution This function has a product within a quotient. Instead of multiplying the factors in the numerator first (which is an option), we can use the quotient rule together with the product rule, as follows. Use the quotient rule first to get

$$D_x\left(\dfrac{(3 - 4x)(5x + 1)}{7x - 9}\right) = \dfrac{(7x - 9)[D_x(3 - 4x)(5x + 1)] - [(3 - 4x)(5x + 1)D_x(7x - 9)]}{(7x - 9)^2}.$$

Now use the product rule to find $D_x(3 - 4x)(5x + 1)$ in the numerator.

$$= \dfrac{(7x - 9)[(3 - 4x)5 + (5x + 1)(-4)] - (3 + 11x - 20x^2)(7)}{(7x - 9)^2}$$

$$= \dfrac{(7x - 9)(15 - 20x - 20x - 4) - (21 + 77x - 140x^2)}{(7x - 9)^2}$$

$$= \dfrac{(7x - 9)(11 - 40x) - 21 - 77x + 140x^2}{(7x - 9)^2}$$

$$= \dfrac{-280x^2 + 437x - 99 - 21 - 77x + 140x^2}{(7x - 9)^2}$$

$$= \dfrac{-140x^2 + 360x - 120}{(7x - 9)^2}$$

EXAMPLE 5 Immune System Activity

The activity of an immune system invaded by a parasite can be described by

$$I(n) = \dfrac{an^2}{b + n^2}$$

where n measures the number of larvae in the host, a is the maximum functional activity of the host's immune response, and b is a measure of the sensitivity of the immune system.* Find and interpret $I'(n)$.

Solution Although it appears that there are three independent variables in the function $I(n)$, both a and b are constants which are specific to the particular host. When taking the derivative of a function with unspecified constants we simply treat those constants in the same manner as we treat any other number. Therefore, the derivative of $I(n)$ is given by

$$I'(n) = \dfrac{(b + n^2)(2an) - an^2(2n)}{(b + n^2)^2}$$

$$= \dfrac{2abn + 2an^3 - 2an^3}{(b + n^2)^2}$$

$$= \dfrac{2abn}{(b + n^2)^2}.$$

In this case, $I'(n)$ is the rate of change of the activity of the immune system with respect to the number of larvae.

*Murray, J. D., *Mathematical Biology,* Springer-Verlag, 1989, p. 634.

4.2 EXERCISES

Use the product rule to find the derivative of each of the following. (Hint for Exercises 3–6: Write the quantity as a product.)

1. $y = (3x^2 + 2)(2x - 1)$

2. $y = (5x^2 - 1)(4x + 3)$

3. $y = (2x - 5)^2$

4. $y = (7x - 6)^2$

5. $k(t) = (t^2 - 1)^2$

6. $g(t) = (3t^2 + 2)^2$

7. $y = (x + 1)(\sqrt{x} + 2)$

8. $y = (2x - 3)(\sqrt{x} - 1)$

9. $p(y) = (y^{-1} + y^{-2})(2y^{-3} - 5y^{-4})$

10. $q(x) = (x^{-2} - x^{-3})(3x^{-1} + 4x^{-4})$

Use the quotient rule to find the derivative of each of the following.

11. $f(x) = \dfrac{7x + 1}{3x + 8}$

12. $f(x) = \dfrac{6x - 11}{8x + 1}$

13. $y = \dfrac{5 - 3t}{4 + t}$

14. $y = \dfrac{9 - 7t}{1 - t}$

15. $y = \dfrac{x^2 + x}{x - 1}$

16. $y = \dfrac{x^2 - 4x}{x + 3}$

17. $f(t) = \dfrac{4t + 11}{t^2 - 3}$

18. $y = \dfrac{-x^2 + 6x}{4x^2 + 1}$

19. $g(x) = \dfrac{x^2 - 4x + 2}{x + 3}$

20. $k(x) = \dfrac{x^2 + 7x - 2}{x - 2}$

21. $p(t) = \dfrac{\sqrt{t}}{t - 1}$

22. $r(t) = \dfrac{\sqrt{t}}{2t + 3}$

23. $y = \dfrac{5x + 6}{\sqrt{x}}$

24. $h(z) = \dfrac{z^{2.2}}{z^{3.2} + 5}$

25. $g(y) = \dfrac{y^{1.4} + 1}{y^{2.5} + 2}$

26. $f(x) = \dfrac{(3x^2 + 1)(2x - 1)}{5x + 4}$

27. If $g(3) = 4$, $g'(3) = 5$, $f(3) = 7$, and $f'(3) = 6$, find $h'(3)$ when $h(x) = f(x)g(x)$.

28. If $g(3) = 4$, $g'(3) = 5$, $f(3) = 7$, and $f'(3) = 6$, find $h'(3)$ when $h(x) = f(x)/g(x)$.

29. Find the error in the following work.

$$D_x\left(\frac{2x + 5}{x^2 - 1}\right) = \frac{(2x + 5)(2x) - (x^2 - 1)2}{(x^2 - 1)^2} = \frac{4x^2 + 10x - 2x^2 + 2}{(x^2 - 1)^2}$$
$$= \frac{2x^2 + 10x + 2}{(x^2 - 1)^2}$$

30. Find the error in the following work.

$$D_x\left(\frac{x^2 - 4}{x^3}\right) = x^3(2x) - (x^2 - 4)(3x^2) = 2x^4 - 3x^4 + 12x^2$$
$$= -x^4 + 12x^2$$

31. Find an equation of the line tangent to the graph of $f(x) = x/(x - 2)$ at $(3, 3)$.

32. Following the steps used to prove the product rule for derivatives, prove the quotient rule for derivatives.

33. Use the fact that $f(x) = u(x)/[v(x)]$ can be rewritten as $f(x)v(x) = u(x)$ and the product rule for derivatives to verify the quotient rule for derivatives. (*Hint:* After applying the product rule, substitute $u(x)/v(x)$ for $f(x)$ and simplify.)

For each of the following functions, find the value(s) of x in which $f'(x) = 0$, to three decimal places.

34. $f(x) = (x^2 - 2)(x^2 - \sqrt{2})$

35. $f(x) = \dfrac{x - 2}{x^2 + 4}$

Applications

LIFE SCIENCES

36. *Muscle Reaction* When a certain drug is injected into a muscle, the muscle responds by contracting. The amount of contraction, s, in millimeters, is related to the concentration of the drug, x, in milliliters, by

$$s(x) = \frac{x}{m + nx},$$

where m and n are constants.

a. Find $s'(x)$.

b. Find the rate of contraction when the concentration of the drug is 50 ml, $m = 10$, and $n = 3$.

37. *Growth Models* In Exercise 52 of the section on Polynomial and Rational Functions, the formula for the growth rate of a population in the presence of a quantity x of food was given as

$$f(x) = \frac{Kx}{A + x}.$$

This was referred to as Michaelis-Menten kinetics.

a. Find the rate of change of the growth rate with respect to the amount of food.

b. The quantity A in the formula for $f(x)$ represents the quantity of food for which the growth rate is half of its maximum. Using your answer from part a, find the rate of change of the growth rate when $x = A$.

38. *Bacteria Population* Assume that the total number (in millions) of bacteria present in a culture at a certain time t (in hours) is given by

$$N(t) = (t - 10)^2(2t) + 50.$$

a. Find $N'(t)$.

Find the rate at which the population of bacteria is changing at each of the following times.

b. 8 hr   **c.** 11 hr

d. The answer in part b is negative, and the answer in part c is positive. What does this mean in terms of the population of bacteria?

39. *Work/Rest Cycles* Murrell's formula for calculating the total amount of rest, in minutes, required after performing a particular type of work activity for 30 minutes is given by the formula:

$$R(w) = \frac{30(w - 4)}{w - 1.5},$$

where w is the work expended in kilocalories per min, kcal/min.*

a. A value of 5 for w indicates light work, such as riding a bicycle on a flat surface at 10 mi/hr. Find $R(5)$.

b. A value of 7 for w indicates moderate work, such as mowing grass with a pushmower on level ground. Find $R(7)$.

c. Find $R'(5)$ and $R'(7)$ and compare your answers. Explain whether these answers make sense.

40. *Optimal Foraging* Using data collected by zoologist Reto Zach, the work done by a crow to break open a whelk (large marine snail) can be estimated by the function

$$W = \left(1 + \frac{20}{H - 0.93}\right)H,$$

where H is the height, in meters, of the whelk when it is dropped.[†]

a. Find dW/dH.

b. One can show that the amount of work is minimized when $dW/dH = 0$. Find the value of H that minimizes W.

c. Interestingly, Zach observed the crows dropping the whelks from an average height of 5.23 meters. What does this imply?

41. *Cell Traction Force* In a matrix of cells, the cell traction force per unit mass is given by

$$T(n) = \frac{an}{1 + bn^2},$$

where n is the number of cells, a is the measure of the traction force generated by a cell, and b measures how force is reduced due to neighboring cells.[‡] Find and interpret $T'(n)$.

*Sanders, Mark, and Ernest McCormick, *Human Factors in Engineering and Design,* 7th ed., New York, McGraw Hill, 1993, pp. 243–246.

†Kellar, Brian, and Heather Thompson, "Whelk-come to Mathematics," *Mathematics Teacher,* Vol. 92, No. 6, September 1999, pp. 475–481.

‡Murray, J. D., *Mathematical Biology,* Springer-Verlag, 1989, p. 535.

42. *Calcium Kinetics* A function used to describe the kinetics of calcium in the cytogel (the gel of cell cytoplasm in the epithelium, an external tissue of epidermal cells) is given by

$$R(c) = \frac{ac^2}{1 + bc^2} - kc,$$

where $R(c)$ measures the release of calcium, c is the amount of free calcium outside the vesicles in which it is stored, and a, b, and k are positive constants.* Find and interpret $R'(c)$.

43. *Cellular Chemistry* The release of a chemical neurotransmitter as a function of the amount C of intracellular calcium is given by

$$L(C) = \frac{aC^n}{k + C^n},$$

where n measures the degree of cooperativity, k measures saturation, and a is the maximum of the process.[†] Find and interpret $L'(C)$.

44. *Memory Retention* Some psychologists contend that the number of facts of a certain type that are remembered after t hours is given by

$$f(t) = \frac{90t}{99t - 90}.$$

Find the rate at which the number of facts remembered is changing after the following numbers of hours.

a. 1   **b.** 10

OTHER APPLICATIONS

45. *Vehicle Waiting Time* The average number of vehicles waiting in a line to enter a parking ramp can be modeled by the function

$$f(x) = \frac{x^2}{2(1 - x)},$$

where x is a quantity between 0 and 1 known as the traffic intensity.[‡] Find the rate of change of the waiting time with respect to the traffic intensity for the following values of the intensity.

a. $x = 0.1$ **b.** $x = 0.6$

46. *Employee Training* A company that manufactures bicycles has determined that a new employee can assemble $M(d)$ bicycles per day after d days of on-the-job training, where

$$M(d) = \frac{100d^2}{3d^2 + 10}.$$

a. Find the rate of change function for the number of bicycles assembled with respect to time.

b. Find and interpret $M'(2)$ and $M'(5)$.

■ **4.3   THE CHAIN RULE**

THINK ABOUT IT

Suppose we know how fast the radius of a circular oil slick is growing, and we know how much the area of the oil slick is growing per unit of change in the radius. How fast is the area growing?

The answer to this question involves the chain rule for derivatives. Before discussing the chain rule, we consider the composition of functions. Many of the most useful functions for modeling are created by combining simpler functions. Viewing complex functions as combinations of simpler functions often makes them easier to understand and use.

Composition of Functions   Suppose a function f assigns to each element x in set X some element $y = f(x)$ in set Y. Suppose also that a function g takes each element in set Y and assigns to it a value $z = g[f(x)]$ in set Z. By using both f and

---

*Murray, J. D., *Mathematical Biology,* Springer-Verlag, 1989, p. 569.
[†]Segel, Lee A., "Toward Molecular Sensory Physiology: Mathematical Models," in *Mathematical Topics in Population Biology, Morphogenesis and Neurosciences,* edited by E. Teramoto and M. Yamaguti, Springer-Verlag, 1987, pp. 313–321.
[‡]Mannering, F., and W. Kilareski, *Principles of Highway Engineering and Traffic Control,* John Wiley and Sons, 1990.

g, an element x in X is assigned to an element z in Z, as illustrated in Figure 5. The result of this process is a new function called the *composition* of functions g and f (discussed briefly in the section on Properties of Functions), and is defined as follows.

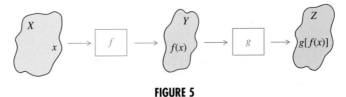

FIGURE 5

COMPOSITE FUNCTION

Let f and g be functions. The **composite function**, or **composition**, of g and f is the function whose values are given by $g[f(x)]$ for all x in the domain of f such that $f(x)$ is in the domain of g. (Read $g[f(x)]$ as "g of f of x".)

EXAMPLE 1 Composite Functions

Let $f(x) = 2x - 1$ and $g(x) = \sqrt{3x + 5}$. Find each of the following.

(a) $g[f(4)]$

Solution Find $f(4)$ first.

$$f(4) = 2 \cdot 4 - 1 = 8 - 1 = 7$$

Then

$$g[f(4)] = g[7] = \sqrt{3 \cdot 7 + 5} = \sqrt{26}.$$

FOR REVIEW

You may want to review how to find the domain of a function. Domain was discussed in the section on Properties of Functions.

(b) $f[g(4)]$

Solution Since $g(4) = \sqrt{3 \cdot 4 + 5} = \sqrt{17}$,

$$f[g(4)] = 2 \cdot \sqrt{17} - 1 = 2\sqrt{17} - 1.$$

(c) $f[g(-2)]$ does not exist since -2 is not in the domain of g.

EXAMPLE 2 Composition of Functions

Let $f(x) = 2x^2 + 5x$ and $g(x) = 4x + 1$. Find each of the following.

(a) $f[g(x)]$

Solution Using the given functions, we have

$$\begin{aligned}
f[g(x)] &= f[4x + 1] \\
&= 2(4x + 1)^2 + 5(4x + 1) \\
&= 2(16x^2 + 8x + 1) + 20x + 5 \\
&= 32x^2 + 16x + 2 + 20x + 5 \\
&= 32x^2 + 36x + 7.
\end{aligned}$$

(b) $g[f(x)]$

Solution By the definition above, with f and g interchanged,

$$g[f(x)] = g[2x^2 + 5x]$$

$$= 4(2x^2 + 5x) + 1$$
$$= 8x^2 + 20x + 1.$$

As Example 2 shows, it is not always true that $f[g(x)] = g[f(x)]$. In fact, it is rare to find two functions f and g such that $f[g(x)] = g[f(x)]$. The domain of both composite functions given in Example 2 is the set of all real numbers.

EXAMPLE 3   Composition of Functions

Write each of the following functions as the composition of two functions f and g so that $h(x) = f[g(x)]$.

(a) $h(x) = 2(4x + 1)^2 + 5(4x + 1)$

Solution   Let $f(x) = 2x^2 + 5x$ and $g(x) = 4x + 1$. Then $f[g(x)] = f(4x + 1) = 2(4x + 1)^2 + 5(4x + 1)$. Notice that $h(x)$ here is the same as $f[g(x)]$ in Example 2(a).

(b) $h(x) = \sqrt{1 - x^2}$

Solution   One way to do this is to let $f(x) = \sqrt{x}$ and $g(x) = 1 - x^2$. Another choice is to let $f(x) = \sqrt{1 - x}$ and $g(x) = x^2$. Verify that with either choice, $f[g(x)] = \sqrt{1 - x^2}$. For the purposes of this section, the first choice is better; it is useful to think of f as being the function on the outer layer and g as the function on the inner layer. With this function h, we see a square root on the outer layer, and when we peel that away we see $1 - x^2$ on the inside.

We now return to the question at the beginning of this section.

 The Chain Rule   A leaking oil well off the Gulf Coast is spreading a circular film of oil over the water surface. At any time t (in minutes) after the beginning of the leak, the radius of the circular oil slick is given by

$$r(t) = 4t, \qquad \text{with} \qquad \frac{dr}{dt} = 4,$$

where dr/dt is the rate of change in radius over time. The area of the oil slick is given by

$$A(r) = \pi r^2, \qquad \text{with} \qquad \frac{dA}{dr} = 2\pi r,$$

where dA/dr is the rate of change in area per unit change in radius.

As these derivatives show, the radius is increasing 4 times as fast as the time t, and the area is increasing $2\pi r$ times as fast as the radius r. It seems reasonable, then, that the area is increasing $2\pi r \cdot 4 = 8\pi r$ times as fast as time. That is,

$$\frac{dA}{dt} = \frac{dA}{dr} \cdot \frac{dr}{dt} = 2\pi r \cdot 4 = 8\pi r.$$

Since $r = 4t$,

$$\frac{dA}{dt} = 8\pi r = 8\pi(4t) = 32\pi t.$$

To check this, use the fact that $r = 4t$ and $A = \pi r^2$ to get the same result:

$$A = \pi(4t)^2 = 16\pi t^2, \qquad \text{with} \qquad \frac{dA}{dt} = 32\pi t.$$

(Notice that because A is a function of r, which is a function of t, A as a function of t is a composition of two functions.)

The product used above,

$$\frac{dA}{dt} = \frac{dA}{dr} \cdot \frac{dr}{dt},$$

is an example of the **chain rule**, which is used to find the derivative of a composite function.

CHAIN RULE

If y is a function of u, say $y = f(u)$, and if u is a function of x, say $u = g(x)$, then $y = f(u) = f[g(x)]$, and

$$\frac{dy}{dx} = \frac{dy}{du} \cdot \frac{du}{dx}.$$

One way to remember the chain rule is to pretend that dy/du and du/dx are fractions, with du "canceling out." The proof of the chain rule requires advanced concepts and therefore is not given here.

EXAMPLE 4 Chain Rule

Find dy/dx if $y = (3x^2 - 5x)^{1/2}$.

Solution Let $y = u^{1/2}$, and $u = 3x^2 - 5x$. Then

$$\frac{dy}{dx} = \frac{dy}{du} \cdot \frac{du}{dx}$$

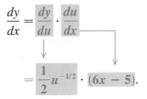

$$= \frac{1}{2} u^{-1/2} \cdot (6x - 5).$$

Replacing u with $3x^2 - 5x$ gives

$$\frac{dy}{dx} = \frac{1}{2}(3x^2 - 5x)^{-1/2}(6x - 5) = \frac{6x - 5}{2(3x^2 - 5x)^{1/2}}.$$

The following alternative version of the chain rule is stated in terms of composite functions.

CHAIN RULE (ALTERNATIVE FORM)

If $y = f[g(x)]$, then

$$\frac{dy}{dx} = f'[g(x)] \cdot g'(x).$$

(To find the derivative of $f[g(x)]$, find the derivative of $f(x)$, replace each x with $g(x)$, and then multiply the result by the derivative of $g(x)$.)

EXAMPLE 5 Chain Rule

Use the chain rule to find $D_x(x^2 + 5x)^8$.

Solution Let $f(x) = x^8$ and $g(x) = x^2 + 5x$. Then $(x^2 + 5x)^8 = f[g(x)]$ and

$$D_x(x^2 + 5x)^8 = f'[g(x)]g'(x).$$

Here $f'(x) = 8x^7$, with $f'[g(x)] = 8[g(x)]^7 = 8(x^2 + 5x)^7$ and $g'(x) = 2x + 5$.

$$D_x(x^2 + 5x)^8 = f'[g(x)]g'(x)$$

$$= 8[g(x)]^7 g'(x)$$

$$= 8(x^2 + 5x)^7(2x + 5)$$

While the chain rule is essential for finding the derivatives of some of the functions discussed later, the derivatives of the algebraic functions discussed so far can be found by the following *generalized power rule,* a special case of the chain rule.

GENERALIZED POWER RULE

Let $g(x)$ be a function of x, and let $y = [g(x)]^n$ for any real number n. Then

$$\frac{dy}{dx} = n \cdot [g(x)]^{n-1} \cdot g'(x).$$

(The derivative of $y = [g(x)]^n$ is found by decreasing the exponent on $g(x)$ by 1 and multiplying the result by the exponent n and by $g'(x)$.)

EXAMPLE 6 Chain Rule

(a) Use the generalized power rule to find the derivative of $y = (3 + 5x)^2$.

Solution Let $g(x) = 3 + 5x$, and $n = 2$. Then $g'(x) = 5$. By the generalized power rule,

$$\frac{dy}{dx} = n \cdot [g(x)]^{n-1}g'(x)$$

$$\overset{n}{\downarrow} \quad \overset{g(x)}{\downarrow} \quad \overset{n-1}{\downarrow} \quad \overset{g'(x)}{\downarrow}$$

$$= 2 \cdot (3 + 5x)^{2-1} \cdot \frac{d}{dx}(3 + 5x)$$

$$= 2(3 + 5x)^{2-1} \cdot 5 = 10(3 + 5x) = 30 + 50x.$$

(b) Find $\frac{dy}{dx}$ if $y = (3 + 5x)^{-3/4}$.

Solution Use the generalized power rule, with $g(x) = 3 + 5x$, $n = -3/4$, and $g'(x) = 5$.

$$\frac{dy}{dx} = -\frac{3}{4}(3 + 5x)^{-3/4-1} \cdot 5 = -\frac{15}{4}(3 + 5x)^{-7/4}$$

This result could not have been found by any of the rules given in previous sections. ▪

CAUTION

(a) A common error is to forget to multiply by $g'(x)$ when using the generalized power rule. Remember, the generalized power rule is an example of the chain rule, and so the derivative must involve a "chain," or product, of derivatives.

(b) Another common mistake is to write the derivative as $n[g'(x)]^{n-1}$. Remember to leave $g(x)$ unchanged, and then to multiply by $g'(x)$.

One way to avoid both of the errors described above is to remember that the chain rule is a two-step process. In Example 6(a), the first step was taking the derivative of the power, and the second step was multiplying by $g'(x)$. Forgetting to multiply by $g'(x)$ would be an erroneous one-step process. The other erroneous one-step process is to take the derivative inside the power, getting $n[g'(x)]^{n-1}$, or $2(5)^1$ in Example 6(a).

Sometimes both the generalized power rule and either the product or quotient rule are needed to find a derivative, as the next examples show.

EXAMPLE 7 Derivative Rules
Find the derivative of $y = 4x(3x + 5)^5$.

Solution Write $4x(3x + 5)^5$ as the product

$$(4x) \cdot (3x + 5)^5.$$

To find the derivative of $(3x + 5)^5$, let $g(x) = 3x + 5$, with $g'(x) = 3$. Now use the product rule and the generalized power rule.

$$\frac{dy}{dx} = 4x\overbrace{[5(3x + 5)^4 \cdot 3]}^{\text{Derivative of }(3x+5)^5} + (3x + 5)^5\overbrace{(4)}^{\text{Derivative of }4x}$$

$$= 60x(3x + 5)^4 + 4(3x + 5)^5$$

$$= 4(3x + 5)^4[15x + (3x + 5)^1] \qquad \text{Factor out the greatest common factor, } 4(3x + 5)^4.$$

$$= 4(3x + 5)^4(18x + 5) \qquad \text{Simplify inside brackets.} \qquad ▪$$

EXAMPLE 8 Derivative Rules
Find $D_x \dfrac{(3x + 2)^7}{x - 1}$.

Solution Use the quotient rule and the generalized power rule.

$$D_x \frac{(3x + 2)^7}{x - 1} = \frac{(x - 1)[7(3x + 2)^6 \cdot 3] - (3x + 2)^7(1)}{(x - 1)^2}$$

$$= \frac{21(x - 1)(3x + 2)^6 - (3x + 2)^7}{(x - 1)^2}$$

$$= \frac{(3x + 2)^6[21(x - 1) - (3x + 2)]}{(x - 1)^2} \qquad \begin{array}{l}\text{Factor out the greatest}\\\text{common factor,}\\(3x + 2)^6.\end{array}$$

$$= \frac{(3x + 2)^6[21x - 21 - 3x - 2]}{(x - 1)^2}$$

Simplify inside brackets.

$$= \frac{(3x + 2)^6(18x - 23)}{(x - 1)^2}$$

Some applications requiring the use of the chain rule or the generalized power rule are illustrated in the next examples.

EXAMPLE 9 Tasmanian Devil

Named for the only place where it is currently known to live, the extremely voracious Tasmanian Devil often preys on animals larger than itself. Researchers have identified the mathematical relationship

$$L(w) = 2.265w^{2.543},$$

where w (in kg) is the weight and $L(w)$ (in mm) is the length of the Tasmanian Devil.* Suppose that the weight of a particular Tasmanian Devil can be estimated by the function

$$w(t) = 0.125 + 0.18t,$$

where $w(t)$ is the weight (in kg) and t is the age, in weeks, of a Tasmanian Devil that is less than one year old. How fast is the length of a 30-week-old Tasmanian Devil changing?

Solution We want to find dL/dt, the rate of change of the length of a Tasmanian Devil. By the chain rule,

$$\frac{dL}{dt} = \frac{dL}{dw} \cdot \frac{dw}{dt}.$$

First find dL/dw, as follows.

$$\frac{dL}{dw} = (2.265)(2.543)w^{2.543-1} \approx 5.76w^{1.543}$$

Also,

$$\frac{dw}{dt} = 0.18.$$

Since the Tasmanian Devil is 30 weeks old, $t = 30$ and

$$w(30) = 0.125 + 0.18(30) = 5.525 \text{ kg}.$$

Putting this all together, we have

$$\frac{dL}{dt} \approx 5.76(5.525)^{1.543}(0.18) \approx 14.5.$$

*Silva, M., "Allometric Scaling of Body Length: Elastic Geometric Similarity in Mammalian Design," *Journal of Mammalogy,* Vol. 79, No. 1, 1998, pp. 20–32.

The length of a 30-week-old Tasmanian Devil is increasing at the rate of about 14.5 mm per week.

Alternatively, we can find dL/dt by first substituting the formula for $w(t)$ into $L(w)$. Then directly take the derivative of $L(t)$ with respect to t.

EXAMPLE 10 Compound Interest

Suppose a sum of $500 is deposited in an account with an interest rate of r percent per year compounded monthly. At the end of 10 years, the balance in the account (as illustrated in Figure 6) is given by

$$A = 500\left(1 + \frac{r}{1,200}\right)^{120}.$$

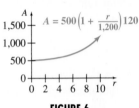

$A = 500\left(1 + \frac{r}{1,200}\right)120$

FIGURE 6

Find the rate of change of A with respect to r if $r = 5$ or 7.*

Solution First find dA/dr using the generalized power rule.

$$\frac{dA}{dr} = (120)(500)\left(1 + \frac{r}{1,200}\right)^{119}\left(\frac{1}{1,200}\right)$$

$$= 50\left(1 + \frac{r}{1,200}\right)^{119}$$

If $r = 5$,

$$\frac{dA}{dr} = 50\left(1 + \frac{5}{1,200}\right)^{119}$$

$$\approx 82.01,$$

or $82.01 per percentage point. If $r = 7$,

$$\frac{dA}{dr} = 50\left(1 + \frac{7}{1,200}\right)^{119}$$

$$\approx 99.90,$$

or $99.90 per percentage point.

> **NOTE** One lesson to learn from this section is that a derivative is always with respect to some variable. In the oil slick example, notice that the derivative of the area with respect to the radius is $2\pi r$, while the derivative of the area with respect to time is $8\pi r$. As another example, consider the velocity of a conductor walking at 2 mph on a train car. Her velocity with respect to the ground may be 50 mph, but the earth on which the train is running is moving about the sun at 1.6 million mph. The derivative of her position function might be 2, 50, or 1.6 million mph, depending upon what variable it is with respect to.

---

*Notice that r is given here as an integer percent, rather than as a decimal, which is why the formula for compound interest has 1,200 where you would expect to see 12. This leads to a simpler interpretation of the derivative.

4.3 EXERCISES

Let $f(x) = 4x^2 - 2x$ and $g(x) = 8x + 1$. Find each of the following.

1. $f[g(2)]$　　　　**2.** $f[g(-5)]$　　　　**3.** $g[f(2)]$　　　　**4.** $g[f(-5)]$　　　　**5.** $f[g(k)]$　　　　**6.** $g[f(5z)]$

Find $f[g(x)]$ and $g[f(x)]$ in each of the following.

7. $f(x) = \dfrac{x}{8} + 12;\ g(x) = 3x - 1$

8. $f(x) = -6x + 9;\ g(x) = \dfrac{x}{5} + 7$

9. $f(x) = \dfrac{1}{x};\ g(x) = x^2$

10. $f(x) = \dfrac{2}{x^4};\ g(x) = 2 - x$

11. $f(x) = \sqrt{x + 2};\ g(x) = 8x^2 - 6$

12. $f(x) = 9x^2 - 11x;\ g(x) = 2\sqrt{x + 2}$

13. $f(x) = \sqrt{x + 1};\ g(x) = \dfrac{-1}{x}$

14. $f(x) = \dfrac{8}{x};\ g(x) = \sqrt{3 - x}$

15. In your own words, explain how to form the composition of two functions.

Write each function as the composition of two functions. (There may be more than one way to do this.)

16. $y = (3x - 7)^{1/3}$

17. $y = (5 - x)^{2/5}$

18. $y = \sqrt{9 - 4x}$

19. $y = -\sqrt{13 + 7x}$

20. $y = (x^{1/2} - 3)^2 + (x^{1/2} - 3) + 5$

21. $y = (x^2 + 5x)^{1/3} - 2(x^2 + 5x)^{2/3} + 7$

22. What is the distinction between the chain rule and the generalized power rule?

Find the derivative of each function defined as follows.

23. $y = (2x^3 + 9x)^5$

24. $y = (8x^4 - 3x^2)^3$

25. $f(x) = -8(3x^4 + 2)^3$

26. $k(x) = -2(12x^2 + 5)^6$

27. $s(t) = 12(2t^4 + 5)^{3/2}$

28. $s(t) = 45(3t^3 - 8)^{3/2}$

29. $f(t) = 8\sqrt{4t^2 + 7}$

30. $g(t) = -3\sqrt{7t^3 - 1}$

31. $r(t) = 4t(2t^5 + 3)^2$

32. $m(t) = -6t(5t^4 - 1)^2$

33. $y = (x^3 + 2)(x^2 - 1)^2$

34. $y = (3x^4 + 1)^2(x^3 + 4)$

35. $p(z) = z(6z + 1)^{4/3}$

36. $q(y) = 4y^2(y^2 + 1)^{5/4}$

37. $y = \dfrac{1}{(3x^2 - 4)^5}$

38. $y = \dfrac{-5}{(2x^3 + 1)^2}$

39. $p(t) = \dfrac{(2t + 3)^3}{4t^2 - 1}$

40. $r(t) = \dfrac{(5t - 6)^4}{3t^2 + 4}$

41. $y = \dfrac{x^2 + 4x}{(3x^3 + 2)^4}$

42. $y = \dfrac{3x^2 - x}{(2x - 1)^5}$

Consider the following table of values of the functions f and g and their derivatives at various points:

x	1	2	3	4
$f(x)$	2	4	1	3
$f'(x)$	−6	−7	−8	−9
$g(x)$	2	3	4	1
$g'(x)$	2/7	3/7	4/7	5/7

Find each of the following using the table on the previous page.

43. a. $D_x(f[g(x)])$ at $x = 1$ **b.** $D_x(f[g(x)])$ at $x = 2$

44. a. $D_x(g[f(x)])$ at $x = 1$ **b.** $D_x(g[f(x)])$ at $x = 2$

In Exercises 45–48, find the equation of the tangent line to the graph of the given function at the given value of x.

45. $f(x) = \sqrt{x^2 + 16}$; $x = 3$

46. $f(x) = (x^3 + 7)^{2/3}$; $x = 1$

47. $f(x) = x(x^2 - 4x + 5)^4$; $x = 2$

48. $f(x) = x^2\sqrt{x^4 - 12}$; $x = 2$

In Exercises 49 and 50, find all values of x for the given function where the tangent line is horizontal.

49. $f(x) = \sqrt{x^3 - 6x^2 + 9x + 1}$

50. $f(x) = \dfrac{x}{(x^2 + 4)^4}$

51. Katie and Sarah are working on taking the derivative of

$$f(x) = \frac{2x}{3x + 4}.$$

Katie uses the quotient rule to get

$$f'(x) = \frac{(3x + 4)2 - 2x(3)}{(3x + 4)^2} = \frac{8}{(3x + 4)^2}.$$

Sarah converts it into a product and uses the product rule and the chain rule:

$$f(x) = 2x(3x + 4)^{-1}$$
$$f'(x) = 2x(-1)(3x + 4)^{-2}(3) + 2(3x + 4)^{-1}$$
$$= 2(3x + 4)^{-1} - 6x(3x + 4)^{-2}.$$

Explain the discrepancies between the two answers. Which procedure do you think is preferable?

52. Margy and Nate are working on taking the derivative of

$$f(x) = \frac{2}{(3x + 1)^4}.$$

Margy uses the quotient rule as follows:

$$f'(x) = \frac{(3x + 1)^4 \cdot 0 - 2 \cdot 4(3x + 1)^3 \cdot 3}{(3x + 1)^8}$$
$$= \frac{-24(3x + 1)^3}{(3x + 1)^8} = \frac{-24}{(3x + 1)^5}.$$

Nate rewrites the function and uses the generalized power rule as follows:

$$f(x) = 2(3x + 1)^{-4}$$
$$f'(x) = (-4)2(3x + 1)^{-5} \cdot 3 = \frac{-24}{(3x + 1)^5}.$$

Compare the two procedures. Which procedure do you think is preferable?

Applications

LIFE SCIENCES

53. *Fish Population* Suppose the population P of a certain species of fish depends on the number x (in hundreds) of a smaller fish that serves as its food supply, so that

$$P(x) = 2x^2 + 1.$$

Suppose, also, that the number of the smaller species of fish depends upon the amount a (in appropriate units) of its food supply, a kind of plankton. Specifically,

$$x = f(a) = 3a + 2.$$

A biologist wants to find the relationship between the population P of the large fish and the amount a of plankton available, that is, $P[f(a)]$. What is the relationship?

54. *Oil Pollution* An oil well off the Gulf Coast is leaking, with the leak spreading oil over the surface as a circle. At any time t (in minutes) after the beginning of the leak, the radius of the circular oil slick on the surface is $r(t) = t^2$ feet. Let $A(r) = \pi r^2$ represent the area of a circle of radius r.

a. Find and interpret $A[r(t)]$.

b. Find and interpret $D_t A[r(t)]$ when $t = 100$.

55. *African Wild Dog* The African wild dog is currently one of the most endangered carnivores in the world. After years of attempts to rescue the dog from extinction, scientists have collected a lot of data on the habits of this nomadic creature. The collected data show the following mathematical relationship between the weight and the length of the dog:

$$L(w) = 2.472w^{2.571},$$

where w (in kg) is the weight and $L(w)$ (in mm) is the length of the mammal.* Suppose that the weight of a particular wild dog can be estimated by

$$w(t) = 0.265 + 0.21t,$$

where $w(t)$ is the weight (in kg) and t is the age, in weeks, of an African wild dog that is less than one year old. How fast is the length of a 25-week-old African wild dog changing?

56. *Thermal Inversion* When there is a thermal inversion layer over a city (as happens often in Los Angeles), pollutants cannot rise vertically but are trapped below the layer and must disperse horizontally. Assume that a factory smokestack begins emitting a pollutant at 8 AM. Assume that the pollutant disperses horizontally, forming a circle. If t represents the time (in hours) since the factory began emitting pollutants ($t = 0$ represents 8 AM), assume that the radius of the circle of pollution is $r(t) = 2t$ miles. Let $A(r) = \pi r^2$ represent the area of a circle of radius r.

a. Find and interpret $A[r(t)]$.

b. Find and interpret $D_t A[r(t)]$ when $t = 4$.

57. *Bacteria Population* The total number of bacteria (in millions) present in a culture is given by

$$N(t) = 2t(5t + 9)^{1/2} + 12,$$

where t represents time (in hours) after the beginning of an experiment. Find the rate of change of the population of bacteria with respect to time for each of the following numbers of hours.

a. 0   b. 7/5   c. 8

58. *Calcium Usage* To test an individual's use of calcium, a researcher injects a small amount of radioactive calcium into the person's bloodstream. The calcium remaining in the bloodstream is measured each day for several days. Suppose the amount of the calcium remaining in the bloodstream (in milligrams per cubic centimeter) t days after the initial injection is approximated by

$$C(t) = \frac{1}{2}(2t + 1)^{-1/2}.$$

Find the rate of change of the calcium level with respect to time for each of the following numbers of days.

a. 0   b. 4   c. 7.5

d. Is C always increasing or always decreasing? How can you tell?

59. *Drug Reaction* The strength of a person's reaction to a certain drug is given by

$$R(Q) = Q\left(C - \frac{Q}{3}\right)^{1/2},$$

where Q represents the quantity of the drug given to the patient and C is a constant.

a. The derivative $R'(Q)$ is called the *sensitivity* to the drug. Find $R'(Q)$.

b. Find the sensitivity to the drug if $C = 59$ and a patient is given 87 units of the drug.

c. Is the patient's sensitivity to the drug increasing or decreasing when $Q = 87$?

60. *Extinction* The probability of a population going extinct by time t when the birth and death rates are the same can be estimated by

$$p(t) = \left(\frac{at}{at + 1}\right)^N,$$

where a is the birth and death rate, and N is the number of individuals in the population at time $t = 0$.[†] Find and interpret $p'(t)$.

OTHER APPLICATIONS

61. *Candy* The volume and surface area of a "jawbreaker" for any radius is given by the formulas

$$V(r) = \frac{4}{3}\pi r^3 \quad \text{and} \quad S(r) = 4\pi r^2,$$

respectively. Roger Guffey estimates the radius of a jawbreaker while in a person's mouth to be

$$r(t) = 6 - \frac{3}{17}t,$$

where $r(t)$ is in mm and t is in minutes.[‡]

a. What is the life expectancy of a jawbreaker?

b. Find dV/dt and dS/dt when $t = 17$ and interpret your answer.

---

*Silva, M., "Allometric Scaling of Body Length: Elastic Geometric Similarity in Mammalian Design," *Journal of Mammalogy*, Vol. 79, No. 1, 1998, pp. 20–32.

[†]Bailey, Norman T., *The Mathematical Approach to Biology and Medicine*, Wiley, 1967, p. 161.

[‡]Guffey, Roger, "The Life Expectancy of a Jawbreaker: An Application of the Composition of Functions," *Mathematics Teacher*, Vol. 92, No. 2, Feb. 1999, pp. 125–127.

c. Construct an analogous experiment using some other type of food or verify the results of this experiment.

62. *Interest* A sum of $1,500 is deposited in an account with an interest rate of r percent per year, compounded daily. At the end of 5 yr, the balance in the account is given by

$$A = 1,500\left(1 + \frac{r}{36,500}\right)^{1,825}.$$

Find the rate of change of A with respect to r for the following interest rates.

a. 6%   **b.** 8%   **c.** 9%

63. *Depreciation* A certain ultrasound machine depreciates according to the formula

$$V = \frac{6,000}{1 + 0.3t + 0.1t^2},$$

where t is time measured in years and $t = 0$ represents the time of purchase (in years). Find the rate at which the value of the machine is changing at the following times.

a. 2 yr   **b.** 4 yr

■ 4.4 DERIVATIVES OF EXPONENTIAL FUNCTIONS

THINK ABOUT IT   What is the relationship between the age and weight of a cow?

We will use a derivative to answer this question in Example 5 at the end of this section.

We can find the derivative of exponential functions by using the definition of the derivative.

$$\frac{d(a^x)}{dx} = \lim_{h \to 0} \frac{a^{x+h} - a^x}{h} \quad \text{(Assume } a > 0\text{)}$$

$$= \lim_{h \to 0} \frac{a^x a^h - a^x}{h} \quad \text{Property 1 of exponents}$$

$$= a^x \lim_{h \to 0} \frac{a^h - 1}{h} \quad \text{Property 1 of limits}$$

In this last step, since a^x does not involve h, we were able to bring a^x in front of the limit. The result says that the derivative of a^x is a^x times a constant that depends on a, namely $\lim_{h \to 0}(a^h - 1)/h$. To find out what this limit is, let us approximate the limit by letting $h = 0.0001$ and graph it as a function of a. The result is shown in Figure 7(a). You may recognize this graph; it looks like the graph of $y = \ln x$. To confirm this, we use the calculator to create the table in Figure 7(b),

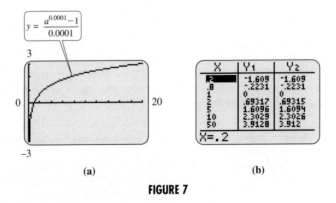

(a)

(b)

FIGURE 7

in which Y_1 represents $\lim\limits_{h \to 0}(a^h - 1)/h$ and Y_2 represents $\ln a$. Notice that the numbers in the two columns are very close. It turns out that $\lim\limits_{h \to 0}(a^h - 1)/h = \ln a$, and the slight discrepancies between the two columns occur because we are only approximating the limit, using $h = 0.0001$. This result is proven rigorously in more advanced courses.

Thus we have the following formula.

DERIVATIVE OF a^x

$$D_x[a^x] = (\ln a)a^x$$

This formula becomes particularly simple when we let $a = e$, because of the fact that $\ln e = 1$.

DERIVATIVE OF e^x

$$D_x[e^x] = e^x$$

We now see why e is the best base to work with: it has the simplest derivative of all the exponential functions. Even if we choose a different base, e appears in the derivative anyway through the $\ln a$ term. (Recall that $\ln a$ is the logarithm of a to the base e.) In fact, of all the functions we have studied, e^x is the simplest to differentiate, because its derivative is just itself.*

The chain rule can be used to find the derivative of the more general exponential function $y = a^{g(x)}$. Let $y = f(u) = a^u$ and $u = g(x)$, so that $f[g(x)] = a^{g(x)}$. Then

$$f'[g(x)] = f'(u) = (\ln a)a^u = (\ln a)a^{g(x)},$$

and by the chain rule,

$$\frac{dy}{dx} = f'[g(x)] \cdot g'(x)$$

$$= (\ln a)a^{g(x)} \cdot g'(x).$$

As before, this formula becomes simpler when we use natural logarithms because $\ln e = 1$. We summarize these results next.

DERIVATIVE OF $a^{g(x)}$ AND $e^{g(x)}$

$$D_x[a^{g(x)}] = (\ln a)a^{g(x)}g'(x)$$

and

$$D_x[e^{g(x)}] = e^{g(x)}g'(x)$$

CAUTION Notice the difference between the derivative of a variable to a constant power, such as $D_x x^3 = 3x^2$, and a constant to a variable power, like $D_x e^x = e^x$. Remember, $D_x e^x \neq xe^{x-1}$.

---

*There is a joke about a deranged mathematician who frightened other inmates at an insane asylum by screaming at them, "I'm going to differentiate you!" But one inmate remained calm and simply responded, "I don't care; I'm e^x."

EXAMPLE 1 Derivatives of Exponential Functions

Find derivatives of the following functions.

(a) $y = e^{5x}$

> **Solution** Let $g(x) = 5x$, with $g'(x) = 5$. Then
>
> $$\frac{dy}{dx} = 5e^{5x}.$$

(b) $s = 3^t$

> **Solution**
>
> $$\frac{ds}{dt} = (\ln 3)3^t$$

(c) $y = 10e^{3x^2}$

> **Solution**
>
> $$\frac{dy}{dx} = 10(e^{3x^2})(6x) = 60xe^{3x^2}$$

(d) $s = 8 \cdot 10^{1/t}$

> **Solution**
>
> $$\frac{ds}{dt} = 8(\ln 10)10^{1/t}\left(\frac{-1}{t^2}\right)$$
> $$= \frac{-8(\ln 10)10^{1/t}}{t^2}$$

EXAMPLE 2 Derivatives of Exponential Functions

Let $y = e^x\sqrt{x}$. Find $\frac{dy}{dx}$.

Solution Use the product rule.

$$\frac{dy}{dx} = e^x \cdot \frac{1}{2\sqrt{x}} + \sqrt{x}e^x$$

$$= \frac{e^x}{2\sqrt{x}} + e^x\sqrt{x} \cdot \frac{2\sqrt{x}}{2\sqrt{x}}$$

$$= \frac{e^x}{2\sqrt{x}}(1 + 2x)$$

EXAMPLE 3 Derivatives of Exponential Functions

Let $f(x) = \dfrac{100{,}000}{1 + 100e^{-0.3x}}$. Find $f'(x)$.

Solution Use the quotient rule.

$$f'(x) = \frac{(1 + 100e^{-0.3x})(0) - 100{,}000(-30e^{-0.3x})}{(1 + 100e^{-0.3x})^2}$$

$$= \frac{3{,}000{,}000e^{-0.3x}}{(1 + 100e^{-0.3x})^2}$$

NOTE In the previous example, we could also have taken the derivative by writing $f(x) = 100,000(1 + 100e^{-0.3x})^{-1}$, from which we have $f'(x) = -100,000(1 + 100e^{-0.3x})^{-2}100e^{-0.3x}(-0.3)$. This simplifies to the same expression as in Example 3.

EXAMPLE 4   Radioactivity

The amount in grams of a radioactive substance present after t years is given by

$$A(t) = 100e^{-0.12t}.$$

Find the rate of change of the amount present after 3 years.

Solution The rate of change is given by the derivative dA/dt.

$$\frac{dA}{dt} = 100(e^{-0.12t})(-0.12) = -12e^{-0.12t}$$

After 3 years ($t = 3$), the rate of change is

$$\frac{dA}{dt} = -12e^{-0.12(3)} = -12e^{-0.36} \approx -8.4$$

grams per year.

Frequently a population, or the size of a living creature, will start growing slowly, then grow more rapidly, and then gradually level off. Such growth can often be approximated by a variety of mathematical models. One such model is illustrated in the next example.

EXAMPLE 5   Age-Weight Relationship in Cattle

$\boxed{?}$ The mathematical relationship between the weight and age of cattle has been studied in Spain. The results of this study conclude that the mathematical function,

$$W(t) = 650.1(1 - e^{-0.038t})^3,$$

can be used to provide an estimate of the weight of Retinta beef cows of various ages, where $W(t)$ is the weight (in kg) at time t (in months).*

(a) Find the rate of growth of a 48-month-old cow.

Solution The derivative of this function, which gives the rate of change of growth, can be determined using the chain rule.

$$W'(t) = (650.1)(3)(1 - e^{-0.038t})^{3-1}(0.038e^{-0.038t})$$
$$= 74.1114e^{-0.038t}(1 - e^{-0.038t})^2.$$

From this, we can determine that $W'(60) \approx 6.11$ kg/month. Thus, the rate of change of weight after 60 months is about 6.11 kg per month.

(b) What happens to the weight of the average cow over time?

Solution As time increases, $t \to \infty$, and

*de Torre, G., J. Candotti, et al., "Effects of Growth Curve Parameters on Cow Efficiency," *Journal of Animal Science,* Vol. 70, 1992, pp. 2668–2672.

$$e^{-0.038t} = \frac{1}{e^{0.038t}} \to 0.$$

Thus,

$$\lim_{t \to \infty} W(t) = 650.1(1 - 0)^3 = 650.1.$$

The weight of an average Retinta cow approaches a horizontal asymptote of 650.1 kg as illustrated by the graph of the function in Figure 8.

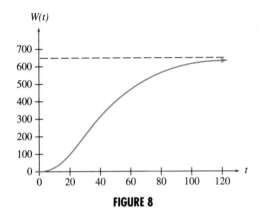

FIGURE 8

4.4 EXERCISES

Find derivatives of the functions defined as follows.

1. $y = e^{4x}$

2. $y = e^{-2x}$

3. $y = -8e^{2x}$

4. $y = 0.2e^{5x}$

5. $y = -16e^{x+1}$

6. $y = -4e^{-0.1x}$

7. $y = e^{x^2}$

8. $y = e^{-x^2}$

9. $y = 3e^{2x^2}$

10. $y = -5e^{4x^3}$

11. $y = 4e^{2x^2-4}$

12. $y = -3e^{3x^2+5}$

13. $y = xe^x$

14. $y = x^2 e^{-2x}$

15. $y = (x - 3)^2 e^{2x}$

16. $y = (3x^2 - 4x)e^{-3x}$

17. $y = \dfrac{x^2}{e^x}$

18. $y = \dfrac{e^x}{2x + 1}$

19. $y = \dfrac{e^x + e^{-x}}{x}$

20. $y = \dfrac{e^x - e^{-x}}{x}$

21. $p = \dfrac{10{,}000}{9 + 4e^{-0.2t}}$

22. $p = \dfrac{500}{12 + 5e^{-0.5t}}$

23. $f(z) = (2z + e^{-z^2})^2$

24. $y = 8^{5x}$

25. $y = 2^{-x}$

26. $y = 3 \cdot 4^{x^2+2}$

27. $y = -10^{3x^2-4}$

28. $s = 2 \cdot 3^{\sqrt{t}}$

29. $s = 5 \cdot 7^{\sqrt{t-2}}$

30. Prove that if $y = y_0 e^{kt}$, then $dy/dt = ky$. (This says that for exponential growth and decay, the rate of change of the population is proportional to the size of the population, and the constant of proportionality is the growth or decay constant.)

31. Use a graphing calculator to sketch the graph of $y = (f(x + h) - f(x))/h$ using $f(x) = e^x$ and $h = 0.0001$. Compare it with the graph of $y = e^x$ and discuss what you observe.

Applications

LIFE SCIENCES

32. *Population Growth* In Exercise 77 of the section on Logarithmic Functions, we found that the population of Florida (in millions) could be approximated by

$$p(t) = 14.0(1.017)^t,$$

where $t = 0$ (in years) corresponds to July 1994. Find the instantaneous rate of change in the population of Florida at the following times.

a. July 1998   **b.** January 2005

33. *Insect Growth* The growth of a population of rare South American beetles is given by the function

$$G(t) = \frac{10,000}{1 + 49e^{-0.1t}},$$

where t is time in months since the initial count.

a. Find the initial count and maximum population of the beetles.

Find the population and rate of growth of the population after the following times.

b. 6 months   **c.** 3 years   **d.** 7 years

e. What happens to the rate of growth over time?

34. *Clam Population* The population of a bed of clams in the Great South Bay off Long Island is described by the function

$$G(t) = \frac{5,200}{1 + 12e^{-0.52t}},$$

where t is time in years since the initial count.

a. Find the initial count and maximum population of the clams.

Find the population and rate of growth of the population after the following times.

b. 1 year   **c.** 4 years   **d.** 10 years

e. What happens to the rate of growth over time?

35. *Pollution Concentration* The concentration of pollutants (in grams per liter) in the east fork of the Big Weasel River is approximated by

$$P(x) = 0.04e^{-4x},$$

where x is the number of miles downstream from a paper mill that the measurement is taken. Find each of the following values.

a. The concentration of pollutants 0.5 mi downstream

b. The concentration of pollutants 1 mi downstream

c. The concentration of pollutants 2 mi downstream

Find the rate of change of concentration with respect to distance for each of the following distances.

d. 0.5 mi   **e.** 1 mi   **f.** 2 mi

36. *Breast Cancer* It has been observed that the following formula accurately models the relationship between the size of a breast tumor and the amount of time that it has been growing.

$$V(t) = 1,100[1,023e^{-0.02415t} + 1]^{-4},$$

where t is in months and $V(t)$ is measured in cubic centimeters.*

a. Find the tumor volume at 240 months.

b. Assuming that the shape of a tumor is spherical, find the radius of the tumor from part a. (*Hint:* The volume of a sphere is given by the formula $V = (4/3)\pi r^3$.)

c. If a tumor of size 0.5 cm³ is detected, according to the formula, how long has it been growing? What does this imply?

d. Find $\lim_{t \to \infty} V(t)$ and interpret this value. Explain whether this makes sense.

e. Calculate the rate of change of tumor volume at 240 months and interpret.

37. *Mortality* The percentage of people of any particular age group that will die in a given year may be approximated by the formula

$$P(t) = 0.00239e^{0.0957t},$$

where t is the age of the person in years.†

a. Find $P(25)$, $P(50)$, and $P(75)$.

b. Find $P'(25)$, $P'(50)$, and $P'(75)$.

c. Interpret your answers for parts a and b. Are there any limitations of this formula?

38. *Dialysis* One measure of whether a dialysis patient has been adequately dialyzed is by the urea reduction ratio (URR). It is generally agreed that a patient has been adequately dialyzed when URR exceeds a value of 0.65. The value of URR can be calculated for a particular patient using the following formula by Gotch:

$$URR = 1 - \left\{(0.96)^{0.14t-1} + \frac{8t}{126t + 900}[1 - (0.96)^{0.14t-1}]\right\},$$

---

*Spratt, John, et al., "Decelerating Growth and Human Breast Cancer," *Cancer,* Vol. 71, No. 6, 1993, pp. 2013–2019.
†U.S. Vital Statistics, 1995.

where t is measured in minutes.*

a. Find the value of URR after a patient receives dialysis for 180 minutes. Has the patient received adequate dialysis?

b. Find the value of URR after a patient receives dialysis for 240 minutes. Has the patient received adequate dialysis?

c. Calculate the instantaneous rate of change of URR when time on dialysis is 240 minutes, and interpret.

39. *Medical Literature* It has been observed that there has been an increase in the proportion of medical research papers that use the word "novel" in the title or abstract, and that this proportion can be accurately modeled by the function

$$p(x) = 0.001131e^{0.1268x},$$

where x is the number of years since 1970.[†]

a. Find $p(25)$.

b. If this phenomenon continues, estimate the year in which every medical article will contain the word "novel" in its title or abstract.

c. Estimate the rate of increase in the proportion of medical papers using this word in the year 2002.

d. Explain some factors that may be contributing to researchers using this word.

40. *Arctic Foxes* The age-weight relationship of female Arctic foxes caught in Svalbard, Norway, can be estimated by the function

$$M(t) = 3,102e^{-e^{0.022(t-56)}},$$

where t is the age of the fox in days and $M(t)$ is the weight of the fox in grams.[‡]

a. Estimate the weight of a female fox that is 200 days old.

b. Use $M(t)$ to estimate the largest size that a female fox can attain. (*Hint:* Find $\lim_{t \to \infty} M(t)$.)

c. Estimate the age of a female fox when it has reached 80% of its maximum weight.

d. Estimate the rate of change in weight of an Arctic fox that is 200 days old. (*Hint:* Recall that $D_t e^{f(t)} = f'(t)e^{f(t)}$.)

e. Use a graphing calculator to graph $M(t)$ and then describe the growth pattern.

f. Use the table function on a graphing calculator or a spreadsheet to develop a chart that shows the estimated weight and growth rate of female foxes for days 50, 100, 150, 200, 250, and 300.

41. *Cutlassfish* The cutlassfish is one of the most important resources of the commercial marine fishing industry in China. Researchers have developed a von Bertalanffy growth model that uses the age of a certain species of cutlassfish to estimate length such that

$$L(t) = 589\{1 - e^{(-0.168(t+2.682))}\},$$

where $L(t)$ is the length of the fish (in mm) at time t (yr).[§]

a. What happens to the length of the average cutlassfish of this species over time?

b. Determine the age of a fish that has grown to 95% of its maximum length.

c. Find $L'(4)$ and interpret the result.

d. Graph the function on $[0, 20]$ by $[0, 600]$.

42. *Beef Cattle* Researchers have compared two models that are used to predict the weight of beef cattle of various ages,

$$W_1(t) = 509.7(1 - 0.941e^{-0.00181t}) \quad \text{and}$$
$$W_2(t) = 498.4(1 - 0.889e^{-0.00219t})^{1.25},$$

where $W_1(t)$ and $W_2(t)$ represent the weight (in kg) of a t-day-old beef cow.[‖]

a. What is the maximum weight predicted by each function for the average beef cow? Is this difference significant?

b. According to each function, find the age that the average beef cow reaches 90% of its maximum weight.

c. Find $W_1'(750)$ and $W_2'(750)$. Compare your results.

d. Graph the two functions on $[0, 2,500]$ by $[0, 525]$ and comment on the differences in growth patterns for each of these functions.

e. Graph the derivative of these two functions on $[0, 2,500]$ by $[0, 1]$ and comment on any differences you notice between these functions.

43. *Holstein Dairy Cattle* Researchers have used the Richardson model to develop the following function that can be

---

*Kessler, Edward, and Nathan Ritchey, et al., "Urea Reduction Ratio and Urea Kinetic Modeling: A Mathematical Analysis of Changing Dialysis Parameters," *American Journal of Nephrology,* Vol. 18, 1998, pp. 471–477.

[†]Friedman, Simon H., and Jens O. Karlsson, "A Novel Paradigm," *Nature,* Vol. 385, No. 6616, Feb. 6, 1997, p. 480.

[‡]Prestrud, Pal, and Kjell Nilssen, "Growth, Size, and Sexual Dimorphism in Arctic Foxes," *Journal of Mammalogy,* Vol. 76, No. 2, May 1995, pp. 522–530.

[§]Kwok, K., and I-Hsun Ni, "Age and Growth of Cutlassfishes, *Trichiurus* spp., from the South China Sea," *Fish. Bulletin,* Vol. 98, No. 4, Oct. 2000, pp. 748–758.

[‖]DeNise, R., and J. Brinks, "Genetic and Environmental Aspects of the Growth Curve Parameters in Beef Cows," *Journal of Animal Science,* Vol. 61, No. 6, 1985, pp. 1431–1440.

used to accurately predict the weight of Holstein cows (females) of various ages:

$$W(t) = 619(1 - 0.905e^{-0.002t})^{1.2386},*$$

where $W(t)$ is the weight of a Holstein cow (in kg) that is t days old.

a. What is the maximum weight of the average Holstein cow?

b. At what age does a cow reach 90% of this maximum weight?

c. Find $W'(1,000)$ and interpret.

44. *Ayrshire Dairy Cattle* The researchers from the previous exercise also developed a function that can be used to accurately predict the weight of Ayrshire cows (females). In this case,

$$W(t) = 532(1 - 0.911e^{-0.0021t})^{1.2466}.*$$

a. What is the maximum weight of the average Ayrshire cow?

b. At what age does a cow reach 90% of this maximum weight?

c. Find $W'(1,000)$ and interpret.

d. Graph the function in this exercise along with the function from the previous exercise on $[0, 2,500]$ by $[0, 700]$. Describe the growth pattern of each cow.

45. *Nervous System* In a model of the nervous system, the intensity of excitation, I, of a nerve pathway is given by

$$I = E[1 - e^{-a(S-h)/E}],$$

where E is the maximum possible excitation, S is the intensity of a stimulus, h is a threshold stimulus, and a is a constant.[†] Find the rate of change of the intensity of the excitation with respect to the intensity of the stimulus.

46. *Extinction* The probability of a population going extinct by time t can be estimated by

$$p(t) = \left(\frac{a[e^{(b-a)t} - 1]}{be^{(b-a)t} - a}\right)^N,$$

where a is the death rate, b is the birth rate ($b \neq a$), and N is the number of individuals in the population at time $t = 0$.[‡] Find and interpret $p'(t)$.

47. *Track and Field* In 1958, L. Lucy developed a method for predicting the world record time in which a human could run the distance of one mile in any given year. His formula is given as follows:

$$t(n) = 218 + 31(0.933)^n,$$

where $t(n)$ is the world record, in seconds, for the mile run in year $1950 + n$. Thus, $n = 5$ corresponds to the year 1955.[§]

a. Find the estimate for the world record in the year 2001.

b. Calculate the instantaneous rate of change for the world record at the end of year 2001 and interpret.

c. Find $\lim_{n \to \infty} t(n)$ and interpret. How does this compare with the current world record?

OTHER APPLICATIONS

48. *Habit Strength* According to work by the psychologist C. L. Hull, the strength of a habit is a function of the number of times the habit is repeated. If N is the number of repetitions and $H(N)$ is the strength of the habit, then

$$H(N) = 1,000(1 - e^{-kN}),$$

where k is a constant. Find $H'(N)$ if $k = 0.1$ and the number of times the habit is repeated is as follows.

a. 10   **b.** 100   **c.** 1,000

d. Show that $H'(N)$ is always positive. What does this mean?

49. *Population* A recent report by the U.S. Census Bureau predicts that the Latino-American population will increase from 26.7 million in 1995 to 96.5 million in 2050.[ǁ] Assuming an exponential growth pattern, the population can be approximated by

$$P(t) = 26.7e^{0.023t},$$

where t is the number of years since 1995.

a. Estimate the Latino-American population for the year 2000.

b. What is the instantaneous rate of change of the Latino-American population when $t = 5$? Interpret your answer.

50. *Radioactive Decay* Assume that the amount (in grams) of a radioactive substance present after t years is given by

*Perotto, D., R. Cue, and A. Lee, "Comparison of Nonlinear Functions Describing the Growth Curve of Three Genotypes of Dairy Cattle," *Canadian Journal of Animal Science,* Vol. 73, Dec. 1992, pp. 773–782.

†Rashevsky, Nicolas, *Mathematical Biology of Social Behavior,* rev. ed., Chicago, The University of Chicago Press, 1959, p. 7.

‡Bailey, Norman T. J., *The Mathematical Approach to Biology and Medicine,* Wiley, 1967, p. 161.

§Bennett, Jay, "Statistical Modeling in Track and Field," *Statistics in Sports,* Arnold, 1998, p. 179.

ǁPopulation Projections of the United States by Age, Race, and Hispanic Origin: 1995–2050, U.S. Census Bureau.

$$A(t) = 500e^{-0.25t}.$$

Find the rate of change of the quantity present after each of the following years.

a. 4   **b.** 6   **c.** 10

d. What is happening to the rate of change of the amount present as the number of years increases?

e. Will the substance ever be gone completely?

51. *Electricity* In a series resistance-capacitance *DC* circuit, the instantaneous charge Q on the capacitor as a function of time (where $t = 0$ is the moment the circuit is energized by closing a switch) is given by the equation

$$Q(t) = CV(1 - e^{-t/RC}),$$

where C, V, and R are constants. Further, the instantaneous charging current I_C is the rate of change of charge on the capacitor, or $I_C = dQ/dt$.*

a. Find the expression for I_C as a function of time.

b. If $C = 10^{-5}$ farads, $R = 10^7$ ohms, and $V = 10$ volts, what is the charging current after 200 seconds? (*Hint:* When placed into the function in part a the units can be combined into amps.)

4.5 DERIVATIVES OF LOGARITHMIC FUNCTIONS

? **THINK ABOUT IT**   How does the average velocity of pedestrians in a city vary with the population size?

In an exercise from an earlier chapter, we found a logarithmic relationship between the average velocity of pedestrians and the population of the city. In this section, we will find the derivative of logarithmic functions and use the result to answer the question above.

Using the definition of the derivative to find the derivative of $\log_a x$ presents an immediate difficulty. When we try to compute

$$\lim_{h \to 0} \frac{\log_a(x + h) - \log_a x}{h},$$

we do not have a way of breaking up $\log_a(x + h)$. Instead, let us see what graphical differentiation tells us. Figure 9 shows a graph of $y = \ln x$, with the derivative estimated at various points. (Recall that $\ln x = \log_a x$ with $a = e$.) In Figure 10, we have plotted those points and connected them with a smooth curve. The result looks remarkably like the positive side of the graph of $y = 1/x$. Could it be that $D_x(\ln x) = 1/x$?

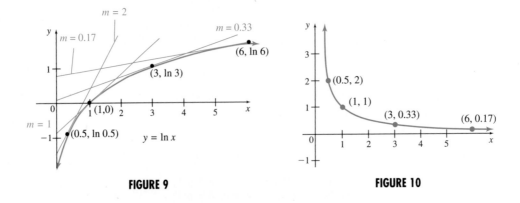

FIGURE 9   **FIGURE 10**

---

*Problem submitted by Kevin Friedrich, Sharon, PA.

This observation is indeed the case and can be further justified, geometrically. Notice what happens to the slope of the line $y = 2x + 4$ if the x-axis and y-axis are switched. That is, if we replace x with y and y with x, then the resulting line $x = 2y + 4$ or $y = x/2 - 2$ is a reflection of the line $y = 2x + 4$ across the line $y = x$, as seen in Figure 11. Furthermore, the slope of the new line is the reciprocal of the original line. In fact, the reciprocal property holds for all lines.

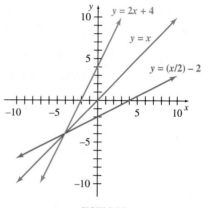

FIGURE 11

In the section on Logarithmic Functions, we showed that switching the x and y variables changes the exponential graph into a logarithmic graph, a defining property of functions that are inverses of each other. We also showed in the previous section that the slope of the tangent line of e^x at any point is e^x—that is, the y coordinate itself. So, if we switch the x and y variables, the new slope of the tangent line will be $1/y$, except that it is no longer y: it is x. Thus, the slope of the tangent line of $y = \ln x$ must be $1/x$ and hence $D_x \ln x = 1/x$.

We can also prove this fact rigorously by using a technique for finding the derivative of a function when you already know the derivative of its inverse function. Returning to the general logarithmic function, $f(x) = \log_a x$, we solve for x.

$$f(x) = \log_a x$$
$$a^{f(x)} = x \qquad \text{Definition of the logarithm}$$

Now consider the left and right sides of the last equation as functions of x that are equal, so their derivatives with respect to x should also be equal. Notice in the first step that we need to use the chain rule when differentiating $a^{f(x)}$.

$$(\ln a)a^{f(x)}f'(x) = 1 \qquad \text{Derivative of the exponential function}$$
$$(\ln a)xf'(x) = 1 \qquad \text{Substitute } a^{f(x)} = x.$$

Finally, divide both sides of this equation by $(\ln a)x$ to get

$$f'(x) = \frac{1}{(\ln a)x}.$$

DERIVATIVE OF $\log_a x$

$$D_x[\log_a x] = \frac{1}{(\ln a)x}$$

As with the exponential function, this formula becomes particularly simple when we let $a = e$, because of the fact that $\ln e = 1$.

DERIVATIVE OF ln x

$$D_x[\ln x] = \frac{1}{x}$$

EXAMPLE 1 Derivatives of Logarithmic Functions
Find the derivative of each function.

(a) $f(x) = \ln 6x$

Solution Use the properties of logarithms and the rules for derivatives.

$$f'(x) = \frac{d}{dx}(\ln 6x)$$

$$= \frac{d}{dx}(\ln 6 + \ln x)$$

$$= \frac{d}{dx}(\ln 6) + \frac{d}{dx}(\ln x) = 0 + \frac{1}{x} = \frac{1}{x}$$

(b) $y = \log x$

Solution Recall that when the base is not specified, we assume that the logarithm is a common logarithm, which has a base of 10.

$$\frac{dy}{dx} = \frac{1}{(\ln 10)x}$$

Applying the chain rule to the formulas for the derivative of logarithmic functions gives us

$$\frac{d}{dx} \log_a g(x) = \frac{1}{\ln a} \cdot \frac{g'(x)}{g(x)}$$

and

$$\frac{d}{dx} \ln g(x) = \frac{g'(x)}{g(x)}.$$

EXAMPLE 2 Derivatives of Logarithmic Functions
Find the derivative of each function.

(a) $f(x) = \ln(x^2 + 1)$

Solution Here $g(x) = x^2 + 1$ and $g'(x) = 2x$. Thus,

$$f'(x) = \frac{g'(x)}{g(x)} = \frac{2x}{x^2 + 1}.$$

(b) $y = \log_2(3x^2 - 4x)$

Solution

$$\frac{dy}{dx} = \frac{1}{\ln 2} \cdot \frac{6x - 4}{3x^2 - 4x}$$

$$= \frac{6x - 4}{(\ln 2)(3x^2 - 4x)}$$

If $y = \ln(-x)$, where $x < 0$, the chain rule with $g(x) = -x$ and $g'(x) = -1$ gives

$$\frac{dy}{dx} = \frac{g'(x)}{g(x)} = \frac{-1}{-x} = \frac{1}{x}.$$

The derivative of $y = \ln(-x)$ is the same as the derivative of $y = \ln x$. For this reason, these two results can be combined into one rule using the absolute value of x. A similar situation holds true for $y = \ln[g(x)]$ and $y = \ln[-g(x)]$, as well as for $y = \log_a[g(x)]$ and $y = \log_a[-g(x)]$. These results are summarized as follows.

DERIVATIVE OF $\log_a |x|$, $\log_a |g(x)|$, $\ln |x|$, **AND** $\ln |g(x)|$

$$D_x\big[\log_a |x|\big] = \frac{1}{(\ln a)x} \qquad D_x\big[\log_a |g(x)|\big] = \frac{1}{\ln a} \cdot \frac{g'(x)}{g(x)}$$

$$D_x\big[\ln |x|\big] = \frac{1}{x} \qquad D_x\big[\ln |g(x)|\big] = \frac{g'(x)}{g(x)}$$

You need not remember four separate formulas. If you know the derivative of $y = \log_a x$ as well as the chain rule, all the other formulas follow. An absolute value inside of a logarithm has no effect on the derivative, other than making the result valid for more values of x.

EXAMPLE 3 Derivatives of Logarithmic Functions
Find the derivative of each function.

(a) $y = \ln |5x|$

Solution Let $g(x) = 5x$, so that $g'(x) = 5$. From the formula above,

$$\frac{dy}{dx} = \frac{g'(x)}{g(x)} = \frac{5}{5x} = \frac{1}{x}.$$

Notice that the derivative of $\ln |5x|$ is the same as the derivative of $\ln |x|$. Also, in Example 1, the derivative of $\ln 6x$ was the same as that for $\ln x$. This suggests that for any constant a,

$$\frac{d}{dx} \ln |ax| = \frac{d}{dx} \ln |x|$$

$$= \frac{1}{x}.$$

Exercise 41 asks for a proof of this result.

(b) $f(x) = 3x \ln x^2$

Solution This function is the product of the two functions $3x$ and $\ln x^2$, so use the product rule.

$$f'(x) = (3x)\left[\frac{d}{dx} \ln x^2\right] + (\ln x^2)\left[\frac{d}{dx} 3x\right]$$

$$= 3x\left(\frac{2x}{x^2}\right) + (\ln x^2)(3)$$

$$= 6 + 3 \ln x^2$$

By the power rule for logarithms,

$$f'(x) = 6 + \ln(x^2)^3$$
$$= 6 + \ln x^6.$$

Alternatively, write the answer as $f'(x) = 6 + 6 \ln x$.

(c) $s(t) = 6 \log_8 t^{3/2}$

Solution

$$s'(t) = 6 \cdot \frac{1}{\ln 8} \cdot \frac{(3/2)t^{1/2}}{t^{3/2}}$$

$$= \frac{9}{(\ln 8)t} \qquad \text{Simplify with algebra.}$$

Alternatively, we could have used one of the rules of logarithms to simplify the function to $s = 6(3/2) \log_8 t = 9 \log_8 t$ and then taken the derivative.

EXAMPLE 4 Pedestrian Speed

In review exercise 107 for the chapter on Exponential, Logarithmic, and Trigonometric Functions, we found that the average speed of pedestrians depends on the population (x) of the city in which they are walking by the function

$$f(x) = 0.873 \log x - 0.0255.$$

Find and interpret $f(1{,}000{,}000)$ and $f'(1{,}000{,}000)$.

Solution Recognizing this function as a common (base 10) logarithm, we have $f(1{,}000{,}000) = 0.873 \log 1{,}000{,}000 - 0.0255 = 5.2125$. In a city of 1,000,000 people, pedestrians walk, on average, about 5.2 feet per second.

$$f'(x) = \frac{0.873}{(\ln 10)x},$$

so $f'(1{,}000{,}000) \approx 3.791 \cdot 10^{-7}$. This means that for each additional person in a city of 1,000,000, the average pedestrian speed increases by about $3.8 \cdot 10^{-7}$ feet per second. Perhaps a simpler interpretation is that for a city of 1,000,000, every additional 10,000 people causes an increase in the average pedestrian velocity of about $3.8 \cdot 10^{-7} \cdot 10{,}000 = 0.0038$ feet per second, a very small quantity. The increase in speed with larger populations is only noticeable when comparing cities that vary greatly in size.

EXAMPLE 5 Black Crappie

Scientists in Florida have collected data which reveal that there is a logarithmic relationship between the length of a black crappie at age three and the total number of crappie in a lake. This relationship can be represented by

$$l(x) = 171.5 + 943.2 \log(x) - 458.15 \log(x^2),$$

where x is the number of black crappie per hectare (ha—2.47 acres) and $l(x)$ (in mm) is the predicted length at age three.*

---

*Allen, B., M. Hoyer, and D. Canfield, Jr., "Factors Related to Black Crappie Occurrence, Density, and Growth in Florida Lakes," *North American Journal of Fisheries Management*, Vol. 18, No. 4, 1998, pp. 864–871.

(a) Use the properties of logarithms to simplify this function and predict the length of three-year-old black crappie when the lake density is 100 crappie/ha.

Solution Since $\log(x^2) = 2\log(x)$, we can rewrite $l(x)$ as

$$l(x) = 171.5 + 943.2\log(x) - 458.15\log(x^2)$$
$$= 171.5 + 943.2\log(x) - 2(458.15)\log(x)$$
$$= 171.5 + 26.9\log(x).$$

We can see that the properties of logarithms can be useful to simplify a function. Also,

$$l(100) = 171.5 + 26.9\log(100) = 225.3.$$

Thus, the predicted length of black crappie at age three when the density of the lake is 100 crappie/ha is 225.3 mm.

(b) Using the simplified function, calculate $l'(100)$ and comment on the results.

Solution The derivative of $l(x)$ is

$$l'(x) = \frac{26.9}{\ln(10)x},$$

so $l'(100) = 26.9/[\ln(10) \cdot 100] \approx 0.1168$. This means that for each additional black crappie per ha in a lake with 100 crappie per ha, the average length at age three will increase by approximately 0.1168 mm. As one can see in Figure 12, the graph of $l(x)$ is always rising. We would certainly not expect this trend to continue for high densities of black crappie since crowding tends to decrease the size of a population of any species of animal. Thus, we should be very cautious in using this function to predict black crappie lengths for high densities of crappie.

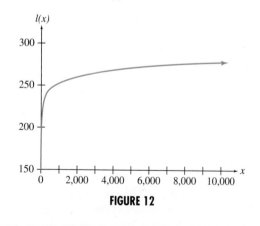

FIGURE 12

4.5 EXERCISES

Find the derivative of each of the following functions.

1. $y = \ln(8x)$

2. $y = \ln(-4x)$

3. $y = \ln(3 - x)$

4. $y = \ln(1 + x^2)$

5. $y = \ln|2x^2 - 7x|$

6. $y = \ln|-8x^2 + 6x|$

7. $y = \ln\sqrt{x + 5}$

8. $y = \ln\sqrt{2x + 1}$

9. $y = \ln(x^4 + 5x^2)^{3/2}$

10. $y = \ln(5x^3 - 2x)^{3/2}$

11. $y = -3x\ln(x + 2)$

12. $y = (3x + 1)\ln(x - 1)$

13. $s = t^2 \ln |t|$

14. $y = x \ln |2 - x^2|$

15. $y = \dfrac{2 \ln(x + 3)}{x^2}$

16. $v = \dfrac{\ln u}{u^3}$

17. $y = \dfrac{\ln x}{4x + 7}$

18. $y = \dfrac{-2 \ln x}{3x - 1}$

19. $y = \dfrac{3x^2}{\ln x}$

20. $y = \dfrac{x^3 - 1}{2 \ln x}$

21. $y = \left(\ln |x + 1| \right)^4$

22. $y = \sqrt{\ln |x - 3|}$

23. $y = \ln |\ln x|$

24. $y = (\ln 4)\left(\ln |3x| \right)$

25. $y = e^{x^2} \ln x$

26. $y = e^{2x-1} \ln(2x - 1)$

27. $y = \dfrac{e^x}{\ln x}$

28. $p(y) = \dfrac{\ln y}{e^y}$

29. $g(z) = (e^{2z} + \ln z)^3$

30. $y = \log(6x)$

31. $y = \log(2x - 3)$

32. $y = \log |1 - x|$

33. $y = \log |3x|$

34. $y = \log_5 \sqrt{5x + 2}$

35. $y = \log_7 \sqrt{2x - 3}$

36. $y = \log_3(x^2 + 2x)^{3/2}$

37. $y = \log_2(2x^2 - x)^{5/2}$

38. $w = \log_8(6^p - 1)$

39. $z = 10^y \log y$

40. Why do we use the absolute value of x or of $g(x)$ in the derivative formulas for the natural logarithm?

41. Prove $\dfrac{d}{dx} \ln |ax| = \dfrac{d}{dx} \ln |x|$ for any constant a.

42. A friend concludes that because $y = \ln 6x$ and $y = \ln x$ have the same derivative, namely $dy/dx = 1/x$, these two functions must be the same. Explain why this is incorrect.

43. Use a graphing calculator to sketch the graph of $y = (f(x + h) - f(x))/h$ using $f(x) = \ln |x|$ and $h = 0.0001$. Compare it with the graph of $y = 1/x$ and discuss what you observe.

Applications

LIFE SCIENCES

44. *Body Suface Area* As we saw in Chapter 2, Boyd found that there is a mathematical relationship between an infant's weight and total body surface area (BSA). The relationship is

$$A(w) = 4.688 w^{(0.8168 - 0.0154 \log_{10} w)},$$

where w is the weight (in g) and $A(w)$ is the BSA in cm^2.*

a. Find the BSA for an infant who weighs 4,000 g.

b. Find $A'(4,000)$ and interpret your answer.

c. Use a graphing calculator to graph $A(w)$ on $[2,000, 10,000]$ by $[0, 6,000]$.

45. *Bologna Sausage* As indicated in Chapter 2, scientists in Italy have developed a modified Gompertz model to predict the growth of *Enterococcus faecium* in bologna sausage at 32°C. Another way of representing the number of bacteria

is given by

$$\ln\left(\frac{N(t)}{N_0} \right) = 9.8901 e^{-e^{(2.54197 - 0.2167t)}},$$

where N_0 is the number of bacteria present at the beginning of the experiment and $N(t)$ is the number of bacteria present at time t (in hours).[†]

a. Use the properties of logarithms to find an expression for $N(t)$. Assume that $N_0 = 1,000$.

b. Use a graphing calculator to estimate the derivative of $N(t)$ when $t = 20$ and interpret.

c. Let $S(t) = \ln(N(t)/N_0)$. Graph $S(t)$ on $[0, 35]$ by $[0, 12]$.

d. Graph $N(t)$ on $[0, 35]$ by $[0, 20,000,000]$ and compare the graphs from parts c and d.

e. Find $\lim\limits_{t \to \infty} S(t)$ and then use this limit to find $\lim\limits_{t \to \infty} N(t)$.

---

*Sharkey, I., et al., "Body Surface Area Estimation in Children Using Weight Alone: Application in Pediatric Oncology," *British Journal of Cancer,* Vol. 85, No. 1, 2001, pp. 23–28.

[†]Zanoni, B., C. Garzaroli, S. Anselmi, and G. Rondinini, "Modeling the Growth of *Enterococcus faecium* in Bologna Sausage," *Applied and Environmental Microbiology,* Vol. 59, No. 10, Oct. 1993, pp. 3411–3417.

46. *Pronghorn Fawns* The field metabolic rate (FMR), or the total energy expenditure per day in excess of growth, can be calculated for pronghorn fawns using Nagy's formula,

$$F(x) = 0.774 + 0.727 \log(x),$$

where x is the mass (in g) of the fawn and $F(x)$ is the energy expenditure (in kJ/day).*

a. Determine the total energy expenditure per day in excess of growth for a pronghorn fawn that weighs 25,000 g.

b. Find $F'(25,000)$ and interpret the result.

c. Graph the function on $[5,000, 30,000]$ by $[3, 5]$.

47. *Fruit Flies* A study of the relation between the rate of reproduction in *Drosophila* (fruit flies) bred in bottles and the density of the mated population found that the number of imagoes (sexually mature adults) per mated female per day (y) can be approximated by

$$\log y = 1.54 - 0.008x - 0.658 \log x,$$

where x is the mean density of the mated population (measured as flies per bottle) over a sixteen-day period.[†]

a. Show that the above equation is equivalent to

$$y = 34.7(1.0186)^{-x}x^{-0.658}.$$

b. Using your answer from part a, find the number of imagoes per mated female per day when the density is

i. 20 flies per bottle;

ii. 40 flies per bottle.

c. Using your answer from part a, find the rate of change in the number of imagoes per mated female per day with respect to the density when the density is

i. 20 flies per bottle;

ii. 40 flies per bottle.

48. *Insect Mating* Consider an experiment in which equal numbers of male and female insects of a certain species are permitted to intermingle. Assume that

$$M(t) = (0.1t + 1) \ln\sqrt{t}$$

represents the number of matings observed among the insects in an hour, where t is the temperature in degrees Celsius. (*Note:* The formula is an approximation at best and holds only for specific temperature intervals.)

a. Find the number of matings when the temperature is 15°C.

b. Find the number of matings when the temperature is 25°C.

c. Find the rate of change of the number of matings when the temperature is 15°C.

49. *Population Growth* Suppose that the population of a certain collection of rare Brazilian ants is given by

$$P(t) = (t + 100) \ln(t + 2),$$

where t represents the time in days. Find the rates of change of the population on the second day and on the eighth day.

50. *Learning* In a model of how animals learn, the number of wrong attempts by an animal is given by

$$w = -\frac{1}{a} \ln\left[1 - \frac{(N - 1)b(1 - e^{-kn})}{N} \right],$$

where n is the number of trials during which an animal attempts to learn a task, N is the number of stimuli, and k, a, and b are constants.[‡] Find dw/dn.

51. *Greenhouse Gas Emissions* The net U.S. greenhouse gas emissions[§] from human activities for the years 1990–1997 can be approximated by

$$E(t) = -15,790.44 + 3,804.6 \ln(t),$$

where t is the year since 1900 and E is measured in millions of metric tons of carbon equivalent.[‖] Assuming that the emissions continue to follow this formula, answer the following questions.

*Miller, Michelle N., and John A. Byers, "Energetic Cost of Locomotor Play in Pronghorn Fawns," *Animal Behavior*, Vol. 41, 1991, pp. 1007–1013.

[†]Pearl, R., and S. Parker, *Proc. Natl. Acad. Sci.*, Vol. 8, 1922, p. 212, quoted in *Elements of Mathematical Biology* by Alfred J. Lotka, Dover Publications, 1956, pp. 308–311.

[‡]Rashevsky, Nicolas, *Mathematical Biology of Social Behavior*, rev. ed., Chicago, The University of Chicago Press, 1959, p. 31.

[§]The interpretation of *net* in this case is total emissions minus the carbon dioxide absorbed by forests and other means.

[‖]*The 2000 World Almanac and Book of Facts*, World Almanac Books, p. 168.

a. Calculate the net U.S. greenhouse gas emissions for the year 2002.

b. Calculate $E'(102)$ and interpret your answer.

OTHER APPLICATIONS

52. *Richter Scale* The Richter Scale provides a measure of the magnitude of an earthquake. In fact, the largest Richter number M ever recorded for an earthquake was 8.9, for the 1933 earthquake in Japan. The following formula shows a relationship between the amount of energy released and the Richter number.

$$M = \frac{2}{3} \log \frac{E}{0.007},$$

where E is measured in kilowatt-hours.*

a. For the 1933 earthquake in Japan, what value of E gives a Richter number $M = 8.9$?

b. If the average household uses 247 kWh/month, how many months would the energy released by an earthquake of this magnitude power 10 million households?

c. Find the rate of change of the Richter number M with respect to energy when $E = 70,000$ kWh.

d. What happens to dM/dE as E increases?

53. *Street Crossing* Consider a child waiting at a street corner for a gap in traffic that is large enough so that he or she can safely cross the street. A mathematical model for traffic shows that if the expected waiting time for the child is to be at most 1 minute, then the maximum traffic flow, in cars per hour, is given by

$$f(x) = \frac{29,000(2.322 - \log x)}{x},$$

where x is the width of the street in feet.[†] Find the maximum traffic flow and the rate of change of the maximum traffic flow with respect to street width for the following values of the street width.

a. 30 ft   **b.** 40 ft

■ **4.6 DERIVATIVES OF TRIGONOMETRIC FUNCTIONS**

? | **THINK ABOUT IT**   How do atmospheric carbon dioxide levels vary over time?

In Example 8 in this section, we will use trigonometry to answer this question. First, we derive formulas for the derivatives of some of the trigonometric functions. All these derivatives can be found from the formula for the derivative of $y = \sin x$.

We will need to use the following identities, which are listed without proof, to find the derivatives of the trigonometric functions.

BASIC IDENTITIES

$$\sin^2 x + \cos^2 x = 1$$

$$\tan x = \frac{\sin x}{\cos x}$$

$$\sin(x + y) = \sin x \cos y + \cos x \sin y$$

$$\sin(x - y) = \sin x \cos y - \cos x \sin y$$

$$\cos(x + y) = \cos x \cos y - \sin x \sin y$$

$$\cos(x - y) = \cos x \cos y + \sin x \sin y$$

---

*Bradley, Christopher, "Media Clips," *Mathematics Teacher,* Vol. 93, No. 4, April 2000, pp. 300–303.
[†]Bender, Edward, *An Introduction to Mathematical Modeling,* John Wiley & Sons, 1978, p. 213.

The derivative of $y = \sin x$ also depends on the value of

$$\lim_{x \to 0} \frac{\sin x}{x}.$$

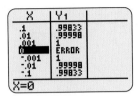

FIGURE 13

To estimate this limit, find the quotient $(\sin x)/x$ for various values of x close to 0. (Be sure that your calculator is set for radian measure.) For example, we used the table feature of the TI-83 calculator to get the values of this quotient shown in Figure 13 as x approaches 0 from either side. Note that, although the calculator shows the quotient equal to 1 for $x = \pm 0.001$, it is an approximation—the value is not exactly 1. Why does the calculator show ERROR when $x = 0$?

These results suggest, and it can be proven, that

$$\lim_{x \to 0} \frac{\sin x}{x} = 1.$$

In Example 1, this limit is used to obtain another limit. Then the derivative of $y = \sin x$ can be found.

EXAMPLE 1 Trigonometric Limit

Find $\displaystyle\lim_{h \to 0} \frac{\cos h - 1}{h}$.

Solution Use the limit above and some trigonometric identities.

$$\lim_{h \to 0} \frac{\cos h - 1}{h} = \lim_{h \to 0} \frac{(\cos h - 1)}{h} \cdot \frac{(\cos h + 1)}{(\cos h + 1)}$$

$$= \lim_{h \to 0} \frac{\cos^2 h - 1}{h(\cos h + 1)}$$

$$= \lim_{h \to 0} \frac{-\sin^2 h}{h(\cos h + 1)} \qquad \cos^2 h = 1 - \sin^2 h$$

$$= \lim_{h \to 0} (-\sin h)\left(\frac{\sin h}{h}\right)\left(\frac{1}{\cos h + 1}\right)$$

$$= (0)(1)\left(\frac{1}{1 + 1}\right)$$

$$\lim_{h \to 0} \frac{\cos h - 1}{h} = 0$$

FOR REVIEW ■

Recall from the section on Limits: When taking the limit of a product, if the limit of each factor exists, the limit of the product is simply the product of the limits.

We can now find the derivative of $y = \sin x$ by using the general definition for the derivative of a function f given in Chapter 3:

$$f'(x) = \lim_{h \to 0} \frac{f(x + h) - f(x)}{h},$$

provided this limit exists. By this definition, the derivative of $f(x) = \sin x$ is

$$f'(x) = \lim_{h \to 0} \frac{\sin(x + h) - \sin x}{h}$$

$$= \lim_{h \to 0} \frac{\sin x \cdot \cos h + \cos x \cdot \sin h - \sin x}{h} \qquad \text{Identity for } \sin(x + h)$$

$$= \lim_{h \to 0} \frac{(\sin x \cdot \cos h - \sin x) + \cos x \cdot \sin h}{h} \qquad \text{Rearrange terms.}$$

$$= \lim_{h \to 0} \frac{\sin x(\cos h - 1) + \cos x \cdot \sin h}{h} \qquad \text{Factor.}$$

$$f'(x) = \lim_{h \to 0}\left(\sin x \frac{\cos h - 1}{h} \right) + \lim_{h \to 0}\left(\cos x \frac{\sin h}{h} \right) \qquad \text{Limit rule for sums}$$

$$= (\sin x)(0) + (\cos x)(1)$$

$$= \cos x.$$

This result is summarized below.

FOR REVIEW ■

Recall that the symbol $D_x[f(x)]$ means the derivative of $f(x)$ with respect to x.

DERIVATIVE OF $\sin x$

$$D_x(\sin x) = \cos x$$

We can use the chain rule to find derivatives of other sine functions, as shown in the following examples.

EXAMPLE 2 Derivatives of $\sin x$

Find the derivative of each function.

(a) $y = \sin 6x$

Solution By the chain rule,

$$\frac{dy}{dx} = (\cos 6x) \cdot D_x(6x)$$

$$= (\cos 6x) \cdot 6$$

$$\frac{dy}{dx} = 6 \cos 6x.$$

(b) $y = 5 \sin(9x^2 + 2) + \cos\left(\frac{\pi}{7}\right)$

Solution

$$\frac{dy}{dx} = [5 \cos(9x^2 + 2)] \cdot D_x(9x^2 + 2) + 0$$

$$= [5 \cos(9x^2 + 2)]18x$$

$$\frac{dy}{dx} = 90x \cos(9x^2 + 2)$$

EXAMPLE 3 Generalized Power Rule

Find $D_x(\sin^4 x)$.

Solution The expression $\sin^4 x$ means $(\sin x)^4$. By the generalized power rule,

$$D_x(\sin^4 x) = 4 \cdot \sin^3 x \cdot D_x(\sin x)$$

$$= 4 \sin^3 x \cos x.$$

The derivative of $y = \cos x$ is found from trigonometric identities and from the fact that $D_x(\sin x) = \cos x$. First, use the identity for $\sin(x - y)$ to get

$$\sin\left(\frac{\pi}{2} - x\right) = \sin\frac{\pi}{2} \cdot \cos x - \cos\frac{\pi}{2} \cdot \sin x$$
$$= 1 \cdot \cos x - 0 \cdot \sin x$$
$$= \cos x.$$

In the same way, $\cos\left(\dfrac{\pi}{2} - x\right) = \sin x$. Therefore,

$$D_x(\cos x) = D_x\left[\sin\left(\frac{\pi}{2} - x\right)\right].$$

By the chain rule,

$$D_x\left[\sin\left(\frac{\pi}{2} - x\right)\right] = \cos\left(\frac{\pi}{2} - x\right) \cdot D_x\left(\frac{\pi}{2} - x\right)$$
$$= \cos\left(\frac{\pi}{2} - x\right) \cdot (-1) \qquad \pi/2 \text{ is constant.}$$
$$= -\cos\left(\frac{\pi}{2} - x\right)$$
$$= -\sin x.$$

DERIVATIVE OF $\cos x$

$$D_x(\cos x) = -\sin x$$

EXAMPLE 4 Derivatives of $\cos x$

Find each derivative.

(a) $D_x[\cos(3x)] = -\sin(3x) \cdot D_x(3x) = -3\sin 3x$

(b) $D_x(\cos^4 x) = 4\cos^3 x \cdot D_x(\cos x) = 4\cos^3 x(-\sin x)$
$$= -4\sin x \cos^3 x$$

(c) $D_x(3x \cdot \cos x)$

Solution Use the product rule.

$$D_x(3x \cdot \cos x) = 3x(-\sin x) + (\cos x)(3)$$
$$= -3x\sin x + 3\cos x$$

As mentioned in the list of basic identities at the beginning of this section, $\tan x = (\sin x)/\cos x$. The derivative of $y = \tan x$ can be found by using the quotient rule to find the derivative of $y = (\sin x)/\cos x$.

$$D_x(\tan x) = D_x\left(\frac{\sin x}{\cos x}\right) = \frac{\cos x \cdot D_x(\sin x) - \sin x \cdot D_x(\cos x)}{\cos^2 x}$$
$$= \frac{\cos x(\cos x) - \sin x(-\sin x)}{\cos^2 x}$$

$$= \frac{\cos^2 x + \sin^2 x}{\cos^2 x}$$

$$= \frac{1}{\cos^2 x} = \sec^2 x$$

The last step follows from the definitions of the trigonometric functions, which could be used to show that $1/\cos x = \sec x$. A similar calculation leads to the derivative of cot x.

DERIVATIVES OF tan x AND cot x

$$D_x(\tan x) = \sec^2 x$$
$$D_x(\cot x) = -\csc^2 x$$

EXAMPLE 5 Derivatives of tan x and cot x
Find each derivative.

(a) $D_x(\tan 9x) = \sec^2 9x \cdot D_x(9x) = 9 \sec^2 9x$

(b) $D_x(\cot^6 x) = 6 \cot^5 x \cdot D_x(\cot x) = -6 \cot^5 x \csc^2 x$

(c) $D_x\big(\ln |6 \tan x|\big) = \dfrac{D_x(6 \tan x)}{6 \tan x} = \dfrac{6 \sec^2 x}{6 \tan x} = \dfrac{\sec^2 x}{\tan x}$

Using the facts that $\sec x = 1/\cos x$ and $\csc x = 1/\sin x$, it is possible to use the quotient rule to find the derivative of each of these functions. In Exercises 34 and 35 at the end of this section, you will be asked to verify the following.

DERIVATIVES OF sec x AND csc x

$$D_x \sec x = \sec x \tan x$$
$$D_x \csc x = -\csc x \cot x$$

EXAMPLE 6 Derivatives of sec x and csc x
Find each derivative.

(a) $D_x(x^2 \sec x) = x^2 \sec x \tan x + 2x \sec x$

(b) $D_x(\csc e^{2x}) = -\csc e^{2x} \cot e^{2x} \cdot D_x(e^{2x}) = -\csc e^{2x} \cot e^{2x} \cdot (2e^{2x})$

$$= -2e^{2x} \csc e^{2x} \cot e^{2x}$$

EXAMPLE 7 Derivatives of Trigonometric Functions
Find the derivative of the function at the specified value of x.

(a) $f(x) = \sin(\pi e^x)$, when $x = 0$

Solution Using the chain rule, the derivative of $f(x)$ is

$$f'(x) = \cos(\pi e^x) \cdot \pi e^x.$$

Thus,

$$f'(0) = \cos(\pi e^0) \cdot \pi e^0 = (-1)\pi(1) = -\pi.$$

(b) $g(x) = e^x \sin(\pi x)$, when $x = 0$

Solution Using the product rule, the derivative of $g(x)$ is

$$g'(x) = e^x \cos(\pi x)\pi + \sin(\pi x)e^x.$$

Thus,

$$g'(0) = e^0 \cos(\pi 0)\pi + \sin(\pi 0)e^0 = \pi.$$

(c) $h(x) = \tan(\cot x)$, when $x = \dfrac{\pi}{4}$

Solution Using the chain rule, the derivative of $h(x)$ is

$$h'(x) = \sec^2(\cot(x)) \cdot (-\csc^2(x)).$$

Thus,

$$h'\left(\frac{\pi}{4}\right) = \sec^2\left(\cot\left(\frac{\pi}{4}\right)\right) \cdot \left(-\csc^2\left(\frac{\pi}{4}\right)\right) = \sec^2(1) \cdot \left(-\frac{1}{2}\right) \approx -1.71.$$

(d) $k(x) = \tan x \cot x$, when $x = \dfrac{\pi}{4}$.

Solution Since $k(x) = \tan x \cot x = \dfrac{\sin x}{\cos x} \cdot \dfrac{\cos x}{\sin x} = 1$, $k'(x) = 0$. In particular, $k'\left(\dfrac{\pi}{4}\right) = 0$.

EXAMPLE 8 Carbon Dioxide Levels

At Mauna Loa, Hawaii, atmospheric carbon dioxide levels in parts per million (ppm) have been measured regularly since 1958. The function defined by

$$L(x) = 0.022x^2 + 0.55x + 316 + 3.5 \sin(2\pi x)$$

can be used to model these levels, where x is in years and $x = 0$ corresponds to 1960.*

(a) Find $L(25)$, $L(35.5)$, and $L(50.2)$.

Solution

$$L(25) = 0.022(25)^2 + 0.55(25) + 316 + 3.5 \sin(50\pi) = 343.5 \text{ ppm},$$
$$L(35.5) = 0.022(35.5)^2 + 0.55(35.5) + 316 + 3.5 \sin(71\pi) = 363.25 \text{ ppm},$$
$$L(50.2) = 0.022(50.2)^2 + 0.55(50.2) + 316 + 3.5 \sin(100.4\pi) = 402.38 \text{ ppm}$$

(b) Find $L'(50.2)$.

Solution Since $L'(x) = 0.044x + 0.55 + 7\pi \cos(2\pi x)$, $L'(50.2)$ is given by

$$L'(50.2) = 0.044(50.2) + 0.55 + 7\pi \cos(100.4\pi) = 9.55 \text{ ppm/yr}.$$

EXAMPLE 9 Piston Velocity

The distance s of a piston from the center of the rotating crankshaft in a 1937 John Deere B engine with respect to the angle θ of the connecting rod, as indicated by Figure 14 on the next page, is given by the formula

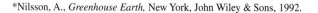

*Nilsson, A., *Greenhouse Earth,* New York, John Wiley & Sons, 1992.

$$s(\theta) = 2.625 \cos \theta + 2.625(15 + \cos^2 \theta)^{1/2},$$

where s is measured in inches and θ in radians.*

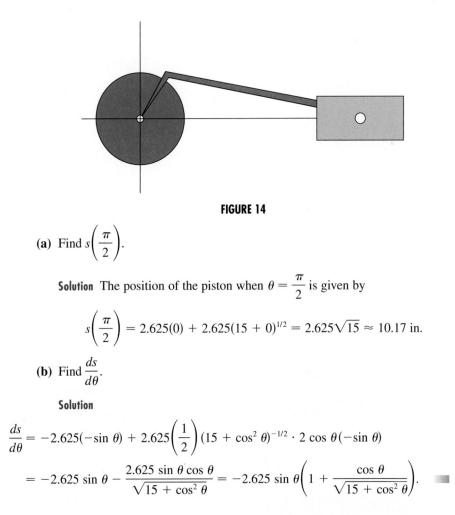

FIGURE 14

(a) Find $s\left(\dfrac{\pi}{2}\right)$.

Solution The position of the piston when $\theta = \dfrac{\pi}{2}$ is given by

$$s\left(\frac{\pi}{2}\right) = 2.625(0) + 2.625(15 + 0)^{1/2} = 2.625\sqrt{15} \approx 10.17 \text{ in.}$$

(b) Find $\dfrac{ds}{d\theta}$.

Solution

$$\frac{ds}{d\theta} = -2.625(-\sin \theta) + 2.625\left(\frac{1}{2}\right)(15 + \cos^2 \theta)^{-1/2} \cdot 2 \cos \theta(-\sin \theta)$$

$$= -2.625 \sin \theta - \frac{2.625 \sin \theta \cos \theta}{\sqrt{15 + \cos^2 \theta}} = -2.625 \sin \theta\left(1 + \frac{\cos \theta}{\sqrt{15 + \cos^2 \theta}}\right).$$

4.6 EXERCISES

Find the derivatives of the functions defined as follows.

1. $y = 2 \sin 6x$

2. $y = -\cos 4x + \cos \dfrac{\pi}{4}$

3. $y = 12 \tan(9x + 1)$

4. $y = -3 \cos(8x^2 + 2)$

5. $y = \cos^4 x$

6. $y = -9 \sin^5 x$

7. $y = \tan^5 x$

8. $y = 2 \cot^4 x$

9. $y = -5x \cdot \sin 4x$

10. $y = 6x \cdot \sec 3x$

11. $y = \dfrac{\csc x}{x}$

12. $y = \dfrac{\tan x}{2x + 4}$

*Drost, John, and Robert Kunferman, "Related Rates Challenge Problem: Calculate the Velocity of a Piston," *The AMATYC Review,* Vol. 21, No. 1, 1999, pp. 17–21.

13. $y = \sin e^{5x}$

14. $y = \cos 4e^{2x}$

15. $y = e^{\sin x}$

16. $y = -8e^{\tan x}$

17. $y = \sin(\ln 4x^2)$

18. $y = \cos(\ln |2x^3|)$

19. $y = \ln |\sin x^2|$

20. $y = \ln |\tan^2 x|$

21. $y = \dfrac{2 \sin x}{3 - 2 \sin x}$

22. $y = \dfrac{4 \cos x}{2 - \cos x}$

23. $y = \sqrt{\dfrac{\sin x}{\sin 3x}}$

24. $y = \sqrt{\dfrac{\cos 4x}{\cos x}}$

25. $y = 2 \csc x - 3 \tan 4x - 7 \cos\left(\dfrac{1}{8}x\right) + e^{3x}$

26. $y = (\sin 3x + \cot(x^3))^8$

In Exercises 27–32, recall that the slope of the tangent line to a graph is given by the derivative of the function. Find the slope of the tangent line to the graph of each equation at the given point. You may wish to use a graphing calculator to support your answers.

27. $y = \sin x; \quad x = 0$

28. $y = \sin x; \quad x = \pi/4$

29. $y = \cos x; \quad x = \pi/2$

30. $y = \cos x; \quad x = -\pi/4$

31. $y = \tan x; \quad x = 0$

32. $y = \cot x; \quad x = \pi/2$

33. Find the derivative of $\cot x$ by using the quotient rule and the fact that $\cot x = \cos x / \sin x$.

34. Verify that the derivative of $\sec x$ is $\sec x \tan x$. (*Hint:* Use the fact that $\sec x = 1/\cos x$.)

35. Verify that the derivative of $\csc x$ is $-\csc x \cot x$. (*Hint:* Use the fact that $\csc x = 1/\sin x$.)

36. In the discussion of the limit of the quotient $(\sin x)/x$, explain why the calculator gave ERROR for the value of $(\sin x)/x$ when $x = 0$.

Applications

LIFE SCIENCES

37. *Swing of a Runner's Arm* A runner's arm swings rhythmically according to the equation

$$y(t) = \frac{\pi}{8} \cos\left[3\pi\left(t - \frac{1}{3}\right)\right],$$

where y denotes the angle between the actual position of the upper arm and the downward vertical position (as shown in the figure*) and where t denotes time in seconds.

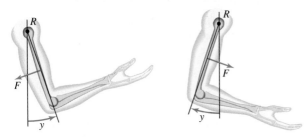

a. Graph y as a function of t.

b. Calculate the velocity, $v(t) = y'(t)$, and the acceleration,

$a(t) = v'(t)$, of the arm. (In the next chapter, we will refer to $a(t)$ as the second derivative of $y(t)$.)

c. Verify that the angle $y(t)$ and the acceleration $a(t)$ are related by the differential equation

$$a(t) + 9\pi^2 y(t) = 0.$$

d. Apply the fact that the force exerted by the muscle as the arm swings is proportional to the acceleration of y, with a positive constant of proportionality, to find the direction of the force (counterclockwise or clockwise) at $t = 1$ sec, $t = 4/3$ sec, and $t = 5/3$ sec. What is the position of the arm at each of these times?

38. *Swing of a Jogger's Arm* A jogger's arm swings according to the equation

$$y(t) = \frac{1}{5} \sin\left[\pi(t - 1)\right].$$

Proceed as directed in parts a–d of the preceding exercise, with the following exceptions: in part c, replace the differential equation with

$$a(t) + \pi^2 y(t) = 0,$$

---

*Art for Exercises 37 and 38 from De Sapio, Rodolfo, *Calculus for the Life Sciences,* Copyright © 1976, 1978 by W. H. Freeman and Company. Reprinted by permission.

and in part d, consider the times $t = 1.5$ seconds, $t = 2.5$ seconds, and $t = 3.5$ seconds.

39. *Carbon Dioxide Levels* At Barrow, Alaska, atmospheric carbon dioxide levels in parts per million can be modeled using the function defined by

$$C(x) = 0.04x^2 + 0.6x + 330 + 7.5 \sin(2\pi x),$$

where x is in years and $x = 0$ corresponds to 1960.*

a. Graph C on $[0, 25]$.

b. Find $C(25)$, $C(35.5)$, and $C(50.2)$.

c. Find $C'(50.2)$ and interpret.

d. C is the sum of a quadratic function and a sine function. What is the significance of each of these functions? Discuss what physical phenomena may be responsible for each function.

40. *Population Growth* Many biological populations, both plant and animal, experience seasonal growth. For example, an animal population might flourish during the spring and summer and die back in the fall. The population, $f(t)$, at time t, is often modeled by

$$f(t) = f(0)e^{c \sin(t)},$$

where $f(0)$ is the size of the population when $t = 0$. Suppose that $f(0) = 1,000$ and $c = 2$. Find the functional values in parts a–d.

a. $f(0.2)$ **b.** $f(1)$ **c.** $f'(0)$ **d.** $f'(0.2)$

e. Graph $f(t)$.

f. Find the maximum and minimum values of $f(t)$ and the values of t where they occur.

OTHER APPLICATIONS

41. *Engine Velocity* In Example 9 a formula that can be used to determine the distance of a piston with respect to the crankshaft for a 1937 John Deere B engine was introduced as

$$s(\theta) = 2.625 \cos(\theta) + 2.625(15 + \cos^2 \theta)^{1/2},$$

where s is measured in inches and θ in radians.[†]

a. Given that the angle θ is changing with respect to time, that is, it is a function of t, use the chain rule to find the derivative of s with respect to t, ds/dt.

b. Use part a, with $\theta = 4.944$ and $d\theta/dt = 1,340$ rev/min, to find the maximum velocity of the engine. Express your answer in miles per hour. (*Hint:* 1,340 rev/min = 505,168.1 rad/hr. Use this value and then convert your answer from inches to miles, where 1 mile = 5,280 ft.)

42. *Sound* If a string with a fundamental frequency of 110 hertz is plucked in the middle, it will vibrate at the odd harmonics of 110, 330, 550,... hertz but not at the even harmonics of 220, 440, 660,... hertz. The resulting pressure P on the eardrum caused by the string can be approximated using the equation

$$P(t) = 0.003 \sin(220\pi t) + \frac{0.003}{3} \sin(660\pi t)$$

$$+ \frac{0.003}{5} \sin(1,100\pi t) + \frac{0.003}{7} \sin(1,540\pi t),$$

where P is in pounds per square foot at a time of t seconds after the string is plucked.[‡]

a. Graph $P(t)$ on $[0, 0.01]$.

b. Use a graphing calculator to find $P'(0.002)$.

c. Now use the rules of differentiation to find $P'(0.002)$. Compare your answers to b and c.[§]

43. *Motion of a Particle* A particle moves along a straight line. The distance of the particle from the origin at time t is given by

$$s(t) = \sin t + 2\cos t.$$

Find the velocity $[v(t) = s'(t)]$ at each of the following times.

a. $t = 0$ **b.** $t = \pi/2$ **c.** $t = \pi$

---

*Zeilik, M., S. Gregory, and E. Smith, *Introductory Astronomy and Astrophysics,* Saunders College Publishing, 1992.
†Drost, John, and Robert Kunferman, "Related Rates Challenge Problem: Calculate the Velocity of a Piston," *The AMATYC Review,* Vol. 21, No. 1, all 1999, pp. 17–21.
‡Benade, Arthur, *Fundamentals of Musical Acoustics,* New York, Oxford University Press, 1976, and Juan Roederer, *The Physics and Psychophysics of Music: An Introduction,* Springer-Verlag, 1995.
§This exercise illustrates the fact that graphing calculators are not always accurate.

Find the acceleration $[a(t) = v'(t)]$ at each of the following times.

d. $t = 0$ **e.** $t = \pi/2$ **f.** $t = \pi$

44. *Revenue from Seasonal Merchandise* The revenue received from the sale of electric fans is seasonal, with maximum revenue in the summer. Let the revenue received from the sale of fans be approximated by

$$R(x) = 100 \cos 2\pi x + 120,$$

where x is time in years, measured from July 1.

a. Find $R'(x)$.

b. Find $R'(x)$ for August 1. (*Hint:* August 1 is 1/12 of a year from July 1.)

c. Find $R'(x)$ for January 1.

d. Find $R'(x)$ for June 1.

e. Discuss whether the answers in parts b–d are reasonable for this model.

■ CHAPTER SUMMARY

In this chapter we used the definition of the derivative to develop techniques for finding derivatives of several types of functions. With the help of the rules that were developed, such as the power rule, product rule, quotient rule, and chain rule, we can now directly compute the derivative of a large variety of functions. In particular, we developed rules for finding derivatives of exponential, logarithmic, and trigonometric functions. We also began to see the usefulness that these functions have in the life sciences and other applications. In the next chapter, we will apply these techniques to study the behavior of certain functions and we will learn that differentiation can be used to find maximum and minimum values of continuous functions.

Rules for Derivatives Summary

Assume all indicated derivatives exist.

Constant Function	If $f(x) = k$, where k is any real number, then $f'(x) = 0$.
Power Rule	If $f(x) = x^n$, for any real number n, then $f'(x) = n \cdot x^{n-1}$.
Constant Times a Function	Let k be a real number. Then the derivative of $y = k \cdot f(x)$ is $dy/dx = k \cdot f'(x)$.
Sum or Difference Rule	If $y = u(x) \pm v(x)$, then $\dfrac{dy}{dx} = u'(x) \pm v'(x)$.

Product Rule   If $f(x) = u(x) \cdot v(x)$, then

$$f'(x) = u(x) \cdot v'(x) + v(x) \cdot u'(x).$$

Quotient Rule   If $f(x) = \dfrac{u(x)}{v(x)}$, then

$$f'(x) = \frac{v(x) \cdot u'(x) - u(x) \cdot v'(x)}{[v(x)]^2}.$$

Chain Rule   If y is a function of u, say $y = f(u)$, and if u is a function of x, say $u = g(x)$, then $y = f(u) = f[g(x)]$, and

$$\frac{dy}{dx} = \frac{dy}{du} \cdot \frac{du}{dx}.$$

Chain Rule (Alternative Form)   Let $y = f[g(x)]$. Then $\dfrac{dy}{dx} = f'[g(x)] \cdot g'(x)$.

Generalized Power Rule	Let $g(x)$ be a function of x, and let $y = [g(x)]^n$ for any real number n. Then $$\frac{dy}{dx} = n \cdot [g(x)]^{n-1} \cdot g'(x).$$

Exponential Functions

$$D_x(e^x) = e^x \qquad\qquad D_x(a^x) = (\ln a)a^x$$
$$D_x(e^{g(x)}) = e^{g(x)}g'(x) \qquad D_x(a^{g(x)}) = (\ln a)a^{g(x)}g'(x)$$

Logarithmic Functions

$$D_x(\ln |x|) = \frac{1}{x} \qquad\qquad D_x(\log_a |x|) = \frac{1}{(\ln a)x}$$
$$D_x(\ln |g(x)|) = \frac{g'(x)}{g(x)} \qquad D_x(\log_a |g(x)|) = \frac{1}{\ln a} \cdot \frac{g'(x)}{g(x)}$$

Trigonometric Functions

$$D_x(\sin x) = \cos x$$
$$D_x(\cos x) = -\sin x$$
$$D_x(\tan x) = \sec^2 x$$
$$D_x(\cot x) = -\csc^2 x$$
$$D_x(\sec x) = \sec x \tan x$$
$$D_x(\csc x) = -\csc x \cot x$$

KEY TERMS

To understand the concepts presented in this chapter, you should know the meaning and use of the following words. For easy reference, the section in the chapter where a word (or expression) was first used is given with each item.

4.3 composite function   chain rule

CHAPTER 4 REVIEW EXERCISES

Use the rules for derivatives to find the derivative of each function defined as follows.

1. $y = 5x^2 - 7x - 9$

2. $y = x^3 - 4x^2$

3. $y = 6x^{7/3}$

4. $y = -3x^{-2}$

5. $f(x) = x^{-3} + \sqrt{x}$

6. $f(x) = 6x^{-1} - 2\sqrt{x}$

7. $k(x) = \dfrac{3x}{x + 5}$

8. $r(x) = \dfrac{-8}{2x + 1}$

9. $y = \dfrac{x^2 - x + 1}{x - 1}$

10. $y = \dfrac{2x^3 - 5x^2}{x + 2}$

11. $f(x) = (3x - 2)^4$

12. $k(x) = (5x - 1)^6$

13. $y = \sqrt{2t - 5}$

14. $y = -3\sqrt{8t - 1}$

15. $y = 3x(2x + 1)^3$

16. $y = 4x^2(3x - 2)^5$

17. $r(t) = \dfrac{5t^2 - 7t}{(3t + 1)^3}$

18. $s(t) = \dfrac{t^3 - 2t}{(4t - 3)^4}$

19. $p(t) = t^2(t^2 + 1)^{5/2}$

20. $g(t) = t^3(t^4 + 5)^{7/2}$

21. $y = -6e^{2x}$

22. $y = 8e^{0.5x}$

23. $y = e^{-2x^3}$

24. $y = -4e^{x^2}$

25. $y = 5xe^{2x}$

26. $y = -7x^2e^{-3x}$

27. $y = \ln(2 + x^2)$

28. $y = \ln(5x + 3)$

29. $y = \dfrac{\ln |3x|}{x - 3}$

30. $y = \dfrac{\ln |2x - 1|}{x + 3}$

31. $y = \dfrac{xe^x}{\ln(x^2 - 1)}$

32. $y = \dfrac{(x^2 + 1)e^{2x}}{\ln x}$

33. $s = (t^2 + e^t)^2$

34. $q = (e^{2p+1} - 2)^4$

35. $y = 3 \cdot 10^{-x^2}$

36. $y = 10 \cdot 2^{\sqrt{x}}$

37. $g(z) = \log_2(z^3 + z + 1)$

38. $h(z) = \log(1 + e^z)$

✎ **39.** Why is e a convenient base for exponential and logarithmic functions?

Find the derivative of each function.

40. $y = -4 \sin 7x$

41. $y = 6 \tan 3x$

42. $y = \tan(4x^2 + 3)$

43. $y = \cot(9 - x^2)$

44. $y = 3 \cos^6 x$

45. $y = 2 \sin^4(4x^2)$

46. $y = \cot(4x^5)$

47. $y = \cos(1 + x^2)$

48. $y = x^2 \csc x$

49. $y = e^{-x} \sin x$

50. $y = \dfrac{\sin x - 1}{\sin x + 1}$

51. $y = \dfrac{\cos^2 x}{1 - \cos x}$

52. $y = \dfrac{x - 2}{\sec x}$

53. $y = \dfrac{\tan x}{1 + x}$

54. $y = \ln |\cos x|$

55. $y = \ln |5 \sin x|$

Find the slope of the tangent line to the given curve at the given value of x. Find the equation of each tangent line.

56. $y = x^2 - 6x; \quad x = 2$

57. $y = 8 - x^2; \quad x = 1$

58. $y = \dfrac{3}{x - 1}; \quad x = -1$

59. $y = \dfrac{x}{x^2 - 1}; \quad x = 2$

60. $y = \sqrt{6x - 2}; \quad x = 3$

61. $y = -\sqrt{8x + 1}; \quad x = 3$

62. $y = e^x; \quad x = 0$

63. $y = xe^x; \quad x = 1$

64. $y = \ln x; \quad x = 1$

65. $y = x \ln x; \quad x = e$

66. $y = x \cos x; \quad x = 0$

67. $y = \tan(\pi x); \quad x = 1$

The following exercise is from the 1991 examination for applicants to the Economics Division of Shiga University in Japan.*

68. Consider the graphs of the function $y = \sqrt{2x - 1}$ and the straight line $y = x + k$. Discuss the number of points of intersection versus the change in the value of k.

▨ Applications

LIFE SCIENCES

69. *Walleye* Scientists in Wisconsin have collected data which reveals that there is a linear relationship between the total number of walleye per acre and an angler's catch rate per hour such that

$$c(x) = 0.19x,$$

where x is the number of walleye per acre and $c(x)$ is the predicted number of fish that an angler will catch per hour.†

a. Use the function above to estimate the number of fish that an angler will catch in two hours if the density of walleye is 20 walleye per acre.

b. Calculate the derivative of this function and interpret the results.

70. *Holstein Cows* A common practice among farmers is to use heart-girth (distance around the animal just behind the front legs) measurements to estimate the weight of a cow.

---

*"Japanese University Entrance Examination Problems in Mathematics," edited by Ling-Erl Eileen T. Wu, published by the Mathematical Association of America, copyright 1993, pp. 18–19.
†Beard, T., S. Hewett, Q. Yang, R. King, and S. Gilbert, "Prediction of Angler Catch Rates Based on Walleye Population Density," *North American Journal of Fisheries Management*, Vol. 17, 1997, pp. 621–627.

Scientists have found that there is a very high linear correlation between heart-girth and weight for Holstein cows such that

$$W(x) = 570 + 5.6(x - 190), \quad x \geq 170,$$

where x is the heart-girth measurement (in cm) and $W(x)$ is the body weight (in kg).[*]

a. Estimate the body weight of a cow that has a heart-girth measurement of 200 cm.

b. Develop a measuring tape that estimates the weight of a Holstein cow.

c. Discuss the relationship between the slope of this line and the derivative of the function $W(x)$.

71. *Shad* Gill-nets were used by scientists to estimate the number of various species of shad in Lake Texoma, located on the border between Texas and Oklahoma. During this investigation a quadratic relationship was found between the total number of fish and the time to lift vertical gill nets, remove the fish, and record data, such that

$$T(n) = 0.25 + (1.93 \times 10^{-2})n - (4.78 \times 10^{-5})n^2,$$

where $T(n)$ is the processing time (in hr) and n is the total number of fish caught in the net.[†]

a. Graph this function on $[0, 100]$ by $[0, 2]$.

b. Find an n-value beyond which this function would no longer make sense. Why?

c. Find the derivative of $T(n)$ when $n = 40$ and interpret.

72. *Exponential Growth* Suppose a population is growing exponentially with an annual growth constant $k = 0.05$. How fast is the population growing when it is 1,000,000? Use the derivative to calculate your answer, and then explain how the answer can be obtained without using the derivative.

73. *Fish* The length of the monkeyface prickleback, a West Coast game fish, can be approximated by

$$L = 71.5(1 - e^{-0.1t})$$

and the weight by

$$W = 0.01289 \cdot L^{2.9},$$

where L is the length in cm, t is the age in years, and W is the weight in g.[‡] (See the Extended Application for the chapter on Nonlinear Functions.)

a. Find the approximate length of a 5-year-old monkeyface.

b. Find how fast the length of a 5-year-old monkeyface is growing.

c. Find the approximate weight of a 5-year-old monkeyface. (*Hint:* Use your answer from part a.)

d. Find the rate of change of the weight with respect to length for a 5-year-old monkeyface.

e. Using the chain rule and your answers to parts b and d, find how fast the weight of a 5-year-old monkeyface is growing.

74. *Arctic Foxes* The age-weight relationship of male Arctic foxes caught in Svalbard, Norway, can be estimated by the function

$$M(t) = 3{,}583e^{-e^{-0.020(t-66)}},$$

where t is the age of the fox in days and $M(t)$ is the weight of the fox in grams.[§]

a. Estimate the weight of a male fox that is 250 days old.

b. Estimate the rate of change in weight of a male Arctic fox that is 250 days old. (*Hint:* Recall that $D_t e^{f(t)} = f'(t)e^{f(t)}$.)

c. Use a graphing calculator to graph $M(t)$ and then describe the growth pattern.

d. Use the table function on a graphing calculator or a spreadsheet to develop a chart that shows the estimated weight and growth rate of male foxes for days 50, 100, 150, 200, 250, and 300.

75. *Blood Pressure* A person's blood pressure at time t (in seconds) is given by

$$P(t) = 90 + 15 \sin 144\pi t.$$

Find the maximum and minimum values of P on the interval $[0, 1/72]$. Graph one period of $y = P(t)$.

Blood Vessel System The body's system of blood vessels is made up of arteries, arterioles, capillaries, and veins. The transport of blood from the heart through all organs of the body and back to the heart should be as efficient as possible. One way this can be done is by having large enough blood vessels to avoid turbulence, with blood cells small enough to minimize viscosity.

In Exercises 76–89, we will find the value of angle θ (see the following figure) such that total resistance to the flow of

---

[*]Warwick, E., and J. Legates, *Breeding and Improvement of Farm Animals*, 7th ed., New York, McGraw-Hill Book Company, 1979, p. 167.

[†]Van Den Avyle, M., G. Ploskey, and P. Bettoli, "Evaluation of Gill-Net Sampling for Estimating Abundance and Length Frequency of Reservoir Shad Populations," *North American Journal of Fisheries Management*, Vol. 15, 1995, pp. 898–917.

[‡]Marshall, William H., and Tina Wylie Echeverria, "Characteristics of the Monkeyface Prickleback," *California Fish and Game*, Vol. 78, No. 2, Spring 1992. Reprinted with permission.

[§]Pestrud, Pal, and Kjell Nilssen, "Growth, Size, and Sexual Dimorphism in Arctic Foxes," *Journal of Mammalogy*, Vol. 76, No. 2, May 1995, pp. 522–530.

blood is minimized. Assume that a main vessel of radius r_1 runs along the horizontal line from A to B. A side artery, of radius r_2, heads for a point C. Choose point B so that CB is perpendicular to AB. Let CB = s and let D be the point where the axis of the branching vessel cuts the axis of the main vessel.

According to Poiseuille's law, the resistance R in the system is proportional to the length L of the vessel and inversely proportional to the fourth power of the radius r. That is,

$$R = k \cdot \frac{L}{r^4},$$

where k is a constant determined by the viscosity of the blood. Let AB = L_0, AD = L_1, and DC = L_2. *

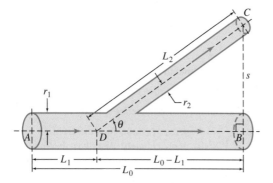

76. Use right triangle *BDC* and find sin θ.

77. Solve the result of Exercise 76 for L_2.

78. Find cot θ in terms of s and $L_0 - L_1$.

79. Solve the result of Exercise 78 for L_1.

80. Write an expression similar to the equation for resistance given above for the resistance R_1 along *AD*.

81. Write a formula for the resistance along *DC*.

82. The total resistance R is given by the sum of the resistances along *AD* and *DC*. Use your answers to Exercises 80 and 81 to write an expression for R.

83. In your formula for R, replace L_1 with the result of Exercise 79 and L_2 with the result of Exercise 77. Simplify your answer.

84. Find $dR/d\theta$. Simplify your answer. (Remember that k, L_1, L_0, s, r_1, and r_2 are constants.)

85. Set $dR/d\theta$ equal to 0.

86. Multiply through by $(\sin^2 \theta)/s$.

87. Solve for cos θ.

88. Suppose $r_1 = 1$ cm and $r_2 = 1/4$ cm. Find cos θ and then find θ.

89. Find θ if $r_1 = 1.4$ cm and $r_2 = 0.8$ cm.

90. *Atmospheric Carbon Dioxide* In Example 7 in the section on Exponential Functions, we found that the concentration (in ppm) of carbon dioxide in the atmosphere could be approximated by

$$C(x) = 353(1.0061)^{x-1990},$$

where x is the year. Find and interpret $C'(1998)$.

91. *Chlorofluorocarbons* In Exercise 37 in the section on Exponential Functions, we found that the concentration (in ppb) of the chlorofluorocarbon CFC-11 in the atmosphere could be approximated by

$$C(x) = 0.05(1.040)^{x-1950},$$

where x is the year. Find and interpret $C'(1998)$.

OTHER APPLICATIONS

92. *Tennis* It is possible to model the flight of a tennis ball that has just been served down the center of the court by the equation

$$y = x \tan \alpha - \frac{16x^2}{V^2} \sec^2 \alpha + h,$$

where y is the height in feet of a tennis ball that is being served at an angle α relative to the horizontal axis, x is the horizontal distance that the ball has traveled in feet, h is the height of the ball when it leaves the server's racket, and V is the velocity of the tennis ball when it leaves the server's racket.[†]

a. If a tennis ball is served from a height of 9 feet and the net is 3 feet high and 39 feet away from the server, does the tennis ball that is hit with a velocity of 50 miles per hour (approximately 73 ft/sec) make it over the net if it is served at an angle of $\pi/24$?

b. When $y = 0$, the corresponding value of x gives the total distance that the tennis ball has traveled while in flight (provided that it cleared the net). For a serving height of 9 feet, the equation for calculating the distance traveled is given by

$$x = \frac{V^2 \sin \alpha \cos \alpha \pm V^2 \cos^2 \alpha \sqrt{\tan^2 \alpha + \frac{576}{V^2} \sec^2 \alpha}}{32}.$$

Use the table function on a graphing calculator or a spreadsheet to determine a range of angles for which

---

*Art from Batschelet, Edward, *Introduction to Mathematics for Life Scientists*. Copyright © 1971 by Springer-Verlag New York, Inc. Reprinted by permission.

[†]Bolt, Brian, "Tennis, Golf, and Loose Gravel: Insight from Easy Math Models," *UMAP Journal*, Vol. 4, No. 1, Spring 1983, pp. 6–18.

the tennis ball will clear the net and travel between 39 and 60 feet when it is hit with an initial velocity of 44 ft/sec.

c. Because calculating $dx/d\alpha$ is so complicated analytically, use a graphing calculator to estimate this derivative when the initial velocity is 44 ft/sec and $\alpha = \pi/8$. Interpret your answer.

93. *Dating a Language* Over time, the number of original basic words in a language tends to decrease as words become obsolete or are replaced with new words. Linguists have used calculus to study this phenomenon and have developed a methodology for dating a language, called *glottochronology*. Experiments have indicated that a good estimate of the number of words that remain in use at a given time is given by

$$N(t) = N_0 e^{-0.217t},$$

where $N(t)$ is the number of words in a particular language, t is measured in the number of millennia, and N_0 is the original number of words in the language.*

a. In 1950, C. Feng and M. Swadesh established that of the original 210 basic ancient Chinese words from 950 A.D. 167 were still being used. Estimate the number of words which still remain in 1950 and compare it to the actual number of words in use.

b. Letting $t = 0$ correspond to 1950, with $N_0 = 167$, estimate the number of words that will remain in the year 2050.

c. Using the formula for $N(t)$ found in part b, find $N'(2)$ and interpret your answer.

94. *Cats* The distance from Lisa Wunderle's cat, Belmar, to a piece of string he is stalking is given in feet by

$$f(t) = \frac{8}{t+1} + \frac{20}{t^2+1},$$

where t is the time in seconds since he began.

a. Find Belmar's average velocity between 1 sec and 3 sec.

b. Find Belmar's instantaneous velocity at 3 sec.

95. *Simple Harmonic Motion* The differential equation $a(t) = -B^2 s(t)$ approximately describes the motion of a pendulum, known as *simple harmonic motion*. Verify that

$$s(t) = A \cos(Bt + C)$$

satisfies this differential equation, where $a(t) = v'(t)$ and $v(t) = s'(t)$.

---

*Lo Bello, Anthony, and Maurice Weir, "Glottochronology: An Application of Calculus to Linguistics," *The UMAP Journal*, Vol. 3, No. 1, Spring 1982, pp. 85–99.

5

Graphs and the Derivative

Derivatives provide useful information about the behavior of functions and the shapes of their graphs. The first derivative describes the rate of increase or decrease, while the second derivative indicates the degree of *nonlinearity* in the function. In an exercise at the end of this chapter we will see what changes in the sign of the second derivative tell us about the shape of a weightlifter's age versus performance graph.

■ **5.1** Increasing and Decreasing Functions

■ **5.2** Relative Extrema

■ **5.3** Higher Derivatives, Concavity, and the Second Derivative Test

■ **5.4** Curve Sketching

Review Exercises

The graph in Figure 1 shows the relationship between the number of sleep-related accidents and traffic density during a 24-hour period.* The blue line indicates the hourly distribution of sleep-related accidents. The green line indicates the hourly distribution of traffic density. The red line indicates the relative risk of sleep-related accidents. For example, the relative risk graph shows us that a person is nearly seven times as likely to have an accident at 4:00 A.M. than at 10:00 P.M.

Given a graph like the one in Figure 1, we can often locate maximum and minimum values simply by looking at the graph. It is difficult to get *exact* values or *exact* locations of maxima and minima from a graph, however, and many functions are difficult to graph. In Chapter 1 we saw how to find exact maximum and minimum values for quadratic functions by identifying the vertex. A more general approach is to use the derivative of a function to determine precise maximum and minimum values of the function. The procedure for doing this is described in this chapter, which begins with a discussion of increasing and decreasing functions.

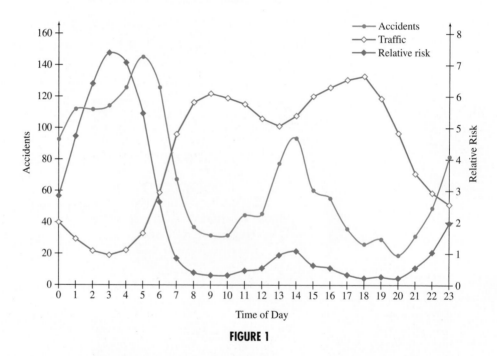

Time of Day

FIGURE 1

5.1 INCREASING AND DECREASING FUNCTIONS

THINK ABOUT IT   What is the yearly horn growth rate of bighorn rams?

We will answer this question in Example 3 after further investigating increasing and decreasing functions.

---

*Garbarino, S., L. Nobili, M. Beelke, F. Phy, and F. Ferrillo, "The Contribution Role of Sleepiness in Highway Vehicle Accidents," *Sleep,* Vol. 24, No. 2, 2001, pp. 203–206.

The graph of a typical function may increase on some intervals and decrease on others as shown in Figure 2. How can we tell from the equation that defines a function where the graph increases and where it decreases? The derivative can be used to answer this question. Remember that the derivative of a function at a point gives the slope of the line tangent to the function at that point. Recall also that a line with a positive slope rises from left to right and a line with a negative slope falls from left to right.

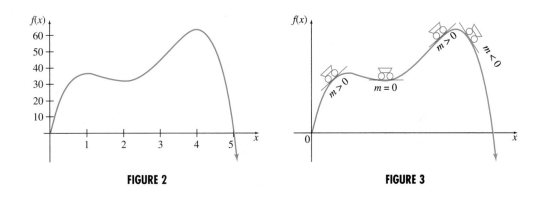

FIGURE 2   **FIGURE 3**

Think of the graph of f in Figure 2 as a roller coaster track moving from left to right along the graph. Now, picture one of the cars on the roller coaster. As shown in Figure 3, when the car is on level ground or parallel to level ground, its floor is horizontal, but as the car moves up the slope, its floor tilts upward. When the car reaches a peak, its floor is again horizontal, but it then begins to tilt downward (very steeply) as the car rolls downhill. The floor of the car as it moves from left to right along the track represents the tangent line at each point. Using this analogy, we can see that the slope of the tangent line will be *positive* when the car travels uphill and f is *increasing,* and the slope of the tangent line will be *negative* when the car travels downhill and f is *decreasing.* (In this case it is also true that the slope of the tangent line will be zero at "peaks" and "valleys.")

Thus, on intervals where $f'(x) > 0$, $f(x)$ will increase, and on intervals where $f'(x) < 0$, $f(x)$ will decrease. We can determine where $f(x)$ peaks by finding the intervals on which it increases and decreases.

Summarizing, a function is *increasing* if the graph goes *up* from left to right and *decreasing* if its graph goes *down* from left to right. Examples of increasing functions are shown in Figures 4(a)–(c) below, and examples of decreasing functions in Figures 4(d)–(f) on the next page.

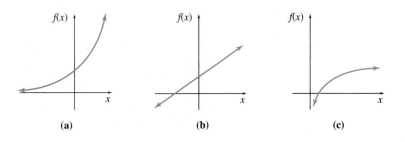

(a)   (b)   (c)

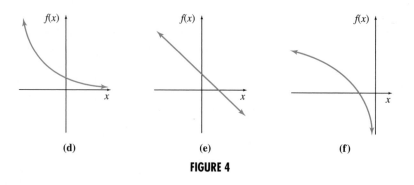

(d)   **(e)**   **(f)**

FIGURE 4

INCREASING AND DECREASING FUNCTIONS

Let f be a function defined on some interval. Then for any two numbers x_1 and x_2 in the interval, f is **increasing** on the interval if

$$f(x_1) < f(x_2) \quad \text{whenever} \quad x_1 < x_2,$$

and f is **decreasing** on the interval if

$$f(x_1) > f(x_2) \quad \text{whenever} \quad x_1 < x_2.$$

EXAMPLE 1   Increasing and Decreasing

Where is the function graphed in Figure 5 increasing? Where is it decreasing?

Solution Moving from left to right, the function is increasing up to $x = -4$, then decreasing from $x = -4$ to $x = 0$, constant (neither increasing nor decreasing) from $x = 0$ to $x = 4$, increasing from $x = 4$ to $x = 6$, and decreasing from $x = 6$ onward. In interval notation, the function is increasing on $(-\infty, -4)$ and $(4, 6)$, decreasing on $(-4, 0)$ and $(6, \infty)$, and constant on $(0, 4)$. ▬

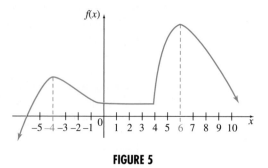

FIGURE 5

Our discussion suggests the following test.

TEST FOR INTERVALS WHERE $f(x)$ IS INCREASING AND DECREASING

Suppose a function f has a derivative at each point in an open interval; then
if $f'(x) > 0$ for each x in the interval, f is *increasing* on the interval;
if $f'(x) < 0$ for each x in the interval, f is *decreasing* on the interval;
if $f'(x) = 0$ for each x in the interval, f is *constant* on the interval.

The derivative $f'(x)$ can change signs from positive to negative (or negative to positive) at points where $f'(x) = 0$, and also at points where $f'(x)$ does not exist. The values of x where this occurs are called *critical numbers* or *critical values*.

CRITICAL NUMBERS

The **critical numbers** for a function f are those numbers c in the domain of f for which $f'(c) = 0$ or $f'(c)$ does not exist. A **critical point** is a point whose x-coordinate is the critical number c, and whose y-coordinate is $f(c)$.

It is shown in more advanced classes that if the critical numbers of a polynomial function are used to determine open intervals on a number line, then the sign of the derivative at any point in an interval will be the same as the sign of the derivative at any other point in the interval. This suggests that the test for increasing and decreasing functions be applied as follows (assuming that no open intervals exist where the function is constant).

FOR REVIEW ■

The method for finding where a function is increasing and decreasing is similar to the method introduced in Section R.5 for solving quadratic inequalities.

APPLYING THE TEST

1. Locate the critical numbers for f on a number line, as well as any points where f is undefined. These points determine several open intervals.

2. Choose a value of x in each of the intervals determined in Step 1. Use these values to decide whether $f'(x) > 0$ or $f'(x) < 0$ in that interval.

3. Use the test above to decide whether f is increasing or decreasing on the interval.

EXAMPLE 2   Increasing and Decreasing

Find the intervals in which the following functions are increasing or decreasing. Locate all points where the tangent line is horizontal. Graph the function.

FOR REVIEW ■

In this chapter you will need all of the rules for derivatives you learned in the previous chapter. If any of these are still unclear, go over the Derivative Summary at the end of that chapter and practice some of the Review Exercises before proceeding.

(a) $f(x) = x^3 + 3x^2 - 9x + 4$

Solution Here $f'(x) = 3x^2 + 6x - 9$. To find the critical numbers, set this derivative equal to 0 and solve the resulting equation by factoring.

$$3x^2 + 6x - 9 = 0$$

$$3(x^2 + 2x - 3) = 0$$

$$3(x + 3)(x - 1) = 0$$

$$x = -3 \quad \text{or} \quad x = 1$$

The tangent line is horizontal at $x = -3$ or $x = 1$. Since there are no values of x where $f'(x)$ fails to exist, the only critical numbers are -3 and 1. To determine where the function is increasing or decreasing, locate -3 and 1 on a number line, as in Figure 6 on the next page. (Be sure to place the values on the number line in numerical order.) These points determine three intervals: $(-\infty, -3)$, $(-3, 1)$, and $(1, \infty)$.

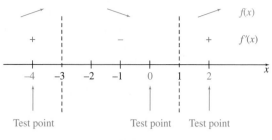

FIGURE 6

Now choose any value of x in the interval $(-\infty, -3)$. Choosing $x = -4$ gives

$$f'(-4) = 3(-4 + 3)(-4 - 1)$$
$$= 3(-1)(-5) = 15,$$

which is positive. Since one value of x in this interval makes $f'(x) > 0$, all values will do so, and therefore f is increasing on $(-\infty, -3)$. Selecting 0 from the middle interval gives $f'(0) = -9$, so f is decreasing on $(-3, 1)$. Finally, choosing 2 in the right-hand region gives $f'(2) = 15$, with f increasing on $(1, \infty)$. The arrows in each interval in Figure 6 indicate where f is increasing or decreasing.

Up to now our only method of graphing most functions has been by plotting points that lie on the graph, either by hand or using a graphing calculator or computer. Now an additional tool is available: the test for determining where a function is increasing or decreasing. (Other tools are discussed in the next few sections.) To graph the function, plot a point at each of the critical numbers by finding $f(-3) = 31$ and $f(1) = -1$. Also plot points for $x = -4, 0,$ and 2, the test values of each interval. Use these points along with the information about where the function is increasing and decreasing to get the graph in Figure 7.

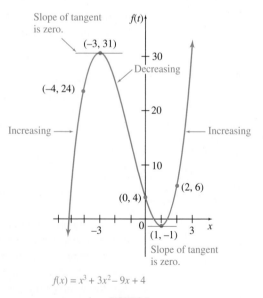

$$f(x) = x^3 + 3x^2 - 9x + 4$$

FIGURE 7

CAUTION Be careful to use $f(x)$, not $f'(x)$, to find the y-values of the points to plot.

(b) $f(x) = \dfrac{x - 1}{x + 1}$

Solution Use the quotient rule to find $f'(x)$.

$$f'(x) = \frac{(x + 1)(1) - (x - 1)(1)}{(x + 1)^2}$$

$$= \frac{x + 1 - x + 1}{(x + 1)^2} = \frac{2}{(x + 1)^2}$$

This derivative is never 0, but it fails to exist at $x = -1$, where the function is undefined. This divides the number line into two intervals: $(-\infty, -1)$ and $(-1, \infty)$. Draw a number line for f', and use a test point in each of these intervals to find that $f'(x) > 0$ for all x except -1. (This can also be seen by observing that $f'(x)$ is the quotient of 2, which is positive, and $(x + 1)^2$, which is always positive or 0.) This means that the function f is increasing on both $(-\infty, -1)$ and $(-1, \infty)$.

We also note that this function has a horizontal asymptote:

$$\lim_{x \to \infty} \frac{x - 1}{x + 1} = \lim_{x \to \infty} \frac{1 - 1/x}{1 + 1/x} \quad \text{Divide numerator and denominator by } x.$$

$$= 1.$$

We get the same limit as x approaches $-\infty$, so the graph has the line $y = 1$ as a horizontal asymptote. Verify that at $x = -1$ the graph has a vertical asymptote. Using this information, as well as the intercept $y = 0$ when $x = 1$, gives the graph in Figure 8.

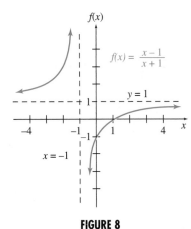

$f(x) = \dfrac{x-1}{x+1}$

$y = 1$

$x = -1$

FIGURE 8

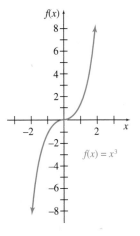

$f(x) = x^3$

FIGURE 9

CAUTION It is important to note that the reverse of the test for increasing and decreasing functions is not true—it is possible for a function to be increasing (or decreasing) on an interval even though the derivative is not positive (or negative) at every point in the interval. A good example is given by $f(x) = x^3$, which is increasing on every interval, even though $f'(x) = 0$ when $x = 0$. See Figure 9.

Similarly, it is incorrect to assume that the sign of the derivative in regions separated by critical numbers must alternate between $+$ and $-$. If this were always so, it would lead to a simple rule for finding the sign of the derivative: just check one test point, and then make the other regions alternate in sign. But this is not true if one of the factors in the derivative is raised to an even power. In the function $f(x) = x^3$ just considered, $f'(x) = 3x^2$ is positive on both sides of the critical number $x = 0$.

A graphing calculator can be used to find the derivative of a function at a particular x-value. The screen in Figure 10 supports our results in Example 2(a) for the test values, -4 and 2. The results are not exact because the calculator uses a numerical method to approximate the derivative at the given x-value.

Some graphing calculators can find where a function changes from increasing to decreasing by finding a maximum or minimum. The calculator windows in

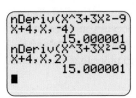

```
nDeriv(X^3+3X²-9
X+4,X,-4)
          15.000001
nDeriv(X^3+3X²-9
X+4,X,2)
          15.000001
■
```

FIGURE 10

Figure 11 show this feature for the function in Example 2(a). Notice that these, too, are approximations. This concept will be explored further in the next section.

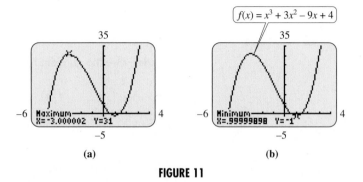

(a)   **(b)**

FIGURE 11

Knowing the intervals where a function is increasing or decreasing can be important in applications, as shown by the next examples.

EXAMPLE 3   Bighorn Sheep

?   The growth rate of horn volume for rams of a certain group can be described by the function

$$v(x) = -1{,}717 + 1{,}499x - 263.7x^2 + 13.53x^3, \qquad 2 \le x \le 9,$$

where x is the age (in yr) of the ram and $v(x)$ is the horn volume growth rate in cm^3/yr.[*]

(a) Determine the interval(s) on which the growth rate of horn volume is increasing. Where is it decreasing?

Solution To find any intervals where this function is increasing, set $v'(x) = 0$.

$$v'(x) = 1{,}499 - 527.4x + 40.59x^2 = 0$$

Solving this with the quadratic formula gives the approximate solutions of $x = 4.20$ and $x = 8.79$. Use these two numbers to determine the three intervals on a number line, as shown in Figure 12 on the next page. Choose $x = 3$, $x = 5$ and $x = 8.9$ as test points.

$$v'(3) = 1{,}499 - 527.4(3) + 40.59(3^2) = 282.11$$
$$v'(5) = 1{,}499 - 527.4(5) + 40.59(5^2) = -123.25$$
$$v'(8.9) = 1{,}499 - 527.4(8.9) + 40.59(8.9)^2 = 20.2739$$

This means that the growth rate of horn volume is increasing at a rate of 282.11 cm^3/yr per year for rams of age three. When the ram is five years old, the growth rate of horn volume is decreasing at a rate of 123.25 cm^3/yr per year. When the ram is almost nine years old the derivative implies that the growth rate begins to increase again. Thus, the analysis of this function indicates that the rate of horn volume is increasing in rams that are less than 4.2 years old, the rate is decreasing for rams that are between 4.2 and 8.79 years old, and it increases for rams that are older than this, as indicated

---

[*]Fitzsimmons, N., S. Buskirk, and M. Smith, "Population History, Genetic Variability, and Horn Growth in Bighorn Sheep," *Conservation Biology*, Vol. 9, No. 2, April 1995, pp. 314–323.

in Figure 13. Note: Although this function indicates that the rate begins to increase for rams that are older than 8.79 years, one must be very skeptical as to the accuracy of the function at this point. It is more likely the case that the function, although accurate for most of the interval $[2, 9]$, begins to lose its accuracy near the right endpoint.

(b) How do these intervals relate to the cumulative horn volume over time?

Solution The function $v(x)$ gives the growth rate of horn volume, which is itself the derivative of horn volume. It may appear surprising that the rate of growth is negative for years 4.2 through 8.79. This, however, does not imply that the total volume of the horns of a ram is decreasing during these years. It simply implies that the rate at which the horns are growing is decreasing. Similarly, in years 2 through 4.2, the yearly horn growth is increasing. This means that the horns are growing faster each successive year in this interval. Thus, a bighorn ram experiences increased horn growth over the first four years of life. After this, the function suggests that the horns continue to grow but the amount of growth decreases with age.

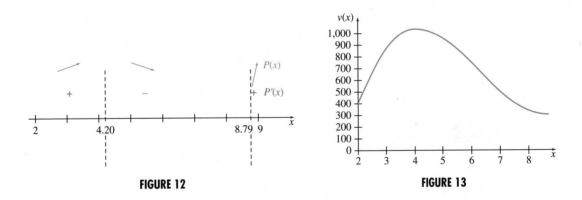

FIGURE 12   **FIGURE 13**

EXAMPLE 4   Recollection of Facts

In the exercises in the previous chapter, the function

$$f(t) = \frac{90t}{99t - 90}$$

gave the number of facts recalled after t hours for $t > 10/11$. Find the intervals in which $f(t)$ is increasing or decreasing.

Solution First find the derivative, $f'(t)$. Using the quotient rule,

$$f'(t) = \frac{(99t - 90)(90) - 90t(99)}{(99t - 90)^2}$$

$$= \frac{8,910t - 8,100 - 8,910t}{(99t - 90)^2} = \frac{-8,100}{(99t - 90)^2}.$$

Since $(99t - 90)^2$ is positive everywhere in the domain of the function and since the numerator is a negative constant, $f'(t) < 0$ for all t in the domain of $f(t)$. Thus $f(t)$ always decreases and, as expected, the number of words recalled decreases steadily over time.

The next example illustrates the case where a function has a critical number at c because the derivative does not exist at c.

EXAMPLE 5 Increasing and Decreasing

Find the critical numbers and decide where f is increasing and decreasing if $f(x) = (x - 1)^{2/3}$.

Solution We find $f'(x)$ first, using the generalized power rule.

$$f'(x) = \frac{2}{3}(x - 1)^{-1/3}(1) = \frac{2}{3(x - 1)^{1/3}}$$

We need to find any values of x that make $f'(x) = 0$, but here $f'(x)$ is never 0. To find the numbers where $f'(x)$ does not exist, set the denominator equal to 0 and solve.

$$3(x - 1)^{1/3} = 0$$
$$x - 1 = 0$$
$$x = 1$$

Since $f'(1)$ does not exist but $f(1)$ is defined, 1 is a critical number, the only critical number. Now use the first derivative test to find where f is increasing and decreasing.

$$f'(0) = \frac{2}{3(0 - 1)^{1/3}} = \frac{2}{-3} = -\frac{2}{3}$$

$$f'(2) = \frac{2}{3(2 - 1)^{1/3}} = \frac{2}{3}$$

Since f is defined for all x, these results show that f is decreasing on $(-\infty, 1)$ and increasing on $(1, \infty)$. A calculator graph of f is shown in Figure 14.

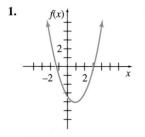

FIGURE 14

5.1 EXERCISES

Find the largest open intervals where the functions graphed as follows are
(a) *increasing, or* **(b)** *decreasing.*

1.

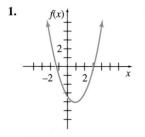

2.

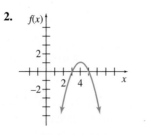

3.

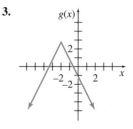

4.

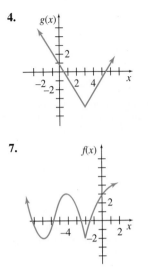

5.

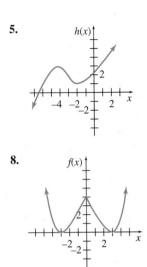

6.

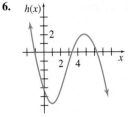

7.

8.

For each function, find **(a)** *the critical numbers;* **(b)** *the largest open intervals where the function is increasing; and* **(c)** *the largest open intervals where it is decreasing.*

9. $y = 2 + 3.6x - 1.2x^2$

10. $y = 0.3 + 0.4x - 0.2x^2$

11. $f(x) = \dfrac{2}{3}x^3 - x^2 - 24x - 4$

12. $f(x) = \dfrac{2}{3}x^3 - x^2 - 4x + 2$

13. $f(x) = 4x^3 - 15x^2 - 72x + 5$

14. $f(x) = 4x^3 - 9x^2 - 30x + 6$

15. $f(x) = x^4 + 4x^3 + 4x^2 + 1$

16. $f(x) = 3x^4 + 8x^3 - 18x^2 + 5$

17. $y = -3x + 6$

18. $y = 6x - 9$

19. $f(x) = \dfrac{x + 2}{x + 1}$

20. $f(x) = \dfrac{x + 3}{x - 4}$

21. $y = \sqrt{x^2 + 1}$

22. $y = x\sqrt{9 - x^2}$

23. $f(x) = x^{2/3}$

24. $f(x) = (x + 1)^{4/5}$

25. $y = x - 4\ln(3x - 9)$

26. $y = xe^{x^2 - 3x}$

27. $y = x^{2/3} - x^{5/3}$

28. $y = x^{1/3} + x^{4/3}$

29. $y = \sin x$

30. $y = 5\tan x$

31. $y = 3\sec x$

32. A friend looks at the graph of $y = x^2$ and observes that if you start at the origin, the graph increases whether you go to the right or the left, so the graph is increasing everywhere. Explain why this reasoning is incorrect.

33. Use the techniques of this chapter to find the vertex and intervals where f is increasing and decreasing, given

$$f(x) = ax^2 + bx + c,$$

where we assume $a > 0$. Verify that this agrees with what we found in Chapter 1.

34. Repeat Exercise 33 under the assumption $a < 0$.

35. Where is the function defined by $f(x) = e^x$ increasing? decreasing? Where is the tangent line horizontal?

36. Repeat Exercise 35 with the function defined by $f(x) = \ln x$.

Applications

LIFE SCIENCES

37. *Air Pollution* The graph shows the amount of air pollution removed by trees in the Chicago urban region for each month of the year.* From the graph we see, for example, that the ozone level starting in May increases up to June, and then abruptly decreases.

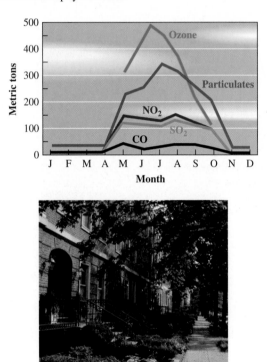

a. Are these curves the graphs of functions?

b. Look at the graph for particulates. Where is the function increasing? decreasing? constant?

c. On what intervals do all four lower graphs indicate that the corresponding functions are constant? Why do you think the functions are constant on those intervals?

38. *Spread of Infection* The number of people $P(t)$ (in hundreds) infected t days after an epidemic begins is approximated by

$$P(t) = \frac{10 \ln(0.19t + 1)}{0.19t + 1}.$$

When will the number of people infected start to decline?

39. *Alcohol Concentration* In an earlier exercise set we gave the function defined by

$$A(x) = -0.015x^3 + 1.058x$$

as the approximate alcohol concentration (in tenths of a percent) in an average person's bloodstream x hr after drinking 8 oz of 100-proof whiskey. The function applies only for the interval $[0, 8]$.

a. On what time intervals is the alcohol concentration increasing?

b. On what intervals is it decreasing?

40. *Drug Concentration* The percent of concentration of a drug in the bloodstream x hr after the drug is administered by mouth is given by

$$K(x) = \frac{4x}{3x^2 + 27}.$$

a. On what time intervals is the concentration of the drug increasing?

b. On what intervals is it decreasing?

41. *Drug Concentration* Suppose a certain drug is administered to a patient by injection, with the percent of concentration of the drug in the bloodstream t hr later given by

$$K(t) = \frac{5t}{t^2 + 1}.$$

a. On what time intervals is the concentration of the drug increasing?

b. On what intervals is it decreasing?

42. *Cardiology* As we saw in Chapter 4, the aortic pressure-diameter relation in a single patient who underwent cardiac catheterization can be modeled by the polynomial

$$D(p) = 0.000002p^3 - 0.0008p^2 + 0.1141p + 16.683,$$
$$55 \le p \le 130,$$

where $D(p)$ is the aortic diameter (in mm) and p is the aortic pressure (in mmHg).[†] Determine where this function is increasing and where it is decreasing within the interval given above.

43. *Thermic Effect of Food* As we saw in Chapter 3, the metabolic rate of a person who has just eaten a meal tends to go up and then, after some time has passed, returns to a resting metabolic rate. This phenomenon is known as the

---

*National Arbor Day Foundation, 100 Arbor Ave., Nebraska City, NE 68410. Ad in *Chicago Tribune,* Feb. 4, 1996, Sec. 2, p. 11. Used with permission of the National Arbor Day Foundation.
†Stefanadis, C., J. Dernellis, et al., "Assessment of Aortic Line of Elasticity Using Polynomial Regression Analysis," *Circulation,* Vol. 101, No. 15, April 18, 2000, pp. 1819–1825.

thermic effect of food. Researchers have indicated that the thermic effect of food for one particular person is

$$F(t) = -10.28 + 175.9te^{-t/1.3},$$

where $F(t)$ is the thermic effect of food, measured in kJ/h, and t is the number of hours that have elapsed since eating a meal.[*]

a. Find $F'(t)$.

b. Determine where this function is increasing and where it is decreasing. Interpret your answers.

44. *Holstein Dairy Cattle* As we saw in Chapter 4, researchers have used the Richardson model to develop the following function that can be used to accurately predict the weight of Holstein cows (females) of various ages:

$$W_1(t) = 619(1 - 0.905e^{-0.002t})^{1.2386},$$

where $W_1(t)$ is the weight of the Holstein cow (in kg) that is t days old.[†] Where is this function increasing?

45. *Ayrshire Dairy Cattle* As we saw in Chapter 4, researchers have developed a function that can be used to accurately predict the weight of Ayrshire cows (females). In this case,

$$W_2(t) = 532(1 - 0.911e^{-0.0021t})^{1.2466}.[‡]$$

a. Where is this function increasing?

b. Use a graphing calculator to graph this function with the function from the previous exercise on $[0, 2,000]$ by $[0, 620]$.

c. Compare the two functions, indicating which type of cow grows the fastest and which type reaches the greatest weight.

46. *Vectorial Capacity* An organism (such as a mosquito) that carries pathogens from one host to another is called a vector. The vectorial capacity, defined as the capacity of the vector population to transmit the disease in terms of the potential number of secondary inoculations originating per unit time from an infective person, is given by

$$C = \frac{ma^2p^n}{-\ln p},$$

where m is the number of vectors per host, a is the number of blood meals taken on a host per vector per day (biting rate), p is the proportion of vectors surviving per day, and n is the length (in days) of the parasite incubation period in the vectors.[§]

a. Calculate dC/dp. (*Hint:* Assume that m, a, and n are constants.) Is vectorial capacity increasing or decreasing with respect to p?

b. Let u be the death rate for the vectors, defined as $u = -\ln p$. Show that

$$C = \frac{ma^2e^{-un}}{u}.$$

c. Using your answer from part b, calculate dC/du. Is vectorial capacity increasing or decreasing with respect to u?

d. Explain how you could conclude that vectorial capacity is decreasing with respect to u by the definition of u and the fact that vectorial capacity is increasing with respect to p. Explain why both of these answers make sense biologically.

47. *Alzheimer's Disease* A study on the circadian rhythms of patients with Alzheimer's disease found that the body temperature of patients can be modeled by the function

$$C(t) = 37.29 + 0.46 \cos\left(\frac{2\pi(t - 16.37)}{24}\right),$$

where t is the time in hours since midnight.[‖] During a 24-hour period, where is this function increasing and where is it decreasing?

OTHER APPLICATIONS

48. *Population* The standard normal probability function is used to describe many different populations. Its graph is the well-known normal curve. This function is defined by

$$f(x) = \frac{1}{\sqrt{2\pi}}e^{-x^2/2}.$$

Give the intervals where the function is increasing and decreasing. Is your answer geometrically obvious?

---

[*]Reed, George, and James Hill, "Measuring the Thermic Effect of Food," *American Journal of Clinical Nutrition,* Vol. 63, 1996, pp. 164–169.

[†]Perotto, D., R. Cue, and A. Lee, "Comparison of Nonlinear Functions of Describing the Growth Curve of Three Genotypes of Dairy Cattle," *Canadian Journal of Animal Science,* Vol. 73, Dec. 1992, pp. 773–782.

[‡]Ibid.

[§]Castillo-Chavez, Carlos, Jorge X. Velasco-Hernandez, and Samuel Fridman, "Modeling Contact Structures in Biology," in *Frontiers in Mathematical Biology,* Simon A. Levin, ed., Springer-Verlag, 1994, pp. 454–491.

[‖]Volicer, L., et al., "Sundowning and Circadian Rhythms in Alzheimer's Disease," *American Journal of Psychiatry,* Vol. 158, No. 5, May 2001, pp. 704–711.

49. *Sports Cars* The following graph shows the horsepower and torque as a function of the engine speed for a 1991 Porsche 928 GT.

928 GT
Performance and Torque

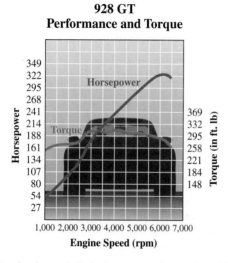

Horsepower (left axis): 27, 54, 80, 107, 134, 161, 188, 214, 241, 268, 295, 322, 349

Torque (in ft. lb) (right axis): 148, 184, 221, 258, 295, 332, 369

Engine Speed (rpm): 1,000 2,000 3,000 4,000 5,000 6,000 7,000

a. On what intervals is the horsepower increasing with engine speed?

b. On what intervals is the horsepower decreasing with engine speed?

c. On what intervals is the torque increasing with engine speed?

d. On what intervals is the torque decreasing with engine speed?

50. *Automobile Mileage* As a mathematics professor loads more weight in the back of his Mazda, the mileage goes down. Let x be the amount of weight (in lb) that he adds, and let $y = f(x)$ be the mileage (in mpg).

a. Is $f'(x)$ positive or negative? Explain.

b. What are the units of $f'(x)$?

51. *Housing Starts* A county realty group estimates that the number of housing starts per year over the next three years will be

$$H(r) = \frac{300}{1 + 0.03r^2},$$

where r is the mortgage rate (in percent).

a. Where is $H(r)$ increasing?

b. Where is $H(r)$ decreasing?

■ 5.2 RELATIVE EXTREMA

THINK ABOUT IT   How much time does it take for a flu epidemic to peak?

Epidemiologists have discovered that a flu epidemic is overtaking Youngstown, Ohio. Based on preliminary data, the number of people (in thousands) who have the flu can be described by the equation

$$f(t) = -\frac{3}{20}t^2 + 6t + 20, \qquad 0 \le t \le 30,$$

where t is the number of days since the epidemic was first discovered. Assuming that this trend continues, on what day is the epidemic expected to peak? How many people are expected to have the flu on that day?

The epidemic will peak when the number of people who have the flu is at a maximum. To find this time, find $f'(t)$.

$$f'(t) = -\frac{3}{10}t + 6 = -0.3t + 6$$

The derivative $f'(t)$ is greater than 0 when $-0.3t + 6 > 0$, $-3t > -60$, or $t < 20$. Similarly, $f'(t) < 0$ when $-0.3t + 6 < 0$, $-3t < -60$, or $t > 20$. Thus, the epidemic increases for the first 20 days and decreases for the last 10 days. The epidemic peaks on day 20. At that time there will be $f(20) = 80$ or 80,000 people who have the flu.

The maximum level of the epidemic (80,000) in the example above is a *relative maximum,* defined as follows.

RELATIVE MAXIMUM OR MINIMUM

Let c be a number in the domain of a function f. Then $f(c)$ is a **relative** (or **local**) **maximum** for f if there exists an open interval (a, b) containing c such that

$$f(x) \le f(c)$$

for all x in (a, b), and $f(c)$ is a **relative** (or **local**) **minimum** for f if there exists an open interval (a, b) containing c such that

$$f(x) \ge f(c)$$

for all x in (a, b).

A function has a **relative** (or **local**) **extremum** (plural: **extrema**) at c if it has either a relative maximum or a relative minimum there.

If c is an endpoint of the domain of f, we only consider x in the half-open interval that is in the domain.*

The intuitive idea is that a relative maximum is the greatest value of the function in some region right around the point, although there may be greater values elsewhere. For example, the highest value of the Dow Jones Industrial Average this week is a relative maximum, although the Dow may have reached a higher value earlier this year. Similarly, a relative minimum is the least value of a function in some region around the point.

EXAMPLE 1   Relative Extrema
Identify the x-values of all points where the graph in Figure 15 has relative extrema.

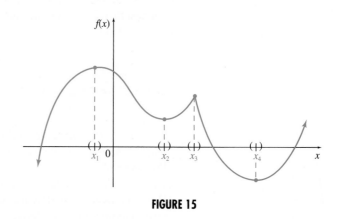

FIGURE 15

---

*There is disagreement on calling an endpoint a maximum or minimum. We define it this way because this is an applied calculus book, and in an application it would be considered a maximum or minimum value of the function.

Solution The parentheses around x_1 show an open interval containing x_1 such that $f(x) \leq f(x_1)$, so there is a relative maximum of $f(x_1)$ at $x = x_1$. Notice that many other open intervals would work just as well. Similar intervals around x_2, x_3, and x_4 can be used to find a relative maximum of $f(x_3)$ at $x = x_3$ and relative minima of $f(x_2)$ at $x = x_2$ and $f(x_4)$ at $x = x_4$. ▪

The function graphed in Figure 16 has relative maxima when $x = x_1$ or $x = x_3$ and relative minima when $x = x_2$ or $x = x_4$. The tangent lines at the points having x-values x_1 and x_2 are shown in the figure. Both tangent lines are horizontal and have slope 0. There is no single tangent line at the point where $x = x_3$.

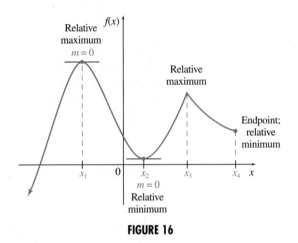

FIGURE 16

Since the derivative of a function gives the slope of a line tangent to the graph of the function, to find relative extrema we first identify all critical numbers and endpoints. A relative extremum *may* exist at a critical number. (A rough sketch of the graph of the function near a critical number often is enough to tell whether an extremum has been found.) These facts about extrema are summarized below.

If a function f has a relative extremum at c, then c is a critical number or c is an endpoint of the domain.

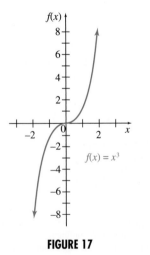

FIGURE 17

CAUTION Be very careful not to get this result backward. It does *not* say that a function has relative extrema at all critical numbers of the function. For example, Figure 17 shows the graph of $f(x) = x^3$. The derivative, $f'(x) = 3x^2$, is 0 when $x = 0$, so that 0 is a critical number for that function. However, as suggested by the graph of Figure 17, $f(x) = x^3$ has neither a relative maximum nor a relative minimum at $x = 0$ (or anywhere else, for that matter). A critical number is a candidate for the location of a relative extremum, but *only* a candidate.

First Derivative Test Suppose all critical numbers have been found for some function f. How is it possible to tell from the equation of the function whether these critical numbers produce relative maxima, relative minima, or neither? One way is suggested by the graph in Figure 18.

As shown in Figure 18, on the left of a relative maximum the tangent lines to the graph of a function have positive slopes, indicating that the function is increasing. At the relative maximum, the tangent line is horizontal. On the right of the relative maximum the tangent lines have negative slopes, indicating that the function is decreasing. Around a relative minimum the opposite occurs. As shown by the tangent lines in Figure 18, the function is decreasing on the left of the relative minimum, has a horizontal tangent at the minimum, and is increasing on the right of the minimum.

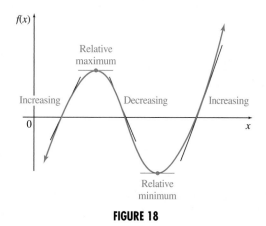

FIGURE 18

Putting this together with the methods from Section 1 for identifying intervals where a function is increasing or decreasing gives the following **first derivative test** for locating relative extrema.

FIRST DERIVATIVE TEST

Let c be a critical number for a function f. Suppose that f is differentiable on (a, b) except possibly at c, and that c is the only critical number for f in (a, b).

1. $f(c)$ is a relative maximum of f if the derivative $f'(x)$ is positive in the interval (a, c) and negative in the interval (c, b).

2. $f(c)$ is a relative minimum of f if the derivative $f'(x)$ is negative in the interval (a, c) and positive in the interval (c, b).

The sketches in the table on the following page show how the first derivative test works. Assume the same conditions on a, b, and c for the table as those given for the first derivative test.

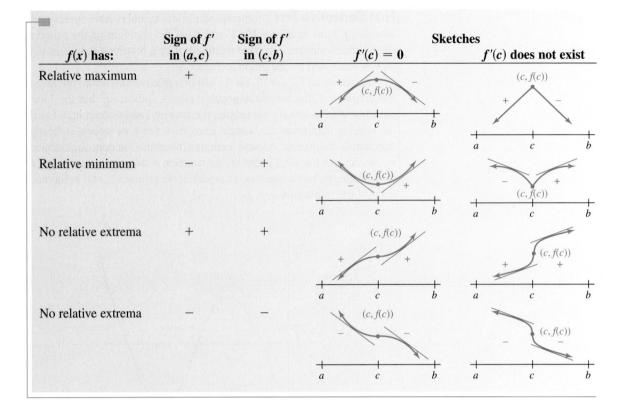

$f(x)$ has:	Sign of f' in (a,c)	Sign of f' in (c,b)	Sketches $f'(c) = 0$	$f'(c)$ does not exist
Relative maximum	+	−		
Relative minimum	−	+		
No relative extrema	+	+		
No relative extrema	−	−		

EXAMPLE 2   Relative Extrema

Find all relative extrema for the following functions, as well as where each function is increasing and decreasing.

(a) $f(x) = 2x^3 - 3x^2 - 72x + 15$

Method 1: First Derivative Test

Solution The derivative is $f'(x) = 6x^2 - 6x - 72$. There are no points where $f'(x)$ fails to exist, so the only critical numbers will be found where the derivative equals 0. Setting the derivative equal to 0 gives

$$6x^2 - 6x - 72 = 0$$
$$6(x^2 - x - 12) = 0$$
$$6(x - 4)(x + 3) = 0$$
$$x - 4 = 0 \quad \text{or} \quad x + 3 = 0$$
$$x = 4 \quad \text{or} \quad x = -3.$$

As in the previous section, the critical numbers 4 and −3 are used to determine the three intervals $(-\infty, -3)$, $(-3, 4)$, and $(4, \infty)$ shown on the number line in Figure 19 on the next page. Any number from each of the three intervals can be used as a test point to find the sign of f' in each interval. Using −4, 0, and 5 gives the following information.

$$f'(-4) = 6(-8)(-1) > 0$$
$$f'(0) = 6(-4)(3) < 0$$
$$f'(5) = 6(1)(8) > 0$$

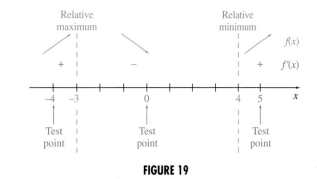

FIGURE 19

Thus, the derivative is positive on $(-\infty, -3)$, negative on $(-3, 4)$, and positive on $(4, \infty)$. By Part 1 of the first derivative test, this means that the function has a relative maximum of $f(-3) = 150$ when $x = -3$; by Part 2, f has a relative minimum of $f(4) = -193$ when $x = 4$.

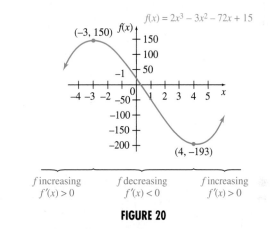

FIGURE 20

The function is increasing on $(-\infty, -3)$ and $(4, \infty)$ and decreasing on $(-3, 4)$. The graph is shown in Figure 20.

Method 2: Graphing Calculator   Many graphing calculators can locate a relative extremum when supplied with an interval containing the extremum. For example, after graphing the function $f(x) = 2x^3 - 3x^2 - 72x + 15$ on a TI-83, we selected "maximum" from the Calc menu and entered a left bound of -4 and a right bound of 0. The calculator asks for an initial guess, but in this example it doesn't matter what we enter. The result of this process, as well as a similar process for finding the relative minimum, is shown in Figure 21 on the next page.

Another way to verify the extrema with a graphing calculator is to graph $y = f'(x)$ and find where the graph crosses the x-axis. Figure 22 on the next page shows the result of this approach for finding the relative minimum of the previous function.

(b) $f(x) = 6x^{2/3} - 4x$

Solution Find $f'(x)$:

$$f'(x) = 4x^{-1/3} - 4 = \frac{4}{x^{1/3}} - 4.$$

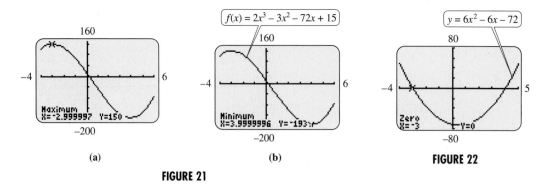

(a)

(b)

FIGURE 21

FIGURE 22

The derivative fails to exist when $x = 0$, but the function itself is defined when $x = 0$, making 0 a critical number for f. To find other critical numbers, set $f'(x) = 0$.

$$f'(x) = 0$$

$$\frac{4}{x^{1/3}} - 4 = 0$$

$$\frac{4}{x^{1/3}} = 4$$

$$4 = 4x^{1/3}$$

$$1 = x^{1/3}$$

$$1 = x$$

The critical numbers 0 and 1 are used to locate the intervals $(-\infty, 0)$, $(0, 1)$, and $(1, \infty)$ on a number line as in Figure 23. Evaluating $f'(x)$ at the test points -1, $1/2$, and 2 and using the first derivative test shows that f has a relative maximum at $x = 1$; the value of this relative maximum is $f(1) = 2$. Also, f has a relative minimum at $x = 0$; this relative minimum is $f(0) = 0$. The function is increasing on $(0, 1)$ and decreasing on $(-\infty, 0)$ and $(1, \infty)$. Notice that the graph, shown in Figure 24, has a sharp point at the critical number where the derivative does not exist. In the last section of this chapter we will show how to verify other features of the graph.

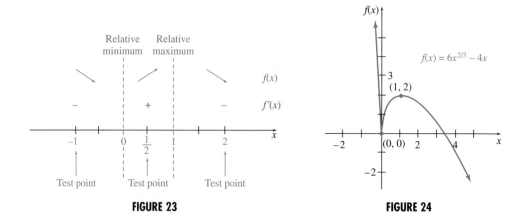

FIGURE 23

FIGURE 24

(c) $f(x) = xe^{2-x^2}$

Solution The derivative is

$$f'(x) = x(-2x)e^{2-x^2} + e^{2-x^2}$$
$$= e^{2-x^2}(-2x^2 + 1).$$

This expression exists for all x in the domain of f. The derivative is 0 when

$$-2x^2 + 1 = 0$$
$$1 = 2x^2$$
$$\frac{1}{2} = x^2$$
$$x = \pm\sqrt{1/2}$$
$$x = \pm\frac{1}{\sqrt{2}} \approx \pm 0.707.$$

There are two critical points, $-1/\sqrt{2}$ and $1/\sqrt{2}$. Using test points of $-1, 0$, and 1 gives the results shown in Figure 25.

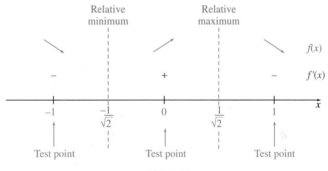

FIGURE 25

The function has a relative minimum at $-1/\sqrt{2}$ of $f(-1/\sqrt{2}) \approx -3.17$, and a relative maximum at $1/\sqrt{2}$ of $f(1/\sqrt{2}) \approx 3.17$. It is decreasing on the interval $(-\infty, -1/\sqrt{2})$, increasing on the interval $(-1/\sqrt{2}, 1/\sqrt{2})$, and decreasing on the interval $(1/\sqrt{2}, \infty)$. The graph is shown in Figure 26.

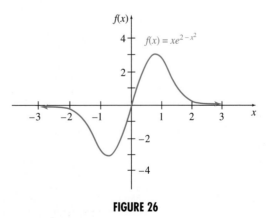

FIGURE 26

> **CAUTION** A critical number must be in the domain of the function. For example, the derivative of $f(x) = x/(x - 4)$ is $f'(x) = -4/(x - 4)^2$, which fails to exist when $x = 4$. But $f(4)$ does not exist, so 4 is not a critical number, and the function has no relative extrema. Even though $f(4)$ does not exist, the function changes from decreasing to increasing at $x = 4$.

As mentioned at the beginning of this section, finding the maximum or minimum value of a quantity is important in applications of mathematics. The final example gives a further illustration.

EXAMPLE 3 Milk Consumption

The daily milk consumption for Aberdeen Angus and Hereford calves can be modeled by the function

$$M(t) = 5.955t^{0.247}e^{-0.027t}, \qquad 1 \le t \le 26,$$

where $M(t)$ is the milk consumption (in kg) and t is the age of the calf (in weeks).*

Find the time in which the maximum daily consumption occurs.

Solution To maximize $M(t)$, find $M'(t)$. Then find the critical numbers.

$$M'(t) = 5.955\{t^{0.247}(-0.027e^{-0.027t}) + 0.247t^{-0.753}e^{-0.247}\}$$
$$= 5.966e^{-0.027t}(-0.027t^{0.247} + 0.247t^{-0.753})$$
$$= 5.966t^{0.247}e^{-0.027t}(-0.027 + 0.247t^{-1}) = 0$$

Since the expression in the parentheses above is the only term that can force the derivative to be zero, we set it equal to zero. That is,

$$-0.027 + 0.247t^{-1} = 0.$$

Solving for t, we find that $M'(t) = 0$ when $t \approx 9.15$. Since $M'(t)$ exists for all t in this interval, 9.15 is the only critical number. To verify that $t = 9.15$ is a maximum, evaluate the derivative on either side of $t = 9.15$.

$$M'(5) \approx 0.1738 \quad \text{and} \quad M'(25) \approx -0.1152$$

As illustrated by these calculations and Figure 27, $M(t)$ is increasing up to $t = 9.15$, then decreasing, so there is a maximum value at $t \approx 9.15$ of $M(9.15) \approx 8.05$. The maximum daily milk consumption will be approximately 8.05 kg and will occur during the ninth week of a calf's life.

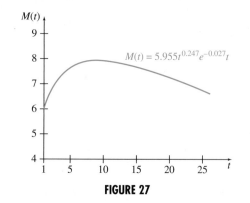

FIGURE 27

---

*Mezzadra, C., R. Paciaroni, S. Vulich, E. Villarreal, and L. Melucci, "Estimation of Milk Consumption Curve Parameters for Different Genetic Groups of Bovine Calves," *Animal Production,* Vol. 49, 1989, pp. 83–87.

CAUTION   Be careful to give the y-value of the point where an extremum occurs. Although we solve the equation $f'(x) = 0$ for x to find the extremum, the maximum or minimum value of the function is the corresponding y-value. Thus, in Example 3, we found that at $t = 9.15$, the maximum daily milk consumption is 8.05 (not 9.15).

Examples that involve the maximization of a quadratic function, such as the opening example, could be solved by the methods described in Chapter 1. But those involving more complicated functions, such as the milk consumption example, are difficult to analyze without the tools of calculus.

Finding extrema for realistic problems requires an accurate mathematical model of the problem. In particular, it is important to be aware of restrictions on the values of the variables. For example, if $N(t)$ closely approximates the number of cells in a tumor at time t, $N(t)$ must certainly be restricted to the positive integers, or perhaps to a few common fractional values.

On the other hand, to apply the tools of calculus to obtain an extremum for some function, the function must be defined and be meaningful at every real number in some interval. Because of this, the answer obtained from a mathematical model might be a number that is not feasible in the actual problem.

Usually, the requirement that a continuous function be used, rather than one that can take on only certain selected values, is of theoretical interest only. In most cases, the methods of calculus give acceptable results as long as the assumptions of continuity and differentiability are not totally unreasonable. If they lead to the conclusion, say, that a tumor has $80\sqrt{2}$ cells, it is usually only necessary to investigate acceptable values close to $80\sqrt{2}$.

5.2 EXERCISES

Find the locations and values of all relative extrema for the functions with the graphs that follow. Compare with Exercises 1–8 in the preceding section.

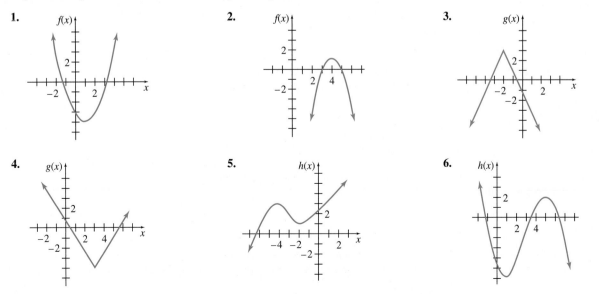

7.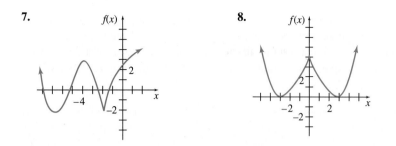

8.

Find the x-value of all points where the functions defined as follows have any relative extrema. Find the value(s) of any relative extrema.

9. $f(x) = x^2 + 12x - 8$

10. $f(x) = x^2 - 4x + 6$

11. $f(x) = x^3 + 6x^2 + 9x - 8$

12. $f(x) = x^3 + 3x^2 - 24x + 2$

13. $f(x) = -\dfrac{4}{3}x^3 - \dfrac{21}{2}x^2 - 5x + 8$

14. $f(x) = -\dfrac{2}{3}x^3 - \dfrac{1}{2}x^2 + 3x - 4$

15. $f(x) = x^4 - 18x^2 - 4$

16. $f(x) = x^4 - 8x^2 + 9$

17. $f(x) = -(8 - 5x)^{2/3}$

18. $f(x) = (2 - 9x)^{2/3}$

19. $f(x) = 2x + 3x^{2/3}$

20. $f(x) = 3x^{5/3} - 15x^{2/3}$

21. $f(x) = x - \dfrac{1}{x}$

22. $f(x) = x^2 + \dfrac{1}{x}$

23. $f(x) = \dfrac{x^2 - 2x + 1}{x - 3}$

24. $f(x) = \dfrac{x^2 - 6x + 9}{x + 2}$

25. $f(x) = x^2 e^x - 3$

26. $f(x) = 3xe^x + 2$

27. $f(x) = 2x + \ln x$

28. $f(x) = \dfrac{x^2}{\ln x}$

29. $f(x) = \sin(\pi x)$

30. $f(x) = \cos(2x)$

Use the derivative to find the vertex of each parabola.

31. $y = -2x^2 + 8x - 1$

32. $y = ax^2 + bx + c$

Graph each of the following functions on a graphing calculator, and then use the graph to find all relative extrema (to three decimal places) of the functions defined as follows. Then confirm your answer by finding the derivative and using the calculator to solve the equation $f'(x) = 0$.

33. $f(x) = x^5 - x^4 + 4x^3 - 30x^2 + 5x + 6$

34. $f(x) = -x^5 - x^4 + 2x^3 - 25x^2 + 9x + 12$

35. Graph $f(x) = 2|x + 1| + 4|x - 5| - 20$ with a graphing calculator in the window $[-10, 10]$ by $[-15, 30]$. Use the graph and the function to determine the x-values of all extrema.

36. Consider the function*

$$g(x) = \frac{1}{x^{12}} - 2\left(\frac{1,000}{x}\right)^6.$$

a. Using a graphing calculator, try to find any local minima, or tell why finding a local minimum is difficult for this function.

---

*Dubinsky, Ed, "Is Calculus Obsolete?" *The Mathematics Teacher*, Vol. 88, No. 2, Feb. 1995, pp. 146–148.

b. Find any local minima using the techniques of calculus.

✎ **c.** Based on your results in parts a and b, describe circumstances under which relative extrema are easier to find using the techniques of calculus than using a graphing calculator.

Applications

LIFE SCIENCES

37. *Activity Level* In the summer the activity level of a certain type of lizard varies according to the time of day. A biologist has determined that the activity level is given by the function

$$a(t) = 0.008t^3 - 0.288t^2 + 2.304t + 7,$$

where t is the number of hours after 12 noon. When is the activity level highest? When is it lowest?

38. *Milk Consumption* The average individual daily milk consumption for herds of Charolais, Angus, and Hereford calves can be described by the function

$$M(t) = 6.281t^{0.242}e^{-0.025t}, \qquad 1 \le t \le 26,$$

where $M(t)$ is the milk consumption (in kg) and t is the age of the calf (in weeks).*

a. Find the time in which the maximum daily consumption occurs and the maximum daily consumption.

b. If the general formula for this model is given by

$$M(t) = at^b e^{-ct},$$

find the time where the maximum consumption occurs and the maximum consumption. (*Hint:* Express your answer in terms of a, b, and c.)

39. *Alaskan Moose* As we saw in Chapter 3, the mathematical relationship between the age of a captive female moose and its weight can be described by the function

$$M(t) = 369(0.93)^t t^{0.36}, \qquad t \le 12,$$

where $M(t)$ is the weight of the moose (in kg) and t is the age (in years) of the moose.† Find the age at which the weight of a female moose is maximized. What is the maximum weight?

40. *Thermic Effect of Food* As we saw in the last section, the metabolic rate after a person eats a meal tends to go up and

then, after some time has passed, returns to a resting metabolic rate. This phenomenon is known as the thermic effect of food and can be described for a particular individual as

$$F(t) = -10.28 + 175.9te^{-t/1.3},$$

where $F(t)$ is the thermic effect of food, measured in kJ/h, and t is the number of hours that have elapsed since eating a meal.‡ Find the time after the meal when the thermic effect of the food is maximized.

41. *Epidemic* The rate of change of the number of individuals infected in an epidemic as a function of time is given by the epidemic curve, which has the form

$$f(t) = \frac{k(n-1)n^2 e^{nt}}{[n-1+e^{nt}]^2},$$

where the constant n is the size of the population and the constant k is the contact rate between individuals of the group.§ Find the time at which the rate of change in the number of infected individuals is a maximum.

OTHER APPLICATIONS

42. *Training Accidents* Far more American troops have died in training accidents than in combat since the Vietnam War. Recently the military has been redoubling its efforts to increase safety standards and prevent violence in the ranks. The number of casualties (in hundreds) each year is approximated by

$$f(x) = 0.198x^3 - 1.56x^2 + 2.03x + 14.8,$$

where x is the number of years since 1990.‖ The function is accurate only on the finite domain $\{0, 1, 2, 3, 4, 5\}$. Find the location and value of any extrema for this function. (*Hint:* Refer to the last two paragraphs in the section.) Interpret your results.

43. *Attitude Change* Social psychologists have found that as the discrepancy between the views of a speaker and

---

*Mezzadra, C., R. Paciaroni, S. Vulich, E. Villarreal, and L. Melucci, "Estimation of Milk Consumption Curve Parameters for Different Genetic Groups of Bovine Calves," *Animal Production,* Vol. 49, 1989, pp. 83–87.

†Schwartz, C., and Kris Hundertmark, "Reproductive Characteristics of Alaskan Moose," *Journal of Wildlife Management,* Vol. 57, No. 3, July 1993, pp. 454–468.

‡Reed, George, and James Hill, "Measuring the Thermic Effect of Food," *American Journal of Clinical Nutrition,* Vol. 63, 1996, pp. 164–169.

§Bailey, Norman T. J., *The Mathematical Approach to Biology and Medicine,* Wiley, 1967, pp. 185–186.

‖*The New York Times,* Dec. 25, 1995, p. 8.

those of an audience increases, the attitude change in the audience also increases to a point, but decreases when the discrepancy becomes too large, particularly if the communicator is viewed by the audience as having low credibility.* Suppose that the degree of change can be approximated by the function

$$D(x) = -x^4 + 8x^3 + 80x^2,$$

where x is the discrepancy between the views of the speaker and those of the audience, as measured by scores on a questionnaire. Find the amount of discrepancy the speaker should aim for to maximize the attitude change in the audience.

44. *Film Length* A group of researchers found that people prefer training films of moderate length; shorter films contain too little information, while longer films are boring. For a training film on the care of exotic birds, the researchers determined that the ratings people gave for the film could be approximated by

$$R(t) = \frac{20t}{t^2 + 100},$$

where t is the length of the film (in minutes). Find the film length that received the highest rating.

45. *Height* After a great deal of experimentation, two Atlantic Institute of Technology senior physics majors determined that when a bottle of French champagne is shaken several times, held upright, and uncorked, its cork travels according to

$$s(t) = -16t^2 + 64t + 3,$$

where s is its height (in feet) above the ground t seconds after being released. How high will it go?

5.3 HIGHER DERIVATIVES, CONCAVITY, AND THE SECOND DERIVATIVE TEST

THINK ABOUT IT   If a diet plan guarantees weight loss, is it always possible to reach your weight loss goal?

The following discussion addresses this question.

To understand the behavior of a function on an interval, it is important to know the *rate* at which the function is increasing or decreasing. For example, suppose that your friend, who is concerned that you are 100 pounds overweight, has studied the diet plan you started two months ago and is trying to get you to continue with the diet. He shows you the following function, which represents your weight $w(t)$ (in pounds):

$$w(t) = 300 - 15 \ln(t + 1),$$

where t is the number of months since starting the diet. He points out that the derivative of the function is always negative, so your weight is always decreasing. He claims that you cannot help but lose all the weight you desire. Should you take his advice and use this diet plan?

It is true that the weight function decreases for all t. The derivative is

$$w'(t) = \frac{-15}{t + 1},$$

---

*Eagly, A. H., and K. Telaak, "Width of the Latitude of Acceptance as a Determinant of Attitude Change," *Journal of Personality and Social Psychology,* Vol. 23, 1972, pp. 388–397.

which is always negative for $t > 0$. The catch lies in *how fast* the function is decreasing. The derivative $w'(t) = -15/(t + 1)$ tells how fast your weight is decreasing at any number of months, t, since you began the diet. For example, when $t = 2$ months, $w'(2) = -5$, and the weight is decreasing at the rate of 5 lbs per month. When $t = 4$ months, $w'(4) = -3$; the weight is decreasing at 3 lbs per month. At $t = 14$ months, $w'(14) = -1$, or 1 lb per month. By the time you've been on the diet for $t = 29$ months, the weight decreases at 0.5 lb per month, and the *rate of decrease* looks as though it will continue to increase.

The rate of decrease in $w'(t)$ is given by the derivative of $w'(t)$, called the **second derivative** of $w(t)$ and denoted by $w''(t)$. Since $w'(t) = -15/(t + 1)$,

$$w''(t) = \frac{15}{(t + 1)^2}.$$

The function $w''(t)$ is positive for all positive values of t and therefore confirms the suspicion that the rate of decrease in weight does indeed increase for all $t \geq 0$. Your weight will not increase, but the amount of weight you lose will certainly not be what your friend predicts. For example, at 30 months, your weight will be about 249 lbs. A year later, it would be about 244 lbs. It would take you approximately 786 months to lose 100 lbs. The only people who could benefit from this diet are those who don't have a lot of weight to lose.

As mentioned earlier, the second derivative of a function f, f'', gives the rate of change of the derivative of f. Before continuing to discuss applications of the second derivative, we need to introduce some additional terminology and notation.

Higher Derivatives   If a function f has a derivative f', then the derivative of f', if it exists, is the second derivative of f, written f''. The derivative of f'', if it exists, is called the **third derivative** of f, and so on. By continuing this process, we can find **fourth derivatives** and other higher derivatives. For example, if $f(x) = x^4 + 2x^3 + 3x^2 - 5x + 7$, then

$$f'(x) = 4x^3 + 6x^2 + 6x - 5, \quad \text{First derivative of } f$$
$$f''(x) = 12x^2 + 12x + 6, \quad \text{Second derivative of } f$$
$$f'''(x) = 24x + 12, \quad \text{Third derivative of } f$$

and
$$f^{(4)}(x) = 24. \quad \text{Fourth derivative of } f$$

NOTATION FOR HIGHER DERIVATIVES

The second derivative of $y = f(x)$ can be written using any of the following notations:

$$f''(x), \qquad \frac{d^2y}{dx^2}, \qquad \text{or} \qquad D_x^2[f(x)].$$

The third derivative can be written in a similar way. For $n \geq 4$, the nth derivative is written $f^{(n)}(x)$ or d^ny/dx^n or $D_x^n[f(x)]$.

■ **CAUTION** Notice the difference in notation between $f^{(4)}$, which indicates the fourth derivative of f, and f^4, which indicates f raised to the fourth power.

EXAMPLE 1 Second Derivative

Let $f(x) = x^3 + 6x^2 - 9x + 8$.

(a) Find $f''(x)$.

Solution To find the second derivative of $f(x)$, find the first derivative, and then take its derivative.

$$f'(x) = 3x^2 + 12x - 9$$
$$f''(x) = 6x + 12$$

(b) Find $f''(0)$.

Solution Since $f''(x) = 6x + 12$,

$$f''(0) = 6(0) + 12 = 12.$$

EXAMPLE 2 Second Derivative

Find the second derivative for the functions defined as follows.

(a) $f(x) = (x^2 - 1)^2$

Solution Here

$$f'(x) = 2(x^2 - 1)(2x) = 4x(x^2 - 1).$$

Use the product rule to find $f''(x)$.

$$\begin{aligned} f''(x) &= 4x(2x) + (x^2 - 1)(4) \\ &= 8x^2 + 4x^2 - 4 \\ &= 12x^2 - 4 \end{aligned}$$

(b) $g(x) = 4x(\ln x)$

Solution Use the product rule.

$$g'(x) = 4x \cdot \frac{1}{x} + (\ln x) \cdot 4 = 4 + 4(\ln x)$$

$$g''(x) = 0 + 4 \cdot \frac{1}{x} = \frac{4}{x}$$

(c) $h(x) = \dfrac{x}{e^x}$

Solution Here, we need the quotient rule.

$$h'(x) = \frac{e^x - xe^x}{(e^x)^2} = \frac{e^x(1 - x)}{(e^x)^2} = \frac{1 - x}{e^x}$$

$$h''(x) = \frac{e^x(-1) - (1 - x)e^x}{(e^x)^2} = \frac{e^x(-1 - 1 + x)}{(e^x)^2} = \frac{-2 + x}{e^x}$$

Earlier, we saw that the first derivative of a function represents the rate of change of the function. The second derivative, then, represents the rate of change of the first derivative. If a function describes the position of a vehicle (along a straight line) at time t, then the first derivative gives the velocity of the vehicle.

That is, if $y = s(t)$ describes the position (along a straight line) of the vehicle at time t, then $v(t) = s'(t)$ gives the velocity at time t.

We also saw that *velocity* is the rate of change of distance with respect to time. Recall, the difference between velocity and speed is that velocity may be positive or negative, whereas speed is always positive. A negative velocity indicates travel in a negative direction (backing up) with regard to the starting point; positive velocity indicates travel in the positive direction (going forward) from the starting point.

The instantaneous rate of change of velocity is called **acceleration.** Since instantaneous rate of change is the same as the derivative, acceleration is the derivative of velocity. Thus, if $a(t)$ represents the acceleration at time t, then

$$a(t) = \frac{d}{dt}v(t) = s''(t).$$

If the velocity is positive and the acceleration is positive, the velocity is increasing, so the vehicle is speeding up. If the velocity is positive and the acceleration is negative, the vehicle is slowing down. A negative velocity and a positive acceleration mean the vehicle is backing up and slowing down. If both the velocity and acceleration are negative, the vehicle is speeding up in the negative direction.

EXAMPLE 3 Velocity and Acceleration

Suppose a car is moving in a straight line, with its position from a starting point (in feet) at time t (in seconds) given by

$$s(t) = t^3 - 2t^2 - 7t + 9.$$

Find the following.

(a) The velocity at any time t

Solution The velocity is given by

$$v(t) = s'(t) = 3t^2 - 4t - 7$$

feet per second.

(b) The acceleration at any time t

Solution Acceleration is given by

$$a(t) = v'(t) = s''(t) = 6t - 4$$

feet per second per second.

(c) The time intervals (for $t \geq 0$) when the car is going forward or backing up

Solution We first find when the velocity is 0, that is, when the car is stopped.

$$v(t) = 3t^2 - 4t - 7 = 0$$
$$(3t - 7)(t + 1) = 0$$
$$t = 7/3 \quad \text{or} \quad t = -1$$

We are interested in $t \geq 0$. Choose a value of t in each of the intervals $(0, 7/3)$ and $(7/3, \infty)$ to see that the velocity is negative in $(0, 7/3)$ and positive in $(7/3, \infty)$. The car is backing up for the first $7/3$ seconds, then going forward.

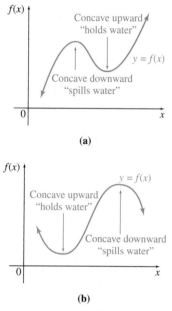

(a)

(b)

FIGURE 28

(d) The time intervals (for $t \geq 0$) when the car is speeding up or slowing down

Solution The car will speed up when the velocity and acceleration are the same sign and slow down when they have opposite signs. Here, the acceleration is positive when $6t - 4 > 0$, that is, $t > 2/3$ seconds, and negative for $t < 2/3$ seconds. Since the velocity is negative in $(0, 7/3)$ and positive in $(7/3, \infty)$, the car is speeding up for $0 \leq t \leq 2/3$ seconds, slowing down for $2/3 \leq t \leq 7/3$ seconds, and speeding up again for $t > 7/3$ seconds. See the sign graphs.

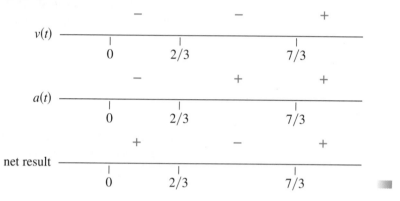

Concavity of a Graph The first derivative has been used to show where a function is increasing or decreasing and where the extrema occur. The second derivative gives the rate of change of the first derivative; it indicates *how fast* the function is increasing or decreasing. The rate of change of the derivative (the second derivative) affects the *shape* of the graph. Intuitively, we say that a graph is *concave upward* on an interval if it "holds water" and *concave downward* if it "spills water." See Figure 28.

More precisely, a function is **concave upward** on an interval (a, b) if the graph of the function lies above its tangent line at each point of (a, b). A function is **concave downward** on (a, b) if the graph of the function lies below its tangent line at each point of (a, b). A point where a graph changes **concavity** is called a **point of inflection.** See Figure 29.

Users of soft contact lenses recognize concavity as the way to tell if a lens is inside out. As Figure 30 shows, a correct contact lens has a profile that is entirely

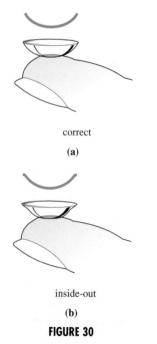

correct

(a)

inside-out

(b)

FIGURE 30

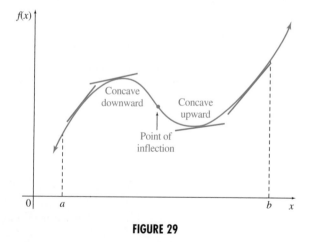

FIGURE 29

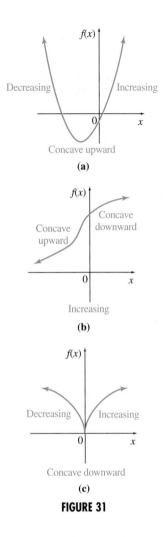

FIGURE 31

concave upward. The profile of an inside-out lens has points of inflection near the edges, where the profile begins to turn concave downward very slightly.

Just as a function can be either increasing or decreasing on an interval, it can be either concave upward or concave downward on an interval. Examples of various combinations are shown in Figure 31.

Figure 32 shows two functions that are concave upward on an interval (a, b). Several tangent lines are also shown. In Figure 32(a), the slopes of the tangent lines (moving from left to right) are first negative, then 0, and then positive. In Figure 32(b), the slopes are all positive, but they get larger.

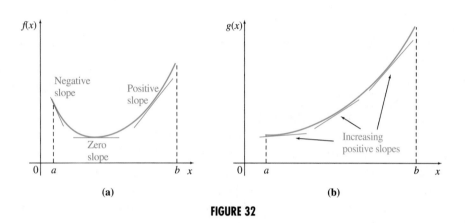

FIGURE 32

In both cases, the slopes are *increasing*. The slope at a point on a curve is given by the derivative. Since a function is increasing if its derivative is positive, its slope is increasing if the derivative of the slope function is positive. Since the derivative of a derivative is the second derivative, a function is concave upward on an interval if its second derivative is positive at each point of the interval.

A similar result is suggested by Figure 33 for functions whose graphs are concave downward. In both graphs, the slopes of the tangent lines are *decreasing* as we move from left to right. Since a function is decreasing if its derivative is negative, a function is concave downward on an interval if its second derivative is negative at each point of the interval. These observations suggest the following test.

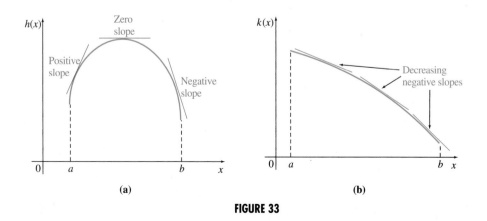

FIGURE 33

TEST FOR CONCAVITY

Let f be a function with derivatives f' and f'' existing at all points in an interval (a, b). Then f is concave upward on (a, b) if $f''(x) > 0$ for all x in (a, b), and concave downward on (a, b) if $f''(x) < 0$ for all x in (a, b).

An easy way to remember this test is by the faces shown in Figure 34. When the second derivative is positive at a point $(+\ +)$, the graph is concave upward $(\smile)$. When the second derivative is negative at a point $(-\ -)$, the graph is concave downward $(\frown)$.

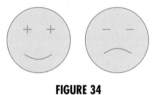

FIGURE 34

EXAMPLE 4 Concavity

Find all intervals where $f(x) = x^4 - 8x^3 + 18x^2$ is concave upward or downward, and find all inflection points.

Solution The first derivative is $f'(x) = 4x^3 - 24x^2 + 36x$, and the second derivative is $f''(x) = 12x^2 - 48x + 36$. We factor $f''(x)$ as $12(x - 1)(x - 3)$, and then create a number line for $f''(x)$ as we did in the previous two sections for $f'(x)$.

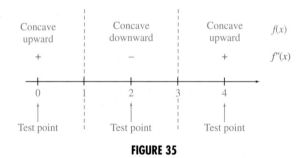

FIGURE 35

We see from Figure 35 that $f''(x) > 0$ on the intervals $(-\infty, 1)$ and $(3, \infty)$, so f is concave upward on these intervals. $f''(x) < 0$ on the interval $(1, 3)$, so f is concave downward on this interval.

Finally, we have inflection points where f'' changes sign, namely at $x = 1$ and $x = 3$. Since $f(1) = 11$ and $f(3) = 27$, the inflection points are $(1, 11)$ and $(3, 27)$. The function is graphed in Figure 36.

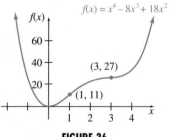

FIGURE 36

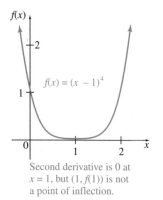

$f(x) = (x - 1)^4$

Second derivative is 0 at
$x = 1$, but $(1, f(1))$ is not
a point of inflection.

FIGURE 37

Example 4 suggests the following result.

> At a point of inflection for a function f, the second derivative is 0 or does not exist.

| **CAUTION** Be careful with the previous statement. Finding a value of x where $f''(x) = 0$ does not mean that a point of inflection has been located. For example, if $f(x) = (x - 1)^4$, then $f''(x) = 12(x - 1)^2$, which is 0 at $x = 1$. The graph of $f(x) = (x - 1)^4$ is always concave upward, however, so it has no point of inflection. See Figure 37.

Second Derivative Test   The idea of concavity can be used to decide whether a given critical number produces a relative maximum or a relative minimum. This test, an alternative to the first derivative test, is based on the fact that a curve with a horizontal tangent at a point c and concave downward on an open interval containing c also has a relative maximum at c. A relative minimum occurs when a graph has a horizontal tangent at a point d and is concave upward on an open interval containing d. See Figure 38.

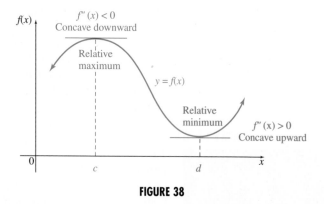

FIGURE 38

A function f is concave upward on an interval if $f''(x) > 0$ for all x in the interval, while f is concave downward on an interval if $f''(x) < 0$ for all x in the interval. These ideas lead to the **second derivative test** for relative extrema.

SECOND DERIVATIVE TEST

Let f'' exist on some open interval containing c, and let $f'(c) = 0$.

1. If $f''(c) > 0$, then $f(c)$ is a relative minimum.
2. If $f''(c) < 0$, then $f(c)$ is a relative maximum.
3. If $f''(c) = 0$, then the test gives no information about extrema.

EXAMPLE 5   Second Derivative Test
Find all relative extrema for

$$f(x) = 4x^3 + 7x^2 - 10x + 8.$$

Solution First, find the points where the derivative is 0. Here $f'(x) = 12x^2 + 14x - 10$. Solve the equation $f'(x) = 0$ to get

$$12x^2 + 14x - 10 = 0$$
$$2(6x^2 + 7x - 5) = 0$$
$$2(3x + 5)(2x - 1) = 0$$
$$3x + 5 = 0 \quad \text{or} \quad 2x - 1 = 0$$
$$3x = -5 \qquad\qquad 2x = 1$$
$$x = -\frac{5}{3} \qquad\qquad x = \frac{1}{2}.$$

Now use the second derivative test. The second derivative is $f''(x) = 24x + 14$. Evaluate $f''(x)$ first at $-5/3$, getting

$$f''\left(-\frac{5}{3}\right) = 24\left(-\frac{5}{3}\right) + 14 = -40 + 14 = -26 < 0,$$

so that by Part 2 of the second derivative test, $-5/3$ leads to a relative maximum of $f(-5/3) = 691/27$. Also, when $x = 1/2$,

$$f''\left(\frac{1}{2}\right) = 24\left(\frac{1}{2}\right) + 14 = 12 + 14 = 26 > 0,$$

with $1/2$ leading to a relative minimum of $f(1/2) = 21/4$. ▪

The second derivative test works only for those critical numbers c that make $f'(c) = 0$. This test does not work for critical numbers c for which $f'(c)$ does not exist (since $f''(c)$ would not exist either). Also, the second derivative test does not work for critical numbers c that make $f''(c) = 0$. In both of these cases, use the first derivative test.

EXAMPLE 6 Catfish Farming

The graph in Figure 39 shows the population of catfish in a commercial catfish farm as a function of time. As the graph shows, the population increases rapidly up to a point and then increases at a slower rate. We saw several types of functions that produce such a graph in earlier sections. The horizontal dashed line shows that the population will approach some upper limit determined by the capacity of the farm. The point at which the rate of population growth starts to slow is the point of inflection for the graph.

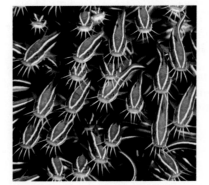

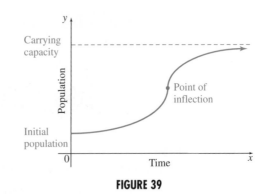

FIGURE 39

To produce the maximum yield of catfish, harvesting should take place at the point of fastest possible growth of the population; here, this is at the point of inflection. The rate of change of the population, given by the first derivative, is increasing up to the inflection point (on the interval where the second derivative is positive) and decreasing past the inflection point (on the interval where the second derivative is negative).

The *law of diminishing returns* in economics is related to the idea of concavity. The function graphed in Figure 40 gives the output y from a given input x. If the input were advertising costs for some product, for example, the output might be the corresponding revenue from sales.

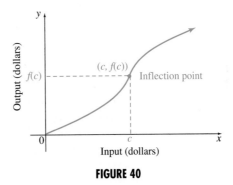

FIGURE 40

The graph in Figure 40 shows an inflection point at $(c, f(c))$. For $x < c$, the graph is concave upward, so the rate of change of the slope is increasing. This indicates that the output y is increasing at a faster rate with each additional dollar spent. When $x > c$, however, the graph is concave downward, the rate of change of the slope is decreasing, and the increase in y is smaller with each additional dollar spent. Thus, further input beyond c dollars produces diminishing returns. The point of inflection at $(c, f(c))$ is called the **point of diminishing returns.** Beyond this point there is a smaller and smaller return for each dollar invested.

As another example of diminishing returns from agriculture, with a fixed amount of land, machinery, fertilizer, and so on, adding workers increases production a lot at first, then less and less with each additional worker.

EXAMPLE 7   Point of Diminishing Returns

The revenue $R(x)$ generated from sales of a certain drug is related to the amount x spent on advertising by

$$R(x) = \frac{1}{15,000}(600x^2 - x^3), \quad 0 \le x \le 600,$$

where x and $R(x)$ are in thousands of dollars. Is there a point of diminishing returns for this function? If so, what is it?

Solution Since a point of diminishing returns occurs at an inflection point, look for an x-value that makes $R''(x) = 0$. Write the function as

$$R(x) = \frac{600}{15,000}x^2 - \frac{1}{15,000}x^3 = \frac{1}{25}x^2 - \frac{1}{15,000}x^3.$$

Now find $R'(x)$ and then $R''(x)$.

$$R'(x) = \frac{2x}{25} - \frac{3x^2}{15,000} = \frac{2}{25}x - \frac{1}{5,000}x^2$$

$$R''(x) = \frac{2}{25} - \frac{1}{2,500}x$$

Set $R''(x)$ equal to 0 and solve for x.

$$\frac{2}{25} - \frac{1}{2,500}x = 0$$

$$-\frac{1}{2,500}x = -\frac{2}{25}$$

$$x = \frac{5,000}{25} = 200$$

Test a number in the interval $(0, 200)$ to see that $R''(x)$ is positive there. Then test a number in the interval $(200, 600)$ to find $R''(x)$ negative in that interval. Since the sign of $R''(x)$ changes from positive to negative at $x = 200$, the graph changes from concave upward to concave downward at that point, and there is a point of diminishing returns at the inflection point $(200, 1,066 \,^2/_3)$. Investments in advertising beyond \$200,000 return less and less for each dollar invested. Verify that $R'(200) = 8$. This means that when \$200,000 is invested, another \$1,000 invested returns approximately \$8,000 in additional revenue. Thus it may still be economically sound to invest in advertising beyond the point of diminishing returns.

5.3 EXERCISES

Find $f''(x)$ for each of the following. Then find $f''(0)$ and $f''(2)$.

1. $f(x) = 3x^3 - 4x + 5$

2. $f(x) = x^3 + 4x^2 + 2$

3. $f(x) = 3x^4 - 5x^3 + 2x^2$

4. $f(x) = -x^4 + 2x^3 - x^2$

5. $f(x) = 3x^2 - 4x + 8$

6. $f(x) = 8x^2 + 6x + 5$

7. $f(x) = \dfrac{x^2}{1 + x}$

8. $f(x) = \dfrac{-x}{1 - x^2}$

9. $f(x) = \sqrt{x + 4}$

10. $f(x) = \sqrt{2x + 9}$

11. $f(x) = 5x^{3/5}$

12. $f(x) = -2x^{2/3}$

13. $f(x) = 5e^{-x^2}$

14. $f(x) = 0.5e^{x^2}$

15. $f(x) = \dfrac{\ln x}{4x}$

16. $f(x) = \ln x + \dfrac{1}{x}$

Find $f'''(x)$, the third derivative of f, and $f^{(4)}(x)$, the fourth derivative of f, for each of the following.

17. $f(x) = -x^4 + 2x^2 + 8$

18. $f(x) = 2x^4 - 3x^3 + x^2$

19. $f(x) = 4x^5 + 6x^4 - x^2 + 2$

20. $f(x) = 3x^5 - x^4 + 2x^3 - 7x$

21. $f(x) = \dfrac{x-1}{x+2}$

22. $f(x) = \dfrac{x+1}{x}$

23. $f(x) = \dfrac{3x}{x-2}$

24. $f(x) = \dfrac{x}{2x+1}$

25. Let $f(x) = \ln x$.

   a. Compute $f'(x)$, $f''(x)$, $f'''(x)$, $f^{(4)}(x)$, and $f^{(5)}(x)$.

   b. Guess a formula for $f^{(n)}(x)$, where n is any positive integer.

26. For $f(x) = e^x$, find $f''(x)$ and $f'''(x)$. What is the nth derivative of f with respect to x?

In Exercises 27–46, find the largest open intervals where the functions are concave upward or concave downward. Find any points of inflection.

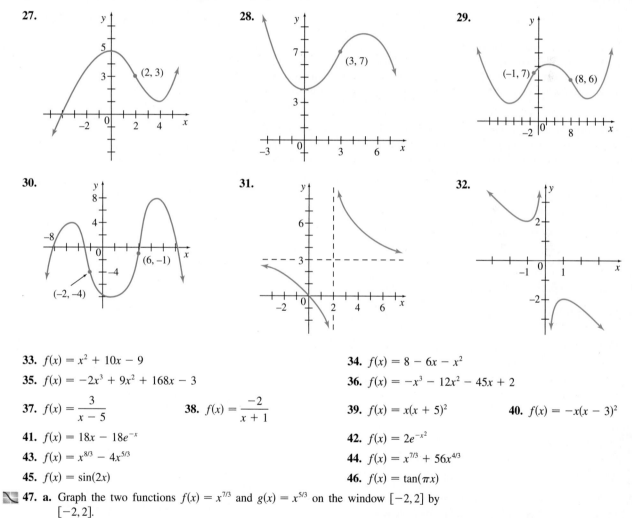

33. $f(x) = x^2 + 10x - 9$

34. $f(x) = 8 - 6x - x^2$

35. $f(x) = -2x^3 + 9x^2 + 168x - 3$

36. $f(x) = -x^3 - 12x^2 - 45x + 2$

37. $f(x) = \dfrac{3}{x-5}$

38. $f(x) = \dfrac{-2}{x+1}$

39. $f(x) = x(x+5)^2$

40. $f(x) = -x(x-3)^2$

41. $f(x) = 18x - 18e^{-x}$

42. $f(x) = 2e^{-x^2}$

43. $f(x) = x^{8/3} - 4x^{5/3}$

44. $f(x) = x^{7/3} + 56x^{4/3}$

45. $f(x) = \sin(2x)$

46. $f(x) = \tan(\pi x)$

47. a. Graph the two functions $f(x) = x^{7/3}$ and $g(x) = x^{5/3}$ on the window $[-2, 2]$ by $[-2, 2]$.

   b. Verify that both f and g have an inflection point at $(0, 0)$.

   c. How is the value of $f''(0)$ different from $g''(0)$?

   d. Based on what you have seen so far in this exercise, is it always possible to tell the difference between a point where the second derivative is 0 or undefined based on the graph? Explain.

48. Describe the slope of the tangent line to the graph of $f(x) = e^x$ for each of the following.

    a. $x \to -\infty$ **b.** $x \to 0$

49. What is true about the slope of the tangent line to the graph of $f(x) = \ln x$ as $x \to \infty$? as $x \to 0$?

Find any critical numbers for f in Exercises 50–55 and then use the second derivative test to decide whether the critical numbers lead to relative maxima or relative minima. If $f''(c) = 0$ for a critical number c, then the second derivative test gives no information. In this case, use the first derivative test instead.

50. $f(x) = -x^2 - 10x - 25$ **51.** $f(x) = x^2 - 12x + 36$ **52.** $f(x) = 3x^3 - 3x^2 + 1$

53. $f(x) = 2x^3 - 4x^2 + 2$ **54.** $f(x) = (x + 3)^4$ **55.** $f(x) = x^3$

56. Suppose a friend makes the following argument. A function f is increasing and concave downward. Therefore, f' is positive and decreasing, so it eventually becomes 0 and then negative, at which point f decreases. Show that your friend is wrong by giving an example of a function that is always increasing and concave downward.

 *Sometimes the derivative of a function is known, but not the function. We will see more of this later in the book. For each function f' defined as shown, find $f''(x)$, then use a graphing calculator to graph f' and f'' in the indicated window. Use the graph to do the following.*

  a. *Give the (approximate) x-values where f has a maximum or minimum.*

  b. *By considering the sign of $f'(x)$, give the (approximate) intervals where $f(x)$ is increasing and decreasing.*

  c. *Give the (approximate) x-values of any inflection points.*

  d. *By considering the sign of $f''(x)$, give the intervals where f is concave upward or concave downward.*

57. $f'(x) = x^3 - 6x^2 + 7x + 4;$ $[-5, 5]$ by $[-5, 15]$

58. $f'(x) = 10x^2(x - 1)(5x - 3);$ $[-1, 1.5]$ by $[-20, 20]$

59. $f'(x) = \dfrac{1 - x^2}{(x^2 + 1)^2};$ $[-3, 3]$ by $[-1.5, 1.5]$

Applications

LIFE SCIENCES

60. *Population Growth* When a hardy new species is introduced into an area, the population often increases as shown. Explain the significance of the following function values on the graph.

  a. f_0

  b. $f(a)$

  c. f_M

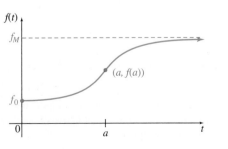

61. *Bacteria Population* Assume that the number of bacteria $R(t)$ (in millions) present in a certain culture at time t (in hours) is given by

$$R(t) = t^2(t - 18) + 96t + 1{,}000.$$

a. At what time before 8 hr will the population be maximized?

b. Find the maximum population.

62. *Ozone Depletion* According to an article in *The New York Times,* "Government scientists reported last week that they had detected a slowdown in the rate at which chemicals that deplete the earth's protective ozone layer are accumulating in the atmosphere."* Letting $c(t)$ be the amount of ozone-depleting chemicals at time t, what does this statement tell you about $c(t)$, $c'(t)$, and $c''(t)$?

63. *Drug Concentration* The percent of concentration of a certain drug in the bloodstream x hr after the drug is administered by mouth is given by

$$K(x) = \frac{3x}{x^2 + 4}.$$

For example, after 1 hr the concentration is given by

$$K(1) = \frac{3(1)}{1^2 + 4} = \frac{3}{5}\% = 0.6\% = 0.006.$$

a. Find the time at which concentration is a maximum.

b. Find the maximum concentration.

64. *Thoroughbred Horses* The association between velocity during exercise and blood lactate concentration (a measure of anaerobic exercise) after submaximal 800-m exercise of thoroughbred racehorses on sand and grass tracks has been studied. Researchers in Australia have determined that the lactate-velocity relationship can be modeled by the functions

$$l_1(v) = 0.08e^{0.33v} \quad \text{and}$$
$$l_2(v) = -0.87v^2 + 28.17v - 211.41,$$

where $l_1(v)$ and $l_2(v)$ are the lactate concentrations (in mmol/L) and v is the velocity (in m/s) of the horse during workout on sand and grass tracks, respectively.†

a. What is the concavity of each function on this interval?

b. Comment on why the concavity might be different for different types of tracks.

65. *Popcorn* As we saw in Chapter 4, researchers have determined that the amount of moisture present in a kernel of popcorn affects the volume of the popped corn and can be modeled for certain sizes of kernels by the function

$$v(x) = -35.98 + 12.09x - 0.4450x^2,$$

where x is moisture content (%, wet basis) and $v(x)$ is the expansion volume (in cm³/g).‡ Describe the concavity of this function.

66. *Drug Concentration* The percent of concentration of a drug in the bloodstream x hours after the drug is administered is given by

$$K(x) = \frac{4x}{3x^2 + 27}.$$

a. Find the time at which the concentration is a maximum.

b. Find the maximum concentration.

The next two exercises are a continuation of exercises first given in the section on Derivatives of Exponential Functions. Find the inflection point of the graph of each function. This is the point at which the growth rate begins to decline.

67. *Insect Growth* The growth function for a population of beetles is given by

$$G(t) = \frac{10{,}000}{1 + 49e^{-0.1t}}.$$

68. *Clam Population Growth* The population of a bed of clams is described by

$$G(t) = \frac{5{,}200}{1 + 12e^{-0.5t}}.$$

Hints for Exercises 69 and 70: Leave B, c, and k as constants until you are ready to calculate your final answer. Recall that $d(e^{g(t)})/dt = g'(t)e^{g(t)}$.

69. *Clam Growth* Researchers used a version of the Gompertz curve to model the growth of razor clams during the first seven years of the clams' lives with the equation

$$L(t) = Be^{-ce^{-kt}},$$

where $L(t)$ gives the length in cm after t years, $B = 14.3032$, $c = 7.267963$, and $k = 0.670840$.§ Find the inflection point and describe what it signifies.

The New York Times, Aug. 29, 1993, p. E2.
†Davie, A., and D. Evans, "Blood Lactate Responses to Submaximal Field Exercise Tests in Thoroughbred Horses," *The Veterinary Journal,* Vol. 159, 2000, pp. 252–258.
‡Song, A., and S. Eckhoff, "Optimum Popping Moisture Content for Popcorn Kernels of Different Sizes," *Cereal Chemistry,* Vol. 71, No. 5, 1994, pp. 458–460.
§Weymouth, F. W., H. C. McMillin, and Willis H. Rich, "Latitude and Relative Growth in the Razor Clam," *Journal of Experimental Biology,* Vol. 8, 1931, pp. 228–249.

70. *Breast Cancer Growth* Researchers used a version of the Gompertz curve to model the growth of breast cancer tumors with the equation

$$N(t) = e^{c(1-e^{-kt})},$$

where $N(t)$ is the number of cancer cells after t days, $c = 27.3$, and $k = 0.011$.* Find the inflection point and describe what it signifies.

71. *Alligator Teeth* Researchers have developed a mathematical model that can be used to estimate the number of teeth $N(t)$ at time t (days of incubation) for *Alligator mississippiensis*,[†] where

$$N(t) = 71.8e^{-8.96e^{-0.0685t}}.$$

Find the inflection point and describe its importance to this research.

OTHER APPLICATIONS

72. *Crime* In 1995, the rate of violent crimes in New York City continued to decrease, but at a slower rate than in previous years.[‡] Letting $f(t)$ be the rate of violent crime as a function of time, what does this tell you about $f(t)$, $f'(t)$, and $f''(t)$?

73. *Chemical Reaction* An autocatalytic chemical reaction is one in which the product being formed causes the rate of formation to increase. The rate of a certain autocatalytic reaction is given by

$$V(x) = 12x(100 - x),$$

where x is the quantity of the product present and 100 represents the quantity of chemical present initially. For what value of x is the rate of the reaction a maximum?

74. *Velocity and Acceleration* When an object is dropped straight down, the distance in feet that it travels in t sec is given by

$$s(t) = -16t^2.$$

Find the velocity at each of the following times.

a. After 3 sec   **b.** After 5 sec   **c.** After 8 sec

d. Find the acceleration. (The answer here is a constant— the acceleration due to the influence of gravity alone near the surface of the earth.)

75. *Baseball* Roger Clemens, ace pitcher of the New York Yankees, is standing on top of the 37-foot-high "Green Monster" left-field wall in Fenway Park, where he used to

pitch for the Boston Red Sox, and to which he has returned for a visit. We have asked him to fire his famous 95 mph (140 ft/sec) fastball straight up. The position equation, which gives the height of the ball at any time t, in seconds, is given by $s(t) = -16t^2 + 140t + 37$.[§] Find the following.

a. The maximum height of the ball

b. The time and velocity when the ball hits the ground

76. *Height of a Ball* If a cannonball is shot directly upward with a velocity of 256 ft/sec, its height above the ground after t sec is given by $s(t) = 256t - 16t^2$. Find the velocity and the acceleration after t sec. What is the maximum height the cannonball reaches? When does it hit the ground?

77. *Velocity and Acceleration of a Car* A car rolls down a hill. Its distance in feet from its starting point is given by $s(t) = 1.5t^2 + 4t$, where t is in seconds.

a. How far will the car move in 10 seconds?

b. What is the velocity at 5 seconds? at 10 seconds?

c. How can you tell from $v(t)$ that the car will not stop?

d. What is the acceleration at 5 seconds? at 10 seconds?

e. What is happening to the velocity and the acceleration as t increases?

78. *Flying Gravel* The grooves or tread in a tire occasionally pick up small pieces of gravel, which then are often thrown into the air as they work loose from the tire. When following behind a vehicle on a highway with loose gravel, it is possible to determine a safe distance to travel behind the vehicle so that your automobile is not hit with flying debris by analyzing the function

$$y = x \tan \alpha - \frac{16x^2}{V^2} \sec^2 \alpha,$$

where y is the height in feet of a piece of gravel that leaves the bottom of a tire at an angle α relative to the roadway, x

---

*Speer, John F., et al., "A Stochastic Numerical Model of Breast Cancer Growth That Simulates Clinical Data," *Cancer Research,* Vol. 44, Sept. 1984, pp. 4124–4130.

[†]Kulesa, P., et al., "On a Model Mechanism for the Spatial Patterning of Teeth Primordia in the Alligator," *Journal of Theoretical Biology,* Vol. 180, 1996, pp. 287–296.

[‡]*The New York Times,* Dec. 17, 1995, p. 49.

[§]This exercise provided by Frederick Russell of Charles County Community College.

is the horizontal distance in feet of the gravel, and V is the velocity of the automobile in feet per second.*

a. If a car is traveling 30 miles per hour (44 ft/sec), find the height of a piece of gravel thrown from a car tire at an angle of $\pi/4$, when the stone is 40 feet from the car.

b. Putting $y = 0$ and solving for x gives the distance that the gravel will fly. Show that the function that gives a relationship between x and the angle α for $y = 0$ is given by

$$x = \frac{V^2}{32} \sin(2\alpha).$$

(*Hint:* $2 \sin \alpha \cos \alpha = \sin(2\alpha)$.)

c. Using part b, if the gravel is thrown from the car at an angle of $\pi/3$ and initial velocity of 44 ft/sec, determine how far the gravel will travel.

d. Find $dx/d\alpha$ and use it to determine the value of α that gives the maximum distance that a stone could fly.

e. Find the maximum distance that a stone can fly from a car that is traveling 60 miles per hour.

79. *Product Life Cycle* The accompanying figure shows the *product life cycle* graph, with typical products marked on it. It illustrates the fact that a new product is often purchased at a faster and faster rate as people become familiar with it. In time, saturation is reached and the purchase rate stays constant until the product is made obsolete by newer products, after which it is purchased less and less.[†]

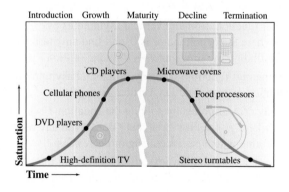

a. Which products on the left side of the graph are closest to the left-hand point of inflection? What does the point of inflection mean here?

b. Which product on the right side of the graph is closest to the right-hand point of inflection? What does the point of inflection mean here?

c. Discuss where home computers, fax machines, and other new technologies should be placed on the graph.

Point of Diminishing Returns In Exercises 80–83, find the point of diminishing returns (x, y) for the given functions, where $R(x)$ represents revenue in thousands of dollars and x represents the amount spent on advertising in thousands of dollars.

80. $R(x) = 10,000 - x^3 + 42x^2 + 800x, \quad 0 \le x \le 20$

81. $R(x) = \dfrac{4}{27}(-x^3 + 66x^2 + 1,050x - 400), 0 \le x \le 25$

82. $R(x) = -0.3x^3 + x^2 + 11.4x, \quad 0 \le x \le 6$

83. $R(x) = -0.6x^3 + 3.7x^2 + 5x, \quad 0 \le x \le 6$

84. *Risk Aversion* In economics, an index of *absolute risk aversion* is defined as

$$I(M) = \frac{-U''(M)}{U'(M)},$$

where M measures how much of a commodity is owned and $U(M)$ is a *utility function*, which measures the ability of quantity M of a commodity to satisfy a consumer's wants. Find $I(M)$ for $U(M) = \sqrt{M}$ and for $U(M) = M^{2/3}$, and determine which indicates a greater aversion to risk.

---

*Bolt, Brian, "Tennis, Golf, and Loose Gravel: Insight from Easy Math Models," *UMAP Journal,* Vol. 4, No. 1, Spring 1983, pp. 6–18.
[†]"The Product Life Cycle: A Key to Strategic Marketing Planning," *MSU Business Topics,* Winter 1973, p. 30. Reprinted by permission of the publisher, Graduate School of Business Administration, Michigan State University.

5.4 CURVE SKETCHING

The test for concavity, the test for increasing and decreasing functions, and the concept of limits at infinity help us sketch the graphs and describe the behavior of a variety of functions. This process, called *curve sketching,* has decreased somewhat in importance in recent years due to the widespread use of graphing calculators. We believe, however, that this topic is worth studying for the following reasons.

For one thing, a graphing calculator picture can be misleading, particularly if important points lie outside the viewing window. Even if all important features are within the viewing windows, there is still the problem that the calculator plots and connects points, and misses what goes on between those points. As an example of the difficulty in choosing an appropriate window without a knowledge of calculus, see Exercise 36 in the second section of this chapter.

Furthermore, curve sketching may be the best way to learn the material in the previous three sections. You may feel confident that you understand what increasing and concave upward mean, but using those concepts in a graph will put your understanding to the test.

Curve sketching may be done with the following steps.

CURVE SKETCHING

To sketch the graph of a function f:

1. Consider the domain of the function, and note any restrictions. (Avoid dividing by zero or taking a square root of a negative number.)
2. Find the y-intercept (if it exists) by substituting $x = 0$ into $f(x)$. Find any x-intercepts by solving $f(x) = 0$ if this is not too difficult.
3. **a.** If f is a rational function, find any vertical asymptotes by investigating where the denominator is 0, and find any horizontal asymptotes by finding the limits as $x \to \infty$ and $x \to -\infty$.
   b. If f is an exponential function, find any horizontal asymptotes; if f is a logarithmic function, find any vertical asymptotes.
4. Find f'. Locate any critical points by solving the equation $f'(x) = 0$ and determining where f' does not exist, but f does. Find any relative extrema and determine where f is increasing or decreasing.
5. Find f''. Locate potential points of inflection by solving the equation $f''(x) = 0$ and determining where $f''(x)$ does not exist. Determine where f is concave upward or concave downward.
6. Plot the intercepts, the critical points, the inflection points, the asymptotes, and other points as needed.
7. Connect the points with a smooth curve using the correct concavity.
8. Check your graph using a graphing calculator. If the picture looks very different from what you've drawn, see in what ways the picture differs and use that information to help find your mistake.

There are four possible combinations for a function to be increasing or decreasing and concave up or concave down, as shown in the chart on the following page.

$f''(x)$	$f'(x)$ + (function is increasing)	− (function is decreasing)
+ (function is concave up)		
− (function is concave down)		

EXAMPLE 1   Polynomial Graph

Graph $f(x) = 2x^3 - 3x^2 - 12x + 1$.

Solution The domain is $(-\infty, \infty)$. The y-intercept is located at $y = f(0) = 1$. Finding the x-intercepts requires solving the equation $f(x) = 0$. But this is a third-degree equation; since we have not covered a procedure for solving such equations, we will skip this step. (One could, however, use a graphing calculator to find estimates of the x-intercepts.) This is not a rational function, so we also skip Step 3.

To find the intervals where the function is increasing or decreasing, find the first derivative.

$$f'(x) = 6x^2 - 6x - 12$$

This derivative is 0 when

$$6(x^2 - x - 2) = 0$$
$$6(x - 2)(x + 1) = 0$$
$$x = 2 \quad \text{or} \quad x = -1.$$

These critical numbers divide the number line in Figure 41 into three regions. Testing a number from each region in $f'(x)$ shows that f is increasing on $(-\infty, -1)$ and $(2, \infty)$ and decreasing on $(-1, 2)$. This is shown with the arrows in Figure 41. By the first derivative test, f has a relative maximum when $x = -1$ and a relative minimum when $x = 2$. The relative maximum is $f(-1) = 8$, while the relative minimum is $f(2) = -19$.

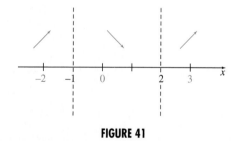

FIGURE 41

Now use the second derivative to find the intervals where the function is concave upward or downward. Here

$$f''(x) = 12x - 6,$$

which is 0 when $x = 1/2$. Testing a point with x less than $1/2$, and one with x greater than $1/2$, shows that f is concave downward on $(-\infty, 1/2)$ and concave upward on $(1/2, \infty)$. The graph has an inflection point at $(1/2, f(1/2))$, or $(1/2, -11/2)$. This information is summarized in the chart on the next page.

Interval	$(-\infty, -1)$	$(-1, 1/2)$	$(1/2, 2)$	$(2, \infty)$
Sign of f'	$+$	$-$	$-$	$+$
Sign of f''	$-$	$-$	$+$	$+$
f Increasing or Decreasing	Increasing	Decreasing	Decreasing	Increasing
Concavity of f	Downward	Downward	Upward	Upward
Shape of Graph	⌒	⌒	⌣	⌣

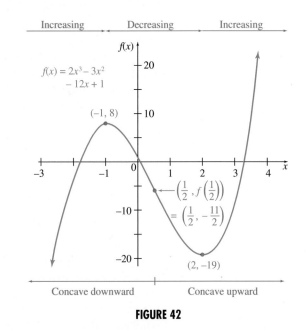

FIGURE 42

Use this information and the critical points to get the graph shown in Figure 42.

A graphing calculator picture of the function in Figure 42 on the arbitrarily chosen window $[-3, 3]$ by $[-7, 7]$ gives a misleading picture, as Figure 43(a) shows. Knowing where the turning points lie tells us that a better window would be $[-3, 4]$ by $[-20, 20]$, with the results shown in Figure 43(b).

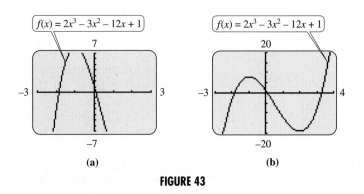

(a)   (b)

FIGURE 43

EXAMPLE 2 Rational Function Graph

Graph $f(x) = x + 1/x$.

Solution Notice that $x = 0$ is not in the domain of the function, so there is no y-intercept. To find the x-intercept, solve $f(x) = 0$.

$$x + \frac{1}{x} = 0$$

$$x = -\frac{1}{x}$$

$$x^2 = -1$$

Since x^2 is always positive, there is also no x-intercept.

The function is a rational function, but it is not written in the usual form of one polynomial over another. It can be rewritten in that form:

$$f(x) = x + \frac{1}{x} = \frac{x^2 + 1}{x}.$$

FOR REVIEW ■

Asymptotes were discussed in the section on Polynomial and Rational Functions. You may wish to refer back to that section to review. To review limits, refer to the first section in the chapter titled The Derivative.

Because $x = 0$ makes the denominator (but not the numerator) 0, the line $x = 0$ is a vertical asymptote. To find any horizontal asymptotes, we investigate

$$\lim_{x \to \infty} \frac{x^2 + 1}{x} = \lim_{x \to \infty} \frac{1 + \frac{1}{x^2}}{\frac{1}{x}} = \frac{1 + 0}{0} = \frac{1}{0}.$$

Division by 0 is undefined, so the limit does not exist. Verify that $\lim_{x \to -\infty} f(x)$ also does not exist, so there are no horizontal asymptotes.

Observe that as x gets very large, the second term $(1/x)$ in $f(x)$ gets very small, so $f(x) = x + (1/x) \approx x$. The graph gets closer and closer to the straight line $y = x$ as x becomes larger and larger. This is what is known as an **oblique asymptote.**

Here $f'(x) = 1 - (1/x^2)$, which is 0 when

$$\frac{1}{x^2} = 1$$

$$x^2 = 1$$

$$x = 1 \quad \text{or} \quad x = -1.$$

The derivative fails to exist at 0, where the vertical asymptote is located. Evaluating $f'(x)$ in each of the regions determined by the critical numbers and the asymptote shows that f is increasing on $(-\infty, -1)$ and $(1, \infty)$ and decreasing on $(-1, 0)$ and $(0, 1)$. By the first derivative test, f has a relative maximum of $y = f(-1) = -2$ when $x = -1$, and a relative minimum of $y = f(1) = 2$ when $x = 1$.

The second derivative is

$$f''(x) = \frac{2}{x^3},$$

which is never equal to 0 and does not exist when $x = 0$. (The function itself also does not exist at 0.) Because of this, there may be a change of concavity, but not an inflection point, when $x = 0$. The second derivative is negative when x is negative, making f concave downward on $(-\infty, 0)$. Also, $f''(x) > 0$ when $x > 0$, making f concave upward on $(0, \infty)$.

Interval	$(-\infty, -1)$	$(-1, 0)$	$(0, 1)$	$(1, \infty)$
Sign of f'	$+$	$-$	$-$	$+$
Sign of f''	$-$	$-$	$+$	$+$
f Increasing or Decreasing	Increasing	Decreasing	Decreasing	Increasing
Concavity of f	Downward	Downward	Upward	Upward
Shape of Graph	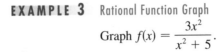			

Use this information, the asymptotes, and the critical points to get the graph shown in Figure 44.

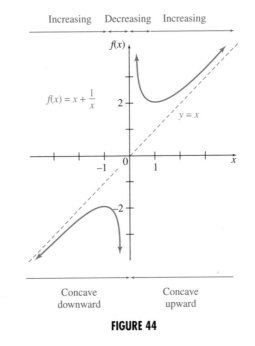

FIGURE 44

EXAMPLE 3 Rational Function Graph

Graph $f(x) = \dfrac{3x^2}{x^2 + 5}$.

Solution The y-intercept is located at $y = f(0) = 0$. Verify that this is also the only x-intercept. There is no vertical asymptote, because $x^2 + 5 \neq 0$ for any value of x. Find any horizontal asymptote by calculating $\lim_{x \to \infty} f(x)$ and $\lim_{x \to -\infty} f(x)$. First, divide both the numerator and the denominator of $f(x)$ by x^2.

$$\lim_{x \to \infty} \frac{3x^2}{x^2 + 5} = \lim_{x \to \infty} \frac{\dfrac{3x^2}{x^2}}{\dfrac{x^2}{x^2} + \dfrac{5}{x^2}} = \frac{3}{1 + 0} = 3$$

Verify that the limit of $f(x)$ as $x \to -\infty$ is also 3. Thus, the horizontal asymptote is $y = 3$.

We now compute $f'(x)$:

$$f'(x) = \frac{(x^2 + 5)(6x) - (3x^2)(2x)}{(x^2 + 5)^2}.$$

Notice that $6x$ can be factored out of each term in the numerator:

$$f'(x) = \frac{(6x)[(x^2 + 5) - x^2]}{(x^2 + 5)^2}$$

$$= \frac{(6x)(5)}{(x^2 + 5)^2} = \frac{30x}{(x^2 + 5)^2}.$$

From the numerator, $x = 0$ is a critical number. The denominator is always positive. (Why?) Evaluating $f'(x)$ in each of the regions determined by $x = 0$ shows that f is decreasing on $(-\infty, 0)$ and increasing on $(0, \infty)$. By the first derivative test, f has a relative minimum when $x = 0$.

The second derivative is

$$f''(x) = \frac{(x^2 + 5)^2(30) - (30x)(2)(x^2 + 5)(2x)}{(x^2 + 5)^4}.$$

Factor $30(x^2 + 5)$ out of the numerator:

$$f''(x) = \frac{30(x^2 + 5)[(x^2 + 5) - (x)(2)(2x)]}{(x^2 + 5)^4}.$$

Divide a factor of $(x^2 + 5)$ out of the numerator and denominator, and simplify the numerator:

$$f''(x) = \frac{30[(x^2 + 5) - (x)(2)(2x)]}{(x^2 + 5)^3}$$

$$= \frac{30[(x^2 + 5) - (4x^2)]}{(x^2 + 5)^3}$$

$$= \frac{30(5 - 3x^2)}{(x^2 + 5)^3}.$$

The numerator of $f''(x)$ is 0 when $x = \pm\sqrt{5/3} \approx \pm1.29$. Testing a point in each of the three intervals defined by these points shows that f is concave downward on $(-\infty, -1.29)$ and $(1.29, \infty)$, and concave upward on $(-1.29, 1.29)$. The graph has inflection points at $\left(\pm\sqrt{5/3}, f\left(\pm\sqrt{5/3}\right)\right) \approx (\pm1.29, 0.75)$.

Interval	$(-\infty, -1.29)$	$(-1.29, 0)$	$(0, 1.29)$	$(1.29, \infty)$
Sign of f'	$-$	$-$	$+$	$+$
Sign of f''	$-$	$+$	$+$	$-$
f Increasing or Decreasing	Decreasing	Decreasing	Increasing	Increasing
Concavity of f	Downward	Upward	Upward	Downward
Shape of Graph	$\frown$	$\smile$	$\smile$	$\frown$

Use this information, the asymptote, the critical point, and the inflection points to get the graph shown in Figure 45 on the following page.

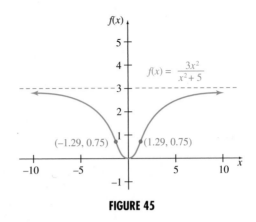

FIGURE 45

EXAMPLE 4 Graph with Logarithm

Graph $f(x) = \dfrac{\ln x}{x^2}$.

Solution The domain is $x > 0$, so there is no y-intercept. The x-intercept is 1, because $\ln 1 = 0$. We know that $y = \ln x$ has a vertical asymptote at $x = 0$, because $\lim\limits_{x \to 0^+} \ln x = -\infty$. Dividing by x^2 when x is small makes $\ln x / x^2$ even more negative than $\ln x$. The first derivative is

$$f'(x) = \frac{x^2 \cdot \dfrac{1}{x} - 2x \ln x}{(x^2)^2} = \frac{x(1 - 2 \ln x)}{x^4} = \frac{1 - 2 \ln x}{x^3}$$

by the quotient rule. Setting the numerator equal to 0 and solving for x gives

$$1 - 2 \ln x = 0$$
$$1 = 2 \ln x$$
$$\ln x = 0.5$$
$$x = e^{0.5} \approx 1.65.$$

Since $f'(1)$ is positive and $f'(2)$ is negative, f increases on $(0, 1.65)$, then decreases on $(1.65, \infty)$, with a maximum value of $f(1.65) \approx 0.18$.

To find any inflection points, we set $f''(x) = 0$.

$$f''(x) = \frac{x^3\left(-2 \cdot \dfrac{1}{x}\right) - (1 - 2 \ln x) \cdot 3x^2}{(x^3)^2}$$

$$= \frac{-2x^2 - 3x^2 + 6x^2 \ln x}{x^6} = \frac{-5 + 6 \ln x}{x^4}$$

$$\frac{-5 + 6 \ln x}{x^4} = 0$$

$$-5 + 6 \ln x = 0$$

$$6 \ln x = 5$$

$$\ln x = 5/6$$

$$x = e^{5/6} \approx 2.3$$

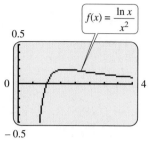

FIGURE 46

There is an inflection point at $(2.3, f(2.3)) \approx (2.3, 0.16)$. Verify that $f''(1)$ is negative and $f''(3)$ is positive, so the graph is concave downward on $(1, 2.3)$ and upward on $(2.3, \infty)$. This information is summarized in the chart and could be used to sketch the graph. A calculator graph of the function is shown in Figure 46.

Interval	$(0, 1.65)$	$(1.65, 2.3)$	$(2.3, \infty)$
Sign of f'	+	−	−
Sign of f''	−	−	+
f Increasing or Decreasing	Increasing	Decreasing	Decreasing
Concavity of f	Downward	Downward	Upward
Shape of Graph	⌒	⌒	⌣

As we saw earlier, a graphing calculator, when used with care, can be helpful in studying the behavior of functions. This section has illustrated that calculus is also a great help. The techniques of calculus show where the important points of a function, such as the relative extrema and the inflection points, are located. Furthermore, they tell how the function behaves between and beyond the points that are graphed, something a graphing calculator cannot always do.

5.4 EXERCISES

1. By sketching a graph of the function or by investigating values of the function near 0, find $\lim_{x \to 0} x \ln |x|$. (This result will be useful in Exercise 21.)

2. Describe how you would find the equation of the horizontal asymptote for the graph of $f(x) = \dfrac{3x^2 - 2x}{2x^2 + 5}$.

Graph each function, considering the domain, critical points, regions where the function is increasing or decreasing, points of inflection, regions where the function is concave upward or concave downward, intercepts where possible, and asymptotes where applicable. (Hint: In Exercise 21, use the result of Exercise 1. In Exercises 25–27, recall from Exercise 60 in the section on Limits that $\lim_{x \to \infty} x^n e^{-x} = 0$.)

3. $f(x) = -2x^3 - 9x^2 + 108x - 10$

4. $f(x) = x^3 - \dfrac{15}{2}x^2 - 18x - 1$

5. $f(x) = -3x^3 + 6x^2 - 4x - 1$

6. $f(x) = x^3 - 6x^2 + 12x - 11$

7. $f(x) = x^4 - 18x^2 + 5$

8. $f(x) = x^4 - 8x^2$

9. $f(x) = x^4 - 2x^3$

10. $f(x) = x^5 - 15x^3$

11. $f(x) = x + \dfrac{2}{x}$

12. $f(x) = 2x + \dfrac{8}{x}$

13. $f(x) = \dfrac{x - 1}{x + 1}$

14. $f(x) = \dfrac{x}{1 + x}$

15. $f(x) = \dfrac{1}{x^2 + x - 2}$

16. $f(x) = \dfrac{-2}{x^2 - x - 6}$

17. $f(x) = \dfrac{x}{x^2 + 1}$

18. $f(x) = \dfrac{1}{x^2 + 1}$

19. $f(x) = \dfrac{1}{x^2 - 4}$

20. $f(x) = \dfrac{x}{x^2 - 1}$

21. $y = x \ln |x|$

22. $y = x - \ln |x|$

23. $y = \dfrac{\ln x}{x}$

24. $y = \dfrac{\ln x^2}{x^2}$

25. $y = xe^{-x}$

26. $y = x^2 e^{-x}$

27. $y = (x - 1)e^{-x}$

28. $y = e^x + e^{-x}$

29. $y = x^{2/3} - x^{5/3}$

30. $y = x^{1/3} + x^{4/3}$

31. $f(x) = x + \cos x$

32. $f(x) = x + \sin x$

33. The default window on many calculators is $[-10, 10]$ by $[-10, 10]$. For the odd exercises between 3 and 15, tell which would give a poor representation in this window. (*Note:* Your answers may differ from ours, depending on what you consider "poor.")

34. Repeat Exercise 33 for the even exercises between 4 and 16.

35. Repeat Exercise 33 for the odd exercises between 17 and 31.

36. Repeat Exercise 33 for the even exercises between 18 and 32.

In Exercises 37–41, sketch the graph of a single function that has all of the properties listed.

37. a. Continuous and differentiable everywhere except at $x = 1$, where it has a vertical asymptote

    b. $f'(x) < 0$ everywhere it is defined

    c. A horizontal asymptote at $y = 2$

    d. $f''(x) < 0$ on $(-\infty, 1)$ and $(2, 4)$

    e. $f''(x) > 0$ on $(1, 2)$ and $(4, \infty)$

38. a. Continuous for all real numbers

    b. $f'(x) < 0$ on $(-\infty, -6)$ and $(1, 3)$

    c. $f'(x) > 0$ on $(-6, 1)$ and $(3, \infty)$

    d. $f''(x) > 0$ on $(-\infty, -6)$ and $(3, \infty)$

    e. $f''(x) < 0$ on $(-6, 3)$

    f. A y-intercept at $(0, 2)$

39. a. Continuous and differentiable for all real numbers

    b. $f'(x) > 0$ on $(-\infty, -3)$ and $(1, 4)$

    c. $f'(x) < 0$ on $(-3, 1)$ and $(4, \infty)$

    d. $f''(x) < 0$ on $(-\infty, -1)$ and $(2, \infty)$

    e. $f''(x) > 0$ on $(-1, 2)$

    f. $f'(-3) = f'(4) = 0$

    g. $f''(x) = 0$ at $(-1, 3)$ and $(2, 4)$

40. a. Continuous for all real numbers

    b. $f'(x) > 0$ on $(-\infty, -2)$ and $(0, 3)$

    c. $f'(x) < 0$ on $(-2, 0)$ and $(3, \infty)$

    d. $f''(x) < 0$ on $(-\infty, 0)$ and $(0, 5)$

    e. $f''(x) > 0$ on $(5, \infty)$

    f. $f'(-2) = f'(3) = 0$

    g. $f'(0)$ doesn't exist

h. Differentiable everywhere except at $x = 0$

i. An inflection point at $(5, 1)$

41. a. Continuous for all real numbers

b. Differentiable everywhere except at $x = 4$

c. $f(1) = 5$

d. $f'(1) = 0$ and $f'(3) = 0$

e. $f'(x) > 0$ on $(-\infty, 1)$ and $(4, \infty)$

f. $f'(x) < 0$ on $(1, 3)$ and $(3, 4)$

g. $\lim\limits_{x \to 4^-} f'(x) = -\infty$ and $\lim\limits_{x \to 4^+} f'(x) = \infty$

h. $f''(x) > 0$ on $(2, 3)$

i. $f''(x) < 0$ on $(-\infty, 2)$, $(3, 4)$, and $(4, \infty)$

42. On many calculators, graphs of rational functions produce lines at vertical asymptotes. For example, graphing $y = (x - 1)/(x + 1)$ on the window $[-4.9, 4.9]$ by $[-4.9, 4.9]$ produces such a line at $x = -1$ on the TI-83 and TI-86. But with the window $[-4.7, 4.7]$ by $[-4.7, 4.7]$ on a TI-83, or $[-6.3, 6.3]$ by $[-6.3, 6.3]$ on a TI-86, the spurious line does not appear. Experiment with this function on your calculator, trying different windows, and try to figure out an explanation for this phenomenon. (*Hint:* Consider the number of pixels on the calculator screen.)

Applications

LIFE SCIENCES

43. *Thoroughbred Horses* The association between velocity during exercise and blood lactate concentration after submaximal 800-m exercise of thoroughbred racehorses on sand and grass tracks has been studied. As we saw in the previous section, the lactate-velocity relationship can be described by the functions

$$l_1(v) = 0.08e^{0.33v} \quad \text{and}$$
$$l_2(v) = -0.87v^2 + 28.17v - 211.41,$$

where $l_1(v)$ and $l_2(v)$ are the lactate concentrations (in mmol/L) and v is the velocity (in m/s) of the horse during workout on sand and grass tracks, respectively.[*] Sketch the graph of both functions for $13 \leq v \leq 17$.

44. *Neuron Communication* In the FitzHugh-Nagumo model of how neurons communicate, the rate of change of the electric potential v with respect to time is given as a function of v by $f(v) = v(a - v)(v - 1)$, where a is a positive

constant.[†] Sketch a graph of this function when $a = 0.25$ and $0 \leq v \leq 1$.

45. *Immune System Activity* As we saw in Chapter 4, the activity of an immune system invaded by a parasite can be described by

$$I(n) = \frac{an^2}{b + n^2}$$

where n measures the number of larvae in the host, a is the maximum functional activity of the host's immune response, and b is a measure of the sensitivity of the immune system.[‡] Sketch a graph of this function when $a = 0.5$ and $b = 10$.

46. *Neuron Membrane Potential* A model of neuron firing involves the function

$$J(z) = \frac{1}{1 + e^{-1.3(z-4)}} - \frac{1}{1 + e^{5.2}},$$

where z is the neuron membrane potential, and the net threshold for a neuron to fire is at $z = 4$.[§] Sketch a graph of this function.

---

[*]Davie, A., and D. Evans, "Blood Lactate Responses to Submaximal Field Exercise Tests in Thoroughbred Horses," *The Veterinary Journal,* Vol. 159, 2000, pp. 252–258.

[†]Murray, J. D., *Mathematical Biology,* Springer-Verlag, 1989, p. 163.

[‡]Murray, J. D., *Mathematical Biology,* Springer-Verlag, 1989, p. 634.

[§]Jinghua, Xu, and Li Wei, "The Dynamics of a Glia-Modulated Neural Network and its Relation to Brain Functions," in *Mathematical Topics in Population Biology, Morphogenesis and Neurosciences,* edited by E. Teramoto and M. Yamaguti, Springer-Verlag, 1987, pp. 292–300.

47. *Cellular Chemistry* As we saw in Chapter 4, the release of a chemical neurotransmitter as a function of the amount C of intracellular calcium is given by

$$L(C) = \frac{aC^n}{k + C^n},$$

where n measures the degree of cooperativity, k measures saturation, and a is the maximum of the process.* Sketch a graph of this function when $n = 4$, $k = 3$, and $a = 100$.

48. *Fruit Flies* In Chapter 4, we showed that the number of imagoes (sexually mature adult fruit flies) per mated female per day (y) can be approximated by

$$y = 34.7(1.0186)^{-x}x^{-0.658},$$

where x is the mean density of the mated population (measured as flies per bottle) over a sixteen-day period.[†] Sketch the graph of the function.

CHAPTER SUMMARY

In this chapter we have explored various concepts related to the graph of a function: increasing and decreasing, relative maxima and minima, concavity, critical points, and inflection points. The first and second derivative tests provide ways to locate relative extrema. The last section brings all these concepts together. Also, we investigated two applications of the second derivative: acceleration and the point of diminishing returns.

KEY TERMS

5.1 increasing function
decreasing function
critical number
critical point
5.2 relative (or local)
  maximum

relative (or local)
  minimum
relative (or local)
  extremum
first derivative test
5.3 second derivative

third derivative
fourth derivative
acceleration
concavity
concave upward and
  downward

point of inflection
second derivative test
point of diminishing
  returns
5.4 oblique asymptote

CHAPTER 5 REVIEW EXERCISES

1. When given the equation for a function, how can you determine where it is increasing and where it is decreasing?

2. When given the equation for a function, how can you determine where the relative extrema are located? Give two ways to test whether a relative extremum is a minimum or a maximum.

3. Does a relative maximum of a function always have the largest y-value in the domain of the function? Explain your answer.

4. What information about a graph can be found from the second derivative?

---

*Segel, Lee A., "Toward Molecular Sensory Physiology: Mathematical Models," in *Mathematical Topics in Population Biology, Morphogenesis and Neurosciences,* edited by E. Teramoto and M. Yamaguti, Springer-Verlag, 1987, pp. 313–321.
[†]Pearl, R., and S. Parker, *Proc. Natl. Acad. Sci.,* Vol. 8, 1922, p. 212, quoted in *Elements of Mathematical Biology* by Alfred J. Lotka, Dover Publications, 1956, pp. 308–311.

Find the largest open intervals where f is increasing or decreasing.

5. $f(x) = x^2 - 5x + 3$

6. $f(x) = -2x^2 - 3x + 4$

7. $f(x) = -x^3 - 5x^2 + 8x - 6$

8. $f(x) = 4x^3 + 3x^2 - 18x + 1$

9. $f(x) = \dfrac{6}{x - 4}$

10. $f(x) = \dfrac{5}{2x + 1}$

11. $f(x) = \ln(x^2 - 1)$

12. $f(x) = 3xe^{2x}$

Find the locations and values of all relative maxima and minima.

13. $f(x) = -x^2 + 4x - 8$

14. $f(x) = x^2 - 6x + 4$

15. $f(x) = 2x^2 - 8x + 1$

16. $f(x) = -3x^2 + 2x - 5$

17. $f(x) = 2x^3 + 3x^2 - 36x + 20$

18. $f(x) = 2x^3 + 3x^2 - 12x + 5$

19. $f(x) = \dfrac{xe^x}{x - 1}$

20. $f(x) = \dfrac{\ln(3x)}{2x^2}$

Find the second derivative of each of the following, and then find $f''(1)$ and $f''(-3)$.

21. $f(x) = 3x^4 - 5x^2 - 11x$

22. $f(x) = 9x^3 + \dfrac{1}{x}$

23. $f(x) = \dfrac{5x - 1}{2x + 3}$

24. $f(x) = \dfrac{4 - 3x}{x + 1}$

25. $f(t) = \sqrt{t^2 + 1}$

26. $f(t) = -\sqrt{5 - t^2}$

Graph each of the following functions, considering the domain, critical points, regions where the function is increasing or decreasing, points of inflection, regions where the function is concave up or concave down, intercepts where possible, and asymptotes where applicable.

27. $f(x) = -2x^3 - \dfrac{1}{2}x^2 + x - 3$

28. $f(x) = -\dfrac{4}{3}x^3 + x^2 + 30x - 7$

29. $f(x) = x^4 - \dfrac{4}{3}x^3 - 4x^2 + 1$

30. $f(x) = -\dfrac{2}{3}x^3 + \dfrac{9}{2}x^2 + 5x + 1$

31. $f(x) = \dfrac{x - 1}{2x + 1}$

32. $f(x) = \dfrac{2x - 5}{x + 3}$

33. $f(x) = -4x^3 - x^2 + 4x + 5$

34. $f(x) = x^3 + \dfrac{5}{2}x^2 - 2x - 3$

35. $f(x) = x^4 + 2x^2$

36. $f(x) = 6x^3 - x^4$

37. $f(x) = \dfrac{x^2 + 4}{x}$

38. $f(x) = x + \dfrac{8}{x}$

39. $f(x) = \dfrac{2x}{3 - x}$

40. $f(x) = \dfrac{-4x}{1 + 2x}$

41. $f(x) = \dfrac{\ln x}{x^2}$

42. $f(x) = x - 2\ln|x|$

43. $f(x) = x - \sin x$

44. $f(x) = x - \cos x$

In Exercises 45 and 46, sketch the graph of a single function that has all of the properties listed.

45. a. Continuous everywhere except at $x = -4$, where there is a vertical asymptote

    b. A y-intercept at $y = -2$

    c. x-intercepts at $x = -3$, 1, and 4

    d. $f'(x) < 0$ on $(-\infty, -5)$, $(-4, -1)$, and $(2, \infty)$

e. $f'(x) > 0$ on $(-5, -4)$ and $(-1, 2)$

f. $f''(x) > 0$ on $(-\infty, -4)$ and $(-4, -3)$

g. $f''(x) < 0$ on $(-3, -1)$ and $(-1, \infty)$

h. Differentiable everywhere except at $x = -4$ and $x = -1$

46. a. Continuous and differentiable everywhere except at $x = -3$, where it has a vertical asymptote

b. A horizontal asymptote at $y = 1$

c. An x-intercept at $x = -2$

d. A y-intercept at $y = 4$

e. $f'(x) > 0$ on the intervals $(-\infty, -3)$ and $(-3, 2)$

f. $f'(x) < 0$ on the interval $(2, \infty)$

g. $f''(x) > 0$ on the intervals $(-\infty, -3)$ and $(4, \infty)$

h. $f''(x) < 0$ on the interval $(-3, 4)$

i. $f'(2) = 0$

j. An inflection point at $(4, 3)$

Applications

LIFE SCIENCES

47. *Weightlifting* An abstract for an article states, "We tentatively conclude that Olympic weightlifting ability in trained subjects undergoes a nonlinear decline with age, in which the second derivative of the performance versus age curve repeatedly changes sign."[*]

a. What does this quote tell you about the first derivative of the performance versus age curve?

b. Describe what you know about the performance versus age curve based on the information in the quote.

48. *Scaling Laws* Many biological variables depend on body mass, with a functional relationship of the form

$$Y = Y_0 M^b,$$

where M represents body mass, b is a multiple of $1/4$, and Y_0 is a constant.[†] For example, when Y represents metabolic rate, $b = 3/4$. When Y represents heartbeat, $b = -1/4$. When Y represents life span, $b = 1/4$.

a. Determine which of metabolic rate, heartbeat, and life span are increasing or decreasing functions of mass.

Also determine which have graphs that are concave upward and which have graphs that are concave downward.

b. Verify that all functions of the form given above satisfy the equation

$$\frac{\partial Y}{\partial M} = \frac{b}{M} Y.$$

This means that the rate of change of Y is proportional to Y and inversely proportional to body mass.

49. *Blood Volume* In Chapter 1 we saw that a formula proposed by Hurley[‡] for the red cell volume (RCV) in milliliters for males is

$$RCV = 1,486S^2 - 4,106S + 4,514,$$

where S is the surface area in square meters. A formula given by Pearson, et al.,[§] is

$$RCV = 1,486S - 825.$$

a. For the value of S for which the RCV values given by the two formulas are closest, find the rate of change of RCV with respect to S for both formulas. What does this number represent?

---

[*]Meltzer, David E., "Age Dependence of Olympic Weightlifting," *Medicine and Science in Sports and Exercise*, Vol. 26, No. 8, Aug. 1994, p. 1053.

[†]West, Geoffrey B., James H. Brown, and Brian J. Enquist, "A General Model for the Origin of Allometric Scaling Laws in Biology," *Science*, Vol. 276, April 4, 1997, pp. 122–126.

[‡]Hurley, Peter J., "Red Cell and Plasma Volumes in Normal Adults," *Journal of Nuclear Medicine*, Vol. 16, 1975, pp. 46–52.

[§]Pearson, T. C., et al., "Interpretation of Measured Red Cell Mass and Plasma Volume in Adults," *British Journal of Haematology*, Vol. 89, 1995, pp. 748–756.

b. The formula for plasma volume for males given by Hurley is

$$PV = 995e^{0.6085S},$$

while the formula given by Pearson, et al., is

$$PV = 1,578S,$$

where PV is measured in milliliters and S in square meters. Find the value of S for which the PV values given by the two formulas are the closest. Then find the value of PV that each formula gives for this value of S.

c. For the value of S found in part b, find the rate of change of PV with respect to S for both formulas. What does this number represent?

d. Notice in parts a and c that both formulas give the same instantaneous rate of change at the value of S for which the function values are closest. Prove that if two functions f and g are differentiable and never cross but are closest together when $x = x_0$, then $f'(x_0) = g'(x_0)$.

50. *Cell Surface Receptors* In an article on the local clustering of cell surface receptors, the researcher sought to verify that the function

$$y(S) = S[1 + C(1 + kS)^{f-1}]$$

is equal to 1 for a unique value of S (the concentration of free receptors), where C is the concentration of free ligand in the medium, f is the number of functional groups of cells, and k is a positive constant.*

a. Verify that $y(0) = 0$ and $y(1) > 1$.

b. Show that

$$y'(S) = 1 + C(1 + fkS)(1 + kS)^{f-2}.$$

Notice from this that $y'(S) > 1$.

c. Explain why the results of parts a and b imply that there is a unique value of S where $y(S) = 1$.

d. If $k = 0$, what is the value of S such that $y(S) = 1$?

e. Explain why, if $k > 0$, the value of S such that $y(S) = 1$ must be between 0 and $1/(1 + C)$.

f. The researcher also calculated $f''(S)$. Show that

$$y''(S) = \begin{cases} 2Ck & \text{if } f = 2 \\ (f - 1)kC(2 + fkS)(1 + kS)^{f-3} & \text{if } f \geq 3. \end{cases}$$

g. Show that solving $y(S) = 1$ for C yields the equation

$$C = \frac{1 - S}{S(1 + kS)^{f-1}},$$

which the researcher denotes $C = F(S)$.

h. Show that

$$F'(S) = -\frac{1 + kS[S + f(1 - S)]}{S^2(1 + kS)^f}.$$

i. For $0 < S < 1$, is F increasing or decreasing? Why?

51. *Cell Surface Receptors* In the same article on the local clustering of cell surface receptors (see previous exercise), the researcher sought to maximize the function

$$z = (1 - S)[1 - (1 + kS)^{-(f-1)}],$$

where z is the fraction of receptor sites bound to multiply bound ligand, S is the concentration of free receptors, f is the number of functional groups of cells, and k is a positive constant.[†]

a. To maximize z, the researcher calculated dz/dS. Show that

$$\frac{dz}{dS} = -1 + (1 + kS)^{-f}[1 + kS + k(f - 1)(1 - S)].$$

b. Show that when $f = 2$, $dz/dS = 0$ when

$$S = \left(-1 + \sqrt{1 + k}\right)/k.$$

52. *Cell Traction Force* In Chapter 4 we saw that, in a matrix of cells, the cell traction force per unit of mass is given by

$$T(n) = \frac{an}{1 + bn^2},$$

where n is the number of cells, a is the measure of the traction force generated by a cell, and b measures how force is reduced due to neighboring cells.[‡] Sketch a graph of this function when $a = 20$ and $b = 1$.

53. *Calcium Kinetics* In Chapter 4 we saw that a function used to describe the kinetics of calcium in the cytogel (the gel of cell cytoplasm in the epithelium, an external tissue of epidermal cells) is given by

$$R(c) = \frac{ac^2}{1 + bc^2} - kc,$$

where $R(c)$ measures the release of calcium, c is the amount of free calcium outside the vesicles in which it is stored, and a, b, and k are positive constants.[§]

a. Sketch a graph of this function when $a = 10$, $b = 0.08$, and $k = 7$.

b. Since R cannot be negative, for what values of c is this function valid?

---

*Perelson, Alan S., "Receptor Clustering on a Cell Surface," *Mathematical Biosciences,* Vol. 53, No. 1/2, Feb. 1981, pp. 1–39.

[†]Ibid.

[‡]Murray, J. D., *Mathematical Biology,* Springer-Verlag, 1989, p. 535.

[§]Murray, J. D., *Mathematical Biology,* Springer-Verlag, 1989, p. 569.

OTHER APPLICATIONS

54. *Murder Rate* The number of murders in Chicago (in hundreds) from 1986–1995 is approximated by $f(x) = -0.028x^3 + 0.413x^2 - 1.39x + 8.26$, where x corresponds to the number of years after 1985—that is, 1986 corresponds to $x = 1$, 1987 corresponds to $x = 2$, and so on.*

  a. In what years did a relative maximum or a relative minimum number of murders occur? How many murders occurred in those years?

  b. In what year did the rate of increase in murders start to slow down?

55. *Learning* Researchers used a version of the Gompertz curve to model the rate that children learn with the equation

$$y(t) = A^{c^t},$$

where $y(t)$ is the portion of children of age t years passing a certain mental test, $A = 0.3982 \cdot 10^{-291}$, and $c = 0.4152$.[†] Find the inflection point and describe what it signifies. (*Hint:* Leave A and c as constants until you are ready to calculate your final answer. If A is too small for your calculator to handle, use common logarithms and properties of logarithms to calculate $(\log A)/(\log e)$. Recall that $d(a^{g(t)})/dt = (\ln a)g'(t)a^{g(t)}$.)

56. *Nuclear Weapons* The graph shows the stockpile of nuclear weapons held by the United States and by the Soviet Union and its successor states from 1945 to 1993.[‡]

Warhead Stockpiles—A New Look

a. In what years was the United States stockpile of weapons at a relative maximum?

b. When the United States stockpile of weapons was at the largest relative maximum, is the graph for the Soviet stockpile concave up or concave down? What does this mean?

57. *Velocity and Acceleration* A projectile is shot straight up with an initial velocity of 512 ft/sec. Its height above the ground after t seconds is given by $s(t) = 512t - 16t^2$.

  a. Find the velocity and acceleration after t seconds.

  b. What is the maximum height attained?

  c. When does the projectile hit the ground and with what velocity?

Stock Prices In Exercises 58 and 59, $P(t)$ is the price of a certain stock at time t during a particular day.

58. a. If the price of the stock is falling faster and faster, are $P'(t)$ and $P''(t)$ positive or negative?

  b. Explain your answer.

59. a. When the stock reaches its highest price of the day, are $P'(t)$ and $P''(t)$ positive, zero, or negative?

  b. Explain your answer.

---

*Chicago Tribune, Dec. 31, 1995, Sec. 4, pp. 1 and 3.

†Courtis, S. A., "Maturation Units for the Measurement of Growth," *School and Society,* Vol. 30, 1929, pp. 683–690.

‡*The New York Times,* Sept. 26, 1993.

CHAPTER

6

Applications of the Derivative

When several variables are related by a single equation, their rates of change are also related. For example, the height and horizontal distance of a kite are related to the length of the string holding the kite. In an exercise in Section 4 we differentiate this relationship to discover how fast the kite flier must let out the string to maintain the kite at a constant height and constant horizontal speed.

6.1 Absolute Extrema

6.2 Applications of Extrema

6.3 Implicit Differentiation

6.4 Related Rates

6.5 Differentials: Linear Approximation

Review Exercises

Extended Application: A Total Cost Model for a Training Program

What do aluminum cans, shipments of antibiotics, and a melting icicle have in common? All involve applications of the derivative. The previous chapter included examples in which we used the derivative to find the maximum or minimum value of a function. This problem is ubiquitous; consider the efforts people expend trying to maximize their income, or to minimize their costs or the time required to complete a task. In this chapter we will treat the topic of optimization in greater depth.

The derivative is applicable in far wider circumstances, however. In roughly 500 B.C., Heraclitus said, "Nothing endures but change," and his observation has relevance here. If change is continuous, rather than in sudden jumps, the derivative can be used to describe the rate of change. This explains why calculus has been applied to so many fields.

6.1 ABSOLUTE EXTREMA

THINK ABOUT IT How can the greatest amount of illegal drug use by young adults in a given period be found?

We will answer this question later in this section.

If a function has more than one relative maximum, in a practical situation it is often important to know if one function value is larger than any other. In other cases we may want to know whether one function value is smaller than any other. For example, in Figure 1, $f(x_1) \geq f(x)$ for all x in the domain. There is no function value that is smaller than all others, however, because $f(x) \to -\infty$ as $x \to \infty$ or as $x \to -\infty$.

The largest possible value of a function is called the *absolute maximum* and the smallest possible value of a function is called the *absolute minimum*. As Figure 1 shows, one or both of these may not exist on the domain of the function, $(-\infty, \infty)$ here. Absolute extrema often coincide with relative extrema, as with $f(x_1)$ in Figure 1. Although a function may have several relative maxima or relative minima, it never has more than one *absolute maximum* or *absolute minimum*.

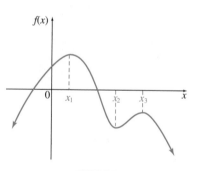

FIGURE 1

ABSOLUTE MAXIMUM OR MINIMUM

Let f be a function defined on some interval. Let c be a number in the interval. Then $f(c)$ is the **absolute maximum** of f on the interval if

$$f(x) \leq f(c)$$

for every x in the interval, and $f(c)$ is the **absolute minimum** of f on the interval if

$$f(x) \geq f(c)$$

for every x in the interval.

A function has an **absolute extremum** (plural: **extrema**) at c if it has either an absolute maximum or an absolute minimum there.

CAUTION Notice that, just like a relative extremum, an absolute extremum is a *y*-value, not an *x*-value.

Now look at Figure 2, which shows three functions defined on closed intervals. In each case there is an absolute maximum value and an absolute minimum value. These absolute extrema may occur at the endpoints or at relative extrema. As the graphs in Figure 2 show, an absolute extremum is either the largest or the smallest function value occurring on a closed interval, while a relative extremum is the largest or smallest function value in some (perhaps small) open interval.

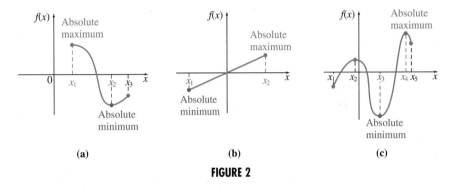

FIGURE 2

Although a function can have only one absolute minimum value and only one absolute maximum value, it can have many points where these values occur. (Note that the absolute maximum value and absolute minimum value are numbers, not points.) As an extreme example, consider the function $f(x) = 2$. The absolute minimum value of this function is clearly 2, as is the absolute maximum value. Both the absolute minimum and the absolute maximum occur at every real number x.

One of the main reasons for the importance of absolute extrema is given by the **extreme value theorem** (which is proved in more advanced courses).

EXTREME VALUE THEOREM

A function f that is continuous on a closed interval $[a, b]$ will have both an absolute maximum and an absolute minimum on the interval.

As Figure 2 shows, an absolute extremum *may* occur on an open interval. See x_2 in part (a) and x_3 in part (c), for example. The extreme value theorem says that a function *must* have both an absolute maximum and an absolute minimum on a closed interval. The conditions that f be *continuous* and on a *closed* interval are very important. For example, in Figure 3, f is discontinuous at $x = b$. Since $f(x)$ gets larger and larger as $x \to b$, there is no absolute (or relative) maximum on (a, b). Also, as $x \to a^+$, the values of $f(x)$ get smaller and smaller approaching $f(a)$, but there is no smallest value of $f(x)$ on the *open* interval (a, b), since there is no endpoint.

The extreme value theorem guarantees the existence of absolute extrema. To find these extrema, use the following steps.

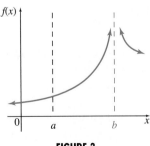

FIGURE 3

FINDING ABSOLUTE EXTREMA

To find absolute extrema for a function f continuous on a closed interval $[a, b]$:

1. Find all critical numbers for f in (a, b).
2. Evaluate f for all critical numbers in (a, b).
3. Evaluate f for the endpoints a and b of the interval $[a, b]$.
4. The largest value found in Step 2 or 3 is the absolute maximum for f on $[a, b]$, and the smallest value found is the absolute minimum for f on $[a, b]$.

EXAMPLE 1 Absolute Extrema

Find the absolute extrema of the function

$$f(x) = x^{8/3} - 16x^{2/3}$$

on the interval $[-1, 8]$.

Solution First look for critical numbers in the interval $(-1, 8)$.

$$f'(x) = \frac{8}{3}x^{5/3} - \frac{32}{3}x^{-1/3}$$

$$= \frac{8}{3}\left[x^{5/3} - \frac{4}{x^{1/3}}\right]$$

$$= \frac{8}{3}\left[x^{5/3} \cdot \frac{x^{1/3}}{x^{1/3}} - \frac{4}{x^{1/3}}\right] \quad \text{Factor and find a common denominator.}$$

$$= \frac{8}{3}\left[\frac{x^2 - 4}{x^{1/3}}\right]$$

Set $f'(x) = 0$ and solve for x. Notice that $f'(x) = 0$ at $x = 2$ and $x = -2$, but -2 is not in the interval $(-1, 8)$, so we ignore it. The derivative is undefined at $x = 0$, but the function is defined there, so 0 is also a critical number.

Evaluate the function at the critical numbers and the endpoints.

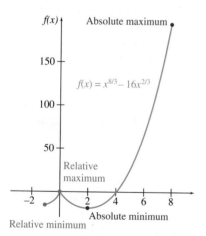

x-value	Value of Function
-1	-15
0	0
2	-19.05
8	192

$f(x) \uparrow$ Absolute maximum

150

$f(x) = x^{8/3} - 16x^{2/3}$

100

50

Relative maximum

-2 2   4   6   8

Absolute minimum

Relative minimum

FIGURE 4

The absolute maximum, 192, occurs when $x = 8$, and the absolute minimum, approximately -19.05, occurs when $x = 2$. A graph of f is shown in Figure 4.

In Example 1, a graphing calculator that gives the maximum and minimum values of a function on an interval, such as the fMax or fMin feature of the TI-83, could replace the table. Alternatively, we could first graph the function on the given interval and then select the feature that gives the maximum or minimum value of the graph of the function instead of completing the table.

EXAMPLE 2 Absolute Extrema

Find the locations of all absolute extrema, if they exist, for the function

$$f(x) = 3x^4 - 4x^3 - 12x^2 + 2.$$

Solution In this example, the extreme value theorem does not apply since the domain is an open interval, $(-\infty, \infty)$, rather than a closed interval. Begin as before by finding any critical numbers.

$$f'(x) = 12x^3 - 12x^2 - 24x = 0$$
$$12x(x^2 - x - 2) = 0$$
$$12x(x + 1)(x - 2) = 0$$
$$x = 0 \quad \text{or} \quad x = -1 \quad \text{or} \quad x = 2$$

There are no values of x where $f'(x)$ does not exist. Evaluate the function at the critical numbers.

x-value	Value of Function
-1	-3
0	2
2	-30

For an open interval, rather than evaluating the function at the endpoints, we evaluate the limit of the function when the endpoints are approached. Because the positive x^4 term dominates the other terms as x becomes large,

$$\lim_{x \to \infty} 3x^4 - 4x^3 - 12x^2 + 2 = \infty.$$

The limit is also ∞ as x approaches $-\infty$. Since the function can be made arbitrarily large, it has no absolute maximum. The absolute minimum, -30, occurs at $x = 2$. This result can be confirmed with a graphing calculator, as shown in Figure 5.

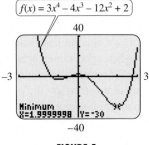

$f(x) = 3x^4 - 4x^3 - 12x^2 + 2$

Minimum
X=1.9999998 Y=-30

FIGURE 5

EXAMPLE 3 Hallucinogen Usage

Based on information provided by the U.S. National Institute on Drug Abuse, the percent y of 18- to 25-year-olds who had used hallucinogens between 1974 and 1991 can be modeled by the function

$$y = 0.025x^3 - 0.70x^2 + 4.43x + 16.77$$

where $x = 0$ corresponds to the year 1974. In what year during this period did this type of drug use reach its absolute maximum? Based on this model, what percent of 18- to 25-year-olds had used hallucinogens?

Solution The function is defined on the interval $[0, 17]$. We look first for critical numbers in this interval. Here $f'(x) = 0.075x^2 - 1.40x + 4.43$. We set this derivative equal to zero and use the quadratic formula to solve for x.

$$0.075x^2 - 1.40x + 4.43 = 0$$
$$x = \frac{1.4 \pm \sqrt{1.4^2 - 4(0.075)(4.43)}}{2(0.075)}$$
$$x = 4.04 \quad \text{or} \quad x = 14.6$$

Both x-values are in the interval $[0, 17]$. Now we evaluate the function at the critical numbers and the endpoints, 0 and 17.

x-value	Value of Function	
0	16.8	
4.04	24.9	←— Absolute maximum
14.6	10.0	
17.0	12.6	

During the fourth year, that is during 1978, such drug use reached its absolute maximum in the given period, with almost 25% of persons in this age group having reported use of hallucinogens.

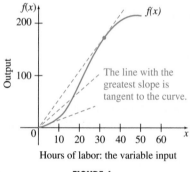

The line with the greatest slope is tangent to the curve.

Hours of labor: the variable input

FIGURE 6

Graphical Optimization   Figure 6, from an economics textbook, shows how the total production output for a product might vary with the hours of labor used.* A manager may want to know how many hours of labor to use in order to maximize the output per hour of labor. For any point on the curve, the y-coordinate measures the output and the x-coordinate measures the hours of labor, so the y-coordinate divided by the x-coordinate gives the output per hour of labor. This quotient is also the slope of the line through the origin and the point on the curve. Therefore, to maximize the output per hour of labor, we need to find where this slope is greatest. As is shown in Figure 6, this occurs when approximately 30 hours of labor are used. Notice that this is also where the line from the origin to the curve is tangent to the curve. Another way of looking at this is to say the point on the curve where the tangent line passes through the origin is the point that maximizes the output per hour of labor.

We can show that, in general, when $y = f(x)$ represents the output as a function of input, the maximum output per unit input occurs when the line from the origin to a point on the graph of the function is tangent to the function. Our goal is to maximize

$$g(x) = \frac{\text{output}}{\text{input}} = \frac{f(x)}{x}.$$

Taking the derivative and setting it equal to 0 gives

$$g'(x) = \frac{xf'(x) - f(x)}{x^2} = 0$$

$$xf'(x) = f(x)$$

$$f'(x) = \frac{f(x)}{x}.$$

Notice that $f'(x)$ gives the slope of the tangent line at the point, and $f(x)/x$ gives the slope of the line from the origin to the point. When these are equal, as in

*Browning, Edgar K., and Jacquelene M. Browning, *Microeconomic Theory and Applications,* 4th ed., New York: HarperCollins, 1992, p. 208.

the function in Figure 6, the output per input is maximized. In other examples, the point on the curve where the tangent line passes through the origin gives a minimum. For a life science example of this, see Exercise 45.

6.1 EXERCISES

Find the locations of any absolute extrema for the functions with graphs as follows.

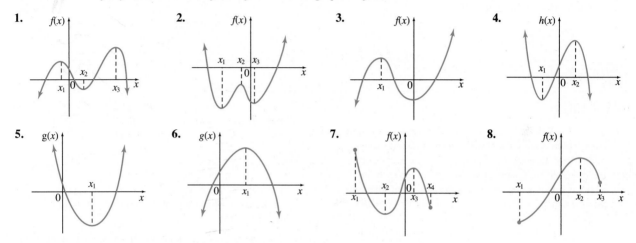

1.   **2.**   **3.**   **4.**

5.   **6.**   **7.**   **8.**

9. What is the difference between a relative extremum and an absolute extremum? Can a relative extremum be an absolute extremum? Is a relative extremum necessarily an absolute extremum?

Find the locations of all absolute extrema for the functions defined as follows, with the specified domains. If you have one, use a graphing calculator to verify your answers.

10. $f(x) = x^3 - 3x^2 - 24x + 5$; $[-3, 6]$

11. $f(x) = x^3 - 6x^2 + 9x - 8$; $[0, 5]$

12. $f(x) = \frac{1}{3}x^3 - \frac{1}{2}x^2 - 6x + 3$; $[-4, 4]$

13. $f(x) = \frac{1}{3}x^3 + \frac{3}{2}x^2 - 4x + 1$; $[-5, 2]$

14. $f(x) = x^4 - 32x^2 - 7$; $[-5, 6]$

15. $f(x) = x^4 - 18x^2 + 1$; $[-4, 4]$

16. $f(x) = \frac{8 + x}{8 - x}$; $[4, 6]$

17. $f(x) = \frac{1 - x}{3 + x}$; $[0, 3]$

18. $f(x) = \frac{x}{x^2 + 2}$; $[0, 4]$

19. $f(x) = \frac{x - 1}{x^2 + 1}$; $[1, 5]$

20. $f(x) = (x^2 + 18)^{2/3}$; $[-3, 3]$

21. $f(x) = (x^2 + 4)^{1/3}$; $[-2, 2]$

22. $f(x) = (x + 1)(x + 2)^2$; $[-4, 0]$

23. $f(x) = (x - 3)(x - 1)^3$; $[-2, 3]$

24. $f(x) = xe^{-x}$; $[0, 10]$

25. $f(x) = x \ln x$; $[1/4, e]$

Graph each function on the indicated domain, and use the capabilities of your calculator to find the location of the absolute extrema.

26. $f(x) = \dfrac{x^3 + 2x + 5}{x^4 + 3x^3 + 10}$; $[-3, 0]$

27. $f(x) = \dfrac{-5x^4 + 2x^3 + 3x^2 + 9}{x^4 - x^3 + x^2 + 7}$; $[-1, 1]$

28. $f(x) = x \sin \pi x$; $[0, 2]$

29. $f(x) = x \cos \dfrac{\pi}{2}x$; $[0, 4]$

In Exercises 30–35, find the locations of all absolute extrema if they exist.

30. $f(x) = 12 - x - 9/x, x > 0$

31. $f(x) = 2x + 8/x^2 + 1, x > 0$

32. $f(x) = x^4 - 4x^3 + 4x^2 + 1$

33. $f(x) = -3x^4 + 8x^3 + 18x^2 + 2$

34. $f(x) = \dfrac{x}{x^2 + 1}$

35. $f(x) = \dfrac{x - 1}{x^2 + 2x + 6}$

36. Find the absolute maximum and minimum of $f(x) = 2x - 3x^{2/3}$ **(a)** on the interval $[-1, 0.5]$; **(b)** on the interval $[0.5, 2]$.

Applications

LIFE SCIENCES

37. *Pollution* A marshy region used for agricultural drainage has become contaminated with selenium. It has been determined that flushing the area with clean water will reduce the selenium for a while, but it will then begin to build up again. A biologist has found that the percent of selenium in the soil x months after the flushing begins is given by

$$f(x) = \frac{x^2 + 36}{2x}, \quad 1 \le x \le 12.$$

When will the selenium be reduced to a minimum? What is the minimum percent?

38. *Salmon Spawning* The number of salmon swimming upstream to spawn is approximated by

$$S(x) = -x^3 + 3x^2 + 360x + 5{,}000, \quad 6 \le x \le 20,$$

where x represents the temperature of the water in degrees Celsius. Find the water temperature that produces the maximum number of salmon swimming upstream.

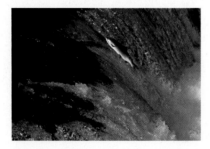

39. *Molars* As we saw in Exercise 23 of Section 3.3, the mesiodistal crown length of deciduous mandibular first molars is related to the postconception age of the tooth as

$$L(t) = -0.01t^2 + 0.788t - 7.048,$$

where $L(t)$ is the crown length, in millimeters, of the molar t weeks after conception.[*] Find the maximum length in mesiodistal crown of mandibular first molars during weeks 22 through 28.

40. *Fungal Growth* Because of the time that many people spend indoors, there is a concern about the health risk of being exposed to harmful fungi that thrive in buildings. The risk appears to increase in damp environments. Researchers have discovered that by controlling both the temperature and the relative humidity in a building, the growth of the fungus *A. versicolor* can be limited. The relationship between temperature and relative humidity, which limits growth, can be described by

$$R(T) = -0.00007T^3 + 0.0401T^2 - 1.6572T + 97.086,$$
$$15 \le T \le 46,$$

where $R(T)$ is the relative humidity (in %) and T is the temperature (in °C).[†] Find the temperature at which the relative humidity is minimized.

41. *Dentin Growth* The growth of dentin in the molars of mice has been studied by researchers in Copenhagen. They determined that the growth curve that best fits dentinal formation for the second molar is

$$M(t) = -0.5035295 + 0.0229883t + 0.0108021t^2$$
$$- 0.0003139t^3 + 0.0000025t^4, \quad 6 \le t \le 48,$$

*Harris, Edward F., Joseph D. Hicks, and Betsy D. Barcroft, "Tissue Contributions to Sex and Race: Differences in Tooth Crown Size of Deciduous Molars," *American Journal of Physical Anthropology,* Vol. 115, 2001, pp. 223–237.
†Rowan, N., C. Johnstone, R. McLean, J. Anderson, and J. Clarke, "Prediction of Toxigenic Fungal Growth in Buildings by Using a Novel Modelling System," *Applied and Environmental Microbiology,* Vol. 65, No. 11, Nov. 1999, pp. 4814–4821.

where t is the age of the mouse (in days), and $M(t)$ is the cumulative dentin volume (in 10^{-1} mm^3).*

a. Use a graphing calculator to sketch the graph of this function on $[6, 51]$ by $[0, 4.5]$.

b. Find the time at which the dentin formation is growing most rapidly. (*Hint:* Find the maximum value of the derivative of this function.)

42. *Dentin Growth* The researchers mentioned in the previous exercise also determined that the growth curve that best fits dentinal formation for the third molar is

$$M(t) = -0.3347806 + 0.0236529t + 0.0012409t^2$$
$$- 0.0000207t^3, \quad 9 \leq t \leq 51,$$

where t is the age of the mouse (in days), and $M(t)$ is the cumulative dentin volume (in 10^{-1} mm^3).[†]

a. Use a graphing calculator to sketch the graph of this function on $[9, 51]$ by $[0, 1.5]$.

b. Find the time at which the dentin formation is growing most rapidly. (*Hint:* Find the maximum value of the derivative of this function.)

c. Discuss whether this function should be used for mice that are older than 48 days.

43. *Human Skin Surface* The surface of the skin is made up of a network of intersecting lines which form polygons. Researchers have discovered a functional relationship between the age of a male and the number of polygons per area of skin according to

$$P(t) = 241.75 - 10.372t + 0.31247t^2 - 0.0044474t^3$$
$$+ 0.000022195t^4, \quad 0 \leq t \leq 95,$$

where t is the age of the person (in years) and $P(t)$ is the number of polygons for a particular surface area of skin.[‡]

a. Use a graphing calculator to sketch the graph of $P(t)$ on $[0, 95]$ by $[0, 300]$.

b. Find the maximum and minimum number of polygons per area predicted by the model.

c. Discuss the accuracy of this model for older people.

44. *Satisfaction* Suppose some substance (such as a preferred food) gives satisfaction to an individual, but the substance requires effort to obtain, so that after a while the individual is no longer interested in expending more effort to obtain the substance. A mathematical model of this situation is given by

$$S = a \ln kx - bx,$$

where S is the amount of satisfaction, x is the amount of the substance, and a, b, and k are constants.[§] Find the amount of the substance that gives the maximum amount of satisfaction.

45. *Flight Speed* The graph on the next page shows the relationship between the speed of the Arctic tern in flight and the required power expended by its flight muscles.[||] Several significant flight speeds are indicated on the curve.

a. The speed V_{mp} minimizes energy costs per unit of time. What is the slope of the line tangent to the curve at the point corresponding to V_{mp}? What is the physical significance of the slope at that point?

b. The speed V_{mr} minimizes the energy costs per unit of distance covered. Estimate the slope of the curve at the point corresponding to V_{mr}. Give the significance of the slope at that point.

c. The speed V_{opt} minimizes the total duration of the migratory journey. Estimate the slope of the curve at the point corresponding to V_{opt}. Relate the significance of this slope to the slopes found in parts a and b.

d. By looking at the shape of the curve, describe how the power level decreases and increases for various speeds.

---

*Matsumoto, B., K. Nonaka, and M. Nakata, "A Genetic Study of Dentin Growth in the Mandibular Second and Third Molars of Male Mice," *Journal of Craniofacial Genetics and Developmental Biology,* Vol. 16, No. 3, Jul.–Sept. 1996, pp. 137–147.

[†]Ibid.

[‡]Voros, E., C. Robert, and A. Robert, "Age-Related Changes of the Human Skin Surface Microrelief," *Gerontology,* Vol. 36, 1990, pp. 276–285.

[§]Rashevsky, Nicolas, *Mathematical Biology of Social Behavior,* rev. ed., Chicago, The University of Chicago Press, 1959, p. 40.

[||]Alerstam, Thomas, "Bird Flight and Optimal Migration," *Trends in Ecology and Evolution,* July 1991. Copyright © 1991 by Elsevier Trends Journals. Reprinted by permission of Elsevier Trends Journals and Thomas Alerstam.

e. Notice that the slope of the lines found in parts a–c represents the power divided by speed. Power is energy per unit time, and speed is distance per unit time, so the slope represents energy per unit distance. If a line is drawn from the origin to a point on the graph, at which point is the slope of the line (representing energy per unit distance) smallest? How does this compare with your answers to parts a–c?

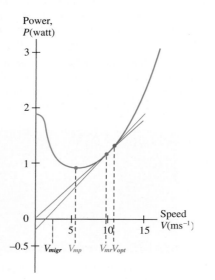

OTHER APPLICATIONS

46. *Gasoline Mileage* From information given in a recent business publication, we constructed the mathematical model

$$M(x) = -\frac{1}{45}x^2 + 2x - 20, \quad 30 \le x \le 65,$$

to represent the miles per gallon used by a certain car at a speed of x mph. Find the absolute maximum miles per gallon and the absolute minimum.

47. *Gasoline Mileage* For a certain compact car,

$$M(x) = -0.018x^2 + 1.24x + 6.2, \quad 30 \le x \le 60,$$

represents the miles per gallon obtained at a speed of x mph. Find the absolute maximum miles per gallon and the absolute minimum.

Area A piece of wire 12 ft long is cut into two pieces. (See the figure.) One piece is made into a circle and the other piece is made into a square. Let the piece of length x be formed into a circle. We allow x to equal 0 or 12, so all the wire can be used for the square or for the circle.

$$\text{Radius of circle} = \frac{x}{2\pi} \qquad \text{Area of circle} = \pi\left(\frac{x}{2\pi}\right)^2$$

$$\text{Side of square} = \frac{12 - x}{4} \qquad \text{Area of square} = \left(\frac{12 - x}{4}\right)^2$$

48. Where should the cut be made in order to minimize the sum of the areas enclosed by both figures?

49. Where should the cut be made in order to make the sum of the areas maximum? (*Hint:* Remember to use the end-points of a domain when looking for absolute maxima and minima.)

50. *Information Content* Suppose dots and dashes are transmitted over a telegraph line so that dots occur a fraction p of the time (where $0 \le p \le 1$) and dashes occur a fraction $1 - p$ of the time. The **information content** of the telegraph line is given by $I(p)$, where

$$I(p) = -p \ln p - (1 - p) \ln(1 - p).$$

a. Show that $I'(p) = -\ln p + \ln(1 - p)$.

b. Set $I'(p) = 0$ and find the value of p that maximizes the information content.

c. How might the result in part b be used?

Cost In Exercises 51 and 52, each graph gives the cost as a function of production level. Use the method of graphical optimization to estimate the production level that results in the minimum cost per item produced.

51.

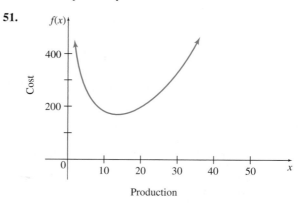

52.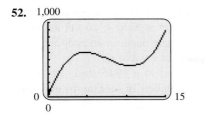

Profit In Exercises 53 and 54, each graph gives the profit as a function of production level. Use graphical optimization to estimate the production level that gives the maximum profit per item produced.

53.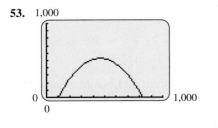

54.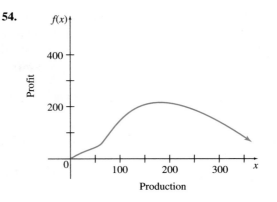

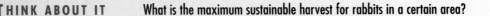

■ 6.2 APPLICATIONS OF EXTREMA

THINK ABOUT IT   What is the maximum sustainable harvest for rabbits in a certain area?

In Example 5 we will use the techniques of calculus to find an answer to this question.

In this section, we give several examples showing applications of calculus to maximum and minimum problems. To solve these examples, go through the following steps.

SOLVING APPLIED EXTREMA PROBLEMS

1. Read the problem carefully. Make sure you understand what is given and what is unknown.

2. If possible, sketch a diagram. Label the various parts.

3. Decide on the variable that must be maximized or minimized. Express that variable as a function of *one* other variable. Be sure to find the domain of the function.

4. Find the critical points for the function from Step 3.

5. If the domain is a closed interval, evaluate the function at the endpoints and the critical points to see which yields the absolute maximum or minimum. If the domain is an open interval, find the limit as the endpoints are approached, as in Example 2 of the previous section, to determine if an absolute maximum or minimum exists at one of the critical points.

> **CAUTION** Do not skip Step 5 in the preceding box. If a problem asks you to maximize a quantity and you find a critical point at Step 4, do not automatically assume the maximum occurs there, for it may occur at an endpoint, as in Exercise 49 of the previous section, or it may not exist at all.
>
> An infamous case of such an error occurred in a 1945 study of "flying wing" aircraft designs similar to the Stealth bomber. In seeking to maximize the range of the aircraft (how far it can fly on a tank of fuel), the study's authors found that a critical point occurred when almost all of the volume of the plane was in the wing. They claimed that this critical point was a maximum. But another engineer later found that this critical point, in fact, *minimized* the range of the aircraft!*

EXAMPLE 1 Maximization

Find two nonnegative numbers x and y for which $2x + y = 30$, such that xy^2 is maximized.

Solution First we must decide what is to be maximized and assign a variable to that quantity. Here, xy^2 is to be maximized, so let

$$M = xy^2.$$

Now, express M in terms of just *one* variable. Use the equation $2x + y = 30$ to do that. Solve $2x + y = 30$ for either x or y. Solving for y gives

$$2x + y = 30$$
$$y = 30 - 2x.$$

Substitute for y in the expression for M to get

$$M = x(30 - 2x)^2$$
$$= x(900 - 120x + 4x^2)$$
$$= 900x - 120x^2 + 4x^3.$$

Note that x must be at least 0. Since y must also be at least 0, $30 - 2x \geq 0$, so $x \leq 15$. Thus x is confined to the interval $[0, 15]$.

Find the critical points for M by finding dM/dx, then solving the equation $dM/dx = 0$ for x.

$$\frac{dM}{dx} = 900 - 240x + 12x^2 = 0$$
$$12(75 - 20x + x^2) = 0$$
$$(5 - x)(15 - x) = 0$$
$$x = 5 \quad \text{or} \quad x = 15$$

Find M for the critical numbers $x = 5$ and $x = 15$ as well as for $x = 0$, one endpoint of the domain. (The other endpoint has already been included as a critical number.)

---

*Biddle, Wayne, "Skeleton Alleged in the Stealth Bomber's Closet," *Science,* Vol. 244, May 12, 1989.

x	M	
0	0	
5	2,000	← Maximum
15	0	

We see in the table that the maximum value of the function occurs when $x = 5$. Since $y = 30 - 2(5) = 20$, the values that maximize xy^2 are $x = 5$ and $y = 20$.

EXAMPLE 2   Minimizing Time

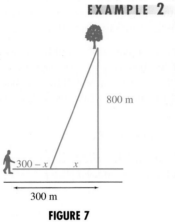

800 m

300 − x   **x**

300 m

FIGURE 7

A math professor participating in the sport of orienteering must get to a specific tree in the woods as fast as possible. He can get there by traveling east along the trail for 300 meters, and then north through the woods for 800 meters. He can run 160 meters per minute along the trail, but only 70 meters per minute through the woods. Running directly through the woods toward the tree minimizes the distance, but he will be going slowly the whole time. He could instead run 300 meters along the trail before entering the woods, maximizing the total distance but minimizing the time in the woods. Perhaps the fastest route is a combination, as shown in Figure 7. Find the path that will get him to the tree in the minimum time.

Solution Let x be the distance shown in Figure 7, so the distance he runs on the trail is $300 - x$. By the Pythagorean theorem, the distance he runs through the woods is $\sqrt{800^2 + x^2}$. The total time is the sum of the time on the trail and the time through the woods. Since time = distance/speed, the total time is

$$T(x) = \frac{\sqrt{800^2 + x^2}}{70} + \frac{300 - x}{160},$$

where $0 \le x \le 300$. Set the derivative equal to 0. Since $\sqrt{800^2 + x^2} = (800^2 + x^2)^{1/2}$,

$$T'(x) = \frac{1}{70}\left(\frac{1}{2}\right)(800^2 + x^2)^{-1/2}(2x) - \frac{1}{160} = 0.$$

$$\frac{x}{70\sqrt{800^2 + x^2}} = \frac{1}{160}$$

$$16x = 7\sqrt{800^2 + x^2} \qquad \text{Cross multiply.}$$

$$256x^2 = 49(800^2 + x^2) = (49 \cdot 800^2) + 49x^2 \qquad \text{Square both sides.}$$

$$207x^2 = 49 \cdot 800^2$$

$$x^2 = \frac{49 \cdot 800^2}{207}$$

$$x = \frac{7 \cdot 800}{\sqrt{207}} \approx 389$$

389 is not in the interval $(0, 300)$, so the minimum time must occur at one of the endpoints.

x	$T(x)$	
0	13.93	
300	12.83	← Minimum

From the table we see that the time is minimized when $x = 300$, that is, when the professor heads straight for the tree.

EXAMPLE 3 Maximizing Volume

An open box is to be made by cutting a square from each corner of a 12-inch by 12-inch piece of metal and then folding up the sides. What size square should be cut from each corner in order to produce a box of maximum volume?

Solution Let x represent the length of a side of the square that is cut from each corner, as shown in Figure 8(a). The width of the box is $12 - 2x$, with the length also $12 - 2x$. As shown in Figure 8(b), the depth of the box will be x inches. The volume of the box is given by the product of the length, width, and height. In this example, the volume, $V(x)$, depends on x:

$$V(x) = x(12 - 2x)(12 - 2x) = 144x - 48x^2 + 4x^3.$$

Clearly, $0 \leq x$, and since neither the length nor the width can be negative, $0 \leq 12 - 2x$, so $x \leq 6$. Thus, the domain of V is the interval $[0, 6]$.

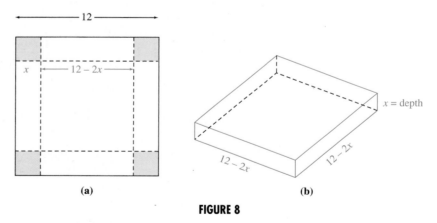

(a)   **(b)**

FIGURE 8

The derivative is $V'(x) = 144 - 96x + 12x^2$. Set this derivative equal to 0.

$$12x^2 - 96x + 144 = 0$$
$$12(x^2 - 8x + 12) = 0$$
$$12(x - 2)(x - 6) = 0$$
$$x - 2 = 0 \quad \text{or} \quad x - 6 = 0$$
$$x = 2 \qquad\qquad x = 6$$

x	$V(x)$	
0	0	
2	128	← Maximum
6	0	

Find $V(x)$ for x equal to 0, 2, and 6 to find the depth that will maximize the volume. The table indicates that the box will have maximum volume when $x = 2$ and that the maximum volume will be 128 cubic inches.

EXAMPLE 4 Minimizing Area

A company wants to manufacture cylindrical aluminum cans with a volume of 1,000 cubic centimeters (one liter). What should the radius and height of the can be to minimize the amount of aluminum used?

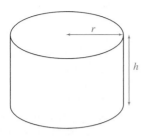

FIGURE 9

Solution The two variables in this problem are the radius and the height of the can, which we shall label r and h, as in Figure 9. Minimizing the amount of aluminum used requires minimizing the surface area of the can, which we will designate S. The surface area consists of a top and a bottom, each of which is a circle with an area πr^2, plus the side. If the side was sliced vertically and unrolled, it would form a rectangle with height h and width equal to the circumference of the can, which is $2\pi r$. Thus the surface area is given by

$$S = 2\pi r^2 + 2\pi rh.$$

The right side of the equation involves two variables. We need to get a function of a single variable. We can do this by using the information about the volume of the can:

$$V = \pi r^2 h = 1,000.$$

(Here we have used the formula for the volume of a cylinder.) Solve this for h:

$$h = \frac{1,000}{\pi r^2}.$$

(Solving for r would have involved a square root and a more complicated function.)

We now substitute this expression for h into the equation for S to get

$$S = 2\pi r^2 + 2\pi r \frac{1,000}{\pi r^2} = 2\pi r^2 + \frac{2,000}{r}.$$

There are no restrictions on r other than that it be a positive number, so the domain of S is $(0, \infty)$.

Find the critical numbers for S by finding dS/dr, then solving the equation $dS/dr = 0$ for r.

$$\frac{dS}{dr} = 4\pi r - \frac{2,000}{r^2} = 0$$

$$4\pi r^3 = 2,000$$

$$r^3 = \frac{500}{\pi}$$

Take the cube root of both sides to get

$$r = \left(\frac{500}{\pi}\right)^{1/3} \approx 5.419$$

centimeters. Substitute this expression into the equation for h to get

$$h = \frac{1,000}{\pi(5.419)^2} \approx 10.84$$

centimeters. Notice that the height of the can is twice its radius. But we are not yet assured that the critical number produces the minimum surface area. Since the domain is an open interval, find the limit as the endpoints are approached.

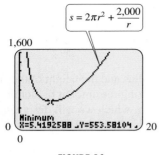

FIGURE 10

$$\lim_{r \to 0} S = \lim_{r \to \infty} S = \infty$$

The surface area becomes arbitrarily large as r approaches the endpoints of the domain, so the critical number must give the absolute minimum surface area. The graphing calculator screen in Figure 10 confirms this result.

Notice that if the previous example had asked for the height and radius that maximize the amount of aluminum used, the problem would have no answer. There is no maximum for a function that can be made arbitrarily large.

In Example 4, we could use the first derivative test to observe that since $dS/dr < 0$ when $0 < r < 5.419$, and $dS/dr > 0$ when $r > 5.419$, there is a relative minimum at $r = 5.419$. The second derivative test also leads to this conclusion because $d^2S/dr^2 = 4\pi + 4{,}000/r^3 > 0$ when $r > 0$. Because there is only one relative extremum, the relative minimum must be the absolute minimum in this case. Be careful when using the first or second derivative test, however—in general, a relative extremum is not necessarily an absolute extremum.

For most living things, reproduction is *seasonal*—it can take place only at selected times of the year.* Large whales, for example, reproduce every two years during a relatively short time span of about two months. Shown on the time axis in Figure 11 are the reproductive periods. Let S = number of adults present during the reproductive period and let R = number of adults that return the next season to reproduce.

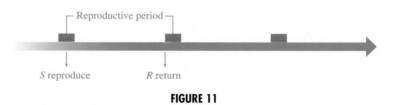

FIGURE 11

If we find a relationship between R and S, $R = f(S)$, then we have formed a **spawner-recruit** function or **parent-progeny** function. These functions are notoriously hard to develop because of the difficulty of obtaining accurate counts and because of the many hypotheses that can be made about the life stages. We will simply suppose that the function f takes various forms.

If $R > S$, we can presumably harvest

$$H = R - S = f(S) - S$$

individuals, leaving S to reproduce. Next season, $R = f(S)$ will return and the harvesting process can be repeated, as shown in Figure 12 on the next page.

Let S_0 be the number of spawners that will allow as large a harvest as possible without threatening the population with extinction. Then $H(S_0)$ is called the **maximum sustainable harvest.**

---

*From Cullen, Michael R., *Mathematics for the Biosciences.* Copyright © 1983 PWS Publishers. Reprinted by permission.

FIGURE 12

EXAMPLE 5 Maximum Sustainable Harvest

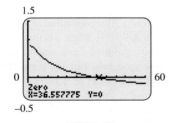

Suppose the spawner-recruit function for Idaho rabbits is $f(S) = 2.17\sqrt{S}\ln(S + 1)$, where S is measured in thousands of rabbits. Find S_0 and the maximum sustainable harvest, $H(S_0)$.

Solution S_0 is the value of S that maximizes H. Since

$$H(S) = f(S) - S$$
$$= 2.17\sqrt{S}\ln(S + 1) - S,$$
$$H'(S) = 2.17\left(\frac{\ln(S + 1)}{2\sqrt{S}} + \frac{\sqrt{S}}{S + 1}\right) - 1.$$

Now we want to set this derivative equal to 0 and solve for S.

$$0 = 2.17\left(\frac{\ln(S + 1)}{2\sqrt{S}} + \frac{\sqrt{S}}{S + 1}\right) - 1.$$

This equation is very difficult to solve analytically, so we will graph $H'(S)$ with a graphing calculator and find any S-values where $H'(S)$ is 0. (An alternative approach is to use the equation solver some graphing calculators have.) The graph with the value where $H'(S)$ is 0 is shown in Figure 13.

From the graph we see that $H'(S) = 0$ when $S = 36.557775$, so the number of rabbits needed to sustain the population is about 36,600. A graph of H will show that this is a maximum. From the graph, using the capability of the calculator, we find that the harvest is $H(36.557775) \approx 11.015504$. These results indicate that after one reproductive season, a population of 36,600 rabbits will have increased to 47,600. Of these, 11,000 may be harvested, leaving 36,600 to regenerate the population. Any harvest larger than 11,000 will threaten the future of the rabbit population, while a harvest smaller than 11,000 will allow the population to grow larger each season. Thus 11,000 is the maximum sustainable harvest for this population.

1.5

0   60

Zero
X=36.557775 Y=0

–0.5

FIGURE 13

6.2 EXERCISES

In Exercises 1–4, use the steps shown in Exercise 1 to find nonnegative numbers x and y that satisfy the given requirements. Give the optimum value of the indicated expression.

1. $x + y = 100$ and the product $P = xy$ is as large as possible.

  a. Solve $x + y = 100$ for y.

b. Substitute the result from part a into $P = xy$, the equation for the variable that is to be maximized.

c. Find the domain of the function P found in part b.

d. Find dP/dx. Solve the equation $dP/dx = 0$.

e. Evaluate P at any solutions found in part d, as well as the endpoints of the domain found in part c.

f. Give the maximum value of P, as well as the two numbers x and y whose product is that value.

2. The sum of x and y is 200 and the sum of the squares of x and y is minimized.

3. $x + y = 150$ and x^2y is maximized.

4. $x + y = 45$ and xy^2 is maximized.

Applications

LIFE SCIENCES

5. *Disease* Epidemiologists have found a new communicable disease running rampant in College Station, Texas. They estimate that t days after the disease is first observed in the community, the percent of the population infected by the disease is approximated by

$$p(t) = \frac{20t^3 - t^4}{1,000}$$

for $0 \le t \le 20$.

a. After how many days is the percent of the population infected a maximum?

b. What is the maximum percent of the population infected?

6. *Pollution* A lake polluted by bacteria is treated with an antibacterial chemical. After t days, the number N of bacteria per ml of water is approximated by

$$N(t) = 20\left(\frac{t}{12} - \ln\left(\frac{t}{12}\right)\right) + 30$$

for $1 \le t \le 15$.

a. When during this time will the number of bacteria be a minimum?

b. What is the minimum number of bacteria during this time?

c. When during this time will the number of bacteria be a maximum?

d. What is the maximum number of bacteria during this time?

7. *Disease* Another disease hits the chronically ill town of College Station, Texas. This time the percent of the popu-

lation infected by the disease t days after it hits town is approximated by $p(t) = 10te^{-t/8}$ for $0 \le t \le 40$.

a. After how many days is the percent of the population infected a maximum?

b. What is the maximum percent of the population infected?

Maximum Sustainable Harvest Find the maximum sustainable harvest in Exercises 8 and 9. See Example 5.

8. $f(S) = 12S^{0.25}$ **9.** $f(S) = \dfrac{25S}{S + 2}$

10. *Maximum Sustainable Harvest* The population of salmon next year is given by $f(S) = Se^{r(1 - S/P)}$, where S is this year's salmon population, P is the equilibrium population, and r is a constant that depends upon how fast the population grows.* The number of salmon that can be fished next year while keeping the population the same is $H(S) = f(S) - S$. The maximum value of $H(S)$ is the maximum sustainable harvest.

a. Show that the maximum sustainable harvest occurs when $f'(S) = 1$. (*Hint:* To maximize, set $H'(S) = 0$.)

b. Let the value of S found in part a be denoted by S_0. Show that the maximum sustainable harvest is given by

$$S_0\left(\frac{1}{1 - rS_0/P} - 1\right).$$

(*Hint:* Set $f'(S_0) = 1$ and solve for $e^{r(1 - S_0/P)}$. Then find $H(S_0)$ and substitute the expression for $e^{r(1 - S_0/P)}$.)

Maximum Sustainable Harvest In Exercises 11 and 12, refer to Exercise 10. Find $f'(S_0)$ and solve the equation $f'(S_0) = 1$, *using a calculator to find the intersection of the graphs of* $f'(S_0)$ *and* $y = 1$.

11. Find the maximum sustainable harvest if $r = 0.1$ and $P = 100$.

12. Find the maximum sustainable harvest if $r = 0.4$ and $P = 500$.

13. *Pigeon Flight* Homing pigeons avoid flying over large bodies of water, preferring to fly around them instead. (One possible explanation is the fact that extra energy is required to fly over water because air pressure drops over water in the daytime.) Assume that a pigeon released from a boat 1 mi from the shore of a lake (point B in the figure) flies first to point P on the shore and then along the straight edge of the lake to reach its home at L. If L is 2 mi from point A, the point on the shore closest to the boat, and if a pigeon needs $4/3$ as much energy per mile to fly over water as over land, find the location of point P, which minimizes energy used.

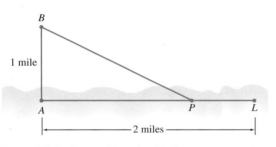

14. *Pigeon Flight* Repeat Exercise 13, but assume a pigeon needs $10/9$ as much energy to fly over water as over land.

15. *Biomechanics* Researchers have determined that the distance that a shot-putter can throw a shot depends on the height, angle, initial velocity, and release angle of the shot. For some shot-putters, the researchers determined that the distance a shot will travel can be estimated by

$$d(\theta) = 5.1 \sin(2\theta)\left[1 + \sqrt{1 + \frac{0.41}{\sin^2 \theta}}\right],$$

where θ is the release angle (in degrees) and $d(\theta)$ is the distance the shot travels (in m).[*] Determine the release angle that maximizes the distance traveled by the shot and the corresponding maximum distance.

16. *Harvesting Cod* A recent article described the population $f(S)$ of cod in the North Sea next year as a function of this year's population S (measured in thousands of tons) by various mathematical models.[†]

$$\text{Shepherd:} \quad f(S) = \frac{aS}{1 + (S/b)^c};$$
$$\text{Ricker:} \quad f(S) = aSe^{-bS};$$
$$\text{Beverton-Holt:} \quad f(S) = \frac{aS}{1 + (S/b)},$$

where a, b, and c are constants. The Shepherd model is the same as the model described in Exercise 51 of Section 1.5, $f(x) = \lambda x/[1 + (ax)^b]$, if we replace S with x, a with λ, b with $1/a$, and c with b.

a. Find a replacement of variables in the Ricker model above that will make it the same as the Ricker model described in Exercise 10 of this section, $f(S) = Se^{r(1-S/P)}$.

b. Find $f'(S)$ for all three models.

c. Find $f'(0)$ for all three models. From your answer, describe in words the geometric meaning of the constant a.

d. The values of a, b, and c reported in the article for the Shepherd model are 3.026, 248.72, and 3.24, respectively. Find the value of this year's population that maximizes next year's population using the Shepherd model.

e. The values of a and b reported in the article for the Ricker model are 4.151 and 0.0039, respectively. Find the value of this year's population that maximizes next year's population using the Ricker model.

f. Explain why, for the Beverton-Holt model, there is no value of this year's population that maximizes next year's population.

g. In Exercise 10, we defined the harvest as $H(S) = f(S) - S$, and showed that the maximum sustainable harvest occurred when $f'(S) = 1$. Find the value of S for the Shepherd model that gives the maximum sustainable harvest, using the values of a, b, and c given in part d. (*Hint:* Use a graphing calculator to find the intersection of the graphs of $y = f'(S)$ and $y = 1$.)

17. *Bird Migration* Suppose a migrating bird flies at a velocity v, and suppose the amount of time the bird can fly depends on its velocity according to the function $T(v)$.[‡]

a. If E is the bird's initial energy, then the bird's effective power is given by kE/T, where k is the fraction of the

*Linthorne, N., "Optimum Release Angle in the Shot Put," *Journal of Sports Sciences,* Vol. 19, 2001, pp. 359–372.

†Cook, R. M., A. Sinclair, and G. Stefánsson, "Potential Collapse of North Sea Cod Stocks," *Nature,* Vol. 385, Feb. 6, 1997, pp. 521–522.

‡This exercise is based on an example in *A Concrete Approach to Mathematical Modelling* by Michael Mesterton-Gibbons, Wiley-Interscience, 1995, pp. 93–96.

power that can be converted into mechanical energy. According to principles of aerodynamics,

$$\frac{kE}{T} = aSv^3 + I,$$

where a is a constant, S is the wind speed, and I is the induced power, or rate of working against gravity. Using this result and the fact that distance is velocity multiplied by time, show that the distance that the bird can fly is given by

$$D(v) = \frac{kEv}{aSv^3 + I}.$$

b. Show that the migrating bird can fly a maximum distance by flying at a velocity

$$v = \left(\frac{I}{2aS}\right)^{1/3}.$$

OTHER APPLICATIONS

18. *Postal Regulations* The U.S. Postal Service stipulates that any boxes sent through the mail must have a length plus girth totaling no more than 108 inches. (See the figure.) Find the dimensions of the box with maximum volume that can be sent through the U.S. mail, assuming that the width and the height of the box are equal.

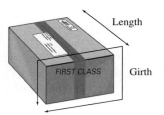

19. *Travel Time* A hunter is at a point on a river bank. He wants to get to his cabin, located 3 mi north and 8 mi west. (See the figure.) He can travel 5 mph on the river but only 2 mph on this very rocky land. How far upriver should he go in order to reach the cabin in minimum time?

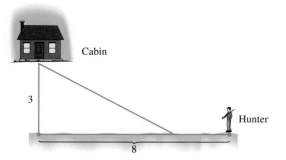

20. *Travel Time* Repeat Exercise 19, but assume the cabin is 19 mi north and 8 mi west.

⬛ *Average Cost* In Exercises 21 and 22, determine the average cost function. Use a graphing calculator to find where the

average cost is smallest by taking the derivative, then finding where the derivative is 0. Check your work by finding the minimum from the graph of the function.

21. $C(x) = \frac{1}{2}x^3 + 2x^2 - 3x + 35$

22. $C(x) = 10 + 20x^{1/2} + 16x^{3/2}$

23. *Area* A farmer has 1,200 m of fencing. He wants to enclose a rectangular field bordering a river, with no fencing needed along the river. (See the sketch.) Let x represent the width of the field.

a. Write an expression for the length of the field.

b. Find the area of the field (area = length × width).

c. Find the value of x leading to the maximum area.

d. Find the maximum area.

24. *Area* Find the dimensions of the rectangular field of maximum area that can be made from 200 m of fencing material. (This fence has four sides.)

25. *Area* An ecologist is conducting a research project on breeding pheasants in captivity. She first must construct suitable pens. She wants a rectangular area with two additional fences across its width, as shown in the sketch. Find the maximum area she can enclose with 3,600 m of fencing.

26. *Cost with Fixed Area* A fence must be built to enclose a rectangular area of 20,000 ft². Fencing material costs $3 per foot for the two sides facing north and south, and $6 per foot for the other two sides. Find the cost of the least expensive fence.

27. *Cost with Fixed Area* A fence must be built in a large field to enclose a rectangular area of 15,625 m². One side of the area is bounded by an existing fence; no fence is needed there. Material for the fence costs $2 per meter for the two ends, and $4 per meter for the side opposite the existing fence. Find the cost of the least expensive fence.

28. *Timing Income* A local group of scouts has been collecting aluminum cans for recycling. The group has already collected 12,000 lb of cans, for which they could currently receive $4 per hundred pounds. The group can continue to collect cans at the rate of 400 lb per day. However, a glut in the aluminum market has caused the recycling company to announce that it will lower its price, starting immediately, by $0.10 per hundred pounds per day. The scouts can make only one trip to the recycling center. Find the best time for the trip. What total income will be received?

29. *Packaging Design* A television manufacturing firm needs to design an open-topped box with a square base. The box must hold 32 in³. Find the dimensions of the box that can be built with the minimum amount of materials. (See the figure.)

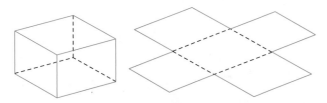

30. *Revenue* A local club is arranging a charter flight to Hawaii. The cost of the trip is $425 each for 75 passengers, with a refund of $5 per passenger for each passenger in excess of 75.

  a. Find the number of passengers that will maximize the revenue received from the flight.

  b. Find the maximum revenue.

31. *Packaging Design* A company wishes to manufacture a box with a volume of 36 ft³ that is open on top and is twice as long as it is wide. Find the dimensions of the box produced from the minimum amount of material.

32. *Packaging Cost* A closed box with a square base is to have a volume of 16,000 cm³. The material for the top and bottom of the box costs $3 per square centimeter, while the material for the sides costs $1.50 per square centimeter. Find the dimensions of the box that will lead to minimum total cost. What is the minimum total cost?

33. *Packaging Design* A cylindrical box will be tied up with ribbon as shown in the figure. The longest piece of ribbon available is 130 cm long, and 10 cm of that are required for the bow. Find the radius and height of the box with the largest possible volume.

34. *Can Design* **a.** For the can problem in Example 4, the minimum surface area required that the height be twice

the radius. Show that this is true for a can of arbitrary volume V.

  b. Do many cans in grocery stores have a height that is twice the radius? If not, discuss why this may be so.

35. *Container Design* Your company needs to design cylindrical metal containers with a volume of 16 ft³. The top and bottom will be made of a sturdy material that costs $2 per square foot, while the material for the sides costs $1 per square foot. Find the radius, height, and cost of the least expensive container.

36. *Container Design* An open box will be made by cutting a square from each corner of a 3-ft by 8-ft piece of cardboard and then folding up the sides. What size square should be cut from each corner in order to produce a box of maximum volume?

37. *Container Design* Consider the problem of cutting corners out of a rectangle and folding up the sides to make a box. Specific examples of this problem are discussed in Example 3 and Exercise 36.

  a. In the solution to Example 3, compare the area of the base of the box with the area of the walls.

  b. Repeat part a for the solution to Exercise 36.

  c. Make a conjecture about the area of the base compared with the area of the walls for the box with the maximum volume.

38. *Use of Materials* A mathematics book is to contain 36 in² of printed matter per page, with margins of 1 in. along the sides and 1½ in. along the top and bottom. Find the dimensions of the page that will require the minimum amount of paper. (See the figure.)

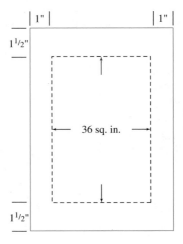

39. *Cost* A company wishes to run a utility cable from point A on the shore (see the figure on the next page) to an installation at point B on the island. The island is 6 mi from the shore. It costs $400 per mile to run the cable on land and $500 per mile underwater. Assume that the cable starts at A

and runs along the shoreline, then angles and runs underwater to the island. Find the point at which the line should begin to angle in order to yield the minimum total cost.

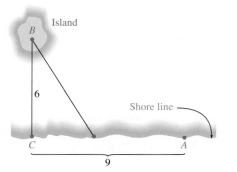

40. *Cost* Repeat Exercise 39, but make point *A* 7 mi from point *C*.

41. *Pricing* Decide what you would do if your assistant presented the following contract for your signature:

Your firm offers to deliver 300 tables to a dealer, at $90 per table, and to reduce the price per table on the entire order by 25¢ for each additional table over 300.

Find the dollar total involved in the largest possible transaction between the manufacturer and the dealer; then find the smallest possible dollar amount.

In Exercises 42–44, use a graphing calculator to determine where the derivative is equal to zero.

42. *Can Design* Modify the can problem in Example 4 so the cost must be minimized. Assume that aluminum costs 3¢ per square centimeter, and that there is an additional cost of 2¢ per centimeter times the perimeter of the top, and a similar cost for the bottom, to seal the top and bottom of the can to the side.

43. *Can Design* In this modification of the can problem in Example 4, the cost must be minimized. Assume that aluminum costs 3¢ per square centimeter, and that there is an additional cost of 1¢ per centimeter times the height of the can to make a vertical seam on the side.

44. *Can Design* This problem is a combination of Exercises 42 and 43. We will again minimize the cost of the can,

assuming that aluminum costs 3¢ per square centimeter. In addition, there is a cost of 2¢ per centimeter to seal the top and bottom of the can to the side, plus 1¢ per centimeter to make a vertical seam.

45. *Ladder* A thief tries to enter a building by placing a ladder over a 9-foot-high fence so it rests against the building, which is 2 ft back from the fence. (See the figure below.) What is the length of the shortest ladder that can be used? (*Hint:* Let θ be the angle between the ladder and the ground. Express the length of the ladder in terms of θ, and then find the value of θ that minimizes the length of the ladder.)

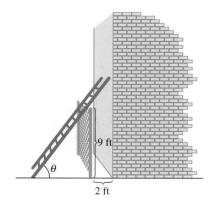

46. *Ladder* A janitor in a hospital needs to carry a ladder around a corner connecting a 10-foot-wide corridor and a 5-foot-wide corridor. (See the figure below.) What is the longest such ladder that can make it around the corner? (*Hint:* Find the narrowest point in the corridor by minimizing the length of the ladder as a function of θ, the angle the ladder makes with the 5-foot-wide corridor.)

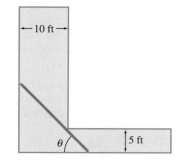

6.3   IMPLICIT DIFFERENTIATION

In almost all of the examples and applications so far, all functions have been defined in the form

$$y = f(x),$$

with y given **explicitly** in terms of x, or as an **explicit function** of x. For example,

$$y = 3x - 2, \qquad y = x^2 + x + 6, \qquad \text{and} \qquad y = -x^3 + 2$$

are all explicit functions of x. The equation $4xy - 3x = 6$ can be expressed as an explicit function of x by solving for y. This gives

$$4xy - 3x = 6$$
$$4xy = 3x + 6$$
$$y = \frac{3x + 6}{4x}.$$

On the other hand, some equations in x and y cannot be readily solved for y, and some equations cannot be solved for y at all. For example, while it would be possible (but tedious) to use the quadratic formula to solve for y in the equation $y^2 + 2yx + 4x^2 = 0$, it is not possible to solve for y in the equation $y^5 + 8y^3 + 6y^2x^2 + 2yx^3 + 6 = 0$. In equations such as these last two, y is said to be given **implicitly** in terms of x.

In such cases, it may still be possible to find the derivative dy/dx by a process called **implicit differentiation.** In doing so, we assume that there exists some function or functions f, which we may or may not be able to find, such that $y = f(x)$ and dy/dx exists. It is useful to use dy/dx here rather than $f'(x)$ to make it clear which variable is independent and which is dependent.

EXAMPLE 1 Implicit Differentiation
Find dy/dx if $3xy + 4y^2 = 10$.

Solution Differentiate with respect to x on both sides of the equation.

$$3xy + 4y^2 = 10$$
$$\frac{d}{dx}(3xy + 4y^2) = \frac{d}{dx}(10) \tag{1}$$

Now differentiate each term on the left side of the equation. Think of $3xy$ as the product $(3x)(y)$ and use the product rule and the chain rule. Since

$$\frac{d}{dx}(3x) = 3 \qquad \text{and} \qquad \frac{d}{dx}(y) = \frac{dy}{dx},$$

the derivative of $(3x)(y)$ is

$$(3x)\frac{dy}{dx} + (y)3 = 3x\frac{dy}{dx} + 3y.$$

To differentiate the second term, $4y^2$, use the generalized power rule, since y is assumed to be some function of x.

$$\frac{d}{dx}(4y^2) = \overset{\text{Derivative of } y^2}{4(2y^1)\frac{dy}{dx}} = 8y\frac{dy}{dx}$$

On the right side of equation (1), the derivative of 10 is 0. Taking the indicated derivatives in equation (1) term by term gives

FOR REVIEW ■

In Chapter 1, we pointed out that when y is given as a function of x, x is the independent variable and y is the dependent variable. We later defined the derivative dy/dx when y is a function of x. In an equation such as $3xy + 4y^2 = 10$, either variable can be considered the independent variable. If a problem asks for dy/dx, consider x the independent variable; if it asks for dx/dy, consider y the independent variable. A similar rule holds when other variables are used.

$$3x\frac{dy}{dx} + 3y + 8y\frac{dy}{dx} = 0.$$

Now solve this result for dy/dx.

$$(3x + 8y)\frac{dy}{dx} = -3y$$

$$\frac{dy}{dx} = \frac{-3y}{3x + 8y}$$

> **NOTE** Because we are treating y as a function of x, notice that each time an expression has y in it, we use the chain rule.

EXAMPLE 2 Implicit Differentiation

Find dy/dx for $x + \sqrt{xy} = y^2$.

Solution Take the derivative on each side.

$$\frac{d}{dx}\left(x + \sqrt{xy}\right) = \frac{d}{dx}(y^2)$$

Since $\sqrt{xy} = \sqrt{x} \cdot \sqrt{y} = x^{1/2} \cdot y^{1/2}$ for nonnegative values of x and y, use the product rule and the chain rule as follows.

$$\overbrace{1}^{\substack{\text{Derivative}\\ \text{of } x}} + \overbrace{x^{1/2}\left(\frac{1}{2}y^{-1/2} \cdot \frac{dy}{dx}\right) + y^{1/2}\left(\frac{1}{2}x^{-1/2}\right)}^{\substack{\text{Derivative}\\ \text{of } x^{1/2}y^{1/2}}} = \overbrace{2y\frac{dy}{dx}}^{\substack{\text{Derivative}\\ \text{of } y^2}}$$

$$1 + \frac{x^{1/2}}{2y^{1/2}} \cdot \frac{dy}{dx} + \frac{y^{1/2}}{2x^{1/2}} = 2y\frac{dy}{dx}$$

Multiply both sides by $2x^{1/2} \cdot y^{1/2}$.

$$2x^{1/2} \cdot y^{1/2} + x\frac{dy}{dx} + y = 4x^{1/2} \cdot y^{3/2} \cdot \frac{dy}{dx}$$

Combine terms and solve for dy/dx.

$$2x^{1/2} \cdot y^{1/2} + y = (4x^{1/2} \cdot y^{3/2} - x)\frac{dy}{dx}$$

$$\frac{dy}{dx} = \frac{2x^{1/2} \cdot y^{1/2} + y}{4x^{1/2} \cdot y^{3/2} - x}$$

EXAMPLE 3 Tangent Line

The graph of $x^3 + y^3 = 9xy$, shown in Figure 14 on the next page, is a *folium of Descartes*.* Find the equation of the tangent line at the point $(2, 4)$, shown in Figure 14.

---

*Information on this curve and others is available on the Famous Curves section of the MacTutor History of Mathematics Archive web site at www-history.mcs.st-and.ac.uk/~history. See Exercises 34–37 for more curves.

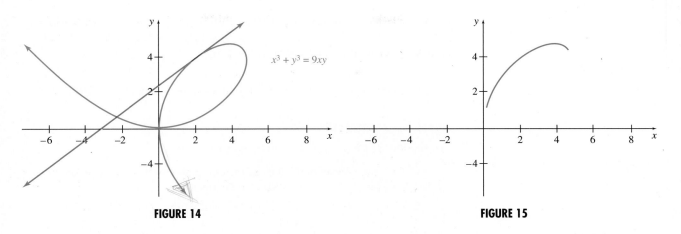

FIGURE 14   **FIGURE 15**

Solution Since this is not the graph of a function, y is not a function of x, and dy/dx is not defined. But if we restrict the curve to the vicinity of $(2, 4)$, as shown in Figure 15, the curve does represent the graph of a function, and we can calculate dy/dx by implicit differentiation.

$$3x^2 + 3y^2 \cdot \frac{dy}{dx} = 9x\frac{dy}{dx} + 9y \qquad \text{Generalized power rule and product rule}$$

$$3y^2 \cdot \frac{dy}{dx} - 9x\frac{dy}{dx} = 9y - 3x^2 \qquad \text{Move all } dy/dx \text{ terms to the same side of the equation.}$$

$$\frac{dy}{dx}(3y^2 - 9x) = 9y - 3x^2 \qquad \text{Factor.}$$

$$\frac{dy}{dx} = \frac{9y - 3x^2}{3y^2 - 9x}$$

$$= \frac{3(3y - x^2)}{3(y^2 - 3x)} = \frac{3y - x^2}{y^2 - 3x}$$

To find the slope of the tangent line at the point $(2, 4)$, let $x = 2$ and $y = 4$. The slope is

$$m = \frac{3y - x^2}{y^2 - 3x} = \frac{3(4) - 2^2}{4^2 - 3(2)} = \frac{8}{10} = \frac{4}{5}.$$

The equation of the tangent line is then found by using the point-slope form of the equation of a line.

$$y - y_1 = m(x - x_1)$$

$$y - 4 = \frac{4}{5}(x - 2)$$

$$y - 4 = \frac{4}{5}x - \frac{8}{5}$$

$$y = \frac{4}{5}x + \frac{12}{5}$$

The tangent line is graphed in Figure 14.

> **NOTE** In Example 3, we could have substituted $x = 2$ and $y = 4$ immediately after taking the derivative implicitly. You may find that such a substitution makes solving the equation for dy/dx easier.

The steps used in implicit differentiation can be summarized as follows.

IMPLICIT DIFFERENTIATION

To find dy/dx for an equation containing x and y:

1. Differentiate on both sides of the equation with respect to x, keeping in mind that y is assumed to be a function of x.
2. Place all terms with dy/dx on one side of the equals sign, and all terms without dy/dx on the other side.
3. Factor out dy/dx, and then solve for dy/dx.

When an applied problem involves an equation that is not given in explicit form, implicit differentiation can be used to locate maxima and minima or to find rates of change.

EXAMPLE 4 Bacteria in Sausage

As we saw in Chapter 4, researchers in Italy have developed a modified Gompertz model to predict that the growth of *Enterococcus faecium* bacteria in bologna sausage at 32°C which can be described by the equation

$$\ln\left(\frac{N(t)}{N_0}\right) = 9.8901 e^{-e^{(2.54197 - 0.2167t)}},$$

where N_0 is the number of bacteria present at the beginning of the experiment and $N(t)$ is the number of bacteria present at time t (in hr).* Use implicit differentiation to determine $N'(3)$. Assume that $N_0 = 1,000$.

Solution We could first solve the above equation for $N(t)$ and then find $N'(t)$. It is also possible and convenient to use implicit differentiation to determine $N'(t)$. Using implicit differentiation, we have

$$\frac{\dfrac{N'(t)}{N_0}}{\dfrac{N(t)}{N_0}} = 9.8901(0.2167)(e^{2.54197 - 0.2167t})(e^{-e^{(2.54197 - 0.2167t)}}).$$

Next, we simplify and solve for $N'(t)$.

$$N'(t) = N(t)(2.14318)(e^{2.54197 - 0.2167t})(e^{-e^{(2.54197 - 0.2167t)}})$$

---

*Zanoni, B., C. Garzaroli, S. Anselmi, and G. Rondinini, "Modeling the Growth of *Enterococcus faecium* in Bologna Sausage," *Applied and Environmental Microbiology,* Vol. 59, No. 10, Oct. 1993, pp. 3411–3417.

To calculate $N'(3)$ we first need to find $N(3)$.

$$\ln\left(\frac{N(3)}{1,000}\right) = 9.8901e^{-e^{(2.54197-0.2167(3))}} \approx 0.013034$$

Thus,

$$N(3) = 1,000e^{0.013034} \approx 1013.1189$$

and

$$N'(3) = 1013.1189(2.14318)(e^{2.54197-0.2167(3)})(e^{-e^{(2.54197-0.2167(3))}}) \approx 19.$$

Thus, we would expect that the number of bacteria will increase by approximately 19 during the third hour.

6.3 EXERCISES

Find dy/dx by implicit differentiation for each of the following.

1. $4x^2 + 3y^2 = 6$ **2.** $2x^2 - 5y^2 = 4$ **3.** $6x^2 + 8xy + y^2 = 6$ **4.** $8x^2 = 6y^2 + 2xy$

5. $x^3 = y^2 + 4$ **6.** $x^3 - 6y^2 = 10$ **7.** $3x^2 = \dfrac{2-y}{2+y}$ **8.** $2y^2 = \dfrac{5+x}{5-x}$

9. $\sqrt{x} + \sqrt{y} = 4$ **10.** $2\sqrt{x} - \sqrt{y} = 1$ **11.** $x^4y^3 + 4x^{3/2} = 6y^{3/2} + 5$ **12.** $(xy)^{4/3} + x^{1/3} = y^6 + 1$

13. $e^{x^2y} = 5x + 4y + 2$ **14.** $x^2e^y + y = x^3$ **15.** $x + \ln y = x^2y^3$ **16.** $y \ln x + 2 = x^{3/2}y^{5/2}$

17. $\sin(xy) = x$ **18.** $\tan y + x = 4$

Find the equation of the tangent line at the given point on each curve in Exercises 19–26.

19. $x^2 + y^2 = 25$; $(-3, 4)$ **20.** $x^2 + y^2 = 100$; $(8, -6)$

21. $x^2y^2 = 1$; $(-1, 1)$ **22.** $x^2y^3 = 8$; $(-1, 2)$

23. $2y^2 - \sqrt{x} = 4$; $(16, 2)$ **24.** $y + \dfrac{\sqrt{x}}{y} = 3$; $(4, 2)$

25. $x - \sin(\pi y) = 1$; $(1, 0)$ **26.** $x^2 + \tan\left(\dfrac{\pi}{4}xy\right) = 2$; $(1, 1)$

In Exercises 27–32, find the equation of the tangent line at the given value of x on each curve.

27. $y^3 + xy - y = 8x^4, x = 1$ **28.** $y^3 + 2x^2y - 8y = x^3 + 19, x = 2$

29. $y^3 + xy^2 + 1 = x + 2y^2, x = 2$ **30.** $y^4(1 - x) + xy = 2, x = 1$

31. $2y^3(x - 3) + x\sqrt{y} = 3, x = 3$ **32.** $\dfrac{y}{18}(x^2 - 64) + x^{2/3}y^{1/3} = 12, x = 8$

33. The graph of $x^2 + y^2 = 100$ is a circle having center at the origin and radius 10.

   a. Write the equations of the tangent lines at the points where $x = 6$.

   b. Graph the circle and the tangent lines.

Information on curves in Exercises 34–37, as well as many other curves, is available on the Famous Curves section of the MacTutor History of Mathematics Archive Web site at www-history.mcs.st-and.ac.uk/~history.

34. The graph of $x^{2/3} + y^{2/3} = 2$, shown in the figure, is an *astroid*. Find the equation of the tangent line at the point $(1, 1)$.

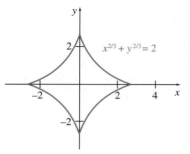

35. The graph of $3(x^2 + y^2)^2 = 25(x^2 - y^2)$, shown in the figure, is a *lemniscate of Bernoulli*. Find the equation of the tangent line at the point $(2, 1)$.

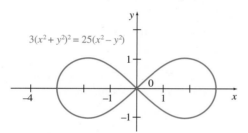

36. The graph of $y^2(x^2 + y^2) = 20x^2$, shown in the figure, is a *kappa curve*. Find the equation of the tangent line at the point $(1, 2)$.

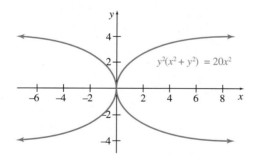

37. The graph of $2(x^2 + y^2)^2 = 25xy^2$, shown in the figure, is a *double folium*. Find the equation of the tangent line at the point $(2, 1)$.

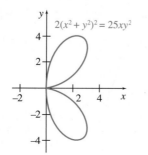

38. Suppose $x^2 + y^2 + 1 = 0$. Use implicit differentiation to find dy/dx. Then explain why the result you got is meaningless. (*Hint:* Can $x^2 + y^2 + 1$ equal 0?)

Let $\sqrt{u} + \sqrt{2v + 1} = 5$. *Find the derivatives in Exercises 39 and 40.*

39. $\dfrac{du}{dv}$

40. $\dfrac{dv}{du}$

▪ Applications

LIFE SCIENCES

41. *Respiratory Rate* Researchers have found a correlation between respiratory rate and body mass in the first three years of life. This correlation can be expressed by the function

$$\log R(w) = 1.83 - 0.43 \log(w),$$

where w is the body weight (in kg) and $R(w)$ is the respiratory rate (in breaths per minute).*

a. Find $R'(w)$ using implicit differentiation.

b. Find $R'(w)$ by first solving the equation for $R(w)$.

c. Discuss the two procedures. Is there a situation when you would want to use one method over another?

42. *Biochemical Reaction* A simple biochemical reaction with three molecules has solutions which oscillate toward a steady state when positive constants a and b are below the curve $b - a = (b + a)^3$.† Find the largest possible value of a for which the reaction has solutions which oscillate toward a steady state. (*Hint:* Find where $da/db = 0$. Derive values for $a + b$ and $a - b$, and then solve the equations in two unknowns.)

43. *Species* The relationship between the number of species in a genus (x) and the number of genera (y) comprising x species is given by

$$xy^a = k,$$

where a and k are constants.‡ Find dy/dx.

OTHER APPLICATIONS

Velocity The position of a particle at time t is given by s. Find the velocity ds/dt.

44. $s^3 - 4st + 2t^3 - 5t = 0$

45. $2s^2 + \sqrt{st} - 4 = 3t$

46. *Cost and Revenue* For a certain product, cost C and revenue R are given as follows, where x is the number of units sold (in hundreds).

$$\text{Cost: } C^2 = x^2 + 100\sqrt{x} + 50$$
$$\text{Revenue: } 900(x - 5)^2 + 25R^2 = 22{,}500$$

a. Find and interpret the marginal cost dC/dx at $x = 5$.

b. Find and interpret the marginal revenue dR/dx at $x = 5$.

▪ 6.4   RELATED RATES

?	**THINK ABOUT IT**	When a skier's blood vessels contract because of the cold, how fast is the velocity of blood changing?

We use related rates to answer this question in Example 5 of this section.

It is common for variables to be functions of time; for example, sales of an item may depend on the season of the year, or a population of animals may be increasing at a certain rate several months after being introduced into an area. Time is often present implicitly in a mathematical model, meaning that derivatives with respect to time must be found by the method of implicit differentiation discussed in the previous section.

---

*Gagliardi, L., and F. Rusconi, "Respiratory Rate and Body Mass in the First Three Years of Life," *Archives of Disease in Children,* Vol. 76, 1997, pp. 151–154.

†Murray, J. D., *Mathematical Biology,* Springer-Verlag, 1989, pp. 156–158.

‡Lotka, Alfred J., *Elements of Mathematical Biology,* Dover Publications, 1956, p. 313.

EXAMPLE 1 Area

A small rock is dropped into a lake. Circular ripples spread over the surface of the water, with the radius of each circle increasing at the rate of 3/2 feet per second. Find the rate of change of the area inside the circle formed by a ripple at the instant the radius is 4 feet.

Solution As shown in Figure 16, the area A and the radius r are related by

$$A = \pi r^2.$$

Take the derivative of each side with respect to time.

$$\frac{d}{dt}(A) = \frac{d}{dt}(\pi r^2)$$

$$\frac{dA}{dt} = 2\pi r \cdot \frac{dr}{dt} \tag{1}$$

Since the radius is increasing at the rate of 3/2 feet per second,

$$\frac{dr}{dt} = \frac{3}{2}.$$

The rate of change of area at the instant $r = 4$ is given by dA/dt evaluated at $r = 4$. Substituting into equation (1) gives

$$\frac{dA}{dt} = 2\pi \cdot 4 \cdot \frac{3}{2}$$

$$\frac{dA}{dt} = 12\pi \approx 37.7 \text{ square feet per second.}$$

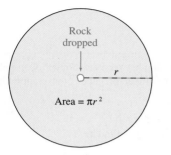

Rock dropped

Area $= \pi r^2$

FIGURE 16

In Example 1, the derivatives (or rates of change) dA/dt and dr/dt are related by equation (1); for this reason they are called **related rates.** As suggested by Example 1, four basic steps are involved in solving problems about related rates.

SOLVING RELATED RATE PROBLEMS

1. Identify all given quantities, as well as the quantities to be found. Draw a sketch when possible.

2. Write an equation relating the variables of the problem.

3. Use implicit differentiation to find the derivative of both sides of the equation in Step 2 with respect to time.

4. Solve for the derivative giving the unknown rate of change and substitute the given values.

CAUTION Differentiate *first,* and *then* substitute values for the variables. If the substitutions were performed first, differentiating would not lead to useful results.

EXAMPLE 2 Sliding Ladder

A 50-foot ladder is placed against a large building. The base of the ladder is resting on an oil spill, and it slips (to the right in Figure 17 on the next page) at the rate of

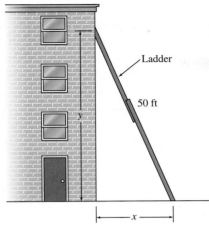

FIGURE 17

3 feet per minute. Find the rate of change of the height of the top of the ladder above the ground at the instant when the base of the ladder is 30 feet from the base of the building.

Solution Let y be the height of the top of the ladder above the ground, and let x be the distance of the base of the ladder from the base of the building. By the Pythagorean theorem,

$$x^2 + y^2 = 50^2. \qquad (2)$$

Both x and y are functions of time t in minutes after the moment that the ladder starts slipping. Take the derivative of both sides of equation (2) with respect to time, getting

$$\frac{d}{dt}(x^2 + y^2) = \frac{d}{dt}(50^2)$$

$$2x\frac{dx}{dt} + 2y\frac{dy}{dt} = 0. \qquad (3)$$

Since the base is sliding at the rate of 3 feet per minute,

$$\frac{dx}{dt} = 3.$$

Also, the base of the ladder is 30 feet from the base of the building. Use this to find y.

$$50^2 = 30^2 + y^2$$
$$2{,}500 = 900 + y^2$$
$$1{,}600 = y^2$$
$$y = 40$$

In summary, $y = 40$ when $x = 30$. Also, the rate of change of x over time t is $dx/dt = 3$. Substituting these values into equation (3) to find the rate of change of y over time gives

$$2(30)(3) + 2(40)\frac{dy}{dt} = 0$$

$$180 + 80\frac{dy}{dt} = 0$$

$$80\frac{dy}{dt} = -180$$

$$\frac{dy}{dt} = \frac{-180}{80} = \frac{-9}{4} = -2.25.$$

At the instant when the base of the ladder is 30 feet from the base of the building, the top of the ladder is sliding down the building at the rate of 2.25 feet per minute. (The minus sign shows that the ladder is sliding *down*, so the distance y is *decreasing*.)*

---

*The model in Example 2 breaks down as the top of the ladder nears the ground. As y approaches 0, dy/dt becomes infinitely large. In reality, the ladder loses contact with the wall before y reaches 0.

EXAMPLE 3   Icicle

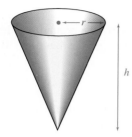

FIGURE 18

A cone-shaped icicle is dripping from the roof. The radius of the icicle is decreasing at a rate of 0.2 centimeter per hour, while the length is increasing at a rate of 0.8 centimeter per hour. If the icicle is currently 4 centimeters in radius and 20 centimeters long, is the volume of the icicle increasing or decreasing, and at what rate?

Solution For this problem we need the formula for the volume of a cone:

$$V = \frac{1}{3}\pi r^2 h, \tag{4}$$

where r is the radius of the cone and h is the height of the cone, which in this case is the length of the icicle, as in Figure 18.

In this problem, both r and h are functions of the time t in hours. Taking the derivative of both sides of equation (4) with respect to time yields

$$\frac{dV}{dt} = \frac{1}{3}\pi\left[r^2\frac{dh}{dt} + (h)(2r)\frac{dr}{dt}\right]. \tag{5}$$

Since the radius is decreasing at a rate of 0.2 centimeter per hour and the length is increasing at a rate of 0.8 centimeter per hour,

$$\frac{dr}{dt} = -0.2 \qquad \text{and} \qquad \frac{dh}{dt} = 0.8.$$

Substituting these, as well as $r = 4$ and $h = 20$, into equation (5) yields

$$\frac{dV}{dt} = \frac{1}{3}\pi[4^2(0.8) + (20)(8)(-0.2)]$$

$$= \frac{1}{3}\pi(-19.2) \approx -20.$$

Because the sign of dV/dt is negative, the volume of the icicle is decreasing at a rate of 20 cubic centimeters per hour.   ▪

EXAMPLE 4   Revenue

A company is increasing production of antibiotics at the rate of 50 cases per day. All cases produced can be sold if the price is set appropriately. The company found that if they wanted to sell q units, the price p (in dollars) needs to be set at $p = 50 - q/200$. Find the rate of change of revenue with respect to time (in days) when the daily production is 200 units.

Solution The revenue function,

$$R = qp = q\left(50 - \frac{q}{200}\right) = 50q - \frac{q^2}{200},$$

relates R and q. The rate of change of q over time (in days) is $dq/dt = 50$. The rate of change of revenue over time, dR/dt, is to be found when $q = 200$. Differentiate both sides of the equation

$$R = 50q - \frac{q^2}{200}$$

with respect to t.

$$\frac{dR}{dt} = 50\frac{dq}{dt} - \frac{1}{100}q\frac{dq}{dt} = \left(50 - \frac{1}{100}q\right)\frac{dq}{dt}$$

Now substitute the known values for q and dq/dt.

$$\frac{dR}{dt} = \left[50 - \frac{1}{100}(200)\right](50) = 2{,}400$$

Thus revenue is increasing at the rate of $2,400 per day.

EXAMPLE 5   Blood Flow

Blood flows faster the closer it is to the center of a blood vessel because of the reduced friction with cell walls. According to Poiseuille's laws, the velocity V of blood is given by

$$V = k(R^2 - r^2),$$

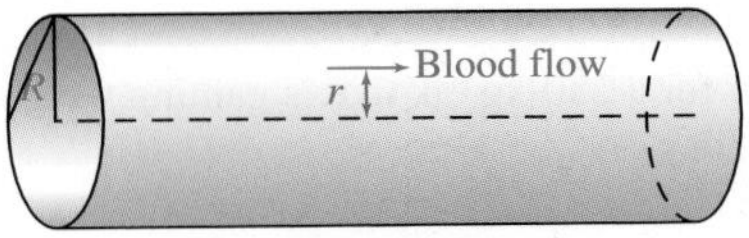

where R is the radius of the blood vessel, r is the distance of a layer of blood flow from the center of the vessel, and k is a constant, assumed here to equal 375. See Figure 19. Suppose a skier's blood vessel has radius $R = 0.08$ millimeter and that cold weather is causing the vessel to contract at a rate of $dR/dt = -0.01$ millimeter per minute. How fast is the velocity of blood changing?

FIGURE 19

Solution Find dV/dt. Treat r as a constant. Assume the given units are compatible.

$$V = 375(R^2 - r^2)$$

$$\frac{dV}{dt} = 375\left(2R\frac{dR}{dt} - 0\right) \qquad r \text{ is a constant.}$$

$$\frac{dV}{dt} = 750R\frac{dR}{dt}$$

Here $R = 0.08$ and $dR/dt = -0.01$, so

$$\frac{dV}{dt} = 750(0.08)(-0.01) = -0.6.$$

That is, the velocity of the blood is decreasing at a rate of -0.6 millimeter per minute each minute. The units (mm/min^2) indicate that this is a deceleration (negative acceleration), since it gives the rate of change of velocity.

6.4 EXERCISES

Assume x and y are functions of t. Evaluate dy/dt for each of the following.

1. $y^2 - 5x^2 = -1$; $\dfrac{dx}{dt} = -3, x = 1, y = 2$

2. $8y^3 + x^2 = 1$; $\dfrac{dx}{dt} = 2, x = 3, y = -1$

3. $xy - 5x + 2y^3 = -70$; $\dfrac{dx}{dt} = -5, x = 2, y = -3$

4. $4x^3 - 9xy^2 + y = -80$; $\dfrac{dx}{dt} = 4, x = -3, y = 1$

5. $\dfrac{x^2 + y}{x - y} = 9$; $\dfrac{dx}{dt} = 2, x = 4, y = 2$

6. $\dfrac{y^3 - x^2}{x + 2y} = \dfrac{17}{7}$; $\dfrac{dx}{dt} = 1, x = -3, y = -2$

7. $xe^y = 1 + \ln x$; $\dfrac{dx}{dt} = 6, x = 1, y = 0$

8. $y \ln x + xe^y = 6$; $\dfrac{dx}{dt} = 5, x = 1, y = 0$

Applications

LIFE SCIENCES

9. *Blood Velocity* A cross-country skier has a history of heart problems. She takes nitroglycerin to dilate blood vessels, thus avoiding angina (chest pain) due to blood vessel contraction. Use Poiseuille's law with $k = 555.6$ to find the rate of change of the blood velocity when $R = 0.02$ mm and R is changing at 0.003 mm per minute. Assume r is constant.

10. *Allometric Growth* Suppose x and y are two quantities that vary with time according to the allometric formula $y = nx^m$. (See Exercise 74 in the section on Logarithmic Functions.) Show that the derivatives of x and y are related by the formula

$$\frac{1}{y}\frac{dy}{dt} = m\frac{1}{x}\frac{dx}{dt}.$$

(*Hint:* Take natural logarithms of both sides before taking the derivatives.)

11. *Brain Weight* The brain weight of a fetus can be estimated using the total weight of the fetus by the function

$$b = 0.22w^{0.87},$$

where w is the weight of the fetus (in g) and b is the brain weight (in g).* Suppose the brain weight of a 25-g fetus is changing at a rate of 0.25 g/day. Use this to estimate the rate of change of the total weight of the fetus dw/dt.

12. *Birds* The energy cost of bird flight as a function of body weight is given by

$$E = 429w^{-0.35},$$

where w is the weight of the bird (in g) and E is the energy expenditure (in cal/g/hr).[†] Suppose that the weight of a bird weighing 10 g is increasing at a rate of 0.001 g/hr. Find the rate at which the energy expenditure is changing with respect to time.

13. *Metabolic Rate* The average daily metabolic rate for rodents can be expressed as a function of weight by

$$m = 85.65w^{0.54},$$

where w is the weight of the rodent (in kg) and m is the metabolic rate (in kcal/day).[‡]

a. Suppose that the weight of the rodent is changing with respect to time at a rate dw/dt. Find dm/dt.

b. Determine dm/dt for a 0.25-kg rodent that is gaining weight at a rate of 0.01 kg/day.

14. *Metabolic Rate* The average daily metabolic rate for captive animals from weasels to elk can be expressed as a function of weight by

$$m = 140.2w^{0.75},$$

where w is the weight of the animal (in kg) and m is the metabolic rate (in kcal/day).[‡]

a. Suppose that the weight of a weasel is changing with respect to time at a rate dw/dt. Find dm/dt.

b. Determine dm/dt for a 250-kg elk that is gaining weight at a rate of 2 kg/day.

15. *Lizards* The energy cost of horizontal locomotion as a function of the body weight of a lizard is given by

$$E = 26.5w^{-0.34},$$

where w is the weight of the lizard (in kg) and E is the energy expenditure (in kcal/kg/km).[§] Suppose that the weight of a 5-kg lizard is increasing at a rate of 0.05 kg/day. Find the rate at which the energy expenditure is changing with respect to time.

16. *Marsupials* The energy cost of horizontal locomotion as a function of the body weight of a marsupial is given by

$$E = 22.8w^{-0.34},$$

where w is the weight of the animal (in kg) and E is the energy expenditure (in kcal/kg/km).[§] Suppose that the weight of a 10-kg marsupial is increasing at a rate of 0.1 kg/day. Find the rate at which the energy expenditure is changing with respect to time.

17. *Crime Rate* Sociologists have found that crime rates are influenced by temperature. In a midwestern town of 100,000 people, the crime rate has been approximated as

$$C = \frac{1}{10}(T - 60)^2 + 100,$$

where C is the number of crimes per month and T is the average monthly temperature in degrees Fahrenheit. The

---

*Wanderley, S., M. Costa-Neves, and R. Rega, "Relative Growth of the Brain in Human Fetuses: First Gestational Trimester," *Archives d'anatomie, d'histologie et d'embryologie*, Vol. 73, 1990, pp. 43–46.
[†]Robbins, C., *Wildlife Feeding and Nutrition*, New York, Academic Press, 1983, p. 119.
[‡]Ibid., p. 133.
[§]Ibid., p. 114.

average temperature for May was 76°, and by the end of May the temperature was rising at the rate of 8° per month. How fast is the crime rate rising at the end of May?

18. *Learning Skills* It is estimated that a person learning a certain assembly-line task takes

$$T(x) = \frac{2 + x}{2 + x^2}$$

minutes to perform the task after x repetitions. Find dT/dt if dx/dt is 4, and 4 repetitions of the task have been performed.

19. *Memorization Skills* Under certain conditions, a person can memorize W words in t minutes, where

$$W(t) = \frac{-0.02t^2 + t}{t + 1}.$$

Find dW/dt when $t = 5$.

OTHER APPLICATIONS*

20. *Sliding Ladder* A 25-ft ladder is placed against a building. The base of the ladder is slipping away from the building at a rate of 4 ft/min. Find the rate at which the top of the ladder is sliding down the building at the instant when the bottom of the ladder is 7 ft from the base of the building.

21. *Distance* **a.** One car leaves a given point and travels north at 30 mph. Another car leaves the same point at the same time and travels west at 40 mph. At what rate is the distance between the two cars changing at the instant when the cars have traveled 2 hr?

  b. Suppose that, in part a, the second car left one hour later than the first car. At what rate is the distance between the two cars changing at the instant when the second car has traveled 1 hr?

22. *Area* A rock is thrown into a still pond. The circular ripples move outward from the point of impact of the rock so that the radius of the circle formed by a ripple increases at the rate of 2 ft per minute. Find the rate at which the area is changing at the instant the radius is 4 ft.

23. *Volume* A spherical snowball is placed in the sun. The sun melts the snowball so that its radius decreases 1/4 inch per hour. Find the rate of change of the volume with respect to time at the instant the radius is 4 inches.

24. *Volume* A sand storage tank used by the highway department for winter storms is leaking. As the sand leaks out, it forms a conical pile. The radius of the base of the pile

increases at the rate of 1 inch per minute. The height of the pile is always twice the radius of the base. Find the rate at which the volume of the pile is increasing at the instant the radius of the base is 5 inches.

25. *Shadow Length* A man 6 ft tall is walking away from a lamp post at the rate of 50 ft per minute. When the man is 8 ft from the lamp post, his shadow is 10 ft long. Find the rate at which the length of the shadow is increasing when he is 25 ft from the lamp post. (See the figure.)

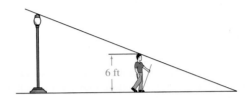

26. *Water Level* A trough has a triangular cross section. The trough is 6 ft across the top, 6 ft deep, and 16 ft long. Water is being pumped into the trough at the rate of 4 cubic feet per minute. Find the rate at which the height of water is increasing at the instant that the height is 4 ft.

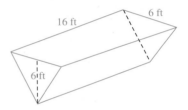

27. *Velocity* A pulley is on the edge of a dock, 8 ft above the water level. (See the figure below.) A rope is being used to pull in a boat. The rope is attached to the boat at water level. The rope is being pulled in at the rate of 1 ft per second. Find the rate at which the boat is approaching the dock at the instant the boat is 8 ft from the dock.

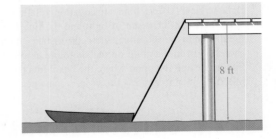

28. *Kite Flying* Christine O'Brien is flying her kite in a wind that is blowing it east at a rate of 50 ft/min. She has already let out 200 ft of string, and the kite is flying 100 ft above

*You may wish to refer to the Appendix of Formulas from Geometry for some of these exercises.

her hand. How fast must she let out string at this moment to keep the kite flying with the same speed and altitude?

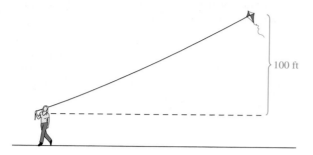

100 ft

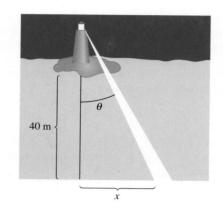

θ

40 m

x

29. *Cost* A manufacturer of handcrafted wine racks has determined that the cost to produce x units per month is given by $C = 0.1x^2 + 10{,}000$. How fast is the cost per month changing when production is changing at the rate of 10 units per month and the production level is 100 units?

30. *Revenue/Cost/Profit* Given the revenue and cost functions $R = 50x - 0.4x^2$ and $C = 5x + 15$, where x is the daily production (and sales), find the following when 40 units are produced daily and the rate of change of production is 10 units per day.

   a. The rate of change of revenue with respect to time

   b. The rate of change of cost with respect to time

   c. The rate of change of profit with respect to time

31. *Revenue/Cost/Profit* Repeat Exercise 30, given that 200 units are produced daily and the rate of change of production is 50 units per day.

32. *Rotating Lighthouse* The beacon on a lighthouse 40 m from a straight shoreline rotates twice per minute. (See the figure.)

   a. How fast is the beam moving along the shoreline at the moment when the light beam and the shoreline are at right angles? (*Hint:* Find an equation relating θ, the angle between the beam of light and the line from the lighthouse to the shoreline, and x, the distance along the shoreline from the point on the shoreline closest to the lighthouse and the point where the beam hits the shoreline. You need to express $d\theta/dt$ in radians per minute.)

   b. In part a, how fast is the beam moving along the shoreline when the beam hits the shoreline 40 m from the point on the shoreline closest to the lighthouse?

33. *Rotating Camera* A television camera on a tripod 60 ft from a road is filming a car carrying the president of the United States. (See the figure.) The car is moving along the road at 600 ft/min.

   a. How fast is the camera rotating (in revolutions per minute) when the car is at the point on the road closest to the camera? (See the hint for Exercise 32.)

   b. How fast is the camera rotating 6 sec after the moment in part a?

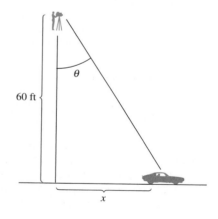

60 ft

θ

x

■ 6.5 DIFFERENTIALS: LINEAR APPROXIMATION

? **THINK ABOUT IT** How can the pulp cavity of the canine tooth of a gray wolf be used to estimate the age of the wolf?

Using differentials, we will answer this question in Exercise 26.

As mentioned earlier, the symbol Δx represents a change in the variable x. Similarly, Δy represents a change in y. An important problem that arises in many applications is to determine Δy given specific values of x and Δx. This quantity is often difficult to evaluate. In this section we show a method of approximating Δy that uses the derivative dy/dx. In essence, we use the tangent line at a particular value of x to approximate $f(x)$ for values close to x.

For values x_1 and x_2,

$$\Delta x = x_2 - x_1.$$

Solving for x_2 gives

$$x_2 = x_1 + \Delta x.$$

For a function $y = f(x)$, the symbol Δy represents a change in y:

$$\Delta y = f(x_2) - f(x_1).$$

Replacing x_2 with $x_1 + \Delta x$ gives

$$\Delta y = f(x_1 + \Delta x) - f(x_1).$$

If Δx is used instead of h, the derivative of a function f at x_1 could be defined as

$$\frac{dy}{dx} = \lim_{\Delta x \to 0} \frac{\Delta y}{\Delta x}.$$

If the derivative exists, then

$$\frac{dy}{dx} \approx \frac{\Delta y}{\Delta x}$$

as long as Δx is close to 0. Multiplying both sides by Δx (assume $\Delta x \neq 0$) gives

$$\Delta y \approx \frac{dy}{dx} \cdot \Delta x.$$

Until now, dy/dx has been used as a single symbol representing the derivative of y with respect to x. In this section, separate meanings for dy and dx are introduced in such a way that their quotient, when $dx \neq 0$, is the derivative of y with respect to x. These meanings of dy and dx are then used to find an approximate value of Δy.

To define dy and dx, look at Figure 20 on the next page, which shows the graph of a function $y = f(x)$. The tangent line to the graph has been drawn at the point P. Let Δx be any nonzero real number (in practical problems, Δx is a small number) and locate the point $x + \Delta x$ on the x-axis. Draw a vertical line through $x + \Delta x$. Let this vertical line cut the tangent line at M and the graph of the function at Q.

Define the new symbol dx to be the same as Δx. Define the new symbol dy to equal the length MR. The slope of PM is $f'(x)$. By the definition of slope, the slope of PM is also dy/dx, so that

$$f'(x) = \frac{dy}{dx},$$

or

$$dy = f'(x)\, dx.$$

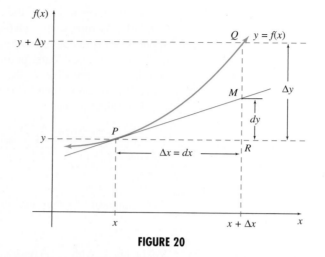

FIGURE 20

In summary, the definitions of the symbols dy and dx are as follows.

DIFFERENTIALS

For a function $y = f(x)$ whose derivative exists, the **differential** of x, written dx, is an arbitrary real number (usually small compared with x); the **differential** of y, written dy, is the product of $f'(x)$ and dx, or

$$dy = f'(x)\,dx.$$

The usefulness of the differential is suggested by Figure 20. As dx approaches 0, the value of dy gets closer and closer to that of Δy, so that for small nonzero values of dx

$$dy \approx \Delta y,$$

or

$$\Delta y \approx f'(x)\,dx.$$

EXAMPLE 1 Differential

Find dy for the following functions.

(a) $y = 6x^2$

> **Solution** The derivative is $dy/dx = 12x$ so
>
> $$dy = 12x\,dx.$$

(b) $y = 800x^{-3/4}$, $x = 16$, $dx = 0.01$

> **Solution**
>
> $$\begin{aligned}
dy &= -600x^{-7/4}\,dx \\
&= -600(16)^{-7/4}(0.01) \\
&= -600\left(\frac{1}{2^7}\right)(0.01) = -0.046875
\end{aligned}$$

Differentials can be used to approximate function values for a given x-value (in the absence of a calculator or computer). As discussed above,

$$\Delta y = f(x + \Delta x) - f(x).$$

For small nonzero values of Δx, $\Delta y \approx dy$, so that

$$dy \approx f(x + \Delta x) - f(x),$$

or
$$f(x) + dy \approx f(x + \Delta x).$$

Replacing dy with $f'(x)\,dx$ gives the following result.

LINEAR APPROXIMATION

Let f be a function whose derivative exists. For small nonzero values of Δx,

$$dy \approx \Delta y,$$

and
$$f(x + \Delta x) \approx f(x) + dy = f(x) + f'(x)\,dx.$$

EXAMPLE 2   Approximation
Approximate $\sqrt{50}$.

Solution We know that $\sqrt{49} = 7$, so we let $f(x) = \sqrt{x}$, $x = 49$, $\Delta x = dx = 1$, and use dy to approximate $\Delta y = \sqrt{50} - \sqrt{49}$. Since $\sqrt{x} = x^{1/2}$,

$$f'(x) = \frac{1}{2} x^{-1/2} = \frac{1}{2x^{1/2}},$$

so
$$dy = \frac{1}{2x^{1/2}}\,dx.$$

Substituting $x = 49$ and $dx = 1$ gives

$$dy = \frac{1}{2 \cdot 49^{1/2}} \cdot 1 = \frac{1}{14}.$$

Thus,
$$\sqrt{50} = f(x + \Delta x) \approx f(x) + dy$$
$$= \sqrt{x} + dy$$
$$= \sqrt{49} + \frac{1}{14} = 7\frac{1}{14}.$$

A calculator gives $7\,1/14 \approx 7.07143$ and $\sqrt{50} \approx 7.07107$. Our approximation of $7\,1/14$ is close to the true answer and does not require a calculator.

While calculators have made differentials less important in solving problems like the one above, the approximation of functions, including linear approximation, is still important in the branch of mathematics known as numerical analysis.

Marginal Analysis   Differentials are used to find an approximation of the change in a value of the dependent variable corresponding to a given change in the independent variable. Marginal analysis is often used in economics to estimate the change in cost (or profit or revenue) for nonlinear functions. These ideas can also be used in the life sciences.

For example, suppose that the weight of a particular animal can be expressed as a function of time, $W(t)$. Then

$$dW = W'(t) \, dt = W'(t) \, \Delta t.$$

Since $\Delta W \approx dW$,

$$\Delta W \approx W'(t) \, \Delta t.$$

If the change in time, Δt, is equal to 1, then

$$W(t + 1) - W(t) = \Delta W$$
$$\approx W'(t) \, \Delta t$$
$$= W'(t),$$

which shows that the marginal weight $W'(t)$ approximates the weight gain during the next unit of time.

EXAMPLE 3 Pigs

Researchers have observed that the weight of a male (boar) pig can be estimated by the function

$$W(t) = 10.1 + 160.1e^{-e^{(-0.0212(t - 103.63))}},$$

where t is the age of the boar (in days) and $W(t)$ is the weight (in kg).* If a particular boar is 70 days old, estimate how much it will weigh when it is 73 days old.

Solution Differentials can be used to find the approximate change in the pig's weight resulting from the extra days. This change can be approximated by $dW = W'(t) \, dt$ where $t = 70$ and $dt = 3$.

$$W'(t) = 160.1(e^{-e^{(-0.0212(t - 103.63))}})(-e^{-0.0212(t - 103.63)})(-0.0212)$$
$$= 3.39412e^{-0.0212(t - 103.63)}e^{-e^{(-0.0212(t - 103.63))}}$$

$$\Delta W \approx dW = 3.39412e^{-0.0212(t - 103.63)}e^{-e^{(-0.0212(t - 103.63))}} \, dt$$
$$= 3.39412e^{-0.0212(100 - 103.63)}e^{-e^{(-0.0212(100 - 103.63))}}(3)$$
$$\approx 3.75$$

Thus a 70-day-old boar is expected to grow approximately 3.75 kg during the next 3 days, which is fairly close to the actual value of about 2.79 kg.

Error Estimation The final example in this section shows how differentials are used to estimate errors that might enter into measurements of a physical quantity.

EXAMPLE 4 Error Estimation

In a precision manufacturing process, ball bearings must be made with a radius of 0.6 millimeter, with a maximum error in the radius of ± 0.015 millimeter. Estimate the maximum error in the volume of the ball bearing.

---

*Van Lunen, T., and D. Cole, "Growth and Body Composition of Highly Selected Boars and Gilts," *Animal Science*, Vol. 67, 1998, pp. 107–116.

Solution The formula for the volume of a sphere is

$$V = \frac{4}{3} \pi r^3.$$

If an error of Δr is made in measuring the radius of the sphere, the maximum error in the volume is

$$\Delta V = \frac{4}{3} \pi (r + \Delta r)^3 - \frac{4}{3} \pi r^3.$$

Rather than calculating ΔV, approximate ΔV with dV, where

$$dV = 4\pi r^2 \, dr.$$

Replacing r with 0.6 and $dr = \Delta r$ with ± 0.015 gives

$$dV = 4\pi(0.6)^2(\pm 0.015)$$
$$\approx \pm 0.0679.$$

The maximum error in the volume is about 0.07 cubic millimeter.

6.5 EXERCISES

For Exercises 1–8, find dy for the given values of x and Δx.

1. $y = 2x^2 - 5x$; $x = -2, \Delta x = 0.2$

2. $y = x^2 - 3x$; $x = 3, \Delta x = 0.1$

3. $y = x^3 - 2x^2 + 3$; $x = 1, \Delta x = -0.1$

4. $y = 2x^3 + x^2 - 4x$; $x = 2, \Delta x = -0.2$

5. $y = \sqrt{3x}$; $x = 1, \Delta x = 0.15$

6. $y = \sqrt{4x - 1}$; $x = 5, \Delta x = 0.08$

7. $y = \dfrac{2x - 5}{x + 1}$; $x = 2, \Delta x = -0.03$

8. $y = \dfrac{6x - 3}{2x + 1}$; $x = 3, \Delta x = -0.04$

Use the differential to approximate each quantity. Then use a calculator to approximate the quantity, and give the absolute value of the difference in the two results to four decimal places.

9. $\sqrt{145}$

10. $\sqrt{23}$

11. $\sqrt{0.99}$

12. $\sqrt{17.02}$

13. $e^{0.01}$

14. $e^{-0.002}$

15. $\ln 1.05$

16. $\ln 0.98$

17. $\sin(0.03)$

18. $\cos(-0.002)$

Applications

LIFE SCIENCES

19. *Alcohol Concentration* The mathematical model

$$y = A(x) = \frac{-7}{480}x^3 + \frac{127}{120}x, \qquad 0 \le x < 9,$$

gives the approximate alcohol concentration (in tenths of a percent) in an average person's bloodstream x hours after drinking 8 ounces of 100-proof whiskey.

a. Approximate the change in alcohol concentration from 3 to 3.5 hours.

b. Approximate the change in alcohol concentration from 6 to 6.25 hours.

20. *Drug Concentration* The concentration of a certain drug in the bloodstream x hr after being administered is approximately

$$C(x) = \frac{5x}{9 + x^2}.$$

Use the differential to approximate the changes in concentration for the following changes in x.

a. 1 to 1.5   **b.** 2 to 2.25

21. *Bacteria Population* The population of bacteria (in millions) in a certain culture x hr after an experimental nutrient is introduced into the culture is

$$P(x) = \frac{25x}{8 + x^2}.$$

Use the differential to approximate the changes in population for the following changes in x.

a. 2 to 2.5   **b.** 3 to 3.25

22. *Area of a Blood Vessel* The radius of a blood vessel is 1.7 mm. A drug causes the radius to change to 1.6 mm. Find the approximate change in the area of a cross section of the vessel.

23. *Volume of a Tumor* A tumor is approximately spherical in shape. If the radius of the tumor changes from 14 mm to 16 mm, find the approximate change in volume.

24. *Area of an Oil Slick* An oil slick is in the shape of a circle. Find the approximate increase in the area of the slick if its radius increases from 1.2 mi to 1.4 mi.

25. *Area of a Bacteria Colony* The shape of a colony of bacteria on a Petri dish is circular. Find the approximate increase in its area if the radius increases from 20 mm to 22 mm.

26. *Gray Wolves* Accurate methods of estimating the age of gray wolves are important to scientists who study wolf population dynamics. One method of estimating the age of a gray wolf is to measure the percent closure of the pulp cavity of a canine tooth and estimate age by

$$A(p) = \frac{1.181p}{94.359 - p},$$

where p is the percent closure and $A(p)$ is the age of the wolf (in yr).*

a. What is a sensible domain for this function?

b. Use differentials to estimate how long it will take for a gray wolf that first measures a 60% closure to obtain a 65% closure. Compare this with the actual value of about 0.55 years.

27. *Pigs* Researchers have observed that the weight of a female (gilt) pig can be estimated by the function

$$W(t) = -3.5 + 197.5e^{-e^{(-0.01394(t - 108.4))}},$$

where t is the age of the pig (in days) and $W(t)$ is the weight of the pig (in kg).†

a. If a particular gilt is 80 days old, use differentials to estimate how much it will gain before it is 90 days old.

b. What is the actual weight gain?

OTHER APPLICATIONS

28. *Volume* A spherical balloon is being inflated. Find the approximate change in volume if the radius increases from 4 cm to 4.2 cm.

29. *Volume* A spherical snowball is melting. Find the approximate change in volume if the radius decreases from 3 cm to 2.8 cm.

30. *Measurement Error* The edge of a square is measured as 3.45 inches, with a possible error of ± 0.002 inch. Estimate the maximum error in the area of the square.

31. *Measurement Error* The radius of a circle is measured as 4.87 inches, with a possible error of ± 0.040 inch. Estimate the maximum error in the area of the circle.

32. *Measurement Error* A sphere has a radius of 5.81 inches, with a possible error of ± 0.003 inch. Estimate the maximum error in the volume of the sphere.

33. *Measurement Error* A cone has a known height of 7.284 inches. The radius of the base is measured as 1.09 inches, with a possible error of ± 0.007 inch. Estimate the maximum error in the volume of the cone. (The volume of the cone is given by $V = (1/3)\pi r^2 h$.)

34. *Average Cost* The average cost to manufacture x dozen syringes is

$$A(x) = 0.04x^3 + 0.1x^2 + 0.5x + 6.$$

Use the differential to approximate the changes in the average cost for the following changes in x.

a. 3 to 4   **b.** 5 to 6

35. *Material Requirement* A cube 4 inches on an edge is given a protective coating 0.1 inch thick. About how much coating should a production manager order for 1,000 such cubes?

36. *Material Requirement* Beach balls 1 ft in diameter have a thickness of 0.03 inch. How much material would be needed to make 5,000 beach balls?

---

*Landon, D., C. Waite, R. Peterson, and L. Mech, "Evaluation of Age Determination Techniques for Gray Wolves," *Journal of Wildlife Management,* Vol. 62, No. 2, 1998, pp. 674–682.
†Van Lunen, T., and D. Cole, "Growth and Body Composition of Highly Selected Boars and Gilts," *Animal Science,* Vol. 67, 1998, pp. 107–116.

■ CHAPTER SUMMARY

In this chapter, after discussing how to find an absolute maximum or minimum, we saw a diversity of applications with that as the goal. Implicit differentiation is more a technique than an application, but it underlies related rate problems, in which one or more rates are given and another is to be found. In the last section of this chapter, we studied the differential as a way to find linear approximations to functions.

KEY TERMS

6.1 absolute maximum
absolute minimum
absolute extremum
(or extrema)

extreme value theorem
graphical optimization
6.2 maximum sustainable
harvest

6.3 explicit function
implicit differentiation

6.4 related rates
6.5 differential

CHAPTER 6 REVIEW EXERCISES

Find the locations and values of all absolute maxima and minima on the given intervals.

1. $f(x) = -x^2 + 5x + 1$; $[1, 4]$

2. $f(x) = 4x^2 - 8x - 3$; $[-1, 2]$

3. $f(x) = x^3 + 2x^2 - 15x + 3$; $[-4, 2]$

4. $f(x) = -2x^3 - 2x^2 + 2x - 1$; $[-3, 1]$

5. When solving applied extrema problems, why is it necessary to check the endpoints of the domain?

6. Find the absolute maximum and minimum of $f(x) = \dfrac{2 \ln x}{x^2}$

a. on the interval $[1, 4]$;

b. on the interval $[2, 5]$.

7. When is it necessary to use implicit differentiation?

8. When a term involving y is differentiated in implicit differentiation, it is multiplied by dy/dx. Why? Why aren't terms involving x multiplied by dx/dx?

Find dy/dx in Exercises 9–14.

9. $x^2y^3 + 4xy = 2$

10. $\dfrac{x}{y} - 4y = 3x$

11. $9\sqrt{x} + 4y^3 = \dfrac{2}{x}$

12. $2\sqrt{y - 1} = 8x^{2/3}$

13. $\dfrac{x + 2y}{x - 3y} = y^{1/2}$

14. $\dfrac{6 + 5x}{2 - 3y} = \dfrac{1}{5x}$

15. Find the equation of the line tangent to the graph of $\sqrt{2x} - 4yx = -22$ at the point $(2, 3)$.

16. What is the difference between a related rate problem and an applied extremum problem?

17. Why is implicit differentiation used in related rate problems?

Find dy/dt in Exercises 18–21.

18. $y = 8x^3 - 7x^2$; $\dfrac{dx}{dt} = 4, x = 2$

19. $y = \dfrac{9 - 4x}{3 + 2x}$; $\dfrac{dx}{dt} = -1, x = -3$

20. $y = \dfrac{1 + \sqrt{x}}{1 - \sqrt{x}}$; $\dfrac{dx}{dt} = -4, x = 4$

21. $\dfrac{x^2 + 5y}{x - 2y} = 2$; $\dfrac{dx}{dt} = 1, x = 2, y = 0$

22. What is a differential? What is it used for?

Evaluate dy in Exercises 23–26.

23. $y = 8 - x^2 + x^3$; $x = -1, \Delta x = 0.02$

24. $y = \dfrac{3x - 7}{2x + 1}$; $x = 2, \Delta x = 0.003$

25. $y = \sin x$; $x = \dfrac{\pi}{3}, \Delta x = -0.01$

26. $y = \tan x$; $x = \dfrac{\pi}{4}, \Delta x = -0.02$

27. a. Suppose x and y are related by the equation

$$-12x + x^3 + y + y^2 = 4.$$

Find all critical points on the curve.

b. Determine whether the critical points found in part a are relative maxima or relative minima by taking values of x nearby and solving for the corresponding values of y.

c. Is there an absolute maximum or minimum for x and y in the relationship given in part a? Why or why not?

28. In Exercise 27, implicit differentiation was used to find the relative extrema. The exercise was contrived to avoid various difficulties that could have arisen. Discuss some of the difficulties that might be encountered in such problems, and how these difficulties might be resolved.

Applications

LIFE SCIENCES

29. *Pollution* A circle of pollution is spreading from a broken underwater waste disposal pipe, with the radius increasing at the rate of 4 ft/min. Find the rate of change of the area of the circle when the radius is 7 ft.

30. *Population Growth* Many populations grow according to the equation

$$\frac{dx}{dt} = rx(N - x),$$

where r is a constant involving the rate of growth and N is the carrying capacity of the environment, beyond which the population decreases. Show that the graph of x has an inflection point where $x = N/2$. (*Hint:* Use implicit differentiation. Then set $d^2x/dt^2 = 0$, and factor.)

31. *Fungal Growth* As we saw earlier in this chapter, because of the time that many people spend indoors, there is a concern about the health risk of being exposed to harmful fungi that thrive in buildings. The risk appears to increase in damp environments. Researchers have discovered that by controlling both the temperature and the relative humidity in a building, the growth of the fungus E. *herbariorum* can be limited. The relationship between the temperature and relative humidity which limits growth can be described by

$$R(T) = -0.0011T^3 + 0.1005T^2 - 2.6876T + 98.171,$$
$$15 \le T \le 46,$$

where $R(T)$ is the relative humidity (in %) and T is the temperature (in °C).* Find the temperature at which the relative humidity is minimized.

*Rowan, N., C. Johnstone, R. McLean, J. Anderson, and J. Clarke, "Prediction of Toxigenic Fungal Growth in Buildings by Using a Novel Modelling System," *Applied and Environmental Microbiology,* Vol. 65, No. 11, Nov. 1999, pp. 4814–4821.

32. *Dentin Growth* As we saw earlier in this chapter, the dentinal formation of molars in mice has been studied by researchers in Copenhagen. They determined that the growth curve that best fits dentinal formation for the first molar is

$$M(t) = 1.3386309 - 0.4321173t + 0.0564512t^2$$
$$- 0.0020506t^3 + 0.0000315t^4 - 0.0000001785t^5,$$
$$5 \le t \le 51,$$

where t is the age of the mouse (in days), and $M(t)$ is the cumulative dentin volume (in 10^{-1} mm^3).*

a. Use a graphing calculator to sketch the graph of this function on $[5, 51]$ by $[0, 7.5]$.

b. Find the time in which the dentin formation is growing most rapidly. (*Hint:* Find the maximum value of the derivative of this function.)

33. *Human Skin Surface* As we saw earlier in this chapter, the surface of the skin is made up of a network of intersecting lines which form polygons. Researchers have discovered a functional relationship between the age of a female and the number of polygons per area of skin according to

$$P(t) = 237.09 - 8.0398t + 0.20813t^2 - 0.0027563t^3$$
$$+ 0.000013016t^4, \qquad 0 \le t \le 95,$$

where t is the age of the person (in yr), and $P(t)$ is the number of polygons for a particular surface area of skin.[†]

a. Use a graphing calculator to sketch a graph of $P(t)$ on $[0, 95]$ by $[0, 300]$.

b. Find the maximum and minimum number of polygons per area predicted by the model.

c. Discuss the accuracy of this model for older people.

34. *Pigs* Researchers have observed that the total nitrogen content of a female (gilt) pig can be estimated by the function

$$W(t) = 0.109 + 4.47e^{-e^{(-0.01708(t-107.4))}},$$

where t is the age of the pig (in days) and $W(t)$ is the weight of the nitrogen in the pig (in kg).[‡]

a. If a particular gilt is 70 days old, estimate how much nitrogen the pig will gain during the next two days.

b. What is the actual nitrogen gain?

OTHER APPLICATIONS

35. *Sliding Ladder* A 50-ft ladder is placed against a building. The top of the ladder is sliding down the building at the rate of 2 ft/min. Find the rate at which the base of the ladder is slipping away from the building at the instant that the base is 30 ft from the building.

36. *Spherical Radius* A large weather balloon is being inflated with air at the rate of 1.2 ft^3/min. Find the rate of change of the radius when the radius is 1.2 ft.

37. *Water Level* A water trough 2 ft across, 4 ft long, and 1 ft deep has ends in the shape of isosceles triangles. (See the figure.) It is being filled with 3.5 ft^3 of water per minute. Find the rate at which the depth of water in the tank is changing when the water is 1/3 ft deep.

38. *Volume* Approximate the volume of coating on a sphere of radius 4 inches if the coating is 0.02 inch thick.

39. *Area* A square has an edge of 9.2 inches, with a possible error in the measurement of ± 0.04 inch. Estimate the possible error in the area of the square.

40. *Package Dimensions* UPS has the following rule regarding package dimensions. The length can be no more than 108 inches, and the length plus the girth (twice the sum of the width and the height) can be no more than 130 inches. If the width of a package is 4 inches more than its height and it has the maximum girth, find the length that produces maximum volume.

---

*Matsumoto, B., K. Nonaka, and M. Nakata, "A Genetic Study of Dentin Growth in the Mandibular Second and Third Molars of Male Mice," *Journal of Craniofacial Genetics and Developmental Biology,* Vol. 16, No. 3, July–Sept. 1996, pp. 137–147.
†Voros, E., C. Robert, and A. Robert, "Age-Related Changes of the Human Skin Surface Microrelief," *Gerontology,* Vol. 36, 1990, pp. 276–285.
‡Van Lunen, T., and D. Cole, "Growth and Body Composition of Highly Selected Boars and Gilts," *Animal Science,* Vol. 67, 1998, pp. 107–116.

41. *Pursuit* A boat moves north at a constant speed. A second boat, moving at the same speed, pursues the first boat in such a way that it always points directly at the first boat. When the first boat is at the point $(0, 1)$, the second boat is at the point $(6, 2.5)$, with the positive y-axis pointing north. It can then be shown that the curve traced by the second boat, known as a pursuit curve, is given by

$$y = \frac{x^2}{16} - 2 \ln x + \frac{1}{4} + 2 \ln 6.*$$

Find the y-coordinate of the southernmost point of the second boat's path.

42. *Playground Area* The city park department is planning an enclosed play area in a new park. One side of the area will be against an existing building, with no fence needed there. Find the dimensions of the maximum rectangular area that can be made with 900 m of fence.

43. *Packaging Design* The packaging department of a corporation is designing a box with a square base and no top. The volume is to be 32 m³. To reduce cost, the box is to have minimum surface area. What dimensions (height, length, and width) should the box have?

44. *Packaging Design* Fruit juice will be packaged in cylindrical cans with a volume of 40 cubic inches each. The top and bottom of the can cost 4¢ per square inch, while the sides cost 3¢ per square inch. Find the radius and height of the can of minimum cost.

45. *Packaging Design* A company plans to package its product in a cylinder that is open at one end. The cylinder is to have a volume of 27π cubic inches. What radius should the circular bottom of the cylinder have to minimize the cost of the material?

---

*For example, see *Differential Equations: Theory and Applications,* Ray Redheffer, Jones and Bartlett Publishers, 1991, pp. 107–108.

EXTENDED APPLICATION: A Total Cost Model for a Training Program*

In this application, we set up a mathematical model for determining the total costs in setting up a training program. Then we use calculus to find the time interval between training programs that produces the minimum total cost. The model assumes that the demand for trainees is constant and that the fixed cost of training a batch of trainees is known. Also, it is assumed that people who are trained, but for whom no job is readily available, will be paid a fixed amount per month while waiting for a job to open up.

The model uses the following variables.

D = demand for trainees per month

N = number of trainees per batch

C_1 = fixed cost of training a batch of trainees

C_2 = marginal cost of training per trainee per month

C_3 = salary paid monthly to a trainee who has not yet been given a job after training

m = time interval in months between successive batches of trainees

t = length of training program in months

$Z(m)$ = total monthly cost of program

The total cost of training a batch of trainees is given by $C_1 + NtC_2$. However, $N = mD$, so that the total cost per batch is $C_1 + mDtC_2$.

After training, personnel are given jobs at the rate of D per month. Thus, $N - D$ of the trainees will not get a job the first month, $N - 2D$ will not get a job the second month, and so on. The $N - D$ trainees who do not get a job the first month produce total costs of $(N - D)C_3$, those not getting jobs during the second month produce costs of $(N - 2D)C_3$, and so on. Since $N = mD$, the costs during the first month can be written as

$$(N - D)C_3 = (mD - D)C_3 = (m - 1)DC_3,$$

while the costs during the second month are $(m - 2)DC_3$, and so on. The total cost for keeping the trainees without a job is thus

$$(m - 1)DC_3 + (m - 2)DC_3$$
$$+ (m - 3)DC_3 + \cdots + 2DC_3 + DC_3,$$

which can be factored to give

$$DC_3[(m - 1) + (m - 2) + (m - 3) + \cdots + 2 + 1].$$

The expression in brackets is the sum of the terms of an arithmetic sequence, discussed in most algebra texts. Using formulas for arithmetic sequences, the expression in brackets can be shown to equal $m(m - 1)/2$, so that we have

$$DC_3\left[\frac{m(m - 1)}{2}\right] \qquad (1)$$

as the total cost for keeping jobless trainees.

The total cost per batch is the sum of the training cost per batch, $C_1 + mDtC_2$, and the cost of keeping trainees without a proper job, given by equation (1). Since we assume that a batch of trainees is trained every m months, the total cost per month, $Z(m)$, is given by

$$Z(m) = \frac{C_1 + mDtC_2}{m} + \frac{DC_3\left[\dfrac{m(m - 1)}{2}\right]}{m}$$

$$= \frac{C_1}{m} + DtC_2 + DC_3\left(\frac{m - 1}{2}\right).$$

Exercises

1. Find $Z'(m)$.

2. Solve the equation $Z'(m) = 0$.

   As a practical matter, it is usually required that m be a whole number. If m does not come out to be a whole number in Exercise 2, then m^+ and m^-, the two whole numbers closest to m, must be chosen. Calculate both $Z(m^+)$ and $Z(m^-)$; the smaller of the two provides the optimum value of Z.

3. Suppose a company finds that its demand for trainees is 3 per month, that a training program requires 12 months, that the fixed cost of training a batch of trainees is $15,000, that the marginal cost per trainee per month is $100, and that trainees are paid $900 per month after training but before going to work. Use your result from Exercise 2 and find m.

4. Since m is not a whole number, find m^+ and m^-.

5. Calculate $Z(m^+)$ and $Z(m^-)$.

6. What is the optimum time interval between successive batches of trainees? How many trainees should be in a batch?

---

*Based on "A Total Cost Model for a Training Program" by P. L. Goyal and S. K. Goyal, Faculty of Commerce and Administration, Concordia University. Used with permission.

Integration

If we know the rate at which a quantity is changing, we can find the total change over a period of time by integrating. An exercise in Section 4 illustrates this process with a model for the growth of the horns in bighorn rams. The integral of the rate function tells us how much the ram's horn increases in size over a given time.

■ 7.1 Antiderivatives

■ 7.2 Substitution

■ 7.3 Area and the Definite Integral

■ 7.4 The Fundamental Theorem of Calculus

■ 7.5 Integrals of Trigonometric Functions

■ 7.6 The Area Between Two Curves

Review Exercises

Extended Application: Estimating Depletion Dates for Minerals

In previous chapters, we studied the mathematical process of finding the derivative of a function, and we considered various applications of derivatives. The material that was covered belongs to the branch of calculus called *differential calculus.* In this chapter and the next we will study another branch of calculus, called *integral calculus.* Like the derivative of a function, the definite integral of a function is a special limit with many diverse applications. Geometrically, the derivative is related to the slope of the tangent line to a curve, while the definite integral is related to the area under a curve. As this chapter will show, differential and integral calculus are connected by the Fundamental Theorem of Calculus.

■ 7.1 ANTIDERIVATIVES

? **THINK ABOUT IT**   How long will it take a population of bacteria to reach a certain size, given that we know its growth rate?

Using *antiderivatives,* we can answer this question.

Functions used in applications in previous chapters have provided information about a *total amount* of a quantity, such as cost, revenue, profit, temperature, gallons of oil, or distance. Working with these functions provided information about the rate of change of these quantities and allowed us to answer important questions about the extrema of the functions. It is not always possible to find ready-made functions that provide information about the total amount of a quantity, but it is often possible to collect enough data to come up with a function that gives the *rate of change* of a quantity. Since we know that derivatives give the rate of change when the total amount is known, is it possible to reverse the process and use a known rate of change to get a function that gives the total amount of a quantity? The answer is yes; this reverse process, called *antidifferentiation,* is the topic of this section. The *antiderivative* of a function is defined as follows.

> **ANTIDERIVATIVE**
>
> If $F'(x) = f(x)$, then $F(x)$ is an **antiderivative** of $f(x)$.

EXAMPLE 1   Antiderivative

(a) If $F(x) = 10x$, then $F'(x) = 10$, so $F(x) = 10x$ is an antiderivative of $f(x) = 10$.

(b) For $F(x) = x^2$, $F'(x) = 2x$, making $F(x) = x^2$ an antiderivative of $f(x) = 2x$.

EXAMPLE 2   Antiderivative
Find an antiderivative of $f(x) = 5x^4$.

Solution To find a function $F(x)$ whose derivative is $5x^4$, work backwards. Recall that the derivative of x^n is nx^{n-1}. If

$$nx^{n-1} \text{ is } 5x^4,$$

then $n - 1 = 4$ and $n = 5$, so x^5 is an antiderivative of $5x^4$.

EXAMPLE 3 Population

Suppose a population is growing at a rate given by $f(x) = e^x$, where x is time in years from some initial date. Find a function giving the population at time x.

Solution Let the population function be $F(x)$. Then

$$f(x) = F'(x) = e^x.$$

The derivative of the function defined by $F(x) = e^x$ is $F'(x) = e^x$, so one possible population function with the given growth rate is $F(x) = e^x$. ■

The function from Example 1(b), defined by $F(x) = x^2$, is not the only function whose derivative is $f(x) = 2x$; for example, both $G(x) = x^2 + 2$ and $H(x) = x^2 - 7$ have $f(x) = 2x$ as a derivative. In fact, for any real number C, the function $F(x) = x^2 + C$ has $f(x) = 2x$ as its derivative. This means that there is a *family* or *class* of functions having $2x$ as an antiderivative. As the next theorem states, if two functions $F(x)$ and $G(x)$ are antiderivatives of $f(x)$, then $F(x)$ and $G(x)$ can differ only by a constant.

If $F(x)$ and $G(x)$ are both antiderivatives of a function $f(x)$ on an interval, then there is a constant C such that

$$F(x) - G(x) = C.$$

(Two antiderivatives of a function can differ only by a constant.) The arbitrary real number C is called an integration constant.

For example,

$$F(x) = x^2 + 2, \qquad G(x) = x^2, \qquad \text{and} \qquad H(x) = x^2 - 4$$

are all antiderivatives of $f(x) = 2x$, and any two of them differ only by a constant. The derivative of a function gives the slope of the tangent line at any x-value. The fact that these three functions have the same derivative, $f(x) = 2x$, means that their slopes at any particular value of x are the same, as shown in Figure 1. Thus, each graph can be obtained from another by a vertical shift of $|C|$ units, where C is any constant. We will represent this family of antiderivatives of $f(x)$ by $F(x) + C$.

The family of all antiderivatives of the function f is indicated by

$$\int f(x) \, dx.$$

The symbol $\int$ is the **integral sign**, $f(x)$ is the **integrand**, and $\int f(x) \, dx$ is called an **indefinite integral**, the most general antiderivative of f.

INDEFINITE INTEGRAL

If $F'(x) = f(x)$, then

$$\int f(x) \, dx = F(x) + C,$$

for any real number C.

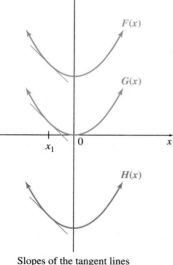

Slopes of the tangent lines at $x = x_1$ are the same.

FIGURE 1

For example, using this notation,

$$\int 2x\,dx = x^2 + C.$$

The dx in the indefinite integral indicates that $\int f(x)\,dx$ is the "integral of $f(x)$ *with respect* to x" just as the symbol dy/dx denotes the "derivative of y with respect to x." For example, in the indefinite integral $\int 2ax\,dx$, dx indicates that a is to be treated as a constant and x as the variable, so that

$$\int 2ax\,dx = \int a(2x)\,dx = ax^2 + C.$$

On the other hand,

$$\int 2ax\,da = a^2x + C = xa^2 + C.$$

The symbol "da" indicates that a is the variable of integration. A more complete interpretation of dx will be discussed later.

The symbol $\int f(x)\,dx$ was created by G. W. Leibniz (1646–1716) in the latter part of the seventeenth century. The $\int$ is an elongated S from *summa,* the Latin word for *sum.* The word *integral* as a term in the calculus was coined by Jakob Bernoulli (1654–1705), a Swiss mathematician who corresponded frequently with Leibniz. The relationship between sums and integrals will be clarified in a later section.

The method of working backwards, as above, to find an antiderivative is not very satisfactory. Some rules for finding antiderivatives are needed. Since the process of finding an indefinite integral is the inverse of the process of finding a derivative, each formula for derivatives leads to a rule for indefinite integrals.

As mentioned in Example 2, the derivative of x^n is found by multiplying x by n and reducing the exponent on x by 1. To find an indefinite integral—that is, to undo what was done—*increase* the exponent by 1 and *divide* by the new exponent, $n + 1$.

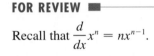

FOR REVIEW

Recall that $\dfrac{d}{dx}x^n = nx^{n-1}$.

POWER RULE

For any real number $n \neq -1$,

$$\int x^n\,dx = \frac{x^{n+1}}{n+1} + C.$$

This rule can be verified by differentiating the expression on the right above:

$$\frac{d}{dx}\left(\frac{x^{n+1}}{n+1} + C\right) = \frac{n+1}{n+1}x^{(n+1)-1} + 0 = x^n.$$

(If $n = -1$, the expression in the denominator is 0, and the above rule cannot be used. Finding an antiderivative for this case is discussed later.)

EXAMPLE 4   Power Rule

Use the power rule to find each indefinite integral.

(a) $\displaystyle\int t^3\,dt$

Solution Use the power rule with $n = 3$.

$$\int t^3 \, dt = \frac{t^{3+1}}{3+1} + C = \frac{t^4}{4} + C$$

(b) $\int \dfrac{1}{t^2} \, dt$

Solution First, write $1/t^2$ as t^{-2}. Then

$$\int \frac{1}{t^2} \, dt = \int t^{-2} \, dt = \frac{t^{-1}}{-1} + C = -\frac{1}{t} + C.$$

(c) $\int \sqrt{u} \, du$

Solution Since $\sqrt{u} = u^{1/2}$,

$$\int \sqrt{u} \, du = \int u^{1/2} \, du = \frac{u^{3/2}}{1/2 + 1} + C = \frac{2}{3} u^{3/2} + C.$$

To check this, differentiate $(2/3)u^{3/2} + C$; the derivative is $u^{1/2}$, the original function.

(d) $\int dx$

Solution Write dx as $1 \cdot dx$ and use the fact that $x^0 = 1$ for any nonzero number x to get

$$\int dx = \int 1 \, dx = \int x^0 \, dx = \frac{x^1}{1} + C = x + C.$$

FOR REVIEW ■

Recall that $\dfrac{d}{dx}[f(x) \pm g(x)] = [f'(x) \pm g'(x)]$ and $\dfrac{d}{dx}[kf(x)] = kf'(x)$.

As shown earlier, the derivative of the product of a constant and a function is the product of the constant and the derivative of the function. A similar rule applies to indefinite integrals. Also, since derivatives of sums or differences are found term by term, indefinite integrals also can be found term by term.

CONSTANT MULTIPLE RULE AND SUM OR DIFFERENCE RULE

If all indicated integrals exist,

$$\int k \cdot f(x) \, dx = k \int f(x) \, dx \qquad \text{for any real number } k,$$

and

$$\int [f(x) \pm g(x)] \, dx = \int f(x) \, dx \pm \int g(x) \, dx.$$

EXAMPLE 5 Rules of Integration

Use the rules to find each integral.

(a) $\int 2v^3 \, dv$

Solution By the constant multiple rule and the power rule,

$$\int 2v^3 \, dv = 2 \int v^3 \, dv = 2\left(\frac{v^4}{4}\right) + C = \frac{v^4}{2} + C.$$

(b) $\int \dfrac{12}{z^5} \, dz$

Solution Use negative exponents.

$$\int \frac{12}{z^5} \, dz = \int 12z^{-5} \, dz = 12 \int z^{-5} \, dz = 12\left(\frac{z^{-4}}{-4}\right) + C$$

$$= -3z^{-4} + C = \frac{-3}{z^4} + C$$

(c) $\int (3z^2 - 4z + 5) \, dz$

Solution Using the rules in this section,

$$\int (3z^2 - 4z + 5) \, dz = 3 \int z^2 \, dz - 4 \int z \, dz + 5 \int dz$$

$$= 3\left(\frac{z^3}{3}\right) - 4\left(\frac{z^2}{2}\right) + 5z + C$$

$$= z^3 - 2z^2 + 5z + C.$$

Only one constant C is needed in the answer; the three constants from integrating term by term are combined. In Example 5(a), C represents any real number, so it is not necessary to multiply it by 2 in the next-to-last step. ◼

NOTE You can always check the result of an integral calculation by taking the derivative of your result and verifying that it matches the original function to be integrated.

EXAMPLE 6 Rules of Integration

Use the rules to find each integral.

(a) $\int \dfrac{x^2 + 1}{\sqrt{x}} \, dx$

Solution First rewrite the integrand as follows.

$$\int \frac{x^2 + 1}{\sqrt{x}} \, dx = \int \left(\frac{x^2}{\sqrt{x}} + \frac{1}{\sqrt{x}}\right) dx$$

$$= \int \left(\frac{x^2}{x^{1/2}} + \frac{1}{x^{1/2}}\right) dx$$

$$= \int (x^{3/2} + x^{-1/2}) \, dx$$

Now find the antiderivative.

$$\int (x^{3/2} + x^{-1/2})\, dx = \frac{x^{5/2}}{5/2} + \frac{x^{1/2}}{1/2} + C$$

$$= \frac{2}{5}x^{5/2} + 2x^{1/2} + C$$

(b) $\displaystyle\int (x^2 - 1)^2\, dx$

Solution Square the binomial first, and then find the antiderivative.

$$\int (x^2 - 1)^2\, dx = \int (x^4 - 2x^2 + 1)\, dx$$

$$= \frac{x^5}{5} - \frac{2x^3}{3} + x + C$$

To check integration, take the derivative of the result. For instance, in Example 5(c) check that $z^3 - 2z^2 + 5z + C$ is the required indefinite integral by taking the derivative

$$\frac{d}{dz}(z^3 - 2z^2 + 5z + C) = 3z^2 - 4z + 5,$$

which agrees with the original information.

The restriction $n \neq -1$ was necessary in the formula for $\int x^n\, dx$ since $n = -1$ made the denominator of $1/(n + 1)$ equal to 0. To find $\int x^n\, dx$ when $n = -1$, that is, to find $\int x^{-1}\, dx$, recall the differentiation formula for the logarithmic function: the derivative of $f(x) = \ln |x|$, where $x \neq 0$, is $f'(x) = 1/x = x^{-1}$. This formula for the derivative of $f(x) = \ln |x|$ gives a formula for $\int x^{-1}\, dx$.

INDEFINITE INTEGRAL OF x^{-1}

$$\int x^{-1}\, dx = \int \frac{1}{x}\, dx = \ln |x| + C$$

CAUTION Don't neglect the absolute value sign in the natural logarithm when integrating x^{-1}. If x can take on a negative value, $\ln x$ will be undefined there. Note, however, that the absolute value is redundant (but harmless) in an expression such as $\ln |x^2 + 1|$, since $x^2 + 1$ can never be negative.

EXAMPLE 7 Integrals

(a) $\displaystyle\int \frac{4}{x}\, dx = 4 \int \frac{1}{x}\, dx = 4 \ln |x| + C$

(b) $\displaystyle\int \left(-\frac{5}{x} + x^2\right) dx = -5 \ln |x| + \frac{x^3}{3} + C$

It was shown earlier that the derivative of $f(x) = e^x$ is $f'(x) = e^x$, and the derivative of $f(x) = a^x$ is $f'(x) = (\ln a)a^x$. Also, the derivative of $f(x) = e^{kx}$ is

$f'(x) = k \cdot e^{kx}$, and the derivative of $f(x) = a^{kx}$ is $f'(x) = k(\ln a)a^{kx}$. These results lead to the following formulas for indefinite integrals of exponential functions.

INDEFINITE INTEGRALS OF EXPONENTIAL FUNCTIONS

$$\int e^x \, dx = e^x + C$$

$$\int e^{kx} \, dx = \frac{e^{kx}}{k} + C, \qquad k \neq 0$$

$$\int a^x \, dx = \frac{a^x}{\ln a} + C$$

$$\int a^{kx} \, dx = \frac{a^{kx}}{k(\ln a)} + C, \qquad k \neq 0$$

EXAMPLE 8   Exponential Functions

(a) $\displaystyle \int 9e^t \, dt = 9 \int e^t \, dt = 9e^t + C$

(b) $\displaystyle \int e^{9t} \, dt = \frac{e^{9t}}{9} + C$

(c) $\displaystyle \int 3e^{(5/4)u} \, du = 3\left(\frac{e^{(5/4)u}}{5/4}\right) + C$

$$= 3\left(\frac{4}{5}\right)e^{(5/4)u} + C$$

$$= \frac{12}{5}e^{(5/4)u} + C$$

(d) $\displaystyle \int 2^{-5x} \, dx = \frac{2^{-5x}}{-5(\ln 2)} + C = -\frac{2^{-5x}}{5(\ln 2)} + C$

In all these examples, the antiderivative family of functions was found. In many applications, however, the given information allows us to determine the value of the integration constant C. The next examples illustrate this idea.

EXAMPLE 9   Slope

Find a function f whose graph has slope $f'(x) = 6x^2 + 4$ and goes through the point $(1, 1)$.

Solution Since $f'(x) = 6x^2 + 4$,

$$f(x) = \int (6x^2 + 4) \, dx = 2x^3 + 4x + C.$$

The graph of f goes through $(1, 1)$, so C can be found by substituting 1 for x and 1 for $f(x)$.

$$1 = 2(1)^3 + 4(1) + C$$
$$1 = 6 + C$$
$$C = -5$$

Finally, $f(x) = 2x^3 + 4x - 5$.

EXAMPLE 10 Population

Suppose that a population of bacteria grows at a rate that slows down with time, so that the growth rate after t days is given by

$$P'(t) = \frac{5{,}000}{\sqrt{t}}$$

for $t \geq 1$. Suppose further that when $t = 1$, the population is 25,000.

(a) Find the function $P(t)$ giving the population at time t.

Solution Write $5{,}000/\sqrt{t}$ as $5{,}000/t^{1/2}$ or $5{,}000t^{-1/2}$, and then use the definite integral rules to integrate the function.

$$\int \frac{5{,}000}{\sqrt{t}}\, dt = \int 5{,}000t^{-1/2}\, dt$$
$$= 5{,}000(2t^{1/2}) + C$$
$$= 10{,}000t^{1/2} + C$$

To find the value of C, use the fact that $P(1) = 25{,}000$.

$$P(t) = 10{,}000t^{1/2} + C$$
$$25{,}000 = 10{,}000 \cdot 1 + C$$
$$C = 15{,}000$$

With this result, the population function is $P(t) = 10{,}000t^{1/2} + 15{,}000$.

(b) How many days will it take for the population to reach 115,000?

Solution Set $P(t)$ equal to 115,000.

$$10{,}000t^{1/2} + 15{,}000 = 115{,}000$$
$$10{,}000t^{1/2} = 100{,}000$$
$$t^{1/2} = 10$$
$$t = 100$$

The population will reach 115,000 after 100 days.

EXAMPLE 11 Velocity and Acceleration

Recall that if the function $s(t)$ gives the position of a particle at time t, then its velocity $v(t)$ and its acceleration $a(t)$ are given by

$$v(t) = s'(t) \qquad \text{and} \qquad a(t) = v'(t) = s''(t).$$

(a) Suppose the velocity of an object is $v(t) = 6t^2 - 8t$ and that the object is at 5 when time is 0. Find $s(t)$.

Solution Since $v(t) = s'(t)$, the function $s(t)$ is an antiderivative of $v(t)$:

$$s(t) = \int v(t)\, dt = \int (6t^2 - 8t)\, dt$$
$$= 2t^3 - 4t^2 + C$$

for some constant C. Find C from the given information that $s = 5$ when $t = 0$.

$$s(t) = 2t^3 - 4t^2 + C$$
$$5 = 2(0)^3 - 4(0)^2 + C$$
$$5 = C$$
$$s(t) = 2t^3 - 4t^2 + 5$$

(b) Many experiments have shown that when an object is dropped, its acceleration (ignoring air resistance) is constant. This constant has been found to be approximately 32 feet per second every second; that is,

$$a(t) = -32.$$

The negative sign is used because the object is falling. Suppose an object is thrown down from the top of the 1,100-foot-tall Sears Tower in Chicago. If the initial velocity of the object is -20 feet per second, find $s(t)$, the distance of the object from the ground at time t.

Solution First find $v(t)$ by integrating $a(t)$:

$$v(t) = \int (-32) \, dt = -32t + k.$$

When $t = 0$, $v(t) = -20$:

$$-20 = -32(0) + k$$
$$-20 = k$$

and

$$v(t) = -32t - 20.$$

Be sure to evaluate the constant of integration k before integrating again to get $s(t)$. Now integrate $v(t)$ to find $s(t)$.

$$s(t) = \int (-32t - 20) \, dt = -16t^2 - 20t + C$$

Since $s(t) = 1{,}100$ when $t = 0$, we can substitute these values into the equation for $s(t)$ to get $C = 1{,}100$ and

$$s(t) = -16t^2 - 20t + 1{,}100$$

as the distance of the object from the ground after t seconds.

(c) Use the equations derived in part (b) to find the velocity of the object when it hit the ground and how long it took to strike the ground.

Solution When the object strikes the ground, $s = 0$, so

$$0 = -16t^2 - 20t + 1{,}100.$$

To solve this equation for t, factor out the common factor of -4 and then use the quadratic formula.

$$0 = -4(4t^2 + 5t - 275)$$
$$t = \frac{-5 \pm \sqrt{25 + 4{,}400}}{8} \approx \frac{-5 \pm 66.5}{8}$$

Only the positive value of t is meaningful here: $t \approx 7.69$. It takes the object about 7.69 seconds to strike the ground. From the velocity equation, with $t = 7.69$, we find

$$v(t) = -32t - 20$$
$$v(7.69) = -32(7.69) - 20 \approx -266,$$

so the object was falling (as indicated by the negative sign) at about 266 feet per second (about 181 mph) when it hit the ground.

7.1 EXERCISES

1. What must be true of $F(x)$ and $G(x)$ if both are antiderivatives of $f(x)$?

2. How is the antiderivative of a function related to the function?

3. In your own words, describe what is meant by an integrand.

4. Explain why the restriction $n \neq -1$ is necessary in the rule $\int x^n \, dx = \dfrac{x^{n+1}}{n+1} + C$.

Find each of the following.

5. $\displaystyle\int 6 \, dk$

6. $\displaystyle\int 2 \, dy$

7. $\displaystyle\int (2z + 3) \, dz$

8. $\displaystyle\int (3x - 5) \, dx$

9. $\displaystyle\int (t^2 - 4t + 5) \, dt$

10. $\displaystyle\int (5x^2 - 6x + 3) \, dx$

11. $\displaystyle\int (4z^3 + 3z^2 + 2z - 6) \, dz$

12. $\displaystyle\int (12y^3 + 6y^2 - 8y + 5) \, dy$

13. $\displaystyle\int 5\sqrt{z} \, dz$

14. $\displaystyle\int t^{1/4} \, dt$

15. $\displaystyle\int x(x^2 - 3) \, dx$

16. $\displaystyle\int x^2(x^4 + 4x + 3) \, dx$

17. $\displaystyle\int \left(4\sqrt{v} - 3v^{3/2}\right) dv$

18. $\displaystyle\int \left(15x\sqrt{x} + 2\sqrt{x}\right) dx$

19. $\displaystyle\int (10u^{3/2} - 14u^{5/2}) \, du$

20. $\displaystyle\int (56t^{5/2} + 18t^{7/2}) \, dt$

21. $\displaystyle\int \left(\frac{1}{z^2}\right) dz$

22. $\displaystyle\int \left(\frac{4}{x^3}\right) dx$

23. $\displaystyle\int \left(\frac{1}{y^3} - \frac{1}{\sqrt{y}}\right) dy$

24. $\displaystyle\int \left(\sqrt{u} + \frac{1}{u^2}\right) du$

25. $\displaystyle\int (-9t^{-2} - 2t^{-1}) \, dt$

26. $\displaystyle\int (8x^{-3} + 4x^{-1}) \, dx$

27. $\displaystyle\int \frac{1}{3x^2} \, dx$

28. $\displaystyle\int \frac{2}{3x^4} \, dx$

29. $\displaystyle\int 3e^{-0.2x} \, dx$

30. $\displaystyle\int -4e^{0.2v} \, dv$

31. $\displaystyle\int \left(\frac{3}{x} + 4e^{-0.5x}\right) dx$

32. $\displaystyle\int \left(\frac{9}{x} - 3e^{-0.4x}\right) dx$

33. $\displaystyle\int \frac{1 + 2t^3}{t} \, dt$

34. $\displaystyle\int \frac{2y^{1/2} - 3y^2}{y} \, dy$

35. $\displaystyle\int (e^{2u} + 4u) \, du$

36. $\displaystyle\int (v^2 - e^{3v}) \, dv$

37. $\displaystyle\int (x + 1)^2 \, dx$

38. $\displaystyle\int (2y - 1)^2 \, dy$

39. $\displaystyle\int \frac{\sqrt{x} + 1}{\sqrt[3]{x}} \, dx$

40. $\displaystyle\int \frac{1 - 2\sqrt[3]{z}}{\sqrt[3]{z}} \, dz$

41. $\displaystyle\int 10^x \, dx$

42. $\displaystyle\int 3^{2x} \, dx$

43. Find an equation of the curve whose tangent line has a slope of

$$f'(x) = x^{2/3},$$

given that the point $(1, 3/5)$ is on the curve.

44. The slope of the tangent line to a curve is given by

$$f'(x) = 6x^2 - 4x + 3.$$

If the point $(0, 1)$ is on the curve, find an equation of the curve.

Applications

LIFE SCIENCES

45. *Flour Beetles* A model for describing the population of adult flour beetles involves evaluating the integral

$$\int \frac{g(x)}{x} \, dx,$$

where $g(x)$ is the per-unit-abundance growth rate for a population of size x.* The researchers consider the simple case in which $g(x) = a - bx$ for positive constants a and b. Find the integral in this case.

46. *Biochemical Excretion* If the rate of excretion of a biochemical compound is given by

$$f'(t) = 0.01e^{-0.01t},$$

the total amount excreted by time t (in minutes) is $f(t)$.

a. Find an expression for $f(t)$.

b. If 0 units are excreted at time $t = 0$, how many units are excreted in 10 minutes?

47. *Concentration of a Solute* According to Fick's law, the diffusion of a solute across a cell membrane is given by

$$c'(t) = \frac{kA}{V}[C - c(t)], \tag{1}$$

where A is the area of the cell membrane, V is the volume of the cell, $c(t)$ is the concentration inside the cell at time t, C is the concentration outside the cell, and k is a constant. If c_0 represents the concentration of the solute inside the cell when $t = 0$, then it can be shown that

$$c(t) = (c_0 - C)e^{-kAt/V} + M. \tag{2}$$

a. Use the last result to find $c'(t)$.

b. Substitute back into equation (1) to show that (2) is indeed the correct antiderivative of (1).

OTHER APPLICATIONS

48. *Seat Belt Usage* In Sacramento County in California, the use of seat belts has risen steadily since the enactment of a

seat belt law in 1986. The rate of change (in %) of drivers using seat belts in year x, where 1985 corresponds to $x = 0$, is modeled by

$$f'(x) = 0.116x^3 - 1.803x^2 + 6.86x + 3.24.^†$$

In 1985 (year 0), 26% of drivers used seat belts.

a. Find the function that gives the percent of drivers using seat belts in year x.

b. According to this function, what percent of drivers used seat belts in 1993?

c. Does the function found in part a give reasonable results past 1996? Explain.

49. *Vehicle-Related Deaths* Vehicle-related deaths in Sacramento County have declined since 1986 when the California seat belt law was enacted.† The rate of change in deaths per 100 million miles of vehicle travel is modeled by

$$g'(x) = -0.00156x^3 + 0.0312x^2 - 0.264x + 0.137,$$

where $x = 0$ corresponds to 1985 and so on. There were 2.4 deaths per 100 million miles driven in 1985.

a. Find the function giving the number of deaths per 100 million miles in year x.

b. How many deaths were there in 1986? in 1990?

c. Does the function found in part a give reasonable results past 1991? Explain.

50. *Telephone Usage* The approximate rate of change in the number of new telephones per 1,000 U.S. population is given by

$$f'(x) = \frac{1,110}{x},$$

where x represents the number of years since 1900.‡ In 1990 ($x = 90$) there were 1,000 phones per 1,000 population.

a. Find the function that gives the total number of phones per 1,000 people in year x.

---

*Dennis, Brian, and Robert F. Costantino, "Analysis of Steady-State Populations with the Gamma Abundance Model: Application to *Tribolium*," *Ecology,* Vol. 69, No. 4, Aug. 1988, pp. 1200–1213.
†California Highway Patrol and National Highway Traffic Safety Administration
‡Bellcore

b. According to this function, how many phones per 1,000 people were there in 1995?

Exercises 51–56 refer to Example 11 in this section.

51. *Velocity* For a particular object, $a(t) = t^2 + 1$ and $v(0) = 6$. Find $v(t)$.

52. *Distance* Suppose $v(t) = 6t^2 - 2/t^2$ and $s(1) = 8$. Find $s(t)$.

53. *Time* An object is dropped from a small plane flying at 6,400 feet. Assume that $a(t) = -32$ feet per second per second and $v(0) = 0$, and find $s(t)$. How long will it take the object to hit the ground?

54. *Distance* Suppose $a(t) = 18t + 8$, $v(1) = 15$, and $s(1) = 19$. Find $s(t)$.

55. *Distance* Suppose $a(t) = (15/2)\sqrt{t} + 3e^{-t}$, $v(0) = -3$, and $s(0) = 4$. Find $s(t)$.

56. *Rocket Science* In the 1999 movie *October Sky,* Homer Hickum was accused of launching a rocket that started a forest fire. Homer proved his innocence by showing that his rocket could not have flown far enough to reach where the fire started. He used the following reasoning.

a. Using the fact that $a(t) = -32$ (see Example 11 (b)), find $v(t)$ and $s(t)$, given $v(0) = v_0$ and $s(0) = 0$. (The initial velocity was unknown, and the initial height was 0 ft.)

b. Homer estimated that the rocket was in the air for 14 sec. Use $s(14) = 0$ to find v_0.

c. If the rocket left the ground at a 45° angle, the velocity in the horizontal direction would be equal to v_0, the velocity in the vertical direction, so the distance traveled horizontally would be $v_0 t$. (The rocket left the ground at a steeper angle, so this would overestimate the distance from starting to landing point.) Find the distance the rocket would travel horizontally during its 14 sec flight.

■ 7.2 SUBSTITUTION

? THINK ABOUT IT

If a pharmaceutical company knows a formula for the marginal revenue for a decongestant, how can they determine a formula for the demand?

Using the method of substitution, this question will be answered in Example 7 of this section.

We saw how to integrate a few simple functions in the previous section. More complicated functions can sometimes be integrated by *substitution.* The substitution technique depends on the idea of a differential, discussed in an earlier chapter. If $u = f(x)$, the *differential* of u, written du, is defined as

$$du = f'(x)\,dx.$$

FOR REVIEW ■

The chain rule, discussed in detail in the chapter on Calculating the Derivative, states that

$$\frac{d}{dx}[f(g(x))] = f'(g(x)) \cdot g'(x).$$

For example, if $u = 6x^4$, then $du = 24x^3\,dx$.

Recall the chain rule for derivatives as used in the following example:

$$\frac{d}{dx}(x^2 - 1)^5 = 5(x^2 - 1)^4(2x) = 10x(x^2 - 1)^4.$$

As in this example, the result of using the chain rule is often a product of two functions. Because of this, functions formed by the product of two functions can sometimes be integrated by using the chain rule in reverse. In the example above, working backwards from the derivative gives

$$\int 10x(x^2 - 1)^4\,dx = (x^2 - 1)^5 + C.$$

To find an antiderivative involving products, it often helps to make a substitution: let $u = x^2 - 1$, so that $du = 2x\,dx$. Now substitute u for $x^2 - 1$ and du for $2x\,dx$ in the indefinite integral above.

$$\int 10x(x^2 - 1)^4 \, dx = \int 5 \cdot 2x(x^2 - 1)^4 \, dx$$

$$= 5 \int (x^2 - 1)^4 (2x \, dx)$$

$$= 5 \int u^4 \, du$$

With substitution we have changed a complicated integral into a simple one. This last integral can now be found by the power rule.

$$5 \int u^4 \, du = 5 \cdot \frac{u^5}{5} + C = u^5 + C$$

Finally, substitute $x^2 - 1$ for u.

$$\int 10x(x^2 - 1)^4 \, dx = (x^2 - 1)^5 + C$$

This method of integration is called **integration by substitution.** As shown above, it is simply the chain rule for derivatives in reverse. The results can always be verified by differentiation.

This discussion leads to the following integration formula, which is sometimes called the general power rule for integrals.

GENERAL POWER RULE FOR INTEGRALS

For $u = f(x)$ and $du = f'(x) \, dx$,

$$\int u^n \, du = \frac{u^{n+1}}{n + 1} + C.$$

EXAMPLE 1 General Power Rule

Find $\int 6x(3x^2 + 4)^4 \, dx$.

Solution A certain amount of trial and error may be needed to decide on the expression to set equal to u. The integrand must be written as two factors, one of which is the derivative of the quantity chosen for u. In this example, if $u = 3x^2 + 4$, then $du = 6x \, dx$ and the integrand can be written as the product of $(3x^2 + 4)^4$ and $6x \, dx$. Now substitute.

$$\int 6x(3x^2 + 4)^4 \, dx = \int (3x^2 + 4)^4 (6x \, dx) = \int u^4 \, du$$

Find this last indefinite integral.

$$\int u^4 \, du = \frac{u^5}{5} + C$$

Now replace u with $3x^2 + 4$.

$$\int 6x(3x^2 + 4)^4 \, dx = \frac{u^5}{5} + C = \frac{(3x^2 + 4)^5}{5} + C$$

To verify this result, find the derivative.

$$\frac{d}{dx}\left[\frac{(3x^2 + 4)^5}{5} + C\right] = \frac{5}{5}(3x^2 + 4)^4(6x) + 0 = (3x^2 + 4)^4(6x)$$

The derivative is the original function (the integrand), as required.

EXAMPLE 2   General Power Rule

Find $\int x^2\sqrt{x^3 + 1}\ dx$.

Solution

Method 1: Solving for du

An expression raised to a power is usually a good choice for u, so because of the square root or $1/2$ power, let $u = x^3 + 1$; then $du = 3x^2\ dx$. The integrand does not contain the constant 3, which is needed for du. To take care of this, multiply by $3/3$, placing 3 inside the integral sign and $1/3$ outside.

$$\int x^2\sqrt{x^3 + 1}\ dx = \frac{1}{3}\int 3x^2\sqrt{x^3 + 1}\ dx = \frac{1}{3}\int \sqrt{x^3 + 1}\,(3x^2\,dx)$$

Now substitute u for $x^3 + 1$ and du for $3x^2\,dx$, and then integrate.

$$\frac{1}{3}\int \sqrt{x^3 + 1}\,(3x^2\,dx) = \frac{1}{3}\int \sqrt{u}\ du = \frac{1}{3}\int u^{1/2}\,du$$

$$= \frac{1}{3}\cdot\frac{u^{3/2}}{3/2} + C = \frac{2}{9}u^{3/2} + C$$

Since $u = x^3 + 1$,

$$\int x^2\sqrt{x^3 + 1}\ dx = \frac{2}{9}(x^3 + 1)^{3/2} + C.$$

Method 2: Solving for the Integrand

As in Method 1, we let $u = x^3 + 1$ and $du = 3x^2\,dx$. Since the integrand does not contain the constant 3, we divide $du = 3x^2\,dx$ by 3 to obtain $(1/3)\,du = x^2\,dx$. Replacing $x^3 + 1$ with u and $x^2\,dx$ with $(1/3)\,du$, we have

$$\int x^2\sqrt{x^3 + 1}\ dx = \int \sqrt{x^3 + 1}\,(x^2\,dx) = \int \sqrt{u}\left(\frac{1}{3}\,du\right)$$

$$= \frac{1}{3}\int u^{1/2}\,du = \frac{1}{3}\cdot\frac{u^{3/2}}{3/2} + C$$

$$= \frac{2}{9}u^{3/2} + C.$$

The answer is the same one we obtained in Method 1; Method 2 is just a slightly different way of performing the steps. In the other examples, we will just show Method 1, but feel free to use Method 2 if it seems simpler to you.

The substitution methods given in the examples above *will not always work*. For example, you might try to find

$$\int x^3\sqrt{x^3 + 1}\ dx$$

by substituting $u = x^3 + 1$, so that $du = 3x^2\,dx$. However, there is no *constant* that can be inserted inside the integral sign to give $3x^2$. This integral, and a great many others, cannot be evaluated by substitution.

With practice, choosing u will become easy if you keep two principles in mind.

1. u should equal some expression in the integral that, when replaced with u, tends to make the integral simpler.

2. u must be an expression whose derivative—disregarding any constant multiplier, such as the 3 in $3x^2$—is also present in the integral.

The substitution should include as much of the integral as possible, as long as its derivative is still present. In Example 1, we could have chosen $u = 3x^2$, but $u = 3x^2 + 4$ is better, because it has the same derivative as $3x^2$ and captures more of the original integral. If we carry this reasoning further, we might try $u = (3x^2 + 4)^4$, but this is a poor choice, for $du = 4(3x^2 + 4)^3(6x)\,dx$, an expression not present in the original integral.

EXAMPLE 3 Substitution

Find $\displaystyle\int \frac{x + 3}{(x^2 + 6x)^2}\,dx.$

Solution Let $u = x^2 + 6x$, so that $du = (2x + 6)\,dx = 2(x + 3)\,dx$. The integral is missing the 2, so multiply by $2 \cdot (1/2)$, putting 2 inside the integral sign and $1/2$ outside.

$$\int \frac{x + 3}{(x^2 + 6x)^2}\,dx = \frac{1}{2}\int \frac{2(x + 3)}{(x^2 + 6x)^2}\,dx$$

$$= \frac{1}{2}\int \frac{du}{u^2} = \frac{1}{2}\int u^{-2}\,du = \frac{1}{2} \cdot \frac{u^{-1}}{-1} + C = \frac{-1}{2u} + C.$$

Substituting $x^2 + 6x$ for u gives

$$\int \frac{x + 3}{(x^2 + 6x)^2}\,dx = \frac{-1}{2(x^2 + 6x)} + C.$$

Recall the formula for $\dfrac{d}{dx}(e^u)$, where $u = f(x)$:

$$\frac{d}{dx}(e^u) = e^u\frac{du}{dx}.$$

For example, if $u = x^2$, then $\dfrac{du}{dx} = \dfrac{d}{dx}(x^2) = 2x$, and

$$\frac{d}{dx}(e^{x^2}) = e^{x^2} \cdot 2x.$$

Working backwards, if $u = x^2$, then $du = 2x\,dx$, so

$$\int e^{x^2} \cdot 2x\,dx = \int e^u\,du = e^u + C = e^{x^2} + C.$$

The work above suggests the following rule for the indefinite integral of e^u, where $u = f(x)$.

INDEFINITE INTEGRAL OF e^u

If $u = f(x)$, then $du = f'(x)\,dx$ and

$$\int e^u\,du = e^u + C.$$

EXAMPLE 4 Substitution

Find $\int x^2 e^{x^3}\,dx$.

Solution Let $u = x^3$, the exponent on e. Then $du = 3x^2\,dx$. Multiplying by $3/3$ gives

$$\int x^2 e^{x^3}\,dx = \frac{1}{3}\int e^{x^3}(3x^2\,dx)$$

$$= \frac{1}{3}\int e^u\,du = \frac{1}{3}e^u + C = \frac{1}{3}e^{x^3} + C.$$

Recall that the antiderivative of $f(x) = 1/x$ is $\ln|x|$. The next example uses the fact that $\int x^{-1}\,dx = \ln|x| + C$, together with the method of substitution.

EXAMPLE 5 Substitution

Find $\displaystyle\int \frac{(2x - 3)\,dx}{x^2 - 3x}$.

Solution Let $u = x^2 - 3x$, so that $du = (2x - 3)\,dx$. Then

$$\int \frac{(2x - 3)\,dx}{x^2 - 3x} = \int \frac{du}{u} = \ln|u| + C = \ln|x^2 - 3x| + C.$$

Generalizing from the results of Example 5 suggests the following rule for finding the indefinite integral of u^{-1}, where $u = f(x)$.

INDEFINITE INTEGRAL OF u^{-1}

If $u = f(x)$, then $du = f'(x)\,dx$ and

$$\int u^{-1}\,du = \int \frac{du}{u} = \ln|u| + C.$$

The next example shows a more complicated integral evaluated by the method of substitution.

EXAMPLE 6 Substitution

Find $\int x\sqrt{1 - x}\,dx$.

Solution Let $u = 1 - x$. To get the x outside the radical in terms of u, solve $u = 1 - x$ for x to get $x = 1 - u$. Then $dx = -du$ and we can substitute as follows.

$$\int x\sqrt{1-x}\,dx = \int (1-u)\sqrt{u}(-du) = \int (u-1)u^{1/2}\,du$$

$$= \int (u^{3/2} - u^{1/2})\,du = \frac{2}{5}u^{5/2} - \frac{2}{3}u^{3/2} + C$$

$$= \frac{2}{5}(1-x)^{5/2} - \frac{2}{3}(1-x)^{3/2} + C$$

The substitution method is useful if the integral can be written in one of the following forms, where u is some function of x.

SUBSTITUTION METHOD

In general, for the types of problems we are concerned with, there are three cases. We choose u to be one of the following:

    1. the quantity under a root or raised to a power;

    2. the exponent on e;

    3. the quantity in the denominator.

Remember that some integrands may need to be rearranged to fit one of these cases.

Some calculators, such as the TI-89 and TI-92, can find indefinite integrals automatically. Many computer algebra systems also do this. Figure 2 shows the integral in Example 6 performed on a TI-92. The answer looks different from that of Example 6. Use algebra to verify that the two answers are equivalent.

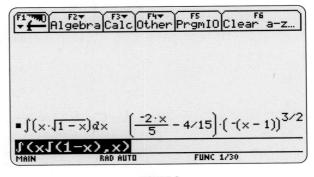

FIGURE 2

EXAMPLE 7 Demand

The marketing department for a pharmaceutical company has studied the number of cases of a decongestant they can sell each week at various prices. When the price is given as a function of the number of cases sold, the function is known as the **demand function.** The research department determined the rate that the price changes with respect to the number of cases sold. Such a function is known as the **marginal demand,*** and in this instance was given by

---

*For more details on demand functions and marginal analysis, see *Calculus with Applications,* 7th ed., Lial, Margaret L., Raymond N. Greenwell, and Nathan P. Ritchey, Addison-Wesley, 2002.

$$p'(x) = \frac{-4,000,000}{(2x + 15)^3}.$$

Find the demand equation if the weekly demand for this type of decongestant is 10 cases when the price of a case of decongestant is $4,000.

Solution To find the demand function, first integrate $p'(x)$ as follows.

$$p(x) = \int p'(x)\, dx = \int \frac{-4,000,000}{(2x + 15)^3}\, dx$$

Let $u = 2x + 15$. Then $du = 2\, dx$, and

$$p(x) = -2,000,000 \int (2x + 15)^{-3}\, 2\, dx$$

$$= -2,000,000 \int u^{-3}\, du$$

$$= (-2,000,000)\frac{u^{-2}}{-2} + C$$

$$= \frac{1,000,000}{u^2} + C$$

$$p(x) = \frac{1,000,000}{(2x + 15)^2} + C. \tag{1}$$

Find the value of C by using the given information that $p = 4,000$ when $x = 10$.

$$4,000 = \frac{1,000,000}{(2 \cdot 10 + 15)^2} + C$$

$$4,000 = \frac{1,000,000}{35^2} + C$$

$$4,000 = 816 + C$$

$$3,184 = C$$

Replacing C with 3,184 in equation (1) gives the demand function,

$$p(x) = \frac{1,000,000}{(2x + 15)^2} + 3,184.$$

With a little practice, you will find you can skip the substitution step for integrals such as that shown in Example 7, in which the derivative of u is a constant. Recall from the chain rule that when you differentiate a function, such as $p'(x) = -4,000,000/(2x + 15)^3$ in the previous example, you multiply by 2, the derivative of $(2x + 15)$. So when taking the antiderivative, simply divide by 2:

$$\int -4,000,000(2x + 15)^{-3}\, dx = \frac{-4,000,000}{2} \cdot \frac{(2x + 15)^{-2}}{-2} + C$$

$$= \frac{1,000,000}{(2x + 15)^2} + C.$$

| **CAUTION** This procedure is valid because of the constant multiple rule presented in the previous section, which says that constant multiples can be brought into or out of integrals, just as they can with derivatives. This procedure is *not* valid with any expression other than a constant.

EXAMPLE 8 Popularity Index

To determine the top 100 popular songs of each year since 1956, Jim Quirin and Barry Cohen developed a function that represents the rate of change on the charts of *Billboard* magazine required for a song to earn a "star" on the *Billboard* "Hot 100" survey.* They developed the function

$$f(x) = \frac{A}{B + x},$$

where $f(x)$ represents the rate of change in position on the charts, x is the position on the "Hot 100" survey, and A and B are constants. The function

$$F(x) = \int f(x)\,dx$$

is defined as the "Popularity Index." Find $F(x)$.

Solution Integrating $f(x)$ gives

$$F(x) = \int f(x)\,dx$$

$$= \int \frac{A}{B + x}\,dx$$

$$= A \int \frac{1}{B + x}\,dx.$$

Let $u = B + x$, so that $du = dx$. Then

$$F(x) = A \int \frac{1}{u}\,du = A \ln u + C$$

$$= A \ln(B + x) + C.$$

(The absolute value bars are not necessary, since $B + x$ is always positive here.)

7.2 EXERCISES

1. Integration by substitution is related to what differentiation method? What type of integrand suggests using integration by substitution?

2. The following integrals may be solved using substitution. Choose a function u that may be used to solve each problem. Then find du.

   a. $\int (3x^2 - 5)^4 2x\,dx$ **b.** $\int \sqrt{1 - x}\,dx$ **c.** $\int \frac{x^2}{2x^3 + 1}\,dx$ **d.** $\int 4x^3 e^{x^4}\,dx$

*Formula for the "Popularity Index" from Quirin, Jim, and Barry Cohen, *Chartmasters' Rock 100,* 4th ed. Copyright 1987 by Chartmasters. Reprinted by permission.

Use substitution to find the following indefinite integrals.

3. $\int 4(2x + 3)^4 \, dx$

4. $\int (-4t + 1)^3 \, dt$

5. $\int \frac{2 \, dm}{(2m + 1)^3}$

6. $\int \frac{3 \, du}{\sqrt{3u - 5}}$

7. $\int \frac{2x + 2}{(x^2 + 2x - 4)^4} \, dx$

8. $\int \frac{6x^2 \, dx}{(2x^3 + 7)^{3/2}}$

9. $\int z\sqrt{z^2 - 5} \, dz$

10. $\int r\sqrt{r^2 + 2} \, dr$

11. $\int -4e^{2p} \, dp$

12. $\int 5e^{-0.3g} \, dg$

13. $\int 3x^2 e^{2x^3} \, dx$

14. $\int re^{-r^2} \, dr$

15. $\int (1 - t)e^{2t - t^2} \, dt$

16. $\int (x^2 - 1)e^{x^3 - 3x} \, dx$

17. $\int \frac{e^{1/z}}{z^2} \, dz$

18. $\int \frac{e^{\sqrt{y}}}{2\sqrt{y}} \, dy$

19. $\int (x^3 + 2x)(x^4 + 4x^2 + 7)^8 \, dx$

20. $\int \frac{t^2 + 2}{t^3 + 6t + 3} \, dt$

21. $\int \frac{2x + 1}{(x^2 + x)^3} \, dx$

22. $\int \frac{B^3 - 1}{(2B^4 - 8B)^{3/2}} \, dB$

23. $\int p(p + 1)^5 \, dp$

24. $\int 4r\sqrt{8 - r} \, dr$

25. $\int \frac{u}{\sqrt{u - 1}} \, du$

26. $\int \frac{2x}{(x + 5)^6} \, dx$

27. $\int \left(\sqrt{x^2 + 12x}\right)(x + 6) \, dx$

28. $\int \left(\sqrt{x^2 - 6x}\right)(x - 3) \, dx$

29. $\int \frac{t}{t^2 + 2} \, dt$

30. $\int \frac{-4x}{x^2 + 3} \, dx$

31. $\int \frac{(1 + \ln x)^2}{x} \, dx$

32. $\int \frac{\sqrt{2 + \ln x}}{x} \, dx$

33. $\int \frac{e^{2x}}{e^{2x} + 5} \, dx$

34. $\int \frac{1}{x(\ln x)} \, dx$

35. Stan and Ollie work on the integral

$$\int 3x^2 e^{x^3} \, dx.$$

Stan lets $u = x^3$ and proceeds to get

$$\int e^u \, du = e^u + C = e^{x^3} + C.$$

Ollie tries $u = e^{x^3}$ and proceeds to get

$$\int du = u + C = e^{x^3} + C.$$

Discuss which procedure you prefer, and why.

36. Stan and Ollie work on the integral

$$\int 2x(x^2 + 2) \, dx.$$

Stan lets $u = x^2 + 2$ and proceeds to get

$$\int u \, du = \frac{u^2}{2} + C = \frac{(x^2 + 2)^2}{2} + C.$$

Ollie multiplies out the function under the integral and gets

$$\int (2x^3 + 4x) \, dx = \frac{x^4}{2} + 2x^2 + C.$$

How can they both be right?

Applications

LIFE SCIENCES

37. *Cell Growth* Under certain conditions, the number of cancer cells $N(t)$ at time t increases at a rate

$$N'(t) = Ae^{kt},$$

where A is the rate of increase at time 0 (in cells per day) and k is a constant.

a. Suppose $A = 50$, and at 5 days, the cells are growing at a rate of 250 per day. Find a formula for the number of cells after t days, given that 300 cells are present at $t = 0$.

b. Use your answer from part a to find the number of cells present after 12 days.

38. *Blood Pressure* The rate of change of the volume $V(t)$ of blood in the aorta at time t is given by

$$V'(t) = -kP(t),$$

where $P(t)$ is the pressure in the aorta at time t and k is a constant that depends upon properties of the aorta. The pressure in the aorta is given by

$$P(t) = P_0 e^{-mt},$$

where P_0 is the pressure at time $t = 0$ and m is another constant. Letting V_0 be the volume at time $t = 0$, find a formula for $V(t)$.

OTHER APPLICATIONS

39. *Border Crossing* Canadians who live near the U.S. border have been shopping in the United States at an increasing rate since 1986. They cross the border to shop because large savings are possible on groceries, gasoline, cigarettes, and children's clothes. The rate of change in the number of Canadian cross-border shoppers (in millions) is given by

$$S'(t) = 4.4e^{0.16t},$$

where t is the number of years since 1986.

a. Find $S(t)$ if there were 27.3 million cross-border shoppers in 1986 (year 0).

b. In how many years will the number of Canadian cross-border shoppers double?

■ 7.3 AREA AND THE DEFINITE INTEGRAL

? THINK ABOUT IT   How might the incidence of foot-and-mouth disease in the United Kingdom have been reduced using more aggressive culling at infected farms?

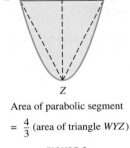

Area of parabolic segment

$$= \frac{4}{3} \text{ (area of triangle } WYZ)$$

FIGURE 3

This section introduces a method for answering questions such as this one, which is answered in one of the exercises.

To calculate the areas of geometric figures such as rectangles, squares, triangles, and circles, we use specific formulas. In this section we consider the problem of finding the area of a figure or region that is bounded by curves, such as the shaded region in Figure 3.

The brilliant Greek mathematician Archimedes (about 287 B.C.–212 B.C.) is considered one of the greatest mathematicians of all time. His development of a rigorous method known as *exhaustion* to derive results was a forerunner of the ideas of integral calculus. Archimedes used a method that would later be verified by the theory of integration. His method involved viewing a geometric figure as a sum of other figures. For example, he thought of a plane surface area as a figure consisting of infinitely many parallel line segments. Among the results established by Archimedes' method was the fact that the area of a segment of a parabola (shown in color in Figure 3) is equal to 4/3 the area of a triangle with the same base and the same height.

Under certain conditions the area of a region can be thought of as a sum of parts. Figure 4 shows the region bounded by the y-axis, the x-axis, and the graph of $f(x) = \sqrt{4 - x^2}$. A very rough approximation of the area of this region can be found by using two inscribed rectangles as in Figure 5. The height of the rectangle on the left is $f(0) = 2$ and the height of the rectangle on the right is $f(1) = \sqrt{3}$. The width of each rectangle is 1, making the total area of the two rectangles

$$1 \cdot f(0) + 1 \cdot f(1) = 2 + \sqrt{3} \approx 3.7321 \text{ square units.}$$

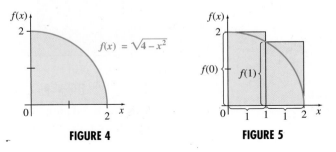

FIGURE 4   **FIGURE 5**

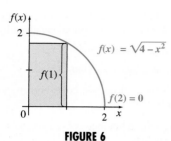

FIGURE 6

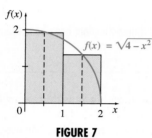

FIGURE 7

In this example, the function is decreasing, and we will overestimate the area when we evaluate the function at the left endpoint to determine the height of the rectangle in that interval. If we use the right endpoint, the answer will be too small. For example, using the right endpoint with $n = 2$, the area of the two rectangles is

$$1 \cdot f(1) + 1 \cdot f(2) = \sqrt{3} + 0 \approx 1.7321 \text{ square units.}$$

See Figure 6.

If the left endpoint gives an answer that is too big, and the right endpoint an answer that is too small, it seems reasonable to average the two answers. This produces the method called the *trapezoidal rule,* discussed in more detail in the next chapter. In this example, we get

$$\frac{3.7321 + 1.7321}{2} = 2.7321.$$

Another way to get an improved answer would be to use the midpoint of each interval, rather than the left endpoint or the right endpoint. This is called the **midpoint rule.** See Figure 7. In this example, it gives

$$1 \cdot f(0.5) + 1 \cdot f(1.5) = \sqrt{3.75} + \sqrt{1.75} \approx 3.2594 \text{ square units.}$$

To improve the accuracy of all of the previous approximations, we could divide the interval from $x = 0$ to $x = 2$ into more parts. The result using the left endpoint again with four equal parts, each of width $1/2$, is shown in Figure 8. This approximation is greater than the actual area. As before, the height of each rectangle is given by the value of f at the left side of the rectangle, and its area is the width, $1/2$, multiplied by the height. The total area of the four rectangles is

$$\frac{1}{2} \cdot f(0) + \frac{1}{2} \cdot f\left(\frac{1}{2}\right) + \frac{1}{2} \cdot f(1) + \frac{1}{2} \cdot f\left(1\frac{1}{2}\right)$$

$$= \frac{1}{2}(2) + \frac{1}{2}\left(\frac{\sqrt{15}}{2}\right) + \frac{1}{2}(\sqrt{3}) + \frac{1}{2}\left(\frac{\sqrt{7}}{2}\right)$$

$$= 1 + \frac{\sqrt{15}}{4} + \frac{\sqrt{3}}{2} + \frac{\sqrt{7}}{4} \approx 3.4957 \text{ square units.}$$

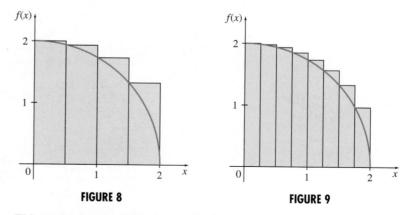

FIGURE 8   **FIGURE 9**

This approximation looks better, but it is still greater than the actual area. To improve the approximation, divide the interval from $x = 0$ to $x = 2$ into 8 parts with equal widths of 1/4. The total area of all these rectangles is

$$\frac{1}{4} \cdot f(0) + \frac{1}{4} \cdot f\left(\frac{1}{4}\right) + \frac{1}{4} \cdot f\left(\frac{1}{2}\right) + \frac{1}{4} \cdot f\left(\frac{3}{4}\right) + \frac{1}{4} \cdot f(1)$$

$$+ \frac{1}{4} \cdot f\left(\frac{5}{4}\right) + \frac{1}{4} \cdot f\left(\frac{3}{2}\right) + \frac{1}{4} \cdot f\left(\frac{7}{4}\right)$$

$$\approx 3.3398 \text{ square units.}$$

This process of approximating the area under a curve by using more and more rectangles to get a better and better approximation can be generalized. To do this, divide the interval from $x = 0$ to $x = 2$ into n equal parts. Each of these n intervals has width

$$\frac{2 - 0}{n} = \frac{2}{n},$$

so each rectangle has width $2/n$ and height determined by the function-value at the left side of the rectangle, or the right side, or the midpoint. We could also average the left and right side values as before. Using a computer or graphing calculator to find approximations to the area for several values of n gives the results in Table 1.

■ Table 1

n	left sum	right sum	trapezoidal	midpoint
2	3.7321	1.7321	2.7321	3.2594
4	3.4957	2.4957	2.9957	3.1839
8	3.3398	2.8398	3.0898	3.1567
10	3.3045	2.9045	3.1045	3.1524
20	3.2285	3.0285	3.1285	3.1454
50	3.1783	3.0983	3.1383	3.1426
100	3.1604	3.1204	3.1404	3.1419
500	3.1455	3.1375	3.1415	3.1416

The numbers in the last four columns of this table represent approximations to the area under the curve, above the x-axis, and between the lines $x = 0$ and $x = 2$. As n becomes larger and larger, all four approximations become better and better, getting closer to the actual area. In this example, the exact area can be found by a formula from plane geometry. Write the given function as

$$y = \sqrt{4 - x^2},$$

then square both sides to get

$$y^2 = 4 - x^2$$
$$x^2 + y^2 = 4,$$

the equation of a circle centered at the origin with radius 2. The region in Figure 4 is the quarter of this circle that lies in the first quadrant. The actual area of this region is one-quarter of the area of the entire circle, or

$$\frac{1}{4}\pi(2)^2 = \pi \approx 3.1416.$$

As the number of rectangles increases without bound, the sum of the areas of these rectangles gets closer and closer to the actual area of the region, π. This can be written as

$$\lim_{n \to \infty}(\text{sum of areas of } n \text{ rectangles}) = \pi.$$

(The value of π was originally found by a process similar to this.)*

Notice in the above example that for a particular value of n, the midpoint rule gave the best answer (the one closest to the true value of 3.1416), followed by the trapezoidal rule, followed by the left and right sums. In fact, the midpoint rule with $n = 20$ gives a value (3.1454) that is slightly more accurate than the left sum with $n = 500$ (3.1455).

Now we can generalize to get a method of finding the area bounded by the curve $y = f(x)$, the x-axis, and the vertical lines $x = a$ and $x = b$, as shown in Figure 10. To approximate this area, we could divide the region under the curve first into 10 rectangles (Figure 10(a)) and then into 20 rectangles (Figure 10(b)). The sum of the areas of the rectangles give approximations to the area under the curve.

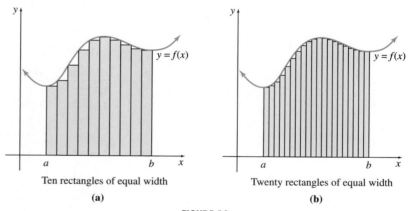

| Ten rectangles of equal width | Twenty rectangles of equal width |
| (a) | (b) |

FIGURE 10

To develop a process that would yield the *exact* area, begin by dividing the interval from a to b into n pieces of equal width, using each of these n pieces as the base of a rectangle (see Figure 11 on the next page). Let x_1 be an arbitrary point in the first interval, x_2 be an arbitrary point in the second interval, and so on, up to the nth interval. In the graph of Figure 11, the symbol Δx is used to represent the width of each of the intervals. The pink rectangle is an arbitrary rectangle called the ith rectangle. Its area is the product of its length and width. Since the width of the ith rectangle is Δx and the length of the ith rectangle is given by the height $f(x_i)$,

$$\text{Area of the } i\text{th rectangle} = f(x_i) \cdot \Delta x.$$

---

*The number π is the ratio of the circumference of a circle to its diameter. It is an example of an *irrational number,* and as such it cannot be expressed as a terminating or repeating decimal. Many approximations have been used for π over the years. A passage in the Bible (I Kings 7:23) indicates a value of 3. The Egyptians used the value 3.16, and Archimedes showed that its value must be between 22/7 and 223/71. A Hindu writer, Brahmagupta, used $\sqrt{10}$ as its value in the seventh century. The search for the digits of π has continued into modern times. Yasumasa Kanada and his coworkers at the University of Tokyo recently computed the value to over 206 billion places.

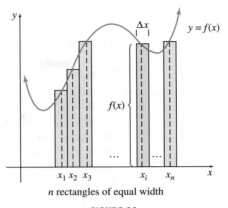

FIGURE 11

Recall from Chapter 1 that the symbol Σ (sigma) indicates "the sum of." Here, we use $\sum\limits_{i=1}^{n}$ to indicate the sum

$f(x_1)\,\Delta x + f(x_2)\,\Delta x + f(x_3)\,\Delta x + \cdots + f(x_n)\,\Delta x$, where we replace i with 1 in the first term, 2 in the second term, and so on, ending with n replacing i in the last term.

The total area under the curve is approximated by the sum of the areas of all n of the rectangles. With sigma notation, the approximation to the total area becomes

$$\text{Area of all } n \text{ rectangles} = \sum_{i=1}^{n} f(x_i) \cdot \Delta x.$$

The exact area is defined to be the limit of this sum (if the limit exists) as the number of rectangles increases without bound:

$$\text{Exact area} = \lim_{n \to \infty} \sum_{i=1}^{n} f(x_i)\,\Delta x.$$

This limit is called the *definite integral* of $f(x)$ from a to b and is written as follows.

THE DEFINITE INTEGRAL

If f is defined on the interval $[a, b]$, the **definite integral** of f from a to b is given by

$$\int_a^b f(x)\,dx = \lim_{n \to \infty} \sum_{i=1}^{n} f(x_i)\,\Delta x,$$

provided the limit exists, where $\Delta x = (b - a)/n$ and x_i is *any* value of x in the ith interval.

▌ **NOTE** In the life sciences literature, the definite integral is often referred to as the area under the curve, abbreviated as **AOC.**

The definite integral can be approximated by

$$\sum_{i=1}^{n} f(x_i)\,\Delta x.$$

If $f(x) \geq 0$ on the interval $[a, b]$, the definite integral gives the area under the curve between $x = a$ and $x = b$. In the midpoint rule, x_i is the midpoint of the ith interval. We may also let x_i be the left endpoint, the right endpoint, or any other point in the ith interval.

Some calculators have a built-in function for evaluating the definite integral. For example, on the TI-83, the command `fnInt(√(4-X²),X,0,2)` gives the answer 3.141593074, with an error of approximately 0.0000004.

As indicated in this definition, although the left endpoint of the *i*th interval has been used to find the height of the *i*th rectangle, any number in the *i*th interval can be used. (A more general definition is possible in which the rectangles do not necessarily all have the same width.) The *b* above the integral sign is called the **upper limit** of integration, and the *a* is the **lower limit** of integration. This use of the word *limit* has nothing to do with the limit of the sum; it refers to the limits, or boundaries, on *x*.

The sum in the definition of the definite integral is an example of a Riemann sum, named for the German mathematician Georg Riemann (1826–1866), who at the age of 20 changed his field of study from theology and the classics to mathematics. Twenty years later he died of tuberculosis while traveling in Italy in search of a cure. The concepts of *Riemann sum* and *Riemann integral* are still studied in rigorous calculus textbooks.

In the example at the beginning of this section, the area bounded by the *x*-axis, the curve $y = \sqrt{4 - x^2}$, and the lines $x = 0$ and $x = 2$ could be written as the definite integral

$$\int_0^2 \sqrt{4 - x^2}\, dx = \pi.$$

NOTE Notice that unlike the indefinite integral, which is a set of *functions,* the definite integral represents a *number.* The next section will show how antiderivatives are used in finding the definite integral and thus the area under a curve.

Keep in mind that finding the definite integral of a function can be thought of as a mathematical process that gives the sum of an infinite number of individual parts (within certain limits). The definite integral represents area only if the function involved is *nonnegative* ($f(x) \geq 0$) at every *x*-value in the interval $[a, b]$. There are many other interpretations of the definite integral, and all of them involve this idea of approximation by appropriate sums.

EXAMPLE 1 Approximation of Area

Approximate $\int_0^4 2x\, dx$, the area of the region under the graph of $f(x) = 2x$, above the *x*-axis, and between $x = 0$ and $x = 4$, by using four rectangles of equal width whose heights are the values of the function at the midpoint of each rectangle.

Solution

Method 1: Calculating by Hand

We want to find the area of the shaded region in Figure 12. The heights of the four rectangles given by $f(x_i)$ for $i = 1, 2, 3,$ and 4 are as follows.

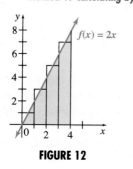

FIGURE 12

Table 2

i	x_i	$f(x_i)$
1	$x_1 = 0.5$	$f(0.5) = 1.0$
2	$x_2 = 1.5$	$f(1.5) = 3.0$
3	$x_3 = 2.5$	$f(2.5) = 5.0$
4	$x_4 = 3.5$	$f(3.5) = 7.0$

The width of each rectangle is $\Delta x = (4 - 0)/4 = 1$. The sum of the areas of the four rectangles is

$$\sum_{i=1}^{4} f(x_i)\, \Delta x = f(x_1)\, \Delta x + f(x_2)\, \Delta x + f(x_3)\, \Delta x + f(x_4)\, \Delta x$$

$$= f(0.5)\, \Delta x + f(1.5)\, \Delta x + f(2.5)\, \Delta x + f(3.5)\, \Delta x$$

$$= 1(1) + 3(1) + 5(1) + 7(1)$$

$$= 16.$$

Using the formula for the area of a triangle, $A = (1/2)bh$, with b, the length of the base, equal to 4 and h, the height, equal to 8, gives

$$A = \frac{1}{2}bh = \frac{1}{2}(4)(8) = 16,$$

the exact value of the area. The approximation equals the exact area in this case because our use of the midpoints of each subinterval distributed the error evenly above and below the graph.

Method 2: Graphing Calculator A graphing calculator can be used to organize the information in this example. For example, the seq feature in the LIST OPS menu of the TI-83 calculator can be used to store the values of i in the list L_1. Using the STAT EDIT menu, the entries for x_i can be generated by entering the formula $-0.5 + L_1$ as the heading of L_2. Similarly, entering the formula for $f(x_i)$, $2 * L_2$, at the top of list L_3 will generate the values of $f(x_i)$ in L_3. (The entries are listed automatically when the formula is entered.) Then the sum feature in the LIST MATH menu can be used to add the values in L_3. The resulting screens are shown in Figure 13.

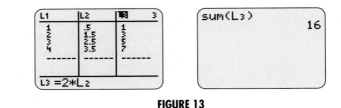

FIGURE 13

Method 3: Spreadsheet The calculations in this example can also be done on a spreadsheet. In Microsoft Excel, for example, store the values of i in column A. Put the command "=A1 − .5" into B1; copying this formula into the rest of column B gives the values of x_i. Similarly, use the formula for $f(x_i)$ to fill column C. Column D is the product of Column C and Δx. Sum column D to get the answer. For more details, see *The Spreadsheet Manual* that is available on this text's Web site.

Total Change Suppose the function $f(x) = x^2 + 20$ gives the rate (in ml/hr) that an intravenous liquid flows into a patient at time x. Then $f(2) = 24$ means that the fluid is flowing at a rate of 24 ml/hr at 2 hours. If the fluid continued to flow at this rate, the patient would receive 24 ml over the next hour. But the rate is changing; because $f(3) = 29$, the fluid is flowing at a rate of 29 ml/hr at 3 hours.

To find the *total* amount of fluid the patient receives as x (the time) changes from 2 to 3 (that is, the total change in the amount of fluid the patient has received), we could divide the interval from 2 to 3 into n equal parts, using each part as the base of a rectangle as we did above. The area of each rectangle would approximate the amount of fluid the patient receives over the time interval indicated by the base of the rectangle. Then the sum of the areas of these rectangles would approximate the net total fluid flow from $x = 2$ to $x = 3$. The limit of this sum as $n \to \infty$ would give the exact total amount of fluid, or the total change in the amount of fluid the patient has received.

This result demonstrates another application of the definite integral. The area of the region that is under the graph of the rate of fluid flow function $f(x)$, and that is above the x-axis and between $x = a$ and $x = b$, gives the *net total change in the fluid* as x goes from a to b.

TOTAL CHANGE IN $F(x)$

If $f(x)$ gives the rate of change of $F(x)$ for x in $[a, b]$, then the **total change** in $F(x)$ as x goes from a to b is given by

$$\lim_{n \to \infty} \sum_{i=1}^{n} f(x_i)\,\Delta x = \int_a^b f(x)\,dx.$$

In other words, the total change in a quantity can be found from the function that gives the rate of change of the quantity, using the same methods used to approximate the area under a curve.

EXAMPLE 2 Total Maintenance Charges

Figure 14 shows the rate of change of the annual maintenance charges for a certain magnetic resonance imaging (MRI) scanner. To approximate the total maintenance charges over the 10-year life of the scanner, use approximating rectangles, dividing the interval from 0 to 10 into ten equal subdivisions. Each rectangle has width 1; using the left endpoint of each rectangle to determine the height of the rectangle, the approximation becomes

$$1 \cdot 0 + 1 \cdot 500 + 1 \cdot 750 + 1 \cdot 1{,}800 + 1 \cdot 1{,}900 + 1 \cdot 3{,}000 + 1 \cdot 3{,}100$$
$$+ 1 \cdot 3{,}400 + 1 \cdot 4{,}200 + 1 \cdot 5{,}200 = 23{,}850.$$

About \$23,850 will be spent on maintenance over the 10-year life of the scanner.

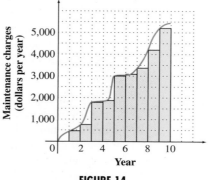

FIGURE 14

Recall, velocity is the rate of change in distance from time a to time b. Thus the area under the velocity function defined by $v(t)$ from $t = a$ to $t = b$ gives the distance traveled in that time period.

EXAMPLE 3   Total Distance

A driver traveling on a business trip checks the speedometer each hour. The table shows the driver's velocity at several times.

Time (hr)	0	1	2	3
Velocity (mph)	0	52	58	60

Approximate the total distance traveled during the three-hour period using the left endpoint of each interval, then the right endpoint.

Solution   Using left endpoints, the total distance is

$$0 \cdot 1 + 52 \cdot 1 + 58 \cdot 1 = 110.$$

With right endpoints, we get

$$52 \cdot 1 + 58 \cdot 1 + 60 \cdot 1 = 170.$$

Again, left endpoints give a total that is too small, while right endpoints give a total that is too large. The average, 140 miles, is a better estimate of the total distance traveled.

Before discussing further applications of the definite integral, we need a more efficient method for evaluating it. This method will be developed in the next section.

7.3 EXERCISES

1. Explain the difference between an indefinite integral and a definite integral.

2. Complete the following statement.

$$\int_0^3 (x^2 + 2)\, dx = \lim_{n \to \infty} \underline{\hspace{1cm}} , \text{ where } \Delta x = \underline{\hspace{1cm}} , \text{ and } x_i \text{ is } \underline{\hspace{1cm}} .$$

3. Let $f(x) = 2x + 1$, $x_1 = 0$, $x_2 = 2$, $x_3 = 4$, $x_4 = 6$, and $\Delta x = 2$.

   a. Find $\sum_{i=1}^{4} f(x_i)\, \Delta x$.

   b. The sum in part a approximates a definite integral using rectangles. The height of each rectangle is given by the value of the function at the left endpoint. Write the definite integral that the sum approximates.

4. Let $f(x) = 1/x$, $x_1 = 1/2$, $x_2 = 1$, $x_3 = 3/2$, $x_4 = 2$, and $\Delta x = 1/2$.

   a. Find $\sum_{i=1}^{4} f(x_i)\, \Delta x$.

   b. The sum in part a approximates a definite integral using rectangles. The height of each rectangle is given by the value of the function at the left endpoint. Write the definite integral that the sum approximates.

5. The booklet *All About Lawns** published by Ortho Books gives the following instructions for measuring the area of an irregularly shaped region.

Irregular Shapes
(within 5% accuracy)
Measure a long (L) axis of the area. Every 10 feet along the length line, measure the width at right angles to the length line. Total widths and multiply by 10.

Area = $(\overline{A_1A_2} + \overline{B_1B_2} + \overline{C_1C_2}$ etc.$) \times 10$
$\quad$ **A** = $(40' + 60' + 32') \times 10$
$\quad$ **A** = $132' \times 10'$
$\quad$ **A** = 1,320 square feet

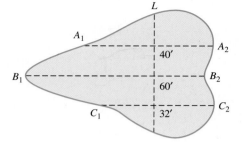

How does this method relate to the discussion in this section?

In Exercises 6–13, approximate the area under the graph of $f(x)$ and above the x-axis using each of the following methods with $n = 4$. **a.** *Use left endpoints.* **b.** *Use right endpoints* **c.** *Average the answers in parts a and b.* **d.** *Use midpoints.*

6. $f(x) = 3x + 2$ from $x = 1$ to $x = 5$

7. $f(x) = x + 5$ from $x = 2$ to $x = 4$

8. $f(x) = x^2$ from $x = 1$ to $x = 5$

9. $f(x) = -x^2 + 4$ from $x = -2$ to $x = 2$

10. $f(x) = e^x - 1$ from $x = 0$ to $x = 4$

11. $f(x) = e^x + 1$ from $x = -2$ to $x = 2$

12. $f(x) = \dfrac{1}{x}$ from $x = 1$ to $x = 5$

13. $f(x) = \dfrac{2}{x}$ from $x = 1$ to $x = 9$

14. Consider the region below $f(x) = x/2$, above the x-axis, and between $x = 0$ and $x = 4$. Let x_i be the midpoint of the ith subinterval.

$\quad$ **a.** Approximate the area of the region using four rectangles.

$\quad$ **b.** Find $\int_0^4 f(x)\,dx$ by using the formula for the area of a triangle.

15. Find $\int_0^5 (5 - x)\,dx$ by using the formula for the area of a triangle.

16. Find $\int_0^4 f(x)\,dx$ for each of the following graphs of $y = f(x)$.

$\quad$ **a.** $\qquad\qquad$ **b.**

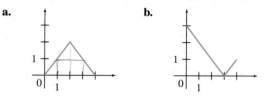

Find the exact value of each of the following integrals using formulas from geometry.

17. $\displaystyle\int_{-3}^{3} \sqrt{9 - x^2}\,dx$

18. $\displaystyle\int_{-4}^{0} \sqrt{16 - x^2}\,dx$

19. $\displaystyle\int_{1}^{3} (5 - x)\,dx$

20. $\displaystyle\int_{2}^{5} (1 + 2x)\,dx$

21. In this exercise, we investigate the value of $\int_0^1 x^2\,dx$ using larger and larger values of n in the definition of the definite integral.

$\quad$ **a.** First let $n = 10$, so $\Delta x = 0.1$. Fill a list on your calculator with values of x^2 as x goes from 0.1 to 1. (On a TI-83, use the command `seq(X^2,X,.1,1,.1)`→L1. On a TI-86, use the command `seq(x^2,x,.1,1,.1)`→A.)

---

*MacLaskey, Michael, *All About Lawns,* ed. by Alice Mace, Ortho Information Services, © 1980, p. 108.

b. Sum the values in the list formed in part a, and multiply by 0.1, to estimate $\int_0^1 x^2\, dx$ with $n = 10$. (On a TI-83, use the command `.1*sum(L1)`. On a TI-86, use the command `.1*sum(A)`.)

c. Repeat parts a and b with $n = 100$.

d. Repeat parts a and b with $n = 500$.

e. Based on your answers to parts b through d, what do you estimate the value of $\int_0^1 x^2\, dx$ to be?

22. Repeat Exercise 21 for $\int_0^1 x^3\, dx$.

Applications

In Exercises 23–28, estimate the area under each curve by summing the area of rectangles. Use the left endpoints and the right endpoints, then give the average of those answers.

LIFE SCIENCES

23. *Cliff Ecology* Cliffs form sheltered environments for a large number of plants and animals. An article on the ecology of cliffs included the following diagram, which gives the daily insolation (calories per square centimeter per hour) on two cliffs, one facing north and one facing south.*

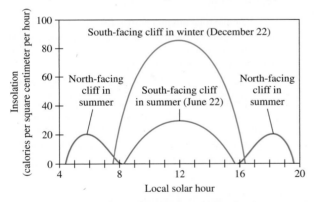

a. Estimate the total calories per square centimeter each day for the south-facing cliff in summer. Use rectangles with widths of 1 hr, starting with 8:15 A.M., and with the last rectangle (ending at 3:45) with a width of 1/2 hr.

b. Repeat part a for the south-facing cliff in winter. Use rectangles with widths of 1 hr, starting with 7:30 A.M.

24. *Foot-and-Mouth Epidemic* In 2001, the United Kingdom suffered an epidemic of foot-and-mouth disease. The graph shows the number of reported cases each day since February 18, as well as the number of cases epidemiologists project would have occurred had they culled all livestock on infected farms within 24 hours, and all livestock on neighboring farms within 48 hours of the infection.†

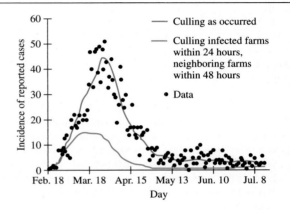

a. Estimate the total number of cases that occurred from February 18 through May 13. Use rectangles with widths of 14 days.

b. Estimate the total number of cases that would have occurred from February 18 through May 13 using the more aggressive culling plan. Use rectangles with widths of 14 days.

25. *Alcohol Concentration* The following graph shows the approximate concentration of alcohol in a person's bloodstream t hr after drinking 2 oz of alcohol. Estimate the total amount of alcohol in the bloodstream by estimating the area under the curve. Use rectangles with widths of 1 hr.

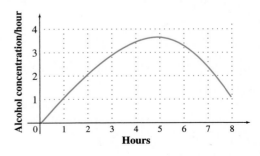

---

*Larson, Douglas W., et al., "Cliffs as Natural Refuges," *American Scientist,* Vol. 87, No. 5, Sept.– Oct. 1999, pp. 410–417.
†*Science,* Vol. 294, Oct. 5, 2001, p. 26.

26. *Oxygen Inhalation* The following graph shows the rate of inhalation of oxygen (in liters per minute) by a person riding a bicycle very rapidly for 10 min. Estimate the total volume of oxygen inhaled in the first 20 min after the beginning of the ride. Use rectangles with widths of 1 min.

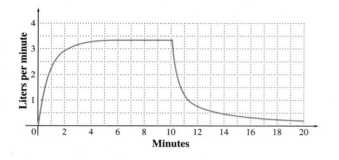

OTHER APPLICATIONS

27. *Oil Consumption* The following graph shows U.S. oil production and consumption rates in millions of barrels per day. The rates for the years beyond 1991 are projected using the policy in place in 1990 (labeled "Current policy base" in the graph) and using a new policy proposed by the Bush administration in 1991 (labeled "With strategy" in the graph.)*

a. Estimate the amount of oil produced between 1990 and 2010 using the policy in place in 1990. Use rectangles with widths of 2 yr.

b. Estimate the amount of oil produced between 1990 and 2010 using the proposed policy. Use rectangles with widths of 2 yr.

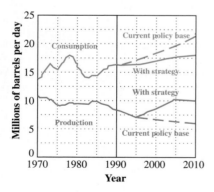

28. *Wages* The graph shows the average manufacturing hourly wage in the United States and Canada for 1987–1991.[†] All figures are in U.S. dollars. Assume that the average employee works 2,000 hours per year.

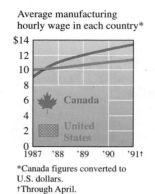

Average manufacturing hourly wage in each country*

*Canada figures converted to U.S. dollars.
†Through April.

a. Estimate the total amount earned by an average U.S. worker during the four-year period from the beginning of 1987 to the beginning of 1991. Use rectangles with widths of 1 yr.

b. Estimate the total amount earned (in U.S. dollars) by an average Canadian worker during the four-year period from the beginning of 1987 to the beginning of 1991. Use rectangles with widths of 1 yr.

Distance The next two graphs on the next page are from *Road & Track* magazine.[‡] The curve shows the velocity at *t* seconds after the car accelerates from a dead stop. To find the total distance traveled by the car in reaching 100 mph, we must estimate the definite integral

$$\int_0^T v(t)\, dt,$$

where *T* represents the number of seconds it takes for the car to reach 100 mph.

Use the graphs to estimate this distance by adding the areas of rectangles with widths of 5 sec. (The last rectangle will have a width of 4 sec or 3 sec.) Use the midpoint rule. To adjust your answer to miles per hour, divide by 3,600 (the number of seconds in an hour). You then have the number of miles that the car traveled in reaching 100 mph. Finally, multiply by 5,280 feet per mile to convert the answer to feet.

---

*Graph, "Uncle Sam's Energy Strategy" from *National Energy Strategy,* Feb. 1991, United States Department of Energy, Washington D.C.
†Graph, "Comparing Wages" from *The New York Times,* 1991. Copyright © 1991 by The New York Times Company. Reprinted with permission.
‡*Road & Track,* April and May, 1978. Reprinted with permission of *Road & Track.*

29. Estimate the distance traveled by the Porsche 928, using the graph below.

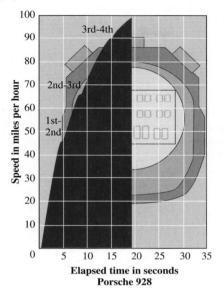

Elapsed time in seconds
Porsche 928

30. Estimate the distance traveled by the BMW 733i, using the graph below.

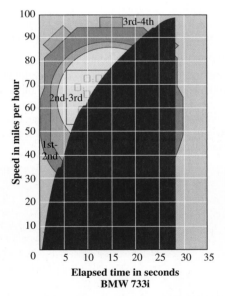

Elapsed time in seconds
BMW 733i

Heat Gain The following graphs* show the typical heat gain, in BTUs per hour per square foot, for a window facing east and one facing south, with plain glass and with a black ShadeScreen. Estimate the total heat gain per square foot by summing the areas of rectangles. Use rectangles with widths of 2 hr, and let the function value at the midpoint of the rectangle give the height of the rectangle.

31. a. Estimate the total heat gain per square foot for a plain glass window facing east.

b. Estimate the total heat gain per square foot for a window facing east with a ShadeScreen.

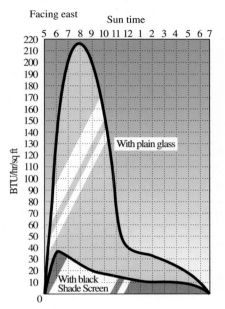

32. a. Estimate the total heat gain per square foot for a plain glass window facing south.

b. Estimate the total heat gain per square foot for a window facing south with a ShadeScreen.

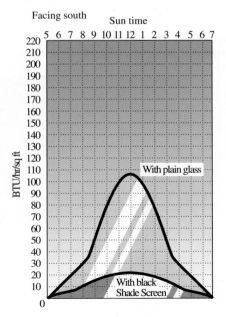

*Graphs courtesy of Phifer Wire Products. Reprinted by permission of Phifer Wire Products.

33. *Automobile Velocity* Two cars start from rest at a traffic light and accelerate for several minutes. The graph shows their velocities (in ft per sec) as a function of time (in sec). Car A is the one that initially has greater velocity.*

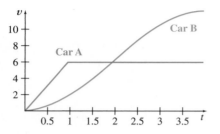

a. How far has car A traveled after 2 seconds? (*Hint:* Use formulas from geometry.)

b. When is car A farthest ahead of car B?

c. Estimate the farthest that car A gets ahead of car B. For car A, use formulas from geometry. For car B, use $n = 4$ and the value of the function at the midpoint of each interval.

d. Give a rough estimate of when car B catches up with car A.

34. *Distance* Musk the friendly pit bull has escaped again! Here is her velocity during the first 4 seconds of her romp.

t (sec)	0	1	2	3	4
v (ft/sec)	0	8	13	17	18

Give two estimates for the total distance Musk traveled during her 4-second trip, one using the left endpoint of each interval and one using the right endpoint.

35. *Distance* The speed of a particle in a test laboratory was noted every second for three seconds. The results are shown in the following table. Use the left endpoints and then the right endpoints to estimate the total distance the particle moved in the first three seconds.

t (sec)	0	1	2	3
v (ft/sec)	10	6.5	6	5.5

36. *Running* In 1987, Canadian Ben Johnson set a world record in the 100-meter sprint. (The record was later taken away when he was found to have used an anabolic steroid to enhance his performance.) His speed at various times in the race is given in the following table.[†]

Time (sec)	Speed (mph)
0	0
1.84	12.9
3.80	23.8
6.38	26.3
7.23	26.3
8.96	26.0
9.83	25.7

a. Use the information in the table and left endpoints to estimate the distance that Johnson ran in miles. You will first need to calculate Δt for each interval. At the end, you will need to divide by 3,600 (the number of seconds in an hour), since the speed is in miles per hour.

b. Repeat part a, using right endpoints.

c. Wait a minute; we know that the distance Johnson ran is 100 meters. Divide this by 1,609, the number of meters in a mile, to find how far Johnson ran in miles. Is your answer from part a or part b closer to the true answer? Briefly explain why you think this answer should be more accurate.

---

*Based on an example given by Stephen Monk of the University of Washington.
[†]Wildbur, Peter, *Information Graphics,* Van Nostrand Reinhold, 1989, pp. 126–127.

■ 7.4 THE FUNDAMENTAL THEOREM OF CALCULUS

THINK ABOUT IT   What is the total energy expenditure of a pronghorn fawn between 1 and 12 weeks of age?

This section introduces a powerful theorem for answering such questions.

The work from the last two sections can now be put together. We have seen that, if $f(x) > 0$,

$$\int_a^b f(x)\, dx$$

gives the area between the graph of $f(x)$ and the x-axis, from $x = a$ to $x = b$. We can find this definite integral by using the antiderivatives discussed earlier. The definite integral was defined and evaluated in the previous section using the limit of a sum. In that section, we also saw that if $f(x)$ gives the rate of change of $F(x)$, the definite integral $\int_a^b f(x)\, dx$ gives the total change of $F(x)$ as x changes from a to b. If $f(x)$ gives the rate of change of $F(x)$, then $F(x)$ is an antiderivative of $f(x)$. Writing the total change in $F(x)$ from $x = a$ to $x = b$ as $F(b) - F(a)$ shows the connection between antiderivatives and definite integrals. This relationship is called the **Fundamental Theorem of Calculus.**

FUNDAMENTAL THEOREM OF CALCULUS

Let f be continuous on the interval $[a, b]$, and let F be *any* antiderivative of f. Then

$$\int_a^b f(x)\, dx = F(b) - F(a) = F(x)\Big|_a^b.$$

The symbol $F(x)\big|_a^b$ is used to represent $F(b) - F(a)$. It is important to note that the Fundamental Theorem does not require $f(x) > 0$. The condition $f(x) > 0$ is necessary only when using the Fundamental Theorem to find area. Also, note that the Fundamental Theorem does not *define* the definite integral; it just provides a method for evaluating it.

EXAMPLE 1   Fundamental Theorem of Calculus

First find $\int 4t^3\, dt$ and then find $\int_1^2 4t^3\, dt$.

Solution By the rules given earlier,

$$\int 4t^3\, dt = t^4 + C.$$

By the Fundamental Theorem, the value of $\int_1^2 4t^3\, dt$ is found by evaluating $t^4\big|_1^2$, with no constant C required.

$$\int_1^2 4t^3\, dt = t^4\Big|_1^2 = 2^4 - 1^4 = 15$$

NOTE No constant C is needed, as it is for the indefinite integral, because if C were added to the antiderivative F, it would be eliminated in the final answer:

$$\int_a^b f(x)\,dx = (F(x) + C)\bigg|_a^b$$
$$= (F(b) + C) - (F(a) + C)$$
$$= F(b) - F(a).$$

In other words, any antiderivative will give the same answer, so for simplicity, we choose the one with $C = 0$.

Example 1 illustrates the difference between the definite integral and the indefinite integral. A definite integral is a real number; an indefinite integral is a family of functions in which all the functions are antiderivatives of a function f.

To see why the Fundamental Theorem of Calculus is true for $f(x) > 0$ when f is continuous, look at Figure 15. Define the function $A(x)$ as the area between the x-axis and the graph of $y = f(x)$ from a to x. We first show that A is an antiderivative of f; that is $A'(x) = f(x)$.

To do this, let h be a small positive number. Then $A(x + h) - A(x)$ is the shaded area in Figure 15. This area can be approximated with a rectangle having width h and height $f(x)$. The area of the rectangle is $h \cdot f(x)$, and

$$A(x + h) - A(x) \approx h \cdot f(x).$$

Dividing both sides by h gives

$$\frac{A(x + h) - A(x)}{h} \approx f(x).$$

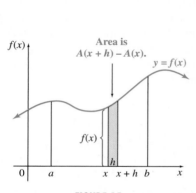

FIGURE 15

This approximation improves as h gets smaller and smaller. Taking the limit on the left as h approaches 0 gives an exact result.

$$\lim_{h \to 0} \frac{A(x + h) - A(x)}{h} = f(x)$$

This limit is simply $A'(x)$, so

$$A'(x) = f(x).$$

This result means that A is an antiderivative of f, as we set out to show.

$A(b)$ is the area under the curve from a to b, and $A(a) = 0$, so the area under the curve can be written as $A(b) - A(a)$. From the previous section, we know that the area under the curve is also given by $\int_a^b f(x)\,dx$. Putting these two results together gives

$$\int_a^b f(x)\,dx = A(b) - A(a)$$
$$= A(x)\bigg|_a^b,$$

where A is an antiderivative of f. From the note after Example 1, we know that any antiderivative will give the same answer, which proves the Fundamental Theorem of Calculus for the case when $f(x) > 0$.

The Fundamental Theorem of Calculus certainly deserves its name, which sets it apart as the most important theorem of calculus. It is the key connection between differential calculus and integral calculus, which originally were developed separately without knowledge of this connection between them. In a rough sense, the theorem says that the area under a curve may be found by a reversal of the process for finding the slope of a curve.

The variable used in the integrand does not matter; each of the following definite integrals represents the number $F(b) - F(a)$.

$$\int_a^b f(x)\, dx = \int_a^b f(t)\, dt = \int_a^b f(u)\, du$$

Key properties of definite integrals are listed below. Some of them are just restatements of properties from Section 1.

PROPERTIES OF DEFINITE INTEGRALS

If all indicated definite integrals exist, then

1. $\displaystyle\int_a^a f(x)\, dx = 0;$

2. $\displaystyle\int_a^b k \cdot f(x)\, dx = k \cdot \int_a^b f(x)\, dx$ for any real constant k
   (constant multiple of a function);

3. $\displaystyle\int_a^b [f(x) \pm g(x)]\, dx = \int_a^b f(x)\, dx \pm \int_a^b g(x)\, dx$
   (sum or difference of functions);

4. $\displaystyle\int_a^b f(x)\, dx = \int_a^c f(x)\, dx + \int_c^b f(x)\, dx$ for any real number c;

5. $\displaystyle\int_a^b f(x)\, dx = -\int_b^a f(x)\, dx.$

For $f(x) > 0$, since the distance from a to a is 0, the first property says that the "area" under the graph of f bounded by $x = a$ and $x = a$ is 0. Also, since $\int_a^c f(x)\, dx$ represents the blue region in Figure 16, and $\int_c^b f(x)\, dx$ represents the pink region,

$$\int_a^b f(x)\, dx = \int_a^c f(x)\, dx + \int_c^b f(x)\, dx,$$

as stated in the fourth property. While the figure shows $a < c < b$, the property is true for any value of c where both $f(x)$ and $F(x)$ are defined.

An algebraic proof is given here for the third property; proofs of the other properties are left for the exercises. If $F(x)$ and $G(x)$ are antiderivatives of $f(x)$ and $g(x)$, respectively,

$$\int_a^b [f(x) + g(x)]\, dx = [F(x) + G(x)]\Big|_a^b$$
$$= [F(b) + G(b)] - [F(a) + G(a)]$$
$$= [F(b) - F(a)] + [G(b) - G(a)]$$
$$= \int_a^b f(x)\, dx + \int_a^b g(x)\, dx.$$

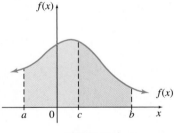

FIGURE 16

EXAMPLE 2 Fundamental Theorem of Calculus
Find $\int_2^5 (6x^2 - 3x + 5)\,dx$.

Solution Use the properties above and the Fundamental Theorem, along with properties from Section 1.

$$\int_2^5 (6x^2 - 3x + 5)\,dx = 6\int_2^5 x^2\,dx - 3\int_2^5 x\,dx + 5\int_2^5 dx$$

$$= 2x^3\Big|_2^5 - \frac{3}{2}x^2\Big|_2^5 + 5x\Big|_2^5$$

$$= 2(5^3 - 2^3) - \frac{3}{2}(5^2 - 2^2) + 5(5 - 2)$$

$$= 2(125 - 8) - \frac{3}{2}(25 - 4) + 5(3)$$

$$= 234 - \frac{63}{2} + 15 = \frac{435}{2}$$

EXAMPLE 3 Fundamental Theorem of Calculus

$$\int_1^2 \frac{dy}{y} = \ln|y|\Big|_1^2 = \ln|2| - \ln|1|$$

$$= \ln 2 - \ln 1 \approx 0.6931 - 0 = 0.6931$$

EXAMPLE 4 Substitution
Evaluate $\int_0^5 x\sqrt{25 - x^2}\,dx$.

Solution Use substitution. Let $u = 25 - x^2$, so that $du = -2x\,dx$. With a definite integral, the limits should be changed, too. The new limits on u are found as follows.

$$\text{If } x = 5, \text{ then } u = 25 - 5^2 = 0.$$

$$\text{If } x = 0, \text{ then } u = 25 - 0^2 = 25.$$

Then

$$\int_0^5 x\sqrt{25 - x^2}\,dx = -\frac{1}{2}\int_0^5 \sqrt{25 - x^2}(-2x\,dx)$$

$$= -\frac{1}{2}\int_{25}^0 \sqrt{u}\,du \qquad \text{Substitute and change limits.}$$

$$= -\frac{1}{2}\int_{25}^0 u^{1/2}\,du$$

$$= -\frac{1}{2}\cdot\frac{u^{3/2}}{3/2}\Big|_{25}^0$$

$$= -\frac{1}{2}\cdot\frac{2}{3}[0^{3/2} - 25^{3/2}]$$

$$= -\frac{1}{3}(-125) = \frac{125}{3}.$$

CAUTION When substitution is used on a definite integral, it is best to not revert from u back to the original variable after antidifferentiation, as we did for the indefinite integral. Notice in Example 4 that after changing from the old limits on x to the new limits on u, we never returned to x or its limits. We recommend the practice of labeling the limits when doing a substitution, so the substitution in Example 4 becomes

$$\int_{x=0}^{x=5} x\sqrt{25 - x^2}\,dx = -\frac{1}{2}\int_{u=25}^{u=0} \sqrt{u}\,du.$$

Area In the previous section we saw that, if $f(x) > 0$ in $[a, b]$, the definite integral $\int_a^b f(x)\,dx$ gives the area below the graph of the function $y = f(x)$, above the x-axis, and between the lines $x = a$ and $x = b$.

To see how to work around the requirement that $f(x) > 0$, look at the graph of $f(x) = x^2 - 4$ in Figure 17. The area bounded by the graph of f, the x-axis, and the vertical lines $x = 0$ and $x = 2$ lies below the x-axis. Using the Fundamental Theorem to find this area gives

$$\int_0^2 (x^2 - 4)\,dx = \left(\frac{x^3}{3} - 4x\right)\Big|_0^2$$

$$= \left(\frac{8}{3} - 8\right) - (0 - 0) = -\frac{16}{3}.$$

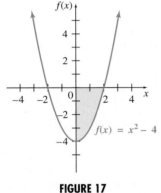

FIGURE 17

The result is a negative number because $f(x)$ is negative for values of x in the interval $[0, 2]$. Since Δx is always positive, if $f(x) < 0$ the product $f(x) \cdot \Delta x$ is negative, so $\int_0^2 f(x)\,dx$ is negative. Since area is nonnegative, the required area is given by $|{-16/3}|$ or $16/3$. Using a definite integral, the area could be written as

$$\left|\int_0^2 (x^2 - 4)\,dx\right| = \left|-\frac{16}{3}\right| = \frac{16}{3}.$$

EXAMPLE 5 Area

Find the area of the region between the x-axis and the graph of $f(x) = x^2 - 3x$ from $x = 1$ to $x = 3$.

Solution The region is shown in Figure 18. Since the region lies below the x-axis, the area is given by

$$\left|\int_1^3 (x^2 - 3x)\,dx\right|.$$

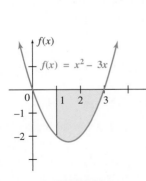

FIGURE 18

By the Fundamental Theorem,

$$\int_1^3 (x^2 - 3x)\,dx = \left(\frac{x^3}{3} - \frac{3x^2}{2}\right)\Big|_1^3 = \left(\frac{27}{3} - \frac{27}{2}\right) - \left(\frac{1}{3} - \frac{3}{2}\right) = -\frac{10}{3}.$$

The required area is $|{-10/3}| = 10/3$.

EXAMPLE 6 Area

Find the area between the x-axis and the graph of $f(x) = x^2 - 4$ from $x = 0$ to $x = 4$.

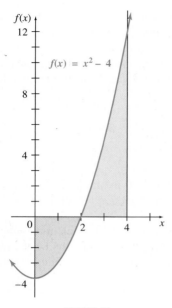

f(x) = x² − 4

FIGURE 19

Solution Figure 19 shows the required region. Part of the region is below the x-axis. The definite integral over that interval will have a negative value. To find the area, integrate the negative and positive portions separately and take the absolute value of the first result before combining the two results to get the total area. Start by finding the point where the graph crosses the x-axis. This is done by solving the equation

$$x^2 - 4 = 0.$$

The solutions of this equation are 2 and −2. The only solution in the interval [0, 4] is 2. The total area of the region in Figure 19 is

$$\left| \int_0^2 (x^2 - 4)\, dx \right| + \int_2^4 (x^2 - 4)\, dx = \left| \left(\frac{1}{3}x^3 - 4x \right)\Big|_0^2 \right| + \left(\frac{1}{3}x^3 - 4x \right)\Big|_2^4$$

$$= \left| \frac{8}{3} - 8 \right| + \left(\frac{64}{3} - 16 \right) - \left(\frac{8}{3} - 8 \right)$$

$$= 16.$$

Incorrectly using one integral over the entire interval to find the area in Example 6 would have given

$$\int_0^4 (x^2 - 4)\, dx = \left(\frac{x^3}{3} - 4x \right)\Big|_0^4 = \left(\frac{64}{3} - 16 \right) - 0 = \frac{16}{3},$$

which is not the correct area. This definite integral does not represent any area, but is just a real number.

For instance, suppose $f(x)$ in Example 6 represents the velocity (in meters per hour) of a duck moving along a straight line after x hours, where positive velocity represents motion to the right. Then 16/3 represents the *displacement* of the duck between $x = 0$ and $x = 4$, that is, the difference (in meters) between where the duck was at the beginning and where it was after 4 hours. The integral between 0 and 2 is −16/3; the negative sign indicates the duck moved to the left for the first two hours. The integral between 2 and 4 is 32/3, indicating the duck moved to the right during this time. The displacement is 32/3 − 16/3 = 16/3 meters. But the *total distance traveled* is represented by the total shaded area of 32/3 + |−16/3| = 16 meters. Since the duck first traveled to the left and then to the right, the total distance traveled is greater than the displacement.

FINDING AREA

In summary, to find the area bounded by a continuous function $f(x)$, $x = a$, $x = b$, and the x-axis, use the following steps.

1. Sketch a graph.
2. Find any x-intercepts of $f(x)$ in $[a, b]$. These divide the total region into subregions.
3. The definite integral will be *positive* for subregions above the x-axis and *negative* for subregions below the x-axis. Use separate integrals to find the (positive) areas of the subregions.
4. The total area is the sum of the areas of all of the subregions.

In the last section, we saw that the area under a rate of change function $f'(x)$ from $x = a$ to $x = b$ gives the total value of $f(x)$ on $[a, b]$. Now we can use the definite integral to solve these problems.

EXAMPLE 7   Energy in Pronghorn Fawns

The total energy expenditure of pronghorn fawns increases at a roughly linear rate with age, from an expenditure of about 2,000 kJ (kilojoules) a day at 1 week to about 9,000 kJ a day at 12 weeks.* Find the total energy expended by a typical pronghorn fawn over this 11 week (77 day) period.

Solution At $t = 7$ days, $E = 2,000$, while at $t = 12 \cdot 7 = 84$ days, $E = 9,000$. To find E as a linear function of t, first find the slope.

$$m = \frac{9,000 - 2,000}{84 - 7} = \frac{7,000}{77} \approx 90.909$$

Using the point slope form of the line,

$$y - 2,000 = 90.909(x - 7)$$
$$y - 2,000 = 90.909x - 636$$
$$y = 90.909x + 1,364$$

Now integrate this function to get the total amount of energy.

$$\int_{7}^{84} (90.909x + 1,364)\, dx = \left[90.909\left(\frac{x^2}{2}\right) + 1,364x \right]\Bigg|_{7}^{84}$$
$$= 435,303 - 11,775 = 423,528$$

A pronghorn fawn uses about 423,500 kJ of energy between 1 and 12 weeks.

7.4   EXERCISES

Evaluate each definite integral.

1. $\displaystyle\int_{-2}^{4} (-1)\, dp$

2. $\displaystyle\int_{-4}^{1} 6x\, dx$

3. $\displaystyle\int_{-1}^{2} (3t - 1)\, dt$

4. $\displaystyle\int_{-2}^{2} (4z + 3)\, dz$

5. $\displaystyle\int_{0}^{2} (5x^2 - 4x + 2)\, dx$

6. $\displaystyle\int_{-2}^{3} (-x^2 - 3x + 5)\, dx$

7. $\displaystyle\int_{0}^{2} 3\sqrt{4u + 1}\, du$

8. $\displaystyle\int_{3}^{9} \sqrt{2r - 2}\, dr$

9. $\displaystyle\int_{0}^{1} 2(t^{1/2} - t)\, dt$

10. $\displaystyle\int_{0}^{4} -(3x^{3/2} + x^{1/2})\, dx$

11. $\displaystyle\int_{1}^{4} \left(5y\sqrt{y} + 3\sqrt{y}\right)\, dy$

12. $\displaystyle\int_{4}^{9} \left(4\sqrt{r} - 3r\sqrt{r}\right)\, dr$

13. $\displaystyle\int_{4}^{6} \frac{2}{(x - 3)^2}\, dx$

14. $\displaystyle\int_{1}^{4} \frac{-3}{(2p + 1)^2}\, dp$

15. $\displaystyle\int_{1}^{5} (5n^{-2} + n^{-3})\, dn$

16. $\displaystyle\int_{2}^{3} (3x^{-2} - x^{-4})\, dx$

17. $\displaystyle\int_{2}^{3} \left(2e^{-0.1A} + \frac{3}{A}\right)\, dA$

18. $\displaystyle\int_{1}^{2} \left(\frac{-1}{B} + 3e^{0.2B}\right)\, dB$

19. $\displaystyle\int_{1}^{2} \left(e^{5u} - \frac{1}{u^2}\right)\, du$

20. $\displaystyle\int_{0.5}^{1} (p^3 - e^{4p})\, dp$

21. $\displaystyle\int_{-1}^{0} y(2y^2 - 3)^5\, dy$

22. $\displaystyle\int_{0}^{3} m^2(4m^3 + 2)^3\, dm$

23. $\displaystyle\int_{1}^{64} \frac{\sqrt{z} - 2}{\sqrt[3]{z}}\, dz$

24. $\displaystyle\int_{1}^{8} \frac{3 - y^{1/3}}{y^{2/3}}\, dy$

*Miller, Michelle N., and John A. Byers, "Energetic Cost of Locomotor Play in Pronghorn Fawns," *Animal Behavior*, Vol. 41, 1991, pp. 1007–1013. In Exercise 46 of Section 4.6, we saw how the energy depends on the mass of the fawn.

25. $\int_1^2 \frac{\ln x}{x} dx$

26. $\int_1^3 \frac{\sqrt{\ln x}}{x} dx$

27. $\int_0^8 x^{1/3}\sqrt{x^{4/3} + 9}\, dx$

28. $\int_1^2 \frac{3}{x(1 + \ln x)} dx$

29. $\int_0^1 \frac{e^t}{(3 + e^t)^2} dt$

30. $\int_0^1 \frac{e^{2z}}{\sqrt{1 + e^{2z}}} dz$

*In each of the following, use the definite integral to find the area between the x-axis and
f(x) over the indicated interval. Check first to see if the graph crosses the x-axis in the
given interval.*

31. $f(x) = 2x + 3;$ $[8, 10]$

32. $f(x) = 4x - 7;$ $[5, 10]$

33. $f(x) = 2 - 2x^2;$ $[0, 5]$

34. $f(x) = 9 - x^2;$ $[0, 6]$

35. $f(x) = x^3;$ $[-1, 3]$

36. $f(x) = x^3 - 2x;$ $[-2, 4]$

37. $f(x) = e^x - 1;$ $[-1, 2]$

38. $f(x) = 1 - e^{-x};$ $[-1, 2]$

39. $f(x) = \frac{1}{x};$ $[1, e]$

40. $f(x) = \frac{1}{x};$ $[e, e^2]$

Find the area of each shaded region.

41.

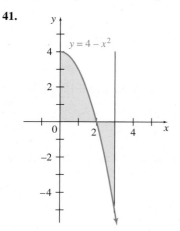

42.

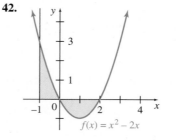

43.

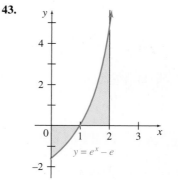

44. Assume $f(x)$ is continuous for $g \le x \le c$ as shown in
the figure. Write an equation relating the three quantities

$$\int_a^b f(x)\, dx, \qquad \int_a^c f(x)\, dx, \qquad \int_b^c f(x)\, dx.$$

45. Is the equation you wrote for Exercise 44 still true

　a. if b is replaced by d?

　b. if b is replaced by g?

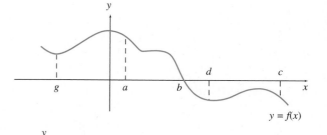

46. The graph of $f(x)$, shown here, consists of two straight line
segments and two quarter circles. Find the value of $\int_0^{16} f(x)\, dx$.

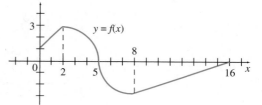

Use the Fundamental Theorem to show that the following are true.

47. $\int_a^b kf(x)\, dx = k\int_a^b f(x)\, dx$

48. $\int_a^b f(x)\, dx = \int_a^c f(x)\, dx + \int_c^b f(x)\, dx$

49. $\int_a^b f(x)\, dx = -\int_b^a f(x)\, dx$

50. Use Exercise 48 to find $\int_{-1}^{4} f(x)\,dx$, given

$$f(x) = \begin{cases} 2x + 3 & \text{if } x \le 0 \\ -\dfrac{x}{4} - 3 & \text{if } x > 0. \end{cases}$$

51. You are given $\int_{0}^{1} e^{x^2}\,dx = 1.46265$ and $\int_{0}^{2} e^{x^2}\,dx = 16.45263$. Use this information to find

a. $\displaystyle\int_{-1}^{1} e^{x^2}\,dx$;   **b.** $\displaystyle\int_{1}^{2} e^{x^2}\,dx$.

52. Let $g(t) = t^4$ and define $f(x) = \displaystyle\int_{c}^{x} g(t)\,dt$ with $c = 1$.

a. Find a formula for $f(x)$.

b. Verify that $f'(x) = g(x)$. The fact that $\dfrac{d}{dx}\displaystyle\int_{c}^{x} g(t)\,dt = g(x)$ is true for all continuous functions g is an alternative version of the Fundamental Theorem of Calculus.

c. Let us verify the result in part b for a function whose antiderivative cannot be found. Let $g(t) = e^{t^2}$ and let $c = 0$. Use the integration feature on a graphing calculator to find $f(x)$ for $x = 1$ and $x = 1.01$. Then use the definition of the derivative with $h = 0.01$ to approximate $f'(1)$, and compare it with $g(1)$.

d. Explain why $f(x) = \displaystyle\int_{a}^{x} g(t)\,dt$ is an antiderivative for g with $f(a) = 0$.

Applications

LIFE SCIENCES

53. *Rams' Horns* The average annual increment in the horn length (in cm) of bighorn rams born since 1986 can be approximated by

$$y = 0.1762x^2 - 3.986x + 22.68,$$

where x is the ram's age in years for x between 3 and 9.* Integrate to find the total increase in the length of a ram's horn during this time.

54. *Beagles* As we saw in the section on Techniques for Finding Derivatives, the daily energy requirements of female beagles who are at least one year old change with respect to time according to the function

$$E(t) = 753t^{-0.1321},$$

where $E(t)$ is the daily energy requirement (in $kJ/W^{0.67}$), where W is the dog's weight (in kg) for a beagle that is t years old.[†]

a. Assuming 365 days in a year, show that the energy requirement for a female beagle that is t days old is given by

$$E(t) = 1{,}642t^{-0.1321}.$$

b. Using the formula from part a, determine the total energy requirements (in $kJ/W^{0.67}$) for a female beagle between her first and third birthday.

55. *Nervous System* In a model of the nervous system, the time required for n circuits of nerves to be excited is given by

$$t = \int_{0}^{n} \frac{du}{K - au},$$

where K and a are constants.[‡]

a. Find a formula for the number of neural circuits that are excited as a function of time.

b. What happens to the number of neural circuits that are excited as time progresses?

56. *Biochemical Reaction* In an example of the Field-Noyes model of a biochemical reaction, where x denotes the concentration of $HBrO_2$ (bromous acid), and T is the period of an oscillation,

$$\int_{2}^{1} \left(x - \frac{1}{x} \right) dx = -\int_{0}^{T/2} dt.[§]$$

Find the value of T.

---

*Jorgenson, Jon T., et al., "Effects of Population Density on Horn Development in Bighorn Rams," *Journal of Wildlife Management,* Vol. 62, No. 3, 1998, pp. 1011–1020.
†Finke, M., "Energy Requirements of Adult Female Beagles," *Journal of Nutrition,* Vol. 124, 1994, pp. 2604s–2608s.
‡Rashevsky, Nicolas, *Mathematical Biology of Social Behavior,* rev. ed., Chicago, The University of Chicago Press, 1959, p. 9.
§Murray, J. D., *Mathematical Biology,* Springer-Verlag, 1989, p. 191.

57. *Drug Concentration* In a model of drug use, the concentration of a drug in a user's bloodstream is given by

$$C(t) = e^{-kt} \int_0^t e^{ky} D(y) \, dy,$$

where k is the proportional rate that the drug is removed from the bloodstream, D is the rate that the drug is introduced into the bloodstream, and t is the time.*

a. Under the simplifying assumption that $D(y) = D$ (a constant), show that

$$C(t) = \frac{D}{k}(1 - e^{-kt}).$$

b. Letting $A(t)$ represent the number of free (active) sites where the drug can cause a response in the individual and N represent the total number of sites, free (active) and bound (inactive),

$$A(t) = N e^{-\int_0^t mC(z) \, dz},$$

where m is a constant. Using the expression for $C(t)$ from part a, show that

$$A(t) = N \exp\left[\frac{-mD}{k}\left(t + \frac{-1 + e^{-kt}}{k}\right)\right].$$

58. *Pollution* Pollution from a factory is entering a lake. The rate of concentration of the pollutant at time t is given by

$$P'(t) = 140t^{5/2},$$

where t is the number of years since the factory started introducing pollutants into the lake. Ecologists estimate that the lake can accept a total level of pollution of 4,850 units before all the fish life in the lake is destroyed. Can the factory operate for 4 years without killing all the fish in the lake?

59. *Spread of an Oil Leak* An oil tanker is leaking oil at the rate given in barrels per hour by

$$L'(t) = \frac{80 \ln(t + 1)}{t + 1},$$

where t is the time in hours after the tanker hits a hidden rock (when $t = 0$).

a. Find the total number of barrels that the ship will leak on the first day.

b. Find the total number of barrels that the ship will leak on the second day.

c. What is happening over the long run to the amount of oil leaked per day?

60. *Tree Growth* After long study, tree scientists conclude that a eucalyptus tree will grow at the rate of $0.2 + 4t^{-4}$ feet per year, where t is time in years.

a. Find the number of feet that the tree will grow in the second year.

b. Find the number of feet the tree will grow in the third year.

61. *Growth of a Fungus* The rate at which a fungus grows is given by

$$R'(x) = 200e^{0.2x},$$

where x is the time in days. What is the total accumulated growth during the first 2.5 days?

62. *Drug Reaction* For a certain drug, the rate of reaction in appropriate units is given by

$$R'(t) = \frac{5}{t} + \frac{2}{t^2},$$

where t is time measured in hours after the drug is administered. Find the total reaction to the drug over the following time periods.

a. From $t = 1$ to $t = 12$ **b.** From $t = 12$ to $t = 24$

63. *Human Mortality* If $f(x)$ is the instantaneous death rate for members of a population at time t, then the number of individuals who survive to age T is given by

$$F(T) = \int_0^T f(x) \, dx.$$

In 1825 the biologist Benjamin Gompertz proposed that $f(x) = kb^x$.[†] Find a formula for $F(T)$.

64. *Cell Division* Let the expected number of cells in a culture that have an x percent probability of undergoing cell division during the next hour be denoted by $n(x)$.

a. Explain why $\int_{20}^{30} n(x) \, dx$ approximates the total number of cells with a 20 to 30% chance of dividing during the next hour.

b. Give an integral representing the number of cells which have less than a 60% chance of dividing during the next hour.

c. Let $n(x) = \sqrt{5x + 1}$ give the expected number of cells (in millions) with x percent probability of dividing during the next hour. Find the number of cells with a 5 to 10% chance of dividing.

65. *Bacterial Growth* A population of *E. coli* bacteria will grow at a rate given by

$$w'(t) = (4t + 1)^{1/3},$$

---

*Hoppensteadt, F. C., and J. D. Murray, "Threshold Analysis of a Drug Use Epidemic Model," *Mathematical Biosciences,* Vol. 53, No. 1/2, Feb. 1981, pp. 79–87.
†Gompertz, Benjamin, "On the Nature of the Function Expressive of the Law of Human Mortality," *Philosophical Transactions of the Royal Society of London,* 1825.

where w is the weight in milligrams after t hours. Find the change in weight of the population from $t = 0$ to $t = 3$.

66. *Blood Flow* In an example from an earlier chapter, the velocity v of the blood in a blood vessel was given as

$$v = k(R^2 - r^2),$$

where R is the (constant) radius of the blood vessel, r is the distance of the flowing blood from the center of the blood vessel, and k is a constant. Total blood flow (in millimeters per minute) is given by

$$Q(R) = \int_0^R 2\pi v r \, dr.$$

a. Find the general formula for Q in terms of R by evaluating the definite integral given above.

b. Evaluate $Q(0.4)$.

67. *Sediment* The density of sediment (in g per cubic cm) at the bottom of Lake Coeur d'Alene, Idaho, is given by

$$p(x) = p_0 e^{0.0133x},$$

where x is the depth in cm and p_0 is the density at the surface.* The total mass of a square-centimeter column of sediment above a depth of h cm is given by

$$\int_0^h p(x) \, dx.$$

If $p_0 = 0.85$ g/cm³, find the total mass above a depth of 100 cm.

68. *Age Distribution* The 1990 U.S. census gives us an age distribution which is approximately given (in millions) by the function

$$f(x) = 32.0 + 4.45x - 0.88x^2,$$

where x varies from 0 to 9 decades.† The population of a given age group can be found by integrating this function over the interval for that age group.

a. Find the integral over the interval $[0, 9]$. What does this integral represent?

b. Baby boomers are those born between 1945 and 1965, that is, those in the range of 2.5 to 4.5 decades in 1990. Find the number of baby boomers.

OTHER APPLICATIONS

69. *Income Distribution* Based on 1990 census data, an approximate income distribution for the U.S. is given by the function

$$f(x) = 15.6 - 0.02x - 0.154x^2,$$

where x is annual income in units of $10,000, $0.5 \leq x \leq 10$.† For example, $x = 0.5$ represents an annual income of

$5,000. (*Note:* This function does not give a good representation for incomes over $100,000.) The percent of the population with an income in a given range can be found by integrating this function over that range. Find the percentage of the population with an income between $25,000 and $50,000.

70. *Oil Consumption* Suppose that the rate of consumption of a natural resource is $c'(t)$, where

$$c'(t) = ke^{rt}.$$

Here t is time in years, r is a constant, and k is the consumption in the year when $t = 0$. In 1998, an oil company sold 1.2 billion barrels of oil. Assume that $r = 0.04$.

a. Write $c'(t)$ for the oil company, letting $t = 0$ represent 1998.

b. Set up a definite integral for the amount of oil that the company will sell in the next ten years.

c. Evaluate the definite integral of part b.

d. The company has about 20 billion barrels of oil in reserve. To find the number of years that this amount will last, solve the equation

$$\int_0^T 1.2e^{0.04t} \, dt = 20.$$

e. Rework part d, assuming that $r = 0.02$.

71. *Oil Consumption* The rate of consumption of oil (in billions of barrels) by the company in Exercise 70 was given as

$$1.2e^{0.04t},$$

where $t = 0$ corresponds to 1998. Find the total amount of oil used by the company from 1998 to year T. At this rate, how much will be used in 5 years?

---

*Nord, Gail, and John Nord, "Sediment in Lake Coeur d'Alene, Idaho," *The Mathematics Teacher,* Vol. 91, No. 4, April 1998, pp. 292–295.

†Exercises 68 and 69 were contributed by Ralph DeMarr, University of New Mexico.

■ 7.5 INTEGRALS OF TRIGONOMETRIC FUNCTIONS

? **THINK ABOUT IT** In a model of an oscillating system in biology, how can we solve for the time that a stimulus is applied to the oscillator?

In Exercise 31 in this section, we will use trigonometry to answer this question.

Any differentiation formula leads to a corresponding formula for integration. In particular, the formulas of the section on Derivatives of Trigonometric Functions lead to the following indefinite integrals.

BASIC TRIGONOMETRIC INTEGRALS

$$\int \sin x \, dx = -\cos x + C \qquad \int \cos x \, dx = \sin x + C$$

$$\int \sec^2 x \, dx = \tan x + C \qquad \int \csc^2 x \, dx = -\cot x + C$$

$$\int \sec x \tan x \, dx = \sec x + C \qquad \int \csc x \cot x \, dx = -\csc x + C$$

EXAMPLE 1 Integrals of $\sin x$ and $\cos x$
Find each integral.

(a) $\displaystyle\int \sin 7x \, dx$

Solution Use substitution. Let $u = 7x$, so that $du = 7 \, dx$. Then

$$\int \sin 7x \, dx = \frac{1}{7} \int \sin 7x (7 \, dx)$$

$$= \frac{1}{7} \int \sin u \, du$$

$$= -\frac{1}{7} \cos u + C$$

$$= -\frac{1}{7} \cos 7x + C.$$

(b) $\displaystyle\int \cos \frac{2}{3} x \, dx = \frac{3}{2} \int \cos \frac{2}{3} x \left(\frac{2}{3} \, dx \right) = \frac{3}{2} \sin \frac{2}{3} x + C$

(c) $\displaystyle\int \sin^2 x \cos x \, dx$

Solution Let $u = \sin x$, with $du = \cos x \, dx$. This gives

$$\int \sin^2 x \cos x \, dx = \int u^2 \, du = \frac{1}{3} u^3 + C.$$

Replacing u with $\sin x$ gives

$$\int \sin^2 x \cos x \, dx = \frac{1}{3} \sin^3 x + C.$$

(d) $\displaystyle\int \frac{\sin x}{\sqrt{\cos x}}\, dx$

Solution Rewrite the integrand as

$$\int (\cos x)^{-1/2}\sin x\, dx.$$

If $u = \cos x$, then $du = -\sin x\, dx$, with

$$\int (\cos x)^{-1/2}\sin x\, dx = -\int (\cos x)^{-1/2}(-\sin x\, dx)$$

$$= -\int u^{-1/2}\, du$$

$$= -2u^{1/2} + C$$

$$= -2\cos^{1/2}x + C.$$

(e) $\displaystyle\int \sec^2 12x\, dx = \frac{1}{12}\int \sec^2 12x\,(12\, dx) = \frac{1}{12}\tan 12x + C$

(f) $\displaystyle\int x^2 \sec x^3 \tan x^3\, dx$

Solution Use substitution. Let $u = x^3$, so that $du = 3x^2\, dx$. Then,

$$\int x^2 \sec x^3 \tan x^3\, dx = \frac{1}{3}\int \sec x^3 \tan x^3\,(3x^2\, dx)$$

$$= \frac{1}{3}\int \sec u \tan u\, du$$

$$= \frac{1}{3}\sec u + C$$

$$= \frac{1}{3}\sec x^3 + C.$$

As in earlier sections of this chapter, we can find the area under a curve by setting up an appropriate definite integral.

EXAMPLE 2 Area Under the Curve
Find the shaded area in Figure 20.

Solution The shaded area in Figure 20 is bounded by $y = \cos x$, $y = 0$, $x = -\pi/2$, and $x = \pi/2$. By the Fundamental Theorem of Calculus, this area is given by

$$\int_{-\pi/2}^{\pi/2} \cos x\, dx = \sin x \Big|_{-\pi/2}^{\pi/2}$$

$$= \sin\frac{\pi}{2} - \sin\left(-\frac{\pi}{2}\right)$$

$$= 1 - (-1)$$

$$= 2.$$

By symmetry, the same area could be found by evaluating

$$2\int_0^{\pi/2} \cos x\, dx.$$

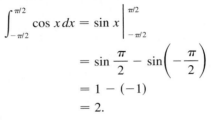

FIGURE 20

The area in Example 2 could also be found using the definite integral feature of a graphing calculator, entering the expression cos x, the variable x, and the limits of integration.

As mentioned earlier, tan $x = (\sin x)/\cos x$, so that

$$\int \tan x\, dx = \int \frac{\sin x}{\cos x}\, dx.$$

To find $\int \tan x\, dx$, let $u = \cos x$, with $du = -\sin x\, dx$. Then

$$\int \tan x\, dx = \int \frac{\sin x}{\cos x}\, dx = -\int \frac{du}{u} = -\ln |u| + C.$$

Replacing u with cos x gives the formula for integrating tan x. The integral for cot x is found in a similar way.

INTEGRALS OF tan x AND cot x

$$\int \tan x\, dx = -\ln |\cos x| + C$$

$$\int \cot x\, dx = \ln |\sin x| + C$$

EXAMPLE 3 Integrals of tan x and cot x

(a) $\displaystyle \int \tan 6x\, dx = \frac{1}{6} \int \tan 6x\, (6\, dx) = -\frac{1}{6} \ln |\cos 6x| + C$

(b) $\displaystyle \int x \cot x^2\, dx = \frac{1}{2} \int (\cot x^2)\, (2x\, dx) = \frac{1}{2} \ln |\sin x^2| + C$

EXAMPLE 4 Natural Gas Consumption

The monthly residential consumption of natural gas in the United States in 1999 is found in Table 3.*

Table 3

Index	Month	Consumption (billion cubic feet)
1	January	903
2	February	680
3	March	660
4	April	417
5	May	234
6	June	155
7	July	128
8	August	117
9	September	135
10	October	227
11	November	362
12	December	648

*Energy Information Administration, *Natural Gas Monthly,* April 2000.

(a) Plot the data. Is it reasonable to assume that the monthly consumption of energy is periodic?

Solution Figure 21 shows a graphing calculator plot of the data. Because of the cyclical nature of the four seasons, it is reasonable to assume that the data are periodic.

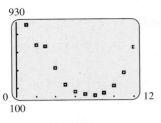

FIGURE 21

(b) Find a trigonometric function of the form $C(x) = a \sin(bx + c) + d$ that models this data when x is the month of the year and $C(x)$ is the natural gas consumption. Use Table 3.

Solution The function $C(x)$, derived by a TI-83 Plus calculator, is given by $C(x) = 467.99 \sin(0.364289x + 1.97066) + 552.579$. Figure 22 shows that this function fits the data well.

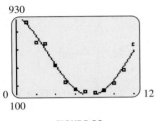

FIGURE 22

(c) Estimate the consumption for the month of September and compare it to the actual value.

Solution $C(9) = 150.43$. The actual value is 135.

(d) Estimate the rate at which the consumption is changing in September.

Solution $C'(x) = (0.364289)467.99 \cos(0.364289x + 1.97066)$
$$= 170.484 \cos(0.364289x + 1.97066), \text{ and}$$
$$C'(9) = 87.1939 \text{ billion cubic feet per month}$$

(e) Estimate the total natural gas consumption for the year for residential customers and compare it to the actual value.

Solution To estimate the total natural gas consumption by residential customers for 1999 we use integration as follows.

$$\int_0^{12} C(x)\,dx = \int_0^{12} 467.99 \sin(0.364289x + 1.97066) + 552.579\,dx$$

$$= -\frac{467.99}{0.364289} \cos(0.364289x + 1.97066) + 552.579x \Big|_0^{12} \approx 4{,}848.4$$

The actual value is 4,666.

(f) What would you expect the period of a function that models annual natural gas consumption to be? What is the period of the function found in part b?

Solution If we assume that the annual natural gas consumption is periodic, we would expect the period to be 12 so that it repeats itself every 12 months. The period for the function given above is $T = 2\pi/0.364289 \approx 17.25$. Although this value is different from 12, we must remember that the derived function was based on finding the trigonometric function that best describes the given set of 12 points. This suggests that although energy usage is seasonal, other factors that affect energy usage are not accounted for in the model. For example, a very cold winter might be followed by a mild one, so that the natural gas consumption 12 months later is significantly lower.

7.5 EXERCISES

Find the following integrals.

1. $\displaystyle\int \cos 5x\,dx$

2. $\displaystyle\int \sin 8x\,dx$

3. $\displaystyle\int (5\cos x + 2\sin x)\,dx$

4. $\displaystyle\int (7\sin x - 8\cos x)\,dx$

5. $\displaystyle\int x\sin x^2\,dx$

6. $\displaystyle\int 2x\cos x^2\,dx$

7. $\displaystyle -\int 6\sec^2 2x\,dx$

8. $\displaystyle -\int 2\csc^2 8x\,dx$

9. $\displaystyle\int \sin^7 x\cos x\,dx$

10. $\displaystyle\int \sin^6 x\cos x\,dx$

11. $\displaystyle\int \sqrt{\sin x}\,(\cos x)\,dx$

12. $\displaystyle\int \frac{\cos x}{\sqrt{\sin x}}\,dx$

13. $\displaystyle\int \frac{\sin x}{1 + \cos x}\,dx$

14. $\displaystyle\int \frac{\cos x}{1 - \sin x}\,dx$

15. $\displaystyle\int x^5 \cos x^6\,dx$

16. $\displaystyle\int (x+2)^4 \sin(x+2)^5\,dx$

17. $\displaystyle\int \tan \frac{1}{4}x\,dx$

18. $\displaystyle\int \cot\left(-\frac{3}{8}x\right)dx$

19. $\displaystyle\int x^2 \cot x^3\,dx$

20. $\displaystyle\int \frac{x}{4}\tan\left(\frac{x}{4}\right)^2 dx$

21. $\displaystyle\int e^x \sin e^x\,dx$

22. $\displaystyle\int e^{-x}\tan e^{-x}\,dx$

23. $\displaystyle\int e^x \csc e^x \cot e^x\,dx$

24. $\displaystyle\int x^4 \sec x^5 \tan x^5\,dx$

Evaluate the following definite integrals. Use the integration feature of a graphing calculator, if you wish, to support your answers.

25. $\displaystyle\int_0^{\pi/4} \sin x\,dx$

26. $\displaystyle\int_{-\pi/2}^0 \cos x\,dx$

27. $\displaystyle\int_0^{\pi/3} \tan x\,dx$

28. $\displaystyle\int_{\pi/4}^{\pi/2} \cot x\,dx$

29. $\displaystyle\int_{\pi/2}^{2\pi/3} \cos x\,dx$

30. $\displaystyle\int_{\pi/4}^{3\pi/4} \sin x\,dx$

Applications

LIFE SCIENCES

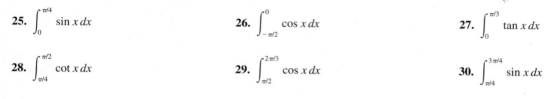

31. *Oscillating System* In a model of an oscillating system in biology,

$$\int_\theta^\phi (1 + I\cos 2x)^{-1}\,dx = \int_0^T dt,$$

where I is a constant, θ is the old phase of the oscillator, ϕ is the new phase, and T is the time that a stimulus is applied to the oscillator.* Solve for T when $I = 1$. (*Hint:* Use the trigonometric identity $\cos 2x = 2\cos^2 x - 1$.)

32. *Migratory Animals* The number of migratory animals (in hundreds) counted at a certain checkpoint is given by

$$T(t) = 50 + 50\cos\left(\frac{\pi}{6}t\right),$$

where t is time in months, with $t = 0$ corresponding to July. The figure in the next column shows a graph of T. Use a definite integral to find the number of animals passing the checkpoint in a year.

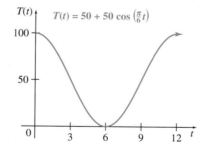

OTHER APPLICATIONS

33. *Energy Consumption* The monthly residential consumption of natural gas in Pennsylvania for 1999 is found in the table on the next page.[†]

---

*Murray, J. D., *Mathematical Biology,* Springer-Verlag, 1989, pp. 204–205.
[†]Energy Information Administration, *Natural Gas Monthly,* April 2000.

Index	Month	Consumption (million cubic feet)
1	January	36,752
2	February	45,967
3	March	37,498
4	April	21,700
5	May	11,260
6	June	6,518
7	July	5,112
8	August	4,808
9	September	5,776
10	October	11,580
11	November	19,778
12	December	34,006

a. Plot the data. Is it reasonable to assume that the monthly consumption of energy is periodic?

b. Find a trigonometric function of the form

$$C(x) = a \sin(bx + c) + d$$

that models this data when x is the month of the year and $C(x)$ is the natural gas consumption.

c. Estimate the total natural gas consumption for the year for residential customers in Pennsylvania and compare it to the actual value.

34. *Voltage* The electrical voltage from a standard wall outlet is given as a function of time t by

$$V(t) = 170 \sin(120\,\pi t).$$

This is an example of alternating current, which is electricity that reverses direction at regular intervals. The common method for measuring the level of voltage from an alternating current is the *root mean square,* which is given by

$$\text{Root mean square} = \sqrt{\frac{\int_0^T V^2(t)\,dt}{T}},$$

where T is one period of the current.

a. Verify that $T = 1/60$ sec for $V(t)$ given above.

b. You may have seen that the voltage from a standard wall outlet is 120 volts. Verify that this is the root mean square value for $V(t)$ given above. (*Hint:* Use the trigonometric identity $\sin^2 x = (1 - \cos 2x)/2$. This identity can be derived by letting $y = x$ in the basic identity for $\cos(x + y)$, and then eliminating $\cos^2 x$ by using the identity $\cos^2 x = 1 - \sin^2 x$.)

35. *Length of Day* The following function can be used to estimate the number of minutes of daylight in Boston for any given day of the year.

$$N(t) = 183.549 \sin(0.0172t - 1.329) + 728.124,$$

where t is the day of the year.* Use this function to estimate the total amount of daylight in a year and compare it to the total amount of daylight, reported to be 4,467.57 hours.

7.6 THE AREA BETWEEN TWO CURVES

THINK ABOUT IT

If a hospital purchasing agent knows how the savings from a new imaging technology will decline over time and how the costs of that technology will increase, how can she determine when the net savings will cease, and what the total savings will be?

This section shows a method for answering such questions.

Many important applications of integrals require finding the area between two graphs. The method used in previous sections to find the area between the graph of a function and the x-axis from $x = a$ to $x = b$ can be generalized to find such an area. For example, the area between the graphs of $f(x)$ and $g(x)$ from $x = a$ to $x = b$ in Figure 23(a) on the next page is the same as the difference of the area from a to b between $f(x)$ and the x-axis, shown in Figure 23(b), and the area from a to b between $g(x)$ and the x-axis (see Figure 23(c)). That is, the area between the graphs is given by

$$\int_a^b f(x)\,dx - \int_a^b g(x)\,dx,$$

---

*Thomas, Robert, *The Old Farmer's Almanac,* 2000.

which can be written as

$$\int_a^b [f(x) - g(x)]\,dx.$$

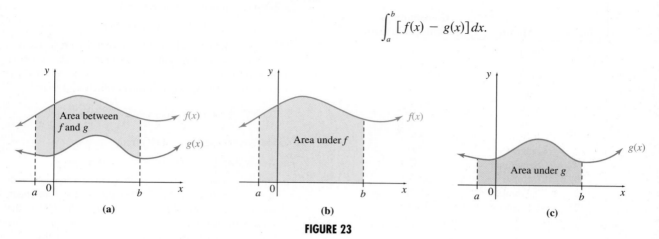

(a)   **(b)**   **(c)**

FIGURE 23

AREA BETWEEN TWO CURVES

If f and g are continuous functions and $f(x) \geq g(x)$ on $[a, b]$, then the area between the curves $f(x)$ and $g(x)$ from $x = a$ to $x = b$ is given by

$$\int_a^b [f(x) - g(x)]\,dx.$$

EXAMPLE 1   Area

Find the area bounded by $f(x) = -x^2 + 1$, $g(x) = 2x + 4$, $x = -1$, and $x = 2$.

Solution A sketch of the four equations is shown in Figure 24. In general, it is not necessary to spend time drawing a detailed sketch, but only to know whether the two functions intersect, and which function is greater between the intersections. To find out, set the two functions equal.

$$-x^2 + 1 = 2x + 4$$
$$0 = x^2 + 2x + 3$$

Verify by the quadratic formula that this equation has no roots. Since the graph of f is a parabola opening downward that does not cross the graph of g (a line), the parabola must be entirely under the line, as shown in Figure 24. Therefore $g(x) \geq f(x)$ for x in the interval $[-1, 2]$, and the area is given by

$$\int_{-1}^2 [g(x) - f(x)]\,dx = \int_{-1}^2 (2x + 4) - (-x^2 + 1)\,dx$$

$$= \int_{-1}^2 (2x + 4 + x^2 - 1)\,dx$$

$$= \int_{-1}^2 (x^2 + 2x + 3)\,dx$$

$$= \frac{x^3}{3} + x^2 + 3x \Big|_{-1}^2$$

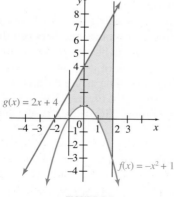

FIGURE 24

$$= \left(\frac{8}{3} + 4 + 6\right) - \left(\frac{-1}{3} + 1 - 3\right)$$

$$= \frac{8}{3} + 10 + \frac{1}{3} + 2$$

$$= 15.$$

NOTE It is not necessary to draw the graphs to determine which function is greater. Since the functions in the previous example do not intersect, we can evaluate them at *any* point to make this determination. For example, $f(0) = 1$ and $g(0) = 4$. Because $g(x) > f(x)$ at $x = 4$, and the two functions are continuous and never intersect, $g(x) > f(x)$ for all x.

EXAMPLE 2   Area

Find the area between the curves $y = x^{1/2}$ and $y = x^3$.

Solution Let $f(x) = x^{1/2}$ and $g(x) = x^3$. As before, set the two equal to find where they intersect.

$$x^{1/2} = x^3$$

$$0 = x^3 - x^{1/2}$$

$$0 = x^{1/2}(x^{5/2} - 1)$$

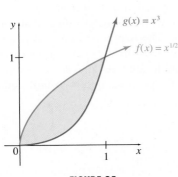

The only solutions are $x = 0$ and $x = 1$. Verify that the graph of f is concave downward, while the graph of g is concave upward, so the graph of f must be greater between 0 and 1. (This may also be verified by taking a point between 0 and 1, such as 0.5, and verifying that $0.5^{1/2} > 0.5^3$.) The graph is shown in Figure 25.

The area between the two curves is given by

$$\int_a^b [f(x) - g(x)]\,dx = \int_0^1 (x^{1/2} - x^3)\,dx.$$

Using the Fundamental Theorem,

$$\int_0^1 (x^{1/2} - x^3)\,dx = \left(\frac{x^{3/2}}{3/2} - \frac{x^4}{4}\right)\Bigg|_0^1$$

$$= \left(\frac{2}{3}x^{3/2} - \frac{x^4}{4}\right)\Bigg|_0^1$$

$$= \frac{2}{3}(1) - \frac{1}{4}$$

$$= \frac{5}{12}.$$

FIGURE 25

A graphing calculator is very useful in approximating solutions of problems involving the area between two curves. First, it can be used to graph the functions and identify any intersection points. Then it can be used to approximate the definite integral that represents the area. (A function that gives the numerical integral is located in the MATH menu of a TI-83 calculator.) Figure 26 on the next page shows the results of using these steps for Example 2. The second window shows that the area closely approximates $5/12$.

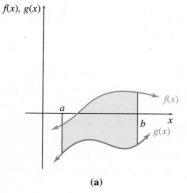

f(x), g(x)

f(x)

a

b x

g(x)

(a)

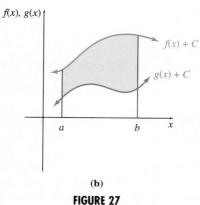

f(x), g(x)

f(x) + C

g(x) + C

a

b x

(b)

FIGURE 27

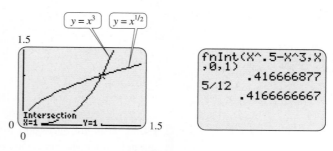

FIGURE 26

The difference between two integrals can be used to find the area between the graphs of two functions even if one graph lies below the x-axis. In fact, if $f(x) \geq g(x)$ for all values of x in the interval $[a, b]$, then the area between the two graphs is always given by

$$\int_a^b [f(x) - g(x)]\, dx.$$

To see this, look at the graphs in Figure 27(a), where $f(x) \geq g(x)$ for x in $[a, b]$. Suppose a constant C is added to both functions, with C large enough so that both graphs lie above the x-axis, as in Figure 27(b). The region between the graphs is not changed. By the work above, this area is given by $\int_a^b [f(x) - g(x)]\, dx$ regardless of where the graphs of $f(x)$ and $g(x)$ are located. As long as $f(x) \geq g(x)$ on $[a, b]$, then the area between the graphs from $x = a$ to $x = b$ will equal $\int_a^b [f(x) - g(x)]\, dx$.

EXAMPLE 3 Area

Find the area of the region enclosed by $y = x^2 - 2x$ and $y = x$ on $[0, 4]$.

Solution Verify that the two graphs cross at $x = 0$ and $x = 3$. Because the first graph is a parabola opening upward, the parabola must be below the line between 0 and 3 and above the line between 3 and 4. See Figure 28. (The greater function could also be identified by checking a point between 0 and 3, such as 1, and a point between 3 and 4, such as 3.5. For each of these values of x, we could calculate the corresponding value of y for the two functions and see which is greater.) Because the graphs cross at $x = 3$, the area is found by taking the sum of two integrals as follows.

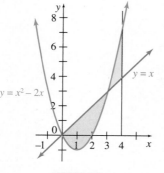

FIGURE 28

$$\text{Area} = \int_0^3 [x - (x^2 - 2x)]\, dx + \int_3^4 [(x^2 - 2x) - x]\, dx$$

$$= \int_0^3 (-x^2 + 3x)\, dx + \int_3^4 (x^2 - 3x)\, dx$$

$$= \left(\frac{-x^3}{3} + \frac{3x^2}{2}\right)\Big|_0^3 + \left(\frac{x^3}{3} - \frac{3x^2}{2}\right)\Big|_3^4$$

$$= \left(-9 + \frac{27}{2} - 0\right) + \left(\frac{64}{3} - 24 - 9 + \frac{27}{2}\right)$$

$$= \frac{19}{3}$$

In the remainder of this section we will consider a typical application that requires finding the area between two curves.

EXAMPLE 4 Savings

A hospital is considering a new magnetic resonance imaging (MRI) machine. The new machine will provide substantial initial savings, with the savings declining with time x (in years) according to the rate-of-savings function

$$S(x) = 100 - x^2,$$

where $S(x)$ is in thousands of dollars per year. At the same time, the cost of operating the new machine increases with time x (in years), according to the rate-of-cost function (in thousands of dollars per year)

$$C(x) = x^2 + \frac{14}{3}x.$$

(a) For how many years will the hospital realize savings?

Solution Figure 29 shows the graphs of the rate-of-savings and rate-of-cost functions. The rate of cost (marginal cost) is increasing, while the rate of savings (marginal savings) is decreasing. The hospital should use this new machine until the difference between these quantities is zero; that is, until the time at which these graphs intersect. The graphs intersect when

$$C(x) = S(x),$$

or

$$x^2 + \frac{14}{3}x = 100 - x^2.$$

Solve this equation as follows.

$$0 = 2x^2 + \frac{14}{3}x - 100$$

$$0 = 3x^2 + 7x - 150 \qquad \text{Multiply by } \tfrac{3}{2}.$$

$$= (x - 6)(3x + 25) \qquad \text{Factor.}$$

Set each factor equal to 0 separately to get

$$x = 6 \qquad \text{or} \qquad x = -25/3.$$

Only 6 is a meaningful solution here. The company should use the new machine for 6 years.

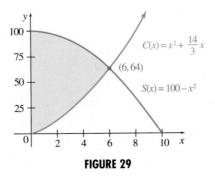

FIGURE 29

(b) What will be the net total savings during this period?

Solution Since the total savings over the 6-year period is given by the area under the rate-of-savings curve and the total additional cost by the area under the rate-of-cost curve, the net total savings over the 6-year period is given by the area between the rate-of-cost and the rate-of-savings curves and the lines $x = 0$ and $x = 6$. This area can be evaluated with a definite integral as follows.

$$
\begin{aligned}
\text{Total savings} &= \int_0^6 \left[(100 - x^2) - \left(x^2 + \frac{14}{3}x \right) \right] dx \\
&= \int_0^6 \left(100 - \frac{14}{3}x - 2x^2 \right) dx \\
&= \left(100x - \frac{7}{3}x^2 - \frac{2}{3}x^3 \right) \Big|_0^6 \\
&= 100(6) - \frac{7}{3}(36) - \frac{2}{3}(216) = 372
\end{aligned}
$$

The hospital will save a total of $372,000 over the 6-year period.

The answer to a problem will not always be an integer. Suppose in solving the quadratic equation in Example 4 we found the solutions to be $x = 6.7$ and $x = -7.3$. It may not be realistic to use a new process for 6.7 years; it may be necessary to choose between 6 years and 7 years. Since the mathematical model produces a result that is not in the domain of the function in this case, it is necessary to find the total savings after 6 years and after 7 years and then select the best result.

7.6 EXERCISES

Find the area between the curves in Exercises 1–26.

1. $x = -2$, $x = 1$, $y = x^2 + 4$, $y = 0$

2. $x = 1$, $x = 2$, $y = x^3$, $y = 0$

3. $x = -3$, $x = 1$, $y = x + 1$, $y = 0$

4. $x = -2$, $x = 0$, $y = 1 - x^2$, $y = 0$

5. $x = -2$, $x = 1$, $y = 2x$, $y = x^2 - 3$

6. $x = 0$, $x = 6$, $y = 5x$, $y = 3x + 10$

7. $y = x^2 - 30$, $y = 10 - 3x$

8. $y = x^2 - 18$, $y = x - 6$

9. $y = x^2$, $y = 2x$

10. $y = x^2$, $y = x^3$

11. $x = 1$, $x = 6$, $y = \dfrac{1}{x}$, $y = -1$

12. $x = 0$, $x = 4$, $y = \dfrac{1}{x + 1}$, $y = \dfrac{x - 1}{2}$

13. $x = -1$, $x = 1$, $y = e^x$, $y = 3 - e^x$

14. $x = -1$, $x = 2$, $y = e^{-x}$, $y = e^x$

15. $x = 1$, $x = 2$, $y = e^x$, $y = \dfrac{1}{x}$

16. $x = 2$, $x = 4$, $y = \dfrac{x}{2} + 3$, $y = \dfrac{1}{x - 1}$

17. $y = x^3 - x^2 + x + 1$, $y = 2x^2 - x + 1$

18. $y = 2x^3 + x^2 + x + 5$, $y = x^3 + x^2 + 2x + 5$

19. $y = x^4 + \ln(x + 10)$, $y = x^3 + \ln(x + 10)$

20. $y = x^5 - 2\ln(x + 5)$, $y = x^3 - 2\ln(x + 5)$

21. $y = x^{4/3}$, $y = 2x^{1/3}$

22. $y = \sqrt{x}$, $y = x\sqrt{x}$

23. $x = 0$, $x = \pi/4$, $y = \cos x$, $y = \sin x$

24. $x = 0$, $x = \pi/4$, $y = \sec^2 x$, $y = \sin 2x$

25. $x = \pi/4$, $y = \tan x$, $y = \sin x$

26. $x = 0$, $x = \pi$, $y = \sin x$, $y = 1 - \sin x$

In Exercises 27 and 28, use a graphing calculator to find the values of x where the curves intersect, and then to find the area between the two curves.

27. $y = e^x$, $y = -x^2 - 2x$

28. $y = \ln x$, $y = x^3 - 5x^2 + 6x - 1$

Applications

LIFE SCIENCES

29. *Pollution* Pollution begins to enter a lake at time $t = 0$ at a rate (in gallons per hour) given by the formula

$$f(t) = 10(1 - e^{-0.5t}),$$

where t is the time in hours. At the same time, a pollution filter begins to remove the pollution at a rate

$$g(t) = 0.4t$$

as long as pollution remains in the lake.

a. How much pollution is in the lake after 12 hours?

b. Use a graphing calculator to find the time when the rate that pollution enters the lake equals the rate the pollution is removed.

c. Find the amount of pollution in the lake at the time found in part b.

d. Use a graphing calculator to find the time when all the pollution has been removed from the lake.

30. *Pollution* Repeat the steps of Exercise 29, using the functions

$$f(t) = 15(1 - e^{-0.05t})$$

and

$$g(t) = 0.3t.$$

OTHER APPLICATIONS

31. *Net Savings* Suppose a company wants to introduce a new machine that will produce a rate of annual savings in dollars given by

$$S(x) = 150 - x^2,$$

where x is the number of years of operation of the machine, while producing a rate of annual costs in dollars of

$$C(x) = x^2 + \frac{11}{4}x.$$

a. For how many years will it be profitable to use this new machine?

b. What are the net total savings during the first year of use of the machine?

c. What are the net total savings over the entire period of use of the machine?

32. *Net Savings* A new smog-control device will reduce the output of sulfur oxides from automobile exhausts. It is estimated that the rate of savings to the community from the use of this device will be approximated by

$$S(x) = -x^2 + 4x + 8,$$

where $S(x)$ is the rate of savings (in millions of dollars) after x years of use of the device. The new device cuts down on the production of sulfur oxides, but it causes an increase in the production of nitrous oxides. The rate of additional costs (in millions) to the community after x years is approximated by

$$C(x) = \frac{3}{25}x^2.$$

a. For how many years will it pay to use the new device?

b. What will be the net savings over this period of time?

33. *Profit* De Win Enterprises had an expenditure rate of $E(x) = e^{0.1x}$ dollars per day and an income rate of $I(x) = 98.8 - e^{0.1x}$ dollars per day on a particular job, where x was the number of days from the start of the job. The company's profit on that job will equal total income less total expenditures. Profit will be maximized if the job ends at the optimum time, which is the point where the two curves meet. Find the following.

a. The optimum number of days for the job to last

b. The total income for the optimum number of days

c. The total expenditures for the optimum number of days

d. The maximum profit for the job

34. *Net Savings* A factory of Marisa Raffaele Industries has installed a new process that will produce an increased rate of revenue (in thousands of dollars per year) of

$$R(t) = 104 - 0.4e^{t/2},$$

where t is time measured in years. The new process produces additional costs (in thousands of dollars per year) at the rate of

$$C(t) = 0.3e^{t/2}.$$

a. When will it no longer be profitable to use this new process?

b. Find the net total savings.

35. *Fuel Economy* In an article in the December 1994 *Scientific American* magazine, the authors estimated future gas use.* Without a change in U.S. policy, auto fuel use is forecasted to rise along the projection shown in the figure below. The shaded band predicts gas use if the technologies for increased fuel economy are phased in by the year 2010. The moderate estimate (center curve) corresponds to an average of 46 miles per gallon for all cars on the road. Discuss the interpretation of the shaded area and other regions of the graph that pertain to the topic in this section.

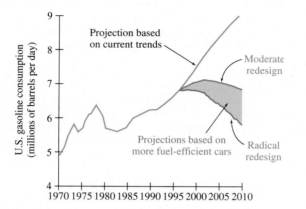

36. *Distribution of Income* Suppose that all the people in a country are ranked according to their incomes, starting at the bottom. Let x represent the fraction of the community

making the lowest income ($0 \le x \le 1$); $x = 0.4$, therefore, represents the lower 40% of all income producers. Let $I(x)$ represent the proportion of the total income earned by the lowest x of all people. Thus, $I(0.4)$ represents the fraction of total income earned by the lowest 40% of the population. The curve described by this function is known as a *Lorenz curve*. Suppose

$$I(x) = 0.9x^2 + 0.1x.$$

Find and interpret the following.

a. $I(0.1)$ **b.** $I(0.4)$

If income were distributed uniformly, we would have $I(x) = x$. The area under this line of complete equality is $1/2$. As $I(x)$ dips further below $y = x$, there is less equality of income distribution. This inequality can be quantified by the ratio of the area between $I(x)$ and $y = x$ to $1/2$. This ratio is called the *Gini index of income inequality* and equals $2 \int_0^1 (x - I(x)) \, dx$.

c. Graph $I(x) = x$ and $I(x) = 0.9x^2 + 0.1x$, for $0 \le x \le 1$, on the same axes.

d. Find the area between the curves.

e. For U.S. families, the Gini index was 0.36 in 1960 and 0.40 in 1992.[†] Describe how the distribution of family incomes has changed over this time.

CHAPTER SUMMARY

Earlier chapters dealt with the derivative, one of the two main ideas of calculus. This chapter deals with integration, the second main idea. There are two aspects of integration. The first is indefinite integration, or finding an antiderivative; the second is definite integration, which can be used to find the area under a curve. The Fundamental Theorem of Calculus unites these two ideas by showing that the way to find the area under a curve is to use the antiderivative. Substitution is a technique for finding antiderivatives. The idea of the definite integral can also be applied to finding the area between two curves.

INTEGRATION SUMMARY

Antidifferentiation Formulas

Power Rule
$$\int x^n \, dx = \frac{x^{n+1}}{n+1} + C, \quad n \ne -1$$

Constant Multiple Rule
$$\int k \cdot f(x) \, dx = k \int f(x) \, dx \quad \text{for any real number } k$$

Sum or Difference Rule
$$\int [f(x) \pm g(x)] \, dx = \int f(x) \, dx \pm \int g(x) \, dx$$

---

*DeCicco, John, and Marc Ross, "Improving Automotive Efficiency," *Scientific American,* Vol. 271, No. 6, Dec. 1994, p. 56. Copyright © 1994 by Scientific American, Inc. All rights reserved.
[†]U.S. Bureau of the Census, *Current Population Reports, Series P60-184, Money Income of Households, Families, and Persons in the United States, 1992,* Sept. 1993.

Integration of Exponential Functions	$\int e^{kx}\,dx = \dfrac{e^{kx}}{k} + C, \quad k \neq 0$
Integration of x^{-1}	$\int x^{-1}\,dx = \ln|x| + C$

Integration of Trigonometric Functions

$$\int \sin x\,dx = -\cos x + C \qquad \int \cos x\,dx = \sin x + C$$

$$\int \sec^2 x\,dx = \tan x + C \qquad \int \csc^2 x\,dx = -\cot x + C$$

$$\int \sec x \tan x\,dx = \sec x + C \qquad \int \csc x \cot x\,dx = -\csc x + C$$

$$\int \tan x\,dx = -\ln|\cos x| + C \qquad \int \cot x\,dx = \ln|\sin x| + C$$

Substitution Method

Choose u to be one of the following:

1. the quantity under a root or raised to a power;
2. the exponent on e;
3. the quantity in the denominator.
4. the quantity in a trigonometric function.

Definite Integrals

Definition of the Definite Integral

$\int_a^b f(x)\,dx = \lim\limits_{n \to \infty} \sum\limits_{i=1}^{n} f(x_i)\,\Delta x$, where $\Delta x = (b - a)/n$ and x_i is any value of x in the ith interval. If $f(x)$ gives the rate of change of $F(x)$ for x in $[a, b]$, then this represents the total change in $F(x)$ as x goes from a to b.

Properties of Definite Integrals

1. $\displaystyle\int_a^a f(x)\,dx = 0$

2. $\displaystyle\int_a^b k \cdot f(x)\,dx = k \int_a^b f(x)\,dx$ for any real number k.

3. $\displaystyle\int_a^b [f(x) \pm g(x)]\,dx = \int_a^b f(x)\,dx \pm \int_a^b g(x)\,dx$

4. $\displaystyle\int_a^b f(x)\,dx = \int_a^c f(x)\,dx + \int_c^b f(x)\,dx$ for any real number c

5. $\displaystyle\int_a^b f(x)\,dx = -\int_b^a f(x)\,dx$

Fundamental Theorem of Calculus

$\int_a^b f(x)\,dx = F(x)\big|_a^b = F(b) - F(a)$, where f is continuous on $[a, b]$ and F is any antiderivative of f.

Area Between Two Curves

$\int_a^b [f(x) - g(x)]\,dx$, where f and g are continuous functions and $f(x) \geq g(x)$ on $[a, b]$

KEY TERMS

7.1 antiderivative
    integral sign
    integrand
    indefinite integral

7.2 integration by
    substitution
    demand function
    marginal demand

7.3 midpoint rule
    definite integral
    AOC (area under the
    curve)

limits of integration
total change
7.4 Fundamental Theorem
    of Calculus

CHAPTER 7 REVIEW EXERCISES

1. Explain the differences between an indefinite integral and a definite integral.

2. Explain under what circumstances substitution is useful in integration.

3. Explain why the limits of integration are changed when u is substituted for an expression in x in a definite integral.

4. Explain how to find the area between two curves.

Find each indefinite integral.

5. $\displaystyle\int (2x + 3)\, dx$

6. $\displaystyle\int (5x - 1)\, dx$

7. $\displaystyle\int (x^2 - 3x + 2)\, dx$

8. $\displaystyle\int (6 - x^2)\, dx$

9. $\displaystyle\int 3\sqrt{x}\, dx$

10. $\displaystyle\int \frac{\sqrt{x}}{2}\, dx$

11. $\displaystyle\int (x^{1/2} + 3x^{-2/3})\, dx$

12. $\displaystyle\int (2x^{4/3} + x^{-1/2})\, dx$

13. $\displaystyle\int \frac{-4}{x^3}\, dx$

14. $\displaystyle\int \frac{5}{x^4}\, dx$

15. $\displaystyle\int -3e^{2x}\, dx$

16. $\displaystyle\int 5e^{-x}\, dx$

17. $\displaystyle\int xe^{3x^2}\, dx$

18. $\displaystyle\int 2xe^{x^2}\, dx$

19. $\displaystyle\int \frac{3x}{x^2 - 1}\, dx$

20. $\displaystyle\int \frac{-x}{2 - x^2}\, dx$

21. $\displaystyle\int \frac{x^2\, dx}{(x^3 + 5)^4}$

22. $\displaystyle\int (x^2 - 5x)^4(2x - 5)\, dx$

23. $\displaystyle\int \frac{x^3}{e^{3x^4}}\, dx$

24. $\displaystyle\int e^{3x^2 + 4}x\, dx$

25. $\displaystyle\int \sin 2x\, dx$

26. $\displaystyle\int \cos 3x\, dx$

27. $\displaystyle\int \tan 9x\, dx$

28. $\displaystyle\int \sec^2 5x\, dx$

29. $\displaystyle\int 5\sec^2 x\, dx$

30. $\displaystyle\int 4\csc^2 x\, dx$

31. $\displaystyle\int x \sin 3x^2\, dx$

32. $\displaystyle\int 5x \sec 2x^2 \tan 2x^2\, dx$

33. $\displaystyle\int \sqrt{\cos x}\, \sin x\, dx$

34. $\displaystyle\int \sin^4 x \cos x\, dx$

35. $\displaystyle\int x \tan 11x^2\, dx$

36. $\displaystyle\int x^2 \cot 8x^3\, dx$

37. $\displaystyle\int (\sin x)^{5/2} \cos x\, dx$

38. $\displaystyle\int (\cos x)^{-4/3} \sin x\, dx$

39. $\displaystyle\int \sec^2 3x \tan 3x\, dx$

40. Let $f(x) = 3x + 1$, $x_1 = -1$, $x_2 = 0$, $x_3 = 1$, $x_4 = 2$, and $x_5 = 3$. Find $\sum\limits_{i=1}^{5} f(x_i)$.

41. Find $\int_0^4 f(x)\, dx$ for each graph of $y = f(x)$.

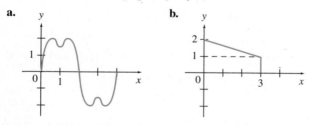

a.

b.

42. Approximate the area under the graph of $f(x) = 2x + 3$ and above the x-axis from $x = 0$ to $x = 4$ using four rectangles. Let the height of each rectangle be the function value on the left side.

43. Find $\int_0^4 (2x + 3)\,dx$ by using the formula for the area of a trapezoid: $A = (1/2)(B + b)h$, where B and b are the lengths of the parallel sides and h is the distance between them. Compare with Exercise 42.

44. In Exercises 29 and 30 of the section on Area and the Definite Integral, you calculated the distance that a car traveled by estimating the integral $\int_0^T v(t)\,dt$.

    a. Let $s(t)$ represent the mileage reading on the odometer. Express the distance traveled between $t = 0$ and $t = T$ using the function $s(t)$.

    b. Since your answer to part a and the original integral both represent the distance traveled by the car, the two can be set equal. Explain why the resulting equation is a statement of the Fundamental Theorem of Calculus.

45. What does the Fundamental Theorem of Calculus state?

Find each definite integral.

46. $\displaystyle\int_1^2 (3x^2 + 5)\,dx$ **47.** $\displaystyle\int_1^6 (2x^2 + x)\,dx$ **48.** $\displaystyle\int_1^5 (3x^{-2} + x^{-3})\,dx$ **49.** $\displaystyle\int_0^1 x\sqrt{5x^2 + 4}\,dx$

50. $\displaystyle\int_1^3 2x^{-1}\,dx$ **51.** $\displaystyle\int_1^6 8x^{-1}\,dx$ **52.** $\displaystyle\int_0^4 2e^x\,dx$ **53.** $\displaystyle\int_1^6 \frac{5}{2}e^{4x}\,dx$

54. $\displaystyle\int_0^{\pi/2} \cos x\,dx$ **55.** $\displaystyle\int_{\pi/2}^{\pi} \sin x\,dx$ **56.** $\displaystyle\int_0^{2\pi} (10 + 10\cos x)\,dx$ **57.** $\displaystyle\int_0^{2\pi} (5 + 5\sin x)\,dx$

58. Use the substitution $u = 4x^2$ and the equation of a semicircle to evaluate

$$\int_0^{1/2} x\sqrt{1 - 16x^4}\,dx.$$

Find the area between the x-axis and $f(x)$ over each of the given intervals.

59. $f(x) = \sqrt{x - 1};\quad [1, 10]$ **60.** $f(x) = (x + 2)^6;\quad [-2, 0]$

61. $f(x) = e^x;\quad [0, 2]$ **62.** $f(x) = 1 + e^{-x};\quad [0, 4]$

Find the area of the region enclosed by each group of curves.

63. $f(x) = 5 - x^2, \quad g(x) = x^2 - 3$ **64.** $f(x) = x^2 - 4x, \quad g(x) = x - 6$

65. $f(x) = x^2 - 4x, \quad g(x) = x + 1, \quad x = 2, \quad x = 4$ **66.** $f(x) = 5 - x^2, \quad g(x) = x^2 - 3, \quad x = 0, \quad x = 4$

Applications

LIFE SCIENCES

67. *Insect Cannibalism* In certain species of flour beetles, the larvae cannibalize the unhatched eggs. In calculating the population cannibalism rate per egg, researchers needed to evaluate the integral

$$\int_0^A c(x)\,dx,$$

where A is the length of the larval stage and $c(x)$ is the cannibalism rate per egg per larva of age x.* The minimum value of A for the flour beetle *Tribolium castaneum* is

---

*Hastings, Alan, and Robert F. Costantino, "Oscillations in Population Numbers: Age-Dependent Cannibalism," *Journal of Animal Ecology,* Vol. 60, No. 2, June 1991, pp. 471–482.

17.6 days, which is the value we will use. The function $c(x)$ starts at day 0 with a value of 0, increases linearly to the value 0.024 at day 12, and then stays constant. Find the value of the integral using

a. formulas from geometry;

b. the Fundamental Theorem of Calculus.

68. *Population Growth* The rate of change of the population of a rare species of Australian spider is given by

$$f(t) = 100 - \sqrt{2.4t + 1},$$

where $f(t)$ is the number of spiders present at time t, measured in months. Find the total number of additional spiders in the first 10 months.

69. *Insulin in Sheep* A research group studied the effect of a large injection of glucose in sheep fed a normal diet compared with sheep that were fasting.* A graph of the plasma insulin levels in pM (pico molars, or 10^{-12} of a molar) for both groups is shown below. The red circles designate the fasting sheep and the green circles the sheep fed a normal diet. The researchers compared the area under the curves for the two groups.

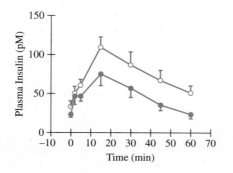

a. For the fasting sheep, estimate the area under the curve using rectangles, first by using the left endpoints, then the right endpoints, and then averaging the two. Note that the width of the rectangles will vary.

b. Repeat part a for the sheep fed a normal diet.

c. How much higher is the area under the curve for the fasting sheep compared with the normal sheep?

70. *Infection Rate* The rate of infection of a disease (in people per month) is given by the function

$$I'(t) = \frac{100t}{t^2 + 1},$$

where t is the time in months since the disease broke out. Find the total number of infected people over the first four months of the disease.

71. *Risk in Fisheries* We saw in the previous chapter that the maximum sustainable yield is an important quantity in managing a fishery. Suppose that θ is the maximum sustainable yield, but that we do not know it exactly, and must instead use an estimator, T, plus a correction factor, x. There is a risk that our estimator is too high or too low. The expected risk of the estimator, $T + x$, is

$$E = \int_{-\infty}^{\infty} \left(\frac{|t + x - \theta|}{\theta} + \frac{a(t + x - \theta)^2}{\theta^2} \right) h(t; \theta)\, dt,$$

where $h(t; \theta)$ is a probability density function of t that depends on the parameter θ and a is a constant.[†] (For more information on estimation, probability density functions, and expected value, see Chapters 12 and 13.) Our goal is to find the value of x that minimizes the risk.

a. For simplicity, let $\theta = 1$. The simplest function h is a constant on an interval, set so that the integral of h over that interval is 1. Let

$$h(t; \theta) = \begin{cases} 1 & \text{if } 0.5 \le t \le 1.5 \\ 0 & \text{otherwise.} \end{cases}$$

Show that

$$E = x^2 + \frac{1}{4} + \frac{a}{3}[(1.5 + x)^3 - (0.5 + x)^3]$$

(*Hint:* To integrate, note that $t + x - 1 < 0$ if $0.5 \le t \le 1 - x$, in which case $|t + x - 1| = -(t + x - 1)$, and $t + x - 1 \ge 0$ if $1 - x \le t \le 1.5$.)

b. We can minimize E by setting $dE/dx = 0$. (How do we know this is a minimum and not a maximum?) Show that

$$x = \frac{-a}{1 + a},$$

*Oliver, M. H., et al., "Maternal Undernutrition During the Periconceptual Period Increases Plasma Taurine Levels and Insulin Response to Glucose but not Arginine in the Late Gestation Fetal Sheep," *Endocrinology,* Vol. 14, No. 10, 2001, pp. 4576–4579.

†Kirkwood, G. P., "Risks in Setting Catch Limits," *Mathematical Biosciences,* Vol. 53, No. 1/2, Feb. 1981, pp. 119–129.

that is, the expected risk is minimized using the estimator $T - a/(1 + a)$.

c. Suppose we leave θ as an unknown parameter in part a, and let

$$h(t; \theta) = \begin{cases} 1/\theta & \text{if } 0.5\theta \le t \le 1.5\theta \\ 0 & \text{otherwise.} \end{cases}$$

Show that in this case, the expected risk is minimized using the estimator $T - a\theta/(1 + a)$.

OTHER APPLICATIONS

72. *Investment* The curve shown below gives the rate that an investment accumulates income (in dollars per year). Use rectangles with widths of 2 units and height determined by the function value at the midpoint to find the total income accumulated over 10 yr.

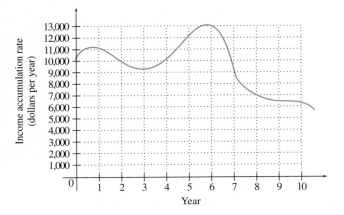

73. *Utilization of Reserves* A manufacturer of electronic equipment requires a certain rare metal. He has a reserve supply of 4,000,000 units that he will not be able to replace. If the rate at which the metal is used is given by

$$f(t) = 100,000e^{0.03t},$$

where t is the time in years, how long will it be before he uses up the supply? (*Hint:* Find an expression for the total amount used in t years and set it equal to the known reserve supply.)

74. *Sales* The rate of change of sales of a new brand of tomato soup (in thousands per month) is given by

$$S'(x) = \sqrt{x} + 2,$$

where x is the time in months that the new product has been on the market. Find the total sales after 9 months.

75. *Productivity* The function defined by $f'(x) = 2.158e^{0.0198x}$ approximates marginal U.S. nonfarm productivity from 1991–1995.* Productivity is measured as total output per hour compared to a measure of 100 for 1982, and x represents the end of the year with 1991 corresponding to $x = 1$, 1992 corresponding to $x = 2$, and so on.

a. Give the function that describes total productivity in year x, if productivity was 115 in 1992.

b. Use your function from part a to find productivity at the end of 1994. In 1994, productivity actually measured 118.6. How does your value using the function compare with this?

76. *Net Savings* A company has installed new machinery that will produce a savings rate (in thousands of dollars per year) of

$$S'(x) = 225 - x^2,$$

where x is the number of years the machinery is to be used. The rate of additional costs (in thousands of dollars per year) to the company due to the new machinery is expected to be

$$C'(x) = x^2 + 25x + 150.$$

For how many years should the company use the new machinery? Find the net savings (in thousands of dollars) over this period.

77. *Crime* Based on data from the New York City Police Department,[†] the homicide rate in New York City between 1988 and 1999 can be approximated by

$$f(t) = 5.45t^3 - 105t^2 + 391t + 1,798,$$

where t is the number of years since 1988. Find the total number of homicides during the 12-year period from the beginning of 1988 to the end of 1999.

78. *Linear Motion* A particle is moving along a straight line with velocity $v(t) = t^2 - 2t$. Its distance from the starting point after 3 sec is 8 cm. Find $s(t)$, the distance of the particle from the starting point after t sec.

79. *Weather* The graph on the next page shows 2000 weather statistics for New York City, as well as the normal high and low temperatures.[‡] The amount of cold weather in a year is measured in heating degree-days, where 1 degree-day is added to the total for each degree that a day's average falls below 65°F. For example, if the average temperature on November 15 is 50°F, 15 degree-days are added to the years' total. Estimate the total number of heating degree-days in an average New York City year, using rectangles of width 1 mo, with the height determined by the average temperature at the middle of the rectangle. Assume that the

Bureau of Labor Statistics
[†]*The New York Times,* Dec. 19, 1999, p. 55.
[‡]*The New York Times,* Jan. 7, 2001, p. 29.

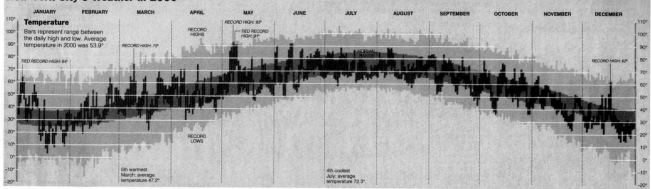

New York City's Weather in 2000

normal average is halfway between the normal high and low (at the center line of the dark band). Use the actual number of days in each month.

80. *Energy Usage* A mathematics textbook author has determined that her monthly gas usage y approximately follows the sine curve

$$y = 12.5 \sin\left(\frac{\pi}{6}(x + 1.2)\right) + 14.7,$$

where y is measured in thousands of cubic feet (MCF) and x is the month of the year ranging from 1 to 12.

   a. Graph this function.

   b. Find the approximate gas usage for the months of February and July.

   c. Find dy/dx, when $x = 7$. Interpret your answer.

81. *Area* A 6-ft board is placed against a wall as shown in the figure, forming a triangular-shaped area beneath it. At what angle θ should the board be placed to make the triangular area as large as possible?

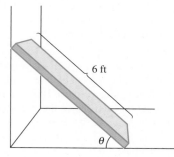

6 ft

θ

82. *Mercator's World Map* Before Gerardus Mercator designed his map of the world in 1569, sailors who traveled in a fixed compass direction could follow a straight line on a map only over short distances. Over long distances, such

a course would be a curve on existing maps, which tried to make area on the map proportional to the actual area. Mercator's map greatly simplified navigation: even over long distances, straight lines on the map corresponded to fixed compass bearings. This was accomplished by distorting distances. On Mercator's map, the distance of an object from the equator to a parallel at latitude θ is given by

$$D(\theta) = k \int_0^\theta \sec x \, dx,$$

where k is a constant of proportionality. Calculus had not yet been discovered when Mercator designed his map; he approximated the distance between parallels of latitude by hand.*

   a. Verify that

   $$\frac{d}{dx} \ln |\sec x + \tan x| = \sec x.$$

   b. Verify that

   $$\frac{d}{dx}\left(-\ln |\sec x - \tan x|\right) = \sec x.$$

   c. Using parts a and b, give two different formulas for $\int \sec x \, dx$. Explain how they can both be correct.

   d. Los Angeles has a latitude of 34°03′N. (The 03′ represents 3 minutes of latitude. Each minute of latitude is 1/60 of a degree.) If Los Angeles is to be 7 inches from the equator on a Mercator map, how far from the equator should we place New York City, which has a latitude of 40°45′N?

   e. Repeat part d for Miami, which has a latitude of 25°46′N.

   f. If you do not live in Los Angeles, New York City, or Miami, repeat part d for your town or city.

*Rickey, V. Frederick, and Philip M. Tuchinsky, "An Application of Geography to Mathematics: History of the Integral of the Secant," *Mathematics Magazine,* Vol. 53, May 1980, pp. 162–66.

EXTENDED APPLICATION: Estimating Depletion Dates for Minerals

It is becoming more and more obvious that the earth contains only a finite quantity of minerals. The "easy and cheap" sources of minerals are being used up, forcing an ever more expensive search for new sources. For example, oil from the North Slope of Alaska would never have been used in the United States during the 1930s because a great deal of Texas and California oil was readily available.

We said in an earlier chapter that population tends to follow an exponential growth curve. Mineral usage also follows such a curve. Thus, if q represents the rate of consumption of a certain mineral at time t, while q_0 represents consumption when $t = 0$, then

$$q = q_0 e^{kt},$$

where k is the growth constant. For example, the world consumption of petroleum in 1970 was 17,100 million barrels. During this period energy use was growing rapidly, and by 1975 annual world consumption had risen to 20,500 million barrels. We can use these two values to make a rough estimate of the constant k, and we find that over this five-year span the average value of k was about 0.037, representing 3.7% annual growth. If we let $t = 0$ correspond to the base year 1970, then

$$q = 17,100 e^{0.037t}$$

is the rate of consumption at time t, assuming that all the trends of the early 1970s have continued. In 1970 a reasonable guess would have put the total amount of oil in provable reserves or likely to be discovered in the future at 1,500,000 million barrels. At the 1970–1975 rate of consumption, in how many years after 1970 would you expect the world's reserves to be depleted? We can use the integral calculus of this chapter to find out.

To begin, we need to know the total quantity of petroleum that would be used between time $t = 0$ and some future time $t = T$. Figure 30 shows a typical graph of the function $q = q_0 e^{kt}$.

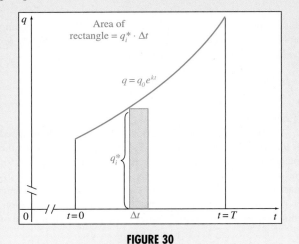

FIGURE 30

Following the work we did in Section 3, divide the time interval from $t = 0$ to $t = T$ into n subintervals. Let each subinterval have width Δt. Let the rate of consumption for the ith subinterval be approximated by q_i^*. Thus, the approximate total consumption for the subinterval is given by

$$q_i^* \cdot \Delta t,$$

and the total consumption over the interval from time $t = 0$ to $t = T$ is approximated by

$$\sum_{i=1}^{n} q_i^* \cdot \Delta t.$$

The limit of this sum as Δt approaches 0 gives the total consumption from time $t = 0$ to $t = T$. That is,

$$\text{Total consumption} = \lim_{\Delta t \to 0} \sum q_i^* \cdot \Delta t.$$

We have seen, however, that this limit is the definite integral of the function $q = q_0 e^{kt}$ from $t = 0$ to $t = T$, or

$$\text{Total consumption} = \int_0^T q_0 e^{kt} \, dt.$$

We can now evaluate this definite integral.

$$\int_0^T q_0 e^{kt} \, dt = q_0 \int_0^T e^{kt} \, dt = q_0 \left(\frac{e^{kt}}{k} \right) \Big|_0^T$$

$$= \frac{q_0}{k} e^{kt} \Big|_0^T = \frac{q_0}{k} e^{kT} - \frac{q_0}{k} e^0$$

$$= \frac{q_0}{k} e^{kT} - \frac{q_0}{k} (1)$$

$$= \frac{q_0}{k} (e^{kT} - 1) \tag{1}$$

Now let us return to the numbers we gave for petroleum. We said that $q_0 = 17,100$ million barrels, where q_0 represents consumption in the base year of 1970. We have $k = 0.037$ with total petroleum reserves estimated at 1,500,000 million barrels. Thus, using equation (1) we have

$$1,500,000 = \frac{17,100}{0.037} (e^{0.037T} - 1).$$

Multiply both sides of the equation by 0.037:

$$55,500 = 17,100(e^{0.037T} - 1).$$

Divide both sides of the equation by 17,100.

$$3.2 = e^{0.037T} - 1$$

Add 1 to both sides.

$$4.2 = e^{0.037T}$$

Take natural logarithms of both sides:

$$\ln 4.2 = \ln e^{0.037T}$$

$$= 0.037T \ln e$$

$$= 0.037T.$$

Finally,

$$T = \frac{\ln 4.2}{0.037} \approx 39.$$

By this result, petroleum reserves should last until 39 years after 1970, that is, until about 2009.

In fact, in the early 1970s some analysts were predicting that reserves would be exhausted before the end of the century, and this was a reasonable guess. But two things have happened since then: The growth in consumption has slowed, and more reserves have been discovered. One way to refine our model is to look at the historical data over a longer time span. Table 4 gives average world annual petroleum consumption in millions of barrels at five-year intervals from 1970 to 1995.

■ **Table 4**

Year	World consumption in millions of barrels
1970	17,100
1975	20,500
1980	23,000
1985	21,900
1990	24,100
1995	25,400

The first step in comparing this data with our exponential model is to estimate a value for the growth constant k. One simple way of doing this is to solve the equation

$$25,400 = 17,100 \cdot e^{k \cdot 25}.$$

Using natural logarithms just as we did in estimating the time to depletion for $k = 0.037$, we find that

$$k = \frac{\ln\left(\dfrac{25,400}{17,100}\right)}{25} \approx 0.016.$$

So the data from the Bureau of Transportation Statistics suggests a growth constant of about 1.6%. We can check the fit by plotting the function $17,100 \cdot e^{0.016t}$ along with a bar graph of the consumption data, shown in Figure 31.

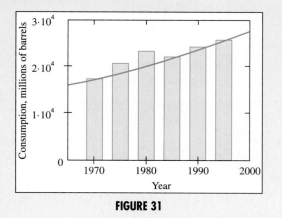

FIGURE 31

The fit looks reasonably good, but over this short range of 25 years, the exponential model is close to a linear model, and the growth in consumption is certainly not smooth.

The exponential model rests on the assumption of a constant growth rate. As already noted, we might expect instead that the growth rate would change as the world comes closer to exhausting its reserves. In particular, scarcity might drive up the price of oil and thus reduce consumption. We can use integration to explore an alternative model in which the factor k changes over time, so that k becomes $k(t)$, a function of time.

As an illustration, we explore a model in which the growth constant k declines toward 0 over time. We'll use 1970 as our base year, so the variable t will count years since 1970. We need a simple positive function $k(t)$ that tends toward 0 as t gets large. To get some numbers to work with, assume that the growth rate was 2% in 1970 and declined to 1% by 1995. There are many possible choices for the function $k(t)$, but a convenient one is

$$k(t) = \frac{0.5}{t + 25}.$$

Using integration to turn the instantaneous rate of consumption into the total consumption up to time T, we can write

$$\text{Total consumption} = 17,100 \int_0^T e^{k(t) \cdot t}\, dt$$

$$= 17,100 \int_0^T e^{0.5t/(t+25)}\, dt$$

We'd like to find out when the world will use up its estimated reserves, but as just noted, the estimates have increased since the 1970s. A paper presented by the U.S. Geological Survey at the 1998 World Petroleum Conference estimates the *current* global petroleum reserves at 2,300,000 million barrels.* So we need to solve

$$2,300,000 = 17,100 \int_0^T e^{0.5t/(t+25)}\, dt \qquad (2)$$

But this problem is much harder to solve than the corresponding problem for constant growth, because *there is no formula for evaluating this definite integral!* The function

$$g(t) = e^{0.5t/(t+25)}$$

doesn't have an antiderivative that we can write down in terms of functions that we know how to compute.

Here the numerical integration techniques to be discussed in the next chapter come to the rescue. We can use one of the integration rules to *approximate* the integral numerically for various values of T, and with some trial and error we can estimate how long the reserves will last. If you have a calculator or computer algebra system that does numerical integration, you can pick some T values and evaluate the right-hand side of

*http://energy.er.usgs.gov/products/papers/WPC/14/.

equation (2). Here are the results produced by one computer algebra system:

> For $T = 90$ the integral is about 2,062,000.
> For $T = 100$ the integral is about 2,316,000.
> For $T = 110$ the integral is about 2,572,000.

So using this model we would estimate that starting in 1970 the petroleum reserves would last for about 100 years, that is, until 2070.

Our integration tools are essential in building and exploring models of resource use, but the difference in our two predictions (39 years vs. 100 years) illustrates the difficulty of making accurate predictions. A model that performs well on historical data may not take the changing dynamics of resource use into account, leading to forecasts that are either unduly gloomy or too optimistic.

Exercises

1. Find the number of years that the estimated petroleum reserves would last if used at the same rate as in the base year.

2. How long would the estimated petroleum reserves last if the growth constant was only 2% instead of 3.7%?

Estimate the length of time until depletion for each of the following minerals.

3. Bauxite (the ore from which aluminum is obtained): estimated reserves in base year 15,000,000 thousand tons; rate of consumption in base year 63,000 thousand tons; growth constant 6%

4. Bituminous coal: estimated world reserves 2,000,000 million tons; rate of consumption in base year 2,200 million tons; growth constant 4%

5. **a.** Verify that the function $k(t)$ defined on the previous page has the right values at $k = 0$ and $k = 25$.

   b. Find a similar function that has $k(0) = 0.03$ and $k(25) = 0.02$.

6. **a.** Use the function you defined in Exercise 5b to write an integral for world petroleum consumption from 1970 until T years after 1970.

   b. If you have access to a numerical integrator, compute some values of your integral and estimate the time required to exhaust the reserve of 2,300,000 million barrels.

Further Techniques and Applications of Integration

It might seem that definite integrals with infinite limits have only theoretical interest, but in fact these *improper* integrals provide answers to many practical questions. An example in Section 4 models an environmental cleanup process in which the amount of pollution entering a stream decreases by a constant fraction each year. An improper integral gives the total amount of pollutant that will ever enter the river.

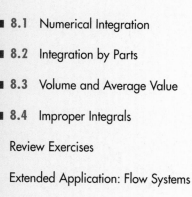

- **8.1** Numerical Integration
- **8.2** Integration by Parts
- **8.3** Volume and Average Value
- **8.4** Improper Integrals

Review Exercises

Extended Application: Flow Systems

In the previous chapter we discussed indefinite and definite integrals, and presented rules for finding the antiderivatives of several types of functions. In this chapter we will show how numerical methods can be used for functions that cannot be integrated by the techniques presented there. We will then develop a method of integrating functions known as integration by parts. We also show how to evaluate an integral that has one or both limits at infinity. These new techniques allow us to consider additional applications of integration, such as volumes of solids of revolution and the average value of a function.

■ 8.1 NUMERICAL INTEGRATION

THINK ABOUT IT   If we know the daily milk consumption by calves as a function of time, how can we calculate the total milk consumed?

Using numerical integration, we will answer this question in Exercise 23 of this section.

Some integrals cannot be evaluated by any technique. One solution to this problem was presented in Section 3 of the previous chapter, in which the area under a curve was approximated by summing the areas of rectangles. This method is seldom used in practice because better methods exist that are more accurate for the same amount of work. These methods are referred to as **numerical integration** methods. We discuss two such methods here: the trapezoidal rule and Simpson's rule.

Trapezoidal Rule   Recall that the trapezoidal rule was mentioned briefly in Section 3 of the previous chapter, where we found approximations with it by averaging the sums of rectangles found by using left endpoints and then using right endpoints. In this section we derive an explicit formula for the trapezoidal rule in terms of function values.* To illustrate the derivation of the trapezoidal rule, consider the integral

$$\int_1^5 \frac{1}{x}\, dx.$$

The shaded region in Figure 1 on the next page shows the area representing that integral, the area under the graph $f(x) = 1/x$, above the x-axis, and between the lines $x = 1$ and $x = 5$. As shown in the figure, if the area under the curve is approximated with trapezoids rather than rectangles, the approximation should be improved. Since $\int (1/x)\, dx = \ln |x| + C$,

$$\int_1^5 \frac{1}{x}\, dx = \ln |x| \Big|_1^5 = \ln 5 - \ln 1 = \ln 5 - 0 = \ln 5 \approx 1.609438.$$

---

*In American English a trapezoid is a four-sided figure with two parallel sides, contrasted with a trapezium, which has no parallel sides. In British English, however, it is just the opposite. What Americans call a trapezoid is called a trapezium in Great Britain.

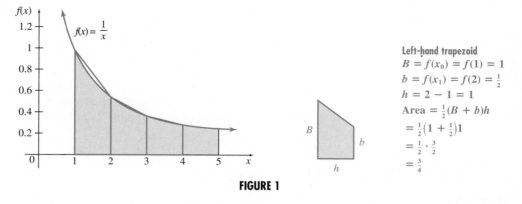

FIGURE 1

As in earlier work, to approximate this area we divide the interval $[1,5]$ into subintervals of equal widths. To get a first approximation to ln 5 by the trapezoidal rule, find the sum of the areas of the four trapezoids shown in Figure 1. From geometry, the area of a trapezoid is half the product of the sum of the bases and the altitude. Each of the trapezoids in Figure 1 has altitude 1. (In this case, the bases of the trapezoid are vertical and the altitudes are horizontal.) Adding the areas gives

$$\ln 5 = \int_1^5 \frac{1}{x}\,dx \approx \frac{1}{2}\left(\frac{1}{1} + \frac{1}{2}\right)(1) + \frac{1}{2}\left(\frac{1}{2} + \frac{1}{3}\right)(1) + \frac{1}{2}\left(\frac{1}{3} + \frac{1}{4}\right)(1) + \frac{1}{2}\left(\frac{1}{4} + \frac{1}{5}\right)(1)$$

$$= \frac{1}{2}\left(\frac{3}{2} + \frac{5}{6} + \frac{7}{12} + \frac{9}{20}\right) \approx 1.68333.$$

To get a better approximation, divide the interval $[1,5]$ into more subintervals. Generally speaking, the larger the number of subintervals, the better the approximation. The results for selected values of n are shown below to 5 decimal places.

■ **Table 1**

n	$\int_1^5 \dfrac{1}{x}\,dx = \ln 5 \approx 1.609438$
6	1.64360
8	1.62897
10	1.62204
20	1.61263
100	1.60957
1,000	1.60944

When $n = 1,000$, the approximation agrees with the true value to five decimal places.

Generalizing from this example, let f be a continuous function on an interval $[a,b]$. Divide the interval from a to b into n equal subintervals by the points $a = x_0, x_1, x_2, \ldots, x_n = b$, as shown in Figure 2 on the next page. Use the subintervals to make trapezoids that approximately fill in the region under the curve. The approximate value of the definite integral $\int_a^b f(x)\,dx$ is given by the sum of the areas of the trapezoids, or

$$\int_a^b f(x)\, dx \approx \frac{1}{2}[f(x_0) + f(x_1)]\left(\frac{b-a}{n}\right) + \frac{1}{2}[f(x_1) + f(x_2)]\left(\frac{b-a}{n}\right)$$

$$+ \cdots + \frac{1}{2}[f(x_{n-1}) + f(x_n)]\left(\frac{b-a}{n}\right)$$

$$= \left(\frac{b-a}{n}\right)\left[\frac{1}{2}f(x_0) + \frac{1}{2}f(x_1) + \frac{1}{2}f(x_1) + \frac{1}{2}f(x_2) + \frac{1}{2}f(x_2) + \cdots + \frac{1}{2}f(x_{n-1}) + \frac{1}{2}f(x_n)\right]$$

$$= \left(\frac{b-a}{n}\right)\left[\frac{1}{2}f(x_0) + f(x_1) + f(x_2) + \cdots + f(x_{n-1}) + \frac{1}{2}f(x_n)\right].$$

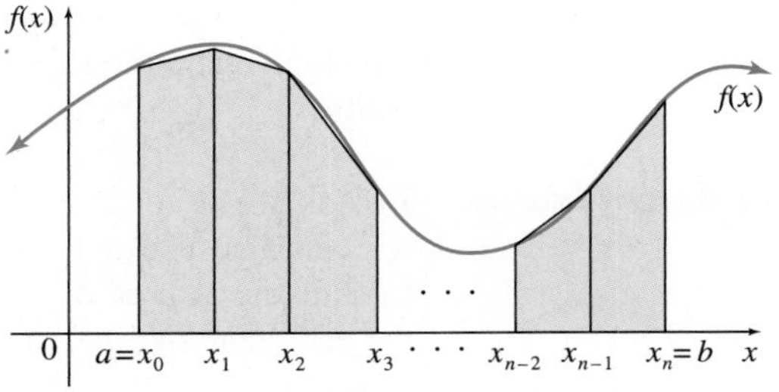

FIGURE 2

This result gives the following rule.

TRAPEZOIDAL RULE

Let f be a continuous function on $[a, b]$ and let $[a, b]$ be divided into n equal subintervals by the points $a = x_0, x_1, x_2, \ldots, x_n = b$. Then, by the **trapezoidal rule**,

$$\int_a^b f(x)\, dx \approx \left(\frac{b-a}{n}\right)\left[\frac{1}{2}f(x_0) + f(x_1) + \cdots + f(x_{n-1}) + \frac{1}{2}f(x_n)\right].$$

EXAMPLE 1 Trapezoidal Rule

Use the trapezoidal rule with $n = 4$ to approximate

$$\int_0^2 \sqrt{x^2 + 1}\, dx.$$

Solution

Method 1: Calculating by Hand

Here $a = 0$, $b = 2$, and $n = 4$, with $(b - a)/n = (2 - 0)/4 = 1/2$ as the altitude of each trapezoid. Then $x_0 = 0$, $x_1 = 1/2$, $x_2 = 1$, $x_3 = 3/2$, and $x_4 = 2$. Now find the corresponding function values. The work can be organized into a table, as follows.

▬ Table 2

i	x_i	$f(x_i)$
0	0	$\sqrt{0^2 + 1} = 1$
1	1/2	$\sqrt{(1/2)^2 + 1} \approx 1.11803$
2	1	$\sqrt{1^2 + 1} \approx 1.41421$
3	3/2	$\sqrt{(3/2)^2 + 1} \approx 1.80278$
4	2	$\sqrt{2^2 + 1} \approx 2.23607$

Substitution into the trapezoidal rule gives

$$\int_0^2 \sqrt{x^2 + 1}\, dx$$

$$\approx \frac{2 - 0}{4}\left[\frac{1}{2}(1) + 1.11803 + 1.41421 + 1.80278 + \frac{1}{2}(2.23607)\right]$$

$$\approx 2.97653.$$

The approximation 2.97653 found above using the trapezoidal rule with $n = 4$ differs from the true value of 2.95789 by 0.01864. As mentioned above, this error would be reduced if larger values were used for n. For example, if $n = 8$, the trapezoidal rule gives an answer of 2.96254, which differs from the true value by 0.00465. Techniques for estimating such errors are considered in more advanced courses.

Method 2: Graphing Calculator Just as we used a graphing calculator to approximate area using rectangles, we can also use it for the trapezoidal rule. As before, put the values of i in L_1 and the values of x_i in L_2. In the heading for L_3, put $\sqrt{(L_2^2 + 1)}$. Using the fact that $(b - a)/n = (2 - 0)/4 = 0.5$, the command `.5*(.5*L₃(1)+sum(L₃,2,4)+.5*L₃(5))` gives the result 2.976528589. For more details, see *The Graphing Calculator Manual* that is available on this text's Web site.

Method 3: Spreadsheet The trapezoidal rule can also be done on a spreadsheet. In Microsoft Excel, for example, store the values of 0 through n in column A. After putting the left endpoint in E1 and Δx in E2, put the command "`=$E$1+A1*$E$2`" into B1; copying this formula into the rest of column B gives the values of x_i. Similarly, use the formula for $f(x_i)$ to fill column C. Using the fact that $n = 5$ in this example, the command "`$E$2*(.5*C1+sum(C2:C4)+.5*C5)`" gives the result 2.976529. For more details, see *The Spreadsheet Manual* that is available on this text's Web site.

Simpson's Rule Another numerical method, *Simpson's rule,* approximates consecutive portions of the curve with portions of parabolas rather than the line segments of the trapezoidal rule. Simpson's rule usually gives a better approximation than the trapezoidal rule for the same number of subintervals. As shown in Figure 3, one parabola is fitted through points A, B, and C, another through C,

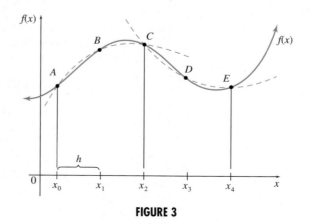

FIGURE 3

D, and E, and so on. Then the sum of the areas under these parabolas will approximate the area under the graph of the function. Because of the way the parabolas overlap, it is necessary to have an even number of intervals, and therefore an odd number of points, to apply Simpson's rule.

If h, the length of each subinterval, is $(b - a)/n$, the area under the parabola through points A, B, and C can be found by a definite integral. The details are omitted; the result is

$$\frac{h}{3}[f(x_0) + 4f(x_1) + f(x_2)].$$

Similarly, the area under the parabola through points C, D, and E is

$$\frac{h}{3}[f(x_2) + 4f(x_3) + f(x_4)].$$

When these expressions are added, the last term of one expression equals the first term of the next. For example, the sum of the two areas given above is

$$\frac{h}{3}[f(x_0) + 4f(x_1) + 2f(x_2) + 4f(x_3) + f(x_4)].$$

This illustrates the origin of the pattern of the terms in the following rule.

SIMPSON'S RULE

Let f be a continuous function on $[a, b]$ and let $[a, b]$ be divided into an even number n of equal subintervals by the points $a = x_0, x_1, x_2, \ldots, x_n = b$. Then by **Simpson's rule,**

$$\int_a^b f(x)\, dx \approx \left(\frac{b - a}{3n}\right)[f(x_0) + 4f(x_1) + 2f(x_2) + 4f(x_3) + \cdots$$
$$+ 2f(x_{n-2}) + 4f(x_{n-1}) + f(x_n)].$$

Thomas Simpson (1710–1761), a British mathematician, wrote texts on many branches of mathematics. Some of these texts went through as many as ten editions. His name became attached to this numerical method of approximating definite integrals even though the method preceded his work.

■ **CAUTION** In Simpson's rule, n (the number of subintervals) must be even.

EXAMPLE 2 Simpson's Rule

Use Simpson's rule with $n = 4$ to approximate

$$\int_0^2 \sqrt{x^2 + 1}\, dx,$$

which was approximated by the trapezoidal rule in Example 1.

Solution As in Example 1, $a = 0$, $b = 2$, and $n = 4$, and the endpoints of the four intervals are $x_0 = 0$, $x_1 = 1/2$, $x_2 = 1$, $x_3 = 3/2$, and $x_4 = 2$. The table of values is also the same.

Table 3

i	x_i	$f(x_i)$
0	0	1
1	1/2	1.11803
2	1	1.41421
3	3/2	1.80278
4	2	2.23607

Since $(b - a)/(3n) = 2/12 = 1/6$, substituting into Simpson's rule gives

$$\int_0^2 \sqrt{x^2 + 1}\, dx \approx \frac{1}{6}[1 + 4(1.11803) + 2(1.41421) + 4(1.80278) + 2.23607] \approx 2.95796.$$

This differs from the true value by 0.00007, which is less than the trapezoidal rule with $n = 8$. If $n = 8$ for Simpson's rule, the approximation is 2.95788, which differs from the true value by only 0.00001.

NOTE

1. Just as we can use a graphing calculator or a spreadsheet for the trapezoidal rule, we can also use such technology for Simpson's rule. For more details, see *The Graphing Calculator Manual* and *The Spreadsheet Manual* that are available on this text's Web site.

2. Let M represent the midpoint rule approximation and T the trapezoidal rule approximation, using n subintervals in each. Then the formula $S = (2M + T)/3$ gives the Simpson's rule approximation with $2n$ subintervals.

Numerical methods make it possible to approximate

$$\int_a^b f(x)\, dx$$

even when $f(x)$ is not known. The next example shows how this is done.

EXAMPLE 3 Total Distance

As mentioned earlier, the velocity $v(t)$ gives the rate of change of distance $s(t)$ with respect to time t. Suppose a vehicle travels an unknown distance. The passengers keep track of the velocity at 10-minute intervals (every 1/6 of an hour) with the following results.

Time in Hours, t	1/6	2/6	3/6	4/6	5/6	1	7/6
Velocity in Miles per Hour, $v(t)$	45	55	52	60	64	58	47

What is the total distance traveled in the 60-minute period from $t = 1/6$ to $t = 7/6$?

Solution The distance traveled in t hours is $s(t)$, with $s'(t) = v(t)$. The total distance traveled between $t = 1/6$ and $t = 7/6$ is given by

$$\int_{1/6}^{7/6} v(t)\, dt.$$

Even though this integral cannot be evaluated since we do not have an expression for $v(t)$, either the trapezoidal rule or Simpson's rule can be used to approximate its value and give the total distance traveled. In either case, let $n = 6$, $a = t_0 = 1/6$, and $b = t_6 = 7/6$. By the trapezoidal rule,

$$\int_{1/6}^{7/6} v(t) \, dt \approx \frac{7/6 - 1/6}{6} \left[\frac{1}{2}(45) + 55 + 52 + 60 + 64 + 58 + \frac{1}{2}(47) \right]$$

$$\approx 55.83.$$

By Simpson's rule,

$$\int_{1/6}^{7/6} v(t) \, dt \approx \frac{7/6 - 1/6}{3(6)} [45 + 4(55) + 2(52) + 4(60) + 2(64) + 4(58) + 47]$$

$$= \frac{1}{18}(45 + 220 + 104 + 240 + 128 + 232 + 47) \approx 56.44.$$

The distance traveled in the 1-hour period was about 56 miles.

As already mentioned, Simpson's rule generally gives a better approximation than the trapezoidal rule. As n increases, the two approximations get closer and closer. For the same accuracy, however, a smaller value of n generally can be used with Simpson's rule so that less computation is necessary. Simpson's rule is the method used by many calculators that have a built-in integration feature.

The branch of mathematics that studies methods of approximating definite integrals (as well as many other topics) is called *numerical analysis*. Numerical integration is useful even with functions whose antiderivatives can be determined if the antidifferentiation is complicated and a computer or calculator programmed with Simpson's rule is handy. You may want to program your calculator for both the trapezoidal rule and Simpson's rule. For some calculators, these programs are in *The Graphing Calculator Manual* that is available on this text's Web site.

8.1 EXERCISES

In Exercises 1–10, use $n = 4$ to approximate the value of each of the given integrals by the following methods: **(a)** the trapezoidal rule and **(b)** Simpson's rule. **(c)** Find the exact value by integration.

1. $\int_0^2 x^2 \, dx$

2. $\int_0^2 (2x + 1) \, dx$

3. $\int_{-1}^3 \frac{1}{4 - x} \, dx$

4. $\int_1^5 \frac{1}{x + 1} \, dx$

5. $\int_{-2}^2 (2x^2 + 1) \, dx$

6. $\int_0^3 (2x^2 + 1) \, dx$

7. $\int_1^5 \frac{1}{x^2} \, dx$

8. $\int_2^4 \frac{1}{x^3} \, dx$

9. $\int_0^1 xe^{x^2} \, dx$

10. $\int_0^4 \sqrt{2x + 1} \, dx$

11. $\int_0^\pi \sin x \, dx$

12. $\int_0^{\pi/4} \sec^2 x \, dx$

13. Find the area under the semicircle $y = \sqrt{4 - x^2}$ and above the x-axis by using $n = 8$ with the following methods:

a. the trapezoidal rule   **b.** Simpson's rule

c. Compare the results with the area found by the formula for the area of a circle. Which of the two approximation techniques was more accurate?

14. Find the area between the x-axis and the ellipse $4x^2 + 9y^2 = 36$ by using $n = 12$ with the following methods:

    a. the trapezoidal rule   **b.** Simpson's rule

    (*Hint:* Solve the equation for y and find the area of the semiellipse.)

    c. Compare the results with the actual area, $3\pi \approx 9.4248$ (which can be found by methods not considered in this text). Which approximation technique was more accurate?

15. Suppose that $f(x) > 0$ and $f''(x) > 0$ for all x between a and b, where $a < b$. Which of the following cases is true of a trapezoidal approximation T for the integral $\int_a^b f(x)\, dx$? Explain.

    a. $T < \displaystyle\int_a^b f(x)\, dx$ **b.** $T > \displaystyle\int_a^b f(x)\, dx$ **c.** Can't say which is larger.

16. Refer to Exercise 15. Which of the three cases applies to these functions?

    a. $f(x) = x^2$, $[0, 3]$ **b.** $f(x) = \sqrt{x}$, $[0, 9]$

    c.

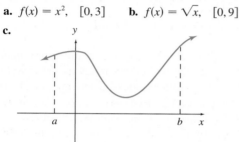

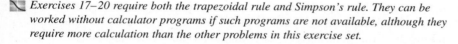 *Exercises 17–20 require both the trapezoidal rule and Simpson's rule. They can be worked without calculator programs if such programs are not available, although they require more calculation than the other problems in this exercise set.*

Error Analysis The difference between the true value of an integral and the value given by the trapezoidal rule or Simpson's rule is known as the error. In numerical analysis, the error is studied to determine how large n must be for the error to be smaller than some specified amount. For both rules, the error is inversely proportional to a power of n, the number of subdivisions. In other words, the error is roughly k/n^p, where k is a constant that depends upon the function and the interval, and p is a power that depends only upon the method used. With a little experimentation, you can find out what the power p is for the trapezoidal rule and for Simpson's rule.

17. a. Find the exact value of $\int_0^1 x^4\, dx$.

    b. Approximate the integral in part a using the trapezoidal rule with $n = 4, 8, 16,$ and 32. For each of these answers, find the absolute value of the error by subtracting the trapezoidal rule answer from the exact answer found in part a.

    c. If the error is k/n^p, then the error times n^p should be approximately a constant. Multiply the errors in part b times n^p for $p = 1, 2,$ etc., until you find a power p yielding the same answer for all four values of n.

18. Based on the results of Exercise 17, what happens to the error in the trapezoidal rule when the number of intervals is doubled?

19. Repeat Exercise 17 using Simpson's rule.

20. Based on the results of Exercise 19, what happens to the error in Simpson's rule when the number of intervals is doubled?

21. For the integral in Exercise 7, apply the midpoint rule (from Section 7.3) with $n = 4$ and Simpson's rule with $n = 8$ to verify the formula $S = (2M + T)/3$.

22. Repeat the instructions of Exercise 21 using the integral in Exercise 8.

Applications

LIFE SCIENCES

23. *Calves* As we saw in the chapter on Graphs and the Derivative, the daily milk consumption (in kilograms) for calves can be approximated by the function

$$y = b_0 w^{b_1} e^{-b_2 w},$$

where w is the age of the calf in weeks and b_0, b_1, and b_2 are constants.[*]

a. The age in days is given by $t = 7w$. Use this fact to convert the function above to a function in terms of t.

b. For a group of Angus calves, $b_0 = 5.955$, $b_1 = 0.233$, and $b_2 = 0.027$. Use the trapezoid rule with $n = 10$, and then Simpson's rule with $n = 10$, to find the total amount of milk consumed by one of these calves over the first 25 weeks of life.

c. For a group of Nelore calves, $b_0 = 8.409$, $b_1 = 0.143$, and $b_2 = 0.037$. Use the trapezoid rule with $n = 10$, and then Simpson's rule with $n = 10$, to find the total amount of milk consumed by one of these calves over the first 25 weeks of life.

24. *Foot-and-Mouth Epidemic* In 2001, the United Kingdom suffered an epidemic of foot-and-mouth disease. The graph below shows the number of reported cases each day since Feb. 18, as well as the number of cases epidemiologists project would have occurred had they culled all livestock on infected farms within 24 hours, and all livestock on neighboring farms within 48 hours of infection.[†] In the section on Area and the Definite Integral, we estimated the number of cases using rectangles.

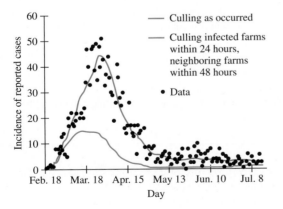

a. Estimate the total number of cases that occurred from Feb. 18 through May 13. Use Simpson's Rule with interval widths of 14 days.

b. Estimate the total number of cases that would have occurred from Feb. 18 through May 13 using the more aggressive culling plan. Use Simpson's Rule with interval widths of 14 days.

25. *Drug Reaction Rate* The reaction rate to a new drug is given by

$$y = e^{-t^2} + \frac{1}{t},$$

where t is time measured in hours after the drug is administered. Find the total reaction to the drug from $t = 1$ to $t = 9$ by letting $n = 8$ and using the following methods:

a. the trapezoidal rule,   **b.** Simpson's rule.

26. *Growth Rate* The growth rate of a certain tree (in feet) is given by

$$y = \frac{2}{t} + e^{-t^2/2},$$

where t is time in years. Find the total growth from $t = 1$ to $t = 7$ by using $n = 12$ with the following methods:

a. the trapezoidal rule,   **b.** Simpson's rule.

Blood Level Curves In the study of bioavailability in pharmacy, a drug is given to a patient. The level of concentration of the drug is then measured periodically, producing blood level curves such as the ones shown in the figure. The areas under the curves give the total amount of the drug available to the patient.[‡] Use the trapezoidal rule with $n = 10$ to find the following areas.

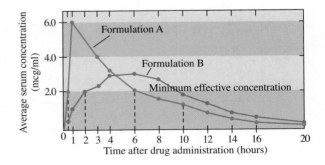

---

[*]Mezzadra, C., R. Paciaroni, S. Vulich, E. Villarreal, and L. Melucci, "Estimation of Milk Consumption Curve Parameters for Different Genetic Groups of Bovine Calves," *Animal Production,* Vol. 49, 1989, pp. 83–87.

[†]*Science,* Vol. 294, Oct. 5, 2001, p. 26.

[‡]These graphs are from D. J. Chodos and A. R. DeSantos, *Basics of Bioavailability.* Copyright 1978 by the Upjohn Company.

27. Find the total area under the curve on the previous page for Formulation A. What does this area represent?

28. Find the total area under the curve on the previous page for Formulation B. What does this area represent?

29. Find the area between the curve on the previous page for Formulation A and the minimum effective concentration line. What does your answer represent?

30. Find the area between the curve on the previous page for Formulation B and the minimum effective concentration line. What does this area represent?

OTHER APPLICATIONS

31. *Educational Psychology* The results from a research study in psychology were as follows.

Number of Hours of Study, x	1	2	3	4	5	6	7
Rate of Extra Points Earned on a Test, f(x)	4	7	11	9	15	16	23

   a. Plot these points. Connect the points with line segments.

   b. Use the trapezoidal rule to find the area bounded by the broken line of part a, the x-axis, the line $x = 1$, and the line $x = 7$.

   c. Find the same area using Simpson's rule.

32. *Chemical Formation* The following table shows the results from a chemical experiment.

Concentration of Chemical A, x	1	2	3	4	5	6	7
Rate of Formation of Chemical B, f(x)	12	16	18	21	24	27	32

   Repeat parts a–c of Exercise 31 for these data.

If you have a program for Simpson's rule in your graphing calculator, use it with n = 20 for Exercises 33–35.

33. *Total Revenue* An electronics company analyst has determined that the rate per month at which revenue comes in from the calculator division is given by

$$R(x) = 105e^{0.01x} + 32,$$

where x is the number of months the division has been in operation. Find the total revenue between the twelfth and thirty-sixth months.

34. *Blood Pressure* Blood pressure in an artery changes rapidly over a very short time for a healthy young adult, from a high of about 120 to a low of about 80. Suppose the blood pressure function over an interval of 1.5 sec is given by

$$f(x) = 0.2x^5 - 0.68x^4 + 0.8x^3 - 0.39x^2 + 0.055x + 100,$$

where x is the time in seconds after a peak reading. The area under the curve for one cycle is important in some blood pressure studies. Find the area under $f(x)$ from 0.1 sec to 1.1 sec.

35. *Probability* The most important function in probability and statistics is the density function for the standard normal distribution, which is the familiar bell-shaped curve. The function is

$$f(x) = \frac{1}{\sqrt{2\pi}}e^{-x^2/2}.$$

   a. The area under this curve between $x = -1$ and $x = 1$ represents the probability that a normal random variable is within 1 standard deviation of the mean. Find this probability.

   b. Find the area under this curve between $x = -2$ and $x = 2$, which represents the probability that a normal random variable is within 2 standard deviations of the mean.

   c. Find the probability that a normal random variable is within 3 standard deviations of the mean.

■ 8.2 INTEGRATION BY PARTS

? THINK ABOUT IT
If we know the rate of growth of a patch of moss, how can we calculate the area the moss covers?

The technique of *integration by parts* often makes it possible to reduce a complicated integral to a simpler integral. We know that if u and v are both differentiable functions, then uv is also differentiable and, by the product rule for derivatives,

$$\frac{d(uv)}{dx} = u\frac{dv}{dx} + v\frac{du}{dx}.$$

This expression can be rewritten, using differentials, as

$$d(uv) = u\,dv + v\,du.$$

Integrating both sides of this last equation gives

$$\int d(uv) = \int u\, dv + \int v\, du,$$

or

$$uv = \int u\, dv + \int v\, du.$$

Rearranging terms gives the following formula.

INTEGRATION BY PARTS

If u and v are differentiable functions, then

$$\int u\, dv = uv - \int v\, du.$$

The process of finding integrals by this formula is called **integration by parts.** There are two ways to do integration by parts: the standard method and column integration. Both methods are illustrated in the following example.

EXAMPLE 1   Integration by Parts
Find $\int xe^{5x}\, dx$.

Solution

Method 1: Standard Method

Although this integral cannot be found by using any method studied so far, it can be found with integration by parts. First write the expression $xe^{5x}\, dx$ as a product of two functions u and dv in such a way that $\int dv$ can be found. One way to do this is to choose the two functions x and e^{5x}. Both x and e^{5x} can be integrated, but $\int x\, dx$, which is $x^2/2$, is more complicated than x itself, while the derivative of x is 1, which is simpler than x. Since e^{5x} remains the same (except for the coefficient) whether it is integrated or differentiated, it is best here to choose

$$dv = e^{5x}\, dx \qquad \text{and} \qquad u = x.$$

Then

$$du = dx,$$

and v is found by integrating dv:

$$v = \int dv = \int e^{5x}\, dx = \frac{e^{5x}}{5}.$$

We need not introduce the constant of integration until the last step, because only one constant is needed. Now substitute into the formula for integration by parts and complete the integration.

$$\int u\, dv = uv - \int v\, du$$

$$\int \underbrace{x}_{u}\,\underbrace{e^{5x}\, dx}_{dv} = \underbrace{x}_{u}\underbrace{\left(\frac{e^{5x}}{5}\right)}_{v} - \int \underbrace{\frac{e^{5x}}{5}}_{v}\underbrace{dx}_{du}$$

$$= \frac{xe^{5x}}{5} - \frac{e^{5x}}{25} + C$$

FOR REVIEW ■

In the section on Substitution, we pointed out that when the chain rule is used to find the derivative of the function e^{kx}, we multiply by k, so when finding the antiderivative of e^{kx}, we divide by k. Thus $\int e^{5x}\, dx = e^{5x}/5 + C$. Keeping this technique in mind makes integration by parts simpler.

The constant C was added in the last step. As before, check the answer by taking its derivative.

Method 2: Column Integration

A technique called **column integration,** or *tabular integration,* is equivalent to integration by parts but helps in organizing the details.* We begin by creating two columns. The first column, labeled D, contains u, the part to be differentiated in the original integral. The second column, labeled I, contains the rest of the integral: that is, the part to be integrated, but without the dx. To create the remainder of the first column, write the derivative of the function in the first row underneath it in the second row. Now write the derivative of the function in the second row underneath it in the third row. Proceed in this manner down the first column, taking derivatives until you get a 0. Form the second column in a similar manner, except take an antiderivative at each row, until the second column has the same number of rows as the first.

To illustrate this process, consider our goal of finding $\int xe^{5x}\,dx$. Here $u = x$, so e^{5x} is left for the second column. Taking derivatives down the first column and antiderivatives down the second column results in the following table.

D	I
x	e^{5x}
1	$e^{5x}/5$
0	$e^{5x}/25$

Next, draw a diagonal line from each term (except the last) in the left column to the term in the row below it in the right column. Label the first such line with "+", the next with "−", and continue alternating the signs as shown. Essentially, the D column corresponds to the u part of the integral in integration by parts, and the I column corresponds to the v part of the integral.

D		I
x	+	e^{5x}
1	−	$e^{5x}/5$
0		$e^{5x}/25$

Then multiply the terms on opposite ends of each diagonal line. Finally, sum up the products just formed, adding the "+" terms and subtracting the "−" terms.

$$\int xe^{5x}\,dx = x(e^{5x}/5) - 1(e^{5x}/25) + C$$

$$= \frac{xe^{5x}}{5} - \frac{e^{5x}}{25} + C$$

Compare these steps with those of Method 1 and convince yourself that the process is the same.

---

*This technique appeared in the 1988 movie *Stand and Deliver.*

CONDITIONS FOR INTEGRATION BY PARTS

Integration by parts can be used only if the integrand satisfies the following conditions.

1. The integrand can be written as the product of two factors, u and dv.

2. It is possible to integrate dv to get v and to differentiate u to get du.

3. The integral $\int v\,du$ can be found.

EXAMPLE 2 Integration by Parts

Find $\int \ln x\,dx$ for $x > 0$.

Solution

Method 1: Standard Method

No rule has been given for integrating $\ln x$, so choose

$$dv = dx \qquad \text{and} \qquad u = \ln x.$$

Then

$$v = x \qquad \text{and} \qquad du = \frac{1}{x}\,dx,$$

and, since $uv = vu$, we have

$$\int u \cdot dv = v \cdot u \; - \; \int v \cdot du$$

$$\int \overbrace{\ln x}\ dx = x\,\overbrace{\ln x} \; - \; \int x \cdot \overbrace{\frac{1}{x}}\,dx$$

$$= x\ln x - \int dx$$

$$= x\ln x - x + C.$$

Method 2: Column Integration

Column integration works a little differently here. As in Method 1, choose $\ln x$ as the part to differentiate. The part to be integrated must be 1. (Think of $\ln x$ as $1 \cdot \ln x$.) No matter how many times $\ln x$ is differentiated, the result is never 0. In this case, stop as soon as the natural logarithm is gone, because continuing to differentiate does not make the function simpler.

D	I
$\ln x$	1
$1/x$	x

Draw diagonal lines with alternating $+$ and $-$ as before. On the last line, because the left column does not contain a 0, draw a horizontal line.

D		I
$\ln x$	$+$	1
$1/x$	$-$	x

The presence of a horizontal line indicates that the product is to be integrated, just as the original integral was represented by the first row of the two columns.

$$\int \ln x \, dx = (\ln x)x - \int \frac{1}{x} \cdot x \, dx$$

$$= x \ln x - \int dx$$

$$= x \ln x - x + C.$$

Note that when setting up the columns, a horizontal line is drawn only when a 0 does not eventually appear in the left column.

Sometimes integration by parts must be applied more than once, as in the next example.

EXAMPLE 3   Integration by Parts
Find $\int 2x^2 e^{-3x} \, dx$.

Solution

Method 1: Standard Method   Choose

$$dv = e^{-3x} \, dx \qquad \text{and} \qquad u = 2x^2.$$

Then

$$v = \frac{-e^{-3x}}{3} \qquad \text{and} \qquad du = 4x \, dx.$$

Substitute these values into the formula for integration by parts.

$$\int u \, dv = uv - \int v \, du$$

$$\int 2x^2 e^{-3x} \, dx = 2x^2 \left(\frac{-e^{-3x}}{3} \right) - \int \left(\frac{-e^{-3x}}{3} \right) 4x \, dx$$

$$= -\frac{2}{3} x^2 e^{-3x} + \frac{4}{3} \int x e^{-3x} \, dx$$

Now apply integration by parts to the last integral, letting

$$dv = e^{-3x} \, dx \qquad \text{and} \qquad u = x,$$

so

$$v = \frac{-e^{-3x}}{3} \qquad \text{and} \qquad du = dx.$$

$$\int 2x^2 e^{-3x} \, dx = -\frac{2}{3} x^2 e^{-3x} + \frac{4}{3} \int x e^{-3x} \, dx$$

$$= -\frac{2}{3} x^2 e^{-3x} + \frac{4}{3} \left[x \left(\frac{-e^{-3x}}{3} \right) - \int \left(\frac{-e^{-3x}}{3} \right) dx \right]$$

$$= -\frac{2}{3} x^2 e^{-3x} + \frac{4}{3} \left[-\frac{x}{3} e^{-3x} - \frac{e^{-3x}}{9} \right] + C$$

$$= -\frac{2}{3} x^2 e^{-3x} - \frac{4}{9} x e^{-3x} - \frac{4}{27} e^{-3x} + C$$

Method 2: Column Integration | Choose $2x^2$ as the part to be differentiated, and put e^{-3x} in the integration column.

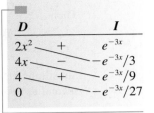

D		I
$2x^2$	$+$	e^{-3x}
$4x$	$-$	$-e^{-3x}/3$
4	$+$	$e^{-3x}/9$
0		$-e^{-3x}/27$

Multiplying and adding as before yields

$$\int 2x^2 e^{-3x}\, dx = 2x^2(-e^{-3x}/3) - 4x(e^{-3x}/9) + 4(-e^{-3x}/27) + C$$

$$= -\frac{2}{3}x^2 e^{-3x} - \frac{4}{9}xe^{-3x} - \frac{4}{27}e^{-3x} + C.$$

With the functions discussed so far in this book, choosing u and dv (or the parts to be differentiated and integrated) is relatively simple. Before trying integration by parts, first see if the integration can be performed using substitution. If substitution does not work, see if $\ln x$ is in the integral. If it is, set $u = \ln x$ and dv equal to the rest of the integral. (Equivalently, put $\ln x$ in the D column and the rest of the function in the I column.) If $\ln x$ is not present, see if x^k is present, where k is any positive integer. If it is present, set $u = x^k$ and dv equal to the rest of the integral. (Equivalently, put x^k in the D column and the rest of the function in the I column.)

EXAMPLE 4 Trigonometric Integral
Find $\int 2x \sin x\, dx$.

Solution Let $u = 2x$ and $dv = \sin x\, dx$. Then $du = 2\, dx$ and $v = -\cos x$. Use the formula for integration by parts,

$$\int u\, dv = uv - \int v\, du,$$

to get

$$\int 2x \sin x\, dx = -2x \cos x - \int (-\cos x)\,(2\, dx)$$

$$= -2x \cos x + 2 \int \cos x\, dx$$

$$= -2x \cos x + 2 \sin x + C.$$

Check the result by differentiating.

EXAMPLE 5 Definite Integral
Find $\displaystyle\int_1^e \frac{\ln x}{x^2}\, dx$.

Solution First we will find the indefinite integral using integration by parts by the standard method. (You may wish to verify this using column integration.) Whenever $\ln x$ is present, it is selected as u, so let

$$u = \ln x \qquad \text{and} \qquad dv = \frac{1}{x^2} \, dx.$$

Then

$$du = \frac{1}{x} \, dx \qquad \text{and} \qquad v = -\frac{1}{x}.$$

Substitute these values into the formula for integration by parts, and integrate the second term on the right.

$$\int u \, dv = uv - \int v \, du$$

$$\int \frac{\ln x}{x^2} \, dx = (\ln x) \frac{-1}{x} - \int \left(-\frac{1}{x} \cdot \frac{1}{x} \right) dx$$

$$= -\frac{\ln x}{x} + \int \frac{1}{x^2} \, dx$$

$$= -\frac{\ln x}{x} - \frac{1}{x} + C$$

$$= \frac{-\ln x - 1}{x} + C$$

Now find the definite integral.

$$\int_1^e \frac{\ln x}{x^2} \, dx = \frac{-\ln x - 1}{x} \bigg|_1^e$$

$$= \left(\frac{-1 - 1}{e} \right) - \left(\frac{0 - 1}{1} \right)$$

$$= \frac{-2}{e} + 1 \approx 0.2642411177$$

Definite integrals can be found with a graphing calculator using the function integral feature or by finding the area under the graph of the function between the limits. For example, using the `fnInt` feature of the TI-83 calculator to find the integral in Example 5 gives 0.2642411177. Using the area under the graph approach gives 0.26424112, the same result rounded.

Many integrals cannot be found by the methods presented so far. For example, consider the integral

$$\int \frac{1}{4 - x^2} \, dx.$$

Substitution of $u = 4 - x^2$ will not help, because $du = -2x \, dx$, and there is no x in the numerator of the integral. We could try integration by parts, using $dv = dx$ and $u = (4 - x^2)^{-1}$. Integration gives $v = x$ and differentiation gives $du = 2x \, dx / (4 - x^2)^2$, with

$$\int \frac{1}{4 - x^2} \, dx = \frac{x}{4 - x^2} - \int \frac{2x^2}{(4 - x^2)^2} \, dx.$$

The integral on the right is more complicated than the original integral, however. A second use of integration by parts on the new integral would only make matters worse. Since we cannot choose $dv = (4 - x^2)^{-1} dx$ because it cannot be integrated by the methods studied so far, integration by parts is not possible for this problem.

This integration can be performed using one of the many techniques of integration beyond the scope of this text.* Tables of integrals can also be used, but technology is rapidly making such tables obsolete, and even reducing the importance of techniques of integration. The following example shows how the table of integrals given in the appendix of this book may be used.

EXAMPLE 6 Tables of Integrals

Find $\displaystyle\int \frac{1}{4 - x^2}\, dx$.

Solution Using formula 7 in the table of integrals in the appendix, with $a = 2$, gives

$$\int \frac{1}{4 - x^2}\, dx = \frac{1}{4} \cdot \ln \left| \frac{2 + x}{2 - x} \right| + C.$$

We mentioned in the previous chapter how computer algebra systems and some calculators can perform integration. Using a TI-92, the answer to the above integral is

$$\frac{\ln\left(\frac{|x + 2|}{|x - 2|}\right)}{4}.$$

(The C is not included.) Verify that this is equivalent to the answer given in Example 6.

If you don't have a calculator or computer program that integrates symbolically, there is a Web site (http://integrals.wolfram.com), as of this writing, that finds indefinite integrals using the computer algebra system Mathematica. It includes instructions on how to enter your function. When the previous integral was entered, it returned the answer

$$-\frac{1}{4} \text{Log}[-2 + x] + \frac{1}{4} \text{Log}[2 + x].$$

Note that Mathematica does not include the C or the absolute value, and that natural logarithms are written as Log. Verify that this answer is equivalent to the answer given by the TI-92 and the answer given in Example 6.

Unfortunately, there are integrals that cannot be antidifferentiated by any technique, in which case numerical integration must be used. (See the previous section.) In this book, for simplicity, all integrals to be antidifferentiated can be done with substitution or by parts, except for Exercises 29–34 in this section.

---

*For example, see Thomas, George B., Ross L. Finney, Maurice D. Weir, and Frank Giordano, *Thomas' Calculus,* 10th ed., Addison Wesley Longman, 2000.

8.2 EXERCISES

Use integration by parts to find the integrals in Exercises 1–16.

1. $\int xe^x\, dx$

2. $\int (x+1)e^x\, dx$

3. $\int (5x-9)e^{-3x}\, dx$

4. $\int (6x+3)e^{-2x}\, dx$

5. $\int_0^1 \dfrac{2x+1}{e^x}\, dx$

6. $\int_0^1 \dfrac{1-x}{3e^x}\, dx$

7. $\int_1^4 \ln 2x\, dx$

8. $\int_1^2 \ln 5x\, dx$

9. $\int x\ln x\, dx$

10. $\int x^2 \ln x\, dx$

11. $\int -6x\cos 5x\, dx$

12. $\int 9x\sin 2x\, dx$

13. $\int 8x\sin x\, dx$

14. $\int -11x\cos x\, dx$

15. $\int -6x^2 \cos 8x\, dx$

16. $\int 10x^2 \sin \dfrac{x}{2}\, dx$

17. Find the area between $y=(x-2)e^x$ and the x-axis from $x=2$ to $x=4$.

18. Find the area between $y=xe^x$ and the x-axis from $x=0$ to $x=1$.

Exercises 19–28 are mixed—some require integration by parts, while others can be integrated by using techniques discussed in the chapter on Integration.

19. $\int x^2 e^{2x}\, dx$

20. $\int_1^2 (1-x^2)e^{2x}\, dx$

21. $\int_0^5 x\sqrt{x^2+2}\, dx$

22. $\int (2x-1)\ln(3x)\, dx$

23. $\int (8x+7)\ln(5x)\, dx$

24. $\int xe^{x^2}\, dx$

25. $\int x^2\sqrt{x+2}\, dx$

26. $\int_0^1 \dfrac{x^2\, dx}{2x^3+1}$

27. $\int_0^1 \dfrac{x^3\, dx}{\sqrt{3+x^2}}$

28. $\int \dfrac{x^2\, dx}{2x^3+1}$

Use the table of integrals, or a computer or calculator with symbolic integration capabilities, to find each indefinite integral.

29. $\int \dfrac{9}{\sqrt{x^2+9}}\, dx$

30. $\int \dfrac{6}{x^2-9}\, dx$

31. $\int \dfrac{3}{x\sqrt{121-x^2}}\, dx$

32. $\int \dfrac{2}{3x(3x-5)}\, dx$

33. $\int \dfrac{-3}{x(4x+3)^2}\, dx$

34. $\int \sqrt{x^2+10}\, dx$

35. What rule of differentiation is related to integration by parts?

36. Explain why the two methods of solving Example 2 are equivalent.

37. Use integration by parts to derive the following formula from the table of integrals.

$$\int x^n \cdot \ln |x|\, dx = x^{n+1}\left[\dfrac{\ln |x|}{n+1} - \dfrac{1}{(n+1)^2}\right] + C, \quad n \neq -1$$

38. Use integration by parts to derive the following formula from the table of integrals.

$$\int x^n e^{ax}\, dx = \dfrac{x^n e^{ax}}{a} - \dfrac{n}{a}\int x^{n-1}e^{ax}\, dx + C, \quad a \neq 0$$

39. a. One way to integrate $\int x\sqrt{x+1}\,dx$ is to use integration by parts. Do so to find the antiderivative.

b. Another way to evaluate the integral in part a is by using the substitution $u = x + 1$. Do so to find the antiderivative.

c. Compare the results from the two methods. If they do not look the same, explain how this can happen. Discuss the advantages and disadvantages of each method.

Applications

LIFE SCIENCES

40. *Thermic Effect of Food* As we saw in Chapters 3 and 5, a person's metabolic rate tends to go up after eating a meal and then, after some time has passed, it returns to a resting metabolic rate. This phenomenon is known as the thermic effect of food, and the effect (in kJ/hr) for one individual is

$$F(t) = -10.28 + 175.9te^{-t/1.3},$$

where t is the number of hours that have elapsed since eating a meal.* Find the total thermic energy of a meal for the next six hours after a meal by integrating the thermic effect function between $t = 0$ and $t = 6$.

41. *Rumen Fermentation* The rumen is the first division of the stomach of a ruminant, or cud-chewing animal. An article on the rumen microbial system reports that the fraction of the soluble material passing from the rumen without being fermented during the first hour after its ingestion could be calculated by the integral

$$\int_0^1 ke^{-kt}(1-t)\,dt,$$

where k measures the rate that the material is fermented.†

a. Determine the above integral, and evaluate it for the following values of k used in the article: $1/12$, $1/24$, and $1/48$ hour.

b. The fraction of intermediate material left in the rumen at 1 hour that escapes digestion by passage between 1 and 6 hours is given by

$$\int_1^6 ke^{-kt}(6-t)/5\,dt.$$

Determine this integral, and evaluate it for the values of k given in part a.

42. *Reaction to a Drug* The rate of reaction to a drug is given by

$$r'(x) = 2x^2e^{-x},$$

where x is the number of hours since the drug was administered. Find the total reaction to the drug from $x = 1$ to $x = 6$.

43. *Growth of a Population* The rate of growth of a microbe population is given by

$$m'(x) = 30xe^{2x},$$

where x is time in days. What is the total accumulated growth during the first 3 days?

44. *Rate of Growth* The area covered by a patch of moss is growing at a rate of

$$A'(t) = \sqrt{t}\,\ln t$$

sq cm per day, for $t \geq 1$. Find the additional amount of area covered by the moss between 4 and 9 days.

OTHER APPLICATIONS

45. *Rate of Change of Revenue* The rate of change of revenue (in dollars per calculator) from the sale of x small desk calculators is

$$R'(x) = (x+1)\ln(x+1).$$

Find the total revenue from the sale of the first 12 calculators. (*Hint:* In this exercise, it simplifies matters to write an antiderivative of $x + 1$ as $(x+1)^2/2$ rather than $x^2/2 + x$.)

*Reed, George, and James Hill, "Measuring the Thermic Effect of Food," *American Journal of Clinical Nutrition,* Vol. 63, 1996, pp. 164–169.
†Hungate, R. E., "The Rumen Microbial Ecosystem," *Annual Review of Ecology and Systematics,* Vol. 6, 1975, pp. 39–66.

8.3 VOLUME AND AVERAGE VALUE

? THINK ABOUT IT   What is the volume of a bird's egg?

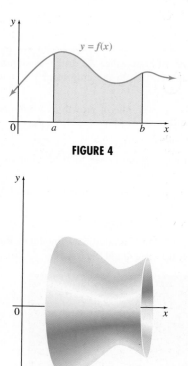

FIGURE 4

In this section, we will answer this and other questions about volume, as well as discover how to find the average value of a function.

Volume   Figure 4 shows the region below the graph of some function $y = f(x)$, above the x-axis, and between $x = a$ and $x = b$. We have seen how to use integrals to find the area of such a region. Now, suppose this region is revolved about the x-axis as shown in Figure 5. The resulting figure is called a **solid of revolution.** In many cases, the volume of a solid of revolution can be found by integration.

To begin, divide the interval $[a, b]$ into n subintervals of equal width Δx by the points $a = x_0, x_1, x_2, \ldots, x_i, \ldots, x_n = b$. Then think of slicing the solid into n slices of equal thickness Δx, as shown in Figure 6. If the slices are thin enough, each slice is very close to being a right circular cylinder. The formula for the volume of a right circular cylinder is $\pi r^2 h$, where r is the radius of the circular base and h is the height of the cylinder. As shown in Figure 7, the height of each slice is Δx. (The height is horizontal here, since the cylinder is on its side.) The radius of the circular base of each slice is $f(x_i)$. Thus, the volume of the slice is closely approximated by $\pi[f(x_i)]^2 \Delta x$. The volume of the solid of revolution will be approximated by the sum of the volumes of the slices:

$$V \approx \sum_{i=1}^{n} \pi[f(x_i)]^2 \Delta x.$$

FIGURE 5

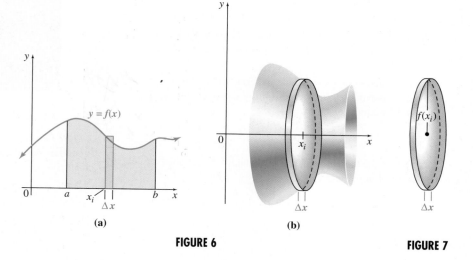

(a)

(b)

FIGURE 6

FIGURE 7

This is a Riemann sum, similar to those found in the previous chapter. By definition, the volume of the solid of revolution is the limit of this sum as the thickness of the slices approaches 0, or

$$V = \lim_{\Delta x \to 0} \sum_{i=1}^{n} \pi[f(x_i)]^2 \Delta x.$$

This limit, like the one discussed earlier for area, is a definite integral.

VOLUME OF A SOLID OF REVOLUTION

If $f(x)$ is nonnegative and R is the region between $f(x)$ and the x-axis from $x = a$ to $x = b$, the volume of the solid formed by rotating R about the x-axis is given by

$$V = \lim_{\Delta x \to 0} \sum_{i=1}^{n} \pi[f(x_i)]^2 \Delta x = \int_a^b \pi[f(x)]^2 dx.$$

The technique of summing disks to approximate volumes was originated by Johannes Kepler (1571–1630), a famous German astronomer who discovered three laws of planetary motion. He estimated volumes of wine casks used at his wedding by means of solids of revolution.

EXAMPLE 1 Volume

Find the volume of the solid of revolution formed by rotating about the x-axis the region bounded by $y = x + 1$, $y = 0$, $x = 1$, and $x = 4$.

Solution The region and the solid are shown in Figure 8. Use the formula given above for the volume, with $a = 1$, $b = 4$, and $f(x) = x + 1$.

$$V = \int_1^4 \pi(x + 1)^2 dx = \pi \left(\frac{(x + 1)^3}{3} \right) \Big|_1^4$$

$$= \frac{\pi}{3}(5^3 - 2^3) = \frac{117\pi}{3} = 39\pi$$

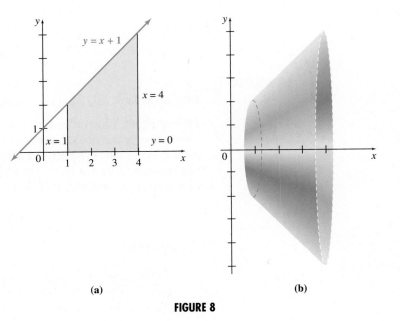

(a)   (b)

FIGURE 8

EXAMPLE 2   Volume

Find the volume of the solid of revolution formed by rotating about the x-axis the area bounded by $f(x) = 4 - x^2$ and the x-axis.

Solution The region and the solid are shown in Figure 9. Find a and b from the x-intercepts. If $y = 0$, then $x = 2$ or $x = -2$, so that $a = -2$ and $b = 2$. The volume is

$$V = \int_{-2}^{2} \pi(4 - x^2)^2 \, dx$$

$$= \int_{-2}^{2} \pi(16 - 8x^2 + x^4) \, dx$$

$$= \pi\left(16x - \frac{8x^3}{3} + \frac{x^5}{5}\right)\bigg|_{-2}^{2} = \frac{512\pi}{15}.$$

A graphing calculator with the `fnInt` feature gives the value as 107.2330292, which agrees with the approximation of $512\pi/15$ to the seven decimal places shown.

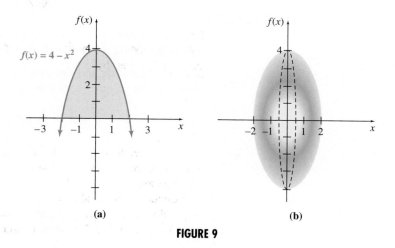

(a)   **(b)**

FIGURE 9

EXAMPLE 3   Volume

Find the volume of a right circular cone with height h and base radius r.

Solution Figure 10(a) on the next page shows the required cone, while Figure 10(b) shows an area that could be rotated about the x-axis to get such a cone. The cone formed by rotation is shown in Figure 10(c). Here $y = f(x)$ is the equation of the line through $(0, 0)$ and (h, r). The slope of this line is h/r, and since the y-intercept is 0, the equation of the line is

$$y = \frac{r}{h}x.$$

Then the volume is

$$V = \int_{0}^{h} \pi\left(\frac{r}{h}x\right)^2 dx = \pi \int_{0}^{h} \frac{r^2 x^2}{h^2} \, dx$$

$$= \pi \frac{r^2 x^3}{3h^2}\bigg|_{0}^{h} \qquad \text{since } r \text{ and } h \text{ are constants}$$

$$= \frac{\pi r^2 h}{3}.$$

This is the familiar formula for the volume of a right circular cone.

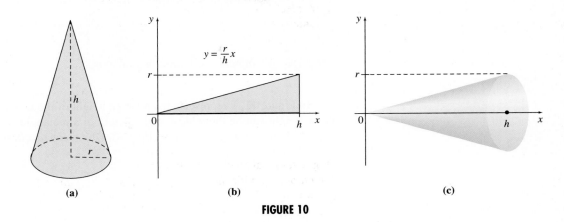

(a)   **(b)**   **(c)**

FIGURE 10

Average Value of a Function   The average of the n numbers $v_1, v_2, v_3, \ldots,$ $v_i, \ldots, v_n$ is given by

$$\frac{v_1 + v_2 + v_3 + \cdots + v_n}{n} = \frac{\sum\limits_{i=1}^{n} v_i}{n}.$$

For example, to compute an average temperature, we could take readings at equally spaced intervals and average the readings.

The average value of a function f on $[a, b]$ can be defined in a similar manner; divide the interval $[a, b]$ into n subintervals, each of width Δx. Then choose an x-value, x_i, in each subinterval, and find $f(x_i)$. The average function value for the n subintervals and the given choices of x_i is

$$\frac{f(x_1) + f(x_2) + \cdots + f(x_n)}{n} = \frac{\sum\limits_{i=1}^{n} f(x_i)}{n}.$$

Since $(b - a)/n = \Delta x$, multiply the expression on the right side of the equation by $(b - a)/(b - a)$ and rearrange the expression to get

$$\frac{b - a}{b - a} \cdot \frac{\sum\limits_{i=1}^{n} f(x_i)}{n} = \frac{b - a}{n} \cdot \frac{\sum\limits_{i=1}^{n} f(x_i)}{b - a} = \Delta x \cdot \frac{\sum\limits_{i=1}^{n} f(x_i)}{b - a}$$

$$= \frac{1}{b - a} \sum\limits_{i=1}^{n} f(x_i) \Delta x.$$

Now, take the limit as $n \to \infty$. If the limit exists, then

$$\lim_{n \to \infty} \frac{1}{b - a} \sum\limits_{i=1}^{n} f(x_i) \Delta x = \frac{1}{b - a} \lim_{n \to \infty} \sum\limits_{i=1}^{n} f(x_i) \Delta x = \frac{1}{b - a} \int_a^b f(x) \, dx.$$

The following definition summarizes this discussion.

> **AVERAGE VALUE OF A FUNCTION**
>
> The **average value of a function** f on the interval $[a, b]$ is
>
> $$\frac{1}{b - a} \int_a^b f(x)\,dx,$$
>
> provided the indicated definite integral exists.

In Figure 11, $\bar{y}$ represents the average height of the irregular region. The average height can be thought of as the height of a rectangle with base $b - a$. For $f(x) \geq 0$, this rectangle has area $\bar{y}(b - a)$, which equals the area under the graph of $f(x)$ from $x = a$ to $x = b$, so that

$$\bar{y}(b - a) = \int_a^b f(x)\,dx.$$

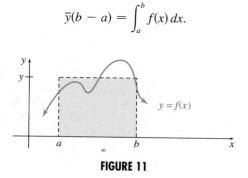

FIGURE 11

EXAMPLE 4 Cholesterol Level

A physician finds that the cholesterol level of a patient on a diet can be approximated by

$$C(t) = 200 + 40e^{-0.1t},$$

where t is the time (in months) since the patient began the diet. Find the average cholesterol level for the patient over the first six months.

Solution Use the formula for average value with $a = 0$ and $b = 6$. The average cholesterol level is

$$\frac{1}{6 - 0} \int_0^6 (200 + 40e^{-0.1t})\,dt = \frac{1}{6}\left(200t + \frac{40}{-0.1}e^{-0.1t} \right)\Bigg|_0^6$$

$$= \frac{1}{6}(200t - 400e^{-0.1t})\Bigg|_0^6$$

$$= \frac{1}{6}(1{,}200 - 400e^{-0.6} + 400)$$

$$\approx 230.$$

Alternatively, using a graphing calculator with the `fnInt` feature to evaluate the integral, we get 1,380.475346. Dividing that result by 6 gives the average cholesterol level as 230.0792243 ≈ 230.

8.3 EXERCISES

Find the volume of the solid of revolution formed by rotating about the x-axis each region bounded by the given curves.

1. $f(x) = x, y = 0, x = 0, x = 2$

2. $f(x) = 2x, y = 0, x = 0, x = 3$

3. $f(x) = 2x + 1, y = 0, x = 0, x = 4$

4. $f(x) = x - 4, y = 0, x = 4, x = 10$

5. $f(x) = \frac{1}{3}x + 2, y = 0, x = 1, x = 3$

6. $f(x) = \frac{1}{2}x + 4, y = 0, x = 0, x = 5$

7. $f(x) = \sqrt{x}, y = 0, x = 1, x = 2$

8. $f(x) = \sqrt{x + 1}, y = 0, x = 0, x = 3$

9. $f(x) = \sqrt{2x + 1}, y = 0, x = 1, x = 4$

10. $f(x) = \sqrt{3x + 2}, y = 0, x = 1, x = 2$

11. $f(x) = e^x, y = 0, x = 0, x = 2$

12. $f(x) = 2e^x, y = 0, x = -2, x = 1$

13. $f(x) = \frac{1}{\sqrt{x}}, y = 0, x = 1, x = 4$

14. $f(x) = \frac{1}{\sqrt{x + 1}}, y = 0, x = 0, x = 2$

15. $f(x) = x^2, y = 0, x = 1, x = 5$

16. $f(x) = \frac{x^2}{2}, y = 0, x = 0, x = 4$

17. $f(x) = 1 - x^2, y = 0$

18. $f(x) = 2 - x^2, y = 0$

19. $f(x) = \sec x, y = 0, x = 0, x = \pi/4$

20. $f(x) = \csc x, y = 0, x = \pi/4, x = \pi/2$

The function defined by $y = \sqrt{r^2 - x^2}$ has as its graph a semicircle of radius r with center at $(0, 0)$ (see the figure). In Exercises 21–23, find the volume that results when each semicircle is rotated about the x-axis. (The result of Exercise 23 gives a formula for the volume of a sphere with radius r.)

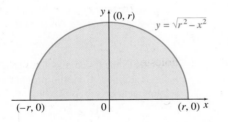

21. $f(x) = \sqrt{1 - x^2}$

22. $f(x) = \sqrt{16 - x^2}$

23. $f(x) = \sqrt{r^2 - x^2}$

24. Find a formula for the volume of an ellipsoid. See Exercises 21–23 and the following figures.

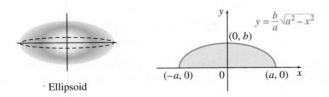

Ellipsoid

25. Use the methods of this section to find the volume of a cylinder with height h and radius r.

Find the average value of each function on the given interval.

26. $f(x) = 3 - 2x^2$; $[1, 9]$

27. $f(x) = x^2 - 2$; $[0, 5]$

28. $f(x) = (2x - 1)^{1/2}$; $[1, 13]$

29. $f(x) = \sqrt{x + 1}$; $[3, 8]$

30. $f(x) = e^{0.1x}$; $[0, 10]$

31. $f(x) = e^{x/5}$; $[0, 5]$

32. $f(x) = x \ln x;$ $[1, e]$ **33.** $f(x) = x^2 e^{2x};$ $[0, 2]$ **34.** $f(x) = \sin x;$ $[0, \pi]$
35. $f(x) = \sec^2 x;$ $[0, \pi/4]$

In Exercises 36 and 37, use the integration feature on a graphing calculator to find the volume of the solid of revolution by rotating about the x-axis each region bounded by the given curves.

36. $f(x) = \dfrac{1}{1 + x^2},\ y = 0,\ x = -1,\ x = 1$ **37.** $f(x) = e^{-x^2},\ y = 0,\ x = -2,\ x = 2$

Applications

LIFE SCIENCES

38. *Bird Eggs* The average length and width of various bird eggs are given in the following table.*

Bird Name	Length (cm)	Width (cm)
Canada goose	8.6	5.8
Robin	1.9	1.5
Turtledove	3.1	2.3
Hummingbird	1.0	1.0
Raven	5.0	3.3

a. Assume for simplicity that a bird's egg is roughly the shape of an ellipsoid. Use the result of Exercise 24 to estimate the volume of an egg of each of the following birds.

   i. Canada goose

   ii. Robin

   iii. Turtledove

   iv. Hummingbird

   v. Raven

b. In Exercise 5 of Sec. 1.2, we showed that the average length in cm of an egg of width w cm is given by

$$l = 1.585w - 0.487.$$

Using this result and the ideas in part a, show that the average volume of an egg of width w cm is given by

$$V = \pi(1.585w^3 - 0.487w^2)/6.$$

Use this formula to calculate the average volume for the bird eggs in part a, and compare with your results from part a.

39. *Blood Flow* The figure in the next column shows the blood flow in a small artery of the body. The flow of blood is *laminar* (in layers), with the velocity very low near the artery walls and highest in the center of the artery. In this model of blood flow, we calculate the total flow in the artery by thinking of the flow as being made up of many layers of concentric tubes sliding one on the other.

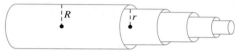

Suppose R is the radius of an artery and r is the distance from a given layer to the center. Then the velocity of blood in a given layer can be shown to equal

$$v(r) = k(R^2 - r^2),$$

where k is a numerical constant.

Since the area of a circle is $A = \pi r^2$, the change in the area of the cross section of one of the layers, corresponding to a small change in the radius, Δr, can be approximated by differentials. For $dr = \Delta r$, the differential of the area A is

$$dA = 2\pi r\, dr = 2\pi r\, \Delta r,$$

where Δr is the thickness of the layer. The total flow in the layer is defined to be the product of velocity and cross-section area, or

$$F(r) = 2\pi rk(R^2 - r^2)\, \Delta r.$$

a. Set up a definite integral to find the total flow in the artery.

b. Evaluate this definite integral.

40. *Drug Reaction* The intensity of the reaction to a certain drug, in appropriate units, is given by

$$R(t) = te^{-0.1t},$$

where t is time (in hours) after the drug is administered. Find the average intensity during each of the following hours.

a. Second hour

b. Twelfth hour

c. Twenty-fourth hour

---
*www.nctm.org/wlme/wlme6/five.htm

OTHER APPLICATIONS

41. *Production Rate* Suppose the number of items a new worker on an assembly line produces daily after t days on the job is given by

$$I(t) = 35 \ln(t + 1).$$

Find the average number of items produced daily by this employee after the following numbers of days.

a. 5   **b.** 10   **c.** 15

42. *Typing Speed* The function $W(t) = -3.75t^2 + 30t + 40$ describes a typist's speed (in words per minute) over a time interval $[0, 5]$.

a. Find $W(0)$.

b. Find the maximum W value and the time t when it occurs.

c. Find the average speed over $[0, 5]$.

■ 8.4 IMPROPER INTEGRALS

? THINK ABOUT IT   If we know the rate at which a pollutant is dumped into a stream, how can we compute the total amount released given that the rate of dumping is decreasing over time?

In this section we will learn how to answer such questions.

Sometimes it is useful to be able to integrate a function over an infinite period of time. For example, we might want to find the total amount of income generated by an apartment building into the indefinite future, or the total amount of pollution into a bay from a source that is continuing indefinitely. In this section we define integrals with one or more infinite limits of integration that can be used to solve such problems.

The graph in Figure 12(a) on the next page shows the area bounded by the curve $f(x) = x^{-3/2}$, the x-axis, and the vertical line $x = 1$. Think of the shaded region below the curve as extending indefinitely to the right. Does this shaded region have an area?

To see if the area of this region can be defined, introduce a vertical line at $x = b$, as shown in Figure 12(b). This vertical line gives a region with both upper and lower limits of integration. The area of this new region is given by the definite integral

$$\int_{1}^{b} x^{-3/2} \, dx.$$

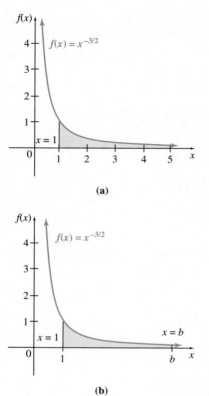

(a)

(b)

FIGURE 12

FOR REVIEW ■

In the section on limits at infinity and curve sketching, we saw that for any positive real number n,

$$\lim_{b \to \infty} \frac{1}{b^n} = 0.$$

By the Fundamental Theorem of Calculus,

$$\int_1^b x^{-3/2}\, dx = (-2x^{-1/2}) \Big|_1^b$$

$$= -2b^{-1/2} - (-2 \cdot 1^{-1/2})$$

$$= -2b^{-1/2} + 2 = 2 - \frac{2}{b^{1/2}}.$$

Suppose we now let the vertical line $x = b$ in Figure 12(b) move farther to the right. That is, suppose $b \to \infty$. The expression $-2/b^{1/2}$ would then approach 0, and

$$\lim_{b \to \infty} \left(2 - \frac{2}{b^{1/2}} \right) = 2 - 0 = 2.$$

This limit is defined to be the *area* of the region shown in Figure 12(a), so that

$$\int_1^\infty x^{-3/2}\, dx = 2.$$

An integral of the form

$$\int_a^\infty f(x)\, dx, \qquad \int_{-\infty}^b f(x)\, dx, \qquad \text{or} \qquad \int_{-\infty}^\infty f(x)\, dx$$

is called an *improper integral*. These **improper integrals** are defined as follows.

IMPROPER INTEGRALS

If f is continuous on the indicated interval and if the indicated limits exist, then

$$\int_a^\infty f(x)\, dx = \lim_{b \to \infty} \int_a^b f(x)\, dx,$$

$$\int_{-\infty}^b f(x)\, dx = \lim_{a \to -\infty} \int_a^b f(x)\, dx,$$

$$\int_{-\infty}^\infty f(x)\, dx = \int_{-\infty}^c f(x)\, dx + \int_c^\infty f(x)\, dx,$$

for real numbers a, b, and c, where c is arbitrarily chosen.

If the expressions on the right side exist, the integrals are **convergent;** otherwise, they are **divergent.** A convergent integral has a value that is a real number. A divergent integral does not, often because the area under the curve is infinitely large.

EXAMPLE 1 Improper Integrals
Find the following integrals.

(a) $\displaystyle \int_1^\infty \frac{dx}{x}$

Solution A graph of this region is shown in Figure 13 on the next page. By the definition of an improper integral,

$$\int_1^\infty \frac{dx}{x} = \lim_{b \to \infty} \int_1^b \frac{dx}{x}.$$

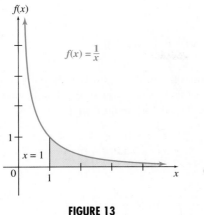

FIGURE 13

Find $\displaystyle\int_1^b \frac{dx}{x}$ by the Fundamental Theorem of Calculus.

$$\int_1^b \frac{dx}{x} = \ln |x| \Big|_1^b = \ln |b| - \ln |1| = \ln |b| - 0 = \ln |b|$$

As $b \to \infty$, $\ln |b| \to \infty$, so $\displaystyle\lim_{b \to \infty} \ln |b|$ does not exist. Since the limit does not exist, $\displaystyle\int_1^\infty \frac{dx}{x}$ is divergent.

(b) $\displaystyle\int_{-\infty}^{-2} \frac{1}{x^2} dx = \lim_{a \to -\infty} \int_a^{-2} \frac{1}{x^2} dx = \lim_{a \to -\infty} \left(\frac{-1}{x}\right) \Big|_a^{-2}$

$$= \lim_{a \to -\infty} \left(\frac{1}{2} + \frac{1}{a}\right) = \frac{1}{2}$$

A graph of this region is shown in Figure 14. Since the limit exists, this integral converges.

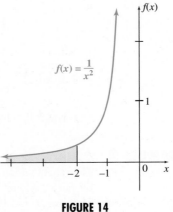

FIGURE 14

It may seem puzzling that the areas under the curves $f(x) = 1/x^{3/2}$ and $f(x) = 1/x^2$ are finite, while $f(x) = 1/x$ has an infinite amount of area. At first glance the graphs of these functions appear similar. The difference is that although all three functions get small as x becomes infinitely large, $f(x) = 1/x$ does not become small enough fast enough.

In the graphing calculator screen in Figure 15, notice how much faster $1/x^2$ becomes small compared with $1/x$.

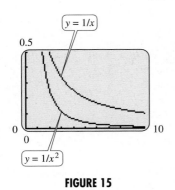

FIGURE 15

CAUTION Since graphing calculators provide only approximations, using them to find improper integrals is tricky and requires skill and care. Although their approximations may be good in some cases, they are wrong in others, and they cannot tell us for certain that an improper integral does not exist. See Exercises 39–41.

EXAMPLE 2 Improper Integral
Find $\int_{-\infty}^{\infty} 4e^{-3x}\, dx$.

Solution Write the integral as

$$\int_{-\infty}^{\infty} 4e^{-3x}\, dx = \int_{-\infty}^{0} 4e^{-3x}\, dx + \int_{0}^{\infty} 4e^{-3x}\, dx$$

and evaluate each of the two improper integrals on the right. If they both converge, the original integral will equal their sum. To show you all the details while maintaining the suspense, we will evaluate the second integral first.

By definition,

$$\int_{0}^{\infty} 4e^{-3x}\, dx = \lim_{b\to\infty} \int_{0}^{b} 4e^{-3x}\, dx = \lim_{b\to\infty}\left(\frac{-4}{3}e^{-3x}\right)\Big|_{0}^{b}$$

$$= \lim_{b\to\infty}\left(\frac{-4}{3e^{3b}} + \frac{4}{3}\right) = 0 + \frac{4}{3} = \frac{4}{3}.$$

Similarly, the second integral is evaluated as

$$\int_{-\infty}^{0} 4e^{-3x}\, dx = \lim_{b\to-\infty} \int_{b}^{0} 4e^{-3x}\, dx = \lim_{b\to-\infty}\left(\frac{-4}{3}e^{-3x}\right)\Big|_{b}^{0}$$

$$= \lim_{b\to-\infty}\left(-\frac{4}{3} + \frac{4}{3e^{3b}}\right) = \infty.$$

Since one of the two improper integrals diverges, the original improper integral diverges.

The following examples describe applications of improper integrals.

EXAMPLE 3 Pollution

The rate at which a pollutant is being dumped into a small stream at time t is given by $P_0 e^{-kt}$, where P_0 is the amount of pollutant initially released into the stream. Suppose $P_0 = 1,000$ and $k = 0.06$. Find the total amount of the pollutant that will be released into the stream into the indefinite future.

Solution Find

$$\int_{0}^{\infty} P_0 e^{-kt}\, dt = \int_{0}^{\infty} 1,000 e^{-0.06t}\, dt.$$

Work as above.

$$\int_{0}^{\infty} 1,000 e^{-0.06t}\, dt = \lim_{b\to\infty} \int_{0}^{b} 1,000 e^{-0.06t}\, dt$$

$$= \lim_{b\to\infty}\left(\frac{1,000}{-0.06}e^{-0.06t}\right)\Big|_{0}^{b}$$

$$= \lim_{b\to\infty}\left(\frac{1,000}{-0.06e^{0.06b}} - \frac{1,000}{-0.06}e^0\right) = \frac{-1,000}{-0.06} = 16,667$$

A total of 16,667 units of the pollutant eventually will be released.

8.4   EXERCISES

Determine whether the improper integrals in Exercises 1–30 converge or diverge, and find the value of each that converges.

1. $\int_2^\infty \frac{1}{x^2} dx$

2. $\int_5^\infty \frac{1}{x^2} dx$

3. $\int_1^\infty \frac{1}{\sqrt{x}} dx$

4. $\int_{16}^\infty \frac{-3}{\sqrt{x}} dx$

5. $\int_{-\infty}^{-1} \frac{2}{x^3} dx$

6. $\int_{-\infty}^{-4} \frac{3}{x^4} dx$

7. $\int_1^\infty \frac{1}{x^{1.0001}} dx$

8. $\int_1^\infty \frac{1}{x^{0.999}} dx$

9. $\int_{-\infty}^{-1} x^{-2} dx$

10. $\int_{-\infty}^{-4} x^{-2} dx$

11. $\int_{-\infty}^{-1} x^{-8/3} dx$

12. $\int_{-\infty}^{-27} x^{-5/3} dx$

13. $\int_0^\infty 4e^{-4x} dx$

14. $\int_0^\infty 10e^{-10x} dx$

15. $\int_{-\infty}^0 4e^x dx$

16. $\int_{-\infty}^0 3e^{4x} dx$

17. $\int_{-\infty}^{-1} \ln |x| \, dx$

18. $\int_1^\infty \ln |x| \, dx$

19. $\int_0^\infty \frac{dx}{(x+1)^2}$

20. $\int_0^\infty \frac{dx}{(2x+1)^3}$

21. $\int_{-\infty}^{-1} \frac{2x-1}{x^2-x} dx$

22. $\int_1^\infty \frac{2x+3}{x^2+3x} dx$

23. $\int_2^\infty \frac{1}{x \ln x} dx$

24. $\int_2^\infty \frac{1}{x(\ln x)^2} dx$

25. $\int_0^\infty xe^{2x} dx$

26. $\int_{-\infty}^0 xe^{3x} dx$

27. $\int_{-\infty}^\infty x^3 e^{-x^4} dx$ (*Hint:* Recall that $\lim_{x \to \infty} x^n e^{-x} = 0$.)

28. $\int_{-\infty}^\infty e^{-|x|} dx$ (*Hint:* Recall that when $x < 0$, $|x| = -x$.)

29. $\int_{-\infty}^\infty \frac{x}{x^2+1} dx$

30. $\int_{-\infty}^\infty \frac{2x+1}{x^2+x+4} dx$

For Exercises 31–34, find the area between the graph of the given function and the x-axis over the given interval, if possible.

31. $f(x) = \dfrac{1}{x-1}$ for $(-\infty, 0]$

32. $f(x) = e^{-x}$ for $(-\infty, e]$

33. $f(x) = \dfrac{1}{(x-1)^2}$ for $(-\infty, 0]$

34. $f(x) = \dfrac{1}{(x-1)^3}$ for $(-\infty, 0]$

35. Find $\int_{-\infty}^\infty xe^{-x^2} dx$.

36. Find $\int_{-\infty}^\infty \dfrac{x}{(1+x^2)^2} dx$.

37. Show that $\int_1^\infty 1/x^p \, dx$ converges if $p > 1$ and diverges if $p \le 1$.

38. Example 1(b) leads to a paradox. On the one hand, the unbounded region in that example has an area of $1/2$, so theoretically it could be colored with ink. On the other hand, the boundary of that region is infinite, so it cannot be drawn with a finite amount of ink. This seems impossible, because coloring the region automatically colors the boundary. Explain why it is possible to color the region.

39. Consider the functions $f(x) = 1/\sqrt{1+x^2}$ and $g(x) = 1/\sqrt{1+x^4}$.

   a. Use your calculator to approximate $\int_1^b f(x) \, dx$ for $b = 20$, 50, 100, 1,000, and 10,000.

   b. Based on your answers from part a, would you guess that $\int_1^\infty f(x) \, dx$ is convergent or divergent?

   c. Use your calculator to approximate $\int_1^b g(x) \, dx$ for $b = 20$, 50, 100, 1,000, and 10,000.

   d. Based on your answers from part c, would you guess that $\int_1^\infty g(x) \, dx$ is convergent or divergent?

e. Show how the answer to parts b and d might be guessed by comparing the integrals with others whose convergence or divergence is known. (*Hint:* For large x, the difference between $1 + x^2$ and x^2 is relatively small.)

Note: The first integral is indeed divergent, and the second convergent, with an approximate value of 0.9270.

40. a. Use your calculator to approximate $\int_0^b e^{-x^2}\, dx$ for $b = 1, 5, 10$, and 20.

b. Based on your answer to part a, does $\int_0^\infty e^{-x^2}\, dx$ appear to be convergent or divergent? If convergent, what seems to be its approximate value?

c. Explain why this integral should be convergent by comparing e^{-x^2} with e^{-x} for $x > 1$.

Note: The integral is convergent, with a value of $\sqrt{\pi}/2$.

41. a. Use your calculator to approximate $\int_0^b e^{-0.00001x}\, dx$ for $b = 10, 50, 100$, and 1,000.

b. Based on your answer to part a, does $\int_0^\infty e^{-0.00001x}\, dx$ appear to be convergent or divergent?

c. To what value does the integral actually converge?

For Exercises 42 and 43 use the integration feature on a graphing calculator and successively larger values of b to estimate $\int_0^\infty f(x)$.

42. $\displaystyle\int_0^b e^{-x} \sin x\, dx$

43. $\displaystyle\int_0^b e^{-x} \cos x\, dx$

▊Applications

LIFE SCIENCES

44. *Drug Epidemic* In an epidemiological model used to study the spread of drug use, a single drug user is introduced into a population of N non-users. Under certain assumptions, the number of people expected to use drugs as a result of direct influence from each drug user is given by

$$S = N \int_0^\infty \frac{a(1 - e^{-kt})}{k} e^{-bt}\, dt,$$

where a, b, and k are constants.* Find the value of S.

45. *Present Value* When harvesting a population, such as fish, the present value of the resource is given by

$$P = \int_0^\infty e^{-rt} n(t) y(t)\, dt,$$

where r is a discount factor, $n(t)$ is the net revenue at time t, and $y(t)$ is the harvesting effort.[†] Suppose $y(t) = K$ and $n(t) = at + b$. Find the present value.

46. *Drug Reaction* The rate of reaction to a drug is given by

$$r'(x) = 2x^2 e^{-x},$$

where x is the number of hours since the drug was administered. Find the total reaction to the drug over all the time since it was administered, assuming this is an infinite time interval. (*Hint:* $\lim\limits_{x\to\infty} x^k e^{-x} = 0$ for all real numbers k.)

OTHER APPLICATIONS

Radioactive Waste Radioactive waste is entering the atmosphere over an area at a decreasing rate. Use the improper integral

$$\int_0^\infty P e^{-kt}\, dt$$

with $P = 50\,mg/day$ to find the total amount of the waste that will enter the atmosphere for each of the following values of k.

47. $k = 0.06$

48. $k = 0.04$

---

*Murray, J. D., *Mathematical Biology,* Springer-Verlag, 1989, p. 642, 648.
†Ludwig, Donald, "An Unusual Free Boundary Problem from the Theory of Optimal Harvesting," in *Lectures on Mathematics in the Life Sciences, Vol. 12: Some Mathematical Questions in Biology,* The American Mathematical Society, 1979, pp. 173–209.

■ CHAPTER SUMMARY

This chapter has covered several topics related to integration. In the first two sections we explored numerical integration and integration by parts as ways to find integrals. The next section covered two applications of the integral: volume and average value. Finally, we learned to evaluate improper integrals, which have upper or lower limits of ∞ or $-\infty$.

KEY TERMS

8.1 numerical integration	**8.2 integration by parts**	**average value of**	**convergent integral**
trapezoidal rule	**column integration**	**a function**	**divergent integral**
Simpson's rule	**8.3 solid of revolution**	**8.4 improper integral**	

CHAPTER 8 REVIEW EXERCISES

1. Describe the type of integral for which numerical integration is useful.

2. Describe the type of integral for which integration by parts is useful.

3. Compare finding the average value of a function with finding the average of n numbers.

4. What is an improper integral? Explain why improper integrals must be treated in a special way.

In Exercises 5–7, use the trapezoidal rule with $n = 4$ to approximate the value of each integral. Then find the exact value and compare the two answers.

5. $\displaystyle\int_{1}^{3} \frac{\ln x}{x}\, dx$

6. $\displaystyle\int_{2}^{10} \frac{x\, dx}{x - 1}$

7. $\displaystyle\int_{0}^{1} e^{x}\sqrt{e^{x} + 1}\, dx$

In Exercises 8–10, use Simpson's rule with $n = 4$ to approximate the value of each integral. Compare your answers with the answers to Exercises 5–7.

8. $\displaystyle\int_{1}^{3} \frac{\ln x}{x}\, dx$

9. $\displaystyle\int_{2}^{10} \frac{x\, dx}{x - 1}$

10. $\displaystyle\int_{0}^{1} e^{x}\sqrt{e^{x} + 1}\, dx$

11. Find the area of the region between the graphs of $y = \sqrt{x - 1}$ and $2y = x - 1$ from $x = 1$ to $x = 5$ in three ways.

 a. Use antidifferentiation.   b. Use the trapezoidal rule with $n = 4$.

 c. Use Simpson's rule with $n = 4$.

12. Let $y = [x(x - 1)(x + 1)(x - 2)(x + 2)]^{2}$.

 a. Find $\int_{-2}^{2} f(x)\, dx$ using the trapezoidal rule with $n = 4$.

 b. Find $\int_{-2}^{2} f(x)\, dx$ using Simpson's rule with $n = 4$.

 c. Without evaluating $\int_{-2}^{2} f(x)\, dx$, explain why your answers to parts a and b cannot possibly be correct.

 d. Explain why the trapezoidal rule and Simpson's rule with $n = 4$ give incorrect answers for $\int_{-2}^{2} f(x)\, dx$ with this function.

Find each of the following integrals, using techniques from this or the previous chapter.

13. $\displaystyle\int x(8 - x)^{3/2}\, dx$

14. $\displaystyle\int \frac{3x}{\sqrt{x - 2}}\, dx$

15. $\displaystyle\int x e^{x}\, dx$

16. $\int (x + 2)e^{-3x}\,dx$

17. $\int \ln|2x + 3|\,dx$

18. $\int (x - 1)\ln|x|\,dx$

19. $\int \dfrac{x}{9 - 4x^2}\,dx$

20. $\int \dfrac{x}{\sqrt{25 + 9x^2}}\,dx$

21. $\int (x + 2)\sin x\,dx$

22. $\int x^2 \cos 2x\,dx$

23. $\int_1^e x^3 \ln x\,dx$

24. $\int_0^1 x^2 e^{x/2}\,dx$

25. Find the area between $y = (3 + x^2)e^{2x}$ and the x-axis from $x = 0$ to $x = 1$.

26. Find the area between $y = x^3(x^2 - 1)^{1/3}$ and the x-axis from $x = 1$ to $x = 3$.

Find the volume of the solid of revolution formed by rotating each of the following bounded regions about the x-axis.

27. $f(x) = 2x - 1$, $y = 0$, $x = 3$

28. $f(x) = \sqrt{x - 2}$, $y = 0$, $x = 11$

29. $f(x) = e^{-x}$, $y = 0$, $x = -2$, $x = 1$

30. $f(x) = \dfrac{1}{\sqrt{x - 1}}$, $y = 0$, $x = 2$, $x = 4$

31. $f(x) = 4 - x^2$, $y = 0$, $x = -1$, $x = 1$

32. $f(x) = \dfrac{x^2}{4}$, $y = 0$, $x = 4$

33. A frustum is what remains of a cone when the top is cut off by a plane parallel to the base. Suppose a right circular frustum (that is, one formed from a right circular cone) has a base with radius r, a top with radius $r/2$, and a height h. (See the figure.) Find the volume of this frustum by rotating about the x-axis the region below the line segment from $(0, r)$ to $(h, r/2)$.

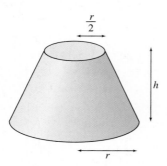

34. How is the average value of a function found?

35. Find the average value of $f(x) = \sqrt{x + 1}$ over the interval $[0, 8]$.

36. Find the average value of $f(x) = x^2(x^3 + 1)^5$ over the interval $[0, 1]$.

Find the value of each integral that converges.

37. $\int_1^{\infty} x^{-1}\,dx$

38. $\int_{-\infty}^{-2} x^{-2}\,dx$

39. $\int_0^{\infty} \dfrac{dx}{(5x + 2)^2}$

40. $\int_1^{\infty} 6e^{-x}\,dx$

41. $\int_{-\infty}^0 \dfrac{x}{x^2 + 3}\,dx$

42. $\int_{10}^{\infty} \ln(2x)\,dx$

Find the area between the graph of each function and the x-axis over the given interval, if possible.

43. $f(x) = \dfrac{3}{(x - 2)^2}$ for $(-\infty, 1]$

44. $f(x) = 3e^{-x}$ for $[0, \infty)$

Applications

LIFE SCIENCES

45. *Milk Production* Researchers report that the average amount of milk produced (in kg/day) by a 4- to 5-year-old cow weighing 700 kg can be approximated by

$$y = 1.87t^{1.49}e^{-0.189(\ln t)^2},$$

where t is the number of days into lactation.*

    a. Approximate the total amount of milk produced from $t = 1$ to $t = 321$ using the trapezoidal rule with $n = 8$.

    b. Repeat part a using Simpson's rule with $n = 8$.

    c. Repeat part a using the integration feature of a graphing calculator, and compare your answer with the answers to parts a and b.

46. *Complimentary Error Function* An article on how household chemicals are transported through septic systems used the complimentary error function

$$\text{erfc}(x) = \frac{2}{\sqrt{\pi}} \int_x^\infty e^{-t^2}\, dt.^\dagger$$

    a. To investigate the value of erfc(0), use Simpson's rule with $n = 20$ to evaluate

$$\frac{2}{\sqrt{\pi}} \int_x^k e^{-t^2}\, dt$$

    for $k = 2$, 3, 4, and 5. Based on these answers, what does erfc(0) seem to equal?

    b. The error function is defined as

$$\text{erf}(x) = \frac{2}{\sqrt{\pi}} \int_0^x e^{-t^2}\, dt.$$

    Show that $\text{erf}(x) = 1 - \text{erfc}(x)$.

47. *Drug Reaction* The reaction rate to a new drug x hours after the drug is administered is

$$r'(x) = 0.5xe^{-x}.$$

Find the total reaction over the first 5 hours.

48. *Oil Leak Pollution* An oil leak from an uncapped well is polluting a bay at a rate of $f(x) = 100e^{-0.05x}$ gallons per year. Use an improper integral to find the total amount of oil that will enter the bay, assuming the well is never capped.

OTHER APPLICATIONS

49. *Average Temperatures* Suppose the temperature in a river at a point x meters downstream from a factory that is discharging hot water into the river is given by

$$T(x) = 400 - 0.25x^2.$$

Find the average temperature over each of the following intervals.

    a. $[0, 10]$ **b.** $[10, 40]$ **c.** $[0, 40]$

50. *Total Revenue* The rate of change of revenue from the sale of x toaster ovens is

$$R'(x) = x(x - 50)^{1/2}.$$

Find the total revenue from the sale of the 50th to the 75th ovens.

---

*Freeze, Brian S., and Timothy J. Richards, "Lactation Curve Estimation for Use in Economic Optimization Models in the Dairy Industry," *Journal of Dairy Science,* Vol. 75, 1992, pp. 2984–2989.
†Lee, Sijin, et al., "Modeling the Fate and Transport of Household Chemicals in Septic Systems," *Ground Water,* Vol. 36, No. 1, Jan.–Feb. 1998, pp. 123–132.

EXTENDED APPLICATION: Flow Systems*

One of the most important biological applications of integration is the determination of properties of flow systems. Among the significant characteristics of a flow system are its flow rate (a heart's output in liters per minute, for example), its volume (for blood, the vascular volume), and the quantity of a substance flowing through it (a pollutant in a stream, perhaps).

A flow system can be anything from a river to an oil pipeline to an artery. We assume that the system has some well-defined volume between two points, and a more or less constant flow rate between them. In arteries the blood pulses, but we shall measure the flow over long enough intervals for the flow rate to be regarded as constant.

The standard method for analyzing flow system behavior is the so-called indicator dilution method, but we shall also discuss the more recent thermodilution technique for determining cardiac output. In the indicator dilution method, a known or unknown quantity of a substance such as a dye, a pollutant, or a radioactive tracer is injected or seeps into a flow system at some entry point. The substance is assumed to mix well with the fluid, and at some downstream measuring point its concentration is sampled either serially or continuously. The concentration function $c(t)$ is then integrated over an appropriate time interval by any of several methods: numerical approximation, a planimeter, electronic calculation, or by curve-fitting and the fundamental theorem of calculus. As we shall see, the integral

$$\int_0^T c(t)\, dt$$

(whose units are mass × time/volume) is extremely useful for finding unknown flow system properties.

The basis for the analysis of flow systems is the principle of mass conservation, or "what goes in must come out." We are going to assume that the indicator does not leak from the stream (for example, by diffusion into tissue), and that the time interval is over before any of the indicator recirculates (less than a minute in the case of blood). What goes *into* the system is a (known or unknown) quantity Q_{in} of indicator. What comes *out* of the system, or more precisely, what passes by the downstream measuring point at a flow rate F, is a volume of fluid with a variable concentration $c(t)$ of the indicator. The product of F and $c(t)$ is called the **mass rate** of the indicator flowing past the measuring point. The total quantity Q_{out} of indicator flowing by during the time interval from 0 to T is the integral of the mass rate from 0 to T:

$$Q_{out} = \int_0^T Fc(t)\, dt.$$

By our assumption of mass conservation (no leakage and no recirculation), we have $Q_{in} = Q_{out}$,

$$Q_{in} = \int_0^T Fc(t)\, dt = F\int_0^T c(t)\, dt, \qquad \text{(1)}$$

provided of course that F is essentially constant.

Equation (1) is the basic relationship of indicator dilution. We shall illustrate its use with an example.

Example 1

Suppose that an industrial plant is discharging toxic kryptonite into the Cahaba River at an unknown variable rate. Downriver, a field station of the Environmental Protection Agency measures the following concentrations:

Time	Concentration
2 A.M.	3.0 mg/m^3
6 A.M.	3.5 mg/m^3
10 A.M.	2.5 mg/m^3
2 P.M.	1.0 mg/m^3
6 P.M.	0.5 mg/m^3
10 P.M.	1.0 mg/m^3

The EPA has found the flow rate of the Cahaba in the vicinity to be 1,000 ft^3/sec = 1.02×10^5 m^3/hr. How much kryptonite is being discharged?

We use equation (1) to estimate the total amount of kryptonite discharged during a 24-hour period by approximating the concentration integral. Dividing the day into six 4-hour subintervals (so that $\Delta t = 4$) and using the EPA data as our function values $c(t)$, we compute as follows:

$$\int_0^{24} c(t)\, dt \approx \sum_{i=1}^6 c(t_1)\, \Delta t = \Delta t \sum_{i=1}^6 c(t_1)$$
$$= 4(3.0 + 3.5 + 2.5 + 1.0 + 0.5 + 1.0)$$
$$= 4(11.5) = 46,$$

or 46 mg × hrs/m^3. Hence, by equation (1), the total daily discharge is

$$Q_{in} = F\int_0^{24} c(t)\, dt$$
$$\approx \left(1.02 \times 10^5\, \frac{m^3}{hour}\right)\left(46\, \frac{mg \times hour}{m^3}\right)$$
$$= 4.7 \times 10^6\, mg = 4.7 \text{ kilograms of kryptonite.}$$

---

*From "Flow System Integrals," by Arthur C. Segal, *The UMAP Journal,* Vol. III, No. 1, 1982. Copyright © 1982 Educational Development Center, Inc. Reprinted by permission of COMAP, Inc.

Determining Cardiac Output by Thermodilution

Knowledge of a patient's cardiac output is a valuable aid to the diagnosis and treatment of heart damage. By cardiac output we mean the volume flow rate of venous blood pumped by the right ventricle through the pulmonary artery to the lungs (to be oxygenated). Normal resting cardiac output is 4 to 5 liters per minute, but in critically ill patients it may fall well below 3 liters per minute.

Unfortunately, cardiac output is difficult to determine by dye dilution because of the need to calibrate the dye and to sample the blood repeatedly. There are also problems with dye instability, recirculation, and slow dissipation. With the development in 1970 of a suitable pulmonary artery catheter, it became possible to make rapid and routine bedside measurements of cardiac output by thermodilution, and to do so with very few patient complications.

During a typical application in the coronary intensive care unit, a balloon-tipped catheter is inserted in an arm vein and guided to the superior vena cava or right atrium (the upper chamber of the heart). A temperature-sensing device called a thermistor is moved through the right side of the heart and then positioned a few centimeters into the pulmonary artery. Then, 10 ml of cold D5W (5% dextrose in water) is injected into the vena cava or atrium. As the cold D5W mixes with blood in the right heart, temperature variations in the flow are detected by the thermistor. Figure 16 shows a typical record of temperature change within a minute after injection. Since the cold D5W is warmed by the body before it recirculates, repeated measurements can be made reliably.

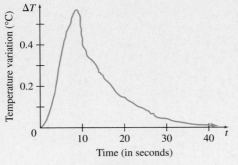

FIGURE 16

This curve shows the temperature variation in the pulmonary artery that resulted from injecting a few milliliters of cold dextrose solution into the right atrium of the heart.

We now derive an equation for thermodilution comparable to equation (1). First, we need to define the thermal equivalent of mass. We define the "quantity of cold" injected to be

$$Q_{in} = V_i(T_b - T_i),$$

where V_i is the volume of injectate, T_i is its temperature, and T_b is the body temperature. Thus, the "quantity of cold" injected is jointly proportional to how cold the injectate is (below body heat) and to how much is injected.

As with indicator dilution, we shall assume "conservation of cold," that is, no loss and no recirculation. (Actually, there is some loss of cold as the injectate travels up through the catheter to the injection point, but that can be corrected for.) The quantity of cold passing by the thermistor in a short time interval dt can be calculated by multiplying the flow volume $F\,dt$ and the temperature variation ΔT during the interval. The total quantity of cold flowing by is

$$Q_{out} = \int_0^\infty F\Delta T\, dt$$

$$= F\int_0^\infty \Delta T\, dt.$$

Equating Q_{in} to Q_{out} and solving for the flow rate F produces the analog of equation (1):

$$Q_{out} = Q_{in} = V_i(T_b - T_i)$$

$$F\int_0^\infty \Delta T\, dt = V_i(T_b - T_i)$$

$$F = \frac{V_i(T_b - T_i)}{\displaystyle\int_0^\infty \Delta T\, dt} \tag{2}$$

If time is measured in seconds and temperature in degrees Celsius, then the denominator of equation (2) has units of (°C)(sec) and the numerator has units of (ml) (°C). Hence, F has units ml/sec, which is converted to liters/minute by multiplying by $60/1,000 = 0.06$.

Although thermistor signals are usually calibrated and integrated electronically, we shall illustrate equation (2) with an example.

Example 2

Suppose $T_b = 37°C$, $T_i = 0°C$, and $V_i = 10$ ml. Suppose that the change in temperature with respect to time is represented by the curve $\Delta T = 0.1t^2 e^{-0.31t}$, shown in Figure 17 on the next page. Finally, suppose that the net effect of all the correction factors is to multiply the right side of equation (2) by 0.891. Then integration by parts gives

$$F = \frac{(10)(37 - 0)(0.891)}{\displaystyle\int_0^\infty (0.1)t^2 e^{-0.31t}\, dt}$$

$$= \frac{330}{(0.1)\dfrac{2}{(0.31)^3}}$$

$$= 49.16 \text{ ml/sec}$$

$$= 2.95 \text{ liters/minute.}$$

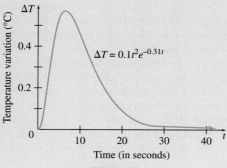

FIGURE 17

The illustration of equation (2) uses this smooth curve in place of the thermodilution curve in Figure 16.

Exercises

1. Approximate the integral in Example 1 using (a) the trapezoidal rule and (b) Simpson's rule. Let t represent the number of hours after 2 A.M., so that 2 A.M. $= 0$, 6 A.M. $= 4$, and so on. Assume the reading after 24 hours is 3.0 mg/m³ again.

2. A dye dilution technique for measuring cardiac output involves injection of a dye into a main vein near the heart and measurement by a catheter in the aorta at regular intervals. Suppose the results from such measurement are as given in the table.

Time (seconds)	Concentration (ml/liter)
0	0
4	0.6
8	2.7
12	4.1
16	2.9
20	0.9
24	0

If 8 ml of dye is injected, approximate the cardiac output F in the equation

$$Q_{in} = F \int_0^t c(t)\, dt$$

by (a) the trapezoidal rule and (b) Simpson's rule.

3. In Example 2 let $T_b = 37.1°C$, $T_i = 5°C$, and $v_i = 8$ ml. Find the cardiac output F.

Multivariable Calculus

Safe diving requires an understanding of how the increased pressure below the surface affects the body's intake of nitrogen. An exercise in Section 2 of this chapter investigates a formula for nitrogen pressure as a function of two variables, depth and dive time. Partial derivatives tell us how this function behaves when one variable is held constant as the other changes. Dive tables based on the formula help divers to choose a safe time for a given depth, or a safe depth for a given time.

- **9.1** Functions of Several Variables

- **9.2** Partial Derivatives

- **9.3** Maxima and Minima

- **9.4** Total Differentials and Approximations

- **9.5** Double Integrals

Review Exercises

Extended Application: Optimization for a Predator

Many of the ideas developed for functions of one variable also apply to functions of more than one variable. In particular, the fundamental idea of derivative generalizes in a very natural way to functions of more than one variable.

9.1 FUNCTIONS OF SEVERAL VARIABLES

? **THINK ABOUT IT**

How is the carbon dioxide released from the lungs related to the total output of blood from the heart?

We will investigate this question in Example 2 of this section. In this application, as with many life science applications, more than one independent variable is required to adequately describe the situation.

If a person takes x tablets of Vitamin C, for instance, then the total amount of the vitamin that is ingested is given by

$$C(x) = 500x,$$

where each tablet contains 500 mg of Vitamin C, and $C(x)$ is total amount (in mg). This is a function of one variable, the number of tablets taken. If the person also drinks y glasses of orange juice, with each glass of juice providing 100 mg of Vitamin C, then the total amount of Vitamin C ingested is a function of two independent variables, x and y. By generalizing $f(x)$ notation, the total amount of Vitamin C can be written as $C(x, y)$, where

$$C(x, y) = 500x + 100y.$$

When $x = 2$ and $y = 3$ the total amount of Vitamin C ingested is written $C(2, 3)$, with

$$C(2, 3) = 500 \cdot 2 + 100 \cdot 3 = 1{,}300.$$

A general definition follows.

FUNCTION OF TWO VARIABLES

$z = f(x, y)$ is a **function of two variables** if a unique value of z is obtained from each ordered pair of real numbers (x, y). The variables x and y are **independent variables,** and z is the **dependent variable.** The set of all ordered pairs of real numbers (x, y) such that $f(x, y)$ exists is the **domain** of f; the set of all values of $f(x, y)$ is the **range.** Similar definitions could be given for functions of three, four, or more independent variables.

EXAMPLE 1 Evaluating Functions
Let $f(x, y) = 4x^2 + 2xy + 3/y$ and find each of the following.

(a) $f(-1, 3)$

Solution Replace x with -1 and y with 3.

$$f(-1, 3) = 4(-1)^2 + 2(-1)(3) + \frac{3}{3} = 4 - 6 + 1 = -1$$

(b) $f(2,0)$

Solution Because of the quotient $3/y$, it is not possible to replace y with 0, so $f(2,0)$ is undefined. By inspection, we see that the domain of the function is the set of all (x, y) such that $y \neq 0$.

EXAMPLE 2 Carbon Dioxide

Let x represent the number of milliliters (ml) of carbon dioxide released by the lungs in one minute. Let y be the change in the carbon dioxide content of the blood as it leaves the lungs (y is measured in ml of carbon dioxide per 100 ml of blood). The total output of blood from the heart in one minute (measured in ml) is given by C, where C is a function of x and y such that

$$C = C(x, y) = \frac{100x}{y}.$$

Find $C(320, 6)$.

Solution Replace x with 320 and y with 6 to get

$$C(320, 6) = \frac{100(320)}{6}$$

$$\approx 5,333 \text{ ml of blood per minute.}$$

EXAMPLE 3 Evaluating Functions

Let $f(x, y, z) = 4xz - 3x^2y + 2z^2$. Find $f(2, -3, 1)$.

Solution Replace x with 2, y with -3, and z with 1.

$$f(2, -3, 1) = 4(2)(1) - 3(2)^2(-3) + 2(1)^2 = 8 + 36 + 2 = 46$$

Graphing Functions of Two Independent Variables

Functions of one independent variable are graphed by using an x-axis and a y-axis to locate points in a plane. The plane determined by the x- and y-axes is called the *xy-plane*. A third axis is needed to graph functions of two independent variables—the z-axis, which goes through the origin in the xy-plane and is perpendicular to both the x-axis and the y-axis.

Figure 1 shows one possible way to draw the three axes. In Figure 1, the yz-plane is in the plane of the page, with the x-axis perpendicular to the plane of the page.

FOR REVIEW

Graph the following lines. Refer to Section 1.1 if you need review.

1. $2x + 3y = 6$

2. $x = 4$

3. $y = 2$

Answers

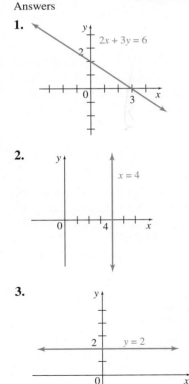

1.

2.

3.

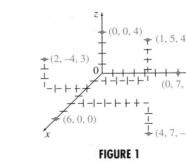

FIGURE 1

Just as we graphed ordered pairs earlier we can now graph **ordered triples** of the form (x, y, z). For example, to locate the point corresponding to the ordered triple $(2, -4, 3)$, start at the origin and go 2 units along the positive x-axis. Then go 4 units in a negative direction (to the left) parallel to the y-axis. Finally, go up 3 units parallel to the z-axis. The point representing $(2, -4, 3)$ is shown in Figure 1, together with several other points. The region of three-dimensional space where all coordinates are positive is called the **first octant.**

In Chapter 1 we saw that the graph of $ax + by = c$ (where a and b are not both 0) is a straight line. This result generalizes to three dimensions.

> **PLANE**
>
> The graph of
>
> $$ax + by + cz = d$$
>
> is a **plane** if a, b, and c are not all 0.

EXAMPLE 4 Graphing Planes

Graph $2x + y + z = 6$.

Solution The graph of this equation is a plane. Earlier, we graphed straight lines by finding x- and y-intercepts. A similar idea helps in graphing a plane. To find the x-intercept, which is the point where the graph crosses the x-axis, let $y = 0$ and $z = 0$.

$$2x + 0 + 0 = 6$$
$$x = 3$$

The point $(3, 0, 0)$ is on the graph. Letting $x = 0$ and $z = 0$ gives the point $(0, 6, 0)$, while $x = 0$ and $y = 0$ lead to $(0, 0, 6)$. The plane through these three points includes the triangular surface shown in Figure 2. This surface is the first-octant part of the plane that is the graph of $2x + y + z = 6$. The surface does not stop at the axes, but extends without bound.

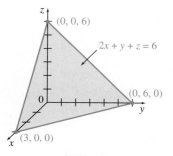

FIGURE 2

EXAMPLE 5 Graphing Planes

Graph $x + z = 6$.

Solution To find the x-intercept, let $z = 0$, giving $(6, 0, 0)$. If $x = 0$, we get the point $(0, 0, 6)$. Because there is no y in the equation $x + z = 6$, there can be no y-intercept. A plane that has no y-intercept is parallel to the y-axis. The first-octant portion of the graph of $x + z = 6$ is shown in Figure 3.

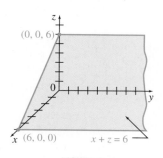

FIGURE 3

EXAMPLE 6 Graphing Planes

Graph each of the following functions in two variables.

(a) $x = 3$

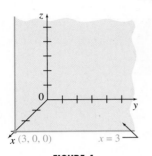

FIGURE 4

Solution This graph, which goes through $(3, 0, 0)$, can have no y-intercept and no z-intercept. It is therefore a plane parallel to the y-axis and the z-axis and, therefore, to the yz-plane. The first-octant portion of the graph is shown in Figure 4.

(b) $y = 4$

Solution This graph goes through $(0, 4, 0)$ and is parallel to the xz-plane. The first-octant portion of the graph is shown in Figure 5.

(c) $z = 1$

Solution The graph is a plane parallel to the xy-plane, passing through $(0, 0, 1)$. Its first-octant portion is shown in Figure 6.

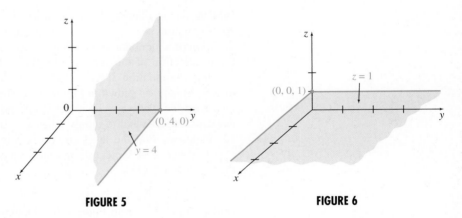

FIGURE 5   **FIGURE 6**

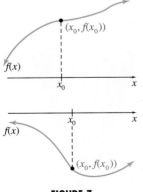

FIGURE 7

The graph of a function of one variable, $y = f(x)$, is a curve in the plane. If x_0 is in the domain of f, the point $(x_0, f(x_0))$ on the graph lies directly above or below the number x_0 on the x-axis, as shown in Figure 7.

The graph of a continuous function of two variables, $z = f(x, y)$, is a **surface** in three-dimensional space. If (x_0, y_0) is in the domain of f, the point $(x_0, y_0, f(x_0, y_0))$ lies directly above or below the point (x_0, y_0) in the xy-plane, as shown in Figure 8 on the next page.

Although computer software is available for drawing the graphs of functions of two independent variables, you can often get a good picture of the graph without it by finding various **traces**—the curves that result when a surface is cut by a plane. The **xy-trace** is the intersection of the surface with the xy-plane. The **yz-trace** and **xz-trace** are defined similarly. You can also determine the intersection of the surface with planes parallel to the xy-plane. Such planes are of the form $z = k$, where k is a constant, and the curves that result when they cut the surface are called **level curves.**

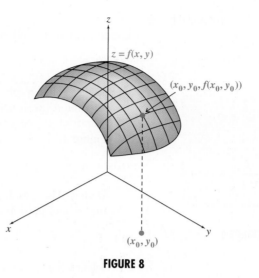

FIGURE 8

EXAMPLE 7 Graphing Functions

Graph $z = x^2 + y^2$.

Solution The yz-plane is the plane in which every point has a first coordinate of 0, so its equation is $x = 0$. When $x = 0$, the equation becomes $z = y^2$, which is the equation of a parabola in the yz-plane, as shown in Figure 9(a). Similarly, to find the intersection of the surface with the xz-plane (whose equation is $y = 0$), let $y = 0$ in the equation. It then becomes $z = x^2$, which is the equation of a parabola in the xz-plane, as shown in Figure 9(a). The xy-trace (the intersection of the surface with the plane $z = 0$) is the single point $(0, 0, 0)$ because $x^2 + y^2$ is never negative, and is equal to 0 only when $x = 0$ and $y = 0$.

Next, we find the level curves by intersecting the surface with the planes $z = 1$, $z = 2$, $z = 3$, etc. (all of which are parallel to the xy-plane). In each case, the result is a circle:

$$x^2 + y^2 = 1, \qquad x^2 + y^2 = 2, \qquad x^2 + y^2 = 3,$$

and so on, as shown in Figure 9(b). Drawing the traces and level curves on the same set of axes suggests that the graph of $z = x^2 + y^2$ is the bowl-shaped figure, called a **paraboloid,** that is shown in Figure 9(c).

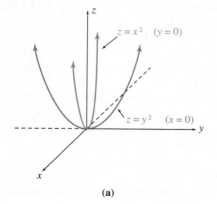

(a)

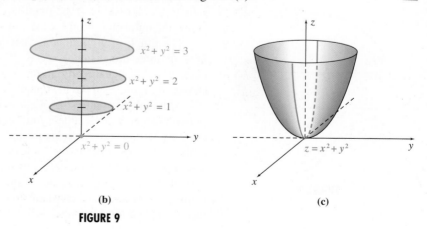

(b)   (c)

FIGURE 9

Figure 10 shows the level curves from Example 7 plotted in the xy-plane. The picture can be thought of as a topographical map which describes the surface generated by $z = x^2 + y^2$, just as the topographical map in Figure 11 describes the surface of the land in a part of New York state. The numbers in Figure 11 indicate the elevation of the land.

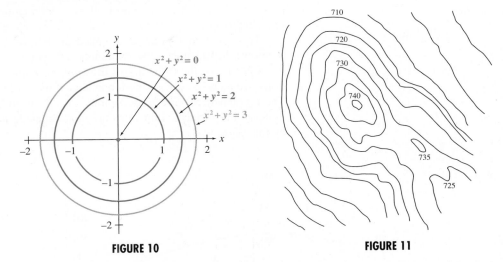

FIGURE 10　　　　　　　　　　　　　　**FIGURE 11**

Applications of level curves occur in economics with production functions. A **production function** $z = f(x, y)$ is a function that gives the quantity z of an item produced as a function of x and y, where x is the amount of labor and y is the amount of capital (in appropriate units) needed to produce z units. For production functions, level curves are used to indicate combinations of the values of x and y that produce the same value of production z. Similarly, level curves can be used in the life sciences where various combinations of the independent variables produce the same value of a function.

EXAMPLE 8　Polar Bears

Researchers in Alaska have developed models that can be used to estimate the mass of a polar bear. For bears that are caught in the spring of the year,

$$M(x, y) = 0.000199x^{1.2823}y^{1.4874},$$

where $M(x, y)$ is the mass of the bear (in kg), x is the straight-line body length (in cm), measured from the tip of the nose to the base of the tail, and y is the girth (in cm), measured directly behind the shoulders.* Find the level curve for a polar bear with a mass of 250 kg.

Solution　Let $M(x, y) = 250$ to get

$$250 = 0.000199x^{1.2823}y^{1.4874}$$

$$\frac{250}{0.000199x^{1.2823}} = y^{1.4874}$$

$$\frac{1{,}256{,}281.41}{x^{1.2823}} = y^{1.4874}.$$

---

*Durner, G., and S. Amstrup, "Mass and Body-Dimension Relationships of Polar Bears in Northern Alaska," *Wildlife Society Bulletin,* Vol. 24, No. 3, Fall 1996, pp. 480–484.

Now, raise both sides of the equation to the $1/1.4874$ power to express y as a function of x: $y = 12,603.82/x^{0.8622}$.

The level curve of mass 250 found in Example 8 is shown graphed in three dimensions in Figure 12(a) and on the familiar xy-plane in Figure 12(b). The points of the graph correspond to those combinations of body length and girth that lead to a mass of 250 kg.

The curve in Figure 12 is called an *isoquant,* for *iso* (equal) and *quant* (amount). In Example 8, the "amounts" all "equal" 250.

Because of the difficulty of drawing the graphs of more complicated functions, we merely list some common equations and their graphs on the next page. These graphs were drawn by computer, a very useful method of depicting three-dimensional surfaces. Equations are sometimes defined implicitly, as illustrated by the ellipsoid.

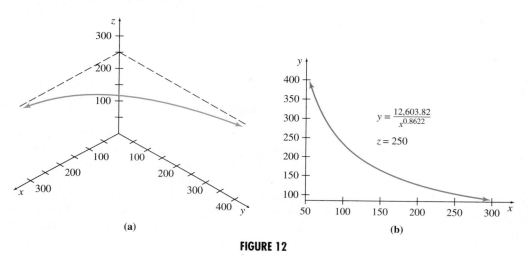

(a)   (b)

FIGURE 12

Instead of computers, we can use graphing calculators to draw the graphs of functions of two variables. Using a program entered into a TI-85, we generated Figure 13, the graph of $z = x^2 + y^2$. Although the graph took several minutes to generate, it is of lesser quality than the computer-generated graph. Figure 14 shows the same graph drawn by a TI-92, which generates graphs as fast as some computers. The graph still has less detail than what a computer can produce. Figure 15 shows the same graph drawn by the computer program Maple™.

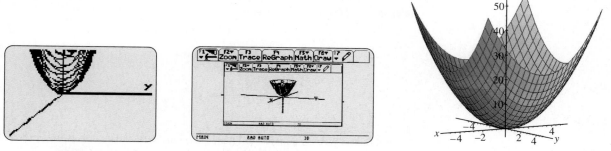

FIGURE 13   **FIGURE 14**   **FIGURE 15**

Paraboloid, $z = x^2 + y^2$

xy-trace: point
yz-trace: parabola
xz-trace: parabola

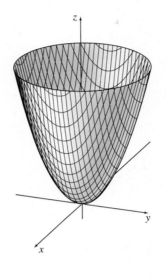

Ellipsoid, $\dfrac{x^2}{a^2} + \dfrac{y^2}{b^2} + \dfrac{z^2}{c^2} = 1$

xy-trace: ellipse
yz-trace: ellipse
xz-trace: ellipse

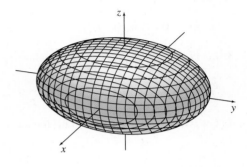

Hyperbolic Paraboloid,
$x^2 - y^2 = z$
(sometimes called a **saddle**)

xy-trace: two intersecting lines
yz-trace: parabola
xz-trace: parabola

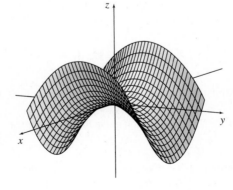

Hyperboloid of Two Sheets,
$-x^2 - y^2 + z^2 = 1$

xy-trace: none
yz-trace: hyperbola
xz-trace: hyperbola

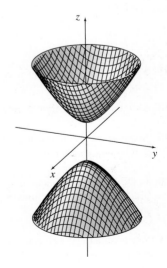

9.1 EXERCISES

1. Let $f(x, y) = 4x + 5y + 3$. Find the following.

    a. $f(2, -1)$ **b.** $f(-4, 1)$ **c.** $f(-2, -3)$ **d.** $f(0, 8)$

2. Let $g(x, y) = -x^2 - 4xy + y^3$. Find the following.

    a. $g(-2, 4)$ **b.** $g(-1, -2)$ **c.** $g(-2, 3)$ **d.** $g(5, 1)$

3. Let $h(x, y) = \sqrt{x^2 + 2y^2}$. Find the following.

    a. $h(5, 3)$ **b.** $h(2, 4)$ **c.** $h(-1, -3)$ **d.** $h(-3, -1)$

4. Let $f(x, y) = \dfrac{\sqrt{9x + 5y}}{\log x}$. Find the following.

    a. $f(10, 2)$ **b.** $f(100, 1)$ **c.** $f(1{,}000, 0)$ **d.** $f\left(\dfrac{1}{10}, 5\right)$

5. Let $f(x, y) = x \sin(x^2 y)$. Find the following.

    a. $f\left(1, \dfrac{\pi}{2}\right)$ **b.** $f\left(\dfrac{1}{2}, \pi\right)$ **c.** $f\left(\sqrt{\pi}, \dfrac{1}{2}\right)$ **d.** $f\left(-1, -\dfrac{\pi}{2}\right)$

Graph the first-octant portion of each plane.

6. $x + y + z = 6$ **7.** $x + y + z = 12$ **8.** $2x + 3y + 4z = 12$ **9.** $4x + 2y + 3z = 24$

10. $x + y = 4$ **11.** $y + z = 5$ **12.** $x = 2$ **13.** $z = 3$

Graph the level curves in the first octant at heights of $z = 0$, $z = 2$, and $z = 4$ for the following functions.

14. $3x + 2y + z = 18$ **15.** $x + 3y + 2z = 8$

16. $y^2 - x = -z$ **17.** $2y - \dfrac{x^2}{3} = z$

18. Discuss how a function of three variables in the form $w = f(x, y, z)$ might be graphed.

19. Suppose the graph of a plane $ax + by + cz = d$ has a portion in the first octant. What can be said about a, b, c, and d?

20. In the chapter on Functions, the vertical line test was presented, which tells whether a graph is the graph of a function. Does this test apply to functions of two variables? Explain.

21. A graph that was not shown in this section is the *hyperboloid of one sheet,* described by the equation $x^2 + y^2 - z^2 = 1$. Describe it as completely as you can.

Match each equation in Exercises 22–27 with its graph in a–f below.

22. $z = x^2 + y^2$

23. $z^2 - y^2 - x^2 = 1$

24. $x^2 - y^2 = z$

25. $z = y^2 - x^2$

26. $\dfrac{x^2}{16} + \dfrac{y^2}{25} + \dfrac{z^2}{4} = 1$

27. $z = 5(x^2 + y^2)^{-1/2}$

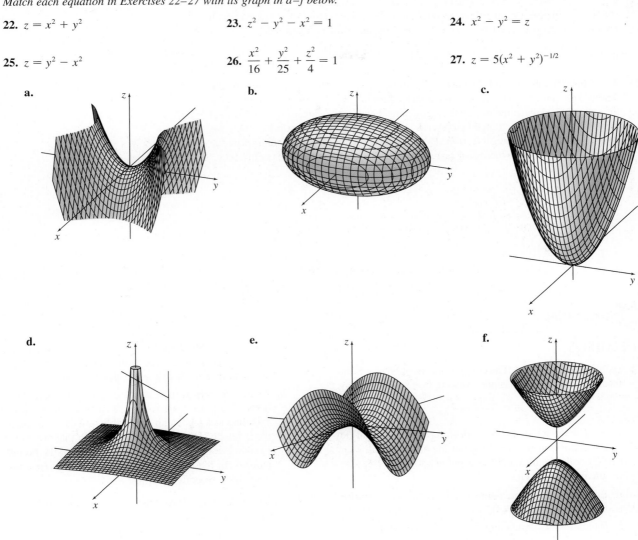

a.

b.

c.

d.

e.

f.

28. Let $f(x, y) = 9x^2 - 3y^2$, and find each of the following.

    a. $\dfrac{f(x + h, y) - f(x, y)}{h}$

    b. $\dfrac{f(x, y + h) - f(x, y)}{h}$

    c. $\lim\limits_{h \to 0} \dfrac{f(x + h, y) - f(x, y)}{h}$

    d. $\lim\limits_{h \to 0} \dfrac{f(x, y + h) - f(x, y)}{h}$

29. Let $f(x, y) = 7x^3 + 8y^2$, and find each of the following.

    a. $\dfrac{f(x + h, y) - f(x, y)}{h}$

    b. $\dfrac{f(x, y + h) - f(x, y)}{h}$

    c. $\lim\limits_{h \to 0} \dfrac{f(x + h, y) - f(x, y)}{h}$

    d. $\lim\limits_{h \to 0} \dfrac{f(x, y + h) - f(x, y)}{h}$

30. Let $f(x, y) = xye^{x^2+y^2}$. Use a graphing calculator or spreadsheet to find each of the following and give a geometric interpretation of the results. (*Hint:* First factor e^2 from the limit and then evaluate the quotient at smaller and smaller values of h.)

    a. $\lim\limits_{h \to 0} \dfrac{f(1 + h, 1) - f(1, 1)}{h}$ **b.** $\lim\limits_{h \to 0} \dfrac{f(1, 1 + h) - f(1, 1)}{h}$

31. The following table provides values of the function $f(x, y)$. However, because of potential errors in measurement, the functional values may be slightly inaccurate. Using the statistical package included with a graphing calculator or spreadsheet and critical thinking skills, find the function $f(x, y) = a + bx + cy$ that best estimates the table where a, b, and c are integers. (*Hint:* Do a linear regression on each column with the value of y fixed and use these four regression equations to determine the coefficient c.)

x \ y	0	1	2	3
0	4.02	7.04	9.98	13.00
1	6.01	9.06	11.98	14.96
2	7.99	10.95	14.02	17.09
3	9.99	13.01	16.01	19.02

Applications

LIFE SCIENCES

32. *Oxygen Consumption* The oxygen consumption of a well-insulated mammal that is not sweating is approximated by

$$m = \frac{2.5(T - F)}{w^{0.67}},$$

where T is the internal body temperature of the animal (in °C), F is the temperature of the outside of the animal's fur (in °C), and w is the animal's weight in kilograms.* Find m for the following data.

a. Internal body temperature = 38°C; outside temperature = 6°C; weight = 32 kg

b. Internal body temperature = 40°C; outside temperature = 20°C; weight = 43 kg

33. *Dinosaur Running* An article entitled "How Dinosaurs Ran" explains that the locomotion of different sized animals can be compared when they have the same Froude number, defined as

$$F = \frac{v^2}{gl},$$

where v is the velocity, g is the acceleration of gravity (9.81 m/sec²), and l is the leg length.[†]

a. One result described in the article is that different animals change from a trot to a gallop at the same Froude

number, roughly 2.56. Find the velocity at which this change occurs for a ferret, with a leg length of 0.09 m, and a rhinoceros, with a leg length of 1.2 m.

b. Ancient footprints in Texas of a sauropod, a large herbivorous dinosaur, are roughly 1 m in diameter, corresponding to a leg length of roughly 4 m. By comparing the stride divided by the leg length with that of various modern creatures, it can be determined that the Froude number for these dinosaurs is roughly 0.025. How fast were the sauropods traveling?

34. *Body Surface Area* The surface area of a human (in square meters) is approximated by

$$S = 0.202W^{0.425}H^{0.725},$$

where W is the weight of the person in kilograms and H is the height in meters.* Find A for the following data.

a. Weight, 72 kg; height, 1.78 m

b. Weight, 65 kg; height, 1.40 m

c. Weight, 70 kg; height, 1.60 m

d. Using your weight and height, find your own surface area. (*Hint:* 1 in = 0.0254 m and 1 lb = 0.4555 kg[‡])

e. Suppose H is the height in centimeters. Verify that the formula, $A = 0.007184W^{0.425}H^{0.725}$, gives the same results in parts a and b.

---

*Exercises 32 and 34 from Clow, Duane J., and N. Scott Urquhart, *Mathematics in Biology.* Copyright © 1974 by W. W. Norton & Company, Inc. Used by permission.
†Alexander, R. McNeill, "How Dinosaurs Ran," *Scientific American,* Vol. 264, April 1991, p. 4.
‡Technically, this equation does not make sense, because pounds are a measure of weight and kg are a measure of mass, but as we pointed out earlier, weight and mass are often used interchangeably.

35. *Cholesterol* Researchers have developed a mathematical relationship between the concentrations of low-density lipoprotein cholesterol (LDL-C), high-density lipoprotein cholesterol (HDL-C), triglyceride (TG), and fasting total cholesterol (TC) as follows:

$$LDL\text{-}C = TC - (HDL\text{-}C + 0.16TG),$$

where each measurement is in milligrams per liter (mg/l).*

a. Express this mathematical relationship using the functional notation, $L(F, H, T)$, where L is the low-density concentration, F is the total fasting concentration, H is the high-density concentration, and T is the triglyceride concentration.

b. Find $L(2{,}000, 1{,}300, 2{,}500)$.

36. *Total Body Water* Accurate prediction of total body water is critical in determining adequate dialysis doses for patients with renal disease. For white males, total body water can be estimated by the function

$$T(A, W, B) = 23.04 - 0.03A + 0.50W - 0.62B,$$

where T is the total body water (in liters), A is age (in yr), W is weight (in kg), and B is the body mass index.[†] Find $T(55, 75, 30)$.

37. *Snake River Salmon* Researchers have determined that the date D (day of the year) in which a Chinook salmon passes the Lower Granite Dam is related to the day of release R (day of the year) of the tagged salmon, the length of the salmon L (in mm), and the mean daily flow of the water F (in m³/s) such that

$$\ln(D) = 4.890179 + 0.005163R - 0.004345L$$
$$- 0.000019F.[‡]$$

a. Solve this expression for D.

b. Estimate the date of passage for a 75-mm salmon that was released on day 150 with an average flow rate of 3,500 m³/s. What day of the year would this be, assuming that the year is not a leap year?

38. *Agriculture* Pregnant sows which are tethered in stalls often show high levels of repetitive behavior, such as bar biting and chain chewing, indicating chronic stress. Researchers from Great Britain have developed a function that estimates the relationship between repetitive behavior, the behavior of sows in adjacent stalls, and food allowances such that

$$\ln(T) = 5.49 - 3.00 \ln(F) + 0.18 \ln(C),$$

where T is the percent of time spent in repetitive behavior, F is the amount of food given to the sow (in kg/day), and C is the percent of time that neighboring sows spent bar biting and chain chewing.[§]

a. Solve the above expression for T.

b. Find and interpret T when $F = 2$ and $C = 40$.

39. *Optimal Foraging* Researchers have developed a function that is used to determine optimal foraging strategies for animals. The function measures the profitability of an animal searching for certain types of prey and is given by

$$P = \frac{\lambda_1 E_1 + \lambda_2 E_2 + \cdots + \lambda_k E_k}{1 + \lambda_1 h_1 + \lambda_2 h_2 + \cdots + \lambda_k h_k},$$

where E_i is the caloric value (in kcal), λ_i is the number of prey encountered per minute, and h_i is the handling time (in min) for the ith prey type. A hypothetical blue jay may eat worms, moths, and grubs. The values of the variables are given in the table below.[‖]

Item	E	λ	h
Worm	162	0.2	3.6
Moth	24	3.0	0.6
Grub	40	3.0	1.6

a. Find the profitability P for a blue jay to eat worms only.

b. Find the profitability for a blue jay to eat worms and moths.

c. What is the combination of prey that provides maximum profitability for the blue jay?

40. *Deer–Vehicle Accidents* Using data collected by the U.S. Forest Service, the annual number of deer–vehicle accidents for any given county in Ohio can be estimated by the function

$$A(L, T, U, C) = 53.02 + 0.383L + 0.0015T + 0.0028U$$
$$- 0.0003C,$$

*DeLong, D., E. DeLong, P. Wood, K. Lippel, B. Rifkind, "A Comparison of Methods for the Estimation of Plasma Low- and Very Low-Density Lipoprotein Cholesterol," *JAMA,* Vol. 256, No. 17, Nov. 7, 1986, pp. 2372–2377.

[†]Chumlea, W., S. Guo, C. Zellar, et al., "Total Body Water Reference Values and Prediction Equations for Adults," *Kidney International,* Vol. 59, 2001, pp. 2250–2258.

[‡]Conner, W., R. Steinhorst, and H. Burge, "Forecasting Survival and Passage of Migratory Juvenile Salmonids," *North American Journal of Fisheries Management,* Vol. 20, 2000, pp. 651–660.

[§]Appleby, M., A. Lawrence, and A. Illius, "Influence of Neighbours on Stereotypic Behaviour of Tethered Sows," *Applied Animal Behaviour Science,* Vol. 24, 1989, pp. 137–146.

[‖]Kolmes, S., K. Mitchell, and J. Ryan, "A Simple Model of Foraging," *The UMAP Journal,* Vol. 18.1, 1997, pp. 6–11.

where A is the estimated number of accidents, L is the road length in kilometers, T is the total county land area in hundred-acres (Ha), U is the urban land area in hundred-acres, and C is the number of hundred-acres of crop land.*

a. Use this formula to estimate the number of deer–vehicle accidents for Mahoning County, where $L = 266$ km, $T = 107{,}484$ Ha, $U = 31{,}697$ Ha, and $C = 24{,}870$ Ha. The actual value was 396.

b. Given the magnitude and nature of the input numbers, which of the variables have the greatest potential to influence the number of deer–vehicle accidents? Explain your answer.

41. *Deer Harvest* Using data collected by the U.S. Forest Service, the annual number of deer that are harvested for any given county in Ohio can be estimated by the function

$$N(R, C) = 329.32 + 0.0377R - 0.0171C,$$

where N is the estimated number of harvested deer, R is the rural land area in hundred acres, and C is the number of acres (in Ha) of crop land.*

a. Use this formula to estimate the number of harvested deer for Tuscarawas County, where $R = 141{,}319$ Ha and $C = 37{,}960$ Ha. The actual value in 1995 was 4,925 deer harvested.

b. Sketch the graph of this function in the first octant.

OTHER APPLICATIONS

42. *Postage Rates* Extra postage is charged for parcels sent by U.S. mail that are more than 84 inches in length and girth combined. (Girth is the distance around the parcel perpendicular to its length. See the figure in the next column.) Express the combined length and girth as a function of L, W, and H.

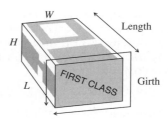

43. *Required Material* Refer to the figure for Exercise 42. Assume L, W, and H are in feet. Write a function in terms of L, W, and H that gives the total material required to build the box.

44. *Elliptical Templates* The holes cut in a roof for vent pipes require elliptical templates. A formula for determining the length of the major axis of the ellipse is $L = f(H, D) = \sqrt{H^2 + D^2}$, where D is the (outside) diameter of the pipe and H is the "rise" of the roof per D units of "run"; that is, the slope of the roof is H/D. (See the figure.) The width of the ellipse (minor axis) equals D. Find the length and width of the ellipse required to produce a hole for a vent pipe with a diameter of 3.75 inches in roofs with the following slopes.

a. 3/4   **b.** 2/5

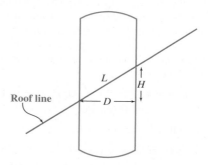

45. *Production* Production of a precision camera is given by

$$P(x, y) = 100\left(\frac{3}{5}x^{-2/5} + \frac{2}{5}y^{-2/5}\right)^{-5},$$

where x is the amount of labor in work-hours and y is the amount of capital. Find the following.

a. What is the production when 32 work-hours and 1 unit of capital are provided?

b. Find the production when 1 work-hour and 32 units of capital are provided.

c. If 32 work-hours and 243 units of capital are used, what is the production output?

---

*Iverson, Aaron, and Louis Iverson, "Spatial and Temporal Trends of Deer Harvest and Deer–Vehicle Accidents in Ohio," *The Ohio Journal of Science*, 99, 1999, pp. 84–94.

Individual Retirement Accounts The multiplier function

$$M = \frac{(1 + i)^n(1 - t) + t}{[1 + (1 - t)i]^n}$$

compares the growth of an Individual Retirement Account (IRA) with the growth of the same deposit in a regular savings account. The function M depends on the three variables n, i, and t, where n represents the number of years an amount is left at interest, i represents the interest rate in both types of accounts, and t represents the income tax rate. Values of

$M > 1$ indicate that the IRA grows faster than t account. Let $M = f(n, i, t)$ and find the followin

46. Find the multiplier when funds are left for 25 years at 5% interest and the income tax rate is 28%. Which account grows faster?

47. What is the multiplier when money is invested for 25 years at 6% interest and the income tax rate is 33%? Which account grows faster?

9.2 PARTIAL DERIVATIVES

? THINK ABOUT IT   What is the relationship between the water pressure on body tissues for various water depths and dive times?

FOR REVIEW ■
You may want to review the chapter on Calculating the Derivative for methods used to find some of the derivatives in this section.

In Chapter 3, we found that the derivative dy/dx gives the rate of change of y with respect to x. In this section, we show how derivatives are found and interpreted for multivariable functions, and we will use that information to answer the question posed above.

The surface area in square meters of a human is dependent upon both height and weight and is given by

$$S(x, y) = 0.007184x^{0.425}y^{0.725},$$

where x is the weight of the person (in kg) and y is the height in centimeters.* How will a change in x or y affect S?

Suppose that the height of famous baseball player Mark McGuire, who is 196-cm tall, is fairly constant but his weight tends to vary. His trainer would like to find the change in surface area with respect to changes in weight, x. Here, y is fixed at 196. Using this information, we begin by finding a new function, $f(x) = S(x, 196)$. Let $y = 196$ to get

$$f(x) = S(x, 196) = 0.007184x^{0.425}(196)^{0.725} = 0.3298x^{0.425}.$$

The function $f(x)$ shows the surface area for a person who weighs x kg, assuming that y is fixed at 196 cm. Find the derivative $f'(x)$ to get the marginal value with respect to x.

$$f'(x) = (0.3298)(0.425)x^{-0.575}$$

In this example, the derivative of the function $f(x)$ was taken with respect to x only; we assumed that y was fixed. To generalize, let $z = f(x, y)$. An intuitive definition of the partial derivatives of f with respect to x and y follows.

---

*Hurley, P., "Red cell and plasma volumes in normal adults," *Journal of Nuclear Medicine,* Vol. 16, 1975, pp. 46–52.

> **PARTIAL DERIVATIVES (INFORMAL DEFINITION)**
>
> The **partial derivative of f with respect to x** is the derivative of f obtained by treating x as a variable and y as a constant.
>
> The **partial derivative of f with respect to y** is the derivative of f obtained by treating y as a variable and x as a constant.

The symbols $f_x(x, y)$ (no prime is used), $\partial z/\partial x$, z_x, and $\partial f/\partial x$ are used to represent the partial derivative of $z = f(x, y)$ with respect to x, with similar symbols used for the partial derivative with respect to y.

Generalizing from the definition of the derivative given earlier, partial derivatives of a function $z = f(x, y)$ are formally defined as follows.

> **PARTIAL DERIVATIVES (FORMAL DEFINITION)**
>
> Let $z = f(x, y)$ be a function of two independent variables. Let all indicated limits exist. Then the partial derivative of f with respect to x is
>
> $$f_x(x, y) = \frac{\partial f(x, y)}{\partial x} = \lim_{h \to 0} \frac{f(x + h, y) - f(x, y)}{h},$$
>
> and the partial derivative of f with respect to y is
>
> $$f_y(x, y) = \frac{\partial f(x, y)}{\partial y} = \lim_{h \to 0} \frac{f(x, y + h) - f(x, y)}{h}.$$
>
> If the indicated limits do not exist, then the partial derivatives do not exist.

Similar definitions could be given for functions of more than two independent variables.

EXAMPLE 1 Partial Derivatives

Let $f(x, y) = 4x^2 - 9xy + 6y^3$. Find $f_x(x, y)$ and $f_y(x, y)$.

Solution To find $f_x(x, y)$, treat y as a constant and x as a variable. The derivative of the first term, $4x^2$, is $8x$. In the second term, $-9xy$, the constant coefficient of x is $-9y$, so the derivative with x as the variable is $-9y$. The derivative of $6y^3$ is zero, since we are treating y as a constant. Thus,

$$f_x(x, y) = 8x - 9y.$$

Now, to find $f_y(x, y)$, treat y as a variable and x as a constant. Since x is a constant, the derivative of $4x^2$ is zero. In the second term, the coefficient of y is $-9x$ and the derivative of $-9xy$ is $-9x$. The derivative of the third term is $18y^2$. Thus,

$$f_y(x, y) = -9x + 18y^2.$$

The next example shows how the chain rule can be used to find partial derivatives.

EXAMPLE 2 Partial Derivatives

Let $f(x, y) = \ln|x^2 + y|$. Find $f_x(x, y)$ and $f_y(x, y)$.

Solution Recall the formula for the derivative of a natural logarithm function. If $g(x) = \ln|x|$, then $g'(x) = 1/x$. Using this formula and the chain rule,

$$f_x(x,y) = \frac{1}{x^2+y} \cdot D_x(x^2+y) = \frac{1}{x^2+y} \cdot 2x = \frac{2x}{x^2+y},$$

and

$$f_y(x,y) = \frac{1}{x^2+y} \cdot D_y(x^2+y) = \frac{1}{x^2+y} \cdot 1 = \frac{1}{x^2+y}.$$

The notation

$$f_x(a,b) \qquad \text{or} \qquad \frac{\partial f}{\partial x}(a,b)$$

represents the value of a partial derivative when $x = a$ and $y = b$, as shown in the next example.

EXAMPLE 3 Evaluating Partial Derivatives

Let $f(x,y) = 2x^2 + 3xy^3 + 2y + 5$. Find the following.

(a) $f_x(-1,2)$

Solution First, find $f_x(x,y)$ by holding y constant.

$$f_x(x,y) = 4x + 3y^3$$

Now let $x = -1$ and $y = 2$.

$$f_x(-1,2) = 4(-1) + 3(2)^3 = -4 + 24 = 20$$

(b) $\dfrac{\partial f}{\partial y}(-4,-3)$

Solution Since $\partial f/\partial y = 9xy^2 + 2$,

$$\frac{\partial f}{\partial y}(-4,-3) = 9(-4)(-3)^2 + 2 = 9(-36) + 2 = -322.$$

The partial derivative can also be approximated using a small value of h in the difference quotient $(f(x+h,y) - f(x,y))/h$ or $(f(x,y+h) - f(x,y))/h$, as we did with functions of single variables. For an application of this, see Exercise 55. To find the derivatives in Example 3(a) numerically, we would calculate

$$\frac{f(-1+h,2) - f(-1,2)}{h} = \frac{[2(-1+h)^2 + 3(-1+h)(2^3) + 2(2) + 5] - [2(-1)^2 + 3(-1)(2^3) + 2(2) + 5]}{h}$$

using small values of h. With $h = 10^{-5}$ and 10^{-6}, the results are 20.00002 and 20.000002, respectively. In this example, the difference quotient above can be reduced algebraically to $20 + 2h$, from which the limit is easily seen to be 20. Review parts (c) and (d) of Exercises 28 and 29 of the previous section for examples in which the partial derivative can be computed using the formal definition.

The derivative of a function of one variable can be interpreted as the tangent line to the graph at that point. With some modification, the same is true of partial derivatives of functions of two variables. At a point on the graph of a function of two variables, $z = f(x,y)$, there may be many tangent lines, all of which lie in the same tangent plane, as shown in Figure 16 on the next page.

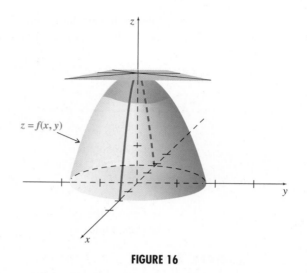

FIGURE 16

In any particular direction, however, there will be only one tangent line. We use partial derivatives to find the slope of the tangent lines in the x- and y-directions as follows.

Figure 17 shows a surface $z = f(x, y)$ and a plane that is parallel to the xz-plane. The equation of the plane is $y = a$. (This corresponds to holding y fixed.) Since $y = a$ for points on the plane, any point on the curve that represents the intersection of the plane and the surface must have the form $(x, y, z) = (x, a, f(x, a))$. Thus, this curve can be described as $z = f(x, a)$. Since a is constant, $z = f(x, a)$ is a function of one variable. When the derivative of $z = f(x, a)$ is evaluated at $x = b$, it gives the slope of the line tangent to this curve at the point $(b, a, f(b, a))$, as shown in Figure 17. Thus, the partial derivative of f with respect to x, $f_x(b, a)$, gives the rate of change of the surface $z = f(x, y)$ in the x-direction at the point $(b, a, f(b, a))$. In the same way, the partial derivative with respect to y will give the slope of the line tangent to the surface in the y-direction at the point $(b, a, f(b, a))$.

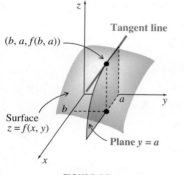

FIGURE 17

Rate of Change   The derivative of $y = f(x)$ gives the rate of change of y with respect to x. In the same way, if $z = f(x, y)$, then $f_x(x, y)$ gives the rate of change of z with respect to x, if y is held constant.

EXAMPLE 4   Water Temperature

Suppose that the temperature of the water at the point on a river where a nuclear power plant discharges its hot waste water is approximated by

$$T(x, y) = 2x + 5y + xy - 40,$$

where x represents the temperature of the river water (in degrees Celsius) before it reaches the power plant and y is the number of megawatts (in hundreds) of electricity being produced by the plant.

(a) Find and interpret $T_x(9, 5)$.

Solution   First, find the partial derivative $T_x(x, y)$.

$$T_x(x, y) = 2 + y$$

This partial derivative gives the rate of change of T with respect to x. Replacing x with 9 and y with 5 gives

$$T_x(9,5) = 2 + 5 = 7.$$

Just as marginal cost is the approximate cost of one more item, this result, 7, is the approximate change in temperature of the output water if input water temperature changes by 1 degree, from $x = 9$ to $x = 9 + 1 = 10$, while y remains constant at 5 (500 megawatts of electricity produced).

(b) Find and interpret $T_y(9,5)$.

Solution The partial derivative $T_y(x, y)$ is

$$T_y(x, y) = 5 + x.$$

This partial derivative gives the rate of change of T with respect to y as

$$T_y(9, 5) = 5 + 9 = 14.$$

This result, 14, is the approximate change in temperature resulting from a 1-unit increase in production of electricity from $y = 5$ to $y = 5 + 1 = 6$ (from 500 to 600 megawatts), while the input water temperature x remains constant at 9°C.

Second-Order Partial Derivatives

The second derivative of a function of one variable is very useful in determining relative maxima and minima. **Second-order partial derivatives** (partial derivatives of a partial derivative) are used in a similar way for functions of two or more variables. The situation is somewhat more complicated, however, with more independent variables. For example, $f(x, y) = 4x + x^2y + 2y$ has two first-order partial derivatives,

$$f_x(x, y) = 4 + 2xy \qquad \text{and} \qquad f_y(x, y) = x^2 + 2.$$

Since each of these has two partial derivatives, one with respect to y and one with respect to x, there are *four* second-order partial derivatives of function f. The notations for these four second-order partial derivatives are given below.

SECOND-ORDER PARTIAL DERIVATIVES

For a function $z = f(x, y)$, if the indicated partial derivative exists, then

$$\frac{\partial}{\partial x}\left(\frac{\partial z}{\partial x}\right) = \frac{\partial^2 z}{\partial x^2} = f_{xx}(x, y) = z_{xx} \qquad \frac{\partial}{\partial y}\left(\frac{\partial z}{\partial y}\right) = \frac{\partial^2 z}{\partial y^2} = f_{yy}(x, y) = z_{yy}$$

$$\frac{\partial}{\partial y}\left(\frac{\partial z}{\partial x}\right) = \frac{\partial^2 z}{\partial y \partial x} = f_{xy}(x, y) = z_{xy} \qquad \frac{\partial}{\partial x}\left(\frac{\partial z}{\partial y}\right) = \frac{\partial^2 z}{\partial x \partial y} = f_{yx}(x, y) = z_{yx}$$

NOTE For most functions found in applications and for all of the functions in this book, the second-order partial derivatives $f_{xy}(x, y)$ and $f_{yx}(x, y)$ are equal. Therefore, it is not necessary to be particular about the order in which these derivatives are found.

EXAMPLE 5 Second-Order Partial Derivatives

Find all second-order partial derivatives for

$$f(x, y) = -4x^3 - 3x^2y^3 + 2y^2.$$

Solution First find $f_x(x, y)$ and $f_y(x, y)$.

$$f_x(x, y) = -12x^2 - 6xy^3 \qquad \text{and} \qquad f_y(x, y) = -9x^2y^2 + 4y$$

To find $f_{xx}(x, y)$, take the partial derivative of $f_x(x, y)$ with respect to x.

$$f_{xx}(x, y) = -24x - 6y^3$$

Take the partial derivative of $f_y(x, y)$ with respect to y; this gives f_{yy}.

$$f_{yy}(x, y) = -18x^2y + 4$$

Find $f_{xy}(x, y)$ by starting with $f_x(x, y)$, then taking the partial derivative of $f_x(x, y)$ with respect to y.

$$f_{xy}(x, y) = -18xy^2$$

Finally, find $f_{yx}(x, y)$ by starting with $f_y(x, y)$; take its partial derivative with respect to x.

$$f_{yx}(x, y) = -18xy^2$$

EXAMPLE 6 Second-Order Partial Derivatives

Let $f(x, y) = 2e^x - 8x^3y^2$. Find all second-order partial derivatives.

Solution Here $f_x(x, y) = 2e^x - 24x^2y^2$ and $f_y(x, y) = -16x^3y$. [Recall: If $g(x) = e^x$, then $g'(x) = e^x$.] Now find the second-order partial derivatives.

$$f_{xx}(x, y) = 2e^x - 48xy^2 \qquad f_{xy}(x, y) = -48x^2y$$
$$f_{yy}(x, y) = -16x^3 \qquad f_{yx}(x, y) = -48x^2y$$

Partial derivatives of functions with more than two independent variables are found in a similar manner. For instance, to find $f_z(x, y, z)$ for $f(x, y, z)$, hold x and y constant and differentiate with respect to z.

EXAMPLE 7 Second-Order Partial Derivatives

Let $f(x, y, z) = 2x^2yz^2 + 3xy^2 - 4yz$. Find $f_x(x, y, z)$, $f_y(x, y, z)$, $f_{xz}(x, y, z)$, and $f_{yz}(x, y, z)$.

Solution

$$f_x(x, y, z) = 4xyz^2 + 3y^2$$
$$f_y(x, y, z) = 2x^2z^2 + 6xy - 4z$$

To find $f_{xz}(x, y, z)$, differentiate $f_x(x, y, z)$ with respect to z.

$$f_{xz}(x, y, z) = 8xyz$$

Differentiate $f_y(x, y, z)$ with respect to z to get $f_{yz}(x, y, z)$.

$$f_{yz}(x, y, z) = 4x^2z - 4$$

9.2 EXERCISES

1. Let $z = f(x, y) = 12x^2 - 8xy + 3y^2$. Find each of the following using the formal definition of the partial derivative.

   a. $\dfrac{\partial z}{\partial x}$ **b.** $\dfrac{\partial z}{\partial y}$ **c.** $\dfrac{\partial f}{\partial x}(2, 3)$ **d.** $f_y(1, -2)$

2. Let $z = g(x, y) = 5x + 9x^2y + y^2$. Find each of the following using the formal definition of the partial derivative.

   a. $\dfrac{\partial g}{\partial x}$ **b.** $\dfrac{\partial g}{\partial y}$ **c.** $\dfrac{\partial z}{\partial y}(-3, 0)$ **d.** $g_x(2, 1)$

In Exercises 3–22, find $f_x(x, y)$ and $f_y(x, y)$. Then find $f_x(2, -1)$ and $f_y(-4, 3)$. Leave the answers in terms of e in Exercises 7–10, 15–16, and 19–20.

3. $f(x, y) = -2xy + 6y^3 + 2$

4. $f(x, y) = 4x^2y - 9y^2$

5. $f(x, y) = 3x^3y^2$

6. $f(x, y) = -2x^2y^4$

7. $f(x, y) = e^{x+y}$

8. $f(x, y) = 3e^{2x+y}$

9. $f(x, y) = -5e^{3x-4y}$

10. $f(x, y) = 8e^{7x-y}$

11. $f(x, y) = \dfrac{x^2 + y^3}{x^3 - y^2}$

12. $f(x, y) = \dfrac{3x^2y^3}{x^2 + y^2}$

13. $f(x, y) = \ln|1 + 3x^2y^3|$

14. $f(x, y) = \ln|2x^5 - xy^4|$

15. $f(x, y) = xe^{x^2y}$

16. $f(x, y) = y^2e^{x+3y}$

17. $f(x, y) = \sqrt{x^4 + 3xy + y^4 + 10}$

18. $f(x, y) = (7x^2 + 18xy^2 + y^3)^{1/3}$

19. $f(x, y) = \dfrac{3x^2y}{e^{xy} + 2}$

20. $f(x, y) = (7e^{x+2y} + 4)(e^{x^2} + y^2 + 2)$

21. $f(x, y) = x \sin(\pi y)$

22. $f(x, y) = x \tan(\pi y^2)$

Find all second-order partial derivatives for the following.

23. $f(x, y) = 6x^3y - 9y^2 + 2x$

24. $g(x, y) = 5xy^4 + 8x^3 - 3y$

25. $R(x, y) = 4x^2 - 5xy^3 + 12y^2x^2$

26. $h(x, y) = 30y + 5x^2y + 12xy^2$

27. $r(x, y) = \dfrac{4x}{x + y}$

28. $k(x, y) = \dfrac{-5y}{x + 2y}$

29. $z = 4xe^y$

30. $z = -3ye^x$

31. $r = \ln|x + y|$

32. $k = \ln|5x - 7y|$

33. $z = x \ln|xy|$

34. $z = (y + 1) \ln|x^3y|$

For the functions defined as follows, find values of x and y such that both $f_x(x, y) = 0$ and $f_y(x, y) = 0$.

35. $f(x, y) = 6x^2 + 6y^2 + 6xy + 36x - 5$

36. $f(x, y) = 50 + 4x - 5y + x^2 + y^2 + xy$

37. $f(x, y) = 9xy - x^3 - y^3 - 6$

38. $f(x, y) = 2{,}200 + 27x^3 + 72xy + 8y^2$

Find $f_x(x, y, z)$, $f_y(x, y, z)$, $f_z(x, y, z)$, and $f_{yz}(x, y, z)$ for the following.

39. $f(x, y, z) = x^2 + yz + z^4$

40. $f(x, y, z) = 3x^5 - x^2 + y^5$

41. $f(x, y, z) = \dfrac{6x - 5y}{4z + 5}$

42. $f(x, y, z) = \dfrac{2x^2 + xy}{yz - 2}$

43. $f(x, y, z) = \ln|x^2 - 5xz^2 + y^4|$

44. $f(x, y, z) = \ln|8xy + 5yz - x^3|$

In Exercises 45 and 46, approximate the indicated derivative for each function by using the definition of the derivative with small values of h.

45. $f(x, y) = (x + y/2)^{x+y/2}$

   a. $f_x(1, 2)$ **b.** $f_y(1, 2)$

46. $f(x, y) = (x + y^2)^{2x+y}$

   a. $f_x(2, 1)$ **b.** $f_y(2, 1)$

Applications

LIFE SCIENCES

47. *Grasshopper Matings* The total number of matings per day between individuals of a certain species of grasshopper is approximated by

$$M(x, y) = 2xy + 10xy^2 + 30y^2 + 20,$$

where x represents the temperature in °C and y represents the number of days since the last rainfall. Find the following.

a. The approximate change in matings when the temperature increases from 20°C to 21°C and the number of days since rain is constant at 4.

b. The approximate change in matings when the temperature is constant at 24°C and the number of days since rain is changed from 10 to 11.

c. Would an increase in temperature or an increase in days since rain cause more of an increase in matings?

48. *Oxygen Consumption* As we saw in the previous section, the oxygen consumption of a well-insulated mammal that is not sweating is approximated by

$$m = m(T, F, w) = \frac{2.5(T - F)}{w^{0.67}} = 2.5(T - F)w^{-0.67},$$

where T is the internal body temperature of the animal (in °C), F is the temperature of the outside of the animal's fur (in °C), and w is the animal's weight (in kilograms). Find the approximate change in oxygen consumption under the following conditions.

a. The internal temperature increases from 38°C to 39°C, while the outside temperature remains at 12°C and the weight remains at 30 kg.

b. The internal temperature is constant at 36°C, the outside temperature increases from 14°C to 15°C, and the weight remains at 25 kg.

49. *Blood Flow* In one method of computing the quantity of blood pumped through the lungs in one minute, a researcher first finds each of the following (in milliliters).

b = quantity of oxygen used by body in one minute

a = quantity of oxygen per liter of blood that has just gone through the lungs

v = quantity of oxygen per liter of blood that is about to enter the lungs

In one minute,

$$\text{Amount of oxygen used} = \text{Amount of oxygen per liter} \\ \times \text{Liters of blood pumped.}$$

If C is the number of liters of blood pumped through the lungs in one minute, then

$$b = (a - v) \cdot C \quad \text{or} \quad C = \frac{b}{a - v}.$$

a. Find the number of liters of blood pumped through the lungs in one minute if $a = 160$, $b = 200$, and $v = 125$.

b. Find the approximate change in C when a changes from 160 to 161, $b = 200$, and $v = 125$.

c. Find the approximate change in C when $a = 160$, b changes from 200 to 201, and $v = 125$.

d. Find the approximate change in C when $a = 160$, $b = 200$, and v changes from 125 to 126.

e. A change of 1 unit in which quantity of oxygen produces the greatest change in the liters of blood pumped?

50. *Health* A weight-loss counselor has prepared a program of diet and exercise for a client. If the client sticks to the program, the weight loss that can be expected (in pounds per week) is given by

$$\text{Weight loss} = f(n, c) = \frac{1}{8}n^2 - \frac{1}{5}c + \frac{1,937}{8},$$

where c is the average daily calorie intake for the week and n is the number of 40-min aerobic workouts per week.

a. How many pounds can the client expect to lose by eating an average of 1,200 cal per day and participating in four 40-min workouts in a week?

b. Find and interpret $\partial f / \partial n$.

c. The client currently averages 1,100 cal per day and does three 40-minute workouts each week. What would be the approximate impact on weekly weight loss of adding a fourth workout per week?

51. *Health* The body-mass index is a number that can be calculated for any individual as follows: Multiply a person's weight by 703 and divide by the person's height squared. That is,

$$B = \frac{703w}{h^2},$$

where w is in pounds and h is in inches.* The National Heart, Lung and Blood Institute uses the body-mass index to determine whether a person is "overweight" ($25 \leq B < 30$) or "obese" ($B \geq 30$).

a. Calculate the body-mass index for a person who weighs 220 pounds and is 74 inches tall.

b. Calculate $\dfrac{\partial B}{\partial w}$ and $\dfrac{\partial B}{\partial h}$ and interpret.

c. Using the fact that $1 \text{ in} = 0.0254 \text{ m}$ and $1 \text{ lb} = 0.4555 \text{ kg}$, transform this formula to handle metric units.†

---

*The National Institutes of Health

†Technically, grams are a measure of mass, not weight. Weight is a measure of the force of gravity, which varies with the distance from the center of the earth. For objects on the surface of the earth, weight and mass are often used interchangeably, and we will do so here.

52. *Drug Reaction* The reaction to x units of a drug t hr after it was administered is given by

$$R(x, t) = x^2(a - x)t^2 e^{-t},$$

for $0 \leq x \leq a$ (where a is a constant). Find the following.

a. $\dfrac{\partial R}{\partial x}$ **b.** $\dfrac{\partial R}{\partial t}$ **c.** $\dfrac{\partial^2 R}{\partial x^2}$ **d.** $\dfrac{\partial^2 R}{\partial x \partial t}$

e. Interpret your answers to parts a and b.

53. *Scuba Diving* In 1908, J. Haldane constructed diving tables that provide a relationship between the water pressure on body tissues for various water depths and dive times. The tables were successfully used by divers to virtually eliminate decompression sickness. The pressure in atmospheres for a no-stop dive is given by the following formula:*

$$p(l, t) = 1 + \frac{l}{33}(1 - 2^{-t/5}),$$

where t is in minutes, l is in feet, and p is in atmospheres (atm).[†]

a. Find the pressure at 33 ft for a 10-minute dive.

b. Find $p_l(33, 10)$ and $p_t(33, 10)$ and interpret. (*Hint:* $D_t(a^t) = \ln(a)a^t$.)

c. Haldane estimated that decompression sickness for no-stop dives could be avoided if the diver's tissue pressure did not exceed 2.15 atm. Find the maximum amount of time that a diver could stay down (time includes going down and coming back up) if he or she wants to dive to a depth of 66 feet.

54. *Wind Chill* In 1941, explorers Paul Siple and Charles Passel discovered that the amount of heat lost when an object is exposed to cold air depends on both the temperature of the air and the velocity of the wind. They developed the *Wind Chill Index* as a way to measure the danger of frostbite while doing outdoor activities. The wind chill can be calculated as follows:

$$W(V, T) = 91.4 - \frac{(10.45 + 6.69\sqrt{V} - 0.447V)(91.4 - T)}{22},$$

where V is the wind speed in miles per hour and T is the temperature in Fahrenheit for wind speeds between 4 and 45 mph.[‡]

a. Find the wind chill for a wind speed of 20 mph and 10°F.

b. If a weather report indicates that the wind chill is -25°F and the actual outdoor temperature is 5°F, use a graphing calculator to find the corresponding wind speed to the nearest mile per hour.

c. Find $W_V(20, 10)$ and $W_T(20, 10)$ and interpret.

d. Using the table command on a graphing calculator or a spreadsheet, develop a wind chill chart for various wind speeds and temperatures.

55. *Heat Index* The following chart shows the heat index, which combines the effects of temperature with humidity to give a measure of the apparent temperature, or how hot it feels to the body.[§] For example, when the outside temperature is 90°F and the relative humidity is 40%, then the apparent temperature is approximately 93°F. Let $I = f(T, H)$ give the heat index, I, as a function of the temperature T in degrees Fahrenheit and the percent humidity H. Estimate each of the following.

a. $f(90, 30)$ **b.** $f(90, 75)$ **c.** $f(80, 75)$

Heat Index

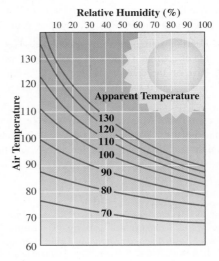

*These estimates are conservative. Please consult modern dive tables before making a dive.
[†]Westbrook, David, "The Mathematics of Scuba Diving," *The UMAP Journal,* Vol. 18, No. 2, 1997, pp. 2–19.
[‡]Bosch, William, and L. Cobb, "Windchill," *The UMAP Journal,* Vol. 13, No. 3, 1990, pp. 481–489.
[§]The Weather Channel, www.weather.com (May 9, 2000).

Estimate each of the following by approximating the partial derivative using a value of $h = 5$ in the difference quotient.

d. $f_T(90, 30)$ **e.** $f_H(90, 30)$ **f.** $f_T(90, 75)$

g. $f_H(90, 75)$

h. Describe in words what your answers in parts d–g mean.

56. *Calorie Expenditure* The average energy expended for an animal to walk or run 1 km can be estimated by the function

$$f(w, v) = 25.92w^{0.68} + \frac{3.62w^{0.75}}{v},$$

where $f(w, v)$ is the energy used (in kcal/hr), w is the weight (in g), and v is the speed of movement (in km/hr) of the animal.*

a. Find $f(300, 10)$.

b. Find $f_w(300, 10)$ and interpret.

c. If a mouse could run at the same speed that an elephant walks, which animal would expend more energy? How can partial derivatives be used to explore this question?

57. *Snake River Salmon* As we saw in the last section, researchers have determined that the date D (day of the year) in which a Chinook salmon passes the Lower Granite Dam can be calculated using the function

$$\ln(D) = 4.890179 + 0.005163R - 0.004345L$$
$$- 0.000019F,$$

where R is the release date (day of the year) of the tagged salmon, L is the length of the salmon L (in mm), and F is the mean daily flow of the water (in m³/s).†

a. Use the result of part a of Exercise 37 of the previous section to find $D_F(R, L, F)$. Interpret your answer.

b. Use the partial derivative calculated above to estimate the effect that a 1,000 m³/s increase of water flow will have on the date of passage for a 75-mm salmon that was released on day 150 with an average flow rate of 3,500 m³/s.

58. *Brown Trout* Researchers from New Zealand have determined that the length of a brown trout depends on both its weight and age, and that the length can be estimated by

$$L(w, t) = (0.00082t + 0.0955)e^{(\ln w + 10.49)/2.842},$$

where $L(w, t)$ is the length of the trout (in cm), w is the weight of the trout (in g), and t is the age of the trout (in yr).‡

a. Find $L(600, 7)$.

b. Find $L_w(600, 7)$ and $L_t(600, 7)$ and interpret.

59. *Learning* In the exercises of Chapter 4 we saw that for one model of how animals learn, the number of wrong attempts by an animal is given by

$$w = -\frac{1}{a} \ln\left[1 - \frac{(N - 1)b(1 - e^{-kn})}{N} \right],$$

where n is the number of trials an animal makes while attempting to learn a task, N is number of stimuli, and k, a, and b are constants.§

a. Find and interpret $\partial w/\partial n$.

b. Find and interpret $\partial w/\partial N$.

60. *Insect Dispersal* A model for the dispersal of insects due to population pressure is the partial differential equation

$$\frac{\partial n}{\partial t} = D_0 \frac{\partial}{\partial x} \left[\left(\frac{n}{n_0}\right)^m \frac{\partial n}{\partial x} \right],$$

where n is the number of insects and D_0, n_0, and m are positive constants.‖ The exact solution of this differential equation is

$$n(x, t) = n_0 [g(t)]^{-1} \left[1 - \left\{ \frac{x}{r_0 g(t)} \right\}^2 \right]^{1/m},$$

where

$$g(t) = \left(\frac{2D_0(m + 2)t}{r_0^2 m} \right)^{1/(2+m)},$$

and r_0 is a positive constant. Show that this solution indeed satisfies the differential equation for the special case where $n_0 = r_0 = m = 1$ and $D_0 = 1/6$.

OTHER APPLICATIONS

61. *Education* A developmental mathematics instructor at a large university has determined that a student's probability of success in the university's pass/fail remedial algebra course is a function of s, n, and a, where s is the student's score on the departmental placement exam, n is the number of semesters of mathematics passed in high school, and a is

---

*Robbins, C., *Wildlife Feeding and Nutrition,* New York, Academic Press, 1983, p. 114.

†Conner, W., R. Steinhorst, and H. Burge, "Forecasting Survival and Passage of Migratory Juvenile Salmonids," *North American Journal of Fisheries Management,* Vol. 20, 2000, pp. 651–660.

‡Hayes, J., J. Stark, and K. Shearer, "Development and Test of a Whole-Lifetime Foraging and Bioenergetics Growth Model for Drift-Feeding Brown Trout," *Transactions of the American Fisheries Society,* Vol. 129, 2000, pp. 315–332.

§Rashevsky, Nicolas, *Mathematical Biology of Social Behavior,* rev. ed., Chicago, The University of Chicago Press, 1959, p. 31.

‖Murray, J. D., *Mathematical Biology,* Springer-Verlag, 1989, p. 238.

the student's mathematics SAT score. She estimates that p, the probability of passing the course (in percent), will be

$$p = f(s, n, a) = 0.003a + 0.1(sn)^{1/2}$$

for $200 \leq a \leq 800$, $0 \leq s \leq 10$, and $0 \leq n \leq 8$. Assuming that the above model has some merit, find the following.

a. If a student scores 8 on the placement exam, has taken 6 semesters of high school math, and has an SAT score of 450, what is the probability of passing the course?

b. Find p for a student with 3 semesters of high school mathematics, a placement score of 3, and an SAT score of 320.

c. Find and interpret $f_n(3, 3, 320)$ and $f_a(3, 3, 320)$.

62. Gravitational Attraction The gravitational attraction F on a body a distance r from the center of the earth, where r is greater than the radius of the earth, is a function of its mass m and the distance r as follows:

$$F = \frac{mgR^2}{r^2},$$

where R is the radius of the earth and g is the force of gravity—about 32 feet per second per second (ft/sec^2).

a. Find and interpret F_m and F_r.

b. Show that $F_m > 0$ and $F_r < 0$. Why is this reasonable?

63. Velocity In 1931, Albert Einstein developed the following formula for adding two velocities, x and y:

$$w(x, y) = \frac{x + y}{1 + \dfrac{xy}{c^2}},$$

where x and y are in miles per second and c represents the speed of light, 186,282 miles per second.*

a. Suppose that, relative to a stationary observer, a new super space shuttle is capable of traveling at 50,000 miles per second and that, while traveling at this speed, it launches a rocket that travels at 150,000 miles per second. How fast is the rocket traveling relative to the stationary observer?

b. What is the instantaneous rate of change of w with respect to the speed of the space shuttle, x, when the space shuttle is traveling at 50,000 miles per second and the rocket is traveling at 150,000 miles per second?

c. Hypothetically, if a person is driving at the speed of light, c, and she turns on the headlights, what is the velocity of the light coming from the headlights, relative to a stationary observer?

64. Movement Time Fitts' law is used to estimate the amount of time it takes for a person, using his or her arm, to pick up a light object, move it, and then place it in a designated target area. Mathematically, Fitts' law for a particular individual is given by

$$T(s, w) = -50 + 105 \log_2 \left(\frac{2s}{w} \right),$$

where s is the distance (in feet) the object is moved, w is the width of the area in which the object is being placed, and T is the time (in msec).[†]

a. Calculate $T(3, 0.5)$.

b. Find $T_s(3, 0.5)$ and $T_w(3, 0.5)$ and interpret these values. (*Hint:* $\log_2 x = \ln x / \ln 2$.)

65. Revenue The revenue from the sale of x units of a tranquilizer and y units of an antibiotic is given by

$$R(x, y) = 5x^2 + 9y^2 - 4xy.$$

Suppose 9 units of tranquilizer and 5 units of antibiotic are sold.

a. What is the approximate effect on revenue if 10 units of tranquilizer and 5 units of antibiotic are sold?

b. What is the approximate effect on revenue if the amount of antibiotic sold is increased to 6 units, while tranquilizer sales remain constant?

66. Ground Temperature Mathematical models of ground temperature variation usually involve Fourier series or other sophisticated methods. However, the elementary model

$$u(x, t) = T_0 + A_0 e^{-ax} \cos \left(\frac{\pi}{6} t - ax \right)$$

has been developed for temperature $u(x, t)$ at a given location at a variable time t (in months) and a variable depth x (in centimeters) beneath the Earth's surface. T_0 is the annual average surface temperature, and A_0 is the amplitude of the seasonal surface temperature variation.[‡]

Assume that $T_0 = 16°C$ and $A_0 = 11°C$ at a certain location. Also assume that $a = 0.00706$ in cgs (centimeter-gram-second) units.

*Fiore, Greg, "An Out-of-Math Experience: Einstein, Relativity, and the Developmental Mathematics Student," *The Mathematics Teacher,* Vol. 93, No. 3, 2000, pp. 194–199.

†Sanders, Mark, and Ernest McCormick, *Human Factors in Engineering Design,* 7th edition, McGraw Hill, New York, 1993, pp. 290–291.

‡Corbitt, Mary Kay, and C. Edwards, "Mathematical Modeling and Cool Buttermilk in the Summer," *Applications in School Mathematics 1979 Yearbook,* ed. by Sidney Sharron and Robert Hays, Reston, VA, NCTM, 1979, p. 221.

a. At what minimum depth x is the amplitude of $u(x, t)$ at most 1°C?

b. Suppose we wish to construct a cellar to keep wine at a temperature between 14°C and 18°C. What minimum depth will accomplish this?

c. At what minimum depth x does the ground temperature model predict that it will be winter when it is summer at

the surface, and vice versa? That is, when will the phase shift correspond to 1/2 year?

d. Show that the ground temperature model satisfies the *heat equation*

$$\frac{\partial u}{\partial t} = k\frac{\partial^2 u}{\partial x^2},$$

where k is a constant.

■ 9.3 MAXIMA AND MINIMA

? THINK ABOUT IT How many scientists does it take to develop a new antibiotic? What is the minimum cost?

FOR REVIEW ■

It may be helpful to review the section on relative extrema in Chapter 5, Graphs and the Derivative, at this point. The concepts presented there are basic to what will be done in this section.

One of the most important applications of calculus is finding maxima and minima of functions. Earlier, we studied this idea extensively for functions of a single independent variable; now we will see that extrema can be found for functions of two variables. In particular, an extension of the second derivative test can be defined and used to identify maxima or minima. We begin with the definitions of relative maxima and minima.

RELATIVE MAXIMA AND MINIMA

Let (a, b) be the center of a circular region contained in the xy-plane. Then, for a function $z = f(x, y)$ defined for every (x, y) in the region, $f(a, b)$ is a **relative maximum** if

$$f(a, b) \geq f(x, y)$$

for all points (x, y) in the circular region, and $f(a, b)$ is a **relative minimum** if

$$f(a, b) \leq f(x, y)$$

for all points (x, y) in the circular region.

As before, the word *extremum* is used for either a relative maximum or a relative minimum. Examples of a relative maximum and a relative minimum are given in Figures 18 and 19.

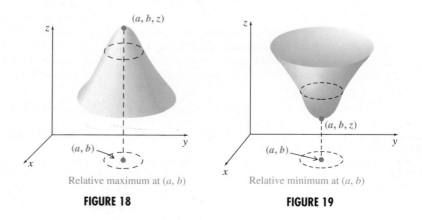

Relative maximum at (a, b)

FIGURE 18

Relative minimum at (a, b)

FIGURE 19

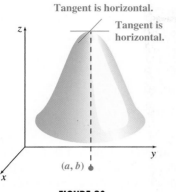

Tangent is horizontal.

Tangent is horizontal.

(a, b)

FIGURE 20

NOTE When functions of a single variable were discussed, a distinction was made between relative extrema and absolute extrema. The methods for finding absolute extrema are quite involved for functions of two variables, so we will discuss only relative extrema here. In some practical applications the relative extrema coincide with the absolute extrema. Also, in this brief discussion of extrema for multivariable functions, we omit cases where an extremum occurs on a boundary of the domain.

As suggested by Figure 20, at a relative maximum the tangent line parallel to the xz-plane has a slope of 0, as does the tangent line parallel to the yz-plane. (Notice the similarity to functions of one variable.) That is, if the function $z = f(x, y)$ has a relative extremum at (a, b), then $f_x(a, b) = 0$ and $f_y(a, b) = 0$, as stated in the next theorem.

LOCATION OF EXTREMA

Let a function $z = f(x, y)$ have a relative maximum or relative minimum at the point (a, b). Let $f_x(a, b)$ and $f_y(a, b)$ both exist. Then

$$f_x(a, b) = 0 \quad \text{and} \quad f_y(a, b) = 0.$$

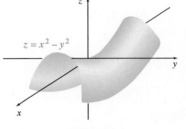

$z = x^2 - y^2$

FIGURE 21

Just as with functions of one variable, the fact that the slopes of the tangent lines are 0 is no guarantee that a relative extremum has been located. For example, Figure 21 shows the graph of $z = f(x, y) = x^2 - y^2$. Both $f_x(0, 0) = 0$ and $f_y(0, 0) = 0$, and yet $(0, 0)$ leads to neither a relative maximum nor a relative minimum for the function. The point $(0, 0, 0)$ on the graph of this function is called a **saddle point;** it is a minimum when approached from one direction but a maximum when approached from another direction. A saddle point is neither a maximum nor a minimum.

The theorem on location of extrema suggests a useful strategy for finding extrema. First, locate all points (a, b) where $f_x(a, b) = 0$ and $f_y(a, b) = 0$. Then test each of these points separately, using the test given after the next example. For a function $f(x, y)$, the points (a, b) such that $f_x(a, b) = 0$ and $f_y(a, b) = 0$ are called **critical points.**

EXAMPLE 1   Critical Points

Find all critical points for

$$f(x, y) = 6x^2 + 6y^2 + 6xy + 36x - 5.$$

Solution Find all points (a, b) such that $f_x(a, b) = 0$ and $f_y(a, b) = 0$. Here

$$f_x(x, y) = 12x + 6y + 36 \quad \text{and} \quad f_y(x, y) = 12y + 6x.$$

Set each of these two partial derivatives equal to 0.

$$12x + 6y + 36 = 0 \quad \text{and} \quad 12y + 6x = 0$$

These two equations make up a system of linear equations. We can use the substitution method to solve this system. First, rewrite $12y + 6x = 0$ as follows:

$$12y + 6x = 0$$
$$6x = -12y$$
$$x = -2y.$$

Now substitute $-2y$ for x in the other equation and solve for y.

$$12x + 6y + 36 = 0$$
$$12(-2y) + 6y + 36 = 0$$
$$-24y + 6y + 36 = 0$$
$$-18y + 36 = 0$$
$$-18y = -36$$
$$y = 2$$

From the equation $x = -2y$, $x = -2(2) = -4$. The solution of the system of equations is $(-4, 2)$. Since this is the only solution of the system, $(-4, 2)$ is the only critical point for the given function. By the theorem above, if the function has a relative extremum, it will occur at $(-4, 2)$.

The results of the next theorem can be used to decide whether $(-4, 2)$ in Example 1 leads to a relative maximum, a relative minimum, or neither.

TEST FOR RELATIVE EXTREMA

For a function $z = f(x, y)$, let f_{xx}, f_{yy}, and f_{xy} all exist in a circular region contained in the xy-plane with center (a, b). Further, let

$$f_x(a, b) = 0 \qquad \text{and} \qquad f_y(a, b) = 0.$$

Define the number D by

$$D = f_{xx}(a, b) \cdot f_{yy}(a, b) - [f_{xy}(a, b)]^2.$$

Then

a. $f(a, b)$ is a relative maximum if $D > 0$ and $f_{xx}(a, b) < 0$;

b. $f(a, b)$ is a relative minimum if $D > 0$ and $f_{xx}(a, b) > 0$;

c. $f(a, b)$ is a saddle point (neither a maximum nor a minimum) if $D < 0$;

d. if $D = 0$, the test gives no information.

This test is comparable to the second derivative test for extrema of functions of one independent variable. The following chart summarizes the conclusions of the theorem.

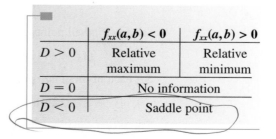

	$f_{xx}(a, b) < 0$	$f_{xx}(a, b) > 0$
$D > 0$	Relative maximum	Relative minimum
$D = 0$	No information	
$D < 0$	Saddle point	

EXAMPLE 2   Relative Extrema

The previous example showed that the only critical point for the function

$$f(x, y) = 6x^2 + 6y^2 + 6xy + 36x - 5$$

is $(-4, 2)$. Does $(-4, 2)$ lead to a relative maximum, a relative minimum, or neither?

Solution Find out by using the test above. From Example 1,

$$f_x(-4, 2) = 0 \quad \text{and} \quad f_y(-4, 2) = 0.$$

Now find the various second partial derivatives used in finding D. From $f_x(x, y) = 12x + 6y + 36$ and $f_y(x, y) = 12y + 6x$,

$$f_{xx}(x, y) = 12, \quad f_{yy}(x, y) = 12, \quad \text{and} \quad f_{xy}(x, y) = 6.$$

(If these second-order partial derivatives had not all been constants, they would have had to be evaluated at the point $(-4, 2)$.) Now

$$D = f_{xx}(-4, 2) \cdot f_{yy}(-4, 2) - [f_{xy}(-4, 2)]^2 = 12 \cdot 12 - 6^2 = 108.$$

Since $D > 0$ and $f_{xx}(-4, 2) = 12 > 0$, part b of the theorem applies, showing that $f(x, y) = 6x^2 + 6y^2 + 6xy + 36x - 5$ has a relative minimum at $(-4, 2)$. This relative minimum is $f(-4, 2) = -77$. ▪

EXAMPLE 3 Saddle Point
Find all points where the function

$$f(x, y) = 9xy - x^3 - y^3 - 6$$

has any relative maxima or relative minima.

Solution First find any critical points. Here

$$f_x(x, y) = 9y - 3x^2 \quad \text{and} \quad f_y(x, y) = 9x - 3y^2.$$

Set each of these partial derivatives equal to 0.

$$
\begin{array}{ll}
f_x(x, y) = 0 & f_y(x, y) = 0 \\
9y - 3x^2 = 0 & 9x - 3y^2 = 0 \\
9y = 3x^2 & 9x = 3y^2 \\
3y = x^2 & 3x = y^2
\end{array}
$$

The substitution method can be used again to solve the system of equations

$$3y = x^2$$
$$3x = y^2.$$

The first equation, $3y = x^2$, can be rewritten as $y = x^2/3$. Substitute this into the second equation to get

$$3x = y^2 = \left(\frac{x^2}{3}\right)^2 = \frac{x^4}{9}.$$

Solve this equation as follows.

$$
\begin{array}{lll}
27x = x^4 & & \\
x^4 - 27x = 0 & & \\
x(x^3 - 27) = 0 & & \text{Factor.} \\
x = 0 \quad \text{or} \quad x^3 - 27 = 0 & & \text{Set each factor equal to 0.} \\
x = 0 \quad \text{or} \quad x^3 = 27 & & \\
x = 0 \quad \text{or} \quad x = 3 & & \text{Take the cube root on each side.}
\end{array}
$$

Use these values of x, along with the equation $3y = x^2$, rewritten as $y = x^2/3$, to find y. If $x = 0$, $y = 0^2/3 = 0$. If $x = 3$, $y = 3^2/3 = 3$. The critical points are $(0, 0)$ and $(3, 3)$. To identify any extrema, use the test. Here

$$f_{xx}(x, y) = -6x, \qquad f_{yy}(x, y) = -6y, \qquad \text{and} \qquad f_{xy}(x, y) = 9.$$

Test each of the possible critical points.

For $(0, 0)$:

$f_{xx}(0, 0) = -6(0) = 0$

$f_{yy}(0, 0) = -6(0) = 0$

$f_{xy}(0, 0) = 9$

$D = 0 \cdot 0 - 9^2 = -81.$

Since $D < 0$, there is a saddle point at $(0, 0)$.

For $(3, 3)$:

$f_{xx}(3, 3) = -6(3) = -18$

$f_{yy}(3, 3) = -6(3) = -18$

$f_{xy}(3, 3) = 9$

$D = -18(-18) - 9^2 = 243.$

Here $D > 0$ and $f_{xx}(3, 3) = -18 < 0$; there is a relative maximum at $(3, 3)$.

Notice that these values are in accordance with the graph generated by the computer program Maple™ shown in Figure 22.

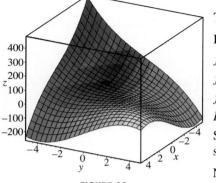

FIGURE 22

EXAMPLE 4 Scientific Development

A pharmaceutical company is developing a new antibiotic. The cost in thousands of dollars to develop an antibiotic is approximated by

$$C(x, y) = 2{,}200 + 27x^3 - 72xy + 8y^2,$$

where x is the number of employees working in quality assurance and y is the number of scientists working in the laboratory. Find the number of employees in each area that results in the minimum cost to develop the antibiotic. What is the minimum cost?

Method 1: Calculation by Hand

Solution Start with the following partial derivatives.

$$C_x(x, y) = 81x^2 - 72y \qquad \text{and} \qquad C_y(x, y) = -72x + 16y$$

Set each of these equal to 0 and solve for y.

$81x^2 - 72y = 0$

$-72y = -81x^2$

$y = \dfrac{9}{8}x^2$

$-72x + 16y = 0$

$16y = 72x$

$y = \dfrac{9}{2}x$

Since $(9/8)x^2$ and $(9/2)x$ both equal y, they are equal to each other. Set them equal, and solve the resulting equation for x.

$$\frac{9}{8}x^2 = \frac{9}{2}x$$

$$9x^2 = 36x$$

$$9x^2 - 36x = 0 \qquad \text{Get 0 on one side.}$$

$$9x(x - 4) = 0 \qquad \text{Factor.}$$

$$9x = 0 \qquad \text{or} \qquad x - 4 = 0 \qquad \text{Set each factor equal to 0.}$$

The equation $9x = 0$ leads to $x = 0$ and $y = 0$, which cannot be a minimizer of $C(x, y)$ since, for example, $C(1, 1) < C(0, 0)$. This fact can also be verified by the test for relative extrema. Substitute $x = 4$, the solution of $x - 4 = 0$, into $y = (9/2)x$ to find y.

$$y = \frac{9}{2}x = \frac{9}{2}(4) = 18$$

Now check to see whether the critical point $(4, 18)$ leads to a relative minimum. For $(4, 18)$,

$$C_{xx}(x, y) = 162x = 162(4) = 648, \quad C_{yy}(x, y) = 16, \quad \text{and} \quad C_{xy}(x, y) = -72.$$

Also,

$$D = (648)(16) - (-72)^2 = 5{,}184.$$

Since $D > 0$ and $C_{xx}(4, 18) > 0$, the cost at $(4, 18)$ is a minimum.

To find the minimum cost, go back to the cost function and evaluate $C(4, 18)$.

$$C(x, y) = 2{,}200 + 27x^3 - 72xy + 8y^2$$
$$C(4, 18) = 2{,}200 + 27(4)^3 - 72(4)(18) + 8(18)^2 = 1{,}336$$

The minimum cost for developing a new antibiotic is \$1,336,000 when four employees work in quality assurance and 18 scientists work in the laboratory.

Method 2: Spreadsheets | Finding the maximum or minimum of a function of one or more variables can be done using a spreadsheet. The most widely used spreadsheets have built-in solvers that are able to optimize complicated functions. These solvers employ more advanced techniques than we have learned thus far, and they are very efficient and practical to use.

The Solver included with Excel is located in the Tools menu and requires that cells be identified ahead of time for each variable in the problem. It also requires that another cell be identified where the function, in terms of the variable cells, is placed. For example, to solve the above problem, we could identify cells A1 and B1 to represent the variables x and y, respectively. The Solver requires that we place a guess for the answer in these cells. Thus, our initial value or guess will be to place the number 5 in each of these cells. The function must be placed in a cell in terms of cells A1 and B1. If we choose cell A3 to represent the function, in cell A3 we would type "=2200+27*A1^3−72*A1*B1+8*B1^2."

We now click on the Tools menu and choose Solver. This solver will attempt to find a solution that either maximizes or minimizes the value of cell A3. Figure 23 illustrates the Solver box and the items placed in it.

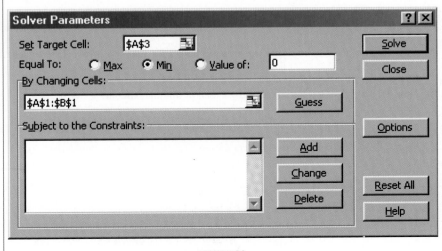

FIGURE 23

To obtain a solution, click on Solve. The rounded solution $x = 4$ and $y = 18$ is located in cells A1 and B1. The minimum cost $C(4, 18) = 1,336$ is located in cell A3.

NOTE One must be careful when using Solver because it will not find a maximizer or minimizer of a function if the initial guess is the exact place in which a saddle point occurs. For example, in the problem above, if our initial guess was $(0, 0)$, the Solver would have returned the value of $(0, 0)$ as the place where a minimum occurs. But $(0, 0)$ is a saddle point. Thus, it is always a good idea to run the Solver for two different initial values and compare the solutions.

9.3 EXERCISES

Find all points where the functions defined as follows have any relative extrema. Identify any saddle points.

1. $f(x, y) = xy + x - y$

2. $f(x, y) = 4xy + 8x - 9y$

3. $f(x, y) = x^2 - 2xy + 2y^2 + x - 5$

4. $f(x, y) = x^2 + xy + y^2 - 6x - 3$

5. $f(x, y) = x^2 - xy + y^2 + 2x + 2y + 6$

6. $f(x, y) = x^2 + xy + y^2 + 3x - 3y$

7. $f(x, y) = x^2 + 3xy + 3y^2 - 6x + 3y$

8. $f(x, y) = 5xy - 7x^2 - y^2 + 3x - 6y - 4$

9. $f(x, y) = 4xy - 10x^2 - 4y^2 + 8x + 8y + 9$

10. $f(x, y) = x^2 + xy + 3x + 2y - 6$

11. $f(x, y) = x^2 + xy - 2x - 2y + 2$

12. $f(x, y) = x^2 + xy + y^2 - 3x - 5$

13. $f(x, y) = 2x^3 + 3y^2 - 12xy + 4$

14. $f(x, y) = 5x^3 + 2y^2 - 60xy - 3$

15. $f(x, y) = x^2 + 4y^3 - 6xy - 1$

16. $f(x, y) = 3x^2 + 7y^3 - 42xy + 5$

17. $f(x, y) = e^{xy}$

18. $f(x, y) = x^2 + e^y$

19. Describe the procedure for finding critical points of a function in two independent variables.

20. How are second-order partial derivatives used in finding extrema?

Figures a–f on the next page show the graphs of the functions defined in Exercises 21–26. Find all relative extrema for each function, and then match the equation to its graph.

21. $z = -3xy + x^3 - y^3 + \dfrac{1}{8}$

22. $z = \dfrac{3}{2}y - \dfrac{1}{2}y^3 - x^2y + \dfrac{1}{16}$

23. $z = y^4 - 2y^2 + x^2 - \dfrac{17}{16}$

24. $z = -2x^3 - 3y^4 + 6xy^2 + \dfrac{1}{16}$

25. $z = -x^4 + y^4 + 2x^2 - 2y^2 + \dfrac{1}{16}$

26. $z = -y^4 + 4xy - 2x^2 + \dfrac{1}{16}$

a.

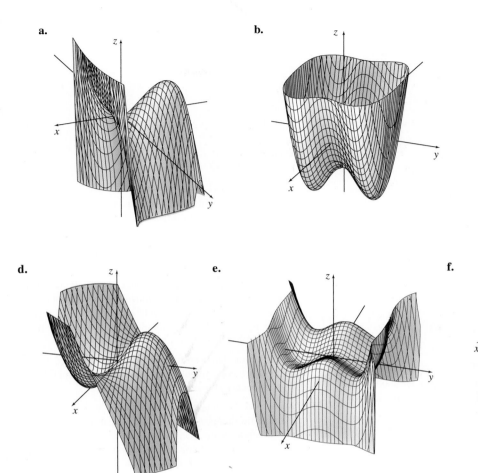

b.

c.

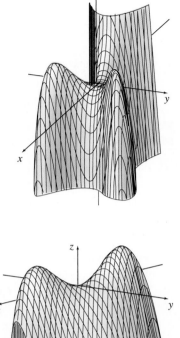

d.

e.

f.

27. Show that $f(x, y) = 1 - x^4 - y^4$ has a relative maximum, even though D in the theorem is 0.

28. Show that $D = 0$ for $f(x, y) = x^3 + (x - y)^2$ and that the function has no relative extrema.

29. A friend taking calculus is puzzled. She remembers that for a function of one variable, if the first derivative is zero at a point and the second derivative is positive, then there must be a relative minimum at the point. She doesn't understand why that isn't true for a function of two variables—that is, why $f_x(x, y) = 0$ and $f_{xx}(x, y) > 0$ doesn't guarantee a relative minimum. Provide an explanation.

30. In Section 1.2, we found the least squares line through a set of n points (x_1, y_1), $(x_2, y_2), \ldots, (x_n, y_n)$ by choosing the slope of the line m and the y-intercept b to minimize the quantity

$$S(m, b) = \sum_{i=1}^{n} (mx_i + b - y_i)^2,$$

where the summation symbol $\sum_{i=1}^{n}$ means that we sum over all the data points. Minimize S by setting $S_m(m, b) = 0$ and $S_b(m, b) = 0$, and then rearrange the results to derive the equations from Section 1.2

$$\left(\sum_{i=1}^{n} x_i\right)b + \left(\sum_{i=1}^{n} x_i^2\right)m = \sum_{i=1}^{n} x_i y_i$$

$$nb + \left(\sum_{i=1}^{n} x_i\right)m = \sum_{i=1}^{n} y_i.$$

31. Consider the function $f(x, y) = ax^2 e^y + y^2 e^x$.*

    a. For what values of a is the point $x = 0$ and $y = 0$ a critical point?

    b. For what values of a does this point produce a relative minimum of the function?

▮ Applications

LIFE SCIENCES

32. *Satisfaction* Suppose two animals get satisfaction from making a sound, which is pleasant to the individuals but only to a point. Let S_1 and S_2 be the satisfaction to the first and second individuals, respectively, and let x_1 and x_2 be the intensity of the sounds made by the two individuals, respectively. A model of the satisfaction of the two animals is given by

$$S_1 = a_1(x_1 + x_2) - b_1(x_1 + x_2)^2 \quad \text{and}$$
$$S_2 = a_2(x_1 + x_2) - b_2(x_1 + x_2)^2,$$

where a_1, a_2, b_1, and b_2 are constants.[†]

    a. Show that if each animal tries to adjust its sound to maximize its own satisfaction, there is no solution unless $a_1/b_1 = a_2/b_2$.

    b. If the animals are altruistic, their only interest is the total satisfaction, namely,

$$S = S_1 + S_2.$$

Show that there are an infinite number of ways to maximize the total satisfaction, namely, all values of x_1 and x_2 satisfying

$$x_1 + x_2 = \frac{a_1 + a_2}{2(b_1 + b_2)}.$$

OTHER APPLICATIONS

33. *Political Science* The probability that a three-person jury will make a correct decision is given by

$$P(\alpha, r, s) = \alpha(3r^2(1 - r) + r^3) + (1 - \alpha)(3s^2(1 - s) + s^3),$$

where $0 < \alpha < 1$ is the probability that the person is guilty of the crime, r is the probability that a given jury member

will vote "guilty" when the defendant is indeed guilty of the crime, and s is the probability that a given jury member will vote "innocent" when the defendant is indeed innocent.[‡]

    a. Calculate $P(0.9, 0.5, 0.6)$ and $P(0.1, 0.8, 0.4)$ and interpret your answers.

    b. Using common sense and without using calculus, what value of r and s would maximize the jury's probability of making the correct verdict? Do these values depend on α in this problem? Should they? What is the maximum probability?

    c. Verify your answer for part b using calculus. (*Hint:* There are two critical points. Argue that the maximum value occurs at one of these points.)

34. *Computer Chips* The following table, which illustrates the dramatic increase in the number of transistors in personal computers since 1971, was given in the chapter on Exponential, Logarithmic, and Trigonometric Functions, Section 1, Exercise 40.[§]

Year (since 1971)	Chip	Transistors
0	4004	2,300
15	386DX	275,000
18	486DX	1,200,000
22	Pentium	3,300,000
24	P6	5,500,000
26	Pentium III	9,500,000

    a. Since the natural logarithm of both sides of the equation $y = ab^x$ yields $\ln y = \ln a + x \ln b$, we could let $w = \ln y$, $r = \ln a$, and $s = \ln b$ to form $w = r + sx$.

---

*This problem was adapted from Exercise 9.8, page 336, *Introduction to Operations Research,* by J. Ecker and M. Kupferschmid, Krieger Pub., 1991.

[†]Rashevsky, Nicolas, *Mathematical Biology of Social Behavior,* rev. ed., Chicago, The University of Chicago Press, 1959, p. 41.

[‡]Grofman, Bernard, "A Preliminary Model of Jury Decision Making as a Function of Jury Size, Effective Jury Decision Rule, and Mean Juror Judgmental Competence," *Frontiers of Economics,* Vol. 3, 1980, pp. 98–110.

[§]Data provided by Intel.

Using linear regression find values for r and s that will fit the data above. (*Hint:* Take the natural logarithm of the values in the transistors column and then use linear regression to find values of r and s that fit the data. Once you know r and s, you can determine the values of a and b by calculating $a = e^r$ and $b = e^s$.)

b. Use the Solver capability of a spreadsheet to find a function of the form $y = ab^x$ that fits the data above. (*Hint:* Using the ideas from part a, find values for a and b that minimize the function

$$f(a, b) = (\ln(2,300) - 0 \ln b - \ln a)^2$$
$$+ (\ln(275,000) - 15 \ln b - \ln a)^2$$
$$+ (\ln(1,200,000) - 18 \ln b - \ln a)^2$$
$$+ (\ln(3,300,000) - 22 \ln b - \ln a)^2$$
$$+ (\ln(5,500,000) - 24 \ln b - \ln a)^2$$
$$+ (\ln(9,500,000) - 26 \ln b - \ln a)^2.)$$

c. Compare your answer to this problem with the one found with a graphing calculator in the chapter on Exponential, Logarithmic, and Trigonometric Functions, Section 1, Exercise 40.

35. *Profit* Suppose that the profit of a certain firm (in hundreds of dollars) is approximated by

$$P(x, y) = 1,000 + 24x - x^2 + 80y - y^2,$$

where x is the cost of a unit of labor and y is the cost of a unit of goods. Find values of x and y that maximize profit. Find the maximum profit.

36. *Labor Costs* Suppose the labor cost (in dollars) for manufacturing a precision camera can be approximated by

$$L(x, y) = \frac{3}{2}x^2 + y^2 - 2x - 2y - 2xy + 68,$$

where x is the number of hours required by a skilled craftsperson and y is the number of hours required by a semiskilled person. Find values of x and y that minimize the labor cost. Find the minimum labor cost.

37. *Cost* The total cost (in dollars) to produce x units of electrical tape and y units of packing tape is given by

$$C(x, y) = 2x^2 + 3y^2 - 2xy + 2x - 126y + 3,800.$$

Find the number of units of each kind of tape that should be produced so that the total cost is a minimum. Find the minimum total cost.

38. *Revenue* The total revenue (in hundreds of dollars) from the sale of x spas and y solar heaters is approximated by

$$R(x, y) = 12 + 74x + 85y - 3x^2 - 5y^2 - 5xy.$$

Find the number of each that should be sold to produce maximum revenue. Find the maximum revenue.

■ 9.4 TOTAL DIFFERENTIALS AND APPROXIMATIONS

? THINK ABOUT IT

How do errors in measuring the length and radius of a blood vessel affect the calculation of its volume?

In the second section of this chapter we used partial derivatives to find the rate of change of a function with respect to one of the variables while the other variables were held constant. To estimate the change of a function for small changes in all the variables, we can extend the concept of differential, introduced in an earlier chapter for functions of one variable, to the concept of *total differential.*

TOTAL DIFFERENTIAL FOR TWO VARIABLES

Let $z = f(x, y)$ be a function of x and y. Let dx and dy be real numbers. Then the **total differential** of z is

$$dz = f_x(x,y) \cdot dx + f_y(x,y) \cdot dy.$$

(Sometimes dz is written df.)

Recall that the differential for a function of one variable $y = f(x)$ is used to approximate the function by its tangent line. This works because a differentiable function appears very much like a line when viewed closely. Similarly, the differential for a function of two variables $z = f(x, y)$ is used to approximate a function by its tangent plane. A differentiable function of two variables looks like a plane when viewed closely, which is why the earth looks flat when you are standing on it.

EXAMPLE 1 Total Differentials

Consider the function $z = f(x, y) = 9x^3 - 8x^2y + 4y^3$.

FOR REVIEW ■──────

In the chapter on Applications of the Derivative, we introduced the differential. Recall that the differential of a function defined by $y = f(x)$ is

$$dy = f'(x) \cdot dx,$$

where dx, the differential of x, is any real number (usually small). We saw that the differential dy is often a good approximation of Δy, where $\Delta y = f(x + \Delta x) - f(x)$ and $\Delta x = dx$.

(a) Find dz.

Solution First find $f_x(x, y)$ and $f_y(x, y)$.

$$f_x(x, y) = 27x^2 - 16xy \qquad \text{and} \qquad f_y(x, y) = -8x^2 + 12y^2$$

By the definition,

$$dz = (27x^2 - 16xy)\,dx + (-8x^2 + 12y^2)\,dy.$$

(b) Evaluate dz when $x = 1$, $y = 3$, $dx = 0.01$, and $dy = -0.02$.

Solution Putting these values into the result from part (a) gives

$$dz = [27(1)^2 - 16(1)(3)](0.01) + [-8(1)^2 + 12(3)^2](-0.02)$$
$$= (-21)(0.01) + (100)(-0.02)$$
$$= -2.21.$$

This result indicates that an increase of 0.01 in x and a decrease of 0.02 in y, when $x = 1$ and $y = 3$, will produce an approximate *decrease* of 2.21 in $f(x, y)$. ■

Approximations Recall that with a function of one variable, $y = f(x)$, the differential dy approximates the change in y, Δy, corresponding to a change in x, Δx or dx. The approximation for a function of two variables is similar.

> **APPROXIMATIONS**
>
> For small values of Δx and Δy,
>
> $$dz \approx \Delta z,$$
>
> where $\Delta z = f(x + \Delta x, y + \Delta y) - f(x, y)$.

EXAMPLE 2 Approximations

Approximate $\sqrt{2.98^2 + 4.01^2}$.

Solution Notice that $2.98 \approx 3$ and $4.01 \approx 4$, and we know that $\sqrt{3^2 + 4^2} = \sqrt{25} = 5$. We therefore let $f(x, y) = \sqrt{x^2 + y^2}$, $x = 3$, $dx = -0.02$, $y = 4$, and $dy = 0.01$. We then use dz to approximate $\Delta z = \sqrt{2.98^2 + 4.01^2} - \sqrt{3^2 + 4^2}$.

$$dz = f_x(x, y) \cdot dx + f_y(x, y) \cdot dy$$

$$= \left(\frac{1}{2\sqrt{x^2 + y^2}} \cdot 2x \right) dx + \left(\frac{1}{2\sqrt{x^2 + y^2}} \cdot 2y \right) dy$$

$$= \left(\frac{x}{\sqrt{x^2 + y^2}} \right) dx + \left(\frac{y}{\sqrt{x^2 + y^2}} \right) dy$$

$$= \frac{3}{5}(-0.02) + \frac{4}{5}(0.01)$$

$$= -0.004$$

Thus, $\sqrt{2.98^2 + 4.01^2} \approx 5 + (-0.004) = 4.996$. A calculator gives $\sqrt{2.98^2 + 4.01^2} \approx 4.996048$. The error is approximately 0.000048. ▬

For small values of dx and dy, the values of Δz and dz are approximately equal. Since $\Delta z = f(x + dx, y + dy) - f(x, y)$,

$$f(x + dx, y + dy) = f(x, y) + \Delta z$$

or

$$f(x + dx, y + dy) \approx f(x, y) + dz.$$

Replacing dz with the expression for the total differential gives the following result.

APPROXIMATIONS BY DIFFERENTIALS

For a function f having all indicated partial derivatives, and for small values of dx and dy,

$$f(x + dx, y + dy) \approx f(x, y) + dz,$$

or

$$f(x + dx, y + dy) \approx f(x, y) + f_x(x, y) \cdot dx + f_y(x, y) \cdot dy.$$

The idea of a total differential can be extended to include functions of three or more independent variables.

TOTAL DIFFERENTIAL FOR THREE VARIABLES

If $w = f(x, y, z)$, then the total differential dw is

$$dw = f_x(x, y, z)\, dx + f_y(x, y, z)\, dy + f_z(x, y, z)\, dz,$$

provided all indicated partial derivatives exist.

EXAMPLE 3   Blood Vessels

A short length of blood vessel is in the shape of a right circular cylinder (see Figure 24).

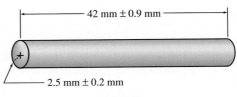

42 mm ± 0.9 mm

2.5 mm ± 0.2 mm

FIGURE 24

(a) The length of the vessel is measured as 42 millimeters, and the radius is measured as 2.5 millimeters. Suppose the maximum error in the measurement of the length is 0.9 millimeter, with an error of no more than 0.2 millimeter in the measurement of the radius. Find the maximum possible error in calculating the volume of the blood vessel.

Solution   The volume of a right circular cylinder is given by $V = \pi r^2 h$. To approximate the error in the volume, find the total differential, dV.

$$dV = (2\pi r h) \cdot dr + (\pi r^2) \cdot dh$$

Here, $r = 2.5$, $h = 42$, $dr = 0.2$, and $dh = 0.9$. Substitution gives

$$dV = [(2\pi)(2.5)(42)(0.2)] + [\pi(2.5)^2](0.9) \approx 149.6.$$

The maximum possible error in calculating the volume is approximately 149.6 cubic millimeters.

(b) Suppose that the errors in measuring the radius and length of the vessel are at most 1% and 3%, respectively. Estimate the maximum percent error in calculating the volume.

Solution To find the percent error, calculate dV/V.

$$\frac{dV}{V} = \frac{(2\pi rh)\,dr + (\pi r^2)\,dh}{\pi r^2 h} = 2\frac{dr}{r} + \frac{dh}{h}$$

Because $dr/r = 0.01$ and $dh/h = 0.03$,

$$\frac{dV}{V} = 2(0.01) + 0.03 = 0.05.$$

The maximum percent error in calculating the volume is approximately 5%.

EXAMPLE 4   Volume of a Can of Beer

The formula for the volume of a cylinder given in Example 3 also applies to cans of beer, for which $r \approx 1$ inch and $h \approx 5$ inches. How sensitive is the volume to changes in the radius compared with changes in the height?

Solution Using the formula for dV from the previous example with $r = 1$ and $h = 5$ gives

$$dV = (2\pi)(1)(5)\,dr + \pi(1)^2\,dh = \pi(10\,dr + dh).$$

The factor of 10 in front of dr in this equation shows that a small change in the radius has 10 times the effect on the volume as a small change in the height. One author argues that this is the reason that beer cans are so tall and thin.* The brewers can reduce the radius by a tiny amount, and compensate by making the can taller. The resulting can appears larger in volume than the shorter, wider can. (Others have argued that a shorter, wider can does not fit as easily in the hand.)

9.4 EXERCISES

Use the total differential to approximate each quantity. Then use a calculator to approximate the quantity, and give the absolute value of the difference in the two results to four decimal places.

1. $\sqrt{3.04^2 + 4.06^2}$

2. $\sqrt{6.07^2 + 7.95^2}$

3. $(1.92^2 + 2.1^2)^{1/3}$

4. $(2.93^2 - 0.94^2)^{1/3}$

5. $0.97e^{0.02}$

6. $1.02e^{-0.03}$

7. $1.04 \ln 0.95$

8. $0.95 \ln 1.04$

Evaluate dz using the given information.

9. $z = x^2 + 3xy + y^2$; $x = 4, y = -2, dx = 0.02, dy = -0.03$

10. $z = 8x^3 + 2x^2y - y$; $x = 1, y = 3, dx = 0.01, dy = 0.02$

---

*Colley, Susan Jane, "Calculus in the Brewery," *The College Mathematics Journal*, Vol. 25, No. 3, May 1994, p. 227.

11. $z = \dfrac{y^2 + 3x}{y^2 - x}$; $\quad x = 4, y = -4, dx = 0.01, dy = 0.03$

12. $z = \ln(x^2 + y^2)$; $\quad x = 2, y = 3, dx = 0.02, dy = -0.03$

Evaluate dw using the given information.

13. $w = \dfrac{5x^2 + y^2}{z + 1}$; $\quad x = -2, y = 1, z = 1, dx = 0.02, dy = -0.03, dz = 0.02$

14. $w = x \ln(yz) - y \ln \dfrac{x}{z}$; $\quad x = 2, y = 1, z = 4, dx = 0.03, dy = 0.02, dz = -0.01$

Applications

LIFE SCIENCES

15. *Bone Preservative Volume* A piece of bone in the shape of a right circular cylinder is 7 cm long and has a radius of 1.4 cm. It is coated with a layer of preservative 0.09 cm thick. Estimate the volume of preservative used.

16. *Blood Vessel Volume* A portion of a blood vessel is measured as having length 7.9 cm and radius 0.8 cm. If each measurement could be off by as much as 0.15 cm, estimate the maximum possible error in calculating the volume of the vessel.

17. *Blood Volume* In Exercise 49 of Section 2 in this chapter, we found that the number of liters of blood pumped through the lungs in one minute is given by

$$C = \frac{b}{a - v}.$$

Suppose $a = 160, b = 200, v = 125$. Estimate the change in C if a becomes 145, b becomes 190, and v changes to 130.

18. *Oxygen Consumption* In Exercise 48 of Section 2 of this chapter, we found that the oxygen consumption of a mammal is

$$m = \frac{2.5(T - F)}{w^{0.67}}.$$

Suppose T is 38°C, F is 12°C, and w is 30 kg. Approximate the change in m if T changes to 36°C, F changes to 13°C, and w becomes 31 kg.

19. *Dialysis* A model that estimates the concentration of urea in the body for a particular dialysis patient, following a dialysis session, is given by

$$C(t, g) = 0.6(0.96)^{7t/50 - 1}$$
$$+ \frac{gt}{126t - 900}(1 - (0.96)^{7t/50 - 1}),$$

where t represents the number of minutes of the dialysis session and g represents the rate at which the body generates urea in mg/min.[*]

a. Find $C(180, 8)$.

b. Using the total differential, estimate the urea concentration if the dialysis session of part a was cut short by 10 minutes and the urea generation rate was 9 mg/min. Compare this with the actual concentration. (*Hint:* First, replace the variable g with the number 8, thus reducing the function to one variable. Then use your graphing calculator to calculate the partial derivative $C_t(180, 8)$. A similar procedure can be done for $C_y(180, 8)$.)

20. *Horn Volume* The volume of the horns from bighorn sheep were estimated by researchers using the equation

$$V = \frac{h\pi}{3}(r_1^2 + r_1 r_2 + r_2^2),$$

where h is the length of a horn segment (in cm) and r_1 and r_2 are the radii of the two ends of the horn segment (in cm).[†]

a. Determine the volume of a segment of horn that is 40 cm long with radii of 5 cm and 3 cm, respectively.

b. Use the total differential to estimate the volume of the segment of horn if the horn segment from part a was actually 42 cm long with radii of 5.1 cm and 2.9 cm, respectively. Compare this with the actual volume.

OTHER APPLICATIONS

21. *Swimming* The amount of time in seconds it takes for a swimmer to hear a single, hand-held, starting signal is given by the formula

$$t(x, y, p, C) = \frac{\sqrt{x^2 + (y - p)^2}}{331.45 + 0.6C},$$

[*]Gotch, Frank, "Clinical Dialysis: Kinetic Modeling in Hemodialysis," *Clinical Dialysis,* 3rd ed., Norwalk, Appleton & Lange, 1995, pp. 156–186.
[†]Fitzsimmons, N., S. Buskirk, and M. Smith, "Population History, Genetic Variability, and Horn Growth in Bighorn Sheep," *Conservation Biology,* Vol. 9, No. 2, April 1995, pp. 314–323.

$x, y)$ is the location of the starter in meters, $(0, p)$ is
...ation of the swimmer in meters, and C is the air temperature in Celsius.*

Assume that the starter is located at the point $(x, y) = (5, -2)$. See the diagram.

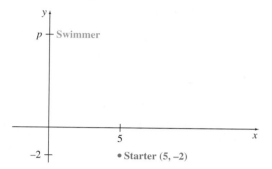

a. Calculate $t(5, -2, 20, 20)$ and $t(5, -2, 10, 20)$. Could the difference in time change the outcome of a race?

b. Calculate the total differential for t if the starter remains stationary, the swimmer moves from 20 m to 20.5 m away from the starter in the y direction, and the temperature decreases from 20°C to 15°C. Interpret your answer.

22. **Estimating Area** The height of a triangle is measured as 42.6 cm, with the base measured as 23.4 cm. The measurement of the height can be off by as much as 1.2 cm, and that of the base by no more than 0.9 cm. Estimate the maximum possible error in calculating the area of the triangle.

23. **Estimating Volume** The height of a cone is measured as 8.4 cm and the radius as 2.9 cm. Each measurement could be off by as much as 0.1 cm. Estimate the maximum possible error in calculating the volume of the cone.

24. **Estimating Volume** Suppose that in measuring the length, width, and height of a box, there is a maximum 1% error in each measurement. Estimate the maximum error in calculating the volume of the box.

25. **Estimating Volume** Suppose there is a maximum error of $a\%$ in measuring the radius of a cone and a maximum error of $b\%$ in measuring the height. Estimate the maximum percent error in calculating the volume of the cone, and compare this value with the maximum percent error in calculating the volume of a cylinder.

26. **Manufacturing** Approximate the amount of aluminum needed for a beverage can of radius 2.5 cm and height 14 cm. Assume the walls of the can are 0.08 cm thick.

27. **Manufacturing** Approximate the amount of material needed to make a water tumbler of diameter 3 cm and height 9 cm. Assume the walls of the tumbler are 0.2 cm thick.

28. **Volume of a Coating** An industrial coating 0.2 inch thick is applied to all sides of a box of dimensions 10 inches by 9 inches by 14 inches. Estimate the volume of the coating used.

9.5 DOUBLE INTEGRALS

? THINK ABOUT IT   How can we find the volume of a bottle with curved sides?

In an earlier chapter, we saw how integrals of functions with one variable may be used to find area. In this section, this idea is extended and used to find volume. We found partial derivatives of functions of two or more variables at the beginning of this chapter by holding constant all variables except one. A similar

*Walker, Anita, "Mathematics Makes a Splash: Evaluating Hand Timing Systems," *The HiMAP Pull-Out Section*, Spring 1992, COMAP.

FOR REVIEW ■▬

You may wish to review the key
ideas of indefinite and definite
integrals from the chapter on
Integration before continuing
with this section. See the review
problems at the end of that
chapter.

process is used in this section to find antiderivatives of functions of two or more
variables. For example, in

$$\int (5x^3y^4 - 6x^2y + 2)\, dy$$

the notation dy indicates integration with respect to y, so we treat y as the variable
and x as a constant. Using the rules for antiderivatives gives

$$\int (5x^3y^4 - 6x^2y + 2)\, dy = 5x^3 \int y^4\, dy - 6x^2 \int y\, dy + 2 \int dy$$

$$= x^3y^5 - 3x^2y^2 + 2y + C(x).$$

The constant C used earlier must be replaced with $C(x)$ to show that the "constant
of integration" here can be any function involving only the variable x. Just as before, check this work by taking the derivative (actually the partial derivative) of
the answer:

$$\frac{\partial}{\partial y}[x^3y^5 - 3x^2y^2 + 2y + C(x)] = 5x^3y^4 - 6x^2y + 2 + 0,$$

which shows that the antiderivative is correct.

EXAMPLE 1 Indefinite Integrals
Find each indefinite integral.

(a) $\displaystyle\int x(x^2 + y)\, dx$

Solution Multiply x and $x^2 + y$. Then (because of the dx) integrate each term
with x as the variable and y as a constant.

$$\int x(x^2 + y)\, dx = \int (x^3 + xy)\, dx$$

$$= \frac{x^4}{4} + \frac{x^2}{2} \cdot y + f(y) = \frac{1}{4}x^4 + \frac{1}{2}x^2y + f(y)$$

(b) $\displaystyle\int x(x^2 + y)\, dy$

Solution Since y is the variable and x is held constant,

$$\int x(x^2 + y)\, dy = \int (x^3 + xy)\, dy = x^3y + \frac{1}{2}xy^2 + g(x).$$

The analogy to integration of functions of one variable can be continued for
evaluating definite integrals. We do this by holding one variable constant and
using the Fundamental Theorem of Calculus with the other variable.

EXAMPLE 2 Definite Integrals
Evaluate each definite integral.

(a) $\displaystyle\int_3^5 (6xy^2 + 12x^2y + 4y)\, dx$

Solution First, find an antiderivative:

$$\int (6xy^2 + 12x^2y + 4y)\, dx = 3x^2y^2 + 4x^3y + 4xy + h(y).$$

Now replace each x with 5, and then with 3, and subtract the results.

$$[3x^2y^2 + 4x^3y + 4xy + h(y)]\Big|_3^5 = [3 \cdot 5^2 \cdot y^2 + 4 \cdot 5^3 \cdot y + 4 \cdot 5 \cdot y + h(y)]$$
$$-[3 \cdot 3^2 \cdot y^2 + 4 \cdot 3^3 \cdot y + 4 \cdot 3 \cdot y + h(y)]$$
$$= 75y^2 + 500y + 20y + h(y)$$
$$-[27y^2 + 108y + 12y + h(y)]$$
$$= 48y^2 + 400y$$

The *function of integration*, $h(y)$, drops out, just as the constant of integration does with definite integrals of functions of one variable. Thus, the function of integration is not included for definite integrals of more than one variable.

(b) $\displaystyle\int_1^2 (6xy^2 + 12x^2y + 4y)\, dy$

Solution Integrate with respect to y; then substitute 2 and 1 for y and subtract.

$$\int_1^2 (6xy^2 + 12x^2y + 4y)\, dy = (2xy^3 + 6x^2y^2 + 2y^2)\Big|_1^2$$
$$= (2x \cdot 2^3 + 6x^2 \cdot 2^2 + 2 \cdot 2^2)$$
$$-(2x \cdot 1^3 + 6x^2 \cdot 1^2 + 2 \cdot 1^2)$$
$$= 16x + 24x^2 + 8 - (2x + 6x^2 + 2)$$
$$= 14x + 18x^2 + 6$$

As Example 2 suggests, an integral of the form

$$\int_a^b f(x, y)\, dy$$

produces a result that is a function of x, while

$$\int_a^b f(x, y)\, dx$$

produces a function of y. These resulting functions of one variable can themselves be integrated, as in the next example.

EXAMPLE 3 Definite Integrals

Evaluate each integral.

(a) $\displaystyle\int_1^2 \left[\int_3^5 (6xy^2 + 12x^2y + 4y)\, dx \right] dy$

Solution In Example 2(a), we found the quantity in brackets to be $48y^2 + 400y$. Thus,

$$\int_1^2 \left[\int_3^5 (6xy^2 + 12x^2y + 4y)\, dx \right] dy = \int_1^2 (48y^2 + 400y)\, dy$$

$$= (16y^3 + 200y^2) \Big|_1^2$$

$$= 16 \cdot 2^3 + 200 \cdot 2^2 - (16 \cdot 1^3 + 200 \cdot 1^2)$$

$$= 128 + 800 - (16 + 200)$$

$$= 712.$$

(b) $\displaystyle \int_3^5 \left[\int_1^2 (6xy^2 + 12x^2y + 4y)\, dy \right] dx$

Solution (This is the same integrand, with the same limits of integration as in part (a), but the order of integration is reversed.)
Use the result from Example 2(b).

$$\int_3^5 \left[\int_1^2 (6xy^2 + 12x^2y + 4y)\, dy \right] dx = \int_3^5 (14x + 18x^2 + 6)\, dx$$

$$= (7x^2 + 6x^3 + 6x) \Big|_3^5$$

$$= 7 \cdot 5^2 + 6 \cdot 5^3 + 6 \cdot 5 - (7 \cdot 3^2 + 6 \cdot 3^3 + 6 \cdot 3)$$

$$= 175 + 750 + 30 - (63 + 162 + 18) = 712 \quad \blacksquare$$

The brackets we have used for the inner integral in Example 3 are not essential because the order of integration is indicated by the order of $dx\,dy$ or $dy\,dx$. For example, if the integral is written as

$$\int_1^2 \int_3^5 (6xy^2 + 12x^2y + 4y)\, dx\, dy,$$

we first integrate with respect to x, letting x vary from 3 to 5, and then with respect to y, letting y vary from 1 to 2, as in Example 3(a).

The answers in the two parts of Example 3 are equal. It can be proved that for a large class of functions, including most functions that occur in applications, the following equation holds true.

$$\int_a^b \int_c^d f(x,y)\, dx\, dy = \int_c^d \int_a^b f(x,y)\, dy\, dx$$

Either of these integrals is called an **iterated integral** since it is evaluated by integrating twice, first using one variable and then using the other. The fact that the iterated integrals above are equal makes it possible to define a *double integral*. First, the set of points (x, y), with $c \le x \le d$ and $a \le y \le b$, defines a rectangular region R in the plane, as shown in Figure 25. Then, the *double integral over R* is defined as follows.

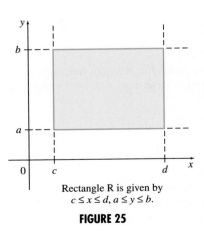

Rectangle R is given by
$c \le x \le d, a \le y \le b.$

FIGURE 25

> **DOUBLE INTEGRAL**
>
> The **double integral** of $f(x, y)$ over a rectangular region R is written
>
> $$\iint_R f(x,y)\, dx\, dy \qquad \text{or} \qquad \iint_R f(x,y)\, dy\, dx,$$
>
> and equals
>
> $$\int_a^b \int_c^d f(x,y)\, dx\, dy \qquad \text{or} \qquad \int_c^d \int_a^b f(x,y)\, dy\, dx.$$

Extending earlier definitions, $f(x, y)$ is the **integrand** and R is the **region of integration.**

EXAMPLE 4 Double Integrals

Find $\displaystyle\iint_R \sqrt{x} \cdot \sqrt{y-2}\, dx\, dy$ over the rectangular region R defined by $0 \le x \le 4,\, 3 \le y \le 11$.

Solution Integrate first with respect to x; then integrate the result with respect to y.

$$\iint_R \sqrt{x} \cdot \sqrt{y-2}\, dx\, dy = \int_3^{11} \int_0^4 \sqrt{x} \cdot \sqrt{y-2}\, dx\, dy$$

$$= \int_3^{11} \left(\frac{2}{3} x^{3/2}\sqrt{y-2} \right)\bigg|_0^4 dy$$

$$= \int_3^{11} \left[\frac{2}{3}(4^{3/2})\sqrt{y-2} - \frac{2}{3}(0^{3/2})\sqrt{y-2} \right] dy$$

$$= \int_3^{11} \left(\frac{16}{3}\sqrt{y-2} - 0 \right) dy = \int_3^{11} \left(\frac{16}{3}\sqrt{y-2} \right) dy$$

$$= \frac{32}{9}(y-2)^{3/2}\bigg|_3^{11} = \frac{32}{9}(9)^{3/2} - \frac{32}{9}(1)^{3/2}$$

$$= 96 - \frac{32}{9} = \frac{832}{9}$$

As a check, integrate with respect to y first. The answer should be the same.

> **NOTE** In the second step of the previous example, it might help you avoid confusion as to whether to put the limits of 0 and 4 into x or y by writing the integral as
>
> $$\int_3^{11} \left(\frac{2}{3}x^{3/2}\sqrt{y-2} \right)\bigg|_{x=0}^{x=4} dy.$$

Volume As shown earlier, the definite integral $\int_a^b f(x)\, dx$ can be used to find the area under a curve. In a similar manner, double integrals are used to find the *volume under a surface.* Figure 26 shows that portion of a surface $f(x, y)$ directly over a rectangle R in the xy-plane. Just as areas were approximated by a large

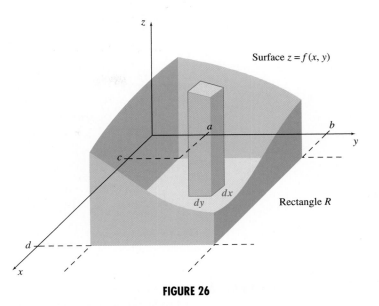

FIGURE 26

number of small rectangles, volume could be approximated by adding the volumes of a large number of properly drawn small boxes. The height of a typical box would be $f(x, y)$ with the length and width given by dx and dy. The formula for the volume of a box would then suggest the following result.

VOLUME

Let $z = f(x, y)$ be a function that is never negative on the rectangular region R defined by $c \le x \le d, a \le y \le b$. The volume of the solid under the graph of f and over the region R is

$$\iint\limits_{R} f(x,y)\,dx\,dy.$$

EXAMPLE 5 Volume

Find the volume under the surface $z = x^2 + y^2$ shown in Figure 27.

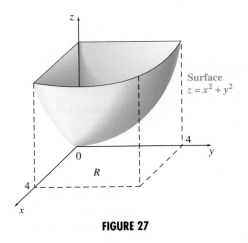

FIGURE 27

Solution By the equation just given, the volume is

$$\iint\limits_{R} f(x, y) \, dx \, dy,$$

where $f(x, y) = x^2 + y^2$ and R is the region $0 \leq x \leq 4, 0 \leq y \leq 4$. By definition,

$$\iint\limits_{R} f(x, y) \, dx \, dy = \int_{0}^{4} \int_{0}^{4} (x^2 + y^2) \, dx \, dy$$

$$= \int_{0}^{4} \left(\frac{1}{3} x^3 + xy^2 \right) \Big|_{0}^{4} \, dy$$

$$= \int_{0}^{4} \left(\frac{64}{3} + 4y^2 \right) dy = \left(\frac{64}{3} y + \frac{4}{3} y^3 \right) \Big|_{0}^{4}$$

$$= \frac{64}{3} \cdot 4 + \frac{4}{3} \cdot 4^3 - 0 = \frac{512}{3}.$$

EXAMPLE 6 Perfume Bottle

A product design consultant for a cosmetics company has been asked to design a bottle for the company's newest perfume. The thickness of the glass is to vary so that the outside of the bottle has straight sides and the inside has curved sides, with flat ends shaped like parabolas on the 4-cm sides, as shown in Figure 28. Before presenting the design to management, the consultant needs to make a reasonably accurate estimate of the amount each bottle will hold. If the base of the bottle is to be 4 centimeters by 3 centimeters, and if a cross section of its interior is to be a parabola of the form $z = -y^2 + 4y$, what is its internal volume?

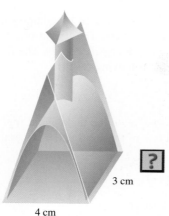

3 cm

4 cm

FIGURE 28

Solution The interior of the bottle can be graphed in three-dimensional space, as shown in Figure 29, where $z = 0$ corresponds to the base of the bottle. Its volume is simply the volume above the region R in the xy-plane and below the graph of $f(x, y) = -y^2 + 4y$. This volume is given by the double integral

$$\int_{0}^{3} \int_{0}^{4} (-y^2 + 4y) \, dy \, dx = \int_{0}^{3} \left(\frac{-y^3}{3} + \frac{4y^2}{2} \right) \Big|_{0}^{4} \, dx$$

$$= \int_{0}^{3} \left(\frac{-64}{3} + 32 - 0 \right) dx$$

$$= \frac{32}{3} x \Big|_{0}^{3}$$

$$= 32 - 0 = 32.$$

The bottle holds 32 cubic centimeters.

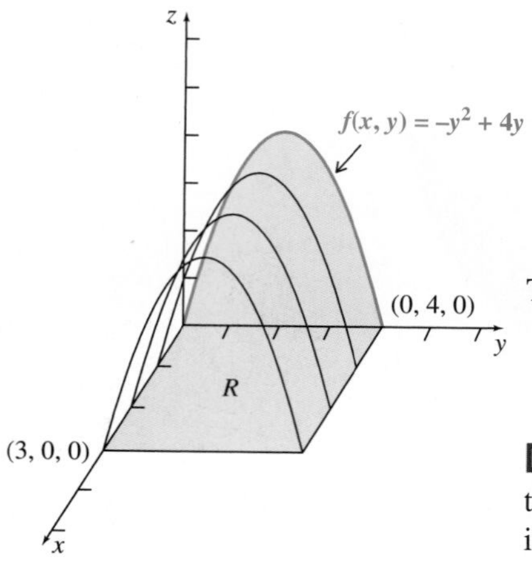

$f(x, y) = -y^2 + 4y$

$(0, 4, 0)$

R

$(3, 0, 0)$

FIGURE 29

Double Integrals Over Other Regions In this section, we found double integrals over rectangular regions by evaluating iterated integrals with constant limits of integration. We can also evaluate iterated integrals with *variable* limits of integration. (Notice in the following examples that the variable limits always go on the *inner* integral sign.)

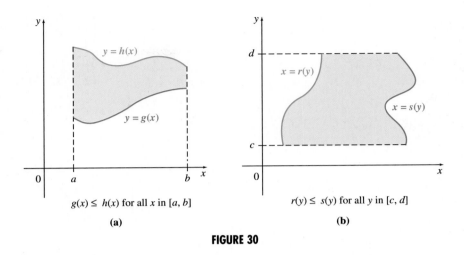

$g(x) \le h(x)$ for all x in $[a, b]$

(a)

$r(y) \le s(y)$ for all y in $[c, d]$

(b)

FIGURE 30

The use of variable limits of integration permits evaluation of double integrals over the types of regions shown in Figure 30. Double integrals over more complicated regions are discussed in more advanced books. Integration over regions such as those in Figure 30 is done with the results of the following theorem.

DOUBLE INTEGRALS OVER VARIABLE REGIONS

Let $z = f(x, y)$ be a function of two variables. If R is the region (in Figure 30(a)) defined by $a \le x \le b$ and $g(x) \le y \le h(x)$, then

$$\iint\limits_{R} f(x,y)\, dy\, dx = \int_a^b \left[\int_{g(x)}^{h(x)} f(x,y)\, dy \right] dx.$$

If R is the region (in Figure 30(b)) defined by $r(y) \le x \le s(y)$ and $c \le y \le d$, then

$$\iint\limits_{R} f(x,y)\, dx\, dy = \int_c^d \left[\int_{r(y)}^{s(y)} f(x,y)\, dx \right] dy.$$

EXAMPLE 7 Double Integrals

Evaluate $\displaystyle\int_1^2 \int_y^{y^2} xy\, dx\, dy$.

Solution The region of integration is shown in Figure 31. Integrate first with respect to x, then with respect to y.

$$\int_1^2 \int_y^{y^2} xy\, dx\, dy = \int_1^2 \left[\int_y^{y^2} xy\, dx \right] dy = \int_1^2 \left(\frac{1}{2} x^2 y \right) \Big|_y^{y^2} dy$$

Replace x first with y^2 and then with y, and subtract.

$$\int_1^2 \int_y^{y^2} xy\, dx\, dy = \int_1^2 \left[\frac{1}{2}(y^2)^2 y - \frac{1}{2}(y)^2 y \right] dy$$

$$= \int_1^2 \left(\frac{1}{2} y^5 - \frac{1}{2} y^3 \right) dy = \left(\frac{1}{12} y^6 - \frac{1}{8} y^4 \right) \Big|_1^2$$

$$= \left(\frac{1}{12} \cdot 2^6 - \frac{1}{8} \cdot 2^4 \right) - \left(\frac{1}{12} \cdot 1^6 - \frac{1}{8} \cdot 1^4 \right)$$

$$= \frac{64}{12} - \frac{16}{8} - \frac{1}{12} + \frac{1}{8} = \frac{27}{8}$$

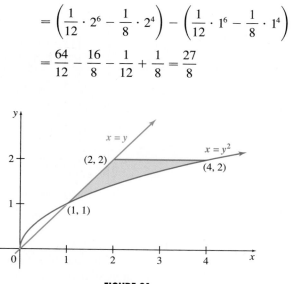

FIGURE 31

EXAMPLE 8 Double Integrals

Let R be the shaded region in Figure 32, and evaluate

$$\iint_R (x + 2y)\, dy\, dx.$$

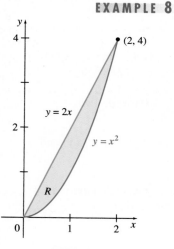

FIGURE 32

Solution Region R is bounded by $h(x) = 2x$ and $g(x) = x^2$, with $0 \le x \le 2$. By the first result in the previous theorem,

$$\iint_R (x + 2y)\, dy\, dx = \int_0^2 \int_{x^2}^{2x} (x + 2y)\, dy\, dx$$

$$= \int_0^2 (xy + y^2) \Big|_{x^2}^{2x} dx$$

$$= \int_0^2 (x(2x) + (2x)^2 - [x \cdot x^2 + (x^2)^2])\, dx$$

$$= \int_0^2 [2x^2 + 4x^2 - (x^3 + x^4)]\, dx$$

$$= \int_0^2 (6x^2 - x^3 - x^4)\, dx$$

$$= \left(2x^3 - \frac{1}{4}x^4 - \frac{1}{5}x^5 \right) \Big|_0^2$$

$$= 2 \cdot 2^3 - \frac{1}{4} \cdot 2^4 - \frac{1}{5} \cdot 2^5 - 0$$

$$= 16 - 4 - \frac{32}{5} = \frac{28}{5}.$$

In Example 8, the same result would be found if we evaluated the double integral first with respect to x, and then with respect to y. In that case, we would need to define the equations of the boundaries in terms of y rather than x, so R would be defined by $y/2 \leq x \leq \sqrt{y}, 0 \leq y \leq 4$. The resulting integral is

$$\int_0^4 \int_{y/2}^{\sqrt{y}} (x + 2y)\, dx\, dy = \int_0^4 \left(\frac{x^2}{2} + 2xy \right) \Bigg|_{y/2}^{\sqrt{y}} dy$$

$$= \int_0^4 \left[\left(\frac{y}{2} + 2y\sqrt{y} \right) - \left(\frac{y^2}{8} + 2(y/2)y \right) \right] dy$$

$$= \int_0^4 \left(\frac{y}{2} + 2y^{3/2} - \frac{9}{8} y^2 \right) dy$$

$$= \left(\frac{y^2}{4} + \frac{4}{5} y^{5/2} - \frac{3}{8} y^3 \right) \Bigg|_0^4$$

$$= 4 + \frac{4}{5} \cdot 4^{5/2} - 24$$

$$= \frac{28}{5}.$$

Interchanging Limits of Integration Sometimes it is easier to integrate first with respect to x, and then y, while with other integrals the reverse process is easier. The limits of integration can be reversed whenever the region R is like the region in Figure 32, which has the property that it can be viewed as either type of region shown in Figure 30. The next example shows how this process works.

EXAMPLE 9 Interchanging Limits of Integration

Evaluate

$$\int_0^{16} \int_{\sqrt{y}}^4 \sqrt{x^3 + 4}\, dx\, dy.$$

Solution Notice that it is impossible to first integrate this function with respect to x. Thus, we attempt to interchange the limits of integration.

For this integral, region R is given by $\sqrt{y} \leq x \leq 4, 0 \leq y \leq 16$. A graph of R is shown in Figure 33.

The same region R can be written in an alternate way. As Figure 33 shows, one boundary of R is $x = \sqrt{y}$. Solving for y gives $y = x^2$. Also, Figure 33 shows that $0 \leq x \leq 4$. Since R can be written as $0 \leq y \leq x^2, 0 \leq x \leq 4$, the double integral above can be written

$$\int_0^4 \int_0^{x^2} \sqrt{x^3 + 4}\, dy\, dx = \int_0^4 y\sqrt{x^3 + 4} \Bigg|_0^{x^2} dx$$

$$= \int_0^4 x^2 \sqrt{x^3 + 4}\, dx$$

$$= \frac{1}{3} \int_0^4 3x^2 \sqrt{x^3 + 4}\, dx \quad \text{Let } u = x^3 + 4.$$

$$= \frac{1}{3} \int_4^{68} u^{1/2}\, du$$

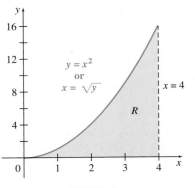

FIGURE 33

$$= \frac{2}{9} u^{3/2} \Big|_4^{68}$$

$$= \frac{2}{9} [68^{3/2} - 4^{3/2}]$$

$$\approx 122.83.$$

9.5 EXERCISES

Evaluate the following integrals.

1. $\int_0^3 (x^3y + y)\, dx$

2. $\int_1^4 (xy^2 - x)\, dy$

3. $\int_4^8 \sqrt{6x + y}\, dx$

4. $\int_3^7 \sqrt{x + 5y}\, dy$

5. $\int_4^5 x\sqrt{x^2 + 3y}\, dy$

6. $\int_3^6 x\sqrt{x^2 + 3y}\, dx$

7. $\int_4^9 \frac{3 + 5y}{\sqrt{x}}\, dx$

8. $\int_2^7 \frac{3 + 5y}{\sqrt{x}}\, dy$

9. $\int_{-1}^1 e^{x+4y}\, dy$

10. $\int_2^6 e^{x+4y}\, dx$

11. $\int_0^5 xe^{x^2+9y}\, dx$

12. $\int_1^6 xe^{x^2+9y}\, dy$

Evaluate the following iterated integrals. (Many of these use results from Exercises 1–12.)

13. $\int_1^2 \int_0^3 (x^3y + y)\, dx\, dy$

14. $\int_0^3 \int_1^4 (xy^2 - x)\, dy\, dx$

15. $\int_0^1 \int_3^6 x\sqrt{x^2 + 3y}\, dx\, dy$

16. $\int_0^3 \int_4^5 x\sqrt{x^2 + 3y}\, dy\, dx$

17. $\int_1^2 \int_4^9 \frac{3 + 5y}{\sqrt{x}}\, dx\, dy$

18. $\int_{16}^{25} \int_2^7 \frac{3 + 5y}{\sqrt{x}}\, dy\, dx$

19. $\int_1^2 \int_1^2 \frac{dx\, dy}{xy}$

20. $\int_1^4 \int_2^5 \frac{dy\, dx}{x}$

21. $\int_2^4 \int_3^5 \left(\frac{x}{y} + \frac{y}{3}\right) dx\, dy$

22. $\int_3^4 \int_1^2 \left(\frac{6x}{5} + \frac{y}{x}\right) dx\, dy$

Find each double integral over the rectangular region R with the given boundaries.

23. $\iint_R (x + 3y^2)\, dx\, dy; \quad 0 \le x \le 2, 1 \le y \le 5$

24. $\iint_R (4x^3 + y^2)\, dx\, dy; \quad 1 \le x \le 4, 0 \le y \le 2$

25. $\iint_R \sqrt{x + y}\, dy\, dx; \quad 1 \le x \le 3, 0 \le y \le 1$

26. $\iint_R x^2\sqrt{x^3 + 2y}\, dx\, dy; \quad 0 \le x \le 2, 0 \le y \le 3$

27. $\iint_R \frac{2}{(x + y)^2}\, dy\, dx; \quad 2 \le x \le 3, 1 \le y \le 5$

28. $\iint_R \frac{y}{\sqrt{6x + 5y^2}}\, dx\, dy; \quad 0 \le x \le 3, 1 \le y \le 2$

29. $\iint_R ye^{x+y^2}\, dx\, dy; \quad 2 \le x \le 3, 0 \le y \le 2$

30. $\iint_R x^2 e^{x^3+2y}\, dx\, dy; \quad 1 \le x \le 2, 1 \le y \le 3$

31. $\iint_R x \cos(xy)\, dy\, dx; \quad \frac{\pi}{2} \le x \le \pi, 0 \le y \le 1$

32. $\iint_R x \sec^2 y\, dy\, dx; \quad 0 \le x \le 1, 0 \le y \le \frac{\pi}{4}$

Find the volume under the given surface $z = f(x, y)$ and above the rectangle with the given boundaries.

33. $z = 6x + 2y + 5; \quad -1 \le x \le 1, 0 \le y \le 3$

34. $z = 9x + 5y + 12; \quad 0 \le x \le 3, -2 \le y \le 1$

35. $z = x^2; \quad 0 \le x \le 1, 0 \le y \le 4$

36. $z = \sqrt{y}; \quad 0 \le x \le 4, 0 \le y \le 9$

37. $z = x\sqrt{x^2 + y}; \quad 0 \le x \le 1, 0 \le y \le 1$

38. $z = yx\sqrt{x^2 + y^2}; \quad 0 \le x \le 4, 0 \le y \le 1$

39. $z = \frac{xy}{(x^2 + y^2)^2}; \quad 1 \le x \le 2, 1 \le y \le 4$

40. $z = e^{x+y}; \quad 0 \le x \le 1, 0 \le y \le 1$

Although it is true that a double integral can be evaluated by using either dx or dy first, sometimes one choice over the other makes the work easier. Evaluate the double integrals in Exercises 41 and 42 in the easiest way possible.

41. $\iint\limits_R xe^{xy}\,dx\,dy;\quad 0\le x\le 2,\,0\le y\le 1$

42. $\iint\limits_R 2x^3 e^{x^2y}\,dx\,dy;\quad 0\le x\le 1,\,0\le y\le 1$

Evaluate each double integral.

43. $\int_2^4\int_2^{x^2}(x^2+y^2)\,dy\,dx$

44. $\int_0^5\int_0^{2y}(x^2+y)\,dx\,dy$

45. $\int_0^4\int_0^x\sqrt{xy}\,dy\,dx$

46. $\int_1^4\int_0^x\sqrt{x+y}\,dy\,dx$

47. $\int_1^2\int_y^{3y}\frac{1}{x}\,dx\,dy$

48. $\int_1^4\int_x^{x^2}\frac{1}{y}\,dy\,dx$

49. $\int_0^4\int_1^{e^x}\frac{x}{y}\,dy\,dx$

50. $\int_0^1\int_{2x}^{4x}e^{x+y}\,dy\,dx$

Evaluate each double integral. If the function seems too difficult to integrate, try interchanging the limits of integration, as in Exercises 41 and 42.

51. $\int_0^{\ln 2}\int_{e^y}^2\frac{1}{\ln x}\,dx\,dy$

52. $\int_0^2\int_{y/2}^1 e^{x^2}\,dx\,dy$

Use the region R with the indicated boundaries to evaluate each of the following double integrals.

53. $\iint\limits_R(4x+7y)\,dy\,dx;\quad 1\le x\le 3,\,0\le y\le x+1$

54. $\iint\limits_R(3x+9y)\,dy\,dx;\quad 2\le x\le 4,\,2\le y\le 3x$

55. $\iint\limits_R(4-4x^2)\,dy\,dx;\quad 0\le x\le 1,\,0\le y\le 2-2x$

56. $\iint\limits_R\frac{dy\,dx}{x};\quad 1\le x\le 2,\,0\le y\le x-1$

57. $\iint\limits_R e^{x/y^2}\,dy\,dx;\quad 1\le y\le 2,\,0\le x\le y^2$

58. $\iint\limits_R(x^2-y)\,dy\,dx;\quad -1\le x\le 1,\,-x^2\le y\le x^2$

59. $\iint\limits_R x^3y\,dy\,dx;\quad R$ bounded by $y=x^2,\,y=2x$

60. $\iint\limits_R x^2y^2\,dx\,dy;\quad R$ bounded by $y=x,\,y=2x,\,x=1$

61. $\iint\limits_R\frac{dy\,dx}{y};\quad R$ bounded by $y=x,\,y=\frac{1}{x},\,x=2$

62. Recall from the section in the previous chapter, "Volume and Average Value," that volume could be found with a single integral. In this section volume is found using a double integral. Explain when volume can be found with a single integral, and when a double integral is needed.

63. Give an example of a region that cannot be expressed by either of the forms shown in Figure 30. One example is the disk with a hole in the middle between the graphs of $x^2+y^2=1$ and $x^2+y^2=2$ in Figure 10 of this chapter.

*The idea of the average value of a function, discussed earlier for functions of the form $y=f(x)$, can be extended to functions of more than one independent variable. For a function $z=f(x,y)$, the **average value** of f over a region R is defined as*

$$\frac{1}{A}\iint\limits_R f(x,y)\,dx\,dy,$$

where A is the area of the region R. Find the average value for each of the following functions over the regions R having the given boundaries.

64. $f(x,y)=5xy+2y;\quad 1\le x\le 4,\,1\le y\le 2$

65. $f(x,y)=x^2+y^2;\quad 0\le x\le 2,\,0\le y\le 3$

66. $f(x,y)=e^{-5y+3x};\quad 0\le x\le 2,\,0\le y\le 2$

67. $f(x,y)=e^{2x+y};\quad 1\le x\le 2,\,2\le y\le 3$

Applications

68. *Packaging* The manufacturer of a fruit juice drink has decided to try innovative packaging in order to revitalize sagging sales. The fruit juice drink is to be packaged in containers in the shape of tetrahedra in which three edges are perpendicular, as shown in the figure. Two of the perpendicular edges will be 3 inches long, and the third edge will be 6 inches long. Find the volume of the container. (*Hint:* The equation of the plane shown in the figure is $z = f(x, y) = 6 - 2x - 2y$.)

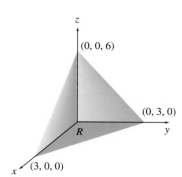

69. *Average Cost* A company's total cost for operating its two warehouses is

$$C(x, y) = \frac{1}{9}x^2 + 2x + y^2 + 5y + 100$$

dollars, where x represents the number of units stored at the first warehouse and y represents the number of units stored at the second. Find the average cost to store a unit if the first warehouse has between 48 and 75 units, and the second has between 20 and 60 units.

70. *Average Profit* The profit (in dollars) from selling x units of one product and y units of a second product is

$$P = -(x - 100)^2 - (y - 50)^2 + 2,000.$$

The weekly sales for the first product vary from 100 units to 150 units, and the weekly sales for the second product vary from 40 units to 80 units. Estimate average weekly profit for these two products.

71. *Average Revenue* A company sells two products. The demand functions of the products are given by

$$q_1 = 300 - 2p_1 \quad \text{and} \quad q_2 = 500 - 1.2p_2,$$

where q_1 units of the first product are demanded at price p_1 and q_2 units of the second product are demanded at price p_2. The total revenue will be given by

$$R = q_1 p_1 + q_2 p_2.$$

Find the average revenue if the price p_1 varies from \$25 to \$50 and the price p_2 varies from \$50 to \$75.

CHAPTER SUMMARY

In this chapter we extended our study of functions to include functions of several variables. The concept of differentiation was generalized to functions of several variables, and partial derivatives were used to determine points where a function of several variables was either maximized or minimized. The concept of differential, introduced in an earlier chapter for functions of one variable, was generalized to the concept of total differential. Total differentials were then used to approximate the value of a function by its tangent plane. Double integrals were introduced and used to find volume.

KEY TERMS

9.1 function of two
  variables
ordered triple
first octant
plane
surface
trace

level curves
paraboloid
production function
ellipsoid
hyperbolic paraboloid
hyperboloid of two
  sheets

9.2 partial derivative
second-order partial
  derivative
9.3 relative maximum
relative minimum
saddle point
critical point

9.4 total differential
9.5 iterated integral
double integral
integrand
region of integration

CHAPTER 9 REVIEW EXERCISES

1. Describe in words how to take a partial derivative.

2. Describe what a partial derivative means geometrically.

3. Describe what a total differential is and how it is useful.

Find $f(-1, 2)$ and $f(6, -3)$ for each of the following.

4. $f(x, y) = -4x^2 + 6xy - 3$

5. $f(x, y) = 3x^2y^2 - 5x + 2y$

6. $f(x, y) = \dfrac{x - 3y}{x + 4y}$

7. $f(x, y) = \dfrac{\sqrt{x^2 + y^2}}{x - y}$

Graph the first-octant portion of each plane.

8. $x + y + z = 4$

9. $x + y + 4z = 8$

10. $5x + 2y = 10$

11. $3x + 5z = 15$

12. $x = 3$

13. $y = 2$

14. Let $z = f(x, y) = -5x^2 + 7xy - y^2$. Find each of the following.

  a. $\dfrac{\partial z}{\partial x}$ **b.** $\left(\dfrac{\partial z}{\partial y}\right)(-1, 4)$ **c.** $f_{xy}(2, -1)$

15. Let $z = f(x, y) = \dfrac{x + y^2}{x - y^2}$. Find each of the following.

  a. $\dfrac{\partial z}{\partial y}$ **b.** $\left(\dfrac{\partial z}{\partial x}\right)(0, 2)$ **c.** $f_{xx}(-1, 0)$

Find $f_x(x, y)$ and $f_y(x, y)$.

16. $f(x, y) = 9x^3y^2 - 5x$

17. $f(x, y) = 6x^5y - 8xy^9$

18. $f(x, y) = \sqrt{4x^2 + y^2}$

19. $f(x, y) = \dfrac{2x + 5y^2}{3x^2 + y^2}$

20. $f(x, y) = x^2e^{2y}$

21. $f(x, y) = (y - 2)^2 e^{x+2y}$

22. $f(x, y) = \ln|2x^2 + y^2|$

23. $f(x, y) = \ln|2 - x^2y^3|$

24. $f(x, y) = \cos(3x^2 + y^2)$

25. $f(x, y) = x \tan(7x^2 + 4y^2)$

Find $f_{xx}(x, y)$ and $f_{xy}(x, y)$.

26. $f(x, y) = 4x^3y^2 - 8xy$

27. $f(x, y) = -6xy^4 + x^2y$

28. $f(x, y) = \dfrac{2x}{x - 2y}$

29. $f(x, y) = \dfrac{3x + y}{x - 1}$

30. $f(x, y) = x^2e^y$

31. $f(x, y) = ye^{x^2}$

32. $f(x, y) = \ln|2 - x^2y|$

33. $f(x, y) = \ln|1 + 3xy^2|$

Find all points where the functions defined below have any relative extrema. Find any saddle points.

34. $z = x^2 + 2y^2 - 4y$

35. $z = x^2 + y^2 + 9x - 8y + 1$

36. $f(x, y) = x^2 + 5xy - 10x + 3y^2 - 12y$

37. $z = x^3 - 8y^2 + 6xy + 4$

38. $z = \dfrac{1}{2}x^2 + \dfrac{1}{2}y^2 + 2xy - 5x - 7y + 10$

39. $f(x, y) = 3x^2 + 2xy + 2y^2 - 3x + 2y - 9$

40. $z = x^3 + y^2 + 2xy - 4x - 3y - 2$

41. $f(x, y) = 7x^2 + y^2 - 3x + 6y - 5xy$

42. Describe the three different types of points that might occur when $f_x(x, y) = f_y(x, y) = 0$.

Evaluate dz using the given information.

43. $z = 2x^2 - 4y^2 + 6xy;$ $x = 2, y = -3, dx = 0.01, dy = 0.05$

44. $z = \dfrac{x + 5y}{x - 2y};$ $x = 1, y = -2, dx = -0.04, dy = 0.02$

Use the total differential to approximate each quantity. Then use a calculator to approximate the quantity, and give the absolute value of the difference in the two results to four decimal places.

45. $\sqrt{5.1^2 + 12.05^2}$

46. $\sqrt{4.06}\,e^{0.04}$

Evaluate each of the following.

47. $\displaystyle\int_4^9 \frac{6y - 8}{\sqrt{x}}\, dx$

48. $\displaystyle\int_3^5 e^{2x-7y}\, dx$

49. $\displaystyle\int_0^5 \frac{6x}{\sqrt{4x^2 + 2y^2}}\, dx$

50. $\displaystyle\int_1^3 \frac{y^2}{\sqrt{7x + 11y^3}}\, dy$

Evaluate each iterated integral.

51. $\displaystyle\int_0^2 \int_0^4 (x^2y^2 + 5x)\, dx\, dy$

52. $\displaystyle\int_0^2 \int_0^3 (x + 5y + y^2)\, dy\, dx$

53. $\displaystyle\int_3^4 \int_2^5 \sqrt{6x + 3y}\, dx\, dy$

54. $\displaystyle\int_1^2 \int_3^5 e^{2x-7y}\, dx\, dy$

55. $\displaystyle\int_2^4 \int_2^4 \frac{dx\, dy}{y}$

56. $\displaystyle\int_1^2 \int_1^2 \frac{dx\, dy}{x}$

Find each double integral over the region R with boundaries as indicated.

57. $\displaystyle\iint_R (x^2 + y^2)\, dx\, dy;$ $0 \le x \le 2, 0 \le y \le 3$

58. $\displaystyle\iint_R \sqrt{2x + y}\, dx\, dy;$ $1 \le x \le 3, 2 \le y \le 5$

59. $\displaystyle\iint_R \sqrt{y + x}\, dx\, dy;$ $0 \le x \le 7, 1 \le y \le 9$

60. $\displaystyle\iint_R ye^{y^2+x}\, dx\, dy;$ $0 \le x \le 1, 0 \le y \le 1$

61. $\displaystyle\iint_R xy \sin(xy^2)\, dy\, dx;$ $0 \le x \le \frac{\pi}{2}, 0 \le y \le 1$

62. $\displaystyle\iint_R ye^x \cos(ye^x)\, dx\, dy;$ $0 \le x \le 1, 0 \le y \le \pi$

Find the volume under the given surface z = f(x, y) and above the given rectangle.

63. $z = x + 9y + 8;$ $1 \le x \le 6, 0 \le y \le 8$

64. $z = x^2 + y^2;$ $3 \le x \le 5, 2 \le y \le 4$

Evaluate each double integral. If the function seems too difficult to integrate, try interchanging the limits of integration.

65. $\displaystyle\int_0^1 \int_0^{2x} xy\, dy\, dx$

66. $\displaystyle\int_0^1 \int_0^{x^3} y\, dy\, dx$

67. $\displaystyle\int_0^1 \int_{x^2}^x x^3y\, dy\, dx$

68. $\displaystyle\int_0^1 \int_y^{\sqrt{y}} x\, dx\, dy$

69. $\displaystyle\int_0^2 \int_{x/2}^1 \frac{1}{y^2 + 1}\, dy\, dx$

70. $\displaystyle\int_0^8 \int_{x/2}^4 \sqrt{y^2 + 4}\, dy\, dx$

Use the region R, with boundaries as indicated, to evaluate the given double integral.

71. $\displaystyle\iint_R (2x + 3y)\, dx\, dy;$ $0 \le y \le 1, y \le x \le 2 - y$

72. $\displaystyle\iint_R (2 - x^2 - y^2)\, dy\, dx;$ $0 \le x \le 1, x^2 \le y \le x$

Applications

LIFE SCIENCES

73. *Blood Vessel Volume* A length of blood vessel is measured as 2.7 cm, with the radius measured as 0.7 cm. If each of these measurements could be off by 0.1 cm, estimate the maximum possible error in the volume of the vessel.

74. *Total Body Water* As we saw earlier in this chapter, accurate prediction of total body water is critical in determining adequate dialysis doses for patients with renal disease. For African-American males, total body water can be estimated by the function

$$T(A, W, S) = -18.37 - 0.09A + 0.34W + 0.25S,$$

where T is the total body water (in liters), A is age (in yr), W is weight (in kg), and S is height (in cm).*

a. Find $T(65, 85, 180)$.

b. Find and interpret $T_A(A, W, S)$, $T_W(A, W, S)$, and $T_S(A, W, S)$.

75. *Brown Trout* As we saw in Exercise 58 of the section on Partial Derivatives, researchers from New Zealand have determined that the length of a brown trout depends on both its weight and age and that the length can be estimated by

$$L(w, t) = (0.00082t + 0.0955)e^{(\ln w + 10.49)/2.842},$$

where $L(w, t)$ is the length of the trout (in cm), w is the weight of the trout (in g), and t is the age of the trout (in yr).†

a. Find $L(450, 4)$.

b. Find $L_w(450, 7)$ and $L_t(450, 7)$ and interpret.

76. *Survival Curves* The figure in the next column shows survival curves (percent surviving as a function of age) for females in the United States in 1900 and 1995.‡ Let $f(x, y)$ give the proportion surviving at age x in year y. Use the graph to estimate each of the following. Interpret each answer in words.

a. $f(60, 1900)$

b. $f(70, 1995)$

c. $f_x(60, 1900)$

d. $f_x(70, 1995)$

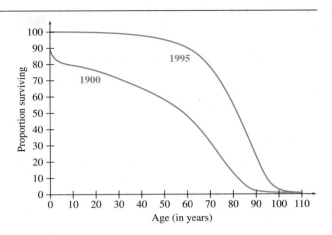

OTHER APPLICATIONS

77. *Area* The bottom of a planter is to be made in the shape of an isosceles triangle, with the two equal sides 3 ft long and the third side 2 ft long. The area of an isosceles triangle with two equal sides of length a and third side of length b is

$$f(a, b) = \frac{1}{4}b\sqrt{4a^2 - b^2}.$$

a. Find the area of the bottom of the planter.

b. The manufacturer is considering changing the shape so that the third side is 2.5 ft long. What would be the approximate effect on the area?

78. *Surface Area* A closed box with square ends must have a volume of 125 cubic inches. Find the dimensions of such a box that has minimum surface area.

79. *Area* Find the maximum rectangular area that can be enclosed with 400 ft of fencing, if no fencing is needed along one side.

80. *Charge for Auto Painting* The charge (in dollars) for painting a sports car is given by

$$C(x, y) = 2x^2 + 4y^2 - 3xy + \sqrt{x},$$

where x is the number of hours of labor needed and y is the number of gallons of paint and sealant used. Find each of the following.

a. The charge for 10 hr and 5 gal of paint and sealant

---

*Chumlea, W., S. Guo, C. Zellar, et al., "Total Body Water Reference Values and Prediction Equations for Adults," *Kidney International,* Vol. 59, 2001, pp. 2250–2258.
†Hayes, J., J. Stark, and K. Shearer, "Development and Test of a Whole-Lifetime Foraging and Bioenergetics Growth Model for Drift-Feeding Brown Trout," *Transactions of the American Fisheries Society,* Vol. 129, 2000, pp. 315–332.
‡*Science,* Vol. 291, Feb. 23, 2001, p. 1491.

b. The charge for 15 hr and 10 gal of paint and sealant

c. The charge for 20 hr and 20 gal of paint and sealant

81. *Manufacturing Costs* The manufacturing cost (in dollars) for a medium-sized business computer is given by

$$c(x, y) = 2x + y^2 + 4xy + 25,$$

where x is the memory capacity of the computer in megabytes (Mb) and y is the number of hours of labor required. For 640Mb and 6 hr of labor, find the following.

a. The approximate change in cost for an additional 1Mb of memory

b. The approximate change in cost for an additional hour of labor

82. *Cost* The cost (in dollars) to manufacture x solar cells and y solar collectors is

$$c(x, y) = x^2 + 5y^2 + 4xy - 70x - 164y + 1,800.$$

a. Find values of x and y that produce minimum total cost.

b. Find the minimum total cost.

83. *Cost* The cost (in dollars) to produce x satellite receiving dishes and y transmitters is given by

$$C(x, y) = \ln(x^2 + y) + e^{xy/20}.$$

Production schedules now call for 15 receiving dishes and 9 transmitters. Use differentials to approximate the change in costs if 1 more dish and 1 fewer transmitter are made.

84. *Production Materials* Approximate the amount of material needed to manufacture a cone of radius 2 cm, height 8 cm, and wall thickness 0.21 cm.

85. *Production Materials* A sphere of radius 2 ft is to receive an insulating coating 1 inch thick. Approximate the volume of the coating needed.

86. *Production Error* The height of a sample cone from a production line is measured as 11.4 cm, while the radius is measured as 2.9 cm. Each of these measurements could be off by 0.2 cm. Approximate the maximum possible error in the volume of the cone.

87. *Profit* The total profit from 1 acre of a certain crop depends on the amount spent on fertilizer, x, and on hybrid seed, y, according to the model

$$P(x, y) = 0.01(-x^2 + 3xy + 160x - 5y^2 + 200y + 2,600).$$

The budget for fertilizer and seed is limited to $280.

a. Use the budget constraint to express one variable in terms of the other. Then substitute into the profit function to get a function with one independent variable. Use the method shown in the chapter on Applications of the Derivative to find the amounts spent on fertilizer and seed that will maximize profit. What is the maximum profit per acre? (*Hint:* Throughout this exercise you may ignore the coefficient of 0.01 until you need to find the maximum profit.)

b. Find the amounts spent on fertilizer and seed that will maximize profit using the first method shown in this chapter. (*Hint:* You will not need to use the budget constraint.)

EXTENDED APPLICATION: Optimization for a Predator*

A predator is an animal that feeds upon another animal. Foxes, coyotes, wolves, weasels, lions, and tigers are well-known predators, but many other animals also fall into this category. In this case, we set up a mathematical model for predation, and then use partial derivatives to minimize the difference between the desired and the actual levels of food consumption. There are several research studies which show that animals *do* control their activities so as to maximize or minimize variables— lobsters orient their bodies by minimizing the discharge rate from certain organs, for example.

In this case, we assume that the predator has a diet consisting of only two foods, food 1 and food 2. We also assume that the predator will hunt only in two locations, location 1 and location 2. To make this mathematical model meaningful, we would have to gather data on the predators and the prey that we wished to study. For example, suppose that we have gathered the following data:

$u_{11} = 0.4$ = rate of feeding on food 1 in location 1;
$u_{12} = 0.1$ = rate of feeding on food 1 in location 2;
$u_{21} = 0.3$ = rate of feeding on food 2 in location 1;
$u_{22} = 0.3$ = rate of feeding on food 2 in location 2.

Let x_1 = proportion of time spent feeding in location 1;
  x_2 = proportion of time spent feeding in location 2;
  x_3 = proportion of time spent on nonfeeding activities.

Using these variables, the total quantity of food 1 consumed is given by Y_1, where

$$Y_1 = u_{11}x_1 + u_{12}x_2$$
$$= 0.4x_1 + 0.1x_2. \tag{1}$$

The total quantity of food 2 consumed is given by Y_2, where

$$Y_2 = u_{21}x_1 + u_{22}x_2$$
$$= 0.3x_1 + 0.3x_2. \tag{2}$$

The total amount of food consumed is thus given by

$$Y_1 + Y_2 = 0.4x_1 + 0.1x_2 + 0.3x_1 + 0.3x_2$$
$$= 0.7x_1 + 0.4x_2. \tag{3}$$

Again, by gathering experimental data, suppose that

$z_t = 0.4$ = desired level of total food consumption;
$z_1 = 0.15$ = desired level of consumption of food 1;
$z_2 = 0.25$ = desired level of consumption of food 2.

The predator wishes to find values of x_1 and x_2 such that the difference between the desired level of consumption (the z's) and the actual level of consumption (the Y's) is minimized. However, much experience has shown that the difference between desired food consumption and actual consumption is not

perceived by the animal as linear, but rather perhaps as a square. That is, a 5% shortfall in the actual consumption of food, as compared to the desired consumption, may be perceived by the predator as a 20–25% shortfall. (This phenomenon is known as *Weber's Law*—many psychology books discuss it.) Thus, to have our model approximate reality, we must find values of x_1 and x_2 that will minimize

$$G = [z_t - (Y_1 + Y_2)]^2 + w_1[z_1 - Y_1]^2$$
$$+ w_2[z_2 - Y_2]^2 + w_3[1 - x_3]^2. \tag{4}$$

The term $w_3[1 - x_3]^2$ represents the fact that the predator does not wish to spend the total available time in searching for food—recall that x_3 represents the proportion of available time that is spent on nonfeeding activities. The variables w_1, w_2, and w_3 represent the relative importance, or weights, assigned by the animal to food 1, food 2, and nonfood activities, respectively. Again, it is necessary to gather experimental data: reasonable values of w_1, w_2, and w_3 are as follows:

$$w_1 = 0.5 \qquad w_2 = 0.4 \qquad w_3 = 0.03.$$

If we substitute 0.4 for z_t, 0.15 for z_1, 0.25 for z_2, 0.5 for w_1, 0.4 for w_2, and 0.03 for w_3, the results of equation (1) for Y_1, (2) for Y_2, and (3) for $Y_1 + Y_2$ into equation (4), we have

$$G = [0.4 - 0.7x_1 - 0.4x_2]^2 + 0.5[0.15 - 0.4x_1 - 0.1x_2]^2$$
$$+ 0.4[0.25 - 0.3x_1 - 0.3x_2]^2 + 0.03[1 - x_3]^2.$$

We want to minimize G, subject to the constraint $x_1 + x_2 + x_3 = 1$, or $x_1 + x_2 + x_3 - 1 = 0$. The original article did this with a mathematical technique known as Lagrange multipliers.[†] Here, we will instead solve the constraint for x_3: $x_3 = 1 - x_1 - x_2$. Substituting this into the equation for G gives

$$G = [0.4 - 0.7x_1 - 0.4x_2]^2 + 0.5[0.15 - 0.4x_1 - 0.1x_2]^2$$
$$+ 0.4[0.25 - 0.3x_1 - 0.3x_2]^2 + 0.03[x_1 + x_2]^2.$$

Now we must find the partial derivatives of G with respect to x_1 and x_2. Doing this we have

$$\frac{\partial G}{\partial x_1} = 2(-0.7)(0.4 - 0.7x_1 - 0.4x_2)$$
$$+ 2(0.5)(-0.4)(0.15 - 0.4x_1 - 0.1x_2)$$
$$+ 2(0.4)(-0.3)(0.25 - 0.3x_1 - 0.3x_2)$$
$$+ 2(0.03)(x_1 + x_2)$$

$$\frac{\partial G}{\partial x_2} = 2(-0.4)(0.4 - 0.7x_1 - 0.4x_2)$$
$$+ 2(0.5)(-0.1)(0.15 - 0.4x_1 - 0.1x_2)$$
$$+ 2(0.4)(-0.3)(0.25 - 0.3x_1 - 0.3x_2)$$
$$+ 2(0.03)(x_1 + x_2)$$

*Based on Gerald G. Martens, "An Optimization Equation for Predation," *Ecology,* Winter 1973, pp. 92–101.

†Lial, Margaret L., Raymond N. Greenwell, Charles D. Miller, *Calculus with Applications,* 6th ed., Addison-Wesley, 1998, pp. 521–522.

Both $\partial G/\partial x_1$ and $\partial G/\partial x_2$ can be simplified, using some rather tedious algebra. Doing this, and placing each partial derivative equal to 0, we get the following system of equations:

$$\frac{\partial G}{\partial x_1} = -0.68 + 1.272x_1 + 0.732x_2 = 0$$

$$\frac{\partial G}{\partial x_2} = -0.395 + 0.732x_1 + 0.462x_2 = 0$$

Although we shall not go through the details of the solution here, it can be found from the system above that the values of the x's that minimize G are given by

$$x_1 = 0.48 \qquad x_2 = 0.09 \qquad x_3 = 0.43.$$

(These values have been rounded to the nearest hundredth.) Recall that $x_3 = 1 - x_1 - x_2$. Thus, the predator should spend

about 0.48 of the available time searching in location 1, and about 0.09 of the time searching in location 2. This will leave about 0.43 of the time free for nonfeeding activities.

Exercises

1. Verify the simplification of $\partial G/\partial x_1$ and $\partial G/\partial x_2$.

2. Verify the solution given in the text by solving the system of equations.

3. Suppose $w_1 = 0.4$ and $w_2 = 0.5$ and all other values remain the same. What proportion of time should the predator spend on feeding in each of locations 1 and 2? What proportion of time will be spent on nonfeeding activities?

10

Matrices

When stocking a lake with various species of fish, wildlife biologists need to take into account the nutrient requirements of each fish, as well as the amounts of the various nutrients available in the lake. An exercise in the first section of this chapter solves such a problem using systems of linear equations and matrices.

■ **10.1**   Solution of Linear Systems

■ **10.2**   Addition and Subtraction of Matrices

■ **10.3**   Multiplication of Matrices

■ **10.4**   Matrix Inverses

■ **10.5**   Eigenvalues and Eigenvectors

Review Exercises

Extended Application: Contagion

Many mathematical models require finding the solutions of two or more equations. The solutions must satisfy *all* of the equations in the model. A set of equations related in this way is called a **system of equations.** In this chapter we will discuss systems of linear equations, introduce the idea of a *matrix*, and then show how matrices are used to solve systems of linear equations. We will also see how matrices provide answers to important questions about population growth.

■ 10.1 SOLUTION OF LINEAR SYSTEMS

? **T**HINK ABOUT IT   How much of each ingredient should be used in an animal feed to meet dietary requirements?

Suppose that an animal feed is made from three ingredients: corn, soybeans, and cottonseed. One gram of each ingredient provides the number of grams of protein, fat, and fiber shown in Table 1. For example, the entries in the first column, 0.25, 0.4, and 0.3, indicate that one gram of corn provides twenty-five hundredths (one-fourth) of a gram of protein, four-tenths of a gram of fat, and three-tenths of a gram of fiber.

■ Table 1

	Corn	Soybeans	Cottonseed
Protein	0.25	0.4	0.2
Fat	0.4	0.2	0.3
Fiber	0.3	0.2	0.1

? Now suppose we need to know the number of grams of each ingredient that should be used to make a feed that contains 22 grams of protein, 28 grams of fat, and 18 grams of fiber. To find out, we let x represent the required number of grams of corn, y the number of grams of soybeans, and z the number of grams of cottonseed. Since each gram of corn provides 0.25 grams of protein, the amount of protein provided by x grams of corn is $0.25x$. Similarly, the amount of protein provided by y grams of soybeans is $0.4y$ and the amount of protein provided by z grams of cottonseed is $0.2z$. Since the total amount of protein is to be 22 grams,

$$0.25x + 0.4y + 0.2z = 22.$$

The feed must supply 28 grams of fat, so

$$0.4x + 0.2y + 0.3z = 28,$$

and 18 grams of fiber, so

$$0.3x + 0.2y + 0.1z = 18.$$

To solve this problem, we must find values of x, y, and z that satisfy this system of equations. Verify that $x = 40$ grams, $y = 15$ grams, and $z = 30$ grams is a solution of the system by substituting these numbers into all three equations. In fact, this is the only solution of this system. Many practical problems lead to such systems of *linear equations*.

A **linear equation in n unknowns** is any equation of the form

$$a_1x_1 + a_2x_2 + \cdots + a_nx_n = k,$$

where $a_1, a_2, \ldots, a_n$ and k are all real numbers. Each of the three equations from the animal feed problem is a linear equation in three unknowns. A *solution* of the linear equation

$$a_1x_1 + a_2x_2 + \cdots + a_nx_n = k$$

is a sequence of numbers $s_1, s_2, \ldots, s_n$ such that

$$a_1s_1 + a_2s_2 + \cdots + a_ns_n = k.$$

A solution of an equation is usually written between parentheses as $(s_1, s_2, \ldots, s_n)$. For example, $(1, 6, 2)$ is a solution of the equation $3x_1 + 2x_2 - 4x_3 = 7$, since $3(1) + 2(6) - 4(2) = 7$. This is an extension of the idea of an ordered pair, which was introduced in Chapter 1. A linear equation in two unknowns is the type of linear equation studied in Chapter 1, and its graph is a straight line. In this section we develop a method for solving a system of linear equations. Although the discussion will be confined to equations with only a few variables, the method of solution can be extended to systems with many variables.

The equations for the animal feed problem form the system

$$0.25x + 0.4y + 0.2z = 22$$
$$0.4x + 0.2y + 0.3z = 28$$
$$0.3x + 0.2y + 0.1z = 18,$$

which can be written in abbreviated form as

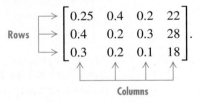

Such a rectangular array of numbers enclosed by brackets is called a **matrix** (plural: **matrices**).* Each number in the array is an **element** or **entry**. To separate the constants in the last column of the matrix from the coefficients of the variables, we use a vertical line, producing the following **augmented matrix.**

$$\left[\begin{array}{ccc|c} 0.25 & 0.4 & 0.2 & 22 \\ 0.4 & 0.2 & 0.3 & 28 \\ 0.3 & 0.2 & 0.1 & 18 \end{array}\right]$$

Any algebraic operation done to the equations of a linear system is equivalent to performing the same operation on the rows of the augmented matrix, since the matrix is just a shorter way of writing the system. The following **row operations** on the augmented matrix transform the system of equations into an **equivalent system,** that is, one that has the same solutions as the original system.

---

*The word matrix, which is Latin for "womb," was coined by James Joseph Sylvester (1814–1897) and made popular by his friend Arthur Cayley (1821–1895). Both mathematicians were English, although Sylvester spent much of his life in the United States.

ROW OPERATIONS

For any augmented matrix of a system of equations, the following operations produce the augmented matrix of an equivalent system:

  1. interchanging any two rows;

  2. multiplying the elements of a row by any nonzero real number;

  3. adding a nonzero multiple of the elements of one row to the corresponding elements of a nonzero multiple of some other row.

In steps 2 and 3, we are replacing a row with a new, modified row which the old row helped to form. This is equivalent to replacing an equation with a new, modified equation.

Row operations, like the transformations of systems of equations, are reversible. If they are used to change matrix A to matrix B, then it is possible to use row operations to transform B back into A.

By the first row operation, we can interchange rows 1 and 2, for example, which we will indicate by the notation $R_2 \rightarrow R_1$ and $R_1 \rightarrow R_2$. (R stands for row.) Row 3 is left unchanged. Thus

$$\begin{bmatrix} 0.25 & 0.4 & 0.2 & | & 22 \\ 0.4 & 0.2 & 0.3 & | & 28 \\ 0.3 & 0.2 & 0.1 & | & 18 \end{bmatrix} \text{ becomes } \begin{matrix} R_2 \rightarrow R_1 \\ R_1 \rightarrow R_2 \end{matrix} \begin{bmatrix} 0.4 & 0.2 & 0.3 & | & 28 \\ 0.25 & 0.4 & 0.2 & | & 22 \\ 0.3 & 0.2 & 0.1 & | & 18 \end{bmatrix}.$$

The second row operation, multiplying a row by a number, allows us to change

$$\begin{bmatrix} 0.25 & 0.4 & 0.2 & | & 22 \\ 0.4 & 0.2 & 0.3 & | & 28 \\ 0.3 & 0.2 & 0.1 & | & 18 \end{bmatrix} \text{ to } \overset{100R_1 \rightarrow R_1}{} \begin{bmatrix} 25 & 40 & 20 & | & 2{,}200 \\ 0.4 & 0.2 & 0.3 & | & 28 \\ 0.3 & 0.2 & 0.1 & | & 18 \end{bmatrix}$$

by multiplying the elements of row 1 of the original matrix by 100. Note that rows 2 and 3 are left unchanged. This operation is helpful in converting all entries in a matrix to integers, which can simplify the calculations. It is also useful for dividing a common factor out of a row, which makes the numbers smaller and hence easier to work with.

Using the third row operation, adding a multiple of one row to a multiple of another, we change

$$\begin{bmatrix} 0.25 & 0.4 & 0.2 & | & 22 \\ 0.4 & 0.2 & 0.3 & | & 28 \\ 0.3 & 0.2 & 0.1 & | & 18 \end{bmatrix} \text{ to } {}_{0.4R_1 - 0.25R_2 \rightarrow R_2} \begin{bmatrix} 0.25 & 0.4 & 0.2 & | & 22 \\ 0 & 0.11 & 0.005 & | & 1.8 \\ 0.3 & 0.2 & 0.1 & | & 18 \end{bmatrix}.$$

This operation is very useful for converting an element in a matrix to 0. In the above operation, notice that the first element in row 2 became $0.4(0.25) - 0.25(0.40) = 0$. Similarly, the second element in row 2 became $0.4(0.4) - 0.25(0.2) = 0.11$. Verify the third and fourth elements in row 2. Also notice that rows 1 and 3 are left unchanged, *even though the elements of row 1 were used to transform row 2.*

The Gauss-Jordan Method

The **Gauss-Jordan method** uses the row operations just described to transform a system of equations into one from which the solution can be immediately read.* For example, in a linear system with two equations in two unknowns, we would like to get the system into the form

$$\begin{bmatrix} 1 & 0 & \Big| & a \\ 0 & 1 & \Big| & b \end{bmatrix},$$

from which we read that $x = a$ and $y = b$. We will see how to do this in Example 1. The idea is to first use the element in row 1, column 1, to get a 0 in every other space in column 1. We then use the element in row 2, column 2, to get a 0 in every other space in column 2, and continue this way if there are more rows. After getting a 0 in the places we want, we then divide each row by an appropriate constant (if necessary) to convert each element on the main diagonal into a 1.

Before the Gauss-Jordan method can be used, the system must be in proper form: the terms with variables should be on the left and the constants on the right in each equation, with the variables in the same order in each equation. The following example illustrates the uses of the Gauss-Jordan method to solve a system of equations.

EXAMPLE 1   Gauss-Jordan
Solve the system

$$3x - 4y = 1 \qquad\qquad (1)$$
$$5x + 2y = 19. \qquad\qquad (2)$$

Solution The system is already in the proper form for use of the Gauss-Jordan method. To begin, change the 5 in row 2 to 0 by using the 3 in row 1, column 1 to clear out the rest of column 1.

$$\begin{bmatrix} 3 & -4 & \Big| & 1 \\ 5 & 2 & \Big| & 19 \end{bmatrix} \quad \text{Augmented matrix representing the system.}$$

$$5R_1 + (-3)R_2 \rightarrow R_2 \quad \begin{bmatrix} 3 & -4 & \Big| & 1 \\ 0 & -26 & \Big| & -52 \end{bmatrix}$$

Row 2 can be simplified by dividing by -26.

$$-\tfrac{1}{26}R_2 \rightarrow R_2 \quad \begin{bmatrix} 3 & -4 & \Big| & 1 \\ 0 & 1 & \Big| & 2 \end{bmatrix}$$

Change the -4 in row 1 to 0.

$$4R_2 + R_1 \rightarrow R_1 \quad \begin{bmatrix} 3 & 0 & \Big| & 9 \\ 0 & 1 & \Big| & 2 \end{bmatrix}$$

Change the first nonzero number in each row to 1.

$$\tfrac{1}{3}R_1 \rightarrow R_1 \quad \begin{bmatrix} 1 & 0 & \Big| & 3 \\ 0 & 1 & \Big| & 2 \end{bmatrix}$$

---

*The great German mathematician Carl Friedrich Gauss (1777–1855), sometimes referred to as the "Prince of Mathematicians," originally developed his elimination method for use in finding least squares coefficients. (See Section 1.2.) The German geodesist Wilhelm Jordan (1842–1899) improved his method and used it in surveying problems. Gauss's method had been known to the Chinese at least 1,800 years earlier and was described in the *Jiuahang Suanshu* (*Nine Chapters on the Mathematical Art*).

The last matrix corresponds to the system

$$x = 3$$
$$y = 2,$$

so the solution can be read directly from the last column of the final matrix. Check the solution by substitution in the equations of the original system. ∎

> **NOTE** If your solution does not check, the most efficient way to find the error is to substitute back through the equations that correspond to each matrix, starting with the last matrix. When you find a system that is not satisfied by your (incorrect) answers, you have probably reached the matrix just before the error occurred. Look for the error in the transformation to the next matrix.

When the Gauss-Jordan method is used to solve a system, the final matrix usually will have zeros above and below the diagonal of 1's on the left of the vertical bar. To transform the matrix, it is best to work column by column from left to right. Such an orderly method avoids confusion and going around in circles. For each column, first perform the steps that give the zeros. When all columns have zeros in place, multiply each row by the reciprocal of the coefficient of the remaining nonzero number in that row to get the required 1's. With certain types of systems, illustrated in Examples 3, 4, and 5, it will not be possible to get the complete diagonal of 1's.

EXAMPLE 2 Gauss-Jordan

Use the Gauss-Jordan method to solve the system

$$x + 5z = -6 + y$$
$$3x + 3y = 10 + z$$
$$x + 3y + 2z = 5.$$

Solution

Method 1: Calculating by Hand   First, rewrite the system in proper form as follows.

$$x - y + 5z = -6$$
$$3x + 3y - z = 10$$
$$x + 3y + 2z = 5$$

Begin to find the solution by writing the augmented matrix of the linear system.

$$\begin{bmatrix} 1 & -1 & 5 & | & -6 \\ 3 & 3 & -1 & | & 10 \\ 1 & 3 & 2 & | & 5 \end{bmatrix}$$

Row operations will be used to transform this matrix (if possible) to the form

$$\begin{bmatrix} 1 & 0 & 0 & | & m \\ 0 & 1 & 0 & | & n \\ 0 & 0 & 1 & | & p \end{bmatrix},$$

where m, n, and p are real numbers. From this final form of the matrix, the solution can be read: $x = m$, $y = n$, $z = p$, or (m, n, p).

In the first column, we need zeros in the second and third rows. Multiply the first row by -3 and add to the second row to get a zero there. Then multiply the first row by -1 and add to the third row to get that zero.

$$\begin{array}{c} \\ -3R_1 + R_2 \rightarrow R_2 \\ -1R_1 + R_3 \rightarrow R_3 \end{array} \left[\begin{array}{ccc|c} 1 & -1 & 5 & -6 \\ 0 & 6 & -16 & 28 \\ 0 & 4 & -3 & 11 \end{array} \right]$$

Now get zeros in the second column in a similar way. We want zeros in the first and third rows. Row two will not change.

$$\begin{array}{c} R_2 + 6R_1 \rightarrow R_1 \\ \\ 2R_2 + (-3)R_3 \rightarrow R_3 \end{array} \left[\begin{array}{ccc|c} 6 & 0 & 14 & -8 \\ 0 & 6 & -16 & 28 \\ 0 & 0 & -23 & 23 \end{array} \right]$$

In transforming the third row, you may have used the operation $4R_2 + (-6)R_3 \rightarrow R_3$ instead of $2R_2 + (-3)R_3 \rightarrow R_3$. This is perfectly fine; the last row would then have -46 and 46 in place of -23 and 23. To avoid errors, it helps to keep the numbers as small as possible. We observe at this point that all of the numbers can be reduced in size by multiplying each row by an appropriate constant. This next step is not essential, but it simplifies the arithmetic.

$$\begin{array}{c} \frac{1}{2}R_1 \rightarrow R_1 \\ \frac{1}{2}R_2 \rightarrow R_2 \\ -\frac{1}{23}R_3 \rightarrow R_3 \end{array} \left[\begin{array}{ccc|c} 3 & 0 & 7 & -4 \\ 0 & 3 & -8 & 14 \\ 0 & 0 & 1 & -1 \end{array} \right]$$

Next, we want zeros in the first and second rows of the third column. Row three will not change.

$$\begin{array}{c} -7R_3 + R_1 \rightarrow R_1 \\ -8R_3 + R_2 \rightarrow R_2 \\ \\ \end{array} \left[\begin{array}{ccc|c} 3 & 0 & 0 & 3 \\ 0 & 3 & 0 & 6 \\ 0 & 0 & 1 & -1 \end{array} \right]$$

Finally, get 1's in each row by multiplying the row by the reciprocal of (or dividing the row by) the number in the diagonal position.

$$\begin{array}{c} \frac{1}{3}R_1 \rightarrow R_1 \\ \frac{1}{3}R_2 \rightarrow R_2 \\ \\ \end{array} \left[\begin{array}{ccc|c} 1 & 0 & 0 & 1 \\ 0 & 1 & 0 & 2 \\ 0 & 0 & 1 & -1 \end{array} \right]$$

The linear system associated with the final augmented matrix is

$$\begin{aligned} x &= 1 \\ y &= 2 \\ z &= -1, \end{aligned}$$

and the solution is $(1, 2, -1)$. Verify that this is the solution to the original system of equations.

CAUTION Notice that we have performed two or three operations on the same matrix in one step. This is permissible as long as we do not use a row that we are changing as part of another row operation. For example, when we changed row 2 in the first step, we could not use row 2 to transform row 3 in the same step. To avoid difficulty, use *only* row 1 to get zeros in column 1, row 2 to get zeros in column 2, and so on.

Method 2: Graphing Calculators

The row operations of the Gauss-Jordan method can also be done on a graphing calculator. For example, Figure 1 shows the result when the augmented matrix is entered into a TI-83. Figures 2 and 3 show how row operations can be used to get zeros in rows 2 and 3 of the first column.

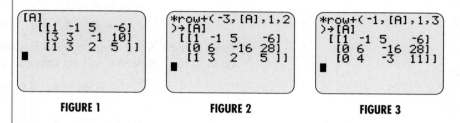

FIGURE 1   **FIGURE 2**   **FIGURE 3**

Calculators typically do not allow any multiple of a row to be added to any multiple of another row, such as in the operation $2R_2 + 6R_1 \rightarrow R_1$. They normally allow a multiple of a row to be added only to another unmodified row. To get around this restriction, we can convert the diagonal element to a 1 before changing the other elements in the column to 0. In this example, we change the 6 in row 2, column 2, to a 1 by dividing by 6. The result is shown in Figure 4. (The right side of the matrix is not visible, but can be seen by pressing the right arrow key.) Notice that this operation introduces decimals. Converting to fractions is preferable on calculators that have that option; $1/3$ is certainly more concise than 0.3333333333. Figure 5 shows such a conversion on the TI-83.

FIGURE 4   **FIGURE 5**

When performing row operations without a graphing calculator, it is best to avoid fractions and decimals, because these make the operations more difficult and more prone to error. A calculator, on the other hand, encounters no such difficulties.

Continuing in the same manner, the solution $(1, 2, -1)$ is found as shown in Figure 6.

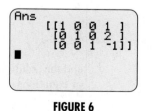

FIGURE 6

Some calculators can do the entire Gauss-Jordan process with a single command; on the TI-83, for example, this is done with the `rref` command. This is very useful in practice, although it does not show any of the intermediate steps.

Method 3: Spreadsheets

The Gauss-Jordan method can be done using a spreadsheet either by using a macro or by developing the row operations using formulas with the copy and paste commands. However, spreadsheets also have built-in methods to solve systems of equations. Although these solvers do not usually employ the Gauss-Jordan method for solving systems of equations, they are, nonetheless, efficient and practical to use.

The Solver that is included with Excel can solve systems of equations that are both linear and nonlinear. The Solver is located in the tools menu and requires that cells be identified ahead of time for each variable in the problem. It also requires that the left-hand side of each equation be placed in the spreadsheet as a formula. For example, to solve the above problem, we could identify cells A1, B1, and C1 for the variables x, y, and z, respectively. The Solver requires that we place a guess for the answer in these cells. Thus, it is convenient to place a zero in each of these cells. The left-hand side of each equation must be placed in a cell. Thus, we could choose A3, A4, and A5 to hold each of these formulas. Thus, in cell A3, we would type "=A1 − B1 + 5*C1" and put the other two equations in cells A4 and A5.

We now click on the Tools menu and choose Solver. Since this solver attempts to find a solution that is best in some way, we are required to identify a cell with a formula in it that we want to optimize. In this case, it is convenient to use the cell with the left-hand side of the first constraint in it, A3. Figure 7 illustrates the Solver box and the items placed in it.

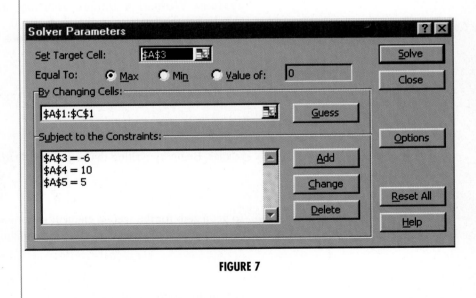

FIGURE 7

To obtain a solution, click on Solve. The solution is located in cells A1, B1, and C1, and these correspond to x, y, and z, respectively.

In summary, the Gauss-Jordan method of solving a linear system requires the following steps.

THE GAUSS-JORDAN METHOD OF SOLVING A LINEAR SYSTEM

1. Write each equation so that variable terms are in the same order on the left side of the equals sign and constants are on the right.

2. Write the augmented matrix that corresponds to the system.

3. Use row operations to transform the first column so that all elements except the element in the first row are zero.

4. Use row operations to transform the second column so that all elements except the element in the second row are zero.

5. Use row operations to transform the third column so that all elements except the element in the third row are zero.

6. Continue in this way, when possible, until the last row is written in the form

$$[0 \quad 0 \quad 0 \quad \cdots \quad 0 \quad j \mid k],$$

where j and k are constants with $j \neq 0$. When this is not possible, continue until every row has more zeros on the left than the previous row (except possibly for any rows of all zero at the bottom of the matrix), and the first nonzero entry in each row is the only nonzero entry in its column.

7. Multiply each row by the reciprocal of the nonzero element in that row.

Systems Without a Unique Solution   In the previous examples, we were able to get the last row in the form $[0 \quad 0 \quad 0 \quad \cdots \quad 0 \quad j \mid k]$, where j and k are constants with $j \neq 0$. We will now look at some examples where this is not the case.

EXAMPLE 3   Systems of Equations with No Solution
Use the Gauss-Jordan method to solve the system

$$\begin{aligned} x + y &= 2 \\ 2x + 2y &= 5. \end{aligned}$$

Solution   Begin by writing the augmented matrix.

$$\begin{bmatrix} 1 & 1 & \bigm| & 2 \\ 2 & 2 & \bigm| & 5 \end{bmatrix}$$

To get a zero for the second element in column 1, multiply the numbers in row 1 by -2 and add the results to the corresponding elements in row 2.

$$-2R_1 + R_2 \to R_2 \quad \begin{bmatrix} 1 & 1 & \bigm| & 2 \\ 0 & 0 & \bigm| & 1 \end{bmatrix}$$

This matrix corresponds to the system

$$\begin{aligned} x + y &= 2 \\ 0x + 0y &= 1. \end{aligned}$$

Since the second equation is $0 = 1$, it is impossible to solve the system. Such a system is said to be **inconsistent** and has no solution. The row $[0 \quad 0 \mid 1]$ is a signal that the given system is inconsistent.

EXAMPLE 4 Systems of Equations with an Infinite Number of Solutions
Use the Gauss-Jordan method to solve the system

$$x + 2y - z = 0$$
$$3x - y + z = 6$$
$$-2x - 4y + 2z = 0.$$

Solution The augmented matrix is

$$\begin{bmatrix} 1 & 2 & -1 & | & 0 \\ 3 & -1 & 1 & | & 6 \\ -2 & -4 & 2 & | & 0 \end{bmatrix}.$$

Get zeros in the second and third rows of column 1.

$$\begin{matrix} -3R_1 + R_2 \to R_2 \\ 2R_1 + R_3 \to R_3 \end{matrix} \begin{bmatrix} 1 & 2 & -1 & | & 0 \\ 0 & -7 & 4 & | & 6 \\ 0 & 0 & 0 & | & 0 \end{bmatrix}$$

To continue, get a zero in the first row of column 2 using the second row, as usual.

$$2R_2 + 7R_1 \to R_1 \begin{bmatrix} 7 & 0 & 1 & | & 12 \\ 0 & -7 & 4 & | & 6 \\ 0 & 0 & 0 & | & 0 \end{bmatrix}$$

We cannot get a zero for the first-row, third-column element without changing the form of the first two columns. We must multiply each of the first two rows by the reciprocal of the first nonzero number.

$$\begin{matrix} \frac{1}{7}R_1 \to R_1 \\ -\frac{1}{7}R_2 \to R_2 \end{matrix} \begin{bmatrix} 1 & 0 & \frac{1}{7} & | & \frac{12}{7} \\ 0 & 1 & -\frac{4}{7} & | & -\frac{6}{7} \\ 0 & 0 & 0 & | & 0 \end{bmatrix}$$

To complete the solution, write the equations that correspond to the first two rows of the matrix.

$$x + \frac{1}{7}z = \frac{12}{7}$$

$$y - \frac{4}{7}z = -\frac{6}{7}$$

The third equation of this system, $0 = 0$, gives us no information. Usually the third equation gives us the value of z. In this case, z can be any real number. Such a system is said to be **dependent** and has an infinite number of solutions, corresponding to the infinite number of values of z. The variable z in this case is called a **parameter.** Solve the first equation for x and the second for y to get

$$x = \frac{12 - z}{7} \quad \text{and} \quad y = \frac{4z - 6}{7}.$$

The general solution is written

$$\left(\frac{12 - z}{7}, \frac{4z - 6}{7}, z \right),$$

where z is any real number. For example, $z = 2$ and $z = 12$ lead to the particular solutions $(10/7, 2/7, 2)$ and $(0, 6, 12)$, respectively.

EXAMPLE 5 Systems of Equations with an Infinite Number of Solutions
Consider the following system of equations.

$$x + 2y + 3z - w = 4$$
$$2x + 3y + w = -3$$
$$3x + 5y + 3z = 1$$

(a) Set this up as an augmented matrix, and verify that the result after the Gauss-Jordan method is

$$\begin{bmatrix} 1 & 0 & -9 & 5 & | & -18 \\ 0 & 1 & 6 & -3 & | & 11 \\ 0 & 0 & 0 & 0 & | & 0 \end{bmatrix}$$

(b) Find the solution to this system of equations.

Solution To complete the solution, write the equations that correspond to the first two rows of the matrix.

$$x \qquad - 9z + 5w = -18$$
$$y + 6z - 3w = \quad 11$$

Because both equations involve both z and w, let z and w be parameters. There are an infinite number of solutions, corresponding to the infinite number of values of z and w. Solve the first equation for x and the second for y to get

$$x = -18 + 9z - 5w \qquad \text{and} \qquad y = 11 - 6z + 3w.$$

In an analogous manner to problems with a single parameter, the general solution is written

$$(-18 + 9z - 5w, \ 11 - 6z + 3w, \ z, \ w),$$

where z and w are any real numbers. For example, $z = 1$ and $w = -2$ leads to the particular solution $(1, -1, 1, -2)$.

Although the examples have used only systems with two equations in two unknowns, three equations in three unknowns, or three equations in four unknowns, the Gauss-Jordan method can be used for any system with n equations and m unknowns. The method becomes tedious with more than three equations in three unknowns; on the other hand, it is very suitable for use by graphing calculators and computers, which can solve fairly large systems quickly. Sophisticated computer programs modify the method to reduce round-off error. Other methods used for special types of large matrices are studied in a course on numerical analysis.

EXAMPLE 6 Rats
A lab has a group of young rats, adult female rats, and adult male rats to be used for experiments. There are 23 rats in all. The rats consume 376 g of food and 56.9 ml of water each day.

(a) Suppose that a young rat consumes 12 g of food and 2 ml of water each day. The corresponding values for an adult female are 16 and 2.5, and for an adult male are 20 and 2.8. How many young rats, adult females, and adult males are there?

Solution We have organized the information in Table 2.

Table 2

	Young	Female	Male	Total
Number	x	y	z	23
Food	12	16	20	376
Water	2.0	2.5	2.8	56.9

The three rows of the table lead to three equations: one for the total number of rats, one for the food, and one for the water.

$$\begin{aligned} x + y + z &= 23 \\ 12x + 16y + 20z &= 376 \\ 2.0x + 2.5y + 2.8z &= 56.9 \end{aligned}$$

Represent this system by an augmented matrix, and verify that the result after the Gauss-Jordan method is

$$\begin{bmatrix} 1 & 0 & 0 & | & 6 \\ 0 & 1 & 0 & | & 9 \\ 0 & 0 & 1 & | & 8 \end{bmatrix}.$$

There are 6 young rats, 9 adult females, and 8 adult males.

(b) Suppose the amounts of water that young rats, adult females, and adult males consume in a new experiment are 2 ml, 4 ml, and 6 ml, respectively, but all other information remains the same. How many young rats, adult females, and adult males are there?

Solution Change the third equation to

$$2x + 4y + 6z = 56.9$$

and go through the Gauss-Jordan method again. The result is

$$\begin{bmatrix} 1 & 0 & -1 & | & 2 \\ 0 & 1 & 2 & | & 25 \\ 0 & 0 & 0 & | & -37.1 \end{bmatrix}.$$

(If you do the row operations in a different order in this example, you will have different numbers in the last column.) The last row of this matrix says that $0 = -37.1$, so the system is inconsistent and has no solution. (In retrospect, this is clear, because each rat consumes a whole number of ml of water, and the total amount of water consumed is not a whole number. In general, however, it is not easy to tell whether a system of equations has a solution or not by just looking at it.)

(c) Suppose the amounts of water consumed by each rat are the same as in part (b), but the total amount of water consumed is 96 ml. Now how many young rats, adult females, and adult males are there?

Solution The third equation becomes

$$2x + 4y + 6z = 96,$$

and the Gauss-Jordan method leads to

$$\begin{bmatrix} 1 & 0 & -1 & | & -2 \\ 0 & 1 & 2 & | & 25 \\ 0 & 0 & 0 & | & 0 \end{bmatrix}.$$

The system is dependent, similar to Example 4. Let z be the parameter, and solve the first two equations for x and y, yielding

$$x = z - 2 \quad \text{and} \quad y = 25 - 2z.$$

Since x, y, and z represent numbers of rats, they must be nonnegative integers. From the equation for x, we must have

$$z \geq 2,$$

and from the equation for y, we must have

$$25 - 2z \geq 0,$$

from which we find

$$z \leq 12.5.$$

We therefore have 11 solutions, corresponding to $z = 2, 3, \ldots, 12$.

(d) Give the solutions from part (c) that have the smallest and the largest number of adult males.

Solution For the smallest number of adult males, let $z = 2$, giving $x = 2 - 2 = 0$ and $y = 25 - 2(2) = 21$. There are 0 young rats, 21 adult females, and 2 adult males.

For the largest number of adult males, let $z = 12$, giving $x = 12 - 2 = 10$ and $y = 25 - 2(12) = 1$. There are 10 young rats, 1 adult female, and 12 adult males.

10.1 EXERCISES

Write the augmented matrix for each of the following systems. **Do not solve.**

1. $2x + 3y = 11$
$\quad x + 2y = 8$

2. $3x + 5y = -13$
$\quad 2x + 3y = -9$

3. $2x + y + z = 3$
$\quad 3x - 4y + 2z = -7$
$\quad x + y + z = 2$

4. $4x - 2y + 3z = 4$
$\quad 3x + 5y + z = 7$
$\quad 5x - y + 4z = 7$

Write the system of equations associated with each of the following augmented matrices.

5. $\begin{bmatrix} 1 & 0 & | & 2 \\ 0 & 1 & | & 3 \end{bmatrix}$

6. $\begin{bmatrix} 1 & 0 & | & 5 \\ 0 & 1 & | & -3 \end{bmatrix}$

7. $\begin{bmatrix} 1 & 0 & 0 & | & 2 \\ 0 & 1 & 0 & | & 3 \\ 0 & 0 & 1 & | & -2 \end{bmatrix}$

8. $\begin{bmatrix} 1 & 0 & 0 & | & 4 \\ 0 & 1 & 0 & | & 2 \\ 0 & 0 & 1 & | & 3 \end{bmatrix}$

9. _____ on a matrix correspond to transformations of a system of equations.

10. Describe in your own words what $2R_1 + R_3 \rightarrow R_3$ means.

Use the indicated row operations to change each matrix.

11. Replace R_2 by $R_1 + (-2)R_2$.

$$\begin{bmatrix} 2 & 3 & 8 & | & 20 \\ 1 & 4 & 6 & | & 12 \\ 0 & 3 & 5 & | & 10 \end{bmatrix}$$

12. Replace R_3 by $-1R_1 + 3R_3$.

$$\begin{bmatrix} 3 & 2 & 6 & | & 18 \\ 2 & -2 & 5 & | & 7 \\ 1 & 0 & 5 & | & 20 \end{bmatrix}$$

13. Replace R_1 by $-4R_2 + R_1$.

$$\begin{bmatrix} 1 & 4 & 2 & | & 9 \\ 0 & 1 & 5 & | & 14 \\ 0 & 3 & 8 & | & 16 \end{bmatrix}$$

14. Replace R_1 by $R_3 + (-3)R_1$.

$$\begin{bmatrix} 1 & 0 & 4 & | & 21 \\ 0 & 6 & 5 & | & 30 \\ 0 & 0 & 12 & | & 15 \end{bmatrix}$$

15. Replace R_1 by $\frac{1}{3}R_1$.

$$\begin{bmatrix} 3 & 0 & 0 & | & 18 \\ 0 & 5 & 0 & | & 9 \\ 0 & 0 & 4 & | & 8 \end{bmatrix}$$

16. Replace R_3 by $\frac{1}{4}R_3$.

$$\begin{bmatrix} 1 & 0 & 0 & | & 6 \\ 0 & 1 & 0 & | & 5 \\ 0 & 0 & 4 & | & 12 \end{bmatrix}$$

Use the Gauss-Jordan method to solve the following systems of equations.

17. $\begin{aligned} x + y &= 5 \\ x - y &= -1 \end{aligned}$

18. $\begin{aligned} x + 2y &= 5 \\ 2x + y &= -2 \end{aligned}$

19. $\begin{aligned} 2x - 5y &= 10 \\ 4x - 5y &= 15 \end{aligned}$

20. $\begin{aligned} 4x - 2y &= 3 \\ -2x + 3y &= 1 \end{aligned}$

21. $\begin{aligned} 2x - 3y &= 2 \\ 4x - 6y &= 1 \end{aligned}$

22. $\begin{aligned} x + 2y &= 1 \\ 2x + 4y &= 3 \end{aligned}$

23. $\begin{aligned} 6x - 3y &= 1 \\ -12x + 6y &= -2 \end{aligned}$

24. $\begin{aligned} x - y &= 1 \\ -x + y &= -1 \end{aligned}$

25. $\begin{aligned} y &= x - 1 \\ y &= 6 + z \\ z &= -1 - x \end{aligned}$

26. $\begin{aligned} x &= 1 - y \\ 2x &= z \\ 2z &= -2 - y \end{aligned}$

27. $\begin{aligned} 2x - 2y &= -2 \\ y + z &= 4 \\ x + z &= 1 \end{aligned}$

28. $\begin{aligned} x - z &= -3 \\ y + z &= 9 \\ -x + z &= 3 \end{aligned}$

29. $\begin{aligned} 4x + 4y - 4z &= 24 \\ 2x - y + z &= -9 \\ x - 2y + 3z &= 1 \end{aligned}$

30. $\begin{aligned} x + 3y - 6z &= 7 \\ 2x - y + 2z &= 0 \\ x + y + 2z &= -1 \end{aligned}$

31. $\begin{aligned} 3x + 5y - z &= 0 \\ 4x - y + 2z &= 1 \\ -6x - 10y + 2z &= 0 \end{aligned}$

32. $\begin{aligned} 3x - 6y + 3z &= 15 \\ 2x + y - z &= 2 \\ -2x + 4y - 2z &= 2 \end{aligned}$

33. $\begin{aligned} 5x - 4y + 2z &= 4 \\ 5x + 3y - z &= 17 \\ 15x - 5y + 3z &= 25 \end{aligned}$

34. $\begin{aligned} 3x + 2y - z &= -16 \\ 6x - 4y + 3z &= 12 \\ 3x + 3y + z &= -11 \end{aligned}$

35. $\begin{aligned} 2x + 3y + z &= 9 \\ 4x + 6y + 2z &= 18 \\ -\frac{1}{2}x - \frac{3}{4}y - \frac{1}{4}z &= -\frac{9}{4} \end{aligned}$

36. $\begin{aligned} 4x - 2y - 3z &= -23 \\ -4x + 3y + z &= 11 \\ 8x - 5y + 4z &= 6 \end{aligned}$

37. $\begin{aligned} x + 2y - w &= 3 \\ 2x + 4z + 2w &= -6 \\ x + 2y - z &= 6 \\ 2x - y + z + w &= -3 \end{aligned}$

38. $\begin{aligned} x + 3y - 2z - w &= 9 \\ 2x + 4y + 2w &= 10 \\ -3x - 5y + 2z - w &= -15 \\ x - y - 3z + 2w &= 6 \end{aligned}$

39. $\begin{aligned} x + y - z + 2w &= -20 \\ 2x - y + z + w &= 11 \\ 3x - 2y + z - 2w &= 27 \end{aligned}$

40. $\begin{aligned} 4x - 3y + z + w &= 21 \\ -2x - y + 2z + 7w &= 2 \\ 10x - 5z - 20w &= 15 \end{aligned}$

41. $\begin{aligned} 10.47x + 3.52y + 2.58z - 6.42w &= 218.65 \\ 8.62x - 4.93y - 1.75z + 2.83w &= 157.03 \\ 4.92x + 6.83y - 2.97z + 2.65w &= 462.3 \\ 2.86x + 19.10y - 6.24z - 8.73w &= 398.4 \end{aligned}$

42. $\begin{aligned} 28.6x + 94.5y + 16.0z - 2.94w &= 198.3 \\ 16.7x + 44.3y - 27.3z + 8.9w &= 254.7 \\ 12.5x - 38.7y + 92.5z + 22.4w &= 562.7 \\ 40.1x - 28.3y + 17.5z - 10.2w &= 375.4 \end{aligned}$

Applications

LIFE SCIENCES

43. *Birds* The date of the first sighting of robins has been occurring earlier each spring over the past 25 years at the Rocky Mountain Biological Laboratory. Scientists from this laboratory have developed two linear equations that estimate the date of the first sighting of robins:

$$y = 759 - 0.338x$$
$$y = 1,637 - 0.779x,$$

where x is the year and y is the estimated number of days into the year when a robin can be expected.*

a. Compare the date of first sighting in 1980 for each of these equations.

b. Solve this system of equations to find the year in which the two estimates agree.

44. *Dietetics* A hospital dietician is planning a special diet for a certain patient. The total amount per meal of food groups A, B, and C must equal 400 grams. The diet should include one-third as much of group A as of group B, and the sum of the amounts of group A and group C should equal twice the amount of group B.

a. How many grams of each food group should be included?

b. Suppose we drop the requirement that the diet include one-third as much of group A as of group B. Describe the set of all possible solutions.

c. Suppose that, in addition to the conditions given in the original problem, foods A and B cost 2 cents per gram and food C costs 3 cents per gram, and that a meal must cost $8. Is a solution possible?

45. *Animal Feed* Solve the animal feed problem posed at the beginning of this section.

46. *Animal Breeding* An animal breeder can buy four types of food for Vietnamese pot-bellied pigs. Each case of Brand A contains 25 units of fiber, 30 units of protein, and 30 units of fat. Each case of Brand B contains 50 units of fiber, 30 units of protein, and 20 units of fat. Each case of Brand C contains 75 units of fiber, 30 units of protein, and 20 units of fat. Each case of Brand D contains 100 units of fiber, 60 units of protein, and 30 units of fat. How many cases of each should the breeder mix together to obtain a food that provides 1,200 units of fiber, 600 units of protein, and 400 units of fat?

47. *Mixing Plant Foods* Natural Brand plant food is made from three chemicals. The mix must include 10.8% of the first chemical, and the other two chemicals must be in a ratio of 4 to 3 as measured by weight. How much of each chemical is required to make 750 kg of the plant food?

48. *Bacterial Food Requirements* Three species of bacteria are fed three foods: I, II, and III. A bacterium of the first species consumes 1.3 units each of foods I and II and 2.3 units of food III each day. A bacterium of the second species consumes 1.1 units of food I, 2.4 units of food II, and 3.7 units of food III each day. A bacterium of the third species consumes 8.1 units of I, 2.9 units of II, and 5.1 units of III each day. If 16,000 units of I, 28,000 units of II, and 44,000 units of III are supplied each day, how many of each species can be maintained in this environment?

49. *Fish Food Requirements* A lake is stocked each spring with three species of fish, A, B, and C. Three foods, I, II, and III, are available in the lake. Each fish of species A requires an average of 1.32 units of food I, 2.9 units of food II, and 1.75 units of food III each day. Species B fish each require 2.1 units of food I, 0.95 unit of food II, and 0.6 unit of food III daily. Species C fish require 0.86, 1.52, and 2.01 units of I, II, and III per day, respectively. If 490 units of food I, 897 units of food II, and 653 units of food III are available daily, how much of each species should be stocked?

50. *Agriculture* According to data from a 1984 Texas agricultural report, the amount of nitrogen (in lb/acre), phosphate (in lb/acre), and labor (in hr/acre) needed to grow honeydews, yellow onions, and lettuce is given by the following table.[†]

	Honeydews	Yellow Onions	Lettuce
Nitrogen	120	150	180
Phosphate	180	80	80
Labor	4.97	4.45	4.65

a. If the farmer has 220 acres, 29,100 lb of nitrogen, 32,600 lb of phosphate, and 480 hours of labor, is it possible to use all resources completely? If so, how many acres should he allot for each crop?

b. Suppose everything is the same as in part a, except that 1,061 hours of labor are available. Is it possible to use

---

*Inouye, David, Billy Barr, Kenneth Armitage, and Brian Inouye, "Climate Change Is Affecting Altitudinal Migrants and Hibernating Species," *Proceedings of the National Academy of Science,* Vol. 97, No. 4, Feb. 15, 2000, pp. 1630–1633.

†Paredes, Miguel, Mohammad Fatehi, and Richard Hinthorn, "The Transformation of an Inconsistent Linear System into a Consistent System," *The AMATYC Review,* Vol. 13, No. 2, Spring 1992.

all resources completely? If so, how many acres should he allot for each crop?

51. *Archimedes' Problem Bovinum* Archimedes is credited with the authorship of a famous problem involving the number of cattle of the sun god. A simplified version of the problem has been stated as follows:*

> *The sun god had a herd of cattle consisting of bulls and cows, one part of which was white, a second black, a third spotted, and a fourth brown.*
> *Among the bulls, the number of white ones was one half plus one third the number of the black greater than the brown; the number of the black, one quarter plus one fifth the number of the spotted greater than the brown; the number of the spotted, one sixth and one seventh the number of the white greater than the brown.*
> *Among the cows, the number of white ones was one third plus one quarter of the total black cattle; the number of the black, one quarter plus one fifth the total of the spotted cattle; the number of the spotted, one fifth plus one sixth the total of the brown cattle; the number of the brown, one sixth plus one seventh the total of the white cattle.*
> *What was the composition of the herd?*

The problem can be solved by converting the statements into two systems of equations, using X, Y, Z, and T for the number of white, black, spotted, and brown bulls, respectively, and x, y, z, and t for the number of white, black, spotted, and brown cows, respectively. For example, the first statement can be written as $X = (1/2 + 1/3)Y + T$ and then reduced. The result is the following two systems of equations:

$$
\begin{aligned}
6X - 5Y &= 6T \\
20Y - 9Z &= 20T \qquad \text{and} \\
42Z - 13X &= 42T
\end{aligned}
\qquad
\begin{aligned}
12x - 7y &= 7Y \\
20y - 9z &= 9Z \\
30z - 11t &= 11T \\
-13x + 42t &= 13X
\end{aligned}
$$

a. Show that these two systems of equations represent Archimedes' Problem Bovinum.

b. If it is known that the number of brown bulls, T, is 4,149,387, use Gauss-Jordan Elimination to first find a solution to the 3×3 system and then use these values and Gauss-Jordan elimination to find a solution to the 4×4 system of equations.

52. *Lead Emissions* Estimates of the total amount of lead emissions in the United States are made by the Environmental Protection Agency's Office of Air Quality Planning and Standards each year. The lead estimates decreased in

the early 1990s but did not continue to fall. The table lists lead emissions estimates for four years.[†]

Year	Lead Emissions (in short tons)
1988	7,053
1992	3,808
1994	4,043
1997	3,915

a. If the relationship between the lead emissions L and the year t is expressed as $L = at^2 + bt + c$, where $t = 0$ corresponds to 1988, use data from 1988, 1992, and 1997 and a linear system of equations to determine the constants a, b, and c.

b. Use the equation from part a to predict the emissions in 1994, and compare the result with the actual data.

c. If the relationship between the lead emissions L and the year t is expressed as $L = at^3 + bt^2 + ct + d$, where $t = 0$ corresponds to 1988, use all four data points and a linear system of equations to determine the constants a, b, c, and d.

d. Discuss the appropriateness of the functions used in parts a and b to model this data.

OTHER APPLICATIONS

53. *Manufacturing* Fred's Furniture Factory has 1,950 machine hours available each week in the cutting department, 1,490 hours in the assembly department, and 2,160 in the finishing department. Manufacturing a chair requires 0.2 hours of cutting, 0.3 hours of assembly, and 0.1 hours of finishing. A cabinet requires 0.5 hours of cutting, 0.4 hours of assembly, and 0.6 hours of finishing. A buffet requires 0.3 hours of cutting, 0.1 hours of assembly, and 0.4 hours of finishing. How many chairs, cabinets, and buffets should be produced in order to use all the available production capacity?

54. *Transportation* An electronics company produces three models of stereo speakers, models A, B, and C, and can deliver them by truck, van, or station wagon. A truck holds 2 boxes of model A, 1 of model B, and 3 of model C. A van holds 1 box of model A, 3 boxes of model B, and 2 boxes of model C. A station wagon holds 1 box of model A, 3 boxes of model B, and 1 box of model C.

a. If 15 boxes of model A, 20 boxes of model B, and 22 boxes of model C are to be delivered, how many

---

*Dorrie, Heinrich, *100 Great Problems of Elementary Mathematics, Their History and Solution,* Dover Publications, New York, 1965, pp. 3–7.
†*The World Almanac and Book of Facts 2000,* World Almanac Books, p. 169.

vehicles of each type should be used so that all operate at full capacity?

b. Model C has been discontinued. Each kind of delivery vehicle can now carry one more box of model B than previously and the same number of boxes of model A. If 16 boxes of model A and 22 boxes of model B are to be delivered, how many vehicles of each type should be used so that all operate at full capacity?

55. *Loans* To get the necessary funds for a planned expansion, a small company took out three loans totaling $25,000. Company owners were able to borrow some of the money at 13%. They borrowed $2,000 more than one-half the amount of the 13% loan at 14%, and the rest at 12%. The total annual interest on the loans was $3,240.

a. How much did they borrow at each rate?

b. Suppose we drop the condition that the amount borrowed at 14% is $2,000 more than one-half the amount borrowed at 13%. What can you say about the amount borrowed at 12%? What is the solution if the amount borrowed at 12% is $5,000?

c. Suppose the company can borrow only $6,000 at 12%. Is a solution possible that still meets the conditions given in part a?

56. *Transportation* A manufacturer purchases a part for use at both of its plants—one at Roseville, California, the other at Akron, Ohio. The part is available in limited quantities from two suppliers. Each supplier has 75 units available. The Roseville plant needs 40 units, and the Akron plant requires 75 units. The first supplier charges $70 per unit delivered to Roseville and $90 per unit delivered to Akron. Corresponding costs from the second supplier are $80 and $120. The manufacturer wants to order a total of 75 units from the first, less expensive supplier, with the remaining 40 units to come from the second supplier. If the company spends $10,750 to purchase the required number of units for the two plants, find the number of units that should be purchased from each supplier.

57. *Broadway Economics* When Neil Simon opens a new play, he has to decide whether to open the show on Broadway or Off Broadway. For example, in his play *London Suite,* he decided to open it Off Broadway. From information provided by Emanuel Azenberg, his producer, the following equations were developed:

$$43,500x - y = 1,295,000$$
$$27,000x - y = 440,000,$$

where x represents the number of weeks that the show has run and y represents the profit or loss from the show (the first equation is for Broadway and the second equation is for Off Broadway).*

a. Solve this system of equations to determine when the profit/loss from the show will be equal for each venue. What is the profit at that point?

b. Discuss which venue is favorable for the show.

58. *Traffic Control* At rush hours, substantial traffic congestion is encountered at the traffic intersections shown in the figure. (The streets are one-way, as shown by the arrows.)

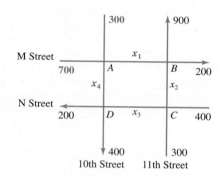

The city wishes to improve the signals at these corners so as to speed the flow of traffic. The traffic engineers first gather data. As the figure shows, 700 cars per hour come down M Street to intersection A, and 300 cars per hour come down 10th Street to intersection A. A total of x_1 of these cars leave A on M Street, and x_4 cars leave A on 10th Street. The number of cars entering A must equal the number leaving, so that

$$x_1 + x_4 = 700 + 300$$

or

$$x_1 + x_4 = 1,000.$$

For intersection B, x_1 cars enter on M Street and x_2 on 11th Street. The figure shows that 900 cars leave B on 11th and 200 on M. We have

$$x_1 + x_2 = 900 + 200$$
$$x_1 + x_2 = 1,100.$$

a. Write two equations representing the traffic entering and leaving intersections C and D.

b. Use the four equations to set up an augmented matrix, and solve the system by the Gauss-Jordan method, using x_4 as the parameter.

c. Based on your solution to part b, what are the largest and smallest possible values for the number of cars leaving intersection A on 10th Street?

---

*Goetz, Albert, "Basic Economics: Calculating Against Theatrical Disaster," *The Mathematics Teacher,* Vol. 89, No. 1, Jan. 1996, pp. 30–32.

d. Answer the question in part c for the other three variables.

e. Verify that you could have discarded any one of the four original equations without changing the solution. What does this tell you about the original problem?

59. *Modeling War* One of the factors that contribute to the success or failure of a particular army during war is its ability to get new troops ready for service. It is possible to analyze the rate of change in the number of troops of two hypothetical armies with the following simplified model,

Rate of increase (RED ARMY) $= 200,000 - 0.5r - 0.3b$

Rate of increase (BLUE ARMY) $= 350,000 - 0.5r - 0.7b$,

where r is the number of soldiers in the Red Army at a given time and b is the number of soldiers in the Blue Army at a given time. The factors 0.5 and 0.7 represent each army's efficiency of bringing new soldiers to the fight.*

a. Solve this system of equations to determine the number of soldiers in each army when the rate of increase for each is zero.

b. Describe what might be going on in a war when the rate of increase is zero.

60. *Toys* 100 toys are to be given out to a group of children. A ball costs \$2, a doll costs \$3, and a car costs \$4. A total of \$295 was spent.

a. A ball weighs 12 oz, a doll 16 oz, and a car 18 oz. The total weight of all the toys was 1,542 oz. Find how many of each toy there are.

b. Now suppose the weight of a ball, doll, and car are 11, 15, and 19 oz, respectively. If the total weight is still 1,542 oz, how many solutions are there now?

c. Keep the weights as in part b, but change the total weight to 1,480 oz. How many solutions are there?

d. Give the solution to part c that has the smallest number of cars.

e. Give the solution to part c that has the largest number of cars.

61. *Ice Cream* Researchers have determined that the amount of sugar contained in ice cream helps to determine the overall "degree of like" that a consumer has toward that particular flavor. They have also determined that too much or too little sugar will have the same negative affect on the "degree of like" and that this relationship follows a quadratic function. In an experiment conducted at Pennsylvania State University, the following condensed table was obtained.[†]

Percentage of Sugar	Degree of Like
8	5.4
13	6.3
18	5.6

a. Use this information and Gauss-Jordan elimination to determine the coefficients a, b, c, of the quadratic equation

$$y = ax^2 + bx + c,$$

where y is the degree of like and x is the percentage of sugar in the ice cream mix.

b. Repeat part a by using the quadratic regression feature on a graphing calculator. Compare your answers.

62. *Lights Out* The Tiger Electronics' game, Lights Out, consists of five rows of five lighted buttons. When a button is pushed, it changes the on/off status of it and the status of all of its vertical and horizontal neighbors. For any given situation where some of the lights are on and some are off, the goal of the game is to push buttons until all of the lights are turned off. Thus, for any given array of lights, solving a system of equations can be used to develop a strategy for turning the lights out.[‡] The following system of equations can

*Bellany, Ian, "Modeling War," *Journal of Peace Research*, Vol. 36, No. 6, 1999, pp. 729–739.
†Guinard, J., C. Zoumas-Morse, L. Mori, B. Uatoni, D. Panyam, and A. Kilar, "Sugar and Fat Effects on Sensory Properties of Ice Cream," *Journal of Food Science,* Vol. 62, No. 4, Sept./Oct. 1997, pp. 1087–1094.
‡Anderson, Marlow, and Todd Feil, "Turning Lights Out with Linear Algebra," *Mathematics Magazine,* Vol. 71, No. 4, 1998, pp. 300–303.

be used to solve the problem for a simplified, 2×2, version of the game where all of the lights are initially turned on.

$$x_{11} + x_{12} + x_{21} = 1$$
$$x_{11} + x_{12} + x_{22} = 1$$
$$x_{11} + x_{21} + x_{22} = 1$$
$$x_{12} + x_{21} + x_{22} = 1,$$

where $x_{ij} = 1$ if the light in row i, column j is on and $x_{ij} = 0$ when it is off. The order in which the buttons are pushed does not matter, so we are only seeking which buttons should be pushed.

a. Solve this system of equations and determine a strategy to turn the lights out. (*Hint:* While doing row operations, if an odd number is found, immediately replace this value with a 1; if an even number is found, then immediately replace that number with a zero. This is called modulo 2 arithmetic, and it is necessary in problems dealing with on/off switches.)

b. Resolve the equation with the right side changed to $(0, 1, 1, 0)$.

■ 10.2 ADDITION AND SUBTRACTION OF MATRICES

? **THINK ABOUT IT**

A pharmaceutical company manufactures different products at its plants in several cities. How might the company keep track of its production most efficiently?

In the previous section, matrices were used to store information about systems of linear equations. In this section, we begin a study of matrices and show additional uses of matrix notation that will answer the question posed above. The use of matrices has gained increasing importance in the fields of management, social science, and the life and other natural sciences because matrices provide a convenient way to organize data, as Example 1 demonstrates.

EXAMPLE 1 Pharmaceutical Manufacturing

The Wellbeing Pharmaceutical Company manufactures decongestants and pain relievers in three forms: capsules, softgels, and tablets. The company has regional manufacturing plants in New York, Chicago, and San Francisco. In August, the New York plant manufactured 10 cases of decongestant capsules, 12 cases of decongestant softgels, and 5 cases of decongestant tablets, as well as 15 cases of pain reliever capsules, 20 cases of pain reliever softgels, and 8 cases of pain reliever tablets. Organize this information in a more readable format.

Solution To organize this data, we might first list it as follows:

Decongestants:	10 capsules	12 softgels	5 tablets
Pain relievers:	15 capsules	20 softgels	8 tablets

Alternatively, we might tabulate the data in a chart, as in Table 3.

■ Table 3

	Capsules	**Softgels**	**Tablets**
Decongestants	10	12	5
Pain relievers	15	20	8

With the understanding that the numbers in each row refer to the pharmaceutical type (decongestants, pain relievers), and the numbers in each column refer to the form (capsule, softgel, tablet), the same information can be given by a matrix, as follows.

$$M = \begin{bmatrix} 10 & 12 & 5 \\ 15 & 20 & 8 \end{bmatrix}$$

Matrices often are named with capital letters, as in Example 1. Matrices are classified by **size;** that is, by the number of rows and columns they contain. For example, matrix M above has two rows and three columns. This matrix is a 2×3 (read "2 by 3") matrix, since it is two rows tall and three columns wide. By definition, a matrix with m rows and n columns is an $m \times n$ matrix. The number of rows is always given first.

EXAMPLE 2   Matrix Size

(a) The matrix $\begin{bmatrix} 6 & 5 \\ 3 & 4 \\ 5 & -1 \end{bmatrix}$ is a 3×2 matrix.

(b) $\begin{bmatrix} 5 & 8 & 9 \\ 0 & 5 & -3 \\ -4 & 0 & 5 \end{bmatrix}$ is a 3×3 matrix.

(c) $\begin{bmatrix} 1 & 6 & 5 & -2 & 5 \end{bmatrix}$ is a 1×5 matrix.

(d) $\begin{bmatrix} 3 \\ -5 \\ 0 \\ 2 \end{bmatrix}$ is a 4×1 matrix.

A matrix with the same number of rows as columns is called a **square matrix.** The matrix in Example 2(b) is a square matrix.

A matrix containing only one row is called a **row matrix** or a **row vector.** The matrix in Example 2(c) is a row matrix, as are

$$\begin{bmatrix} 5 & 8 \end{bmatrix}, \qquad \begin{bmatrix} 6 & -9 & 2 \end{bmatrix}, \qquad \text{and} \qquad \begin{bmatrix} -4 & 0 & 0 & 0 \end{bmatrix}.$$

A matrix of only one column, as in Example 2(d), is a **column matrix** or a **column vector.**

Equality for matrices is defined as follows.

MATRIX EQUALITY

Two matrices are equal if they are the same size and if each pair of corresponding elements is equal.

By this definition,

$$\begin{bmatrix} 2 & 1 \\ 3 & -5 \end{bmatrix} \qquad \text{and} \qquad \begin{bmatrix} 1 & 2 \\ -5 & 3 \end{bmatrix}$$

are not equal (even though they contain the same elements and are the same size) since the corresponding elements differ.

EXAMPLE 3 Matrix Equality

(a) From the definition of matrix equality given above, the only way that the statement

$$\begin{bmatrix} 2 & 1 \\ p & q \end{bmatrix} = \begin{bmatrix} x & y \\ -1 & 0 \end{bmatrix}$$

can be true is if $2 = x$, $1 = y$, $p = -1$, and $q = 0$.

(b) The statement

$$\begin{bmatrix} x \\ y \end{bmatrix} = \begin{bmatrix} 1 \\ 4 \\ 0 \end{bmatrix}$$

can never be true, since the two matrices are different sizes. (One is 2×1 and the other is 3×1.) ■

Addition The matrix given in Example 1,

$$M = \begin{bmatrix} 10 & 12 & 5 \\ 15 & 20 & 8 \end{bmatrix},$$

shows the number of cases of decongestants and pain relievers in each of the three forms manufactured by the New York plant of the Wellbeing Pharmaceutical Company in August. If matrix N below gives the amounts manufactured in September, what are the total amounts manufactured over these two months?

$$N = \begin{bmatrix} 45 & 35 & 30 \\ 65 & 40 & 35 \end{bmatrix}$$

If 10 cases of decongestant capsules were manufactured in August and 45 in September, then altogether $10 + 45 = 55$ cases were manufactured in the two months. The other corresponding entries can be added in a similar way to get a new matrix Q, which represents the total amounts manufactured over the two months.

$$Q = \begin{bmatrix} 55 & 47 & 25 \\ 80 & 60 & 43 \end{bmatrix}$$

It is convenient to refer to Q as the sum of M and N.

The way these two matrices were added illustrates the following definition of addition of matrices.

ADDITION OF MATRICES

The sum of two $m \times n$ matrices X and Y is the $m \times n$ matrix $X + Y$ in which each element is the sum of the corresponding elements of X and Y.

CAUTION It is important to remember that only matrices that are the same size can be added.

EXAMPLE 4 Addition of Matrices

Find each sum if possible.

Solution

(a) $\begin{bmatrix} 5 & -6 \\ 8 & 9 \end{bmatrix} + \begin{bmatrix} -4 & 6 \\ 8 & -3 \end{bmatrix} = \begin{bmatrix} 5 + (-4) & -6 + 6 \\ 8 + 8 & 9 + (-3) \end{bmatrix} = \begin{bmatrix} 1 & 0 \\ 16 & 6 \end{bmatrix}$

(b) The matrices

$$A = \begin{bmatrix} 5 & 8 \\ 6 & 2 \end{bmatrix} \quad \text{and} \quad B = \begin{bmatrix} 3 & 9 & 1 \\ 4 & 2 & 5 \end{bmatrix}$$

are different sizes. Therefore, the sum $A + B$ does not exist.　■

EXAMPLE 5 Pharmaceutical Manufacturing
The number of cases manufactured by the Wellbeing Pharmaceutical Company in September at the New York, San Francisco, and Chicago manufacturing plants are given in matrices N, S, and C below.

$$N = \begin{bmatrix} 45 & 35 & 20 \\ 65 & 40 & 35 \end{bmatrix} \quad S = \begin{bmatrix} 30 & 32 & 28 \\ 43 & 47 & 30 \end{bmatrix} \quad C = \begin{bmatrix} 22 & 25 & 38 \\ 31 & 34 & 35 \end{bmatrix}$$

What was the total amount manufactured at the three warehouses in September?

Solution The total amount manufactured in September is represented by the sum of the three matrices N, S, and C.

$$N + S + C = \begin{bmatrix} 45 & 35 & 20 \\ 65 & 40 & 35 \end{bmatrix} + \begin{bmatrix} 30 & 32 & 28 \\ 43 & 47 & 30 \end{bmatrix} + \begin{bmatrix} 22 & 25 & 38 \\ 31 & 34 & 35 \end{bmatrix}$$

$$= \begin{bmatrix} 97 & 92 & 86 \\ 139 & 121 & 100 \end{bmatrix}$$

For example, this sum shows that the total number of cases of decongestants in tablet form manufactured at the three plants in September was 86.　■

The additive inverse of the real number a is $-a$; a similar definition applies to matrices.

ADDITIVE INVERSE

The **additive inverse** (or **negative**) of a matrix X is the matrix $-X$ in which each element is the additive inverse of the corresponding element of X.

If

$$A = \begin{bmatrix} 1 & 2 & 3 \\ 0 & -1 & 5 \end{bmatrix} \quad \text{and} \quad B = \begin{bmatrix} -2 & 3 & 0 \\ 1 & -7 & 2 \end{bmatrix},$$

then by the definition of the additive inverse of a matrix,

$$-A = \begin{bmatrix} -1 & -2 & -3 \\ 0 & 1 & -5 \end{bmatrix} \quad \text{and} \quad -B = \begin{bmatrix} 2 & -3 & 0 \\ -1 & 7 & -2 \end{bmatrix}.$$

By the definition of matrix addition, for each matrix X the sum $X + (-X)$ is a **zero matrix,** O, whose elements are all zeros. For the matrix A above,

$$A - A = \begin{bmatrix} 0 & 0 & 0 \\ 0 & 0 & 0 \end{bmatrix}.$$

FOR REVIEW ■

Compare this with the identity property for real numbers: for any real number a,
$a + 0 = 0 + a = a$.
Exercises 34–37 give other properties of matrices that are parallel to the properties of real numbers.

There is an $m \times n$ zero matrix for each pair of values of m and n. Such a matrix serves as the $m \times n$ **additive identity,** similar to the additive identity 0 for any real number. Zero matrices have the following identity property.

ZERO MATRIX

If O is the $m \times n$ zero matrix, and A is any $m \times n$ matrix, then
$$A + O = O + A = A.$$

Subtraction   The subtraction of matrices is defined in a manner comparable to subtraction of real numbers.

SUBTRACTION OF MATRICES

For two $m \times n$ matrices X and Y, the difference $X - Y$ is the $m \times n$ matrix defined by
$$X - Y = X + (-Y).$$

This definition means that matrix subtraction can be performed by subtracting corresponding elements. For example, with A and B as defined above,

$$A - B = \begin{bmatrix} 1 & 2 & 3 \\ 0 & -1 & 5 \end{bmatrix} - \begin{bmatrix} -2 & 3 & 0 \\ 1 & -7 & 2 \end{bmatrix}$$
$$= \begin{bmatrix} 1 - (-2) & 2 - 3 & 3 - 0 \\ 0 - 1 & -1 - (-7) & 5 - 2 \end{bmatrix}$$
$$= \begin{bmatrix} 3 & -1 & 3 \\ -1 & 6 & 3 \end{bmatrix}.$$

Matrix operations are easily performed on a graphing calculator. Figure 8 shows the previous operation; the matrices A and B were already entered into the calculator.

Spreadsheet programs are designed to effectively organize data that can be represented in rows and columns. Accordingly, matrix operations are also easily performed on spreadsheets. See *The Spreadsheet Manual* that is available on this text's Web site for details.

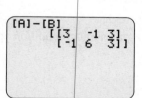

FIGURE 8

EXAMPLE 6   Subtraction of Matrices

(a) $[8 \quad 6 \quad -4] - [3 \quad 5 \quad -8] = [5 \quad 1 \quad 4]$

(b) The matrices
$$\begin{bmatrix} -2 & 5 \\ 0 & 1 \end{bmatrix} \quad \text{and} \quad \begin{bmatrix} 3 \\ 5 \end{bmatrix}$$
are different sizes and cannot be subtracted.

EXAMPLE 7   Pharmaceutical Manufacturing

During September the Chicago manufacturing plant of the Wellbeing Pharmaceutical Company shipped out the following number of cases of each pharmaceutical.

$$K = \begin{bmatrix} 5 & 10 & 8 \\ 11 & 14 & 15 \end{bmatrix}.$$

How many cases were left at the plant on October 1, taking into account only the cases manufactured and sent out during the month?

Solution The number of cases of each pharmaceutical manufactured during September is given by matrix C from Example 5; the number of cases of each pharmaceutical shipped out during September is given by matrix K. The October 1 inventory will be represented by the matrix $C - K$:

$$\begin{bmatrix} 22 & 25 & 38 \\ 31 & 34 & 35 \end{bmatrix} - \begin{bmatrix} 5 & 10 & 8 \\ 11 & 14 & 15 \end{bmatrix} = \begin{bmatrix} 17 & 15 & 30 \\ 20 & 20 & 20 \end{bmatrix}.$$

10.2 EXERCISES

Mark each statement as true or false. If false, tell why.

1. $\begin{bmatrix} 1 & 3 \\ 5 & 7 \end{bmatrix} = \begin{bmatrix} 1 & 5 \\ 3 & 7 \end{bmatrix}$

2. $\begin{bmatrix} 1 \\ 2 \\ 3 \end{bmatrix} = \begin{bmatrix} 1 & 2 & 3 \end{bmatrix}$

3. $\begin{bmatrix} x \\ y \end{bmatrix} = \begin{bmatrix} 3 \\ 5 \end{bmatrix}$ if $x = 3$ and $y = 5$.

4. $\begin{bmatrix} 3 & 5 & 2 & 8 \\ 1 & -1 & 4 & 0 \end{bmatrix}$ is a 4×2 matrix.

5. $\begin{bmatrix} 1 & 9 & -4 \\ 3 & 7 & 2 \\ -1 & 1 & 0 \end{bmatrix}$ is a square matrix.

6. $\begin{bmatrix} 2 & 4 & -1 \\ 3 & 7 & 5 \\ 0 & 0 & 0 \end{bmatrix} = \begin{bmatrix} 2 & 4 & -1 \\ 3 & 7 & 5 \end{bmatrix}$

Find the size of each matrix. Identify any square, column, or row matrices. Give the additive inverse of each matrix.

7. $\begin{bmatrix} -4 & 8 \\ 2 & 3 \end{bmatrix}$

8. $\begin{bmatrix} -9 & 6 & 2 \\ 4 & 1 & 8 \end{bmatrix}$

9. $\begin{bmatrix} -6 & 8 & 0 & 0 \\ 4 & 1 & 9 & 2 \\ 3 & -5 & 7 & 1 \end{bmatrix}$

10. $\begin{bmatrix} 8 & -2 & 4 & 6 & 3 \end{bmatrix}$

11. $\begin{bmatrix} 2 \\ 4 \end{bmatrix}$

12. $\begin{bmatrix} -9 \end{bmatrix}$

13. The sum of an $n \times m$ matrix and its additive inverse is _____.

14. If A is a 5×2 matrix and $A + K = A$, what do you know about K?

Find the values of the variables in each equation.

15. $\begin{bmatrix} 2 & 1 \\ 4 & 8 \end{bmatrix} = \begin{bmatrix} x & 1 \\ y & z \end{bmatrix}$

16. $\begin{bmatrix} -5 \\ y \end{bmatrix} = \begin{bmatrix} -5 \\ 8 \end{bmatrix}$

17. $\begin{bmatrix} x + 6 & y + 2 \\ 8 & 3 \end{bmatrix} = \begin{bmatrix} -9 & 7 \\ 8 & k \end{bmatrix}$

18. $\begin{bmatrix} 9 & 7 \\ r & 0 \end{bmatrix} = \begin{bmatrix} m - 3 & n + 5 \\ 8 & 0 \end{bmatrix}$

19. $\begin{bmatrix} -7 + z & 4r & 8s \\ 6p & 2 & 5 \end{bmatrix} + \begin{bmatrix} -9 & 8r & 3 \\ 2 & 5 & 4 \end{bmatrix} = \begin{bmatrix} 2 & 36 & 27 \\ 20 & 7 & 12a \end{bmatrix}$

20. $\begin{bmatrix} a + 2 & 3z + 1 & 5m \\ 4k & 0 & 3 \end{bmatrix} + \begin{bmatrix} 3a & 2z & 5m \\ 2k & 5 & 6 \end{bmatrix} = \begin{bmatrix} 10 & -14 & 80 \\ 10 & 5 & 9 \end{bmatrix}$

Perform the indicated operations where possible.

21. $\begin{bmatrix} 1 & 2 & 5 & -1 \\ 3 & 0 & 2 & -4 \end{bmatrix} + \begin{bmatrix} 8 & 10 & -5 & 3 \\ -2 & -1 & 0 & 0 \end{bmatrix}$

22. $\begin{bmatrix} 1 & 5 \\ 2 & -3 \\ 3 & 7 \end{bmatrix} + \begin{bmatrix} 2 & 3 \\ 8 & 5 \\ -1 & 9 \end{bmatrix}$

23. $\begin{bmatrix} 1 & 3 & -2 \\ 4 & 7 & 1 \end{bmatrix} + \begin{bmatrix} 3 & 0 \\ 6 & 4 \\ -5 & 2 \end{bmatrix}$

24. $\begin{bmatrix} 1 & 3 & -2 \\ 4 & 7 & 1 \end{bmatrix} - \begin{bmatrix} 3 & 6 & -5 \\ 0 & 4 & 2 \end{bmatrix}$

25. $\begin{bmatrix} 2 & 8 & 12 & 0 \\ 7 & 4 & -1 & 5 \\ 1 & 2 & 0 & 10 \end{bmatrix} - \begin{bmatrix} 1 & 3 & 6 & 9 \\ 2 & -3 & -3 & 4 \\ 8 & 0 & -2 & 17 \end{bmatrix}$

26. $\begin{bmatrix} 2 & 1 \\ 5 & -3 \\ -7 & 2 \\ 9 & 0 \end{bmatrix} + \begin{bmatrix} 1 & -8 & 0 \\ 5 & 3 & 2 \\ -6 & 7 & -5 \\ 2 & -1 & 0 \end{bmatrix}$

27. $\begin{bmatrix} 2 & 3 \\ -2 & 4 \end{bmatrix} + \begin{bmatrix} 4 & 3 \\ 7 & 8 \end{bmatrix} - \begin{bmatrix} 3 & 2 \\ 1 & 4 \end{bmatrix}$

28. $\begin{bmatrix} 4 & 3 \\ 1 & 2 \end{bmatrix} - \begin{bmatrix} 1 & 1 \\ 1 & 0 \end{bmatrix} + \begin{bmatrix} 1 & 1 \\ 1 & 4 \end{bmatrix}$

29. $\begin{bmatrix} 1 & 5 \\ -3 & 7 \end{bmatrix} - \begin{bmatrix} 6 & 3 \\ 2 & 4 \end{bmatrix} + \begin{bmatrix} 8 & 10 \\ -1 & 0 \end{bmatrix}$

30. $\begin{bmatrix} -1 & -1 \\ -1 & 0 \end{bmatrix} + \begin{bmatrix} 4 & 3 \\ 1 & 2 \end{bmatrix} + \begin{bmatrix} 1 & 1 \\ 1 & 4 \end{bmatrix}$

31. $\begin{bmatrix} -4x + 2y & -3x + y \\ 6x - 3y & 2x - 5y \end{bmatrix} + \begin{bmatrix} -8x + 6y & 2x \\ 3y - 5x & 6x + 4y \end{bmatrix}$

32. $\begin{bmatrix} 4k - 8y \\ 6z - 3x \\ 2k + 5a \\ -4m + 2n \end{bmatrix} - \begin{bmatrix} 5k + 6y \\ 2z + 5x \\ 4k + 6a \\ 4m - 2n \end{bmatrix}$

33. For matrix X given below, find the matrix $-X$.

Using matrices $O = \begin{bmatrix} 0 & 0 \\ 0 & 0 \end{bmatrix}$, $P = \begin{bmatrix} m & n \\ p & q \end{bmatrix}$, $T = \begin{bmatrix} r & s \\ t & u \end{bmatrix}$, *and* $X = \begin{bmatrix} x & y \\ z & w \end{bmatrix}$, *verify the statements in Exercises 34–37.*

34. $X + T = T + X$ (commutative property of addition of matrices)

35. $X + (T + P) = (X + T) + P$ (associative property of addition of matrices)

36. $X + (-X) = O$ (inverse property of addition of matrices)

37. $P + O = P$ (identity property of addition of matrices)

38. Which of the above properties are valid for matrices that are not square? What must be true about O, P, T, and X for these statements to make sense?

Applications

LIFE SCIENCES

39. *Dietetics* A dietician prepares a diet specifying the amounts a patient should eat of four basic food groups: group I, meats; group II, fruits and vegetables; group III, breads and starches; group IV, milk products. Amounts are given in "exchanges" that represent 1 oz (meat), 1/2 cup (fruits and vegetables), 1 slice (bread), 8 oz (milk), or other suitable measurements.

  a. The number of "exchanges" for breakfast for each of the four food groups, respectively, are 2, 1, 2, and 1; for

lunch, 3, 2, 2, and 1; and for dinner, 4, 3, 2, and 1. Write a 3 × 4 matrix using this information.

  b. The amounts of fat, carbohydrates, and protein (in appropriate units) in each food group, respectively, are as follows.

  Fat: 5, 0, 0, 10

  Carbohydrates: 0, 10, 15, 12

  Protein: 7, 1, 2, 8

  Use this information to write a 4 × 3 matrix.

c. There are 8 calories per exchange of fat, 4 calories per exchange of carbohydrates, and 5 calories per exchange of protein; summarize this data in a 3×1 matrix.

40. *Animal Growth* At the beginning of a laboratory experiment, five baby rats measured 5.6, 6.4, 6.9, 7.6, and 6.1 centimeters in length, and weighed 144, 138, 149, 152, and 146 grams, respectively.

a. Write a 2×5 matrix using this information.

b. At the end of two weeks, their lengths in centimeters were 10.2, 11.4, 11.4, 12.7, and 10.8 and their weights (in grams) were 196, 196, 225, 250, and 230. Write a 2×5 matrix with this information.

c. Use matrix subtraction and the matrices found in parts a and b to write a matrix that gives the amount of change in length and weight for each rat.

d. During the third week, the rats grew by the amounts shown in the matrix below.

$$\begin{array}{c} \text{Length} \\ \text{Weight} \end{array} \begin{bmatrix} 1.8 & 1.5 & 2.3 & 1.8 & 2.0 \\ 25 & 22 & 29 & 33 & 20 \end{bmatrix}$$

What were their lengths and weights at the end of this week?

41. *Testing Medication* A drug company is testing 200 patients to see if Painfree (a new headache medicine) is effective. Half the patients receive Painfree and half receive a placebo. The data on the first 50 patients is summarized in this matrix:

$$\begin{array}{c} \\ \text{Painfree} \\ \text{Placebo} \end{array} \overset{\overset{\textstyle\text{Pain Relief Obtained}}{\text{Yes \quad No}}}{\begin{bmatrix} 22 & 3 \\ 8 & 17 \end{bmatrix}}.$$

a. Of those who took the placebo, how many got relief?

b. Of those who took the new medication, how many got no relief?

c. The test was repeated on three more groups of 50 patients each, with the results summarized by these matrices.

$$\begin{bmatrix} 21 & 4 \\ 6 & 19 \end{bmatrix} \quad \begin{bmatrix} 19 & 6 \\ 10 & 15 \end{bmatrix} \quad \begin{bmatrix} 23 & 2 \\ 3 & 22 \end{bmatrix}$$

Find the total results for all 200 patients.

d. On the basis of these results, does it appear that Painfree is effective?

OTHER APPLICATIONS

42. *Car Accidents* The following tables give the death rates, per million person trips, for male and female drivers for various ages and number of passengers.*

Male Drivers	Number of Passengers			
Age	0	1	2	≥ 3
16	2.61	4.39	6.29	9.08
17	1.63	2.77	4.61	6.92
30–59	0.92	0.75	0.62	0.54

Female Drivers	Number of Passengers			
Age	0	1	2	≥ 3
16	1.38	1.72	1.94	3.31
17	1.26	1.48	2.82	2.28
30–59	0.41	0.33	0.27	0.40

a. Write a matrix for the death rate of male drivers.

b. Write a matrix for the death rate of female drivers.

c. Use the matrices from parts a and b to write a matrix showing the difference between the death rates of males and females.

d. Analyze the results of part c and make some suggestions on how to reduce the rates.

43. *Life Expectancy* The table on the next page gives the life expectancy for African-American males and females and

---

*Chen, Li-Hui, Susan Baker, Elisa Braver, and Guohua Li, "Carrying Passengers as a Risk Factor for Crashes Fatal to 16- and 17-Year-Old Drivers," *JAMA*, Vol. 283, No. 12, Mar. 22/29, 2000, pp. 1578–1582.

white American males and females for various times in the last 30 years.*

	Black		White	
Year	M	F	M	F
1970	60.0	68.3	68.0	75.6
1980	63.8	72.5	70.7	78.1
1990	64.5	73.6	72.7	79.4
1997	67.2	74.7	74.3	79.9

a. Write a matrix for the life expectancy of African Americans.

b. Write a matrix for the life expectancy of white Americans.

c. Use the matrices from parts a and b to write a matrix showing the difference between the life expectancy between the two groups.

d. Analyze the results of part c and make some suggestions on how to decrease these numbers.

44. *Educational Attainment* The following table gives the educational attainment of the U.S. population 25 years and older.†

	Male		Female	
	Percent with Four Years of High School or More	Percent with Four Years of College or More	Percent with Four Years of High School or More	Percent with Four Years of College or More
1960	39.5	9.7	42.5	5.8
1970	51.9	13.5	52.8	8.1
1980	67.3	20.1	65.8	12.8
1990	77.7	24.4	77.5	18.4
1995	81.7	26.0	81.6	20.2
1998	82.8	26.5	82.9	22.4

a. Write a matrix for the educational attainment of males.

b. Write a matrix for the educational attainment of females.

c. Use the matrices from parts a and b to write a matrix showing how much more (or less) education males have attained than females.

45. *Educational Attainment* The following table gives the educational attainment of African Americans 25 years and older.†

	Male		Female	
	Percent with Four Years of High School or More	Percent with Four Years of College or More	Percent with Four Years of High School or More	Percent with Four Years of College or More
1960	18.2	2.8	21.8	3.3
1970	30.1	4.2	32.5	4.6
1980	50.8	8.4	51.5	8.3
1990	65.8	11.9	66.5	10.8
1995	73.4	13.6	74.1	12.9
1998	75.2	13.9	76.7	15.4

a. Write a matrix for the educational attainment of African-American males.

b. Write a matrix for the educational attainment of African-American females.

c. Use the matrices from parts a and b to write a matrix showing how more (or less) education African-American males have attained than African-American females.

46. *Animal Interactions* When two kittens named Cauchy and Cliché were introduced into a household with Jamie (an older cat) and Musk (a dog), the interactions between animals were complicated. The two kittens liked each other and Jamie, but didn't like Musk. Musk liked everybody, but Jamie didn't like any of the other animals.

a. Write a 4 × 4 matrix in which rows (and columns) 1, 2, 3, and 4 refer to Musk, Jamie, Cauchy, and Cliché. Make an element a 1 if the animal for that row likes the animal for that column, and otherwise make the element a 0. Assume every animal likes herself.

b. Within a few days, Cauchy and Cliché decided that they liked Musk after all. Write a 4 × 4 matrix, as you did in part a, representing the new situation.

47. *Management* A toy company has plants in Boston, Chicago, and Seattle that manufacture toy phones and calculators. The following matrix gives the production costs (in dollars) for each item at the Boston plant:

$$\begin{array}{cc} & \text{Phones} \quad \text{Calculators} \\ \begin{array}{c} \text{Material} \\ \text{Labor} \end{array} & \begin{bmatrix} 4.27 & 6.94 \\ 3.45 & 3.65 \end{bmatrix} \end{array}$$

a. In Chicago, a phone costs $4.05 for material and $3.27 for labor; a calculator costs $7.01 for material and $3.51 for labor. In Seattle, material costs are $4.40 for a phone and $6.90 for a calculator; labor costs are $3.54 for a phone and $3.76 for a calculator. Write the production cost matrices for Chicago and Seattle.

b. Suppose labor costs increase by $0.11 per item in Chicago and material costs there increase by $0.37 for a phone and $0.42 for a calculator. What is the new production cost matrix for Chicago?

48. *Management* There are three convenience stores in Folsom. This week, Store I sold 88 loaves of bread, 48 quarts of milk, 16 jars of peanut butter, and 112 pounds of cold cuts. Store II sold 105 loaves of bread, 72 quarts of milk, 21 jars of peanut butter, and 147 pounds of cold cuts. Store III sold 60 loaves of bread, 40 quarts of milk, no peanut butter, and 50 pounds of cold cuts.

a. Use a 4 × 3 matrix to express the sales information for the three stores.

b. During the following week, sales on these products at Store I increased by 25%; sales at Store II increased by 1/3; and sales at Store III increasec by 10%. Write the sales matrix for that week.

c. Write a matrix that represents total sales over the two week period.

10.3 MULTIPLICATION OF MATRICES

? THINK ABOUT IT   How can the long-term survival of the northern spotted owl be predicted?

Matrix multiplication will be used to answer this question in one of the exercises.

We begin by defining the product of a real number and a matrix. In work with matrices, a real number is called a **scalar.**

PRODUCT OF A MATRIX AND A SCALAR

The product of a scalar k and a matrix X is the matrix kX, each of whose elements is k times the corresponding element of X.

For example,

$$(-3)\begin{bmatrix} 2 & -5 \\ 1 & 7 \end{bmatrix} = \begin{bmatrix} -6 & 15 \\ -3 & -21 \end{bmatrix}.$$

Finding the product of two matrices is more involved, but such multiplication is important in solving practical problems. To understand the reasoning behind matrix multiplication, it may be helpful to consider another example concerning Wellbeing Pharmaceutical Company discussed in the previous section. Suppose Wellbeing makes three antibiotics, which we will label A, B, and C. Matrix W shows the number of cases of each antibiotic manufactured in a month at each plant.

$$\begin{array}{c} \\ \text{New York} \\ \text{Chicago} \\ \text{San Francisco} \end{array} \begin{array}{ccc} \text{A} & \text{B} & \text{C} \\ \begin{bmatrix} 10 & 7 & 3 \\ 5 & 9 & 6 \\ 4 & 8 & 2 \end{bmatrix} \end{array} = W$$

If the selling price for a case of antibiotic A is $800, for a case of antibiotic B is $1,000, and for a case of antibiotic C is $1,200, the total value of the antibiotics manufactured at the New York plant is found in Table 4.

Table 4

Antibiotic	Number of Cases		Price of Case		Total
A	10	$\times$	$800	=	$8,000
B	7	$\times$	$1,000	=	$7,000
C	3	$\times$	$1,200	=	$3,600
		(Total for New York)			$18,600

The total value of the three kinds of antibiotics manufactured in New York is $18,600.

The work done in Table 4 is summarized as follows:

$$10(\$800) + 7(\$1,000) + 3(\$1,200) = \$18,600.$$

In the same way, we find that the Chicago antibiotics have a total value of

$$5(\$800) + 9(\$1,000) + 6(\$1,200) = \$20,200,$$

and in San Francisco, the total value of the antibiotics is

$$4(\$800) + 8(\$1,000) + 2(\$1,200) = \$13,600.$$

The selling prices can be written as a column matrix P, and the total value in each location as another column matrix, V.

$$\begin{bmatrix} 800 \\ 1,000 \\ 1,200 \end{bmatrix} = P \qquad \begin{bmatrix} 18,600 \\ 20,200 \\ 13,600 \end{bmatrix} = V$$

Look at the elements of W and P below; multiplying the first, second, and third elements of the first row of W by the first, second, and third elements, respectively, of the column matrix P and then adding these products gives the first element in V. Doing the same thing with the second row of W gives the second element of V; the third row of W leads to the third element of V, suggesting that it is reasonable to write the product of matrices

$$W = \begin{bmatrix} 10 & 7 & 3 \\ 5 & 9 & 6 \\ 4 & 8 & 2 \end{bmatrix} \quad \text{and} \quad P = \begin{bmatrix} 800 \\ 1,000 \\ 1,200 \end{bmatrix}$$

as

$$WP = \begin{bmatrix} 10 & 7 & 3 \\ 5 & 9 & 6 \\ 4 & 8 & 2 \end{bmatrix} \begin{bmatrix} 800 \\ 1,000 \\ 1,200 \end{bmatrix} = \begin{bmatrix} 18,600 \\ 20,200 \\ 13,600 \end{bmatrix} = V.$$

The product was found by multiplying the elements of *rows* of the matrix on the left and the corresponding elements of the *column* of the matrix on the right, and then adding these separate products. Notice that the product of a 3×3 matrix and a 3×1 matrix is a 3×1 matrix. Notice also that this matrix product is equivalent to the three equations following Table 4. The product of any two matrices can be defined this way as long as the number of columns in the first matrix is the same as the number of rows in the second matrix.

The product AB of an $m \times n$ matrix A and an $n \times k$ matrix B is found as follows. Multiply each element of the *first row* of A by the corresponding element of the *first column* of B. The sum of these n products is the *first-row, first-column* element of AB. Similarly, the sum of the products found by multiplying the elements of the *first row* of A by the corresponding elements of the *second column* of B gives the *first-row, second-column* element of AB, and so on.

> **PRODUCT OF TWO MATRICES**
>
> Let A be an $m \times n$ matrix and let B be an $n \times k$ matrix. To find the element in the ith row and jth column of the **product matrix** AB, multiply each element in the ith row of A by the corresponding element in the jth column of B, and then add these products. The product matrix AB is an $m \times k$ matrix.

EXAMPLE 1   Matrix Product

Find the product AB of matrices

$$A = \begin{bmatrix} 2 & 3 & -1 \\ 4 & 2 & 2 \end{bmatrix} \quad \text{and} \quad B = \begin{bmatrix} 1 \\ 8 \\ 6 \end{bmatrix}.$$

Solution

Step 1 Multiply the elements of the first row of A and the corresponding elements of the column of B.

$$\begin{bmatrix} \mathbf{2} & \mathbf{3} & \mathbf{-1} \\ 4 & 2 & 2 \end{bmatrix} \begin{bmatrix} \mathbf{1} \\ \mathbf{8} \\ \mathbf{6} \end{bmatrix} \quad \mathbf{2 \cdot 1 + 3 \cdot 8 + (-1) \cdot 6} = 20$$

Thus, 20 is the first-row entry of the product matrix AB.

Step 2 Multiply the elements of the second row of A and the corresponding elements of B.

$$\begin{bmatrix} 2 & 3 & -1 \\ \mathbf{4} & \mathbf{2} & \mathbf{2} \end{bmatrix} \begin{bmatrix} \mathbf{1} \\ \mathbf{8} \\ \mathbf{6} \end{bmatrix} \quad \mathbf{4 \cdot 1 + 2 \cdot 8 + 2 \cdot 6} = 32$$

The second-row entry of the product matrix AB is 32.

Step 3 Write the product as a column matrix using the two entries found above.

$$AB = \begin{bmatrix} 2 & 3 & -1 \\ 4 & 2 & 2 \end{bmatrix} \begin{bmatrix} 1 \\ 8 \\ 6 \end{bmatrix} = \begin{bmatrix} 20 \\ 32 \end{bmatrix}$$

EXAMPLE 2   Matrix Product

Find the product CD of matrices

$$C = \begin{bmatrix} -3 & 4 & 2 \\ 5 & 0 & 4 \end{bmatrix} \quad \text{and} \quad D = \begin{bmatrix} -6 & 4 \\ 2 & 3 \\ 3 & -2 \end{bmatrix}.$$

Solution

Step 1

$$\begin{bmatrix} -3 & 4 & 2 \\ 5 & 0 & 4 \end{bmatrix} \begin{bmatrix} -6 & 4 \\ 2 & 3 \\ 3 & -2 \end{bmatrix} \qquad (-3) \cdot (-6) + 4 \cdot 2 + 2 \cdot 3 = 32$$

Step 2

$$\begin{bmatrix} -3 & 4 & 2 \\ 5 & 0 & 4 \end{bmatrix} \begin{bmatrix} -6 & 4 \\ 2 & 3 \\ 3 & -2 \end{bmatrix} \qquad (-3) \cdot 4 + 4 \cdot 3 + 2 \cdot (-2) = -4$$

Step 3

$$\begin{bmatrix} -3 & 4 & 2 \\ 5 & 0 & 4 \end{bmatrix} \begin{bmatrix} -6 & 4 \\ 2 & 3 \\ 3 & -2 \end{bmatrix} \qquad 5 \cdot (-6) + 0 \cdot 2 + 4 \cdot 3 = -18$$

Step 4

$$\begin{bmatrix} -3 & 4 & 2 \\ 5 & 0 & 4 \end{bmatrix} \begin{bmatrix} -6 & 4 \\ 2 & 3 \\ 3 & -2 \end{bmatrix} \qquad 5 \cdot 4 + 0 \cdot 3 + 4 \cdot (-2) = 12$$

Step 5 The product is

$$CD = \begin{bmatrix} -3 & 4 & 2 \\ 5 & 0 & 4 \end{bmatrix} \begin{bmatrix} -6 & 4 \\ 2 & 3 \\ 3 & -2 \end{bmatrix} = \begin{bmatrix} 32 & -4 \\ -18 & 12 \end{bmatrix}.$$

Here the product of a 2 × 3 matrix and a 3 × 2 matrix is a 2 × 2 matrix. ■

NOTE One way to avoid errors in matrix multiplication is to lower the first matrix so it is below and to the left of the second matrix, and then write the product in the space between the two matrices. For example, to multiply the matrices in Example 2, we could rewrite the product as shown below.

$$\begin{bmatrix} \downarrow & \\ -6 & 4 \\ 2 & 3 \\ 3 & -2 \end{bmatrix}$$
$$\rightarrow \begin{bmatrix} -3 & 4 & 2 \\ 5 & 0 & 4 \end{bmatrix} \begin{bmatrix} * & \\ & \end{bmatrix}$$

To find the entry where the ∗ is, for example, multiply the row and the column indicated by the arrows: $5 \cdot (-6) + 0 \cdot 2 + 4 \cdot 3 = -18$.

As the definition of matrix multiplication shows,

the product AB of two matrices A and B can be found only if the number of columns of A is the same as the number of rows of B.

The final product will have as many rows as *A* and as many columns as *B*.

EXAMPLE 3   Matrix Product

Suppose matrix *A* is 2×2 and matrix *B* is 2×4. Can the products *AB* and *BA* be calculated? If so, what is the size of each product?

Solution The following diagram helps decide the answers to these questions.

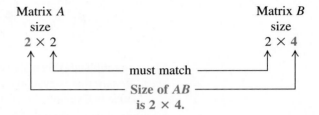

The product of *A* and *B* can be found because *A* has two columns and *B* has two rows. The size of the product is 2×4.

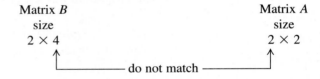

The product *BA* cannot be found because *B* has 4 columns and *A* has 2 rows.

EXAMPLE 4   Comparing Matrix Products *AB* and *BA*

Find *AB* and *BA*, given

$$A = \begin{bmatrix} 1 & -3 \\ 7 & 2 \\ -2 & 5 \end{bmatrix} \quad \text{and} \quad B = \begin{bmatrix} 1 & 0 & -1 \\ 3 & 1 & 4 \end{bmatrix}.$$

Solution

Method 1: Calculating by Hand

$$AB = \begin{bmatrix} 1 & -3 \\ 7 & 2 \\ -2 & 5 \end{bmatrix} \begin{bmatrix} 1 & 0 & -1 \\ 3 & 1 & 4 \end{bmatrix}$$

$$= \begin{bmatrix} -8 & -3 & -13 \\ 13 & 2 & 1 \\ 13 & 5 & 22 \end{bmatrix}$$

$$BA = \begin{bmatrix} 1 & 0 & -1 \\ 3 & 1 & 4 \end{bmatrix} \begin{bmatrix} 1 & -3 \\ 7 & 2 \\ -2 & 5 \end{bmatrix}$$

$$= \begin{bmatrix} 3 & -8 \\ 2 & 13 \end{bmatrix}$$

Method 2: Graphing Calculators

Matrix multiplication is easily performed on a graphing calculator. Figure 9 shows the results. The matrices A and B were already entered into the calculator.

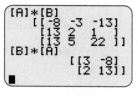

FIGURE 9

Matrix multiplication can also be easily done with a spreadsheet. See *The Spreadsheet Manual* that is available on this text's Web site for details.

Notice in Example 4 that $AB \neq BA$; AB and BA aren't even the same size. In Example 3, we showed that they may not both exist. This means that matrix multiplication is *not* commutative. Even if both A and B are square matrices, in general, matrices AB and BA are not equal. (See Exercise 31.) Of course, there may be special cases in which they are equal, but this is not true in general.

CAUTION Since matrix multiplication is not commutative, always be careful to multiply matrices in the correct order.

Matrix multiplication *is* associative, however. For example, if

$$C = \begin{bmatrix} 2 & 1 \\ 3 & 4 \\ 1 & 5 \end{bmatrix},$$

then $(AB)C = A(BC)$, where A and B are the matrices given in Example 4. (Verify this.) Also, there is a distributive property of matrices such that, for appropriate matrices A, B, and C,

$$A(B + C) = AB + AC.$$

(See Exercises 32 and 33.) Other properties of matrix multiplication involving scalars are included in the exercises. Multiplicative inverses and multiplicative identities are defined in the next section.

EXAMPLE 5 Bird Feeding
A wildlife center finds that the number of cardinals and finches that visit a bird feeder on a regular basis varies with the season, as shown in the matrix P below.

$$\begin{array}{c} \\ \text{Spring} \\ \text{Summer} \\ \text{Fall} \\ \text{Winter} \end{array} \begin{array}{cc} \overset{\text{Cardinals}}{} & \overset{\text{Finches}}{} \\ \begin{bmatrix} 25 & 31 \\ 20 & 35 \\ 22 & 29 \\ 36 & 20 \end{bmatrix} \end{array} = P$$

The wildlife center has also found that cardinals and finches eat different amounts of sunflower seed, corn, and millet (in grams) when they visit the feeder. The matrix Q shows how much one bird of each species eats over an entire season.

$$\begin{array}{c} \text{Sunflower} \quad \text{Corn} \quad \text{Millet} \\ \begin{array}{l} \text{Cardinals} \\ \text{Finches} \end{array} \left[\begin{array}{ccc} 32 & 10 & 30 \\ 10 & 18 & 27 \end{array} \right] = Q \end{array}$$

The cost (in cents) per gram of each type of seed is shown in the matrix R.

$$\begin{array}{c} \text{Cost} \\ \begin{array}{l} \text{Sunflower} \\ \text{Corn} \\ \text{Millet} \end{array} \left[\begin{array}{c} 3 \\ 2 \\ 1 \end{array} \right] = R \end{array}$$

(a) What is the total cost of seed for each season of the year to feed the cardinals and finches that regularly visit the feeder?

Solution To find the total cost for each season, first find PQ, which shows how much of each type of seed is eaten each season of the year.

$$PQ = \begin{bmatrix} 25 & 31 \\ 20 & 35 \\ 22 & 29 \\ 36 & 20 \end{bmatrix} \begin{bmatrix} 32 & 10 & 30 \\ 10 & 18 & 27 \end{bmatrix}$$

$$= \begin{array}{c} \text{Sunflower} \quad \text{Corn} \quad \text{Millet} \\ \begin{bmatrix} 1{,}110 & 808 & 1{,}587 \\ 990 & 830 & 1{,}545 \\ 994 & 742 & 1{,}443 \\ 1{,}352 & 720 & 1{,}620 \end{bmatrix} \begin{array}{l} \text{Spring} \\ \text{Summer} \\ \text{Fall} \\ \text{Winter} \end{array} \end{array}$$

Now multiply PQ by R, the cost matrix, to get the total cost of seed for each season.

$$\begin{bmatrix} 1{,}110 & 808 & 1{,}587 \\ 990 & 830 & 1{,}545 \\ 994 & 742 & 1{,}443 \\ 1{,}352 & 720 & 1{,}620 \end{bmatrix} \begin{bmatrix} 3 \\ 2 \\ 1 \end{bmatrix} = \begin{array}{c} \text{Cost} \\ \begin{bmatrix} 6{,}533 \\ 6{,}175 \\ 5{,}909 \\ 7{,}116 \end{bmatrix} \begin{array}{l} \text{Spring} \\ \text{Summer} \\ \text{Fall} \\ \text{Winter} \end{array} \end{array}$$

These costs are in cents, so multiply by 0.01 to convert to dollars. The total cost of seed is \$65.33 in spring, \$61.75 in summer, \$59.09 in fall, and \$71.16 in winter.

(b) How many grams of each kind of seed are used over an entire year?

Solution The totals of the columns of matrix PQ give a matrix whose elements represent the total number of grams of each seed used over an entire year. Call this matrix T, and write it as a row matrix.

$$T = [4{,}446 \quad 3{,}100 \quad 6{,}195]$$

Thus, 4,446 grams of sunflower seed, 3,100 grams of corn, and 6,195 grams of millet are used over an entire year.

(c) What is the total cost of seed in a year?

Solution For the total cost of seed in a year, find the product of the matrix T, the matrix showing the total amount of each seed, and matrix R, the cost matrix. (To multiply these and get a 1×1 matrix representing total cost, we need a 1×4

matrix multiplied by a 4×1 matrix. This is why T was written as a row matrix in (b) above.)

$$TR = [4{,}446 \quad 3{,}100 \quad 6{,}195] \begin{bmatrix} 3 \\ 2 \\ 1 \end{bmatrix} = [25{,}733]$$

The total cost is $257.33. Of course, we could have calculated this same answer by adding the amounts found in part (a).

(d) Suppose the wildlife center puts bird feeders in four more locations, for a total of five feeders. The locations are very distant from each other, but have similar habitats, so the number of birds and the amount they eat is expected to be the same as at the original feeder. Calculate the total amount of each type of seed expected to be used in each season.

Solution Multiply PQ by the scalar 5, as follows.

$$5(PQ) = 5 \begin{bmatrix} 1{,}110 & 808 & 1{,}587 \\ 990 & 830 & 1{,}545 \\ 994 & 742 & 1{,}443 \\ 1{,}352 & 720 & 1{,}620 \end{bmatrix} = \begin{bmatrix} 5{,}550 & 4{,}040 & 7{,}935 \\ 4{,}950 & 4{,}150 & 7{,}725 \\ 4{,}970 & 3{,}710 & 7{,}215 \\ 6{,}760 & 3{,}600 & 8{,}100 \end{bmatrix}$$

The total amount of corn used in fall, for example, is 3,710 grams.

Choosing Matrix Notation It is helpful to use a notation that keeps track of the quantities a matrix represents. We will use the notation

meaning of the rows/meaning of the columns,

that is, writing the meaning of the rows first, followed by the meaning of the columns. In Example 5, we would use the notation seasons/birds for matrix P, birds/seeds for matrix Q, and seeds/cost for matrix R. In multiplying PQ, we are multiplying seasons/birds by birds/seeds. The result is seasons/seeds. Notice that birds, the common quantity in both P and Q, was eliminated in the product PQ. By this method, the product $(PQ)R$ represents seasons/cost.

In practical problems this notation helps us decide in which order to multiply matrices so that the results are meaningful. In Example 5(c) either RT or TR can be calculated. Since T represents year/seeds and R represents seeds/cost, the product TR gives year/cost, while the product RT is meaningless.

10.3 EXERCISES

Let $A = \begin{bmatrix} -2 & 4 \\ 0 & 3 \end{bmatrix}$ *and* $B = \begin{bmatrix} -6 & 2 \\ 4 & 0 \end{bmatrix}$. *Find each of the following.*

1. $2A$

2. $-3B$

3. $-4B$

4. $5A$

5. $-4A + 5B$

6. $3A - 10B$

In Exercises 7–12, the sizes of two matrices A and B are given. Find the sizes of the product AB and the product BA, whenever these products exist.

7. *A* is 2 × 2, and *B* is 2 × 2.   **8.** *A* is 3 × 3, and *B* is 3 × 3.   **9.** *A* is 3 × 5, and *B* is 5 × 2.

10. *A* is 4 × 3, and *B* is 3 × 6.   **11.** *A* is 4 × 2, and *B* is 3 × 4.   **12.** *A* is 7 × 3, and *B* is 2 × 7.

13. To find the product matrix *AB*, the number of _____ of *A* must be the same as the number of _____ of *B*.

14. The product matrix *AB* has the same number of _____ as *A* and the same number of _____ as *B*.

Find each of the matrix products in Exercises 15–30 if possible.

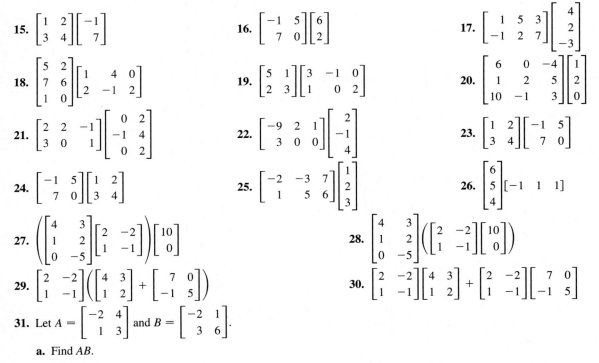

15. $\begin{bmatrix} 1 & 2 \\ 3 & 4 \end{bmatrix} \begin{bmatrix} -1 \\ 7 \end{bmatrix}$

16. $\begin{bmatrix} -1 & 5 \\ 7 & 0 \end{bmatrix} \begin{bmatrix} 6 \\ 2 \end{bmatrix}$

17. $\begin{bmatrix} 1 & 5 & 3 \\ -1 & 2 & 7 \end{bmatrix} \begin{bmatrix} 4 \\ 2 \\ -3 \end{bmatrix}$

18. $\begin{bmatrix} 5 & 2 \\ 7 & 6 \\ 1 & 0 \end{bmatrix} \begin{bmatrix} 1 & 4 & 0 \\ 2 & -1 & 2 \end{bmatrix}$

19. $\begin{bmatrix} 5 & 1 \\ 2 & 3 \end{bmatrix} \begin{bmatrix} 3 & -1 & 0 \\ 1 & 0 & 2 \end{bmatrix}$

20. $\begin{bmatrix} 6 & 0 & -4 \\ 1 & 2 & 5 \\ 10 & -1 & 3 \end{bmatrix} \begin{bmatrix} 1 \\ 2 \\ 0 \end{bmatrix}$

21. $\begin{bmatrix} 2 & 2 & -1 \\ 3 & 0 & 1 \end{bmatrix} \begin{bmatrix} 0 & 2 \\ -1 & 4 \\ 0 & 2 \end{bmatrix}$

22. $\begin{bmatrix} -9 & 2 & 1 \\ 3 & 0 & 0 \end{bmatrix} \begin{bmatrix} 2 \\ -1 \\ 4 \end{bmatrix}$

23. $\begin{bmatrix} 1 & 2 \\ 3 & 4 \end{bmatrix} \begin{bmatrix} -1 & 5 \\ 7 & 0 \end{bmatrix}$

24. $\begin{bmatrix} -1 & 5 \\ 7 & 0 \end{bmatrix} \begin{bmatrix} 1 & 2 \\ 3 & 4 \end{bmatrix}$

25. $\begin{bmatrix} -2 & -3 & 7 \\ 1 & 5 & 6 \end{bmatrix} \begin{bmatrix} 1 \\ 2 \\ 3 \end{bmatrix}$

26. $\begin{bmatrix} 6 \\ 5 \\ 4 \end{bmatrix} \begin{bmatrix} -1 & 1 & 1 \end{bmatrix}$

27. $\left(\begin{bmatrix} 4 & 3 \\ 1 & 2 \\ 0 & -5 \end{bmatrix} \begin{bmatrix} 2 & -2 \\ 1 & -1 \end{bmatrix} \right) \begin{bmatrix} 10 \\ 0 \end{bmatrix}$

28. $\begin{bmatrix} 4 & 3 \\ 1 & 2 \\ 0 & -5 \end{bmatrix} \left(\begin{bmatrix} 2 & -2 \\ 1 & -1 \end{bmatrix} \begin{bmatrix} 10 \\ 0 \end{bmatrix} \right)$

29. $\begin{bmatrix} 2 & -2 \\ 1 & -1 \end{bmatrix} \left(\begin{bmatrix} 4 & 3 \\ 1 & 2 \end{bmatrix} + \begin{bmatrix} 7 & 0 \\ -1 & 5 \end{bmatrix} \right)$

30. $\begin{bmatrix} 2 & -2 \\ 1 & -1 \end{bmatrix} \begin{bmatrix} 4 & 3 \\ 1 & 2 \end{bmatrix} + \begin{bmatrix} 2 & -2 \\ 1 & -1 \end{bmatrix} \begin{bmatrix} 7 & 0 \\ -1 & 5 \end{bmatrix}$

31. Let $A = \begin{bmatrix} -2 & 4 \\ 1 & 3 \end{bmatrix}$ and $B = \begin{bmatrix} -2 & 1 \\ 3 & 6 \end{bmatrix}$.

    a. Find *AB*.

    b. Find *BA*.

    c. Did you get the same answer in parts a and b?

    d. In general, for matrices *A* and *B* such that *AB* and *BA* both exist, does *AB* always equal *BA*?

Given matrices $P = \begin{bmatrix} m & n \\ p & q \end{bmatrix}$, $X = \begin{bmatrix} x & y \\ z & w \end{bmatrix}$, *and* $T = \begin{bmatrix} r & s \\ t & u \end{bmatrix}$, *verify that the statements in Exercises 32–35 are true. The statements are valid for any matrices whenever matrix multiplication and addition can be carried out. This, of course, depends on the size of the matrices.*

32. $(PX)T = P(XT)$ (associate property: see Exercises 27 and 28)

33. $P(X + T) = PX + PT$ (distributive property: see Exercises 29 and 30)

34. $k(X + T) = kX + kT$ for any real number *k*.

35. $(k + h)P = kP + hP$ for any real numbers *k* and *h*.

36. Let I be the matrix $I = \begin{bmatrix} 1 & 0 \\ 0 & 1 \end{bmatrix}$, and let matrices P, X, and T be defined as for Exercises 32–35.

   a. Find IP, PI, and IX.

   b. Without calculating, guess what the matrix IT might be.

   c. Suggest a reason for naming a matrix such as I an *identity matrix*.

37. Show that the system of linear equations

$$2x_1 + 3x_2 + x_3 = 5$$
$$x_1 - 4x_2 + 5x_3 = 8$$

can be written as the matrix equation

$$\begin{bmatrix} 2 & 3 & 1 \\ 1 & -4 & 5 \end{bmatrix} \begin{bmatrix} x_1 \\ x_2 \\ x_3 \end{bmatrix} = \begin{bmatrix} 5 \\ 8 \end{bmatrix}.$$

38. Let $A = \begin{bmatrix} 1 & 2 \\ -3 & 5 \end{bmatrix}$, $X = \begin{bmatrix} x_1 \\ x_2 \end{bmatrix}$, and $B = \begin{bmatrix} -4 \\ 12 \end{bmatrix}$. Show that the equation $AX = B$ represents a linear system of two equations in two unknowns. Solve the system and substitute into the matrix equation to check your results.

Use a computer or graphing calculator and the following matrices to find the matrix products and sums in Exercises 39–41.

$$A = \begin{bmatrix} 2 & 3 & -1 & 5 & 10 \\ 2 & 8 & 7 & 4 & 3 \\ -1 & -4 & -12 & 6 & 8 \\ 2 & 5 & 7 & 1 & 4 \end{bmatrix} \quad B = \begin{bmatrix} 9 & 3 & 7 & -6 \\ -1 & 0 & 4 & 2 \\ -10 & -7 & 6 & 9 \\ 8 & 4 & 2 & -1 \\ 2 & -5 & 3 & 7 \end{bmatrix}$$

$$C = \begin{bmatrix} -6 & 8 & 2 & 4 & -3 \\ 1 & 9 & 7 & -12 & 5 \\ 15 & 2 & -8 & 10 & 11 \\ 4 & 7 & 9 & 6 & -2 \\ 1 & 3 & 8 & 23 & 4 \end{bmatrix} \quad D = \begin{bmatrix} 5 & -3 & 7 & 9 & 2 \\ 6 & 8 & -5 & 2 & 1 \\ 3 & 7 & -4 & 2 & 11 \\ 5 & -3 & 9 & 4 & -1 \\ 0 & 3 & 2 & 5 & 1 \end{bmatrix}$$

39. a. Find AC.   **b.** Find CA.   **c.** Does $AC = CA$?

40. a. Find CD.   **b.** Find DC.   **c.** Does $CD = DC$?

41. a. Find $C + D$.   **b.** Find $(C + D)B$.   **c.** Find CB.

   d. Find DB.   **e.** Find $CB + DB$.   **f.** Does $(C + D)B = CB + DB$?

42. Which property of matrices does Exercise 41 illustrate?

Applications

LIFE SCIENCES

43. *Dietetics* In Exercise 39 from the previous section, label the matrices

$$\begin{bmatrix} 2 & 1 & 2 & 1 \\ 3 & 2 & 2 & 1 \\ 4 & 3 & 2 & 1 \end{bmatrix}, \quad \begin{bmatrix} 5 & 0 & 7 \\ 0 & 10 & 1 \\ 0 & 15 & 2 \\ 10 & 12 & 8 \end{bmatrix}, \quad \text{and} \quad \begin{bmatrix} 8 \\ 4 \\ 5 \end{bmatrix}$$

found in parts a, b, and c, respectively, X, Y, and Z. Matrix X represents the number of exchanges for each food group in each meal. Matrix Y represents the amount of fat, carbohydrates, and protein in each food group. Matrix Z represents the number of calories per exchange of fat, carbohydrates, and protein.

   a. Find the product matrix XY. What do the entries of this matrix represent?

b. Find the product matrix *YZ*. What do the entries represent?

c. Find the products (*XY*)*Z* and *X*(*YZ*) and verify that they are equal. What do the entries represent?

44. *Car Accidents* In Exercise 42 of the previous section, you constructed matrices that represent the death rates, per million person trips, for both male and female drivers for various ages and number of passengers. Use matrix operations to combine these two matrices to form one matrix that represents the combined death rates for males and females, per million person trips, of drivers of various ages and number of passengers. (*Hint:* Add the two matrices together and then multiply the resulting matrix by the scalar 1/2.)

45. *Life Expectancy* In Exercise 43 of the previous section, you constructed matrices that represent the life expectancy of African American and white American males and females. Use matrix operations to combine these two matrices to form one matrix that represents the combined life expectancy for African American and white Americans from 1970 through 1997. See the hint for Exercise 44.

46. *Northern Spotted Owl Population** In an attempt to save the endangered northern spotted owl, the U.S. Fish and Wildlife Service imposed strict guidelines for the use of 12 million acres of Pacific Northwest forest. This decision led to a national debate between the logging industry and environmentalists. Mathematical ecologists have created a mathematical model to analyze population dynamics of the northern spotted owl by dividing the female owl population into three categories: juvenile (up to 1 year old), subadult (1 to 2 years), and adult (over 2 years old). By analyzing these three subgroups, it is possible to use the number of females in each subgroup at time *n* to estimate the number of females in each group at any time *n* + 1 with the following matrix equation:

$$\begin{bmatrix} j_{n+1} \\ s_{n+1} \\ a_{n+1} \end{bmatrix} = \begin{bmatrix} 0 & 0 & 0.33 \\ 0.18 & 0 & 0 \\ 0 & 0.71 & 0.94 \end{bmatrix} \begin{bmatrix} j_n \\ s_n \\ a_n \end{bmatrix},$$

where j_n is the number of juveniles, s_n is the number of subadults, and a_n is the number of adults at time n.[†]

a. If there are currently 4,000 female northern spotted owls made up of 900 juveniles, 500 subadults, and 2,600 adults, use a graphing calculator or spreadsheet and matrix operations to determine the total number of

female owls for each of the next 5 years. (*Hint:* Round each answer to the nearest whole number after each matrix multiplication.)

b. With advanced techniques from linear algebra, it is possible to show that in the long run, the following holds:

$$\begin{bmatrix} j_{n+1} \\ s_{n+1} \\ a_{n+1} \end{bmatrix} \approx 0.98359 \begin{bmatrix} j_n \\ s_n \\ a_n \end{bmatrix}.$$

What can we conclude about the long-term survival of the northern spotted owl?

c. Notice that only 18 percent of the juveniles become subadults. Assuming that, through better habitat management, this number could be increased to 40 percent, rework part a. Discuss possible reasons why only 18 percent of the juveniles become subadults. Under the new assumption, what can you conclude about the long-term survival of the northern spotted owl?

47. *World Population* The 1998 birth and death rates per million for several regions and the world population (in millions) by region are given below and on the next page.[‡]

	Births	Deaths
Africa	0.038	0.014
Asia	0.022	0.008
Latin America	0.023	0.007
North America	0.014	0.009
Europe	0.010	0.011

---

*This problem was created by David I. Schneider, University of Maryland.

†Lamberson, R., R. McKelvey, B. Noon, and C. Voss, "A Dynamic Analysis of Northern Spotted Owl Viability in a Fragmented Forest Landscape," *Conservation Biology,* Vol. 6, No. 4, Dec. 1992, pp. 505–512.

‡"Vital Events and Rates by Region and Development Category, 1998" and "World Population by Region and Development Category, 1950–2050" from U.S. Bureau of the Census, *World Population Profile: 1998.*

	Africa	Asia	Latin America	North America	Europe
1960	283	1,627	218	199	425
1970	360	2,038	286	227	460
1980	468	2,498	362	252	484
1990	621	2,987	443	278	498
2000	798	3,451	523	306	508

a. Write the information in each table as a matrix.

b. Use the matrices from part a to find the total number (in millions) of births and deaths in each year.

c. Using the results of part b, compare the number of births in 1960 and in 2000. Also compare the birth rates from part a. Which gives better information?

d. Using the results of part b, compare the number of deaths in 1980 and in 2000. Discuss how this comparison differs from comparing death rates from part a.

48. *Bird Feeding* Consider the matrices P, Q, and R given in Example 5.

a. Find and interpret the matrix product QR.

b. Verify that $P(QR)$ is equal to $(PQ)R$ calculated in Example 5.

OTHER APPLICATIONS

49. *Cost Analysis* The four departments of Spangler Enterprises need to order the following amounts of the same products.

	Paper	Tape	Printer Ribbon	Memo Pads	Pens
Department 1	10	4	3	5	6
Department 2	7	2	2	3	8
Department 3	4	5	1	0	10
Department 4	0	3	4	5	5

The unit price (in dollars) of each product is given below for two suppliers.

	Supplier A	Supplier B
Paper	2	3
Tape	1	1
Printer Ribbon	4	3
Memo Pads	3	3
Pens	1	2

a. Use matrix multiplication to get a matrix showing the comparative costs for each department for the products from the two suppliers.

b. Find the total cost over all departments to buy products from each supplier. From which supplier should the company make the purchase?

50. *Cost Analysis* The Mundo Candy Company makes three types of chocolate candy: Cheery Cherry, Mucho Mocha, and Almond Delight. The company produces its products in San Diego, Mexico City, and Managua using two main ingredients: chocolate and sugar.

a. Each kilogram of Cheery Cherry requires 0.5 kg of sugar and 0.2 kg of chocolate; each kilogram of Mucho Mocha requires 0.4 kg of sugar and 0.3 kg of chocolate; and each kilogram of Almond Delight requires 0.3 kg of sugar and 0.3 kg of chocolate. Put this information into a 2 × 3 matrix, labeling the rows and columns.

b. The cost of 1 kg of sugar is $4 in San Diego, $2 in Mexico City, and $1 in Managua. The cost of 1 kg of chocolate is $3 in San Diego, $5 in Mexico City, and $7 in Managua. Put this information into a matrix in such a way that when you multiply it with your matrix from part a, you get a matrix representing the ingredient cost of producing each type of candy in each city.

c. Multiply the matrices in parts a and b, labeling the product matrix.

d. From part c, what is the combined sugar-and-chocolate cost to produce 1 kg of Mucho Mocha in Managua?

e. Mundo Candy needs to quickly produce a special shipment of 100 kg of Cheery Cherry, 200 kg of Mucho Mocha, and 500 kg of Almond Delight, and it decides to select one factory to fill the entire order. Use matrix multiplication to determine in which city the total sugar-and-chocolate cost to produce the order is the smallest.

51. *Shoe Sales* Sal's Shoes and Fred's Footwear both have outlets in California and Arizona. Sal's sells shoes for $80, sandals for $40, and boots for $120. Fred's prices are $60, $30, and $150 for shoes, sandals, and boots, respectively. Half of all sales in California stores are shoes, 1/4 are sandals, and 1/4 are boots. In Arizona the fractions are 1/5 shoes, 1/5 sandals, and 3/5 boots.

a. Write a 2 × 3 matrix called P representing prices for the two stores and three types of footwear.

b. Write a 3 × 2 matrix called F representing the fraction of each type of footwear sold in each state.

c. Only one of the two products PF and FP is meaningful. Determine which one it is, calculate the product, and describe what the entries represent.

■ 10.4 MATRIX INVERSES

THINK ABOUT IT

How can we determine the amount of various kinds of fertilizers to use on a farm, given the requirements for the soil?

This question is answered in Example 4 of this section by solving a matrix equation. In this section, we introduce the idea of a matrix inverse, which is comparable to the reciprocal of a real number. This will allow us to solve a matrix equation.

Earlier, we defined a zero matrix as an additive identity matrix with properties similar to those of the real number 0, the additive identity for real numbers. The real number 1 is the *multiplicative* identity for real numbers: for any real number a, $a \cdot 1 = 1 \cdot a = a$. In this section, we define a *multiplicative identity matrix I* that has properties similar to those of the number 1. We then use the definition of matrix I to find the *multiplicative inverse* of any square matrix that has an inverse.

If I is to be the identity matrix, both of the products AI and IA must equal A. This means that an identity matrix exists only for square matrices. The 2×2 **identity matrix** that satisfies these conditions is

$$I = \begin{bmatrix} 1 & 0 \\ 0 & 1 \end{bmatrix}.$$

To check that I, as defined above, is really the 2×2 identity, let

$$A = \begin{bmatrix} a & b \\ c & d \end{bmatrix}.$$

Then AI and IA should both equal A.

$$AI = \begin{bmatrix} a & b \\ c & d \end{bmatrix}\begin{bmatrix} 1 & 0 \\ 0 & 1 \end{bmatrix} = \begin{bmatrix} a(1) + b(0) & a(0) + b(1) \\ c(1) + d(0) & c(0) + d(1) \end{bmatrix} = \begin{bmatrix} a & b \\ c & d \end{bmatrix} = A$$

$$IA = \begin{bmatrix} 1 & 0 \\ 0 & 1 \end{bmatrix}\begin{bmatrix} a & b \\ c & d \end{bmatrix} = \begin{bmatrix} 1(a) + 0(c) & 1(b) + 0(d) \\ 0(a) + 1(c) & 0(b) + 1(d) \end{bmatrix} = \begin{bmatrix} a & b \\ c & d \end{bmatrix} = A$$

This verifies that I has been defined correctly.

It is easy to verify that the identity matrix I is unique. Suppose there is another identity; call it J. Then IJ must equal I, because J is an identity, and IJ must also equal J, because I is an identity. Thus $I = J$.

The identity matrices for 3×3 matrices and 4×4 matrices, respectively, are

$$I = \begin{bmatrix} 1 & 0 & 0 \\ 0 & 1 & 0 \\ 0 & 0 & 1 \end{bmatrix} \quad \text{and} \quad I = \begin{bmatrix} 1 & 0 & 0 & 0 \\ 0 & 1 & 0 & 0 \\ 0 & 0 & 1 & 0 \\ 0 & 0 & 0 & 1 \end{bmatrix}.$$

By generalizing, we can find an $n \times n$ identity matrix for any value of n.

Recall that the multiplicative inverse of the nonzero real number a is $1/a$. The product of a and its multiplicative inverse $1/a$ is 1. Given a matrix A, can a **multiplicative inverse matrix A^{-1} (read "A-inverse")** be found satisfying both

$$AA^{-1} = I \quad \text{and} \quad A^{-1}A = I?$$

For a given matrix, we often can find an inverse matrix by using the row operations of the first section in this chapter.

> **NOTE** A^{-1} does not mean $1/A$; here, A^{-1} is just the notation for the multiplicative inverse of matrix A. Also, only square matrices can have inverses because both $A^{-1}A$ and AA^{-1} must exist and be equal to I.

If an inverse exists, it is unique. That is, any given square matrix has no more than one inverse. The proof of this is left to Exercise 50 in this section.

As an example, let us find the inverse of

$$A = \begin{bmatrix} 2 & 4 \\ 1 & -1 \end{bmatrix}.$$

Let the unknown inverse matrix be

$$A^{-1} = \begin{bmatrix} x & y \\ z & w \end{bmatrix}.$$

By the definition of matrix inverse, $AA^{-1} = I$, or

$$AA^{-1} = \begin{bmatrix} 2 & 4 \\ 1 & -1 \end{bmatrix} \begin{bmatrix} x & y \\ z & w \end{bmatrix} = \begin{bmatrix} 1 & 0 \\ 0 & 1 \end{bmatrix}.$$

By matrix multiplication,

$$\begin{bmatrix} 2x + 4z & 2y + 4w \\ x - z & y - w \end{bmatrix} = \begin{bmatrix} 1 & 0 \\ 0 & 1 \end{bmatrix}.$$

Setting corresponding elements equal gives the system of equations

$$2x + 4z = 1 \tag{1}$$
$$2y + 4w = 0 \tag{2}$$
$$x - z = 0 \tag{3}$$
$$y - w = 1. \tag{4}$$

Since equations (1) and (3) involve only x and z, while equations (2) and (4) involve only y and w, these four equations lead to two systems of equations,

$$2x + 4z = 1 \quad \text{and} \quad 2y + 4w = 0$$
$$x - z = 0 \qquad\qquad y - w = 1.$$

Representing the two systems as augmented matrices gives

$$\begin{bmatrix} 2 & 4 & | & 1 \\ 1 & -1 & | & 0 \end{bmatrix} \quad \text{and} \quad \begin{bmatrix} 2 & 4 & | & 0 \\ 1 & -1 & | & 1 \end{bmatrix}.$$

Each of these systems can be solved by the Gauss-Jordan method. Notice, however, that the elements to the left of the vertical bar are identical. The two systems can be combined into the single matrix

$$\begin{bmatrix} 2 & 4 & | & 1 & 0 \\ 1 & -1 & | & 0 & 1 \end{bmatrix}.$$

This is of the form $[A\,|\,I]$. It is solved simultaneously as follows.

$$R_1 + (-2)R_2 \rightarrow R_2 \qquad \begin{bmatrix} 2 & 4 & | & 1 & 0 \\ 0 & 6 & | & 1 & -2 \end{bmatrix} \qquad \text{Get 0 in the second row, first-column position.}$$

$$-2R_2 + 3R_1 \rightarrow R_1 \qquad \begin{bmatrix} 6 & 0 & | & 1 & 4 \\ 0 & 6 & | & 1 & -2 \end{bmatrix} \qquad \text{Get 0 in the first-row, second-column position.}$$

$$\begin{array}{c} \frac{1}{6}R_1 \rightarrow R_1 \\ \frac{1}{6}R_2 \rightarrow R_2 \end{array} \qquad \begin{bmatrix} 1 & 0 & | & \frac{1}{6} & \frac{2}{3} \\ 0 & 1 & | & \frac{1}{6} & -\frac{1}{3} \end{bmatrix} \qquad \text{Get ones down the diagonal.}$$

The numbers in the first column to the right of the vertical bar give the values of x and z. The second column gives the values of y and w. That is,

$$\begin{bmatrix} 1 & 0 & | & x & y \\ 0 & 1 & | & z & w \end{bmatrix} = \begin{bmatrix} 1 & 0 & | & \frac{1}{6} & \frac{2}{3} \\ 0 & 1 & | & \frac{1}{6} & -\frac{1}{3} \end{bmatrix}$$

so that

$$A^{-1} = \begin{bmatrix} x & y \\ z & w \end{bmatrix} = \begin{bmatrix} \frac{1}{6} & \frac{2}{3} \\ \frac{1}{6} & -\frac{1}{3} \end{bmatrix}.$$

To check, multiply A by A^{-1}. The result should be I.

$$AA^{-1} = \begin{bmatrix} 2 & 4 \\ 1 & -1 \end{bmatrix} \begin{bmatrix} \frac{1}{6} & \frac{2}{3} \\ \frac{1}{6} & -\frac{1}{3} \end{bmatrix} = \begin{bmatrix} \frac{1}{3} + \frac{2}{3} & \frac{4}{3} - \frac{4}{3} \\ \frac{1}{6} - \frac{1}{6} & \frac{2}{3} + \frac{1}{3} \end{bmatrix} = \begin{bmatrix} 1 & 0 \\ 0 & 1 \end{bmatrix} = I$$

Verify that $A^{-1}A = I$, also. Finally,

$$A^{-1} = \begin{bmatrix} \frac{1}{6} & \frac{2}{3} \\ \frac{1}{6} & -\frac{1}{3} \end{bmatrix}.$$

FINDING A MULTIPLICATIVE INVERSE MATRIX

To obtain A^{-1} for any $n \times n$ matrix A for which A^{-1} exists, follow these steps.

1. Form the augmented matrix $[A\,|\,I]$, where I is the $n \times n$ identity matrix.
2. Perform row operations on $[A\,|\,I]$ to get a matrix of the form $[I\,|\,B]$.
3. Matrix B is A^{-1}.

EXAMPLE 1 Inverse Matrix

Find A^{-1} if $A = \begin{bmatrix} 1 & 0 & 1 \\ 2 & -2 & -1 \\ 3 & 0 & 0 \end{bmatrix}$.

Solution

Method 1: Calculating by Hand | Write the augmented matrix $[A\,|\,I]$.

$$[A\,|\,I] = \begin{bmatrix} 1 & 0 & 1 & | & 1 & 0 & 0 \\ 2 & -2 & -1 & | & 0 & 1 & 0 \\ 3 & 0 & 0 & | & 0 & 0 & 1 \end{bmatrix}$$

Begin by selecting the row operation that produces a zero for the first element in row 2.

$$\begin{array}{c} \\ -2R_1 + R_2 \to R_2 \\ -3R_1 + R_3 \to R_3 \end{array} \left[\begin{array}{ccc|ccc} 1 & 0 & 1 & 1 & 0 & 0 \\ 0 & -2 & -3 & -2 & 1 & 0 \\ 0 & 0 & -3 & -3 & 0 & 1 \end{array}\right]$$ **Get zeros in the first column.**

Column 2 already has zeros in the required positions, so we work on column 3.

$$\begin{array}{c} R_3 + 3R_1 \to R_1 \\ R_3 + (-1)R_2 \to R_2 \\ \\ \end{array} \left[\begin{array}{ccc|ccc} 3 & 0 & 0 & 0 & 0 & 1 \\ 0 & 2 & 0 & -1 & -1 & 1 \\ 0 & 0 & -3 & -3 & 0 & 1 \end{array}\right]$$ **Get zeros in the third column.**

Now get 1's down the main diagonal.

$$\begin{array}{c} \frac{1}{3}R_1 \to R_1 \\ \frac{1}{2}R_2 \to R_2 \\ -\frac{1}{3}R_3 \to R_3 \end{array} \left[\begin{array}{ccc|ccc} 1 & 0 & 0 & 0 & 0 & \frac{1}{3} \\ 0 & 1 & 0 & -\frac{1}{2} & -\frac{1}{2} & \frac{1}{2} \\ 0 & 0 & 1 & 1 & 0 & -\frac{1}{3} \end{array}\right]$$ **Get ones down the diagonal.**

From the last transformation, the desired inverse is

$$A^{-1} = \begin{bmatrix} 0 & 0 & \frac{1}{3} \\ -\frac{1}{2} & -\frac{1}{2} & \frac{1}{2} \\ 1 & 0 & -\frac{1}{3} \end{bmatrix}.$$

Confirm this by forming the products $A^{-1}A$ and AA^{-1}, both of which should equal I.

Method 2: Graphing Calculators The inverse of A can also be found with a graphing calculator, as shown in Figure 10. (The matrix A had previously been entered into the calculator.) The entire answer can be viewed by pressing the right and left arrow keys on the calculator.

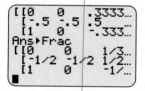

FIGURE 10

Spreadsheets also have the capability of calculating the inverse of a matrix with a simple command. See *The Spreadsheet Manual* that is available on this text's Web site for details.

EXAMPLE 2 Matrix Without an Inverse

Find A^{-1} if $A = \begin{bmatrix} 2 & -4 \\ 1 & -2 \end{bmatrix}$.

Solution Using row operations to transform the first column of the augmented matrix

$$\left[\begin{array}{cc|cc} 2 & -4 & 1 & 0 \\ 1 & -2 & 0 & 1 \end{array}\right]$$

gives the following results.

$$R_1 + (-2)R_2 \to R_2 \quad \left[\begin{array}{cc|cc} 2 & -4 & 1 & 0 \\ 0 & 0 & 1 & -2 \end{array}\right]$$

Because the last row has all zeros to the left of the vertical bar, there is no way to complete the process of finding the inverse matrix. What is wrong? Just as the real number 0 has no multiplicative inverse, some matrices do not have inverses. Matrix A is an example of a matrix that has no inverse: there is no matrix A^{-1} such that $AA^{-1} = A^{-1}A = I$.

NOTE As is shown in Exercise 46, there is a quick way to find the inverse of a 2 × 2 matrix:

$$\text{When } A = \begin{bmatrix} a & b \\ c & d \end{bmatrix} \text{ and } ad - bc \neq 0, \ A^{-1} = \frac{1}{ad - bc} \begin{bmatrix} d & -b \\ -c & a \end{bmatrix}.$$

Solving Systems of Equations with Inverses

We used matrices to solve systems of linear equations by the Gauss-Jordan method in the first section of this chapter. Another way to use matrices to solve linear systems is to write the system as a matrix equation $AX = B$, where A is the matrix of the coefficients of the variables of the system, X is the matrix of the variables, and B is the matrix of the constants. Matrix A is called the **coefficient matrix.**

To solve the matrix equation $AX = B$, first see if A^{-1} exists. Assuming A^{-1} exists and using the facts that $A^{-1}A = I$ and $IX = X$ gives

$$\begin{aligned} AX &= B \\ A^{-1}(AX) &= A^{-1}B \qquad \text{Multiply both sides by } A^{-1}. \\ (A^{-1}A)X &= A^{-1}B \qquad \text{Associative property} \\ IX &= A^{-1}B \qquad \text{Multiplicative inverse property} \\ X &= A^{-1}B. \qquad \text{Identity property} \end{aligned}$$

CAUTION When multiplying by matrices on both sides of a matrix equation, be careful to multiply in the same order on both sides of the equation, since multiplication of matrices is not commutative (unlike multiplication of real numbers).

The work above leads to the following method of solving a system of equations written as a matrix equation.

SOLVING A SYSTEM $AX = B$ USING MATRIX INVERSES

To solve a system of equations $AX = B$, where A is the matrix of coefficients, X is the matrix of variables, and B is the matrix of constants, first find A^{-1}. Then $X = A^{-1}B$.

This method is most practical in solving several systems that have the same coefficient matrix but different constants, as in Example 4 in this section. Then just one inverse matrix must be found.

EXAMPLE 3 Inverse Matrices and Systems of Equations

Use the inverse of the coefficient matrix to solve the linear system

$$\begin{aligned} 2x - 3y &= 4. \\ x + 5y &= 2. \end{aligned}$$

Solution To represent the system as a matrix equation, use the coefficient matrix of the system together with the matrix of variables and the matrix of constants:

$$A = \begin{bmatrix} 2 & -3 \\ 1 & 5 \end{bmatrix}, \qquad X = \begin{bmatrix} x \\ y \end{bmatrix}, \qquad \text{and} \qquad B = \begin{bmatrix} 4 \\ 2 \end{bmatrix}.$$

The system can now be written in matrix form as the equation $AX = B$ since

$$AX = \begin{bmatrix} 2 & -3 \\ 1 & 5 \end{bmatrix} \begin{bmatrix} x \\ y \end{bmatrix} = \begin{bmatrix} 2x - 3y \\ x + 5y \end{bmatrix} = \begin{bmatrix} 4 \\ 2 \end{bmatrix} = B.$$

To solve the system, first find A^{-1}. Do this by using row operations on matrix $[A \,|\, I]$ to get

$$\begin{bmatrix} 1 & 0 & \Big| & \frac{5}{13} & \frac{3}{13} \\ 0 & 1 & \Big| & -\frac{1}{13} & \frac{2}{13} \end{bmatrix}.$$

From this result,

$$A^{-1} = \begin{bmatrix} \frac{5}{13} & \frac{3}{13} \\ -\frac{1}{13} & \frac{2}{13} \end{bmatrix}.$$

Next, find the product $A^{-1}B$.

$$A^{-1}B = \begin{bmatrix} \frac{5}{13} & \frac{3}{13} \\ -\frac{1}{13} & \frac{2}{13} \end{bmatrix} \begin{bmatrix} 4 \\ 2 \end{bmatrix} = \begin{bmatrix} 2 \\ 0 \end{bmatrix}$$

Since $X = A^{-1}B$,

$$X = \begin{bmatrix} x \\ y \end{bmatrix} = \begin{bmatrix} 2 \\ 0 \end{bmatrix}.$$

The solution of the system is (2,0).

EXAMPLE 4   Fertilizer

Three brands of fertilizer are available that provide nitrogen, phosphoric acid, and soluble potash to the soil. One bag of each brand provides the following units of each nutrient.

■ Table 5

Nutrient	Fertifun	Big Grow	Soakem
Nitrogen	1	2	3
Phosphoric Acid	3	1	2
Potash	2	0	1

For ideal growth, the soil on a Michigan farm needs 18 units of nitrogen, 23 units of phosphoric acid, and 13 units of potash per acre. The corresponding numbers for a California farm are 31, 24, and 11, and for a Kansas farm are 20, 19, and 15. How many bags of each brand of fertilizer should be used per acre for ideal growth on each farm?

Solution Rather than solve three separate systems, consider the single system

$$x + 2y + 3z = a$$
$$3x + y + 2z = b$$
$$2x \qquad + z = c,$$

where a, b, and c represent the units of nitrogen, phosphoric acid, and potash needed for the different farms. The system of equations is then of the form $AX = B$, where

$$A = \begin{bmatrix} 1 & 2 & 3 \\ 3 & 1 & 2 \\ 2 & 0 & 1 \end{bmatrix} \quad \text{and} \quad X = \begin{bmatrix} x \\ y \\ z \end{bmatrix}.$$

Matrix B has different values for the different farms. Find A^{-1} first, then use it to solve all three systems.

To find A^{-1}, we start with the matrix

$$[A \,|\, I] = \begin{bmatrix} 1 & 2 & 3 & | & 1 & 0 & 0 \\ 3 & 1 & 2 & | & 0 & 1 & 0 \\ 2 & 0 & 1 & | & 0 & 0 & 1 \end{bmatrix}$$

and use row operations to get $[I \,|\, A^{-1}]$. The result is

$$A^{-1} = \begin{bmatrix} -\frac{1}{3} & \frac{2}{3} & -\frac{1}{3} \\ -\frac{1}{3} & \frac{5}{3} & -\frac{7}{3} \\ \frac{2}{3} & -\frac{4}{3} & \frac{5}{3} \end{bmatrix}.$$

Now we can solve each of the three systems by using $X = A^{-1}B$.

For the Michigan farm, $B = \begin{bmatrix} 18 \\ 23 \\ 13 \end{bmatrix}$, and

$$X = \begin{bmatrix} -\frac{1}{3} & \frac{2}{3} & -\frac{1}{3} \\ -\frac{1}{3} & \frac{5}{3} & -\frac{7}{3} \\ \frac{2}{3} & -\frac{4}{3} & \frac{5}{3} \end{bmatrix} \begin{bmatrix} 18 \\ 23 \\ 13 \end{bmatrix} = \begin{bmatrix} 5 \\ 2 \\ 3 \end{bmatrix}.$$

Therefore, $x = 5$, $y = 2$, and $z = 3$. Buy 5 bags of Fertifun, 2 bags of Big Grow, and 3 bags of Soakem.

For the California farm, $B = \begin{bmatrix} 31 \\ 24 \\ 11 \end{bmatrix}$, and

$$X = \begin{bmatrix} -\frac{1}{3} & \frac{2}{3} & -\frac{1}{3} \\ -\frac{1}{3} & \frac{5}{3} & -\frac{7}{3} \\ \frac{2}{3} & -\frac{4}{3} & \frac{5}{3} \end{bmatrix} \begin{bmatrix} 31 \\ 24 \\ 11 \end{bmatrix} = \begin{bmatrix} 2 \\ 4 \\ 7 \end{bmatrix}.$$

Buy 2 bags of Fertifun, 4 bags of Big Grow, and 7 bags of Soakem.

For the Kansas farm, $B = \begin{bmatrix} 20 \\ 19 \\ 15 \end{bmatrix}$. Verify that this leads to $x = 1$, $y = -10$, and $z = 13$. We cannot have a negative number of bags, so this solution is impossible. In buying enough bags to meet all of the nutrient requirements the farmer must purchase an excess of some nutrients. ∎

In Example 4, using the matrix inverse method of solving the systems involved considerably less work than using row operations for each of the three systems.

EXAMPLE 5   Inconsistent Systems of Equations

Use the inverse of the coefficient matrix to solve the system

$$2x - 4y = 13$$
$$x - 2y = 1.$$

Solution   We say in Example 2 that the coefficient matrix $\begin{bmatrix} 2 & -4 \\ 1 & -2 \end{bmatrix}$ does not have an inverse. This means that the given system either has no solution or has an infinite number of solutions. Verify that this system is inconsistent and has no solution.

EXAMPLE 6   Cryptography

Throughout the Cold War and as the Internet has grown and developed, the need for sophisticated methods of coding and decoding messages has increased. Although there are many methods of encrypting messages, one fairly sophisticated method uses matrix operations. This method first assigns a number to each letter of the alphabet. The simplest way to do this is to assign the number 1 to A, 2 to B, and so on, with the number 27 used to represent a space between words.

For example, the message *math is cool* can be divided into groups of three letters each and then converted into numbers as follows

$$\begin{bmatrix} m \\ a \\ t \end{bmatrix} = \begin{bmatrix} 13 \\ 1 \\ 20 \end{bmatrix}.$$

The entire message would then consist of four 3×1 columns of numbers:

$$\begin{bmatrix} 13 \\ 1 \\ 20 \end{bmatrix}, \quad \begin{bmatrix} 8 \\ 27 \\ 9 \end{bmatrix}, \quad \begin{bmatrix} 19 \\ 27 \\ 3 \end{bmatrix}, \quad \begin{bmatrix} 15 \\ 15 \\ 12 \end{bmatrix}.$$

This code is easy to break, so we further complicate the code by choosing a matrix that has an inverse (in this case a 3×3 matrix) and calculate the products of the matrix and each of the column vectors above.

If we choose the coding matrix

$$A = \begin{bmatrix} 1 & 3 & 4 \\ 2 & 1 & 3 \\ 4 & 2 & 1 \end{bmatrix},$$

then the products of A with each of the column vectors above produce a new set of vectors

$$\begin{bmatrix} 96 \\ 87 \\ 74 \end{bmatrix}, \quad \begin{bmatrix} 125 \\ 70 \\ 95 \end{bmatrix}, \quad \begin{bmatrix} 112 \\ 74 \\ 133 \end{bmatrix}, \quad \begin{bmatrix} 108 \\ 81 \\ 102 \end{bmatrix}.$$

This set of vectors represents our coded message and it will be transmitted as 96, 87, 74, 125 and so on.

When the intended person receives the message, it is divided into groups of three numbers, and each group is formed into a column matrix. The message is easily decoded if the receiver knows the inverse of the original matrix. The inverse of matrix A is

$$A^{-1} = \begin{bmatrix} -0.2 & 0.2 & 0.2 \\ 0.4 & -0.6 & 0.2 \\ 0 & 0.4 & -0.2 \end{bmatrix}.$$

Thus, the message is decoded by taking the product of the inverse matrix with each column vector of the received message. For example,

$$A^{-1} \begin{bmatrix} 96 \\ 87 \\ 74 \end{bmatrix} = \begin{bmatrix} 13 \\ 1 \\ 20 \end{bmatrix}.$$

Unless the original matrix or its inverse is known, this type of code can be difficult to break. In fact, very large matrices can be used to encrypt data. It is interesting to note that many mathematicians are employed by the National Security Agency to develop encryption methods that are virtually unbreakable. ■

10.4 EXERCISES

In Exercises 1–8, decide whether the given matrices are inverses of each other. (Check to see if their product is the identity matrix I.)

1. $\begin{bmatrix} 2 & 3 \\ 1 & 1 \end{bmatrix}$ and $\begin{bmatrix} -1 & 3 \\ 1 & -2 \end{bmatrix}$

2. $\begin{bmatrix} 5 & 7 \\ 2 & 3 \end{bmatrix}$ and $\begin{bmatrix} 3 & -7 \\ -2 & 5 \end{bmatrix}$

3. $\begin{bmatrix} 2 & 1 \\ 3 & 2 \end{bmatrix}$ and $\begin{bmatrix} 2 & 1 \\ -3 & 2 \end{bmatrix}$

4. $\begin{bmatrix} -1 & 2 \\ 3 & -5 \end{bmatrix}$ and $\begin{bmatrix} -5 & -2 \\ -3 & -1 \end{bmatrix}$

5. $\begin{bmatrix} 1 & 2 & 0 \\ 0 & 1 & 0 \\ 0 & 1 & 0 \end{bmatrix}$ and $\begin{bmatrix} 1 & -2 & 0 \\ 0 & 1 & 0 \\ 0 & -1 & 1 \end{bmatrix}$

6. $\begin{bmatrix} 0 & 1 & 0 \\ 0 & 0 & -2 \\ 1 & -1 & 0 \end{bmatrix}$ and $\begin{bmatrix} 1 & 0 & 1 \\ 1 & 0 & 0 \\ 0 & -1 & 0 \end{bmatrix}$

7. $\begin{bmatrix} 1 & 3 & 3 \\ 1 & 4 & 3 \\ 1 & 3 & 4 \end{bmatrix}$ and $\begin{bmatrix} 7 & -3 & -3 \\ -1 & 1 & 0 \\ -1 & 0 & 1 \end{bmatrix}$

8. $\begin{bmatrix} -1 & 0 & 2 \\ 3 & 1 & 0 \\ 0 & 2 & -3 \end{bmatrix}$ and $\begin{bmatrix} -\frac{1}{5} & \frac{4}{15} & -\frac{2}{15} \\ \frac{3}{5} & \frac{1}{5} & \frac{2}{5} \\ \frac{2}{5} & \frac{2}{15} & -\frac{1}{15} \end{bmatrix}$

9. Does a matrix with a row of all zeros have an inverse? Why?

10. Matrix A has A^{-1} as its inverse. What does $(A^{-1})^{-1}$ equal? (*Hint:* Experiment with a few matrices to see what you get.)

Find the inverse, if it exists, for each matrix.

11. $\begin{bmatrix} 1 & -1 \\ 2 & 0 \end{bmatrix}$

12. $\begin{bmatrix} -1 & 2 \\ -2 & -1 \end{bmatrix}$

13. $\begin{bmatrix} 3 & -1 \\ -5 & 2 \end{bmatrix}$

14. $\begin{bmatrix} -1 & -2 \\ 3 & 4 \end{bmatrix}$

15. $\begin{bmatrix} -6 & 4 \\ -3 & 2 \end{bmatrix}$

16. $\begin{bmatrix} 5 & 10 \\ -3 & -6 \end{bmatrix}$

17. $\begin{bmatrix} 1 & 0 & 0 \\ 0 & -1 & 0 \\ 1 & 0 & 1 \end{bmatrix}$

18. $\begin{bmatrix} 1 & 0 & 1 \\ 0 & -1 & 0 \\ 2 & 1 & 1 \end{bmatrix}$

19. $\begin{bmatrix} -1 & -1 & -1 \\ 4 & 5 & 0 \\ 0 & 1 & -3 \end{bmatrix}$

20. $\begin{bmatrix} 2 & 0 & 4 \\ 3 & 1 & 5 \\ -1 & 1 & -2 \end{bmatrix}$

21. $\begin{bmatrix} 1 & 2 & 3 \\ -3 & -2 & -1 \\ -1 & 0 & 1 \end{bmatrix}$

22. $\begin{bmatrix} 2 & 0 & 4 \\ 1 & 0 & -1 \\ 3 & 0 & -2 \end{bmatrix}$

23. $\begin{bmatrix} 2 & 4 & 6 \\ -1 & -4 & -3 \\ 0 & 1 & -1 \end{bmatrix}$ **24.** $\begin{bmatrix} 2 & 2 & -4 \\ 2 & 6 & 0 \\ -3 & -3 & 5 \end{bmatrix}$ **25.** $\begin{bmatrix} 1 & -2 & 3 & 0 \\ 0 & 1 & -1 & 1 \\ -2 & 2 & -2 & 4 \\ 0 & 2 & -3 & 1 \end{bmatrix}$ **26.** $\begin{bmatrix} 1 & 1 & 0 & 2 \\ 2 & -1 & 1 & -1 \\ 3 & 3 & 2 & -2 \\ 1 & 2 & 1 & 0 \end{bmatrix}$

Solve each system of equations by using the inverse of the coefficient matrix.

27. $\begin{aligned} 2x + 3y &= 10 \\ x - y &= -5 \end{aligned}$ **28.** $\begin{aligned} -x + 2y &= 15 \\ -2x - y &= 20 \end{aligned}$ **29.** $\begin{aligned} 2x + y &= 5 \\ 5x + 3y &= 13 \end{aligned}$ **30.** $\begin{aligned} -x - 2y &= 8 \\ 3x + 4y &= 24 \end{aligned}$

31. $\begin{aligned} -x + y &= 1 \\ 2x - y &= 1 \end{aligned}$ **32.** $\begin{aligned} 3x - 6y &= 1 \\ -5x + 9y &= -1 \end{aligned}$ **33.** $\begin{aligned} -x - 8y &= 12 \\ 3x + 24y &= -36 \end{aligned}$ **34.** $\begin{aligned} x + 3y &= -14 \\ 2x - y &= 7 \end{aligned}$

Solve each system of equations by using the inverse of the coefficient matrix. (The inverses for the first four problems were found in Exercises 19, 20, 23, and 24.)

35. $\begin{aligned} -x - y - z &= 1 \\ 4x + 5y &= -2 \\ y - 3z &= 3 \end{aligned}$ **36.** $\begin{aligned} 2x + 4z &= -8 \\ 3x + y + 5z &= 2 \\ -x + y - 2z &= 4 \end{aligned}$ **37.** $\begin{aligned} 2x + 4y + 6z &= 4 \\ -x - 4y - 3z &= 8 \\ y - z &= -4 \end{aligned}$

38. $\begin{aligned} 2x + 2y - 4z &= 12 \\ 2x + 6y &= 16 \\ -3x - 3y + 5z &= -20 \end{aligned}$ **39.** $\begin{aligned} 2x - 2y &= 5 \\ 4y + 8z &= 7 \\ x + 2z &= 1 \end{aligned}$ **40.** $\begin{aligned} x + z &= 3 \\ y + 2z &= 8 \\ -x + y &= 4 \end{aligned}$

Solve the systems of equations in Exercises 41 and 42 by using the inverse of the coefficient matrix. (The inverses were found in Exercises 25 and 26.)

41. $\begin{aligned} x - 2y + 3z &= 4 \\ y - z + w &= -8 \\ -2x + 2y - 2z + 4w &= 12 \\ 2y - 3z + w &= -4 \end{aligned}$ **42.** $\begin{aligned} x + y + 2w &= 3 \\ 2x - y + z - w &= 3 \\ 3x + 3y + 2z - 2w &= 5 \\ x + 2y + z &= 3 \end{aligned}$

Let $A = \begin{bmatrix} a & b \\ c & d \end{bmatrix}$ *in Exercises 43–48.*

43. Show that $IA = A$.   **44.** Show that $AI = A$.   **45.** Show that $A \cdot 0 = 0$.

46. Find A^{-1}.   **47.** Show that $A^{-1}A = I$.   **48.** Show that $AA^{-1} = I$.
(Assume $ad - bc \neq 0$.)

49. Using the definition and properties listed in this section, show that for square matrices A and B of the same size, if $AB = 0$ and if A^{-1} exists, then $B = O$.

50. Prove that, if it exists, the inverse of a matrix is unique. (*Hint:* Assume there are two inverses B and C for some matrix A, so that $AB = BA = I$ and $AC = CA = I$. Multiply the first equation by C and the second by B.)

Use matrices C and D in Exercises 51–55.

$$C = \begin{bmatrix} -6 & 8 & 2 & 4 & -3 \\ 1 & 9 & 7 & -12 & 5 \\ 15 & 2 & -8 & 10 & 11 \\ 4 & 7 & 9 & 6 & -2 \\ 1 & 3 & 8 & 23 & 4 \end{bmatrix}, \quad D = \begin{bmatrix} 5 & -3 & 7 & 9 & 2 \\ 6 & 8 & -5 & 2 & 1 \\ 3 & 7 & -4 & 2 & 11 \\ 5 & -3 & 9 & 4 & -1 \\ 0 & 3 & 2 & 5 & 1 \end{bmatrix}$$

51. Find C^{-1}.   **52.** Find $(CD)^{-1}$.   **53.** Find D^{-1}.   **54.** Is $C^{-1}D^{-1} = (CD)^{-1}$?   **55.** Is $D^{-1}C^{-1} = (CD)^{-1}$?

Solve the matrix equation $AX = B$ for X by finding A^{-1}, given A and B as follows.

56. $A = \begin{bmatrix} 2 & 3 & 5 \\ 1 & 7 & 9 \\ -3 & 2 & 10 \end{bmatrix}$, $B = \begin{bmatrix} 3 \\ 4 \\ 1 \end{bmatrix}$

57. $A = \begin{bmatrix} 2 & 5 & 7 & 9 \\ 1 & 3 & -4 & 6 \\ -1 & 0 & 5 & 8 \\ 2 & -2 & 4 & 10 \end{bmatrix}$, $B = \begin{bmatrix} 3 \\ 7 \\ -1 \\ 5 \end{bmatrix}$

58. $A = \begin{bmatrix} 3 & 2 & -1 & -2 & 6 \\ -5 & 17 & 4 & 3 & 15 \\ 7 & 9 & -3 & -7 & 12 \\ 9 & -2 & 1 & 4 & 8 \\ 1 & 21 & 9 & -7 & 25 \end{bmatrix}$, $B = \begin{bmatrix} -2 \\ 5 \\ 3 \\ -8 \\ 25 \end{bmatrix}$

Applications

Solve the following exercises by using the inverse of the coef-ficient matrix to solve a system of equations.

LIFE SCIENCES

59. *Vitamins* Greg Tobin mixes together three types of vitamin tablets. Each Super Vim tablet contains, among other things, 15 mg of niacin and 12 I.U. of Vitamin E. The figures for a Multitab tablet are 20 mg and 15 I.U., and for a Mighty Mix are 25 mg and 35 I.U. How many of each tablet are there if the total number of tablets, total amount of niacin, and total amount of Vitamin E are as follows?

a. 225 tablets, 4,750 mg of niacin, and 5,225 I.U. of Vitamin E

b. 185 tablets, 3,625 mg of niacin, and 3,750 I.U. of Vitamin E

c. 230 tablets, 4,450 mg of niacin, and 4,210 I.U. of Vitamin E

OTHER APPLICATIONS

60. *Analysis of Orders* The Bread Box Bakery sells three types of cakes, each requiring the amounts of the basic ingredients shown in the following matrix.

		Type of Cake		
		I	II	III
	Flour (in cups)	2	4	2
Ingredient	Sugar (in cups)	2	1	2
	Eggs	2	1	3

To fill its daily orders for these three kinds of cake, the bakery uses 72 cups of flour, 48 cups of sugar, and 60 eggs.

a. Write a 3 × 1 matrix for the amounts used daily.

b. Let the number of daily orders for cakes be a 3 × 1 matrix X with entries x_1, x_2, and x_3. Write a matrix equa-

tion that can be solved for X, using the given matrix and the matrix from part a.

c. Solve the equation from part b to find the number of daily orders for each type of cake.

61. *Production Requirements* An electronics company produces transistors, resistors, and computer chips. Each transistor requires 3 units of copper, 1 unit of zinc, and 2 units of glass. Each resistor requires 3, 2, and 1 units of the three materials, and each computer chip requires 2, 1, and 2 units of these materials, respectively. How many of each product can be made with the following amounts of materials?

a. 810 units of copper, 410 units of zinc, and 490 units of glass

b. 765 units of copper, 385 units of zinc, and 470 units of glass

c. 1,010 units of copper, 500 units of zinc, and 610 units of glass

62. *Investments* An investment firm recommends that a client invest in AAA, A, and B rated bonds. The average yield on AAA bonds is 6%, on A bonds 7%, and on B bonds 10%. The client wants to invest twice as much in AAA bonds as in B bonds. How much should be invested in each type of bond under the following conditions?

a. The total investment is $25,000, and the investor wants an annual return of $1,810 on the three investments.

b. The values in part a are changed to $30,000 and $2,150, respectively.

c. The values in part a are changed to $40,000 and $2,900, respectively.

63. *Production* Pretzels cost $3 per pound, dried fruit is $4 per pound, and nuts $8 per pound. The three ingredients are to be combined in a trail mix containing twice the weight of pretzels as dried fruit. How many pounds of each should be used to produce the following amounts at the given cost?

a. 140 lb at \$6 per lb

b. 112 lb at \$6.50 per lb

c. 126 lb at \$5 per lb

64. *Encryption* Use the matrices presented in Example 6 of this section to do the following:

a. Encode the message, "All is fair in love and war."

b. Decode the message 94, 48, 91, 118, 71, 112, 71, 47, 69.

65. *Music* During a marching band's half-time show, the band members generally line up in such a way that a common shape is recognized by the fans. For example, as illustrated in the figure, a band might form a letter T, where an X represents a member of the band. As the music is played the band will either create a new shape or rotate the original shape. In doing this, each member of the band will need to move from one point on the field to another. For larger bands, keeping track of who goes where can be a daunting task. However, it is possible to use matrix inverses to make the process a bit easier.* The entire process is calculated by knowing how three band members, all of whom cannot be in a straight line, will move from the current position to a new position. For example, in the figure, we can see that there are band members at $(50, 0)$, $(50, 15)$, and $(45, 20)$. We will assume that these three band members move to $(40, 10)$, $(55, 10)$, and $(60, 15)$, respectively. (We have chosen to assign the 40-yard line to the right of the 50-yard line an x-coordinate of 60.)

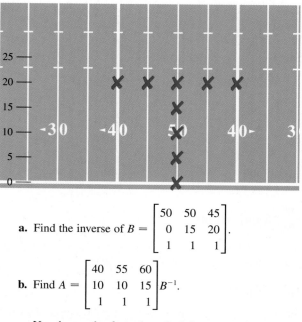

a. Find the inverse of $B = \begin{bmatrix} 50 & 50 & 45 \\ 0 & 15 & 20 \\ 1 & 1 & 1 \end{bmatrix}$.

b. Find $A = \begin{bmatrix} 40 & 55 & 60 \\ 10 & 10 & 15 \\ 1 & 1 & 1 \end{bmatrix} B^{-1}$.

c. Use the result of part b to find the new position of the other band members. What is the shape of the new position? (*Hint:* Multiply the matrix A by a 3×1 column vector with the first two components equal to the original position of each band member and the third component equal to 1. The new position of the band member is in the first two components of the product.)

■ 10.5 EIGENVALUES AND EIGENVECTORS

? THINK ABOUT IT How can we determine a population of insects that grows at a constant rate while keeping the same proportion of adults and juveniles?

This kind of question is of fundamental interest in mathematical biology. To answer it, suppose there is an insect population consisting of juveniles (insects in their first year of life) and adults (insects in their second year of life).[†] We will assume that no insects of this species live more than two years.

Let $x_1(t)$ and $x_2(t)$ be the number of juveniles and adults in year t. We will assume that a proportion s of the juveniles survive to become adults the following year. We will also assume that a proportion p of the juveniles have offspring that are juveniles the following year, and that a proportion q of the adults have

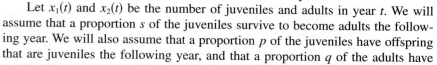

---

*Isaksen, Daniel, "Linear Algebra on the Gridiron," *The College Mathematics Journal,* Vol. 26, No. 5, Nov. 1995, pp. 358–360.

[†]The model here is described in *Population Biology: Concepts and Models,* by Alan Hastings, Springer-Verlag, 1997, pp. 16–31.

offspring that are juveniles the following year. Then the number of juveniles and adults next year can be calculated from the number this year using the equation

$$x_1(t + 1) = px_1(t) + qx_2(t)$$
$$x_2(t + 1) = sx_1(t).$$

In matrix form, this is

$$\begin{bmatrix} x_1(t+1) \\ x_2(t+1) \end{bmatrix} = \begin{bmatrix} p & q \\ s & 0 \end{bmatrix}\begin{bmatrix} x_1(t) \\ x_2(t) \end{bmatrix},$$

which can also be written as

$$X(t + 1) = MX(t),$$

where

$$M = \begin{bmatrix} p & q \\ s & 0 \end{bmatrix}$$

and

$$X(t) = \begin{bmatrix} x_1(t) \\ x_2(t) \end{bmatrix}.$$

The advantage of this last notation is that it can be extended to any number of groups within the population. The matrix M is known as a **Leslie matrix** after P. H. Leslie, who first wrote about them in 1945.*

For example, suppose

$$M = \begin{bmatrix} 0.6 & 0.8 \\ 0.9 & 0 \end{bmatrix}$$

and that

$$X(0) = \begin{bmatrix} 1{,}000 \\ 2{,}000 \end{bmatrix}.$$

Then

$$\begin{bmatrix} x_1(1) \\ x_2(1) \end{bmatrix} = \begin{bmatrix} 0.6 & 0.8 \\ 0.9 & 0 \end{bmatrix}\begin{bmatrix} 1{,}000 \\ 2{,}000 \end{bmatrix}$$
$$= \begin{bmatrix} 2{,}200 \\ 900 \end{bmatrix}$$

and

$$\begin{bmatrix} x_1(2) \\ x_2(2) \end{bmatrix} = \begin{bmatrix} 0.6 & 0.8 \\ 0.9 & 0 \end{bmatrix}\begin{bmatrix} 2{,}200 \\ 900 \end{bmatrix}$$
$$= \begin{bmatrix} 2{,}040 \\ 1{,}980 \end{bmatrix}.$$

Notice that the population of juveniles has gone from 1,000 in year 0 to 2,200 in year 1 to 2,040 in year 2. Meanwhile, the population of adults has gone from 2,000 in year 0 to 900 in year 1 to 1,980 in year 2.

---

*P. H. Leslie, "On the Use of Matrices in Certain Population Mathematics," *Biometrika,* Vol. 33, 1945, pp. 183–212.

In many actual populations, the proportion of the population in each age group stays the same from year to year, even as the population grows. For example, if

$$X(0) = \begin{bmatrix} 400 \\ 300 \end{bmatrix},$$

then

$$X(1) = \begin{bmatrix} 0.6 & 0.8 \\ 0.9 & 0 \end{bmatrix} \begin{bmatrix} 400 \\ 300 \end{bmatrix} = \begin{bmatrix} 480 \\ 360 \end{bmatrix} = 1.2 \begin{bmatrix} 400 \\ 300 \end{bmatrix}.$$

Both the population of juveniles and adults have grown by 20%. If we denote the growth factor by the Greek letter λ (pronounced lambda), then

$$MX = \lambda X.$$

A number λ that satisfies the above equation is called an **eigenvalue.** (Eigen is the German word meaning "characteristic.") The matrix (or vector) X in the above equation is called an **eigenvector.** Eigenvalues and eigenvectors have a wide variety of applications in mathematics in addition to solving problems about populations. In the next chapter, we will see how they are used in solving systems of differential equations. The important thing to notice here is that multiplying an eigenvector X by the matrix M has the same result as multiplying X by the scalar λ.

Now we must investigate how to find eigenvalues and eigenvectors. By subtracting λX from both sides of the equation above, we get

$$MX - \lambda X = O,$$

where O represents the zero matrix. This equation can be rewritten as

$$(M - \lambda I)X = O,$$

where I is the identity matrix. This equation has the trivial solution $X = 0$, but we are interested in nontrivial solutions. In the language of Section 10.1, the above equation would have more than one solution (in fact, an infinite number of solutions) if it describes a dependent system. So we need to know for what values of λ the above system is dependent.

We now need to stretch a little to some mathematics that is explained more fully in a text on linear algebra.* It can be shown that the above system has a nontrivial system if something called the **determinant** of $M - \lambda I$ is zero. (For a proof of this fact in the case where M is a 2×2 matrix, see Exercise 9.) Computing the determinant of a 2×2 matrix is easy:

$$\det \begin{bmatrix} a & b \\ c & d \end{bmatrix} = ad - bc.$$

Recall from the previous section that the expression $ad - bc$ is part of the formula for the inverse of a 2×2 matrix. You can think of calculating the determinant by starting with the product of the numbers along the diagonal from upper

---

*For example, see Lay, David, *Linear Algebra and Its Applications*, 2nd ed., Addison Wesley, 2000.

left to lower right, and subtracting the product of the numbers along the diagonal from upper right to lower left:

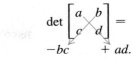

$$\det \begin{bmatrix} a & b \\ c & d \end{bmatrix} = -bc + ad.$$

For example,

$$\det \begin{bmatrix} 2 & 3 \\ 4 & 5 \end{bmatrix} =$$

$$-3 \times 4 + 2 \times 5 = -12 + 10 = -2.$$

To find the determinant of a 3×3 matrix, take the first two columns of the matrix and duplicate them on the right, as we will show in a moment. Then add the products of the numbers along each of the three diagonals going from upper left to lower right, and subtract the products of the numbers along each of the three diagonals from upper right to lower left. Here we demonstrate the process for a general 3×3 matrix.

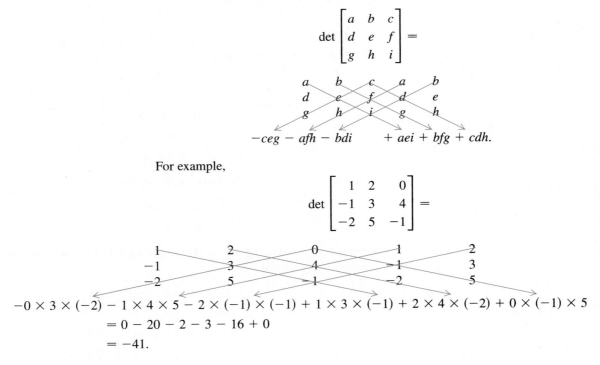

$$\det \begin{bmatrix} a & b & c \\ d & e & f \\ g & h & i \end{bmatrix} =$$

$$-ceg - afh - bdi + aei + bfg + cdh.$$

For example,

$$\det \begin{bmatrix} 1 & 2 & 0 \\ -1 & 3 & 4 \\ -2 & 5 & -1 \end{bmatrix} =$$

$$-0 \times 3 \times (-2) - 1 \times 4 \times 5 - 2 \times (-1) \times (-1) + 1 \times 3 \times (-1) + 2 \times 4 \times (-2) + 0 \times (-1) \times 5$$

$$= 0 - 20 - 2 - 3 - 16 + 0$$

$$= -41.$$

CAUTION The determinant of any square matrix can be calculated, but the procedure is more involved for matrices larger than 3×3, so we will not go into it in this book.

Now we return to the problem of solving

$$\det(M - \lambda I) = 0$$

for the matrix

$$M = \begin{bmatrix} 0.6 & 0.8 \\ 0.9 & 0 \end{bmatrix}.$$

$$\det(M - \lambda I) = \det\begin{bmatrix} 0.6 - \lambda & 0.8 \\ 0.9 & 0 - \lambda \end{bmatrix}$$

$$= (0.6 - \lambda)(0 - \lambda) - 0.8 \times 0.9$$

$$= \lambda^2 - 0.6\lambda - 0.72$$

We can solve $\lambda^2 - 0.6\lambda - 0.72 = 0$ with the quadratic formula.

$$\lambda = \frac{0.6 \pm \sqrt{0.6^2 - 4(-0.72)}}{2}$$

$$= \frac{0.6 \pm \sqrt{3.24}}{2}$$

$$= \frac{0.6 \pm 1.8}{2}$$

$$= 1.2, \quad -0.6$$

To find the corresponding eigenvectors, substitute each eigenvalue back into the equation $(M - \lambda I)X = 0$ and solve the dependent system. First try this with $\lambda = 1.2$.

$$(M - \lambda I)X = \begin{bmatrix} 0.6 - \lambda & 0.8 \\ 0.9 & 0 - \lambda \end{bmatrix}\begin{bmatrix} x_1 \\ x_2 \end{bmatrix} \qquad \text{Let } \lambda = 1.2.$$

$$= \begin{bmatrix} -0.6 & 0.8 \\ 0.9 & -1.2 \end{bmatrix}\begin{bmatrix} x_1 \\ x_2 \end{bmatrix} = \begin{bmatrix} 0 \\ 0 \end{bmatrix}$$

Now create the augmented matrix for this system.

$$\left[\begin{array}{cc|c} -0.6 & 0.8 & 0 \\ 0.9 & -1.2 & 0 \end{array}\right]$$

Before row reducing, let's simplify by getting rid of the decimals, and then dividing out any common factors.

$$\begin{array}{c} -10R_1 \to R_1 \\ 10R_2 \to R_2 \end{array}\left[\begin{array}{cc|c} 6 & -8 & 0 \\ 9 & -12 & 0 \end{array}\right]$$

$$\begin{array}{c} R_1/2 \to R_1 \\ R_2/3 \to R_2 \end{array}\left[\begin{array}{cc|c} 3 & -4 & 0 \\ 3 & -4 & 0 \end{array}\right]$$

Next, get a zero in row 2 of column 1.

$$-R_1 + R_2 \to R_2 \left[\begin{array}{cc|c} 3 & -4 & 0 \\ 0 & 0 & 0 \end{array}\right]$$

The zeros in row 2 indicate that we have a dependent system. Since the first row indicates that $3x_1 - 4x_2 = 0$, the simplest solution is $x_1 = 4$ and $x_2 = 3$. There are, of course, an infinite number of solutions, described by letting x_2 be an arbitrary parameter and $x_1 = 4x_2/3$, but we only need one solution. An eigenvector corresponding to $\lambda = 1.2$ is therefore

$$\begin{bmatrix} x_1 \\ x_2 \end{bmatrix} = \begin{bmatrix} 4 \\ 3 \end{bmatrix}.$$

Any multiple of this matrix is also an eigenvector.

Now find the eigenvector corresponding to $\lambda = -0.6$.

$$(M - \lambda I)X = \begin{bmatrix} 0.6 - \lambda & 0.8 \\ 0.9 & 0 - \lambda \end{bmatrix} \begin{bmatrix} x_1 \\ x_2 \end{bmatrix} \qquad \text{Let } \lambda = -0.6.$$

$$= \begin{bmatrix} 1.2 & 0.8 \\ 0.9 & 0.6 \end{bmatrix} \begin{bmatrix} x_1 \\ x_2 \end{bmatrix} = \begin{bmatrix} 0 \\ 0 \end{bmatrix}$$

Next create the augmented matrix for this system.

$$\begin{bmatrix} 1.2 & 0.8 & \bigg| & 0 \\ 0.9 & 0.6 & \bigg| & 0 \end{bmatrix}$$

Verify that one solution to this system is

$$\begin{bmatrix} x_1 \\ x_2 \end{bmatrix} = \begin{bmatrix} 2 \\ -3 \end{bmatrix}.$$

Since a population cannot have a negative value, a population with a stable structure cannot be a multiple of this eigenvector. This eigenvector can, however, be a component of a population whose age structure is changing with time, as we shall see in Example 2. Meanwhile, the population in which the proportion of each age group stays the same from one year to the next must be a multiple of the eigenvector

$$\begin{bmatrix} x_1 \\ x_2 \end{bmatrix} = \begin{bmatrix} 4 \\ 3 \end{bmatrix},$$

for which each year's population is 1.2 times the previous year's population. In other words, the population grows 20%, and the proportion of juveniles and adults stays the same.

For example, if we wanted to start in year 0 with a population of size 14,000, then we would take a multiple of the above eigenvector, say

$$\begin{bmatrix} x_1 \\ x_2 \end{bmatrix} = k \begin{bmatrix} 4 \\ 3 \end{bmatrix} = \begin{bmatrix} 4k \\ 3k \end{bmatrix}.$$

The size of the population is then $4k + 3k$, or $7k$. Setting this equal to 14,000 makes $k = 2{,}000$, so the population in year 0 is

$$\begin{bmatrix} x_1 \\ x_2 \end{bmatrix} = 2{,}000 \begin{bmatrix} 4 \\ 3 \end{bmatrix} = \begin{bmatrix} 8{,}000 \\ 6{,}000 \end{bmatrix}.$$

The population the next year is

$$1.2 \begin{bmatrix} 8{,}000 \\ 6{,}000 \end{bmatrix} = \begin{bmatrix} 9{,}600 \\ 7{,}200 \end{bmatrix},$$

and the population the following year is

$$1.2 \begin{bmatrix} 9{,}600 \\ 7{,}200 \end{bmatrix} = \begin{bmatrix} 11{,}520 \\ 8{,}640 \end{bmatrix}.$$

Verify that these results are exactly what you would get by multiplying the original population vector by the matrix M, and then multiplying that result by M. It is, of course, much simpler to multiply a vector by a scalar than by a matrix, which is a nice feature of eigenvectors and eigenvalues.

EXAMPLE 1 Eigenvectors

Find the eigenvalues and their corresponding eigenvectors for the matrix

$$M = \begin{bmatrix} 5 & 2 & 3 \\ 0 & 8 & 3 \\ 0 & 0 & 4 \end{bmatrix}.$$

Solution Calculate $\det(M - \lambda I)$.

$$\det(M - \lambda I) = \det \begin{bmatrix} 5 - \lambda & 2 & 3 \\ 0 & 8 - \lambda & 3 \\ 0 & 0 & 4 - \lambda \end{bmatrix} =$$

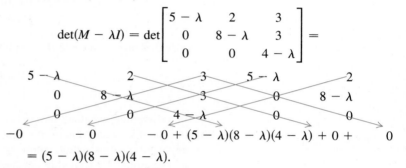

$$\quad -0 \qquad\qquad -0 \qquad\qquad -0 + (5 - \lambda)(8 - \lambda)(4 - \lambda) + 0 + \qquad 0$$

$$= (5 - \lambda)(8 - \lambda)(4 - \lambda).$$

We now see that $\det(M - \lambda I) = 0$ when $\lambda = 5, 8,$ and 4. Notice that these are the values along the main diagonal of the original matrix. This will always be the case when all the values above the main diagonal or all the values below the main diagonal are 0.

To find an eigenvector corresponding to the eigenvalue $\lambda = 5$, set $\lambda = 5$ in the augmented matrix corresponding to the system $(M - \lambda I)X = 0$.

$$\begin{bmatrix} 5 - \lambda & 2 & 3 & | & 0 \\ 0 & 8 - \lambda & 3 & | & 0 \\ 0 & 0 & 4 - \lambda & | & 0 \end{bmatrix}$$

$$= \begin{bmatrix} 0 & 2 & 3 & | & 0 \\ 0 & 3 & 3 & | & 0 \\ 0 & 0 & -1 & | & 0 \end{bmatrix}$$

Row reduction is simplified by the fact that the first column consists entirely of zeros.

$$\begin{array}{r} \\ R_2/3 \to R_2 \\ -R_3 \to R_3 \end{array} \begin{bmatrix} 0 & 2 & 3 & | & 0 \\ 0 & 1 & 1 & | & 0 \\ 0 & 0 & 1 & | & 0 \end{bmatrix} \quad \text{Simplify the rows.}$$

$$-2R_2 + R_1 \to R_1 \begin{bmatrix} 0 & 0 & 1 & | & 0 \\ 0 & 1 & 1 & | & 0 \\ 0 & 0 & 1 & | & 0 \end{bmatrix} \quad \text{Get zeros in column 2.}$$

$$\begin{matrix} R_1 - R_3 \rightarrow R_1 \\ R_2 - R_3 \rightarrow R_1 \end{matrix} \left[\begin{array}{ccc|c} 0 & 0 & 0 & 0 \\ 0 & 1 & 0 & 0 \\ 0 & 0 & 1 & 0 \end{array}\right] \quad \text{Get zeros in column 3.}$$

The second row tells us that $x_2 = 0$, and the third row tells us that $x_3 = 0$. Since x_1 is arbitrary, we will let $x_1 = 1$. Therefore, an eigenvector corresponding to $\lambda = 5$ is

$$\begin{bmatrix} 1 \\ 0 \\ 0 \end{bmatrix}.$$

Similarly, when $\lambda = 8$, the augmented matrix corresponding to the system $(M - \lambda I)X = 0$ is

$$\left[\begin{array}{ccc|c} -3 & 2 & 3 & 0 \\ 0 & 0 & 3 & 0 \\ 0 & 0 & -4 & 0 \end{array}\right].$$

Verify that this augmented matrix reduces to

$$\left[\begin{array}{ccc|c} -3 & 2 & 0 & 0 \\ 0 & 0 & 1 & 0 \\ 0 & 0 & 0 & 0 \end{array}\right]$$

and that an eigenvector is

$$\begin{bmatrix} 2 \\ 3 \\ 0 \end{bmatrix}.$$

We leave it as an exercise to show that an eigenvector corresponding to $\lambda = 4$ is

$$\begin{bmatrix} 6 \\ 3 \\ -4 \end{bmatrix}.$$

The following result, mentioned in Example 1, is worth noting.

> If all the values above the main diagonal or all the values below the main diagonal of a matrix are 0, the eigenvalues are the values along the main diagonal.

NOTE The eigenvalues of a matrix can be complex numbers. There can also be eigenvalues with a multiplicity greater than 1, in the sense that, rather than having two distinct eigenvalues for a 2×2 matrix or three distinct eigenvalues for a 3×3 matrix, some of the eigenvalues may be identical. In the examples and exercises in this text, the eigenvalues will be real and distinct, with the exception of Exercises 10–13 in this section.

Earlier in this section, we showed how a population described by an eigenvector has a stable distribution from one generation to the next. In the next example, we will see how eigenvalues and eigenvectors help us analyze an arbitrary population that is not an eigenvector. The key is to represent the initial population in terms of the eigenvectors.

EXAMPLE 2   Long-term Population

In the insect example with which we started this section, recall that the Leslie matrix was $M = \begin{bmatrix} 0.6 & 0.8 \\ 0.9 & 0 \end{bmatrix}$ and that the eigenvectors $\lambda_1 = 1.2$ and $\lambda_2 = -0.6$ corresponded to the eigenvectors $V_1 = \begin{bmatrix} 4 \\ 3 \end{bmatrix}$ and $V_2 = \begin{bmatrix} 2 \\ -3 \end{bmatrix}$, respectively. Recall also that the initial population was $X(0) = \begin{bmatrix} 1,000 \\ 2,000 \end{bmatrix}$.

(a) Find values a and b such that $aV_1 + bV_2 = X(0)$.

Solution The equation

$$a\begin{bmatrix} 4 \\ 3 \end{bmatrix} + b\begin{bmatrix} 2 \\ -3 \end{bmatrix} = \begin{bmatrix} 1,000 \\ 2,000 \end{bmatrix}$$

corresponds to the augmented matrix

$$\begin{bmatrix} 4 & 2 & | & 1,000 \\ 3 & -3 & | & 2,000 \end{bmatrix}.$$

Use the Gauss-Jordan method to reduce this matrix to

$$\begin{bmatrix} 1 & 0 & | & 3,500/9 \\ 0 & 1 & | & -2,500/9 \end{bmatrix}.$$

The solution is $a = 3,500/9$ and $b = -2,500/9$.

(b) Use the result from part (a), as well as properties of eigenvalues and eigenvectors, to find the population distribution the following year.

Solution The population in the next year is given by

$$X(1) = MX(0) = M\left(\frac{3,500}{9}V_1 - \frac{2,500}{9}V_2\right) \qquad \text{Use the results of part (a).}$$

$$= \frac{3,500}{9}MV_1 - \frac{2,500}{9}MV_2$$

$$= \frac{3,500}{9}\lambda_1V_1 - \frac{2,500}{9}\lambda_2V_2 \qquad \text{Use properties of eigenvalues and eigenvectors.}$$

$$= \frac{3,500}{9}(1.2)\begin{bmatrix} 4 \\ 3 \end{bmatrix} - \frac{2,500}{9}(-0.6)\begin{bmatrix} 2 \\ -3 \end{bmatrix}$$

$$= \begin{bmatrix} 2,200 \\ 900 \end{bmatrix}.$$

(c) Find a formula for the long-term population distribution.

Solution By continuing to multiply by M, the population after n years is given by

$$X(n) = M^nX(0) = M^n\left(\frac{3,500}{9}V_1 - \frac{2,500}{9}V_2\right)$$

$$= \frac{3,500}{9}M^nV_1 - \frac{2,500}{9}M^nV_2$$

$$= \frac{3,500}{9} \lambda_1^n V_1 - \frac{2,500}{9} \lambda_2^n V_2$$

$$= \frac{3,500}{9} (1.2)^n \begin{bmatrix} 4 \\ 3 \end{bmatrix} - \frac{2,500}{9} (-0.6)^n \begin{bmatrix} 2 \\ -3 \end{bmatrix}.$$

Since $\lim_{n \to \infty} (0.6)^n = 0$, the second term becomes smaller and smaller over time. Therefore, when n is large,

$$X(n) \approx \frac{3,500}{9} (1.2)^n \begin{bmatrix} 4 \\ 3 \end{bmatrix}.$$

Example 2 demonstrates an important result. Even if the initial population is not an eigenvector, the population over time approaches a multiple of one of the eigenvectors. Since any multiple of an eigenvector is itself an eigenvector, this shows that, at least in the case with two eigenvalues, one greater than 1 in magnitude and one less than 1 in magnitude, the population will approach an eigenvector over time, and that each year's population can be approximately found by multiplying the previous year's population by an eigenvalue. This behavior is typical of a large range of applications in the life sciences.

10.5 EXERCISES

For Exercises 1–8, find the eigenvalues and their corresponding eigenvectors.

1. $\begin{bmatrix} 5 & 0 \\ 2 & 1 \end{bmatrix}$

2. $\begin{bmatrix} 7 & 4 \\ -3 & -1 \end{bmatrix}$

3. $\begin{bmatrix} 3 & 2 \\ 3 & 8 \end{bmatrix}$

4. $\begin{bmatrix} 1 & 0 \\ 6 & -1 \end{bmatrix}$

5. $\begin{bmatrix} 4 & -3 \\ 2 & -1 \end{bmatrix}$

6. $\begin{bmatrix} 2 & 3 \\ 4 & 1 \end{bmatrix}$

7. $\begin{bmatrix} 4 & 0 & 0 \\ 3 & -1 & 0 \\ 2 & 5 & -3 \end{bmatrix}$

8. $\begin{bmatrix} -5 & 0 & 0 \\ 1 & 2 & 0 \\ -1 & 4 & 4 \end{bmatrix}$

9. Show that the system of equations

$$ax + by = 0$$
$$cx + dy = 0$$

is dependent whenever $ac - bd = 0$, that is, when the determinant of the coefficient matrix is 0. (*Hint:* Use the Gauss-Jordan method.)

In Exercises 10 and 11, each 2 × 2 matrix has only one eigenvalue. Find the eigenvalue and the corresponding eigenvector.

10. $\begin{bmatrix} 7 & -2 \\ 2 & 3 \end{bmatrix}$

11. $\begin{bmatrix} 10 & -9 \\ 4 & -2 \end{bmatrix}$

In Exercises 12 and 13, each matrix has complex numbers as eigenvalues. Find the eigenvalues and their corresponding eigenvectors. (Note: These exercises require a knowledge of complex numbers. Complex eigenvalues are important in biology applications.)

12. $\begin{bmatrix} 1 & -2 \\ 1 & 3 \end{bmatrix}$

13. $\begin{bmatrix} 1 & 5 \\ -2 & 3 \end{bmatrix}$

Applications

LIFE SCIENCES

Leslie Matrices For each of the following Leslie matrices, **(a)** find a population of size 10,000 for which the proportion of the population in each age group stays the same from one year to the next, and **(b)** tell by what factor the population grows or declines each year.

14. $\begin{bmatrix} 0.7 & 0.9 \\ 1.6 & 0 \end{bmatrix}$

15. $\begin{bmatrix} 0.5 & 0.9 \\ 1.4 & 0 \end{bmatrix}$

16. $\begin{bmatrix} 0.95 & 0.03 \\ 0.05 & 0.97 \end{bmatrix}$

17. $\begin{bmatrix} 0.2 & 0.7 \\ 0.7 & 0.2 \end{bmatrix}$

18. $\begin{bmatrix} 0.7 & 0.2 \\ 0.2 & 0.4 \end{bmatrix}$

19. $\begin{bmatrix} 0.1 & 1.2 \\ 0.4 & 0.3 \end{bmatrix}$

20. *Long-term Population* For the matrix in Exercise 14, find an approximate expression for the population distribution after n years, given that the initial population distribution is given by $X(0) = \begin{bmatrix} 3{,}000 \\ 1{,}000 \end{bmatrix}$.

21. *Long-term Population* Repeat Exercise 20 using the matrix in Exercise 15.

22. *Northern Spotted Owl Population* In Exercise 46 of the section on Multiplication of Matrices, we saw that the number of female northern spotted owls at time $n + 1$ could be estimated by the equation

$$\begin{bmatrix} j_{n+1} \\ s_{n+1} \\ a_{n+1} \end{bmatrix} = \begin{bmatrix} 0 & 0 & 0.33 \\ 0.18 & 0 & 0 \\ 0 & 0.71 & 0.94 \end{bmatrix} \begin{bmatrix} j_n \\ s_n \\ a_n \end{bmatrix},$$

where j_n is the number of juveniles, s_n is the number of subadults, and a_n is the number of adults at time n. In part b of that exercise, it was stated that in the long run, the following approximation holds:

$$\begin{bmatrix} j_{n+1} \\ s_{n+1} \\ a_{n+1} \end{bmatrix} \approx 0.98359 \begin{bmatrix} j_n \\ s_n \\ a_n \end{bmatrix}.$$

Use the concept of eigenvalues to show that this approximation is exactly true if $\begin{bmatrix} j_n \\ s_n \\ a_n \end{bmatrix}$ is an eigenvector. (*Hint:* After calculating $\det(M - \lambda I)$, use a graphing calculator to plot $y = \det(M - \lambda I)$ as a function of λ, and find where $y = 0$.)

CHAPTER SUMMARY

In this chapter we studied matrices and systems of linear equations. The Gauss-Jordan method was developed and used to solve such systems. We saw that matrices can be combined and manipulated using addition, subtraction, scalar multiplication, and matrix multiplication. The concept of multiplicative inverse of real numbers was then extended to matrices, and matrix inverses were explored. The fact that matrix inverses can be used to solve systems of equations was introduced and then applied to solve a wide range of problems. Eigenvalues and eigenvectors were introduced to solve certain types of matrix equations that arise in applications.

KEY TERMS

system of equations
10.1 linear equation in
    n unknowns
    matrix (matrices)

element (entry)
augmented matrix
row operations
equivalent system

Gauss-Jordan method
inconsistent system
dependent system
parameter

10.2 size
square matrix
row matrix (row
    vector)

column matrix	zero matrix	multiplicative inverse	eigenvector
(column vector)	additive identity	matrix	determinant
additive inverse	10.3 scalar	coefficient matrix	
(negative) of a	product matrix	10.5 Leslie matrix	
matrix	10.4 identity matrix	eigenvalue	

CHAPTER 10 REVIEW EXERCISES

1. What is true about the number of solutions to a system of m linear equations in n unknowns if $m = n$? If $m < n$? If $m > n$?

2. Suppose someone says that a more reasonable way to multiply two matrices than the method presented in the text is to multiply corresponding elements. For example, the result of

$$\begin{bmatrix} 1 & 2 \\ 3 & 4 \end{bmatrix} \cdot \begin{bmatrix} 3 & 5 \\ 7 & 11 \end{bmatrix} \text{ should be } \begin{bmatrix} 3 & 10 \\ 21 & 44 \end{bmatrix},$$

according to this person. How would you respond?

Solve each system by the Gauss-Jordan method.

3. $\begin{aligned} 2x + 3y &= 10 \\ -3x + y &= 18 \end{aligned}$

4. $\begin{aligned} \dfrac{x}{2} + \dfrac{y}{4} &= 3 \\ \dfrac{x}{4} - \dfrac{y}{2} &= 4 \end{aligned}$

5. $\begin{aligned} 2x - 3y + z &= -5 \\ x + 4y + 2z &= 13 \\ 5x + 5y + 3z &= 14 \end{aligned}$

6. $\begin{aligned} x - y &= 3 \\ 2x + 3y + z &= 13 \\ 3x - 2z &= 21 \end{aligned}$

7. $\begin{aligned} 2x + 4y &= -6 \\ -3x - 5y &= 12 \end{aligned}$

8. $\begin{aligned} x + 2y &= -9 \\ 4x + 9y &= 41 \end{aligned}$

9. $\begin{aligned} x - y + 3z &= 13 \\ 4x + y + 2z &= 17 \\ 3x + 2y + 2z &= 1 \end{aligned}$

10. $\begin{aligned} x - 2z &= 5 \\ 3x + 2y &= 8 \\ -x + 2z &= 10 \end{aligned}$

11. $\begin{aligned} 3x - 6y + 9z &= 12 \\ -x + 2y - 3z &= -4 \\ x + y + 2z &= 7 \end{aligned}$

Find the size of each matrix, find the values of any variables, and identify any square, row, or column matrices.

12. $\begin{bmatrix} 2 & 3 \\ 5 & q \end{bmatrix} = \begin{bmatrix} a & b \\ c & 9 \end{bmatrix}$

13. $\begin{bmatrix} 2 & x \\ y & 6 \\ 5 & z \end{bmatrix} = \begin{bmatrix} a & -1 \\ 4 & 6 \\ p & 7 \end{bmatrix}$

14. $[m \quad 4 \quad z \quad -1] = [12 \quad k \quad -8 \quad r]$

15. $\begin{bmatrix} a+5 & 3b & 6 \\ 4c & 2+d & -3 \\ -1 & 4p & q-1 \end{bmatrix} = \begin{bmatrix} -7 & b+2 & 2k-3 \\ 3 & 2d-1 & 4l \\ m & 12 & 8 \end{bmatrix}$

Given the matrices

$$A = \begin{bmatrix} 4 & 10 \\ -2 & -3 \\ 6 & 9 \end{bmatrix}, \quad B = \begin{bmatrix} 2 & 3 & -2 \\ 2 & 4 & 0 \\ 0 & 1 & 2 \end{bmatrix}, \quad C = \begin{bmatrix} 5 & 0 \\ -1 & 3 \\ 4 & 7 \end{bmatrix},$$

$$D = \begin{bmatrix} 6 \\ 1 \\ 0 \end{bmatrix}, \quad E = [1 \quad 3 \quad -4], \quad F = \begin{bmatrix} -1 & 4 \\ 3 & 7 \end{bmatrix}, \quad G = \begin{bmatrix} 2 & 5 \\ 1 & 6 \end{bmatrix},$$

find each of the following, if it exists.

16. $A + C$

17. $2G - 4F$

18. $3C + 2A$

19. $B - A$

20. $2A - 5C$

21. AF **22.** AC **23.** DE **24.** ED **25.** BD

26. EA **27.** F^{-1} **28.** B^{-1} **29.** $(A + C)^{-1}$

Find the inverse of each of the following matrices that has an inverse.

30. $\begin{bmatrix} 2 & 1 \\ 5 & 3 \end{bmatrix}$
31. $\begin{bmatrix} -4 & 2 \\ 0 & 3 \end{bmatrix}$
32. $\begin{bmatrix} 2 & 0 \\ -1 & 5 \end{bmatrix}$
33. $\begin{bmatrix} 6 & 4 \\ 3 & 2 \end{bmatrix}$

34. $\begin{bmatrix} 2 & -1 & 0 \\ 1 & 0 & 1 \\ 1 & -2 & 0 \end{bmatrix}$
35. $\begin{bmatrix} 2 & 0 & 4 \\ 1 & -1 & 0 \\ 0 & 1 & -2 \end{bmatrix}$
36. $\begin{bmatrix} 1 & 3 & 6 \\ 4 & 0 & 9 \\ 5 & 15 & 30 \end{bmatrix}$
37. $\begin{bmatrix} 2 & 3 & 5 \\ -2 & -3 & -5 \\ 1 & 4 & 2 \end{bmatrix}$

Solve the matrix equation $AX = B$ for X using the given matrices.

38. $A = \begin{bmatrix} 2 & 4 \\ -1 & -3 \end{bmatrix}, \quad B = \begin{bmatrix} 8 \\ 3 \end{bmatrix}$
39. $A = \begin{bmatrix} 1 & 2 \\ 2 & 4 \end{bmatrix}, \quad B = \begin{bmatrix} 5 \\ 10 \end{bmatrix}$

40. $A = \begin{bmatrix} 1 & 0 & 2 \\ -1 & 1 & 0 \\ 3 & 0 & 4 \end{bmatrix}, \quad B = \begin{bmatrix} 8 \\ 4 \\ -6 \end{bmatrix}$
41. $A = \begin{bmatrix} 2 & 4 & 0 \\ 1 & -2 & 0 \\ 0 & 0 & 3 \end{bmatrix}, \quad B = \begin{bmatrix} 72 \\ -24 \\ 48 \end{bmatrix}$

Solve each of the following systems of equations by inverses.

42. $2x + y = 5$
    $3x - 2y = 4$

43. $5x + 10y = 80$
    $3x - 2y = 120$

44. $x + y + z = 1$
    $2x + y = -2$
    $3y + z = 2$

45. $x + 4y - z = 6$
    $2x - y + z = 3$
    $3x + 2y + 3z = 16$

For Exercises 46–49, find the eigenvalues and their corresponding eigenvectors.

46. $\begin{bmatrix} -3 & 12 \\ -2 & 7 \end{bmatrix}$
47. $\begin{bmatrix} 5 & 3 \\ 3 & 5 \end{bmatrix}$
48. $\begin{bmatrix} 1 & 0 & 0 \\ 2 & 2 & 0 \\ 2 & 1 & -3 \end{bmatrix}$
49. $\begin{bmatrix} -2 & 0 & 0 \\ 1 & 3 & 0 \\ -1 & 1 & 2 \end{bmatrix}$

50. The following system of equations is given.

$$x + 2y + z = 7$$
$$2x - y - z = 2$$
$$3x - 3y + 2z = -5$$

    a. Solve by the Gauss-Jordan method.

    b. Write the system as a matrix equation, $AX = B$.

    c. Find the inverse of matrix A from part b.

    d. Solve the system using A^{-1} from part c.

Applications

LIFE SCIENCES

51. *Animal Activity* The activities of a grazing animal can be classified roughly into three categories: grazing, moving, and resting. Suppose horses spend 8 hr grazing, 8 moving, and 8 resting; cattle spend 10 grazing, 5 moving, and 9 resting; sheep spend 7 grazing, 10 moving, and 7 resting; and goats spend 8 grazing, 9 moving, and 7 resting. Write this information as a 4×3 matrix.

52. *CAT Scans* Computer Aided Tomograph (CAT) scanners take X-rays of a part of the body from different directions, and put the information together to create a picture of a cross section of the body.* The amount by which the energy

---

*Exercises 52 and 53 are based on the article "Medical Applications of Linear Equations" by Jabon, David, Gail Nord, Bryce W. Wilson, and Penny Coffman, *The Mathematics Teacher,* Vol. 89, No. 5, May 1996, p. 398.

of the X-ray decreases, measured in linear-attenuation units, tells whether the X-ray has passed through healthy tissue, tumorous tissue, or bone (because of the different densities), based on the following table.

Type of Tissue	Linear-Attenuation Values
Healthy tissue	0.1625–0.2977
Tumorous tissue	0.2679–0.3930
Bone	0.3857–0.5108

The part of the body to be scanned is divided into cells. If an X-ray passes through more than one cell, the total linear-attenuation value is the sum of the values for the cells. For example, in the figure, let a, b, and c be the values for cells A, B, and C. The attenuation value for beam 1 is $a + b$ and for beam 2 is $a + c$.

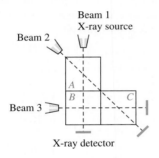

Beam 1
X-ray source

Beam 2

A
B   C

Beam 3

X-ray detector

a. Find the attenuation value for beam 3.

b. Suppose that the attenuation values are 0.8, 0.55, and 0.65 for beams 1, 2, and 3, respectively. Set up and solve the system of three equations for a, b, and c. What can you conclude about cells A, B, and C?

c. Find the inverse of the coefficient matrix from part b to find a, b, and c for the following three cases, and make conclusions about cells A, B, and C for each.

Patient	Linear-Attenuation Values		
	Beam 1	Beam 2	Beam 3
X	0.54	0.40	0.52
Y	0.65	0.80	0.75
Z	0.51	0.49	0.44

53. *CAT Scans* (Refer to Exercise 52.)* Four X-ray beams are aimed at four cells, as shown in the following figure.

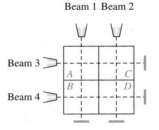

Beam 1 Beam 2

Beam 3

A   C
B   D

Beam 4

a. Suppose the attenuation values for beams 1, 2, 3, and 4 are 0.60, 0.75, 0.65, and 0.70, respectively. Do we have enough information to determine the values of a, b, c, and d? Explain.

b. Suppose we have the data from part a, as well as the following values for d. Find the values for a, b, and c, and make conclusions about cells A, B, C, and D in each case.

   (i) 0.33   **(ii)** 0.43

c. Two X-ray beams are added, as shown in the figure. In addition to the data in part a, we now have attenuation values for beams 5 and 6 of 0.85 and 0.50. Find the values for a, b, c, and d, and make conclusions about cells A, B, C, and D.

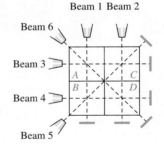

Beam 1 Beam 2

Beam 6

Beam 3

A   C
B   D

Beam 4

Beam 5

d. Six X-ray beams are not necessary because four appropriately chosen beams are sufficient. Give two examples of four beams (chosen from beams 1–6 in part c) that will give the solution. (*Note:* There are 12 possible solutions.)

e. Discuss what properties the four beams selected in part d must have in order to provide a unique solution.

54. *Hockey* In a recent study, the number of head and neck injuries among hockey players wearing full face shields and half face shields were compared. The following table provides the rates per 1,000 athlete-exposures for specific injuries that caused a player wearing either shield to miss one or more events.*

	Half Shield	Full Shield
Head and Face Injuries (Excluding Concussions)	3.54	1.41
Concussions	1.53	1.57
Neck Injuries	0.34	0.29
Other	7.53	6.21

If an equal number of players in a large league wear each type of shield and the total number of athlete-exposures for the league in a season is 8,000, use matrix operations to estimate the total number of injuries of each type.

Leslie Matrices For each of the following Leslie matrices, **a.** *find a population of size 10,000 for which the proportion of the population in each age group stays the same from one year to the next and* **b.** *tell by what factor the population grows or declines each year.*

55. $\begin{bmatrix} 0.3 & 0.2 \\ 0.3 & 0.8 \end{bmatrix}$ **56.** $\begin{bmatrix} 0.2 & 0.3 \\ 0.4 & 0.1 \end{bmatrix}$

57. *Buffalo* A mathematical model for a herd of buffalo is given by the Leslie matrix

$$\begin{bmatrix} 0.95 & 0 & 0.75 & 0 & 0 & 0 \\ 0 & 0.95 & 0 & 0.75 & 0 & 0 \\ 0 & 0 & 0 & 0 & 0.6 & 0 \\ 0 & 0 & 0 & 0 & 0 & 0.6 \\ 0 & 0.48 & 0 & 0 & 0 & 0 \\ 0 & 0.42 & 0 & 0 & 0 & 0 \end{bmatrix},$$

where the first row corresponds to the number of adult males, the second to the number of adult females, the third to the number of male yearlings, the fourth to the number of female yearlings, the fifth to the number of male calves, and the sixth to the number of female calves.† The matrix is used to give next year's herd as a function of this year's.

a. Explain what each row of the matrix means.

b. Verify that $\lambda = 0.95$ is an eigenvalue corresponding to the eigenvector $\begin{bmatrix} k \\ 0 \\ 0 \\ 0 \\ 0 \\ 0 \end{bmatrix}$ for any constant k. Explain what this tells us about the herd.

c. The manager of a buffalo herd wants to harvest some of the herd each year. Suppose that Q is the vector describing a harvest that leaves the population for next year the same as this year's. Show that the population X must satisfy the equation

$$X = (M - I)^{-1}Q.$$

d. Suppose the manager in part c wants a harvest given by

$$Q = \begin{bmatrix} 100 \\ 100 \\ 10 \\ 10 \\ 0 \\ 0 \end{bmatrix}.$$

Find the matrix X describing the herd.

e. Suppose the manager wants the herd to increase by 10% each year while still yielding a harvest Q. Show that the population X must satisfy the equation

$$X = (M - 1.1I)^{-1}Q.$$

f. Suppose the manager in part e wants a harvest given by

$$Q = \begin{bmatrix} 100 \\ 100 \\ 10 \\ 10 \\ 0 \\ 0 \end{bmatrix}.$$

Find the matrix X describing the herd.

58. *Carbon Dioxide* Determining the amount of carbon dioxide in the atmosphere is important because carbon dioxide

*Benson, B., N. Mohtadi, S. Rose, and W. Meeuwisse, "Head and Neck Injuries Among Ice Hockey Players Wearing Full Face Shields vs. Half Face Shields," *JAMA*, Vol. 282, No. 24, Dec. 22/29 1999, pp. 2328–2332.
†Tuchinsky, Philip M., "Management of a Buffalo Herd," *UMAP Module 207*, 1977.

is known to be a greenhouse gas. Carbon dioxide concentrations (in parts per million) have been measured at Mauna Loa, Hawaii, over the past 30 years. The concentrations have increased quadratically.* The table lists readings for three years.

Year	CO_2
1959	316
1979	337
1999	368

a. If the relationship between the carbon dioxide concentration C and the year t is expressed as $C = at^2 + bt + c$, where $t = 0$ corresponds to 1959, use a linear system of equations to determine the constants a, b, and c.

b. Predict the year when the amount of carbon dioxide in the atmosphere will double from its 1959 level. (*Hint:* This requires solving a quadratic equation. For review on how to do this, see Section R.4.)

OTHER APPLICATIONS

59. *Chemistry* When carbon monoxide (CO) reacts with oxygen (O_2), carbon dioxide (CO_2) is formed. This can be written as $CO + (1/2)O_2 = CO_2$ and as a matrix equation.[†] If we form a 2×1 column matrix by letting the first element be the number of carbon atoms and the second element be the number of oxygen atoms, then CO would have the column matrix

$$\begin{bmatrix} 1 \\ 1 \end{bmatrix}.$$

Similarly, O_2 and CO_2 would have the column matrices $\begin{bmatrix} 0 \\ 2 \end{bmatrix}$ and $\begin{bmatrix} 1 \\ 2 \end{bmatrix}$, respectively.

a. Use the Gauss-Jordan method to find numbers x and y (known as *stoichiometric numbers*) that solve the system of equations

$$\begin{bmatrix} 1 \\ 1 \end{bmatrix} x + \begin{bmatrix} 0 \\ 2 \end{bmatrix} y = \begin{bmatrix} 1 \\ 2 \end{bmatrix}.$$

Compare your answers to the equations written above.

b. Repeat the process for $xCO_2 + yH_2 + zCO = H_2O$, where H_2 is hydrogen, and H_2O is water. In words, what does this mean?

60. *Roof Trusses* Linear systems occur in the design of roof trusses for new homes and buildings. The simplest type of roof truss is a triangle. The truss shown in the figure below is used to frame roofs of small buildings. If a 100-lb force is applied at the peak of the truss, then the forces or weights W_1 and W_2 exerted parallel to each rafter of the truss are determined by the following linear system of equations.

$$\frac{\sqrt{3}}{2}(W_1 + W_2) = 100$$

$$W_1 - W_2 = 0$$

Solve the system to find W_1 and W_2.[‡]

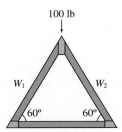

61. *Roof Trusses* (Refer to Exercise 60.) Use the following system of equations to determine the force or weights W_1 and W_2 exerted on each rafter for the truss shown in the figure.

$$\frac{1}{2}W_1 + \frac{\sqrt{2}}{2}W_2 = 150$$

$$\frac{\sqrt{3}}{2}W_1 - \frac{\sqrt{2}}{2}W_2 = 0$$

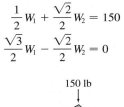

62. *Students* Suppose 20% of the boys and 30% of the girls in a high school like tennis, and 60% of the boys and 90% of the girls like math. If 500 students like tennis, and 1,500 like math, how many boys and girls are in the school? Find all possible solutions.

63. *Scheduling Production* An office supply manufacturer makes two kinds of paper clips, standard and extra large. To make 1,000 standard paper clips requires 1/4 hr on a cutting machine and 1/2 hr on a machine that shapes the clips. One thousand extra large paper clips require 1/3 hr on each

---

*Atmospheric Carbon Dioxide Record from Mauna Loa, University of California, La Jolla, http://cdiac.esd.ornl.gov/trends/co2/slo-mlo.htm.

[†]Alberty, Robert, "Chemical Equations Are Actually Matrix Equations," *Journal of Chemical Education,* Vol. 68, No. 12, Dec. 1991, p. 984.

[‡]Hibbeler, R., *Structural Analysis,* Prentice Hall, 1995.

machine. The manager of paper clip production has 4 hr per day available on the cutting machine and 6 hr per day on the shaping machine. How many of each kind of clip can he make?

64. *Investment* Gretchen Schmidt plans to buy shares of two stocks. One costs $32 per share and pays dividends of $1.20 per share. The other costs $23 per share and pays dividends of $1.40 per share. She has $10,100 to spend and wants to earn dividends of $540. How many shares of each stock should she buy?

65. *Production Requirements* The Waputi Indians make woven blankets, rugs, and skirts. Each blanket requires 24 hr for spinning the yarn, 4 hr for dyeing the yarn, and 15 hr for weaving. Rugs require 30, 5, and 18 hr and skirts 12, 3, and 9 hr, respectively. If there are 306, 59, and 201 hr available for spinning, dyeing, and weaving, respectively, how many of each item can be made? (*Hint:* Simplify the equations you write, if possible, before solving the system.)

66. *Distribution* An oil refinery in Tulsa sells 50% of its production to a Chicago distributor, 20% to a Dallas distributor, and 30% to an Atlanta distributor. Another refinery in New Orleans sells 40% of its production to the Chicago distributor, 40% to the Dallas distributor, and 20% to the Atlanta distributor. A third refinery in Ardmore sells the same distributors 30%, 40%, and 30% of its production.

The three distributors received 219,000, 192,000, and 144,000 gallons of oil, respectively. How many gallons of oil were produced at each of the three plants?

67. *Filling Orders* A printer has three orders for pamphlets that require three kinds of paper, as shown in the following matrix.

$$
\begin{array}{c}
 \\
\text{Paper}
\end{array}
\begin{array}{c}
 \\
\text{High-grade} \\
\text{Medium-grade} \\
\text{Coated}
\end{array}
\begin{array}{c}
\text{Order} \\
\begin{array}{ccc}
\text{I} & \text{II} & \text{III}
\end{array} \\
\begin{bmatrix}
10 & 5 & 8 \\
12 & 0 & 4 \\
0 & 10 & 5
\end{bmatrix}
\end{array}
$$

The printer has on hand 3,170 sheets of high-grade paper, 2,360 sheets of medium-grade paper, and 1,800 sheets of coated paper. All the paper must be used in preparing the order.

a. Write a 3 × 1 matrix for the amounts of paper on hand.

b. Write a matrix of variables to represent the number of pamphlets that must be printed in each of the three orders.

c. Write a matrix equation using the given matrix and your matrices from parts a and b.

d. Solve the equation from part c.

EXTENDED APPLICATION: Contagion

Suppose that three people have contracted a contagious disease.* A second group of five people may have been in contact with the three infected persons. A third group of six people may have been in contact with the second group. We can form a 3×5 matrix P with rows representing the first group of three and columns representing the second group of five. We enter a one in the corresponding position if a person in the first group has contact with a person in the second group. These direct contacts are called *first-order contacts*. Similarly, we form a 5×6 matrix Q representing the first-order contacts between the second and third group. For example, suppose

$$P = \begin{bmatrix} 1 & 0 & 0 & 1 & 0 \\ 0 & 0 & 1 & 1 & 0 \\ 1 & 1 & 0 & 0 & 0 \end{bmatrix} \text{ and}$$

$$Q = \begin{bmatrix} 1 & 1 & 0 & 1 & 1 & 1 \\ 0 & 0 & 0 & 0 & 1 & 0 \\ 0 & 0 & 0 & 0 & 0 & 0 \\ 0 & 1 & 0 & 1 & 0 & 0 \\ 1 & 0 & 0 & 0 & 1 & 0 \end{bmatrix}.$$

From matrix P we see that the first person in the first group had contact with the first and fourth persons in the second group. Also, none of the first group had contact with the last person in the second group.

A *second-order contact* is an indirect contact between persons in the first and third groups through some person in the second group. The product matrix PQ indicates these contacts. Verify that the second-row, fourth-column entry of PQ is 1. That is, there is one second-order contact between the second person in group one and the fourth person in group three. Let a_{ij} denote the element in the i-th row and j-th col-

umn of the matrix PQ. By looking at the products that form a_{24} below, we see that the common contact was the fourth individual in group two. (The p_{ij} are entries in P, and the q_{ij} are entries in Q.)

$$a_{24} = p_{21}q_{14} + p_{22}q_{24} + p_{23}q_{34} + p_{24}q_{44} + p_{25}q_{54}$$
$$= 0 \cdot 1 + 0 \cdot 0 + 1 \cdot 0 + 1 \cdot 1 + 0 \cdot 1$$
$$= 1$$

The second person in group one and the fourth person in group three both had contact with the fourth person in group two.

This idea could be extended to third-, fourth-, and larger-order contacts. It indicates a way to use matrices to trace the spread of a contagious disease. It could also pertain to the dispersal of ideas or anything that might pass from one individual to another.

Exercises

1. Find the second-order contact matrix PQ mentioned in the text.

2. How many second-order contacts were there between the second contagious person and the third person in the third group?

3. Is there anyone in the third group who has had no contacts at all with the first group?

4. The totals of the columns in PQ give the total number of second-order contacts per person, while the column totals in P and Q give the total number of first-order contacts per person. Which person(s) in the third group had the most contacts, counting first- and second-order contacts?

---

*Grossman, Stanley, "First and Second Order Contact to a Contagious Disease," *Finite Mathematics with Applications to Business, Life Sciences, and Social Sciences*, WCB/McGraw-Hill, 1993.

CHAPTER

11

Differential Equations

Since 1963, when the bald eagle was put on the endangered species list, the number of bald eagles has risen: slowly at first, then more rapidly, and more recently at a slower rate. Such changes in the rate of growth are typical of the logistic growth model, a differential equation we will study in the first section of this chapter.

■ **11.1** Solutions of Elementary and Separable Differential Equations

■ **11.2** Linear First-Order Differential Equations

■ **11.3** Euler's Method

■ **11.4** Linear Systems of Differential Equations

■ **11.5** Nonlinear Systems of Differential Equations

■ **11.6** Applications of Differential Equations

Review Exercises

Extended Application: Pollution of the Great Lakes

Suppose a population biologist wants to develop an equation that will forecast the population of an endangered wildcat. By studying data on the population in the past, she hopes to find a relationship between the population and its rate of change. A function giving the rate of change of the population would be the derivative of the function describing the population itself. A **differential equation** is an equation that involves an unknown function $y = f(x)$ and a finite number of its derivatives. Solving the differential equation for y would give the unknown function to be used for forecasting the wildcat population.

Differential equations have been important in the study of physical science and engineering since the eighteenth century. Among the pioneers in the field of differential equations was the French mathematician Alexis Claude Clairaut (1713–1765). A particular type of equation studied in elementary courses on differential equations bears his name.

More recently, differential equations have become useful in social sciences, life sciences, and economics for solving problems about population growth, ecological balance, and interest rates. In this chapter, we will introduce some methods for solving differential equations and give examples of their applications.

■ 11.1 SOLUTIONS OF ELEMENTARY AND SEPARABLE DIFFERENTIAL EQUATIONS

? THINK ABOUT IT   How can we predict the future population of a flock of mountain goats?

Using differential equations, we will learn to answer such questions.

Usually a solution of an algebraic equation is a *number*. A solution of a differential equation, however, is a *function*. For example, the solutions of a differential equation such as

$$\frac{dy}{dx} = 3x^2 - 2x \qquad (1)$$

consist of all expressions for y that satisfy the equation. Since the left side of the equation is the derivative of y with respect to x, we can solve the equation for y by finding an antiderivative on each side. On the left, the antiderivative is $y + C_1$. On the right side,

$$\int (3x^2 - 2x)\,dx = x^3 - x^2 + C_2.$$

The solutions of equation (1) are given by

$$y + C_1 = x^3 - x^2 + C_2$$

or $$y = x^3 - x^2 + C_2 - C_1.$$

Replacing the constant $C_2 - C_1$ with the single constant C gives

$$y = x^3 - x^2 + C. \qquad (2)$$

(From now on we will add just one constant, with the understanding that it represents the difference between the two constants obtained in the two integrations.)

Each different value of C in equation (2) leads to a different solution of equation (1), showing that a differential equation can have an infinite number of solutions. Equation (2) is the **general solution** of the differential equation (1). Some of the solutions of equation (1) are graphed in Figure 1.

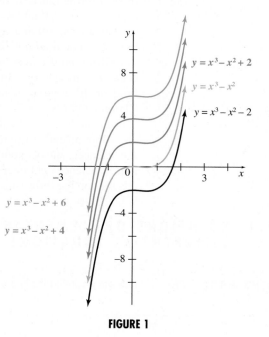

FIGURE 1

The simplest kind of differential equation has the form

$$\frac{dy}{dx} = f(x).$$

Since equation (1) has this form, the solution of equation (1) suggests the following generalization.

GENERAL SOLUTION OF $\dfrac{dy}{dx} = f(x)$

The general solution of the differential equation $dy/dx = f(x)$ is

$$y = \int f(x)\,dx.$$

EXAMPLE 1 Population

The population P of a flock of birds is growing exponentially so that

$$\frac{dP}{dx} = 20e^{0.05x},$$

where x is time in years. Find P in terms of x if there were 20 birds in the flock initially.

Solution Solve the differential equation:

$$P = \int 20e^{0.05x}\,dx = \frac{20}{0.05}e^{0.05x} + C = 400e^{0.05x} + C.$$

Since P is 20 when x is 0,

$$20 = 400e^0 + C$$
$$-380 = C,$$

and

$$P = 400e^{0.05x} - 380.$$

In Example 1, the given information was used to produce a solution with a specific value of C. Such a solution is called a **particular solution** of the given differential equation. The given information, $P = 20$ when $x = 0$, is called an **initial condition.** An **initial value problem** is a differential equation with a value of y given at $x = x_0$, where x_0 is any real number.

Sometimes a differential equation must be rewritten in the form

$$\frac{dy}{dx} = f(x)$$

before it can be solved.

EXAMPLE 2   Initial Value Problem

Find the particular solution of

$$\frac{dy}{dx} - 2x = 5,$$

given that $y = 2$ when $x = -1$.

Solution To get this equation into the proper form, add $2x$ to both sides, yielding

$$\frac{dy}{dx} = 2x + 5.$$

The general solution is

$$y = \frac{2x^2}{2} + 5x + C = x^2 + 5x + C.$$

Substituting 2 for y and -1 for x gives

$$2 = (-1)^2 + 5(-1) + C$$
$$C = 6.$$

The particular solution is $y = x^2 + 5x + 6$.

So far in this section, we have used a method that is essentially the same as that used in the section on antiderivatives, when we first started the topic of integration. But not all differential equations can be solved so easily. For example, a simple model of the growth of a bacteria population assumes that the population grows at a rate proportional to the number of bacteria present. If P is the population at time t, then for some constant k, the differential equation

$$\frac{dP}{dt} = kP$$

gives the rate of growth of P with respect to t. This differential equation is different from those discussed previously, which had the form

$$\frac{dy}{dx} = f(x).$$

CAUTION Since the right-hand side of the differential equation for bacterial growth is a function of P, rather than a function of t, it would be completely invalid to simply integrate both sides as we did before. The previous method only works when the side opposite the derivative is simply a function of the independent variable.

The differential equation for bacterial growth is an example of a more general differential equation we will now learn to solve; namely, those that can be written in the form

$$\frac{dy}{dx} = \frac{f(x)}{g(y)}.$$

Suppose we think of dy/dx as a fraction dy over dx. This is incorrect, of course; the derivative is actually the limit of a small change in y over a small change in x, but the notation is chosen so that this interpretation gives a correct answer, as we shall see. Multiply on both sides by $g(y)\,dx$ to get

$$g(y)\,dy = f(x)\,dx.$$

In this form all terms involving y (including dy) are on one side of the equation and all terms involving x (and dx) are on the other side. A differential equation that can be put into this form is said to be *separable*, since the variables x and y can be separated. After separation, a **separable differential equation** may be solved by integrating each side. This method is known as **separation of variables.**

$$\int g(y)\,dy = \int f(x)\,dx$$

$$G(y) = F(x) + C,$$

where F and G are antiderivatives of f and g. To show that this answer is correct, differentiate implicitly with respect to x.

$$G'(y)\frac{dy}{dx} = F'(x) \qquad \text{The chain rule is used on the left side.}$$

$$g(y)\frac{dy}{dx} = f(x)$$

$$\frac{dy}{dx} = \frac{f(x)}{g(y)}$$

This last equation is the one we set out to solve.

EXAMPLE 3   Separation of Variables
Find the general solution of

$$y\frac{dy}{dx} = x^2.$$

Solution Begin by separating the variables to get

$$y\,dy = x^2\,dx.$$

The general solution is found by taking antiderivatives on each side.

$$\int y\,dy = \int x^2\,dx$$

$$\frac{y^2}{2} = \frac{x^3}{3} + C$$

$$y^2 = \frac{2}{3}x^3 + 2C = \frac{2}{3}x^3 + K$$

The constant K was substituted for $2C$ in the last step. The solution is left in implicit form, not solved explicitly for y.

EXAMPLE 4   Separation of Variables

Find the general solution of $dy/dx = ky$, where k is a constant.

Solution Separating variables leads to

$$\frac{1}{y}\,dy = k\,dx.$$

To solve this equation, take antiderivatives on each side.

$$\int \frac{1}{y}\,dy = \int k\,dx$$

$$\ln|y| = kx + C$$

Use the definition of logarithm to write the equation in exponential form as

$$|y| = e^{kx+C}.$$

By properties of exponents,

$$|y| = e^{kx}e^{C}.$$

Finally, use the definition of absolute value to get

$$y = e^{kx}e^{C} \qquad \text{or} \qquad y = -e^{kx}e^{C}.$$

Since e^C and $-e^C$ are constants, replace them with the constant M, which may have any nonzero real-number value, to get the single equation

$$y = Me^{kx}.$$

This equation, $y = Me^{kx}$, defines the exponential growth or decay function that was discussed previously (in the chapter on Exponential, Logarithmic, and Trigonometric Functions).

CAUTION Notice that $y = 0$ is also a solution to the differential equation in Example 4, but after we divide by y (which is not possible if $y = 0$) and integrate, the resulting equation $|y| = e^{kx+C}$ does not allow y to equal 0. In this example, the lost solution can be recovered in the final answer if we allow M to equal 0, a value that was previously excluded. When dividing by an expression in separation of variables, look for solutions that would make this expression 0 and may be lost.

In the chapter on Exponential, Logarithmic, and Trigonometric Functions, we saw that the amount of money in an account with interest compounded continuously is given by

$$A = Pe^{rt},$$

where P is the initial amount in the account, r is the annual interest rate, and t is the time in years. Observe that this is the same as the equation for the amount of money in an account derived here, where A, P, and r have been replaced with P, P_0, and k, respectively.

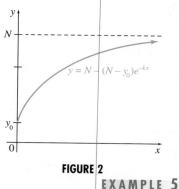

FIGURE 2

Recall that equations of the form $y = Me^{kx}$ arise in situations where the rate of change of a quantity is proportional to the amount present at time x; that is, where

$$\frac{dy}{dx} = ky.$$

The constant k is called the **growth rate constant,** while M represents the initial condition, the amount present at time $x = 0$. (A positive value of k indicates growth, while a negative value of k indicates decay.)

Applying the results of Example 4 to the equation discussed earlier,

$$\frac{dP}{dt} = kP,$$

shows that the bacteria population at time t is

$$P = P_0 e^{kt},$$

where P_0 is the initial population.

As a model of population growth, the equation $y = Me^{kx}$ is not realistic over the long run for most populations. As shown by graphs of functions of the form $y = Me^{kx}$, with both M and k positive, growth would be unbounded. Additional factors, such as space restrictions or a limited amount of food, tend to inhibit growth of populations as time goes on. In an alternative model that assumes a maximum population of size N, the rate of growth of a population is proportional to how close the population is to that maximum, that is, to the difference between N and x. These assumptions lead to the differential equation

$$\frac{dy}{dx} = k(N - y),$$

which has as its solution the limited growth function mentioned in the chapter on Exponential, Logarithmic, and Trigonometric Functions. Graphs of limited growth functions look like the graph in Figure 2, where y_0 is the initial population.

EXAMPLE 5 Population

A certain nature reserve can support no more than 4,000 mountain goats. Assume that the rate of growth is proportional to how close the population is to this maximum, with a growth rate of 20 percent. There are currently 1,000 goats in the area.

(a) Write a differential equation for the rate of growth of this population and solve it to obtain a function y describing the population at time x.

Solution Let $N = 4,000$ and $k = 0.20$. The rate of growth of the population is given by

$$\frac{dy}{dx} = 0.20(4,000 - y).$$

To solve for y, first separate the variables.

$$\frac{dy}{4,000 - y} = 0.2\, dx$$

$$\int \frac{dy}{4,000 - y} = \int 0.2\, dx$$

$$-\ln(4,000 - y) = 0.2x + C$$

$$\ln(4,000 - y) = -0.2x - C$$

$$4,000 - y = e^{-0.2x - C} = (e^{-0.2x})(e^{-C})$$

The absolute value bars are not needed for $\ln(4{,}000 - y)$ since y must be less than 4,000 for this population, so that $4{,}000 - y$ is always nonnegative. Let $e^{-C} = B$. Then

$$4{,}000 - y = Be^{-0.2x}$$
$$y = 4{,}000 - Be^{-0.2x}.$$

Find B by using the fact that $y = 1{,}000$ when $x = 0$.

$$1{,}000 = 4{,}000 - B$$
$$B = 3{,}000$$

Notice that the value of B is the difference between the maximum population and the initial population. Substituting 3,000 for B in the equation for y gives

$$y = 4{,}000 - 3{,}000e^{-0.2x}.$$

(b) What will the goat population be in 5 years?

Solution In 5 years, the population will be

$$y = 4{,}000 - 3{,}000e^{-(0.2)(5)} = 4{,}000 - 3{,}000e^{-1}$$
$$= 4{,}000 - 1{,}103.6 = 2{,}896.4,$$

or about 2,900 goats.

Logistic Growth Let y be the size of a certain population at time x. In the standard model for unlimited growth,

$$\frac{dy}{dx} = ky, \tag{3}$$

the rate of growth is proportional to the current population size. The constant k, the growth rate constant, is the difference between the birth and death rates of the population. The unlimited growth model predicts that the population's growth rate is a constant, k.

Growth usually is not unlimited, however, and the population's growth rate is usually not constant because the population is limited by environmental factors to a maximum size N, called the **carrying capacity** of the environment for the species. In the limited growth model already given,

$$\frac{dy}{dx} = k(N - y),$$

the rate of growth is proportional to the remaining room for growth, $N - y$. In the **logistic growth model**

$$\frac{dy}{dx} = k\left(1 - \frac{y}{N}\right)y \tag{4}$$

the rate of growth is proportional to both the current population size y and a factor $(1 - y/N)$ that is equal to the remaining room for growth, $N - y$, divided by N. Equation (4) is called the **logistic equation.** As $y \to 0$, $(1 - y/N) \to 1$, and the differential equation can be approximated as

$$\frac{dy}{dx} = k\left(1 - \frac{y}{N}\right)y \approx k(1)y = ky.$$

In other words, when y is small, the growth of the population behaves as if it were unlimited. On the other hand, as $y \to N$, $(1 - y/N) \to 0$, so

$$\frac{dy}{dx} = k\left(1 - \frac{y}{N}\right)y \approx k(0)y = 0.$$

That is, population growth levels off as y nears the maximum population N. Thus, the logistic equation (4) is the unlimited growth equation (3) with a damping factor $(1 - y/N)$ to account for limiting environmental factors when y nears N. Let y_0 denote the initial population size. Under the assumption $0 < y < N$, the general solution of equation (4) is

$$y = \frac{N}{1 + be^{-kx}}, \tag{5}$$

where $b = (N - y_0)/y_0$ (see Exercise 33). This solution, called a **logistic curve,** is shown in Figure 3.

As expected, the logistic curve begins exponentially and subsequently levels off. Another important feature is the point of inflection $((\ln b)/k, N/2)$, where dy/dx is a maximum (see Exercise 35).

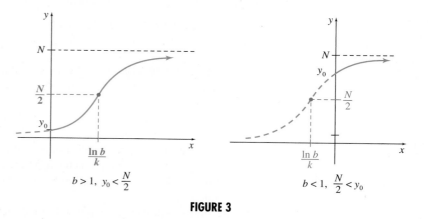

$$b > 1, \quad y_0 < \frac{N}{2}$$

$$b < 1, \quad \frac{N}{2} < y_0$$

FIGURE 3

Logistic equations arise frequently in the study of populations. In about 1840 the Belgian sociologist P. F. Verhulst fitted a logistic curve to U.S. census figures and made predictions about the population that were subsequently proven to be quite accurate. American biologist Raymond Pearl (circa 1920) found that the growth of a population of fruit flies in a limited space could be modeled by the logistic equation

$$\frac{dy}{dx} = 0.2y - \frac{0.2}{1,035}y^2.$$

Some calculators can fit a logistic curve to a set of data points. For example, the TI-83 has this capability, listed as LOGISTIC in the STAT CALC function menu, along with other types of regression. See Exercises 45 and 48.

Logistic growth is an example of how a model is modified over time as new insights occur. The model for population growth changed from the early exponential curve $y = Me^{kx}$ to the logistic curve

$$y = \frac{N}{1 + be^{-kx}}.$$

Many other quantities besides population grow logistically. That is, their initial rate of growth is slow, but as time progresses, their rate of growth increases to a maximum value and subsequently begins to decline and to approach zero.

EXAMPLE 6   Logistic Curve

Rapid technological advancements in the last 20 years have made many products obsolete practically overnight. J. C. Fisher and R. H. Pry* successfully described the phenomenon of a technically superior new product replacing another product by the logistic equation

$$\frac{dz}{dx} = k(1 - z)z, \tag{6}$$

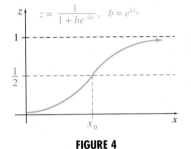

FIGURE 4

where z is the market share of the new product and $1 - z$ is the market share of the other product. The new product will initially have little or no market share; that is, $z_0 \approx 0$. Thus, the constant b in equation (5) will have to be determined in a different way. Let x_0 be the time at which $z = 1/2$. Under the assumption $0 < z < 1$, the general solution of equation (6) is

$$z = \frac{1}{1 + be^{-kx}},$$

where $b = e^{kx_0}$ (see Exercise 34). This solution is shown in Figure 4.

The market share of the new product will be growing most rapidly when the new product has captured exactly half the market, and the market share of the older product will be shrinking most rapidly at the same time. Notice that the logistic equation (4) can be transformed into the simpler logistic equation (6) by the change of variable $z = y/N$.

11.1 EXERCISES

Find general solutions for the following differential equations.

1. $\dfrac{dy}{dx} = -2x + 3x^2$

2. $\dfrac{dy}{dx} = 3e^{-2x}$

3. $3x^3 - 2\dfrac{dy}{dx} = 0$

4. $3x^2 - 3\dfrac{dy}{dx} = 2$

5. $y\dfrac{dy}{dx} = x$

6. $y\dfrac{dy}{dx} = x^2 - 1$

7. $\dfrac{dy}{dx} = 2xy$

8. $\dfrac{dy}{dx} = x^2y$

9. $\dfrac{dy}{dx} = 3x^2y - 2xy$

10. $(y^2 - y)\dfrac{dy}{dx} = x$

11. $\dfrac{dy}{dx} = \dfrac{y}{x}, x > 0$

12. $\dfrac{dy}{dx} = \dfrac{y}{x^2}$

13. $\dfrac{dy}{dx} = y - 5$

14. $\dfrac{dy}{dx} = 3 - y$

15. $\dfrac{dy}{dx} = y^2e^x$

16. $\dfrac{dy}{dx} = \dfrac{e^x}{e^y}$

17. $\dfrac{dy}{dx} = \dfrac{\cos x}{\sin y}$

18. $\dfrac{dy}{dx} = \tan x \cos^2 y$

Find particular solutions for the following initial value problems.

19. $\dfrac{dy}{dx} + 2x = 3x^2;\quad y = 2$ when $x = 0$

20. $x\dfrac{dy}{dx} = x^2e^{3x};\quad y = \dfrac{8}{9}$ when $x = 0$

---

*Fisher, J. C., and R. H. Pry, "A Simple Substitution Model of Technological Change," *Technological Forecasting and Social Change*, Vol. 3, 1971–1972. Copyright © 1972 by Elsevier Science Publishing Co., Inc. Reprinted by permission of the publisher.

21. $2\dfrac{dy}{dx} = 4xe^{-x}$; $\quad y = 42$ when $x = 0$

22. $x\dfrac{dy}{dx} - y\sqrt{x} = 0$; $\quad y = 1$ when $x = 0$

23. $\dfrac{dy}{dx} = \dfrac{x^2}{y}$; $\quad y = 3$ when $x = 0$

24. $\dfrac{dy}{dx} = \dfrac{x^2 + 5}{2y - 1}$; $\quad y = 11$ when $x = 0$

25. $(2x + 3)y = \dfrac{dy}{dx}$; $\quad y = 1$ when $x = 0$

26. $\dfrac{dy}{dx} = \dfrac{2x + 1}{y - 3}$; $\quad y = 4$ when $x = 0$

Find particular solutions for the following equations.

27. $\dfrac{dy}{dx} = 4x^3 - 3x^2 + x$; $\quad y = 0$ when $x = 1$

28. $x^2\dfrac{dy}{dx} = y$; $\quad y = -1$ when $x = 1$

29. $\dfrac{dy}{dx} = \dfrac{y^2}{x}$; $\quad y = 5$ when $x = e$

30. $\dfrac{dy}{dx} = x^{1/2}y^2$; $\quad y = 12$ when $x = 4$

31. $\dfrac{dy}{dx} = (y - 1)^2 e^{x-1}$; $\quad y = 2$ when $x = 1$

32. $\dfrac{dy}{dx} = (x + 2)^2 e^y$; $\quad y = 0$ when $x = 1$

33. a. Solve the logistic equation (4) in this section by observing that

$$\frac{1}{y} + \frac{1}{N - y} = \frac{N}{(N - y)y}.$$

b. Assume $0 < y < N$. Verify that $b = (N - y_0)/y_0$ in equation (5), where y_0 is the initial population size.

c. Assume $0 < N < y$ for all y. Verify that $b = (y_0 - N)/y_0$.

34. Suppose that $0 < z < 1$ for all z. Solve the logistic equation (6) as in Exercise 33. Verify that $b = e^{kx_0}$, where x_0 is the time at which $z = 1/2$.

35. Suppose that $0 < y_0 < N$. Let $b = (N - y_0)/y_0$, and let $y(x) = N/(1 + be^{-kx})$ for all x. Show the following.

a. $0 < y(x) < N$ for all x.

b. The lines $y = 0$ and $y = N$ are horizontal asymptotes of the graph.

c. $y(x)$ is an increasing function.

d. $((\ln b)/k, N/2)$ is a point of inflection of the graph.

e. dy/dx is a maximum at $x_0 = (\ln b)/k$.

36. Suppose that $0 < N < y_0$. Let $b = (y_0 - N)/y_0$ and let

$$y(x) = \frac{N}{1 - be^{-kx}} \text{ for all } x \neq \frac{\ln b}{k}.$$

See the figure. Show the following.

a. $0 < b < 1$

b. The lines $y = 0$ and $y = N$ are horizontal asymptotes of the graph.

c. The line $x = (\ln b)/k$ is a vertical asymptote of the graph.

d. $y(x)$ is decreasing on $((\ln b)/k, \infty)$ and on $(-\infty, (\ln b)/k)$.

e. $y(x)$ is concave upward on $((\ln b)/k, \infty)$ and concave downward on $(-\infty, (\ln b)/k)$.

$y(x) = \dfrac{N}{1 - be^{-kx}}$, $\quad b = \dfrac{y_0 - N}{y_0}$

Applications

LIFE SCIENCES

37. *Bald Eagles* In 1963, the bald eagle was put on the endangered species list. Since then, its numbers have increased, as shown by the following table that gives the estimated number of bald eagle pairs in the world for various years.*

Year	Estimated Number of Bald Eagle Pairs
1963	400
1974	800
1981	1,200
1984	1,800
1986	1,900
1988	2,500
1989	2,700
1990	3,000
1991	3,400
1992	3,700
1993	4,000
1994	4,500
1996	5,000
1999	5,800

Use a calculator with logistic regression capability to complete the following.

a. Letting x represent the years since 1900, plot the number of bald eagle pairs on the y-axis against the year on the x-axis. Discuss the appropriateness of fitting a logistic function to these data.

b. Use the logistic regression function on your calculator to determine the logistic equation that best fits the data.

c. Plot the logistic equation from part b on the same graph as the data points. Discuss how well the logistic equation fits the data.

d. Assuming that the logistic equation found in part b continues to be accurate, what seems to be the limiting size of the number of bald eagle pairs in the world?

38. *Guernsey Growth* The growth of Guernsey cows can be approximated by the equation

$$\frac{dW}{dt} = 0.0152(486 - W),$$

where W is the weight in kg after t weeks.[†]

a. Find the solution to the initial value problem with $W(0) = 32$ kg.

b. What limit does the weight of a Guernsey cow approach as t goes to infinity?

39. *Flea Beetles* A study of flea beetles found that the change in the rate of flea beetles moving in and out of a patch of beetles could be described by the differential equation

$$\frac{dN}{dt} = mN + i,$$

where N is the number of beetles in a patch, m is the rate at which beetles move out of the patch, and i is the rate at which they move in.[‡]

a. Solve the differential equation above with the initial condition $N(0) = N_0$.

b. After the researchers cleared a patch of beetles, so that $N_0 = 0$, they would return 8 hours later and count the number of beetles. Show that the parameter i can then be estimated by the equation

$$i = \frac{mN(8)}{e^{8m} - 1}.$$

c. The researchers estimated m using the equation $m = \ln F_{sd}$, where F_{sd} is the fraction of beetles who remained in the patch in which they were released. For the beetles *P. striolata* released in July in the lush interior of patches 5 meters apart, the average values of F_{sd} and $N(8)$ were 0.709 and 4.5, respectively. Find the values of m and i.

40. *Tracer Dye* The amount of a tracer dye injected into the bloodstream decreases exponentially, with a decay constant of 3% per minute. If 6 cc are present initially, how many cubic centimeters are present after 10 min? (Here k will be negative.)

41. *Soil Moisture* The evapotranspiration index I is a measure of soil moisture. An article on 10- to 14-year-old heath

*www.nctm.org/wlme/wlme6/twelve.htm
[†]France, J., J. Kijkstra, and M. S. Dhanoa, "Growth Functions and Their Application in Animal Science," *Annales de Zootechnie*, Vol. 45 (Supplement), 1996, pp. 165–174.
[‡]Kareiva, Peter, "Experimental and Mathematical Analyses of Herbivore Movement: Quantifying the Influence of Plant Spacing and Quality on Foraging Discrimination," *Ecological Monographs*, Vol. 52, No. 3, Sept. 1982, pp. 261–282.

vegetation described the rate of change of I with respect to W, the amount of water available, by the equation

$$\frac{dI}{dW} = 0.088(2.4 - I).*$$

a. According to the article, I has a value of 1 when $W = 0$. Solve the initial value problem.

b. What happens to I as W becomes larger and larger?

42. *Fish Population* An isolated fish population is limited to 5,000 by the amount of food available. If there are now 150 fish and the population is growing with a growth constant of 1% a year, find the expected population at the end of 5 yr.

Dieting *A person's weight depends both on the daily rate of energy intake, say C calories per day, and on the daily rate of energy consumption, typically between 15 and 20 calories per pound per day. Using an average value of 17.5 calories per pound per day, a person weighing w pounds uses 17.5w calories per day. If C = 17.5w, then weight remains constant, and weight gain or loss occurs according to whether C is greater or less than 17.5w.*[†]

43. To determine how fast a change in weight will occur, a plausible assumption is that dw/dt is proportional to the net excess (or deficit) $C - 17.5w$ in the number of calories per day.

a. Assume C is constant and write a differential equation to express this relationship. Use k to represent the constant of proportionality. What does C being constant imply?

b. The units of dw/dt are pounds/day, and the units of $C - 17.5w$ are calories/day. What units must k have?

c. Use the fact that 3,500 calories is equivalent to one pound to rewrite the differential equation in part a.

d. Solve the differential equation.

e. Let w_0 represent the initial weight and use it to express the coefficient of $e^{-0.005t}$ in terms of w_0 and C.

44. (Refer to Exercise 43.) Suppose someone initially weighing 180 pounds adopts a diet of 2,500 calories per day.

a. Write the weight function for this individual.

b. Graph the weight function on the window $[0, 300]$ by $[120, 200]$. What is the asymptote? This value of w is the equilibrium weight w_{eq}. According to the model, can a person ever achieve this weight?

c. How long will it take a dieter to reach a weight just 2 pounds more than w_{eq}?

45. *Toronto's Jewish Population* The table below gives the population of Toronto's Jewish community at various times.[‡]

Year	Population
1901	3,103
1911	18,294
1921	34,770
1931	46,751
1941	52,798
1951	66,773
1961	85,000
1971	97,000
1981	128,650
1991	162,605

Use a calculator with logistic regression capability to complete the following.

a. Letting x represent the years since 1900, plot the population on the y-axis against the year on the x-axis. Discuss the appropriateness of fitting a logistic function to these data.

b. Use the logistic regression function on your calculator to determine the logistic equation that best fits the data.

c. Plot the logistic equation from part b on the same graph as the data points. Discuss how well the logistic equation fits the data.

d. Assuming that the logistic equation found in part b continues to be accurate, what seems to be the limiting size of the Jewish population in Toronto?

e. See Exercise 78 in the section on Logarithmic Functions (in the chapter on Exponential, Logarithmic, and Trigonometric Functions), where we fit a linear and an exponential function to these data. Compare the results of that exercise with the results of this exercise, and discuss which type of function best fits the data.

46. *U.S. Latino Population* A recent report by the U.S. Census Bureau predicts that the Latino American population will

---

*Specht, R. L., "Dark Island Heath (Ninety-Mile Plain, South Australia) V: The Water Relationships in Heath Vegetation and Pastures on the Makin Sand," *Australian Journal of Botany*, Vol. 5, No. 2, Sept. 1957, pp. 151–172.

[†]Segal, Arthur C., "A Linear Diet Model," *College Mathematics Journal*, Vol. 18, No. 1, Jan. 1987.

[‡]*The Globe and Mail*, Feb. 17, 1995. The data were quoted and fitted to a logistic curve by Ron Lancaster and Charlie Marion, *The Mathematics Teacher*, Vol. 90, No. 2, Feb. 1997.

increase from 31.4 million in 2000 to 96.5 million in 2050.*
Assuming the unlimited growth model $dy/dt = ky$ fits this
population growth, express the population y as a function
of the year t. Let 2000 correspond to $t = 0$.

47. *U.S. African American Population* (Refer to Exercise 46.)
The report also predicted that the African American population of the U.S. would increase from 35.5 million in 2000 to
60.6 million in 2050.* Repeat Exercise 46 using this data.

48. *World Population* The following table gives the population of the world at various times over the last two centuries, plus projections for the next century.†

Year	Population (billions)
1804	1
1927	2
1960	3
1974	4
1987	5
1999	6
2011	7
2025	8
2041	9
2071	10

Use a calculator with logistic regression capability to complete the following.

a. Use the logistic regression function on your calculator
to determine the logistic equation that best fits the data.

b. Plot the logistic function found in part a and the original data in the same window. Does the logistic function
seem to fit the data from 1927 on? Before 1927?

c. To get a better fit, subtract 0.99 from each value of the
population in the table. (This makes the population in
1804 small, but not 0 or negative.) Find a logistic function that fits the new data.

d. Plot the logistic function found in part c and the modified data in the same window. Does the logistic function
now seem to be a better fit than in part b?

e. Based on the results from parts c and d, predict the limiting value of the world's population as time increases.
For comparison, the *New York Times* article predicts a
value of 10.73 billion. (*Hint:* After taking the limit, remember to add the 0.99 that was removed earlier.)

f. Based on the results from parts c and d, predict the limiting value of the world population as you go further and
further back in time. Does that seem reasonable? Explain.

g. Look up population projections for your own state at
www.census.gov and determine the logistic equation that
best fits the data. You may have to subtract some constant
from each value of the population, as we did in part c.

OTHER APPLICATIONS

49. *Spread of a Rumor* Suppose the rate at which a rumor
spreads—that is, the number of people who have heard the
rumor over a period of time—increases with the number of
people who have heard it. If y is the number of people who
have heard the rumor, then

$$\frac{dy}{dt} = ky,$$

where t is the time in days.

a. If y is 1 when $t = 0$, and y is 5 when $t = 2$, find k.

Using the value of k from part a, find y for each of the following times.

b. $t = 3$ **c.** $t = 5$ **d.** $t = 10$

50. *Worker Productivity* A company has found that the rate at
which a person new to the assembly line produces items is

$$\frac{dy}{dx} = 7.5e^{-0.3y},$$

where x is the number of days the person has worked on the
line. How many items can a new worker be expected to
produce on the eighth day if he produces none when $x = 0$?

51. *Radioactive Decay* The amount of a radioactive substance decreases exponentially, with a decay constant of
5% per month.

a. Write a differential equation to express the rate of
change.

b. Find a general solution to the differential equation from
part a.

c. If there are 90 g at the start of the decay process, find a
particular solution for the differential equation from
part a.

d. Find the amount left after 10 mo.

52. *Snowplow* One morning snow began to fall at a heavy
and constant rate. A snowplow started out at 8:00 A.M.
At 9:00 A.M. it had traveled 2 mi. By 10:00 A.M. it had

Statistical Abstracts of the United States, 1999, U.S. Census Bureau, U.S. Dept. of Commerce,
Table 12, p. 14.
†*The New York Times,* Nov. 17, 1996, p. 3.

traveled 3 mi. Assuming that the snowplow removes a constant volume of snow per hour, determine the time at which it started snowing. (*Hint:* Let t denote the time since the snow started to fall, and let T be the time when the snowplow started out. Let x, the distance the snowplow has traveled, and h, the height of the snow, be functions of t. The assumption that a constant volume of snow per hour is removed implies that the speed of the snowplow times the height of the snow is a constant. Set up and solve differential equations involving dx/dt and dh/dt.)*

■ 11.2 LINEAR FIRST-ORDER DIFFERENTIAL EQUATIONS

? THINK ABOUT IT   What happens over time to the glucose level in a patient's bloodstream?

The solution to a linear differential equation gives us a function as an answer.

Recall that $f^{(n)}(x)$ represents the nth derivative of $f(x)$, and that $f^{(n)}(x)$ is called an *nth-order* derivative. By this definition, the derivative $f'(x)$ is first-order, $f''(x)$ is second-order, and so on. The *order of a differential equation* is that of the highest-order derivative in the equation. In this section only first-order differential equations are discussed.

A **linear first-order differential equation** is an equation of the form

$$\frac{dy}{dx} + P(x)y = Q(x).$$

Many useful models produce such equations. In this section we develop a general method for solving first-order linear differential equations.

For example, to solve the equation

$$x\frac{dy}{dx} + 6y + 2x^4 = 0, \tag{1}$$

we need to get it in the form of a linear first-order differential equation. Thus, dy/dx should have a coefficient of 1. To accomplish this, we divide both sides of the equation by x and rearrange the terms to get the linear differential equation

$$\frac{dy}{dx} + \frac{6}{x}y = -2x^3.$$

This equation does not have separable variables and cannot be solved by the methods discussed so far. (Verify this.) Instead, multiply both sides of the equation by x^6 (the reason will be explained shortly) to get

$$x^6\frac{dy}{dx} + 6x^5y = -2x^9. \tag{2}$$

---

*This problem first appeared in the *American Mathematical Monthly,* Vol. 44, Dec. 1937.

On the left, $6x^5$, the coefficient of y, is the derivative of x^6, the coefficient of dy/dx. Recall the product rule for derivatives:

$$D_x(uv) = u\frac{dv}{dx} + \frac{du}{dx}v.$$

If $u = x^6$ and $v = y$, the product rule gives

$$D_x(x^6y) = x^6\frac{dy}{dx} + 6x^5y,$$

which is the left side of equation (2). Substituting $D_x(x^6y)$ for the left side of equation (2) gives

$$D_x(x^6y) = -2x^9.$$

Assuming $y = f(x)$, as usual, both sides of this equation can be integrated with respect to x and the result solved for y to get

$$x^6y = \int -2x^9\,dx = -2\left(\frac{x^{10}}{10}\right) + C = -\frac{x^{10}}{5} + C$$

$$y = -\frac{x^4}{5} + \frac{C}{x^6}. \tag{3}$$

Equation (3) is the general solution of equation (2) and therefore of equation (1).

This procedure has given us a solution, but what motivated our choice of the multiplier x^6? To see where x^6 came from, let $I(x)$ represent the multiplier, and multiply both sides of the general equation

$$\frac{dy}{dx} + P(x)y = Q(x)$$

by $I(x)$:

$$I(x)\frac{dy}{dx} + I(x)P(x)y = I(x)Q(x). \tag{4}$$

The method illustrated above will work only if the left side of the equation is the derivative of the product function $I(x) \cdot y$, which is

$$I(x)\frac{dy}{dx} + I'(x)y. \tag{5}$$

Comparing the coefficients of y in equations (4) and (5) shows that $I(x)$ must satisfy

$$I'(x) = I(x)P(x),$$

or

$$\frac{I'(x)}{I(x)} = P(x).$$

Integrating both sides of this last equation gives

$$\ln |I(x)| = \int P(x)\,dx + C$$

$$|I(x)| = e^{\int P(x)\,dx + C}$$

or

$$I(x) = \pm e^C e^{\int P(x)\,dx}.$$

Only one value of $I(x)$ is needed, so let $C = 0$, so that $e^C = 1$, and use the positive result, giving

$$I(x) = e^{\int P(x)\, dx}.$$

In summary, choosing $I(x)$ as $e^{\int P(x)\, dx}$ and multiplying both sides of a linear first-order differential equation by $I(x)$ puts the equation in a form that can be solved by integration.

INTEGRATING FACTOR

$I(x) = e^{\int P(x)\, dx}$ is called an **integrating factor** for the differential equation $\dfrac{dy}{dx} + P(x)y = Q(x)$.

For equation (1), written as the linear differential equation

$$\frac{dy}{dx} + \frac{6}{x}\, y = -2x^3,$$

$P(x) = 6/x$, and the integrating factor is

$$I(x) = e^{\int (6/x)\, dx} = e^{6 \ln |x|} = e^{\ln |x|^6} = e^{\ln x^6} = x^6.$$

This last step used the fact that, for all positive a, $e^{\ln a} = a$.

In summary, we solve a linear first-order differential equation with the following steps.

SOLVING A LINEAR FIRST-ORDER DIFFERENTIAL EQUATION

1. Put the equation in the linear form $\dfrac{dy}{dx} + P(x)y = Q(x)$.
2. Find the integrating factor $I(x) = e^{\int P(x)\, dx}$.
3. Multiply each term of the equation from Step 1 by $I(x)$.
4. Replace the sum of terms on the left with $D_x[I(x)y]$.
5. Integrate both sides of the equation.
6. Solve for y.

EXAMPLE 1 Linear Equation

Give the general solution of $dy/dx + 2xy = x$.

Solution

Step 1 This equation is already in the required form.

Step 2 The integrating factor is

$$I(x) = e^{\int 2x\, dx} = e^{x^2}.$$

Step 3 Multiplying each term by e^{x^2} gives

$$e^{x^2}\frac{dy}{dx} + 2xe^{x^2}y = xe^{x^2}.$$

Step 4 The sum of terms on the left can now be replaced with $D_x(e^{x^2}y)$, to get

$$D_x(e^{x^2}y) = xe^{x^2}.$$

Step 5 Integrating on both sides leads to the general solution

$$e^{x^2}y = \int xe^{x^2}\,dx = \frac{1}{2}e^{x^2} + C.$$

Step 6 Divide both sides by e^{x^2}.

$$y = \frac{1}{2} + Ce^{-x^2}.$$

EXAMPLE 2 Linear Equation

Solve the initial value problem $2(dy/dx) - 6y - e^x = 0$ if $y = 5$ when $x = 0$.

Solution Write the equation in the required form:

$$\frac{dy}{dx} - 3y = \frac{1}{2}e^x.$$

The integrating factor is

$$I(x) = e^{\int(-3)\,dx} = e^{-3x}.$$

Multiplying each term by $I(x)$ gives

$$e^{-3x}\frac{dy}{dx} - 3e^{-3x}y = \frac{1}{2}e^x e^{-3x}$$

$$e^{-3x}\frac{dy}{dx} - 3e^{-3x}y = \frac{1}{2}e^{-2x}.$$

The left side can now be replaced by $D_x(e^{-3x}y)$ to get

$$D_x(e^{-3x}y) = \frac{1}{2}e^{-2x}.$$

Integrating on both sides gives

$$e^{-3x}y = \int \frac{1}{2}e^{-2x}\,dx = \frac{1}{2}\left(\frac{e^{-2x}}{-2}\right) + C.$$

Now, multiply both sides by e^{3x} to get

$$y = -\frac{e^x}{4} + Ce^{3x},$$

the general solution. Find the particular solution by substituting 0 for x and 5 for y:

$$5 = -\frac{e^0}{4} + Ce^0 = -\frac{1}{4} + C$$

or

$$\frac{21}{4} = C,$$

which leads to the particular solution

$$y = -\frac{e^x}{4} + \frac{21}{4}e^{3x}.$$

> **NOTE** We can write the condition $y = 5$ when $x = 0$ as $y(0) = 5$. We will use this notation from now on.

EXAMPLE 3 Glucose

Suppose glucose is infused into a patient's bloodstream at a constant rate of a grams per minute. At the same time, glucose is removed from the bloodstream at a rate proportional to the amount of glucose present. Then the amount of glucose, $G(t)$, present at time t satisfies

$$\frac{dG}{dt} = a - KG$$

for some constant K. Solve this equation for G. Does the glucose concentration eventually reach a constant? That is, what happens to G as $t \to \infty$?*

 Solution The equation can be written in the form of the linear first-order differential equation

$$\frac{dG}{dt} + KG = a. \tag{6}$$

The integrating factor is

$$I(t) = e^{\int K\,dt} = e^{Kt}.$$

Multiply both sides of equation (6) by $I(t) = e^{Kt}$.

$$e^{Kt}\frac{dG}{dt} + Ke^{Kt}G = ae^{Kt}$$

Write the left side as $D_t(e^{Kt}G)$ and solve for G by integrating on each side.

$$D_t(e^{Kt}G) = ae^{Kt}$$

$$e^{Kt}G = \int ae^{Kt}\,dt = \frac{a}{K}e^{Kt} + C$$

Multiply both sides by e^{-Kt} to get

$$G = \frac{a}{K} + Ce^{-Kt}.$$

As $t \to \infty$,

$$\lim_{t\to\infty} G = \lim_{t\to\infty}\left(\frac{a}{K} + Ce^{-Kt}\right) = \lim_{t\to\infty}\left(\frac{a}{K} + \frac{C}{e^{Kt}}\right) = \frac{a}{K}.$$

Thus, the glucose concentration stabilizes at a/K.

> **NOTE** The equation in Example 3 can also be solved by separation of variables. You are asked to do this in Exercise 26.

---

*Andrews, Larry C., *Ordinary Differential Equations with Applications,* Scott, Foresman and Company, 1982, p. 79.

11.2 EXERCISES

Find general solutions for the following differential equations.

1. $\dfrac{dy}{dx} + 2y = 5$

2. $\dfrac{dy}{dx} + 4y = 10$

3. $\dfrac{dy}{dx} + xy = 3x$

4. $\dfrac{dy}{dx} + 2xy = x$

5. $x\dfrac{dy}{dx} - y - x = 0, \quad x > 0$

6. $x\dfrac{dy}{dx} + 2xy - x^2 = 0$

7. $2\dfrac{dy}{dx} - 2xy - x = 0$

8. $3\dfrac{dy}{dx} + 6xy + x = 0$

9. $x\dfrac{dy}{dx} + 2y = x^2 + 3x, \quad x > 0$

10. $x^2\dfrac{dy}{dx} + xy = x^3 - 2x^2, \quad x > 0$

11. $y - x\dfrac{dy}{dx} = x^3, \quad x > 0$

12. $2xy + x^3 = x\dfrac{dy}{dx}$

13. $\dfrac{dy}{dx} + y \cot x = x$

14. $\dfrac{dy}{dx} + y \tan x = \cos^2 x$

Solve each differential equation, subject to the given initial condition.

15. $\dfrac{dy}{dx} + y = 2e^x; \quad y(0) = 100$

16. $\dfrac{dy}{dx} + 2y = e^{3x}; \quad y(0) = 50$

17. $\dfrac{dy}{dx} - xy - x = 0; \quad y(1) = 10$

18. $x\dfrac{dy}{dx} - 3y + 2 = 0; \quad y(1) = 8$

19. $x\dfrac{dy}{dx} + 5y = x^2; \quad y(2) = 12$

20. $2\dfrac{dy}{dx} - 4xy = 5x; \quad y(1) = 10$

21. $x\dfrac{dy}{dx} + (1 + x)y = 3; \quad y(4) = 50$

22. $\dfrac{dy}{dx} + 2xy - e^{-x^2} = 0; \quad y(0) = 100$

Applications

LIFE SCIENCES

23. *Drug Use* The rate of change in the concentration of a drug with respect to time in a user's blood is given by

$$\frac{dC}{dt} = -kC + D(t),$$

where $D(t)$ is dosage at time t and k is the rate that the drug leaves the bloodstream.*

a. Solve this linear equation to show that, if $C(0) = 0$, then

$$C(t) = e^{-kt}\int_0^t e^{ky}D(y)\,dy.$$

(*Hint:* To integrate both sides of the equation in Step 5 of "Solving a Linear First-Order Differential Equation," integrate from 0 to t, and change the variable of integration to y.)

b. Show that if $D(y)$ is a constant D, then

$$C(t) = \frac{D(1 - e^{-kt})}{k}.$$

24. *Mouse Infection* A model for the spread of an infectious disease among mice is

$$\frac{dN}{dt} = rN - \frac{\alpha r(\alpha + b + v)}{\beta\left[\alpha - r\left(1 + \dfrac{v}{b + \gamma}\right)\right]},$$

where N is the size of the population of mice, α is the mortality rate due to infection, b is the mortality rate due to natural causes for infected mice, β is a transmission coefficient for the rate that infected mice infect susceptible mice, v is the rate the mice recover from infection, and γ is the rate that mice lose immunity.[†] Show that the solution to this

*Hoppensteadt, F. C., and J. D. Murray, "Threshold Analysis of a Drug Use Epidemic Model," *Mathematical Biosciences,* Vol. 53, No. 1/2, Feb. 1981, pp. 79–87.

†Anderson, Roy M., "The Persistence of Direct Life Cycle Infectious Diseases Within Populations of Hosts," in *Lectures on Mathematics in the Life Sciences, Vol. 12: Some Mathematical Questions in Biology,* the American Mathematical Society, 1979, pp. 1–67.

equation, with the initial condition $N(0) = (\alpha + b + v)/\beta$, can be written as

$$N(t) = \frac{(\alpha + b + v)}{\beta R}[(R - \alpha)e^{rt} + \alpha],$$

where

$$R = \alpha - r\left(1 + \frac{v}{b + \gamma}\right).$$

25. *Population Growth* The logistic equation introduced in Section 1,

$$\frac{dy}{dx} = k\left(1 - \frac{y}{N}\right)y \tag{7}$$

can be written as

$$\frac{dy}{dx} = cy - py^2, \tag{8}$$

where c and p are positive constants. Although this is a nonlinear differential equation, it can be reduced to a linear equation by a suitable substitution for the variable y.

a. Letting $y = 1/z$ and $dy/dx = (-1/z^2)\, dz/dx$, rewrite equation (8) in terms of z. Solve for z and then for y.

b. Let $z(0) = 1/y_0$ in part a and find a particular solution for y.

c. Find the limit of y as $x \to \infty$. This is the saturation level of the population.

26. *Glucose Level* Solve the glucose level example (Example 3) using separation of variables.

OTHER APPLICATIONS

Immigration and Emigration If a population is changed either by immigration or emigration, the population model discussed in Section 1 is modified to

$$\frac{dy}{dt} = ky + f(t),$$

where y is the population at time t and f(t) is some (other) function of t that describes the net effect of the emigration/immigration. Assume k = 0.02 and y(0) = 10,000. Solve this differential equation for y, given the following functions f(t).

27. $f(t) = e^t$

28. $f(t) = e^{-t}$

29. $f(t) = -t$

30. $f(t) = t$

Newton's Law of Cooling Newton's law of cooling states that the rate of change of temperature of an object is proportional to the difference in temperature between the object and the surrounding medium. Thus, if T is the temperature of the object after t hours and T_0 is the (constant) temperature of the surrounding medium, then

$$\frac{dT}{dt} = -k(T - T_0),$$

where k is a constant. Use this equation in Exercises 31–35.

31. Use the method of this section to show that the solution of this differential equation is

$$T = ce^{-kt} + T_0,$$

where c is a constant.

32. Solve the differential equation for Newton's law of cooling using separation of variables.

33. According to the solution of the differential equation for Newton's law of cooling, what happens to the temperature of an object after it has been in a surrounding medium with constant temperature for a long period of time? How well does this agree with reality?

Newton's Law of Cooling When a dead body is discovered, one of the first steps in the ensuing investigation is for a medical examiner to determine the time of death as closely as possible. Have you ever wondered how this is done? If the temperature of the medium (air, water, or whatever) has been fairly constant and less than 48 hours have passed since the death, Newton's law of cooling can be used. The medical examiner does not actually solve the equation for each case. Instead, a table based on the formula is consulted. Use Newton's law of cooling to work the following exercises.*

34. Assume the temperature of a body at death is 98.6°F, the temperature of the surrounding air is 68°F, and at the end of one hour the body temperature is 90°F.

---

*Callas, Dennis, and David J. Hildreth, "Snapshots of Applications in Mathematics," *The College Mathematics Journal,* Vol. 26, No. 2, March 1995.

a. What is the temperature of the body after 2 hours?

b. When will the temperature of the body be 75°F?

c. Approximately when will the temperature of the body be within 0.01° of the surrounding air?

35. Suppose the air temperature surrounding a body remains at a constant 10°F, $c = 88.6$, and $k = 0.24$.

a. Determine a formula for the temperature at any time t.

b. Use a graphing calculator to graph the temperature T as a function of time t on the window $[0, 30]$ by $[0, 100]$.

c. When does the temperature of the body decrease more rapidly: just after death, or much later? How do you know?

d. What will the temperature of the body be after 4 hours?

e. How long will it take for the body temperature to reach 40°F? Use your calculator graph to verify your answer.

11.3 EULER'S METHOD

? **THINK ABOUT IT**

How can we predict the weight of a goat at 5 weeks given its weight at birth and a formula for its growth rate?

This question will be answered in the exercises at the end of this section.

Applications sometimes involve differential equations such as

$$\frac{dy}{dx} = \frac{x + y}{y}$$

that cannot be solved by the methods discussed so far, but approximate solutions to these equations often can be found by numerical methods. For many applications, these approximations are quite adequate. In this section we introduce Euler's method, which is only one of numerous mathematical contributions made by Leonhard Euler (1707–1783) of Switzerland. (His name is pronounced "oiler.") He also introduced the $f(x)$ notation used throughout this text. Despite becoming blind during his later years, he was the most prolific mathematician of his era. He published in all mathematical fields.

Euler's method of solving differential equations gives approximate solutions to differential equations involving $y = f(x)$ where the initial values of x and y are known; that is, equations of the form

$$\frac{dy}{dx} = g(x, y), \quad \text{with} \quad y(x_0) = y_0.$$

Geometrically, Euler's method approximates the graph of the solution $y = f(x)$ with a polygonal line whose first segment is tangent to the curve at the point (x_0, y_0), as shown in Figure 5 on the next page.

To use Euler's method, divide the interval from x_0 to another point x_n into n subintervals of equal width (see Figure 5.) The width of each subinterval is $h = (x_n - x_0)/n$.

Recall from Section 6.5 on Differentials: Linear Approximation, that if Δx is a small change in x, then the corresponding small change in y, Δy, is approximated by

$$\Delta y \approx dy = \frac{dy}{dx} \cdot \Delta x.$$

In Section 6.5 on Differentials: Linear Approximation, we defined Δy to be the actual change in y as x changed by an amount Δx:

$$\Delta y = f(x + \Delta x) - f(x).$$

The differential dy is an approximation to Δy. We find dy by following the tangent line from the point $(x, f(x))$, rather than by following the actual function. Then dy is found by using the formula $dy = (dy/dx) \, dx$ where $dx = \Delta x$. For example, let $f(x) = x^3$, $x = 1$, and $dx = \Delta x = 0.2$. Then $dy = f'(x) \, dx = 3x^2 \, dx = 3(1^2)(0.2) = 0.6$. The actual change in y as x changes from 1 to 1.2 is

$$
\begin{aligned}
f(x + \Delta x) &- f(x) \\
&= f(1.2) - f(1) \\
&= 1.2^3 - 1 \\
&= 0.728.
\end{aligned}
$$

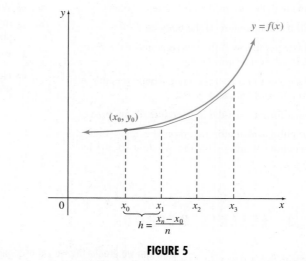

FIGURE 5

The differential dy is the change in y along the tangent line. On the interval from x_i to x_{i+1}, dy is just $y_{i+1} - y_i$, where y_i is the approximate solution at x_i. We also have $dy/dx = g(x_i, y_i)$ and $\Delta x = h$. Putting these into the above equation yields

$$y_{i+1} - y_i = g(x_i, y_i)h$$
$$y_{i+1} = y_i + g(x_i, y_i)h.$$

Because y_0 is given, we can use the equation just derived with $i = 0$ to get y_1. We can then use y_1 and the same equation with $i = 1$ to get y_2, and continue in this manner until we get y_n. A summary of Euler's method follows.

EULER'S METHOD

Let $y = f(x)$ be the solution of the differential equation

$$\frac{dy}{dx} = g(x, y), \quad \text{with} \quad y(x_0) = y_0,$$

for $x_0 \le x \le x_n$. Let $x_{i+1} - x_i = h$ and

$$y_{i+1} = y_i + g(x_i, y_i)h,$$

for $0 \le i \le n$. Then

$$f(x_{i+1}) \approx y_{i+1}.$$

As the following examples will show, the accuracy of the approximation varies for different functions. As h gets smaller, however, the approximation improves, although making h too small can make things worse. (See the discussion at the end of this section.) Euler's method is not difficult to program; it then becomes possible to try smaller and smaller values of h to get better and better approximations. Graphing calculator programs for Euler's method are included in *The Graphing Calculator Manual* that is available on this text's Web site.

EXAMPLE 1 Euler's Method

Use Euler's method to approximate the solution of $dy/dx + 2xy = x$, with $y(0) = 1.5$, for $[0, 1]$. Use $h = 0.1$.

Solution

Method 1: Calculating by Hand

The general solution of this equation was found in Example 1 of the last section, so the results using Euler's method can be compared with the actual solution. Begin by writing the differential equation in the required form as

$$\frac{dy}{dx} = x - 2xy, \qquad \text{so that} \qquad g(x, y) = x - 2xy.$$

Since $x_0 = 0$ and $y_0 = 1.5$, $g(x_0, y_0) = 0 - 2(0)(1.5) = 0$, and

$$y_1 = y_0 + g(x_0, y_0)h = 1.5 + 0(0.1) = 1.5.$$

Now $x_1 = 0.1$, $y_1 = 1.5$, and $g(x_1, y_1) = 0.1 - 2(0.1)(1.5) = -0.2$. Then

$$y_2 = 1.5 + (-0.2)(0.1) = 1.48.$$

The eleven values for x_i and y_i for $0 \le i \le 10$ are shown in Table 1, together with the actual values using the result from Example 1 in the last section. (Since the result was only a general solution, replace x with 0 and y with 1.5 to get the particular solution $y = 1/2 + e^{-x^2}$.)

■ **Table 1**

	Euler's Method	Actual Solution	Difference
x_i	y_i	$f(x_i)$	$y_i - f(x_i)$
0	1.5	1.5	0
0.1	1.5	1.4900498	0.0099502
0.2	1.48	1.4607894	0.0192106
0.3	1.4408	1.4139312	0.0268688
0.4	1.384352	1.3521438	0.0322082
0.5	1.3136038	1.2788008	0.034803
0.6	1.232243	1.1976763	0.0345667
0.7	1.1443742	1.1126264	0.0317102
0.8	1.0541619	1.0272924	0.0268695
0.9	0.96549595	0.94485807	0.02063788
1	0.88170668	0.86787944	0.01382724

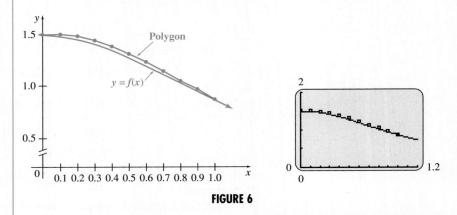

FIGURE 6

The results in Table 1 look quite good. The graphs in Figure 6 show that the polygonal line follows the actual graph of $f(x)$ quite closely.

Method 2: Graphing Calculator

A graphing calculator can readily implement Euler's method. To do this example on a TI-83, put X − 2X*Y into the function Y₁. Start x with the value $x_0 - h$ by storing −0.1 in X. Store y_0, or 1.5, in Y. Then the command X + .1 → X : Y + Y₁* .1 → Y gives the next value of y, which is still 1.5. Continue to press the ENTER key to get subsequent values of y. For more details, see *The Graphing Calculator Manual* that is available on this text's Web site.

Method 3: Spreadsheet

Euler's method can also be performed on a spreadsheet. In Microsoft Excel, for example, store the values of x in column A and the initial value of y in B1. Then put the command "=B1+(A1−2*A1*B1)*.1" into B2 to get the next value of y, using the formula for $g(x, y)$ in this example. Copy this formula into the rest of column B to get subsequent values of y. For more details, see *The Spreadsheet Manual* that is available on this text's Web site. ▪

Euler's method produces a very good approximation for this differential equation because the slope of the solution $f(x)$ is not steep in the interval under investigation. The next example shows that such good results cannot always be expected.

EXAMPLE 2 Euler's Method

Use Euler's method to solve $dy/dx = 3y + (1/2)e^x$, with $y(0) = 5$, for $[0, 1]$, using 10 subintervals.

Solution This is the differential equation of Example 2 in the last section. The general solution found there, with the initial condition given above, leads to the particular solution

$$y = -\frac{1}{4}e^x + \frac{21}{4}e^{3x}.$$

To solve by Euler's method, start with $g(x, y) = 3y + (1/2)e^x$, $x_0 = 0$, and $y_0 = 5$. For $n = 10$, $h = (1 - 0)/10 = 0.1$ again, and

$$y_{i+1} = y_i + g(x_i, y_i)h = y_i + \left(3y_i + \frac{1}{2}e^{x_i}\right)h.$$

For y_1, this gives
$$y_1 = y_0 + \left(3y_0 + \frac{1}{2}e^{x_0}\right)h$$

$$= 5 + \left[3(5) + \frac{1}{2}(e^0)\right](0.1)$$

$$= 5 + (15.5)(0.1) = 6.55.$$

Similarly,
$$y_2 = 6.55 + \left[3(6.55) + \frac{1}{2}e^{0.1}\right](0.1)$$

$$= 6.55 + (19.65 + 0.55258546)(0.1)$$

$$= 8.57025855.$$

These and the remaining values for the interval $[0, 1]$ are shown in Table 2 on the next page.

In this example the absolute value of the differences grows very rapidly as x_i gets farther from x_0. See Figure 7. These large differences come from the term e^{3x} in the solution; this term grows very quickly as x increases. ▪

Table 2 Approximate Solution Using $h = 0.1$

x_i	Euler's Method y_i	Actual Solution $f(x_i)$	Difference $y_i - f(x_i)$
0	5	5	0
0.1	6.55	6.8104660	−0.260466
0.2	8.570259	9.2607730	−0.690514
0.3	11.20241	12.575452	−1.373042
0.4	14.63062	17.057658	−2.427038
0.5	19.09440	23.116687	−4.022287
0.6	24.90516	31.305120	−6.399960
0.7	32.46781	42.368954	−9.901144
0.8	42.30884	57.315291	−15.006451
0.9	55.11277	77.503691	−22.390921
1	71.76958	104.76950	−32.999920

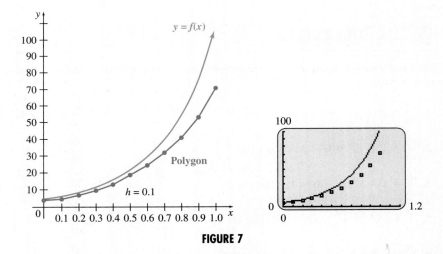

FIGURE 7

As these examples show, numerical methods may produce large errors. The error often can be reduced by using more subintervals of smaller width—letting $n = 100$ or 1,000, for example. Approximations for the function in Example 2 with $n = 100$ and $h = (1 - 0)/100 = 0.01$ are shown in Table 3 on the next page. The approximations are considerably improved.

We could improve the accuracy of Euler's method by using a smaller h, but there are two difficulties. First, this requires more calculations, and consequently more time. Such calculations are usually done by computer, so the increased time may not matter. But this introduces a second difficulty: The increased number of calculations causes more round-off error, so there is a limit to how small we can make h and still get improvement. The preferred way to get greater accuracy is to use a more sophisticated procedure, such as the Runge-Kutta method. Such methods are beyond the scope of this book but are discussed in numerical analysis and differential equations courses.*

*For example, see Nagel, R. K., E. B. Saff, and A. D. Snider, *Fundamentals of Differential Equations*, 5th ed., Addison Wesley Longman, 2000.

■ Table 3 Approximate Solution Using $h = 0.01$

x_i	Euler's Method y_i	Actual Solution $f(x_i)$	Difference $y_i - f(x_i)$
0	5	5	0
0.1	6.779419	6.810466	−0.031047
0.2	9.177102	9.260773	−0.083671
0.3	12.40634	12.575452	−0.169112
0.4	16.75386	17.057658	−0.303798
0.5	22.60505	23.116687	−0.511637
0.6	30.47795	31.305120	−0.827170
0.7	41.06884	42.368954	−1.30011
0.8	55.31358	57.315291	−2.001711
0.9	74.46998	77.503691	−3.033711
1	100.22860	104.769498	−4.540898

11.3   EXERCISES

Use Euler's method to approximate the indicated function value to three decimal places, using $h = 0.1$.

1. $\dfrac{dy}{dx} = x^2 + y^2$; $y(0) = 1$;   find $y(0.5)$

2. $\dfrac{dy}{dx} = xy + 2$; $y(0) = 0$;   find $y(0.5)$

3. $\dfrac{dy}{dx} = 1 + y$; $y(0) = 2$;   find $y(0.6)$

4. $\dfrac{dy}{dx} = x + y^2$; $y(0) = 0$;   find $y(0.6)$

5. $\dfrac{dy}{dx} = x + \sqrt{y}$; $y(0) = 1$;   find $y(0.4)$

6. $\dfrac{dy}{dx} = 1 + \dfrac{y}{x}$; $y(1) = 0$;   find $y(1.4)$

7. $\dfrac{dy}{dx} = x\sqrt{1 + y^2}$; $y(1) = 1$;   find $y(1.5)$

8. $\dfrac{dy}{dx} = e^{-y} + x$; $y(0) = 0$;   find $y(0.5)$

9. $\dfrac{dy}{dx} = y + \cos x$; $y(0) = 0$;   find $y(0.5)$

10. $\dfrac{dy}{dx} = \sin(x + y)$; $y(1) = 3$;   find $y(1.5)$

Use Euler's method to approximate the indicated function value to three decimal places, using $h = 0.1$. Next, solve the differential equation and find the indicated function value to three decimal places. Compare the result with the approximation.

11. $\dfrac{dy}{dx} = -4 + x$; $y(0) = 1$;   find $y(0.4)$

12. $\dfrac{dy}{dx} = 4x + 3$; $y(1) = 0$;   find $y(1.5)$

13. $\dfrac{dy}{dx} = x^2$; $y(0) = 2$;   find $y(0.5)$

14. $\dfrac{dy}{dx} = \dfrac{1}{x}$; $y(1) = 1$;   find $y(1.4)$

15. $\dfrac{dy}{dx} = 2xy$; $y(1) = 1$;   find $y(1.6)$

16. $\dfrac{dy}{dx} = x^2y$; $y(0) = 1$;   find $y(0.6)$

17. $\dfrac{dy}{dx} = ye^x$; $y(0) = 1$;   find $y(0.3)$

18. $\dfrac{dy}{dx} = \dfrac{x}{y}$; $y(0) = 2$;   find $y(0.3)$

19. Repeat Exercise 17 with $h = 0.05$. How does the error (the difference between this answer and the true answer) compare with the error in Exercise 17?

20. Repeat Exercise 18 with $h = 0.05$. How does the error (the difference between this answer and the true answer) compare with the error in Exercise 18?

In each of the following, use the program for Euler's method in The Graphing Calculator Manual that is available on this text's Web site to construct a table like the ones in the examples for $0 \le x \le 1$, with $h = 0.2$.

21. $\dfrac{dy}{dx} = \sqrt[3]{x};\quad y(0) = 0$

22. $\dfrac{dy}{dx} = y;\quad y(0) = 1$

23. $\dfrac{dy}{dx} = 1 - y;\quad y(0) = 0$

24. $\dfrac{dy}{dx} = x - xy;\quad y(0) = 0.5$

Solve each differential equation and graph the function $y = f(x)$ and the polygonal approximation on the same axes. (The approximations were found in Exercises 21–24.)

25. $\dfrac{dy}{dx} = \sqrt[3]{x};\quad y(0) = 0$

26. $\dfrac{dy}{dx} = y;\quad y(0) = 1$

27. $\dfrac{dy}{dx} = 1 - y;\quad y(0) = 0$

28. $\dfrac{dy}{dx} = x - xy;\quad y(0) = 0.5$

29. a. Use Euler's method with $h = 0.2$ to approximate $f(1)$, where $f(x)$ is the solution to the differential equation

$$\frac{dy}{dx} = y^2;\quad y(0) = 1.$$

b. Solve the differential equation in part a using separation of variables, and discuss what happens to $f(x)$ as x approaches 1.

c. Based on what you learned from parts a and b, discuss what might go wrong when using Euler's method. (More advanced courses on differential equations discuss the question of whether a differential equation has a solution for a given interval in x.)

Applications

LIFE SCIENCES

30. *Goat Growth* The growth of male Saanen goats can be approximated by the equation

$$\frac{dW}{dt} = -0.01189W + 0.92389W^{0.016},$$

where W is the weight in kg after t weeks.* Find the weight of a goat at 5 weeks, given that the weight at birth is 3.65 kg. Use Euler's method with $h = 1$.

31. *Growth of Algae* The phosphate compounds found in many detergents are highly water soluble and are excellent fertilizers for algae. Assume that there are 3,000 algae present at time $t = 0$ and conditions will support at most 100,000 algae. Assume that the rate of growth of algae, in the presence of sufficient phosphates, is proportional both to the number present (in thousands) and to the difference

between 100,000 and the number present (in thousands).

a. Write a differential equation using the given information. Use 0.01 for the constant of proportionality.

b. Approximate the number present when $t = 2$, using Euler's method with $h = 0.5$.

32. *Immigration* An island is colonized by immigration from the mainland, where there are 100 species. Let the number of species on the island at time t (in years) equal y, where $y = f(t)$. Suppose the rate at which new species immigrate to the island is

$$\frac{dy}{dt} = 0.02(100 - y^{1/2}).$$

Use Euler's method with $h = 0.5$ to approximate y when $t = 5$ if there were 10 species initially.

*France, J., J. Kijkstra, and M. S. Dhanoa, "Growth Functions and Their Application in Animal Science," *Annales de Zootechnie*, Vol. 45 (Supplement), 1996, pp. 165–174.

33. *Insect Population* A population of insects, y, living in a circular colony grows at a rate

$$\frac{dy}{dt} = 0.05y - 0.1y^{1/2},$$

where t is time in weeks. If there were 60 insects initially, use Euler's method with $h = 1$ to approximate the number of insects after 6 wk.

34. *Whale Population* Under certain conditions a population may exhibit a polynomial growth rate function. A population of blue whales is growing according to the function

$$\frac{dy}{dt} = -y + 0.02y^2 + 0.003y^3.$$

Here y is the population in thousands and t is measured in years. Use Euler's method with $h = 1$ to approximate the population in 4 yr if the initial population is 15,000.

35. *Learning* In an early article describing how people learn, the rate of change of the probability that a person performs a task correctly (p) with respect to time (t) is given by

$$\frac{dp}{dt} = \frac{2k}{\sqrt{m}}(p - p^2)^{3/2},$$

where k and m are constants related to the rate that the person learns the task.* For this exercise, let $m = 4$ and $k = 0.5$.

a. Letting $p = 0.1$ when $t = 0$, use the program for Euler's method in *The Graphing Calculator Manual* that is available on this text's Web site to construct a table for t_i and p_i like the ones in the examples for $0 \le x \le 30$, with $h = 5$.

b. Based on your answer to part a, what does p seem to approach as t increases? Explain why this answer makes sense.

36. *Spread of Rumors* A rumor spreads through a community of 500 people at the rate

$$\frac{dN}{dt} = 0.2(500 - N)N^{1/2},$$

where N is the number of people who have heard the rumor at time t (in hours). Use Euler's method with $h = 0.5$ to find the number who have heard the rumor after 3 hr, if only 2 people heard it initially.

■ 11.4   LINEAR SYSTEMS OF DIFFERENTIAL EQUATIONS

? THINK ABOUT IT   How can we describe the way a drug flows from one part of the body to another?

We will answer this question using a linear system of differential equations.

    In the previous three sections, we have studied differential equations involving one function of one variable. Some of the most important applications of calculus to the life sciences, however, involve more than one function of one variable. In an area of mathematical physiology known as *compartmental analysis,* different organs of the body are considered as different compartments, with material traveling from one compartment to another.[†] Figure 8 on the next page illustrates a simplified version of such a system involving two components.

    For the system in Figure 8, let V_1 and V_2 be the volumes of the two compartments, and let x_1 and x_2 be the concentration of the material under consideration (such as a pharmaceutical drug) in each of the two compartments, so that $V_i x_i$ represents the amount of the substance in compartment i for $i = 1$ or 2. The rate at which the substance flows from one compartment to another depends on the amount of the substance in each compartment as well as the fractional rate at

---

*Thurstone, L. L., "The Learning Function," *The Journal of General Psychology,* Vol. 3, No. 4, Oct. 1930, pp. 469–493.

[†]For an example of the type of problem solved in compartmental analysis, see Greenwell, Raymond N., "Generalization of Results on Almost Closed Compartment Systems," *Mathematical Biosciences,* Vol. 46, 1979, pp. 125–129.

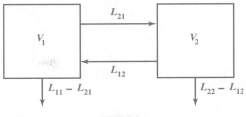

FIGURE 8

which the substance leaves and enters a compartment. To describe these rates, which vary from one compartment to another, we create the matrix

$$L = \begin{bmatrix} -L_{11} & L_{12} \\ L_{21} & -L_{22} \end{bmatrix},$$

where L_{ij} is the fractional rate at which the substance flows from compartment i to compartment j (for $i = 1$ and $j = 2$, or $i = 2$ and $j = 1$), and L_{ii} is the fractional rate at which the substance leaves compartment i (for $i = 1$ or 2). Some of the substance will move to the other compartment (at a rate L_{ji}) and the remainder will move outside the system (at a rate $L_{ii} - L_{ji}$). For example, if

$$L = \begin{bmatrix} -2 & 2 \\ 2 & -3 \end{bmatrix},$$

then

$$\frac{d(V_1 x_1)}{dt} = -2(V_1 x_1) + 2(V_2 x_2),$$

where $V_1 x_1$ is the amount of the substance in compartment 1, and $V_2 x_2$ is the amount in compartment 2.

If we define

$$V = \begin{bmatrix} V_1 & 0 \\ 0 & V_2 \end{bmatrix} \quad \text{and} \quad X = \begin{bmatrix} x_1 \\ x_2 \end{bmatrix},$$

then the equations describing how the amount of the drug changes in the compartments over time can be written as

$$\frac{d(VX)}{dt} = LVX.$$

We can simplify further by noting that

$$\frac{d(VX)}{dt} = V\frac{dX}{dt},$$

because V is a matrix of constants. Also, because V is a diagonal matrix, we know V^{-1} exists. Multiplying both sides of the equation above by V^{-1} gives

$$\frac{dX}{dt} = V^{-1}LVX$$

$$= MX,$$

where

$$M = V^{-1}LV.$$

For example, suppose $V_1 = 5$ ml and $V_2 = 10$ ml. Then

$$V = \begin{bmatrix} 5 & 0 \\ 0 & 10 \end{bmatrix}.$$

Verify that

$$V^{-1} = \begin{bmatrix} 1/5 & 0 \\ 0 & 1/10 \end{bmatrix}$$

and that using the matrix L given above yields

$$
\begin{aligned}
M &= V^{-1}LV \\
&= \begin{bmatrix} 1/5 & 0 \\ 0 & 1/10 \end{bmatrix} \begin{bmatrix} -2 & 2 \\ 2 & -3 \end{bmatrix} \begin{bmatrix} 5 & 0 \\ 0 & 10 \end{bmatrix} \\
&= \begin{bmatrix} -2 & 2 \\ 1 & -3 \end{bmatrix}.
\end{aligned}
$$

Our original problem boils down to solving

$$\frac{dX}{dt} = MX.$$

This equation represents the **linear system of differential equations**

$$\frac{dx_1}{dt} = M_{11}x_1 + M_{12}x_2$$

$$\frac{dx_2}{dt} = M_{21}x_1 + M_{22}x_2.$$

This system is a generalization of the linear differential equations of Section 11.2 because it involves more than one differential equation. This particular system is simpler than many because the matrix M is constant, and because we don't have another matrix added to the right-hand side, as in the equation

$$\frac{dX}{dt} = MX + Q,$$

where the entries in the matrix Q might be functions of t. (Example 1 shows how to handle the case when the matrix Q is present.) What complicates solving the system is that we can't solve the first equation for x_1 without knowing what x_2 is, and we can't solve the second equation for x_2 without knowing what x_1 is. The method for solving this system involves disentangling the equations using ideas from the previous chapter.

To begin, diagonalize the matrix M by finding the eigenvalues and eigenvectors, as we did in the previous chapter. Recall that to find the eigenvalues, set $\det(M - \lambda I) = 0$.

$$M - \lambda I = \begin{bmatrix} -2 - \lambda & 2 \\ 1 & -3 - \lambda \end{bmatrix}$$

$$
\begin{aligned}
\det(M - \lambda I) &= (-2 - \lambda)(-3 - \lambda) - 2(1) \\
&= \lambda^2 + 5\lambda + 6 - 2 \\
&= \lambda^2 + 5\lambda + 4
\end{aligned}
$$

Set this expression equal to 0 and solve.

$$0 = \lambda^2 + 5\lambda + 4$$
$$= (\lambda + 4)(\lambda + 1)$$
$$\lambda = -4 \quad \text{and} \quad \lambda = -1$$

We must next find the eigenvectors. For each value of λ, find a solution to the equation $(M - \lambda I)X = O$, where O is the zero matrix. For $\lambda = -4$,

$$(M - \lambda I)X = \begin{bmatrix} 2 & 2 \\ 1 & 1 \end{bmatrix} \begin{bmatrix} x \\ y \end{bmatrix} = \begin{bmatrix} 0 \\ 0 \end{bmatrix}.$$

This system has an infinite number of solutions; let us arbitrarily choose the solution $\begin{bmatrix} 1 \\ -1 \end{bmatrix}$. For $\lambda = -1$,

$$(M - \lambda I)X = \begin{bmatrix} -1 & 2 \\ 1 & -2 \end{bmatrix} \begin{bmatrix} x \\ y \end{bmatrix} = \begin{bmatrix} 0 \\ 0 \end{bmatrix}.$$

This system also has an infinite number of solutions; let us arbitrarily choose the solution $\begin{bmatrix} 2 \\ 1 \end{bmatrix}$.

We next form the matrix P whose columns are the eigenvectors:

$$P = \begin{bmatrix} 1 & 2 \\ -1 & 1 \end{bmatrix}.$$

Verify that

$$P^{-1} = \begin{bmatrix} 1/3 & -2/3 \\ 1/3 & 1/3 \end{bmatrix} \quad \text{and that} \quad P^{-1}MP = \begin{bmatrix} -4 & 0 \\ 0 & -1 \end{bmatrix}.$$

Notice that $P^{-1}MP$ is the diagonal matrix with the eigenvalues of M on the diagonal.

> ■ **NOTE** Since there are an infinite number of choices for the eigenvectors, there are also an infinite number of choices for the matrix P.

Now let $X = PY$, so

$$\frac{dX}{dt} = MX$$

becomes

$$\frac{d(PY)}{dt} = MPY.$$

Using the fact that P is a constant, so that $d(PY)/dt = P\,dY/dt$, and multiplying both sides by P^{-1}, we have

$$\frac{dY}{dt} = P^{-1}MPY,$$

or

$$\begin{bmatrix} dy_1/dt \\ dy_2/dt \end{bmatrix} = \begin{bmatrix} -4 & 0 \\ 0 & -1 \end{bmatrix} \begin{bmatrix} y_1 \\ y_2 \end{bmatrix}.$$

This yields two differential equations:

$$\frac{dy_1}{dt} = -4y_1 \quad \text{and} \quad \frac{dy_2}{dt} = -y_2,$$

each of which can be solved by separation of variables or with the use of an integrating factor, yielding

$$\begin{bmatrix} y_1 \\ y_2 \end{bmatrix} = \begin{bmatrix} C_1 e^{-4t} \\ C_2 e^{-t} \end{bmatrix}$$

where C_1 and C_2 are two arbitrary constants. Finally, calculate $X = PY$.

$$\begin{bmatrix} x_1 \\ x_2 \end{bmatrix} = \begin{bmatrix} 1 & 2 \\ -1 & 1 \end{bmatrix} \begin{bmatrix} C_1 e^{-4t} \\ C_2 e^{-t} \end{bmatrix}$$

$$= \begin{bmatrix} C_1 e^{-4t} + 2C_2 e^{-t} \\ -C_1 e^{-4t} + C_2 e^{-t} \end{bmatrix}$$

The values of C_1 and C_2 are determined by the initial conditions. For example, if $x_1(0) = 5$ and $x_2(0) = 1$, then

$$C_1 + 2C_2 = 5$$
$$-C_1 + C_2 = 1.$$

This system of equations can be solved by the Gauss-Jordan method, or in this case by simply adding the equations together. The result is $C_1 = 1$ and $C_2 = 2$. Therefore,

$$x_1 = e^{-4t} + 4e^{-t} \quad \text{and} \quad x_2 = -e^{-4t} + 2e^{-t}.$$

Let's now look at a case where $Q \neq 0$, and then summarize the steps.

EXAMPLE 1 Receptor Dynamics

The following linear system of differential equations describes a model of how the body responds when a drug is introduced into the bloodstream:

$$\frac{dx_1}{dt} = -\frac{bc}{k}x_1 + \frac{a}{k}x_2 + g(t)$$

$$\frac{dx_2}{dt} = \frac{bc}{k}x_1 - \frac{a}{k}x_2.$$

Here x_1 is the number of free (or active) sites at time t; x_2 is the number of bound (or inactive) sites; k measures the ratio of the binding rate time scale to the dosage time scale; bcx_1/k gives the binding rate of active sites; ax_2/k gives the rate at which bound sites are freed; and $g(t)$ gives the rate at which new active sites are produced in the body.* The initial conditions are $x_1(0) = p$ and $x_2(0) = 0$.

---

*Hoppensteadt, F. C., and J. D. Murray, "Threshold Analysis of a Drug Use Epidemic Model," *Mathematical Biosciences,* Vol. 53, No. 1/2, Feb. 1981, pp. 79–87.

Suppose $a = 3$, $b = 2$, $c = 2$, $k = 1$, $p = 6$, and $g(t) = e^{-t}$. Solve the initial value problem.

Solution This system of differential equations can be represented in matrix form as

$$\begin{bmatrix} dx_1/dt \\ dx_2/dt \end{bmatrix} = \begin{bmatrix} -4 & 3 \\ 4 & -3 \end{bmatrix} \begin{bmatrix} x_1 \\ x_2 \end{bmatrix} + \begin{bmatrix} e^{-t} \\ 0 \end{bmatrix}.$$

Find the eigenvalues by setting $\det(M - \lambda I) = 0$.

$$M - \lambda I = \begin{bmatrix} -4 - \lambda & 3 \\ 4 & -3 - \lambda \end{bmatrix}$$

$$0 = \det(M - \lambda I) = (-4 - \lambda)(-3 - \lambda) - 3(4)$$
$$0 = \lambda^2 + 7\lambda$$
$$\lambda = 0 \quad \text{and} \quad \lambda = -7$$

For $\lambda = 0$,

$$(M - \lambda I)X = \begin{bmatrix} -4 & 3 \\ 4 & -3 \end{bmatrix} \begin{bmatrix} x \\ y \end{bmatrix} = \begin{bmatrix} 0 \\ 0 \end{bmatrix}.$$

One solution of this system is $\begin{bmatrix} 3 \\ 4 \end{bmatrix}$. For $\lambda = -7$,

$$(M - \lambda I)X = \begin{bmatrix} 3 & 3 \\ 4 & 4 \end{bmatrix} \begin{bmatrix} x \\ y \end{bmatrix} = \begin{bmatrix} 0 \\ 0 \end{bmatrix}.$$

One solution of this system is $\begin{bmatrix} 1 \\ -1 \end{bmatrix}$. The matrix of eigenvectors is then

$$P = \begin{bmatrix} 3 & 1 \\ 4 & -1 \end{bmatrix}$$

and we can calculate

$$P^{-1} = \begin{bmatrix} 1/7 & 1/7 \\ 4/7 & -3/7 \end{bmatrix} \quad \text{and} \quad P^{-1}MP = \begin{bmatrix} 0 & 0 \\ 0 & -7 \end{bmatrix}.$$

Now let $X = PY$ and multiply both sides of the differential equation by P^{-1}, yielding

$$\frac{dY}{dt} = P^{-1}MPY + P^{-1} \begin{bmatrix} e^{-t} \\ 0 \end{bmatrix},$$

or

$$\begin{bmatrix} dy_1/dt \\ dy_2/dt \end{bmatrix} = \begin{bmatrix} 0 & 0 \\ 0 & -7 \end{bmatrix} \begin{bmatrix} y_1 \\ y_2 \end{bmatrix} + \begin{bmatrix} e^{-t}/7 \\ 4e^{-t}/7 \end{bmatrix}.$$

This yields two differential equations:

$$\frac{dy_1}{dt} = \frac{e^{-t}}{7} \quad \text{and} \quad \frac{dy_2}{dt} = -7y_2 + \frac{4e^{-t}}{7}.$$

The first equation is integrated to yield

$$y_1 = -\frac{e^{-t}}{7} + C_1.$$

The second is a linear equation which can be solved by multiplying by the integrating factor $e^{\int 7\,dt} = e^{7t}$.

$$\frac{dy_2}{dt} + 7y_2 = \frac{4e^{-t}}{7}$$

$$e^{7t}\frac{dy_2}{dt} + 7e^{7t}y_2 = \frac{4e^{-t}}{7}e^{7t}$$

$$D_x(e^{7t}y_2) = \frac{4e^{6t}}{7}$$

$$e^{7t}y_2 = \int \frac{4e^{6t}}{7}\,dt = \frac{2e^{6t}}{21} + C_2$$

$$y_2 = \frac{2e^{-t}}{21} + C_2 e^{-7t}$$

Finally, calculate $X = PY$.

$$\begin{bmatrix} x_1 \\ x_2 \end{bmatrix} = \begin{bmatrix} 3 & 1 \\ 4 & -1 \end{bmatrix}\begin{bmatrix} -e^{-t}/7 + C_1 \\ 2e^{-t}/21 + C_2 e^{-7t} \end{bmatrix}$$

$$= \begin{bmatrix} -3e^{-t}/7 + 3C_1 + 2e^{-t}/21 + C_2 e^{-7t} \\ -4e^{-t}/7 + 4C_1 - 2e^{-t}/21 - C_2 e^{-7t} \end{bmatrix}$$

Setting $x_1(0) = 6$ and $x_2(0) = 0$ gives

$$-3/7 + 3C_1 + 2/21 + C_2 = 6$$
$$-4/7 + 4C_1 - 2/21 - C_2 = 0.$$

Adding these two equations eliminates C_2, allowing us to solve for C_1. Verify that $C_1 = 1$ and $C_2 = 10/3$. Therefore,

$$x_1 = -\frac{3}{7}e^{-t} + 3 + \frac{2}{21}e^{-t} + \frac{10}{3}e^{-7t}$$

and

$$x_2 = -\frac{4}{7}e^{-t} + 4 - \frac{2}{21}e^{-t} - \frac{10}{3}e^{-7t}.$$

SOLVING A LINEAR SYSTEM OF DIFFERENTIAL EQUATIONS

$\dfrac{dX}{dt} = MX + Q$ where M is constant.

1. Find the eigenvalues of the matrix M by solving the equation $\det(M - \lambda I) = 0$.

2. For each eigenvalue, find the corresponding eigenvector by solving the equation $(M - \lambda I)X = O$.

3. Create the matrix P whose columns are the eigenvectors of M.

4. Solve the system $dY/dt = (P^{-1}MP)Y + P^{-1}Q$, where $P^{-1}MP$ is the diagonal matrix with the eigenvalues of M on the diagonal. Each equation in this system is linear and involves only one of the functions of Y.

5. The solution to the original system is given by $X = PY$.

6. The solution at step 5 will have arbitrary constants in it. For an initial value problem, substitute the values of the variables and solve the resulting system of linear equations to find the constants.

NOTE In this section, we have only considered cases in which the eigenvalues are distinct real numbers. In actual applications, the eigenvalues are often complex numbers. See Exercise 15 for an example of this.

11.4 EXERCISES

For Exercises 1–8, solve the system of differential equations.

1. $\dfrac{dx_1}{dt} = 3x_1 - 2x_2$

   $\dfrac{dx_2}{dt} = x_1$

2. $\dfrac{dx_1}{dt} = x_1 + 6x_2$

   $\dfrac{dx_2}{dt} = 5x_1 + 2x_2$

3. $\dfrac{dx_1}{dt} = 3x_1 + 2x_2$

   $\dfrac{dx_2}{dt} = 3x_1 + 8x_2$

4. $\dfrac{dx_1}{dt} = 7x_1 + 3x_2$

   $\dfrac{dx_2}{dt} = 3x_1 - x_2$

5. $\dfrac{dx_1}{dt} = 3x_1$

   $\dfrac{dx_2}{dt} = 5x_1 + 2x_2$

   $\dfrac{dx_3}{dt} = 2x_1 + 4x_2 - x_3$

6. $\dfrac{dx_1}{dt} = 2x_1$

   $\dfrac{dx_2}{dt} = 3x_1 + 5x_2$

   $\dfrac{dx_3}{dt} = 6x_1 + 5x_2 - 4x_3$

7. $\dfrac{dx_1}{dt} = 4x_1 - 2x_2 + e^t$

   $\dfrac{dx_2}{dt} = -3x_1 + 9x_2 + e^{2t}$

8. $\dfrac{dx_1}{dt} = 2x_1 + 3x_2 + e^{2t}$

   $\dfrac{dx_2}{dt} = 3x_1 - 6x_2 + e^{-4t}$

For Exercises 9–14, find the particular solution for the initial value problem.

9. Exercise 1, $x_1(0) = 7$, $x_2(0) = 5$

10. Exercise 2, $x_1(0) = 12$, $x_2(0) = 1$

11. Exercise 3, $x_1(0) = 3$, $x_2(0) = 23$

12. Exercise 4, $x_1(0) = 9$, $x_2(0) = 23$

13. Exercise 5, $x_1(0) = 6$, $x_2(0) = 15$, $x_3(0) = -1$

14. Exercise 6, $x_1(0) = -6$, $x_2(0) = -3$, $x_3(0) = 0$

Applications

LIFE SCIENCES

15. *Population Modeling* An article* describing the mathematical modeling of populations considers the example

$$\frac{dx_1}{dt} = x_2 - 2$$

$$\frac{dx_2}{dt} = -x_1 + 2$$

a. Show that the characteristic equation is $\lambda^2 + 1 = 0$.

b. The equation in part a has no real roots. Show, however, that if we let $i = \sqrt{-1}$, then the equation has two roots, i and $-i$. Notice for later use that $1/i = -i$.

c. Show that $\begin{bmatrix} 1 \\ i \end{bmatrix}$ and $\begin{bmatrix} 1 \\ -i \end{bmatrix}$ are eigenvectors corresponding to the eigenvalues found in part b, so that $P = \begin{bmatrix} 1 & 1 \\ i & -i \end{bmatrix}$.

d. Verify that $P^{-1} = \begin{bmatrix} 1/2 & -i/2 \\ 1/2 & i/2 \end{bmatrix}$.

e. Show that the matrix P in part c leads to the equations

$$\frac{dy_1}{dt} = iy_1 + (-1 - i)$$

$$\frac{dy_2}{dt} = -iy_2 + (-1 + i).$$

f. Show that the solutions to the equations in part e are

$$y_1 = C_1 e^{it} + (1 - i)$$

and

$$y_2 = C_2 e^{-it} + (1 + i).$$

g. Show that the solutions to the original equations are

$$x_1 = C_1 e^{it} + C_2 e^{-it} + 2$$

and

$$x_2 = iC_1 e^{it} - iC_2 e^{-it} + 2.$$

h. The article considered the initial condition $x_1(0) = 2$, $x_2(0) = 3$. Show that $C_1 = -i/2$ and $C_2 = i/2$.

i. Use identity of Euler,

$$e^{it} = \cos t + i \sin t,$$

and its equivalent form

$$e^{-it} = \cos t - i \sin t,$$

to show that the solution to the initial value problem is given by

$$x_1 = \sin t + 2$$

and

$$x_2 = \cos t + 2.$$

Notice that the final answer has no complex numbers.

j. Why is the answer to part i biologically reasonable?

16. *Pollution* Compartmental models can be used to describe the dispersal of pollutants. Researchers have studied the model

$$\frac{dX}{dt} = FX,$$

where x_i represents the amount of the pollutant in compartment i. The matrix F is the *fate matrix*, in which the negative elements on the diagonal represent the rate at which the pollutant leaves each compartment, and the other terms represent the rate at which the pollutant is transferred from one compartment to another.*

a. For the simple two-compartment model, let $F = \begin{bmatrix} L_1 & T_{21} \\ T_{12} & L_2 \end{bmatrix}$. Show that the eigenvalues are given by $\lambda = \left(L_1 + L_2 \pm \sqrt{(L_1 - L_2)^2 + 4T_{12}T_{21}}\right)/2$.

b. The researchers studied a seven-compartment model. For simplicity, we have reduced this to four compartments: air, plants, surface soil, and root-zone soil, for which

$$F = \begin{bmatrix} -1.7982 & 0.00191 & 4.8 \times 10^{-5} & 0 \\ 1.740748 & -0.00768 & 4.25 \times 10^{-17} & 5.44 \times 10^{-13} \\ 0.027803 & 0.005767 & -0.00044 & 2.57 \times 10^{-7} \\ 0 & 6.12 \times 10^{-12} & 5.19 \times 10^{-5} & -2.6 \times 10^{-7} \end{bmatrix}.$$

The matrix of eigenvectors is given by the matrix P shown below. Verify that the columns of the matrix P are the eigenvectors corresponding to the eigenvalues -0.005873653661, $-2.260432550 \times 10^{-7}$, -1.80005664, and $-3.897392958 \times 10^{-4}$ by multiplying F by P, multiplying each eigenvalue by P, and verifying that F times each eigenvector equals the corresponding eigenvalue times that eigenvector.

c. Write out the solution to x_1 in the linear system of differential equations corresponding to the matrix F in part b.

d. According to the answer in part c, what happens to the amount of the pollutant in the air after a long time?

$$P = \begin{bmatrix} -0.001034833155 & -0.9978255773 & 1.064336125 & -0.009404958855 \\ -2.300721130 \times 10^{-6} & -5.218032278 \times 10^{-4} & -0.06542733470 & -100.0001234 \\ 0.0995311655 & -0.09671974321 & -0.001227752248 & 3.548283458 \times 10^{-8} \\ -3.56625132 \times 10^{-4} & -0.8520286615 & -9.966809274 & 1.328125544 \end{bmatrix}$$

---

*Hertwich, Edgar G., "Fugacity Superposition: A New Approach to Dynamic Multimedia Fate Modeling," *Chemosphere*, Vol. 84, 2001, pp. 843–853.

■ 11.5 NONLINEAR SYSTEMS OF DIFFERENTIAL EQUATIONS

? THINK ABOUT IT   How do the populations of a predator and its prey change over time?

We will answer this question using a nonlinear system of differential equations.

In the previous section we studied linear systems of differential equations. If all systems of differential equations were linear, the world would be a simpler but less interesting place. Most differential equation models in the life sciences are nonlinear. There are no general methods for finding the solution to nonlinear systems, but there are vast amounts of literature on methods for analyzing their behavior. In this section we will touch on a few basic ideas.

Consider the following model of the interaction between baleen whales and Antarctic krill described by a group of researchers in 1979.* Krill tend to grow according to the logistic model described in Sec. 11.1, so if x represents the population of krill at time t, k_1 is the growth rate, and N is the carrying capacity, then

$$\frac{dx_1}{dt} = k_1\left(1 - \frac{x_1}{N}\right)x_1. \tag{1}$$

The population of krill is reduced by the whales, which eat the krill. If the population of whales is denoted by x_2, and if the rate at which the whales consume the krill is denoted by c, then

$$\frac{dx_1}{dt} = k_1\left(1 - \frac{x_1}{N}\right)x_1 - cx_1x_2. \tag{2}$$

A similar equation describes how the population of whales varies with time, but their carrying capacity is affected by how many krill there are to eat. We will therefore let the carrying capacity of the whales be qx_1, where q is a constant describing how many krill are needed to sustain a whale. Thus

$$\frac{dx_2}{dt} = k_2\left(1 - \frac{x_2}{qx_1}\right)x_2. \tag{3}$$

Equations (2) and (3) form a nonlinear system of differential equations. We cannot explicitly solve this system for x_1 and x_2, even if we knew the values of the constants k_1, k_2, c, and q. Nevertheless, we will explore this system to see what conclusions we can make.

One point of interest is the **equilibrium point,** that is, the values of x_1 and x_2 at which $dx_1/dt = dx_2/dt = 0$. This means that if the populations ever reached these values, they would stay there forever, or at least until something changed. Let's start with equation (3), since it's simpler than equation (2). Setting

---

*May, Robert M., et al., "Management of Multispecies Fisheries," *Science,* Vol. 205, No. 4403, July 20, 1979. More details on the discussion in this section can be found in Greenwell, Raymond N., "Whales and Krill: A Mathematical Model," *UMAP Module 610,* 1982. This unit also appeared in *The UMAP Journal,* Vol. 3, 1982, pp. 165–183.

$dx_2/dt = 0$ implies either $x_2 = 0$ (that is, there are no whales, an undesirable situation of little interest) or

$$1 - \frac{x_2}{qx_1} = 0,$$

from which we conclude that

$$x_2 = qx_1. \tag{4}$$

Now set $dx_1/dt = 0$ in Equation (2) and substitute the value for x_2 from Equation (4).

$$
\begin{aligned}
0 &= k_1\left(1 - \frac{x_1}{N}\right)x_1 - cx_1(qx_1) \\
&= x_1\left[k_1\left(1 - \frac{x_1}{N}\right) - cqx_1\right] \qquad \text{Factor out the } x_1. \\
&= x_1\left[k_1 - \frac{k_1}{N}x_1 - cqx_1\right] \\
&= x_1\left[k_1 - \left(\frac{k_1}{N} + cq\right)x_1\right]
\end{aligned}
$$

This means either that $x_1 = 0$ (no more whales) or

$$
\begin{aligned}
x_1 &= \frac{k_1}{\dfrac{k_1}{N} + cq} \\
&= \frac{k_1 N}{k_1 + cqN}. \tag{5}
\end{aligned}
$$

From this we find

$$
\begin{aligned}
x_2 &= qx_1 \\
&= \frac{qk_1 N}{k_1 + cqN}. \tag{6}
\end{aligned}
$$

Together, equations (5) and (6) give the equilibrium point for the system of differential equations given by equations (2) and (3).

Our mathematical model has not yet taken into account the effect of humans harvesting both the whales and the krill for food. Let us denote by F_1 and F_2 the fishing effort for krill and whales, respectively, scaled so that $F_1 = 1$ and $F_2 = 1$ corresponds to an effort which yields at a rate proportional to the natural growth rate. Then the system of differential equations (2) and (3) can be modified to

$$
\begin{aligned}
\frac{dx_1}{dt} &= k_1\left(1 - \frac{x_1}{N}\right)x_1 - cx_1x_2 - k_1F_1x_1 \\
\frac{dx_2}{dt} &= k_2\left(1 - \frac{x_2}{qx_1}\right)x_2 - k_2F_2x_2. \tag{7}
\end{aligned}
$$

To simplify the system in (7), let $y_1 = x_1/N$ and $y_2 = x_2/(qN)$. Then $y_1 = 1$ corresponds to the krill population being equal to the carrying capacity, and

$y_2 = 1$ corresponds to the whale population being equal to the carrying capacity when the krill are at their carrying capacity. Then

$$\frac{d(Ny_1)}{dt} = k_1(1 - y_1)Ny_1 - c(Ny_1)(qNy_2) - k_1F_1Ny_1$$

$$\frac{d(qNy_2)}{dt} = k_2\left(1 - \frac{y_2}{y_1}\right)qNy_2 - k_2F_2qNy_2. \tag{8}$$

Dividing the first equation in (8) by N and the second by qN, and factoring k_1y_1 from the right-hand side of the first equation and k_2y_2 from the right-hand side of the second equation gives

$$\frac{dy_1}{dt} = k_1y_1\left(1 - y_1 - \frac{cqNy_2}{k_1} - F_1\right)$$

$$\frac{dy_2}{dt} = k_2y_2\left(1 - \frac{y_2}{y_1} - F_2\right). \tag{9}$$

Denoting cqN/k_1 by b and rearranging slightly, (9) becomes

$$\frac{dy_1}{dt} = k_1y_1(1 - F_1 - y_1 - by_2)$$

$$\frac{dy_2}{dt} = k_2y_2\left(1 - F_2 - \frac{y_2}{y_1}\right). \tag{10}$$

Note that the constants c, q, and N have disappeared from the system of differential equations. This means that these constants by themselves do not affect the behavior of the system; it is only their product, which becomes the numerator of the constant b, that appears in (9). This is an interesting conclusion that is not at all apparent in the original mathematical model.

EXAMPLE 1 Predator-Prey

The Austrian mathematician A. J. Lotka (1880–1949) and the Italian mathematician Vito Volterra (1860–1940) proposed the following simple model for the way in which the fluctuations of populations of a predator and its prey affect each other.* Let $x_1 = f(t)$ denote the population of the predator and $x_2 = g(t)$ denote the population of the prey at time t. The predator might be a wolf and its prey a moose, or the predator might be a ladybug and the prey an aphid.

Assume that if there were no predators present, the population of prey would increase at a rate px_2 proportional to their number, but that in the presence of predators, the predators consume the prey at a rate qx_1x_2 proportional to the product of the number of prey and the number of predators. The net rate of change dx_2/dt of x_2 is the rate of increase of the prey minus the rate at which the prey are eaten, that is,

$$\frac{dx_2}{dt} = px_2 - qx_1x_2, \tag{11}$$

with positive constants p and q.

Assume that if there were no prey, the predators would starve and their population would *decrease* at a rate rx_1 proportional to their number, but that in the

*Lotka, A. J., *Elements of Mathematical Biology*, Dover, 1956.

presence of prey, the rate of growth of the population of predators is increased by an amount sx_1x_2. These assumptions give a second differential equation,

$$\frac{dx_1}{dt} = -rx_1 + sx_1x_2,\tag{12}$$

with additional positive constants r and s.

Equations (11) and (12) form a system of differential equations known as the **Lotka-Volterra equations.**

(a) Find the equilibrium point for the Lotka-Volterra equations.

Solution Equations (11) and (12) factor as

$$\frac{dx_1}{dt} = -rx_1 + sx_1x_2 = x_1(-r + sx_2) \quad \text{and}$$

$$\frac{dx_2}{dt} = px_2 - qx_1x_2 = x_2(p - qx_1).$$

Assuming that neither x_1 nor x_2 equals 0 (a trivial equilibrium point),

$$-r + sx_2 = 0 \quad \text{and} \quad p - qx_1 = 0,$$

so

$$x_2 = \frac{r}{s} \quad \text{and} \quad x_1 = \frac{p}{q}.$$

(b) By considering the sign of dx_1/dt and dx_2/dt at points other than the equilibrium point, determine the behavior of any solution to the system of differential equations.

Solution In Figure 9, we have shown the x_1- and x_2-axes, with the equilibrium point $(p/q, r/s)$ marked. This divides the first quadrant of the plane into four regions as shown.

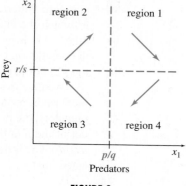

FIGURE 9

In region 1, $x_1 > p/q$ and $x_2 > r/s$. Then $dx_1/dt = x_1(-r + sx_2) > 0$ and $dx_2/dt = x_2(p - qx_1) < 0$. Since x_1 is increasing and x_2 is decreasing, the graph will go down and to the right, as shown.

In region 2, $x_1 < p/q$ and $x_2 > r/s$. Then $dx_1/dt > 0$ and $dx_2/dt > 0$. Since x_1 and x_2 are both increasing, the graph will go up and to the right, as shown.

Similarly, in region 3 $dx_1/dt < 0$ and $dx_2/dt > 0$, so the graph will go up and to the left, as shown. Verify the behavior of the solution in region 4.

We can see from this analysis that any solution will move about the equilibrium point in a clockwise direction. The population of prey tends to rise when the population of predators is low. Then, with the resulting abundance of food, the population of predators increases. As a result, the population of prey decreases again, and this forces the population of predators to decline because of the lack of food. The pattern repeats indefinitely. Figure 9 is an example of a **phase plane diagram.** ∎

We can say more about the graph of the solution by dividing equation (11) by equation (12), giving the separable differential equation

$$\frac{dx_2}{dx_1} = \frac{\dfrac{dx_2}{dt}}{\dfrac{dx_1}{dt}} = \frac{px_2 - qx_1x_2}{-rx_1 + sx_1x_2}$$

$$\frac{dx_2}{dx_1} = \frac{x_2(p - qx_1)}{x_1(sx_2 - r)} \tag{13}$$

Equation (13) is solved for specific values of the constants p, q, r, and s in the next example.

EXAMPLE 2 Predator-Prey
Suppose that $x_1 = f(t)$ (hundreds of predators) and $x_2 = g(t)$ (thousands of prey) satisfy the Lotka-Volterra equations (11) and (12) with $p = 3$, $q = 1$, $r = 4$, and $s = 2$. Suppose that at a time when there are 100 predators ($x_1 = 1$), there are 1,000 prey ($x_2 = 1$). Find an equation relating x_1 and x_2.

Solution With the given values of the constants, p, q, r, and s, equation (13) reads

$$\frac{dx_2}{dx_1} = \frac{x_2(3 - x_1)}{x_1(2x_2 - 4)}.$$

Separating the variables yields

$$\frac{2x_2 - 4}{x_2} dx_2 = \frac{3 - x_1}{x_1} dx_1,$$

or

$$\int \left(2 - \frac{4}{x_2} \right) dx_2 = \int \left(\frac{3}{x_1} - 1 \right) dx_1.$$

Evaluating the integrals gives

$$2x_2 - 4 \ln x_2 = 3 \ln x_1 - x_1 + C.$$

(It is not necessary to use absolute value for the logarithms since x_1 and x_2 are positive.) Use the initial conditions $x_1 = 1$ and $x_2 = 1$ to find C.

$$2 - 4(0) = 3(0) - 1 + C$$
$$C = 3$$

The desired equation is

$$x_1 + 2x_2 - 3 \ln x_1 - 4 \ln x_2 = 3. \tag{14}$$

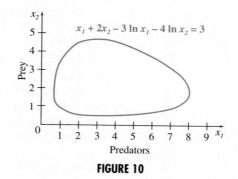

FIGURE 10

A graph of equation (14), in Figure 10, shows that the variations in the populations of the predator and prey are cyclic. We know from Example 1 that the solution moves about this cycle in a clockwise direction. So even though we cannot solve the nonlinear system of differential equations, we can say quite a bit about the solution.

11.5 EXERCISES

For Exercises 1–4, sketch a phase plane diagram, similar to Figure 9.

1. $\dfrac{dx_1}{dt} = 4x_1 - 2x_1x_2$

$\dfrac{dx_2}{dt} = 3x_2 - x_1x_2$

2. $\dfrac{dx_1}{dt} = 3x_1^2x_2 - 6x_1^2$

$\dfrac{dx_2}{dt} = x_1x_2^2 - 2x_2^2$

3. $\dfrac{dx_1}{dt} = 3x_1x_2 - x_1x_2^2$

$\dfrac{dx_2}{dt} = 2x_1^2x_2 - 2x_1x_2$

4. $\dfrac{dx_1}{dt} = x_1x_2^3 - 2x_1x_2^2$

$\dfrac{dx_2}{dt} = 9x_1^2x_2 - 3x_1^3x_2$

*In Exercises 5–8, **(a)** find the nontrivial equilibrium point and **(b)** sketch a phase plane diagram, similar to Figure 9. Note that in these exercises, the lines where $dx_1/dt = 0$ and $dx_2/dt = 0$ are neither vertical nor horizontal.*

5. $\dfrac{dx_1}{dt} = x_1^2 + x_1x_2 - 2x_1$

$\dfrac{dx_2}{dt} = 2x_1x_2 - x_2^2 - x_2$

6. $\dfrac{dx_1}{dt} = x_2^2 - x_1x_2 - x_2$

$\dfrac{dx_2}{dt} = 2x_1^2 + x_1x_2 - 7x_1$

7. $\dfrac{dx_1}{dt} = x_1^2x_2 + x_1x_2^2 - 3x_1x_2$

$\dfrac{dx_2}{dt} = 2x_1x_2^2 - x_1^2x_2 - 3x_1x_2$

8. $\dfrac{dx_1}{dt} = x_1^2x_2 - x_1^3$

$\dfrac{dx_2}{dt} = 6x_2^2 + x_2^3 - 3x_1x_2^2$

 Applications

LIFE SCIENCES

9. *Competing Species* The system of equations

$$\frac{dx_1}{dt} = -2x_1 + 3x_1x_2$$

$$\frac{dx_2}{dt} = 3x_2 - 2x_1x_2$$

describes the influence of the populations of two competing species (such as two birds with similar diets) on their growth rates.

a. Following Example 1 in the text, find an equation relating x and y, assuming $x_2 = 2$ when $x_1 = 1$.

b. Find the equilibrium point.

c. Determine what happens to the species as time progresses.

10. *Symbiotic Species* When two species, such as the rhinoceros and birds pictured below, coexist in a symbiotic (dependent) relationship, they either increase together or decrease together. Typical equations for the growth rates of two such species might be

$$\frac{dx_1}{dt} = -3x_1 + 4x_1x_2$$

$$\frac{dx_2}{dt} = -3x_2 + x_1x_2.$$

a. Find an equation relating x_1 and x_2 if $x_1 = 3$ when $x_2 = 1$.

b. Find the equilibrium point.

c. If both populations are greater than their values at the equilibrium point, what happens to the populations?

d. If both populations are less than their values at the equilibrium point, what happens to the populations?

11. *Whales and Krill* For the system of differential equations (10), suppose that $F_1 < 1$ and $F_2 < 1$. Show that the equilibrium point is given by

$$y_1 = \frac{1 - F_1}{1 + b(1 - F_2)} \quad \text{and} \quad y_2 = \frac{(1 - F_1)(1 - F_2)}{1 + b(1 - F_2)}.$$

12. *Whales and Krill* For the system of differential equations (10), consider the sign of dy_1/dt and dy_2/dt to determine what would happen to the whales and krill if

a. $F_1 < 1$ and $F_2 > 1$;

b. $F_1 > 1$ and $F_2 > 1$.

13. *Mouse Infection* A more detailed model than the one in Exercise 24 of Sec. 11.2 for the spread of an infectious disease among mice is given by the system of differential equations

$$\frac{dx_1}{dt} = a - bx_1 - \beta x_1x_2 + \gamma x_3$$

$$\frac{dx_2}{dt} = \beta x_1x_2 - (b + \alpha + \nu)x_2$$

$$\frac{dx_3}{dt} = \nu x_2 - (\beta + \gamma)x_3,$$

where x_1 is the number of susceptible mice, x_2 is the number of infected mice, x_3 is the number of immune mice, α is the mortality rate due to infection, b is the mortality rate due to natural causes for infected mice, β is a transmission coefficient for the rate that infected mice infect susceptible mice, ν is the rate at which the mice recover from infection, and γ is the rate at which mice lose immunity.[*] In the article describing this model, the author states that an equilibrium solution only exists when

$$\frac{a}{b} > \frac{b + \alpha + \nu}{\beta}.$$

Verify this claim by finding the equilibrium solution and determining that the above inequality must be satisfied for the equilibrium values of x_1, x_2, and x_3 to be positive.

14. *Competing Species* A Lotka-Volterra model in which two species compete for the same source of food (as in Exercise 9) can be described by the system of differential equations

$$\frac{dx_1}{dt} = r_1x_1\left(1 - \frac{x_1}{k_1} - b_1\frac{x_2}{k_1}\right)$$

$$\frac{dx_2}{dt} = r_2x_2\left(1 - \frac{x_2}{k_2} - b_2\frac{x_1}{k_2}\right),$$

where r_1 and r_2 are growth rates for the two species, k_1 and k_2 are the carrying capacity for each species in the absence of the other, and b_1 and b_2 measure the competitive effect of each species on the other.[†] Suppose that $k_1 = 6$, $k_2 = 4$, $b_1 = 2$, and $b_2 = 1$.

a. Find the equilibrium point.

b. Sketch a phase plane diagram, similar to Figure 9.

c. Based on your sketch in part b, what do you expect the populations of the two species to do in the long run?

---

[*]Anderson, Roy M., "The Persistence of Direct Life Cycle Infectious Diseases Within Populations of Hosts," in *Lectures on Mathematics in the Life Sciences, Vol. 12: Some Mathematical Questions in Biology,* the American Mathematical Society, 1979, pp. 1–67.
[†]Murray, J. D., *Mathematical Biology,* Springer-Verlag, 1989, pp. 78–79.

■ 11.6 APPLICATIONS OF DIFFERENTIAL EQUATIONS

 THINK ABOUT IT   During an epidemic, how does the infection rate of a disease vary with time?

Epidemics   Under certain conditions, the spread of a contagious disease can be described with the logistic growth model, as in the next example.

EXAMPLE 1   Spread of an Epidemic

Consider a population of size N that satisfies the following conditions.

1. Initially there is only one infected individual.
2. All uninfected individuals are susceptible, and infection occurs when an uninfected individual contacts an infected individual.
3. Contact between any two individuals is just as likely as contact between any other two individuals.
4. Infected individuals remain infectious.

Let $t =$ the time (in days) and $y =$ the number of individuals infected at time t. At any moment there are $(N - y)y$ possible contacts between an uninfected individual and an infected individual. Thus, it is reasonable to assume that the rate of spread of the disease satisfies the following logistic equation (discussed in Section 11.1):

$$\frac{dy}{dt} = k\left(1 - \frac{y}{N}\right)y. \tag{1}$$

As shown in Section 11.1, the general solution of equation (1) is

$$y = \frac{N}{1 + be^{-kt}}, \tag{2}$$

where $b = (N - y_0)/y_0$. Since just one individual is infected initially, $y_0 = 1$ here. Substituting these values into equation (2) gives

$$y = \frac{N}{1 + (N - 1)e^{-kt}} \tag{3}$$

as the general solution of equation (1).

The infection rate will be a maximum when its derivative is 0, that is, when $d^2y/dt^2 = 0$. Since

$$\frac{dy}{dt} = k\left(1 - \frac{y}{N}\right)y,$$

we have
$$\frac{d^2y}{dt^2} = k\left[\left(1 - \frac{y}{N}\right)(1) + y\left(-\frac{1}{N}\right)\right]$$

$$= k\left(1 - \frac{2y}{N}\right).$$

Set $d^2y/dt^2 = 0$ to get

$$0 = 1 - \frac{2y}{N}$$

$$\frac{2y}{N} = 1$$

$$y = \frac{N}{2}.$$

The maximum infection rate occurs when exactly half the total population is still uninfected and equals

$$\frac{dy}{dt} = k\left(1 - \frac{N/2}{N}\right)\frac{N}{2} = \frac{kN}{4}.$$

Letting $y = N/2$ in equation (3) and solving for t shows that the maximum infection rate occurs at time

$$t_m = \frac{\ln(N - 1)}{k}.$$

The graph of dy/dt, shown in Figure 11, is called the *epidemic curve*. It is symmetric about the line $t = t_m$.

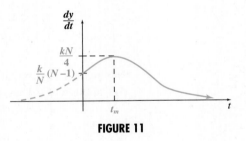

FIGURE 11

Arms Race Models   Lewis F. Richardson* proposed a pair of linear first-order differential equations as a model of an arms race between two rival countries X and Y. In order to simplify the situation, the following assumptions are made.

1. X and Y are roughly equal in size.
2. T_0, the annual trade flow from X to Y (in dollars), is a constant.
3. U_0, the annual trade flow from Y to X (in dollars), is a constant.
4. The arms race between X and Y is not affected by the policies of other countries.

Let

$t =$ the time in years,

$x =$ the armaments budget of X in year t (in dollars) minus T_0,

$y =$ the armaments budget of Y in year t (in dollars) minus U_0.

---

*Richardson, Lewis F., *Arms and Insecurity*, The Boxwood Press, 1960.

Richardson assumed that x stimulates the growth of y, but the sheer size of x would eventually inhibit its own growth. He made similar assumptions about y. These assumptions led him to the following model:

$$\frac{dx}{dt} = ky - mx + g \quad \text{and} \quad \frac{dy}{dt} = kx - my + h, \tag{4}$$

where k, m, g, and h are constants. The constants g and h account for the flow of trade between the two nations and their underlying attitudes toward each other. For example, if the two countries have long-standing grievances and ideological differences that outweigh the value of the constant trade flow, then $g > 0$, $h > 0$, and the armament budgets will grow even when $x = 0 = y$. On the other hand, if the constant trade flow is the dominant factor, then $g < 0$, $h < 0$, and the armament budgets will shrink when $x = 0 = y$. Richardson found that the constant k was roughly proportional to the size of the country, but the constant m was roughly the same for all countries.

To model the mutual stimulation for the growth of armaments, Richardson set $z = x + y$ and added the two equations in (4), obtaining

$$\frac{dz}{dt} = \frac{dx}{dt} + \frac{dy}{dt}$$
$$= ky - mx + g + kx - my + h$$
$$= k(x + y) - m(x + y) + g + h$$
$$= kz - mz + g + h$$
$$\frac{dz}{dt} = g + h - (m - k)z. \tag{5}$$

Solving this differential equation as in Example 3 in Section 11.2 gives the general solution of (6) as

$$z = \frac{g + h}{m - k} + Ce^{-(m-k)t}. \tag{6}$$

Richardson tested his model on the growth of armaments in the years preceding World War I for

    X = the bloc consisting of Russia and France,

    Y = the bloc consisting of Germany and Austro-Hungary.

He found a remarkably good fit of equations (5) and (6) to the data.

R. Taagepera, et al.,* proposed the following model for the arms race (within a finite time span) between Israel and some Arab countries:

$$\frac{dx}{dt} = kx \quad \text{and} \quad \frac{dy}{dt} = my, \tag{7}$$

where k and m are constants. By Example 4 in Section 11.1 the general solution of (7) is

$$x = x_0 e^{kt} \quad \text{and} \quad y = y_0 e^{mt}. \tag{8}$$

---

*Taagepera, R., G. M. Shiffler, R. T. Perkins, and D. L. Wagner, "Soviet-American and Israeli-Arab Arms Races and the Richardson Model," *General Systems,* Vol. 20, 1975, pp. 151–158.

This model asserts that x stimulates the growth of x, but not of y, and vice versa. Here we are supposing that self-stimulation is the controlling factor, whereas the Richardson model presupposes mutual stimulation. Taagepera, et al., found that the self-stimulation model (7)–(8) fit the Israeli-Arab arms race better than the Richardson model (4)–(6). This result is very surprising, for it challenges the common assumption that the driving force in any arms race is mutual stimulation.

Mixing Problems
The mixing of two solutions can lead to a first-order differential equation, as the next example shows.

EXAMPLE 2   Salt Concentration

Suppose a tank contains 100 gallons of a solution of dissolved salt and water, which is kept uniform by stirring. If pure water is allowed to flow into the tank at the rate of 4 gallons per minute, and the mixture flows out at the rate of 3 gallons per minute (see Figure 12), how much salt will remain in the tank after t minutes if 15 pounds of salt are in the mixture initially?*

Solution Let the amount of salt present in the tank at any specific time be $y = f(t)$. The net rate at which y changes is given by

$$\frac{dy}{dt} = (\text{Rate of salt in}) - (\text{Rate of salt out}).$$

Since pure water is coming in, the rate of salt entering the tank is zero. The rate at which salt is leaving the tank is the product of the amount of salt per gallon (in V gallons) and the number of gallons per minute leaving the tank:

$$\text{Rate of salt out} = \left(\frac{y}{V}\,\text{lb/gal}\right)(3\text{ gal/min}).$$

The differential equation, therefore, can be written as

$$\frac{dy}{dt} = -\frac{3y}{V}; \qquad y(0) = 15,$$

where $y(0)$ is the initial amount of salt in the solution. We must take into account the fact that the volume, V, of the mixture is not constant but is determined by

$$\frac{dV}{dt} = (\text{Rate of liquid in}) - (\text{Rate of liquid out}) = 4 - 3 = 1,$$

or

$$\frac{dV}{dt} = 1,$$

from which

$$V(t) = t + C_1.$$

Because the volume is known to be 100 at time $t = 0$, we have $C_1 = 100$, and

$$\frac{dy}{dt} = \frac{-3y}{t + 100}; \qquad y(0) = 15,$$

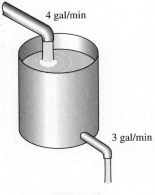

4 gal/min

3 gal/min

FIGURE 12

---

*Andrews, Larry C., *Ordinary Differential Equations with Boundary Value Problems,* HarperCollins Publishers, Inc., 1991, pp. 85–86.

a separable equation with solution

$$\frac{dy}{y} = \frac{-3}{t + 100} \, dt$$

$$\ln y = -3 \ln(t + 100) + C.$$

Since $y = 15$ when $t = 0$,

$$\ln 15 = -3 \ln 100 + C$$

$$\ln 15 + 3 \ln 10^2 = C$$

$$\ln(15 \cdot 10^6) = C.$$

Finally,

$$\ln y = \ln(t + 100)^{-3} + \ln(15 \cdot 10^6)$$

$$= \ln[(t + 100)^{-3}(15 \cdot 10^6)]$$

$$y = \frac{15 \cdot 10^6}{(t + 100)^3}.$$

11.6 EXERCISES

Applications

LIFE SCIENCES

1. *Spread of an Epidemic* The native Hawaiians lived for centuries in isolation from other peoples. When foreigners finally came to the islands they brought with them diseases such as measles, whooping cough, and smallpox, which decimated the population. Suppose such an island has a native population of 5,000, and a sailor from a visiting ship introduces measles, which has an infection rate of 0.00005. Also suppose that the model for spread of an epidemic described in Example 1 applies.

a. Write an equation for the number of natives who remain uninfected. Let t represent time in days.

b. How many are uninfected after 30 days?

c. How many are uninfected after 50 days?

d. When will the maximum infection rate occur?

2. *Spread of an Epidemic* In Example 1, the number of infected individuals is given by equation (3).

a. Show that the number of uninfected individuals is given by

$$N - y = \frac{N(N - 1)}{N - 1 + e^{kt}}.$$

b. Graph the equation in part a and equation (3) on the same axes when $N = 100$ and $k = 1$.

c. Find the common inflection point of the two graphs.

d. What is the significance of the common inflection point?

e. What are the limiting values of y and $N - y$?

3. *Spread of an Epidemic* An influenza epidemic spreads at a rate proportional to the product of the number of people infected and the number not yet infected. Assume that 50 people are infected at the beginning of the epidemic in a community of 10,000 people, and 300 are infected 10 days later.

a. Write an equation for the number of people infected, y, after t days.

b. When will half the community be infected?

4. *Spread of an Epidemic* The Gompertz growth law,

$$\frac{dy}{dt} = kye^{-at},$$

for constants k and a, is another model used to describe the growth of an epidemic. Repeat Exercise 3, using this differential equation with $a = 0.02$.

5. *Spread of Gonorrhea* Gonorrhea is spread by sexual contact, takes 3 to 7 days to incubate, and can be treated with antibiotics. There is no evidence that a person ever develops immunity. One model proposed for the rate of change in the number of men infected by this disease is

$$\frac{dy}{dt} = -ay + b(f - y)Y,$$

where y is the fraction of men infected, f is the fraction of men who are promiscuous, Y is the fraction of women infected, and a and b are appropriate constants.*

a. Assume $a = 1$, $b = 1$, and $f = 0.5$. Choose $Y = 0.01$, and solve for y using $y = 0.02$ when t is 0 as an initial condition. Round your answer to three decimal places.

b. A comparable model for women is

$$\frac{dY}{dt} = -AY + B(F - Y)y,$$

where F is the fraction of women who are promiscuous and A and B are constants. Assume $A = 1$, $B = 1$, and $F = 0.03$. Choose $y = 0.1$ and solve for Y, using $Y = 0.01$ as an initial condition.

OTHER APPLICATIONS

Spread of a Rumor The equation developed in the text for the spread of an epidemic also can be used to describe diffusion of information. In a population of size N, let y be the number of people who have heard a particular piece of information. Then

$$\frac{dy}{dt} = k\left(1 - \frac{y}{N}\right)y$$

for a positive constant k. Use this model in Exercises 6–8.

6. Suppose a rumor starts among 3 people in a certain office building. That is, $y_0 = 3$. Suppose 500 people work in the building and 50 people have heard the rumor in 2 days. Using equation (2), write an equation for the number of people who have heard the rumor in t days. How many people will have heard the rumor in 5 days?

7. A rumor spreads at a rate proportional to the product of the number of people who have heard it and the number who have not heard it. Assume that 5 people in an office with 50 employees heard the rumor initially, and 15 people have heard it 3 days later.

a. Write an equation for the number, y, of people who have heard the rumor in t days.

b. When will 30 employees have heard the rumor?

8. A news item is heard on the late news by 5 of the 100 people in a small community. By the end of the next day 20 people have heard the news. Using equation (2), write an equation for the number of people who have heard the news in t days. How many have heard the news after 3 days?

9. Repeat Exercise 7 using the Gompertz growth law,

$$\frac{dy}{dt} = kye^{-at},$$

for constants k and a, with $a = 0.1$.

10. *Salt Concentration* A tank holds 100 gal of water that contains 20 lb of dissolved salt. A brine (salt) solution is flowing into the tank at the rate of 2 gal/min while the solution flows out of the tank at the same rate. The brine solution entering the tank has a salt concentration of 2 lb/gal.

a. Find an expression for the amount of salt in the tank at any time.

b. How much salt is present after 1 hr?

c. As time increases, what happens to the salt concentration?

11. Solve Exercise 10 if the brine solution is introduced at the rate of 3 gal/min while the rate of outflow remains the same.

12. Solve Exercise 10 if the brine solution is introduced at the rate of 1 gal/min while the rate of outflow stays the same.

13. Solve Exercise 10 if pure water is added instead of brine.

14. *Chemical in a Solution* Five grams of a chemical is dissolved in 100 liters of alcohol. Pure alcohol is added at the rate of 2 liters/min and at the same time the solution is being drained at the rate of 1 liter/min.

a. Find an expression for the amount of the chemical in the mixture at any time.

b. How much of the chemical is present after 30 min?

15. Solve Exercise 14 if a 10% solution of the same mixture is added instead of pure alcohol.

16. *Soap Concentration* A prankster puts 4 lb of soap in a fountain that contains 200 gal of water. To clean up the mess a city crew runs clear water into the fountain at the rate of 8 gal/min allowing the excess solution to drain off at the same rate. How long will it be before the amount of soap in the mixture is reduced to 1 lb?

*Bender, Edward A., *An Introduction to Mathematical Modeling.* Copyright © 1978 by John Wiley and Sons, Inc. Reprinted by permission.

■ CHAPTER SUMMARY

In this chapter we solved differential equations by finding functions that satisfy a particular equation involving the derivative of that function. We developed the methods of separation of variables and linear first-order equations to solve a wide range of differential equations. We also introduced a numerical method, Euler's method, that employs technology to solve problems in which the other methods cannot be applied. We then studied linear and nonlinear systems of differential equations. Applications in the chapter included the logistic equation for populations, a predator-prey model, an arms race model, and mixing problems.

■ KEY TERMS

	separation of variables	**11.2** linear first-order	**11.5** equilibrium point
differential equation	growth rate constant	differential equation	Lotka-Volterra
11.1 general solution	carrying capacity	integrating factor	equations
particular solution	logistic growth model	**11.3** Euler's method	phase plane diagram
initial condition	logistic equation	**11.4** linear system of	
initial value problem	logistic curve	differential	
separable differential		equations	
equation			

■ CHAPTER 11 REVIEW EXERCISES

1. What is a differential equation? What is it used for?

2. What is the difference between a particular solution and a general solution to a differential equation?

3. How can you tell that a differential equation is separable? That it is linear?

4. Can a differential equation be both separable and linear? Explain why not, or give an example of an equation that is both.

In Exercises 5–12, classify each equation as separable, linear, both, or neither.

5. $y\dfrac{dy}{dx} = x + y$

6. $\dfrac{dy}{dx} + y^2 = xy^2$

7. $\sqrt{x}\,\dfrac{dy}{dx} = \dfrac{1 + e^x}{y}$

8. $\dfrac{dy}{dx} = xy + \ln x$

9. $\dfrac{dy}{dx} + x = xy$

10. $\dfrac{x}{y}\dfrac{dy}{dx} = 1 + x^{3/2}$

11. $x\dfrac{dy}{dx} + y = e^x(1 + y)$

12. $\dfrac{dy}{dx} = x^2 + y^2$

Find general solutions for the following differential equations.

13. $\dfrac{dy}{dx} = 2x^3 + 6x$

14. $\dfrac{dy}{dx} = x \cos x$

15. $\dfrac{dy}{dx} = 4e^x$

16. $\dfrac{dy}{dx} = \dfrac{1}{2x + 3}$

17. $\dfrac{dy}{dx} = \dfrac{3x + 1}{y}$

18. $\dfrac{dy}{dx} = \dfrac{e^x + x}{y - 1}$

19. $\dfrac{dy}{dx} = \dfrac{2y + 1}{x}$

20. $\dfrac{dy}{dx} = \dfrac{3 - y}{e^x}$

21. $\dfrac{dy}{dx} + 5y = 12$

22. $\dfrac{dy}{dx} + xy = 4x$

23. $3\dfrac{dy}{dx} + xy - x = 0$

24. $x\dfrac{dy}{dx} + y - e^x = 0$

Find particular solutions for the following initial value problems. (Some solutions may give y implicitly.)

25. $\dfrac{dy}{dx} = 2x \cos x^2; \quad y(0) = 1$

26. $\dfrac{dy}{dx} = (3x + 2)^2 e^y; \quad y(0) = 0$

27. $\dfrac{dy}{dx} = 5(e^{-x} - 1); \quad y(0) = 17$

28. $e^x \dfrac{dy}{dx} - e^x y = x^2 - 1; \quad y(0) = 42$

29. $(5 - 2x)y = \dfrac{dy}{dx}; \quad y(0) = 2$

30. $\dfrac{dy}{dx} + x^2 y = x^2; \quad y(0) = 8$

31. $\dfrac{dy}{dx} = \dfrac{1 - 2x}{y + 3}; \quad y(0) = 16$

32. $x \dfrac{dy}{dx} - 2x^2 y + 3x^2 = 0; \quad y(0) = 15$

Find particular solutions for the following differential equations.

33. $\dfrac{dy}{dx} = 4x^3 + 2; \quad y(1) = 3$

34. $\dfrac{dy}{dx} = \dfrac{x}{x^2 - 3}; \quad y(2) = 52$

35. $\sqrt{x} \dfrac{dy}{dx} = xy; \quad y(1) = 4$

36. $x^2 \dfrac{dy}{dx} + 4xy - e^{2x^3} = 0; \quad y(1) = e^2$

37. When is Euler's method useful?

Use Euler's method to approximate the indicated function value for $y = f(x)$ to three decimal places, using $h = 0.2$.

38. $\dfrac{dy}{dx} = x + y^{-1}; \quad y(0) = 1; \quad$ find $y(1)$

39. $\dfrac{dy}{dx} = e^x + y; \quad y(0) = 1; \quad$ find $y(0.6)$

40. Let $y = f(x)$ and $dy/dx = (x/2) + 4$, with $y(0) = 0$. Use Euler's method with $h = 0.1$ to approximate $y(0.3)$ to three decimal places. Then solve the differential equation and find $f(0.3)$ to three decimal places. Also, find $y_3 - f(x_3)$.

41. Let $y = f(x)$ and $dy/dx = 3 + \sqrt{y}$, with $y(0) = 0$. Construct a table for x_i and y_i like the one in Section 11.3, Example 2, for $[0, 1]$, with $h = 0.2$. Then graph the polygonal approximation of the graph of $y = f(x)$.

42. What is the logistic equation? Why is it useful?

Solve each of the following systems of differential equations.

43. $\dfrac{dx_1}{dt} = 2x_1 + 7x_2 + t$

$\dfrac{dx_2}{dt} = 7x_1 + 2x_2 + 2t$

44. $\dfrac{dx_1}{dt} = 7x_1 + 2x_2 + 3t$

$\dfrac{dx_2}{dt} = -4x_1 + x_2 - t$

In Exercises 45 and 46, (a) find the nontrivial equilibrium point and (b) sketch a phase plane diagram, similar to Figure 9.

45. $\dfrac{dx_1}{dt} = 4x_1 - x_1^2 - x_1 x_2$

$\dfrac{dx_2}{dt} = 2x_1^2 - x_1 x_2 - 2x_1$

46. $\dfrac{dx_1}{dt} = x_1 x_2 - x_2^2 - x_2$

$\dfrac{dx_2}{dt} = 4x_1 x_2 - x_1^2 x_2 - 2x_1 x_2^2$

Applications

LIFE SCIENCES

47. *Effect of Insecticide* After use of an experimental insecticide, the rate of decline of an insect population is

$$\frac{dy}{dt} = \frac{-10}{1 + 5t},$$

where t is the number of hours after the insecticide is applied. Assume that there were 50 insects initially.

a. How many are left after 24 hours?

b. How long will it take for the entire population to die?

48. *Growth of a Mite Population* A population of mites grows at a rate proportional to the number present, y. If the growth constant is 10% and 120 mites are present at time $t = 0$ (in weeks), find the number present after 6 weeks.

49. *Competing Species* Find an equation relating x_1 to x_2 given the following equations, which describe the interaction of two competing species and their growth rates.

$$\frac{dx_1}{dt} = 0.2x_1 - 0.5x_1x_2$$

$$\frac{dx_2}{dt} = -0.3x_2 + 0.4x_1x_2$$

Find the values of x_1 and x_2 for which both growth rates are 0.

50. *Smoke Content in a Room* The air in a meeting room of 15,000 cubic feet has a smoke content of 20 parts per million (ppm). An air conditioner is turned on, which brings fresh air (with no smoke) into the room at a rate of 1,200 cubic feet per minute and forces the smoky air out at the same rate. How long will it take to reduce the smoke content to 5 ppm?

51. In Exercise 50, how long will it take to reduce the smoke content to 10 ppm if smokers in the room are adding smoke at the rate of 5 ppm per minute?

52. *Spread of Influenza* A small, isolated mountain community with a population of 700 is visited by an outsider who carries influenza. After 6 weeks, 300 people are uninfected.

a. Write an equation for the number of people who remain uninfected at time t (in weeks).

b. Find the number still uninfected after 7 weeks.

c. When will the maximum infection rate occur?

53. *Population Growth* Let

$$y = \frac{N}{1 + be^{-kx}}.$$

If y is y_1, y_2, and y_3 at times x_1, x_2, and $x_3 = 2x_2 - x_1$ (that is, at three equally spaced times), then prove that

$$N = \frac{1/y_1 + 1/y_3 - 2/y_2}{1/(y_1y_3) - 1/y_2^2}.$$

Population Growth In the following table of United States census figures, y is population in millions.*

Year	y	Year	y
1790	3.9	1900	76.0
1800	5.3	1910	92.0
1810	7.2	1920	105.7
1820	9.6	1930	122.8
1830	12.9	1940	131.7
1840	17.1	1950	150.7
1850	23.2	1960	179.3
1860	31.4	1970	203.3
1870	39.8	1980	226.5
1880	50.2	1990	248.7
1890	62.9	2000	281.4

54. Use Exercise 53 and the table to find the following.

a. Find N using the years 1800, 1850, and 1900.

b. Find N using the years 1850, 1900, and 1950.

c. Find N using the years 1870, 1920, and 1970.

d. Explain why different values of N were obtained in parts a–c. What does this suggest about the validity of this model and others?

55. Let $x = 0$ correspond to 1870, and let every decade correspond to an increase in x of 1.

a. Use 1870, 1920, and 1970 to find N, 1870 to find b, and 1920 to find k in the equation

$$y = \frac{N}{1 + be^{-kx}}.$$

b. Estimate the population of the United States in 1990 and compare your estimate to the actual population in 1990.

c. Predict the populations of the United States in 2010 and 2050.

56. Let $x = 0$ correspond to 1790, and let every decade correspond to an increase in x of 1. Use a calculator with logistic regression capability to complete the following.

a. Plot the data points. Do the points suggest that a logistic function is appropriate here?

---

*U. S. Bureau of the Census

b. Use the logistic regression function on your calculator to determine the logistic equation that best fits the data.

c. Plot the logistic equation from part a on the same graph as the data points. How well does the logistic equation seem to fit the data?

d. What seems to be the limiting size of the United States population?

OTHER APPLICATIONS

57. *Education* Researchers have proposed that the amount a full-time student is educated (x) changes with respect to the student's age t according to the differential equation

$$\frac{dx}{dt} = 1 - kx,$$

where k is a constant measuring the rate that education depreciates due to forgetting or technological obsolescence.*

a. Solve the equation using the method of separation of variables.

b. Solve the equation using an integrating factor.

c. What does x approach over time?

58. *Spread of a Rumor* A rumor spreads through the offices of a company with 100 employees, starting in a meeting with 4 people. After 3 days, 15 people have heard the rumor.

a. Write an equation for the number of people who have heard the rumor in x days.

b. How many people have heard the rumor in 5 days?

59. *Newton's Law of Cooling* A roast at a temperature of 40°F is put in a 300°F oven. After 1 hr the roast has reached a temperature of 150°F. Newton's law of cooling states that

$$\frac{dT}{dt} = k(T - T_F),$$

where T is the temperature of an object, the surrounding medium has temperature T_F at time t, and k is a constant. Use Newton's law to find the temperature of the roast after 2 hr.

60. In Exercise 59, how long does it take for the roast to reach a temperature of 250°F?

61. *Air Resistance* In the section on antiderivatives, we saw that the acceleration of gravity is a constant if air resistance is ignored. But air resistance cannot always be ignored, or

parachutes would be of little use. In the presence of air resistance, the equation for acceleration also contains a term roughly proportional to the velocity squared. Since acceleration forces a falling object downward and air resistance pushes it upward, the air resistance term is opposite in sign to the acceleration of gravity. Thus,

$$a(t) = \frac{dv}{dt} = g - kv^2,$$

where g and k are positive constants. Future calculations will be simpler if we replace g and k by the squared constants G^2 and K^2, giving

$$\frac{dv}{dt} = G^2 - K^2v^2.$$

a. Use separation of variables and the fact that

$$\frac{1}{G^2 - K^2v^2} = \frac{1}{2G}\left(\frac{1}{G - Kv} + \frac{1}{G + Kv}\right)$$

to solve the differential equation above. Assume $v < G/K$, which is certainly true when the object starts falling (with $v = 0$). Write your solution in the form of v as a function of t.

b. Find $\lim_{t \to \infty} v(t)$, where $v(t)$ is the solution you found in part a. What does this tell you about a falling object in the presence of air resistance?

c. According to *Harper's Index,* the terminal velocity of a cat falling from a tall building is 60 mph.† Use your answers from part b, plus the fact that 60 mph = 88 ft per sec and g, the acceleration of gravity, is 32 ft per sec², to find a formula for the velocity of a falling cat (in ft per sec) as a function of time (in sec). (*Hint:* Find K in terms of G. Then substitute into the answer from part a.)

---

*Southwick, Lawrence, Jr., and Stanley Zionts, "An Optimal-Control-Theory Approach to the Education-Investment Decision," *Operations Research,* Vol. 22, 1974, pp. 1156–1174.
†*Harper's,* Oct. 1994, p. 13.

EXTENDED APPLICATION: Pollution of the Great Lakes*

Industrial nations are beginning to face the problems of water pollution. Lakes present a problem, because a polluted lake contains a considerable amount of water that must somehow be cleaned. The main cleanup mechanism is the natural process of gradually replacing the water in the lake. This application deals with pollution in the Great Lakes. The basic idea is to regard the flow in the Great Lakes as a mixing problem.

We make the following assumptions.

1. Rainfall and evaporation balance each other, so the average rates of inflow and outflow are equal.

2. The average rates of inflow and outflow do not vary much seasonally.

3. When water enters the lake, perfect mixing occurs, so that the pollutants are uniformly distributed.

4. Pollutants are not removed from the lake by decay, sedimentation, or in any other way except outflow.

5. Pollutants flow freely out of the lake; they are not retained (as DDT is).

(The first two are valid assumptions; however, the last three are questionable.)

We will use the following variables in the discussion to follow.

V = volume of the lake
P_L = pollution concentration in the lake at time t
P_i = pollution concentration in the inflow
    to the lake at time t
r = rate of flow
t = time in years

By the assumptions stated above, the net change in total pollutants during the time interval Δt is (approximately)

$$V \cdot \Delta P_L = (P_i - P_L)(r \cdot \Delta t),$$

where ΔP_L is the change in the pollution concentration. Dividing this equation by Δt and by V and taking the limit as $\Delta t \to 0$, we get the differential equation

$$\frac{dP_L}{dt} = \frac{(P_i - P_L)r}{V}.$$

Since we are treating V and r as constants, we replace r/V with k, so the equation can be written as the first-order linear equation

$$\frac{dP_L}{dt} + kP_L = kP_i.$$

The solution is

$$P_L(t) = e^{-kt}\left[P_L(0) + k \int_0^t P(x)e^{kx}\,dx \right]. \tag{1}$$

Figure 13 shows values of $1/k$ for each lake (except Huron) measured in years. *If the model is reasonable*, the numbers in the figure can be used in equation (1) to determine the effect of various pollution abatement schemes. Lake Ontario is excluded from the discussion because about 84% of its inflow comes from Erie and can be controlled only indirectly.

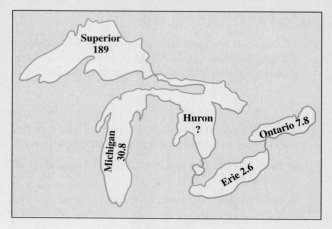

FIGURE 13

The fastest possible cleanup will occur if all pollution inflow ceases. This means that $P_i = 0$. In this case, equation (1) leads to

$$t = \frac{1}{k} \ln\left(\frac{P_L(0)}{P_L(t)} \right).$$

From this we can tell the length of time necessary to reduce pollution to a given percentage of its present level. For example, from the figure, for Lake Superior $1/k = 189$. Thus, to reduce pollution to 50% of its present level, $P_L(0)$, we want

$$\frac{P_L(t)}{P_L(0)} = 0.5 \quad \text{or} \quad \frac{P_L(0)}{P_L(t)} = 2,$$

from which

$$t = 189 \ln 2 \approx 131.$$

The figures in Table 4, representing years, were found in this way.

Table 4

Lake	50%	20%	10%	5%
Erie	2	4	6	8
Michigan	21	50	71	92
Superior	131	304	435	566

Fortunately, the pollution in Lake Superior is quite low at present.

---

*Bender, Edward A., *An Introduction to Mathematical Modeling.* Copyright © 1978 by John Wiley & Sons, Inc. Reprinted by permission.

As mentioned before, assumptions 3, 4, and 5 are questionable. For persistent pollutants like DDT, the estimated cleanup times may be too low. For other pollutants, how assumptions 4 and 5 affect cleanup times is unclear. However, the values of $1/k$ given in the figure probably provide rough lower bounds for the cleanup times of persistent pollutants.

Exercises

1. Calculate the number of years to reduce pollution in Lake Erie to each of the following levels.

   a. 40%   **b.** 30%

2. Repeat Exercise 1 for Lake Michigan.

3. Repeat Exercise 1 for Lake Superior.

4. We claim that equation (1) is a solution of the differential equation

$$\frac{dP_L}{dt} + kP_L(t) = kP_i(t),$$

   where t measures time from the present. The constant $k = r/V$ measures how quickly the water in the lake is replaced through inflow and corresponding outflow. The constant $P_L(0)$ is the current pollution level.

   a. To show that equation (1) does define a solution of the differential equation, multiply both sides of equation (1) by e^{kt} and then differentiate both sides with respect to t. Remember from Exercise 52 in the section on the Fundamental Theorem of Calculus that you can differentiate an integral by using the version of the Fundamental Theorem that says

$$\frac{d}{dt}\int_a^t f(x)\, dx = f(t).$$

   b. When you substitute $t = 0$ into the right-hand side of equation (1), you should get $P_L(0)$. Do you? What happens to the integral? What happens to the factor of e^{-kt}?

   c. The map indicates a value of 30.8 for Lake Michigan. What value of k does this correspond to? What percent of the water in Lake Michigan is replaced each year by inflow? Which lake has the biggest annual water turnover?

5. Suppose that instead of assuming that all pollution inflow immediately ceases, we model $P_i(t)$ by a decaying exponential of the form $a \cdot e^{-pt}$, where p is a constant that tells us how fast the inflow is being cleaned of pollution. To simplify things, we'll also assume that initially the inflow and the lake have the same pollution concentration, so $a = P_L(0)$. Now substitute $P_L(0)e^{-px}$ for $P_i(x)$ in equation (1), and evaluate the integral as a function of t.

6. When you simplify the right-hand side of equation (1) using your new expression for the integral, and then factor out and divide by $P_L(0)$, you'll get the following nice expression for the ratio $P_L(t)/P_L(0)$:

$$\frac{P_L(t)}{P_L(0)} = \frac{1}{k - p}(ke^{-pt} - pe^{-kt}).$$

   a. Suppose that for Lake Michigan the constant p is equal to 0.02. Use a graph of the ratio $P_L(t)/P_L(0)$ to estimate how long it will take to reduce pollution to 50% of its current value. How does this compare with the time, assuming pollution-free inflow?

   b. If the constant p has the value 0 for Lake Michigan, what does that tell you about the pollution level in the inflow? In this case, what happens to the ratio $P_L(t)/P_L(0)$ over time?

CHAPTER

12

Probability

Physicians are required to determine the best treatment strategy for a particular patient with a particular disease. Although it is generally impossible to know beforehand exactly how a patient will respond to certain types of medical procedures or pharmacological agents, probability can help physicians and researchers compare various options. In this chapter, some of the basic ideas from probability theory provide the foundation from which powerful decision tools are constructed.

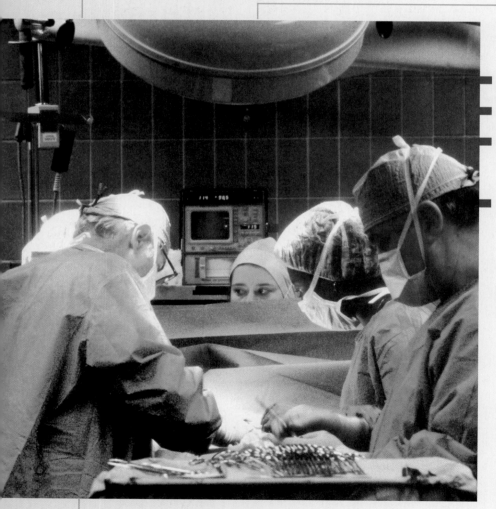

12.1 Sets

12.2 Introduction to Probability

12.3 Conditional Probability; Independent Events; Bayes' Theorem

12.4 Discrete Random Variables; Applications to Decision Making

Review Exercises

Extended Application: Medical Diagnosis

In this chapter we introduce the basic ideas of probability theory, a branch of mathematics that has become increasingly important in the biological and social sciences. Probability theory is valuable because it provides a way to deal with uncertainty. Since the language of sets and set operations is used in the study of probability, we begin there.

■ 12.1 SETS

| ? | THINK ABOUT IT | How many hockey players sustained head and face injuries while wearing a full shield? |

Using knowledge of sets, we will answer this question in one of the examples.

Think of a **set** as a well-defined collection of objects in which it is possible to determine if a given object is included in the collection. A set of bones might include one of each bone found in the human body. Another set might be made up of all the students in your biology class. In mathematics, sets are often made up of numbers. The set consisting of the numbers 3, 4, and 5 is written

$$\{3, 4, 5\},$$

with set braces, { }, enclosing the numbers belonging to the set. The numbers 3, 4, and 5 are called the **elements** or **members** of this set. To show that 4 is an element of the set $\{3, 4, 5\}$, we use the symbol $\in$ and write

$$4 \in \{3, 4, 5\},$$

read "4 is an element of the set containing 3, 4, and 5." Also, $5 \in \{3, 4, 5\}$. To show that 8 is *not* an element of this set, place a slash through the symbol:

$$8 \notin \{3, 4, 5\}.$$

Sets often are named with capital letters, so that if

$$B = \{5, 6, 7\},$$

then, for example, $6 \in B$ and $10 \notin B$.

A collection of people called "young adults" does not constitute a set unless the words "young adults" are defined so that membership in the set can be determined.

It is possible to have a set with no elements. Some examples are the set of counting (or natural) numbers less than one, the set of female presidents of the United States, and the set of men more than 10 feet tall. A set with no elements is called the **empty set** and is written $\emptyset$.

▌**CAUTION** Be careful to distinguish between the symbols, 0, $\emptyset$, $\{0\}$, and $\{\emptyset\}$. The symbol 0 represents a *number;* $\emptyset$ represents a *set* with 0 elements; $\{0\}$ represents a set with one element, 0; and $\{\emptyset\}$ represents a set with one element, $\emptyset$.

Two sets are *equal* if they contain the same elements. The sets $\{5, 6, 7\}, \{7, 6, 5\}$, and $\{6, 5, 7\}$ all contain exactly the same elements and are equal. In symbols,

$$\{5, 6, 7\} = \{7, 6, 5\} = \{6, 5, 7\}.$$

Sets that do not contain exactly the same elements are *not equal.* For example, the sets $\{5, 6, 7\}$ and $\{7, 8, 9\}$ do not contain exactly the same elements and thus are not equal. To indicate that these sets are not equal, we write

$$\{5, 6, 7\} \neq \{7, 8, 9\}.$$

The sets $\{1, 2, 3\}$ and $\{1, 1, 2, 3, 3\}$, however, are equal because both sets contain the same elements.

Sometimes we are interested in a common property of the elements in a set, rather than a list of the elements. This common property can be expressed by using **set-builder notation,** for example,

$$\{x \,|\, x \text{ has property } P\}$$

(read "the set of all elements x such that x has property P") represents the set of all elements x having some stated property P.

EXAMPLE 1 Sets

Write the elements belonging to each of the following sets.

(a) $\{x \,|\, x \text{ is a natural number less than } 5\}$

Solution The natural numbers less than 5 make up the set $\{1, 2, 3, 4\}$.

(b) $\{x \,|\, x \text{ is a bone in the middle ear}\}$

Solution The bones in the middle ear make up the set {malleus, incus, stapes}.

The **universal set** for a particular discussion is a set that includes all the objects being discussed. In elementary school arithmetic, for instance, the set of whole numbers might be the universal set, while in a college algebra class the universal set might be the set of real numbers. The universal set will be specified when necessary, or it will be clearly understandable from the context of the problem.

Subsets Sometimes every element of one set also belongs to another set. For example, if

$$A = \{3, 4, 5, 6\}$$

and

$$B = \{2, 3, 4, 5, 6, 7, 8\},$$

then every element of A is also an element of B. This is an example of the following definition.

SUBSET

Set A is a **subset** of set B (written $A \subseteq B$) if every element of A is also an element of B. Set A is a *proper subset* (written $A \subset B$) if $A \subseteq B$ and $A \neq B$.

To indicate that A is *not* a subset of B, we write $A \nsubseteq B$.

EXAMPLE 2 Sets

Decide whether the following statements are true or false.

(a) $\{3, 4, 5, 6\} = \{4, 6, 3, 5\}$

Solution Both sets contain exactly the same elements, so the sets are equal and the given statement is true. (The fact that the elements are listed in a different order does not matter.)

(b) $\{5, 6, 9, 12\} \subseteq \{5, 6, 7, 8, 9, 10, 11\}$

Solution The first set is not a subset of the second because it contains an element, 12, that does not belong to the second set. Therefore, the statement is false. ▪

By the definition of subset, the empty set (which contains no elements) is a subset of every set. That is, if A is any set, and the symbol $\emptyset$ represents the empty set, then $\emptyset \subseteq A$. Also, the definition of subset can be used to show that every set is a subset of itself; that is, if A is any set, then $A \subseteq A$.

For any set A,

$$\emptyset \subseteq A \quad \textbf{and} \quad A \subseteq A.$$

EXAMPLE 3 Subsets

List all possible subsets for each of the following sets.

(a) $\{7, 8\}$

Solution There are 4 subsets of $\{7, 8\}$:

$$\emptyset, \quad \{7\}, \quad \{8\}, \quad \text{and} \quad \{7, 8\}.$$

(b) $\{a, b, c\}$

Solution There are 8 subsets of $\{a, b, c\}$:

$$\emptyset, \quad \{a\}, \quad \{b\}, \quad \{c\}, \quad \{a, b\}, \quad \{a, c\}, \quad \{b, c\}, \quad \text{and} \quad \{a, b, c\}. \quad ▪$$

In Example 3, all the subsets of $\{7, 8\}$ and all the subsets of $\{a, b, c\}$ were found by trial and error. An alternative method uses a **tree diagram,** a systematic way of listing all the subsets of a given set. Figure 1 shows tree diagrams for finding the subsets of $\{7, 8\}$ and $\{a, b, c\}$.

As Figure 1 shows, there are two possibilities for each element (either it's in the subset or it's not), so a set with 2 elements has $2 \cdot 2 = 2^2 = 4$ subsets, and a set with 3 elements has $2^3 = 8$ subsets. This idea can be extended to a set with any finite number of elements, which leads to the following conclusion.

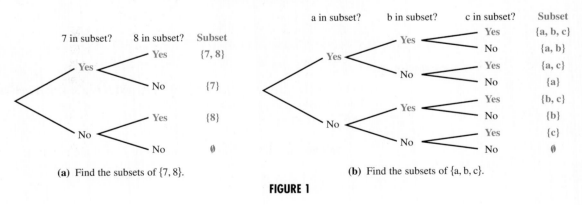

(a) Find the subsets of $\{7, 8\}$. **(b)** Find the subsets of $\{a, b, c\}$.

FIGURE 1

A set of n distinct elements has 2^n subsets.

EXAMPLE 4 Subsets

Find the number of subsets for each of the following sets.

(a) $\{3, 4, 5, 6, 7\}$

Solution This set has 5 elements; thus, it has 2^5 or 32 subsets.

(b) $\{-1, 2, 3, 4, 5, 6, 12, 14\}$

Solution This set has 8 elements and therefore has 2^8 or 256 subsets.

(c) $\emptyset$

Solution Since the empty set has 0 elements, it has $2^0 = 1$ subset—itself.

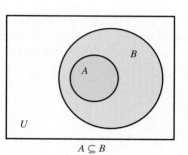

$A \subseteq B$

FIGURE 2

Figure 2 shows a set A that is a subset of set B. The rectangle represents the universal set, U. Such diagrams, called **Venn diagrams**—after the English logician John Venn (1834–1923), who invented them in 1876—are used to help illustrate relationships among sets.

Set Operations It is possible to form new sets by combining or manipulating one or more existing sets. Given a set A and a universal set U, the set of all elements of U that do *not* belong to A is called the **complement** of set A. For example, if set A is the set of all the female students in a class, and U is the set of all students in the class, then the complement of A would be the set of all male students in the class. The complement of set A is written A', read "A-prime."

COMPLEMENT OF A SET

Let A be any set, with U representing the universal set. Then the complement of A, colored red in the figure is

$$A' = \{x \mid x \notin A \text{ and } x \in U\}.$$

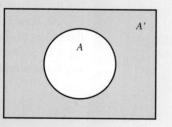

EXAMPLE 5 Set Operations

Let $U = \{1, 2, 3, 4, 5, 6, 7\}$, $A = \{1, 3, 5, 7\}$, and $B = \{3, 4, 6\}$. Find each of the following sets.

(a) A'

Solution Set A' contains the elements of U that are not in A.

$$A' = \{2, 4, 6\}$$

(b) $B' = \{1, 2, 5, 7\}$

(c) $\emptyset' = U$ and $U' = \emptyset$

(d) $(A')' = A$

Given two sets A and B, the set of all elements belonging to *both* set A and set B is called the **intersection** of the two sets, written $A \cap B$. For example, the elements that belong to both set $A = \{1, 2, 4, 5, 7\}$ and set $B = \{2, 4, 5, 7, 9, 11\}$ are 2, 4, 5, and 7, so that

$$A \cap B = \{1, 2, 4, 5, 7\} \cap \{2, 4, 5, 7, 9, 11\} = \{2, 4, 5, 7\}.$$

INTERSECTION OF TWO SETS

The intersection of sets A and B, shown in green in the figure, is

$$A \cap B = \{x \mid x \in A \text{ and } x \in B\}.$$

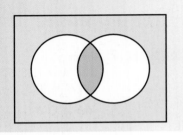

EXAMPLE 6 Set Operations

Let $A = \{3, 6, 9\}$, $B = \{2, 4, 6, 8\}$, and the universal set $U = \{0, 1, 2, \ldots, 10\}$. Find each of the following.

(a) $A \cap B$

Solution

$$A \cap B = \{3, 6, 9\} \cap \{2, 4, 6, 8\} = \{6\}$$

(b) $A \cap B'$

Solution

$$A \cap B' = \{3, 6, 9\} \cap \{0, 1, 3, 5, 7, 9, 10\} = \{3, 9\}$$

Two sets that have no elements in common are called **disjoint sets.** For example, there are no elements common to both $\{50, 51, 54\}$ and $\{52, 53, 55, 56\}$, so these two sets are disjoint, and

$$\{50, 51, 54\} \cap \{52, 53, 55, 56\} = \emptyset.$$

This result can be generalized.

DISJOINT SETS

For any sets A and B, if A and B are disjoint sets, then $A \cap B = \emptyset$.

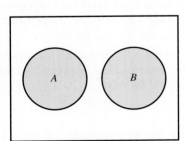

A and B are disjoint sets.

FIGURE 3

Figure 3 shows a pair of disjoint sets.

The set of all elements belonging to set A, to set B, or to both sets is called the **union** of the two sets, written $A \cup B$. For example,

$$\{1, 3, 5\} \cup \{3, 5, 7, 9\} = \{1, 3, 5, 7, 9\}.$$

UNION OF TWO SETS

The union of sets A and B, shown in blue in the figure, is

$$A \cup B = \{x \mid x \in A \text{ or } x \in B\}.$$

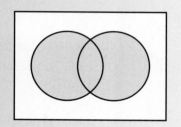

EXAMPLE 7 Union of Sets

Let $A = \{1, 3, 5, 7, 9, 11\}$, $B = \{3, 6, 9, 12\}$, $C = \{1, 2, 3, 4, 5\}$, and the universal set $U = \{0, 1, 2, \ldots, 12\}$. Find each of the following.

(a) $A \cup B$

Solution Begin by listing the elements of the first set, $\{1, 3, 5, 7, 9, 11\}$. Then include any elements from the second set *that are not already listed.* Doing this gives

$$A \cup B = \{1, 3, 5, 7, 9, 11\} \cup \{3, 6, 9, 12\} = \{1, 3, 5, 7, 9, 11, 6, 12\}$$
$$= \{1, 3, 5, 6, 7, 9, 11, 12\}.$$

(b) $(A \cup B) \cap C'$

Solution Begin with the expression in parentheses, which we calculated in part (a), and then intersect this with C'.

$$(A \cup B) \cap C' = \{1, 3, 5, 6, 7, 9, 11, 12\} \cap \{0, 6, 7, 8, 9, 10, 11, 12\}$$
$$= \{6, 7, 9, 11, 12\}$$

NOTE 1. As Example 7 shows, when forming sets, do not list the same element more than once. In our final answer, we listed the elements in numerical order to make it easier to see what elements are in the set, but the set is the same, regardless of the order of the elements.

2. As shown in the definitions, an element is in the *intersection* of sets A and B if it is in A *and* B. On the other hand, an element is in the *union* of sets A and B if it is in A *or* B (or both).

Applications of Venn Diagrams   As we will see, Venn diagrams can be used to efficiently represent a variety of complicated sets. The rectangular region of a Venn diagram represents the universal set U. Including only a single set A inside

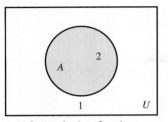

One set leads to 2 regions
(numbering is arbitrary).

FIGURE 4

the universal set, as in Figure 4, divides U into two regions. Region 1 represents those elements of U outside set A (that is, the elements in A'), and region 2 represents those elements belonging to set A. (Our numbering of these regions is arbitrary.)

The Venn diagram in Figure 5(a) shows two sets inside U. These two sets divide the universal set into four regions. As labeled in Figure 5(a), region 1 represents the set whose elements are outside both set A and set B. Region 2 shows the set whose elements belong to A and not to B. Region 3 represents the set whose elements belong to both A and B. Which set is represented by region 4? (Again, the labeling is arbitrary.) Two other situations can arise when representing sets by Venn diagrams. If it is known that $A \cap B = \emptyset$, then the Venn diagram is drawn as in Figure 5(b). If it is known that $A \subseteq B$, then the Venn diagram is drawn as in Figure 5(c). For the material presented throughout this chapter we will only refer to Venn diagrams like the one in Figure 5(a), and note that some of the regions of the Venn diagram may be equal to the null set.

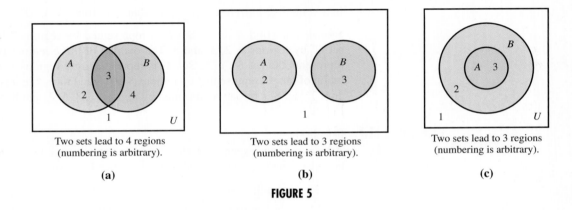

Two sets lead to 4 regions
(numbering is arbitrary).

(a)

Two sets lead to 3 regions
(numbering is arbitrary).

(b)

Two sets lead to 3 regions
(numbering is arbitrary).

(c)

FIGURE 5

EXAMPLE 8   Venn Diagrams

Draw Venn diagrams similar to Figure 5(a) and shade the regions representing the following sets.

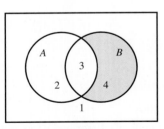

FIGURE 6

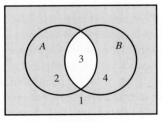

FIGURE 7

(a) $A' \cap B$

Solution Set A' contains all the elements outside set A. As labeled in Figure 5(a), A' is represented by regions 1 and 4. Set B is represented by regions 3 and 4. The intersection of sets A' and B, the set $A' \cap B$, is given by the region common to the combined regions 1 and 4 and the combined regions 3 and 4. The result is the set represented by region 4, which is blue in Figure 6. When looking for the intersection, remember to choose the area that is in one region *and* the other region.

(b) $A' \cup B'$

Solution Again, set A' is represented by regions 1 and 4, and set B' by regions 1 and 2. To find $A' \cup B'$, identify the region that represents the set of all elements in A', B', or both. The result, which is blue in Figure 7, includes regions 1, 2, and 4. When looking for the union, remember to choose the area that is in one region *or* the other region (or both).

Venn diagrams also can be drawn with three sets inside U. These three sets divide the universal set into eight regions, which can be numbered (arbitrarily) as in Figure 8.

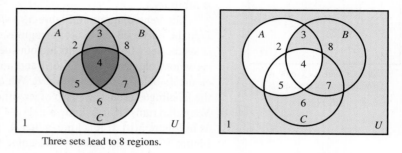

Three sets lead to 8 regions.

FIGURE 8   **FIGURE 9**

EXAMPLE 9   Venn Diagrams

In a Venn diagram, shade the region that represents $A' \cup (B \cap C')$.

Solution First find $B \cap C'$. Set B is represented by regions 3, 4, 7, and 8, and set C' by regions 1, 2, 3, and 8. The overlap of these regions (regions 3 and 8) represents the set $B \cap C'$. Set A' is represented by regions 1, 6, 7, and 8. The union of the set represented by regions 3 and 8 and the set represented by regions 1, 6, 7, and 8 is the set represented by regions 1, 3, 6, 7, and 8, which are blue in Figure 9.

EXAMPLE 10   Genetics

After a genetics experiment with 50 pea plants, the number of plants having certain characteristics was tallied, with the following results:

22 were tall;

25 had green peas (as opposed to yellow);

39 had smooth peas;

14 had smooth peas and were tall;

9 were tall and had green peas;

20 had green peas and smooth peas;

6 had all three characteristics.

Use this information to answer the following questions.

(a) How many plants had none of the characteristics?

(b) How many plants had smooth peas, but were neither tall nor had green peas?

(c) How many plants had exactly one of the above characteristics?

Solution A Venn diagram like the one in Figure 10 on the next page will help sort out the information. To solve problems like this, it is usually best to begin with the innermost region, the intersection of all categories. Thus, in Figure 10, we put the number 6 in the region common to all three characteristics, because 6 of the plants were tall and had smooth, green peas. Notice that, of the 20 plants that had both green and smooth peas, $20 - 6 = 14$ are not tall, so in Figure 11 we put 14 in the region that is common to green and smooth but not tall. Similarly, $9 - 6 = 3$ plants were tall and had green peas that were not smooth, so we put a

3 in that region. Also, $14 - 6 = 8$ of the plants had smooth peas and were tall so 8 is placed in that region.

The data show that 39 of the plants were smooth. However, $8 + 6 + 14 = 28$ of the plants have already been placed in the region representing smooth peas. The balance of this region will contain only $39 - 28 = 11$ plants. These plants only had the characteristic of having smooth peas—they were not tall and did not have green peas. In the same way, 2 plants had only green peas and 5 plants were only tall, as indicated by Figure 12.

A total of $2 + 14 + 6 + 3 + 11 + 8 + 5 = 49$ plants were placed in the three circles of Figure 12. Since 50 plants were in the experiment, $50 - 49 = 1$ plant had none of the three characteristics, and a 1 is placed outside all three regions.

Now Figure 12 can be used to answer the questions asked above.

(a) Only one plant had none of the three characteristics.

(b) There were 11 plants with smooth peas that were neither tall nor had green peas.

(c) There were 18 plants that had exactly one of the characteristics.

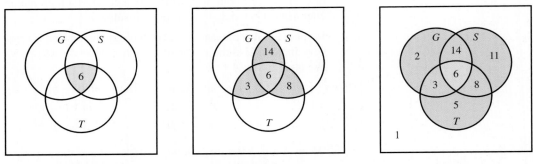

| FIGURE 10 | FIGURE 11 | FIGURE 12 |

CAUTION A common error in solving problems of this type is to make a circle represent one set and another circle represent its complement. In Example 10, with one circle representing those plants that were smooth, we did not draw another for those plants that were not smooth. An additional circle is not only unnecessary (because those not in one set are automatically in the other) but very confusing, because the region outside or inside both circles must be empty. Similarly, if a problem involves men and women, do not draw one circle for men and another for women. Draw one circle; if you label it "women," for example, then men are automatically those outside the circle.

We use the symbol $n(A)$ to indicate the *number* of elements in a set A. For example, if $A = \{a, b, c, d, e\}$, then $n(A) = 5$. The following statement about the number of elements in the union of two sets will be used later in our study of probability.

UNION RULE FOR SETS

$$n(A \cup B) = n(A) + n(B) - n(A \cap B)$$

To prove this statement, let $x + y$ represent $n(A)$, y represent $n(A \cap B)$, and $y + z$ represent $n(B)$, as shown in Figure 13. Then

$$n(A \cup B) = x + y + z,$$
$$n(A) + n(B) - n(A \cap B) = (x + y) + (y + z) - y = x + y + z,$$

so
$$n(A \cup B) = n(A) + n(B) - n(A \cap B).$$

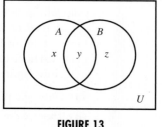

FIGURE 13

EXAMPLE 11 School Activities

A group of 10 students meet to plan a school function. All are majoring in biology or chemistry or both. Five of the students are biology majors and 7 are majors in chemistry. How many major in both subjects?

Solution Let A represent the set of biology majors and B represent the set of chemistry majors. Use the union rule, with $n(A) = 5$, $n(B) = 7$, and $n(A \cup B) = 10$. We must find $n(A \cap B)$.

$$n(A \cup B) = n(A) + n(B) - n(A \cap B)$$
$$10 = 5 + 7 - n(A \cap B),$$

so
$$n(A \cap B) = 5 + 7 - 10 = 2.$$

When A and B are disjoint, $n(A \cap B) = 0$ and the union rule simplifies.

UNION RULE FOR DISJOINT SETS

If A and B are disjoint sets,

$$n(A \cup B) = n(A) + n(B).$$

EXAMPLE 12 Hockey

The number of head and neck injuries for 319 ice hockey players wearing either a full shield or half shield in the Canadian Inter-University Athletics Union during the 1997–1998 season are listed below.*

Table 1

	Half Shield (H)	Full Shield (F)	Totals
Head and Face Injuries (A)	95	34	129
Concussions (B)	41	38	79
Neck Injuries (C)	9	7	16
Other Injuries (D)	202	150	352
Totals	347	229	

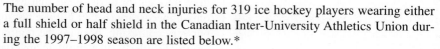

*Benson, B., N. Mohtadi, M. Rose, and W. Meeuwisse, "Head and Neck Injuries Among Ice Hockey Players Wearing Full Face Shields vs. Half Face Shields," *JAMA*, Vol. 282, No. 24, Dec. 22/29, 1999, pp. 2328–2332.

Using the letters given in the table, find the number of injuries in each set.

(a) $A \cap F$

Solution The set $A \cap F$ consists of all head and face injuries that were sustained by players wearing a full shield. From the table, we see that there were 34 such injuries.

(b) $C \cup H$

Solution The set $C \cup H$ consists of all those who had neck injuries or wore a half shield. We include all 347 people who wore a half shield, plus the 7 who sustained a neck injury and wore a full shield, for a total of 354. Alternatively, we could use the formula $n(C \cup H) = n(C) + n(H) - n(C \cap H) = 16 + 347 - 9 = 354$.

(c) $H' \cap (B \cup C)$

Solution Begin with the set $B \cup C$, which includes all concussions and neck injuries. This consists of the two categories, half shield and full shield, with 41, 38, 9, and 7 people. Of this set, include only those who did not wear a half shield, for a total of $38 + 7 = 45$ people. This is the number of people who did not wear a half shield and sustained either a neck injury or a concussion.

12.1 EXERCISES

In Exercises 1–5, write true or false for each statement.

1. $3 \in \{2, 5, 7, 9, 10\}$

2. $9 \notin \{2, 1, 5, 8\}$

3. $\{3, 7, 12, 14\} = \{3, 7, 12, 14, 0\}$

4. $\{x \mid x \text{ is an odd integer}; 6 \le x \le 18\} = \{7, 9, 11, 15, 17\}$

5. $0 \in \emptyset$

6. What is set-builder notation? Give an example.

Let $A = \{2, 4, 6, 8, 10, 12\}$; $B = \{4, 8, 10\}$; $C = \{2, 14\}$; and $U = \{2, 4, 6, 8, 10, 12, 14\}$.
Insert $\subseteq$ or $\nsubseteq$ to make the statement true.

7. A ___ U

8. B ___ A

9. A ___ B

10. $\emptyset$ ___ A

11. $\{0, 2\}$ ___ C

Insert a number in each blank to make the statement true, using the sets for Exercises 7—11.

12. There are exactly ___ subsets of A.

13. There are exactly ___ subsets of B.

14. Describe the intersection and union of sets. How do they differ?

Insert $\cap$ or $\cup$ to make each statement true.

15. $\{5, 7, 9, 19\}$ ___ $\{7, 9, 11, 15\} = \{7, 9\}$

16. $\{8, 11, 15\}$ ___ $\{8, 11, 19, 20\} = \{8, 11\}$

17. $\{3, 5, 9, 10\}$ ___ $\emptyset = \emptyset$

18. $\{3, 5, 9, 10\}$ ___ $\emptyset = \{3, 5, 9, 10\}$

19. $\{1, 2, 4\}$ ___ $\{1, 2, 4\} = \{1, 2, 4\}$

20. Is it possible for two nonempty sets to have the same intersection and union? If so, give an example.

Let $U = \{2, 3, 4, 5, 7, 9\}$; $X = \{2, 3, 4, 5\}$; $Y = \{3, 5, 7, 9\}$; and $Z = \{2, 4, 5, 7, 9\}$. List the members of each of the following sets, using set braces.

21. $X \cap Y$

22. $X \cup Y$

23. X'

24. $X' \cap Y'$

25. $Y \cap (X \cup Z)$

26. $X' \cap (Y' \cup Z)$

27. $(X \cap Y') \cup Z'$

28. a. In Example 6, what set do you get when you calculate $(A \cap B) \cup (A \cap B')$?

b. Explain in words why $(A \cap B) \cup (A \cap B') = A$.

29. Refer to the sets listed for Exercises 7–11. Which pairs of sets are disjoint?

30. Let $A = \{1, 2, 3, \{3\}, \{1, 4, 7\}\}$. Answer each of the following as true or false.

a. $1 \in A$ **b.** $\{3\} \in A$ **c.** $\{2\} \in A$ **d.** $4 \in A$ **e.** $\{\{3\}\} \subset A$ **f.** $\{1, 4, 7\} \in A$ **g.** $\{1, 4, 7\} \subseteq A$

Sketch a Venn diagram like the one in the figure, and use shading to show each of the following sets.

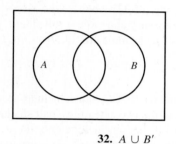

31. $B \cap A'$

32. $A \cup B'$

33. $A' \cap B'$

34. $B' \cup (A' \cap B')$

35. $(A \cap B) \cup B'$

36. Three sets divide the universal set into at most _____ regions.

37. What does the notation $n(A)$ represent?

Sketch a Venn diagram like the one shown, and use shading to show each of the following sets.

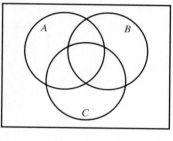

38. $(A \cap B) \cap C$

39. $(A \cap C') \cup B$

40. $A \cap (B \cup C')$

41. $A' \cap (B \cap C)$

42. $A' \cap (B' \cup C)$

Use the union rule to answer the following questions.

43. If $n(A) = 5$, $n(B) = 8$, and $n(A \cap B) = 4$, what is $n(A \cup B)$?

44. If $n(A) = 12$, $n(B) = 27$, and $n(A \cup B) = 30$, what is $n(A \cap B)$?

45. Suppose $n(B) = 7$, $n(A \cap B) = 3$, and $n(A \cup B) = 20$. What is $n(A)$?

46. Suppose $n(A \cap B) = 5$, $n(A \cup B) = 35$, and $n(A) = 13$. What is $n(B)$?

In Exercises 47–50, show that the statement is true by drawing Venn diagrams and shading the regions representing the sets on each side of the equals sign. *

47. $(A \cup B)' = A' \cap B'$

48. $(A \cap B)' = A' \cup B'$

---

*The statements in Exercises 47 and 48 are known as DeMorgan's laws. They are named for the English mathematician Augustus DeMorgan (1806–1871).

49. $A \cap (B \cup C) = (A \cap B) \cup (A \cap C)$

50. $A \cup (B \cap C) = (A \cup B) \cap (A \cup C)$

51. Use the union rule of sets to prove that $n(A \cup B \cup C) = n(A) + n(B) + n(C) - n(A \cap B) - n(A \cap C) - n(B \cap C) + n(A \cap B \cap C)$. (*Hint:* Write $A \cup B \cup C$ as $A \cup (B \cup C)$ and use the formula from Exercise 49.)

■ Applications

LIFE SCIENCES

Health The table below shows some symptoms of an overactive thyroid and an underactive thyroid.*

Underactive Thyroid	Overactive Thyroid
Sleepiness, s	Insomnia, i
Dry hands, d	Moist hands, m
Intolerance of cold, c	Intolerance of heat, h
Goiter, g	Goiter, g

Let U be the smallest possible set that includes all the symptoms listed, N be the set of symptoms for an underactive thyroid, and O be the set of symptoms for an overactive thyroid. Find each of the following sets.

52. O' **53.** N' **54.** $N \cap O$

55. $N \cup O$ **56.** $N \cap O'$

57. *Blood Antigens* Human red blood cells can contain the A antigen, the B antigen, both the A and B antigens, or neither antigen. A third antigen, called the Rh antigen, is important in human reproduction, and again may or may not be present in an individual. Blood is called type A-positive if the individual has the A and Rh, but not the B antigen. A person having only the A and B antigens is said to have type AB-negative blood. A person having only the Rh antigen has type O-positive blood. Other blood types are defined in a similar manner. Identify the blood types on the individuals in regions (a)–(g) below. (One region has no letter assigned to it.)

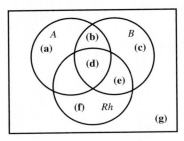

58. *Blood Antigens* (Use the diagram from Exercise 57.) In a certain hospital, the following data were recorded.

25 patients had the A antigen;

17 had the A and B antigens;

27 had the B antigen;

22 had the B and Rh antigens;

30 had the Rh antigen;

12 had none of the antigens;

16 had the A and Rh antigens;

15 had all three antigens.

How many patients

a. were represented?

b. had exactly one antigen?

c. had exactly two antigens?

d. had O-positive blood?

e. had AB-positive blood?

f. had B-negative blood?

g. had O-negative blood?

h. had A-positive blood?

59. *Mortality* The table below lists the number of deaths in the U.S. during 1997 according to race and sex.[†] Use this information and the letters given to find the number of people in each set.

	White (W)	Black (B)	American Indian (I)	Asian or Pacific Islander (A)
Female (F)	1,009,509	132,410	4,591	13,696
Male (M)	986,884	144,110	5,985	17,060

a. F **b.** $F \cap (I \cup A)$

c. $M \cup B$ **d.** $W' \cup I' \cup A'$

e. In words, describe the set in part b.

---

*The Merck Manual of Diagnosis and Therapy, 16th ed., Merck Research Laboratories, 1992, pp. 1075 and 1080.

†National Vital Statistics Reports, Vol. 47, No. 19, June 30, 1999, Table I.

	Non-Hispanic White (A)	Hispanic (B)	African American (C)	Asian American (D)	American Indian (E)	Totals
Under 45 (F)	120.1	25.7	24.4	7.6	1.5	179.3
45–64 (G)	47.5	4.9	6.3	2.2	0.4	61.3
65 and over (H)	29.1	1.9	2.8	0.8	0.2	34.8
Totals	196.7	32.5	33.5	10.6	2.1	275.4

U.S. Population The U.S. population by age and race in 2000 (in millions) is given in the table above. Use this information in Exercises 60–65.*

Using the letters given in the table, find the number of people in each set.

60. $A \cap F$

61. $G \cup B$

62. $G \cup (C \cap H)$

63. $F \cap (B \cup H)$

64. $H \cup D$

65. $G' \cap (A' \cap C')$

OTHER APPLICATIONS

66. *Military* The number of female military personnel on active duty in September 1998 is given in the table below.[†] Use this information and the letters given to find the number of female military personnel in each of the following sets.

    a. $A \cup B$ **b.** $E \cup (C \cup D)$ **c.** $O' \cap M'$

67. *Living Arrangements* In 1998, the percent of white children under 18 years of age who lived with both parents was 74; the percent of white children under 18 years of age who lived with their father only was 5; and the percent of white children under 18 years of age that lived with neither parent was 3.[‡] What percent of children under 18 lived with their mother only?

68. *Living Arrangements* In 1998, there were 3,989 (in thousands) children under the age of 18 living with their grandparents. Of these children, 503 had both parents also living with them, 1,827 had only their mother living with them, and 241 had only their father living with them.[‡] How many children lived with their grandparents only?

Mutual Funds The table on the next page shows the top five holdings of four major mutual funds on July 12, 2000.[§]

Let U be the smallest possible set that includes all the corporations listed, and V, J, F, and T be the set of top-five holdings for each mutual fund, respectively. Find each of the following sets.

69. $V \cap J$ **70.** $V \cap (F \cup T)$

71. $(J \cup F)'$ **72.** $J' \cap T'$

	Army (A)	Air Force (B)	Navy (C)	Marines (D)	Total
Officers (O)	10,367	11,971	7,777	854	30,969
Enlisted (E)	60,787	53,542	42,261	8,928	165,518
Cadets & Midshipmen (M)	624	653	656	n.a.	1,933
Total	71,778	66,166	50,694	9,782	198,420

---

*"Projections of the Total Resident Population by 5-Year Age Groups, Race, and Hispanic Origin, 1999 to 2000," from U.S. Bureau of the Census, *Current Population Reports.*

[†]Brunner, Borga, ed., *Time Almanac 2000*, p. 395.

[‡]*The World Almanac and Book of Facts 2000*, p. 392.

[§]Top 5 holdings found for each fund at www.vanguard.com, ww4.janus.com, www.fidelity.com, www.troweprice.com, respectively.

Vanguard 500	Janus Worldwide	Fidelity Magellan	T. Rowe Price Blue Chip Growth Fund
General Electric Inc.	China Telecom Ltd.	General Electric Inc.	Microsoft Corp.
Intel Corp.	Cisco Systems Inc.	Microsoft Corp.	Cisco Systems Inc.
Cisco Systems Inc.	Nokia Oyj	Cisco Systems Inc.	Citigroup Inc.
Microsoft Corp.	Nortel Networks	Home Depot Inc.	Intel Corp.
Exxon Mobil Corp.	NTT Mobile Corp.	Intel Corp.	Tyco International

73. *Harvesting Fruit* Toward the middle of the harvesting season, peaches for canning come in three types, early, late, and extra late, depending on the expected date of ripening. During a certain week, the following data were recorded at a fruit delivery station:

> 34 trucks went out carrying early peaches;
>
> 61 carried late peaches;
>
> 50 carried extra late;
>
> 25 carried early and late;
>
> 30 carried late and extra late;
>
> 8 carried early and extra late;
>
> 6 carried all three;
>
> 9 carried only figs (no peaches at all).

a. How many trucks carried only late variety peaches?

b. How many carried only extra late?

c. How many carried only one type of peach?

d. How many trucks (in all) went out during the week?

74. *Poultry Analysis* A chicken farmer surveyed his flock with the following results. The farmer had

> 9 fat red roosters;
>
> 2 fat red hens;
>
> 37 fat chickens;
>
> 26 fat roosters;
>
> 7 thin brown hens;
>
> 18 thin brown roosters;
>
> 6 thin red roosters;
>
> 5 thin red hens.

Answer the following questions about the flock. (Assume all chickens are thin or fat, red or brown, and hens or roosters.) How many chickens were

a. fat?   **b.** red?

c. male?   **d.** fat, but not male?

e. brown, but not fat?   **f.** red and fat?

75. *Chinese New Year* A survey of people attending a Lunar New Year celebration in Chinatown yielded the following results:

> 120 were women;
>
> 150 spoke Cantonese;
>
> 170 lit firecrackers;
>
> 108 of the men spoke Cantonese;
>
> 100 of the men did not light firecrackers;
>
> 18 of the non-Cantonese-speaking women lit firecrackers;
>
> 78 non-Cantonese-speaking men did not light firecrackers;
>
> 30 of the women who spoke Cantonese lit firecrackers.

a. How many attended?

b. How many of those who attended did not speak Cantonese?

c. How many women did not light firecrackers?

d. How many of those who lit firecrackers were Cantonese-speaking men?

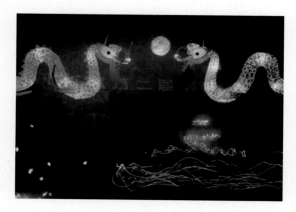

76. *Games* In David Gale's game of Subset Takeaway, the object is for each player, at his or her turn, to pick a nonempty proper subset of a given set subject to the condition that no subset chosen earlier by either player can be a subset of the newly chosen set.* Consider the set $A = \{1, 2, 3\}$. Suppose

Joe and Dorothy are playing the game and Dorothy goes first. If she chooses the proper subset $\{1\}$, then Joe cannot choose any subset that includes the element 1. Joe can, however, choose $\{2\}$ or $\{3\}$ or $\{2, 3\}$. Develop a strategy for Joe so that he can always win the game if Dorothy goes first.

77. *Electoral College* U.S. presidential elections are decided by the Electoral College, in which each of the 50 states, plus the District of Columbia, gives all of its votes to a candidate.† Ignoring the number of votes each state has in the Electoral College, but including all possible combinations of states that could be won by either candidate, how many outcomes are possible in the Electoral College if there are two candidates? (*Hint:* The states that can be won by a candidate form a subset of all the states.)

78. *Musicians* A concert featured a cellist, a flutist, a harpist, and a vocalist. Throughout the concert, different subsets of the four musicians performed together, with at least two musicians playing each piece. How many subsets of at least two are possible?

79. *Cat Food* Suppose 9 flavors of cat food are available in a store. Euclid, the mathematical cat, could like all 9 flavors, or none, or any combination of selected flavors. How many possibilities are there for the set of flavors that Euclid likes? (*Hint:* Each set of flavors is a subset of the original 9 flavors.)

■ 12.2 INTRODUCTION TO PROBABILITY

[?] THINK ABOUT IT   What fraction of all deaths in the United States is caused by heart disease or cancer?

After introducing probability, we will answer this question in one of the examples.

When a physician prescribes an antibiotic, she specifies the *exact* amount of the drug that you should receive. On the other hand, the managing pharmacist is faced with the problem of ordering the medicine. The pharmacist may have a good estimate of the amount of an antibiotic that will be sold during the day, but it is impossible to predict the *exact* amount. The quantity of the antibiotic that customers will purchase during a day is *random:* it cannot be predicted exactly. There are many problems that come up in life science applications of mathematics that involve random phenomena—those for which *exact* prediction is impossible. The best that we can do is to determine the *probability* of the possible outcomes.

---

*Stewart, Ian, "Mathematical Recreations: A Strategy for Subsets," *Scientific American,* Mar. 2000, pp. 96–98.
†The exceptions are Maine and Nebraska, which allocate their electoral college votes according to the winner in each congressional district.

Sample Spaces   In probability, an **experiment** is an activity or occurrence with an observable result. Each repetition of an experiment is called a **trial.** The possible results of each trial are called **outcomes.** The set of all possible outcomes for an experiment is the **sample space** for that experiment. A sample space for the experiment of tossing a coin is made up of the outcomes heads (h) and tails (t). If S represents this sample space, then

$$S = \{h, t\}.$$

EXAMPLE 1   Sample Space

Give the sample space for each experiment.

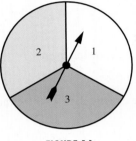

FIGURE 14

(a) A spinner like the one in Figure 14 is spun.

> **Solution** The three outcomes are 1, 2, or 3, so the sample space is
>
> $$\{1, 2, 3\}.$$

(b) For the purposes of a public opinion poll, respondents are classified as young, middle-aged, or senior, and as male or female.

> **Solution** A sample space for this poll could be written as a set of ordered pairs:
>
> {(young, male), (young, female), (middle-aged, male),
> (middle-aged, female), (senior, male), (senior, female)}.

(c) An experiment consists of studying the numbers of boys and girls in families with exactly 3 children. Let b represent *boy* and g represent *girl*.

> **Solution** A three-child family can have 3 boys, written *bbb*, 3 girls, *ggg*, or various combinations, such as *bgg*. A sample space with four outcomes (not equally likely) is
>
> $$S_1 = \{3 \text{ boys}, 2 \text{ boys and } 1 \text{ girl}, 1 \text{ boy and } 2 \text{ girls}, 3 \text{ girls}\}.$$

Notice that a family with 3 boys or 3 girls can occur in just one way, but a family of 2 boys and 1 girl or 1 boy and 2 girls can occur in more than one way. If the *order* of the births is considered, so that *bgg* is different from *gbg* or *ggb*, for example, another sample space is

$$S_2 = \{bbb, bbg, bgb, gbb, bgg, gbg, ggb, ggg\}.$$

The second sample space, S_2, has equally likely outcomes if we assume that boys and girls are equally likely. This assumption, while not quite true, is approximately true, so we will use it throughout this book. The outcomes in S_1 are not equally likely, since there is more than one way to get a family with 2 boys and 1 girl or a family with 2 girls and 1 boy, but only one way to get 3 boys or 3 girls.

CAUTION   An experiment may have more than one sample space, as shown in Example 1(c). The most convenient sample spaces have equally likely outcomes, but it is not always possible to choose such a sample space.

Events An **event** is a subset of a sample space. If the sample space for tossing a coin is $S = \{h, t\}$, then one event is $E = \{h\}$, which represents the outcome "heads."

An ordinary die is a cube whose six different faces show the following numbers of dots: 1, 2, 3, 4, 5, and 6. If the die is fair (not "loaded" to favor certain faces over others), then any one of the faces is equally likely to come up when the die is rolled. The sample space for the experiment of rolling a single fair die is $S = \{1, 2, 3, 4, 5, 6\}$. Some possible events are listed below.

The die shows an even number: $E_1 = \{2, 4, 6\}$.

The die shows a 1: $E_2 = \{1\}$.

The die shows a number less than 5: $E_3 = \{1, 2, 3, 4\}$.

The die shows a multiple of 3: $E_4 = \{3, 6\}$.

EXAMPLE 2 Events

For the sample space S_2 in Example 1(c), write the following events.

(a) Event H: the family has exactly 2 girls

Solution Families with three children can have exactly 2 girls with either *bgg*, *gbg*, or *ggb*, so event H is

$$H = \{bgg, gbg, ggb\}.$$

(b) Event K: the three children are the same sex

Solution Two outcomes satisfy this condition: all boys or all girls.

$$K = \{bbb, ggg\}$$

(c) Event J: the family has three girls

Solution Only *ggg* satisfies this condition, so

$$J = \{ggg\}.$$

In Example 2(c), event J had only one possible outcome, *ggg*. Such an event, with only one possible outcome, is a **simple event.** If event E equals the sample space S, then E is called a **certain event.** If event $E = \emptyset$, then E is called an **impossible event.**

EXAMPLE 3 Events

Suppose a fair die is rolled. As shown above, the sample space is $\{1, 2, 3, 4, 5, 6\}$.

(a) The event "the die shows a 4," $\{4\}$, has only one possible outcome. It is a simple event.

(b) The event "the number showing is less than 10" equals the sample space. $S = \{1, 2, 3, 4, 5, 6\}$. This event is a certain event; if a die is rolled, the number showing (either 1, 2, 3, 4, 5, or 6) must be less than 10.

(c) The event "the die shows a 7" is the empty set $\emptyset$; this is an impossible event.

Since events are sets, we can use set operations to find unions, intersections, and complements of events. A summary of the set operations for events is given on the next page.

SET OPERATIONS FOR EVENTS

Let E and F be events for a sample space S.

$E \cap F$ occurs when both E and F occur;

$E \cup F$ occurs when E or F or both occur;

E' occurs when E does not occur.

Two events that cannot both occur at the same time, such as getting both a head and a tail on the same toss of a coin, are called **mutually exclusive events.**

MUTUALLY EXCLUSIVE EVENTS

Events E and F are mutually exclusive events if $E \cap F = \emptyset$.

For any event E, E and E' are mutually exclusive. By definition, mutually exclusive events are disjoint sets.

EXAMPLE 4 Mutually Exclusive Events

Let $S = \{1, 2, 3, 4, 5, 6\}$, the sample space for tossing a single die. Let $E = \{4, 5, 6\}$, and let $G = \{1, 2\}$. Then E and G are mutually exclusive events since they have no outcomes in common: $E \cap G = \emptyset$. See Figure 15.

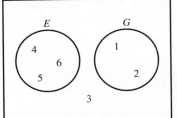

$E \cap G = \emptyset$

FIGURE 15

Probability   For sample spaces with *equally likely* outcomes, the probability of an event is defined as follows.

BASIC PROBABILITY PRINCIPLE

Let S be a sample space of equally likely outcomes, and let event E be a subset of S. Then the **probability that event E occurs** is

$$P(E) = \frac{n(E)}{n(S)}.$$

By this definition, the probability of an event is a number that indicates the relative likelihood of the event.

CAUTION The basic probability principle only applies when the outcomes are equally likely.

EXAMPLE 5 Basic Probabilities

Suppose a single fair die is rolled. Use the sample space $S = \{1, 2, 3, 4, 5, 6\}$ and give the probability of each of the following events.

(a) E: the die shows an even number

Solution Here, $E = \{2, 4, 6\}$, a set with three elements. Since S contains six elements,

$$P(E) = \frac{3}{6} = \frac{1}{2}.$$

(b) *F*: the die shows a number less than 10

Solution Event *F* is a certain event, with

$$F = \{1, 2, 3, 4, 5, 6\},$$

so that

$$P(F) = \frac{6}{6} = 1.$$

(c) *G*: the die shows an 8

Solution This event is impossible, so

$$P(G) = 0.$$

A standard deck of 52 cards has four suits: hearts (♥), clubs (♣), diamonds (♦), and spades (♠), with 13 cards in each suit. The hearts and diamonds are red, and the spaces and clubs are black. Each suit has an ace (A), a king (K), a queen (Q), a jack (J), and cards numbered from 2 to 10. The jack, queen, and king are called *face cards* and for many purposes can be thought of as having values 11, 12, and 13, respectively. The ace can be thought of as the low card (value 1) or the high card (value 14). See Figure 16. We will refer to this standard deck of cards often in our discussion of probability.

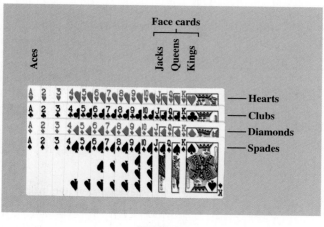

FIGURE 16

EXAMPLE 6 Playing Cards

If a single playing card is drawn at random from a standard 52-card deck, find the probability of each of the following events.

(a) Drawing an ace

Solution There are 4 aces in the deck. The event "drawing an ace" is

$$\{\text{heart ace, diamond ace, club ace, spade ace}\}.$$

Therefore,

$$P(\text{ace}) = \frac{4}{52} = \frac{1}{13}.$$

(b) Drawing a face card

Solution Since there are 12 face cards (three in each of the four suits),

$$P(\text{face card}) = \frac{12}{52} = \frac{3}{13}.$$

(c) Drawing a spade

Solution The deck contains 13 spades, so

$$P(\text{spade}) = \frac{13}{52} = \frac{1}{4}.$$

(d) Drawing a spade or a heart

Solution Besides the 13 spades, the deck contains 13 hearts, so

$$P(\text{spade or heart}) = \frac{26}{52} = \frac{1}{2}.$$

In the preceding examples, the probability of each event was a number between 0 and 1. The same thing is true in general. Any event E is a subset of the sample space S, so $0 \leq n(E) \leq n(S)$. Since $P(E) = n(E)/n(S)$, it follows that $0 \leq P(E) \leq 1$.

> For any event E, $0 \leq P(E) \leq 1$.

We now need to develop additional rules for calculating probability, beginning with the probability of a union of two events.

To determine the probability of the union of two events E and F in a sample space S, use the union rule for sets,

$$n(E \cup F) = n(E) + n(F) - n(E \cap F),$$

which was proved in Section 12.1. Dividing both sides by $n(S)$ shows that

$$\frac{n(E \cup F)}{n(S)} = \frac{n(E)}{n(S)} + \frac{n(F)}{n(S)} - \frac{n(E \cap F)}{n(S)}$$

$$P(E \cup F) = P(E) + P(F) - P(E \cap F).$$

This result is called the **union rule for probability.**

> **UNION RULE FOR PROBABILITY**
>
> For any events E and F from a sample space S,
>
> $$P(E \cup F) = P(E) + P(F) - P(E \cap F).$$

(Although the union rule applies to any events E and F from any sample space, the derivation we have given is valid only for sample spaces with equally likely simple events.)

EXAMPLE 7 Probabilities with Dice

Suppose two fair dice are rolled. Find each of the following probabilities.

(a) The first die shows a 2, or the sum of the results is 6 or 7.

Solution The sample space for the throw of two dice is shown in Figure 17, where 1-1 represents the event "the first die shows a 1 and the second die shows a 1," 1-2 represents "the first die shows a 1 and the second die shows a 2," and so on. Let A represent the event "the first die shows a 2," and B represent the event "the sum of the results is 6 or 7." These events are indicated in Figure 17. From the diagram, event A has 6 elements, B has 11 elements, and the sample space has 36 elements. Thus,

$$P(A) = \frac{6}{36}, \quad P(B) = \frac{11}{36}, \quad \text{and} \quad P(A \cap B) = \frac{2}{36}.$$

By the union rule,

$$P(A \cup B) = P(A) + P(B) - P(A \cap B)$$

$$P(A \cup B) = \frac{6}{36} + \frac{11}{36} - \frac{2}{36} = \frac{15}{36} = \frac{5}{12}.$$

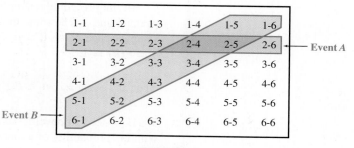

FIGURE 17

(b) The sum of the results is 11, or the second die shows a 5.

Solution $P(\text{sum is } 11) = 2/36$, $P(\text{second die shows a 5}) = 6/36$, and $P(\text{sum is 11 and second die shows a 5}) = 1/36$, so

$$P(\text{sum is 11 or second die shows a 5}) = \frac{2}{36} + \frac{6}{36} - \frac{1}{36} = \frac{7}{36}.$$

CAUTION You may wonder why we did not use $S = \{2, 3, 4, 5, \ldots, 12\}$ as the sample space in Example 7. Remember, we prefer to use a sample space with equally likely outcomes. The outcomes in set S above are not equally likely—a sum of 2 can occur in just one way, a sum of 3 in two ways, a sum of 4 in three ways, and so on, as shown in Figure 17.

If events E and F are mutually exclusive, then $E \cap F = \emptyset$ by definition; hence, $P(E \cap F) = 0$. Applying the union rule yields this useful fact.

UNION RULE FOR MUTUALLY EXCLUSIVE EVENTS

For mutually exclusive events E and F,

$$P(E \cup F) = P(E) + P(F).$$

By the definition of E', for any event E from a sample space S,

$$E \cup E' = S \quad \text{and} \quad E \cap E' = \emptyset.$$

Since $E \cap E' = \emptyset$, events E and E' are mutually exclusive, so that

$$P(E \cup E') = P(E) + P(E').$$

However, $E \cup E' = S$, the sample space, and $P(S) = 1$. Thus

$$P(E \cup E') = P(E) + P(E') = 1.$$

Rearranging these terms gives the following useful rule.

> **COMPLEMENT RULE**
>
> $$P(E) = 1 - P(E') \quad \text{and} \quad P(E') = 1 - P(E).$$

EXAMPLE 8   Complement Rule
In Example 7, find the probability that the sum of the numbers rolled is greater than 3.

Solution To calculate this probability directly, we must find the probabilities that the sum is 4, 5, 6, 7, 8, 9, 10, 11, or 12 and then add them. It is much simpler to first find the probability of the complement, the event that the sum is less than or equal to 3.

$$P(\text{sum} \le 3) = P(\text{sum is 2}) + P(\text{sum is 3})$$

$$= \frac{1}{36} + \frac{2}{36}$$

$$= \frac{3}{36} = \frac{1}{12}$$

Now use the fact that $P(E) = 1 - P(E')$ to get

$$P(\text{sum} > 3) = 1 - P(\text{sum} \le 3)$$

$$= 1 - \frac{1}{12} = \frac{11}{12}.$$

Odds   Sometimes probability statements are given in terms of **odds**, a comparison of $P(E)$ with $P(E')$. For example, suppose $P(E) = 4/5$. Then $P(E') = 1 - 4/5 = 1/5$. These probabilities predict that E will occur 4 out of 5 times and E' will occur 1 out of 5 times. Then we say the **odds in favor** of E are 4 to 1.

> **ODDS**
>
> The **odds in favor** of an event E are defined as the ratio of $P(E)$ to $P(E')$, or
>
> $$\frac{P(E)}{P(E')}, \quad P(E') \ne 0.$$

EXAMPLE 9   Odds of Flat Feet

About 1 in 7 children in the United States has persistent flat feet, which do not usually require treatment unless they cause pain. Find the odds that a child will have flat feet.

Solution Let E be the event "flat feet." Then E' is the event "no flat feet." Since $P(E) = 1/7$, $P(E') = 6/7$. By the definition of odds, the odds in favor of flat feet are

$$\frac{1/7}{6/7} = \frac{1}{6}, \qquad \text{written} \qquad 1 \text{ to } 6, \text{ or } 1:6.$$

On the other hand, the odds that a child will not have flat feet, or the *odds against* flat feet, are

$$\frac{6/7}{1/7} = \frac{6}{1}, \qquad \text{written} \qquad 6 \text{ to } 1, \text{ or } 6:1.$$

If the odds in favor of an event are, say, 3 to 5, then the probability of the event is $3/8$, while the probability of the complement of the event is $5/8$. (Odds of 3 to 5 indicate 3 outcomes in favor of the event out of a total of 8 possible outcomes.) This example suggests the following generalization.

If the odds favoring event E are m to n, then

$$P(E) = \frac{m}{m + n} \qquad \text{and} \qquad P(E') = \frac{n}{m + n}.$$

EXAMPLE 10   Motorcycle Accidents

Suppose the odds of dying while riding on a motorcycle are 1 to 1,250. What is the probability of death while riding on a motorcycle?

Solution The odds indicate chances of 1 out of 1,251 $(1 + 1,250 = 1,251)$ that one will die, so

$$P(\text{death}) = \frac{1}{1,251}.$$

Empirical Probability   In many real-life problems, it is not possible to establish exact probabilities for events. Instead, useful approximations are often found by drawing on past experience as a guide to the future. The next example shows one approach to such **empirical probabilities.**

EXAMPLE 11   Causes of Death

There were 2,329,520 deaths in the United States in 1998 and they are listed, according to cause, in Table 2 on the next page.*

---

*Brunner, Borga, ed., *Time Almanac 2000,* p. 809.

Table 2

Cause	Number of Deaths
Heart disease	727,624
Cancer	540,702
Cerebrovascular disease	159,059
Abstructive pulmonary disease	111,823
Accidents	136,959
Pneumonia and influenza	92,048
Diabetes mellitus	63,813
All other causes	497,492

We can find the empirical probability that a death was caused by each category by dividing the number of deaths for each cause by the total number of deaths. Verify that the amounts in the table sum to 2,329,520. The probability that a randomly selected death was caused by an accident, for example, is P(death by accident) = 136,959/2,329,520 ≈ 0.059. Similarly, we could divide each amount by 2,329,520, with the results (rounded to three decimal places) shown in Table 3.

Table 3

Cause	Probabilities
Heart disease	0.312
Cancer	0.232
Cerebrovascular disease	0.068
Abstructive pulmonary disease	0.048
Accidents	0.059
Pneumonia and influenza	0.040
Diabetes mellitus	0.027
All other causes	0.214

The numbers in Table 3 sum to 1.000. In theory, they should always total 1.000, but this sometimes does not occur with rounded numbers.

The categories in the table are mutually exclusive simple events. Thus, to find the probability that a death is caused by heart disease or cancer, we use the union rule to calculate

P(death by heart disease or cancer) = 0.312 + 0.232 = 0.544.

We could get this same result by summing the number of deaths caused by heart disease and cancer, and dividing the total by 2,329,520.

Thus, a little more than half of all deaths were caused by heart disease or cancer.

A table of probabilities, as in Example 11, sets up a **probability distribution;** that is, for each possible outcome of an experiment, a number, called the probability of that outcome, is assigned. This assignment may be done in any reasonable way (on an empirical basis, as in Example 11, or by theoretical reasoning), provided that it satisfies the following conditions.

> **PROPERTIES OF PROBABILITY**
>
> Let S be a sample space consisting of n distinct outcomes, $s_1, s_2, \ldots, s_n$. An acceptable probability assignment consists of assigning to each outcome s_i a number p_i (the probability of s_i) according to these rules.
>
> **1.** The probability of each outcome is a number between 0 and 1.
>
> $$0 \le p_1 \le 1, \quad 0 \le p_2 \le 1, \ldots, \quad 0 \le p_n \le 1$$
>
> **2.** The sum of the probabilities of all possible outcomes is 1.
>
> $$p_1 + p_2 + p_3 + \cdots + p_n = 1$$

12.2 EXERCISES

1. What is meant by a "fair" coin or die?

2. Explain the concept of sample space for an experiment.

Write sample spaces for the experiments in Exercises 3–8.

3. A month of the year is chosen for a wedding.

4. A student is asked how many points she earned on a recent 80-point test.

5. A patient must decide whether to have surgery, treat the problem with medicine, or wait six months to see if the problem resolves.

6. A record is kept each day for three days about whether a patient's blood pressure goes up or goes down.

7. A coin is tossed, and a die is rolled.

8. A box contains five balls, numbered 1, 2, 3, 4, and 5. A ball is drawn at random, the number on it is recorded, and the ball replaced. The box is shaken, a second ball is drawn, and its number is recorded.

9. Define an event.

10. What is a simple event?

For the experiments in Exercises 11–14 write out the sample space, and then write the indicated events in set notation.

11. A committee of 2 people is selected from 5 doctors: Alam, Bartolini, Chinn, Dickson, and Ellsberg.

   a. Chinn is on the committee.

   b. Dickson and Ellsberg are not both on the committee.

   c. Both Alam and Chinn are on the committee.

12. Slips of paper marked with the numbers 1, 2, 3, 4, and 5 are placed in a box. After being mixed, two slips are drawn simultaneously.

   a. Both slips are marked with even numbers.

   b. One slip is marked with an odd number and the other is marked with an even number.

   c. Both slips are marked with the same number.

13. An unprepared student takes a three-question, true/false quiz in which he guesses the answers to all three questions, so each answer is equally likely to be correct or wrong.

   a. The student gets three answers wrong.

   b. The student gets exactly two answers correct.

   c. The student gets only the first answer correct.

14. A coin is flipped at most four times, or until two heads occur, whichever comes first.

   a. The coin is tossed four times.

   b. Exactly two heads are tossed.

   c. No heads are tossed.

A single fair die is rolled. Find the probabilities of the following events.

15. Getting a 2

16. Getting an odd number

17. Getting a number less than 5

18. Getting a number greater than 2

19. Getting a 3 or a 4

20. Getting any number except 3

A card is drawn from a well-shuffled deck of 52 cards. Find the probability of drawing each of the following.

21. A 9

22. A black card

23. A black 9

24. A 2 or a queen

25. A black 7 or a red 8

26. A red face card

A jar contains 2 white, 3 orange, 5 yellow, and 8 black marbles. If a marble is drawn at random, find the probability that it is the following.

27. Orange

28. Not black

29. Orange or yellow

30. The student sitting next to you in class concludes that the probability of the ceiling falling down on both of you before class ends is 1/2, because there are two possible outcomes—the ceiling will fall or not fall. What is wrong with this reasoning?

31. Define mutually exclusive events in your own words.

Decide whether the events in Exercises 32–35 are mutually exclusive.

32. Owning a car and owning a truck

33. Being married and being over 30 years old

34. Being a teenager and being over 30 years old

35. Rolling a die once and getting a 4 and an odd number

Two dice are rolled. Find the probabilities of the following events.

36. The first die is 3 or the sum is 8.

37. The second die is 5 or the sum is 10.

38. Three unusual dice, *A*, *B*, and *C*, are constructed such that die *A* has the numbers 3, 3, 4, 4, 8, 8; die *B* has the numbers 1, 1, 5, 5, 9, 9; and die *C* has the numbers 2, 2, 6, 6, 7, 7.

   a. If dice *A* and *B* are rolled, find the probability that *B* beats *A*, that is, the number that appears on die *B* is greater than the number that appears on die *A*.

   b. If dice *B* and *C* are rolled, find the probability that *C* beats *B*.

   c. If dice *A* and *C* are rolled, find the probability that *A* beats *C*.

   d. Which die is better? Explain.

39. Laurie Rosatone, a game show contestant, could win one of two prizes: a shiny new Porsche or a shiny new penny. Laurie is given two boxes of marbles. The first box has 50 pink marbles in it and the second box has 50 blue marbles in it. The game show host will pick someone from the audience to blindfold and then draw a marble from one of the two boxes. If a pink marble is drawn, she wins the Porsche. Otherwise Laurie wins the penny.* Can Laurie increase her chances of winning by redistributing some of the marbles from one box to the other? Explain.

One card is drawn from an ordinary deck of 52 cards. Find the probabilities of drawing the following cards.

40. a. A 9 or 10

   b. A red card or a 3

   c. A 9 or a black 10

   d. A heart or a black card

   e. A face card or a diamond

41. a. Less than a 4 (count aces as ones)

   b. A diamond or a 7

   c. A black card or an ace

   d. A heart or a jack

   e. A red card or a face card

Ms. Elliott invites 10 relatives to a party: her mother, 2 aunts, 3 uncles, 2 brothers, 1 male cousin, and 1 female cousin. If the chances of any 1 guest arriving first are equally likely, find the probabilities that the first guest to arrive is as follows.

42. a. A brother or an uncle   **b.** A brother or a cousin   **c.** A brother or her mother

43. a. An uncle or a cousin   **b.** A male or a cousin   **c.** A female or a cousin

44. Define what is meant by "odds in favor."

A single fair die is rolled. Find the odds in favor of getting the results in Exercises 45–48.

45. 5

46. 3, 4, or 5

47. 1, 2, 3, or 4

48. Some number less than 2

49. In the "Ask Marilyn" column of *Parade* magazine, a reader wrote about the following game: You and I each roll a die. If your die is higher than mine, you win. Otherwise, I win. The reader thought that the probability that each player wins is 1/2. Is this correct? If not, what is the probability that each player wins?†

50. On page 134 of Roger Staubach's autobiography, *First Down, Lifetime to Go,* Staubach makes the following statement regarding his experience in Vietnam:

   "Odds against a direct hit are very low but when your life is in danger, you don't worry too much about the odds."

Is this wording consistent with our definition of odds, for and against? How could it have been said so as to be technically correct?

---

*This problem is based on the "Puzzler of the Week: Prison Marbles" from the week of Sept. 7, 1996, on National Public Radio's *Car Talk.*
†*Parade* Magazine, Nov. 6, 1994, p. 10. Reprinted by permission of the William Morris Agency, Inc. on behalf of the author. Copyright © 1994 by Marilyn vos Savant.

51. The following table gives the odds that a particular event will occur.* Convert each odd to the probability that the event will occur.

Event	Odds for the Event
You will eat out today.	1 to 2
The next bottled water you buy will be nothing more than tap water.	1 to 4
The Earth will be struck by a huge meteor during your lifetime.	1 to 9,000
You will go to Disney World this year.	1 to 9
You'll regain weight you lost by dieting.	9 to 10

Which of Exercises 52–56 are examples of empirical probability?

52. The probability of heads on 5 consecutive tosses of a coin

53. The probability that a person is allergic to penicillin

54. The probability of drawing an ace from a standard deck of 52 cards

55. The probability that a person will get lung cancer from smoking cigarettes

56. A surgeon's prediction that a patient has a 90% chance of full recovery

57. What is a probability distribution?

An experiment is conducted for which the sample space is $S = \{s_1, s_2, s_3, s_4, s_5\}$. Which of the probability assignments in Exercises 58 and 59 is possible for this experiment? If an assignment is not possible, tell why.

58.

Outcomes	s_1	s_2	s_3	s_4	s_5
Probabilities	0.09	0.32	0.21	0.25	0.13

59.

Outcomes	s_1	s_2	s_3	s_4	s_5
Probabilities	0.64	-0.08	0.30	0.12	0.02

One way to solve a probability problem is to repeat the experiment many times, keeping track of the results. Then the probability can be approximated using the basic definition of the probability of an event E: $P(E) = n(E)/n(S)$, where E occurs $n(E)$ times out of $n(S)$ trials of an experiment. This is called the Monte Carlo method of finding probabilities. If physically repeating the experiment is too tedious, it may be simulated using a random number generator, available on most computers and scientific or graphing calculators. To simulate a coin toss or the roll of a die on the TI-83, change the setting to fixed decimal mode with 0 digits displayed, and enter rand *or* rand*6+.5, *respectively. For a coin toss, interpret 0 as a head and 1 as a tail. In either case, the* ENTER *key can be pressed repeatedly to perform multiple simulations.*

60. Suppose two dice are rolled. Use the Monte Carlo method with at least 50 repetitions to approximate the following probabilities.

    a. P(the sum is not more than 5)   **b.** P(the sum is not less than 8)

61. Suppose two dice are rolled. Use the Monte Carlo method with at least 50 repetitions to approximate the following probabilities.

    a. P(the sum is 9 or more)   **b.** P(the sum is less than 7)

---

*The Forum for Investor Advice; Krantz, Les, *What the Odds Are,* Harper Perennial, 1992; and Laudan, Larry, *Danger Ahead: The Risks You Really Face on Life's Highway,* John Wiley & Sons, 1997.

62. Suppose three dice are rolled. Use the Monte Carlo method with at least 100 repetitions to approximate the following probabilities.

  a. P(the sum is 5 or less)   **b.** P(neither a 1 nor a 6 is rolled)

63. Suppose a coin is tossed 5 times. Use the Monte Carlo method with at least 50 repetitions to approximate the following probabilities.

  a. P(exactly 4 heads)   **b.** P(2 heads and 3 tails)

Applications

LIFE SCIENCES

64. *Medical Survey* For a medical experiment, people are classified as to whether they smoke, have a family history of heart disease, or are overweight. Define events E, F, and G as follows.

  E: person smokes

  F: person has a family history of heart disease

  G: person is overweight

Describe each of the following events in words.

  a. G' **b.** $F \cap G$ **c.** $E \cup G'$

65. *Medical Survey* Refer to Exercise 64. Describe each of the following events in words.

  a. $E \cup F$ **b.** $E' \cap F$ **c.** $F' \cup G'$

66. *Body Types* A study on body types gave the following results: 45% were short; 25% were short and overweight; and 24% were tall and not overweight. Find the probabilities that a person is the following.

  a. Overweight

  b. Short, but not overweight

  c. Tall and overweight

67. *Color Blindness* Color blindness is an inherited characteristic that is much more common in males than in females. If M represents male and C represents red-green color blindness, we use the relative frequencies of the incidences of males and red-green color blindness as probabilities to get

$P(C) = 0.039$, $P(M \cap C) = 0.035$, $P(M \cup C) = 0.491$.*

Find the following probabilities.

  a. $P(C')$ **b.** $P(M)$ **c.** $P(M')$

  d. $P(M' \cap C')$ **e.** $P(C \cap M')$ **f.** $P(C \cup M')$

68. *Genetics* Gregor Mendel, an Austrian monk, was the first to use probability in the study of genetics. In an effort to

understand the mechanism of character transmittal from one generation to the next in plants, he counted the number of occurrences of various characteristics. Mendel found that the flower color in certain pea plants obeyed this scheme:

Pure purple crossed with pure white produces purple.

From its parents, the purple offspring received genes for both purple (P) and white (W), but in this case purple is *dominant* and white *recessive,* so the offspring exhibits the color purple. However, the offspring still carries both genes, and when two such offspring are crossed, several things can happen in the third generation. The table in the next column, which is called a *Punnet square,* shows the equally likely outcomes.

Use the fact that purple is dominant over white to find each of the following. Assume that there are an equal number of purple and white genes in the population.

  a. P(a flower is purple)   **b.** P(a flower is white)

		Second Parent	
		P	W
First Parent	P	PP	PW
	W	WP	WW

69. *Genetics* Mendel found no dominance for color in snapdragons, with one red gene and one white gene producing pink-flowered offspring. These second-generation pinks, however, still carry one red and one white gene, and when they are crossed, the next generation still yields the Punnet square shown above (where P represents a red gene and W represents a white gene). Find each of the following probabilities.

  a. P(red)   **b.** P(pink)   **c.** P(white)

(Mendel verified these probability ratios experimentally and did the same for many characteristics other than flower color. His work, published in 1866, was not recognized until 1890.)

---

*The probabilities of a person being male or female are from *The World Almanac and Book of Facts,* 1995. The probabilities of a male and female being color-blind are from *Parsons' Diseases of the Eye* (18th ed.) by Stephen J.H. Miller, Churchill Livingstone, 1990, p. 269. This reference gives a range of 3 to 4% for the probability of gross color blindness in men; we used the midpoint of this range.

70. *Genetics* In most animals and plants, it is very unusual for the number of main parts of the organism (such as arms, legs, toes, or flower petals) to vary from generation to generation. Some species, however, have *meristic variability,* in which the number of certain body parts varies from generation to generation. One researcher studied the front feet of certain guinea pigs and produced the following probabilities.*

$$P(\text{only four toes, all perfect}) = 0.77$$
$$P(\text{one imperfect toe and four good ones}) = 0.13$$
$$P(\text{exactly five good toes}) = 0.10$$

Find the probability of each of the following events.

a. No more than four good toes

b. Five toes, whether perfect or not

71. *U.S. Population* The population of the United States by race in 2000 and the projected population by race for the year 2025 are given below (in thousands).[†]

Race	2000	2025
White	196,700	209,900
Hispanic	32,500	56,900
African-American	33,500	44,700
Asian-American	10,600	24,000
Other	2,100	2,800

Find the probability of a randomly selected person being each of the following.

a. Hispanic in 2000

b. Hispanic in 2025

c. African-American in 2000

d. African-American in 2025

OTHER APPLICATIONS

72. *Civil War* Estimates of the Union Army's strength and losses for the battle of Gettysburg are given in the following table, where *strength* is the number of soldiers immediately preceding the battle and *loss* indicates a soldier who was killed, wounded, captured, or missing.[‡]

Unit	Strength	Loss
I Corps (Reynolds)	12,222	6,059
II Corps (Hancock)	11,347	4,369
III Corps (Sickles)	10,675	4,211
V Corps (Sykes)	10,907	2,187
VI Corps (Sedgwick)	13,596	242
XI Corps (Howard)	9,188	3,801
XII Corps (Slocum)	9,788	1,082
Cavalry (Pleasonton)	11,851	610
Artillery (Tyler)	2,376	242
Total	91,950	22,803

a. Find the probability that a randomly selected union soldier was from the XI Corps.

b. Find the probability that a soldier was lost in the battle.

c. Find the probability that a I Corps soldier was lost in the battle.

d. Which group had the highest probability of not being lost in the battle?

e. Which group had the highest probability of loss?

f. Explain why these probabilities vary.

73. *Civil War* Estimates of the Confederate Army's strength and losses for the battle of Gettysburg are given in the table on the next page, where *strength* is the number of soldiers

---

*Wright, J.R., "An Analysis of Variability in Guinea Pigs," *Genetics,* Vol. 19, pp. 506–536. Reprinted by permission.
[†]"Projections of the Total Resident Population by 5-Year Age Groups, Race, and Hispanic Origin, 1999 to 2000," from U.S. Bureau of the Census, *Current Population Reports.*
[‡]Busey, John, and David Martin, *Regimental Strengths and Losses at Gettysburg,* Hightstown, N.J., Longstreet House, 1986, p. 270.

immediately preceding the battle and *loss* indicates a soldier who was killed, wounded, captured, or missing.[*]

Unit	Strength	Loss
I Corps (Longstreet)	20,706	7,661
II Corps (Ewell)	20,666	6,603
III Corps (Hill)	22,083	8,007
Cavalry (Stuart)	6,621	286
Total	70,076	22,557

a. Find the probability that a randomly selected confederate soldier was from the III Corps.

b. Find the probability that a confederate soldier was lost in the battle.

c. Find the probability that a I Corps soldier was lost in the battle.

d. Which group had the highest probability of not being lost in the battle?

e. Which group had the highest probability of loss?

74. *Research Funding* According to an article in *Business Week*, in 1989 funding for university research in the United States totaled $15 billion. Support came from various sources, as shown in the table below.[†]

Source	Amount (in billions of dollars)
Federal government	9.0
State and local government	1.2
Institutional	2.7
Industry	1.0
Other	1.1

Find the probability that funds for a particular project came from each of the following sources.

a. Federal government

b. Industry

c. The institution

75. *Military* There were 198,420 female military personnel on active duty in September 1998 in various ranks and military branches, as listed in the table below.[‡]

	Army (A)	Air Force (B)	Navy (C)	Marines (D)
Officers (O)	10,367	11,971	7,777	854
Enlisted (E)	60,787	53,542	42,261	8,928
Cadets & Midshipmen (M)	624	653	656	0

a. Convert the numbers in the table to probabilities.

b. Find the probability that a randomly selected female soldier is enlisted in the Army.

c. Find the probability that a randomly selected female soldier is an officer in the Navy or Marine Corps.

d. $P(A \cup B)$

e. $P(E \cup (C \cup D))$

76. *Perceptions of Threat* Research has been carried out to measure the amount of intolerance that citizens of Russia have for left-wing Communists and right-wing Fascists, as indicated in the table below. Note that the numbers are given as percents and each row sums to 100 (except for rounding).[§]

	None at All	Don't Know	Not Very Much	Somewhat	Extremely
Left-wing Communists	47.8	6.7	31.0	10.5	4.1
Right-wing Fascists	3.0	3.2	7.1	27.1	59.5

---

[*]Busey, John, and David Martin, *Regimental Strengths and Losses at Gettysburg,* Hightstown, N.J., Longstreet House, 1986, p. 270.

[†]"University Research: The Squeeze Is On," *Business Week,* May 20, 1991. Reprinted by special permission. Copyright © 1991 by McGraw-Hill, Inc.

[‡]Brunner, Borga, ed., *Time Almanac 2000,* p. 395.

[§]Gibson, J. L., "Putting Up with Fellow Russians: An Analysis of Political Tolerance in the Fledgling Russian Democracy," *Political Research Quarterly,* Vol. 51, No. 1, Mar. 1998, pp. 37–68.

a. Find the probability that a randomly chosen citizen of Russia would be somewhat or extremely intolerant of right-wing Fascists.

b. Find the probability that a randomly chosen citizen of Russia would be completely tolerant of left-wing Communists.

✎ **c.** Compare your answers to parts a and b and provide possible reasons for these numbers.

77. *Perceptions of Threat* Research has been carried out to measure the amount of intolerance that U.S. citizens have for left-wing Communists and right-wing Fascists, as indicated in the table below. Note that the numbers are given as percents and each row sums to 100 (except for rounding).*

a. Find the probability that a randomly chosen U.S. citizen would have at least some intolerance of right-wing Fascists.

b. Find the probability that a randomly chosen U.S. citizen would have at least some intolerance of left-wing Communists.

✎ **c.** Compare your answers to parts a and b and provide possible reasons for these numbers.

✎ **d.** Compare these answers to the answers to Exercise 76.

	None at All	Don't Know	Not Very Much	Somewhat	Extremely
Left-wing Communists	13.0	2.7	33.0	34.2	17.1
Right-wing Fascists	10.1	3.3	20.7	43.1	22.9

■ 12.3 CONDITIONAL PROBABILITY; INDEPENDENT EVENTS; BAYES' THEOREM

? THINK ABOUT IT

What is the probability that a player who wears a full shield will sustain a head and face injury?

As we saw in section 1 of this chapter, the number of head and neck injuries for 319 ice hockey players wearing either a full shield or half shield in the Canadian Inter-University Athletics Union during the 1997–1998 season are listed below.[†]

■ Table 4

	Half Shield (H)	Full Shield (F)	Totals
Head and Face Injuries (A)	95	34	129
Concussions (B)	41	38	79
Neck Injuries (C)	9	7	16
Other Injuries (D)	202	150	352
Totals	347	229	576

*Gibson, J. L., "Putting Up with Fellow Russians: An Analysis of Political Tolerance in the Fledgling Russian Democracy," *Political Research Quarterly,* Vol. 51, No. 1, Mar. 1998, pp. 37–68.
†Benson, B., N. Mohtadi, M. Rose, and W. Meeuwisse, "Head and Neck Injuries Among Ice Hockey Players Wearing Full Face Shields vs. Half Face Shields," *JAMA,* Vol. 282, No. 24, Dec. 22/29, 1999, pp. 2328–2332.

Researchers can use this information to determine if there is a difference in the number and type of injuries that occur when a hockey player wears a half shield versus a full shield.

Using the letters above, we can find the following probabilities.

$$P(A) = \frac{129}{576} \approx 0.224 \qquad P(A') = \frac{447}{576} \approx 0.776$$

$$P(F) = \frac{229}{576} \approx 0.398 \qquad P(F') = \frac{347}{576} \approx 0.602$$

 Suppose we want to find the probability that an injury sustained by a player while wearing a full shield will be a head and face injury. From the table on the previous page, of the 229 injuries sustained in players who wore a full shield, 34 injuries were to the head and face, with

$$P\begin{pmatrix}\text{injury sustained while wearing full} \\ \text{shield was a head and face injury}\end{pmatrix} = \frac{34}{229} \approx 0.148.$$

This is a different number than the probability of a head and face injury, 0.224, since we have additional information (the player wore a full shield) that has *reduced the sample space*. In other words, we found the probability that an injured player has sustained a head and face injury, A, given that the player was wearing a full shield, F. This is called the *conditional probability* of event A, given that event F has occurred, written $P(A|F)$. The probability $P(A|F)$ is read as "the probability of A given F." In the example above,

$$P(A|F) = \frac{34}{229},$$

which can also be written as

$$P(A|F) = \frac{34/576}{229/576} = \frac{P(A \cap F)}{P(F)},$$

where $P(A \cap F)$ represents, as usual, the probability that both A and F will occur.

To generalize this result, assume that E and F are two events for a particular experiment. Assume that the sample space S for this experiment has n possible equally likely outcomes. Suppose event F has m elements, and $E \cap F$ has k elements ($k \le m$). Using the fundamental principle of probability,

$$P(F) = \frac{m}{n} \qquad \text{and} \qquad P(E \cap F) = \frac{k}{n}.$$

We now want $P(E|F)$, the probability that E occurs given that F has occurred. Since we assume F has occurred, reduce the sample space to F: look only at the m elements inside F. See Figure 18. Of these m elements, there are k elements where E also occurs, since $E \cap F$ has k elements. This makes

$$P(E|F) = \frac{k}{m}.$$

Divide numerator and denominator by n to get

$$P(E|F) = \frac{k/n}{m/n} = \frac{P(E \cap F)}{P(F)}.$$

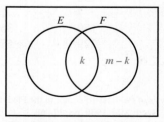

Event F has a total of m elements.

FIGURE 18

This last result motivates the definition of conditional probability.

DEFINITION OF CONDITIONAL PROBABILITY

The **conditional probability** of event E given event F, written $P(E|F)$, is

$$P(E|F) = \frac{P(E \cap F)}{P(F)}, \quad \text{where } P(F) \neq 0.$$

This definition tells us that, for equally likely outcomes, conditional probability is found by *reducing the sample space to event F*, and then finding the number of outcomes in F that are also in event E. Thus,

$$P(E|F) = \frac{n(E \cap F)}{n(F)}.$$

Although the definition of conditional probability was motivated by an example with equally likely outcomes, it is valid in all cases. For an intuitive explanation, think of the formula as giving the probability that both E and F occur compared with the entire probability of F.

EXAMPLE 1 Hockey

Use the information given in the chart at the beginning of this section to find the following probabilities.

(a) $P(F|A)$

Solution This represents the probability that a player was wearing a full shield given that a head and face injury occurred. Reduce the sample space to A. Then find $n(A \cap F)$ and $n(A)$.

$$P(F|A) = \frac{n(A \cap F)}{n(A)} = \frac{34}{129} \approx 0.264$$

If a head and face injury occurred, then the probability is about 0.264 that the player wore a full shield. This suggests that fewer head and face injuries occur in players who wear a full shield.

(b) $P(A'|F)$

Solution In words, this is the probability that an injury will not be to the head and face for players who wear full shields.

$$P(A'|F) = \frac{n(A' \cap F)}{n(F)} = \frac{195}{229} \approx 0.852$$

(c) $P(F'|A')$

Solution Here, we want the probability that an injured player was not wearing a full shield even though the injury was not to the head and face.

$$P(F'|A') = \frac{n(A' \cap F')}{n(A')} = \frac{252}{447} \approx 0.564$$

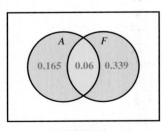

FIGURE 19

Venn diagrams are useful for illustrating problems in conditional probability. A Venn diagram for Example 1, in which the probabilities are used to indicate the number in the set defined by each region, is shown in Figure 19. In the diagram,

$P(F|A)$ is found by reducing the sample space to just set A. Then $P(F|A)$ is the ratio of the number in that part of set A which is also in F to the number in set A, or $0.059/0.224 \approx 0.263$. The difference between this value and the one calculated in Example 1(a) is due to rounding.

EXAMPLE 2 Conditional Probabilities
Given $P(E) = 0.4$, $P(F) = 0.5$, and $P(E \cup F) = 0.7$, find $P(E|F)$.

Solution Find $P(E \cap F)$ first. By the union rule,

$$P(E \cup F) = P(E) + P(F) - P(E \cap F)$$
$$0.7 = 0.4 + 0.5 - P(E \cap F)$$
$$P(E \cap F) = 0.2.$$

$P(E|F)$ is the ratio of the probability of that part of E which is in F to the probability of F, or

$$P(E|F) = \frac{P(E \cap F)}{P(F)} = \frac{0.2}{0.5} = \frac{2}{5}.$$

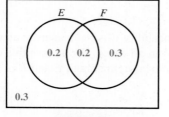

FIGURE 20

The Venn diagram in Figure 20 illustrates Example 2.

EXAMPLE 3 Tossing Coins
Two fair coins were tossed, and it is known that at least one was a head. Find the probability that both were heads.

Solution At first glance the answer to this question appears to be $1/2$. Using mathematics, however, we will see otherwise. The sample space has four equally likely outcomes, $S = \{hh, ht, th, tt\}$. Define two events:

$$E_1 = \text{at least 1 head} = \{hh, ht, th\},$$

and

$$E_2 = \text{2 heads} = \{hh\}.$$

Since there are four equally likely outcomes, $P(E_1) = 3/4$. Also, $P(E_1 \cap E_2) = 1/4$. We want the probability that both were heads, given that at least one was a head; that is, we want to find $P(E_2|E_1)$. Because of the condition that at least one coin was a head, the reduced sample space is

$$\{hh, ht, th\}.$$

Since only one outcome in this reduced sample space is 2 heads,

$$P(E_2|E_1) = \frac{1}{3}.$$

Alternatively, use the definition given above.

$$P(E_2|E_1) = \frac{P(E_2 \cap E_1)}{P(E_1)} = \frac{1/4}{3/4} = \frac{1}{3}$$

It is important not to confuse $P(A|B)$ with $P(B|A)$. For example, in a criminal trial, a prosecutor may point out to the jury that the probability of the defendant's DNA profile matching that of a sample taken at the scene of the crime, given that the defendant is innocent, is very small. What the jury must decide, however, is the probability that the defendant is innocent, given that the defendant's DNA profile matches the sample. Confusing the two is an error sometimes called "the prosecutor's fallacy," and the 1990 conviction of a rape suspect in England was overturned by a panel of judges, who ordered a retrial, because the fallacy made the original trial unfair.*

Later in this section, we will see how to compute $P(A|B)$ when we know $P(B|A)$.

Product Rule   If $P(E) \neq 0$ and $P(F) \neq 0$, then the definition of conditional probability shows that

$$P(E|F) = \frac{P(E \cap F)}{P(F)} \qquad \text{and} \qquad P(F|E) = \frac{P(F \cap E)}{P(E)}.$$

Using the fact that $P(E \cap F) = P(F \cap E)$, and solving each of these equations for $P(E \cap F)$, we obtain the following rule.

PRODUCT RULE OF PROBABILITY

If E and F are events, then $P(E \cap F)$ may be found by either of these formulas.

$$P(E \cap F) = P(F) \cdot P(E|F) \qquad \text{or} \qquad P(E \cap F) = P(E) \cdot P(F|E)$$

The product rule gives a method for finding the probability that events E and F both occur, as illustrated by the next few examples.

EXAMPLE 4   Biology Majors

In a class with 2/5 women and 3/5 men, 25% of the women are biology majors. Find the probability that a student chosen from the class at random is a female biology major.

Solution Let B and W represent the events "biology major" and "woman," respectively. We want to find $P(B \cap W)$. By the product rule,

$$P(B \cap W) = P(W) \cdot P(B|W).$$

Using the given information, $P(W) = 2/5 = 0.4$ and $P(B|W) = 0.25$. Thus,

$$P(B \cap W) = 0.4(0.25) = 0.10.$$

The next examples show how a tree diagram is used with the product rule to find the probability of a sequence of events.

---

*Pringle, David, "Who's the DNA Fingerprinting Pointing At?" *New Scientist*, Jan. 29, 1994, pp. 51–52.

EXAMPLE 5 Tree Diagram

From a box containing 3 white, 2 green, and 1 red marble, 2 marbles are drawn one at a time without replacing the first before the second is drawn. Find the probability that 1 white and 1 green marble are drawn.

Solution A tree diagram showing the various possible outcomes is given in Figure 21. In this diagram, W represents the event "drawing a white marble" and G represents "drawing a green marble." On the first draw, $P(W \text{ first}) = 3/6 = 1/2$ because 3 of the 6 marbles in the box are white. On the second draw $P(G \text{ second} \,|\, W \text{ first}) = 2/5$. One white marble has been removed, leaving 5, of which 2 are green.

Now we want to find the probability of drawing exactly one white marble and one green marble. This event can occur in two ways: drawing a white marble first and then a green one (branch 2 of the tree diagram), or drawing a green marble first and then a white one (branch 4). For branch 2,

$$P(W \text{ first}) \cdot P(G \text{ second} \,|\, W \text{ first}) = \frac{1}{2} \cdot \frac{2}{5} = \frac{1}{5}.$$

For branch 4, where the green marble is drawn first,

$$P(G \text{ first}) \cdot P(W \text{ second} \,|\, G \text{ first}) = \frac{1}{3} \cdot \frac{3}{5} = \frac{1}{5}.$$

Since the two events are mutually exclusive, the final probability is the sum of these two probabilities, or

$$P(1\ W, 1\ G) = P(W \text{ first}) \cdot P(G \text{ second} \,|\, W \text{ first})$$

$$+ P(G \text{ first}) \cdot P(W \text{ second} \,|\, G \text{ first}) = \frac{2}{5}.$$

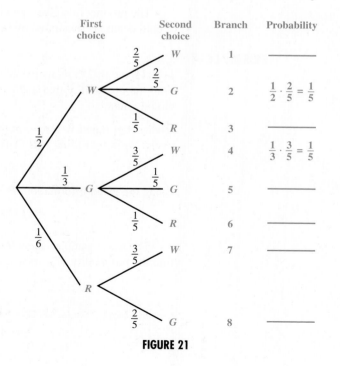

FIGURE 21

Independent Events   Suppose, in Example 5, that we draw the two marbles *with* replacement rather than without replacement (that is, we put the first marble back before drawing the second marble). If the first marble is white, then the probability of drawing a green marble on the second pick is 2/6, rather than 2/5, because there are still 6 marbles in the box, 2 of them green. In this case, $P($green second $|$ white first$)$ is the same as $P($green second$)$. The color of the second marble is not affected by the color of the first marble. We say that the event that the second marble is green is *independent* of the event that the first marble is white since the knowledge of the first marble does not influence what happens to the second marble. On the other hand, when we draw without replacement, the events that the first marble is white and that the second marble is green are *dependent* events. The fact that the first marble is white means there is one less white marble in the box, influencing the probability that the second marble is green.

As another example, consider tossing a fair coin twice. If the first toss shows heads, the probability that the next toss is heads is still 1/2. Coin tosses are independent events, since the outcome of one toss does not influence the outcome of the next toss. Similarly, rolls of a fair die are independent events. On the other hand, the events "the milk is old" and "the milk is sour" are dependent events; if the milk is old, there is an increased chance that it is sour. Also, in the example at the beginning of this section, the events A (player sustained a head or face injury) and F (player wore a full shield) are dependent events, because information about the use of shields affected the probability of a head or face injury. That is, $P(A \,|\, F)$ is different from $P(A)$.

If events E and F are independent, then the knowledge that E has occurred gives no (probability) information about the occurrence or nonoccurrence of event F. That is $P(F)$ is exactly the same as $P(F \,|\, E)$, or

$$P(F \,|\, E) = P(F).$$

This, in fact, is the formal definition of independent events.

INDEPENDENT EVENTS

E and F are **independent events** if

$$P(F \,|\, E) = P(F) \quad \text{or} \quad P(E \,|\, F) = P(E).$$

If the events are not independent, they are **dependent events.**

When E and F are independent events, then $P(F \,|\, E) = P(F)$ and the product rule becomes

$$P(E \cap F) = P(E) \cdot P(F \,|\, E) = P(E) \cdot P(F).$$

Conversely, if this equation holds, then it follows that $P(F) = P(F \,|\, E)$. Consequently, we have this useful fact:

PRODUCT RULE FOR INDEPENDENT EVENTS

E and F are independent events if and only if

$$P(E \cap F) = P(E) \cdot P(F).$$

EXAMPLE 6 Product Rule

For a particular couple expecting their first child, the probability the child will have blue eyes is 0.25 and the probability the child will be a boy is 0.5. Find the probability that the child will be a blue-eyed boy.

Solution If the child having blue eyes and the child being a boy are independent events, then

$$P(\text{blue-eyed boy}) = P(\text{child has blue eyes}) \cdot P(\text{child is a boy})$$
$$= (0.25)(0.5) = 0.125.$$

CAUTION It is common for students to confuse the ideas of *mutually exclusive* events and *independent* events. Events E and F are mutually exclusive if $E \cap F = \emptyset$. For example, if a family has exactly one child, the only possible outcomes are $B = \{\text{boy}\}$ and $G = \{\text{girl}\}$. These two events are mutually exclusive. The events are *not* independent, however, since $P(G|B) = 0$ (if a family with only one child has a boy, the probability it has a girl is then 0). Since $P(G|B) \neq P(G)$, the events are not independent.

Of all the families with exactly two children, the events $G_1 = \{\text{first child is a girl}\}$ and $G_2 = \{\text{second child is a girl}\}$ are independent, since $P(G_2|G_1)$ equals $P(G_2)$. However, G_1 and G_2 are not mutually exclusive, since $G_1 \cap G_2 = \{\text{both children are girls}\} \neq \emptyset$.

To show that two events E and F are independent, show that $P(F|E) = P(F)$ or that $P(E|F) = P(E)$ or that $P(E \cap F) = P(E) \cdot P(F)$. Another way is to observe that knowledge of one outcome does not influence the probability of the other outcome, as we did for coin tosses.

NOTE In some cases, it may not be apparent from the physical description of the problem whether two events are independent or not. For example, it is not obvious whether the event that a baseball player gets a hit tomorrow is independent of the event that he got a hit today. In such cases, it is necessary to calculate whether $P(F|E) = P(F)$, or, equivalently, whether $P(E \cap F) = P(E) \cdot P(F)$.

EXAMPLE 7 Snow in Manhattan

On a typical January day in Manhattan the probability of snow is 0.10, the probability of a traffic jam is 0.80, and the probability of snow or a traffic jam (or both) is 0.82. Are the event "it snows" and the event "a traffic jam occurs" independent?

Solution Let S represent the event "it snows" and T represent the event "a traffic jam occurs." We must determine whether

$$P(T|S) = P(T) \qquad \text{or} \qquad P(S|T) = P(S).$$

We know $P(S) = 0.10$, $P(T) = 0.8$, and $P(S \cup T) = 0.82$. We can use the union rule (or a Venn diagram) to find $P(S \cap T) = 0.08$, $P(T|S) = 0.8$, and $P(S|T) = 0.1$. Since

$$P(T|S) = P(T) = 0.8 \qquad \text{and} \qquad P(S|T) = P(S) = 0.1,$$

the events "it snows" and "a traffic jam occurs" are independent.

Although we showed $P(T|S) = P(T)$ and $P(S|T) = P(S)$ in Example 7, only one of these results is needed to establish independence. It is also important to note that independence of events does not necessarily follow intuition; it is established from the mathematical definition of independence.

Bayes' Theorem   Suppose the probability that a person gets lung cancer, given that the person smokes a pack or more of cigarettes daily, is known. For a research project, it might be necessary to know the probability that a person smokes a pack or more of cigarettes daily, given that the person has lung cancer. More generally, if $P(E|F)$ is known for two events E and F, can $P(F|E)$ be found? The answer is yes, we can find $P(F|E)$ using the formula to be developed in this section. To develop this formula, we can use a tree diagram to find $P(F|E)$. Since $P(E|F)$ is known, the first outcome is either F or F'. Then for each of these outcomes, either E or E' occurs, as shown in Figure 22. The four cases have the probabilities shown on the right. By the definition of conditional probability,

$$P(F|E) = \frac{P(F \cap E)}{P(E)} = \frac{P(F) \cdot P(E|F)}{P(F) \cdot P(E|F) + P(F') \cdot P(E|F')}.$$

Notice that $P(E)$ is the sum of the first and third cases. This result is a special case of Bayes' theorem, which is generalized later in this section.

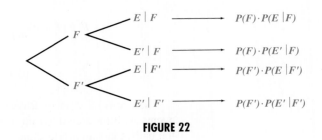

FIGURE 22

BAYES' THEOREM (SPECIAL CASE)

$$P(F|E) = \frac{P(F) \cdot P(E|F)}{P(F) \cdot P(E|F) + P(F') \cdot P(E|F')}$$

EXAMPLE 8   Worker Errors

For a fixed length of time, the probability of a worker error on a certain production line is 0.1, the probability that an accident will occur when there is a worker error is 0.3, and the probability that an accident will occur when there is no worker error is 0.2. Find the probability of a worker error if there is an accident.

Solution Let E represent the event of an accident, and let F represent the event of worker error. From the information above,

$$P(F) = 0.1, \qquad P(E|F) = 0.3, \qquad \text{and} \qquad P(E|F') = 0.2.$$

These probabilities are shown on the tree diagram in Figure 23 on the next page.

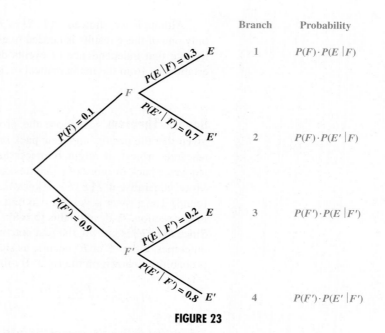

	Branch	Probability
	1	$P(F) \cdot P(E\,|\,F)$
	2	$P(F) \cdot P(E'\,|\,F)$
	3	$P(F') \cdot P(E\,|\,F')$
	4	$P(F') \cdot P(E'\,|\,F')$

FIGURE 23

Find $P(F\,|\,E)$ by dividing the probability that both E and F occur, given by branch 1, by the probability that E occurs, given by the sum of branches 1 and 3.

$$P(F\,|\,E) = \frac{P(F) \cdot P(E\,|\,F)}{P(F) \cdot P(E\,|\,F) + P(F') \cdot P(E\,|\,F')}$$

$$= \frac{(0.1)(0.3)}{(0.1)(0.3) + (0.9)(0.2)} = \frac{1}{7} \approx 0.143$$

The special case of Bayes' theorem can be generalized to more than two events with the tree diagram in Figure 24 on the next page. This diagram shows the paths that can produce an event E. We assume that the events $F_1, F_2, \ldots, F_n$ are mutually exclusive events (that is, disjoint events) whose union is the sample space, and that E is an event that has occurred. See Figure 25 on the next page.

The probability $P(F_i\,|\,E)$, where $1 \leq i \leq n$, can be found by dividing the probability for the branch containing $P(E\,|\,F_i)$ by the sum of the probabilities of all the branches producing event E.

BAYES' THEOREM

$$P(F_i\,|\,E) = \frac{P(F_i) \cdot P(E\,|\,F_i)}{P(F_1) \cdot P(E\,|\,F_1) + P(F_2) \cdot P(E\,|\,F_2) + \cdots + P(F_n) \cdot P(E\,|\,F_n)}$$

This result is known as **Bayes' theorem,** after the Reverend Thomas Bayes (1702–1761), whose paper on probability was published about three years after his death.

The statement of Bayes' theorem can be daunting. Actually, it is easier to remember the formula by thinking of the tree diagram that produced it. Go through the following steps.

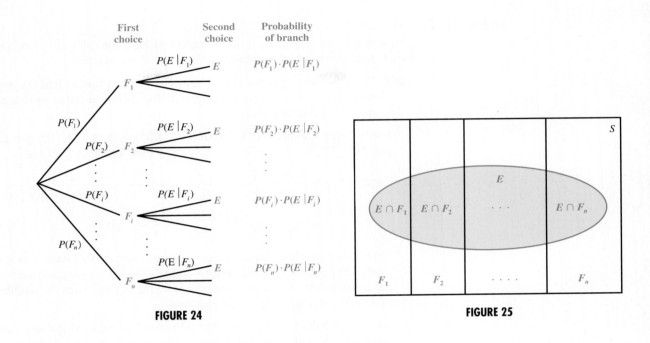

FIGURE 24

FIGURE 25

USING BAYES' THEOREM

1. Start a tree diagram with branches representing $F_1, F_2, \ldots, F_n$. Label each branch with its corresponding probability.

2. From the end of each of these branches, draw a branch for event E. Label this branch with the probability of getting to it, $P(E \mid F_i)$.

3. You now have n different paths that result in event E. Next to each path, put its probability—the product of the probabilities that the first branch occurs, $P(F_i)$, and that the second branch occurs, $P(E \mid F_i)$; that is, the product $P(F_i) \cdot P(E \mid F_i)$, which equals $P(F_i \cap E)$.

4. $P(F_i \mid E)$ is found by dividing the probability of the branch for F_i by the sum of the probabilities of all the branches producing event E.

Sensitivity and Specificity   The *sensitivity* of a medical test is defined as the probability that a test will be positive given that a person has a disease, written $P(T^+ \mid D^+)$. The *specificity* of a test is defined as the probability that a test will be negative given that the person does not have the disease, written $P(T^- \mid D^-)$. Both of these values provide important information with regard to the accuracy of the test. It is important to note that sensitivity and specificity are not dependent upon disease prevalence, or how common the disease is in a given population. Notice that these tests are a measure of the accuracy of the test and not of whether a given patient has the disease. For patients, the important probability is $P(D^+ \mid T^+)$, that is, the probability that the disease is present given that a positive test has occurred. We can use Bayes' Theorem, test accuracy, and disease prevalence to find this value, as the next example illustrates.

EXAMPLE 9 Breast Cancer

The sensitivity and specificity for breast cancer during a clinical breast examination by a trained expert are approximately 0.54 and 0.94, respectively.*

(a) If 2% of the women in the United States have breast cancer, find the probability that a woman who tests positive during a clinical breast examination actually has breast cancer.

Solution We know that $P(D^+) = 0.02$, $P(T^+|D^+) = 0.54$, $P(T^-|D^-) = 0.94$, and $P(T^+|D^-) = 1 - P(T^-|D^-)$. Using Bayes' Theorem, we have

$$P(D^+|T^+) = \frac{P(D^+ \cap T^+)}{P(T^+)} = \frac{P(D^+) \cdot P(T^+|D^+)}{P(D^+) \cdot P(T^+|D^+) + P(D^-) \cdot P(T^+|D^-)}$$

$$= \frac{(0.02)(0.54)}{(0.02)(0.54) + (0.98)(1 - 0.94)} \approx 0.155.$$

Thus, a positive clinical exam suggests an approximate 16% chance of breast cancer. Notice that, although the probability of breast cancer given a positive test is fairly low, the information gained by the test increases the likelihood of actually having breast cancer eightfold.

(b) Suppose that the positive clinical examination has occurred and the woman is directed to receive a mammogram. Suppose that for this particular woman, the sensitivity and specificity for the mammogram are 0.89 and 0.93, respectively. Given that both the clinical exam and the mammogram result in a positive test, what is the probability that the woman has breast cancer? Assume that the mammogram and the clinical examination are independent events.

Solution First construct a tree diagram using the given information, as in Figure 26 on the next page. The steps leading to this event are shown in blue type. Let M^+ and M^- indicate a positive and negative mammogram, respectively. We can calculate the probability of this event from the highlighted branches of the tree and Bayes' Theorem.

$$P(D^+|T^+ \cap M^+) = \frac{P(D^+ \cap T^+ \cap M^+)}{P(T^+ \cap M^+)}$$

$$= \frac{P(D^+) \cdot P(T^+|D^+) \cdot P(M^+|D^+ \cap T^+)}{P(D^+) \cdot P(T^+|D^+) \cdot P(M^+|D^+ \cap T^+) + P(D^-) \cdot P(T^+|D^-) \cdot P(M^+|D^- \cap T^+)}$$

$$= \frac{P(D^+) \cdot P(T^+|D^+) \cdot P(M^+|D^+)}{P(D^+) \cdot P(T^+|D^+) \cdot P(M^+|D^+) + P(D^-) \cdot P(T^+|D^-) \cdot P(M^+|D^-)} \quad \text{Assume that events } T \text{ and } M \text{ are independent.}$$

$$= \frac{(0.02)(0.54)(0.89)}{(0.02)(0.54)(0.89) + (0.98)(0.06)(0.07)} = 0.70$$

Thus, for this woman, who has a positive clinical exam and a positive mammogram, there is a 70% chance that she has breast cancer.

---

*Barton, M., R. Harris, and S. Fletcher, "Does This Patient Have Breast Cancer? The Screening Clinical Breast Examinations: Should it Be Done? How?," *JAMA*, Vol. 282, No. 13, October 6, 1999, 1270–1280.

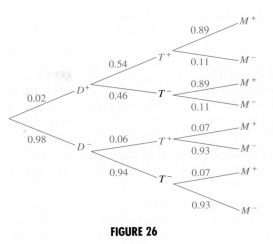

FIGURE 26

12.3 EXERCISES

If a single fair die is rolled, find the probabilities of the following results.

1. A 2, given that the number rolled was odd

2. A 4, given that the number rolled was even

3. An even number, given that the number rolled was 6

If two fair dice are rolled, find the probabilities of the following results.

4. A sum of 8, given that the sum is greater than 7

5. A sum of 6, given that the roll was a "double" (two identical numbers)

6. A double, given that the sum was 9

If two cards are drawn without replacement from an ordinary deck, find the probabilities of the following results.

7. The second is a heart, given that the first is a heart.

8. The second is black, given that the first is a spade.

9. The second is a face card, given that the first is a jack.

10. The second is an ace, given that the first is not an ace.

11. A jack and a 10 are drawn.

12. An ace and a 4 are drawn.

13. Two black cards are drawn.

14. Two hearts are drawn.

15. In your own words, explain the meaning of conditional probability.

16. Your friend asks you to explain how the product rule for independent events differs from the product rule for dependent events. How would you respond?

17. Another friend asks you to explain how to tell whether two events are dependent or independent. How would you reply? (Use your own words.)

18. A student reasons that the probability in Example 3 of both coins being heads is just the probability that the other coin is a head, that is, 1/2. Explain why this reasoning is wrong.

19. The following problem, submitted by Daniel Hahn of Blairstown, Iowa, appeared in the "Ask Marilyn' column of *Parade* magazine.*

"You discover two booths at a carnival. Each is tended by an honest man with a pair of covered coin shakers. In each shaker is a single coin, and you are allowed to bet upon the chance that both coins in that booth's shakers are heads after the man in the booth shakes them, does an inspection and can tell you that at least one of the shakers contains a head. The difference is that the man in the first booth always looks inside both of his shakers, whereas the man in the second booth looks inside only one of the shakers. Where will you stand the best chance?"

20. A student has two examinations on the same day. If the probability of passing the first test is 0.90 and the probability of passing the second test is 0.85, argue that these two events are not independent.

21. Suppose a male defendant in a court trial has a mustache, beard, tattoo, and an earring. Suppose, also, that an eyewitness has identified the perpetrator as someone with these characteristics. If the respective probabilities for the male population in this region are 0.35, 0.30, 0.10, and 0.05, is it fair to multiply these probabilities together to conclude that the probability that a person having these characteristics is 0.000525, or 21 in 40,000, and thus decide that the defendant must be guilty?

22. In a two-child family, if we assume that the probabilities of a male child and a female child are each 0.5, are the events *each child is the same sex* and *at most one male* independent? Are they independent for a three-child family?

23. Let A and B be independent events with $P(A) = \dfrac{1}{4}$ and $P(B) = \dfrac{1}{5}$. Find $P(A \cap B)$ and $P(A \cup B)$.

24. If A and B are events such that $P(A) = 0.5$ and $P(A \cup B) = 0.80$, find $P(B)$ when

a. A and B are mutually exclusive.

b. A and B are independent.

For two events M and N, $P(M) = 0.4$, $P(N|M) = 0.3$, and $P(N|M') = 0.4$. Find each of the following.

25. $P(M|N)$　　　　　　　　　　　　　　　　**26.** $P(M'|N)$

For mutually exclusive events R_1, R_2, and R_3, we have $P(R_1) = 0.05$, $P(R_2) = 0.6$, and $P(R_3) = 0.35$. Also, $P(Q|R_1) = 0.40$, $P(Q|R_2) = 0.30$, and $P(Q|R_3) = 0.60$. Find each of the following.

27. $P(R_1|Q)$　　　　**28.** $P(R_2|Q)$　　　　**29.** $P(R_3|Q)$　　　　**30.** $P(R_1'|Q)$

Suppose you have three jars with the following contents: 2 black balls and 1 white ball in the first, 1 black ball and 2 white balls in the second, and 1 black ball and 1 white ball in the third. One jar is to be selected, and then 1 ball is to be drawn from the selected jar. If the probabilities of selecting the first, second, or third jar are 1/2, 1/3, and 1/6, respectively, find the probabilities that if a white ball is drawn, it came from the following jars.

31. The second jar　　　　　　　　　　　　　　**32.** The third jar

Applications

LIFE SCIENCES

33. *Genetics* Both of a certain pea plant's parents had a gene for purple and a gene for white flowers. (See Exercise 68 in Section 12.2.) If the offspring has purple flowers, find the probability that it combined a gene for purple and a gene for white (rather than 2 for purple).

Genetics Assuming that boy and girl babies are equally likely, fill in the remaining probabilities on the tree diagram and use the following information to find the probability that a family with three children has all girls, given the following.

34. The first is a girl.

35. The third is a girl.

36. At least 2 are girls.

37. At least 1 is a girl.

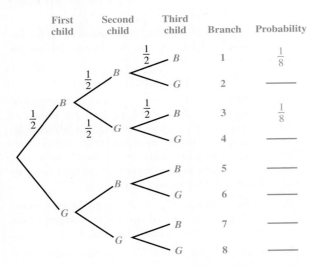

38. *AIDS* The table in the next column, based on data from the Centers for Disease Control, gives the number of new cases of the AIDS virus for men and women in the United States in 1998 by method of transmission.*

  a. Find the probability that a male resident who was newly diagnosed with AIDS in 1998 contracted it via homosexual contact.

  b. Find the probability that a female resident who was newly diagnosed with AIDS in 1998 contracted it via intravenous drug use.

Method of Transmission	Male	Female	Total
Homosexual contact	8,388	0	8,388
Heterosexual contact	1,146	1,806	2,952
Intervenous drug use	3,652	1,541	5,193
Other	5,237	2,054	7,291
Total	18,423	5,401	23,824

  c. Find the probability that a newly diagnosed person with AIDS is a female.

  d. Find the probability that a newly diagnosed person who contracted AIDS via heterosexual contact was a female.

  e. Are the events that a person with AIDS is a female and that the person contracted AIDS via heterosexual contact independent? Explain your answer.

39. *Medical Experiment* A medical experiment showed that the probability that a new medicine is effective is 0.75, the probability that a patient will have a certain side effect is 0.4, and the probability that both events occur is 0.3. Decide whether these events are dependent or independent.

Color Blindness The following table shows frequencies for red-green color blindness, where M represents "person is male" and C represents "person is color-blind." Use this table to find the following probabilities. (See Exercise 67, Section 12.2.)

	M	M'	Totals
C	0.035	0.004	0.039
C'	0.452	0.509	0.961
Totals	0.487	0.513	1.000

40. $P(M)$

41. $P(C)$

42. $P(M \cap C)$

43. $P(M \cup C)$

44. $P(M \mid C)$

45. $P(C \mid M)$

46. Are the events C and M, described above, dependent? What does this mean?

47. *Color Blindness* A scientist wishes to determine whether there is a relationship between color blindness (C) and deafness (D).

---

*Health United States, 1999, CDC, National Center for HIV, STD, and TB Prevention, Division of HIV/AIDS Prevention and *The World Almanac and Book of Facts 2000*, p. 903.

a. Suppose the scientist found the probabilities listed in the table. What should the findings be? (See Exercises 40–46.)

b. Explain what your answer tells us about color blindness and deafness.

	D	**D'**	**Totals**
C	0.0008	0.0392	0.0400
C'	0.0192	0.9408	0.9600
Totals	0.0200	0.9800	1.0000

48. *Breast Cancer* To explain why the chance of a woman getting breast cancer in the next year goes up each year, while the chance of a woman getting breast cancer in her lifetime goes down, Ruma Falk made the following analogy.* Suppose you are looking for a letter that you may have lost. You have 8 drawers in your desk. There is a probability of 0.1 that the letter is in any one of the 8 drawers, and a probability of 0.2 that the letter is not in any of the drawers.

a. What is the probability that the letter is in drawer 1?

b. Given that the letter is not in drawer 1, what is the probability that the letter is in drawer 2?

c. Given that the letter is not in drawer 1 or 2, what is the probability that the letter is in drawer 3?

d. Given that the letter is not in drawers 1–7, what is the probability that the letter is in drawer 8?

e. Based on your answers to parts a–d, what is happening to the probability that the letter is in the next drawer?

f. What is the probability that the letter is in some drawer?

g. Given that the letter is not in drawer 1, what is the probability that the letter is in some drawer?

h. Given that the letter is not in drawer 1 or 2, what is the probability that the letter is in some drawer?

i. Given that the letter is not in drawers 1–7, what is the probability that the letter is in some drawer?

j. Based on your answers to parts f–i, what is happening to the probability that the letter is in some drawer?

49. *Rain Forecasts* In a letter to the journal *Nature,* Robert A. J. Matthews gives the following table of outcomes of forecast and weather over 1,000 1-hour walks, based on the United Kingdom's Meteorological Office's 83% accuracy in 24-hour forecasts.†

	Rain	**No Rain**	**Sum**
Forecast of rain	66	156	222
Forecast of no rain	14	764	778
Sum	80	920	1,000

a. Verify that the probability that the forecast called for rain, given that there was rain, is indeed 83%. Also verify that the probability that the forecast called for no rain, given that there was no rain, is also 83%.

b. Calculate the probability that there was rain, given that the forecast called for rain.

c. Calculate the probability that there was no rain, given that the forecast called for no rain.

d. Observe that your answer to part c is higher than 83%, and that your answer to part b is much lower. Discuss which figure best describes the accuracy of the weather forecast in recommending whether or not you should carry an umbrella.

50. *Drug Screening* In searching for a new drug with commercial possibilities, drug company researchers use the ratio

$$N_S : N_A : N_P : 1.$$

That is, if the company gives preliminary screening to N_S substances, if may find that N_A of them are worthy of further study, with N_P of these surviving into full-scale development. Finally, 1 of the substances will result in a marketable drug. Typical numbers used by Smith, Kline, and French Laboratories in planning research budgets might be $2,000 : 30 : 8 : 1$.‡ Use this ratio for parts a–f.

a. Suppose a compound has been chosen for preliminary screening. Find the probability that the compound will survive and become a marketable drug.

b. Find the probability that the compound will not lead to a marketable drug.

c. Suppose the number of such compounds receiving preliminary screening is *a*. Set up the probability that none of them produces a marketable drug. (Assume independence throughout these exercises.)

d. Use your results from part c to find the probability that at least one of the drugs will prove marketable.

e. Suppose now that *N* scientists are employed in the preliminary screening, and that each scientist can screen

---

*Falk, Ruma, *Chance News,* July 23, 1995.
†Matthews, Robert A. J., *Nature,* Vol. 382, Aug. 29, 1996, p. 3.
‡Pyle, E. B., III, B. Douglas, G. W. Ebright, W. J. Westlake, and A. B. Bender, "Scientific Manpower Allocation to New Drug Screening Programs," *Management Science,* Vol. 19, No. 12, August 1973. Copyright © 1973 by The Institute of Management Sciences. Reprinted by permission.

c compounds per year. Find the probability that no marketable drugs will be discovered in a year.

f. Find the probability that at least one marketable drug will be discovered.

51. *Toxemia Test* A magazine article described a new test for toxemia, a disease that affects pregnant women. To perform this test, the woman lies on her left side and then rolls onto her back. The test is considered positive if there is a 20 millimeter rise in her blood pressure within one minute. The article gives the following probabilities, where T represents having toxemia at some time during the pregnancy, and N represents a negative test.

$$P(T' \mid N) = 0.90 \quad \text{and} \quad P(T \mid N') = 0.75$$

Assume that $P(N') = 0.11$, and find each of the following.

a. $P(N \mid T)$ **b.** $P(N' \mid T)$

52. *AIDS Testing* In 1987, it was found that one woman in 10,000 was infected with the AIDS virus. The most commonly used test always detected an infected woman, but one healthy woman out of 20,000 was incorrectly identified as being infected with the AIDS virus.*

a. Find the probability that a woman who tested positive in 1987 was not infected.

b. It has been argued that everyone should be tested for AIDS. Based on the results in part a, how useful would the results of such testing be?

53. *Hepatitis Blood Test* The probability that a person with certain symptoms has hepatitis is 0.8. The blood test used to confirm this diagnosis gives positive results for 90% of people with the disease and 5% of those without the disease. What is the probability that an individual who has the symptoms and who reacts positive to the test actually has hepatitis?

54. *Test for HIV* A new test for the virus that causes AIDS, developed by Octopus Diagnostics Research of Hantsport, Nova Scotia, shows the presence or absence of HIV in a drop of blood in two minutes, compared with five days for a conventional test.[†] Preliminary results indicate a false positive rate (an indication that the HIV virus is present when it is not) of less than 2%, and a false negative rate (a failure to detect the presence of the HIV virus) of up to 5%. Assume for this exercise that these rates are exactly 2% and 5%. In 1996, there were

780,000 people in North America with the HIV virus, out of a population of 295 million.[‡] Suppose a resident of North America is chosen at random and given this test. If the result is positive, what is the probability that the person actually has the HIV virus?

55. *Binge Drinking* A 1995 study by the Harvard School of Public Health reported that 86% of male students who live in a fraternity house are binge drinkers. The figure for fraternity members who are not residents of a fraternity house is 71%, while the figure for men who do not belong to a fraternity is 45%.[§] Suppose that 10% of U.S. male students live in a fraternity house, 15% belong to a fraternity but do not live in a fraternity house, and 75% do not belong to a fraternity.

a. What is the probability that a randomly selected male student is a binge drinker?

b. If a randomly selected male student is a binge drinker, what is the probability that he lives in a fraternity house?

56. *Murder* During the murder trial of O. J. Simpson, Alan Dershowitz, an advisor to the defense team, stated on television that only about 0.1% of men who batter their wives actually murder them. Statistician I. J. Good observed that even if, given that a husband is a batterer, the probability he is guilty of murdering his wife is 0.001, what we really want to know is the probability that the husband is guilty, given that the wife was murdered.[‖] Good estimates the probability of a battered wife being murdered, given that her husband is not guilty, as 0.001. The probability that she is murdered if her husband is guilty is 1, of course. Using these numbers and Dershowitz's 0.001 probability of the husband being guilty, find the probability that the husband is guilty, given that the wife was murdered.

OTHER APPLICATIONS

57. *Working Women* A survey has shown that 52% of the women in a certain community work outside the home. Of these women, 64% are married, while 86% of the women who do not work outside the home are married. Find the probabilities that a woman in that community can be categorized as follows.

a. Married

b. A single woman working outside the home

The New York Times, Sept. 5, 1987.
[†]*Maclean's,* Feb. 17, 1997, p. 70.
[‡]*The World Almanac and Book of Facts 1997.*
[§]*The New York Times,* Dec. 6, 1995, p. B16.
[‖]Good, I. J., "When Batterer Turns Murderer," *Nature,* Vol. 375, No. 15, June 15, 1995, p. 541.

58. *Titanic* The following table lists the number of passengers who were on the *Titanic* and the number of passengers who survived, according to class of ticket.*

	Children		Women		Men		Totals	
	On	Survived	On	Survived	On	Survived	On	Survived
First class	6	6	144	140	175	57	325	203
Second class	24	24	165	76	168	14	357	114
Third class	79	27	93	80	462	75	634	182
Total	109	57	402	296	805	146	1,316	499

Use this information to determine the following (round answers to three decimal places).

a. What is the probability that a randomly selected passenger was from second class?

b. What is the overall probability of surviving?

c. What is the probability of a first-class passenger surviving?

d. What is the probability of a child who was also in the third class surviving?

e. Given that the survivor is from first class, what is the probability that she was a woman?

f. Given that a male has survived, what is the probability that he was in third class?

g. Are the events third-class survival and male survival independent events? What does this imply?

Never-Married Adults by Age Groups The following tables give the proportions of men and women in the U.S. population, and the proportions of men and women who have never married in a recent year.[†]

	Men	
Age	Proportion in Population	Proportion Never Married
18–24	0.133	0.874
25–34	0.205	0.397
35–44	0.232	0.187
45–64	0.287	0.075
65 or over	0.142	0.038

	Women	
Age	Proportion in Population	Proportion Never Married
18–24	0.123	0.775
25–34	0.194	0.298
35–44	0.219	0.121
45–64	0.284	0.062
65 or over	0.181	0.047

59. Find the probability that a randomly selected man who has never married is between 35 and 44 years old (inclusive).

60. Find the probability that a randomly selected woman who has been married is between 18 and 24 (inclusive).

61. Find the probability that a randomly selected woman who has never been married is between 45 and 64 (inclusive).

Seat-Belt Effectiveness A federal study showed that in 1990, 49% of all those involved in a fatal car crash wore seat belts. Of those in a fatal crash who wore seat belts, 44% were injured and 27% were killed. For those not wearing seat belts, the comparable figures were 41% and 50%, respectively.[‡]

62. Find the probability that a randomly selected person who was killed in a car crash was wearing a seat belt.

63. Find the probability that a randomly selected person who was unharmed in a fatal crash was not wearing a seat belt.

---

*Takis, Sandra L., "Titanic: A Statistical Exploration," *Mathematics Teacher*, Vol. 92, No. 8, Nov. 1999, pp. 660–664.

[†]"Marital Status and Living Arrangements of Adults 18 Years and Over," Mar. 1998, Table B, U.S. Bureau of the Census.

[‡]National Highway Traffic Safety Administration, Office of Driver and Pedestrian Research: "Occupant Protection Trends in 19 Cities," Nov. 1989, and "Use of Automatic Safety Belt Systems in 19 Cities," Feb. 1991.

Voting *The following table shows the proportion of people over 18 who are in various age categories, along with the probabilities that a person in a given age category will vote in a general election.*

Age	Percent of Voting Age Population	Probability of a Person of This Age Voting
18–21	11.0%	0.48
22–24	7.6%	0.53
25–44	37.6%	0.68
45–64	28.3%	0.64
65 or over	15.5%	0.74

Suppose a voter is picked at random. Find the probabilities that the voter is in the following age categories.

64. 18–21

65. 65 or over

66. Find the probability that a person who did not vote was in the age category 45–64.

12.4 DISCRETE RANDOM VARIABLES; APPLICATIONS TO DECISION MAKING

? THINK ABOUT IT   Should patients with malaria-related anemia receive blood transfusions?

In this section we shall see that the *expected value* of a probability distribution is a type of average. We will also see how this concept can be used to analyze a wide variety of decision problems.

Probability distributions were introduced briefly earlier in this chapter. Now we take a more complete look at probability distributions. A probability distribution depends on the idea of a *random variable*, so we begin with that.

Random Variables   Suppose that 100 patients are asked to rate the services provided by a local hospital. The rating scale ranges from 1 (Outstanding), 2 (Good), 3 (Okay), 4 (Poor), and 5 (Terrible). The value that each patient provides, which we will label X, is one of the numbers 1, 2, 3, 4, 5. For this experiment, X is called a **random variable.**

RANDOM VARIABLE

A random variable X is a function that assigns a real number to each simple event of an experiment.

It is important to note that random variables are functions. They are not random and they are not considered variables. One explanation for the use of these words comes from the observation that there are usually many different random variables that could be used to represent outcomes of an experiment. For example, in the rating scale above, the number 5 could have been assigned to "Outstanding," the number 4 could have been assigned to "Good," etc. Of course many other assignments could also be made. For this reason the words "random variable" have been used to describe the function.

Random variables are generally categorized as either discrete or continuous. A **discrete random variable** is used in experiments where there is either a finite or countably infinite number of simple events. **Continuous random variables** are used to describe situations where the random variable can be assigned any number in a given interval. We will study continuous random variables in the next chapter.

Probability Distribution In the example above, we can use the survey results to construct a probability distribution. Suppose the results of the survey show that 15 people listed the services as Outstanding, 38 people listed the services as Good, 25 people listed the services as Okay, 13 people listed the services as Poor, and 9 people listed the services as Terrible. We can place these results in a table as illustrated in Table 5.

Table 5

x	1	2	3	4	5
$P(x)$	15/100	38/100	25/100	13/100	9/100

Such a table that lists the possible values of a discrete random variable, together with the corresponding probabilities, is called a **probability distribution.** The function, $P(x) = P(X = x)$, is called a **probability function.** The sum of the probabilities in a probability distribution must always equal 1. (The sum in some distributions may appear to vary slightly from 1 because of rounding.)

Instead of writing the probability distribution as a table, we could write the same information as a set of ordered pairs:

$$\{(1, 15/100), (2, 38/100), (3, 25/100), (4, 13/100), (5, 9/100)\}.$$

There is just one probability for each value of the random variable. Thus, a probability distribution defines a function, the probability function.

The information in a probability distribution is often displayed graphically as a special kind of bar graph called a **histogram.** The bars of a histogram all have the same width, usually 1. The heights of the bars are determined by the probabilities. A histogram for the data in Table 5 is given in Figure 27 on the next page. A histogram shows important characteristics of a distribution that may not be readily apparent in tabular form, such as the relative sizes of the probabilities and any symmetry in the distribution.

The area of the bar above $x = 1$ in Figure 27 is the product of 1 and 15/100, or $1 \cdot 15/100 = 0.15$. Since each bar has a width of 1, its area is equal to the probability that corresponds to that value of x. Thus, the probability that a particular value will occur is given by the area of the appropriate bar of the graph. For example, the probability that a person rated the hospital as Outstanding, Good, or Okay is the sum of the areas for $x = 1$, $x = 2$, and $x = 3$. This area, shown in red in Figure 28, corresponds to 78/100, which corresponds to 78% of the total area, since

$$P(x \leq 3) = P(x = 1) + P(x = 2) + P(x = 3).$$

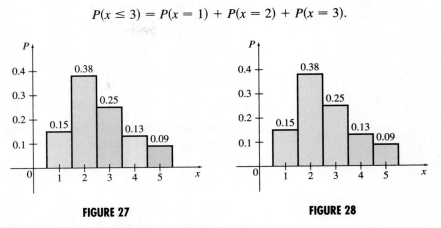

FIGURE 27   **FIGURE 28**

EXAMPLE 1   Probability Distributions

(a) Give the probability distribution for the number of heads showing when two coins are tossed.

Solution Let x represent the random variable "number of heads." Then x can take on the values 0, 1, or 2. Now find the probability of each outcome. The results are shown in Table 6.

(b) Draw a histogram for the distribution in Table 6. Find the probability that at least one coin comes up heads.

Solution The histogram is shown in Figure 29. The portion in red represents

$$P(x \geq 1) = P(x = 1) + P(x = 2)$$

$$= \frac{3}{4}.$$

Table 6

x	0	1	2
$P(x)$	1/4	1/2	1/4

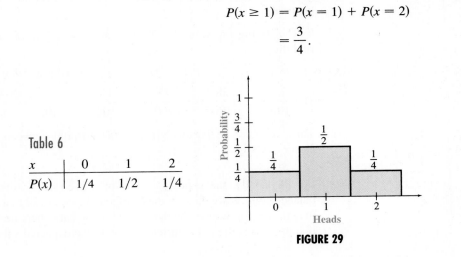

FIGURE 29

Expected Value   In working with probability distributions, it is useful to have a concept of the typical or average value that the random variable takes on. In Example 1, for instance, it seems reasonable that, on the average, one head shows when two coins are tossed. This does not tell what will happen the next time we toss two coins; we may get two heads, or we may get none. If we tossed two coins many times, however, we would expect that, in the long run, we would average about one head for each toss of two coins.

A way to solve such problems in general is to imagine flipping two coins 4 times. Based on the probability distribution in Example 1, we would expect that 1 of the 4 times we would get 0 heads, 2 of the 4 times we would get 1 head, and 1 of the 4 times we would get 2 heads. The total number of heads we would get, then, is

$$0 \cdot 1 + 1 \cdot 2 + 2 \cdot 1 = 4.$$

The expected numbers of heads per toss is found by dividing the total number of heads by the total number of tosses, or

$$\frac{0 \cdot 1 + 1 \cdot 2 + 2 \cdot 1}{4} = 0 \cdot \frac{1}{4} + 1 \cdot \frac{1}{2} + 2 \cdot \frac{1}{4} = 1.$$

Notice that the expected number of heads turns out to be the sum of the three values of the random variable X multiplied by their corresponding probabilities. We can use this idea to define the **expected value** of a discrete random variable as follows.

EXPECTED VALUE

The expected value of a discrete random variable X is defined as

$$E(X) = \sum_x xP(X = x) = \sum_x xP(x),$$

where the sum is taken over all possible values of the random variable X.

EXAMPLE 2 Hospital Satisfaction

In the example with hospital ratings, find the expected satisfaction rating of the hospital.

Solution Using the values in Table 5 and the definition of expected value, we find that

$$E(X) = 1 \cdot P(X = 1) + 2 \cdot P(X = 2) + 3 \cdot P(X = 3) + 4 \cdot P(X = 4) + 5 \cdot P(X = 5)$$

$$= 1 \cdot \frac{15}{100} + 2 \cdot \frac{38}{100} + 3 \cdot \frac{25}{100} + 4 \cdot \frac{13}{100} + 5 \cdot \frac{9}{100} = \frac{263}{100} = 2.63.$$

Thus, the average person will rate the hospital between Good and Okay, but closer to Okay.

Physically, the expected value of a probability distribution represents a balance point. If we think of the histogram in Figure 27 as a series of weights with magnitudes represented by the heights of the bars, then the system would balance if supported at the point corresponding to the expected value.

EXAMPLE 3 Friendly Wager

Each day Donna and Mary toss a coin to see who buys coffee (80 cents a cup). One tosses and the other calls the outcome. If the person who calls the outcome is correct, the other buys the coffee; otherwise the caller pays. Find Donna's expected winnings.

Solution Assume that an honest coin is used, that Mary tosses the coin, and that Donna calls the outcome. The possible results and corresponding probabilities are shown in Table 7.

Table 7

	Possible Results			
Result of toss	Heads	Heads	Tails	Tails
Call	Heads	Tails	Heads	Tails
Caller wins?	Yes	No	No	Yes
Probability	1/4	1/4	1/4	1/4

Donna wins an 80¢ cup of coffee whenever the results and calls match, and she loses an 80¢ cup when there is no match. Her expected winnings are

$$(0.80)\left(\frac{1}{4}\right) + (-0.80)\left(\frac{1}{4}\right) + (-0.80)\left(\frac{1}{4}\right) + (0.80)\left(\frac{1}{4}\right) = 0.$$

On the average, over the long run, Donna neither wins nor loses.

A game with an expected value of 0 (such as the one in Example 3) is called a **fair game.** Casinos do not offer fair games. If they did, they would win (on the average) $0, and have a hard time paying the help! Casino games have expected winnings for the house that vary from 1.5 cents per dollar to 60 cents per dollar. Exercises 25–27 at the end of the section ask you to find the expected winnings for certain games of chance.

The idea of expected value can be very useful in decision making, as shown by the next example.

EXAMPLE 4   Life Insurance

At age 50, you receive a letter from Mutual of Mauritania Insurance Company. According to the letter, you must tell the company immediately which of the following two options you will choose: take $20,000 at age 60 (if you are alive, $0 otherwise) or $30,000 at age 70 (again, if you are alive, $0 otherwise). Based *only* on the idea of expected value, which should you choose?

Solution Life insurance companies have constructed elaborate tables showing the probability of a person living a given number of years into the future. From a recent such table, the probability of living from age 50 to 60 is 0.88, while the probability of living from age 50 to 70 is 0.64. The expected values of the two options are given below.

First option: $(20,000)(0.88) + (0)(0.12) = 17,600$

Second option: $(30,000)(0.64) + (0)(0.36) = 19,200$

Based strictly on expected values, choose the second option.

Variance   Although the expected value provides a measure of central tendency, it does not provide a measure of how much the data varies from the expected value. For example, the three sets of numbers in Table 8 on the next page all have an average of 50. But, as you can see, the values in the third distribution vary significantly from the mean.

■ Table 8

List	Values	Mean
Set 1	15, 25, 70, 90	50
Set 2	45, 46, 47, 48, 49, 51, 52, 53, 54, 55	50
Set 3	1, 17, 45, 70, 117	50

The **variance** of a probability distribution is a measure of the *spread* of the values of a probability distribution. For a discrete distribution, the variance is found by taking the expected value of the squares of the differences of the values of the random variable and the mean. Thus,

$$\text{Var}(X) = \sum_x (x - \mu)^2 P(x),$$

where the sum is taken over all possible values of the random variable X, and μ is the expected value of X.

Think of the variance as the expected value of $(X - \mu)^2$, which measures how far X is from the mean μ. The **standard deviation** of X is defined as

$$\sigma = \sqrt{\text{Var}(X)}$$

EXAMPLE 5 Hospital Satisfaction

Find the variance and standard deviation of the hospital satisfaction ratings data.

Solution

$$\text{Var}(X) = (1 - 2.63)^2 \frac{15}{100} + (2 - 2.63)^2 \frac{38}{100} + (3 - 2.63)^2 \frac{25}{100} + (4 - 2.63)^2 \frac{13}{100} + (5 - 2.63)^2 \frac{9}{100}$$

$$= 1.3331.$$

Using the definition of standard deviation we find that $\sigma = \sqrt{1.3331} \approx 1.15$.

Decision Analysis   The concept of expected value is used extensively in medicine to help physicians determine diagnostic and treatment strategies that provide the best care for the patients they serve. This area of research is commonly called **decision analysis.** The following example illustrates the power of expected value and decision analysis.

EXAMPLE 6 Blood Tranfusions for Malaria-Related Anemia

? Blood transfusions are commonly used to treat childhood malarial anemia in Africa. Although the transfusions are effective in treating the anemia, much of the blood is unscreened and there is a significant risk of transmitting the HIV virus with the transfusion. Given the risk of HIV transmission, what type of transfusion policy would maximize childhood survival? Researchers have used the concept of decision analysis to study this problem.*

---

*Obonyo, C., E. Steyerberg, A. Oloo, and J. Habbema, "Blood Transfusions for Severe Malaria-Related Anemia in Africa: A Decision Analysis," *Am. J. Trop. Med. Hyg.,* Vol. 59, No. 5, 1998, pp. 808–812.

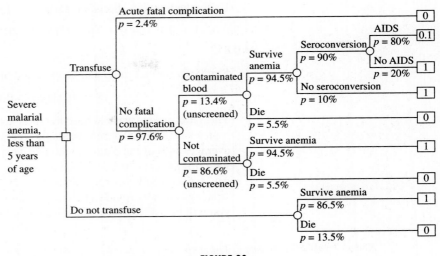

FIGURE 30

The decision tree in Figure 30 illustrates the two choices of transfusion or no transfusion and the possible effects of both choices. The square ☐ represents a decision node and the circles ○ represent a chance node. The choice of transfusion includes the possibility of immediate fatal complications. If a fatal complication from the transfusion does not occur, then several outcomes are possible: the blood may or may not be contaminated with the HIV virus, the patient may or may not die from anemia, seroconversion (HIV antigens appearing in the blood) may or may not occur, and HIV may or may not convert to AIDS. If the patient does not receive the transfusion, then he or she will either live or die from the anemia. The values at the end of the tree indicate that the patient has died (0), fully recovered (1), or has AIDS (0.1) The value of 0.1 indicates that this patient's life expectancy will be one-tenth of that for a patient without AIDS.

To determine the strategy that gives the highest life expectancy, we will calculate the expected value for each of the two choices.

Solution

Method 1: Probability Distribution Table

The first seven branches of the decision tree in Figure 30 are related to the choice of giving the transfusion. For each branch, we can calculate the probability that the outcome for a particular patient will follow that branch. For example, the probability that a transfused patient will not have a fatal complication due to the blood transfusion, will receive noncontaminated blood, and survive the anemia (Branch 6) is

$$P = 0.976(0.866)(0.945) = 0.799.$$

We can calculate the probabilities for the other branches in a similar manner, as illustrated in Table 9. Verify that the probabilities in the table are correct.

▨ Table 9

Branch	1	2	3	4	5	6	7
Outcome (x)	0	0.1	1	1	0	1	0
$P(x)$	0.024	0.089	0.022	0.012	0.007	0.799	0.046

| Using the methods discussed earlier, we find that the expected outcome is

$$E_1 = 0(0.024) + 0.1(0.089) + 1(0.022) + 1(0.012) + 0(0.007) + 1(0.799) + 0(0.046)$$
$$\approx 0.842.$$

Similarly, we can calculate the expected outcome of not transfusing.

$$E_2 = 1(0.865) + 0(0.135) = 0.865$$

There are various ways to interpret these expected values. Since 0 is the value that represents death and 1 represents full recovery, the values above indicate that, on average, 84% of the patients who receive a transfusion can expect a full recovery with full life expectancy. Of the patients who do not receive a transfusion, 86.5% can expect a full recovery. Thus, for unscreened blood, it appears that the best strategy is to not provide a transfusion for malaria-related anemia. In the exercises you will be asked to evaluate the same strategies for blood that is screened.

Method 2: Foldback Procedure

The expected value of each strategy can also be calculated using a procedure called **folding back.** In this procedure, we start calculating at the end of a branch and by working our way back, we compress the entire tree to the decision nodes. For example, in the transfusion branches, we start at the top-most endpoint and calculate inward.

The AIDS/No AIDS branches can be replaced by one endpont that has the value $0.1(0.80) + 1(0.2) = 0.28$, as illustrated in Figure 31. The Seroconversion/No Seroconversion branches, with the new value of 0.28, can then be replaced by the value $0.28(0.9) + 1(0.10) = 0.352$, as shown in Figure 32 on the next page. In a similar manner, the Survive anemia/Die branches, with the newly calculated value of 0.352, can be replaced by the value $0.352(0.945) + 0(0.055) = 0.333$. With this new value, the Contaminated blood/Not contaminated blood branches can be replaced by the single value, $0.333(0.134) + (1(0.945) + 0(0.055))(0.866) = 0.863$ as shown in Figure 33 on the next page. Continuing the foldback procedure for all branches, the tree is folded back to the original two strategy branches, representing Transfuse/Do not transfuse, with expected values illustrated in Figure 34 on the next page.

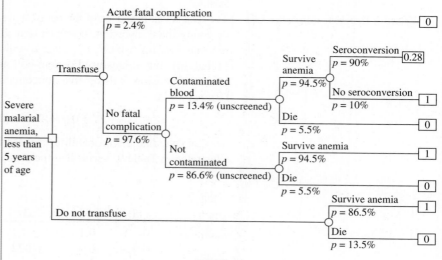

FIGURE 31

The random variables representing outcomes that were used in this example provide an illustration of the types of values that are used to analyze various decision strategies. Quality of life indicators for various outcomes are often used in medicine to distinguish between outcomes. Costs are also often used to evaluate strategies. Occasionally, two indicators—cost and risk—are combined to form a cost-benefit decision analysis.

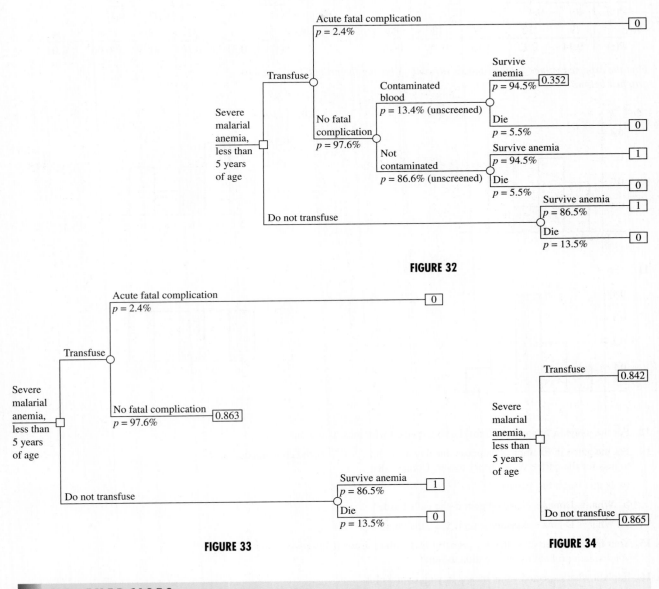

FIGURE 32

FIGURE 33

FIGURE 34

12.4 EXERCISES

For each of the experiments described below, let x determine a random variable, and use your knowledge of probability to prepare a probability distribution.

1. Four coins are tossed, and the number of heads is noted.

2. Two dice are rolled, and the total number of points is recorded.

Draw a histogram for each of the following, and shade the region that gives the indicated probability.

3. Exercise 1; $P(x \le 2)$

4. Exercise 2; $P(x \ge 11)$

Find the expected value and standard deviation for each random variable.

5.

x	2	3	4	5
P(x)	0.1	0.4	0.3	0.2

6.

y	4	6	8	10
P(y)	0.4	0.4	0.05	0.15

7.

z	9	12	15	18	21
P(z)	0.14	0.22	0.36	0.18	0.10

8.

x	30	32	36	38	44
P(x)	0.31	0.30	0.29	0.06	0.04

Find the expected values for the random variables x having the probability functions graphed below.

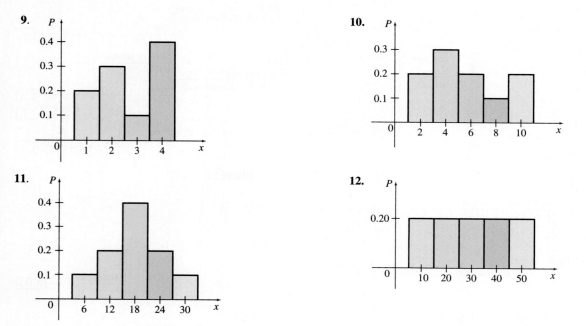

9.

10.

11.

12.

13. For the game in Example 3, find Mary's expected winnings. Is it a fair game?

14. For the game in Example 3, suppose one day Mary brings a 2-headed coin and uses it to toss for the coffee. Since Mary tosses, Donna calls.

   a. Is this still a fair game?

   b. What is Donna's expected gain if she calls heads?

   c. What is Donna's expected gain if she calls tails?

15. Your friend missed class the day probability distributions were discussed. How would you explain probability distribution to him?

16. Explain what expected value means in your own words.

Applications

LIFE SCIENCES

17. *Animal Offspring* In a certain animal species, the probability that a healthy adult female will have no offspring in a given pregnancy is 0.31, while the probabilities of 1, 2, 3, or 4 offspring are, respectively, 0.21, 0.19, 0.17, and 0.12. Find the expected number of offspring and the standard deviation.

18. *Ear Infections* Otitis media, or middle ear infection, is initially treated with an antibiotic. Researchers have compared two antibiotics, amoxicillin and cefaclor, for their cost effectiveness. Amoxicillin is inexpensive, safe and effective. Cefaclor is also safe. However, it is considerably more expensive and it is generally more effective. Use the tree diagram below (where the costs are estimated as the total cost of medication, office visit, ear check, and hours of lost work) to answer the following.*

a. Find the expected cost of using each antibiotic to treat a middle ear infection.

b. To minimize the total expected cost, which antibiotic should be chosen?

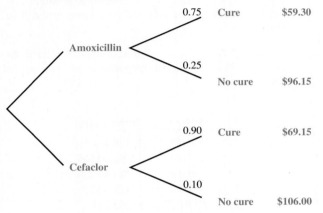

19. *Blood Transfusions* Suppose that in Example 6 the blood was screened for HIV prior to being used and that the screening process reduces the probability that the blood is contaminated to 1%. Calculate the expected survival rate for each strategy. Should blood transfusions be given after the blood has been screened?

20. *Perinatal Mortality* Preterm birth and low birth weight are consistently the primary causes of perinatal mortality. Researchers have determined that treating infections during pregnancy may be one of the most effective ways to prevent such complications and reduce infant mortality.[†] In the study, 870 women from Germany were screened for bacterial vaginosis. The patients were assigned to one of three different treatment strategies, or practices, according to the

clinic which they visited. The decision tree below gives the results of the study. Calculate the expected cost of each strategy. Which strategy yields the lowest expected cost?

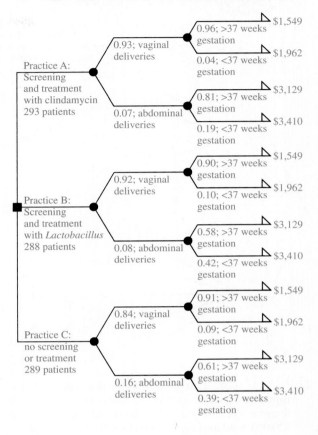

OTHER APPLICATIONS

21. *Seeding Storms* One of the few methods that can be used in an attempt to cut the severity of a hurricane is to *seed* the storm. In this process, silver iodide crystals are dropped into the storm. Unfortunately, silver iodide crystals sometimes cause the storm to *increase* its speed. Wind speeds may also increase or decrease even with no seeding. Use the tree diagram on the next page to answer the following.[‡]

*Weiss, Jeffrey, and Shoshana Melman, "Cost Effectiveness in the Choice of Antibiotics for the Initial Treatment of Otitis Media in Children: A Decision Analysis Approach," *Journal of Pediatric Infectious Disease,* Vol. 7, No. 1, 1988, pp. 23–26.

†Muller, Eva, K. Berger, N. Dennemark, and M. Oleen-Burkey, "Cost of Bacterial Vaginosis in Pregnancy: Decision Analysis and Cost Evaluation of a Clinical Study in Germany," *The Journal of Reproductive Medicine,* Vol. 44, No. 9, September 1999, pp. 807–814.

‡The probabilities and amounts of property damage in the tree diagram for Exercise 21 are from Howard, R. A., J. E. Matheson, and D. W. North, "The Decision to Seed Hurricanes," *Science,* Vol. 176, No. 16, June 1972, pp. 1191–1202. Copyright © 1972 by the American Association for the Advancement of Science. Reprinted by permission.

a. Find the expected amount of damage under each option, "seed" and "do not seed."

b. To minimize total expected damage, what option should be chosen?

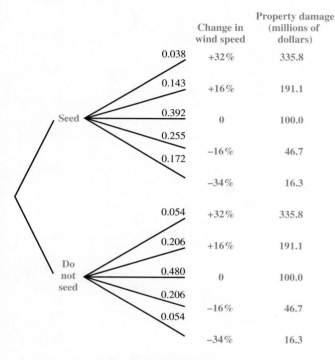

	Change in wind speed	Property damage (millions of dollars)
Seed 0.038	+32%	335.8
0.143	+16%	191.1
0.392	0	100.0
0.255	−16%	46.7
0.172	−34%	16.3
Do not seed 0.054	+32%	335.8
0.206	+16%	191.1
0.480	0	100.0
0.206	−16%	46.7
0.054	−34%	16.3

22. *Golf Tournament* At the end of play in a major golf tournament, two players, an "old pro" and a "new kid," are tied. Suppose the first prize is $80,000 and second prize is $20,000. Find the expected winnings for the old pro if

a. both players are of equal ability;

b. the new kid will freeze up, giving the old pro a 3/4 change of winning.

23. *Raffle* A raffle offers a first prize of $100 and 2 second prizes of $40 each. One ticket costs $1, and 500 tickets are sold. Find the expected winnings for a person who buys 1 ticket. Is this a fair game?

24. *Raffle* A raffle offers a first prize of $1,000, 2 second prizes of $300 each, and 20 third prizes of $10 each. If 10,000 tickets are sold at 50¢ each, find the expected winnings for a person buying 1 ticket. Is this a fair game?

Find the expected winnings for the games of chance described in Exercises 25–27.

25. *Roulette* In one form of roulette, you bet $1 on "even." If 1 of the 18 even numbers comes up, you get your dollar

back, plus another one. If 1 of the 20 noneven (18 odd, 0, and 00) numbers comes up, you lose your dollar.

26. *Roulette* In another form of roulette, there are only 19 noneven numbers (no 00).

27. *Numbers* Numbers is a game in which you bet $1 on any three-digit number from 000 to 999. If you number comes up, you get $500.

28. *Contests* A magazine distributor offers a first prize of $100,000, two second prizes of $40,000 each, and two third prizes of $10,000 each. A total of 2,000,000 entries are received in the contest. Find the expected winnings if you submit one entry to the contest. If it would cost you 50¢ in time, paper, and stamps to enter, would it be worth it?

29. *Contests* A contest at a fast-food restaurant offered the following cash prizes and probabilities of winning on one visit.

Prize	Probability
$100,000	1/176,402,500
$25,000	1/39,200,556
$5,000	1/17,640,250
$1,000	1/1,568,022
$100	1/282,244
$5	1/7,056
$1	1/588

Suppose you spend $1 to buy a bus pass that lets you go to 25 different restaurants in the chain and pick up entry forms. Find your expected value.

30. *The Hog Game* In the hog game, each player states the number of dice that he or she would like to roll. The player then rolls that many dice. If a 1 comes up on any die, the player's score is 0. Otherwise, the player's score is the sum of the numbers rolled.*

a. Find the expected value of the player's score when the player rolls one die.

b. Find the expected value of the player's score when the player rolls two dice.

c. Verify that the expected nonzero score of a single die is 4, so that if a player rolls n dice that do not result in a sore of 0, the expected score is $4n$.

d. Verify that if a player rolls n dice, there are 5^n possible ways to get a nonzero score, and 6^n possible ways to roll

---

*Bohan, James, and John Shultz, "Revisiting and Extending the Hog Game," *The Mathematics Teacher,* Vol. 89, No. 9, Dec. 1996, p. 728–733.

the dice. Explain why the expected value, E, of the player's score when the player rolls n dice is then

$$E = \frac{5^n(4n)}{6^n}.$$

31. *Football* After a team scores a touchdown, it can either attempt to kick an extra point or attempt a two-point conversion. During the 1999–2000 NFL season, two-point conversions were successful 37% of the time and the extra-point kicks were successful 94% of the time.*

a. Calculate the expected value of each strategy.

b. Which strategy, over the long run, will maximize the number of points scored?

c. Using this information, should a team always only use one strategy? Explain.

■ CHAPTER SUMMARY

We began this chapter by introducing sets, set operations, tree diagrams, and Venn diagrams. We used these ideas to define and study probability experiments, sample spaces, mutually exclusive events, and probability distributions. These concepts were further extended to include conditional probability, odds, independent events, and Bayes' theorem. Finally, the concept of expected value was derived and used to illustrate an important use of decision trees, called decision analysis. In the next chapter we will use calculus to extend the ideas and power of probability.

Probability Summary

Basic Probability Principle

Let S be a sample space of equally likely outcomes, and let event E be a subset of S. Then the probability that event E occurs is

$$P(E) = \frac{n(E)}{n(S)}.$$

Union Rule

For any events E and F from a sample space S,

$$P(E \cup F) = P(E) + P(F) - P(E \cap F).$$

---

*Leonhardt, David, "In Football, 6 + 2 Often Equals 6," *The New York Times (Week in Review),* Sunday, Jan. 16, 2000, p. 4–2.

For mutually exclusive events E and F,

$$P(E \cup F) = P(E) + P(F).$$

Complement Rule

$$P(E) = 1 - P(E') \quad \text{and} \quad P(E') = 1 - P(E)$$

Odds

The odds in favor of event E are $\dfrac{P(E)}{P(E')}$, $P(E') \neq 0$.

Properties of Probability

1. For any event E in sample space S, $0 \leq P(E) \leq 1$.
2. The sum of the probabilities of all possible distinct outcomes is 1.

Conditional Probability

The conditional probability of event E, given that event F has occurred, is

$$P(E \mid F) = \frac{P(E \cap F)}{P(F)}, \quad \text{where } P(F) \neq 0.$$

For equally likely outcomes, conditional probability is found by reducing the sample space to event F; then

$$P(E \mid F) = \frac{n(E \cap F)}{n(F)}.$$

Product Rule of Probability

If E and F are events, then $P(E \cap F)$ may be found by either of these formulas.

$$P(E \cap F) = P(F) \cdot P(E \mid F) \quad \text{or} \quad P(E \cap F) = P(E) \cdot P(F \mid E)$$

If E and F are independent events, then $P(E \cap F) = P(E) \cdot P(F)$.

Bayes' Theorem

$$P(F_i \mid E) = \frac{P(F_i) \cdot P(E \mid F_i)}{P(F_1) \cdot P(E \mid F_1) + P(F_2) \cdot P(E \mid F_2) + \cdots + P(F_n) \cdot P(E \mid F_n)}$$

KEY TERMS

12.1 set
element (member)
empty set
set-builder notation
universal set
subset
tree diagram
Venn diagram
set operations
complement
intersection
disjoint sets
union

union rule for sets
union rule for
disjoint sets
12.2 experiment
trial
outcome
sample space
event
simple event
certain event
impossible event
mutually exclusive
events

union rule for
probability
odds
empirical probability
probability distribution
12.3 conditional probability
product rule
independent events
dependent events
Bayes' theorem
12.4 random variable
discrete random
variable

continuous random
variable
probability distribution
probability function
histogram
expected value
fair game
variance
standard deviation
decision analysis
folding back

CHAPTER 12 REVIEW EXERCISES

In Exercises 1–6, let $U = \{a, b, c, d, e, f, g\}$, $K = \{c, d, f, g\}$, *and* $R = \{a, c, d, e, g\}$. *Find the following.*

1. The number of subsets of K

2. K'

3. $K \cap R$

4. $K \cup R$

5. $(K \cap R)'$

6. $(K \cup R)'$

Draw a Venn diagram and shade each set in Exercises 7–10.

7. $A \cup B'$ **8.** $A' \cap B$ **9.** $(A \cap B) \cup C$ **10.** $(A \cup B)' \cap C$

Write the sample space for each of the following experiments.

11. Rolling a die

12. Measuring the weight of a person to the nearest half pound (the scale will not measure more than 300 lb)

13. Tossing a coin 4 times

When a single card is drawn from an ordinary deck, find the probabilities that it will be the following.

14. A heart

15. A face card

16. Black or a face card

17. Red, given that it is a queen

18. A face card, given that it is a king

19. Describe what is meant by disjoint sets.

20. Describe what is meant by mutually exclusive events.

21. How are disjoint sets and mutually exclusive events related?

22. Define independent events.

23. Are independent events always mutually exclusive? Are they ever mutually exclusive?

24. An uproar has raged since September 1990 over the answer to a puzzle* published in *Parade* magazine, a supplement of the Sunday newspaper. In the "Ask Marilyn" column, Marilyn vos Savant answered the following question: "Suppose you're on a game show, and you're given the choice of three doors. Behind one door is a car; behind the others, goats. You pick a door, say number 1, and the host, who knows what's behind the other doors, opens another door, say number 3, which has a goat. He then says to you, 'Do you want to pick door number 2?' Is it to your advantage to take the switch?"

  Ms. vos Savant estimates that she has since received some 10,000 letters; most of them, including many from mathematicians and statisticians, disagreed with her answer. Her answer has been debated by both professionals and amateurs, and tested in classes at all levels, from grade school to graduate school. But by performing the experiment repeatedly, it can be shown that vos Savant's answer was correct. Find the probabilities of getting the car if you switch or do not switch, and then answer the question yourself. (*Hint:* Consider the sample space.)

Find the odds in favor of a card drawn from an ordinary deck being each of the following.

25. A club   **26.** A black jack

Find the probabilities of getting the following sums when 2 fair dice are rolled.

27. 8   **28.** 0   **29.** No more than 5

30. 12, given that the sum is greater than 10   **31.** 7, given that at least one die shows a 4

32. At least 9, given that at least one die shows a 5

---

*Parade Magazine, Sept. 9, 1990, p. 13. Reprinted by permission of the William Morris Agency, Inc. on behalf of the author. Copyright © 1990 by Marilyn vos Savant.

33. Suppose $P(E) = 0.51$, $P(F) = 0.37$, and $P(E \cap F) = 0.22$. Find each of the following.

    a. $P(E \cup F)$ **b.** $P(E \cap F')$ **c.** $P(E' \cup F)$ **d.** $P(E' \cap F')$

34. Box A contains 5 red balls and 1 black ball; box B contains 2 red balls and 3 black balls. A box is chosen, and a ball is selected from it. The probability of choosing box A is 3/8. If the selected ball is black, what is the probability that it came from box A?

35. Find the probability that the ball in Exercise 34 came from box B, given that it is red.

*In Exercises 36 and 37, (**a**) give a probability distribution, (**b**) sketch its histogram, (**c**) find the expected value, and (**d**) find the standard deviation.*

36. A coin is tossed 3 times and the number of heads is recorded.

37. A pair of dice is rolled and the sum of the results for each roll is recorded.

In Exercises 38 and 39, give the probability that corresponds to the shaded region of each histogram.

38.

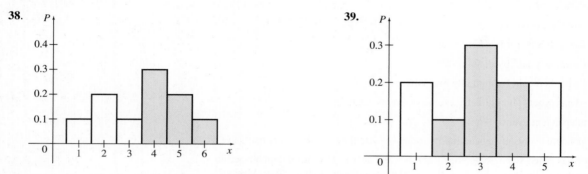

39.

40. You pay $6 to play in a game where you will roll a die, with payoffs as follows: $8 for a 6, $7 for a 5, and $4 for any other results. What are your expected winnings? Is the game fair?

41. Find the expected number of girls in a family of 5 children.

42. Three cards are drawn from a standard deck of 52 cards.

    a. What is the expected number of aces?

    b. What is the expected number of clubs?

Applications

LIFE SCIENCES

43. *Sickle Cell Anemia* The square shows the four possible (equally likely) combinations when both parents are carriers of the sickle cell anemia trait. Each carrier parent has normal cells (N) and trait cells (T).

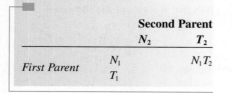

	Second Parent	
	N_2	T_2
First Parent N_1		$N_1 T_2$
T_1		

a. Complete the table.

b. If the disease occurs only when two trait cells combine, find the probability that a child born to these parents will have sickle cell anemia.

c. The child will carry the trait but not have the disease if a normal cell combines with a trait cell. Find this probability.

d. Find the probability that the child is neither a carrier nor has the disease.

OTHER APPLICATIONS

44. *Elections* In the 1994 elections for the U.S. House of Representatives, 46% of men and 54% of women voted Democratic. Of those who voted, 49% were men and 51% were women.* What is the probability that a person who voted Democratic was a man?

45. *Television Viewing Habits* A telephone survey of television viewers revealed the following information.

> 20 watch situation comedies;
>
> 19 watch game shows;
>
> 27 watch movies;
>
> 5 watch both situation comedies and game shows;
>
> 8 watch both game shows and movies;
>
> 10 watch both situation comedies and movies;
>
> 3 watch all three;
>
> 6 watch none of these.

a. How many viewers were interviewed?

b. How many viewers watch comedies and movies but not game shows?

c. How many viewers watch only movies?

d. How many viewers watch comedies and game shows but not movies?

46. *Randomized Response Method for Getting Honest Answers to Sensitive Questions*[†] Basically, this is a method to guarantee that an individual who answers sensitive questions will remain anonymous, thus encouraging a truthful response. This method is, in effect, an application of the formula for finding the probability of an intersection, and operates as follows. Questions A and B are posed, one of which is sensitive and the other not. The probability of receiving a "yes" to the nonsensitive question must be known. For example, one could ask

> A: Does your Social Security number end in an odd digit? (Nonsensitive)
>
> B: Have you ever intentionally cheated on your income tax? (Sensitive)

We know that $P(\text{answer yes} \mid \text{answer } A) = 1/2$. We wish to approximate $P(\text{answer yes} \mid \text{answer } B)$. The subject is asked to flip a coin and answer A if the coin comes up heads and otherwise to answer B. In this way, the interviewer does not know which question the subject is answering. Thus, a "yes" answer is not incriminating. There is no way for the interviewer to know whether the subject is saying "Yes, my Social Security number ends in an odd digit" or "Yes, I have intentionally cheated on my income taxes." The percentage of subjects in the group answering "yes" is used to approximate $P(\text{answer yes})$.

a. Use the fact that the event "answer yes" is the union of the event "answer yes and answer A" with the event "answer yes and answer B" to prove that

$P(\text{answer yes} \mid \text{answer } B)$

$$= \frac{P(\text{answer yes}) - P(\text{answer yes} \mid \text{answer } A) \cdot P(\text{answer } A)}{P(\text{answer } B)}.$$

b. If this technique is tried on 100 subjects and 60 answered "yes," what is the approximate probability that a person randomly selected from the group has intentionally cheated on income taxes?

47. *Earthquake* It has been reported that government scientists have predicted that the odds for a major earthquake occurring in the San Francisco Bay area during the next 30 years are 9 to 1.[‡] What is the probability that a major earthquake will occur during the next 30 years in San Francisco?

48. *Making a First Down* A first down is desirable in football—it guarantees four more plays by the team making it, assuming no score or turnover occurs in the plays. After getting a first down, a team can get another by advancing the ball at least 10 yards. During the four plays given by a first down, a team's position will be indicated by a phrase such as "third and 4," which means that the team has already had two of its four plays, and that 4 more yards are needed to get 10 yards necessary for another first down. An article in a management journal[§] offers the following results for 189 games for a recent National Football League season. "Trials" represents the number of times a team tried to make a first down, given that it was currently playing either a third or a fourth down. Here, *n* represents the number of yards still needed for a first down.

n	Trials	Successes	Probability of Making First Down with *n* Yards to Go
1	543	388	
2	327	186	
3	356	146	
4	302	97	
5	336	91	

a. Complete the table.

b. Why is the sum of the answers in the table not equal to 1?

*The New York Times, Nov. 13, 1994, p. 24.

[†]Milton, J. S., and J. J. Corbet, *Applied Statistics with Probability.* Copyright © 1979 by Litton Educational Publishing, Inc. Reprinted by permission of Brooks/Cole Publishing Company, Monterey, California.

[‡]*The San Francisco Chronicle,* June 8, 1994, p. A1.

[§]Carter Virgil, and Robert Machols, "Optimal Strategies on Fourth Down," *Management Science,* Vol. 24, No. 16, Dec. 1978. Copyright © 1978 by The Institute of Management Sciences. Reprinted by permission.

49. *Missiles* In his novel *Debt of Honor,* Tom Clancy writes the following:*

"There were ten target points—missile silos, the intelligence data said, and it pleased the Colonel [Zacharias] to be eliminating the hateful things, even though the price of that was the lives of other men. There were only three of them [bombers], and his bomber, like the others, carried only eight weapons [smart bombs]. The total number of weapons carried for the mission was only twenty-four, with two designated for each silo, and Zacharias's last four for the last target. Two bombs each. Every bomb had a 95% probability of hitting within four meters of the aim point, pretty good numbers really, except that this sort of mission had precisely no margin for error. Even the paper probability was less than half a percent chance of a double miss, but that number times ten targets meant a 5% chance that [at least] one missile would survive, and that could not be tolerated."

Determine if the calculations in this quote are correct by the following steps.

a. Given that each bomb had a 95% probability of hitting the missile silo on which it was dropped, and that two bombs were dropped on each silo, what is the probability of a double miss?

b. What is the probability that a specific silo was destroyed (that is, that at least one bomb of the two bombs struck the silo)?

c. What is the probability that all ten silos were destroyed?

d. What is the probability that at least one silo survived? Does this agree with the quote?

e. What assumptions need to be made for the calculations in parts a and b to be valid? Discuss whether these assumptions seem reasonable.

Exercises 50 and 51 are taken from actuarial examinations given by the Society of Actuaries.[†]

50. *Product Success* A company is considering the introduction of a new product that is believed to have probability 0.5 of being successful and probability of 0.5 of being unsuccessful. Successful products pass quality control 80% of the time. Unsuccessful products pass quality control 25% of the time. If the product is successful, the net profit to the company will be $40 million; if unsuccessful, the net loss will be $15 million. Determine the expected net profit if the product passes quality control.

a. $23 million   **b.** $24 million   **c.** $25 million

d. $26 million   **e.** $27 million

51. *Sampling Fruit* A merchant buys boxes of fruit from a grower and sells them. Each box of fruit is either Good or Bad. A Good box contains 80% excellent fruit and will earn $200 profit on the retail market. A Bad box contains 30% excellent fruit and will produce a loss of $1,000. The a priori probability of receiving a Good box of fruit is 0.9. Before the merchant decides to put the box on the market, he can sample one piece of fruit to test whether it is excellent. Based on that sample, he has the option of rejecting the box without paying for it. Determine the expected value of the right to sample. (*Hint:* If the merchant samples the fruit, what are the probabilities of accepting a Good box, accepting a Bad box, and not accepting the box? What are these probabilities if he does not sample the fruit?)

a. 0   **b.** $16   **c.** $34   **d.** $72   **e.** $80

52. *Baseball* The number of runs scored in 16,456 half-innings of the 1986 National League Baseball season was analyzed by Hal Stern. Use the following table to answer the following questions.[‡]

Runs	Frequency	Probability
0	12,087	0.7345
1	2,451	0.1489
2	1,075	0.0653
3	504	0.0306
4	225	0.0137
5	66	0.0040
6	29	0.0018
7	12	0.0007
8	5	0.0003
9	2	0.0001

a. What is the probability that a given team scored 5 or more runs in any given one-half inning during the 1986 season?

b. What is the probability that a given team scored fewer than two runs in any given one-half inning of the 1986 season?

c. What is the expected number of runs that a team scored during any given half-inning of the 1986 season? Interpret this number.

---

*Clancy, Tom, *Debt of Honor,* G. P. Putnam's Sons, 1994, pp. 686–687. Reprinted with permission.

[†]Course 130 Examination, Operations Research, Nov. 1989.

[‡]J. Laurie Snell's report of Hal Stern's analysis in *Chance News 7.05,* Apr. 27–May 26, 1998.

EXTENDED APPLICATION: Medical Diagnosis

When a patient is examined, information (typically incomplete) is obtained about his state of health. Probability theory provides a mathematical model appropriate for this situation, as well as a procedure for quantitatively interpreting such partial information to arrive at a reasonable diagnosis.*

To develop a model, we list the states of health that can be distinguished in such a way that the patient can be in one and only one state at the time of the examination. For each state of health H, we associate a number, $P(H)$, between 0 and 1 such that the sum of all these numbers is 1. This number $P(H)$ represents the probability, before examination, that a patient is in the state of health H, and $P(H)$ may be chosen subjectively from medical experience, using any information available prior to the examination. The probability may be most conveniently established from clinical records; that is, a mean probability is established for patients in general, although the number would vary from patient to patient. Of course, the more information that is brought to bear in establishing $P(H)$, the better the diagnosis.

For example, limiting the discussion to the condition of a patient's heart, suppose there are exactly three states of health, with probabilities as follows.

State of Health, H		$P(H)$
H_1	Patient has a normal heart	0.8
H_2	Patient has minor heart irregularities	0.15
H_3	Patient has a severe heart condition	0.05

Having selected $P(H)$, the information from the examination is processed. First, the results of the examination must be classified. The examination itself consists of observing the state of a number of characteristics of the patient. Let us assume that the examination for a heart condition consists of a stethoscope examination and a cardiogram. The outcome of such an examination, C, might be one of the following:

$C_1 =$ stethoscope shows normal heart and cardiogram shows normal heart;

$C_2 =$ stethoscope shows normal heart and cardiogram shows minor irregularities;

and so on.

It remains to assess for each state of health H the conditional probability $P(C|H)$ of each examination outcome C using only the knowledge that a patient is in a given state of health. (This may be based on the medical knowledge and clinical experience of the doctor.) The conditional probabilities $P(C|H)$ will not vary from patient to patient (although they

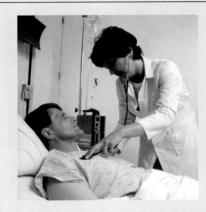

should be reviewed periodically), so that they may be built into a diagnostic system.

Suppose the result of the examination is C_1. Let us assume the following probabilities:

$$P(C_1|H_1) = 0.9,$$
$$P(C_1|H_2) = 0.4,$$
$$P(C_1|H_3) = 0.1.$$

Now, for a given patient, the appropriate probability associated with each state of health H, after examination, is $P(H|C)$, where C is the outcome of the examination. This can be calculated by using Bayes' theorem. For example, to find $P(H_1|C_1)$—that is, the probability that the patient has a normal heart given that the examination showed a normal stethoscope examination and a normal cardiogram—we use Bayes' theorem as follows:

$$P(H_1|C_1)$$
$$= \frac{P(C_1|H_1)P(H_1)}{P(C_1|H_1)P(H_1) + P(C_1|H_2)P(H_2) + P(C_1|H_3)P(H_3)}$$
$$= \frac{(0.9)(0.8)}{(0.9)(0.8) + (0.4)(0.15) + (0.1)(0.05)} \approx 0.92.$$

Hence, the probability is about 0.92 that the patient has a normal heart on the basis of the examination results. This means that in 8 out of 100 patients, some abnormality will be present and not be detected by the stethoscope or the cardiogram.

Exercises

1. Find $P(H_2|C_1)$.
2. Assuming the following probabilities, find $P(H_1|C_2)$.
   $$P(C_2|H_1) = 0.2 \qquad P(C_2|H_2) = 0.8 \qquad P(C_2|H_3) = 0.3$$
3. Assuming the probabilities of Exercise 2, find $P(H_3|C_2)$.

*Wright, Roger, "Probabilistic Medical Diagnosis," *Some Mathematical Models in Biology*, rev. ed., Robert M. Thrall, ed., University of Michigan, 1967. Used by permission of Robert M. Thrall.

Probability and Calculus

Though earthquakes may appear to strike at random, the times between quakes can be modeled with an exponential density function. Such *continuous probability models* have many applications in science, engineering, and medicine. In an exercise in Section 1 of this chapter we'll use an exponential density function to describe the times between major earthquakes in southern California, and in Section 3 we will compute the mean and standard deviation for this distribution.

■ **13.1** Continuous Probability Models

■ **13.2** Expected Value and Variance of Continuous Random Variables

■ **13.3** Special Probability Density Functions

Review Exercises

Extended Application: Exponential Waiting Times

In recent years, probability has become increasingly useful in fields ranging from manufacturing to medicine, as well as in all types of research. The foundations of probability were laid in the seventeenth century by Blaise Pascal (1623–1662) and Pierre de Fermat (1601–1665), who investigated *the problem of the points.* This problem dealt with the fair distribution of winnings in an interrupted game of chance between two equally matched players whose scores were known at the time of the interruption.

Probability has advanced from a study of gambling to a well-developed, deductive mathematical system. In this chapter we give a brief introduction to the use of calculus in probability.

■ 13.1 CONTINUOUS PROBABILITY MODELS

? THINK ABOUT IT   What is the probability that there is a bird's nest within 0.5 kilometers of a given point?

In this section, we show how calculus is used to find the probability of certain events. Later in the section, we will answer the question posed above. Before discussing probability, however, we need to introduce some new terminology.

Suppose an ornithologist is studying the length of time that sparrows spend feeding at a birdfeeder. The lengths of time spent feeding, rounded to the nearest minute, are shown in Table 1.

Table 1

Time	1	2	3	4	5	6	7	8	9	10	
Frequency	3	5	9	12	15	11	10	6	3	1	(Total: 75)

The table shows, for example, that 9 of the 75 sparrows in the study stayed 3 minutes, 15 sparrows stayed 5 minutes, and 1 sparrow stayed 10 minutes. Because the time for any particular feeding is a random event, the number of minutes for a feeding is a random variable, as defined in the previous chapter. The frequencies can be converted to probabilities by dividing each frequency by the total number of feedings (75) to get the results shown in Table 2.*

Table 2

Time	1	2	3	4	5	6	7	8	9	10
Probability	0.04	0.07	0.12	0.16	0.20	0.15	0.13	0.08	0.04	0.01

Because each value of the random variable is associated with just one probability, Table 2 defines a function. As we saw in the last chapter, such a function is called a probability function.

The information in Table 2 can be displayed graphically with a histogram. Recall that the bars of a histogram have the same width, and their heights are determined by the probabilities of the various values of the random variable. See Figure 1 on the following page.

---

*One definition of the *probability of an event* is the number of outcomes that favor the event divided by the total number of equally likely outcomes in an experiment.

As we learned in the previous chapter, the probability function in Table 2 is a discrete probability function because it has a finite domain—the integers from 1 to 10, inclusive. A discrete probability function has a finite domain or an infinite domain that can be listed. For example, if we flip a coin until we get heads, and let the random variables be the number of flips, then the domain is 1, 2, 3, 4, ... On the other hand, the distribution of heights (in inches) of college women includes infinitely many possible measurements, such as 53, 54.2, 66.5, 72.$\overline{3}$, and so on, *within some real number interval.* Probability functions with such domains are called *continuous probability distributions.*

CONTINUOUS PROBABILITY DISTRIBUTION

A **continuous random variable** can take on any value in some interval of real numbers. The distribution of this random variable is called a **continuous probability distribution.**

In the bird feeder example discussed earlier in this section, it would have been possible to time the feedings with greater precision—to the nearest tenth of a minute or even to the nearest second (one-sixtieth of a minute), if desired. Theoretically, at least, t could take on any positive real-number value between, say, 0 and 11 minutes. The graph of the probabilities $f(t)$ of these feeding times can be thought of as the continuous curve shown in Figure 1. As indicated in Figure 1, the curve was derived from Table 2 by connecting the points at the tops of the bars in the corresponding histogram and smoothing the resulting polygon into a curve.

For a discrete probability function, the area of each bar (or rectangle) gives the probability of a particular feeding time. Thus, by considering the possible feeding times t as all the real numbers between 0 and 11, the area under the curve of Figure 2 between any two values of t can be interpreted as the probability that a feeding time will be between those two numbers. For example, the shaded region in Figure 2 corresponds to the probability that t is between a and b, written $P(a \leq t \leq b)$.

FOR REVIEW ■

The connection between area and the definite integral is discussed in the chapter on Integration. For example, in that chapter we solved such problems as the following.

Find the area between the x-axis and the graph of $f(x) = x^2$ from $x = 1$ to $x = 4$.

Answer: $\int_1^4 x^2\,dx = 21$

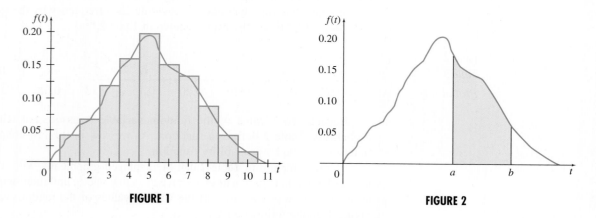

FIGURE 1

FIGURE 2

It was shown earlier that the definite integral of a continuous function f, where $f(x) \geq 0$, gives the area under the graph of $f(x)$ from $x = a$ to $x = b$. If a function f can be found to describe a continuous probability distribution, then the definite integral can be used to find the area under the curve from a to b that represents the probability that x will be between a and b.

> If x is a continuous random variable whose distribution is described by the function f on $[a, b]$, then
>
> $$P(a \leq x \leq b) = \int_a^b f(x)\,dx.$$

Probability Density Functions   A function f that describes a continuous probability distribution is called a *probability density function*. Such a function must satisfy the following conditions.

PROBABILITY DENSITY FUNCTION

The function f is a **probability density function** of a random variable x in the interval $[a, b]$ if

1. $f(x) \geq 0$ for all x in the interval $[a, b]$; and

2. $\int_a^b f(x)\,dx = 1$.

Intuitively, condition 1 says that the probability of a particular event can never be negative. Condition 2 says that the total probability for the interval must be 1; *something* must happen.

EXAMPLE 1   Probability Density Functions

(a) Show that the function defined by $f(x) = (3/26)x^2$ is a probability density function for the interval $[1, 3]$.

Solution First, note that Condition 1 holds; that is, $f(x) \geq 0$ for the interval $[1, 3]$. Next show that Condition 2 holds.

$$\int_1^3 \frac{3}{26}x^2\,dx = \frac{3}{26}\left(\frac{x^3}{3}\right)\Bigg|_1^3 = \frac{3}{26}\left(9 - \frac{1}{3}\right) = 1$$

Since both conditions hold, $f(x)$ is a probability density function.

(b) Find the probability that x will be between 1 and 2.

Solution The desired probability is given by the area under the graph of $f(x)$ between $x = 1$ and $x = 2$, as shown in Figure 3. The area is found by using a definite integral.

$$P(1 \leq x \leq 2) = \int_1^2 \frac{3}{26}x^2\,dx = \frac{3}{26}\left(\frac{x^3}{3}\right)\Bigg|_1^2 = \frac{7}{26}$$

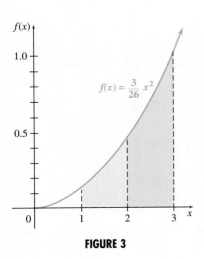

FIGURE 3

Earlier, we noted that determining a suitable function is the most difficult part of applying mathematics to actual situations. Sometimes a function appears to model an application well, but does not satisfy the requirements for a probability density function. In such cases, we may be able to change the function into a probability density function by multiplying it by a suitable constant, as shown in the next example.

EXAMPLE 2 Probability Density Functions

Is there a constant k such that $f(x) = kx^2$ is a probability density function for the interval $[0,4]$?

Solution First,

$$\int_0^4 kx^2\, dx = \frac{kx^3}{3}\Big|_0^4 = \frac{64k}{3}.$$

The integral must be equal to 1 for the function to be a probability density function. To convert it to one, let $k = 3/64$. The function defined by $(3/64)x^2$ for $[0,4]$ will be a probability density function, since $(3/64)x^2 \geq 0$ for all x in $[0,4]$ and

$$\int_0^4 \frac{3}{64}x^2 = 1.$$

An important distinction is made between a discrete probability function and a probability density function (which is continuous). In a discrete distribution, the probability that the random variable, x, will assume a specific value is given in the distribution for every possible value of x. In a probability density function, however, the probability that x equals a specific value, say, c, is

$$P(x = c) = \int_c^c f(x)\, dx = 0.$$

For a probability density function, only probabilities of *intervals* can be found. For example, suppose the random variable is the annual rainfall for a given region. The amount of rainfall in one year can take on any value within some continuous interval that depends on the region; however, the probability that the rainfall in a given year will be some specific amount, say 33.25 inches, is actually zero.

The definition of a probability density function is extended to intervals such as $(-\infty, b]$, $(-\infty, b)$, $[a, \infty)$, (a, ∞), or $(-\infty, \infty)$ by using improper integrals, as follows.

PROBABILITY DENSITY FUNCTIONS ON $(-\infty, \infty)$

If f is a probability density function for a continuous random variable x on $(-\infty, \infty)$, then

$$P(x \leq b) = P(x < b) = \int_{-\infty}^b f(x)\, dx,$$

$$P(x \geq a) = P(x > a) = \int_a^\infty f(x)\, dx,$$

$$P(-\infty < x < \infty) = \int_{-\infty}^\infty f(x)\, dx = 1.$$

The total area under the graph of a probability density function of this type must still equal 1.

EXAMPLE 3 Location of a Bird's Nest

Suppose the random variable x is the distance (in kilometers) from a given point to the nearest bird's nest, with the probability density function of the distribution given by $f(x) = 2xe^{-x^2}$ for $x \geq 0$.

(a) Show that $f(x)$ is a probability density function.

Solution Since $e^{-x^2} = 1/e^{x^2}$ is always positive, and $x \geq 0$,

$$f(x) = 2xe^{-x^2} \geq 0.$$

Use substitution to evaluate the definite integral $\int_0^\infty 2xe^{-x^2}\,dx$. Let $u = -x^2$, so that $du = -2x\,dx$, and

$$\int 2xe^{-x^2}\,dx = -\int e^{-x^2}(-2x\,dx)$$
$$= -\int e^u\,du = -e^u = -e^{-x^2}.$$

Then

$$\int_0^\infty 2xe^{-x^2}\,dx = \lim_{b\to\infty}\int_0^b 2xe^{-x^2}\,dx = \lim_{b\to\infty}(-e^{-x^2})\Big|_0^b$$
$$= \lim_{b\to\infty}\left(-\frac{1}{e^{b^2}} + e^0\right) = 0 + 1 = 1.$$

The function defined by $f(x) = 2xe^{-x^2}$ satisfies the two conditions required of a probability density function.

(b) Find the probability that there is a bird's nest within 0.5 kilometers of the given point.

Solution Find $P(x \leq 0.5)$ where $x \geq 0$. This probability is given by

$$P(0 \leq x \leq 0.5) = \int_0^{0.5} 2xe^{-x^2}\,dx.$$

Now evaluate the integral. The indefinite integral was found in part (a).

$$P(0 \leq x \leq 0.5) = \int_0^{0.5} 2xe^{-x^2}\,dx = (-e^{-x^2})\Big|_0^{0.5}$$
$$= -e^{-(0.5)^2} - (-e^0) = -e^{-0.25} + 1$$
$$\approx -0.78 + 1 = 0.22$$

The probability that a bird's nest will be found within 0.5 kilometers of the given point is about 0.22.

EXAMPLE 4 Computing Mortality

According to the National Center for Health Statistics, if we start with 100,000 people who are 60 years old, we can expect a certain number of them to die within each 5-year time interval, as indicated by Table 3 on the next page.*

National Vital Statistics Reports, Vol. 47, No. 19, June 30, 1999, Table 4.

FOR REVIEW

Improper integrals, those with one or two infinite limits, were discussed in the chapter on Further Techniques and Applications of Integration. The type of improper integral we shall need was defined as

$$\int_a^\infty f(x)\,dx = \lim_{b\to\infty}\int_a^b f(x)\,dx.$$

For example,

$$\int_1^\infty x^{-2}\,dx = \lim_{b\to\infty}\int_1^b x^{-2}\,dx$$
$$= \lim_{b\to\infty}\left(-\frac{1}{x}\Big|_1^b\right)$$
$$= \lim_{b\to\infty}\left(-\frac{1}{b} + \frac{1}{1}\right)$$
$$= 0 + 1 = 1.$$

Table 3

Years from Age 60	Midpoint of Interval	Number Dying in Each Interval
0–5	2.5	6,457
5–10	7.5	8,898
10–15	12.5	12,159
15–20	17.5	15,075
20–25	22.5	18,138
25–30	27.5	18,126
30–35	32.5	13,255
35–40	37.5	6,102
40–45	42.5	1,790

18,500

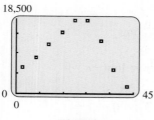

0
0
45

FIGURE 4

18,500

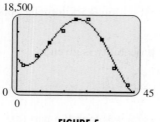

0
0
45

FIGURE 5

(a) Plot the data.

Solution Figure 4 shows that the plot appears to have the shape of a polynomial.

(b) Use the regression feature on a TI-83 calculator to find a quartic equation that models the number of years, x, since age 60 and the number of deaths, $N(x)$. Use the midpoints and the number of deaths in each interval from Table 3.

Solution The function

$$N(x) = 0.0681571x^4 - 6.72318x^3 + 184.397x^2 - 1,091.40x + 8,477.08$$

provided by the calculator models this data quite well, as illustrated by Figure 5.

(c) Convert this function into a probability density function.

Solution The particular function found in part (b) becomes negative at about $x = 44.29$. Thus, we will restrict the domain of the density function to the interval $[0, 44]$. Using the integration feature on our calculator, we find that

$$\int_0^{44} N(x)\,dx = 500,669.$$

Since the interval is not equal to 1, the function is not a probability density function. To convert it to one, multiply $N(x)$ by $1/500,669$. The function defined by

$$S(x) = \frac{1}{500,669} N(x)$$

$$= \frac{1}{500,669}(0.0681571x^4 - 6.72318x^3 + 184.397x^2 - 1,091.40x + 8,477.08)$$

will be a probability density function for $[0, 44]$, because

$$\int_0^{44} S(x)\,dx = 1, \text{ and } S(x) \geq 0 \text{ for all } x \text{ in } [0, 44].$$

(d) Find the probability that a randomly chosen 60-year-old person will live at least until age 80.

Solution Again, using the integration feature on our calculator,

$$P(x > 20) = \int_{20}^{44} S(x)\, dx \approx 0.57.$$

Thus, a 60-year-old person has a 57% chance of living at least until age 80. Notice that this value could also be estimated from the table by finding the number of people who have not died by age 80 and then dividing this number by 100,000. Thus, according to the table, there are 57,411 people still alive at age 80, representing approximately 57% of the original population. As you can see, our estimate agrees quite well with the actual number. ◼

13.1 EXERCISES

Decide whether the functions defined as follows are probability density functions on the indicated intervals. If not, tell why.

1. $f(x) = \dfrac{1}{9}x - \dfrac{1}{18}$; $[2, 5]$

2. $f(x) = \dfrac{1}{3}x - \dfrac{1}{6}$; $[3, 4]$

3. $f(x) = \dfrac{x^2}{21}$; $[1, 4]$

4. $f(x) = \dfrac{3}{98}x^2$; $[3, 5]$

5. $f(x) = 4x^3$; $[0, 3]$

6. $f(x) = \dfrac{x^3}{81}$; $[0, 3]$

7. $f(x) = \dfrac{x^2}{16}$; $[-2, 2]$

8. $f(x) = 2x^2$; $[-1, 1]$

9. $f(x) = \dfrac{5}{3}x^2 - \dfrac{5}{90}$; $[-1, 1]$

10. $f(x) = \dfrac{3}{13}x^2 - \dfrac{12}{13}x + \dfrac{45}{52}$; $[0, 4]$

Find a value of k that will make f a probability density function on the indicated interval.

11. $f(x) = kx^{1/2}$; $[1, 4]$

12. $f(x) = kx^{3/2}$; $[4, 9]$

13. $f(x) = kx^2$; $[0, 5]$

14. $f(x) = kx^2$; $[-1, 2]$

15. $f(x) = kx$; $[0, 3]$

16. $f(x) = kx$; $[2, 3]$

17. $f(x) = kx$; $[1, 5]$

18. $f(x) = kx^3$; $[2, 4]$

19. The total area under the graph of a probability density function always equals _____ .

20. In your own words, define a random variable.

21. What is the difference between a discrete probability function and a probability density function?

22. Why is $P(x = c) = 0$ for any number c in the domain of a probability density function?

Show that each function defined as follows is a probability density function on the given interval; then find the indicated probabilities.

23. $f(x) = \dfrac{1}{2}(1 + x)^{-3/2}$; $[0, \infty)$

   a. $P(0 \le x \le 2)$ **b.** $P(1 \le x \le 3)$

   c. $P(x \ge 5)$

24. $f(x) = e^{-x}$; $[0, \infty)$

   a. $P(0 \le x \le 1)$ **b.** $P(1 \le x \le 2)$

   c. $P(x \le 2)$

25. $f(x) = (1/2)e^{-x/2}$; $[0, \infty)$

   a. $P(0 \le x \le 1)$ **b.** $P(1 \le x \le 3)$

   c. $P(x \ge 2)$

26. $f(x) = \dfrac{20}{(x + 20)^2}$; $[0, \infty)$

   a. $P(0 \le x \le 1)$ **b.** $P(1 \le x \le 5)$

   c. $P(x \ge 5)$

Applications

LIFE SCIENCES

27. *Petal Length* The length of a petal, x, on a certain flower varies from 1 cm to 4 cm and has a probability density function defined by

$$f(x) = \frac{1}{2\sqrt{x}}.$$

Find the probabilities that the length of a randomly selected petal will be as follows.

a. Greater than or equal to 3 cm

b. Less than or equal to 2 cm

c. Between 2 cm and 3 cm

28. *Clotting Time of Blood* The clotting time of blood is a random variable x with values from 1 second to 20 seconds and probability density function defined by

$$f(x) = \frac{1}{(\ln 20)x}.$$

Find the following probabilities for a person selected at random.

a. The probability that the clotting time is between 1 and 5 sec

b. The probability that the clotting time is greater than 10 sec

29. *Flour Beetles* Researchers who study the abundance of the flour beetle, *Tribolium castaneum*, have developed a probability density function that can be used to estimate the abundance of the beetle in a population. The density function, which is a member of the gamma distribution, is

$$f(x) = 1.185 \cdot 10^{-9}x^{4.5222}e^{-0.049846x},$$

where x is the size of the population.*

a. Estimate the probability that a randomly selected flour beetle population is between 0 and 150.

b. Estimate the probability that a randomly selected flour beetle population is between 100 and 200.

30. *Flea Beetles* As we saw in Chapter 11, the mobility of an insect is an important part of its survival. Researchers have determined that the probability that a marked flea beetle, *Phyllotreta cruciferae* and *Phylotreta striolata,* will be recaptured within a certain distance and time after release can be calculated from the probability density function

$$p(x, t) = \frac{e^{-x^2/(4Dt)}}{\int_0^L e^{-u^2/(4Dt)} \, du},$$

where t is the time after release (in hr), x is the distance (in m) from the release point that recaptures occur, L is the maximum distance from the release point that recaptures can occur, and D is the diffusion coefficient.[†]

a. If $t = 12$, $L = 6$, and $D = 38.3$, find the probability that a flea beetle will be recaptured within 3 meters of the release point.

b. Using the same coefficients, find the probability that a flea beetle will be recaptured between 1 and 5 meters of the release point.

31. *Firearms* The number of deaths in the United States caused by firearms for each age group in 1997 is given in the table on the next page.[‡]

a. Plot the data. What type of function appears to best match this data?

b. Use the regression feature on your graphing calculator to find a quartic equation that models the number of years, x, since birth and the number of deaths caused by firearms, $N(x)$. Use the midpoint value to estimate the point in each interval when the person died. Graph the function with the plot of the data. Does the function resemble the data?

c. Convert this function into a probability density function. (*Hint:* Because the function in part b is negative for

*Dennis, Brian, and Robert F. Costantino, "Analysis of Steady-State Populations with the Gamma Abundance Model: Application to *Tribolium*," *Ecology*, Vol. 69, No. 4, Aug. 1988, pp. 1200–1213.
†Kareiva, Peter, "Experimental and Mathematical Analyses of Herbivore Movement: Quantifying the Influence of Plant Spacing and Quality on Foraging Discrimination," *Ecology Monographs*, Vol. 2, No. 3, Sept. 1982, pp. 261–282.
‡*National Vital Statistics Reports*, Vol. 47, No. 19, June 30, 1999, Table 16.

Age interval (years)	Midpoint of interval (year)	Number dying in each interval
0–15	7.5	630
15–25	20	8,173
25–35	30	7,045
35–45	40	5,802
45–55	50	3,872
55–65	60	2,390
65–75	70	2,202
75–85	80	1,740
85 +	90 (est)	555
Total		32,409

values less than 6.8 and greater than 91.5, restrict the domain of the density function to the interval $[6.8, 91.5]$. That is, integrate the function you found in part b from 6.8 to 91.5.)

d. For a randomly chosen person who was killed by a firearm, find the probabilities that the person killed was less than 25 years old, between 45 and 65 years old, and at least 75 years old, and compare these with the actual probabilities.

32. *Annual Rainfall* The annual rainfall in a remote Middle Eastern country varies from 0 to 5 inches and is a random variable with probability density function defined by

$$f(x) = \frac{5.5 - x}{15}.$$

Find the following probabilities for the annual rainfall in a randomly selected year.

a. The probability that the annual rainfall is greater than 3 inches

b. The probability that the annual rainfall is less than 2 inches

c. The probability that the annual rainfall is between 1 inch and 4 inches

33. *Earthquakes* The time between major earthquakes in the Southern California region is a random variable with probability density function

$$f(t) = \frac{1}{960} e^{-t/960},$$

where t is measured in days.*

a. Find the probability that the time between a major earthquake and the next one is less than 365 days.

b. Find the probability that the time between a major earthquake and the next one is more than 960 days.

34. *Earthquakes* The time between major earthquakes in the Taiwan region is a random variable with probability density function

$$f(t) = \frac{1}{3,650.1} e^{-t/3,650.1},$$

where t is measured in days.*

a. Find the probability that the time between a major earthquake and the next one is more than one year but less than three years.

b. Find the probability that the time between a major earthquake and the next one is more than 7,300 days.

35. *Time to Learn a Task* The time x required for a person to learn a certain task is a random variable with probability density function defined by

$$f(x) = \frac{8}{7(x - 2)^2}.$$

The time required to learn the task is between 3 and 10 min. Find the probabilities that a randomly selected person will learn the task in the following lengths of time.

a. Less than 4 min **b.** More than 5 min

36. *Drunk Drivers* According to an article in the journal *Traffic Safety,* the percentage of alcohol-related traffic fatalities for teenagers and young adults has dropped significantly since 1982. Based on data given in the article, the age of a randomly selected, alcohol-impaired driver in a fatal car crash is a random variable with probability density function given by

$$f(x) = \frac{105}{4x^2} \quad \text{for } x \text{ in } [15, 35].^\dagger$$

Find the following probabilities of the age of such a driver.

a. Less than 25 **b.** Less than 21

c. Between 21 and 25

OTHER APPLICATIONS

37. *Length of a Telephone Call* The length of a telephone call (in minutes), x, for a certain town is a continuous random variable with probability density function defined by

---

*Wang, Jeen-Hwa, and Chiao-Hui Kuo, "On the Frequency Distribution of Interoccurrence Times of Earthquakes," *Journal of Seismology,* Vol. 2, 1998, pp. 351–358.
†Castelli, Jim, "Drunk Driving: Not What It Used to Be," *Traffic Safety,* Vol. 91, No. 4, July/Aug. 1991, pp. 14–15.

$$f(x) = 3x^{-4} \quad \text{for } x \text{ in } [1, \infty).$$

Find the probabilities for the following situations.

a. The call lasts between 1 and 2 min.

b. The call lasts between 3 and 5 min.

c. The call lasts longer than 3 min.

38. *Life Span of a Computer Part* The life (in months) of a certain electronic computer part has a probability density function defined by

$$f(x) = \frac{1}{2} e^{-x/2} \quad \text{for } x \text{ in } [0, \infty).$$

Find the probability that a randomly selected component will last each of the following lengths of time.

a. At most 12 mo

b. Between 12 and 20 mo

39. *Machine Life* A machine has a useful life of 4 to 9 years, and its life (in years) has a probability density function defined by

$$f(x) = \frac{1}{11}\left(1 + \frac{3}{\sqrt{x}}\right).$$

Find the probabilities that the useful life of such a machine selected at random will be the following.

a. Longer than 6 yr   **b.** Less than 5 yr

c. Between 4 and 7 yr

13.2 EXPECTED VALUE AND VARIANCE OF CONTINUOUS RANDOM VARIABLES

? THINK ABOUT IT   What is the average age of a drunk driver in a fatal car crash?

It often is useful to have a single number, a typical or "average" number, that represents a random variable. The *mean* or *expected value* for a discrete random variable is found by multiplying each value of the random variable by its corresponding probability.

As we saw in the last chapter, if x is a discrete random variable that takes on the n values, $x_1, x_2, \ldots, x_n$, with respective probabilities $p_1, p_2, \ldots, p_n$, then the mean, or expected value, of the random variable is

$$\mu = x_1 p_1 + x_2 p_2 + x_3 p_3 + \cdots + x_n p_n = \sum_{i=1}^{n} x_i p_i.$$

For the bird feeder example in the previous section, the expected value is given by

$$\mu = 1(0.04) + 2(0.07) + 3(0.12) + 4(0.16) + 5(0.20) + 6(0.15) + 7(0.13)$$
$$+ 8(0.08) + 9(0.04) + 10(0.01)$$
$$= 5.09.$$

Thus, the average time a sparrow spends at a bird feeder is 5.09 minutes.

This definition can be extended to continuous random variables by using definite integrals. Suppose a continuous random variable has probability density function f on $[a, b]$. We can divide the interval from a to b into n subintervals of length Δx, where $\Delta x = (b - a)/n$. In the ith subinterval, the probability that the random variable takes a value close to x_i is approximately $f(x_i) \Delta x$, and so

$$\mu \approx \sum_{i=1}^{n} x_i f(x_i) \Delta x.$$

As $n \to \infty$, the limit of this sum gives the expected value

$$\mu = \int_a^b xf(x)\,dx.$$

As we saw in the previous chapter, the *variance* of a discrete probability distribution is a measure of the *spread* of the values of the distribution. For a discrete distribution, the variance is found by taking the expected value of the squares of the differences of the values of the random variable and the mean. If the random variable x takes the values $x_1, x_2, x_3, \ldots, x_n$, with respective probabilities $p_1, p_2, p_3, \ldots, p_n$ and mean μ, then the variance of x is

$$\text{Var}(x) = \sum_{i=1}^{n} (x_i - \mu)^2 p_i.$$

Think of the variance as the expected value of $(x - \mu)^2$, which measures how far x is from the mean μ. The *standard deviation* of x is defined as

$$\sigma = \sqrt{\text{Var}(x)}.$$

For the bird feeder example in the previous section, the variance and standard deviation are

$$\begin{aligned}
\text{Var}(X) = {} & (1 - 5.09)^2(0.04) + (2 - 5.09)^2(0.07) + (3 - 5.09)^2(0.12) \\
& + (4 - 5.09)^2(0.16) + (5 - 5.09)^2(0.20) + (6 - 5.09)^2(0.15) \\
& + (7 - 5.09)^2(0.13) + (8 - 5.09)^2(0.08) + (9 - 5.09)^2(0.04) \\
& + (10 - 5.09)^2(0.01) \\
= {} & 4.1819
\end{aligned}$$

and,

$$\sigma = \sqrt{\text{Var}(x)} \approx 2.045.$$

Like the mean or expected value, the variance of a continuous random variable is an integral.

$$\text{Var}(x) = \int_a^b (x - \mu)^2 f(x)\,dx$$

To find the standard deviation of a continuous probability distribution, like that of a discrete distribution, we find the square root of the variance. The formulas for the expected value, variance, and standard deviation of a continuous probability distribution are summarized here.

EXPECTED VALUE, VARIANCE, AND STANDARD DEVIATION

If x is a continuous random variable with probability density function f on $[a, b]$, then the expected value of x is

$$E(x) = \mu = \int_a^b xf(x)\,dx.$$

The variance of x is

$$\text{Var}(x) = \int_a^b (x - \mu)^2 f(x)\,dx,$$

and the standard deviation of x is

$$\sigma = \sqrt{\text{Var}(x)}.$$

Geometrically, the expected value (or mean) of a probability distribution represents the balancing point of the distribution. If a fulcrum were placed at μ on the x-axis, the figure would be in balance. See Figure 6.

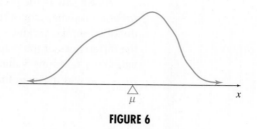

FIGURE 6

The variance and standard deviation of a probability distribution indicate how closely the values of the distribution cluster about the mean. These measures are most useful for comparing different distributions, as in Figure 7.

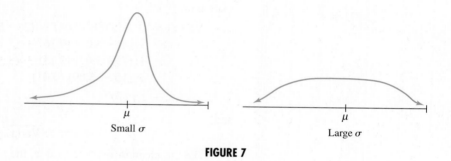

Small σ 　　　　　　　Large σ

FIGURE 7

EXAMPLE 1 Expected Value and Variance

Find the expected value and variance of the random variable x with probability density function defined by $f(x) = (3/26)x^2$ on $[1,3]$.

Solution By the definition of expected value just given,

$$\mu = \int_1^3 xf(x)\,dx$$

$$= \int_1^3 x\left(\frac{3}{26}x^2\right)dx$$

$$= \frac{3}{26}\int_1^3 x^3\,dx$$

$$= \frac{3}{26}\left(\frac{x^4}{4}\right)\Bigg|_1^3 = \frac{3}{104}(81 - 1) = \frac{30}{13},$$

or about 2.31.

The variance is

$$\text{Var}(x) = \int_1^3 \left(x - \frac{30}{13}\right)^2 \left(\frac{3}{26}x^2\right) dx$$

$$= \int_1^3 \left(x^2 - \frac{60}{13}x + \frac{900}{169}\right)\left(\frac{3}{26}x^2\right) dx \quad \text{Square } \left(x - \tfrac{30}{13}\right).$$

$$= \frac{3}{26}\int_1^3 \left(x^4 - \frac{60}{13}x^3 + \frac{900}{169}x^2\right) dx \quad \text{Multiply.}$$

$$= \frac{3}{26}\left(\frac{x^5}{5} - \frac{60}{13}\cdot\frac{x^4}{4} + \frac{900}{169}\cdot\frac{x^3}{3}\right)\Big|_1^3 \quad \text{Integrate.}$$

$$= \frac{3}{26}\left[\left(\frac{243}{5} - \frac{60(81)}{52} + \frac{900(27)}{169(3)}\right) - \left(\frac{1}{5} - \frac{60}{52} + \frac{300}{169}\right)\right]$$

$$\approx 0.259.$$

From the variance, the standard deviation is $\sigma \approx \sqrt{0.259} \approx 0.51$. The expected value and standard deviation are shown on the graph of the probability density function in Figure 8.

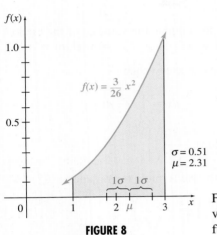

FIGURE 8

Calculating the variance in the last example was a messy job. An alternative version of the formula for the variance is easier to compute. This alternative formula is derived as follows.

$$\text{Var}(x) = \int_a^b (x - \mu)^2 f(x)\, dx$$

$$= \int_a^b (x^2 - 2\mu x + \mu^2) f(x)\, dx$$

$$= \int_a^b x^2 f(x)\, dx - 2\mu \int_a^b x f(x)\, dx + \mu^2 \int_a^b f(x)\, dx \qquad \textbf{(1)}$$

By definition,

$$\int_a^b x f(x)\, dx = \mu,$$

and, since $f(x)$ is a probability density function,

$$\int_a^b f(x)\, dx = 1.$$

Substitute back into equation (1) to get the alternative formula,

$$\text{Var}(x) = \int_a^b x^2 f(x)\, dx - 2\mu^2 + \mu^2 = \int_a^b x^2 f(x)\, dx - \mu^2.$$

ALTERNATIVE FORMULA FOR VARIANCE

If x is a random variable with probability density function f on $[a, b]$, and if $E(x) = \mu$, then

$$\textbf{Var}(x) = \int_a^b x^2 f(x)\, dx - \mu^2.$$

| **CAUTION** Notice that the term μ^2 comes *after* the dx, and so is *not* integrated.

EXAMPLE 2 Variance

Use the alternative formula for variance to compute the variance of the random variable x with probability density function defined by $f(x) = 3/x^4$ for $x \geq 1$.

Solution To find the variance, first find the expected value:

$$\mu = \int_1^\infty xf(x)\,dx = \int_1^\infty x \cdot \frac{3}{x^4}\,dx = \int_1^\infty \frac{3}{x^3}\,dx$$

$$= \lim_{b \to \infty} \int_1^b \frac{3}{x^3}\,dx = \lim_{b \to \infty} \left(\frac{3}{-2x^2}\right)\Bigg|_1^b = \frac{3}{2},$$

or 1.5. Now find the variance by the alternative formula for variance:

$$\text{Var}(x) = \int_1^\infty x^2\left(\frac{3}{x^4}\right)dx - \left(\frac{3}{2}\right)^2$$

$$= \int_1^\infty \frac{3}{x^2}\,dx - \frac{9}{4}$$

$$= \lim_{b \to \infty} \int_1^b \frac{3}{x^2}\,dx - \frac{9}{4}$$

$$= \lim_{b \to \infty} \left(\frac{-3}{x}\right)\Bigg|_1^b - \frac{9}{4}$$

$$= 3 - \frac{9}{4} = \frac{3}{4}, \quad \text{or } 0.75.$$

EXAMPLE 3 Patient Wait

A recent study has shown that the time patients wait at the doctor's office (in hours) after the time of their scheduled appointment is given by the probability density function $f(x) = 6x - 6x^2$ for $0 \leq x \leq 1$.

(a) Find and interpret the expected value for this distribution.

Solution The expected value is

$$\mu = \int_0^1 x(6x - 6x^2)\,dx = \int_0^1 (6x^2 - 6x^3)\,dx$$

$$= \left(2x^3 - \frac{3}{2}x^4\right)\Bigg|_0^1 = \frac{1}{2},$$

or 0.5. This result indicates that patients wait an average of 1/2 hour past the scheduled office appointment time.

(b) Compute the standard deviation.

Solution First compute the variance. We use the alternative formula.

$$\text{Var}(x) = \int_0^1 x^2(6x - 6x^2)\, dx - \left(\frac{1}{2}\right)^2$$

$$= \int_0^1 (6x^3 - 6x^4)\, dx - \left(\frac{1}{2}\right)^2$$

$$= \left(\frac{3}{2}x^4 - \frac{6}{5}x^5\right)\Big|_0^1 - \frac{1}{4}$$

$$= \frac{3}{10} - \frac{1}{4} = \frac{1}{20} = 0.05$$

The standard deviation is $\sigma = \sqrt{0.05} \approx 0.22$.

EXAMPLE 4 Life Expectancy

In the previous section of this chapter we used statistics compiled by the National Center for Health Statistics to determine a probability density function that can be used to study the proportion of all 60-year-olds who will be alive in x years. The function is given by

$$S(x) = \frac{1}{500,669}(0.0681571x^4 - 6.72318x^3 + 184.397x^2 - 1,091.40x + 8,477.08)$$

for $0 \le x \le 44$.

(a) Find the life expectancy of a 60-year-old person.

Solution Since this is a complicated function that is tedious to integrate analytically, we will employ the integration feature on a TI-83 calculator to calculate

$$\mu = \int_0^{44} xS(x)\, dx \approx 21.32 \text{ years.}$$

According to life tables, the life expectancy of a 60-year-old person is 21.4 years. Our estimate is remarkably accurate given the limited number of data points and the function used in our original analysis. Life expectancy is generally calculated with techniques from life table analysis.*

(b) Find the standard deviation of this probability function.

Solution Using the alternate formula, we first calculate the variance.

$$\text{Var}(x) = \int_0^{44} x^2 S(x)\, dx - \mu^2 = 554.6281 - (21.32)^2 \approx 100.42$$

Thus, $\sigma = \sqrt{\text{Var}(x)} \approx 10.02$ years.

---

*Chiang, Chin Long, *The Life Table and Its Applications*, Malabar, FL, Krieger Pub., 1984.

13.2 EXERCISES

In Exercises 1–8, a probability density function of a random variable is defined. Find the expected value, the variance, and the standard deviation. Round answers to the nearest hundredth.

1. $f(x) = \dfrac{1}{4}$; $[3, 7]$

2. $f(x) = \dfrac{1}{10}$; $[0, 10]$

3. $f(x) = \dfrac{x}{8} - \dfrac{1}{4}$; $[2, 6]$

4. $f(x) = 2(1 - x)$; $[0, 1]$

5. $f(x) = 1 - \dfrac{1}{\sqrt{x}}$; $[1, 4]$

6. $f(x) = \dfrac{1}{11}\left(1 + \dfrac{3}{\sqrt{x}}\right)$; $[4, 9]$

7. $f(x) = 4x^{-5}$; $[1, \infty)$

8. $f(x) = 3x^{-4}$; $[1, \infty)$

9. What information does the mean (expected value) of a continuous random variable give?

10. Suppose two random variables have standard deviations of 0.10 and 0.23, respectively. What does this tell you about their distributions?

In Exercises 11–14, the probability density function of a random variable is defined.

a. *Find the expected value to the nearest hundredth.*

b. *Find the variance to the nearest hundredth.*

c. *Find the standard deviation. Round to the nearest hundredth.*

d. *Find the probability that the random variable has a value greater than the mean.*

e. *Find the probability that the value of the random variable is within one standard deviation of the mean.*

11. $f(x) = \dfrac{\sqrt{x}}{18}$; $[0, 9]$

12. $f(x) = \dfrac{x^{-1/3}}{6}$; $(0, 8]$

13. $f(x) = \dfrac{1}{2}x$; $[0, 2]$

14. $f(x) = \dfrac{3}{2}(1 - x^2)$; $[0, 1]$

*If x is a random variable with probability density function f on $[a, b]$, then the **median** of x is the number m such that*

$$\int_a^m f(x)\, dx = \dfrac{1}{2}.$$

a. *Find the median of each random variable for the probability density functions in Exercises 15–20.*

b. *In each case, find the probability that the random variable is between the expected value (mean) and the median. The expected value for each of these functions was found in Exercises 1–8.*

15. $f(x) = \dfrac{1}{4}$; $[3, 7]$

16. $f(x) = \dfrac{1}{10}$; $[0, 10]$

17. $f(x) = \dfrac{x}{8} - \dfrac{1}{4}$; $[2, 6]$

18. $f(x) = 2(1 - x)$; $[0, 1]$

19. $f(x) = 4x^{-5}$; $[1, \infty)$

20. $f(x) = 3x^{-4}$; $[1, \infty)$

Applications

LIFE SCIENCES

21. *Blood Clotting Time* The clotting time of blood (in seconds) is a random variable with probability density function defined by

$$f(x) = \frac{1}{(\ln 20)x} \quad \text{for } x \text{ in } [1, 20].$$

a. Find the mean clotting time.

b. Find the standard deviation.

c. Find the probability that a person's blood clotting time is within one standard deviation of the mean.

22. *Length of a Leaf* The length of a leaf on a tree is a random variable with probability density function defined by

$$f(x) = \frac{3}{32}(4x - x^2) \quad \text{for } x \text{ in } [0, 4].$$

a. What is the expected leaf length?

b. Find σ for this distribution.

c. Find the probability that the length of a given leaf is within one standard deviation of the expected value.

23. *Petal Length* The length (in centimeters) of a petal on a certain flower is a random variable with probability density function defined by

$$f(x) = \frac{1}{2\sqrt{x}} \quad \text{for } x \text{ in } [1, 4].$$

a. Find the expected petal length.

b. Find the standard deviation.

c. Find the probability that a petal selected at random has a length more than two standard deviations above the mean.

24. *Firearms* In Exercise 31 of the previous section, the probability density function for the number of firearm deaths in the United States in 1997 was found to be

$$S(x) = \frac{1}{343,795}(-0.00351465x^4 + 0.792884x^3 - 61.7955x^2 + 1,814.54x - 9,709.20),$$

where x was the number of years since birth on $[6.8, 91.5]$. Calculate the expected age at which a person will be killed with a firearm, as well as the standard deviation.

25. *Drunk Drivers* The age of a randomly selected, alcohol-impaired driver in a fatal car crash is a random variable with probability density function given by

$$f(x) = \frac{105}{4x^2} \quad \text{for } x \text{ in } [15, 35].*$$

a. Find the expected age of a drunk driver in a fatal car crash.

b. Find the standard deviation of the distribution.

c. Find the probability that the age of such a driver will be less than one standard deviation below the mean.

26. *Flour Beetles* As we saw in the previous section, a probability density function has been developed to estimate the abundance of the flour beetle, *Tribolium castaneum*. The density function, which is a member of the gamma distribution, is

$$f(x) = 1.185 \cdot 10^{-9}x^{4.5222}e^{-0.049846x},$$

where x is the size of the population.[†] Calculate the expected size of a flour beetle population. (*Hint:* Use 1,000 as the upper limit of integration.)

27. *Flea Beetles* As we saw in Chapter 11 and in the previous section, the probability that a marked flea beetle, *Phyllotreta cruciferae* and *Phylotreta striolata,* will be recaptured within a certain distance and time after release can be calculated from the probability density function

$$p(x, t) = \frac{e^{-x^2/(4Dt)}}{\int_0^L e^{-u^2/(4Dt)} \, du},$$

where t is the time (in hr) after release, x is the distance (in m) from the release point that recaptures occur, L is the maximum distance from the release point that recaptures

---

*Castelli, Jim, "Drunk Driving: Not What It Used to Be," *Traffic Safety,* Vol. 91, No. 4, July/Aug. 1991, pp. 14–15.

†Dennis, Brian, and Robert F. Costantino, "Analysis of Steady-State Populations with the Gamma Abundance Model: Application to *Tribolium,*" *Ecology,* Vol. 69, No. 4, Aug. 1988, pp. 1200–1213.

can occur, and D is the diffusion coefficient.* If $t = 12$, $L = 6$, and $D = 38.3$, find the expected recapture distance.

28. *Annual Rainfall* The annual rainfall in a remote Middle Eastern country is a random variable with probability density function defined by

$$f(x) = \frac{5.5 - x}{15} \quad \text{for } x \text{ in } [0, 5].$$

   a. Find the mean annual rainfall.

   b. Find the standard deviation.

   c. Find the probability of a year with rainfall less than one standard deviation below the mean.

29. *Earthquakes* As we saw in the previous section, the time between major earthquakes in the Southern California region is a random variable with probability density function defined by

$$f(t) = \frac{1}{960} e^{-t/960},$$

where t is measured in days.† Find the expected value and the standard deviation of this probability density function.

OTHER APPLICATIONS

30. *Life of a Light Bulb* The life (in hours) of a certain kind of light bulb is a random variable with probability density function defined by

$$f(x) = \frac{1}{58\sqrt{x}} \quad \text{for } x \text{ in } [1, 900].$$

   a. What is the expected life of such a bulb?

   b. Find σ.

   c. Find the probability that one of these bulbs lasts longer than one standard deviation above the mean.

31. *Machine Life* The life (in years) of a certain machine is a random variable with probability density function defined by

$$f(x) = \frac{1}{11}\left(1 + \frac{3}{\sqrt{x}}\right) \quad \text{for } x \text{ in } [4, 9].$$

   a. Find the mean life of this machine.

   b. Find the standard deviation of the distribution.

   c. Find the probability that a particular machine of this kind will last longer than the mean number of years.

32. *Life of an Automobile Part* The life span of a certain automobile part (in months) is a random variable with probability density function defined by

$$f(x) = \frac{1}{2} e^{-x/2} \quad \text{for } x \text{ in } [0, \infty).$$

   a. Find the expected life of this part.

   b. Find the standard deviation of the distribution.

   c. Find the probability that one of these parts lasts less than the mean number of months.

■ 13.3 SPECIAL PROBABILITY DENSITY FUNCTIONS

? THINK ABOUT IT   What is the probability that the maximum outdoor temperature will be higher than 24°C? What is the probability that a hearing aid battery will last longer than 400 hours?

These questions can be answered if the probability density function for the maximum temperature and for the life of the battery are known. In practice, however, it is not feasible to construct a probability density function for each experiment. Instead, a researcher uses one of several probability density functions that are well known, matching the shape of the experimental distribution to one of the known distributions. In this section we discuss some of the most commonly used probability distributions.

---

*Kareiva, Peter, "Experimental and Mathematical Analyses of Herbivore Movement: Quantifying the Influence of Plant Spacing and Quality on Foraging Discrimination," *Ecology Monographs,* Vol. 2, No. 3, 1982, pp. 261–282.
†Wang, Jeen-Hwa, and Chiao-Hui Kuo, "On the Frequency Distribution of Interoccurrence Times of Earthquakes," *Journal of Seismology,* Vol. 2, 1998, pp. 351–358.

Uniform Distribution   The simplest probability distribution occurs when the probability density function of a random variable remains constant over the sample space. In this case, the random variable is said to be *uniformly distributed* over the sample space. The probability density function for the **uniform distribution** is defined by

$$f(x) = \frac{1}{b - a} \text{ for } x \text{ in } [a, b],$$

where a and b are constant real numbers. The graph of $f(x)$ is shown in Figure 9.

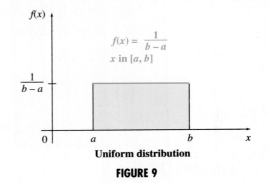

Uniform distribution

FIGURE 9

Since $b - a$ is positive, $f(x) \geq 0$, and

$$\int_a^b \frac{1}{b - a} dx = \frac{1}{b - a} x \Big|_a^b = \frac{1}{b - a}(b - a) = 1.$$

Therefore, the function is a probability density function.

The expected value for the uniform distribution is

$$\mu = \int_a^b \left(\frac{1}{b - a}\right) x \, dx = \left(\frac{1}{b - a}\right) \frac{x^2}{2} \Big|_a^b$$

$$= \frac{1}{2(b - a)}(b^2 - a^2) = \frac{1}{2}(b + a). \qquad b^2 - a^2 = (b - a)(b + a)$$

The variance is given by

$$\text{Var}(x) = \int_a^b \left(\frac{1}{b - a}\right) x^2 \, dx - \left(\frac{b + a}{2}\right)^2$$

$$= \left(\frac{1}{b - a}\right) \frac{x^3}{3} \Big|_a^b - \frac{(b + a)^2}{4}$$

$$= \frac{1}{3(b - a)}(b^3 - a^3) - \frac{1}{4}(b + a)^2$$

$$= \frac{b^2 + ab + a^2}{3} - \frac{b^2 + 2ab + a^2}{4} \qquad b^3 - a^3 = (b - a)(b^2 + ab + a^2)$$

$$= \frac{b^2 - 2ab + a^2}{12}. \qquad \text{Get a common denominator; subtract.}$$

Thus

$$\text{Var}(x) = \frac{1}{12}(b - a)^2, \qquad \text{Factor.}$$

and

$$\sigma = \frac{1}{\sqrt{12}}(b - a).$$

These properties of the uniform distribution are summarized below.

UNIFORM DISTRIBUTION

If x is a random variable with probability density function

$$f(x) = \frac{1}{b - a} \text{ for } x \text{ in } [a, b],$$

then

$$\mu = \frac{1}{2}(b + a) \qquad \text{and} \qquad \sigma = \frac{1}{\sqrt{12}}(b - a).$$

EXAMPLE 1 Daily Temperature

A couple is planning to vacation in San Francisco. They have been told that the maximum daily temperature during the time they plan to be there ranges from 15°C to 27°C. Assume that the probability of any temperature between 15°C and 27°C is equally likely for any given day during the specified time period.

(a) What is the probability that the maximum temperature on the day they arrive will be higher than 24°C?

Solution If the random variable t represents the maximum temperature on a given day, then the uniform probability density function for t is defined by $f(t) = 1/12$ for the interval $[15, 27]$. By definition,

$$P(t > 24) = \int_{24}^{27} \frac{1}{12}\, dt = \frac{1}{12}t \Big|_{24}^{27} = \frac{1}{4}.$$

(b) What average maximum temperature can they expect?

Solution The expected maximum temperature is

$$\mu = \frac{1}{2}(27 + 15) = 21,$$

or 21°C.

(c) What is the probability that the maximum temperature on a given day will be at least one standard deviation below the mean?

Solution First find σ.

$$\sigma = \frac{1}{\sqrt{12}}(27 - 15) = \frac{12}{\sqrt{12}} = \sqrt{12} = 2\sqrt{3} \approx 3.5.$$

One standard deviation below the mean indicates a temperature of $21 - 3.5 = 17.5$°C.

$$P(T \le 17.5) = \int_{15}^{17.5} \frac{1}{12} \, dt = \frac{1}{12} t \Big|_{15}^{17.5} \approx 0.21$$

The probability is 0.21 that the temperature will not exceed 17.5°C. ▪

Exponential Distribution   The next distribution is very important in reliability and survival analysis. When manufactured items and living things have a constant failure rate over a period of time, the exponential distribution is used to describe their probability of failure. In this case, the random variable is said to be *exponentially distributed* over the sample space. The probability density function for the **exponential distribution** is defined by

$$f(x) = ae^{-ax} \quad \text{for } x \text{ in } [0, \infty),$$

where a is a positive constant. The graph of $f(x)$ is shown in Figure 10.

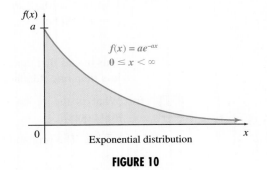

FIGURE 10

Here $f(x) \ge 0$, since e^{-ax} and a are both positive for all values of x. Also,

$$\int_0^\infty ae^{-ax} \, dx = \lim_{b \to \infty} \int_0^b ae^{-ax} \, dx$$

$$= \lim_{b \to \infty} (-e^{-ax}) \Big|_0^b = \lim_{b \to \infty} \left(\frac{-1}{e^{ab}} + \frac{1}{e^0} \right) = 1,$$

so the function is a probability density function.

The expected value and the standard deviation of the exponential distribution can be found using integration by parts. The results are given below.

EXPONENTIAL DISTRIBUTION

If x is a random variable with probability density function

$$f(x) = ae^{-ax} \quad \text{for } x \text{ in } [0, \infty),$$

then

$$\mu = \frac{1}{a} \quad \text{and} \quad \sigma = \frac{1}{a}.$$

EXAMPLE 2 Hearing Aid Battery

Suppose the useful life (in hours) of a hearing aid battery is the random variable t, with probability density function given by the exponential distribution

$$f(t) = \frac{1}{300} e^{-t/300} \quad \text{for } t \geq 0.$$

(a) Find the probability that a particular battery, selected at random, has a useful life of less than 600 hours.

Solution The probability is given by

$$P(t \leq 100) = \int_0^{600} \frac{1}{300} e^{-t/300} \, dt = \frac{1}{300} (-300 e^{-t/300}) \Big|_0^{600}$$

$$= -(e^{-600/300} - e^0) = -(e^{-2} - 1)$$

$$\approx 1 - 0.1353 = 0.8647.$$

(b) Find the expected value and standard deviation of the distribution.

Solution Use the formulas given above. Both μ and σ equal $1/a$, and since $a = 1/300$ here,

$$\mu = 300 \quad \text{and} \quad \sigma = 300.$$

This means that the average life of a battery is 300 hours, and no battery lasts less than one standard deviation below the mean.

 (c) What is the probability that a hearing aid battery will last longer than 400 hours?

Solution The probability is given by

$$P(t > 400) = \int_{400}^{\infty} \frac{1}{300} e^{-t/300} \, dt = \lim_{b \to \infty} (-e^{-t/300}) \Big|_{400}^{b} = \frac{1}{e^{4/3}} \approx 0.2636,$$

or about 26%.

Normal Distribution The **normal distribution,** with its well-known bell-shaped graph, is undoubtedly the most important probability density function. It is widely used in various applications of statistics. The random variables associated with these applications are said to be normally distributed. The probability density function for the normal distribution has the following characteristics.

NORMAL DISTRIBUTION

If μ and σ are real numbers, $\sigma > 0$, and if x is a random variable with probability density function defined by

$$f(x) = \frac{1}{\sigma\sqrt{2\pi}} e^{-(x-\mu)^2/(2\sigma^2)} \quad \text{for } x \text{ in } (-\infty, \infty),$$

then

$$E(x) = \mu \quad \text{and} \quad Var(x) = \sigma^2, \quad \text{with **standard deviation** } \sigma.$$

Notice that the definition of the probability density function includes σ, which is the standard deviation of the distribution.

Advanced techniques can be used to show that

$$\int_{-\infty}^{\infty} \frac{1}{\sigma\sqrt{2\pi}} e^{-(x-\mu)^2/(2\sigma^2)} dx = 1.$$

Deriving the expected value and standard deviation for the normal distribution also requires techniques beyond the scope of this text.

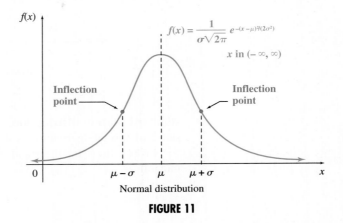

$$f(x) = \frac{1}{\sigma\sqrt{2\pi}} e^{-(x-\mu)^2/(2\sigma^2)}$$

x in $(-\infty, \infty)$

FIGURE 11

Each normal probability distribution has associated with it a bell-shaped curve, called a **normal curve,** such as the one in Figure 11. Each normal curve is symmetric about a vertical line through the mean, μ. Vertical lines at points $+1\sigma$ and -1σ from the mean show the inflection points of the graph. A normal curve never touches the x-axis; it extends indefinitely in both directions.

The development of the normal curve is credited to the Frenchman Abraham De Moivre (1667–1754). Three of his publications dealt with probability and associated topics: *Annuities upon Lives* (which contributed to the development of actuarial studies), *Doctrine of Chances,* and *Miscellanea Analytica.*

Many different normal curves have the same mean. In such cases, a larger value of σ produces a "flatter" normal curve, while smaller values of σ produce more values near the mean, resulting in a "taller" normal curve. See Figure 12.

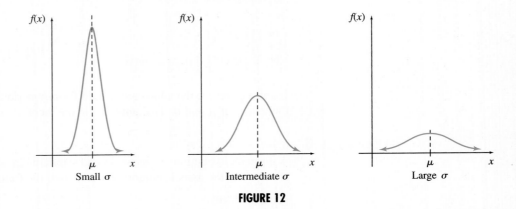

FIGURE 12

It would be far too much work to calculate values for the normal probability distribution for various values of μ and σ. Instead, values are calculated for the

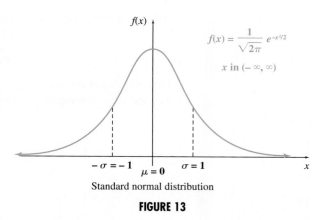

Standard normal distribution

FIGURE 13

standard normal distribution, which has $\mu = 0$ and $\sigma = 1$. The graph of the standard normal distribution is shown in Figure 13.

Probabilities for the standard normal distribution come from the definite integral

$$\int_a^b \frac{1}{\sqrt{2\pi}} e^{-x^2/2}\, dx.$$

Since $f(x) = e^{-x^2/2}$ does not have an antiderivative that can be expressed in terms of functions used in this course, numerical methods are used to find values of this definite integral. A table in the appendix of this book gives areas under the standard normal curve, along with a sketch of the curve. Each value in this table is the total area under the standard normal curve to the left of the number z.

If a normal distribution does not have $\mu = 0$ and $\sigma = 1$, we use the following theorem, which is proven in Exercise 21.

z-SCORES THEOREM

Suppose a normal distribution has mean μ and standard deviation σ. The area under the associated normal curve that is to the left of the value x is exactly the same as the area to the left of

$$z = \frac{x - \mu}{\sigma}$$

for the standard normal curve.

Using this result, the table can be used for *any* normal distribution, regardless of the values of μ and σ. The number z in the theorem is called a **z-score.**

EXAMPLE 3 Normal Distribution

A normal distribution has mean 35 and standard deviation 5.9. By computing the areas under the associated normal curve, find the following probabilities.

(a) $P(x < 40)$

Solution Find the appropriate z-score using $x = 40$, $\mu = 35$, and $\sigma = 5.9$. Round to the nearest hundredth.

$$z = \frac{40 - 35}{5.9} = \frac{5}{5.9} \approx 0.85$$

Look up 0.85 in the normal curve table in the Appendix. The corresponding area is 0.8023. Thus, the shaded area shown in Figure 14 is 0.8023. This area represents 80.23% of the total area under the normal curve and the proportion of observations that are less than 40.

(b) $P(x > 32)$

Solution

$$z = \frac{32 - 35}{5.9} = \frac{-3}{5.9} \approx -0.51$$

The area to the *left* of $z = -0.51$ is 0.3050, so the area to the *right is* $1 - 0.3050 = 0.6950$. See Figure 15. Thus, $P(x > 32) = 0.6950$.

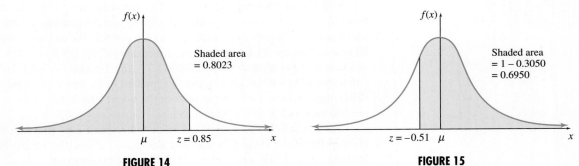

FIGURE 14   **FIGURE 15**

(c) $P(30 \le x \le 33)$

Solution Find z-scores for both values.

$$z = \frac{30 - 35}{5.9} = \frac{-5}{5.9} \approx -0.85 \quad \text{and} \quad z = \frac{33 - 35}{5.9} = \frac{-2}{5.9} \approx -0.34$$

Start with the area to the left of $z = -0.34$ and subtract the area to the left of $z = -0.85$. Thus,

$$P(30 \le x \le 33) = 0.3669 - 0.1977 = 0.1692.$$

The required area is shaded in Figure 16.

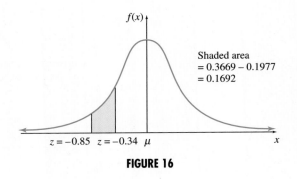

FIGURE 16

As an alternative to using the normal curve table, we can use a graphing calculator. Enter the formula for the normal distribution into the calculator, using $\mu = 35$ and $\sigma = 5.9$. Plot the function in a window that contains at least four standard deviations to the left and right of μ; for this example, we will let $0 \leq x \leq 60$. Then use the integration feature (under CALC on a TI-83) to find the area under the curve.

The result is shown in Figure 17. In place of $-\infty$, we have used $x = 0$ as the left endpoint. This is far enough to the left of $\mu = 35$ that it can be considered as $-\infty$ for all practical purposes. You can verify that choosing a lower limit even further to the left makes little or no difference in the answer. In fact, the answer of 0.80162995 is more accurate than the answer of 0.8023 that we found in Example 3, where we needed to round $5/5.9 \approx 0.8474576$ to 0.85 in order to use the table.

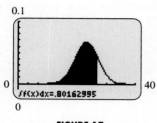

0.1

0   40

$\int f(x)dx=.80162995$

0

FIGURE 17

We could get the answer on a TI-83 without generating a graph using the command `fnInt(Y₁,X,0,40)`, where Y_1 is the formula for the normal distribution with $\mu = 35$ and $\sigma = 5.9$.

The numerical integration method works with any probability density function. In addition, many graphing calculators are programmed with information about specific density functions, such as the normal. We can solve the first part of Example 3 on the TI-83 by entering `normalcdf(-1E99,40,35,5.9)`. The calculator responds with 0.8016300043. ($-1\text{E}99$ stands for $-1 \cdot 10^{99}$, which the calculator uses for $-\infty$.) If you use this method in the exercises, your answers will differ slightly from those in the back of the book, which were generated using the normal curve table in the Appendix.

The z-scores are actually standard deviation multiples; that is, a z-score of 2.5 corresponds to a value 2.5 standard deviations above the mean. For example, looking up $z = 1.00$ and $z = -1.00$ in the table shows that

$$0.8413 - 0.1587 = 0.6826,$$

so that 68.26% of the area under a normal curve is within one standard deviation of the mean. Also, using $z = 2.00$ and $z = -2.00$,

$$0.9772 - 0.0228 = 0.9544,$$

meaning 95.44% of the area is within two standard deviations of the mean. These results, summarized in Figure 18, can be used to get a quick estimate of results when working with normal curves.

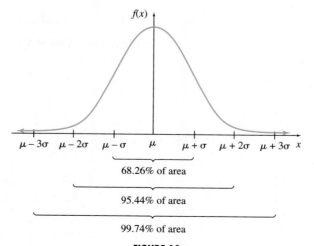

$f(x)$

$\mu - 3\sigma$ $\mu - 2\sigma$ $\mu - \sigma$ μ $\mu + \sigma$ $\mu + 2\sigma$ $\mu + 3\sigma$ x

68.26% of area

95.44% of area

99.74% of area

FIGURE 18

Manufacturers make use of the fact that a normal random variable is almost always within three standard deviations of the mean to design control charts. When a sample of items produced by a machine has a mean further than three standard deviations from the desired specification, the machine is assumed to be out of control, and adjustments are made to ensure that the items produced meet the tolerance required.

EXAMPLE 4   Lead Poisoning

Historians and biographers have collected evidence suggesting that President Andrew Jackson suffered from lead poisoning. Recently, researchers measured the amount of lead in samples of Jackson's hair from 1815. The results of this experiment showed that Jackson had a mean lead level of 130.5 ppm.*

(a) If levels of lead in hair samples from that time period follow a normal distribution with mean 93 and standard deviation 16[†], find the probability that a randomly selected person from this time period would have a lead level of 130.5 ppm or higher. Does this provide evidence that Jackson suffered from lead poisoning during this time period?

Solution

$$P(x > 130.5) = P\left(z > \frac{130.5 - 93}{16}\right) = P(z > 2.34) \approx 0.01.$$

Since this probability is so low, it is likely that Jackson suffered from lead poisoning during this time period.[‡]

(b) Today's normal lead levels follow a normal distribution with approximate mean of 10 ppm and standard deviation of 5 ppm.[§] By today's standards, calculate the probability that a randomly selected person from today would have a lead level of 130.5 or higher. From this, can we conclude that Andrew Jackson had lead poisoning?

Solution

$$P(x > 130.5) = P\left(z > \frac{130.5 - 10}{5}\right) = P(z > 24.1) \approx 0$$

By today's standards, which may not be valid for this experiment, Jackson certainly suffered from lead poisoning.

---

*Deppisch, Lidwig, Jose Centeno, David Gemmel, and Norca Torres, "Andrew Jackson's Exposure to Mercury and Lead," *JAMA,* Vol. 282, No. 6, Aug. 11, 1999, pp. 569–571.
†Weiss, D., B. Whitten, and D. Leddy, "Lead Content of Human Hair (1871–1971)," *Science,* Vol. 178, 1972, pp. 69–70.
‡Although this provides evidence that Andrew Jackson had elevated lead levels, the authors of the paper concluded that Andrew Jackson did not die from lead poisoning.
§Iyengar, V., and J. Woittiez, "Trace Elements in Human Clinical Specimens," *Clinical Chemistry,* Vol. 34, 1988, pp. 474–481.

13.3 EXERCISES

For Exercises 1–6, find the mean of the distribution, the standard deviation of the distribution, and the probability that the random variable is between the mean and one standard deviation above the mean.

1. The length (in centimeters) of the leaf of a certain plant is a continuous random variable with probability density function defined by

$$f(x) = \frac{5}{4} \quad \text{for } x \text{ in } [4, 4.8].$$

2. The price of an item (in dollars) is a continuous random variable with probability density function defined by

$$f(x) = 2 \quad \text{for } x \text{ in } [1.25, 1.75].$$

3. The length of time (in years) until a particular radioactive particle decays is a random variable t with probability density function defined by

$$f(t) = 0.03e^{-0.03t} \quad \text{for } t \text{ in } [0, \infty).$$

4. The length of time (in years) that a seedling tree survives is a random variable t with probability density function defined by

$$f(t) = 0.05e^{-0.05t} \quad \text{for } t \text{ in } [0, \infty).$$

5. The length of time (in days) required to learn a certain task is a random variable t with probability density function defined by

$$f(t) = e^{-t} \quad \text{for } t \text{ in } [0, \infty).$$

6. The distance (in meters) that seeds are dispersed from a certain kind of plant is a random variable x with probability density function defined by

$$f(x) = 0.1e^{-0.1x} \quad \text{for } x \text{ in } [0, \infty).$$

Find the proportion of observations of a standard normal distribution that are between the mean and the number of standard deviations above the mean given in Exercises 7 and 8.

7. 3.50

8. 1.68

Find the proportion of observations of a standard normal distribution that are between the z-scores given in Exercises 9 and 10.

9. 1.28 and 2.05

10. −2.13 and −0.04

Find a z-score satisfying the conditions given in Exercises 11–14. (Hint: Use the table backwards.)

11. 10% of the total area is to the left of z.

12. 2% of the total area is to the left of z.

13. 18% of the total area is to the right of z.

14. 22% of the total area is to the right of z.

15. Describe the standard normal distribution. What are its characteristics?

16. What is a z-score? How is it used?

17. Describe the shape of the graph of each of the following probability distributions.

   a. Uniform   b. Exponential   c. Normal

In the exercises for the second section of this chapter, we defined the median of a probability distribution as an integral. The median also can be defined as the number m such that $P(x \leq m) = P(x \geq m)$.

18. Find an expression for the median of the uniform distribution.

19. Find an expression for the median of the exponential distribution.

20. Verify the expected value and standard deviation of the exponential distribution given in the text.

21. Prove the z-scores theorem. (*Hint:* Write the formula for the normal distribution with mean μ and standard deviation σ, using t instead of x as the variable. Then write the integral representing the area to the left of the value x, and make the substitution $u = (t - \mu)/\sigma$.)

22. Show that a normal random variable has inflection points at $x = \mu - \sigma$ and $x = \mu + \sigma$.

23. Use Simpson's rule with $n = 100$, or use the integration feature on a graphing calculator, to approximate the following integrals.

 a. $\int_0^{50} 0.5e^{-0.5x}\, dx$ **b.** $\int_0^{50} 0.5xe^{-0.5x}\, dx$ **c.** $\int_0^{50} 0.5x^2 e^{-0.5x}\, dx$

24. Use your results from Exercise 23 to verify that, for the exponential distribution with $a = 5$, the total probability is 1, and both the mean and the standard deviation are equal to $1/a$.

25. Use Simpson's rule with $n = 100$, or the integration feature on a graphing calculator, to approximate the following for the standard normal probability distribution. Use limits of -4 and 4 in place of $-\infty$ and ∞.

 a. The mean **b.** The standard deviation

26. A very important distribution for analyzing the reliability of manufactured goods is the Weibull distribution, whose probability density function is defined by

$$f(x) = abx^{b-1}e^{-ax^b} \quad \text{for } x \text{ in } [0, \infty),$$

where a and b are constants. Notice that when $b = 1$, this reduces to the exponential distribution. The Weibull distribution is more general than the exponential, because it applies even when the failure rate is not constant. Use Simpson's rule with $n = 100$, or the integration feature on a graphing calculator, to approximate the following for the Weibull distribution with $a = 4$ and $b = 1.5$. Use a limit of 3 in place of ∞.

 a. The mean **b.** The standard deviation

Applications

LIFE SCIENCES

27. *Insect Life Span* The life span of a certain insect (in days) is uniformly distributed over the interval $[20, 36]$.

 a. What is the expected life of this insect?

 b. Find the probability that one of these insects, randomly selected, lives longer than 30 days.

28. *Location of a Bee Swarm* A swarm of bees is released from a certain point. The proportion of the swarm located at least 2 m from the point of release after 1 hr is a random variable that is exponentially distributed, with $a = 2$.

a. Find the expected proportion under the given conditions.

b. Find the probability that fewer than $1/3$ of the bees are located at least 2 m from the release point after 1 hr.

29. *Digestion Time* The digestion time (in hours) of a fixed amount of food is exponentially distributed, with $a = 1$.

a. Find the mean digestion time.

b. Find the probability that the digestion time is less than 30 min.

30. *Pygmy Heights* The average height of a member of a certain tribe of pygmies is 3.2 ft, with a standard deviation of 0.2 ft. If the heights are normally distributed, what are the largest and smallest heights of the middle 50% of this population?

31. *Finding Prey* H. R. Pulliam found that the time (in minutes) required by a predator to find prey is a random variable that is exponentially distributed, with $\mu = 25$.*

a. According to this distribution, what is the longest time within which the predator will be 90% certain of finding prey?

b. What is the probability that the predator will have to spend more than one hour looking for prey?

32. *Life Expectancy* According to the National Center for Health Statistics, the life expectancy for a 55-year-old African American female is 24.8 years.† Assuming that from age 55, the survival of African American females

follows an exponential distribution, determine each of the following probabilities.

a. The probability that a randomly selected 55-year-old African American woman will live beyond 80 years of age (at least 25 more years)

b. The probability that a randomly selected 55-year-old African American woman will live less than 20 more years

33. *Life Expectancy* According to the National Center for Health Statistics, life expectancy for a 70-year-old African American male is 11.5 years.† Assuming that from age 70, the survival of African American males follows an exponential distribution, determine each of the following probabilities.

a. The probability that a randomly selected 70-year-old African American man will live beyond 90 years of age

b. The probability that a randomly selected 70-year-old African American man will live between 10 and 20 more years

34. *Mercury Poisoning* Historians and biographers have collected evidence that suggests that President Andrew Jackson suffered from mercury poisoning. Recently, researchers measured the amount of mercury in samples of Jackson's hair from 1815. The results of this experiment showed that Jackson had a mean mercury level of 6.0 ppm.‡

a. If levels of mercury in hair samples from that time period followed a normal distribution with mean 6.9 and standard deviation 4.6,§ find the probability that a randomly selected person from that time period would have a mercury level of 6.0 ppm or higher. Discuss whether this provides evidence that Jackson suffered from mercury poisoning during this time period.

b. Today's accepted normal mercury levels follow a normal distribution with approximate mean 0.6 ppm and standard deviation 0.3 ppm.‖ By today's standards, how likely is it that a randomly selected person from today would have a mercury level of 6.0 ppm or higher? Discuss whether we can conclude that Andrew Jackson suffered from mercury poisoning.

c. Compare these answers with the results of Example 4 of this section.

*Pulliam, H. R., "On the Theory of Optimal Diets," *American Naturalist,* Vol. 108, 1974, pp. 59–74.
†*National Vital Statistics Reports,* Vol. 47, No. 19, June 30, 1999, Table 5.
‡Deppisch, Lidwig, Jose Centeno, David Gemmel, and Norca Torres, "Andrew Jackson's Exposure to Mercury and Lead," *JAMA,* Vol. 282, No. 6, August 11, 1999, pp. 569–571.
§Suzuki, T., T. Hongo, M. Morita, and R. Yamamoto, "Elemental Contamination of Japanese Women's Hair from Historical Samples," *Sci. Total Environ.,* Vol. 39, 1984, pp. 81–91.
‖Iyengar, V., and J. Woittiez, "Trace Elements in Human Clinical Specimens," *Clinical Chemistry,* Vol. 34, 1988, pp. 474–481.

35. *Rainfall* The rainfall (in inches) in a certain region is uniformly distributed over the interval $[32, 44]$.

   a. What is the expected number of inches of rainfall?

   b. What is the probability that the rainfall will be between 38 and 40 inches?

36. *Dry Length Days* Researchers have shown that the number of successive dry days that occur after a rainstorm for particular regions of Catalonia, Spain, is a random variable that is distributed exponentially with a mean of 8 days.*

   a. Find the probability that 10 or more successive dry days occur after a rainstorm.

   b. Find the probability that fewer than two dry days occur after a rainstorm.

37. *Earthquakes* The proportion of the times (in days) between major earthquakes in the north-south seismic belt of China is a random variable that is exponentially distributed, with $a = 1/609.5$.[†]

   a. Find the expected number of days and the standard deviation between major earthquakes for this region.

   b. Find the probability that the time between a major earthquake and the next one is more than one year.

OTHER APPLICATIONS

38. *Dating a Language* Over time, the number of original basic words in a language tends to decrease as words become obsolete or are replaced with new words. In 1950, C. Feng and M. Swadesh established that of the original 210 basic ancient Chinese words from 950 A.D., 167 were still being used.[‡] The proportion of words that remain after t millennia is a random variable that is exponentially distributed with $a = 0.229$.

   a. Find the life expectancy and standard deviation of a Chinese word.

   b. Calculate the probability that a randomly chosen Chinese word will remain after 2,000 years.

39. *Insurance Sales* The amount of insurance (in thousands of dollars) sold in a day by a particular agent is uniformly distributed over the interval $[10, 85]$.

   a. What amount of insurance does the agent sell on an average day?

   b. Find the probability that the agent sells more than $50,000 of insurance on a particular day.

40. *Fast-Food Outlets* The number of new fast-food outlets opening during June in a certain city is exponentially distributed, with a mean of 5.

   a. Give the probability density function for this distribution.

   b. What is the probability that the number of outlets opening is between 2 and 6?

41. *Sales Expense* A salesperson's monthly expenses (in thousands of dollars) are exponentially distributed, with an average of 4.25 (thousand dollars).

   a. Give the probability density function for the expenses.

   b. Find the probability that the expenses are more than $10,000.

In Exercises 42–44, assume a normal distribution.

42. *Machine Accuracy* A machine that fills quart bottles with apple juice averages 32.8 oz per bottle, with a standard deviation of 1.1 oz. What are the probabilities that the amount of juice in a bottle is as follows?

   a. Less than 1 qt

   b. At least 1 oz more than a quart

43. *Machine Accuracy* A machine produces screws with a mean length of 2.5 cm and a standard deviation of 0.2 cm. Find the probabilities that a screw produced by this machine has lengths as follows.

   a. Greater than 2.7 cm

   b. Within 1.2 standard deviations of the mean

44. *Customer Expenditures* Customers at a certain pharmacy spend an average of $54.40, with a standard deviation of $13.50. What are the largest and smallest amounts spent by the middle 50% of these customers?

---

*Lana, X., and A. Burgueno, "Daily Dry-Wet Behaviour in Catalonia (NE Spain) from the Viewpoint of Markov Chains," *International Journal of Climatology*, Vol. 18, 1998, pp. 793–815.
[†]Wang, Jeen-Hwa, and Chiao-Hui Kuo, "On the Frequency Distribution of Interoccurrence Times of Earthquakes," *Journal of Seismology*, Vol. 2, 1998, pp. 351–358.
[‡]Lo Bello, Anthony, and Maurice Weir, "Glottochronology: An Application of Calculus to Linguistics," *The UMAP Journal*, Vol. 3., No. 1, Spring 1982, pp. 85–99.

■ CHAPTER SUMMARY

In this chapter, we gave a brief introduction to the use of calculus in the study of probability. In particular, the conceptual framework of a continuous random variable and its connection to a probability function was given. Integration techniques were used to determine probabilities, expected value, and variance of continuous random variables. Three probability density functions that have a wide range of applications—uniform, exponential, and normal—were studied in detail.

■ KEY TERMS

13.1 continuous random
  variable
  continuous probability
  distribution

probability density
  function
13.2 median

13.3 uniform distribution
  exponential
  distribution
  normal distribution

normal curve
standard normal
  distribution
z-score

■ CHAPTER 13 REVIEW EXERCISES

1. In a probability function, the y-values (or function values) represent _____ .

 2. Define a continuous random variable.

3. Give the two conditions that a probability density function for $[a, b]$ must satisfy.

Decide whether each function defined as follows is a probability density function for the given interval.

4. $f(x) = \dfrac{1}{27}(2x + 4);$ $[1, 4]$

5. $f(x) = \sqrt{x};$ $[4, 9]$

6. $f(x) = 0.1;$ $[0, 10]$

7. $f(x) = e^{-x};$ $[0, \infty)$

In Exercises 8 and 9, find a value of k that will make f(x) define a probability density function for the indicated interval.

8. $f(x) = k\sqrt{x};$ $[1, 4]$

9. $f(x) = kx^2;$ $[0, 3]$

10. The probability density function of a random variable x is defined by

$$f(x) = 1 - \frac{1}{\sqrt{x - 1}} \quad \text{for } x \text{ in } [2, 5].$$

Find the following probabilities.

  a. $P(x \geq 3)$ **b.** $P(x \leq 4)$ **c.** $P(3 \leq x \leq 4)$

11. The probability density function of a random variable x is defined by

$$f(x) = \frac{1}{10} \quad \text{for } x \text{ in } [10, 20].$$

Find the following probabilities.

  a. $P(x \leq 12)$ **b.** $P(x \geq 31/2)$ **c.** $P(10.8 \leq x \leq 16.2)$

12. Describe what the expected value or mean of a probability distribution represents geometrically.

13. The probability density functions shown in the graphs below have the same mean. Which has the smallest standard deviation?

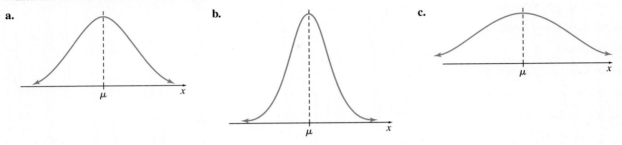

a.   b.   c.

For the probability density functions defined in Exercises 14–17, find the expected value, the variance, and the standard deviation.

14. $f(x) = \dfrac{1}{5}$; $[4, 9]$

15. $f(x) = \dfrac{2}{9}(x - 2)$; $[2, 5]$

16. $f(x) = \dfrac{1}{7}\left(1 + \dfrac{2}{\sqrt{x}}\right)$; $[1, 4]$

17. $f(x) = 5x^{-6}$; $[1, \infty)$

18. The probability density function of a random variable is defined by $f(x) = 4x - 3x^2$ for x in $[0, 1]$. Find each of the following for the distribution.

  a. The mean   **b.** The standard deviation

  c. The probability that the value of the random variable will be less than the mean

  d. The probability that the value of the random variable will be within one standard deviation of the mean

19. Find the median of the random variable of Exercise 18. (See Exercises 15–20 in the second section of this chapter.) Then find the probability that the value of the random variable will lie between the median and the mean of the distribution.

For Exercises 20 and 21, find **(a)** *the mean of the distribution,* **(b)** *the standard deviation of the distribution, and* **(c)** *the probability that the value of the random variable is within one standard deviation of the mean.*

20. $f(x) = \dfrac{5}{112}(1 - x^{-3/2})$ for x in $[1, 25]$

21. $f(x) = 0.01e^{-0.01x}$ for x in $[0, \infty)$

In Exercises 22–27, find the proportion of observations of a standard normal distribution for each of the following regions.

22. The region to the right of $z = 1.53$

23. The region to the left of $z = -0.49$

24. The region between $z = -1.47$ and $z = 1.03$

25. The region between $z = -0.98$ and $z = -0.15$

26. The region that is up to 2.5 standard deviations above the mean

27. The region that is up to 1.2 standard deviations below the mean

28. Find a z-score so that 21% of the area under the normal curve is to the left of z.

29. Find a z-score so that 52% of the area under the normal curve is to the right of z.

The topics in this short chapter involved much of the material studied earlier in this book, including functions, domain and range, exponential functions, area and integration, improper integrals, integration by parts, and numerical integration. For each of the following special probability density functions, give

a. the type of distribution;

b. the domain and range;

c. the graph;

d. the mean and standard deviation;

e. $P(\mu - \sigma \le x \le \mu + \sigma)$.

30. $f(x) = 0.05$ for x in $[10, 30]$

31. $f(x) = e^{-x}$ for x in $[0, \infty)$

32. $f(x) = \dfrac{e^{-x^2}}{\sqrt{\pi}}$ for x in $(-\infty, \infty)$ (*Hint:* $\sigma = 1/\sqrt{2}$.)

Applications

LIFE SCIENCES

33. *Movement of a Released Animal* The distance (in meters) that a certain animal moves away from a release point is exponentially distributed, with a mean of 100 m. Find the probability that the animal will move no farther than 100 m away.

34. *Weight Gain of Rats* The weight gain (in grams) of rats fed a certain vitamin supplement is a continuous random variable with probability density function defined by

$$f(x) = \frac{8}{7}x^{-2} \quad \text{for } x \text{ in } [1, 8].$$

a. Find the mean of the distribution.

b. Find the standard deviation of the distribution.

c. Find the probability that the value of the random variable is within one standard deviation of the mean.

35. *Body Temperature of a Bird* The body temperature (in degrees Celsius) of a particular species of bird is a continuous random variable with probability density function defined by

$$f(x) = \frac{6}{15,925}(x^2 + x) \quad \text{for } x \text{ in } [20, 25].$$

a. What is the expected body temperature of this species?

b. Find the probability of a body temperature below the mean.

36. *Snowfall* The snowfall (in inches) in a certain area is uniformly distributed over the interval $[2, 40]$.

a. What is the expected snowfall?

b. What is the probability of getting more than 20 inches of snow?

37. *Heart Muscle Tension* In a pilot study on tension of the heart muscle in dogs, the mean developed tension was 2.4 g, with a standard deviation of 0.4 g. Find the probability of a tension of less than 1.9 g. Assume a normal distribution.

38. *Average Birth Weight* The average birth weight of infants in the United States is 7.8 lb, with a standard deviation of 1.1 lb. Assuming a normal distribution, what is the probability that a newborn will weigh more than 9 lb?

39. *Suicides* The number of suicides in the U.S. caused by firearms for each age group in 1997 is given in the table below.*

Age Interval (years)	Midpoint of Interval (year)	Number Dying in Each Interval
5–15	10	127
15–25	20	2,587
25–35	30	3,010
35–45	40	3,321
45–55	50	2,647
55–65	60	1,859
65–75	70	1,906
75–85	80	1,608
85 +	90 (est)	494
Total		17,559

a. Plot the data. What type of function appears to best match this data?

---

*National Vital Statistics Reports, Vol. 47, No. 19, June 30, 1999, Table 16.

b. Use the regression feature on your graphing calculator to find a quartic equation that models the number of years, x, since birth and the number of suicides caused by firearms, $N(x)$. Use the midpoint value to estimate the point in each interval when the person died. Graph the function with the plot of the data. Does the function resemble the data?

c. Convert this function into a probability density function. (*Hint:* Integrate the function you found in part b from 9.7 to 93.2 yrs.)

d. For a randomly chosen person who committed suicide with a firearm, find the probabilities that the person was between 25 and 35 years old, between 45 and 65 years old, and at least 55 years old. Compare these with the actual probabilities.

e. Estimate the expected age of suicide.

f. Estimate the median of this distribution. (*Hint:* See Exercises 15–20 in the second section of this chapter.)

40. *Earthquakes* The time between major earthquakes in the Taiwan region is a random variable with probability density function defined by

$$f(t) = \frac{1}{3{,}650.1}\,e^{-t/3{,}650.1},$$

where t is measured in days.* Find the expected value and standard deviation of this probability density function.

41. *Extinction* The probability of a population going extinct by time t when the birth and death rates are the same can be estimated by

$$p(t) = \left(\frac{at}{at+1}\right)^{N},$$

where a is the birth and death rate, and N is the number of individuals in the population at time $t = 0$.[†] Sketch a graph of this function when $a = 50$ and $N = 1{,}000$.

OTHER APPLICATIONS

42. *State-Run Lotteries* The average state "take" on lotteries is 40%, with a standard deviation of 13%. Assuming a normal distribution, what is the probability that a state-run lottery will have a "take" of more than 50%?

43. *Mutual Funds* The price per share (in dollars) of a particular mutual fund is a random variable x with probability density function defined by

$$f(x) = \frac{3}{4}(x^2 - 16x + 65) \quad \text{for } x \text{ in } [8, 9].$$

Find the probability that the price will be less than $8.50.

44. *Machine Repairs* The time (in years) until a certain machine requires repairs is a random variable t with probability density function defined by

$$f(x) = \frac{5}{112}(1 - t^{-3/2}) \quad \text{for } t \text{ in } [1, 25].$$

Find the probability that no repairs are required in the first three years by finding the probability that a repair will be needed in years 4 through 25.

45. *Product Repairs* The number of repairs required by a new product each month is exponentially distributed, with an average of 8.

a. What is the probability density function for this distribution?

b. Find the expected number of repairs per month.

c. Find the standard deviation.

d. What is the probability that the number of repairs per month will be between 5 and 10?

46. *Retail Outlets* The number of new outlets for a clothing manufacturer is an exponential distribution with probability density function defined by

$$f(x) = \frac{1}{6}e^{-x/6} \quad \text{for } x \text{ in } [0, \infty).$$

---

*Wang, Jeen-Hwa, and Chiao-Hui Kuo, "On the Frequency Distribution of Interoccurrence Times of Earthquakes," *Journal of Seismology,* Vol. 2, 1998, pp. 351–358.
[†]Bailey, Norman T. J., *The Mathematical Approach to Biology and Medicine,* Wiley, 1967, p. 161.

Find each of the following for this distribution.

a. The mean

b. The standard deviation

c. The probability that the number of new outlets will be greater than the mean

47. *Useful Life of an Appliance Part* The useful life of a certain appliance part (in hundreds of hours) is 46.2, with a standard deviation of 15.8. Find the probability that one such part would last for at least 6,000 (60 hundred) hr. Assume a normal distribution.

EXTENDED APPLICATION: Exponential Waiting Times

We have seen in this chapter how probabilities that are spread out over continuous time intervals can be modeled by continuous probability density functions. The exponential distribution you met in the last section of this chapter is often used to model *waiting times,* the gaps between events that are randomly distributed in time, such as decays of a radioactive nucleus or arrivals of customers in the waiting line at a bank. In this application we investigate some properties of the exponential family of distributions.

Suppose that in a large hospital system, patients arrive at a waiting room and wait for a member of the transport department to arrive and take them in a wheelchair to their next destination. The times between arrivals of transports are exponentially distributed with a mean of 10 minutes. Sometimes transports arrive very close together, sometimes far apart, but if you keep track over many days, you'll find that the *average* time between transports is 10 minutes. According to the last section of this chapter, the exponential distribution with density function $f(t) = ae^{-at}$ has mean $1/a$, so the probability density function for our interarrival times is

$$f(t) = \frac{1}{10} e^{-t/10}.$$

First let's see what these waiting times look like. We have used a random-number generator from a statistical software package to draw 25 waiting times from this distribution. Figure 19 shows cumulative arrival times, which is what you would observe if you recorded the arrival time of each transport measured in minutes from an arbitrary 0 point.

You can see that 25 transports arrive in a span of about 260 minutes, so the average interarrival time was indeed close to 10 minutes. You may also notice that there are some large gaps and some cases where transports arrived very close together.

To get a better feeling for the distribution of long and short interarrival times, look at the following list, which gives the 25 interarrival times in minutes, sorted from smallest to largest.

0.016	4.398	15.659
0.226	4.573	15.954
0.457	5.415	16.403
0.989	9.570	18.978
1.576	10.413	20.736
1.988	10.916	33.013
2.738	13.109	39.073
3.133	13.317	
3.895	14.622	

You can see that there were some very short waits. (In fact, the shortest time between transports is only 1 second.) The longest time between transports was 39 minutes, almost four times as long as the average! Although the exponential model exaggerates the irregularities of typical service, the problem of pile-ups and long gaps is very real. This is particularly true for public transportation, especially for bus routes which are subject to unpredictable traffic delays. Anyone who works at a customer service job is also familiar with this behavior: The waiting line at a bank may be empty for minutes at a stretch, and then several customers walk in at nearly the same time. In this case, the customer interarrival times are exponentially distributed.

Planners who are involved with scheduling need to understand this "clumping" behavior. One way to explore it is to find probabilities for ranges of interarrival times. Here integrals are the natural tool. For example, if we want to estimate the fraction of interarrival times that will be less than 2 minutes, we compute

$$\frac{1}{10} \int_0^2 e^{-t/10}\, dt = 1 - e^{-1/5} \approx 0.181.$$

So on average, 18% of the interarrival times will be less than 2 minutes, which indicates that clustering of transports will be a problem in our hospital. We can also compute the probability of a gap of 30 minutes or longer. It will be

$$\frac{1}{10} \int_{30}^{\infty} e^{-t/10}\, dt = e^{-3} \approx 0.05.$$

So in a random sample of 25 interarrival times we might expect one or two long waits, and our simulation, which includes times of 33 and 39 minutes, is not a fluke. Of course, the patient's experience depends on when she arrives at the waiting room, which is another random input to our model. If she arrives in the middle of a cluster, she'll get a transport right away, but if she arrives at the beginning of a long gap she may have a half-hour wait. So we would also like to model the patient's *waiting time:* the time between the patient's arrival at the waiting room and the arrival of the next transport.

A remarkable fact about the exponential distribution is that if our patient arrives at the waiting room at a random time, the distribution of the patient's waiting times is *the same* as the distribution of interarrival times (that is, exponential with mean 10 minutes). At first this seems paradoxical; since she usually arrives between transports, she should wait less, on average,

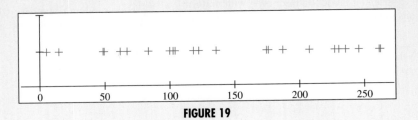

FIGURE 19

than the average time between transports. But remember that she's more likely to arrive at the waiting room in one of those long gaps. In our simulation, 72 out of 160 minutes is taken up with long gaps, and even if the patient arrives at the middle of such a gap she'll still wait longer than 15 minutes. Because of this feature the exponential distribution is often called *memoryless:* If you dip into the process at random, it is as if you were starting all over. If you arrive at the waiting room just as a transport leaves, your waiting time for the next one still has an exponential distribution with mean 10 minutes. The next transport doesn't "know" anything about the one that just left.*

Because the patients' waiting times are exponential, the calculations we have already made tell us what patients will experience: A wait of less than 2 minutes has probability 0.18. The average wait is 10 minutes, but long waits of more than 30 minutes are not all that rare (probability 0.05).

Customers waiting for service care about the average wait, but they may care even more about the *predictability* of the wait. In this chapter we stated that the standard deviation for an exponential distribution is the same as the mean, so in our model the standard deviation of patients' waiting times will be 10 minutes. This indicates that a wait twice the average length is not a rare event. (See Exercise 3.)

Let's compare the experience of patients on our exponential transport service with the experience of patients of a perfectly regulated transport service in which transports arrive *exactly* 10 minutes apart. We'll still assume that the patient arrives at random. But now the waiting time is uniformly distributed on the time interval [0 minutes, 10 minutes]. This uniform distribution has density function

$$f(t) = \begin{cases} \dfrac{1}{10} & \text{for } 0 \le t \le 10 \\ 0 & \text{otherwise} \end{cases}.$$

The mean waiting time is

$$\int_0^{10} \frac{1}{10} \cdot t \, dt = 5 \text{ minutes}$$

and the standard deviation of the waiting times is

$$\sqrt{\int_0^{10} \frac{(t-5)^2}{10} \, dt} = \sqrt{\frac{25}{3}} \approx 2.89 \text{ minutes.}$$

Clearly the patient has a better experience on this system. Even though the same average number of transports is coming per hour as in the exponential transport system, the average wait for the uniform transport system is only 5 minutes with a standard deviation of 2.89 minutes, and no one ever waits longer than 10 minutes!

Any transport system is subject to unpredictable variations, and this random input is always pushing the patients' waiting times toward the exponential model. Indeed, even with uniform scheduling of transports, there will be service bottlenecks be-

cause the exponential distribution is also a reasonable model (over a short time period) for interarrival times of *patients* entering the waiting room. The goal of schedulers is to move patients efficiently in spite of random transport delays and random input of patients.

The transport scheduling problem is part of a branch of statistics called *queueing theory,* the study of any process in which inputs arrive at a service point and wait in a line or queue to be served. Examples include telephone calls arriving at a customer service center, passengers entering the subway station, packets of information traveling through the Internet, and even pieces of code waiting for a processor in a multiprocessor computer. The following Web sites provide a small sampling of work in this very active research area.

- *http://www.research.att.com/~wow/E.html* (articles dealing with scheduling problems in telephone centers and networks)

- *http://faculty.washington.edu/jbs/itrans/ingsim.htm* (an article on scheduling a PRT)

- *http://byte.com/art/9506/sec8/art9.htm* (an article on queueing theory in computer network design)

Exercises

1. If x is a continuous random variable, $P(a \le x \le b)$ is the same as $P(a < x < b)$. Since these are different events, how can they have the same probability?

2. A patient who visits the hospital often for many tests makes about 8 trips a month. On the exponential transport system, how many times a month can this patient expect a wait longer than half an hour?

3. Find the probability that a patient using the exponential transport system waits more than 20 minutes for a transport; that is, find the probability of a wait more than twice as long as the average.

4. On the exponential transport system, what is the probability that a randomly arriving patient has a wait of between 9 and 10 minutes? What is the corresponding probability on the uniform transport system?

5. If our system is aiming for an average interarrival time of 10 minutes, we might set a tolerance of plus or minus 2 minutes and try to keep the interarrival times between 8 and 12 minutes. Under the exponential model, what fraction of interarrival times fall in this range? How about under the uniform model?

6. Most mathematical software includes routines for generating "pseudo-random" numbers (that is, numbers that behave randomly even though they are generated by arithmetic).

---

*See Chapter 1 in Volume 2 of Feller, William, *An Introduction to Probability Theory and Its Applications,* 2nd ed., New York, Wiley, 1971.

That's what we used to simulate the exponential waiting times for our transport system. But a source on the Internet (http://www.fourmilab.ch/hotbits/) delivers random numbers based on the times between decay events in a sample of Krypton-85. As noted above, the waiting times between decay events have an exponential distribution, so we can see what nature's random numbers look like. Here's a short sample:

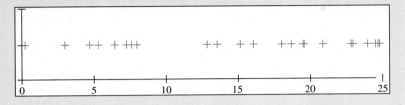

Actually, this source builds its random numbers from random bits, that is, 0s and 1s that occur with equal probability. See if you can think of a way of turning a sequence of exponential waiting times into a random sequence of 0s and 1s.

TABLES

Table 1 Formulas from Geometry

PYTHAGOREAN THEOREM
For a right triangle with legs of lengths a and b and hypotenuse of length c, $a^2 + b^2 = c^2$.

CIRCLE
Area: $A = \pi r^2$
Circumference: $C = 2\pi r$

RECTANGLE
Area: $A = lw$
Perimeter: $P = 2l + 2w$

TRIANGLE
Area: $A = \dfrac{1}{2}bh$

SPHERE
Volume: $V = \dfrac{4}{3}\pi r^3$
Surface area: $A = 4\pi r^2$

CONE
Volume: $V = \dfrac{1}{3}\pi r^2 h$

RECTANGULAR BOX
Volume: $V = lwh$
Surface area: $A = 2lh + 2wh + 2lw$

CIRCULAR CYLINDER
Volume: $V = \pi r^2 h$
Surface area: $A = 2\pi r^2 + 2\pi rh$

TRIANGULAR CYLINDER
Volume: $V = \dfrac{1}{2}bhl$

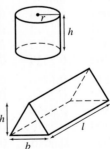

GENERAL INFORMATION ON SURFACE AREA
To find the surface area of a figure, break down the total surface area into the individual components and add up the areas of the components. For example, a rectangular box has six sides, each of which is a rectangle. A circular cylinder has two ends, each of which is a circle, plus the side, which forms a rectangle when opened up.

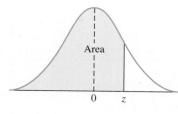

Table 2 Area Under a Normal Curve to the Left of z, Where $z = \dfrac{x - \mu}{\sigma}$

z	0.00	0.01	0.02	0.03	0.04	0.05	0.06	0.07	0.08	0.09
−3.4	0.0003	0.0003	0.0003	0.0003	0.0003	0.0003	0.0003	0.0003	0.0003	0.0002
−3.3	0.0005	0.0005	0.0005	0.0004	0.0004	0.0004	0.0004	0.0004	0.0004	0.0003
−3.2	0.0007	0.0007	0.0006	0.0006	0.0006	0.0006	0.0006	0.0005	0.0005	0.0005
−3.1	0.0010	0.0009	0.0009	0.0009	0.0008	0.0008	0.0008	0.0008	0.0007	0.0007
−3.0	0.0013	0.0013	0.0013	0.0012	0.0012	0.0011	0.0011	0.0011	0.0010	0.0010
−2.9	0.0019	0.0018	0.0017	0.0017	0.0016	0.0016	0.0015	0.0015	0.0014	0.0014
−2.8	0.0026	0.0025	0.0024	0.0023	0.0023	0.0022	0.0021	0.0021	0.0020	0.0019
−2.7	0.0035	0.0034	0.0033	0.0032	0.0031	0.0030	0.0029	0.0028	0.0027	0.0026
−2.6	0.0047	0.0045	0.0044	0.0043	0.0041	0.0040	0.0039	0.0038	0.0037	0.0036
−2.5	0.0062	0.0060	0.0059	0.0057	0.0055	0.0054	0.0052	0.0051	0.0049	0.0048
−2.4	0.0082	0.0080	0.0078	0.0075	0.0073	0.0071	0.0069	0.0068	0.0066	0.0064
−2.3	0.0107	0.0104	0.0102	0.0099	0.0096	0.0094	0.0091	0.0089	0.0087	0.0084
−2.2	0.0139	0.0136	0.0132	0.0129	0.0125	0.0122	0.0119	0.0116	0.0113	0.0110
−2.1	0.0179	0.0174	0.0170	0.0166	0.0162	0.0158	0.0154	0.0150	0.0146	0.0143
−2.0	0.0228	0.0222	0.0217	0.0212	0.0207	0.0202	0.0197	0.0192	0.0188	0.0183
−1.9	0.0287	0.0281	0.0274	0.0268	0.0262	0.0256	0.0250	0.0244	0.0239	0.0233
−1.8	0.0359	0.0352	0.0344	0.0336	0.0329	0.0322	0.0314	0.0307	0.0301	0.0294
−1.7	0.0446	0.0436	0.0427	0.0418	0.0409	0.0401	0.0392	0.0384	0.0375	0.0367
−1.6	0.0548	0.0537	0.0526	0.0516	0.0505	0.0495	0.0485	0.0475	0.0465	0.0455
−1.5	0.0668	0.0655	0.0643	0.0630	0.0618	0.0606	0.0594	0.0582	0.0571	0.0559
−1.4	0.0808	0.0793	0.0778	0.0764	0.0749	0.0735	0.0722	0.0708	0.0694	0.0681
−1.3	0.0968	0.0951	0.0934	0.0918	0.0901	0.0885	0.0869	0.0853	0.0838	0.0823
−1.2	0.1151	0.1131	0.1112	0.1093	0.1075	0.1056	0.1038	0.1020	0.1003	0.0985
−1.1	0.1357	0.1335	0.1314	0.1292	0.1271	0.1251	0.1230	0.1210	0.1190	0.1170
−1.0	0.1587	0.1562	0.1539	0.1515	0.1492	0.1469	0.1446	0.1423	0.1401	0.1379
−0.9	0.1841	0.1814	0.1788	0.1762	0.1736	0.1711	0.1685	0.1660	0.1635	0.1611
−0.8	0.2119	0.2090	0.2061	0.2033	0.2005	0.1977	0.1949	0.1922	0.1894	0.1867
−0.7	0.2420	0.2389	0.2358	0.2327	0.2296	0.2266	0.2236	0.2206	0.2177	0.2148
−0.6	0.2743	0.2709	0.2676	0.2643	0.2611	0.2578	0.2546	0.2514	0.2483	0.2451
−0.5	0.3085	0.3050	0.3015	0.2981	0.2946	0.2912	0.2877	0.2843	0.2810	0.2776

Table 2 Area Under a Normal Curve (continued)

z	0.00	0.01	0.02	0.03	0.04	0.05	0.06	0.07	0.08	0.09
−0.4	0.3446	0.3409	0.3372	0.3336	0.3300	0.3264	0.3228	0.3192	0.3156	0.3121
−0.3	0.3821	0.3783	0.3745	0.3707	0.3669	0.3632	0.3594	0.3557	0.3520	0.3483
−0.2	0.4207	0.4168	0.4129	0.4090	0.4052	0.4013	0.3974	0.3936	0.3897	0.3859
−0.1	0.4602	0.4562	0.4522	0.4483	0.4443	0.4404	0.4364	0.4325	0.4286	0.4247
−0.0	0.5000	0.4960	0.4920	0.4880	0.4840	0.4801	0.4761	0.4721	0.4681	0.4641
0.0	0.5000	0.5040	0.5080	0.5120	0.5160	0.5199	0.5239	0.5279	0.5319	0.5359
0.1	0.5398	0.5438	0.5478	0.5517	0.5557	0.5596	0.5636	0.5675	0.5714	0.5753
0.2	0.5793	0.5832	0.5871	0.5910	0.5948	0.5987	0.6026	0.6064	0.6103	0.6141
0.3	0.6179	0.6217	0.6255	0.6293	0.6331	0.6368	0.6406	0.6443	0.6480	0.6517
0.4	0.6554	0.6591	0.6628	0.6664	0.6700	0.6736	0.6772	0.6808	0.6844	0.6879
0.5	0.6915	0.6950	0.6985	0.7019	0.7054	0.7088	0.7123	0.7157	0.7190	0.7224
0.6	0.7257	0.7291	0.7324	0.7357	0.7389	0.7422	0.7454	0.7486	0.7517	0.7549
0.7	0.7580	0.7611	0.7642	0.7673	0.7704	0.7734	0.7764	0.7794	0.7823	0.7852
0.8	0.7881	0.7910	0.7939	0.7967	0.7995	0.8023	0.8051	0.8078	0.8106	0.8133
0.9	0.8159	0.8186	0.8212	0.8238	0.8264	0.8289	0.8315	0.8340	0.8365	0.8389
1.0	0.8413	0.8438	0.8461	0.8485	0.8508	0.8531	0.8554	0.8577	0.8599	0.8621
1.1	0.8643	0.8665	0.8686	0.8708	0.8729	0.8749	0.8770	0.8790	0.8810	0.8830
1.2	0.8849	0.8869	0.8888	0.8907	0.8925	0.8944	0.8962	0.8980	0.8997	0.9015
1.3	0.9032	0.9049	0.9066	0.9082	0.9099	0.9115	0.9131	0.9147	0.9162	0.9177
1.4	0.9192	0.9207	0.9222	0.9236	0.9251	0.9265	0.9278	0.9292	0.9306	0.9319
1.5	0.9332	0.9345	0.9357	0.9370	0.9382	0.9394	0.9406	0.9418	0.9429	0.9441
1.6	0.9452	0.9463	0.9474	0.9484	0.9495	0.9505	0.9515	0.9525	0.9535	0.9545
1.7	0.9554	0.9564	0.9573	0.9582	0.9591	0.9599	0.9608	0.9616	0.9625	0.9633
1.8	0.9641	0.9649	0.9656	0.9664	0.9671	0.9678	0.9686	0.9693	0.9699	0.9706
1.9	0.9713	0.9719	0.9726	0.9732	0.9738	0.9744	0.9750	0.9756	0.9761	0.9767
2.0	0.9772	0.9778	0.9783	0.9788	0.9793	0.9798	0.9803	0.9808	0.9812	0.9817
2.1	0.9821	0.9826	0.9830	0.9834	0.9838	0.9842	0.9846	0.9850	0.9854	0.9857
2.2	0.9861	0.9864	0.9868	0.9871	0.9875	0.9878	0.9881	0.9884	0.9887	0.9890
2.3	0.9893	0.9896	0.9898	0.9901	0.9904	0.9906	0.9909	0.9911	0.9913	0.9916
2.4	0.9918	0.9920	0.9922	0.9925	0.9927	0.9929	0.9931	0.9932	0.9934	0.9936
2.5	0.9938	0.9940	0.9941	0.9943	0.9945	0.9946	0.9948	0.9949	0.9951	0.9952
2.6	0.9953	0.9955	0.9956	0.9957	0.9959	0.9960	0.9961	0.9962	0.9963	0.9964
2.7	0.9965	0.9966	0.9967	0.9968	0.9969	0.9970	0.9971	0.9972	0.9973	0.9974
2.8	0.9974	0.9975	0.9976	0.9977	0.9977	0.9978	0.9979	0.9979	0.9980	0.9981
2.9	0.9981	0.9982	0.9982	0.9983	0.9984	0.9984	0.9985	0.9985	0.9986	0.9986
3.0	0.9987	0.9987	0.9987	0.9988	0.9988	0.9989	0.9989	0.9989	0.9990	0.9990
3.1	0.9990	0.9991	0.9991	0.9991	0.9992	0.9992	0.9992	0.9992	0.9993	0.9993
3.2	0.9993	0.9993	0.9994	0.9994	0.9994	0.9994	0.9994	0.9995	0.9995	0.9995
3.3	0.9995	0.9995	0.9995	0.9996	0.9996	0.9996	0.9996	0.9996	0.9996	0.9997
3.4	0.9997	0.9997	0.9997	0.9997	0.9997	0.9997	0.9997	0.9997	0.9997	0.9998

Table 3 Integrals

(C is an arbitrary constant.)

1. $\displaystyle\int x^n\,dx = \frac{x^{n+1}}{n+1} + C \quad (\text{if } n \neq -1)$

2. $\displaystyle\int e^{kx}\,dx = \frac{e^{kx}}{k} + C$

3. $\displaystyle\int \frac{a}{x}\,dx = a\ln|x| + C$

4. $\displaystyle\int \ln|ax|\,dx = x(\ln|ax| - 1) + C$

5. $\displaystyle\int \frac{1}{\sqrt{x^2 + a^2}}\,dx = \ln\left|x + \sqrt{x^2 + a^2}\right| + C$

6. $\displaystyle\int \frac{1}{\sqrt{x^2 - a^2}}\,dx = \ln\left|x + \sqrt{x^2 - a^2}\right| + C$

7. $\displaystyle\int \frac{1}{a^2 - x^2}\,dx = \frac{1}{2a}\cdot\ln\left|\frac{a + x}{a - x}\right| + C \quad (a \neq 0)$

8. $\displaystyle\int \frac{1}{x^2 - a^2}\,dx = \frac{1}{2a}\cdot\ln\left|\frac{x - a}{x + a}\right| + C \quad (a \neq 0)$

9. $\displaystyle\int \frac{1}{x\sqrt{a^2 - x^2}}\,dx = -\frac{1}{a}\cdot\ln\left|\frac{a + \sqrt{a^2 - x^2}}{x}\right| + C \quad (a \neq 0)$

10. $\displaystyle\int \frac{1}{x\sqrt{a^2 + x^2}}\,dx = -\frac{1}{a}\cdot\ln\left|\frac{a + \sqrt{a^2 + x^2}}{x}\right| + C \quad (a \neq 0)$

11. $\displaystyle\int \frac{x}{ax + b}\,dx = \frac{x}{a} - \frac{b}{a^2}\cdot\ln|ax + b| + C \quad (a \neq 0)$

12. $\displaystyle\int \frac{x}{(ax + b)^2}\,dx = \frac{b}{a^2(ax + b)} + \frac{1}{a^2}\cdot\ln|ax + b| + C \quad (a \neq 0)$

13. $\displaystyle\int \frac{1}{x(ax + b)}\,dx = \frac{1}{b}\cdot\ln\left|\frac{x}{ax + b}\right| + C \quad (b \neq 0)$

14. $\displaystyle\int \frac{1}{x(ax + b)^2}\,dx = \frac{1}{b(ax + b)} + \frac{1}{b^2}\cdot\ln\left|\frac{x}{ax + b}\right| + C \quad (b \neq 0)$

15. $\displaystyle\int \sqrt{x^2 + a^2}\,dx = \frac{x}{2}\sqrt{x^2 + a^2} + \frac{a^2}{2}\cdot\ln\left|x + \sqrt{x^2 + a^2}\right| + C$

16. $\displaystyle\int x^n\cdot\ln|x|\,dx = x^{n+1}\left[\frac{\ln|x|}{n+1} - \frac{1}{(n+1)^2}\right] + C \quad (n \neq -1)$

17. $\displaystyle\int x^n e^{ax}\,dx = \frac{x^n e^{ax}}{a} - \frac{n}{a}\cdot\int x^{n-1}e^{ax}\,dx + C \quad (a \neq 0)$

Table 4 Integrals Involving Trigonometric Functions

18. $\displaystyle\int \sin u\,du = -\cos u + C$

19. $\displaystyle\int \cos u\,du = \sin u + C$

20. $\displaystyle\int \sec^2 u\,du = \tan u + C$

21. $\displaystyle\int \csc^2 u\,du = -\cot u + C$

22. $\displaystyle\int \sec u \tan u\,du = \sec u + C$

23. $\displaystyle\int \csc u \cot u\,du = -\csc u + C$

24. $\displaystyle\int \tan u\,du = \ln |\sec u| + C$

25. $\displaystyle\int \cot u\,du = \ln |\sin u| + C$

26. $\displaystyle\int \sec u\,du = \ln |\sec u + \tan u| + C$

27. $\displaystyle\int \csc u\,du = \ln |\csc u - \cot u| + C$

28. $\displaystyle\int \sin^n u\,du = -\frac{1}{n} \sin^{n-1} u \cos u + \frac{n-1}{n}\int \sin^{n-2} u\,du \quad (n \neq 0)$

29. $\displaystyle\int \cos^n u\,du = \frac{1}{n} \cos^{n-1} u \sin u + \frac{n-1}{n}\int \cos^{n-2} u\,du \quad (n \neq 0)$

30. $\displaystyle\int \tan^n u\,du = \frac{1}{n-1} \tan^{n-1} u - \int \tan^{n-2} u\,du \quad (n \neq 1)$

31. $\displaystyle\int \sec^n u\,du = \frac{1}{n-1} \tan u \sec^{n-2} u + \frac{n-2}{n-1}\int \sec^{n-2} u\,du \quad (n \neq 1)$

32. $\displaystyle\int \sin au \sin bu\,du = \frac{\sin(a-b)u}{2(a-b)} - \frac{\sin(a+b)u}{2(a+b)} + C, \quad |a| \neq |b|$

33. $\displaystyle\int \cos au \cos bu\,du = \frac{\sin(a-b)u}{2(a-b)} + \frac{\sin(a+b)u}{2(a+b)} + C, \quad |a| \neq |b|$

34. $\displaystyle\int \sin au \cos bu\,du = -\frac{\cos(a-b)u}{2(a-b)} - \frac{\cos(a+b)u}{2(a+b)} + C, \quad |a| \neq |b|$

35. $\displaystyle\int u \sin u\,du = \sin u - u \cos u + C$

36. $\displaystyle\int u^n \sin u\,du = -u^n \cos u + n\int u^{n-1} \cos u\,du$

37. $\displaystyle\int e^{au} \sin bu\,du = \frac{e^{au}}{a^2 + b^2}(a \sin bu - b \cos bu) + C$

38. $\displaystyle\int e^{au} \cos bu\,du = \frac{e^{au}}{a^2 + b^2}(a \cos bu + b \sin bu) + C$

ANSWERS TO SELECTED EXERCISES

Answers to selected writing exercises are provided.

Chapter R Algebra Reference

Exercises R.1 (page xviii)

1. $-x^2 + x + 9$ **2.** $-6y^2 + 3y + 10$ **3.** $-14q^2 + 11q - 14$ **4.** $9r^2 - 4r + 19$ **5.** $-0.327x^2 - 2.805x - 1.458$
6. $-2.97r^2 - 8.083r + 7.81$ **7.** $-18m^3 - 27m^2 + 9m$ **8.** $12k^2 - 20k + 3$ **9.** $25r^2 + 5rs - 12s^2$ **10.** $18k^2 - 7kq - q^2$
11. $(6/25)y^2 + (11/40)yz + (1/16)z^2$ **12.** $(15/16)r^2 - (7/12)rs - (2/9)s^2$ **13.** $144x^2 - 1$ **14.** $36m^2 - 25$
15. $27p^3 - 1$ **16.** $6p^3 - 11p^2 + 14p - 5$ **17.** $8m^3 + 1$ **18.** $12k^4 + 21k^3 - 5k^2 + 3k + 2$
19. $m^2 + mn - 2n^2 - 2km + 5kn - 3k^2$ **20.** $2r^2 - 7rs + 3s^2 + 3rt - 4st + t^2$ **21.** $x^3 + 6x^2 + 11x + 6$
22. $x^3 - 2x^2 - 5x + 6$ **23.** $9a^2 + 6ab + b^2$ **24.** $x^3 - 6x^2y + 12xy^2 - 8y^3$

Exercises R.2 (page xxi)

1. $8a(a^2 - 2a + 3)$ **2.** $3y(y^2 + 8y + 3)$ **3.** $5p^2(5p^2 - 4pq + 20q^2)$ **4.** $10m^2(6m^2 - 12mn + 5n^2)$ **5.** $(m + 7)(m + 2)$
6. $(x + 5)(x - 1)$ **7.** $(z + 4)(z + 5)$ **8.** $(b - 7)(b - 1)$ **9.** $(a - 5b)(a - b)$ **10.** $(s - 5t)(s + 7t)$
11. $(y - 7z)(y + 3z)$ **12.** $6(a - 10)(a + 2)$ **13.** $3m(m + 3)(m + 1)$ **14.** $(2x + 1)(x - 3)$ **15.** $(3a + 7)(a + 1)$
16. $(2a - 5)(a - 6)$ **17.** $(5y + 2)(3y - 1)$ **18.** $(7m + 2n)(3m + n)$ **19.** $2a^2(4a - b)(3a + 2b)$
20. $4z^3(8z + 3a)(z - a)$ **21.** $(x + 8)(x - 8)$ **22.** $(3m + 5)(3m - 5)$ **23.** $(11a + 10)(11a - 10)$ **24.** Prime
25. $(z + 7y)^2$ **26.** $(m - 3n)^2$ **27.** $(3p - 4)^2$ **28.** $(a - 6)(a^2 + 6a + 36)$ **29.** $(2r - 3s)(4r^2 + 6rs + 9s^2)$
30. $(4m + 5)(16m^2 - 20m + 25)$ **31.** $(x - y)(x + y)(x^2 + y^2)$ **32.** $(2a - 3b)(2a + 3b)(4a^2 + 9b^2)$

Exercises R.3 (page xxiv)

1. $z/2$ **2.** $5p/2$ **3.** $8/9$ **4.** $3/(t - 3)$ **5.** $2(x + 2)/x$ **6.** $4(y + 2)$ **7.** $(m - 2)/(m + 3)$ **8.** $(r + 2)/(r + 4)$
9. $(x + 4)/(x + 1)$ **10.** $(z - 3)/(z + 2)$ **11.** $(2m + 3)/(4m + 3)$ **12.** $(2y + 1)/(y + 1)$ **13.** $3k/5$ **14.** $25p^2/9$
15. $6/(5p)$ **16.** 2 **17.** $2/9$ **18.** $3/10$ **19.** $2(a + 4)/(a - 3)$ **20.** $2/(r + 2)$ **21.** $(k + 2)/(k + 3)$
22. $(m + 6)/(m + 3)$ **23.** $(m - 3)/(2m - 3)$ **24.** $(2n - 3)/(2n + 3)$ **25.** 1 **26.** $(6 + p)/(2p)$ **27.** $(8 - y)/(4y)$
28. $137/(30m)$ **29.** $(3m - 2)/[m(m - 1)]$ **30.** $(r - 12)/[r(r - 2)]$ **31.** $14/[3(a - 1)]$ **32.** $23/[20(k - 2)]$
33. $(7x + 9)/[(x - 3)(x + 1)(x + 2)]$ **34.** $y^2/[(y + 4)(y + 3)(y + 2)]$ **35.** $k(k - 13)/[(2k - 1)(k + 2)(k - 3)]$
36. $m(3m - 19)/[(3m - 2)(m + 3)(m - 4)]$ **37.** $(4a + 1)/[a(a + 2)]$ **38.** $(5x^2 + 4x - 4)/[x(x - 1)(x + 1)]$

Exercises R.4 (page xxx)

1. 12 **2.** $-2/7$ **3.** -12 **4.** $3/4$ **5.** $-7/8$ **6.** $-6/11$ **7.** -1 **8.** $-10/19$ **9.** $-3, -2$ **10.** $-1, 3$ **11.** 4
12. $-2, 5/2$ **13.** $-1/2, 4/3$ **14.** $2, 5$ **15.** $-4/3, 4/3$ **16.** $-4, 1/2$ **17.** $0, 4$
18. $(5 + \sqrt{13})/6 \approx 1.434, (5 - \sqrt{13})/6 \approx 0.232$ **19.** $(1 + \sqrt{33})/4 \approx 1.686, (1 - \sqrt{33})/4 \approx -1.186$
20. $(-1 + \sqrt{5})/2 \approx 0.618, (-1 - \sqrt{5})/2 \approx -1.618$ **21.** $5 + \sqrt{5} \approx 7.236, 5 - \sqrt{5} \approx 2.764$
22. $(-6 + \sqrt{26})/2 \approx -0.450, (-6 - \sqrt{26})/2 \approx -5.550$ **23.** $1, 5/2$ **24.** No real number solutions
25. $(-1 + \sqrt{73})/6 \approx 1.257, (-1 - \sqrt{73})/6 \approx -1.591$ **26.** $-1, 0$ **27.** 3 **28.** 12 **29.** $-59/6$ **30.** $-11/5$
31. No real number solutions **32.** $-5/2$ **33.** $2/3$ **34.** 1 **35.** $(-13 - \sqrt{185})/4 \approx -6.650, (-13 + \sqrt{185})/4 \approx 0.150$
36. No solution **37.** $-15/4$

Exercises R.5 (page xxxv)

1. $(-\infty, 0)$ **2.** $[-3, \infty)$ **3.** $[1, 2)$
4. $(-5, -4]$ **5.** $(-\infty, -9)$ **6.** $[6, \infty)$

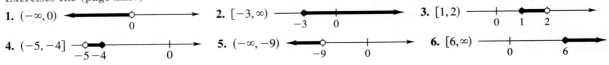

7. $-4 < x < 3$ **8.** $2 \le x < 7$ **9.** $x \le -1$ **10.** $x > 3$ **11.** $-2 \le x < 6$ **12.** $0 < x < 8$ **13.** $x \le -4$ or $x \ge 4$
14. $x < 0$ or $x \ge 3$ **15.** $(-\infty, -1]$ **16.** $(-\infty, 1)$

17. $(-1, \infty)$ **18.** $(-\infty, 1]$ **19.** $(1/5, \infty)$

20. $(1/3, \infty)$ **21.** $(-5, 6)$ **22.** $[7/3, 4]$

23. $[-11/2, 7/2]$ **24.** $[-1, 2]$

25. $[-17/7, \infty)$ **26.** $(-\infty, 50/9]$

27. $(-2, 4)$ **28.** $(-\infty, -6] \cup [1, \infty)$

29. $(1, 2)$ **30.** $(-\infty, -4) \cup (1/2, \infty)$

31. $[1, 6]$ **32.** $[-3/2, 5]$

33. $(-\infty, -1/2) \cup (1/3, \infty)$ **34.** $[-1/2, 2/5]$

35. $[-3, 1/2]$ **36.** $(-\infty, -2) \cup (5/3, \infty)$

37. $[-5, 5]$ **38.** $(-\infty, 0) \cup (16, \infty)$ **39.** $(-5, 3]$

40. $(-\infty, -1) \cup (1, \infty)$ **41.** $(-\infty, -2)$ **42.** $(-2, 3/2)$ **43.** $[-8, 5)$ **44.** $(-\infty, -3/2) \cup [-13/9, \infty)$ **45.** $(-2, \infty)$
46. $(-\infty, -1)$ **47.** $(-\infty, -1) \cup (-1/2, 1) \cup (2, \infty)$ **48.** $(-4, -2) \cup (0, 2)$ **49.** $(1, 3/2]$ **50.** $(-\infty, -2) \cup (-2, 2) \cup [4, \infty)$

Exercises R.6 (page xl)

1. $1/64$ **2.** $1/81$ **3.** 1 **4.** 1 **5.** $-1/9$ **6.** $1/9$ **7.** $49/4$ **8.** $27/64$ **9.** $1/3^6$ **10.** 8^5 **11.** $1/10^8$ **12.** 5
13. x^2 **14.** y^3 **15.** $2^3 k^3$ **16.** $1/(3z^7)$ **17.** $x^2/(2y)$ **18.** $m^3/5^4$ **19.** $a^3 b^6$ **20.** $d^6/(2^2 c^4)$ **21.** x^4/y^4 **22.** b/a^3
23. $(a + b)/(ab)$ **24.** $(1 - ab^2)/b^2$ **25.** $2(m - n)/[mn(m + n^2)]$ **26.** $(3n^2 + 4m)/(mn^2)$ **27.** $xy/(y - x)$
28. $x^4 y^4/(x^2 + y^2)^2$ **29.** 9 **30.** 3 **31.** 4 **32.** -25 **33.** $2/3$ **34.** $4/3$ **35.** $1/32$ **36.** $1/5$ **37.** $4/3$
38. $1{,}000/1{,}331$ **39.** 2^2 **40.** $27^{1/3}$ **41.** 4^2 **42.** 1 **43.** r **44.** $12^3/y^8$ **45.** $1/(2^2 \cdot 3k^{5/2})$ or $1/(12k^{5/2})$ **46.** $1/(2p^2)$
47. $a^{2/3} b^2$ **48.** $y/(x^{4/3} z^{1/2})$ **49.** $h^{1/3} t^{1/5}/k^{2/5}$ **50.** $m^3 p/n$ **51.** $4x(x^2 + 2)(-x^3 + 6x - 1)$ **52.** $6x(x^3 + 7)(-2x^3 - 5x + 7)$
53. $3x^3(x^2 - 1)^{-1/2}$ **54.** $3(6x + 2)^{-1/2}(27x + 5)$ **55.** $(2x + 5)(x^2 - 4)^{-1/2}(4x^2 + 5x - 8)$
56. $(4x^2 + 1)(2x - 1)^{-1/2}(36x^2 - 16x + 1)$

Exercises R.7 (page xliv)

1. 5 **2.** 6 **3.** -5 **4.** $5\sqrt{2}$ **5.** $20\sqrt{5}$ **6.** $4y^2\sqrt{2y}$ **7.** $7\sqrt{2}$ **8.** $9\sqrt{3}$ **9.** $2\sqrt{5}$ **10.** $-2\sqrt{7}$ **11.** $5\sqrt[3]{2}$
12. $7\sqrt[3]{3}$ **13.** $3\sqrt[3]{4}$ **14.** $xyz^2\sqrt{2x}$ **15.** $7rs^2t^5\sqrt{2r}$ **16.** $2x^2yz\sqrt[3]{2x^2yz^2}$ **17.** $x^2yz^2\sqrt[4]{y^3z^3}$ **18.** $ab\sqrt{ab}(b - 2a^2 + b^3)$
19. $p^2\sqrt{pq}(pq - q^4 + p^2)$ **20.** $5\sqrt{7}/7$ **21.** $-2\sqrt{3}/3$ **22.** $-\sqrt{3}/2$ **23.** $\sqrt{2}$ **24.** $-3(1 + \sqrt{5})/4$
25. $-5(2 + \sqrt{6})/2$ **26.** $-2(\sqrt{3} + \sqrt{2})$ **27.** $(\sqrt{10} - \sqrt{3})/7$ **28.** $(\sqrt{r} + \sqrt{3})/(r - 3)$ **29.** $5(\sqrt{m} + \sqrt{5})/(m - 5)$
30. $\sqrt{y} + \sqrt{5}$ **31.** $\sqrt{z} + \sqrt{11}$ **32.** $-2x - 2\sqrt{x(x + 1)} - 1$ **33.** $[p^2 + p + 2\sqrt{p(p^2 - 1)} - 1]/(-p^2 + p + 1)$
34. $-1/[2(1 - \sqrt{2})]$ **35.** $-2/[3(1 + \sqrt{3})]$ **36.** $-1/[2x - 2\sqrt{x(x + 1)} + 1]$
37. $(-p^2 + p + 1)/[p^2 + p - 2\sqrt{p(p^2 - 1)} - 1]$ **38.** $|4 - x|$ **39.** $|2y + 1|$ **40.** Cannot be simplified
41. Cannot be simplified

Chapter 1 Functions

Exercises 1.1 (page 15)

1. $3/5$ **3.** Not defined   **5.** 2   **7.** $5/9$ **9.** Not defined   **11.** 0   **13.** 2   **15.** $y = -2x + 5$ **17.** $y = 1$
19. $y = -(1/3)x + 10/3$ **21.** $y = -(3/5)x + 59/30$ **23.** $x = -8$ **25.** $y = (2/3)x - 2$ **27.** $x = -6$
29. $y = -(1/3)x + 11/3$ **31.** $y = x - 7$ **33.** $y = (2/3)x + 2$ **35.** No   **39.** a   **41.** -4

45.

$y = 2x + 3$

47.

$y = -6x + 12$

49.

$3x - y = -9$

51.

$4y + 5x = 10$

53.

$x = 4$

55.

$y - 4 = 0$

57.

$y = -5x$

59.

$x - 3y = 0$

61. False   **63.** True

65. a. $u = 187 - 0.85x$, $l = 154 - 0.7x$ **b.** 140 to 170 beats per minute   **c.** 126 to 153 beats per minute   **d.** 16 and 52; 143 beats per minute   **67.** About 81 years   **69. a.** $y = 35 + 2.8x$ **b.** $y = 100 - 2.2x$ **c.** 2003   **71. a.** $y = 2.1x - 2.3$
b. -0.2, or roughly 0   **c.** 20 years old   **73.** 88   **75. a.** $T = 0.03t + 15$ **b.** About 2103   **77. a.** $y = 5{,}121x + 86{,}821$
b. About 271,177   **79. a.** $p = 0.31t + 8.4$ **b.** About 21%   **c.** 2043   **81.** $-40°$

Exercises 1.2 (page 27)

3. c.   **5. a.**   **b.** $Y = 1.585x - 0.487$

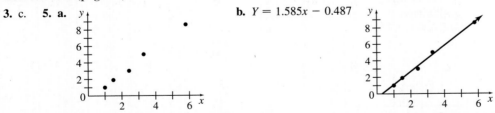

c. No; it gives negative values for small widths.   **d.** 0.999   **7. a.** 0.994; very well   **b.** $Y = 1.3525x - 2.51$
c.   Yes   **d.** 1.3525 cm   **e.** 58.35 cm   **9. a.** $Y = 0.212x - 0.309$ **b.** 15.2   **c.** $86.4°$

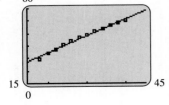

d. 0.835   **11. a.** 3.16 miles per hour   **b.**   110

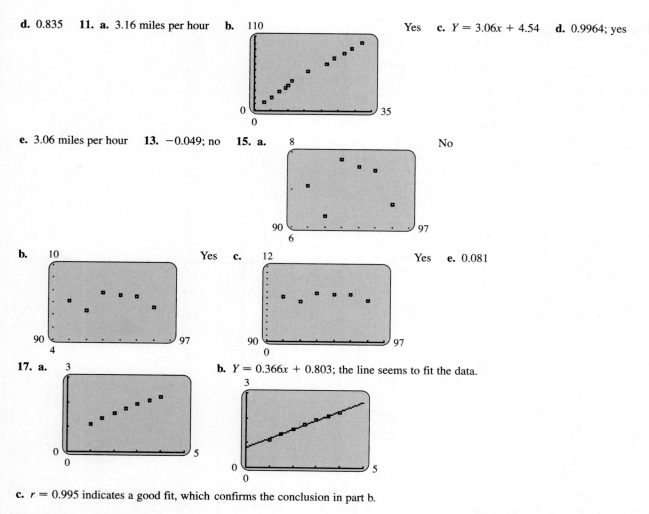

Yes   **c.** $Y = 3.06x + 4.54$ **d.** 0.9964; yes

e. 3.06 miles per hour   **13.** -0.049; no   **15. a.**   8   No

b.   10   Yes   **c.**   12   Yes   **e.** 0.081

17. a.   3   **b.** $Y = 0.366x + 0.803$; the line seems to fit the data.

c. $r = 0.995$ indicates a good fit, which confirms the conclusion in part b.

Exercises 1.3 (page 40)

1. Not a function   **3.** Function   **5.** Function   **7.** Not a function

9. $(-2, -1), (-1, 1),$
$(0, 3), (1, 5), (2, 7),$
$(3, 9)$; Range:
$\{-1, 1, 3, 5, 7, 9\}$

11. $(-2, 3/2), (-1, 2),$
$(0, 5/2), (1, 3), (2, 7/2),$
$(3, 4)$; Range:
$\{3/2, 2, 5/2, 3, 7/2, 4\}$

13. $(-2, 2), (-1, 0),$
$(0, 0), (1, 2), (2, 6),$
$(3, 12)$; Range:
$\{0, 2, 6, 12\}$

15. $(-2, 4), (-1, 1),$
$(0, 0), (1, 1), (2, 4),$
$(3, 9)$; Range:
$\{0, 1, 4, 9\}$

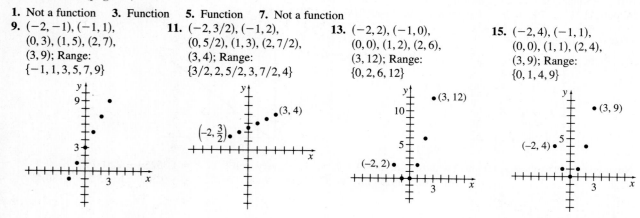

17. $(-2, 1), (-1, 1/2),$
$(0, 1/3), (1, 1/4), (2, 1/5),$
$(3, 1/6)$; Range:
$\{1, 1/2, 1/3, 1/4, 1/5, 1/6\}$

19. $(-2, -3), (-1, -3/2),$
$(0, -3/5), (1, 0), (2, 3/7),$
$(3, 3/4)$; Range:
$\{-3, -3/2, -3/5, 0, 3/7, 3/4\}$

21. $(-\infty, \infty)$ **23.** $(-\infty, \infty)$ **25.** $[-4, 4]$ **27.** $[3, \infty)$

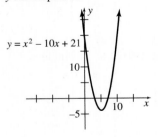

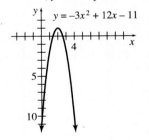

29. $(-\infty, -2) \cup (-2, 2) \cup (2, \infty)$ **31.** $(-\infty, \infty)$ **33.** $(-\infty, -1] \cup [5, \infty)$ **35.** $(-\infty, 2) \cup (4, \infty)$
37. Domain: $[-5, 4)$; range: $[-2, 6]$ **39.** Domain: $(-\infty, \infty)$; range: $(-\infty, 12]$ **41. a.** 5   **b.** $-7/4$ **c.** $-a^2 + 5a + 1$
d. $-4/m^2 + 10/m + 1$ or $(-4 + 10m + m^2)/m^2$ **e.** 0, 5   **43. a.** 9/2   **b.** 0   **c.** $(2a + 1)/(a - 2)$
d. $(4 + m)/(2 - 2m)$ **e.** -3 **45.** domain: $[-2, 4]$; range: $[0, 4]$ **a.** 0   **b.** 4   **c.** 3   **d.** $-1.5, 1.5, 2.5$
47. domain: $[-2, 4]$; range: $[-3, 2]$ **a.** -3 **b.** -2 **c.** -1 **d.** 2.5   **49.** $6m^2 - 36m + 52$
51. $r^2 + 2rh + h^2 - 2r - 2h + 5$ **53.** $9/q^2 - 6/q + 5$ or $(9 - 6q + 5q^2)/q^2$ **55.** Function   **57.** Not a function
59. Function   **61. a.** $x^2 + 2xh + h^2 - 4$ **b.** $2xh + h^2$ **c.** $2x + h$ **63. a.** $2x^2 + 4xh + 2h^2 - 4x - 4h - 5$
b. $4xh + 2h^2 - 4h$ **c.** $4x + 2h - 4$ **65. a.** $1/(x + h)$ **b.** $-h/[x(x + h)]$ **c.** $-1/[x(x + h)]$
67. $18x^2 + 3x - 2; 6x^2 + 15x + 2$ **69. a.** About 140 m   **b.** About 250 m   **71. a. i.** 3.6 kcal/km   **ii.** 61 kcal/km
b. $g(z) = 1{,}000x$ **c.** $(f \circ g)(x) = 4.4x^{0.88}$ gives the energy expenditure in kcal/km where x is the weight in kg.
73. a. The years   **b.** The number of Internet users   **c.** 98 million   **d.** Domain: $1995 \leq x \leq 1999$; range:
$26{,}000{,}000 \leq y \leq 205{,}000{,}000$ **75. a.** $A = (3{,}000 - w)w$ **b.** $0 \leq w \leq 3{,}000$ **c.**

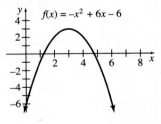

Exercises 1.4 (page 50)
3. D   **5.** B   **7.** E
9. Vertex is $(5, -4)$; axis is $x = 5$;
x-intercepts are 3 and 7;
y-intercept is 21

11. Vertex is $(2, 1)$; axis is $x = 2$;
x-intercepts are $2 \pm \sqrt{3}/3 \approx 2.58$
or 1.42; y-intercept is -11

13. Vertex is $(3, 3)$; axis is $x = 3$;
x-intercepts are $3 \pm \sqrt{3} \approx 4.73$ or 1.27;
y-intercept is -6

15. Vertex is $(4, 12)$; axis is $x = 4$; x-intercepts are 6 and 2; y-intercept is -36

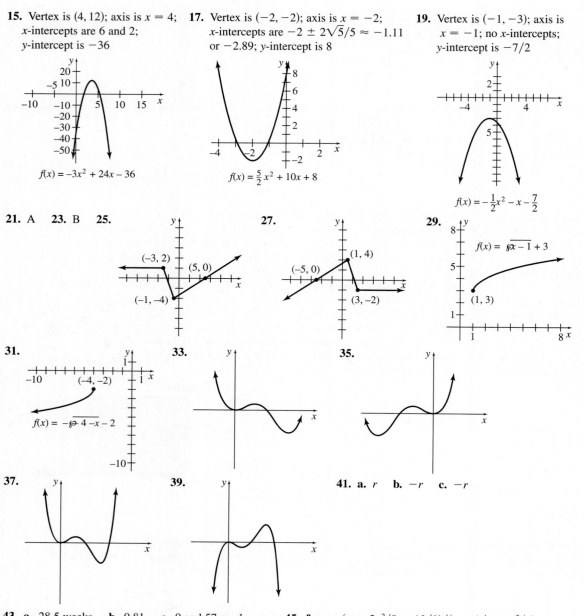

$f(x) = -3x^2 + 24x - 36$

17. Vertex is $(-2, -2)$; axis is $x = -2$; x-intercepts are $-2 \pm 2\sqrt{5}/5 \approx -1.11$ or -2.89; y-intercept is 8

$f(x) = \frac{5}{2}x^2 + 10x + 8$

19. Vertex is $(-1, -3)$; axis is $x = -1$; no x-intercepts; y-intercept is $-7/2$

$f(x) = -\frac{1}{2}x^2 - x - \frac{7}{2}$

21. A   **23.** B   **25.**

$(-3, 2)$ $(5, 0)$ $(-1, -4)$

27.

$(1, 4)$ $(-5, 0)$ $(3, -2)$

29.

$f(x) = \sqrt{x - 1} + 3$

$(1, 3)$

31.

$(-4, -2)$

$f(x) = -\sqrt{4 - x} - 2$

33.

35.

37.

39.

41. a. r **b.** $-r$ **c.** $-r$

43. a. 28.5 weeks   **b.** 0.81   **c.** 0 and 57 weeks; no   **45. f.** $q = (y - 3x^2/8 - 13/8)/(x - 3/4 - x^2/4)$ or $q = (8y - 3x^2 - 13)/(8x - 6 - 2x^2)$ **47.** 1995; $0 \le x \le 6$ **49. a.** 16 ft   **b.** 2 sec   **51. a.** 1993   **b.** \$195 million

c.

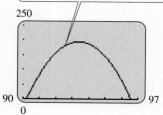

$f(x) = -19.321x^2 + 3,608.7x - 168,310$

250

90   97
0

53. 80 ft by 160 ft

55.

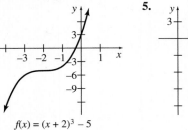

$y = -x^2/15;\ 10\sqrt{3}\text{ m} \approx 17.32\text{ m}$

Exercises 1.5 (page 61)

3.

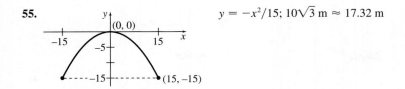

$f(x) = (x + 2)^3 - 5$

5.

$f(x) = -(x - 3)^4 + 1$

7. D   **9.** E   **11.** I   **13.** G   **15.** A   **17.** D   **19.** E

21. 4, 6, etc.; +   **23.** 5, 7, etc.; +   **25.** 7, 9, etc.; −

27. Horizontal asymptote: $y = 0$; vertical asymptote: $x = 3$; no x-intercept; y-intercept $= 4/3$

29. Horizontal asymptote: $y = 0$; vertical asymptote: $x = -3/2$; no x-intercept; y-intercept $= 2/3$

31. Horizontal asymptote: $y = 3$; vertical asymptote: $x = 1$; x-intercept $= 0$; y-intercept $= 0$

33. Horizontal asymptote: $y = 1$; vertical asymptote: $x = 4$; x-intercept $= -1$; y-intercept $= -1/4$

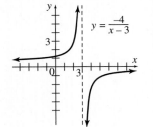

$y = \dfrac{-4}{x - 3}$

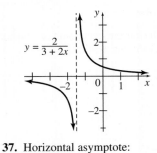

$y = \dfrac{2}{3 + 2x}$

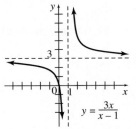

$y = \dfrac{3x}{x - 1}$

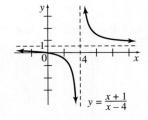

$y = \dfrac{x + 1}{x - 4}$

35. Horizontal asymptote: $y = -2/5$; vertical asymptote: $x = -4$; x-intercept $= 1/2$; y-intercept $= 1/20$

37. Horizontal asymptote: $y = -1/3$; vertical asymptote: $x = -2$; x-intercept $= -4$; y-intercept $= -2/3$

39. One possible answer is $y = 2x/(x - 1)$.

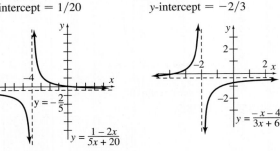

$y = \dfrac{1 - 2x}{5x + 20}$

$y = \dfrac{-x - 4}{3x + 6}$

41. a. 0   **b.** 2, −3   **d.** $(x + 1)(x - 1)(x + 2)$ **e.** $3(x + 1)(x - 1)(x + 2)$ **f.** $(x - a)$

43. a. Two; one at $x = -1.4$ and one at $x = 1.4$ **b.** Three; one at $x = -1.414$, one at $x = 1.414$ and one at $x = 1.442$

45. c. $y = (0.025 + 0.895x)/(0.24 + 0.76x)$ **d.** 0.50   **e.** 0.031

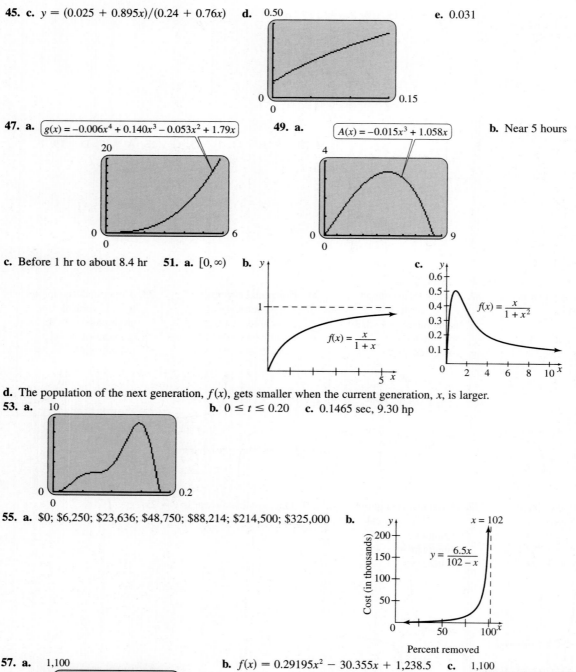

47. a. $g(x) = -0.006x^4 + 0.140x^3 - 0.053x^2 + 1.79x$ **49. a.** $A(x) = -0.015x^3 + 1.058x$ **b.** Near 5 hours

c. Before 1 hr to about 8.4 hr   **51. a.** $[0, \infty)$ **b.**   **c.**

$f(x) = \dfrac{x}{1 + x}$

$f(x) = \dfrac{x}{1 + x^2}$

d. The population of the next generation, $f(x)$, gets smaller when the current generation, x, is larger.

53. a. 10   **b.** $0 \le t \le 0.20$ **c.** 0.1465 sec, 9.30 hp

55. a. \$0; \$6,250; \$23,636; \$48,750; \$88,214; \$214,500; \$325,000   **b.**

$y = \dfrac{6.5x}{102 - x}$, $x = 102$

Cost (in thousands) / Percent removed

57. a. 1,100   **b.** $f(x) = 0.29195x^2 - 30.355x + 1,238.5$ **c.** 1,100

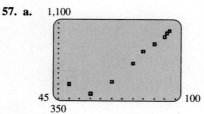

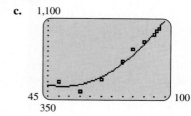

d. $f(x) = -0.01382x^3 + 3.3697x^2 - 252.65x + 6,425.5$ **e.**

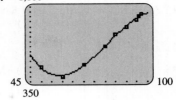

Chapter 1 Review Exercises (page 68)

3. $1/3$ **5.** $-2/11$ **7.** $-2/3$ **9.** 0 **11.** -3 **13.** $y = (2/3)x - 13/3$ **15.** $y = -(5/4)x + 17/4$ **17.** $x = -1$
19. $y = 3x - 7$ **21.** $y = -10$ **23.** $x = -7$

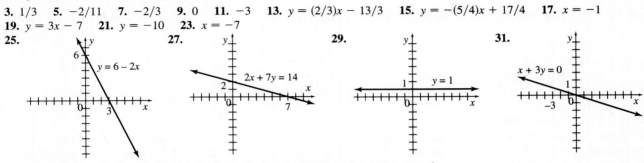

25.   **27.**   **29.**   **31.**

35. $(-3, 20), (-2, 9), (-1, 2), (0, -1), (1, 0), (2, 5), (3, 14)$; Range: $\{-1, 0, 2, 5, 9, 14, 20\}$

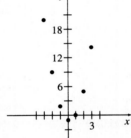

37. a. -28 **b.** -12 **c.** -28 **d.** $-r^2 - 3$ **39. a.** 17 **b.** 4 **c.** $5k^2 - 3$ **d.** $-9m^2 + 12m + 1$
e. $5x^2 + 10xh + 5h^2 - 3$ **f.** $-x^2 - 2xh - h^2 + 4x + 4h + 1$ **g.** $10x + 5h$ **h.** $-2x - h + 4$ **41.** $(-\infty, 0) \cup (0, \infty)$

43.   **45.**   **47.**   **49.**

$y = -\dfrac{1}{4}x^2 + x + 2$ $y = -3x^2 - 12x - 1$ $f(x) = 1 - x^4$ $y = -(x + 2)^4 - 2$

51.   **53.**   **55.** $3x^2 - 18x - 17$; $9x^2 + 6x - 15$

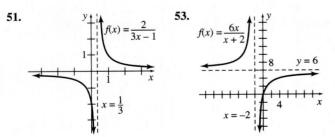

$f(x) = \dfrac{2}{3x - 1}$ $f(x) = \dfrac{6x}{x + 2}$

57. a. 8.06; each additional local species per 0.5 square meters leads to about 8 additional regional species per 0.5 hectare.
b. 39.5   **c.** 9.78   **59. b.** 0 and 1.35 square meters; 0 ml and 1,889 ml   **61. a.** 0.8286; yes, but not as close as in some other examples   **b.**

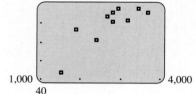

Yes   **c.** $Y = 0.0131x + 33.8$ **d.** About 75 years, which is slightly lower than the actual figure

63. Third day; about 104.2°F   **65. a.** 1/8; 1 over the SPF rating   **b.**

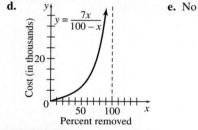

c. UVB $= 1 - 1/$SPF

d. 12.5%   **e.** About 3.3%   **f.** The increase in percent protection decreases to zero.   **67. a.** About 6.640 billion; this is about 644 million more than the estimate in the abstract of 5,996 million.   **b.** About 26.560 billion; 96.320 billion   **69. a.** $28,000
b. $7,000   **c.** $63,000   **d.**

$$y = \frac{7x}{100 - x}$$

Cost (in thousands)   Percent removed

e. No   **71. a.** The general trend has been to decrease from 9.125 to 7.454 parts per million.

b. $g(x) = -0.012053(x - 1982)^2 - 0.046607(x - 1982) + 9.125$ **73. a.** 0.494; not very closely
b.

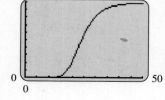

c. $Y = 1.70x + 87.8$ **d.** About $1,700

Chapter 2 Exponential, Logarithmic, and Trigonometric Functions

Exercises 2.1 (page 85)

1. 2, 4, 8, 16, 32,..., 1,024, $1.125899907 \times 10^{15}$ **3.** E   **5.** C   **7.** F   **9.** A   **11.** C   **13.** −3   **15.** −2   **17.** 6
19. 7/4   **21.** −2   **23.** 2, −2   **25.** 0, −1   **33. a.** Approximately 1,006, 432,500, 1,637,000, 1,931,000, 1,969,000
b. 2,000,000   **c.** It levels off at around 2,000,000.   **35. a.** Very closely   **b.** 5,341 million

c. 7,656 million   **37. a.** 2.6, 2.7; yes.   **b.** $C(x) = 0.05(1.040)^{x-1950}$
c.

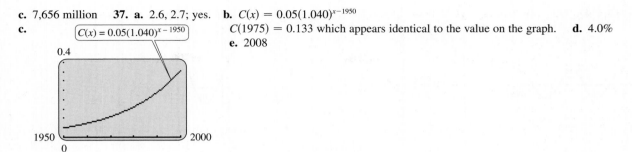

$C(1975) = 0.133$ which appears identical to the value on the graph.   **d.** 4.0%
e. 2008

39. a. $P = 1{,}013e^{-1.34 \cdot 10^{-4}x}$, $P = -0.0748x + 1{,}013$, $P = 1/(2.79 \cdot 10^{-7}x + 9.87 \cdot 10^{-4})$
b. 1,100 $P = 1{,}013e^{-1.34 \cdot 10^{-4}x}$ is the best fit.   **c.** 829 millibars, 232 millibars

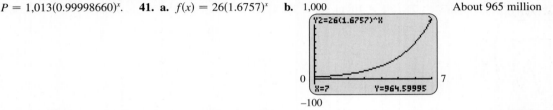

d. $P = 1{,}038(0.99998661)^x$. This is slightly different from the function found in part b, which can be rewritten as
$P = 1{,}013(0.99998660)^x$.   **41. a.** $f(x) = 26(1.6757)^x$ **b.** 1,000   About 965 million   **c.** About 68%
d. 1998

43. a. \$10,528.13   **b.** \$10,881.50   **c.** \$11,069.78   **d.** \$11,199.99   **45. a.** 12.5%   **b.** 11.9%   **47. a.** 8.84%   **b.** 8.75%

Exercises 2.2 (page 98)

1. $\log_2 8 = 3$ **3.** $\log_3 81 = 4$ **5.** $\log_3(1/9) = -2$ **7.** $2^7 = 128$ **9.** $25^{-1} = 1/25$ **11.** $10^4 = 10{,}000$ **13.** 2   **15.** 3
17. -2 **19.** $-2/3$ **21.** 1   **23.** $5/3$ **25.** $\log_3 4$ **27.** $\log_9 7 + \log_9 m$ **29.** $1 + \log_3 p - \log_3 5 - \log_3 k$
31. $\log_3 5 + (1/2)\log_3 2 - (1/4)\log_3 7$ **33.** $3a$ **35.** $2c + 3a + 1$ **37.** 1.86   **39.** -3.28 **41.** $x = 1/5$ **43.** $z = 2/3$
45. $r = 49$ **47.** $x = 1$ **49.** No solution   **51.** $x = 3$ **53.** $x = 1.79$ **55.** $y = 1.24$ **57.** $z = 2.10$ **59.** $x = 7.41$
61. $x < -2$ or $x > 2$ **65. a.** 23.4 yr   **b.** 11.9 yr   **c.** 9.0 yr   **67.** 21.3 years   **69. a.** 2,590 sq cm   **b.** 4,211 sq cm
c. 7,430 gm   **71. a.** About 0.693   **b.** $\ln 2$ **c.** yes   **73. a.** About 1.099   **b.** About 1.386   **75.** About every 7 hours,
$T = 3 \ln 5/\ln 2$ **77.** 2011   **79. a.** 530   **b.** 3,500   **c.** 6,000   **d.** 1,800   **e.** 0.8   **81. a.** 6   **b.** 8   **c.** About
$5{,}000{,}000 I_0$ **d.** About $126{,}000{,}000 I_0$ **e.** The 1985 earthquake had an amplitude more than 25 times that of the 1999
earthquake.   **f.** The 1985 earthquake had an energy about 126 times that of the 1999 earthquake.

Exercises 2.3 (page 107)

1. y_0 represents the initial quantity; k represents the rate of growth or decay.   **3.** The half-life of a quantity is the time period for
the quantity to decay to one-half of the initial amount.   **7.** No; 6.6%   **9. a.** $y = 25{,}000e^{0.047t}$ **b.** 18.6 hr   **11. a.** 17.9 days
b. January 17   **13. a.** 2, 5, 20, 125   **b.** 0.061   **c.** 0.26   **d.** No   **e.** Between 3 and 4   **15.** 87%
17. About 4,100 years old   **19.** About 1,600 years   **21. a.** 1.9 g   **b.** Approximately 7,000 years   **23. a.** $y = 25.0e^{-0.00497t}$
b. 139 days   **25. a.** $y = 10e^{0.0095t}$ **b.** 42.7°C   **27.** About 30 minutes   **29.** No; 17.2%

Exercises 2.4 (page 121)

1. $\pi/3$ **3.** $5\pi/6$ **5.** $7\pi/6$ **7.** $13\pi/6$ **9.** 315°   **11.** 330°   **13.** 288°   **15.** 48°   *Note: In Exercises 17–23 we give the
answers in the following order: sine, cosine, tangent, cotangent, secant, and cosecant.*   **17.** $4/5$; $-3/5$; $-4/3$; $-3/4$; $-5/3$; $5/4$
19. $4/5$; $3/5$; $4/3$; $3/4$; $5/3$; $5/4$ **21.** $+$ $+$ $+$ $+$ $+$ $+$ **23.** $-$ $-$ $+$ $+$ $-$ $-$ **25.** $\sqrt{3}/3$; $\sqrt{3}$; 2   **27.** $\sqrt{3}/2$; $\sqrt{3}/3$;
$2\sqrt{3}/3$ **29.** -1; -1 **31.** $-\sqrt{3}/2$; $-2\sqrt{3}/3$ **33.** $\sqrt{3}/2$ **35.** 1   **37.** $2\sqrt{3}/3$ **39.** -1 **41.** $-\sqrt{3}/2$ **43.** $-\sqrt{2}$

45. $-\sqrt{3}$ **47.** $1/2$ **49.** 0.6293 **51.** 7.1154 **53.** 0.3907 **55.** 0.1565 **57.** $a = 1, T = 2\pi/3$ **59.** $a = 3, T = 1/440$
61.   **63.**   **65.**

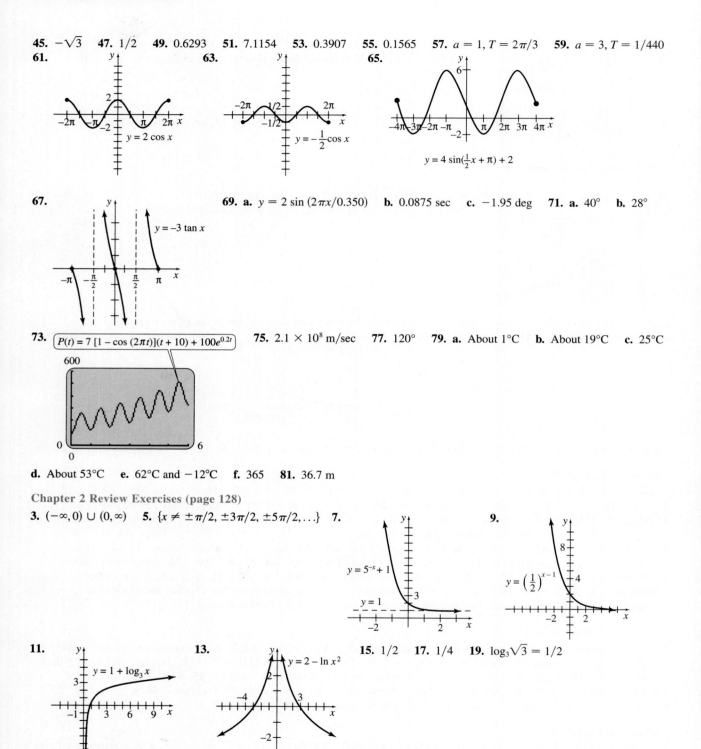

67.   **69. a.** $y = 2 \sin (2\pi x/0.350)$ **b.** 0.0875 sec   **c.** -1.95 deg   **71. a.** $40°$ **b.** $28°$

73. $P(t) = 7\,[1 - \cos (2\pi t)](t + 10) + 100e^{0.2t}$ **75.** 2.1×10^8 m/sec   **77.** $120°$ **79. a.** About $1°C$ **b.** About $19°C$ **c.** $25°C$

d. About $53°C$ **e.** $62°C$ and $-12°C$ **f.** 365 **81.** 36.7 m

Chapter 2 Review Exercises (page 128)
3. $(-\infty, 0) \cup (0, \infty)$ **5.** $\{x \neq \pm\pi/2, \pm3\pi/2, \pm5\pi/2, \ldots\}$ **7.**   **9.**

11.   **13.**   **15.** $1/2$ **17.** $1/4$ **19.** $\log_3\sqrt{3} = 1/2$

21. $\log_{10} 12 = 1.07918$ **23.** $10^2 = 100$ **25.** $10^{1.18921} = 15.46$ **27.** $4/5$ **29.** $3/2$ **31.** $\log_3 (y/4)$ **33.** $-\log_4 r$
35. $(\ln 11)/(\ln 3) + 2 \approx 4.183$ **37.** $-(\ln 9)/(\ln 15) \approx -0.811$ **39.** $(1 + \ln 12)/3 \approx 1.162$
41. $\left(-5 \pm 5\sqrt{3}\right)/2 \approx 1.830, -6.830$ **43.** 2 **45.** 14 **47. a.** $(0, \infty)$ **b.** $(-\infty, \infty)$ **c.** 1 **d.** None   **e.** $x = 0$ **f.** $a > 1$

g. $0 < a < 1$ **53.** $\pi/2$ **55.** $7\pi/6$ **57.** 2π **59.** $1{,}260°$ **61.** $81°$ **63.** $117°$ **65.** $\sqrt{3}/2$ **67.** $\sqrt{3}/2$ **69.** $2\sqrt{3}/3$
71. $1/2$ **73.** $-\sqrt{2}$ **75.** 0.7314 **77.** 6.3138 **79.** 0.9945 **81.** 0.8290 **83.**

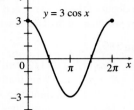

85. 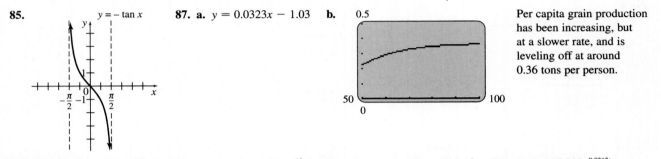   **87. a.** $y = 0.0323x - 1.03$ **b.**

Per capita grain production has been increasing, but at a slower rate, and is leveling off at around 0.36 tons per person.

89. a. About 4.19×10^{-15} cubic meters   **b.** $4.19 \times 10^{-15}2^t$ cubic meters   **c.** About 27 days   **91. a.** $y = 15{,}000e^{0.0313t}$
b. About 35 yrs   **93.** The maximum concentration of 0.25 occurs at $t = 0.69$ minutes   **95. a.** $28{,}000$ **b.** $7{,}000$
c. $63{,}000$ **d.**   **e.** No   **97.** 105; 75   **99.** $1{,}692.28$

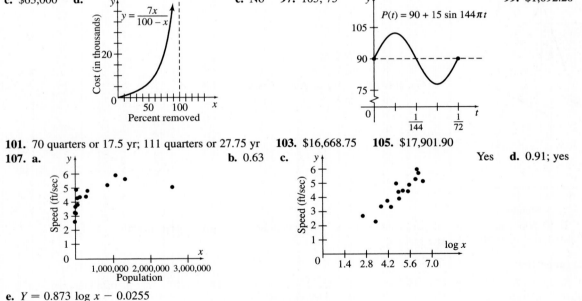

101. 70 quarters or 17.5 yr; 111 quarters or 27.75 yr   **103.** $16{,}668.75$ **105.** $17{,}901.90$
107. a.   **b.** 0.63   **c.**   Yes   **d.** 0.91; yes

e. $Y = 0.873 \log x - 0.0255$

Chapter 3 The Derivative

Exercises 3.1 (page 148)

1. c   **3.** b   **5.** 3   **7.** 0   **9. a.** $-1, -1/2$, does not exist, does not exist   **b.** $-1/2, -1/2, -1/2, -1/2$ **11.** 3   **15.** 2
17. 10   **19.** Does not exist   **21.** 8   **23.** 2   **25.** 4   **27.** 256   **29.** $3/2$ **31.** 1   **33.** 6   **35.** $3/2$ **37.** -5 **39.** $-1/9$
41. $1/10$ **43.** $2x$ **45.** $3/5$ **47.** $1/2$ **49.** 0   **51.** ∞ **53.** 1   **55. a.** Does not exist   **b.** $x = -2$ **c.** If $x = -2$ is an
asymptote for the graph of $F(x)$, then $\lim_{x \to -2} F(x)$ does not exist.   **59. a.** 0   **b.** $y = 0$ **61. a.** $-\infty$ **b.** $x = 0$ **65.** 5

67. 0.3333 or 1/3   **69. a.** 1.5   **71. a.** −2   **73. a.** 8   **77. a.** 96   **b.** 92   **c.** 96   **d.** Does not exist   **e.** 92
79. 0; the concentration of the drug in the bloodstream approaches 0 as the number of hours after injection increases.
81. a. 36.2 cm; the depth of the contaminated sediment layer deposited below the bottom of the lake in 1970 is 37.2 cm.
b. 155 cm; all of the sediment in the lake is within 155 cm of the bottom of the lake.   **83.** *E*

Exercises 3.2 (page 159)

1. $a = -1$ **a.** 1/2   **b.** 1/2   **c.** 1/2   **d.** $f(a)$ does not exist.   **3.** $a = 1$ **a.** −2   **b.** −2   **c.** −2   **d.** 2
5. $a = -5$ **a.** ∞   **b.** −∞   **c.** Limit does not exist.   **d.** $f(a)$ does not exist; $a = 0$ **a.** 0   **b.** 0   **c.** 0
d. $f(a)$ does not exist.   **7.** $a = 0$, limit does not exist; $a = 2$, limit does not exist.   **9.** $a = 2, 4$ **11.** Nowhere   **13.** $a = -2$,
limit does not exist.   **15.** $a = -2$, limit does not exist.   **17.** $a < 1$, limit does not exist.   **19.** $0 \le a \le 1$, limit does not exist.
21. a.   **b.** 2   **c.** 1, 5   **23. a.**   **b.** −1   **c.** 11, 3

25. a.   **b.** None   **29.** Discontinuous at $x = 1.2$ **31. a.** 1.0995×10^{12}

b.   **c.** No, at $t = 40$ **35. a.** About 687 g   **b.** No   **c.** 3,000

37. a. $500   **b.** $1,500   **c.** $1,000   **d.** Does not exist   **e.** Discontinuous at $x = 10$; a change in shifts   **f.** 15
39. a. $96   **b.** $150   **c.** $120   **d.** At $x = 100$

Exercises 3.3 (page 169)

1. 5   **3.** 8   **5.** 1/3   **7.** −1/3   **9.** 17   **11.** 5   **13.** 2   **15.** 2   **17.** 6.7726   **19.** 1.1207   **23. a.** 0.288 mm/wk
b. 0.348 mm/wk   **c.**   9   **25. a.** 6% per day   **b.** 7% per day   **27. a.** The bacteria are

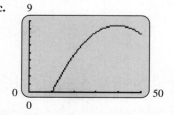

increasing at a rate of 2 million per min.   **b.** The bacteria are decreasing at a rate of −0.8 million per min.   **c.** The bacteria are
decreasing at a rate of −2.2 million per min.   **d.** The bacteria are decreasing at a rate of −1 million per min.   **e.** After 2 min
f. Around 3 min   **29. a.** Opium: 400 tons per year; −400 tons per year. Marijuana: −750 tons per year; −975 tons per year

b. Marijuana   **31. a.** Approximately $6.25 billion   **b.** Approximately $14.58 billion   **c.** Approximately $0 billion
33. a. The ratio of Internet messages to telephone calls is increasing at a faster rate than the ratio of households with PCs to households without PCs.   **b.** The ratio of Internet messages to telephone calls is growing at an increasing rate.   **35. a.** 5 ft/sec
b. 2 ft/sec   **c.** 3 ft/sec   **d.** 5 ft/sec   **e. i.** 2.5 ft/sec   **ii.** 2.5 ft/sec   **f. i.** 4 ft/sec   **ii.** 4 ft/sec

Exercises 3.4 (page 185)

1. a. 0   **b.** 1   **c.** −1   **d.** Does not exist   **e.** m **3.** At $x = -2$ **5.** 2   **7.** 1/4   **9.** 0   **11.** $f'(x) = -8x + 11$;
27; 11; −13   **13.** $f'(x) = 2/x^2$; 1/2; does not exist; 2/9   **15.** $f'(x) = 1/(2\sqrt{x})$; does not exist; does not exist; $1/(2\sqrt{3})$
17. $y = 8x - 9$ **19.** $y = -5x/4 + 5$ **21.** $y = 2x/3 + 6$ **23.** −5; −117; 35   **25.** 8; 8; 8   **27.** 1/2; 1/128; 2/9
29. $1/(2\sqrt{2})$; 1/8; does not exist   **31.** 0   **33.** −3; −1; 0; 2; 3; 5   **35. a.** $(a, 0)$ and (b, c) **b.** $(0, b)$ **c.** $x = 0$ and $x = b$
37. a. Distance   **b.** Velocity   **39.** 56.6626   **41.** −0.0158   **43. a.** 8.3 cm   **b.** 0.17 cm of catfish/cm of snake; longer
snakes eat bigger catfish at the constant rate of 0.17 cm increase in length of prey to each 1 cm increase in size of snake.
c. About 22 cm   **45. a.** 57 g/min   **c.** Range: $0 \le t \le 24$ **47.** 1,000; the population is increasing at a rate of 1,000 shellfish
per time unit. 570; the population is increasing more slowly at 570 shellfish per time unit. 250; the population is increasing at a
much slower rate of 250 shellfish per time unit.   **49. a.** 0.75 and 3   **b.** 1,033; the oven temperature is increasing at 1,033°F per
hour.   **c.** 0; the oven temperature is not changing.   **d.** −1,033; the oven temperature is decreasing at 1,033°F per hour.
51. 0 mph per sec for the hands and 640 mph per sec for the bat. This represents the acceleration of the hands and the bat at the
moment when the velocities are equal.

Exercises 3.5 (page 194)

3. $f: Y_2$; $f': Y_1$ **5.** $f: Y_1$; $f': Y_2$ **7.**   **9.**

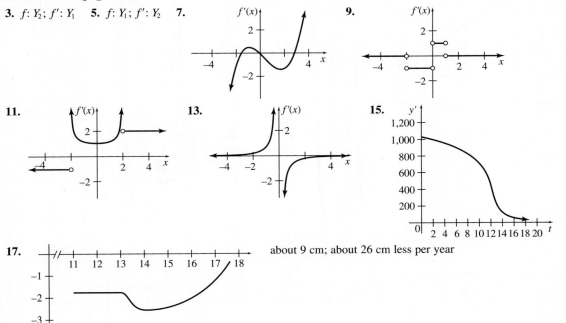

17.   about 9 cm; about 26 cm less per year

Chapter 3 Review Exercises (page 197)

5. a. 4   **b.** 4   **c.** 4   **d.** 4   **7. a.** ∞   **b.** −∞   **c.** Does not exist   **d.** Does not exist   **9.** ∞   **11.** 17/3   **13.** 8
15. −13   **17.** 1/6   **19.** 1/5   **21.** 3/4   **23.** Discontinuous at x_2 and x_4 **25.** 0, does not exist, does not exist; 1/2, does not
exist, does not exist   **27.** −5, does not exist, does not exist   **29.** Continuous everywhere

31. a.

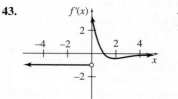

b. 1   **c.** 0, 2   **33.** 2   **35.** 30; 12   **37.** 9/77; 18/49   **39.** $f'(x) = 4$ **41.** 0.5171

43.

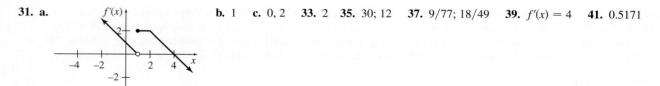

45. a. 0.02; the risk of heart attack is going up at a rate of 0.02 per 1,000 people for each increase in the blood cholesterol of 1 mg/dL

b. 0.15; the risk of heart attack is going up at a rate of 0.15 per 1,000 people for each increase in the blood cholesterol of 1 mg/dL
c. 0.065 per 1,000 people per mg/dL of cholesterol

47. a.

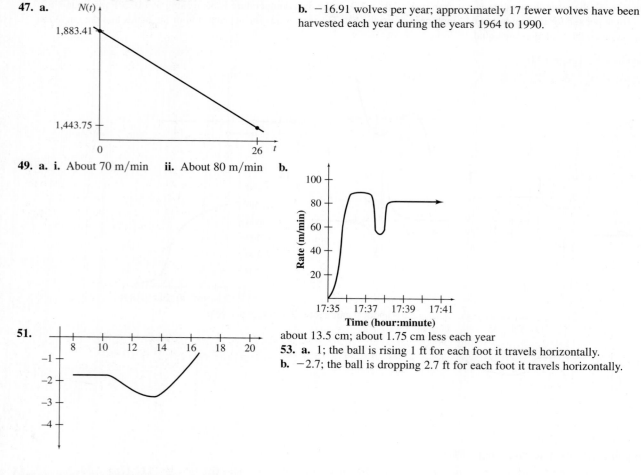

b. -16.91 wolves per year; approximately 17 fewer wolves have been harvested each year during the years 1964 to 1990.

49. a. i. About 70 m/min   **ii.** About 80 m/min   **b.**

about 13.5 cm; about 1.75 cm less each year
53. a. 1; the ball is rising 1 ft for each foot it travels horizontally.
b. -2.7; the ball is dropping 2.7 ft for each foot it travels horizontally.

51.

Chapter 4 Calculating the Derivative

Exercises 4.1 (page 212)

1. $dy/dx = 30x^2 - 18x + 6$ **3.** $dy/dx = 4x^3 - 15x^2 + (2/9)x$ **5.** $f'(x) = 9x^{0.5} - 2x^{-0.5}$ or $9x^{0.5} - 2/x^{0.5}$
7. $dy/dx = 4x^{-1/2} + (9/2)x^{-1/4}$ or $4/x^{1/2} + 9/(2x^{1/4})$ **9.** $g'(x) = -30x^{-6} + x^{-2}$ or $-30/x^6 + 1/x^2$

11. $dy/dx = -5x^{-6} + 2x^{-3} - 5x^{-2}$ or $-5/x^6 + 2/x^3 - 5/x^2$ **13.** $f'(t) = -4t^{-2} - 6t^{-4}$ or $-4/t^2 - 6/t^4$
15. $dy/dx = -18x^{-7} - 5x^{-6} + 14x^{-3}$ or $-18/x^7 - 5/x^6 + 14/x^3$ **17.** $h'(x) = -x^{-3/2}/2 + 21x^{-5/2}$ or $-1/(2x^{3/2}) + 21/x^{5/2}$
19. $dy/dx = 2x^{-4/3}/3$ or $2/(3x^{4/3})$ **21.** $g'(x) = (3/2)\sqrt{x} - 1/\sqrt{x}$ **23.** $h'(x) = 6x^5 - 12x^3 + 6x$ **27.** $-(9/2)x^{-3/2} - 3x^{-5/2}$ or
$-9/(2x^{3/2}) - 3/x^{5/2}$ **29.** $-14/3$ **31.** -28; $y = -28x + 34$ **33.** $5/2$ **35.** $(4/9, 20/9)$ **37.** $-3, -7$ **39.** $(5 \pm \sqrt{7})/3$
41. $(-1, 6), (4, -59)$ **43.** 38   **45. a.** 2   **b.** $1/2$ **c.** $[-1, \infty)$ **d.** $[0, \infty)$ **49. a.** 0.4824   **b.** 2.216   **51. a.** 264
b. 510   **c.** About 97/4 or 24.25 matings per degree   **53. a.** 1,232.62 cm^3 **b.** 948.08 cm^3/yr   **55.** $R = (2/3)R_0$
57. a. $[18, 40]$ **b.** $l'(x) = 0.2356 - 0.005348x$ **c.** 0.1019 cm/wk   **59.** $-38,037$; the number of species is decreasing at a
rate of about 38,037 species per gram of body mass, or about 38 for an increase of 1 microgram.
61. a.

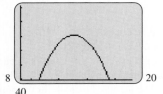

b. About 13.58%   **c.** 0   **63. a.** $v(t) = 22t + 4$ **b.** 4; 114; 224

65. a. $v(t) = 12t^2 + 16t$ **b.** 0; 380; 1,360   **67. a.** -32 ft per sec; -64 ft per sec   **b.** In 3 sec   **c.** -96 ft per sec
69. a. 1.3275 g/cm^3 **b.** -0.43 g/cm^3; when the level of the Dead Sea decreases to 50% of its current level, the density of the
brine is decreasing at the rate of 0.43 g/cm^3.   **71.** $M'(x) = 8.97x^2 - 764.24x + 15,316.35$ **a.** 462.15   **b.** -443.8
c. -901.25 **d.** 417.4   **e.** 11,585.15   **f.** 27,571.56   **g.** The amount of money in circulation was increasing at the rate of
\$462.15/year in 1930 but then began to decrease. The decrease of money continued through the 1950s and then began to increase.
It has continued to increase through the year 2000.

Exercises 4.2 (page 221)

1. $dy/dx = 18x^2 - 6x + 4$ **3.** $dy/dx = 8x - 20$ **5.** $k'(t) = 4t^3 - 4t$ **7.** $dy/dx = (3/2)x^{1/2} + (1/2)x^{-1/2} + 2$ or
$3x^{1/2}/2 + 1/(2x^{1/2}) + 2$ **9.** $-8y^{-5} + 15y^{-6} + 30y^{-7}$ **11.** $f'(x) = 53/(3x + 8)^2$ **13.** $dy/dx = -17/(4 + t)^2$
15. $dy/dx = (x^2 - 2x - 1)/(x - 1)^2$ **17.** $f'(t) = (-4t^2 - 22t - 12)/(t^2 - 3)^2$ **19.** $g'(x) = (x^2 + 6x - 14)/(x + 3)^2$
21. $p'(t) = [-\sqrt{t}/2 - 1/(2\sqrt{t})]/(t - 1)^2$ or $(-t - 1)/[2\sqrt{t}(t - 1)^2]$ **23.** $dy/dx = (5\sqrt{x}/2 - 3/\sqrt{x})/x$ or
$(5x - 6)/(2x\sqrt{x})$ **25.** $g'(y) = (-1.1y^{2.9} - 2.5y^{1.5} + 2.8y^4)/(y^{2.5} + 2)^2$ **27.** 59   **29.** In the first step, the numerator should
be $(x^2 - 1)2 - (2x + 5)(2x)$.   **31.** $y = -2x + 9$ **35.** $x = -0.828, 4.828$ **37. a.** $f'(x) = AK/(A + x)^2$ **b.** $K/(4A)$
39. a. 8.57 min.   **b.** 16.36 min.   **c.** 6.12 min.^{2}/kcal; 2.48 min.^{2}/kcal   **41.** $T'(n) = a(1 - bn^2)/(1 + bn^2)^2$; the rate of change
of the cell traction force with respect to the number of cells.   **43.** $aknC^{n-1}/(k + C^n)^2$; the rate of change of the amount of the
neurotransmitter released with respect to the amount of calcium.   **45. a.** 0.1173   **b.** 2.625

Exercises 4.3 (page 231)

1. 1,122   **3.** 97   **5.** $256k^2 + 48k + 2$ **7.** $(3x + 95)/8$; $(3x + 280)/8$ **9.** $1/x^2$; $1/x^2$ **11.** $2\sqrt{2x^2 - 1}$; $8x + 10$
13. $\sqrt{(x - 1)/x}$; $-1/\sqrt{x + 1}$ **17.** If $f(x) = x^{2/5}$ and $g(x) = 5 - x$, then $y = f[g(x)]$.   **19.** If $f(x) = -\sqrt{x}$ and
$g(x) = 13 + 7x$, then $y = f[g(x)]$.   **21.** If $f(x) = x^{1/3} - 2x^{2/3} + 7$ and $g(x) = x^2 + 5x$, then $y = f[g(x)]$.
23. $dy/dx = 5(2x^3 + 9x)^4(6x^2 + 9)$ **25.** $f'(x) = -288x^3(3x^4 + 2)^2$ **27.** $s'(t) = 144t^3(2t^4 + 5)^{1/2}$
29. $f'(t) = 32t/\sqrt{4t^2 + 7}$ **31.** $r'(t) = 4(2t^5 + 3)(22t^5 + 3)$ **33.** $dy/dx = x(x^2 - 1)(7x^3 - 3x + 8)$
35. $(6z + 1)^{1/3}(14z + 1)$ **37.** $dy/dx = -30x/(3x^2 - 4)^6$ **39.** $p'(t) = [2(2t + 3)^2(4t^2 - 12t - 3)]/(4t^2 - 1)^2$
41. $dy/dx = (-30x^4 - 132x^3 + 4x + 8)/(3x^3 + 2)^5$ **43. a.** -2 **b.** $-24/7$ **45.** $y = (3/5)x + 16/5$ **47.** $y = x$
49. 1, 3   **53.** $P[f(a)] = 18a^2 + 24a + 9$ **55.** About 19.5 mm/wk   **57. a.** 6   **b.** $39/4 = 9.75$ **c.** $138/7 \approx 19.71$
59. a. $R'(Q) = -Q/[6(C - Q/3)^{1/2}] + (C - Q/3)^{1/2}$ **b.** 2.83   **c.** Increasing   **61. a.** 34 minutes
b. $-(108/17)\pi$ mm^3/min., $-(72/17)\pi$ mm^2/min.   **63. a.** $-\$1,050$ **b.** $-\$457.06$

Exercises 4.4 (page 238)

1. $dy/dx = 4e^{4x}$ **3.** $dy/dx = -16e^{2x}$ **5.** $dy/dx = -16e^{x+1}$ **7.** $dy/dx = 2xe^{x^2}$ **9.** $dy/dx = 12xe^{2x^2}$ **11.** $dy/dx = 16xe^{2x^2-4}$
13. $dy/dx = xe^x + e^x = e^x(x + 1)$ **15.** $dy/dx = 2(x - 3)(x - 2)e^{2x}$ **17.** $dy/dx = [2xe^x - x^2e^x]/e^{2x} = [x(2 - x)]/e^x$
19. $dy/dx = [x(e^x - e^{-x}) - (e^x + e^{-x})]/x^2$ **21.** $dp/dx = 8,000e^{-0.2t}/(9 + 4e^{-0.2t})^2$ **23.** $f'(z) = 4(2z + e^{-z})(1 - ze^{-z})$
25. $dy/dx = -(\ln 2)(2^{-x})$ **27.** $dy/dx = -6x(10^{3x^2-4}) \ln 10$ **29.** $ds/dx = (5 \ln 7)7^{\sqrt{t-2}}/(2\sqrt{t - 2})$

31.

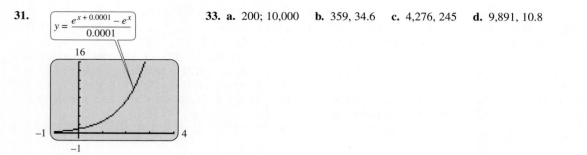

$$y = \frac{e^{x+0.0001} - e^x}{0.0001}$$

33. a. 200; 10,000   **b.** 359, 34.6   **c.** 4,276, 245   **d.** 9,891, 10.8

e. It increases for a while and then gradually decreases to 0.   **35. a.** 0.005   **b.** 0.0007   **c.** 0.000013   **d.** −0.022
e. −0.0029   **f.** −0.000054   **37. a.** 0.026%, 0.286%, 3.130%   **b.** 0.0025%/year, 0.0274%/year, 0.300%/year
c. The percentage of people in each of the age groups that die in a given year is increasing as indicated by the answers in parts a
and b. A person who is 75 years old has a 3% chance of dying during the year and the rate is increasing by almost 0.3%. The
formula implies that everyone will be dead by about age 86.   **39. a.** 0.027   **b.** 2024   **c.** The marginal increase in the population
per year in 2002 is approximately 0.008.   **41. a.** Approaches 589 mm   **b.** About 15 yr   **c.** About 32.2 mm/yr; the rate of
growth of a cutlassfish increases by approximately 32.2 mm during its fourth year of life.   **d.**   600

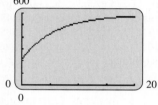

43. a. 619 kg   **b.** About 1,203 days   **c.** About 0.18 kg/day; a Holstein cow will grow about 0.18 kg during day 1,000.
45. $ae^{-a(S-h)/E}$ **47. a.** 218.9 sec   **b.** The record is decreasing by 0.063 seconds per year at the end of 2001.   **c.** 218 sec. If the
estimate is correct, then this is the least amount of time that it will ever take for a human to run a mile.   **49. a.** 29.95 million
b. The Latino-American population will be increasing at the rate of 0.689 million/year at the end of the year 2000.
51. a. $I_C = (V/R)e^{-t/RC}$ **b.** 1.35×10^{-7} amps

Exercises 4.5 (page 247)

1. $dy/dx = 1/x$ **3.** $dy/dx = -1/(3 - x)$ or $1/(x - 3)$ **5.** $dy/dx = (4x - 7)/(2x^2 - 7x)$ **7.** $dy/dx = 1/[2(x + 5)]$
9. $dy/dx = [3(2x^2 + 5)]/[x(x^2 + 5)]$ **11.** $dy/dx = -3x/(x + 2) - 3\ln(x + 2)$ **13.** $ds/dt = t + 2t \ln |t|$
15. $dy/dx = [2x - 4(x + 3)\ln(x + 3)]/[x^3(x + 3)]$ **17.** $dy/dx = (4x + 7 - 4x \ln x)/[x(4x + 7)^2]$
19. $dy/dx = (6x \ln x - 3x)/(\ln x)^2$ **21.** $dy/dx = 4(\ln |x + 1|)^3/(x + 1)$ **23.** $dy/dx = 1/(x \ln x)$
25. $dy/dx = e^{x^2}/x + 2xe^{x^2} \ln x$ **27.** $dy/dx = (xe^x \ln x - e^x)/[x(\ln x)^2]$ **29.** $g'(z) = [3(e^{2z} + \ln z)^2(2ze^{2z} + 1)]/z$
31. $dy/dx = 2/[(\ln 10)(2x - 3)]$ **33.** $dy/dx = 1/(x \ln 10)$ **35.** $dy/dx = 1/[(\ln 7)(2x - 3)]$
37. $dy/dx = 5(4x - 1)/[(2 \ln 2)(2x^2 - x)]$ **39.** $dz/dy = 10^y/[(\ln 10)y] + (\log y)(\ln 10)10^y$
43.

$$y = \frac{\ln |x + 0.0001| - \ln |x|}{0.0001}$$

45. a. $N(t) = 1{,}000e^{9.8901e^{-e[2.54197 - 0.2167t]}}$ **b.** 1,307,414 bacteria/hr; the number of bacteria is
increasing at a rate of 1,307,414 per hour, 20 hours after the experiment began.

c.   12

d. 20,000,000

e. 9.8901; $1{,}000e^{9.8901} \approx 19{,}734{,}033$

47. b. i. 3.343   **ii.** 1.466   **c. i.** -0.172 **ii.** -0.0511 **49.** 26.9; 13.1   **51. a.** 1,805.73 million metric tons
b. The greenhouse gas emissions are increasing at the rate of 37.3 million metric tons per year at the end of 2002.
53. a. 817 vehicles/hour, -41.2 vehicles/hour per foot   **b.** 522 vehicles/hour, -20.9 vehicles/hour per foot

Exercises 4.6 (page 256)

1. $dy/dx = 12 \cos 6x$ **3.** $dy/dx = 108 \sec^2(9x + 1)$ **5.** $dy/dx = -4 \cos^3 x \sin x$ **7.** $dy/dx = 5 \tan^4 x \sec^2 x$
9. $dy/dx = -5(4x \cos 4x + \sin 4x)$ **11.** $dy/dx = -(x \csc x \cot x + \csc x)/x^2$ **13.** $dy/dx = 5e^{5x} \cos e^{5x}$
15. $dy/dx = (\cos x)e^{\sin x}$ **17.** $dy/dx = (2/x) \cos(\ln 4x^2)$ **19.** $dy/dx = (2x \cos x^2)/\sin x^2$ or $2x \cot x^2$
21. $dy/dx = (6 \cos x)/(3 - 2 \sin x)^2$ **23.** $dy/dx = \sqrt{\sin 3x}(\sin 3x \cos x - 3 \sin x \cos 3x)/(2\sqrt{\sin x}(\sin^2 3x))$
25. $dy/dx = -2 \csc x \cot x - 12 \sec^2 4x + (7/8) \sin[(1/8)x] + 3e^{3x}$ **27.** 1   **29.** -1 **31.** 1   **33.** $-\csc^2 x$
37. a.

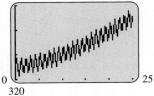

$$y = \frac{\pi}{8} \cos 3\pi \left(t - \frac{1}{3}\right)$$

0.4

0

$\frac{4}{3}$

-0.4

b. $v(t) = dy/dt = -(3\pi^2/8) \sin[3\pi(t - (1/3))]$;
$a(t) = v'(t) = -(9\pi^3/8) \cos[3\pi(t - (1/3))]$

d. At $t = 1$, acceleration is negative, arm is moving clockwise and is at an angle of $\pi/8$ radians from the vertical; at $t = 4/3$, acceleration is positive, arm is moving counterclockwise, and is at an angle of $-\pi/8$ radians from the vertical; at $t = 5/3$, acceleration is negative, arm is moving clockwise and is at an angle of $\pi/8$ radians from the vertical.
39. a.   380

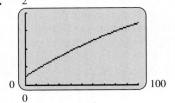

0   25
320

b. 370 ppm, 401.71 ppm, 468.05 ppm   **c.** $C'(50.2) = 19.18$ ppm/yr. The level of carbon dioxide will be increasing at the beginning of 2010 at a rate of 19.18 ppm.

41. a. $dx/dt = -2.625 \sin(\theta)\left(1 + \cos(\theta)/\sqrt{15 + \cos^2(\theta)}\right) d\theta/dt$ **b.** 21.6 mph   **43. a.** 1   **b.** -2 **c.** -1 **d.** -2
e. -1 **f.** 2

Chapter 4 Review Exercises (page 260)

1. $dy/dx = 10x - 7$ **3.** $dy/dx = 14x^{4/3}$ **5.** $f'(x) = -3x^{-4} + (1/2)x^{-1/2}$ or $-3/x^4 + 1/(2x^{1/2})$ **7.** $k'(x) = 15/(x + 5)^2$
9. $dy/dx = (x^2 - 2x)/(x - 1)^2$ **11.** $f'(x) = 12(3x - 2)^3$ **13.** $dy/dx = 1/(2t - 5)^{1/2}$ **15.** $dy/dx = 3(2x + 1)^2(8x + 1)$
17. $r'(t) = (-15t^2 + 52t - 7)/(3t + 1)^4$ **19.** $p'(t) = t(t^2 + 1)^{3/2}(7t^2 + 2)$ **21.** $dy/dx = -12e^{2x}$ **23.** $dy/dx = -6x^2 e^{-2x^3}$
25. $dy/dx = 10xe^{2x} + 5e^{2x} = 5e^{2x}(2x + 1)$ **27.** $dy/dx = 2x/(2 + x^2)$ **29.** $dy/dx = (x - 3 - x \ln|3x|)/[x(x - 3)^2]$
31. $dy/dx = [e^x(x + 1)(x^2 - 1) \ln(x^2 - 1) - 2x^2 e^x]/[(x^2 - 1)(\ln(x^2 - 1))^2]$ **33.** $ds/ds = 2(t^2 + e^t)(2t + e^t)$
35. $dy/dx = -6x(\ln 10) \cdot 10^{-x^2}$ **37.** $q'(z) = (3z^2 + 1)/[(\ln 2)(z^3 + z + 1)]$ **41.** $dy/dx = 18 \sec^2 3x$
43. $dy/dx = 2x \csc^2(9 - x^2)$ **45.** $dy/dx = 64x \sin^3(4x^2) \cos(4x^2)$ **47.** $dy/dx = -2x \sin(1 + x^2)$
49. $dy/dx = e^{-x}(\cos x - \sin x)$ **51.** $dy/dx = (-2 \cos x \sin x + \cos^2 x \sin x)/(1 - \cos x)^2$
53. $dy/dx = (\sec^2 x + x \sec^2 x - \tan x)/(1 + x)^2$ **55.** $dy/dx = (\cos x)/(\sin x)$ or $\cot x$ **57.** -2; $y = -2x + 9$
59. $-5/9$; $y = -(5/9)x + 16/9$ **61.** $-4/5$; $y = -(4/5)x - 13/5$ **63.** $2e$; $y = 2ex - e$ **65.** 2; $y = 2x - e$
67. π; $y = \pi x - \pi$ **69. a.** 7.6 fish   **b.** 0.19 caught walleye/hr/walleye/acre; the catch rate of walleye grows at a constant rate of 0.19 as the density of walleye increases.   **71. a.**   2

b. 201 fish; the processing time begins to decrease.   **c.** 0.015 hr/fish; the rate at which the processing time is increasing when $n = 40$ is 0.015 hr/fish.

0   100
0

73. a. 28.1 cm   **b.** 4.34 cm/year   **c.** 205 g   **d.** 21.2 g/cm   **e.** 91.9 g/year   **75.** 105; 75

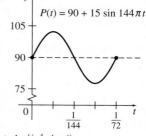

77. $L_2 = s/\sin \theta$ **79.** $L_1 = L_0 - s \cot \theta$ **81.** $R_2 = k \cdot L_2/r_2^4$ **83.** $R = k(L_0 - s \cot \theta)/r_1^4 + ks/(r_2^4 \sin \theta)$

85. $0 = [ks\,(\csc^2 \theta)]/r_1^4 - (ks \cos \theta)/(r_2^4 \sin^2 \theta)$ **87.** $\cos \theta = (r_2^4)/(r_1^4)$ **89.** 84° to the nearest degree   **91.** 0.0129; the concentration of CFC-11 is increasing at a rate of 0.0129 ppb per year in 1998.   **93. a.** 169; the value is close to the number 167. **b.** 163   **c.** In 2950, the language will be decreasing at a rate of 29.5 words per millennium.

Chapter 5 Graphs and the Derivative

Exercises 5.1 (page 274)

1. a. $(1, \infty)$ **b.** $(-\infty, 1)$ **3. a.** $(-\infty, -2)$ **b.** $(-2, \infty)$ **5. a.** $(-\infty, -4), (-2, \infty)$ **b.** $(-4, -2)$
7. a. $(-7, -4), (-2, \infty)$ **b.** $(-\infty, -7), (-4, -2)$ **9. a.** 3/2   **b.** $(-\infty, 3/2)$ **c.** $(3/2, \infty)$ **11. a.** $-3, 4$
b. $(-\infty, -3), (4, \infty)$ **c.** $(-3, 4)$ **13. a.** $-3/2, 4$ **b.** $(-\infty, -3/2), (4, \infty)$ **c.** $(-3/2, 4)$ **15. a.** $-2, -1, 0$
b. $(-2, -1), (0, \infty)$ **c.** $(-\infty, -2), (-1, 0)$ **17. a.** None   **b.** None   **c.** $(-\infty, \infty)$ **19. a.** None   **b.** None
c. $(-\infty, -1), (-1, \infty)$ **21. a.** 0   **b.** $(0, \infty)$ **c.** $(-\infty, 0)$ **23. a.** 0   **b.** $(0, \infty)$ **c.** $(-\infty, 0)$ **25. a.** 7   **b.** $(7, \infty)$
c. $(3, 7)$ **27. a.** $0, 2/5$ **b.** $(0, 2/5)$ **c.** $(-\infty, 0), (2/5, \infty)$ **29. a.** $n\pi/2$ where n is an odd integer
b. $(n\pi, (n + 1/2)\pi) \cup ((n + 3/2)\pi, (n + 2)\pi)$ where n is an even integer   **c.** $((n + 1/2)\pi, (n + 3/2)\pi)$ where n is an even integer   **31. a.** $n\pi$, where n is an integer   **b.** $(n\pi, (n + 1/2)\pi) \cup ((n + 1/2)\pi, (n + 1)\pi)$, where n is an even integer
c. $((n + 1)\pi, (n + 3/2)\pi) \cup ((n + 3/2)\pi, (n + 2)\pi)$, where n is an even integer   **33.** Vertex: $(-b/(2a), (4ac - b^2)/(4a))$;
increasing on $(-b/(2a), \infty)$, decreasing on $(-\infty, -b/(2a))$ **35.** On $(-\infty, \infty)$; nowhere; nowhere   **37. a.** Yes   **b.** From April to July; July to November; January to April; and November to December   **c.** January to April and November and December. In these months most trees do not have leaves.   **39. a.** $(0, 4.8)$ **b.** $(4.8, 8)$ **41. a.** $(0, 1)$ **b.** $(1, \infty)$ **43. a.** $175.9e^{-t/1.3}(1 - 0.769t)$
b. Increasing on $(0, 1.3)$; decreasing on $(1.3, \infty)$ **45. a.** $(0, \infty)$ **b.**

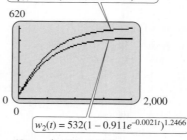

47. Increasing on $(4.37, 16.37)$, decreasing on $(0, 4.37) \cup (16.37, 24)$ **49. a.** $(1,000, 6,100)$
b. $(6,100, 6,500)$ **c.** $(1,000, 3,000), (3,600, 4,200)$ **d.** $(3,000, 3,600), (4,200, 6,500)$ **51. a.** Nowhere   **b.** $(0, \infty)$

Exercises 5.2 (page 287)

1. Relative minimum of -4 at 1   **3.** Relative maximum of 3 at -2 **5.** Relative maximum of 3 at -4; relative minimum of 1 at -2 **7.** Relative maximum of 3 at -4; relative minimum of -2 at -7 and -2 **9.** Relative minimum of -44 at -6
11. Relative maximum of -8 at -3; relative minimum of -12 at -1 **13.** Relative maximum of 827/96 at $-1/4$; relative minimum of $-377/6$ at -5 **15.** Relative maximum of -4 at 0; relative minimum of -85 at 3 and -3 **17.** Relative maximum of 0 at 8/5   **19.** Relative maximum of 1 at -1; relative minimum of 0 at 0   **21.** No relative extrema   **23.** Relative maximum of 0 at 1; relative minimum of 8 at 5   **25.** Relative maximum of -2.46 at -2; relative minimum of -3 at 0   **27.** No relative extrema   **29.** Relative maximum of 1 at $x = \ldots, -7/2, -3/2, 1/2, 5/2, \ldots$; relative minimum of -1 at
$x = \ldots, -5/2, -1/2, 3/2, 7/2, \ldots$ **31.** $(2, 7)$ **33.** Relative maximum of 6.211 at 0.085; relative minimum of -57.607

at 2.161   **35.** Relative minimum at $x = 5$ 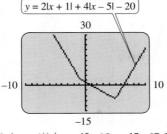   **37.** 5:04 P.M.; 6:56 A.M.

39. About 4.96 yr; about 458.22 kg   **41.** $(\ln(n - 1))/n$ **43.** 10   **45.** 67 ft

Exercises 5.3 (page 300)

1. $f''(x) = 18x$; 0; 36   **3.** $f''(x) = 36x^2 - 30x + 4$; 4; 88   **5.** $f''(x) = 6$; 6; 6   **7.** $f''(x) = 2/(1 + x)^3$; 2; 2/27
9. $f''(x) = -1/[4(x + 4)^{3/2}]$; $-1/32$; $-1/[4(6^{3/2})] \approx -0.0170$ **11.** $f''(x) = -(6/5)x^{-7/5}$ or $-6/(5x^{7/5})$; $f''(0)$ does not exist;
$-6/[5(2^{7/5})] \approx -0.4547$ **13.** $f''(x) = 20x^2 e^{-x^2} - 10e^{-x^2}$; -10; $70e^{-4} \approx 1.282$ **15.** $f''(x) = (-3 + 2\ln x)/(4x^3)$; does not
exist; -0.050 **17.** $f'''(x) = -24x$; $f^{(4)}(x) = -24$ **19.** $f'''(x) = 240x^2 + 144x$; $f^{(4)}(x) = 480x + 144$
21. $f'''(x) = 18(x + 2)^{-4}$ or $18/(x + 2)^4$; $f^{(4)}(x) = -72(x + 2)^{-5}$ or $-72/(x + 2)^5$ **23.** $f'''(x) = -36(x - 2)^{-4}$ or $-36/(x - 2)^4$;
$f^{(4)}(x) = 144(x - 2)^{-5}$ or $144/(x - 2)^5$ **25. a.** $f'(x) = 1/x$; $f''(x) = -1/x^2$; $f'''(x) = 2/x^3$; $f^{(4)}(x) = -6/x^4$; $f^{(5)}(x) = 24/x^5$
b. $f^{(n)}(x) = (-1)^{n-1}[1 \cdot 2 \cdot 3 \cdots (n - 1)]/x^n$ or, using factorial notation, $f^{(n)}(x) = (-1)^{n-1}(n - 1)!/x^n$ **27.** Concave upward on
$(2, \infty)$; concave downward on $(-\infty, 2)$; point of inflection at $(2, 3)$ **29.** Concave upward on $(-\infty, -1)$ and $(8, \infty)$; concave
downward on $(-1, 8)$; points of inflection at $(-1, 7)$ and $(8, 6)$ **31.** Concave upward on $(2, \infty)$; concave downward on $(-\infty, 2)$;
no points of inflection   **33.** Always concave upward; no points of inflection   **35.** Concave upward on $(-\infty, 3/2)$; concave
downward on $(3/2, \infty)$; point of inflection at $(3/2, 525/2)$ **37.** Concave upward on $(5, \infty)$; concave downward on $(-\infty, 5)$; no
points of inflection   **39.** Concave upward on $(-10/3, \infty)$; concave downward on $(-\infty, -10/3)$; point of inflection at
$(-10/3, -250/27)$ **41.** Never concave upward; always concave downward; no inflection points   **43.** Concave upward on
$(-\infty, 0)$ and $(1, \infty)$; concave downward on $(0, 1)$; points of inflection at $(0, 0)$ and $(1, -3)$.   **45.** Concave upward on
$\ldots \cup (-3\pi/2, -\pi) \cup (-\pi/2, 0) \cup (\pi/2, \pi) \cup \ldots$; concave downward on
$\ldots \cup (-\pi, (-\pi)/2) \cup (0, \pi/2) \cup (\pi, (3\pi)/2) \cup \ldots$; points of inflection at $(n\pi/2, 0)$ where n is an integer
47. a.      **c.** $f''(0) = 0$, while $g''(0)$ is undefined.   **d.** No.   **49.** Approaches 0; approaches ∞

51. Relative minimum at 6   **53.** Relative maximum at 0; relative minimum at 4/3   **55.** Critical number at 0, but neither a
maximum nor minimum there   **57. a.** Minimum at about -0.4 and 4.0; maximum at about 2.4   **b.** Increasing on about
$(-0.4, 2.4)$ and about $(4.0, \infty)$; decreasing on about $(-\infty, -0.4)$ and $(2.4, 4.0)$ **c.** About 0.7 and 3.3   **d.** Concave upward on
about $(-\infty, 0.7)$ and $(3.3, \infty)$; concave downward on about $(0.7, 3.3)$ **59. a.** Maximum at -1; minimum at 1   **b.** Increasing on
$(-1, 1)$; decreasing on $(-\infty, -1)$ and $(1, \infty)$ **c.** About 1.7, and -1.7 and 0   **d.** Concave upward on about $(-1.7, 0)$ and $(1.7, \infty)$;
concave downward on about $(-\infty, -1.7)$, $(0, 1.7)$ **61. a.** At 4 hr   **b.** 1,160 million   **63. a.** After 2 hr   **b.** 3/4%
65. Always concave down   **67.** (38.9182, 5,000)   **69.** (2.96, 5.27)   **71.** (32.01, 26.41)   **73.** 50   **75. a.** 343.25 ft
b. About 9 sec, about -148 ft/sec   **77. a.** 190 ft   **b.** 19 ft/sec; 34 ft/sec   **c.** The car stops if $v(t) = 0$, but here $v(t) > 0$ for
all nonnegative t.   **d.** 3 ft/sec^2; 3 ft/sec^2 **e.** Velocity is increasing; acceleration is constant   **79. a.** Cellular phones and DVD
players; the rate of growth of sales will now decline.   **b.** Food processors; the rate of decline of sales is starting to slow.
81. (22, 6,517.9)   **83.** (2.06, 20.8)

Exercises 5.4 (page 313)

1. 0

3.

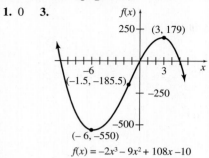

$f(x) = -2x^3 - 9x^2 + 108x - 10$

5.

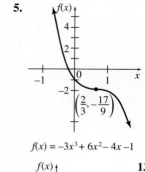

$f(x) = -3x^3 + 6x^2 - 4x - 1$

7.

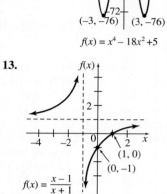

$f(x) = x^4 - 18x^2 + 5$

9.

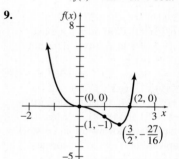

$f(x) = x^4 - 2x^3$

11.

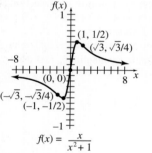

$f(x) = x + \dfrac{2}{x}$

13.

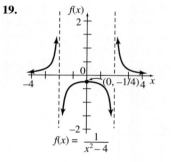

$f(x) = \dfrac{x-1}{x+1}$

15.

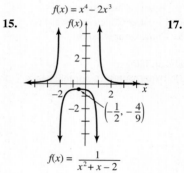

$f(x) = \dfrac{1}{x^2 + x - 2}$

17.

$f(x) = \dfrac{x}{x^2 + 1}$

19.

$f(x) = \dfrac{1}{x^2 - 4}$

21.

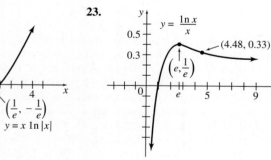

$y = x \ln |x|$

23.

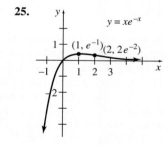

$y = \dfrac{\ln x}{x}$

25.

$y = xe^{-x}$

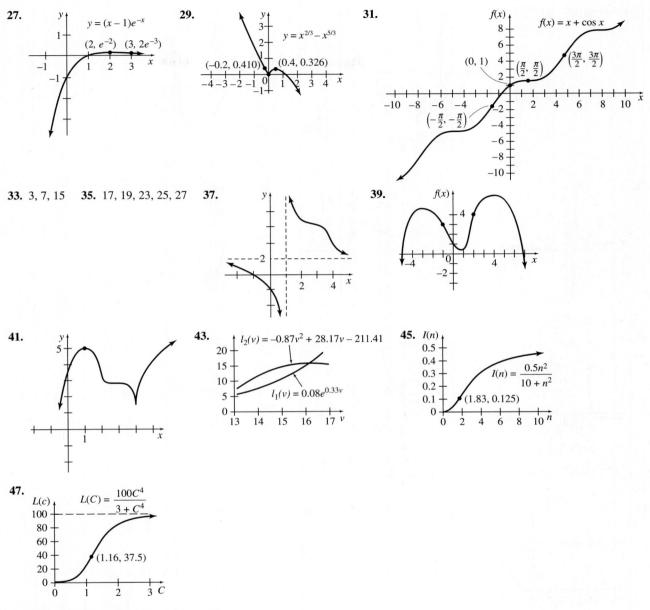

27. $y = (x-1)e^{-x}$ $(2, e^{-2})$ $(3, 2e^{-3})$

29. $y = x^{2/3} - x^{5/3}$ $(-0.2, 0.410)$ $(0.4, 0.326)$

31. $f(x) = x + \cos x$ $(0, 1)$ $\left(\frac{\pi}{2}, \frac{\pi}{2}\right)$ $\left(\frac{3\pi}{2}, \frac{3\pi}{2}\right)$ $\left(-\frac{\pi}{2}, -\frac{\pi}{2}\right)$

33. 3, 7, 15   **35.** 17, 19, 23, 25, 27   **37.**

39. $f(x)$ 4

41.

43. $l_2(v) = -0.87v^2 + 28.17v - 211.41$ $l_1(v) = 0.08e^{0.33v}$

45. $I(n)$ $I(n) = \dfrac{0.5n^2}{10 + n^2}$ $(1.83, 0.125)$

47. $L(c)$ $L(C) = \dfrac{100C^4}{3 + C^4}$ $(1.16, 37.5)$

Chapter 5 Review Exercises (page 316)

5. Increasing on $(5/2, \infty)$; decreasing on $(-\infty, 5/2)$ **7.** Increasing on $(-4, 2/3)$; decreasing on $(-\infty, -4)$ and $(2/3, \infty)$
9. Never increasing; decreasing on $(-\infty, 4)$ and $(4, \infty)$ **11.** Decreasing on $(-\infty, -1)$; increasing on $(1, \infty)$ **13.** Relative
maximum of -4 at 2   **15.** Relative minimum of -7 at 2   **17.** Relative maximum of 101 at -3; relative minimum of -24 at 2
19. Relative maximum at $(-0.618, 0.206)$; relative minimum at $(1.618, 13.203)$ **21.** $f''(x) = 36x^2 - 10$; 26; 314
23. $f''(x) = -68(2x + 3)^{-3}$ or $-68/(2x + 3)^3$; $-68/125$; $68/27$ **25.** $f''(t) = (t^2 + 1)^{-3/2}$ or $1/(t^2 + 1)^{3/2}$; $1/2^{3/2} \approx 0.354$;

$1/10^{3/2} \approx 0.032$ **27.**

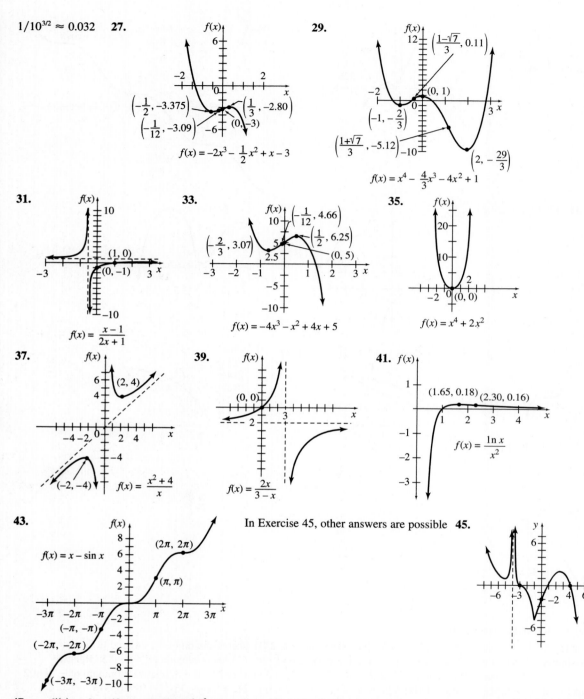

27.
$$f(x) = -2x^3 - \tfrac{1}{2}x^2 + x - 3$$
$\left(-\tfrac{1}{2}, -3.375\right)$ $\left(\tfrac{1}{3}, -2.80\right)$
$\left(-\tfrac{1}{12}, -3.09\right)$ $(0, -3)$

29.
$$f(x) = x^4 - \tfrac{4}{3}x^3 - 4x^2 + 1$$
$\left(\tfrac{1-\sqrt{7}}{3}, 0.11\right)$ $(0, 1)$
$\left(-1, -\tfrac{2}{3}\right)$
$\left(\tfrac{1+\sqrt{7}}{3}, -5.12\right)$ $\left(2, -\tfrac{29}{3}\right)$

31.
$$f(x) = \frac{x-1}{2x+1}$$
$(1, 0)$ $(0, -1)$

33.
$$f(x) = -4x^3 - x^2 + 4x + 5$$
$\left(-\tfrac{1}{12}, 4.66\right)$ $\left(\tfrac{1}{2}, 6.25\right)$
$\left(-\tfrac{2}{3}, 3.07\right)$ $(0, 5)$

35.
$$f(x) = x^4 + 2x^2$$
$(0, 0)$

37.
$$f(x) = \frac{x^2 + 4}{x}$$
$(2, 4)$ $(-2, -4)$

39.
$$f(x) = \frac{2x}{3-x}$$
$(0, 0)$

41. $f(x)$
$$f(x) = \frac{\ln x}{x^2}$$
$(1.65, 0.18)$ $(2.30, 0.16)$

43.
$$f(x) = x - \sin x$$
$(2\pi, 2\pi)$ (π, π) $(-\pi, -\pi)$ $(-2\pi, -2\pi)$ $(-3\pi, -3\pi)$

In Exercise 45, other answers are possible   **45.**

47. a. $f'(x) < 0$ **49. a.** 1,486 ml/m²; for males with 1.88 m² of surface area, the red cell volume increases approximately 1,486 ml for each additional m² of surface area.   **b.** 1.57 m²; 2,593 ml (Hurley); 2,484 ml (Pearson, et al.)   **c.** 1,578 ml/m²; for males with 1.57 m² of surface area, the red cell volume increases approximately 1,578 ml for each additional m² of surface area.

53. a.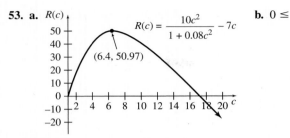

b. $0 \le C \le 17.13$ **55.** 7.405 yrs; the age at which the rate of learning to pass the test begins to slow down.

57. a. $v(t) = 512 - 32t$; $a(t) = -32$ **b.** 4,096 ft   **c.** After 32 sec; -512 ft/sec   **59. a.** $P'(t)$ is 0 and $P''(t)$ is negative

Chapter 6 Applications of the Derivative

Exercises 6.1 (page 327)

1. Absolute maximum at x_3; no absolute minimum   **3.** No absolute extrema   **5.** Absolute minimum at x_1; no absolute maximum   **7.** Absolute maximum at x_1; absolute minimum at x_2 **11.** Absolute maximum at $x = 5$; absolute minimum at $x = 0$ and $x = 3$ **13.** Absolute maximum at $x = -4$; absolute minimum at $x = 1$ **15.** Absolute maximum at $x = 0$; absolute minimum at $x = -3$ and $x = 3$ **17.** Absolute maximum at $x = 0$; absolute minimum at $x = 3$ **19.** Absolute maximum at $x = 1 + \sqrt{2} \approx 2.4$; absolute minimum at $x = 1$ **21.** Absolute maximum at $x = -2$ and $x = 2$; absolute minimum at $x = 0$ **23.** Absolute maximum at $x = -2$; absolute minimum at $x = 5/2$ **25.** Absolute maximum at $x = e$; absolute minimum at $x = e^{-1}$ **27.** Absolute maximum at $x = 0.6085$; absolute minimum at $x = -1$ **29.** Absolute maximum at $x = 4$; absolute minimum at $x = 2.18$ **31.** Absolute minimum at $x = 2$; no absolute maximum   **33.** Absolute maximum at $x = 3$; no absolute minimum   **35.** Absolute maximum at $x = 4$; absolute minimum at $x = -2$ **37.** 6 mo; 6%   **39.** About 8.5 mm   **41. a.** 4.5   **b.** About 15 days   **43. a.** 300   **b.** Maximum: about 242; minimum: about 44

45. a. Approximately 0; the power expended is not changing at that point.   **b.** About 0.1; the power expended is increasing 0.1 unit for each unit increase in speed.   **c.** About 0.12; the power expended increases 0.12 units for each 1 unit increase in speed.   **d.** The power level decreases to V_{mp}, which minimizes energy costs; then it increases at an alarming rate   **e.** V_{mr} **47.** 27.6; 15.8   **49.** Use all the wire to make a circle.   **51.** 20 units   **53.** 300 units

Exercises 6.2 (page 337)

1. a. $y = 100 - x$ **b.** $P = x(100 - x)$ **c.** $[0, 100]$ **d.** $P'(x) = 100 - 2x$; $x = 50$ **e.** $P(0) = 0$; $P(100) = 0$; $P(50) = 2,500$ **f.** 2,500; 50 and 50   **3.** 100; 50; 500,000   **5. a.** 15 days   **b.** 16.875%   **7. a.** 8 days   **b.** 29.4%   **9.** 12.86 thousand   **11.** 49.37   **13.** Point P is $3\sqrt{7}/7 \approx 1.134$ mi from Point A.   **15.** About 40°; about 12 m   **19.** $(56 - 2\sqrt{21})/7 \approx 6.7$ mi   **21.** $A(x) = x^2/2 + 2x - 3 + 35/x$; $x = 2.722$ **23. a.** $1,200 - 2x$ **b.** $A(x) = 1,200x - 2x^2$ **c.** 300 m   **d.** 180,000 m^2 **25.** 405,000 m^2 **27.** \$1,000   **29.** 4 in by 4 in by 2 in   **31.** 3 ft by 6 ft by 2 ft   **33.** 10 cm and 10 cm   **35.** Radius is 1.08 ft; height is 4.34 ft; cost is \$44.11 using the rounded values for the height and radius.   **37. a.** Both are 64 square in.   **b.** Both are 100/9 square ft.   **c.** They are equal. (This conjecture is true.)   **39.** 1 mi from point A **41.** Fire the assistant; when 330 tables are ordered, the revenue is \$27,225; when 660 tables are ordered, each one is free.   **43.** Radius $= 5.454$ cm, height $= 10.70$ cm

Exercises 6.3 (page 347)

1. $dy/dx = -4x/(3y)$ **3.** $dy/dx = (-6x - 4y)/(4x + y)$ **5.** $dy/dx = 3x^2/(2y)$ **7.** $dy/dx = -3x(2 + y)^2/2$ **9.** $dy/dx = -y^{1/2}/x^{1/2}$ **11.** $dy/dx = (4x^3y^3 + 6x^{1/2})/(9y^{1/2} - 3x^4y^2)$ **13.** $dy/dx = (5 - 2xye^{x^2y})/(x^2e^{x^2y} - 4)$ **15.** $dy/dx = y(2xy^3 - 1)/(1 - 3x^2y^3)$ **17.** $dy/dx = \sec(xy)/x - y/x$ **19.** $y = (3/4)x + 25/4$ **21.** $y = x + 2$ **23.** $y = x/64 + 7/4$ **25.** $y = x/\pi - 1/\pi$ **27.** $y = (5/2)x - 1/2$ **29.** $y = 1$ **31.** $y = -2x + 7$ **33. a.** $y = -(3/4)x + 25/2$; $y = (3/4)x - 25/2$

b.

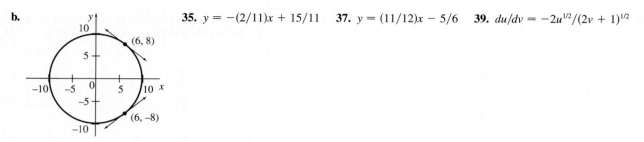

35. $y = -(2/11)x + 15/11$ **37.** $y = (11/12)x - 5/6$ **39.** $du/dv = -2u^{1/2}/(2v + 1)^{1/2}$

41. $R'(w) = -29.0716w^{-1.43}$ **43.** $-y/(ax)$ **45.** $ds/dt = \left(-s + 6\sqrt{st}\right)/\left(8s\sqrt{st} + t\right)$

Exercises 6.4 (page 353)

1. $-15/2$ **3.** $-5/7$ **5.** $1/5$ **7.** 0 **9.** 0.067 mm/min   **11.** About 1.9849 g/day   **13. a.** $dm/dt = 46.251w^{-0.46}\, dw/dt$
b. 0.875 kcal/day^2 **15.** $dE/dt = -0.0521$ kcal/kg/km/day   **17.** 25.6 crimes/month   **19.** 0.008 **21. a.** 50 mph
b. About 47.15 mph   **23.** -16π in^3/hr   **25.** 62.5 ft/min   **27.** $\sqrt{2} \approx 1.41$ ft/sec   **29.** $\$200$ per month   **31. a.** Revenue is
decreasing at a rate of $\$5,500$ per day.   **b.** Cost is increasing at a rate of $\$250$ per day.   **c.** Profit is decreasing at a rate of
$\$5,750$ per day.

Exercises 6.5 (page 361)

1. -2.6 **3.** 0.1 **5.** 0.130 **7.** -0.023 **9.** $12.0417; 12.0416; 0.0001$ **11.** $0.995; 0.9950; 0$ **13.** $1.01; 1.0101; 0.0001$
15. $0.05; 0.0488; 0.0012$ **17.** $0.03; 0.0300; 0$ **19. a.** Alcohol concentration increases by about 0.33 tenths of a percent.
b. Alcohol concentration decreases by about 0.13 tenths of a percent.   **21. a.** 0.347 million   **b.** -0.022 million
23. $1,568$ mm^3 **25.** 80π mm^2 **27. a.** About 9.3 kg   **b.** About 9.5 kg   **29.** -7.2π cm^3 **31.** ± 1.224 in^2
33. ± 0.116 in^3 **35.** About $9,600$ in^3

Chapter 6 Review Exercises (page 363)

1. Absolute maximum of $29/4$ at $5/2$; absolute minimum of 5 at 1 and 4 **3.** Absolute maximum of 39 at -3; absolute minimum
of $-319/27$ at $5/3$ **9.** $dy/dx = (-4y - 2xy^3)/(3x^2y^2 + 4x)$ **11.** $dy/dx = (-4 - 9x^{3/2})/(24x^2y^2)$
13. $dy/dx = (2y - 2y^{1/2})/(4y^{1/2} - x + 9y)$ (This form of the answer was obtained by multiplying both sides of the given function
by $x - 3y$.)   **15.** $y = -(23/16)x + 47/8$ **19.** $10/3$ **21.** $-2/9$ **23.** 0.1 **25.** -0.005 **27. a.** $(2, -5), (2, 4)$ **b.** $(2, 4)$
is a relative maximum and $(2, -5)$ is a relative minimum   **c.** No   **29.** 56π ft^2/min   **31.** About $19.82°C$
33. a.

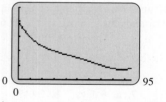

b. Maximum: about 237 at birth; minimum: about 44 **35.** $8/3$ ft/min

37. $21/16 = 1.3125$ ft/min   **39.** ± 0.736 in^2 **41.** $1.25 + 2\ln 1.5$ **43.** 2 m by 4 m by 4 m   **45.** 3 in

Chapter 7 Integration

Exercises 7.1 (page 378)

1. They differ only by a constant.   **5.** $6k + C$ **7.** $z^2 + 3z + C$ **9.** $t^3/3 - 2t^2 + 5t + C$ **11.** $z^4 + z^3 + z^2 - 6z + C$
13. $10z^{3/2}/3 + C$ **15.** $x^4/4 - 3x^2/2 + C$ **17.** $8v^{3/2}/3 - 6v^{5/2}/5 + C$ **19.** $4u^{5/2} - 4u^{7/2} + C$ **21.** $-1/z + C$
23. $-1/(2y^2) - 2y^{1/2} + C$ **25.** $9/t - 2\ln|t| + C$ **27.** $-1/(3x) + C$ **29.** $-15e^{-0.2x} + C$ **31.** $3\ln|x| - 8e^{-0.5x} + C$
33. $\ln|t| + 2t^3/3 + C$ **35.** $e^{2u}/2 + 2u^2 + C$ **37.** $x^3/3 + x^2 + x + C$ **39.** $6x^{7/6}/7 + 3x^{2/3}/2 + C$ **41.** $10^x/(\ln 10) + C$
43. $f(x) = 3x^{5/3}/5$ **45.** $a\ln x - bx + C$ **47. a.** $c'(t) = -(kA/V)(c_0 - C)e^{-kAt/V}$
49. a. $g(x) = -0.00039x^4 + 0.0104x^3 - 0.132x^2 + 0.137x + 2.4$ **b.** About 2.4; about $0.8 \approx 1$ **c.** No
51. $v(t) = t^3/3 + t + 6$ **53.** $s(t) = -16t^2 + 6,400$; 20 sec   **55.** $s(t) = 2t^{5/2} + 3e^{-t} + 1$

Exercises 7.2 (page 387)

3. $2(2x + 3)^5/5 + C$ **5.** $-(2m + 1)^{-2}/2 + C$ **7.** $-(x^2 + 2x - 4)^{-3}/3 + C$ **9.** $(z^2 - 5)^{3/2}/3 + C$ **11.** $-2e^{2p} + C$
13. $e^{2x^3}/2 + C$ **15.** $e^{2t-t^2}/2 + C$ **17.** $-e^{1/z} + C$ **19.** $(x^4 + 4x^2 + 7)^9/36 + C$ **21.** $-1/[2(x^2 + x)^2] + C$
23. $(p + 1)^7/7 - (p + 1)^6/6 + C$ **25.** $2(u - 1)^{3/2}/3 + 2(u - 1)^{1/2} + C$ **27.** $(x^2 + 12x)^{3/2}/3 + C$
29. $[\ln(t^2 + 2)]/2 + C$ **31.** $(1 + \ln x)^3/3 + C$ **33.** $(1/2) \ln(e^{2x} + 5) + C$ **37. a.** $N(t) = 155.3337e^{0.3218873t} + 144.666$
b. 7,537   **39. a.** $S(t) = 27.5e^{0.16t} - 0.2$ **b.** 4.3 years

Exercises 7.3 (page 397)

3. a. 56   **b.** $\int_0^8 (2x + 1)\, dx$ **7. a.** 15.5   **b.** 16.5   **c.** 16   **d.** 16   **9. a.** 10   **b.** 10   **c.** 10   **d.** 11   **11. a.** 8.22
b. 15.48   **c.** 11.85   **d.** 10.96   **13. a.** 6.70   **b.** 3.15   **c.** 4.93   **d.** 4.17   **15.** 12.5   **17.** $9\pi/2$ **19.** 6   **21. b.** 0.385
c. 0.33835   **d.** 0.334334   **e.** 0.333333   **23. a.** About 155 cal/cm^2 **b.** About 564 cal/cm^2 **25.** A concentration of about
20 units   **27. a.** About 52 billion barrels   **b.** About 65 billion barrels   **29.** About 1,900 ft   **31. a.** About 1,230 BTUs
b. About 230 BTUs   **33. a.** 9 ft   **b.** 2 seconds   **c.** 4.6 ft   **d.** Between 3 and 3.5 seconds   **35.** 22.5 and 18 ft

Exercises 7.4 (page 409)

1. -6 **3.** 3/2   **5.** 28/3   **7.** 13   **9.** 1/3   **11.** 76   **13.** 4/3   **15.** 112/25
17. $20e^{-0.2} - 20e^{-0.3} + 3 \ln 3 - 3 \ln 2 \approx 2.775$ **19.** $e^{10}/5 - e^5/5 - 1/2 \approx 4,375.1$ **21.** 91/3   **23.** $447/7 \approx 63.857$
25. $(\ln 2)^2/2 \approx 0.24023$ **27.** 49   **29.** $1/4 - 1/(3 + e) \approx 0.075122$ **31.** 42   **33.** 76   **35.** 41/2
37. $e^2 - 3 + 1/e \approx 4.757$ **39.** 1   **41.** 23/3   **43.** $e^2 - 2e + 1 \approx 2.9525$ **45. a.** Yes   **b.** Yes   **51. a.** 2.92530
b. 14.98998   **53.** 33.8 cm   **55. a.** $(K - Ke^{-at})/a$ **b.** approaches K/a **59. a.** About 414 barrels   **b.** About 191 barrels
c. It is decreasing to 0.   **61.** $1,000(e^{0.5} - 1) \approx 648.72$ **63.** $F(T) = k(b^T - 1)/(\ln b)$
65. $3(13^{4/3} - 1)/16 \approx 5.5439$ mg   **67.** 178 g   **69.** About 33%   **71.** $30(e^{0.04T} - 1)$; 6.64 billion barrels

Exercises 7.5 (page 418)

1. $(1/5) \sin 5x + C$ **3.** $5 \sin x - 2 \cos x + C$ **5.** $(-\cos x^2)/2 + C$ **7.** $-3 \tan 2x + C$ **9.** $(1/8) \sin^8 x + C$
11. $(2/3) \sin^{3/2} x + C$ **13.** $-\ln |1 + \cos x| + C$ **15.** $(1/6) \sin x^6 + C$ **17.** $-4 \ln |\cos (x/4)| + C$
19. $(1/3) \ln |\sin x^3| + C$ **21.** $-\cos e^x + C$ **23.** $-\csc e^x + C$ **25.** $1 - \sqrt{2}/2$ **27.** $-\ln(1/2)$ or $\ln 2$ **29.** $\sqrt{3}/2 - 1$
31. $(\tan \phi - \tan \theta)/2$ **33. a.**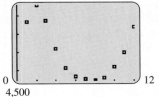

The data appear to be periodic, although there is a strong increase
in February due to an extended period of cold weather.

b. $C(x) = 20,277.8 \sin(0.484742x + 1.02112) + 21,442.2$.   **c.** 243,603 million cubic feet. The actual value is 240,755 million
cubic feet.   **35.** 4,430 hours. This result is relatively close to the actual value.

Exercises 7.6 (page 424)

1. 15   **3.** 4   **5.** 23/3   **7.** 366.1667   **9.** 4/3   **11.** $5 + \ln 6 \approx 6.792$ **13.** $6 \ln(3/2) - 6 + 2e^{-1} + 2e \approx 2.6051$
15. $e^2 - e - \ln 2 \approx 3.978$ **17.** 1/2   **19.** 1/20   **21.** $3(2^{4/3})/2 - 3(2^{7/3})/7 \approx 1.6199$ **23.** $\sqrt{2} - 1$
25. $\sqrt{2}/2 - 1 + (\ln 2)/2$ **27.** $-1.9241, -0.4164, 0.6650$ **29. a.** 71.25 gal   **b.** 25.00 hrs   **c.** 105.00 gal   **d.** 47.91 hrs
31. a. 8 yr   **b.** About $148   **c.** About $771   **33. a.** 39 days   **b.** $3,369.18   **c.** $484.02   **d.** $2,885.16

Chapter 7 Review Exercises (page 428)

5. $x^2 + 3x + C$ **7.** $x^3/3 - 3x^2/2 + 2x + C$ **9.** $2x^{3/2} + C$ **11.** $2x^{3/2}/3 + 9x^{1/3} + C$ **13.** $2x^{-2} + C$ **15.** $-3e^{2x}/2 + C$
17. $e^{2x^2}/6 + C$ **19.** $(3 \ln |x^2 - 1|)/2 + C$ **21.** $-(x^3 + 5)^{-3}/9 + C$ **23.** $-e^{-3x^4}/12 + C$ **25.** $-(1/2) \cos 2x + C$
27. $-(1/9) \ln |\cos 9x| + C$ **29.** $5 \tan x + C$ **31.** $-(1/6) \cos 3x^2 + C$ **33.** $-(2/3)(\cos x)^{3/2} + C$
35. $(-1/22) \ln |\cos 11x^2| + C$ **37.** $2(\sin x)^{7/2}/7 + C$ **39.** $(\tan^2 3x)/6 + C$ **41. a.** 0   **b.** 4.5   **43.** 28
47. $965/6 \approx 160.83$ **49.** 19/15   **51.** $8 \ln 6 \approx 14.334$ **53.** $5(e^{24} - e^4)/8 \approx 1.656 \times 10^{10}$ **55.** 1   **57.** 10π **59.** 18
61. $e^2 - 1 \approx 6.3891$ **63.** 64/3   **65.** 40/3   **67. a. and b.** 0.2784   **69. a.** About 4,600 pM   **b.** About 2,800 pM
c. About 60%   **73.** About 26.3 yr   **75. a.** $f(x) = 109.0e^{0.0198x} + 1.6$ **b.** 119.6, which differs from the actual value by 1.0.
77. About 17,500   **79.** Approximately 4,800 degree-days (the actual value according to the National Weather Service is
4,868 degree-days.)   **81.** $\theta = \pi/4$ or $45°$

Chapter 8 Further Techniques and Applications of Integration

Exercises 8.1 (page 443)

1. a. 2.7500 **b.** 2.6667 **c.** $8/3 \approx 2.6667$ **3. a.** 1.6833 **b.** 1.6222 **c.** $\ln 5 \approx 1.6094$ **5. a.** 16 **b.** 14.6667
c. $44/3 \approx 14.6667$ **7. a.** 0.9436 **b.** 0.8374 **b.** $4/5 = 0.8$ **9. a.** 0.895892 **b.** 0.860997 **c.** $e/2 - 1/2 \approx 0.859141$
11. a. 1.8961 **b.** 2.0046 **c.** 2 **13. a.** 5.9914 **b.** 6.1672 **c.** 6.2832; Simpson's rule **15.** b is true **17. a.** 0.2
b. 0.220703, 0.205200, 0.201302, 0.200325, 0.020703, 0.005200, 0.001302, 0.000325 **c.** $p = 2$ **19. a.** 0.2 **b.** 0.2005208,
0.2000326, 0.2000020, 0.2000001, 0.0005208, 0.0000326, 0.0000020, 0.0000001 **c.** $p = 4$ **21.** $M = 0.7355$; $S = 0.8048$
23. a. $y = b_0(t/7)^{b_1}e^{-b_2 t/7}$ **b.** 1,212 kg, 1,231 kg **c.** 1,224 kg, 1,250 kg **25. a.** 2.4759 **b.** 2.3572 **27.** About
30 mcg/ml; this represents the total amount of drug available to the patient. **29.** About 9 mcg/ml; this represents the total
effective amount of the drug available to the patient. **31. a.**  **b.** 71.5 **c.** 69.0

33. 3,979.24 **35. a.** 0.682690 **b.** 0.954498 **c.** 0.997293

Exercises 8.2 (page 454)

1. $xe^x - e^x + C$ **3.** $-5xe^{-3x}/3 - 5e^{-3x}/9 + 3e^{-3x} + C$ or $-5xe^{-3x}/3 + 22e^{-3x}/9 + C$ **5.** $-5e^{-1} + 3 \approx 1.1606$
7. $11 \ln 2 - 3 \approx 4.6246$ **9.** $(x^2 \ln x)/2 - x^2/4 + C$ **11.** $(-6/5)x \sin 5x - (6/25) \cos 5x + C$
13. $-8x \cos x + 8 \sin x + C$ **15.** $(-3/4)x^2 \sin 8x - (3/16)x \cos 8x + (3/128) \sin 8x + C$ **17.** $e^4 + e^2 \approx 61.9872$
19. $x^2 e^{2x}/2 - xe^{2x}/2 + e^{2x}/4 + C$ **21.** $243/8 - 3\sqrt[3]{2}/4 \approx 29.4301$ **23.** $4x^2 \ln(5x) + 7x \ln(5x) - 2x^2 - 7x + C$
25. $2x^2(x + 2)^{3/2}/3 - 8x(x + 2)^{5/2}/15 + 16(x + 2)^{7/2}/105 + C$ **27.** 0.13077 **29.** $9 \ln \left|x + \sqrt{x^2 + 9}\right| + C$
31. $-(3/11) \ln \left|(11 + \sqrt{121 - x^2})/x\right| + C$ **33.** $-1/(4x + 3) - (1/3) \ln |x/(4x + 3)| + C$
39. a. $(2/3)x(x + 1)^{3/2} - (4/15)(x + 1)^{5/2} + C$ **b.** $(2/5)(x + 1)^{5/2} - (2/3)(x + 1)^{3/2} + C$
41. a. $e^{-k}/k + 1 - 1/k$; 0.0405, 0.0205, 0.0103 **b.** $e^{-6k}/(5k) + e^{-k}(1 - 1/5k)$; 0.1676, 0.0933, 0.0493
43. $15(5e^6 + 1)/2 \approx 15,136$ **45.** $(169/2) \ln 13 - 42 \approx \$174.74$

Exercises 8.3 (page 461)

1. $8\pi/3$ **3.** $364\pi/3$ **5.** $386\pi/27$ **7.** $3\pi/2$ **9.** 18π **11.** $\pi(e^4 - 1)/2 \approx 84.19$ **13.** $\pi \ln 4 \approx 4.36$ **15.** $3,124\pi/5$
17. $16\pi/15$ **19.** π **21.** $4\pi/3$ **23.** $4\pi r^3/3$ **25.** $\pi r^2 h$ **27.** 19/3 **29.** 38/15 **31.** $e - 1 \approx 1.718$
33. $(5e^4 - 1)/8 \approx 33.999$ **35.** $4/\pi$ **37.** 3.9372 **39. a.** $2\pi k \int_0^R r(R^2 - r^2)\,dr$ **b.** $\pi k R^4/2$
41. a. $7(6 \ln 6 - 5) \approx 40.254$ **b.** $7(11 \ln 11 - 10)/2 \approx 57.319$ **c.** $7(16 \ln 16 - 15)/3 \approx 68.510$

Exercises 8.4 (page 467)

1. 1/2 **3.** Divergent **5.** −1 **7.** 10,000 **9.** 1 **11.** 3/5 **13.** 1 **15.** 4 **17.** Divergent **19.** 1 **21.** Divergent
23. Divergent **25.** Divergent **27.** 0 **29.** Divergent **31.** Divergent **33.** 1 **35.** 0 **39. a.** 2.8081, 3.7239, 4.4170,
6.7195, 9.0221 **b.** Divergent **c.** 0.8770, 0.9070, 0.9170, 0.9260, 0.9269 **d.** Convergent **41. a.** 9.9995, 49.9875, 99.9500,
995.0166 **b.** Divergent **c.** 100,000 **43.** 1/2 **45.** $K(a + br)/r^2$ **47.** 833.33 mg

Chapter 8 Review Exercises (page 469)

5. 0.58325; 0.60347 **7.** 2.9223; 2.8943 **9.** 10.27725 **11. a.** 4/3 **b.** 1.14626 **c.** 1.2522
13. $-2x(8 - x)^{5/2}/5 - 4(8 - x)^{7/2}/35 + C$ **15.** $xe^x - e^x + C$ **17.** $(2x + 3)(\ln |2x + 3| - 1)/2 + C$
19. $-(1/8) \ln |9 - 4x^2| + C$ **21.** $-(x + 2) \cos x + \sin x + C$ **23.** $(3e^4 + 1)/16 \approx 10.300$ **25.** $7(e^2 - 1)/4 \approx 11.181$
27. $125\pi/6 \approx 65.45$ **29.** $\pi(e^4 - e^{-2})/2 \approx 85.55$ **31.** $406\pi/15 \approx 85.03$ **33.** $7\pi r^2 h/12$ **35.** 13/6 **37.** Divergent
39. 1/10 **41.** Divergent **43.** 3 **45. a.** 8,208 kg **b.** 8,430 kg **c.** 8,558 kg **47.** 0.480 **49. a.** 391.7 **b.** 225
c. 266.7

Chapter 9 Multivariable Calculus

Exercises 9.1 (page 484)

1. a. 6 **b.** −8 **c.** −20 **d.** 43 **3. a.** $\sqrt{43}$ **b.** 6 **c.** $\sqrt{19}$ **d.** $\sqrt{11}$ **5. a.** 1 **b.** $1/(2\sqrt{2})$ **c.** $\sqrt{\pi}$ **d.** 1
7.   **9.**   **11.**

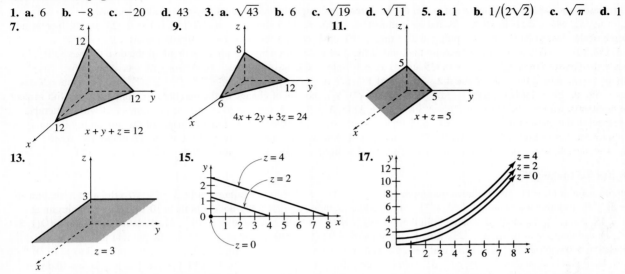

23. f **25.** a **27.** d **29. a.** $21x^2 + 21xh + 7h^2$ **b.** $16y + 8h$ **c.** $21x^2$ **d.** $16y$ **31.** $f(x, y) = 4 + 2x + 3y$
33. a. ferret: 1.5 m/sec, rhinoceros: 5.5 m/sec **b.** 1 m/sec **35. a.** $L(F, H, T) = F − (H + 0.16T)$ **b.** 300 mg/l
37. a. $D = 132.977375e^{0.005163R − 0.004345L − 0.000019F}$ **b.** On day 194; July 13 **39. a.** About 18.8 kcal/min **b.** About
29.7 kcal/min **c.** Worms and moths **41. a.** About 5,008 **b.**

43. $g(L, W, H) = 2LW + 2WH + 2LH$ **45. a.** 1,987 **b.** 595 **c.** 359,768 **47.** $M \approx 1.197$. The IRA account grows faster.

Exercises 9.2 (page 495)

1. a. $dz/dx = 24x − 8y$ **b.** $dz/dy = −8x + 6y$ **c.** 24 **d.** −20 **3.** $f_x(x, y) = −2y$; $f_y(x, y) = −2x + 18y^2$; 2; 170
5. $f_x(x, y) = 9x^2y^2$; $f_y(x, y) = 6x^3y$; 36; −1,152 **7.** $f_x(x, y) = e^{x+y}$; $f_y(x, y) = e^{x+y}$; e; $1/e$ **9.** $f_x(x, y) = −15e^{3x−4y}$;
$f_y(x, y) = 20e^{3x−4y}$; $−15e^{10}$; $20e^{−24}$ **11.** $f_x(x, y) = (−x^4 − 2xy^2 − 3x^2y^3)/(x^3 − y^2)^2$; $f_y(x, y) = (3x^3y^2 − y^4 + 2x^2y)/(x^3 − y^2)^2$;
$−8/49$; $−1,713/5,329$ **13.** $f_x(x, y) = 6xy^3/(1 + 3x^2y^3)$; $f_y(x, y) = 9x^2y^2/(1 + 3x^2y^3)$; 12/11; 1,296/1,297
15. $f_x(x, y) = e^{x^2y}(1 + 2x^2y)$; $f_y(x, y) = x^3e^{x^2y}$; $−7e^{−4}$; $−64e^{48}$ **17.** $f_x(x, y) = (4x^3 + 3y)/(2\sqrt{x^4 + 3xy + y^4 + 10})$;
$f_y(x, y) = (3x + 4y^3)/(2\sqrt{x^4 + 3xy + y^4 + 10})$; $29/(2\sqrt{21})$; $48/\sqrt{311}$
19. $f_x(x, y) = [(e^{xy} + 2)(6xy) − 3x^2y^2e^{xy}]/(e^{xy} + 2)^2$; $f_y(x, y) = [(e^{xy} + 2)(3x^2) − 3x^3ye^{xy}]/(e^{xy} + 2)^2$; $−24(e^{−2} + 1)/(e^{−2} + 2)^2$;
$(624e^{−12} + 96)/(e^{−12} + 2)^2$ **21.** $f_x(x, y) = \sin(\pi y)$; $f_y(x, y) = \pi x \cos(\pi y)$; 0; 4π **23.** $f_{xx}(x, y) = 36xy$; $f_{yy}(x, y) = −18$;
$f_{xy}(x, y) = f_{yx}(x, y) = 18x^2$ **25.** $R_{xx}(x, y) = 8 + 24y^2$; $R_{yy}(x, y) = −30xy + 24x^2$; $R_{xy}(x, y) = R_{yx}(x, y) = −15y^2 + 48yx$
27. $r_{xx}(x, y) = −8y/(x + y)^3$; $r_{yy}(x, y) = 8x/(x + y)^3$; $r_{xy}(x, y) = r_{yx}(x, y) = (4x − 4y)/(x + y)^3$ **29.** $z_{xx} = 0$; $z_{yy} = 4xe^y$;
$z_{xy} = z_{yx} = 4e^y$ **31.** $r_{xx} = −1/(x + y)^2$; $r_{yy} = −1/(x + y)^2$; $r_{xy} = r_{yx} = 2/(x + y)^2$ **33.** $z_{xx} = 1/x$; $z_{yy} = −x/y^2$;
$z_{xy} = z_{yx} = 1/y$ **35.** $x = −4, y = 2$ **37.** $x = 0, y = 0$; or $x = 3, y = 3$ **39.** $f_x(x, y, z) = 2x$; $f_y(x, y, z) = z$;
$f_z(x, y, z) = y + 4z^3$; $f_{yz}(x, y, z) = 1$ **41.** $f_x(x, y, z) = 6/(4z + 5)$; $f_y(x, y, z) = −5/(4z + 5)$; $f_z(x, y, z) = −4(6x − 5y)/(4z + 5)^2$;
$f_{yz}(x, y, z) = 20/(4z + 5)^2$ **43.** $f_x(x, y, z) = (2x − 5z^2)/(x^2 − 5xz^2 + y^4)$; $f_y(x, y, z) = 4y^3/(x^2 − 5xz^2 + y^4)$;
$f_z(x, y, z) = −10xz/(x^2 − 5xz^2 + y^4)$; $f_{yz}(x, y, z) = 40xy^3z/(x^2 − 5xz^2 + y^4)^2$ **45. a.** 6.773 **b.** 3.386 **47. a.** 168 **b.** 5,448

c. An increase in days since rain would cause more of an increase in matings. **49. a.** 5.71 **b.** -0.163 **c.** 0.0286 **d.** 0.163 **e.** Changing a produces the greatest decrease in the liters of blood pumped, while changing v produces the same amount of increase in the liters of blood pumped. **51. a.** 28 **b.** $dB/dw = 703/h^2$, and $db/dh = -1,406w/h^3$. This implies that as w increases, so does the body-mass index. On the other hand, as height increases, the body-mass index decreases. **c.** $B_m \approx w_m/h_m^2$, where w_m is in kilograms and h_m is in meters. **53. a.** 1.75 atm **b.** $(dp/dl)(33, 10) = 0.023$ atm/ft. Increasing the depth of the dive to 34 ft while keeping the dive time at 10 min increases the pressure by approximately 0.023 atm. $(dp/dt)(33, 10) = 0.035$ atm/min. Increasing the time of the dive to 11 min while keeping the depth of the dive time at 33 ft increases the pressure by approximately 0.035 atm. **c.** 6.17 min **55. a.** About 90 **b.** About 109 **c.** About 86 **d.** About 1 **e.** About 0.4 **f.** About 4.2 **g.** About 1 **57. a.** $D_F = -0.002527e^{0.005163R - 0.004345L - 0.000019F}$ **b.** About 3.7 days earlier **59. a.** $(N - 1)bke^{-kn}/\{aN[1 - (N - 1)b(1 - e^{-kn})/N]\}$; the rate of change of the number of wrong attempts with respect to the number of trials. **b.** $b(1 - e^{-kn})/\{aN^2[1 - (N - 1)^b(1 - e^{-kn})/N]\}$; the rate of change of the number of wrong attempts with respect to the number of stimuli. **61. a.** 2.04% **b.** 1.26% **c.** 0.05% is the rate of change of the probability for an additional semester of high school math; 0.003% is the rate of change of the probability per unit of change in the SAT score. **63. a.** 164,456 mi/sec **b.** 0.238 mi/sec² **c.** The speed of light, c **65. a.** Increase of $70 **b.** Increase of $54

Exercises 9.3 (page 506)

1. Saddle point at $(1, -1)$ **3.** Relative minimum at $(-1, -1/2)$ **5.** Relative minimum at $(-2, -2)$ **7.** Relative minimum at $(15, -8)$ **9.** Relative maximum at $(2/3, 4/3)$ **11.** Saddle point at $(2, -2)$ **13.** Saddle point at $(0, 0)$; relative minimum at $(4, 8)$ **15.** Saddle point at $(0, 0)$; relative minimum at $(9/2, 3/2)$ **17.** Saddle point at $(0, 0)$ **21.** Relative maximum of $1(1/8)$ at $(-1, 1)$; saddle point at $(0, 0)$; a. **23.** Relative minima of $-2(1/16)$ at $(0, 1)$ and at $(0, -1)$; saddle point at $(0, 0)$; b. **25.** Relative maxima of $1(1/16)$ at $(1, 0)$ and $(-1, 0)$; relative minima of $-15/16$ at $(0, 1)$ and at $(0, -1)$; saddle points at $(0, 0)$, $(-1, 1)$, $(1, -1)$, $(1, 1)$, and $(-1, -1)$; e. **31. a.** all values of a **b.** $a \geq 0$ **33. a.** 0.5148, 0.4064. The jury is less likely to make the correct decision when in the second situation. **b.** $r = s = 1$, $P(\alpha, 1, 1) = 1$. The values of r and s that give the maximum probability do not depend on α; however, α is likely to influence r and s in a real-life situation. **c.** $r = s = 1$; $P(\alpha, 1, 1) = 1$ **35.** $12 per unit of labor and $40 per unit of goods produce maximum profit of $274,400 **37.** 12 units of electrical tape and 25 units of packing tape give the minimum cost of $2,237.

Exercises 9.4 (page 512)

1. 5.072; 5.0720; 0 **3.** 2.0067; 2.0080; 0.0013 **5.** 0.99; 0.9896; 0.0004 **7.** -0.05; -0.0533; 0.0033 **9.** -0.2 **11.** 0.0311 **13.** -0.335 **15.** 6.65 cm³ **17.** 2.98 liters **19. a.** 0.265 **b.** Actual 0.282; approximation 0.281 **21. a.** 0.066 sec, 0.038 sec. In a close race, this could certainly affect the outcome. **b.** 0.0020 sec. This is the approximate change in time when the temperature decreases 5 degrees and the swimmer stands 0.5 m farther away from the starter in the y direction, while the starter's position is held fixed. **23.** 5.98 cm³ **25.** $2a + b$, the same as for a cylinder **27.** 18.4 cm³

Exercises 9.5 (page 524)

1. $93y/4$ **3.** $(1/9)[(48 + y)^{3/2} - (24 + y)^{3/2}]$ **5.** $(2x/9)[(x^2 + 15)^{3/2} - (x^2 + 12)^{3/2}]$ **7.** $6 + 10y$ **9.** $(1/4)e^{x+4} - (1/4)e^{x-4}$ **11.** $(1/2)e^{25+9y} - (1/2)e^{9y}$ **13.** $279/8$ **15.** $(2/45)(39^{5/2} - 12^{5/2} - 7,533)$ **17.** 21 **19.** $(\ln 2)^2$ **21.** $8\ln 2 + 4$ **23.** 256 **25.** $(4/15)(33 - 2^{5/2} - 3^{5/2})$ **27.** $-2\ln(6/7)$ or $2\ln(7/6)$ **29.** $(1/2)(e^7 - e^6 - e^3 + e^2)$ **31.** 1 **33.** 48 **35.** $4/3$ **37.** $(2/15)(2^{5/2} - 2)$ **39.** $(1/4)\ln(17/8)$ **41.** $e^2 - 3$ **43.** $97,632/105 \approx 929.83$ **45.** $128/9$ **47.** $\ln 3$ **49.** $64/3$ **51.** 1 **53.** 116 **55.** $10/3$ **57.** $7(e - 1)/3$ **59.** $16/3$ **61.** $4\ln 2 - 2$ **65.** $13/3$ **67.** $(e^7 - e^6 - e^5 + e^4)/2$ **69.** $2,583 **71.** $34,833

Chapter 9 Review Exercises (page 527)

5. 21; 936 **7.** $-\sqrt{5}/3$; $\sqrt{5}/3$ **9.** **11.**

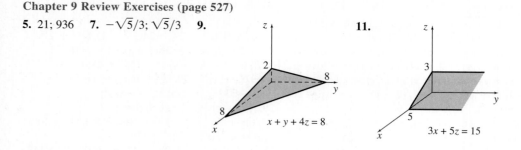

13.

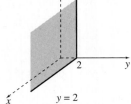

15. a. $4xy/(x - y^2)^2$ **b.** $-1/2$ **c.** 0   **17.** $f_x(x, y) = 30x^4y - 8y^9$; $f_y(x, y) = 6x^5 - 72xy^8$

19. $f_x(x, y) = (-6x^2 + 2y^2 - 30xy^2)/(3x^2 + y^2)^2$; $f_y(x, y) = (30x^2y - 4xy)/(3x^2 + y^2)^2$ **21.** $f_x(x, y) = (y - 2)^2e^{x+2y}$;
$f_y(x, y) = 2(y - 2)(y - 1)e^{x+2y}$ **23.** $f_x(x, y) = -2xy^3/(2 - x^2y^3)$; $f_y(x, y) = -3x^2y^2/(2 - x^2y^3)$
25. $f_x(x, y) = \tan(7x^2 + 4y^2) + 14x^2 \sec^2(7x^2 + 4y^2)$; $f_y(x, y) = 8xy \sec^2(7x^2 + 4y^2)$ **27.** $f_{xx}(x, y) = 2y$; $f_{xy}(x, y) = -24y^3 + 2x$
29. $f_{xx}(x, y) = 2(3 + y)/(x - 1)^3$; $f_{xy}(x, y) = -1/(x - 1)^2$ **31.** $f_{xx}(x, y) = 2ye^{x^2}(2x^2 + 1)$; $f_{xy}(x, y) = 2xe^{x^2}$
33. $f_{xx}(x, y) = -9y^4/(1 + 3xy^2)^2$; $f_{xy}(x, y) = 6y/(1 + 3xy^2)^2$ **35.** Relative minimum at $(-9/2, 4)$ **37.** Relative maximum at
$(-3/4, -9/32)$; saddle point at $(0, 0)$ **39.** Relative minimum at $(4/5, -9/10)$ **41.** Relative minimum at $(-8, -23)$ **43.** 1.7
45. 13.0846; 13.0848; 0.0002   **47.** $12y - 16$ **49.** $(3/2)[(100 + 2y^2)^{1/2} - (2y^2)^{1/2}]$ **51.** 1,232/9
53. $2[(42)^{5/2} - (24)^{5/2} - (39)^{5/2} + (21)^{5/2}]/135$ **55.** 2 ln 2 or ln 4   **57.** 26   **59.** $(4/15)(782 - 8^{5/2})$ **61.** $-1/2 + \pi/4$
63. 1,900   **65.** 1/2   **67.** 1/48   **69.** ln 2   **71.** 3   **73.** 1.3 cm^3 **75. a.** About 33.98 cm   **b.** About 0.027 cm/g; about
0.28 cm/yr; the approximate change in the length of a trout if its weight increases from 450 to 451 g while age is held constant at
7 years is 0.027 cm; the approximate change in the length of a trout if its age increases from 7 to 8 yrs while weight is held
constant at 450 g is 0.28 cm.   **77. a.** About 2.83 ft^2 **b.** An increase of about 0.6187 ft^2 **79.** 20,000 ft^2 with dimensions
100 ft by 200 ft   **81. a.** $26   **b.** $2,572   **83.** A decrease of $256.09   **85.** 4.19 ft^3 **87. a.** $200 spent on fertilizer and $80
spent on seed will produce maximum profit of $266 per acre.   **b.** Spend $200 on fertilizer and $80 on seed.

Chapter 10 Matrices

Exercises 10.1 (page 546)

1. $\begin{bmatrix} 2 & 2 & | & 11 \\ 1 & 3 & | & 8 \end{bmatrix}$ **3.** $\begin{bmatrix} 2 & 1 & 1 & | & 3 \\ 3 & -4 & 2 & | & -7 \\ 1 & 1 & 1 & | & 2 \end{bmatrix}$ **5.** $x = 2, y = 3$ **7.** $x = 2, y = 3, z = -2$ **9.** Row operations

11. $\begin{bmatrix} 2 & 3 & 8 & | & 20 \\ 0 & -5 & -4 & | & -4 \\ 0 & 3 & 5 & | & 10 \end{bmatrix}$ **13.** $\begin{bmatrix} 1 & 0 & -18 & | & -47 \\ 0 & 1 & 5 & | & 14 \\ 0 & 3 & 8 & | & 16 \end{bmatrix}$ **15.** $\begin{bmatrix} 1 & 0 & 0 & | & 6 \\ 0 & 5 & 0 & | & 9 \\ 0 & 0 & 4 & | & 8 \end{bmatrix}$ **17.** $(2, 3)$ **19.** $(5/2, -1)$

21. No solution   **23.** $((3y + 1)/6, y)$ **25.** $(3, 2, -4)$ **27.** No solution   **29.** $(-1, 23, 16)$
31. $((-9z + 5)/23, (10z - 3)/23, z)$ **33.** $((-2z + 80)/35, (3z + 13)/7, z)$ **35.** $((9 - 3y - z)/2, y, z)$ **37.** $(0, 2, -2, 1)$ The
answers are given in the order x, y, z, w.   **39.** $(-w - 3, -4w - 19, -3w - 2, w)$ **41.** $(28.9436, 36.6326, 9.6390, 37.1036)$
43. a. About March 30; about April 3 (1980 was a leap year)   **b.** About 1991   **45.** 40 units of corn, 15 units of soybeans, and
30 units of cottonseed   **47.** 81 kg of the first chemical, 382.286 kg of the second, and 286.714 kg of the third   **49.** 244 of A,
39 of B, and 101 of C (rounded)   **51. b.** 10,366,482 white bulls, 7,460,514 black bulls, 7,358,060 spotted bulls, 7,206,360 white
cows, 4,893,246 black cows, 3,515,820 spotted cows, 5,439,213 brown cows   **53.** 2,000 chairs, 1,600 cabinets, 2,500 buffets
55. a. $10,000 at 13%, $7,000 at 14%, $8,000 at 12%   **b.** It must be between $1,000 and $13,000. $16,000 is borrowed at 13%
and $4,000 at 14%.   **c.** No   **57. a.** About 51.8 weeks; about $959,091   **59. a.** $r = 175,000$; $b = 375,000$
61. a. $y = -0.032x^2 + 0.852x + 0.632$ **b.** Same as a

Exercises 10.2 (page 557)

1. False; not all corresponding elements are equal.   **3.** True   **5.** True   **7.** 2×2; square; $\begin{bmatrix} 4 & -8 \\ -2 & -3 \end{bmatrix}$

9. 3×4; $\begin{bmatrix} 6 & -8 & 0 & 0 \\ -4 & -1 & -9 & -2 \\ -3 & 5 & -7 & -1 \end{bmatrix}$ **11.** 2×1; column; $\begin{bmatrix} -2 \\ -4 \end{bmatrix}$ **13.** The $n \times m$ zero matrix   **15.** $x = 2, y = 4, z = 8$

17. $x = -15, y = 5, k = 3$ **19.** $z = 18, r = 3, s = 3, p = 3, a = 3/4$ **21.** $\begin{bmatrix} 9 & 12 & 0 & 2 \\ 1 & -1 & 2 & -4 \end{bmatrix}$ **23.** Not possible

25. $\begin{bmatrix} 1 & 5 & 6 & -9 \\ 5 & 7 & 2 & 1 \\ -7 & 2 & 2 & -7 \end{bmatrix}$ **27.** $\begin{bmatrix} 3 & 4 \\ 4 & 8 \end{bmatrix}$ **29.** $\begin{bmatrix} 3 & 12 \\ -6 & 3 \end{bmatrix}$ **31.** $\begin{bmatrix} -12x + 8y & -x + y \\ x & 8x - y \end{bmatrix}$ **33.** $\begin{bmatrix} -x & -y \\ -z & -w \end{bmatrix}$

39. a. $\begin{bmatrix} 2 & 1 & 2 & 1 \\ 3 & 2 & 2 & 1 \\ 4 & 3 & 2 & 1 \end{bmatrix}$ **b.** $\begin{bmatrix} 5 & 0 & 7 \\ 0 & 10 & 1 \\ 0 & 15 & 2 \\ 10 & 12 & 8 \end{bmatrix}$ **c.** $\begin{bmatrix} 8 \\ 4 \\ 5 \end{bmatrix}$ **41. a.** 8 **b.** 3 **c.** $\begin{bmatrix} 85 & 15 \\ 27 & 73 \end{bmatrix}$ **d.** Yes **43. a.** $\begin{bmatrix} 60.0 & 68.3 \\ 63.6 & 72.5 \\ 64.5 & 73.6 \\ 67.2 & 74.7 \end{bmatrix}$

b. $\begin{bmatrix} 68.0 & 75.6 \\ 70.7 & 78.1 \\ 72.7 & 79.4 \\ 74.3 & 79.9 \end{bmatrix}$ **c.** $\begin{bmatrix} -8.0 & -7.3 \\ -6.9 & -5.6 \\ -8.2 & -5.8 \\ -7.1 & -5.2 \end{bmatrix}$ **45. a.** $\begin{bmatrix} 18.2 & 2.8 \\ 30.1 & 4.2 \\ 50.8 & 8.4 \\ 65.8 & 11.9 \\ 73.4 & 13.6 \\ 75.2 & 13.9 \end{bmatrix}$ **b.** $\begin{bmatrix} 21.8 & 3.3 \\ 32.5 & 4.6 \\ 51.5 & 8.3 \\ 66.5 & 10.8 \\ 74.1 & 12.9 \\ 76.7 & 15.4 \end{bmatrix}$ **c.** $\begin{bmatrix} -3.6 & -0.5 \\ -2.4 & -0.4 \\ -0.7 & 0.1 \\ -0.7 & 1.1 \\ -0.7 & 0.7 \\ -1.5 & -1.5 \end{bmatrix}$

47. a. Chicago: $\begin{bmatrix} 4.05 & 7.01 \\ 3.27 & 3.51 \end{bmatrix}$, Seattle: $\begin{bmatrix} 4.40 & 6.90 \\ 3.54 & 3.76 \end{bmatrix}$ **b.** $\begin{bmatrix} 4.42 & 7.43 \\ 3.38 & 3.62 \end{bmatrix}$

Exercises 10.3 (page 568)

1. $\begin{bmatrix} -4 & 8 \\ 0 & 6 \end{bmatrix}$ **3.** $\begin{bmatrix} 24 & -8 \\ -16 & 0 \end{bmatrix}$ **5.** $\begin{bmatrix} -22 & -6 \\ 20 & -12 \end{bmatrix}$ **7.** $2 \times 2; 2 \times 2$ **9.** $3 \times 2; BA$ does not exist.

11. AB does not exist; 3×2 **13.** Columns; rows **15.** $\begin{bmatrix} 13 \\ 25 \end{bmatrix}$ **17.** $\begin{bmatrix} 5 \\ -21 \end{bmatrix}$ **19.** $\begin{bmatrix} 16 & -5 & 2 \\ 9 & -2 & 6 \end{bmatrix}$ **21.** $\begin{bmatrix} -2 & 10 \\ 0 & 8 \end{bmatrix}$

23. $\begin{bmatrix} 13 & 5 \\ 25 & 15 \end{bmatrix}$ **25.** $\begin{bmatrix} 13 \\ 29 \end{bmatrix}$ **27.** $\begin{bmatrix} 110 \\ 40 \\ -50 \end{bmatrix}$ **29.** $\begin{bmatrix} 22 & -8 \\ 11 & -4 \end{bmatrix}$ **31. a.** $\begin{bmatrix} 16 & 22 \\ 7 & 19 \end{bmatrix}$ **b.** $\begin{bmatrix} 5 & -5 \\ 0 & 30 \end{bmatrix}$ **c.** No **d.** No

39. a. $\begin{bmatrix} 6 & 106 & 158 & 222 & 28 \\ 120 & 139 & 64 & 75 & 115 \\ -146 & -2 & 184 & 144 & -129 \\ 106 & 94 & 24 & 116 & 110 \end{bmatrix}$ **b.** Cannot be found **c.** No **41. a.** $\begin{bmatrix} -1 & 5 & 9 & 13 & -1 \\ 7 & 17 & 2 & -10 & 6 \\ 18 & 9 & -12 & 12 & 22 \\ 9 & 4 & 18 & 10 & -3 \\ 1 & 6 & 10 & 28 & 5 \end{bmatrix}$

b. $\begin{bmatrix} -2 & -9 & 90 & 77 \\ -42 & -63 & 127 & 62 \\ 413 & 76 & 180 & -56 \\ -29 & -44 & 198 & 85 \\ 137 & 20 & 162 & 103 \end{bmatrix}$ **c.** $\begin{bmatrix} -56 & -1 & 1 & 45 \\ -156 & -119 & 76 & 122 \\ 315 & 86 & 118 & -91 \\ -17 & -17 & 116 & 51 \\ 118 & 19 & 125 & 77 \end{bmatrix}$ **d.** $\begin{bmatrix} 54 & -8 & 89 & 32 \\ 114 & 56 & 51 & -60 \\ 98 & -10 & 62 & 35 \\ -12 & -27 & 82 & 34 \\ 19 & 1 & 37 & 26 \end{bmatrix}$

e. $\begin{bmatrix} -2 & -9 & 90 & 77 \\ -42 & -63 & 127 & 62 \\ 413 & 76 & 180 & -56 \\ -29 & -44 & 198 & 85 \\ 137 & 20 & 162 & 103 \end{bmatrix}$ **f.** Yes **43. a.** $\begin{bmatrix} 20 & 52 & 27 \\ 25 & 62 & 35 \\ 30 & 72 & 43 \end{bmatrix}$ The rows represent the amounts of fat, carbohydrates, and protein,

respectively, in each of the daily meals. **b.** $\begin{bmatrix} 75 \\ 45 \\ 70 \\ 168 \end{bmatrix}$ The rows give the number of calories in one exchange of each of the food groups. **c.** Both equal $\begin{bmatrix} 503 \\ 623 \\ 743 \end{bmatrix}$. The rows give the number of calories in each meal. **45.** $\begin{bmatrix} 64.0 & 72.0 \\ 67.3 & 75.3 \\ 68.6 & 76.5 \\ 70.8 & 77.3 \end{bmatrix}$

47. a. $\begin{bmatrix} 0.038 & 0.014 \\ 0.022 & 0.008 \\ 0.023 & 0.007 \\ 0.014 & 0.009 \\ 0.010 & 0.011 \end{bmatrix}$; $\begin{bmatrix} 283 & 1,627 & 218 & 199 & 425 \\ 360 & 2,038 & 286 & 227 & 460 \\ 468 & 2,498 & 362 & 252 & 484 \\ 621 & 2,987 & 443 & 278 & 498 \\ 798 & 3,451 & 523 & 306 & 508 \end{bmatrix}$ **b.** $\begin{array}{c} \\ 1960 \\ 1970 \\ 1980 \\ 1990 \\ 2000 \end{array}\begin{array}{cc} \text{Births} & \text{Deaths} \\ \begin{bmatrix} 58.598 & 24.97 \\ 72.872 & 30.449 \\ 89.434 & 36.662 \\ 108.373 & 43.671 \\ 127.639 & 50.783 \end{bmatrix} \end{array}$ **49. a.** $\begin{array}{c} \\ \text{Dept.1} \\ \text{Dept. 2} \\ \text{Dept. 3} \\ \text{Dept. 4} \end{array}\begin{array}{cc} \text{A} \quad \text{B} \\ \begin{bmatrix} 57 & 70 \\ 41 & 54 \\ 27 & 40 \\ 39 & 40 \end{bmatrix} \end{array}$

b. Supplier A: \$164; Supplier B: \$204; Supplier A **51. a.** $\begin{bmatrix} 80 & 40 & 120 \\ 60 & 30 & 150 \end{bmatrix}$ **b.** $\begin{bmatrix} 1/2 & 1/5 \\ 1/4 & 1/5 \\ 1/4 & 3/5 \end{bmatrix}$ **c.** $PF = \begin{bmatrix} 80 & 96 \\ 75 & 108 \end{bmatrix}$ The rows

give the average price per pair of footwear sold by each store, and the columns give the state.

Exercises 10.4 (page 581)

1. Yes **3.** No **5.** No **7.** Yes **9.** No; the row of all zeros makes it impossible to get all the 1s in the diagonal of the identity

matrix, no matter what matrix is used as an inverse. **11.** $\begin{bmatrix} 0 & 1/2 \\ -1 & 1/2 \end{bmatrix}$ **13.** $\begin{bmatrix} 2 & 1 \\ 5 & 3 \end{bmatrix}$ **15.** No inverse **17.** $\begin{bmatrix} 1 & 0 & 0 \\ 0 & -1 & 0 \\ -1 & 0 & 1 \end{bmatrix}$

19. $\begin{bmatrix} 15 & 4 & -5 \\ -12 & -3 & 4 \\ -4 & -1 & 1 \end{bmatrix}$ **21.** No inverse **23.** $\begin{bmatrix} 7/4 & 5/2 & 3 \\ -1/4 & -1/2 & 0 \\ -1/4 & -1/2 & -1 \end{bmatrix}$ **25.** $\begin{bmatrix} 1/2 & 1/2 & -1/4 & 1/2 \\ -1 & 4 & -1/2 & -2 \\ -1/2 & 5/2 & -1/4 & -3/2 \\ 1/2 & -1/2 & 1/4 & 1/2 \end{bmatrix}$ **27.** $(-1, 4)$

29. $(2, 1)$ **31.** $(2, 3)$ **33.** No inverse, $(-8y - 12, y)$ **35.** $(-8, 6, 1)$ **37.** $(15, -5, -1)$ **39.** No inverse, no solution for system

41. $(-7, -34, -19, 7)$ **51.** Entries are rounded to four places. $\begin{bmatrix} -0.0447 & -0.0230 & 0.0292 & 0.0895 & -0.0402 \\ 0.0921 & 0.0150 & 0.0321 & 0.0209 & -0.0276 \\ -0.0678 & 0.0315 & -0.0404 & 0.0326 & 0.0373 \\ 0.0171 & -0.0248 & 0.0069 & -0.0003 & 0.0246 \\ -0.0208 & 0.0740 & 0.0096 & -0.1018 & 0.0646 \end{bmatrix}$

53. Entries are rounded to four places. $\begin{bmatrix} 0.0394 & -0.0880 & 0.0033 & 0.0530 & -0.1499 \\ -0.1492 & 0.0289 & 0.0187 & 0.1033 & 0.1668 \\ -0.1330 & -0.0543 & 0.0356 & 0.1768 & 0.1055 \\ 0.1407 & 0.0175 & -0.0453 & -0.1344 & 0.0655 \\ 0.0102 & -0.0653 & 0.0993 & 0.0085 & -0.0388 \end{bmatrix}$ **55.** Yes **57.** $\begin{bmatrix} 1.51482 \\ 0.053479 \\ -0.637242 \\ 0.462629 \end{bmatrix}$

59. a. 50 Super Vim, 75 Multitab, and 100 Mighty Mix **b.** 75 Super Vim, 50 Multitab, and 60 Mighty Mix **c.** 80 Super Vim, 100 Multitab, and 50 Mighty Mix **61. a.** 100 transistors, 110 resistors, and 90 computer chips **b.** 95 transistors, 100 resistors, and 90 computer chips **c.** 140 transistors, 130 resistors, and 100 computer chips **63. a.** 40 lb pretzels, 20 lb dried fruit, and 80 lb nuts **b.** 24 lb pretzels, 12 lb dried fruit, and 76 lb nuts **c.** 54 lb pretzels, 27 lb dried fruit, and 45 lb nuts

65. a. $\begin{bmatrix} -1/15 & -1/15 & 13/3 \\ 4/15 & 1/15 & -40/3 \\ -1/5 & 0 & 10 \end{bmatrix}$ **b.** $\begin{bmatrix} 0 & 1 & 40 \\ -1 & 0 & 60 \\ 0 & 0 & 1 \end{bmatrix}$ **c.** The shape is a sideways T whose vertical and horizontal intersection is at the mark $(60, 10)$.

Exercises 10.5 (page 593)

1. 1, 5; $\begin{bmatrix} 0 \\ 1 \end{bmatrix}$, $\begin{bmatrix} 2 \\ 1 \end{bmatrix}$ **3.** 2, 9; $\begin{bmatrix} 2 \\ -1 \end{bmatrix}$, $\begin{bmatrix} 1 \\ 3 \end{bmatrix}$ **5.** 2, 1; $\begin{bmatrix} 3 \\ 2 \end{bmatrix}$, $\begin{bmatrix} 1 \\ 1 \end{bmatrix}$ **7.** 4, −1, −3; $\begin{bmatrix} 35 \\ 21 \\ 25 \end{bmatrix}$, $\begin{bmatrix} 0 \\ 2 \\ 5 \end{bmatrix}$, $\begin{bmatrix} 0 \\ 0 \\ 1 \end{bmatrix}$ **11.** 4; $\begin{bmatrix} 3 \\ 2 \end{bmatrix}$

13. $2 + 3i, 2 - 3i$; $\begin{bmatrix} 1 - 3i \\ 2 \end{bmatrix}$, $\begin{bmatrix} 1 + 3i \\ 2 \end{bmatrix}$ **15.** $\begin{bmatrix} 5,000 \\ 5,000 \end{bmatrix}$, 1.4 **17.** $\begin{bmatrix} 5,000 \\ 5,000 \end{bmatrix}$, 0.9 **19.** $\begin{bmatrix} 6,000 \\ 4,000 \end{bmatrix}$, 0.9

21. $X(n) \approx (51,000/23)(1.4)^n \begin{bmatrix} 1 \\ 1 \end{bmatrix}$

Chapter 10 Review Exercises (page 595)

3. $(-4, 6)$ **5.** $(-1, 2, 3)$ **7.** $(-9, 3)$ **9.** $(7, -9, -1)$ **11.** $(6 - 7z/3, 1 + z/3, z)$ **13.** 3×2; $a = 2, x = -1, y = 4, p = 5$,

$z = 7$ **15.** 3×3 (square); $a = -12, b = 1, k = 9/2, c = 3/4, d = 3, l = -3/4, m = -1, p = 3, q = 9$ **17.** $\begin{bmatrix} 8 & -6 \\ -10 & -16 \end{bmatrix}$

19. Not possible **21.** $\begin{bmatrix} 26 & 86 \\ -7 & -29 \\ 21 & 87 \end{bmatrix}$ **23.** $\begin{bmatrix} 6 & 18 & -24 \\ 1 & 3 & -4 \\ 0 & 0 & 0 \end{bmatrix}$ **25.** $\begin{bmatrix} 15 \\ 16 \\ 1 \end{bmatrix}$ **27.** $\begin{bmatrix} -7/19 & 4/19 \\ 3/19 & 1/19 \end{bmatrix}$ **29.** No inverse

31. $\begin{bmatrix} -1/4 & 1/6 \\ 0 & 1/3 \end{bmatrix}$ **33.** No inverse **35.** $\begin{bmatrix} 1/4 & 1/2 & 1/2 \\ 1/4 & -1/2 & 1/2 \\ 1/8 & -1/4 & -1/4 \end{bmatrix}$ **37.** No inverse **39.** Matrix A has no inverse.

Solution: $(-2y + 5, y)$ **41.** $X = \begin{bmatrix} 6 \\ 15 \\ 16 \end{bmatrix}$ **43.** $(34, -9)$ **45.** $(1, 2, 3)$ **47.** $8, 2$; $\begin{bmatrix} 1 \\ 1 \end{bmatrix}$, $\begin{bmatrix} 1 \\ -1 \end{bmatrix}$ **49.** $-2, 3, 2$; $\begin{bmatrix} 10 \\ -2 \\ 3 \end{bmatrix}$, $\begin{bmatrix} 0 \\ 1 \\ 1 \end{bmatrix}$, $\begin{bmatrix} 0 \\ 0 \\ 1 \end{bmatrix}$

51. $\begin{bmatrix} 8 & 8 & 8 \\ 10 & 5 & 9 \\ 7 & 10 & 7 \\ 8 & 9 & 7 \end{bmatrix}$ **53. a.** No **b. (i)** 0.23, 0.37, 0.42; A is healthy; B and D are tumorous, C is bone. **(ii)** 0.33, 0.27, 0.32;

A and C are tumorous, B could be healthy or tumorous, D is bone. **c.** 0.2, 0.4, 0.45, 0.3; A is healthy, B and C are bone, D is tumorous. **d.** Any four beams except $(1, 2, 3, 4)$, $(1, 2, 5, 6)$, or $(3, 4, 5, 6)$. One example is to choose beams 1, 2, 3, and 6.

55. $\begin{bmatrix} 2,500 \\ 7,500 \end{bmatrix}$, 0.9 **57. d.** $\begin{bmatrix} 1,191 \\ 773 \\ 213 \\ 185 \\ 371 \\ 325 \end{bmatrix}$ **f.** $\begin{bmatrix} 19,797 \\ 17,233 \\ 4,093 \\ 3,580 \\ 7,520 \\ 6,580 \end{bmatrix}$ **59. a.** $x = 1, y = 1/2$. These are the values that balance the equation.

b. $x = 1, y = 1, z = -1$ **61.** $W_1 = 300/(1 + \sqrt{3}) \approx 110$ lb; $W_2 = 300\sqrt{3}/[(1 + \sqrt{3})\sqrt{2}] \approx 134$ lb **63.** 8,000 standard,

6,000 extra large **65.** 5 blankets, 3 rugs, 8 skirts **67. a.** $\begin{bmatrix} 3,170 \\ 2,360 \\ 1,800 \end{bmatrix}$ **b.** $\begin{bmatrix} x \\ y \\ z \end{bmatrix}$ **c.** $\begin{bmatrix} 10 & 5 & 8 \\ 12 & 0 & 4 \\ 0 & 10 & 5 \end{bmatrix} \begin{bmatrix} x \\ y \\ z \end{bmatrix} = \begin{bmatrix} 3,170 \\ 2,360 \\ 1,800 \end{bmatrix}$ **d.** $\begin{bmatrix} 150 \\ 110 \\ 140 \end{bmatrix}$

Chapter 11 Differential Equations

Exercises 11.1 (page 611)

1. $y = -x^2 + x^3 + C$ **3.** $y = 3x^4/8 + C$ **5.** $y^2 = x^2 + C$ **7.** $y = ke^{x^2}$ **9.** $y = ke^{x^3 - x^2}$ **11.** $y = Mx$
13. $y = Me^x + 5$ **15.** $y = -1/(e^x + C)$ **17.** $\cos y = -\sin x + C$ **19.** $y = x^3 - x^2 + 2$ **21.** $y = -2xe^{-x} - 2e^{-x} + 44$
23. $y^2 = 2x^3/3 + 9$ **25.** $y = e^{x^2 + 3x}$ **27.** $y = x^4 - x^3 + x^2/2 - 1/2$ **29.** $y = 5/(6 - 5 \ln |x|)$
31. $y = (e^{x-1} - 3)(e^{x-1} - 2)$ **37. a.** 6,000 **b.** $y = 11,074/(1 + 151,378e^{-0.1219x})$

c. 6,000

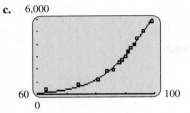

60   100
0

d. 11,074   **39. a.** $N = (N_0 + i/m)e^{mt} - i/m$ **c.** $-0.344, 1.65$

41. a. $I = 2.4 - 1.4e^{-0.088W}$ **b.** I approaches 2.4.   **43. a.** $dw/dt = k(C - 17.5w)$; the calorie intake per day is constant.
b. pounds/calorie   **c.** $dw/dt = 1/3{,}500(C - 17.5w)$ **d.** $w = C/17.5 - e^{-0.005(M+t)}/17.5$
e. $w = C/17.5 + (w_0 - C/17.5)e^{-0.005t}$ **45. a.** 175,000   **b.** $y = 484{,}900/(1 + 28.32e^{-0.02893x})$

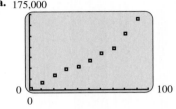

0   100
0

c. 175,000

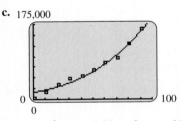

0   100
0

d. 484,900   **47.** $y = 35.5e^{0.01070t}$ **49. a.** $k \approx 0.8$ **b.** 11   **c.** 55   **d.** 2,981

51. a. $dy/dt = -0.05y$ **b.** $y = Me^{-0.05t}$ **c.** $y = 90e^{-0.05t}$ **d.** 55 g

Exercises 11.2 (page 621)

1. $y = 5/2 + Ce^{-2x}$ **3.** $y = 3 + Ce^{-x^2/2}$ **5.** $y = x \ln x + Cx$ **7.** $y = -1/2 + Ce^{x^2/2}$ **9.** $y = x^2/4 + x + Cx^{-2}$
11. $y = -x^3/2 + Cx$ **13.** $y = -x \cot x + 1 + C/\sin x$ **15.** $y = e^x + 99e^{-x}$ **17.** $y = -1 + 11e^{(x^2-1)/2}$
19. $y = x^2/7 + 2{,}560/(7x^5)$ **21.** $y = (3 + 197e^{4-x})/x$ **25. a.** $y = c/(p + Kce^{-cx})$ **b.** $y = cy_0/[py_0 + (c - py_0)e^{-cx}]$
c. c/p **27.** $y = 1.02e^t + 9{,}999e^{0.02t}$ (rounded)   **29.** $y = 50t + 2{,}500 + 7{,}500e^{0.02t}$ **33.** The temperature approaches T_0, the
temperature of the surrounding medium. This agrees with reality.   **35. a.** $T = 88.6e^{-0.24t} + 10$
b. 100

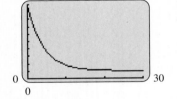

0   30
0

c. Just after death—the graph shows that the most rapid decrease occurs in the first few
hours.

d. About 43.9°F   **e.** About 4.5 hours

Exercises 11.3 (page 628)

1. 1.837   **3.** 4.315   **5.** 1.491   **7.** 2.022   **9.** 0.595   **11.** $-0.540; -0.520$ **13.** 2.030; 2.042   **15.** 3.806; 4.759
17. 1.371; 1.419   **19.** 1.393; the error is about half.

21.

x_i	y_i	$y(x_i)$	$y_i - y(x_i)$
0	0	0	0
0.2	0	0.08772053	-0.08772053
0.4	0.11696071	0.22104189	-0.10408118
0.6	0.26432197	0.37954470	-0.11522273
0.8	0.43300850	0.55699066	-0.12398216
1.0	0.61867206	0.75000000	-0.13132794

23.

x_i	y_i	$y(x_i)$	$y_i - y(x_i)$
0	0	0	0
0.2	0.2	0.1812692	0.0187308
0.4	0.36	0.32967995	0.03032005
0.6	0.488	0.45118836	0.03681164
0.8	0.5904	0.55067104	0.03972896
1.0	0.67232	0.63212056	0.04019944

25.

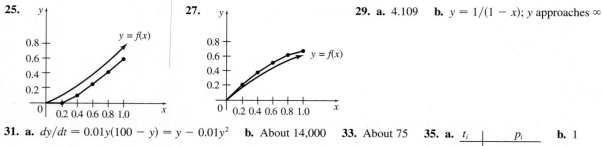

27.

29. a. 4.109   **b.** $y = 1/(1 - x)$; y approaches ∞

31. a. $dy/dt = 0.01y(100 - y) = y - 0.01y^2$ **b.** About 14,000   **33.** About 75   **35. a.**

t_i	p_i
0	0.1
5	0.1675
10	0.297678
15	0.536660
20	0.846644
25	0.963605
30	0.980024

b. 1

Exercises 11.4 (page 637)

1. $x_1 = C_1e^t + 2C_2e^{2t}$, $x_2 = C_1e^t + C_2e^{2t}$ **3.** $x_1 = 2C_1e^{2t} + C_2e^{9t}$, $x_2 = -C_1e^{2t} + 3C_2e^{9t}$
5. $x_1 = 2C_1e^{3t}$, $x_2 = 10C_1e^{3t} + 3C_2e^{2t}$, $x_3 = 11C_1e^{3t} + 4C_2e^{2t} - C_3e^{-t}$
7. $x_1 = 2C_1e^{3t} + C_2e^{10t} - 4e^t/9 - e^{2t}/4$, $x_2 = C_1e^{3t} - 3C_2e^{10t} - e^t/6 - e^{2t}/4$ **9.** $x_1 = 3e^t + 4e^{2t}$, $x_2 = 3e^t + 2e^{2t}$
11. $x_1 = -4e^{2t} + 7e^{9t}$, $x_2 = 2e^{2t} + 21e^{9t}$ **13.** $x_1 = 6e^{3t}$, $x_2 = 30e^{3t} - 15e^{2t}$, $x_3 = 33e^{3t} - 20e^{2t} - 14e^{-t}$

Exercises 11.5 (page 644)

1.   **3.**   **5. a.** $(1, 1)$ **b.**

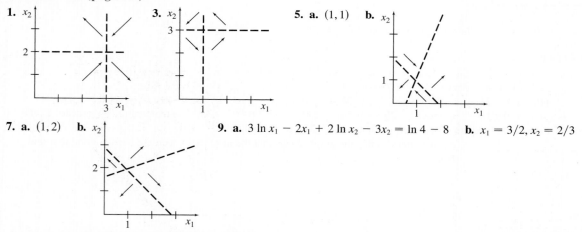

7. a. $(1, 2)$ **b.**   **9. a.** $3 \ln x_1 - 2x_1 + 2 \ln x_2 - 3x_2 = \ln 4 - 8$ **b.** $x_1 = 3/2$, $x_2 = 2/3$

c. They cycle clockwise in the phase plane, as shown in Figure 9.   **13.** $x_1 = (b + \alpha + v)\beta$, $x_2 = (b + \gamma)[\beta a - b(b + \alpha + v)]/$
$\{\beta[\gamma v + (b + \alpha + v)(b + \gamma)]\}$, $x_3 = v[\beta a - b(b + \alpha + v)]/\{\beta[\gamma v + (b + \alpha + v)(b + \gamma)]\}$

Exercises 11.6 (page 650)

1. a. $y = 24{,}995{,}000/(4{,}999 + e^{0.25t})$ **b.** 3,672   **c.** 91   **d.** On the 34th day   **3. a.** $y = 10{,}000e^{0.18t}/(e^{0.18t} + 199)$ **b.** In
about 29 days   **5. a.** $y = 0.005 + 0.015e^{-1.010t}$ **b.** $Y = 0.00273 + 0.00727e^{-1.1t}$ **7. a.** $y = 50/(1 + 9e^{-0.45t})$ **b.** In about
6 days   **9. a.** $y = 347e^{-4.24e^{-0.1t}}$ **b.** In about 5.5 days   **11. a.** $y = [2(t + 100)^3 - 1{,}800{,}000]/(t + 100)^2$ **b.** About 250 lb
c. It continues to increase.   **13. a.** $y = 20e^{-0.02t}$ **b.** About 6 lb   **c.** It continues to decrease.
15. a. $y = [0.1(t + 100)^2 - 500]/(t + 100)$ **b.** About 9.2 g

Chapter 11 Review Exercises (page 652)

5. Neither   **7.** Separable   **9.** Both   **11.** Linear   **13.** $y = x^4/2 + 3x^2 + C$ **15.** $y = 4e^x + C$ **17.** $y^2 = 3x^2 + 2x + C$
19. $y = (Cx^2 - 1)/2$ **21.** $y = 12/5 + Ce^{-5x}$ **23.** $y = 1 + Ce^{-x^2/6}$ **25.** $y = \sin(x^2) + 1$ **27.** $y = -5e^{-x} - 5x + 22$

29. $y = 2e^{5x-x^2}$ **31.** $y^2 + 6y = 2x - 2x^2 + 352$ **33.** $y = x^4 + 2x$ **35.** $y = e^{2x^{3/2}/3 + 0.720}$ or $y = 2.054e^{2x^{3/2}/3}$ **39.** 2.6075

41.

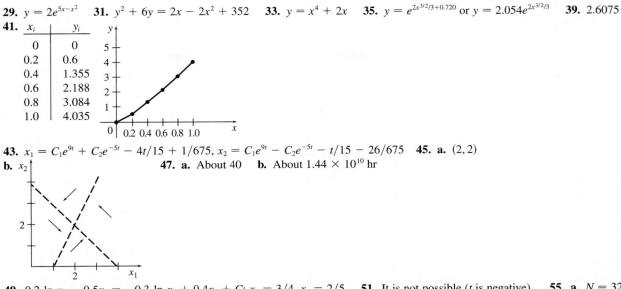

x_i	y_i
0	0
0.2	0.6
0.4	1.355
0.6	2.188
0.8	3.084
1.0	4.035

43. $x_1 = C_1e^{9t} + C_2e^{-5t} - 4t/15 + 1/675$, $x_2 = C_1e^{9t} - C_2e^{-5t} - t/15 - 26/675$ **45. a.** $(2,2)$

b.

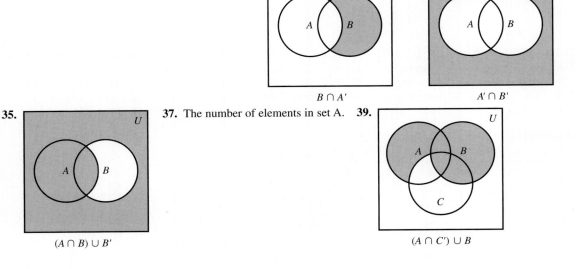

47. a. About 40   **b.** About 1.44×10^{10} hr

49. $0.2 \ln x_2 - 0.5x_2 = -0.3 \ln x_1 + 0.4x_1 + C$; $x_1 = 3/4$, $x_2 = 2/5$ **51.** It is not possible (t is negative).   **55. a.** $N = 326$, $b = 7.20$, $k = 0.248$ **b.** $y \approx 238$ million, which is less than the table value of 248.7 million   **c.** About 266 million for 2010, about 301 million for 2050   **57. a. and b.** $x = 1/k + Ce^{-kt}$ **c.** $1/k$ **59.** 213°   **61. a.** $v = (G/K)(e^{2GKt} - 1)/(e^{2GKt} + 1)$ **b.** G/K **c.** $v = 88(e^{0.727t} - 1)/(e^{0.727t} + 1)$

Chapter 12 Probability

Exercises 12.1 (page 669)

1. False   **3.** False   **5.** False   **7.** $\subseteq$ **9.** $\not\subseteq$ **11.** $\not\subseteq$ **13.** 8   **15.** $\cap$ **17.** $\cap$ **19.** $\cup$ or $\cap$ **21.** $\{3, 5\}$ **23.** $\{7, 9\}$
25. Y or $\{3, 5, 7, 9\}$ **27.** $\{2, 3, 4\}$ **29.** B and C   **31.**

$B \cap A'$

33.

$A' \cap B'$

35.

$(A \cap B) \cup B'$

37. The number of elements in set A.   **39.**

$(A \cap C') \cup B$

41.

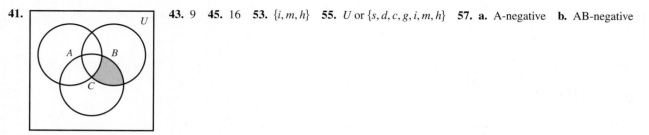

$A' \cap (B \cap C)$

43. 9   **45.** 16   **53.** $\{i, m, h\}$ **55.** U or $\{s, d, c, g, i, m, h\}$ **57. a.** A-negative   **b.** AB-negative

c. B-negative   **d.** AB-positive   **e.** B-positive   **f.** O-positive   **g.** O-negative   **59. a.** 1,160,206   **b.** 18,287   **c.** 1,286,449
d. 2,314,245   **e.** Females who are either American Indian or Asian or Pacific Islander.   **61.** 88.9 million   **63.** 25.7 million
65. 37.7 million   **67.** 18 percent   **69.** {Cisco Systems Inc.}   **71.** {Exxon Mobil Corp., Citigroup Inc., Tyco International}
73. a. 12   **b.** 18   **c.** 37   **d.** 97   **75. a.** 342   **b.** 192   **c.** 72   **d.** 86   **77.** $2^{51} \approx 2.252 \times 10^{15}$ **79.** 512

Exercises 12.2 (page 684)

3. {January, February, March, ... , December}   **5.** {surgery, medicine, wait}
7. $\{(h, 1), (h, 2), (h, 3), (h, 4), (h, 5), (h, 6), (t, 1), (t, 2), (t, 3), (t, 4), (t, 5), (t, 6)\}$ **11.** {AB, AC, AD, AE, BC, BD, BE, CD, CE, DE}
a. {AC, BC, CD, CE}   **b.** {AB, AC, AD, AE, BC, BD, BE, CD, CE}   **c.** {AC}   **13.** {www, wwc, wcw, cww, ccw, cwc, wcc, ccc}
a. {www}   **b.** {ccw, cwc, wcc}   **c.** {cww}   **15.** 1/6   **17.** 2/3   **19.** 1/3   **21.** 1/13   **23.** 1/26   **25.** 1/13   **27.** 1/6   **29.** 4/9
33. No   **35.** Yes   **37.** 2/9   **39.** Yes, Laurie should put 1 pink marble in one box and the other 49 in the box with the 50 blue
marbles.   **41. a.** 3/13   **b.** 4/13   **c.** 7/13   **d.** 4/13   **e.** 8/13   **43. a.** 1/2   **b.** 7/10   **c.** 1/2   **45.** 1 to 5   **47.** 2 to 1
49. No. The probability the reader wins is 5/12, while the probability his opponent wins is 7/12.   **51.** 1/3; 1/5; 1/9,001; 1/10;
9/19   **53.** Empirical   **55.** Empirical   **59.** Not possible; a probability cannot be negative   **61. a.** 0.2778   **b.** 0.4167
63. a. 0.15625   **b.** 0.3125   **65. a.** Person smokes or has a family history of heart disease.   **b.** Person does not smoke and has a
family history of heart disease.   **c.** Person does not have a family history of heart disease or is not overweight.   **67. a.** 0.961
b. 0.487   **c.** 0.513   **d.** 0.509   **e.** 0.004   **f.** 0.548   **69. a.** 1/4   **b.** 1/2   **c.** 1/4   **71. a.** 0.118   **b.** 0.168   **c.** 0.122
d. 0.132   **73. a.** About 0.32   **b.** About 0.32   **c.** About 0.37   **d.** Cavalry   **e.** I Corps
75. a.

	A	B	C	D
O	0.052	0.060	0.039	0.004
E	0.306	0.270	0.213	0.045
M	0.003	0.003	0.003	0

b. 0.306   **c.** About 0.04   **d.** About 0.7   **e.** About 0.88

77. a. 0.867   **b.** 0.843

Exercises 12.3 (page 703)

1. 0   **3.** 1   **5.** 1/6   **7.** 4/17   **9.** 11/51   **11.** 0.012   **13.** 0.245   **19.** The second booth, for which the probability that both are
heads is 1/2. The probability is 1/3 for the man in the first booth.   **21.** No, these events are not independent.   **23.** 1/20; 2/5
25. 1/3   **27.** 2/41   **29.** 21/41   **31.** 8/17   **33.** 2/3   **35.** 1/4   **37.** 1/7   **39.** They are independent.   **41.** 0.039   **43.** 0.491
45. 0.072   **47. a.** Color blindness and deafness are independent events.   **49. b.** 0.30   **c.** 0.98   **51. a.** 0.519   **b.** 0.481
53. $72/73 \approx 0.986$ **55. a.** 0.53   **b.** 0.1623   **57. a.** 0.7456   **b.** 0.1872   **59.** 0.162   **61.** 0.086   **63.** 0.244   **65.** 0.178

Exercises 12.4 (page 717)

1.

Number of Heads	0	1	2	3	4
Probability	1/16	1/4	3/8	1/4	1/16

3.

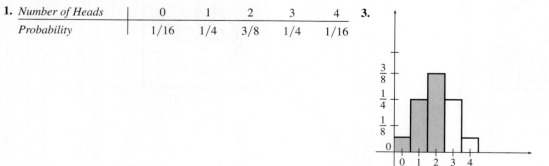

5. 3.6; 0.917   **7.** 14.64; 3.480   **9.** 2.7   **11.** 18   **13.** 0; yes   **17.** 1.58, 1.39   **19.** Transfusion: about 0.92; No Transfusion: about 0.87; yes   **21. a.** $94.0 million for seeding; $116.0 million for not seeding   **b.** Seed   **23.** $-64¢$; no   **25.** $-5.3¢$
27. $-50¢$ **29.** $-87.8¢$ **31. a.** 0.74; 0.94   **b.** Extra-point kick

Chapter 12 Review Exercises (page 722)

1. 16   **3.** $\{c, d, g\}$ **5.** $\{a, b, e, f\}$ **7.**

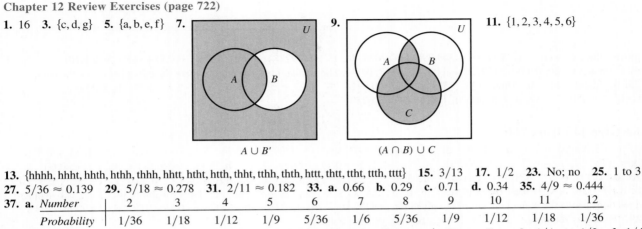

$A \cup B'$ $(A \cap B) \cup C$

11. $\{1, 2, 3, 4, 5, 6\}$

13. {hhhh, hhht, hhth, hthh, thhh, hhtt, htht, htth, thht, tthh, thth, httt, thtt, ttht, ttth, tttt}   **15.** 3/13   **17.** 1/2   **23.** No; no   **25.** 1 to 3
27. $5/36 \approx 0.139$ **29.** $5/18 \approx 0.278$ **31.** $2/11 \approx 0.182$ **33. a.** 0.66   **b.** 0.29   **c.** 0.71   **d.** 0.34   **35.** $4/9 \approx 0.444$
37. a.

Number	2	3	4	5	6	7	8	9	10	11	12
Probability	1/36	1/18	1/12	1/9	5/36	1/6	5/36	1/9	1/12	1/18	1/36

b.

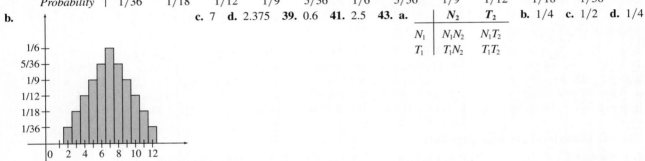

c. 7   **d.** 2.375   **39.** 0.6   **41.** 2.5   **43. a.**

	N_2	T_2
N_1	$N_1 N_2$	$N_1 T_2$
T_1	$T_1 N_2$	$T_1 T_2$

b. 1/4   **c.** 1/2   **d.** 1/4

45. a. 52   **b.** 7   **c.** 12   **d.** 2   **47.** 0.90   **49. a.** 0.0025   **b.** 0.9975   **c.** 0.9753   **d.** 0.0247; no   **e.** The events that each of the two bombs hit their targets are assumed to be independent. The events that each silo is destroyed are assumed to be independent.   **51.** (c)

Chapter 13 Probability and Calculus

Exercises 13.1 (page 735)

1. Yes   **3.** Yes   **5.** No; $\int_0^3 4x^3 \, dx \neq 1$ **7.** No; $\int_{-2}^2 x^2/16 \, dx \neq 1$ **9.** No; $f(x) < 0$ for some x values in $[-1, 1]$.
11. $k = 3/14$ **13.** $k = 3/125$ **15.** $k = 2/9$ **17.** $k = 1/12$ **19.** 1   **23. a.** 0.4226   **b.** 0.2071   **c.** 0.4082
25. a. 0.3935   **b.** 0.3834   **c.** 0.3679   **27. a.** 0.2679   **b.** 0.4142   **c.** 0.3178   **29. a.** About 0.81   **b.** About 0.49
31. a. 8,200   polynomial function

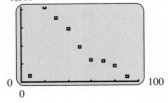

b. $N(x) = -0.00351465x^4 + 0.792884x^3 - 61.7955x^2 + 1,814.54x - 9,709.20;$ the function

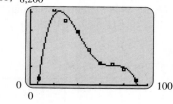

models the data well. **c.** $S(x) = (1/343,795)(-0.00351465x^4 + 0.792884x^3 - 61.7955x^2 + 1,814.54x - 9,709.20)$
d. Estimates 0.300; 0.179; 0.102; Actual 0.272; 0.193; 0.071. The probabilities are closer where the curve better fits the data.
33. a. About 0.32 **b.** About 0.37 **35. a.** $4/7 \approx 0.5714$ **b.** $5/21 \approx 0.2381$ **37. a.** $7/8 = 0.875$ **b.** 0.029
c. $1/27 \approx 0.037$ **39. a.** 0.5730 **b.** 0.2197 **c.** 0.6250

Exercises 13.2 (page 744)

1. $\mu = 5$; Var$(x) \approx 1.33$; $\sigma \approx 1.15$ **3.** $\mu = 14/3 \approx 4.67$; Var$(x) \approx 0.89$; $\sigma \approx 0.94$ **5.** $\mu = 17/6 \approx 2.83$; Var$(x) \approx 0.57$;
$\sigma \approx 0.76$ **7.** $\mu = 4/3 \approx 1.33$; Var$(x) \approx 2/9 \approx 0.22$; $\sigma \approx 0.47$ **11. a.** 5.40 **b.** 5.55 **c.** 2.36 **d.** 0.54 **e.** 0.60
13. a. $4/3 \approx 1.33$ **b.** 0.22 **c.** 0.47 **d.** 0.56 **e.** 0.63 **15. a.** 5 **b.** 0 **17. a.** 4.83 **b.** 0.055 **19. a.** $\sqrt[3]{2} \approx 1.19$
b. 0.18 **21. a.** 6.34 sec **b.** 5.14 sec **c.** 0.75 **23. a.** 2.33 cm **b.** 0.87 cm **c.** The probability is 0, since two standard
deviations above the mean falls out of the given interval $[1, 4]$. **25. a.** 22.2 **b.** 5.51 **c.** 0.18 **27.** About 2.99 m
29. 960 days; 960 days **31. a.** 6.41 yr **b.** 1.45 yr **c.** 0.49

Exercises 13.3 (page 756)

1. a. 4.4 cm **b.** 0.23 cm **c.** 0.29 **3. a.** 33.33 yr **b.** 33.33 yr **c.** 0.23 **5. a.** 1 day **b.** 1 day **c.** 0.23
7. 49.98% **9.** 8.01% **11.** -1.28 **13.** 0.92 **19.** $m = (-\ln 0.5)/a$ or $(\ln 2)/a$ **23. a.** 1.00000 **b.** 1.99999 **c.**
8.00003 **25. a.** $\mu = -1.75 \times 10^{-14} \approx 0$ **b.** $\sigma = 0.999433 \approx 1$ **27. a.** 28 days **b.** 0.375 **29. a.** 1 hr **b.** 0.39 **31.**
a. About 58 min **b.** 0.09 **33. a.** About 0.18 **b.** About 0.24 **35. a.** 38 in **b.** 0.17 **37. a.** 609.5, 609.5
b. About 0.55 **39. a.** \$47,500 **b.** 0.47 **41. a.** $f(x) = 0.235e^{-0.235x}$ on $[0, \infty)$ **b.** 0.095 **43. a.** 0.1587 **b.** 0.7698

Chapter 13 Review Exercises (page 760)

1. probabilities **3.** (1) $f(x) \geq 0$ for all x in $[a, b]$ (2) $\int_a^b f(x)\,dx = 1$ **5.** Not a probability density function **7.** Probability
density function **9.** $k = 1/9$ **11. a.** $1/5 = 0.2$ **b.** $9/20 = 0.45$ **c.** 0.54 **13. b.** **15.** $\mu = 4$; Var$(x) \approx 0.5$; $\sigma \approx 0.71$
17. $\mu = 5/4$; Var$(x) = 5/48 \approx 0.10$; $\sigma \approx 0.32$ **19.** $m = 0.60$; 0.02 **21. a.** 100 **b.** 100 **c.** 0.86 **23.** 31.21%
25. 27.69% **27.** 11.51% **29.** -0.05 **31. a.** Exponential **b.** Domain: $[0, \infty)$, range: $(0, 1]$
c. $f(x)$ **d.** $\mu = 1$; $\sigma = 1$ **e.** 0.86 **33.** 0.63 **35. a.** 22.68°C **b.** 0.48 **37.** 0.1056

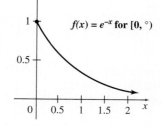

$f(x) = e^{-x}$ for $[0, \infty)$

39. a. polynomial function **b.**

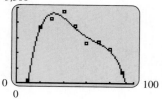

The function models the data well.   **c.** $S(x) = (1/177,514)(-0.00114251x^4 + 0.261752x^3 - 21.6938x^2 + 729.694x - 5,266.33)$
d. Estimates 0.181; 0.266; 0.338; actual 0.171; 0.257; 0.334   **e.** About 46.4 years   **f.** About 43.9 years

41.

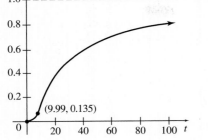

43. 0.406   **45. a.** $f(x) = e^{-x/8}/8$ for x in $[0, \infty)$ **b.** 8   **c.** 8   **d.** 0.249

47. 0.1922

PHOTO ACKNOWLEDGMENTS

INDEX OF APPLICATIONS

AUTOMOTIVE

Abandoned cars, 67
Automobile velocity, 402
Car accidents, 53, 268, 559, 571
Deer-vehicle accidents, 487–488
Flying gravel from a tire tread, 306–307
Gasoline mileage, 280, 330
Highway research, 53
Life of an automobile part, 746
Motorcycle accidents, 682
Seat belts, 379, 708
Sports cars, 280
Stopping distance, 53
Traffic density, 268
Vehicle waiting time, 223
Vehicle-related deaths, 379

BIOLOGY

Allometric growth, 100, 354
Animal offspring, 718
Animals, 508, 559, 560, 596
Bacteria, 107, 170–171, 222, 233, 305, 362
Bacterial food requirements, 548
Biochemicals, 349, 379, 411
Bologna sausage, 248
Botany, 108
Cell growth, 388
Competing species, 644, 645, 654
Dinosaur running, 486
Ecosystem diversity, 70
Extinction, 200, 233, 241, 763
Flow systems, 472–474
Fungal growth, 328, 364
Genetics, 176–177, 666–667, 688, 689, 705
Growth models, 66, 222
Growth of a fungus, 412
Growth of a population, 455
Growth of algae, 629
Growth of bacteria, 87, 107, 412–413
Growth rates, 445, 455
Index of diversity, 97–98, 100
Length of a leaf, 745
Leslie matrices, 594, 598
Logistic growth, 364
Maximum sustainable harvest, 337, 338–339
Mixing plant foods, 548
Mouse infection, 621–622, 645
Optimal foraging, 222, 487
Oscillating system, 418
Oxygen consumption, 82–83, 486, 496, 513

Petal length, 736, 745
Population biology, 66
Rumen fermentation, 455
Scaling laws, 320
Species, 349, 645
Vectorial capacity, 279
Weight gain of rats, 762

BIRDS

Bald eagles, 613
Bird eggs, 27, 215, 462
Bird feeding, 566–568, 572
Bird flight, 354
Bird migration, 339–340
Bird population, 604–605
Body temperature of a bird, 762
Location of a bird's nest, 733
Pigeon flight, 339
Robin sightings, 548
Sparrows at a bird feeder, 729–730, 738

BUSINESS

Average cost, 153, 200–201, 340, 362, 526
Cost, 330–331, 341–342, 349, 356, 509, 530
Cost analysis, 13–14, 18, 161, 572
Cost-benefit analysis, 60–61, 66
Customer expenditures, 759
Distribution, 600
Error estimation, 360–361
Fast-food outlets, 759
Insurance sales, 759
Management, 561
Manufacturing, 514, 530, 549
Net savings, 425, 431
Orders, 583, 600
Packaging, 341, 366, 520, 529
Point of diminishing returns, 301–302, 307
Postal regulations, 340
Pricing, 342
Product life cycle, 307
Product repairs, 763
Product success, 726
Production, 161, 463, 488, 530, 583–584, 599–600
Productivity, 431
Profit, 331, 356, 425, 509, 526, 530
Purchasing parts, 550
Quality control of cheese, 189
Retail outlets, 763–764

Revenue, 260, 341, 349, 352–353, 356, 446, 455, 471, 499, 509, 526
Risk in fisheries, 430–431
Sales expense, 759
Sales, 431, 572
Telephone usage, 379–380
Total maintenance charges, 396–397

CHEMISTRY

Acidity of a solution, 102
Carbon dating, 104–105, 108
Cellular chemistry, 223, 318
Chemicals, 109, 306, 446, 599, 651
Concentration of a solute, 379
Half-life, 108–109
Newton's law of cooling, 109
Radioactive decay, 88, 105, 108, 109, 237, 241–242, 615
Salt concentration, 649–650, 651
Soap concentration, 651
Yeast production, 103

CONSTRUCTION

Elliptical templates for a roof vent, 488
Heat gain in a window, 401
Housing starts, 280
Ladders, 259–260, 342, 350–351, 355, 365
Material requirement, 362, 488
Measurement, 125
Parabolic arch, 53
Roof trusses, 599

CONSUMER INFORMATION

Air fares, 29–30
Charge for auto painting, 529–530
Consumer demand, 152
Electric rates, 109
Life insurance, 713
Postal rates, 143–144, 161–162, 216, 488
Red meat consumption, 73
Rental costs, 109

DISEASE

AIDS, 52–53, 71, 72, 705, 707
Alzheimer's disease, 123, 279
Blood transfusions for malaria-related anemia, 714–717

Cancer, 66, 107–108, 130, 213, 239, 306, 702, 706
Contagion, 601
Disease, 338
Epidemic, 170, 280–281, 291, 646–647, 650
Foot-and-mouth epidemic, 399, 445
HIV infection, 16–17
Hepatitis blood test, 707
Infection, 278, 430
Measles, 58
Melanoma, 107
Sickle cell anemia, 724
Spread of a virus, 199
Spread of gonorrhea, 651
Spread of influenza, 654
Test for HIV, 707
Tumor growth, 99–100, 160

ECONOMICS

Broadway economics, 550
Coal consumption, 67
Demand, 385–386
Depreciation, 234
Energy consumption, 43, 418–419
Energy use, 432
Fuel economy, 425–426
Money in circulation, 216
Natural gas consumption, 416–417
Oil consumption, 400, 413
Stock prices, 322

EDUCATION

Biology majors, 695
Educational attainment, 560
Educational expenditures, 30
Educational psychology, 446
Full-time student, 655
Learning skills, 249, 322, 355, 498, 630
Medical school admissions, 28
Older college students, 19
Research funding, 690
SAT scores, 30–31
School activities, 668
Students, 599
Student's success, 498–499

ENGINEERING

Biomechanics, 339
Electricity, 242
Engine velocity, 258
Piston velocity, 255–256
Rocket science, 380
Rotating camera, 259, 356
Rotating lighthouse, 259, 356

ENVIRONMENT

Air pollution, 124, 278
Atmospheric pressure, 88
Carbon dioxide levels, 83–84, 255, 258, 477, 598–599
Carbon emissions, 18, 72
Carbon monoxide levels, 73
Chlorofluorocarbons (CFCs) in the atmosphere, 87–88, 265
Dead Sea, 216
Dry length days, 759
Effect of insecticide, 654
Giardia, 107
Global warming, 18
Greenhouse gas emissions, 249–250
Ground temperature, 258–259, 499–500
Heat index, 497–498
Icicle, 352
Lead emissions, 47, 549
Oil leak pollution, 232–233, 471
Oxygen concentration in a pond, 144–145
Ozone depletion, 305
Parabolic culvert, 53
Pollution, 73, 131, 239, 328, 338, 364, 412, 425, 466, 638, 656–657
Radioactive waste, 468
Rainfall, 706, 737, 746, 759
Seeding storms, 719–720
Snowfall, 615–616, 698, 762
Spread of an oil leak, 412
Sunrise, 120–121
Sunset, 125
Temperature, 14, 19, 125, 132, 172–173, 189, 191–192, 201, 471, 748–749
Thermal inversion, 233
Tree growth, 412
Water level, 355, 365
Water temperature, 492–493
Weather, 431–432
Wind chill, 497

FARMING

Age-weight relationship in cattle, 237–238
Agriculture, 487, 548
Animal feed, 534, 548–549
Ayrshire dairy cattle, 241, 279
Beef cattle, 240
Calves, 445
Fertilizer, 578–579
Goat growth, 629
Guernsey cow growth, 613
Harvesting fruit, 673
Holstein dairy cattle, 240–241, 263, 279
Insulin in sheep, 430
Milk production, 471
Pigs, 360, 362, 365
Poultry, 160–161, 673

Soil ingestion, 64–65
World grain production, 130

FINANCE

Individual Retirement Account (IRA), 489
Interest, 80, 82, 89, 131, 230, 234
Investment, 431, 583, 600
Loans, 550
Mutual funds, 672–673, 763
Savings, 423–424

FISH

Black crappie, 246–247
Brown trout, 498, 529
Catfish farming, 300–301
Cutlassfish, 240
Fish food requirements, 548
Fish population, 232, 468, 614
Harvesting cod, 339
Salmon spawning, 328
Sculpin growth, 106
Shad, 263
Snake River Salmon, 487, 498
Walleye, 263

GEOLOGY

Cliff ecology, 399
Dating rocks, 131
Depleting dates for minerals, 433–435
Earthquakes, 102, 725, 737, 746, 759, 763
Rare metal reserve supply, 431
Richter scale, 250
Sediment, 153, 413
Soil moisture, 613–614

GEOMETRY

Area, 340, 350, 355, 365, 432, 529
Area of a bacteria colony, 362
Area of a rectangular region, 39–40
Area of an oil slick, 362
Can design, 341, 342
Container design, 341
Cost with fixed area, 340
Estimating area, 514
Estimating volume, 514
Maximizing area, 53
Maximizing volume, 334
Measurement error, 362
Minimizing area, 335–336
Package dimensions, 365
Playground area, 366
Spherical radius, 365
Surface area, 529
Use of materials, 341

Volume, 355, 362, 365
Volume of a can of beer, 512
Volume of a coating, 514

GOVERNMENT

Arms race models, 647–649
Civil War, 689–690
Cryptography, 580–581, 584
Elections, 725
Electoral college, 674
Governor's salaries, 73
Legislative voting, 153–154
Missiles, 726
Modeling war, 551
Nuclear energy, 109
Nuclear weapons, 322
Perceptions of threat, 690–691
Political science, 508
Social Security assets, 190
State-run lotteries, 763
Voting, 709

HEALTH

Alcohol concentration, 65, 278, 361, 399
Average birth weight, 762
Calcium usage, 233
Calorie expenditure, 498
Cholesterol, 71, 199, 460, 487
Contact lenses, 65
Dietetics, 548, 558–559, 570–571
Dieting, 292–293, 614
Drug concentration, 100–101, 130–131, 153, 278, 305, 361, 412
Drug control, 164–165
Drug epidemic, 468
Exercise heart rate, 17
Fever, 71
Hallucinogen usage, 325–326
Human cough, 214
Life expectancy, 52, 75, 559–560, 571, 743, 758
Life span, 17
Mortality, 239–240, 412, 671, 733–735
Oxygen inhalation, 400
Recommended Daily Allowance (RDA) of vitamin C, 2
Sunscreen, 72
Survival curves, 529
Thermic effect of food, 170, 278–279, 291, 455
Thyroid, 671
Tobacco deaths, 17–18
Transylvania hypothesis, 123
Vitamins, 476, 583
Weight, 182, 196, 496
World health, 71

HUMAN BODY

Blood antigens, 671
Blood flow, 353, 413, 462, 496
Blood pressure, 131, 264, 388–389, 446
Blood sugar level, 71, 213
Blood velocity, 354
Blood vessels, 264–265, 362, 511–512, 513, 529
Blood volume, 70, 320–321, 513
Body mass index, 214–215, 496
Body surface area, 100, 248, 486
Body temperature, 18, 32–33
Body types, 688
Bone preservation volume, 513
Brain weight, 64, 214, 354
Cell division, 412
Cell surface receptors, 153, 321
Cell traction force, 222, 321
Dentin growth, 328–329, 365
Digestion time, 758
Femoral angles, 123–124
Fetal stature, 28
Flat feet, 682
Heart, 215, 762
Human growth, 196, 200
Human skin surface, 329, 365
Human strength, 172
Immune system activity, 220, 317
Metabolic rate, 354
Molars, 170, 328
Muscle reaction, 222
Nervous system, 153, 241, 411
Neurons, 317
Oxygen consumption, 101
Plasma volume, 86
Pregnancy, 160
Respiratory rate, 70–71
Skeletal maturity, 27–28
Splenic artery resistance, 51
Surface area of a human, 489
Tooth length, 51–52
Total body water, 487, 529

INSECTS

Ant colonies, 18
Crickets chirping, 28–29
Deer ticks, 17
Flea beetles, 613, 736, 745–746
Flour beetles, 379, 736, 745
Fruit flies, 249, 318, 610
Grasshopper matings, 496
Growth of a mite population, 654
Insect cannibalism, 429–430
Insect dispersal, 498
Insect growth, 239, 305
Insect life span, 757
Insect mating, 214, 249

Insect population, 592–593, 630
Insect species, 100, 215
Location of a bee swarm, 757–758
Population growth of a spider, 430

LABOR

Cost of a training program, 367
Employee training, 223
Income distribution, 413, 426
Labor costs, 509
Military, 672, 690
Timing income, 340–341
Training accidents, 291
Wages, 400
Work/rest cycles, 222
Worker errors, 699–700
Worker productivity, 615
Working women, 707

MARINE LIFE

Characteristics of the monkeyface prickleback, 133
Clam growth, 305
Clam population, 239, 305
Shellfish population, 189, 195
Velocity of a marine organism, 215
Whale population, 630
Whales and krill, 639–641, 645
Whales diving, 43, 200

MEDICAL

Angioplasty, 52
Antibiotic resistance, 13
Blood clotting time, 736, 745
Blood level curves, 445–446
Blood transfusions, 719
CAT scans, 596–597
Calcium kinetics, 321
Cardiac output, 65
Cardiology, 211, 278
Causes of death, 682–683
Chromosomal abnormality, 108
Circadian testosterone, 131
Color blindness, 688, 705–706
Cortisone concentration, 86, 119–120
Dialysis, 153, 239–240, 513
Drug reaction, 233, 412, 445, 455, 462, 468, 471, 497
Drug screening, 706–707
Drug use, 621
Ear infections, 719
Electrocardiogram, 161
Glucose, 131, 620, 622
Heart attack risk, 87
Hospital satisfaction, 712, 714
Lead poisoning, 755

Medical diagnosis, 727
Medical experiment, 705
Medical literature, 240
Medical survey, 688
Medicare trust fund, 171–172
Medicine, 52
Mercury poisoning, 758
Mutation, 130
Patient wait times, 742–743, 765–767
Perinatal mortality, 719
Pharmaceutical manufacturing, 552–553, 555, 556–557, 561–563
Pharmacology, 160
Receptor dynamics, 634–636
Scientific development, 504–506
Silicone implants, 26
Testing medication, 559
Toxemia test, 707
Tracer dye injected into the bloodstream, 613
TYLENOL, 199–200
Volume of a tumor, 362

MISCELLANEOUS

Archimedes' problem bovinum, 549
Bacteria in sausage, 86–87, 346–347
Candy, 233–234
Complimentary error function, 471
Contests, 720
Drug trade, 171
Firearms, 736–737, 745
Friendly wager, 712–713
Hearing aid battery, 750
Ice cream, 551
Information content, 330
Length of a day, 419
Length of a telephone call, 737–738
Library budget decline, 30
Life of a light bulb, 746
Lights Out, 551–552
Milk consumption, 288, 291
Oven temperature, 189
Popcorn, 215, 305
Probability, 446
Raffle, 720
Sampling fruit, 726
Smoke content in a room, 654
Street crossing, 250
Toys, 551
Typing speed, 463

PETS

Animal breeding, 548
Beagles, 208, 411
Cats, 265–266, 322, 674
Dog's human age, 215
Ponies trotting, 17
Rats, 544–546
Thoroughbred horses, 305, 317

PHYSICS

Acceleration, 295–296, 306, 322, 376–378
Air resistance, 655
Gravitational attraction, 499
Height of an object, 292, 306
Intensity of light, 130
Intensity of sound, 102
Length of a pendulum, 31, 66–67
Light emitted by a nova, 185
Linear motion, 431
Motion of a particle, 259
Movement time, 499
Newton's law of cooling, 622–623, 655
Planets, 74
Shadow length, 355
Simple harmonic motion, 266
Sound, 124, 258
Velocity, 166, 168–169, 173, 216, 295–296, 306, 322, 349, 355, 376–378, 380, 499
Voltage, 419

REPTILES

Alligator teeth, 153, 306
Beaked sea snake, 188
Lizards, 291, 354
Water snakes, 188

SOCIAL SCIENCE

Accidental death rate, 19–24
Attitude change, 291–292
Binge drinking, 707
Chinese New Year, 673–674
Cohabitation, 19
Crime, 306, 354–355, 431
Drunk drivers, 737, 745
Eating behavior, 70, 182, 188–189
Habit strength, 241
High school drug use, 171
Immigration, 18–19, 622, 629
Juveniles becoming adults, 584–590
Languages, 102, 265, 759
Living arrangements, 672
Marital status, 73
Memorization skills, 355
Memory retention, 223
Murder, 322, 707
Neediest cases, 25
Never-married adults, 708
Pace of life, 131–132
Poverty levels, 30
Pygmy heights, 758
Randomized response, 725
Recollection of facts, 275
Satisfaction, 329
Spread of a rumor, 615, 630, 651, 655
Suicides, 762–763

Television viewing habits, 725
Time to learn a task, 737

SPORTS AND ENTERTAINMENT

Athletic records, 29
Baseball, 189–190, 201, 306, 726
Batting power, 66
Billboard "Hot 100" survey, 387
Film length, 292
Football, 721
Games, 674, 720–721
Golf tournament, 720
Hockey, 597–598, 668–669, 691–693
Kite flying, 355–356
Making a first down, 725
Music, 584, 674
Roulette, 720
Running, 29, 402
Scuba diving, 497
Swimming, 43, 513–514
Swing of a runner's arm, 257–258
Tennis, 265
Track and field, 214, 241
Weightlifting, 320

STATISTICS AND DEMOGRAPHICS

Age distribution, 413
Centenarians, 210–211
Doubling time of a population, 97
Jewish population, 101, 614
Long-term population, 594
Population growth, 72, 87, 99, 101, 107, 130, 239, 241, 249, 258, 263, 279, 304, 370, 376, 622, 637–638, 654–655
U.S. population, 614–615, 672, 689
World population growth, 170, 571–572, 615

TECHNOLOGY

Computer chips, 88, 508–509
Internet users, 43–44, 88
Life span of a computer part, 738
Machine accuracy, 759
Machine life, 738, 746
Machine repairs, 763
New technology versus old, 172
Rapid technological advancements, 611
Useful life of an appliance part, 764

TRANSPORTATION

Average speed, 73–74
Border crossing, 389
Deliveries, 549–550
Distance, 355, 380, 397, 400–401, 402, 442–443
Flight speed, 195, 329–330
Mercator's world map, 432

Minimizing travel time, 333–334, 340
Pedestrian speed, 246
Pursuit, 366
Speed, 162–163, 165
Time, 380
Titanic, 708
Traffic control, 550–551

WILDLIFE

African wild dog, 233
Alaskan moose, 163–164, 167–168, 291

Arctic foxes, 147–148, 240, 264
Bighorn sheep, 214, 274–275, 513
Buffalo, 598
Deer harvest, 488
Deer population variation, 65
Elk energy, 11–12
Endangered species, 70
Finding prey, 758
Gray wolves, 362
Marsupials, 354
Mass of bighorn yearlings, 166–167, 171
Metabolic rate for anteaters, 43

Migratory animals, 418
Monkey eyes, 123
Mountain goat population, 608–609
Movement of a released animal, 762
Northern spotted owl population, 571, 594
Optimization for a predator, 531
Polar bears, 130, 481–482
Predator-prey, 641–644
Pronghorn fawns, 249, 409
Ram's horns, 411
Tasmanian devil, 229–230
Wolf harvest, 199

I N D E X

Absolute extrema, 322
   steps to find, 324
Absolute maximum, 322
Absolute minimum, 322
Absolute value, xliii
   definition of, xliii
Absolute value equation, xxv
Absolute value function, 184
   derivative of, 184
Acceleration, 293
Acrophase, 118
Acute angle, 111
Addition
   of matrices, 554
   of polynomials, xvi
   of rational expressions, xxii
Addition property
   of equality, xxv
   of inequality, xxxi
Additive identity of a matrix, 556
Additive inverse of a matrix, 555
Agreement on domain, 35
Amplitude of a trigonometric function, 118
Analysis
   compartmental, 630
   cost, 13
   decision, 714
   error, 444
   Fourier, 120
   marginal, 359
   numerical, 443
Angle(s), 110
   initial side of, 110
   measure of, 110
   naming of, 110
   negative, 110
   positive, 110
   standard position of, 110
   terminal side of, 110
   types of, 110
   vertex of, 110
Antiderivatives, 369
Antidifferentiation, 369
Approximately equal to, xxviii, 12
   symbol for, xxviii, 12
Approximation
   area by, 389
   by differentials, 359, 510, 623
   linear, 359
Arc of a circle, 111
Archimedes, 389, 392

Area
   by approximation, 389
   between curves, 419
   between curves by calculator, 421
   by definite integral, 408
   exact, 392
   by integration, 383, 408
   by midpoint rule, 390
   by trapezoidal rule, 390, 437, 439
   under a normal curve, A-2
Area formulas, A-1
Arms race models, 647
Associative properties, xvi
Asteroid, 348
Asymptotes, 59
   horizontal, 309
   oblique, 309
   of a rational function, 59
   vertical, 309
Augmented matrix, 535
   row operations on, 535
Average cost per item, 61
Average rate of change, 162
   definition of, 163
Average speed, 162
Average value of a function, 459, 525
Axes, coordinate, 2
Axis of a parabola, 45

Base of an exponent, xxxvi
Basic probability principle, 677
Basic trigonometric identities, 250
Bayes, Thomas, 700
Bayes' theorem, 699
   special case of, 699
   steps to use, 701
Bernoulli, Jakob, 371
Binomial, xviii
   multiplication of, xviii
Binomial theorem, 205
Brahmagupta, 392
Briggs, Henry, 92

Calculus
   differential, 369
   fundamental theorem of, 403
   integral, 369
   multivariable, 475
Carbon 14 dating, 104
Cardiac output by thermodilution, 473
Carrying capacity, 609

Cartesian coordinate system, 2
   quadrants of, 3
Cauchy, Augustin-Louis, 137
Cayley, Arthur, 535
Celsius, Anders, 14
Celsius to Fahrenheit equation, 14
Certain event, 676
Chain rule for derivatives, 225
   alternative form of, 226
Change in x, 3
Change in y, 3
Change-of-base theorem
   for exponentials, 96
   for logarithms, 92
Change of limits, 406
Characteristics of monkeyface prickleback, 133
Circle, A-1
   arc of, 111
   tangent line to, 173
Circular cylinder, A-1
Clairaut, Alexis Claude, 603
Class of functions, 370
Closed interval, xxxii
   continuity on, 156, 323
Coefficient, xvi, 54
   of correlation, 23
   leading, 54
Coefficient matrix, 577
Column integration, 448
Column matrix, 553
Column vector of a matrix, 553
Combining rational expressions, xxiii
Common denominator, xxiv
Common logarithmic function, 90, 157
   derivative of, 243, 245
Common logarithms, 92
Commutative properties, xvi
Compartmental analysis, 630
Complement of a set, 662
Complement rule for sets, 681
Completing the square, 46
Composite function, 224
Composition of functions, 38, 224
Compound amount, 80
Compound interest, 79, 230, 608
   continuous, 81
Concavity
   downward, 294
   of a graph, 294
   second derivative test for, 297

test for, 296
upward, 294
Conditional probability, 692
definition of, 693
Conditions for integration by parts, 449
Cone, A-1
Conjugate of real numbers, xliii
Constant
decay, 103
growth, 103
growth rate, 608
Constant multiplier rule for integration, 372
Constant of integration, 370
Constant rule for derivatives, 204
Constant times a function rule for derivatives, 207
Contagion, 601
Continuity, 154
on a closed interval, 156, 323
definition of, 155
from the left, 156
from the right, 156
for functions, 157
on an open interval, 156, 323
at a point, 154
Continuous compounding, 81, 608
formula for, 82
Continuous functions, 154, 157
table of, 157
Continuous probability distribution, 730
Continuous probability models, 728
Continuous random variable, 710, 730
Convergent integrals, 464
Coordinate system, Cartesian, 2
Correlation, coefficient of, 23
Cosecant function, 113
derivative of, 254
Cosine function, 113
amplitude of, 118
derivative of, 253
graph of, 117
integral of, 414
translating graphs for, 118
Cost
fixed, 13
marginal, 13, 167, 423
Cost analysis, 13
Cost-benefit model, 60
Cost function, 14
Cost model for training program, 367
Cost per item, 13
Cotangent function, 113
derivative of, 254
integral of, 416
Critical numbers, 269
Critical points, 269, 501
Cryptography, 580
Cube root, xxxix
Cubes
difference of, xxi

sum of, xxi
Curve(s)
area between, 419
level, 479
logistic, 610
normal, 751
secant line to, 174
slope of, 174
tangent line to, 173
Curve sketching, 306
steps for, 306

Decay constant, 103
Decibel, 102
Decision analysis, 714
Decreasing function, 267
definition of, 268
graphing calculator method to find, 271
tests for, 268
Definite integrals, 393
by calculator, 395
finding area by, 408
lower limit of, 394
properties of, 405
upper limit of, 394
Degree measure, 110
to radian measure, 112
Degree of a polynomial, 54
Degree/radian relationship, 111
Delta x, 3
Delta y, 3
Demand, 385
marginal, 167, 385
Demand function, 385
De Moivre, Abraham, 751
DeMorgan, Augustus, 670
DeMorgan's laws, 670
Denominator
common, xxiv
least common, xxiv
rationalizing, xlii
Dependent events, 697
Dependent system, 543
Dependent variable, 11, 476
Depletion dates for minerals, 433
Derivative(s), 177
of absolute value functions, 184
calculator process to find, 210
chain rule for, 225
of common logarithmic functions, 243, 245
constant rule for, 204
constant times a function rule for, 207
definition of, 177, 204
difference rule for, 209
existence of, 183
of exponential functions, 235
finding from definition, 181
finding graphically, 190
first, 291
first-order, 616

four-step process to find, 180
fourth, 291
generalized power rule for, 227
by graphing calculator, 179
higher, 291
interpretations of, 178
Leibniz notation for, 203
of natural logarithmic functions, 244, 245
notations for, 177, 203, 291
nth, 616
partial, 490
power rule for, 205
product rule for, 217
quotient rule for, 219
second, 291
second-order partial, 493
sum rule for, 209
summary of rules for, 261, 262
techniques for finding, 203
third, 291
of trigonometric functions, 252, 253, 254
Descartes, René, 2
Determinant, 586
evaluating, 587
Determining cardiac output, 473
Deviation, standard, 714, 740
Difference of cubes, xxi
factoring of, xxi
Difference of squares, xxi
factoring of, xxi
Difference quotient, 163, 178
Difference rule
for differentiation, 209
for integration, 372
Differential calculus, 369
Differential equations, 603
applications of, 641, 646
Euler's method for solving, 623
general solution of, 604
initial condition of, 605
integrating factor for, 618
linear first-order, 616
linear system of, 630
nonlinear system of, 639
order of, 616
particular solution of, 605
separable, 606
separation of variables method for solving, 606
system of, 630, 639
Differentials, 356
approximation by, 359, 510, 623
definition of, 358
error estimation by, 360
total, 509
of x, 358
of y, 358
Differentiation, 177
implicit, 342
Dimensions of a matrix, 553

Diminishing returns
    law of, 299
    point of, 299
Discontinuity, 154
Discontinuous at a point, 154
Discrete probability function, 730
Discrete random variable, 710
    expected value for, 738
    mean for, 738
Disjoint sets, 663
    union rule for, 668
Distance, rate, time relationship, 162
Distribution
    continuous probability, 730
    exponential, 749
    normal, 750
    probability, 683, 710
    standard normal, 752
    uniform, 747
Distributive property, xvi
Divergent integrals, 464
Division of rational expressions, xxii
Domain of a function, 32, 476
    agreement on, 35
Double folium, 348
Double integrals, 514, 517
    interchanging limits of, 523
    over variable regions, 520
Doubling time, 89, 97
Downward concavity, 294

e, 80
    approximating values of, 81
    definition of, 81
Eigenvalue, 586
Eigenvector, 586
Element
    of a matrix, 535
    of a set, 659
Elementary trigonometric identities, 113
Ellipsoid, 483
Empirical probabilities, 682
Empty set, 659
Endpoint of a ray, 109
Entry of a matrix, 535
Equal, approximately, xxviii, 12
Equal matrices, 553
Equal sets, 659
Equality, properties of, xxv
Equation(s)
    absolute value, xxv
    Celsius to Fahrenheit, 14
    differential, 603
    exponential, 78, 95
    Fahrenheit to Celsius, 14
    with fractions, xxviii
    graph of, 3
    of a horizontal line, 7, 10
    of a line, 5
    linear, xxv, 5

linear in one variable, xxv
linear in two variables, 5
linear system of, 534
logarithmic, 94
logistic, 609
Lotka-Volterra, 642
quadratic, xxvi
rational, xxviii
system of, 534
of a tangent line, 183
of a vertical line, 8
Equilibrium point, 639
Equivalent systems, 535
Error analysis, 444
Error estimation by differentials, 360
Estimating depletion dates for minerals, 433
Euler, Leonhard, 80, 623
Euler's method for solving differential
        equations, 623
Events
    certain, 676
    dependent, 697
    impossible, 676
    independent, 697, 698
    mutually exclusive, 677, 698
    odds in favor of, 681
    of a sample space, 676
    set operations for, 677
    simple, 676
Exact area, 392
Existence
    of the derivative, 183
    of limits, 139
Expected value
    of a discrete random variable, 738
    of a probability density function, 739
    of a random variable, 711
Experiment, 675
    outcomes of, 675
Explicit function, 343
Explicitly in terms of x, 343
Exponential decay function, 103, 607
Exponential distribution, probability density
        function for, 749
Exponential equations, 78, 95
    converting to logarithmic equations, 89
Exponential function rule for integration, 371,
        384
Exponential functions, 77, 157
    continuity on, 157
    derivative of, 235
    extended application of, 133
    graph of, 77
    limited growth, 105
Exponential growth function, 103, 607
Exponential waiting times, 765
Exponentials, change-of-base theorem for, 96
Exponents, xvi, xxxvi
    base of, xxxvi
    definition of, xxxvi

integer, xxxvi
    negative, xxxvi
    properties of, xxxvii
    rational, xxxix
    zero, xxxvi
Expressions, rational, xxii
Extraneous solutions, xxx
Extrapolation to predict life expectancy, 75
Extrema, 500
    absolute, 322
    applications of, 331
    first derivative test for, 281
    location of, 501
    relative, 56, 278
    relative maximum, 279
    relative minimum, 279
    steps to solve problems with, 331
    using spreadsheet, 505
Extreme value theorem, 323

Factor(s), xix
    greatest common, xix
    of real numbers, xix
Factoring, xix
    difference of cubes, xxi
    difference of squares, xxi
    greatest common factor, xix
    of a perfect square trinomial, xxi
    sum of cubes, xxi
    trinomials, xx
Factoring method for solving quadratic
        equations, xxvi
Fahrenheit, Gabriel, 14
Fahrenheit to Celsius equation, 14
Fair game, 713
Family of functions, 370
Fermat, Pierre de, 729
First derivative, 291
    notation for, 291
First derivative test, 281
First octant, 478
First-order derivative, 616
Fisher, J. C., 611
Fixed cost, 13
Flow systems, 472
FOIL, xviii
Folding back, 716
Folium of Descartes, 344
Four-step process to find derivatives, 180
Fourier analysis, 120
Fourth derivative, 291
    notation for, 291
Fractional equation, xxviii
Fractional inequality, xxxiii, xxxiv
Frequency, 118
Frustum, 470
Function(s), 31
    absolute value, 184
    average value of, 459, 525
    class of, 370

composite, 224
composition of, 38, 224
concaved downward, 294
concaved upward, 294
continuous, 154, 157
cost, 14
decay, 103
decreasing, 267
decreasing on an interval, 268
demand, 385
differentiable at x, 177
discrete probability, 730
domain of, 32, 476
evaluating, 37
explicit, 343
exponential, 77, 157
exponential decay, 103, 607
exponential growth, 103, 607
family of, 370
graphs of basic, 127
growth, 103
implicit, 343
increasing, 267
increasing on an interval, 268
input-output, 36
inverse, 90
limit of, 136
limited growth, 105
linear, 10
logarithmic, 90, 157
multivariable, 476
nonlinear, 33
nth derivative of, 616
parent-progeny, 336
period of, 116
periodic, 116, 675
piecewise, 158
polynomial, 54, 157
power, 54
probability, 710
probability density, 731
production, 481
quadratic, 44
range of, 32, 476
rational, 58, 157
real zeros of, 57
reflection of, 45, 55
root, 157
spawner-recruit, 336
square root, 49
summary of translations and reflections of, 48
total change in, 395
transcendental functions, 77
translations of, 45, 55
trigonometric, 113
turning points of, 56
of two variables, 476
vertical line test for, 38
vertical reflection of, 45

zeros of, 57
Functional notation, 11
Fundamental property of rational expressions, xxii
Fundamental Theorem of Calculus, 403
$f(x)$ notation, 11
$f'(x)$ notation, 177

Gauss, Carl Friedrich, 537
Gauss-Jordan method for solving linear systems, 536
   by graphing calculator, 540
   steps to use, 542
General solution of a differential equation, 604
Generalized power rule
   for derivatives, 227
   for integration, 381
Geometric series, infinite, 640
Geometry formulas, A-1
Graph(s)
   of basic functions, 127
   concavity of, 294
   of an equation, 3
   of an exponential function, 77
   of a horizontal line, 10
   of an interval, xxxii
   of a line, 9
   of a linear function, 9
   of a linear inequality, xxxii
   of a multivariable function, 477
   of an ordered pair, 2
   of a parabola, 45
   of a plane, 478
   of a polynomial function, 56
   of a quadratic function, 45
   of a rational function, 58
   of trigonometric functions, 116
   of a two independent variable function, 477
   of a vertical line, 7
Graphical differentiation, 190
Graphical optimization, 326
Graphing calculator method for solving linear systems, 540
Greatest common factor, xix
   factoring out, xix
Growth, logistic, 609
Growth and decay functions, 103
   exponential, 103
Growth constant, 103
Growth curve, von Bertalanffy, 106
Growth rate constant, 608
Gunter, Edmund, 92

Half-life, 104
Half-open interval, xxxii
Higher derivatives, 291
   notation for, 291
Histogram, 710, 730
Horizontal asymptote, 59, 309

Horizontal line
   equation of, 7, 10
   graph of, 10
   slope of, 4
Horizontal translation of a function, 45
Hyperbolic paraboloid, 483
Hyperboloid of two sheets, 483

Implicit differentiation, 342
   steps to find, 346
Implicit function, 343
Implicitly in terms of x, 343
Impossible event, 676
Improper integrals, 464, 733
   convergent, 464
   divergent, 464
Inconsistent system, 542
Increasing function, 267
   definition of, 268
   graphing calculator method to find, 271
   tests for, 268
Indefinite integrals, 370
   of exponential functions, 371, 384
Independent events, 697, 698
   product rule for, 697
Independent variables, 10, 476
Index of a radical, xli
Inequalities, xxxi
   with fractions, xxxiii, xxxiv
   graph of, xxxii
   linear in one variable, xxxi
   polynomial, xxxiii
   properties of, xxxi
   quadratic, xxxii
   rational, xxxiii
   symbols of, xxxi
Infinity
   limits at, 144
   steps to find limits at, 147
Inflection point, 294
   of a normal curve, 751
Initial condition of a differential equation, 605
Initial side of an angle, 110
Initial value problem, 605
Input-output function, 36
Instantaneous rate of change, 165
Integer exponents, xxxvi
Integral calculus, 369
Integral sign, 370
Integrals
   convergent, 464
   definite, 393
   divergent, 464
   double, 514, 517
   improper, 464, 733
   indefinite, 370
   iterated, 517
   Riemann, 394
   tables of, 453, A-4
   of trigonometric functions, 414, 416

7.2 General Power Rule for Integrals

For $u = f(x)$ and $du = f'(x)\,dx$,

$$\int u^n\,du = \frac{u^{n+1}}{n+1} + C.$$

7.2 Indefinite Integral of e^u

If $u = f(x)$, then $du = f'(x)\,dx$ and

$$\int e^u\,du = e^u + C.$$

7.2 Indefinite Integral of u^{-1}

If $u = f(x)$, then $du = f'(x)\,dx$ and

$$\int u^{-1}\,du = \int \frac{du}{u} = \ln|u| + C.$$

7.4 Fundamental Theorem of Calculus

Let f be continuous on the interval $[a, b]$, and let F be *any* antiderivative of f. Then

$$\int_a^b f(x)\,dx = F(b) - F(a) = F(x)\Big|_a^b.$$

7.5 Basic Trigonometric Integrals

$$\int \sin x\,dx = -\cos x + C \qquad \int \cos x\,dx = \sin x + C$$

$$\int \sec^2 x\,dx = \tan x + C \qquad \int \csc^2 x\,dx = -\cot x + C$$

8.1 Trapezoidal Rule

Let f be a continuous function on $[a, b]$ and let $[a, b]$ be divided into n equal subintervals by the points $a = x_0, x_1, x_2, \ldots, x_n = b$. Then, by the trapezoidal rule,

$$\int_a^b f(x)\,dx \approx \left(\frac{b-a}{n}\right)\left[\frac{1}{2}f(x_0) + f(x_1) + \cdots + f(x_{n-1}) + \frac{1}{2}f(x_n)\right].$$

8.1 Simpson's Rule

Let f be a continuous function on $[a, b]$ and let $[a, b]$ be divided into an even number n of equal subintervals by the points $a = x_0, x_1, x_2, \ldots, x_n = b$. Then, by Simpson's rule,

$$\int_a^b f(x)\,dx \approx \frac{b-a}{3n}\left[f(x_0) + 4f(x_1) + 2f(x_2) + 4f(x_3) + \cdots + 2f(x_{n-2}) + 4f(x_{n-1}) + f(x_n)\right].$$

8.2 Integration by Parts

If u and v are differentiable functions, then

$$\int u\,dv = uv - \int v\,du.$$

8.4 Improper Integrals

If f is continuous on the indicated interval and if the indicated limits exist, then

$$\int_a^\infty f(x)\,dx = \lim_{b\to\infty} \int_a^b f(x)\,dx,$$

$$\int_{-\infty}^b f(x)\,dx = \lim_{a\to-\infty} \int_a^b f(x)\,dx,$$

$$\int_{-\infty}^\infty f(x)\,dx = \int_{-\infty}^c f(x)\,dx + \int_c^\infty f(x)\,dx,$$

for real numbers a, b, and c, where c is arbitrarily chosen.